25th International Mining Congress and Exhibition of Turkey (IMCET 2017)

New Trends in Mining

Antalya, Turkey
11-14 April 2017

Editors:

Irfan Bayraktar
Mehmet Karadeniz
Firat Atalay
Elif Torun Bilgic
Mehtap Gulsun Kilic
Oznur Onel
Mehmet Ozyurt

ISBN: 978-1-5108-3974-8

Printed from e-media with permission by:

Curran Associates, Inc.
57 Morehouse Lane
Red Hook, NY 12571

Some format issues inherent in the e-media version may also appear in this print version.

Printed by Curran Associates, Inc. (2017)

For permission requests, please contact Chamber of Mining Engineers of Turkey at the address below.

Chamber of Mining Engineers of Turkey
Selanik Caddesi No: 19/4
Kizilay / Ankara
Turkey

Phone: (+90) 312 425 10 80
Fax: (+90) 312 417 52 90

maden@maden.org.tr

Additional copies of this publication are available from:

Curran Associates, Inc.
57 Morehouse Lane
Red Hook, NY 12571 USA
Phone: 845-758-0400
Fax: 845-758-2633
Email: curran@proceedings.com
Web: www.proceedings.com

IMCET 2017

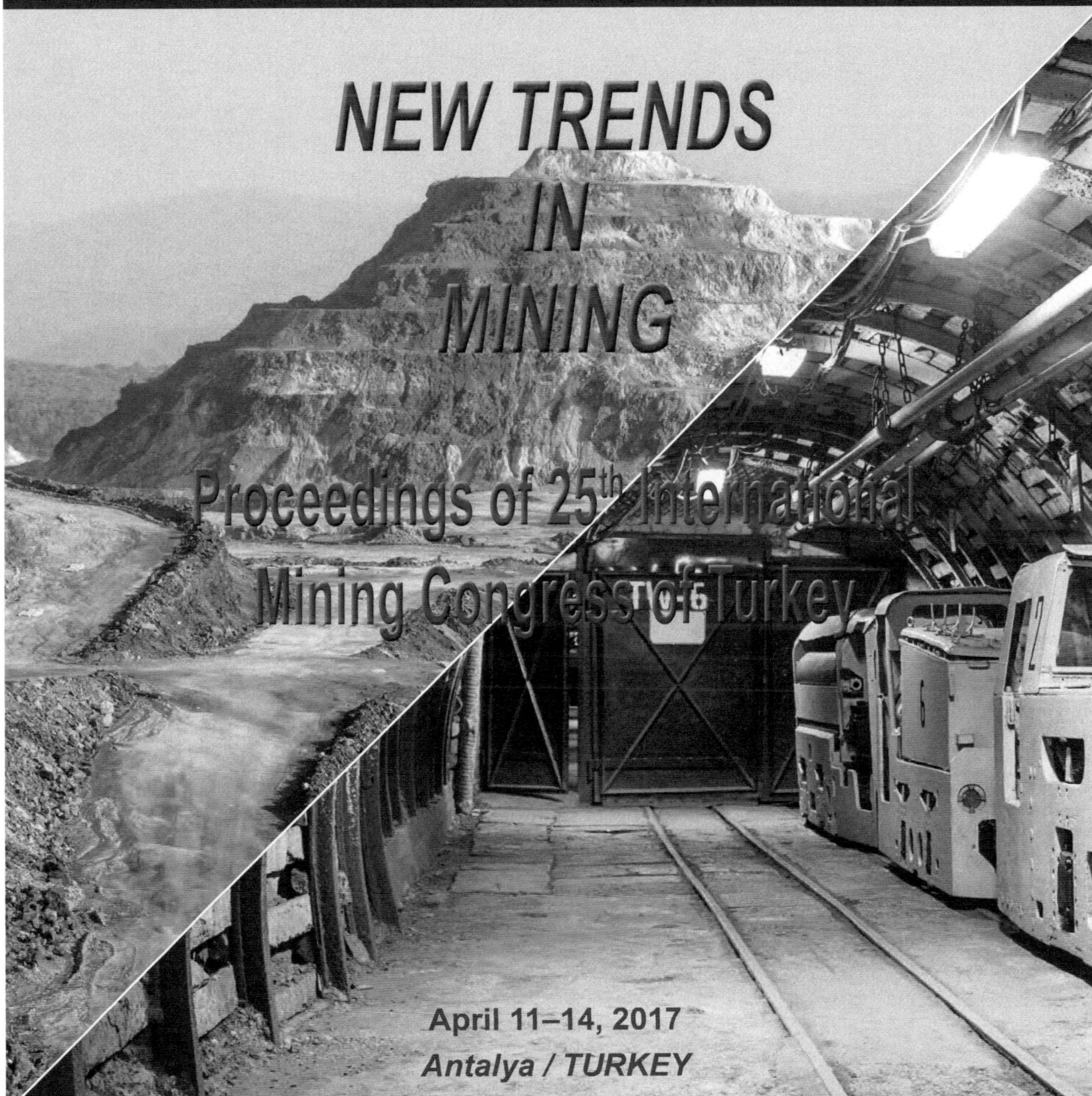

NEW TRENDS IN MINING

Proceedings of 25th International Mining Congress of Turkey

April 11–14, 2017

Antalya / TURKEY

Editors:

Prof. Dr. İrfan BAYRAKTAR

Dr. Mehmet KARADENİZ, Dr. Mehtap GÜLSÜN KILIÇ, Dr. Fırat ATALAY, Öznur ÖNEL, Elif TORUN BİLGİÇ, Mehmet ÖZYURT

IMCET 2017

NEW TRENDS IN MINING

Proceedings of 25th International Mining Congress of Turkey

Editors:

Prof. Dr. İrfan BAYRAKTAR

Dr. Mehmet KARADENİZ

Dr. Mehtap GÜLSÜN KILIÇ

Dr. Fırat ATALAY

Öznur ÖNEL

Elif TORUN BİLGİÇ

Mehmet ÖZYURT

UCTEA
Chamber of Mining Engineers

Executive Committee of the Congress

Chairman	Prof. Dr. İrfan BAYRAKTAR
Vice Chairmen	Ümit Ragıp ÜNCÜ Dr. Mehmet KARADENİZ Dr. Mehtap GÜLSÜN KILIÇ
Secretaries	Niyazi KARADENIZ Fatih TÜTÜNLÜ Dr. Fırat ATALAY
Treasurers	İsmail Fatih ÖZKAN Mehmet ÖZYURT
Members	Necmi ERGİN Nadir AVŞAROĞLU Mehmet Erşat AKYAZILI Ali ÖNEMLİ Davut ÖZLEN Selim ALTUN Elif TORUN BİLGİÇ Öznur ÖNEL İmge TÜMÜKLÜ Pelin KERTMEN

Members of Board of the Chamber

President	Ayhan YÜKSEL
Vice President	Emre DEMİR
Secretary	Necmi ERGİN
Treasurer	Mehmet ÖZYURT
Members	Sinan GİRDAPLI Öznur AKÇA Servet GÜRER

FOREWORD

In the course of the first Mining Congress of Turkey which was organized in 1969, we presume, not many delegates conceived of such a lasting, periodic and successful history of the Congress. It has been able consistently to grow and eventually develop an international identity in 2001.

Needless to say mining engineering has evolved intrinsically, being somewhat parallel to the other engineering subjects, in terms of understanding of underlying fundamental principles and ultimately development of new technologies. Yet, the mining industry has been facing many challenges associated with environmental, labour safety, public scrutiny and civil participation issues in addition to pure technical and economic aspects.

In this context, the 25th International Mining Congress and Exhibition of Turkey was organized to describe, discuss and record the solution recommendations.

The 25th Congress comprises of 11 invited and 182 peer-reviewed papers from authors around the world. This book is the culmination of invaluable contributions by many experts. So,we anticipate that it will be most beneficial to the mining industry.

The editors express their appreciation to them and also gratefully acknowledge the assistance provided by the Organization Committees' members and the management of the Chamber of Mining Engineers of Turkey.

Prof. Dr. İrfan BAYRAKTAR
Dr. Mehmet KARADENİZ
Dr. Mehtap GÜLSÜN KILIÇ
Dr. Fırat ATALAY
Öznur ÖNEL
Elif TORUN BİLGİÇ
Mehmet ÖZYURT

INVITED PAPERS

MINING
ROCK MECHANICS / PRODUCTION / ENVIRONMENT / EDUCATION / LEGISLATION

MINERAL PROCESSING / EXTRACTIVE METALLURGY

POSTER PAPERS

INVITED PAPERS

Mining Economics and Technology

I. C. Runge
Founder and Director, RPMGlobal, Brisbane, Australia

ABSTRACT This paper examines the interaction of economics and technology in mining. It is in three parts, the common link being the influence of economics on decisions. The first part highlights how economics underpins choices to explore for mineral commodities. It shows that popular concerns regarding exhaustion of non-renewable mineral commodities are largely unfounded. The second part examines the impact of technology and its effect on mine economics. Using real-world cases it suggests that the industry application of technology is focussing on evident but less-economically-valuable applications and overlooking less-evident but more-economically-important applications. It uses an example of path-dependent processes to highlight the importance of the *process* of mine planning. The third part examines the impact of cyclical commodity prices on mining company decision-making. It concludes that the current phase of the commodity price cycle presents a significant opportunity for implementation of technology in mines for lasting long-term efficiencies.

Keywords: Economics, Technology, Reserves, Computer, Commodity Prices

1 INTRODUCTION

In competitive free market environments economics underpins every choice - in exploring for mineral deposits; in planning for, developing, and operating mines and processing plants; and in the financing and marketing of mineral commodities. Yet, understandably, few technical professionals have a comprehensive understanding of economics. Mining is a multi-disciplinary industry, and, as with elsewhere in the industry where there is overlap between disciplines simplified proxies from the other field are deployed to incorporate the necessary guidelines or requirements.

A grade control engineer in the mine, for example, might only have a rudimentary understanding of processing plant metallurgy, but will use simplified guidelines given to him or her from the plant manager. Likewise, a geologist looking for the latest mineral commodity of interest will incorporate guidelines such as minimum widths, depths, and grades from mining and processing disciplines to inform his exploration targeting. Sometimes these guidelines from other disciplines prove reliable enough for the task at hand, and sometimes not.

What about the overlap with economics – an overlap that impacts all the technical disciplines?

When a planning engineer optimizes something, how well does the "something" translate into the best economic outcomes? What defines the best economic outcome? If we achieve lower costs or better returns in the short term, is this at the expense of higher costs or worse returns in the long term? Are better returns associated with higher risk?

The sections that follow don't answer these questions explicitly, but look at three areas of industry endeavour through an economic lens to arrive at a subjective report card of success or otherwise from an economic perspective.

2 MINERAL COMMODITIES: LONG TERM RESERVES AND PRICES

2.1 Reserves of Mineral Commodities

It is common perception that all the key mineral commodities – non-renewables – are gradually becoming exhausted. Deposits are becoming deeper, harder to find, and more complicated to extract, whilst at the same time the rate of production is increasing. The implication is that we must conserve existing resources and focus on recycling and on the use of alternatives, especially renewables.

Various authors (see: Baumol and Blackman, 1993; Repetto, 1987; Runge, 1998) have looked at world reserves and resources of non-renewable mineral commodities and how they change through time. Table 1 shows the reserves and production for three selected minerals illustrating the situation since 1950.

Even though production in the 30 years following on from 1950 exceeded, or nearly exceeded, the reserves available in 1950, the reserves by 1980 far outweighed the reserves at the start. Even today, with higher production rates and a much stricter definition of what constitutes a "reserve," the current reserve position translates into 26 years of production (for Iron), and more than this for the other commodities. For most commodities the reserve position is typically increasing both in total terms and in years-of-production terms.

This trend is a consequence of simple economics. Exploring for and proving up reserves is costly. It is uneconomic to spend money *now* to prove up reserves too far into the future, because the return from exploiting them only materializes when they are extracted.

The conclusion to be drawn is that there is no evidence yet to suggest that any of the important mineral commodities are becoming exhausted.

2.2 Long-term Prices of Mineral Commodities

When it comes to sustainable production of mineral commodities sufficient reserves are only half the equation however. The other half is "cost." It is of little value if the world reserves of oil were good for (say) 100 years at current production rates but gasoline was going to cost $100/litre.

In a market economy if there is a shortage of any commodity, the price will change (rise) until supply matches demand. The economics will favour additional exploration and deployment of technologies to exploit deposits previously considered uneconomic. At the same time, demand will also reduce. In this purely economic scenario the world will probably never really run out of any mineral commodity.

Is Table 1 just underpinned by ever-increasing commodity prices?

The answer is: No. There are short term fluctuations in the price of all commodities, but

Table 1 – World Reserves and Production of Three Selected Minerals

Mineral	Reserves* 1950	Production 1950 – 1980	Reserves 1980	Reserves[β] 2016	Years of Production
Aluminum	1,400	1,346	5,200	7,000 (16,000)	>100 years
Copper	100	156	494	720 (2,100)	36 to 112 years
Iron	19,000	11,040	93,466	85,000 (230,000)	26 to 66 years

* Reserves and Production from 1950 and 1980 from Repetto (1987). Reserves and production for 2016 estimated by the author based on information from the U.S. Geological Survey. Units: *millions of tonnes of metal content*

β Definitions for reserves in 1950 and 1980 are not compatible with the definitions in use in 2016, and include mineralization which would now be classified as "resources." For 2016, the first estimate is quoted reserves, the second estimate in parentheses includes resources (insufficiently well-defined to be classified as reserves under currently accepted definitions).

when these changes are excluded *the long term price of most mineral commodities is either constant in inflation-adjusted terms, or declining.*

Figure 1, adapted from Deverell and Yu (2011) shows the price of iron ore since 1885, corrected for inflation and plotted on a relative log scale.

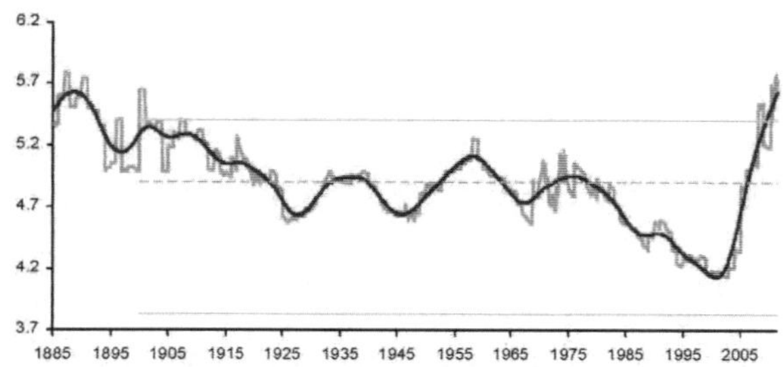

Figure 1 Iron Ore Price since 1885

Apart from the dramatic short-term price spike in the early 2000s, the long-term trend in iron ore price is unambiguously *declining.*

The same trend is evident with the long-term price of copper shown in Figure 2.

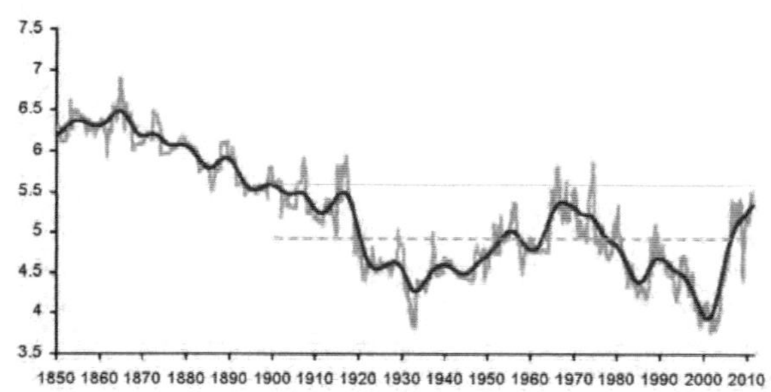

Figure 2 Copper Price since 1850

Despite increased demand, despite deeper more complex and lower-grade orebodies, and despite increased environmental and safety imposts, technology to find and exploit orebodies at competitive commodity prices has more than kept pace with the rate of exhaustion of known deposits over at least 150 years.

The long-term price reflects the fundamental costs of production. This declining trend is almost totally due to advances in technology.

It is technology applied to finding previously unknown deposits or understanding existing deposits better. It is technology applied to the mining of deposits previously considered too difficult, too complex, or perhaps too unsafe to mine. It is technology applied to extracting minerals in ways previously too hard to extract.

Technology finds its way into mining in many ways. It may be widely applicable or narrowly focussed. It might be just different but not necessarily better. For every idea put to practical use there are probably 10 ideas that end up in a dead end. There are probably another 10 ideas that just establish the ground-work – ideas examined years or decades previously in some university or some corporate back-room environment.

This technological advancement is the foundation of how the world advances, however long it sometimes takes to come to fruition in standard-of-living terms, and however unrecognized and unappreciated it often is in the mainstream consumer world.

3 TECHNOLOGY

Technology and economics are integrally linked. Technology and mining are integrally linked.

Runge (1995) examined the growth of technology starting from before the industrial revolution in England, how the mining industry adopted the technology, and how this affected both the industry and the wider community.

Technology is a part of mining now more than ever before. The following sections set out examples from the last 40 years of extensive use of technology in mines around the world to draw some lessons for the mining industry today. The lessons apply to any technology, but the primary focus is on computer technology aimed at planning and operating mines better.

3.1 Computer Applications and Mine Design.

Early computer tools for equipment simulation and mine design were very primitive, but so too would any new technology seem to be when viewed from 40 years in the future. Reliable results from these early applications were only possible if the work was being undertaken by someone well-versed and experienced in the system and aware of its limitations – someone who was wary of the kinds of situations likely to yield unreliable results.

Nevertheless, huge gains were made because the technology allowed things to be done that were previously not possible to do. The analysis of the complex interaction between multiple trucks and loaders was something not hitherto possible. A mine plan that previously took weeks of tedious hard work to schedule just once could be scheduled in less than an hour. For the first time this facilitated analysing alternative mine layouts and alternative schedules that were simply impossible to conceive previously.

Today the tools are immensely more powerful and sophisticated.

Have there been significant *economic* gains? There has been a saving in planning personnel manpower. Complex mines have been commissioned where previously the complexity would have been a barrier to start-up, though this doesn't necessarily mean economic gains over mines from the previous era.

The greatest economic gains come from better decisions, and judging by decision-making in the industry today it is hard to conclude that better decisions are now being made than in previous eras. The problems stem from lesser involvement of experienced personnel with the introduction of technology; from planning personnel distracted by the technology to the detriment of mine economics; and from technology that automates a planning process that is itself flawed or inappropriate to the application.

This is the key conclusion of this paper: *that the advances in computer applications for the analytical aspects of mine planning have not been matched by advancements in the understanding of the process of mine planning, with some of the biggest potential economic benefits not being realized.*

I illustrate this firstly with an example from a study of a large South African dragline mine planned in the late 1980s. A stylized mine layout with the different directions of mining possible is shown in Figure 3.

Dragline mines always progress in strip-by-strip fashion, because the waste from each strip

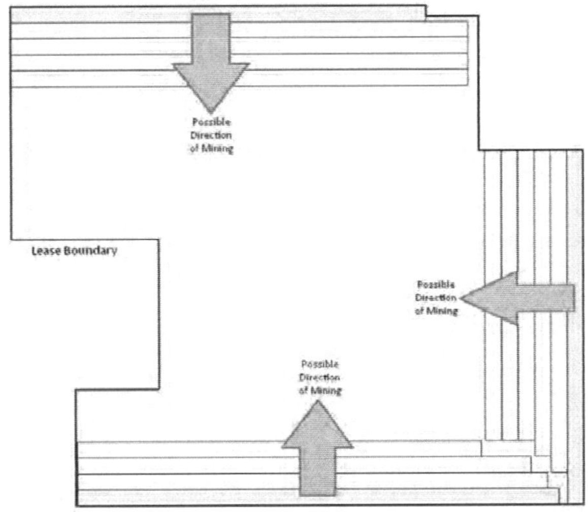

Figure 3 Stylized Plan View of Open Pit Dragline Mine

is placed in the mined-out void from the previous strip. The starting point is usually the coal outcrop or some property boundary. In this deposit the topography was undulating and the coal seam was relatively flat-lying and it extended over almost the whole lease area. There was no outcrop, and no obvious place to start mining. Such a case presented a prime target for the new (at the time) technology. For the first time in such an environment it was possible to try multiple different mine layouts with mining advancing in just about any conceivable direction. The company undertook such a study, analysing scores of layouts and schedules. The optimum mine plan was the layout and schedule that yielded the lowest price of coal when assessed on a discounted cash flow, net present value basis.

Unfortunately, the planning personnel inadvertently fell into the "knowledge problem" trap that was the subject of the paper referred to in Runge (1995). The "knowledge problem" isn't a problem explicitly associated with technology, but technology – in this case by using the power of the computer to examine many hundreds of cases – can deceive us into believing that our assessment has been comprehensive. In this case, although the personnel involved were experienced mining engineers, they had limited knowledge of dragline operations, and simply failed to examine a set of cases that were (as it turned

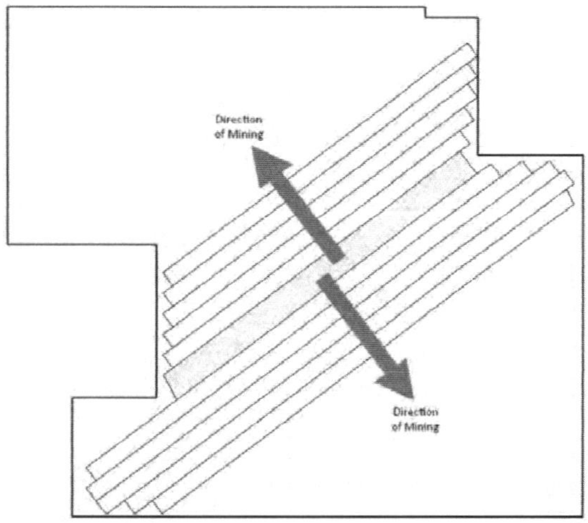

Figure 4 Final Configuration and Mine Layout

out) 30% more cost efficient than the best case previously studied. The selected case started with a boxcut in the centre of the deposit (excavated by shovels and trucks) and progressed in two directions outward as shown in Figure 4.

Technology is a tool, but deciding how to use it and how to rely on it is a task quite different to the task of using it. This is not a case of "garbage in, garbage out." Nor is it a shortcoming in the computer program. It is an example of a shortcoming in the *process* of planning a mine.

"Experience" definitely provides some protection against this shortcoming, but it isn't the only tool that can provide such guidance. Nevertheless, with the advent of advanced technology the value of experience has often been overlooked. Sometimes other computer tools can help make these choices (i.e. to determine if something is worth studying, or not). The lesson, however, is that the task is not something that can be assigned to a lesser experienced person simply because he or she has the requisite computer skills to drive the program. Understanding the *process* is something quite different to undertaking the *tasks* that make up the process.

This example showed a 30% lower cost of production than the case that might otherwise have been chosen. Such huge changes in economics are not uncommon *at the start of mining projects*.

This characteristic of mining, e.g. the inability at the start to define a comprehensive set of alternatives for evaluation and consideration, is something that sets our industry apart from most other industries.

3.2 Early Stage Assessments and Choices Subject to Uncertainty

This section extends the example from above, and again illustrates the importance of correct *process* in early-stage assessments.

Figure 5 shows a classical sub-vertical-trending metalliferous orebody that is subject to possible mine development either as an open

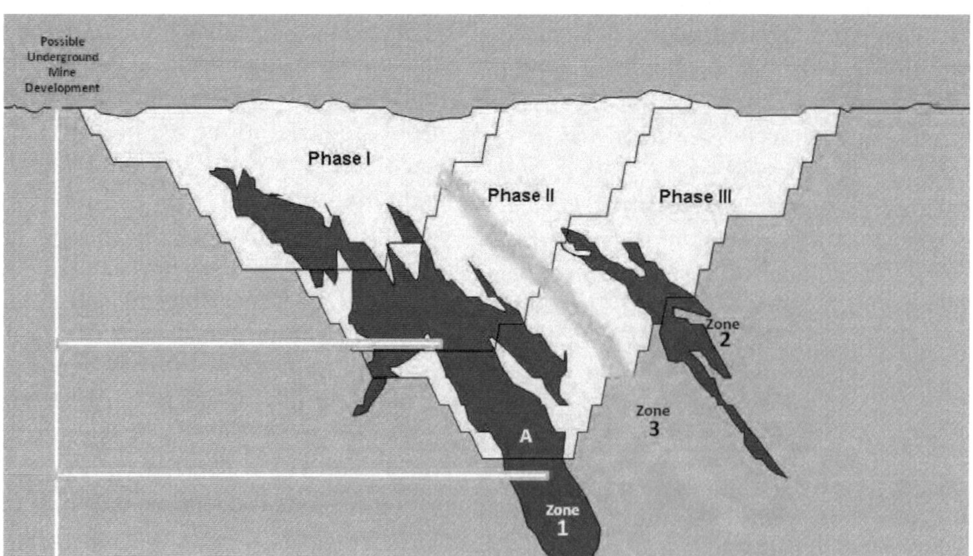

Figure 5 – Example Metalliferous Deposit and Possible Mining Methods

pit mine or as an underground mine. Two high-grade zones (Zone 1, and Zone 2) have been identified, with a third ill-defined region of lower-grade mineralization (Zone 3) also present. The highest grade of ore so far identified in the deposit is located at "A."

Consider now the characteristics of this deposit and the decision-making environment that might lead to an optimum way of exploiting it. Five characteristics are shown in Table 2.

Table 2 only covers a few of the differences between the mining methods. Also, the mining methods are themselves not necessarily exclusive - open pit mining can coexist with underground mining, and frequently does in mines around the world. Choices are not "digital" – hundreds of variations and combinations are possible, and not all of the subtle differences between the variations can be identified in advance, or even reduced to economic criteria. The complexities in this example are similar to the example shown in the previous section.

In the previous section the shortcoming in the process of evaluation was ascribed to the "knowledge" problem. This section introduces another aspect of the process that can lead to sub-optimal results, namely, the risk of *path-dependent evaluation processes*.

For example, at the early stages of evaluation, if a geologist believes that the deposit is likely to be mined by underground methods, then he or she can logically and rationally choose to ignore the uneconomic mineralization in Zone 3. It costs money to drill out and evaluate deep deposits and there is little point in doing so if it will never be mined. This omission will have no bearing on the net present value of any future underground mine.

Consider now any subsequent assessment of the deposit *as an open pit mining proposition.* The mineralization in Zone 3 won't be evident - it likely won't even be shown on geological plans. The net present value of any open pit mine will not benefit from this inclusion. Due to the early-in-the-process assumption by the geologist the comparison between open pit and underground mining options has been biased towards the underground mining option.

Table 2 Deposit Characteristics with Open Pit Mining and Underground Mining Alternatives.

Characteristic	Typical Open Pit Mining Method	Typical Underground Mining Method
Sequence of Mining	The highest grade ore will not be accessible until the last stages of the mine life	There is scope to mine the highest grade ore relatively early in the mine life
Ore Grade for Economic Viability	Ore in zone 3 is viable to mine because the material has to be extracted anyway. Once there is already a processing plant in place and once the material has already been hauled to the surface, the return from this lower grade ore is attractive	Zone 3 is uneconomic when mined using underground methods. Narrow ore zones within Zones 1 and 2 may also not be mineable.
Reserves	Maximum extraction of in-situ mineralization	Proportion of in-situ mineralization that can be extracted is much less.
Development Effort and Timing	Requires extraction of the shallower reserves first. Even the shallow reserves may require a lot of waste to be prestripped before reaching the first ore. Initial development work (prestripping) can be expedited using contract earthmoving	More flexibility in choosing which ore can be mined first. Initial development work (shafts, drives, stope development) constrained (cannot easily be expedited) because of limited access and tasks undertaken in series
Exploration Effort and Data Reliability	Reserves that are mined first are best known (shallowest, easiest to drill out) reducing risk and increasing reliability of plant design and marketing.	Deep reserves are expensive to drill out in advance. Higher cost, up-front geological assessment.

Indeed, this result "confirms" the judgement of the geologist in the first place to exclude the mineralization in Zone 3. The process just followed is a path-dependent one. The wrong result will potentially be arrived at even though all choices leading to that point were logically and systematically made.

This path dependency is an endemic characteristic of any decision process where there is uncertainty that can be resolved only at a cost that itself impacts the viability of the project. Path-dependent processes don't necessarily yield incorrect answers, and even where they do few operators would even be aware of it because the alternative path that was not followed (the opportunity cost) is seldom evident. Runge (2000, p. 128) sets out a number of examples of such processes.

As with the previous example, the evaluation of the various alternatives in this case is definitely one for modern computer tools. But the lesson with use of these tools is the same: the greatest economic value added (or greatest loss of economic value suffered, even if unknowingly) occurs at the start of projects, and is a function of the process followed and the choices as to which cases are to be examined.

4 DECISIONS THROUGH THE COMMODITY PRICE CYCLE

Decision-making in the mining industry extends across a spectrum from the urgent (survival) to choices spanning decades. All the of the previous examples have been on the less urgent part of this spectrum. This section considers decisions on the more shorter-term part of the spectrum.

The impact of shorter term commodity price changes can be dramatic, and any assessment and valuation of the economics of mines in the longer term must also consider how well-equipped the mine is to handle circumstances in the shorter-term. Mines that can readily adapt – either because of deposit characteristics, or because of mining methods selected, or because of some other characteristic - are to be preferred over mines that are less adaptable.

4.1 Technical and Economic Impact of Commodity Price Changes

Figure 6 shows the price of iron ore over the last 8 years.

The price changes from an initial high of US$180/tonne in early 2008, to a low of just one-third of that later in the same year; to a

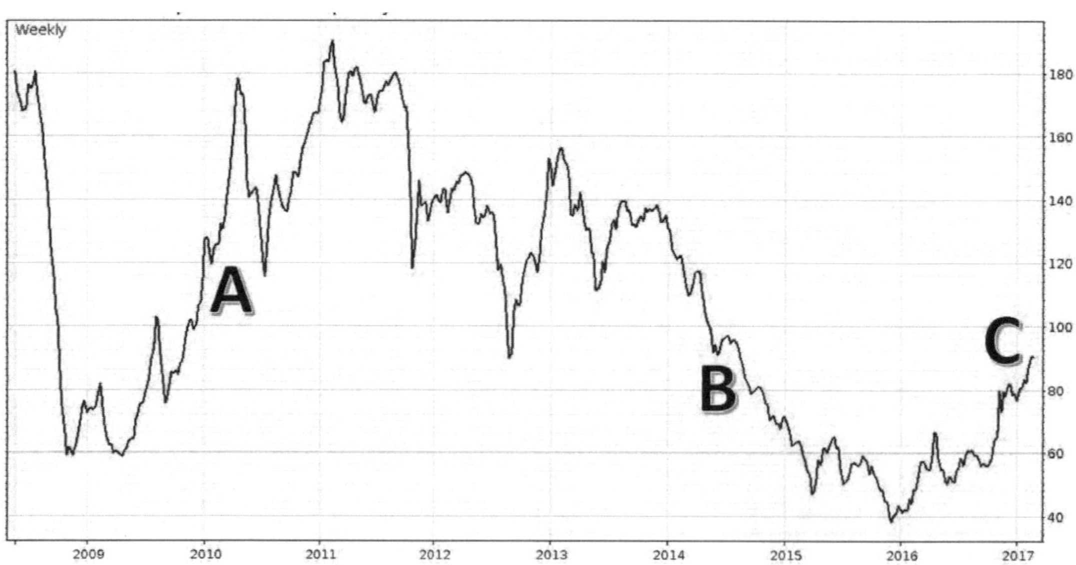

Figure 6 Iron ore price 2008-2016

Import price of Iron Ore fines (62% Fe) into China. Price in US$ per dry metric tonne. Data from www.fullertreacymoney.com

high of more than US$180/tonne in early 2011, and to a low of less than US$40/tonne at the end of 2015. The current price (March, 2017) is around US$90/tonne.

Commodity price changes over the last eight years may have been more dramatic than in most 8-year periods, but nevertheless, these fluctuations are a characteristic of mining much more than most other industries. Imagine the impact on industries such as motor vehicle manufacturing, or house construction if the selling price of their product fell by two-thirds in less than one year (2008 in the above figure) or more than doubled in price in one year (as in 2016 in the above figure)?

How can price changes of this magnitude (and changes in the way that the mine is operated) be reconciled with the long-term trends and application of technology discussed in the first part of this paper?

Assuming you were involved in assessing an iron ore deposit, or in planning a mine, or in managing a mine during this period, how would price changes like the ones shown in Figure 6 impact your decision-making?

Reserves: The tonnage of reserves that are viable at $40/tonne will almost certainly be less than the tonnage when the selling price is $180/tonne. What should be the basis for reporting reserves? How should exploration effort be prioritized over this period with change in selling price? If some mineralization is clearly not viable at current prices, but is likely to become viable under some future envisaged price scenario, should it be examined now, and at what cost?

Mine Design: An "Optimum" pit when the selling price is $40/tonne is surely much smaller and a different shape than an "Optimum" pit when the selling price is $180/tonne. If a mine has been designed, and is in operation, using the "optimum" pit shape based on the "$180/tonne" price, then at some other price how much "less-than-optimum" is it, and what should be the strategy for changing the design to accommodate the changed price?

Management: Anyone can look good managing a mine during periods of high commodity prices. But when selling prices are low many operating mines are unprofitable and require a lot of cash to keep running. Closure might not be a viable option, because high closure costs might require even more cash. Yet low points in the commodity price cycle are when the raising of cash is the most expensive - when the marginal cost of capital is highest. What should be the strategy to avoid this vicious circle? Is there anything that can be done prior to mine start-up, or is it something that can really only be addressed operationally?

These are not just rhetorical questions. It is not sufficient to simply focus on keeping the costs of production in the lowest quartile of the industry. Mining companies have failed because the commodity price remained below the long-term trend price for too long, and they ran out of money waiting for the upturn.

The answer to the questions is one for each specific mining operation, however there are guidelines to be drawn and lessons to be learnt that apply to all mines.

4.2 Change.

Every mine changes throughout its life. It changes because the orebody changes. It changes because the price of the product changes. It changes because demands of the customers change. And it changes because technology changes.

One lesson from the last 8 years is that our ability to adapt to change has been found wanting. This isn't surprising since when mines are being planned few operators plan for adaptability.

Classically mine assessments are based on relatively fixed scenarios, albeit examining multiple alternative mine plans consistent with that scenario. The scenario is initially taken as a given by the mine planners because it involves inputs outside of his or her area of expertise – expected selling price, cost of capital (required return on investment), market characteristics etc. Sensitivity studies are conducted to assess the impact on the net present value of various changes to these starting assumptions.

However in the face of significant change in some fundamental parameter operators don't just accept the change as implied by the sensitivity analysis. The mine plan changes to respond to the external changes. Only then does the ability and resilience of the mine to respond to change become evident.

Could this "ability and resilience of the mine to respond to change" have been understood *before the mine was commenced?* If so, a more robust alternative plan better able to cope with the change might have materialized.

How do mining enterprises value plans that are more resilient over plans that have less ability to change and adapt? How can increased expenditure leading to increased adaptability ever be justified when under any base case (fixed) scenario the less adaptable alternative (with lower capital requirements) will yield a higher return on investment?

4.3 Commodity Price Cycles and Management Decision-Making

Most technological advances and long term cost-of-production efficiencies originate with technical professionals, often in conjunction with operations personnel who have the most knowledge of aspects of the mine that might be done better.

Improving efficiency means change – doing something a "better way." At least initially this takes additional time, effort and investment compared to simply maintaining the status quo. At what stage does implementing change make sense?

Implementing change takes time that may not be available. Short-term commodity price changes exacerbate the problem. As a technical person focussed on improving efficiency how can these constraints be reconciled?

Consider again Figure 6 as a proxy for any short term commodity price cycle, characterized in three phases, labelled "A", "B", and "C."

Commodity Prices Increasing (Phase "A"). During the "up" phase of the cycle, few mining enterprises are interested in efficiency; they are interested in expansion, and production (often "at all costs"). Skilled operational personnel are hard to find, and operational efficiencies suffer. Economics favour expansion and maximization of production because the profit from an additional tonne is more important than increased profit from [more efficient] current production tonnes. This phase is *not* characterised by mine efficiency.

Commodity Prices Decreasing (Phase "B"). During this "down" phase of the cycle company management is focussed on cutting costs. The focus is on reducing any costs where the return is not immediate – exploration and long term planning, for example. The economic driver is survival and protection of cash, meaning reduction of working capital and minimization of development effort. This too is not the phase characterised by mine efficiency.

Stability, and "Reasonable" Returns (Phase "C"). This phase – the current phase of the commodity price cycle – represents the best opportunity for technology professionals to really make a difference to mine efficiency. Commodity prices have risen from the cyclical low point, cash flows have improved, and debt has been reduced. Mining companies are aware than many of the cost savings during the previous phase were of a temporary nature. Also some savings achieved were at the expense of higher costs later in the mine life. Whilst in this phase few mining companies have any appetite for large capital expenditures, modest capital expenditure on technological improvements justified on the savings from these improvements can be supported.

5 CONCLUSION

There is a well-known saying commonly considered to be a Chinese curse about "living in interesting times," and that certainly describes the mining industry over the last decade. Yet for those of us on the technology and economic assessment side of the industry the situation today offers many opportunities for challenging and interesting jobs and improvements across the whole spectrum of mining.

- The industry is now entering an efficiency regime from previous expansion and cost cutting regimes. Opportunities abound.
- Whilst the opportunities to add the greatest economic value present themselves at the

start of projects, the same conditions occur with any major change in the mine – particularly changes that facilitate recapitalization following change of ownership or change of mining method. Judicious use of technology, coupled with more robust evaluation processes, can yield great returns in this environment.

- For mines already in operation there is scope for changes that can also yield great returns. Even without quantum changes, myriads of smaller changes can lead to efficiencies that aggregate into large economic improvements. New mines commence with limited knowledge of many characteristics – the orebody, processing limitations and subtleties, and market requirements. Now, after some time in operation, all of these things are better known. Re-examining all aspects of the mine to refine operations (often termed de-bottlenecking) can yield high marginal returns for relatively small additional investment.

REFERENCES

Baumol, W.J, and Blackman, S.B.B, 1993. Natural Resources, in *The Fortune Encyclopaedia of Economics* (ed: Henderson, D.R.), pp 40-43, Warner Books, Inc: New York

Deverell, R and Yu, M. 2011 *Long Run Commodity Prices: Where Do We Stand?* http://www.credit.suisse.com Commodity Research Report, Credit Suisse.

Repetto, R, 1987. Population, Resources, Environment: An Uncertain Future, *Population Bureau,* 42:2, quoted in Baumol and Blackman (1993)

Runge, I. C. 1995 Economics, Technological Change and the Knowledge Problem. *Proceedings, APCOM XXV, Application of Computers and Operations Research in the Minerals Industries*, AusIMM, Melbourne, Australia.

Runge, I. C. 1998 *Mining Economics and Strategy* Society for Mining, Metallurgy, and Exploration, Inc. Littleton, Colorado

Runge, I. C. 2000 *Capital and Uncertainty – The Capital Investment Process in a Market Economy Edward* Elgar Publishing Limited. Cheltenham, Glos, UK

Current Status of Secondary Raw Material Processing in Japan

T. Fujita, G. Dodbiba, J. Ponou

Graduate School of Engineering, The University of Tokyo, Tokyo, 113-8656, Japan

ABSTRACT Current status of secondary raw materials in east Asia and world is initially described. Next, the trends of secondary raw materials processing, and waste treatment in Japan is shown with material flow and acts. The intermediate treatment companies and smelting companies and cement industries are contributed for the recycling in Japan. Also the recent new technologies for the sustainable development of secondary material processing for recycle in Japan are introduced.

1 INTRODUCTION

Nowadays, in Japan, about 300 calcite mines from small to large scale (Limestone assoc., 2017) are operating, however, there is only one coal mine (Kushiro coal mine Co., Ltd., 2017) for training and only one commercial scale operating metal mine (Hishikari Mine, 2017) producing gold. Therefore, the metal import and recycling to get secondary raw materials are very important for Japanese industry.
METI (Agency for Natural Resources and Energy, 2017) in Japan promotes the cooperation of exploration and environmental operation in foreign countries, recycling, stock, research of alternative materials production, the research of exploiting the bottom of the sea, cultivation of human resources development for minerals and energy.

2 BACKGORUND OF SECONDARY RAW MATERIALS IN EAST ASIA AND WORLD

Printed circuit boards (PCBs), consisting of several mounted parts and motherboard, are used in many kinds of industries. They contain many kinds of metals, which should be recycled. PCBs are produced mainly East Asia, including Japan as shown in Figure 1. Most of electrical and machine parts in the world are produced in those countries like PCBs. The East Asian countries have discussed recycling technologies and policies in the symposium on East Asian resource recycling technology for about 30 years. The eight categories of topics are shown in Figure 2. The largest number of topics depends on

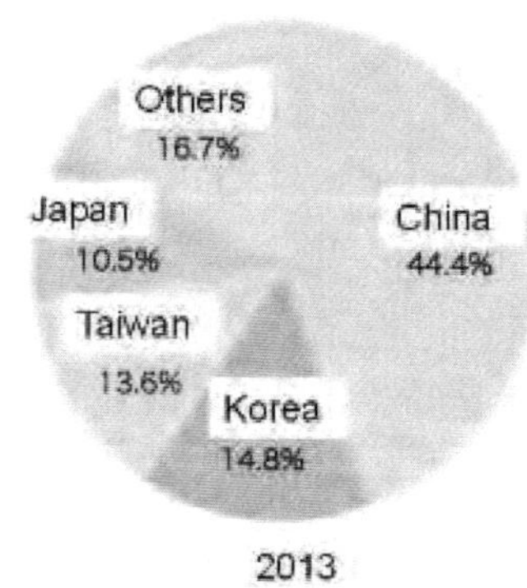

Figure 1. Print circuit board (PCB) production in the world (Economic news 2013)

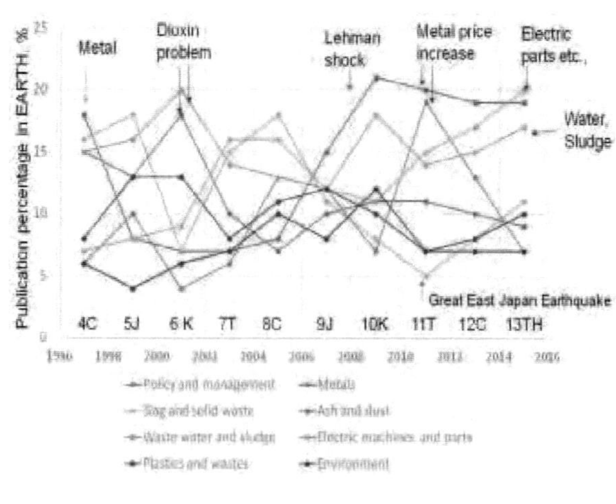

Figure 2. Past publication in eight categories percentage in Symposium on East Asian Resource Recycling Technology (EARTH, 2015)

the importance in each period. Recently, the research topics on the electric parts and metal recycling, as well as on water sludge treatments are increasing.

Metal prices increased 2011, however, the prices decreased and recently they are stable. Figure 3 shows the comparison of world metal market in various metals between 2012 and 2015. In 2012 iron is the largest market and if iron is 100%, copper and gold is 40% and the rare metals are several percent. However, in 2015, the iron price decreases and the copper becomes the largest market. If the copper is 100%, iron is 97%, gold is 90% and aluminum is 76%. Therefore, the copper and gold recycling becomes more important, while the circulation of iron scrap becomes difficult.

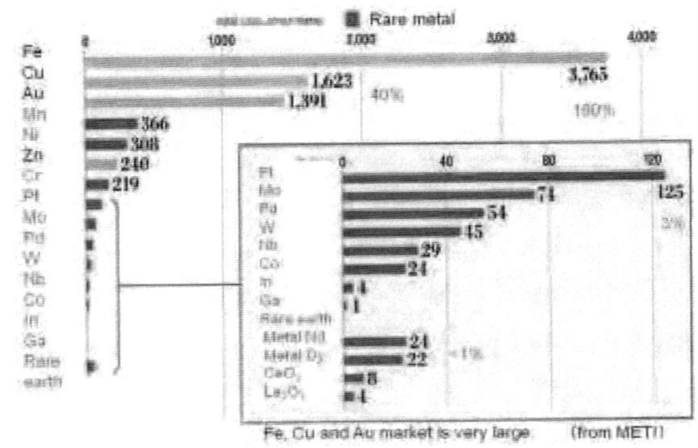

(a) 2012

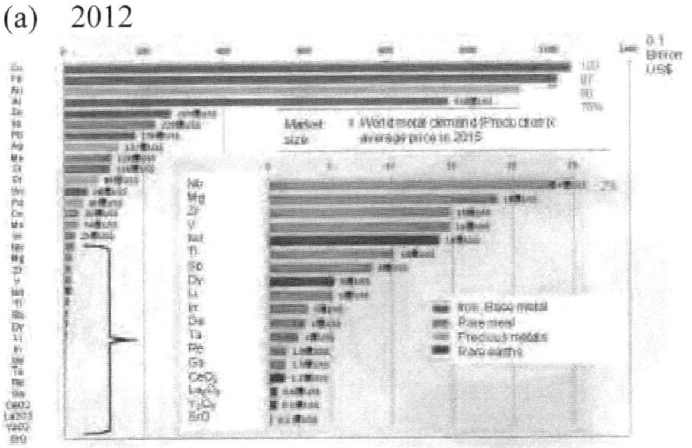

(b) 2015

Figure 3. Comparison of world metal market in various metals between 2012 and 2015 (METI,2016)

The future demand of metals comparing 2020 and 2030 is listed in Table 1. In the base metals demand, aluminum and copper will increase more than 3 times in 10 years. In the rare metals, the demand is much larger than base metals, especially lithium, cobalt and rare earth demand is 8, 6 and 5.9 times, respectively. Rare metal utilizes renewable energy production and to save the energy. The lithium and cobalt is utilized in the secondary buttery and rare earth is necessary for production of strong magnet and LED etc..

On the other hand, according to the UNEP's report published in 2011 the percentage of metal recycling is not so large (Table 2).

Table 1. Forecast of world metal growth rate of supply and demand in 2020 and 2030 (MERI/J, 2015).

Metal	Growth rate of supply and demand per year, % (from 2020 to 2030)	Supply and demand (production) per year in 2020	Supply and demand (production) per year in 2030	Reference
Fe	1.8	1700 Mt	2050 Mt	JFE group
Al	3.7	65 Mt	74 Mt	USGS
Cu	3.4	26 Mt	37 Mt	World metal statics
Zn	2	14.5 Mt	16.5 Mt	ILZSG
Pb	1.2	12.8 Mt	14 Mt	ILZSG
Ni	2.3→1.8	2.2 Mt	2.6 Mt	Wood Mackenzie
Co	6 ← Increase	0.13 Mt		CDI
Mn	3.2→1.7	19 Mt	23 Mt	BHP Billiton
Li	8	0.045 Mt		TRU
Rare earth	5.9	0.14 Mt		Roskill
Pt*	2	125 t (139)	143 t (173)	MERI/J
Pd*	1.4	246 t	280 t	MERI/J

* only for automobiles

Table 2. The recycling rate of lithium and rare earth

Recycling rate	Metal Abundance of metals ←large small→			
	more than 1000ppm	1000-100ppm	100-1ppm	Less than 1ppm
50%以上	Al, Fe, Ti, Mn	Cr	Ni, Zn, Cu, Co, Nb, Pb, Sn	Ag, Pt, Au, Pd, Re, Rh
25-50%	Mg		Mo	Ir
10-25%		W		Cd, Ru
1-10%				Sb, Hg
Less than 1%		Ba, Sr, V, Zr	Li, Ga, B, Hf, Be, Ta, Ge, As Rare earth	Tl, Se, In, Bi, Te, Os

Metal recycling percentage of used materials and metal terrestrial abundance (UNEP, 2011) were less than 1%. The new recycling technology and design to reuse these metals will be necessary.

3 BACKGORUND OF SECONDARY RAW MATERIALS IN JAPAN

In Japan, resource productivity forecast is shown in Figure 4. Resource productivity is GDP / volume of resources. It gradually increases and will reach 460 thousand yen per ton in 2020. The material recycle forecast in Japan is shown in Figure 5. Metal recycling rate of Japan will be 17% in 2020. Final disposal forecast in Japan is shown in Figure 6. The disposal area is not enough in Japan, and the final disposal of Japan will gradually decrease to17million tons in 2020. Figure 7 shows the comparison between 2000 and 2012 for material flow in Japan. Comparing past and present, the import amount is almost same, however, in 2012 the domestic resource amount supply, the final disposal and the total material input decrease more and the volume of recycle use increase, therefore, the material reduction is proceeding.

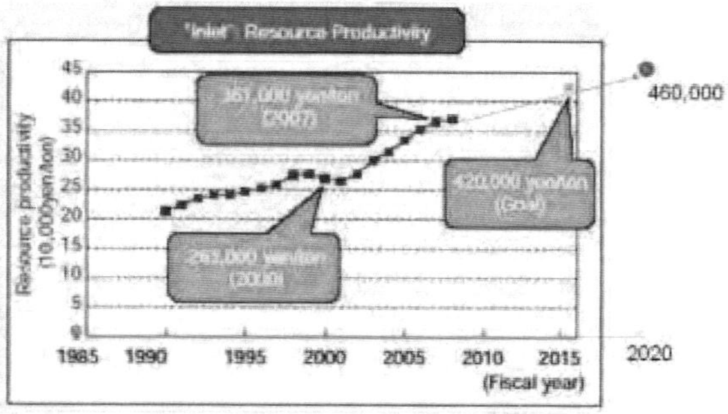

Figure 4. Resource productivity forecast in Japan (Ministry of the Environment, 2015)

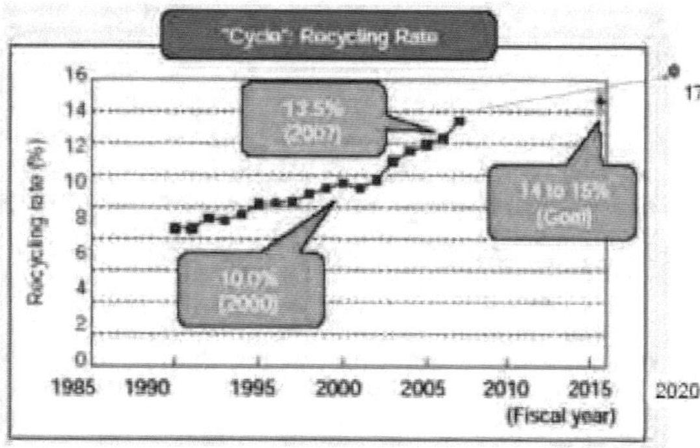

Figure 5. Forecast of Material Recycling in Japan. (Ministry of the Environment, 2015)

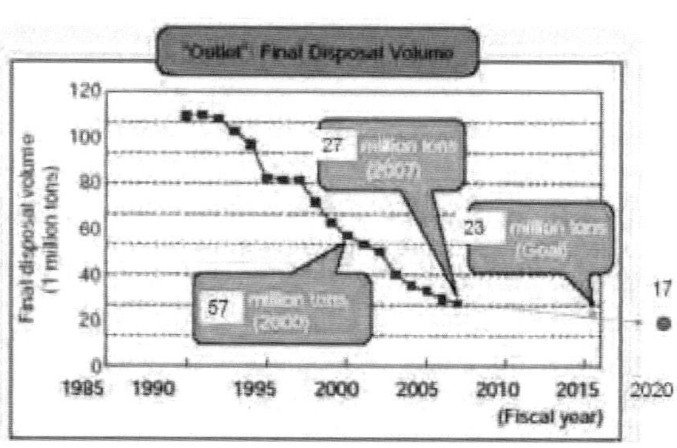

Figure 6. Final disposal forecast in Japan (Ministry of the Environment, 2015)

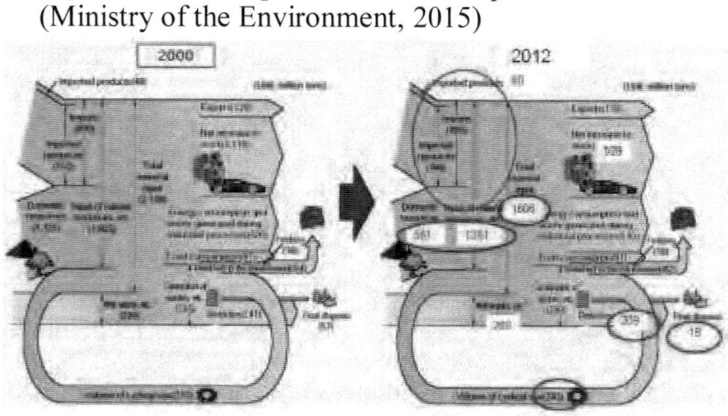

Figure 7. Comparison between 2000 and 2012 for material flow in Japan
(Ministry of the Environment, 2015)

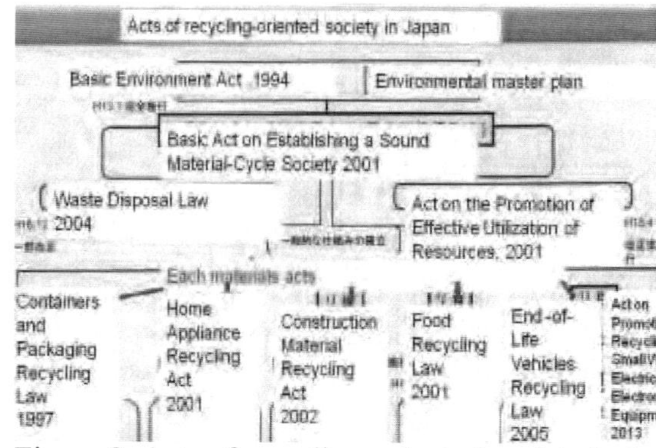

Figure 8. Acts of recycling-oriented society in Japan

Acts of recycling-oriented society in Japan is shown in Figure 8. Basic environment act, the promotion of effective utilization of resources and waste disposal law are established in 1944, 2001 and 2004, respectively. After 1997 to 2013 each material acts are established in six kinds of materials and improved.

4 WASTE TREATMENT AND RECYCLING COMPANIES IN JAPAN

In Japan there are many intermediate treatment contractors. In Table 3, the world enterprises are listed comparing Suzutoku Holdings in Japan. There are many big companies in the world and Suzutoku Holdings is number 8th in USA and Europe.

Table 3. World enterprises for recycling and waste treatment in USA and Europe (Waste age 100,2015)

	Company in USA	Sales revenue Billion $	employee number
1	Waste Management Inc	13.9	43,700
2	Republic Services Inc	8.4	30,000
3	Clean Harbors Inc	3.5	13,180
4	Stericycle Inc	2.1	[illegible]
5	Progressive Waste Solutions Ltd	2.0	7,500
6	Waste Connections Inc	1.9	7,000
7	Covanta Energy Corporation	1.6	3,500
8	Advanced Disposal Services LLC	1.3	[illegible]
9	Recology	0.7	[illegible]
10	Rumpke Consolidated Companies Inc	0.5	[illegible]

Suzutoku Holdings →

	Company in Europe		Sales revenue Billion €
1	Veolia	Veolia	7.0
2	Suez Environnement	GDF-Suez	6.4
3	Remondis	Rethmann	5.3
4	Alba	Alba Group	2.7
5	FCC	FCC	2.6
6	Indaver	Delta NV	2.2
7	Urbaser	ACS	1.7
8	Van Gansewinkel	[illegible]	1.2
9	Cespa	Ferrovial	1.0
10	Biffa Group	Montagu PE	0.9

Suzutoku Holdings →

Market change of resource circulation in Japan is listed in Table 4. Recycling industry produces new business. Japanese market for resources circulation is about 400 billion dollars. Waste treatment and recycling industry is about 40 billion dollars.

Table 4. Market change of resources circulation (Sakai, 2015)

Market change of resources circulation, 0.1 billion yen

Classification	2000	2005	2010	2013
Waste treatment and utilization	394,602	438,217	420,378	437,790
Recycling	39,058	37,858	34,783	37,593
Containers and Packaging	251	546	425	393
Home Appliance Recycling	0	546	1,129	513
automobile recycling	0	207	340	316
Reuse of parts and resources	354,636	387,740	364,560	374,100
Long life design	909	12,619	21,035	26,094

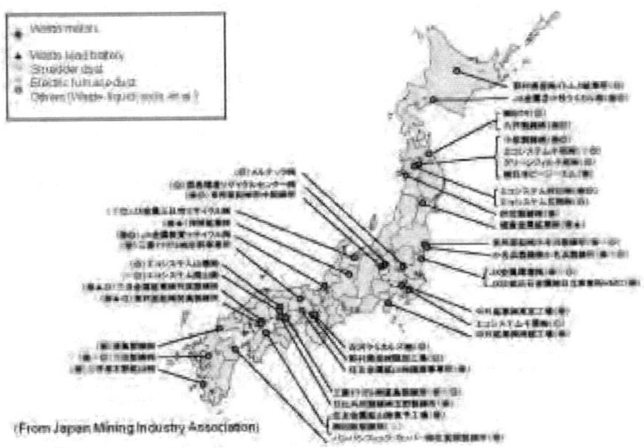

Figure 9. Waste treatment and recycling smelter in Japan (JMIA, 2015)

Table 5 Materials to utilize in the cement industry (Taiheiyo Cement Co., 2007)

Materials	Main Utilization	Year 1	Year 2	Year 3	Year 4	Year 5
Burst furnace slag	R,M	[illegible]	[illegible]	[illegible]	[illegible]	[illegible]
Coal ash	R,M	[illegible]	[illegible]	[illegible]	[illegible]	[illegible]
Sludge	R	[illegible]	[illegible]	[illegible]	[illegible]	[illegible]
Construction soil	R	[illegible]	[illegible]	[illegible]	[illegible]	[illegible]
Gypsum	R	[illegible]	[illegible]	[illegible]	[illegible]	[illegible]
Incinerated ash	R,T	[illegible]	[illegible]	[illegible]	[illegible]	[illegible]
Slag	R	[illegible]	[illegible]	[illegible]	[illegible]	[illegible]
Sand blast	R	[illegible]	[illegible]	[illegible]	[illegible]	[illegible]
Steel making slag	R	[illegible]	[illegible]	[illegible]	[illegible]	[illegible]
Waste plastics	T	[illegible]	[illegible]	[illegible]	[illegible]	[illegible]
Woods	R,T	[illegible]	[illegible]	[illegible]	[illegible]	[illegible]
Refined oil	T	[illegible]	[illegible]	[illegible]	[illegible]	[illegible]
Oil	T	[illegible]	[illegible]	[illegible]	[illegible]	[illegible]
Alumina	R,T	[illegible]	[illegible]	[illegible]	[illegible]	[illegible]
Slagheap	R,T	[illegible]	[illegible]	[illegible]	[illegible]	[illegible]
Waste tire	R,T	[illegible]	[illegible]	[illegible]	[illegible]	[illegible]
Meat-and-bone meal	R,T	[illegible]	[illegible]	[illegible]	[illegible]	[illegible]
Others	-	[illegible]	[illegible]	[illegible]	[illegible]	[illegible]
Total		[illegible]	[illegible]	[illegible]	[illegible]	[illegible]
Utilization kg per ton cement,	kg/t	[illegible]	[illegible]	[illegible]	[illegible]	[illegible]

Many kinds of inorganic materials can be used in Cement

Chlorine and Chromium have to be removed to prepare cement.

R: raw materials
M: Mixture
T: Thermal energy

Waste treatment and recycling smelter in Japan is shown in Figure 9. Waste metals, waste lead battery, shredder residue and electric furnace dust etc. are marked. Approximately 40 companies are operating. After recovery of metals, inorganic waste is used in cement industry. Materials to be utilized in cement industry are listed in Table 5. Municipal solid waste ash treatment in cement industry as one example is shown in Figure 10. Chlorine content in cement has to be reduced. Many kinds of inorganic nonmetal wastes are utilized to produce cement. Therefore, the cement factory is located all over Japan. The cement factory map in Japan is shown in Figure 11.

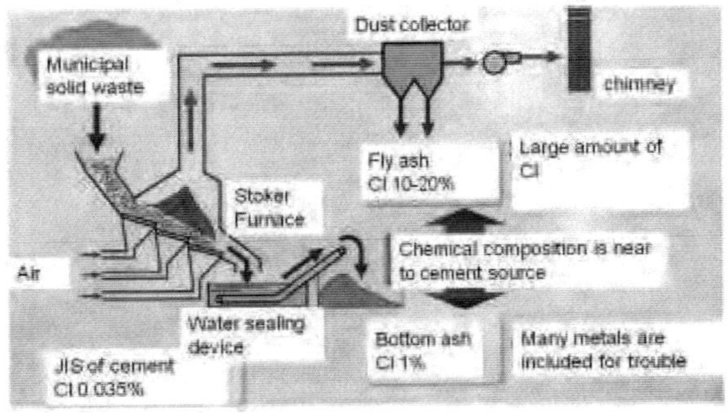

Figure 10. Municipal solid waste ash treatment in cement industry. (Taiheiyo Cement Co., 2016)

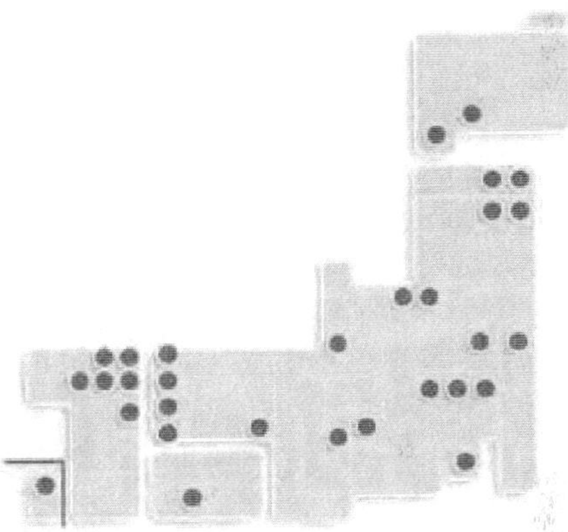

Figure 11. Cement factory map in Japan (Japan Cement Association, 2017)

5 SECONDARY RAW MATERIAL TREATMENT FACILITIES IN JAPAN

Japanese several groups of smelting companies to recycle metals are introduced in this session. Recycle network in JX Nippon Ming & Metals Co. group is shown in Figure 12. Seven places are connected for recycling. Hitachi metal-recycling company (HMC) collects various kinds of metals. Recycle network in DOWA Eco-system is shown in Figure 13. There are seven places to connect the recycling system. The incineration plant in Eco-system Akita treats many kinds of waste materials.

Recycle network in Mitsubishi Materials Co. is shown in Figure 14. The recycle system connects in about 10 places. Naoshima smelter recovers various metals. In Figure 15 the material flow in Mitsubishi Materials Co. is shown. The various base and rare metals and precious metals are recovered and produced. Recycle network in Mitsui Ming & Smelting group is shown in Figure 16.

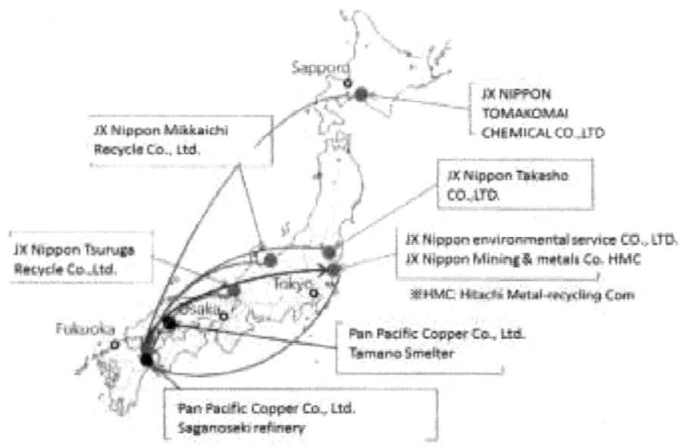

Figure 12. Recycle network in JX Nippon Ming & Metals Co. group. (JX Nippon Mining & Metals, 2017)

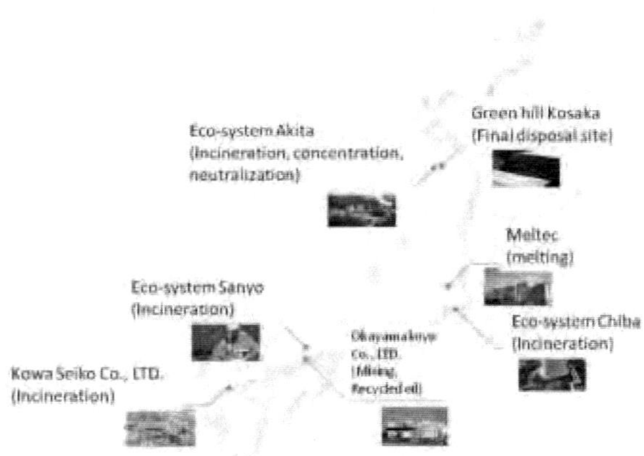

Figure 13. Recycle network in DOWA Eco-system (Dowa Eco-system, 2017)

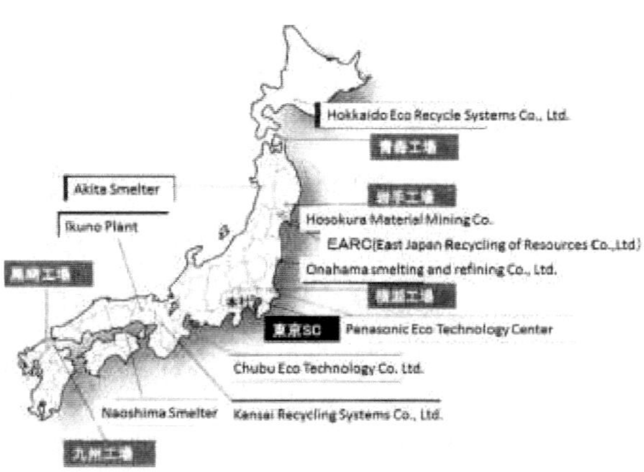

Figure 14. Recycle network in Mitsubishi Materials Co. (Mitsubishi Materials Co., 2017)

Seven places are connected for recycling. Especially Takehara refinery collected various metals. Waste PCB treatment plant in Mitsui Ming & Smelting is shown in Figure17. After crushing of PCBs, iron and aluminum are removed in Takehara refinery. The reminder is incinerated by kiln gasification and copper containing precious metals are recovered.

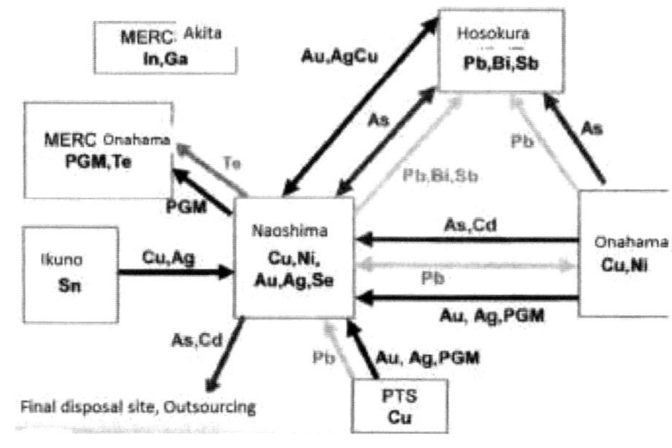

Figure 15. Material flow in Mitsubishi Materials Co. (Mitsubishi Materials Co., 2017)

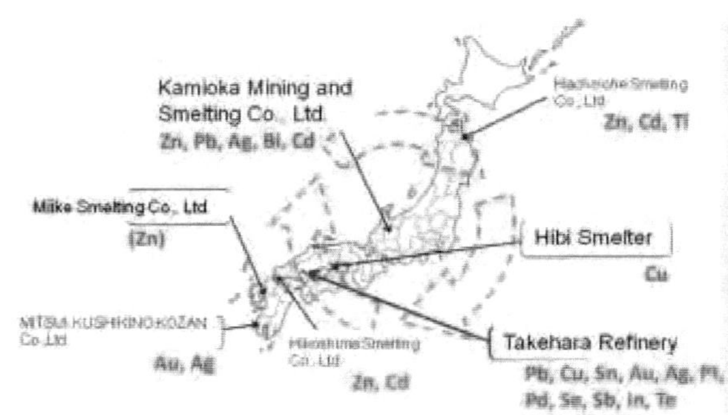

Figure 16. Recycle network in Mitsui Ming & Smelting Co., Ltd. group. (Mitsui Ming & Smelting, 2017)

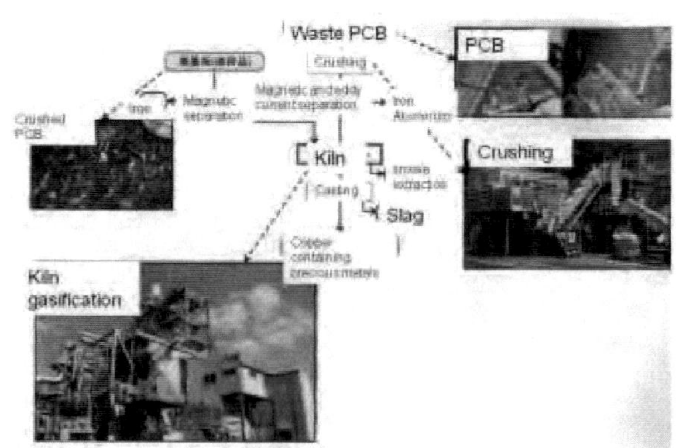

Figure 17. Waste PCB treatment plant in Mitsui Ming & Smelting Co., Ltd. (Mitsui Ming & Smelting, 2017)

Produced recycled metals in Taehara refinery of Mitsui Ming & Smelting plant is shown in Figure 18. The metals are processed in different size and sent.

Precious metals recovery process in Naoshima smelter of Mitsubishi Materials Co. is shown in Figure 19. De-copperized slimes go to flotation process and silver is recovered by cupellation and electro refining, while gold is collected with wet chlorination process.

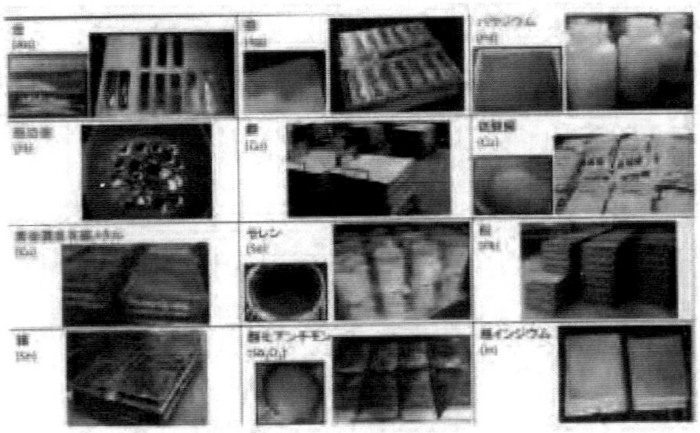

Figure 18. Produced recycled metals in Taehara refinery of Mitsui Ming & Smelting plant. (Mitsui Ming & Smelting, 2017)

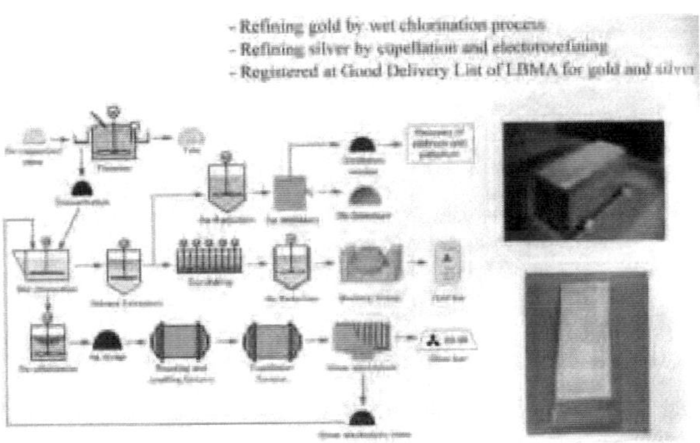

Figure 19. Precious metals recovery process in Naoshima smelter of Mitsubishi Materials Co. (Mitsubishi Materials Co., 2017)

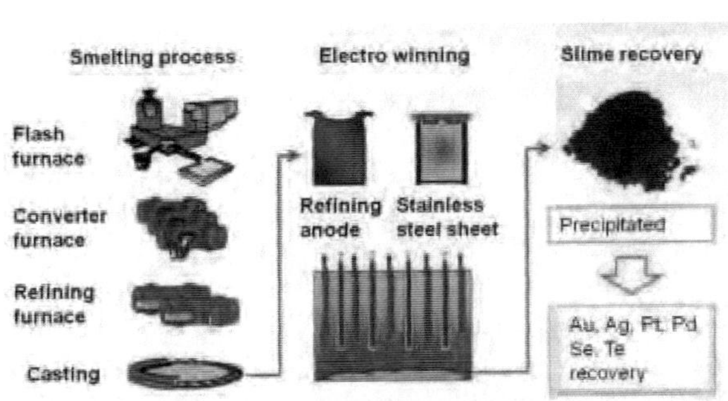

Figure 20. Copper and precious metals recovery process in Pan Pacific Company Co., Ltd. (Pan Pacific Company, 2017)

Copper and precious metals recovery process in Pan Pacific Company is shown in Figure 20. After the copper is recovered by electro winning, precious metals are collected from the slime.

6 SECONDARY RAW MATERIAL PROCESSING AND RECENT TECHNOLOGY

6.1 Crushing as pretreatment of recycle

For the separation to get the secondary raw material processing, the liberation is important, therefore, the large size crushing to liberate to the same materials lumps by preventing the over finer crushing. One of method is electrical crushing in water. The liquid crystal display (LCD) is sandwiched with ITO coated two glasses and films. If the high voltage pulse in more than 50kV is applied to LCD, large current flows in micro second period and the LCD panel can be separated two parts. (Dodbiba, 2012) In Figure 21, the LCD is set in the water and applied high voltage pulse. It is possible to separate 40-inch size of LCD. The crushed ITO coated glass is easy to dissolve indium ion by acid and recovered as hydroxide.

The other method is underwater crushing. The concept of crushing is shown in Figure 22. The explosive hanging in the water is exploded and shockwave and bubbles are periodically appeared. (Dodbiba, 2012) Setting of sample and photos of sample and crushed powder of hard tungsten scraps by underwater explosion and flowsheet to recover APT is shown in Figure 23. (Baik, 2004) Underwater crushing is convenient to crush hard metals and liberate different density composite materials, erase the information and crush the battery to prevent combustion. Metal recovery from lithium ion battery Flowsheet to recycle of lithium ion battery after explosion treatment is shown in Figure 24. After the sieving, magnetic separation, eddy current separation, air table and flotation are used to separate metals, plastics and powders. (Yamaji, 2011)

Figure 21. Electrical crushing apparatus, setting liquid crystal display (LCD) in water and crushed glass. (Dodbiba, 2012)

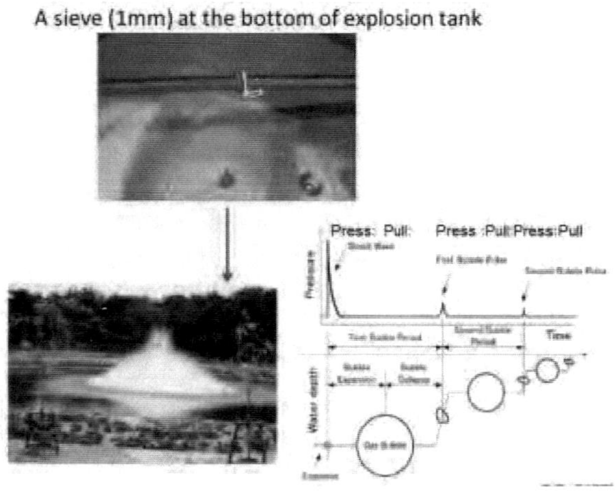

Figure 22. Underwater explosion pound to crush the materials and pressure and bubble size change in water (Dodbiba, 2012)

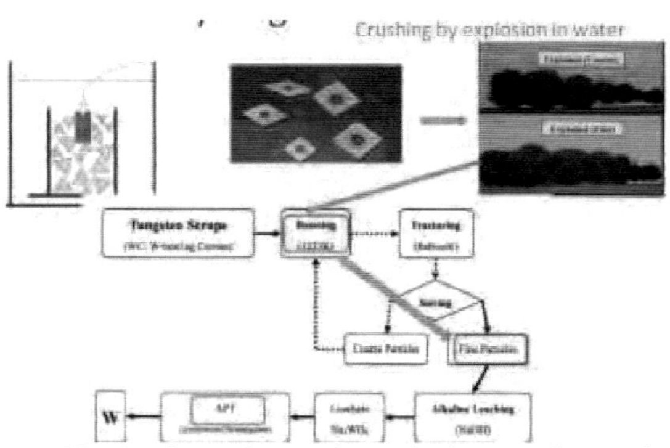

Figure 23. Setting of sample and photos of sample and crushed powder of hard tungsten scraps by underwater explosion and flowsheet to recover APT. (Baik, 2004)

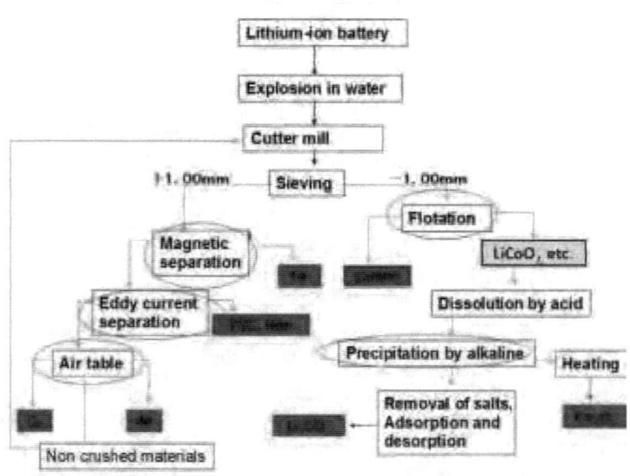

Figure 24. Flowsheet to recycle of lithium ion battery after explosion treatment (Yamaji, 2011)

6.2 Dismantling and sorting for recycle

Recent number of recycled home appliances in Japan is shown in Figure 25. CRT recycle number increased in 2010 because the terrestrial digital broadcasting system was adapted in Japan. Nowadays, washing machine and refrigerator represent the largest number of recycled home appliances. Robots are used to dismantle flat panel TV and after crushing home appliances the copper pipes are separated by using color sorter. Those photos are shown in Figure 26. Aluminum is used as many kinds of alloys, which contains number 1000's to 7000's. If many kinds of alloys are dissolved to recycle, large amount of aluminum dross is produced and discarded. LIBS sorter can fast analyze aluminum composition in alloy by laser and separate aluminum alloy fragments by air spray or puddle. The photos of LIBS and separation results is shown in Figur27.

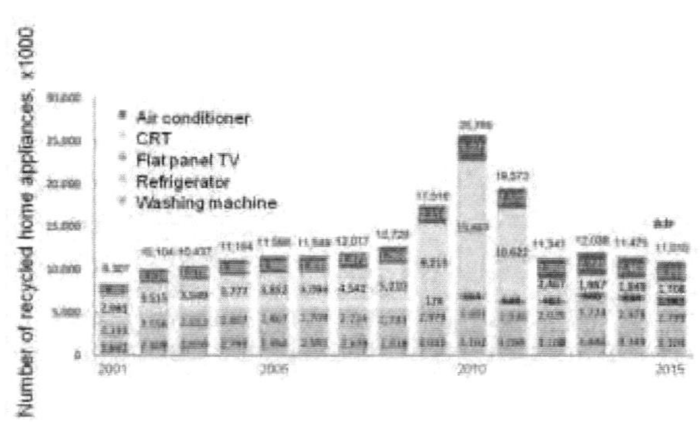

Figure 25. Recent number of recycled home appliances in Japan.
(Association for electric home appliances, 2017)

6.2 Some new separation technologies

Dry fluidized bed separation using fine spherical iron powder like 0.3 mm is shown in Figure 28. The air is blowing from the bottom of iron sand bed. The apparent density is about 3g/cm^3. More than several mm of aluminum solid float on the bed and can be separated from the other sank heavy metals like copper, brass and lead. The continuous dry fluidized bed equipment is also developed. (Nagata Engineering, 2017)

The separation using magnetic fluid is the technique more than 30 years ago. Recently the cheep water based magnetic fluid (about 10$/little) is developed and utilized in the separation of floated zinc alloy particles from sank heavy copper and lead particles in shredded automobile scraps. (Kyushu metal Industry, 2017) The photo of magnetic fluid sink and float separator is shown in Figure 29.

Figure 26. Photographs of TV decomposition by robot and Color sorter to separate copper (Mitsubishi Material Co., 2015)

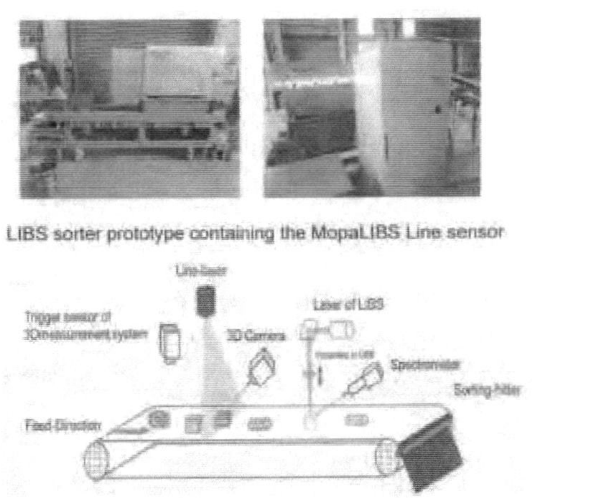

AluminumAlloy	Cast	Wrought Aluminum				
		2000's	3000's	5000's	6000's	7000's
Purity (wt%)	—	50.0	99.2	99.3	100	100
Recovery (wt%)	0	100	99.2	98.4	96.1	96.3

Al-Cu Al-Mn Al-Mg Al-Mg-Si Al-Zn-Mg

Figure 27. LIBS sorter photograph and explanation and aluminum alloy separation result by LIBS (Harita metal Co., Ltd., 2017)

Figure 28. Dry fluidized bed separation using fine iron powder. (Nagata Engineering, 2017)

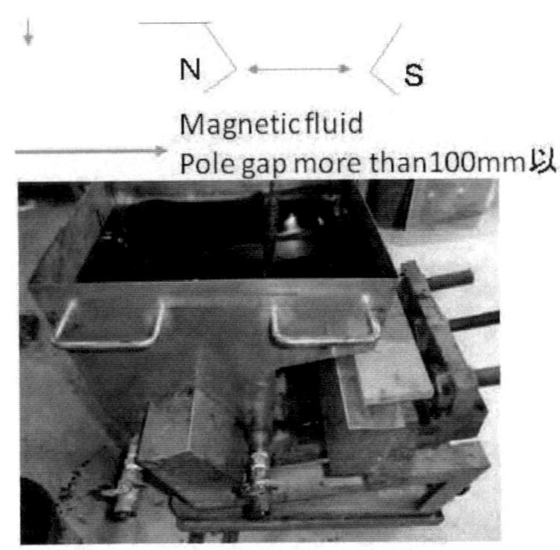

Figure 29. Sink and float separator using magnetic fluid (Kyushumetal Industry Co., Ltd, 2017)

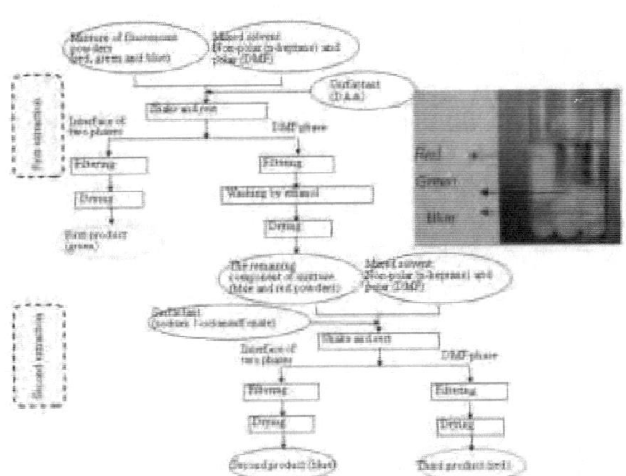

Figure 30. Flowsheet for separating three kinds of fluorescent powders by liquid-liquid powder extraction method. (Otsuki, 2006)

Fine particles mixture can be separated by liquid-liquid powder extraction. One liquid is hydrophobic light oil and the other liquid is heavy hydrophilic liquid like water. If the particles are coated by the surfactant and the particle surface becomes hydrophobic, they are moved to oil phase and dispersed in oil phase. While, non-coated hydrophilic particles are dispersed in water phase. Therefore, the different material particles are separated using liquid-liquid powder extraction by adding optimum surfactant. Flow sheet to separate three kinds of fluorescent powders by liquid-liquid powder extraction is shown in Figure 30. The red, green and blue fluorescent powders have been separated at more than 90% grade and recovery. (Otsuki, 2006)

In the recycling of PCB, the several separation techniques combination is useful to separate different materials after heat treatment like carbonization. The recycling flow sheet for printed circuit board by physical separation and heat treatment and photos of separated copper and silica by carbonization are shown in Figure 31.

The ion exchange resin is useful to separate mixed precious metal ions in nitric acid aqueous solution.

Pt, Ag, Pd and Cu recovery in nitric acid solution by using iminodi-acetic acid type chelate resin is shown in Figure 32. The adsorption and stripping processes separate Ag, Pt, Pd and Cu ions, respectively. (Tanaka Holdings, 2017)

The conventional solvent extraction uses mixer and settler, which need large operation area. The emulsion flow method is developed as shown in Figure 33. Upper and lower fine nozzles can produce fine emulsion that specific surface area is large, therefore, metal ion can be captured by solvent extractant in the high capacity and fast speed. (Yanase, 2011)

In the recycling, large amount of scrap collection is important. If the collected amount is small, the company has to stock the scrap till enough amount collection. The mobile type container for recycling of rare earth magnets is considered as shown in Figure 34. One container contains the crusher, leaching apparatus, emulsion flow solvent extraction cell, crystallization, filtration and drying equipment. (Kawabe, 2017) The Nd and Dy ions leached solution are separated by solvent extraction using PC88A or D2EHPA.

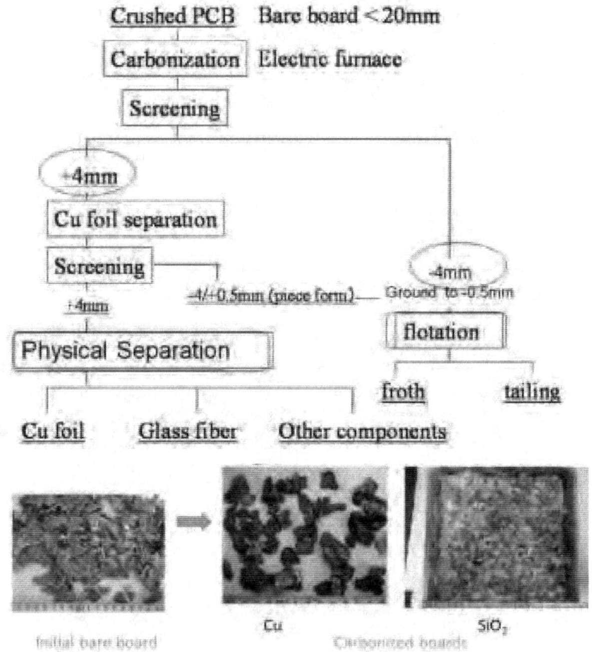

Figure 31. Recycling flowsheet or printed circuit board by physical separation and heat treatment and photos of separated copper and silica by carbonization. (Fujita, 2014)

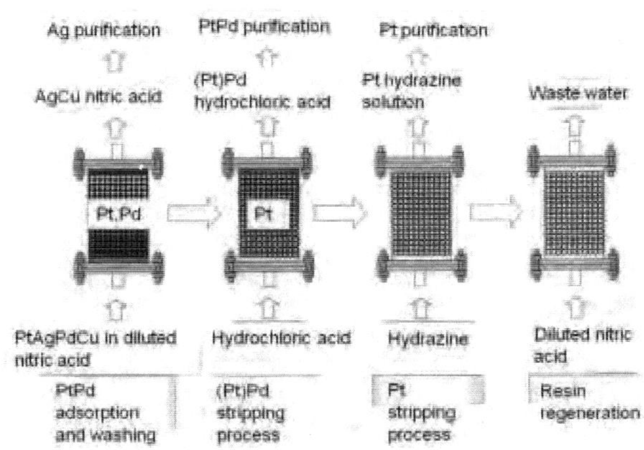

Figure 32. Pt, Ag, Pd and Cu recovery in nitric acid solution by using iminodi-acetic acid type chelate resin. (Tanaka Holdings, 2017)

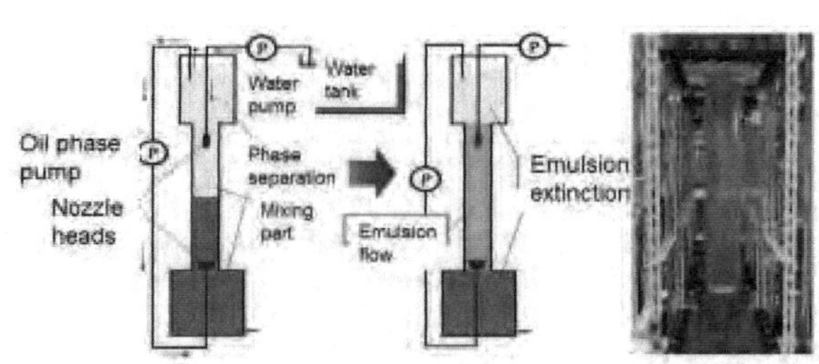

Figure 33. Emulsion flow for rapid solvent extraction method (Yanase, 2011)

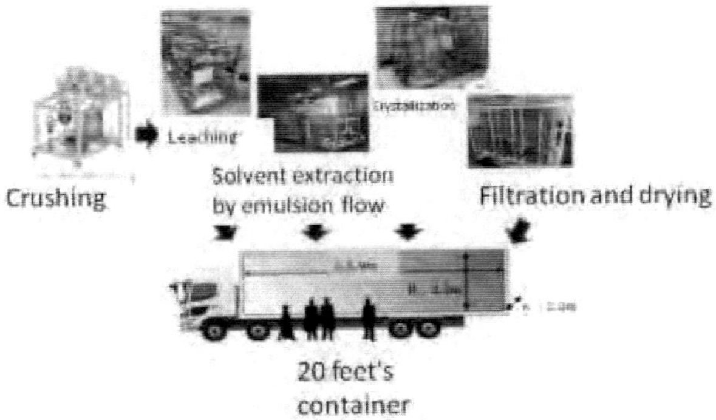

Figure 34. Mobile type container for recycling of rare earth magnets (Kawabe, 2015)

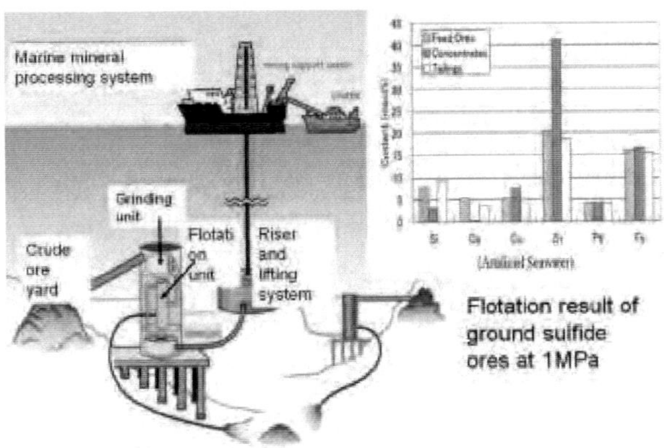

Figure 35. Development to collect seafloor and seabed resources. (Nakajima, 2012)

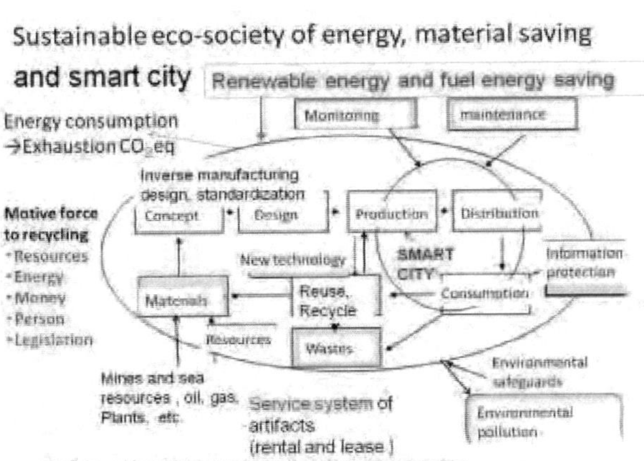

Figure 36. Sustainable eco-society considering circulation oriented society.

7 CONCLUSION

Recently the metal mineral grade is decreasing in the mines all over the world, therefore, it is important to proceed the secondary raw material processing as recycling. In Japan the intermediate treatment companies and smelting companies and cement industries are contributed for the recycling of metals and industrial minerals processing. Some new technologies are developed in the field of crushing, sorting, density separation, adsorption, liquid-liquid powder separation and solvent extraction, etc.. Also, as the future resources in Japan, the metal collection research from seafloor and seabed minerals is studied as shown in Figure 35. The system to construct environmental friendly material circulation society for sustainability is shown in Figure 36. The produced artifacts should be designed, produced, utilized and discarded by considering the conservation of the balances of resource, energy and environment. Though the life cycle only includes from the production to wastes, the total life-cycle contains the reuse and recycle of wastes, environmental burden estimation of artifacts and the cost and relationship between society and artifacts. The circulation society should be considered all over the world as listed in sustainable development goals by United Nations (United Nations, 2015).

References:

Agency for Natural Resources and Energy, 2017 http://www.enecho.meti.go.jp/en/

Association for electric home appliances, 2017 http://www.aeha.or.jp/recycling_report/03.html

Baik, S.W., A. Shibayama, K. Murata, T. Fujita, The Effect of Underwater Explosion on the Kinetics of Alkaline Leaching of Roasted Tungsten Carbide Scraps for Recycling, Int. J. Soc. Mater. Eng. Resour.,Vol. 12, No. 2, 55-59 ,2004

Dodbiba, G., K. Murata, K. Okaya, T. Fujita, Liberation of various types of composite materials by controlled underwater explosion, Minerals Engineering, Vol.89,63-70,2016

Dodbiba, G., H. Nagai, L. P. Wang, K. Okaya, T. Fujita, Leaching of indium from obsolete liquid crystal displays: Comparing grinding with electrical disintegration in context of LCA, Waste Management, Vol. 32, Issue 10, 1937-1944, 2012

DOWA ECO-SYSTEM Co.,Ltd., 2017 http://www.dowa-eco.co.jp/business/waste/

EARTH,2015 https://www.mtec.or.th/earth2015/index.php/component/content/article/9-uncategorised/1-introduction

Economic news, 2013 https://www.ysconsulting.com.tw/news/47310.html

Fujita, T., H. Ono, K. Yamaguchi, G. Dodbiba, Evaluation of a recycling process for printed circuit board by physical separation and heat treatment, Waste management 34, 1264-1273, 2014

Hishikari mine, 2017 http://www.smm.co.jp/E/corp_info/domestic/hishikari/kyoten.html

Harita metal Co., Ltd., 2017 http://www.harita.co.jp/technology/advanced-recycling/libs/

Japan Cement Association, 2017 http://www.jcassoc.or.jp/cement/2eng/index.html

JMIA(Japan mining industry Association),2016 http://www.kogyokyokai.gr.jp/category/1850702.html

JOGMEC, 2017 http://www.jogmec.go.jp/english/index.html

JX Nippon Mining & Metals, 2017 http://www.nmm.jx-group.co.jp/english/

Kawabe, K., Mobile recycle to reduce the capital investment, Eco materials Forum, 23-32, 2015

Kushiro coal mineCo.,Ltd.,2017 http://www.k-coal.co.jp/

Kyushu metal Industry Co., Ltd, 2017 http://www.nishitetsu.ne.jp/kyusyumetal/

Limestone association Japan, 2017 https://www.limestone.gr.jp/introduction/qa.htm

METI, 2016 from world metal statistics (WBMS) and world metal commodity summaries (USGS)

MERI/J, Automobiles and metal, No. II Metal market, No. 214, 2015

Ministry of Environment, 2015 https://www.env.go.jp/policy/hakusyo/h27/pdf/2_3.pdf

Mitsui Ming & Smelting Co., Ltd, 2017 https://www.mitsui-kinzoku.co.jp/en/

Mitsubishi Materials Co., 2017 http://www.mmc.co.jp/corporate/en/

Mitsubishi Material Co., 2015 http://www.mmc.co.jp/corporate/en/

Nakajima, Y., J. Yamamoto, S.Kanada, S. Masanobu, I. Takahashi, J. Sadaki, R. Abe, K. Okaya, S. Matsuo, T.Fujita, Study on seafloor mineral processing for mining of seafloor massive sulfides, Proc. of the ASME 2012, 31st Int. Conf. on Ocean, Offshore and Arctic Eng., OMAE2012-83354, 1-7 ,2012

Nagata Engineering Co., 2017 http://www.nagata-kit.co.jp/j12-003%20dry.pdf

Otsuki, A. et al.: Solid-solid separation of fluorescent powders by liquid-liquid extraction using aqueous and organic phases, Resources Processing, 53(2006),121-133.

Pan Pacific Company Co., Ltd., 2017 http://www.ppcu.co.jp/eng/index.html

Sakai S. Material cycle and waste management research,26, p.430, 2015

Taiheiyo Cement Co.,2007, 2016 http://www.taiheiyocement.co.jp/english/index.html

Tanaka Holdings Co., Ltd., 2017 http://pro.tanaka.co.jp/en/solution/main-product/product02/index.html

UNEP,2011 http://www.unep.org/resourcepanel/portals/24102/pdfs/metals_recycling_rates_110412-1.pdf

United Nations, 2015 http://www.un.org/sustainabledevelopment/sustainable-development-goals/

Waste age 100, 2015 http://www.waste360.com/special-report-waste360-top-100

Yanase, N., H. Naganawa et.al.:New apparatus for liquid-liquid extraction, “Emulsion flow extractor”, ANALYTICAL SCIENCES FEBRUARY, 27,171-176, 2011

Yamaji,Y., G. Dodbiba, S. Matsuo, K. Okaya, A. Shibayama, T. Fujita, A Novel Flow Sheet for Processing of Used Lithium-ion Batteries for Recycling, Resources Processing, Vol. 58, 9-13 ,2011

Can Geometallurgy Add Tangible Value During the Production Phase of Mining Projects?

JW Mann[1,2], M Becker[1,2]

1 Centre for Minerals Research, Department of Chemical Engineering, University of Cape Town, Rondebosch

R Schouwstra[2] , DJ Bradshaw[2]

2 Minerals to Metals Initiative, Department of Chemical Engineering, University of Cape Town, Rondebosch

ABSTRACT A significant body of fundamental geometallurgical research has been conducted in the last two decades. Historically geometallurgy has evolved in the technical sphere mainly through collaboration of geologists and metallurgists. It has been focused on improved sampling and measurement of predictive ore characteristic properties for integration into the resource block model, using information derived primarily from drill core data at the feasibility and prefeasibility stages of mining and processing. Its major value to date for the industry has been the improved ability to establish 3D spatial models, thus providing a more robust methodology to converting a mineral Resource to a mineral Reserve.

Currently however, it is the exception, rather than the norm, for geometallurgy to be actively applied to operational production activities, to manage ore body variation across the full sequence of activities in the mining value chain.

Under the prevailing global financial climate, with continuing volatile commodity prices and stunted world economic growth, mining companies have had to take austere measures to "get their balance sheets right". Sustainable management measures to adapt to uncontrollable changes and volatility in the commodity market are rapidly being exhausted.

Surviving and new operations are called to extract the maximum value from their existing mineral endowments, and deliver consistent, predictable metal production. Operational resilience - the ability of an organisation to prevent disruptions from occurring, as well as the ability to respond quickly to, and recover from, a disruption in the primary business processes, has become an essential operational management focus.

This paper advocates that the implementation of operational geometallurgy offers industry the potential to realise significant tangible value in delivering a higher conversion of their current Resource to Reserve and ultimately improving economic extraction of value.

1 INTRODUCTION

There is some early optimism of recovery in the mining industry, however, the current reality is that the medium term forecast for commodity prices and world economic growth are modest and reinvestment and growth in the sector is going to be challenging. With the severe project development cuts over the last 3 to 4 years, the lead time for new project development and their ramp up will require that surviving operations continue to extract the maximum value from their existing mineral endowments.

The focus of existing mining operations will remain on delivering consistent, predictable production and operational resilience. A real opportunity exists to achieve this using operational geometallurgy in order to limit the extent of unforeseen production responses from ore feed variability at the front end of the mining value chain.

A significant body of fundamental geometallurgical research has been conducted in the last two decades. However there has been limited adoption of this work in the form of a tool to manage ore body variation and systematically change operational planning and scheduling of the full set of day-to-day production activities, spanning in situ exploration, through mining and process activities to final dispatch of a product or products to the market.

The application of advance ore body knowledge in short term geological monitoring (ore control) and mine planning, and associated decision making is a critical frontier of geometallurgy – hard-won knowledge of ore body variability can be exploited to add near-term value (McKay *et al.*, 2016).

This paper reviews the impact of the recent history of the global commodity markets on the mining industry. It discusses the emergence of the enduring need for the concept of "operational resilience" and the potential of operational geometallurgy to limit unforeseen production responses from ore body variability at the feed end of the value chain.

2 REVIEW OF GLOBAL TRENDS IN THE MINING INDUSTRY

The PricewaterhouseCoopers (PwC) annual review of 2015 global trends in the mining industry (PwC, 2016), found that the Top 40 mining companies experienced their first ever collective net loss, their lowest return on capital employed, unprecedented capex containment and the tag team effect of prevailing debt levels plus impairments, sending leverage to new heights. The market capitalisation of the Top 40 was at the lowest level seen since 2004. The June 2016 market capitalisation of the Top 40 was only a third of its value from five years ago, and the combined market capitalisation of the traditional companies went below book value. All gains made during the commodity super-cycle were effectively wiped out.

Capital efficiency hit at an all-time low. In 2015, the adjusted return on capital employed (ROCE) halved from 8% to 4% year-on-year. This compares to the commonly cited 15% hurdle rate for new projects and the rising costs of capital. Without impairments included, the ROCE was negative. The significant drop off in capex will inevitably slow future growth. The industry received broad ratings downgrades from key agencies throughout 2015 and the beginning of 2016. Overall, finance costs were higher in 2015, despite the decrease in borrowings. Interest expenses as a percentage of total debt increased from 5% to 7%. The Top 40 appear to have worked smarter and implemented more productive methods to drive greater volume growth from existing plant and equipment at lower unit costs.

In the face of continued weakening commodity prices, to survive, Major traditional miners began 2016 attempting to manage debt and shedding non-core assets, mothballing marginal projects or curtailing capacity. There was a significant drop-off in capex, signalling an almost stagnant investment environment. At the time of the writing of this paper (March 2017), almost nine months have passed since the publishing of the PwC 2016 annual review of global trends in the mining industry.

The exploration sector remained depressed again in 2016 and, after a record high in 2012, exploration budgets have now fallen for four consecutive years. Exploration budgets for all three stages of project development (grassroots, late stage and minesite exploration) continued to retreat in 2016 from 2012's record highs (SNL Metals & Mining Research, CES - Minesite exploration hits record-high share of Budgets, 22 February 2017).

Not all discoveries turn into mines – historic conversion rates are only 60-80%, depending on the metal, size, quality and location. For the successful projects, there is a lag of 10-15 years between discovery and development. (Schodde. MinEx Consulting, November 2011).

The Scenario however, appears to be improving: According to S&P Global Market Intelligence (Worldwide Mining Exploration Trends, March 2017) the past year was a better one for mining. Nevertheless, concern remains. Commodity markets are delicately poised. The current reality is that the medium term forecast for commodity prices is modest. The International Monetary Fund Commodity price data projects almost flat line movement in most metal commodity prices over the near term i.e. to December 2018.

With the severe project development cuts over the last 3 to 4 years, the lead time for new project development and their ramp up will require that surviving operations continue to extract the maximum value from their existing mineral endowments.

It is clear that the mining industry is still particularly vulnerable. The forecast of prolonged flat-line of the near term metal commodity prices, the paucity of free cash, a lower appetite for expansion (from lenders and shareholders alike), the broad ratings downgrades from key agencies for the industry and the resultant higher finance costs will inevitably result in an extended hiatus in new project development and exploration expenditure.

In this paper we argue that sustainable management measures to adapt to the uncontrollable changes and volatility in the commodity market (variability at the tail end of the value chain) are rapidly being exhausted. The current industry commodities price downturn has decisively swung the emphasis from volume to value.

Management action should now also focus on the limiting the extent of unforeseen production responses from variability on the front end of the value chain. There is growing evidence verifying the economic value of advance knowledge about the variable distribution within an ore body of mineralogy, alteration and structural features and the response their changes cause to equipment and process performance.

Existing mining operations are called to extract the maximum value from their existing mineral endowments, and deliver consistent, predictable production. Operational resilience - the ability of an organization to prevent disruptions from occurring as well as the ability to respond quickly to, and recover from a disruption in the primary business processes, has become an essential operational management focus.

In a geometallurgical context, operational resilience translates explicitly into limiting the extent of unforeseen production responses from variability at the front end of the value chain. Ore variability is an inherent fact of nature. "Neither geologists, mining engineers, nor metallurgists can change the fact that all ore deposits are zoned with respect to metal grades, mineralogy, mineral abundances, mineral textures and other parameters such as physical hardness, all of which impact on the treatment costs and recovery of metals as an ore deposit is progressively mined and depleted." (Woodall, 2007 cited in Johnson & Munro, 2008). Ore feed variability can no longer be tolerated as a natural and unavoidable occurrence that impacts production.

3 OVERIEW OF GEOMETALLURGY

A significant body of fundamental geometallurgical research has been conducted in the last two decades. This was through the extensive effort by a host of research institutions and operating companies in the industry, by their own projects and via collaborative programs like AMIRA P843 & P843A.

Historically geometallurgy has evolved in the technical sphere mainly through collaboration of geologists and metallurgists, and been focused on improved sampling and measurement of predictive ore characteristic properties for integration into the resource block model. Most of these collaborative projects made use of information derived primarily from drill core data at the feasibility and prefeasibility stages of mining and processing. Its major value to date for the industry has been the improved ability to establish 3D spatial models, thus providing a more robust methodology to converting a mineral Resource to a mineral Reserve.

Geometallurgy, its tools and protocols, have to-date predominantly found application in the early phases of project development i.e. pre-feasibility, feasibility design and planning phases. (Bye, 2011; Liebezeit *et al.*, 2016)

The modelling of response variables rather than the simple use of traditional "modifying factors" now can provide a more competent statement of the economically mineable part of a measured resource as required by minimum standards, recommendations and guidelines for public reporting codes *e.g.* JORC, SAMREC codes.

Integration of the 3D spatial models with other independent models – geological, geotechnical, geohydrology *etc.*, and better understanding of process mineralogy, has also enabled improved mine and plant design. More recently increasing application is being found as a "Life of Mine" business evaluation and decision making tool (Macfarlane *et al.,* 2014; Butler *et al.,* 2016; Williams, S, 2017)

Currently however, it is the exception, rather than the norm, for geometallurgy to be actively applied to operational production activities, and manage ore body variation across the value chain i.e. spanning in situ exploration, through mining and process activities to final dispatch of a product or products to the market. Geometallurgy must be translated into a routine approach in mine planning and operations. (McKay *et al.,* 2016).

We argue that it is with the implementation of operational geometallurgy at this level that the industry will stand to realise significant tangible value – *i.e.* deliver a higher conversion of their current Resource to Reserve viz. materially improving ultimate economic extraction of more metal and value.

However, there are many reasons why the implementation of a multi-disciplinary approach to geometallurgy on existing sites has been low (Walters *et al.*, 2006; Alruiz *et al.*, 2009; Jackson *et al.*, 2011). These include the following:

- The existence of a strong Inter-discipline silo mentality in the mining industry.
- The lack of representation from the mining discipline during the evolution of geometallurgy as a concept.
- Little involvement by operational decision makers in the development of geometallurgical programs.
- Prevalence of discipline specific business and technical jargon, and the lack of a consistent approach in the communication of the context and purpose of both general and specific geometallurgical programs.
- The value of "geometallurgy" not clearly seen by individual disciplines, and the resultant difficulty justifying the time required and expense of acquiring relevant ore body characteristics and metallurgical response data.
- Traditional production metrics of the separate disciplines in the value chain present barriers to integrated implementation.

McKay *et al.*, (2016) posit that "this situation arises because of a communication failure in which technical specialists, including resource geologists, mining engineers, mining geologists and mineral processors can envision the value of geometallurgy. The actual delivery plan to achieve this improved value has not yet been crystallised."

This paper offers an approach to allow integrated site teams to synthesise their own site bespoke fit-for-purpose operational geometallurgical management program.

4 ESTABLISHING A SITE SPECIFIC GEOMETALLRGICAL MANAGEMENT PROGRAM.

Any operational program that is expected to deliver on potential improved value must be customised to cater for the unique set of conditions that prevail at the specific site. This includes the constraints imposed by the installed mining and processing infrastructure, the standard operational procedures and production measurement criteria. Importantly, the programme has to cater for their unique orebody and its variability with respect those key ore characteristics that elicit significant responses (both negative and positive) in the installed mining infrastructure and process circuit.

This will be best achieved using a facilitated collaborative approach, involving an integrated team of the relevant multidisciplinary stakeholders and specialists (technical specialists, resource geologists, mining engineers, mining geologists and mineral processors, experienced operators etc. from across the whole value chain). "The key is communicating the right data in the right format at the right time, and having an organisational culture to do so effectively." (Cropp *et al.*, 2014). The team will need clear visibility of the whole production system, access to information that is undisputed, with clear accountabilities for driving improved operational resilience through better, smarter, faster planning, scheduling and decision making. A facilitated approach is deemed necessary to take cognisance of the fact that individual participants will have varying levels of understanding and knowledge of upstream or downstream activities, including the impacts of production disturbance caused by variation of the feed to each particular section of the production chain.

5 ENSURING A COMMON APPROACH TO COMMUNICATION

Jackson *et al.* (2011) presented a multi-dimensional construct for communicating geometallurgy programs consisting of six dimensions. (Figure 1) which can further considered in terms of "operational geometallurgy".

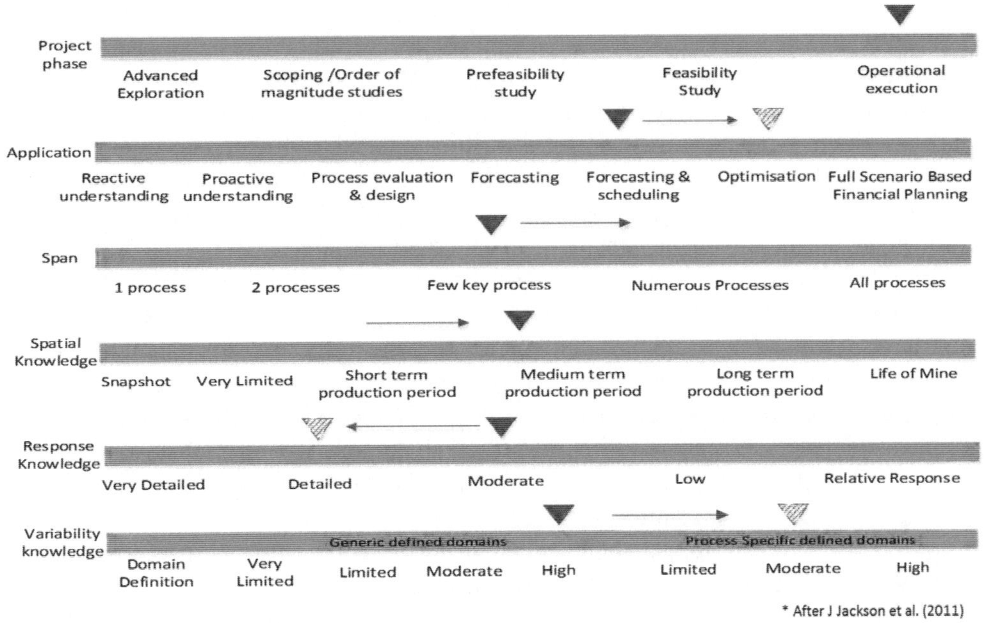

Figure 1: The dimensions of Operational Geometallurgy

The project phase dimension - the stage of the mineral project development life cycle at which the geometallurgical program is being applied. Operational geometallurgy is clearly in the operational execution phase of project development life cycle - it is focused on existing and operating mines, with signed off resource and reserve statements, with an established and approved mining plan, mining schedule and production targets.*The application dimension* relates to how the geometallurgical data is to be used, and ranges from reactive understanding (in hindsight) to full scenario based technical and financial assessment. The application dimension in operational geometallurgy is aimed at exerting management control on forecasting and scheduling, evolving over time towards optimisation.

The process span dimension refers to the number of processes under consideration within the geometallurgical program. Process here is defined in a general sense and can be a mechanical or chemical operation on the feed material, within any one of the value chain disciplines. Operational geometallurgy will initially only span the few key processes whose production performance is materially impacted by ore variability response, but could include more processes as the program matures.

The spatial knowledge dimension provides an indication of the volume of the mineral deposit under consideration, as a function of the production period. Operational geometallurgy is specifically limited to the volume of the mineral deposit that will be mined in the short to medium production period *i.e.* the planned ore feed.

The response knowledge dimension refers to the level of knowledge of the response to a process for the mineral deposit. In operational geometallurgy the degree of response knowledge of the ore feed to the process initially may only be moderate, limited to only those processes included in the process span. As the operational geometallurgy program matures it may become more detailed.

The variability knowledge dimension is the extent of variability of the response knowledge. This gives rise to domain definition. Domains could be either conventional generic defined domains based on traditional geological variables (lithology, grade *etc.*) or a process specific domain based on a specific process response variable or a proxy thereof. In operational geometallurgy, the extent of the variability knowledge is only within the volume of ore that comprises the feed to the operations in the short term, and will be specific to only those measurable ore characteristics or proxies that have material impact on the production performance on those processes included in the process span.

6 DEFINING THE AGREED GOALS OF AN OPERATIONAL GEOMETALLURGICAL MANAGEMENT PROGRAM

For an operational geometallurgical management program (with a short to medium term operational focus) to be cost effective and operationally implementable, it should include only sufficient ore body variability information that can inform proactive production planning and action. The information should be derived at the granularity of the site's ore control drilling programme and complimented by the site's other existing models e.g. geological, lithological, structural, geotechnical, geo-hydrological *etc*. The degree of data required should be sufficient to provide only specific advance knowledge of the variability in the ore body that will allow alternative options to be evaluated, planned and scheduled in order to optimally extract and deliver mining blocks to the correct destination.

We propose a bottom-up stepwise approach to analyse the site's own historical production data (of each of the value chain disciplines) to identify root causes for deviation from their targeted production performance. This will enable them to develop their own fit-for-purpose operational geometallurgical management program that will seek to maximise the value creation from stable processing of the ore being treated.

7 CAN GEOMETALLURGY ADD TANGIBLE VALUE DURING THE PRODUCTION PHASE OF MINING PROJECTS?

Geometallurgy enhances and leverages the spatial knowledge of the orebody into the value-based decision space of planning and operations. This is done at two levels. Strategic geometallurgy which includes the project stage-gate and investment processes and long-term operational mine planning systems. Tactical geometallurgy is the decision mechanism whereby mined blocks are delivered to their optimal destination to maximise value (McKay *et al.*, 2016).

This paper offers on alternative definition - an operational geometallurgical management program is a routine approach to planning and scheduling operations in the short to medium term. Its primary outputs are the optimal allocation of ore blocks to specific destinations, with a dynamic mine schedule that will result in efficient extraction and transport of the delineated blocks to the pre assigned destinations (stockpile - blending, waste dumps, or directly to the processing plant).This should be executed in a time frame and sequence that will allow operational control measures (upstream and / or within the processing plant) to ensure predictable and optimal plant performance and maximum value delivery.

We believe that a stepwise approach as outlined, could produce a pragmatic implementable fit-for-purpose geometallurgical management program with defined objectives and benefits at operational mine sites.

A structured approach to analysing historical production performance across the value chain as advocated, will at the very least:

- Identify the significant causes (due to ore variability or otherwise) of production disruptions, throughout the value chain, and identity actions to improve operational resilience.
- Create common understanding of the impacts of the identified disruptions, which can then be quantified, justifying the need (or not) for proactive management action to prevent their occurrence.
- Clearly identify which (if any) of the ore characteristics (when varying) indeed do cause responses that result in significant production or performance disruptions (whether in mining activities or processing activities), and
- whether there is any feasible proactive action (whether upstream or downstream) that can be taken, and
- whether there is any quantifiable benefit in doing so, and if so
- how, where and how frequently they need to be proactively monitored to prevent disruptions.
- Clearly define the need for (or not) and benefits of a geometallurgical management program.
- Contribute to breaking down the traditional inter-discipline silo mentality - participants will have better of understanding and knowledge of upstream or downstream activities, and the impacts of production disturbances, whether by ore variation or otherwise.

If a geometallurgical management program is then established, it will be site fit-for-purpose in that it will be sampling and acquiring data on only the relevant ore characteristics at appropriate locations along the value chain, at a frequency that will allow decision making within the operational time frame required. Such an approach should contribute tangible value during the production phase of mining projects.

8 CHALLENGES TO THE IMPLEMENTATION OF OPERATIONAL GEOMETALLURGICAL PROGRAMS

Historically geometallurgy has evolved in the technical sphere (head office technical specialists and consultants, and academia). The development of this scientific body of knowledge has been primarily through the collaborative effort of geologists and metallurgists.

This body of geometallurgical knowledge has predominantly found application at the level of strategic geometallurgy in the early phases of project development - providing an information base for developing an optimised reserve, production schedule and cash flows as a time series over the Life of Mine (LOM). It has typically used relevant data (bench-scale metallurgical tests, mineralogy, geochemistry, ore texture *etc.*) to predict processing performance and to model spatial ore variability and metallurgical response throughout the LOM. This has enabled better resource definition, more optimal mine planning and improved process design over the LOM. (McKay et al., 2016).

Conspicuous by its omission, is the inclusion of any consideration for optimal mine design and extraction methodology. Indeed, one of the fundamental flaws in the historic development of "geometallurgy" as a scientific body of knowledge, is the lack of any meaningful participation from the mining discipline.

As one progresses through the early phases of project development, towards the final phases of design and engineering, construction, commissioning and operation, the level of flexibility in mining and processes design rapidly diminishes to the stage where it is fixed with the installed infrastructure and equipment. To continue to deliver optimal value, the level of geometallurgy inevitably has to move from the realm of strategic geometallurgy to operational geometallurgy. Operational geometallurgy by definition is constrained to an operational time frame – looking ahead days to weeks to months. The level of flexibility no longer resides in available alternatives for mining or processing infrastructure, but rather in options of value assignment to ore blocks and the allocation of these blocks to pre assigned destinations. This assumes efficient extraction and transport of the delineated blocks to the pre assigned destinations, in a time frame and sequence that will allow operational control measures that ensure predictable and optimal plant performance and maximum value delivery.

Herein lies the challenge to the implementation of operational geometallurgy – the exclusion of the mining discipline during the advancement of geometallurgy over the last two decades has resulted in little or no concurrent development in mine design/ planning / scheduling or extraction activities that are commensurate with the developments in the geoscience and metallurgical disciplines. This omission did not significantly or detrimentally impact the value delivery of geometallurgical modelling in the strategic time frame - the current mining technology and capability is aligned with the more lenient long term time frame. Optional solutions to infrastructural constraints and known orebody variations or complications could be timeously evaluated and optimal solutions realistically planned and implemented.

In the realm of operational geometallurgy, called to deliver value in an operational time frame varying from days to a few months, this lack of flexibility does pose a significant challenge. With a granularity and density of information that is typical of operational sampling and performance measurement, current state-of-the-art geometallurgical tools, protocols and systems, are more than adequate for the geosciences discipline to do value assignment to ore blocks and allocate these blocks to pre assigned destinations. At the other end of the value chain, with sufficient advance knowledge, and the availability of the current state-of-the-art tools and systems to determine process response to ore variability, and with tested control action options, the metallurgical processing discipline are adequately capable of taking specific control action to ensure predictable and optimal plant performance and maximum value delivery. However, these two ends of the value chain are inextricably linked by mining activities.

We argue that the inability for the mining discipline to dynamically and rapidly re-plan and reschedule their production activities of drilling, blasting, extraction and transportation of delineated blocks of assigned value to pre-assigned destinations in an operational time frame, is the primary hurdle to implementation of operational geometallurgy.

These mining technology constraints are unlikely to be eliminated in the short term – however this should not necessarily exclude the implementation of a bespoke fit-for-use geometallurgical program which still could deliver real value through improved operational resilience.

Another barrier to the more general adoption and implementation of operational geometallurgy within the broader mining industry, is the fact that it is a relatively young scientific body of knowledge – Its development over the last decades has been through the sponsorship of only a handful of major mining companies. We believe that wider exposure and access to the current body of knowledge, available in the public domain, could deliver real value to existing mining operations who have not been able to afford the expense of developing their own strategic geometallurgical models.

We argue that there is only limited benefit in developing a strategic geometallurgical model for established small and medium sized mining operations, particularly if they do not have a long LOM. We do however believe that real value could be delivered to these operations through the implementation of a fit-for-purpose operational geometallurgical management program. The challenge for implementation here is to enhance their capability for implementation, assuming that they have had little or no exposure to geometallurgy.

We believe that the benefits of operational geometallurgy can be realised for these small and medium sized operations, if a generic decision making framework, and a toolbox of available geometallurgical tests, equipment, protocols, models *etc.* was readily accessible. The framework should allow a multidisciplinary site team to conduct the structured and facilitated analysis of their historic production data as we have advocated above, and guide them to identify production responses typical of their own orebody and installed mining and processing infrastructure, supported by a toolbox of available tests, protocols, technologies and technological solutions, with a list of potential service and technology providers who could cost effectively help them implement a fit-for-purpose program that will add value through improved operational resilience.

9 CONCLUSION

The mining industry is still vulnerable after an extended period of challenging global metal commodity market conditions and a forecast of prolonged flat-line near term metal commodity prices. The severe curtailment of capacity, exploration and project development will inevitably result in an extended lag time to bring in new production when the market does eventually improve.

To remain viable, mining operations are called to extract the maximum value from their existing mineral endowments and deliver consistent, predictable production. Operational resilience has become an essential survival strategy.

Limiting unforeseen production responses from variability in the ore feed by implementing site specific, fit-for-purpose operational geometallurgy can improve operational resilience and will add tangible value to existing operations.

A common and consistent approach to communication of geometallurgy (its context, purpose and benefits) and the establishment of shared value and benefit across all the disciplines in the entire mining value chain is a necessity for implementing an integrated operational geometallurgical management program.

Using a bottom-up stepwise approach to analyse the site's own historical production data could produce a pragmatic implementable fit-for-purpose geometallurgical management program with defined objectives and benefits.

The existing constraints of the mining discipline to dynamically and rapidly re-plan and reschedule their production activities of drilling, blasting, extraction and transportation of delineated blocks of assigned value to pre-assigned destinations in an operational time frame, is the primary hurdle to implementation of operational geometallurgy.

The benefits of operational geometallurgy can be realised for small and medium sized operations, if a generic decision making framework, and a toolbox of available geometallurgical tests, equipment, protocols, models *etc*. was developed and made readily accessible.

REFERENCES

Alruiz, O M, Morrell, S, Suazo, C J and Naranjo, A, (2009). A novel approach to the geometallurgical modelling of the Collahuasi grinding circuit, Minerals Engineering, 22:1060-1067.

Bye, A R, (2011). Case studies demonstrating value from geometallurgical initiatives. Proceedings of the 1st AUSIMM International Geometallurgy Conference, Brisbane, 9 – 30.

Butler, C, Dale, R, Robinson, S, Turner, A, (2016). Geometallurgy – Bridging the gap between mine and mill: a case study of the DeGrussa geometallurgy program. Proceedings of the 3rd AUSIMM International Geometallurgy Conference, Perth, 77 – 88.

Cropp, A, Goodall, W, Bradshaw, D, Hunt, J, berry, R, (2014). Communicating and integrating geometallurgical data along the mining value chain. 27th International Mineral Processing Congress, Santiago.

Jackson, J, McFarlane, A J, Olson Hoal, K, (2011). Geometallurgy – Back to the future: Scoping and communicating geomet programs. Proceedings of the 1st AUSIMM International Geometallurgy Conference, Brisbane, 125 – 131.

Johnson, N W, Munro P D, (2008). Methods for Assigning Domains in the Primary Sulfide Zone of a Sulfide Orebody. Ninth International Congress for Applied Mineralogy, Brisbane, 597 – 603.

Liebezeit, V, Ehrig, K, Robertson, A, Grant, D, Smith, M, Bruyn, H, (2016). Embedding Geometallurgy into mine planning practices – Practical examples at Olympic Dam. Proceedings of the 3rd AUSIMM International Geometallurgy Conference, Perth, 135 -143.

Macfarlane, A S, Williams, T P, (2014). Optimizing value on a copper mine by adopting a geometallurgical solution. The Journal of the Southern African Institute of Mining and Metallurgy, 114, 929 – 936.

McKay, N, Vann, J, Ware, W, Morley, C, Hodkiewicz, P (2016). Strategic and tactical geometallurgy – a systematic process to add and sustain resource value. Proceedings of the 3rd AUSIMM International Geometallurgy Conference, Perth, 29 – 36.

PwC, (2016). Mine 2016 – Slower, lower, weaker… but not defeated. Review of global trends in the mining industry.

Schodde, R C, (2011).Recent trends in Mineral Exploration – are we finding enough? Presentation to the RMG 8th Annual Exploration & Mining Investment Conference, Stockholm, November 2011.

S&P Global, (2017). Worldwide mining exploration trends. A special report from S&P Global Market Intelligence for the PDAC international convention.

SNL Metals & Mining Research, (2017). CES – Minesite exploration hits record-high share of budgets. Newsflash 22 February 2017.

Williams, S R, (2017). Reflections on the benefits and tasks ahead for Geometallurgy – from Metallurgist to Junior Miner at Pasinex Resources. Key note address, Process Mineralogy 2017, Cape Town, March 2017

Woodall, R, (2007). Project evaluation: Getting the geology right, The AusIMM Bulletin, 5:67-72.

Spatial Outlier Detection in Mineral Resource Estimation

A.Erhan Tercan
Hacettepe University, Department of Mining Engineering, Ankara, Turkey

ABSTRACT Mineral resource estimation is an important task for geoscientists and mining engineers. It requires detailed consideration of a number of critical issues. This paper addresses spatial outlier detection problem which is not sufficiently examined in mineral industry.

A number of spatial detection algorithms are reviewed and one novel method is suggested. The reviewed algorithms are based on estimating the given target point from a set of observation locations adjacent to it and computing residual between actual and estimated value. The suggested method uses pairwise relations between actual and nearby values. These algorithms are compared on a simulated values lognormally distributed. The results show the value of spatial detection algorithms in resource estimation.

1 INTRODUCTION

A Mineral Resource is an inventory of mineralization and natural material of intrinsic economic interest. The size, geometry, quantity, quality, geological characteristics, continuity and the prospect of economic extraction are key attributes of a mineral resource and resource estimation includes estimation of these attributes from geological, mining and other related data and information. Economic extraction is deliberately put into the list of attributes because a mineral resource is not only the inventory revealed by drill-holes and sampling but also economically extractable one under available economic and technical conditions, as rightfully stated in worldwide resource / reserve estimation reporting standards such as JORC (2012), CIM (2014), SME (2014), PERC (2013), CRIRSCO (2013). Mineral resource estimates are used by mining companies, financial institutions and investors.

Resource modeling process can roughly be divided into the four main stages: database construction, geological solid modeling, block modeling-grade estimation and assessment of economic extraction-resource model (Figure 1). It is an iterative process that is initiated by data collection followed by geological interpretation, block and resource modeling.

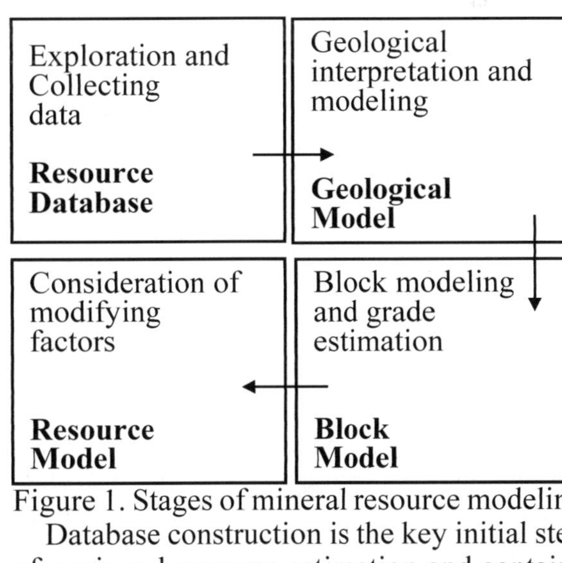

Figure 1. Stages of mineral resource modeling

Database construction is the key initial step of a mineral resource estimation and contains the collection and assessment of (1) primary data which consists of geological, geochemical, geophysical, topographic and assay data and (2) secondary data such as geological projections, block models and cutoff grades derived from the interpretation of primary data. Before its use the database should be debugged and verified.

Geological models are representations of key features of mineralization in a three dimensional computer environment. These models are developed by interpreting lithology, structure, alteration, mineralization and other relevant data on cross sections and plan maps, then extending the interpretations to three-dimensional environments. Once the geological model is complete, estimation domains which are stationary zones in the deposit are defined, based on geological model with support of statistical and geostatistical analysis (Rossi and Deutsch, 2014; Stephenson and Vann, 2001).

Block modeling and estimation of grades proceeds within these domains. Prior to grade estimation it is usual practice to convert domain models into block models which comprise a series of orthogonal cuboid blocks (Glacken and Snowden, 2001). The mean grade of each block is then estimated in some manner by the neighboring data. The final stage of grade estimation includes classification of the resources.

A mineral resource is not everything in the ground but only that part of mineralization for which there are reasonable prospects for economic extraction (Rendu, 2007). Therefore the prospects for economic extraction should be identified for mineral resource by considering the technical and economic factors such as metal prices, recoveries, cut-off grades and the proposed mining and processing methods. During resource estimation these factors are assessed on the basis of generally accepted industry practice and experience.

Resource estimation requires to consider a number of critical issues, including domaing, outliers, compositing, block size, search ellipse, estimation method, resource classification, cutoff grades and so on. In this paper, I address one specific area that I have chosen to call challenge. This challenge is spatial outliers. Outliers in general have strong effects on statistics such as the mean, variance, the variogram and also on block estimates. Therefore they should be identified and treated in a proper manner in a mineral resource estimation study. Depending on the their size of the neighborhood, outliers can be divided into two main categories: global outliers and spatial outliers. Mineral resource industry currently seems to give more attention to the global outliers. However a few high grade values (spatial outliers) in zones of predominantly low grades can lead to overestimation of block grades and affect overall tonnage and grade.

The algorithms proposed for spatial outlier detection in the literature are mainly due to data mining community. The term spatial outlier is first introduced by Shekhar et al. (2001) when studying graph-based outliers. The methods for detecting spatial outliers are used in many application domains including transporation, ecology, public health and climatology.

Hawkins and Cressie (1984) propose robust kriging to solve problems in resource estimation caused by the presence of high outlier values in positive-skewed distribution and Costa (2003) applies it to an exhaustively sampled data set and a gold deposit. Grade capping, i.e., lowering the highest grades in a drill hole data set to some level, is a common practice in mineral resource estimation. Rivoirard et al. (2012) proposed the top-cut model based on decomposing the grade of interest into three components (the truncated grade, a weighted indicator above the top-cut grade, and a zero-mean residual) and jointly estimate the truncated grade and the indicator by cokriging. They demonstrated the application of this top-cut model to blast-hole data from a gold deposit and also to a synthetic example. Nowak et al. (2013) review the capping methods used in the mineral industry and demonstrate the utility of these tools in a gold data set. Leuangthong and Nowak (2015) review a number of methods used to mitigate the impact of high grade data in three stages of resource estimation: domain determination, grade capping and estimation - implementation. All these studies are related to detection and treatment of global outliers. However, research for spatial outlier detection is rare.

Babakhani (2014) studies outliers to mitigate their effects on on histogram, variogram and block estimates and propose a point spatial detection algorithm based on the uniform rank transformation of data and its cross validation estimation. In attempting to detect multivariate spatial outliers Harris et al. (2014) suggest local adaptations of various global detection methods and evaluate them using simulated and real data

Spatial outliers may be more interesting than global outliers because they are likely to be less known and therefore more surprising. The present study attempts to review point spatial outlier detection algorithms recently used and proposes a new detection measure called pairwise relative residual function. These algorithms are then compared in a simulated environment. The multivariate and regional spatial outliers and also treatment of outliers are not considered.

The outline of the paper is as follows: the first section of the paper deals with describing global and spatial outliers in mineral resource estimation. The second section defines the spatial outlier detection problem with its basic constituents: the neighborhood, the neighborhood function and the comparison function. The comparison functions yield spatial outlier detection algortihms. The third section compares these algorithms on a set of simulated data. The final section includes conclusions.

2 GLOBAL AND SPATIAL OUTLIERS

Outliers are observations in a data set, which appear to be inconsistent with the remainder of the set (Barnett and Lewis, 1994) or which deviate from other observations in the same grouping (Hawkins, 1980). These definitions require priorly to group the data together and to measure inconsistency or deviation, which always include some kind of subjectivity as emphasized by several authors such as Costa (2003) and Rossi and Deutsch (2014).

To detect outliers, observations should be compared with each other. Based on the size of the neighborhood used in comparisons, outliers can be studied into two main categories: outliers in a local neighborhood (local or spatial outliers) and outliers in a global neighboorhod (traditional or global outliers). A spatial outlier is an observation that is significantly different from those of other observations in its spatial neighborhood. It is usually viewed as a local anomaly whose attribute value is extreme to its neighbors. In contrast to global outliers, spatial outliers do not necessarily deviate from the remainder of the whole data set (Chen et al., 2008). In other words, spatial outlier is defined in its close spatial neighbor while global outlier is an unusual observation among all the values in a set of data. Therefore descriptive statistics such as mean, variance and variogram may be more affected by global outliers. In both cases they should be identified using raw data rather than composites because compositing smooths the raw data. Figure 2 shows global and spatial outliers for grade measured along a vertical drill-hole.

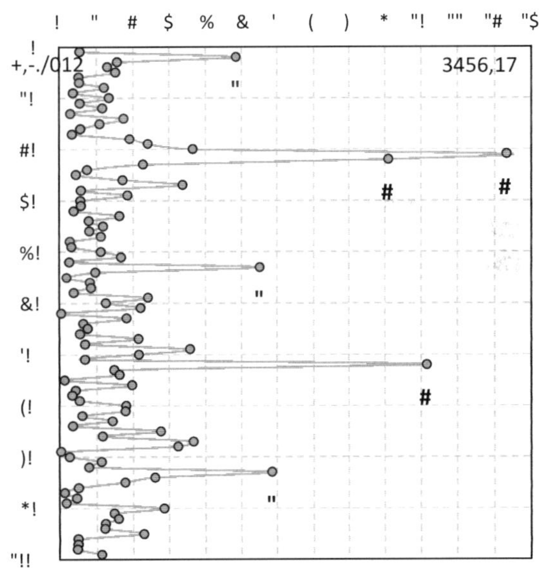

Figure 2. Global (G) and spatial (S) outliers for grade values measured along a vertical drill-hole

The simplest procedures used in identifying global outliers in resource estimation include mean ± two or three standard deviation and box-plots. More complex methods are probability plots, decile analysis, metal at risk, cutting curve plots, cutting statistics plots. Nowak et al. (2013) and Leuangthong and Nowak (2015) describe these techniques and compare them in an application to a West African gold deposit.

3 PROBLEM OF SPATIAL OUTLIER DETECTION

Consider a stationary random function Z(x) defined over point locations x within some region D and let the values $z(x_i)$ at locations x_i, i=1,...,K be a realization of the random function Z(x). For a given point x_j, define $N(x_j)$ the neighborhood of point x_j, $z^*(x_j)$ the neighborhood function summarising $z(x_j)$ values for all the points but except x_j inside $N(x_j)$ and finally $c(x_j)$ the comparison function that compares $z(x_j)$ with $z^*(x_j)$. Then mathematical definition of the spatial outlier detection problem is to obtain the value $z(x_j)$ for which $c(x_j)$ is an extreme value of the set $[c(x_1),...,c(x_K)]$. Note that definition depends on the choices of the neighborhood, the neighborhood function and the comparison function. Many choices are possible for each case.

3.1 The neighborhood

The neighborhood of a given target location x_j, $N(x_j)=(x_i, i=1,...,n; i \neq j$ and $n<<K)$, is a set of observation locations adjacent to x_j. Its definition requires to determine search parameters such as the shape, the size and the number of samples. The neighborhood can be defined by search radius as traditionallly applied in mineral resource estimation or the k nearest neighbours as considered in data mining applications. The latter, being the simplest choice, always guarantees the same number of samples for each x_j while the number of samples and size of the search neighborhood (radii of standard or rotated ellipse / ellipsoid) can vary from point to another, depending on the sampling density and the spatial continuity of grade. The size basically depends on the distance that a sample has effect on another and can be determined from range and nugget effect of grade variogram traditionally (see for example Cressie, 1993, p.134) or from both grade values and spatial coordinates of samples as suggested by Abedini et al. (2012) in which the search capability of a genetic algorithm is exploited. Another option would be to use point madogram functions which can better characterize the range for a target point. It is first suggested by Tutmez (2005) in an attempt to integrate fuzzy sets into resource estimation. By point madogram functions the size of the neighborhood for each x_j can differ from one location to another so that local variability can be handled in a more effective way but it takes too much time and effort.

3.2 The neighborhood functions

This is a function that estimates the attribute value at a target location x_j from measured values $[z(x_i), i=1,...,n; j \neq i]$ at locations in the neighborhood of the target location. The neighborhood function can take any form ranging from the simple arithmetic mean to more complex functions of nearby data. Some of the functions used in practice are listed in the following:

3.2.1 Mean

It is the simple arithmetic mean of the attribute values $[z(x_i), i=1,...,n; j \neq i]$. Every data point gets the same weight (1/n):

$$z_m^*(x_j) = \frac{1}{n}\sum_{\substack{i=1 \\ i \neq j}}^{n} z(x_i),\ (x_i, i = 1, \dots, n) \in N(x_j)$$

3.2.2 Ordinary kriging

The neighborhood function is defined as follows:

$$z_{kr}^*(x_j) = \sum_{\substack{i=1 \\ i \neq j}}^{n} \lambda_i\, z(x_i)$$

The unknown weights λ_i are obtained the solution of the following kriging system:

$$\sum_{k=1}^{n} \lambda_k \gamma(d_{ij}) + \mu = \gamma(d_{ji}),\ i = 1, \dots, n; i \neq j$$
$$\sum_{i=1}^{n} \lambda_i = 1$$

where $\gamma(d_{ij})$ is the variogram value corresponding to distance d_{ij} between locations x_i and x_j and μ is Lagrange multiplier.

3.2.3 Weighted median

It is based on employing a weighted median of λ_i found in ordinary kriging and the nearby attribute values $z(x_i)$:

$$z^*_{wm}(x_j) = weighted\ median\ [z(x_i), \lambda_i;\ i \neq j]$$

3.3 Comparison functions

Spatial outlier detection algorithms considered in this study work on a simple principle: compare the attribute value $z(x_j)$ at each target location x_j against a neighboorhod function $z^*(x_j)$ that summarizes the neighborhood attribute values $[z(x_i), i = 1, \ldots, n, i \neq j]$ (Su, 2011). Examples of comparison functions include residual [difference between $z(x_j)$ and $z^*(x_j)$] or ratio of $z(x_j)$ to $z^*(x_j)$. In practical applications the normalized residuals (division of raw residuals by standart error) are mostly used and the objective of spatial outlier algorithms is to produce the normalized residuals and to consider observations with the highest residuals as spatial outliers. This procedure is similar to cross validation exercise in which the parameters for a variogram model are fine-tuning. The normalized residuals can be calculated in many ways. Some of them are given below:

3.3.1 Mean (ME) residuals

$c_{ME}(x_j) = \frac{|e_m(x_j) - m_e|}{sd_e}, j = 1, \ldots, K$ where $e_m(x_j) = z(x_j) - z^*_m(x_j)$, $m_e = (\frac{1}{K}) \sum_{i=1}^{K} e_m(x_i)$ and $sd_e(x_j) = \sqrt{1/K \sum_{i=1}^{K} (e_m(x_i) - m_e)^2}$

This function is first suggested by Shekhar et al. (2003) to produce the residuals and is named spatial statistics z erroneously.

3.3.2 Local (LO) residuals (Kleinjan et al., 2002)

$c_{LO}(x_j) = \frac{|e_m(x_j)|}{sd_z(x_j)}, j = 1, \ldots, K$ where $e_m(x_j) = z(x_j) - z^*_m(x_j)$ with $z^*_m(x_j) = 1/K \sum_{i=1, i \neq j}^{n} z(x_i)$ and $sd_z(x_j) = \sqrt{1/K \sum_{i=1, i \neq j}^{n} (z(x_i) - z^*_m(x_i))^2}$.

Raw residuals are not normalized over all neighborhoods but for each neighborhood locally.

3.3.3 Regression (RE) residuals (Shekhar et al., 2003)

$c_{RE}(x_j) = \frac{|e_r(x_j) - m_r|}{sd_r}, j = 1, \ldots, K$

where $e_r(x_j) = z^*_m(x_j) - (a \times z(x_j) + b)$. Coefficients a and b are obtained from regressing $z(x_j)$ over $z^*_m(x_j)$. m_r and sd_r are the mean and standard deviation of $e_r(x_j)$ respectively.

3.3.4 Ordinary kriging (OK) residuals

$c_{OK}(x_j) = \frac{|e_{kr}(x_j) - m_{kr}|}{sd_{kr}}, j = 1, \ldots, K$ where $e_{kr}(x_j) = z(x_j) - z^*_{kr}(x_j)$, $m_{kr} = (\frac{1}{K}) \sum_{i=1}^{K} e_{kr}(x_i)$ and $sd_{kr}(x_j) = \sqrt{1/K \sum_{i=1}^{K} (e_{kr}(x_i) - m_{kr})^2}$.

3.3.5 Weighted median (WM) residuals (Cressie, 1993)

$c_{wm}(x_j) = \frac{|e_{wm}(x_j)|}{w\sigma_{kr}}, j = 1, \ldots, K$

where $e_{wm}(x_j) = z(x_j) - z^*_{wm}(x_j)$, $\sigma_{kr} = ordinary\ kriging\ standard\ deviation$ and w is a constant that controls the amount of downweighting the outlier values. Cressie (1993) suggests w values in the 1.5-2.5 interval.

3.3.6 Pairwise relative (PW) residuals

The variance in a set of positively skewed data is a function of the spatial location such that the data show greater variability in high valued areas. Therefore it is necessary to consider detection measures taking into account the local stability around a data point and downweighting outliers in highly unstable areas. The following measure is proposed for this purpose and is closely related to the pairwise relative variograms used in Geostatistics.

$$c_{pw}(x_j) = \frac{1}{n} \sum_{i=1}^{n} [\frac{2(z(x_j) - z(x_i))}{z(x_j) + z(x_i)}]^2, j = 1, \ldots, K$$

4 A SİMULATED EXPERIMENT

To evaluate the spatial outlier detection algorithms under study, a simulation experiment is designed. The experiment consists of (1) generating a spatially dependent and skewed data distribution, (2) contaminating the distribution by local outliers, (3) detecting the spatial outliers and (4) assessing performance of the algortihms.

100 data values are unconditionally simulated on a regular grid of 1 × 1 unit by using sequential Gaussian simulation method and these simulated values are then transformed into a lognormal distribution by simply taking their exponents. Lognormal distribution is chosen because it resembles the skewed data distributions frequently encountered in mineral resource estimation.

Since the simulated values are spatially dependent by design, they are free of spatial outliers and a certain percantage of them (for example 5%) should be contaminated. Global outliers are still possible however and values outside three standard deviation from the mean are identified as global outlier values. This procedure produced two global outliers (Figure 3b).

Contamination of a data set can be achived in many ways: one method is to swap data at locations with high simulated values with data at locations with low simulated values (Harris et al. 2014). Another one includes generating a certain number of error terms (5% of the population) and adding these values to randomly selected locations of the simulated field (Cressie, 1993 and Su, 2011). I chose the former one and a higher (95-98%) and lower (5-2%) intervals of the simulated values are used to determine the swapping locations. This ensured that data at the global outlier locations are not swapped and in this way 6 of 100 simulated values are contaminated. Figure 3 shows the bubble plots of the simulated values before and after contamination together with the histogram and the omnidirectional variogram of the pre-contaminated simulation values.

The size of the search neighborhood is specified by the variogram range (2 units) of the original simulated values to avoid the error incurred in estimating and modelling the variogram of the contaminated data. This choice leads to a varying number of samples in the neighborhood with minimum 5 and maximum 12.

The use of fixed n neighborhood and also varying neighborhoods derived from point madogram functions for each target point could be investigated more deeply. The algorithms are applied to the set of contaminated simulation values. The resulting residuals are put into decreasing number of order and the first 8 residuals are retained as outliers. A well performed algorithm captures the first two as global outliers and the remaing six residuals as spatial outliers. Table 1 shows the results. The first and second numbers in each cell present X and Y coordinates of the location. The underlined and bold numbers show the locations detected correctly by the algorithm. The last column presents the highest 9th residual. It is added to see some of the global or spatial outlier locations missed by the algorithms.

Table 1. Comparison of the results[+]

	GO[1]		HSO[2]			LSO[3]			9th
True	8,7*	8,8	3,4	3,8	9,3	1,1	7,6	7,2	NA
ME	**8,7**	**8,8**	**3,4**	**3,8**	**9,3**	2,1	3,10	8,1	1,1
LO	3,4	9,3	**3,8**	8,1	2,1	1,6	1,9	8,7	8,8
RE	**8,7**	**8,8**	**3,4**	**3,8**	**9,3**	9,4	10,4	**1,1**	7,6
OK	**8,7**	**8,8**	**3,4**	**3,8**	**9,3**	2,1	3,10	**1,1**	7,7
WM	**8,7**	**8,8**	**3,4**	**3,8**	**9,3**	2,1	9,8	3,10	8,1
PW	**1,1**	**3,4**	**7,6**	**3,8**	**9,3**	**7,2**	9,4	8,7	8,8

+) The underlined and bold numbers show the locations detected correctly by the algorithm,

*) The first and second numbers are X and Y coordinates of the location,

1) GO: Global outlier,

2) HSO: High spatial outlier,

3) LSO: Low spatial outlier,

**) The highest 9th residual

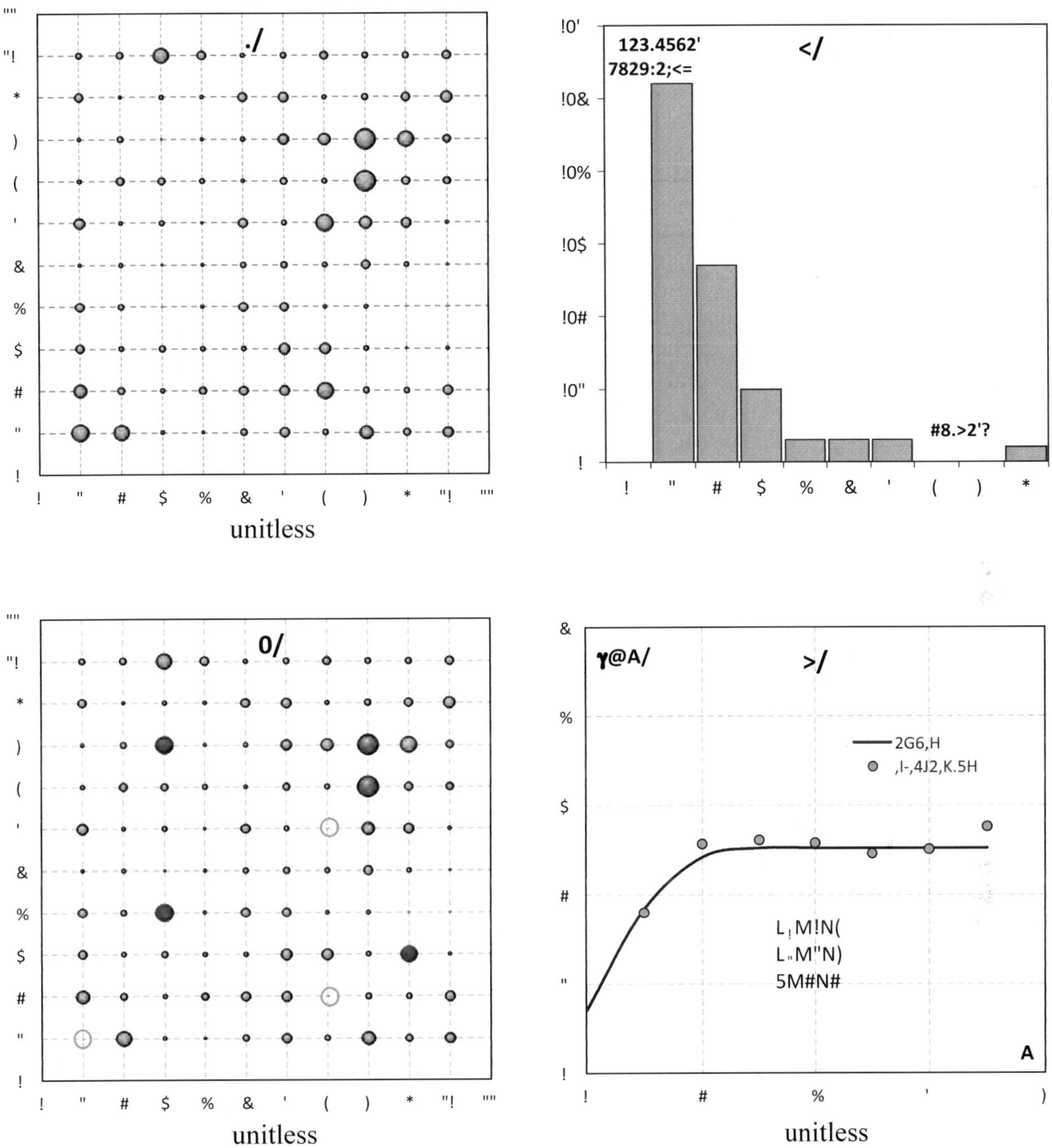

Figure 3. The bubble plots of the simulated values a) before and b) after contamination together with c) the histogram and d) variogram of the non-contaminated simulation values. In b), black balls show global outliers, red balls high spatial outliers, blue balls non-outliers and green circles low spatial outliers.

PW seems to be the best performing detection measure although global outliers are captured as LSO. All the algorithms except LO detect global and high spatial outliers correctly while these algorithms except OK and R fail to identify low spatial outliers. OK and RE detect correctly only one location as low spatial outlier. A poor performance of LO is due to reporting of outliers in highly unstable areas. Note that in the presence of a positively skewed data distribution high spatial outliers are more critical than low spatial outliers although few low grade samples in a zone of predominantly high values lead to underestimation of block grade.

CONCLUSIONS

In mineral resource estimation spatial outliers may be more critical than global outliers because they are likely to be less known and therefore more surprising. In this study I have considered five spatial detection algorithms and proposed the novel one. The algorithms under consideration are very simple to understand and can easily be implemented to resource estimation studies. The algorithms except PW use the normalized residuals derived from some function of the neighboring data.

A performance comparison of the algorithms on a simulated data set lognormally distributed shows that the shape of the data distribution is an important factor in choosing right algorithm. PW detects spatial outliers well since it supresses the effect of outliers in unstable areas of high variance while LO performs worse for not doing so. Most of the algorithms capture high outlier values while they fail to detect low outlier values. To see whether OK would capture low spatial outliers it would be interesting to use interpolation variance instead of kriging variance. Modelling the behaivour of outliers in different spatial scales especially in short and long range remains as a future work.

REFERENCES

Abedini, M.J., Nasseri, M., Burn, D.H., 2012, The use of a genetic algorithm-based search strategy in geostatistics: application to a set of anisotropic piezometric head data, *Computers & Geosciences*, 41, 136-146.

Babakhani, M., 2014, Geostatistical modeling in presence of extreme values, *M.Sc. Thesis*, University of Alberta, 85 pp.

Barnett, V., and Lewis, T., 1994, *Outliers in statistical data*, Wiley, New York.

Chen, D., Lu, C-T., Kou, Y., Chen, F., 2008, On detecting spatial outliers, *GeoInformatica*, 12, 455-475.

CIM, 2014, Canadian fefinition standards on mineral resources and reserves (CIM definition standards).

Costa, J.F., 2003, Reducing the impact of outliers in ore reserves estimation, *Mathematical Geology*, 33, 3, 323-345.

Cressie, N., 1993, Statistics for spatial data, Revised edition, Wiley, New York, 900 pp.

CRIRSCO, 2013, International reporting template for the public reporting of exploration results, mineral resources and mineral reserves, 41 pp.

Glacken, I.M. and Snowden, D.V., 2001, Mineral resource estimation, in *Mineral resource and ore reserve estimation- The AusIMM guide to good practice* (Ed: Edwards, A.C.), 189-198, The Australasian Institute of Mining and Metallurgy, Melbourne.

Harris, P., Brunsdon, C., Charlton, M., Juggins, S., Clarke, A., 2014, Multivariate spatial outlier detection using robust geographically weighted methods, *Mathematical Geosciences*, 46, 1-31.

Hawkins, D., 1980, *Identification of outliers*, Chapman and Hall.

Hawkins, D.M., and Cressie, N., 1984, Robust kriging-a proposal, *Mathematical Geology*, 16, 1, 3-18.

JORC, 2012. Australasian code for reporting of exploration results, mineral resources and ore reserves (The JORC Code).

Kleinjan, J., Chang, J., Wilson., Humburg, D., Carlson, G., Clay, D., and Long, D., 2002, *Cleaning yield data*, SDSU publication.

Nowak, M., Leuangthong, O., and Srivastava, R.M., 2013, Suggestions for good capping practices from historical literature, *Proceedings of the 23rd World Mining Congress 2013*, Montreal, Canada, 11-15 august 2013. Canadian Institute of Mining, Metallurgy and Petroleum, 10 pp.

Leuangthong, O., and Nowak, M., 2015, Dealing with high-grade data in resource estimation, *The Journal of The Southern African Institute of Mining and Metallurgy*, 115, 27-36.

PERC, 2013, Pan-European standard for reporting of exploration results, mineral resources and reserves (The PERC reporting standard).

Rendu, J-M., 2007, Orebody modelling, mine planning, reserve evaluation and the regulatory environment, Mineral resource estimation, in *Orebody Modelling and Strategic Mine Planning-*

Uncertainty and risk management Models (2nd Edition) (Ed: Dimitrakopoulos, R.), 219-226, The Australasian Institute of Mining and Metallurgy, Melbourne.

Rivoirard, J., Demange, C., Freulon, X., Lecureuil, A., and Bellot, N., 2012, A top-cut model for deposits with heavy-tailed grade distribution, *Mathematical Geosciences*, 45, 8, 967-982.

Rossi, M.E., and Deutsch, C.V., 2014, *Mineral Resource Estimation*, Springer, Dordrecht, 332 pp.

Shekhar, S., Lu, C-T., Zhang. P., 2001, Detecting graph-based spatial outlier: algorithms and applications (a summary of results), *Proc. Of the Seventh ACM-SIGKDD International Conference on Knowledge Discovery and Data Mining.*

Shekhar, S., Lu, C-T., and Zhang, P., 2003, A unified approach to detecting spatial outliers, *GeoInformatica*, 7, 2, 139-166.

SME, 2014, The SME guide for reporting exploration results, mineral resources, and mineral reserves (The 2014 SME guide), 65 pp.

Stephenson, R.P., and Vann, J., 2001, Common sense and good communication in mineral resource and ore reserve estimation, *Mineral resource and ore reserve estimation- The AusIMM guide to good practice* (Ed: Edwards, A.C.), 13-20, The Australasian Institute of Mining and Metallurgy, Melbourne.

Su, P.C., 2011, Statistical geocomputing: spatial outlier detection in precision agriculture, *M.Sc. Thesis*, The University of Waterloo, 141 pp.

Tutmez, B., 2005, Bulanık küme yaklaşımı ile rezerv kestirimi (in Turkish), *Ph.D. Thesis*, Hacettepe University.

MINING

ROCK MECHANICS / PRODUCTION / ENVIRONMENT / EDUCATION / LEGISLATION

Comparative Study of ANFO Versus Emulsion Use in Anisotropic Jointed Rock Mass

T. Chikande, T. Zvarivadza
University of the Witwatersrand, Johannesburg, SA

ABSTRACT Appropriate choice of an explosive is critical in blasting to come out with stable excavations at the same time meeting production requirements. Since inception, the studied mine has been using ANFO as the main blasting agent in primary blasting. As mining progresses, there is a general deterioration of ground conditions, hence there is need to find an explosive with energy to just break the rock into the required fragments and at the same time avoiding surrounding rockmass disturbances. This study compared and contrasted Emulsion and ANFO use in bad ground conditions. The following comparisons were made for both types of explosives: mining profiles after blasting, the effects of explosives on the rock mass, advance per blast, fragmentation, re-entry periods and charging time per end. The research approach entailed the following: extensive literature review, data collection, measurements and observations, trials for the selected alternative explosive were conducted and results collected were compared with its target, the mine's target and the mine's actual results using ANFO. Trials for bulk emulsion were carried out in two sections operating in bad ground conditions. The results gathered and analyzed showed that, technically, emulsion explosives are beneficial but the increase of operational cost down-weighs them. However, in solution to the problem which prompted this research, the authors suggest the mine to take up emulsion as it promotes safety at higher productivity in terms of tonnage output.

1 INTRODUCTION

Sound mining practices based on the appropriate choice of explosive energy lead to significantly safer mining operations. The selection of the exact charge mass, explosive type and blasting pattern is imperative in mining the desired slice, thereby reducing stoping overbreak and subsequently minimizing Platinum Group Elements (PGE) dilution. ANFO explosive is used in numerous mines due to its simplicity of use. The area of research is sited on the Great Dyke of Zimbabwe and is infested by faults and sympathetic joints. PGEs are the primary metals exploited and base metals are also recovered as secondary commodities. The Zimbabwean Great Dyke is the second prominent reserve of PGEs following the South African Bushveld complex (Oberthür et al, 2012). The widespread section of the Great Dyke is nearly bowl shaped and encompasses strata that are dipping towards the axis. The Dyke extends for about 550km having a maximum width of 11km (Prendergast, 1989). Zero harm and innovation being some of the core values in the mining industry, there was need to reflect on the suitability of the current generation of explosive to the ground conditions that the mine is now experiencing. The accident statistics gathered as from 2012 shows that the greatest share of accidents is coming from fall of ground incidents (FOG) as shown in Figure 1, thereby making it an area of concern.

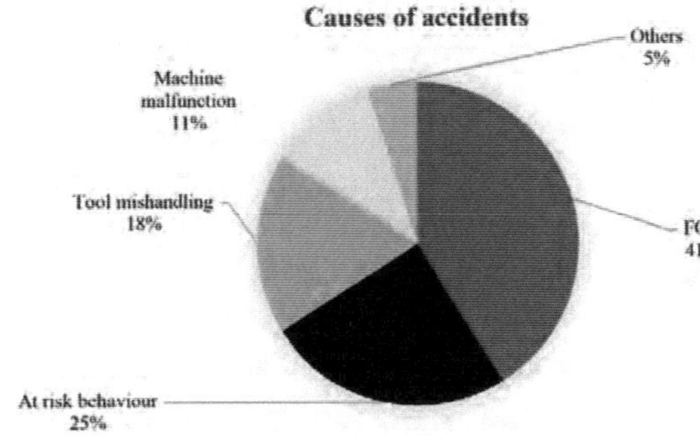

Figure 1. Summary of accidents by cause at the mine

1.1 Mining Operations Overview

The study was carried out at a shallow underground mine with its operations carried out less than 200m from the surface. The mining method and the cycle of drilling, blasting, lashing and supporting are all described in this section.

1.1.1 Mining method

The mining method utilized at the studied mine is room and pillar. Mining operations are carried out from the main decline advancing towards the strike direction. The main decline divides the mine into two regions, the northern part and the southern region. The size of the rooms mined varies with ground quality. Figure 2 shows the standard mining layout used in poor ground conditions. The stoping height is maintained at 2m to avoid PGMs dilution. In case of poor ground conditions, twin gullies which are 6m wide are mined leaving in situ pillars of 3m by 3m. Drives were developed from the main decline to the working areas.

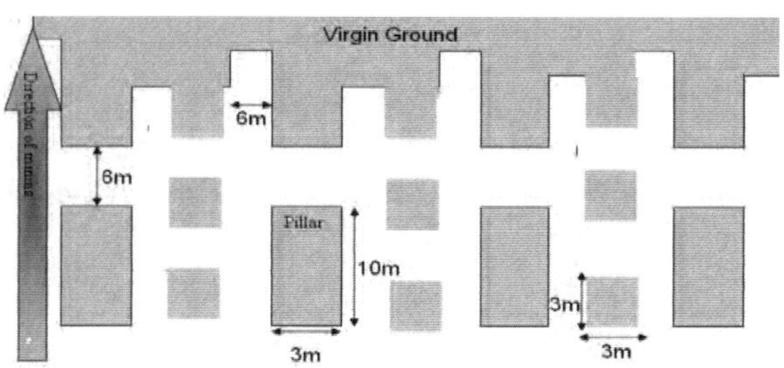

Figure 2. Mining layout for a 6m stope

1.1.2 Drilling

Drilling rigs are used to drill 51 holes with a diameter of 45mm. Three of these holes are enlarged to 102mm to give a second free face. A gulley is first marked after being cleaned up by LHDs and after all support installation has taken place. The length of holes drilled is 3.2m to give an advance of 2.8m to 3m. The average spacing is 0.5m while the burden is 0.67m. Figure 3 shows the drilling pattern followed at the mine. Drilling accuracy is critical to have a good advance and also for maintaining the designed stoping height. Timing is then done by connecting the shock tubes according to the pattern shown.

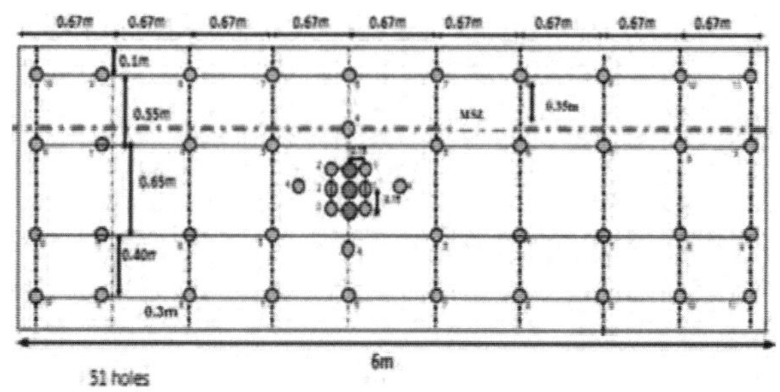

Figure 3. Drilling pattern

1.1.3 Charging and blasting

After drilling the holes, water is pumped out of the holes and the holes are cleaned up. Shock tubes with detonators at both ends called dual dets are used, inserted in megamite cartridges to form a primer. ANFO is then used to charge the drilled holes. ANFO is a high energy explosive and its effects will be described in the next sections. The authors looked at the explosives currently used at the mine. High gas explosives widen the joints leading to the unravelling of rocks which results in a decrease in safety and productivity. Figure 4 presents FOG trend at the mine for years 2005 to 2016.

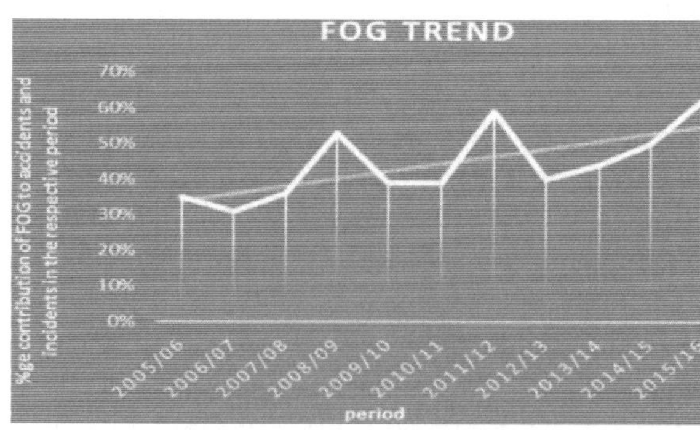

Figure 4. FOG trend at the mine (2005-2016)

Due to the rising FOG incidents (Figure 4), the rock mechanics department came up with number of interventions such as changing regional pillar dimensions, span reduction and installation of 4m spiling bolts after every blast in poor ground conditions to work in conjunction with resin bolts. In as much as an explosive is very essential in a mining set up to break the ground, it can also pose greater risk of worsening/damaging the rockmass thereby creating a hazardous working environment for employees. Due to augmented generation of bad hangings and the unpredictable unravelling of loose rocks, there was need to find an explosive with energy to just break the rock into the required fragments and at the same time avoiding surrounding rockmass disturbances.

1.2 Study Aim and Objectives

The study was undertaken with the main aim to improve the company Key Performance Indicators (KPIs) of safety and production. This was achieved through a comparative study of ANFO versus Emulsion use in anisotropic jointed rock mass at the mine. To achieve the main aim, the following objectives were set:

- To compare mining profiles after blasting (Emulsion vs ANFO)
- To Analyse the effects of explosives on the rock mass
- To measure and compare advance per blast for ANFO (currently in use) against emulsion.
- To compare fragmentation for ANFO relative to Emulsion
- To compare re-entry periods
- To compare charging time per end.

2 RELEVANT LITERATURE SURVEY

Safety, production and costs are the most significant parameters when evaluating the overall mining operation. ANFO is a supreme explosive used in the mining industry than most common bulk explosive because it is cheap, easy and safe to handle (Sellers, 2011). ANFO is unfortunately hygroscopic and despite a lot of efforts being made to add ingredients in order to increase the water resistance, this has only succeeded to a marginal degree.

2.1 Explosives properties

Various explosives have different magnitudes of shock energy as well as gas energy (Bohanek, et al., 2013). The merits of using ANFO include simple production, cheaper and its lack of sensitivity to mechanical impacts during mechanical loading to drill holes (Mather, 1997). ANFO has also some disadvantages which include lack of water resistance and low detonation parameters which reduce their range of use to dry blasting holes in truncated compactness rock masses (Maranda, 2011). In a bid to improve the mining practices, the authors reviewed the other bulk explosives which include bulk emulsion (UG100). The merits of bulk emulsion explosive over ANFO and packaged products include easy transportation, handling, string charging, low gas emissions, water resistant, full coupling, increased velocity of detonation, detonator sensitivity and improved work environment (Maranda, 2011).

A study was carried out to find out the effect of ANFO on the supporting systems used in anisotropic jointed rock mass. ANFO is well-known for generating a lot of gases therefore widening the cracks thereby affecting the half cast factor. The preliminary review of ANFO shows that it will result in a frequent number of keyblocks which will lead to an unstable hangingwall (Bohanek, et al., 2013). In addition, the cut slice will increase due to overbreak and more bad hangings will be formed which require intense barring down.

Bulk emulsion comprises of oxidizer solution droplets suspended in incessant fuel phase (Budin, 2009). Droplet sizes are orders of size smaller than ammonium nitrate prills and are soothed with emulsifiers. Budin (2009) pointed out that emulsion can be sensitized by Microballoons, blending with ANFO or gassing for underground application such as the studied mine.

2.2 Reflections on Explosive Characteristics in Jointed Areas

ANFO is often regarded as a substantially copious gas explosive than bulk emulsions and is thought to cause extensively longer cracks, thereby damaging the hangingwall further (Sellers, 2011). ANFO has a lower detonation pressure and a slower delivery of energy than other bulk explosives. The consequence of this will result in less expansion of the blast hole by the shock wave and leaving energy for driving crack growth and heaving of the fragments that have been created (Szendrei, et al., 2006). The extensive driving force that acts on the borehole widens the fractures. It is imperative to note that the densities of ANFO and bulk emulsion are different, hence they deliver different amounts of energy at different stages during their reaction process. The stability of an excavation is not only determined by the blast induced fractures, but also by the anisotropic jointed rock mass. The formation of argumentative hangingwall conditions in any given mining scenario can be changed to some degree by the correct choice of explosive type, drill holes diameter and round design. Minimum overbreak with good perimeter blasting can be achieved through smoothwall blasting (Lee, et al., 1993). Smoothwall blasting works more efficiently with bulk emulsion hence the need to compare the blasting results. The damage extent depends on the rock characterisation and the in situ geological settings. In smooth wall blasting, the final row of holes contains a lighter than normal charge and should be fired after the main charge is completed in order to limit the confinement of the holes and minimise damage back into the sidewalls. Hustrulid and Iverson (2010) pointed out that the accomplishment of smoothwall blasting pivots on sound design of blasting parameters.

2.3 Half Cast Factor

Half cast factor is defined as the ratio of the total visible drill barrel length in the sidewalls and hangingwall after blast and the total drilling length (Dey and Murthy, 2010). Half cast factor is vital in the determination of stoping overbreak. Singh (1992) pointed out that blasting can be described as a destructive process and the effects of blast damage are deleterious to both safety and productivity. The authors measured the length and number of barrels and determined the half cast factor using equation 1, as given by Mcknown (1984).

$$HCF = \frac{\sum_{i=1}^{n} L_i}{\sum_{r=1}^{n} L_r} \quad (1)$$

Where; HCF = Half cast factor
L_i = Post-blast drill mark length visible (m)
L_r = Pre-blast drilled length (m)

3 STUDY APPROACH

It is the aim of the study to optimize blasting at the studied mine for safety's sake while not neglecting other parameter changes accompanied. Described herein is the research criterion and techniques used in collecting necessary and sufficient data in order to fulfil this study's aim.

3.1 Research Criterion

- The research started off with an extensive literature review related to the study.
- Data collection, measurements and observations
- Trials for the selected alternative were conducted and results were collected and a comprehensive comparison was carried out against the mine's actual results using ANFO.
- Trials for bulk emulsion were carried out in North 1(N1) and North 7 (N7) for the following reasons:

-these are adjacent sections hence easier for the two to share the 1.5t main charging unit (MCU).
- All of them are operating in bad ground conditions.

3.2 Collection of Current Blast Output

Data collection techniques for the study included observations and data capturing pertaining to the blast output which lies in the hands of safety and production. Collected data gave the following:

i. Advance per blast
ii. Powder factor
iii. Fragmentation
iv. Half cast factor

3.3 Measurements and Calculations

3.3.1 Pre-blasting

- Hole diameter (mm) – was measured using a vienier calliper.
- Gulley length (m) – was measured using a distometer which measures to the nearest mm
- Mining width (m) – was measured using a distometer
- Holes depth (m) – was measured using a 5m tape.
- Burden (m) and spacing – were measured using a 5m tape.
- Advance (m) per blast – subtract distance from peg to face before blast from distance from peg to face after blast.
- Blasts per day- were captured on daily blast measurement form.
- Average No of holes/panel – physical counting on the face.
- Average meters drilled/day (m) – was captured on daily blast measurement form.
- Tonnes per blast – calculated from data captured on the blast measurement form.
- Tonnes per day – calculated from data captured on the blast measurement form.
- Half cast factor – calculated from data captured on the blast measurement form.

3.3.2 Blasting

- Stemming length (m) – was measured using a one meter clino rule.
- Polypipes/ round – physical counting of perimeter holes.
- Total primers mass (kg) per day – physical counting of primers used per end.
- Charged column length (m) – subtract stemming length from drilled length.
- Column charge (kg/m^3) – to be calculated from data captured on the blast measurement form.
- Volume of column charge (m^3) per hole.
- Mass of explosives – column charge/hole (kg) recorded on the MCU.
- Total mass of explosive/section – was calculated from the data captured on the blast measurement form.
- Total cost of explosives/gulley – was calculated from cost of explosives by the planning department.
- Explosives cost/t - (USD) was calculated from cost of explosives and the blasted tonnage.

4 RESULTS AND ANALYSIS

The results and analysis of the trial findings are given in this section. Table 1 shows a snapshot of the data collected from north 1 section during the month of May. Figure 5 shows the advance per blast trend during the month of April 2016. The average advance per blast during the first month of trial (thus April 2016) was 2.85m. This was lower than the targeted 3.0m mainly due to calibration fault on the MCU causing holes to be undercharged and also poor charging hose handling as employees were still acclimatizing with the new charging system.

Table 1. Sample of the collected data from one section

Date	Density	Distance before Blast	Distance after Blast	Advance	Height	Width	Tonnes Blasted	Mass Charged	Powder Factor	Number of Perimeter Holes	Expected sum of Barrel Length	Actual Barrel Length	Half Cast Factor (%)
5/23/2016	1.06	2.11	5.208	3.098	2.16	4.86	102.44	196.9	1.92	13	39	2.7	6.92
5/23/2016	1.06	24.015	27.033	3.018	2.02	6.03	115.80	231.2	2.00	13	39	4.896	12.55
5/23/2016	1.06	5.046	8.057	3.011	2.12	6.49	130.50	241.8	1.85	13	39	2.9	7.44
5/23/2016	1.06	10.695	13.762	3.067	2.12	5.019	102.80	200.2	1.95	14	42	6.1	14.52
5/24/2016	1.07	7.92	10.909	2.989	2.11	5.01	99.53	227.7	2.29	12	36	2.876	7.99
5/24/2016	1.07	2.043	5.321	3.278	2.4	5.1	126.39	237.2	1.88	12	36	2.65	7.36
5/24/2016	1.07	3.795	6.752	2.957	2.23	6.41	133.14	286.2	2.15	14	42	6.887	16.40
5/24/2016	1.07	5.267	8.089	2.822	2.11	5.98	112.16	210.5	1.88	11	33	2.548	7.72
5/25/2016	1.12	14.494	17.806	3.312	2.14	5.92	132.17	262	1.98	12	36	6.78	18.83
5/25/2016	1.12	2.962	6.346	3.384	2.033	4.947	107.21	246.1	2.30	12	36	2.467	6.85
5/25/2016	1.12	21.561	24.725	3.164	2.068	4.738	97.65	279.5	2.86	14	42	10.923	26.01
5/25/2016	1.12	6.787	9.838	3.051	2.149	6.631	136.95	196.3	1.43	11	33	6.623	20.07
5/25/2016	1.12	7.853	10.773	2.92	2.13	6.34	124.21	287.3	2.31	14	42	4.36	10.38
5/26/2016	1.12	26.509	29.709	3.2	2.157	6.289	136.74	192.5	1.41	10	30	5.37	17.90
5/26/2016	1.12	18.737	22.098	3.361	2.181	4.86	112.22	255.6	2.28	15	45	7.89	17.53
5/26/2016	1.09	6.752	9.832	3.08	2.116	6.03	123.79	213.4	1.72	14	42	6.21	14.79
5/27/2016	1.09	5.724	8.843	3.119	2.197	5.024	108.44	189.9	1.75	11	33	6.892	20.88
5/27/2016	1.09	26.43	29.546	3.116	2.102	5.564	114.80	272.6	2.37	14	42	5.41	12.88
5/27/2016	1.09	3.833	7.075	3.242	1.871	4.192	80.10	229	2.86	12	36	6.27	17.42
5/27/2016	1.09	3.552	6.892	3.34	2.083	5.705	125.03	270.3	2.16	14	42	4.513	10.75
5/27/2016	1.09	16.328	19.508	3.18	2.235	5.187	116.13	209.5	1.80	13	39	4.915	12.60
5/27/2016	1.09	8.046	11.275	3.229	2.077	5.23	110.49	205.7	1.86	11	33	7.81	23.67

4.1 April 2016 Results and Analysis

4.1.1Advance per blast

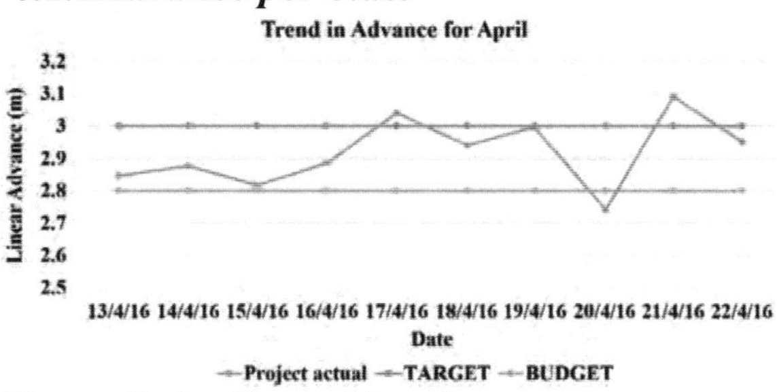

Figure 5. April trend in advance

4.1.2. April half cast factor

Half cast factor is a measure of the blast induced over break. The half cast factors were determined using Equation 1. Figure 6 shows the half cast factor trend for the month of April. From the graph, it can be noted that the average half cast factor for the month was at 23.5%, which was below the target of 50%. Overbreak is predominantly affected by the properties of the host rock, blast design and explosive parameters (Dey and Murthy, 2010). Rock mass characterisation played a crucial part in lowering the half cast factors since operations were carried in geotechnically poor grounds.

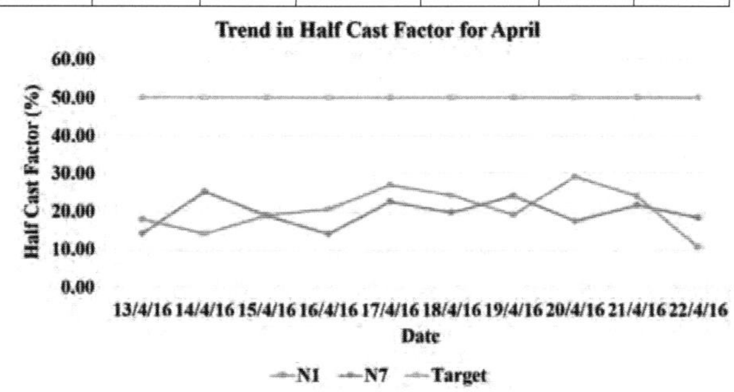

Figure 6. April trend in half cast factor

4.1.3 April powder factor

Powder factor refers to the amount of explosives used to break one tonne of rock. Figure 7 shows the trend in powder factor for the 2 sections in April 2016. The powder factor averaged 1.704kg/t against a target of 1.6kg/t. The powder factor was above the desired as a result of the aforementioned minimum advance.

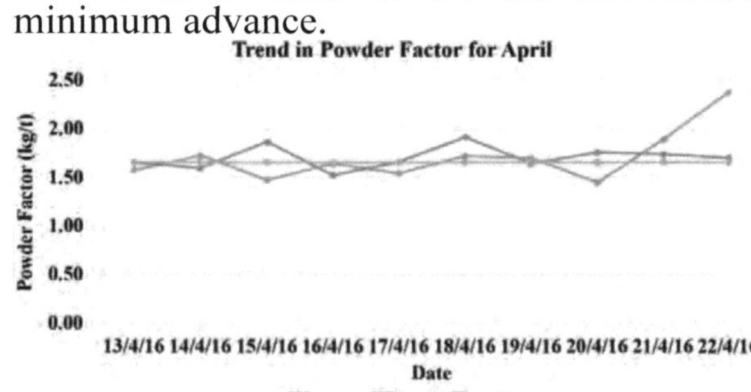

Figure 7. April trend in powder factor

4.2 May 2016 Results and Analysis

4.2.1 Advance per blast

The linear advance increased during the second month of the trial to 2.95m against a budget (minimum linear advance planned for) of 2.8m. Figure 8 shows the advance per blast trend during the month of May.

Figure 8. May trend in linear advance

4.2.2 May 2016 half cast factor

Figure 9 shows the trend in half cast factors during the month of May. The half cast factor figures were again below the targeted 50% as a result of the prevailing geological discontinuities.

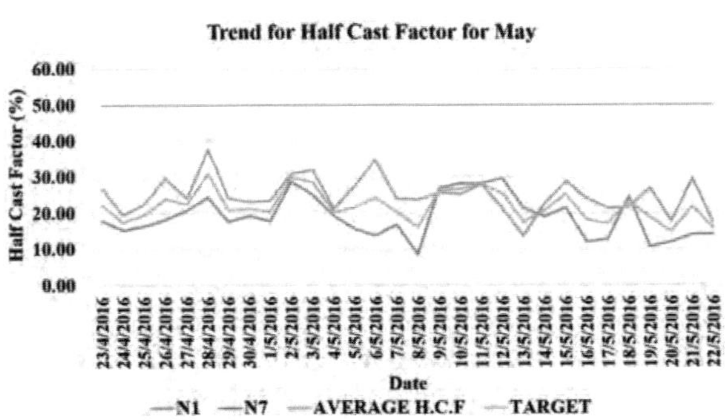

Figure 9. May trend in half cast factor

4.2.3 May 2016 powder factor

Figure 10 shows the trend in powder factor during the month of May against a target of 1.6kg/t. The results increased as a result of the improved advances.

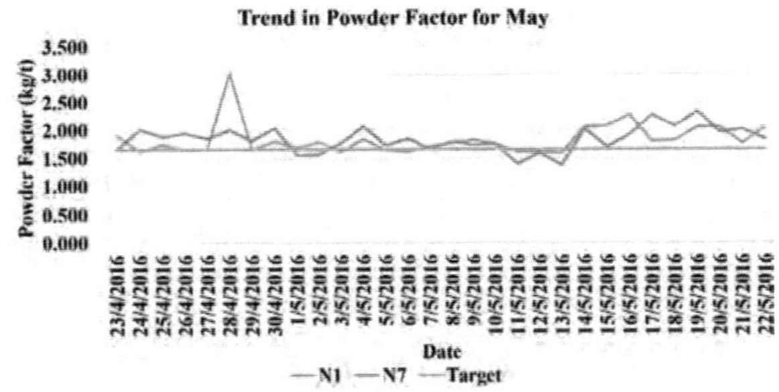

Figure 10. May Trend in powder factor

4.3 June 2016 Results and Analysis

4.3.1 Advance per blast

The linear advance increased during the month of June to 3.01m against a budget of 2.8m. Figure 11 shows the advance per blast trend during the month of June 2016.

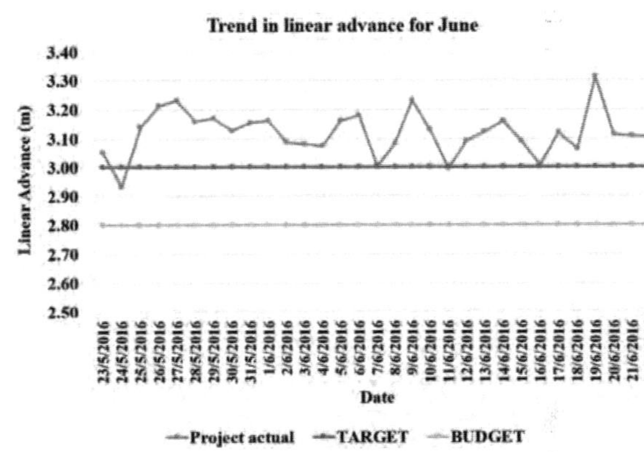

Figure 11. June trend in linear advance

4.3.2 June 2016 half cast factor

Figure 12 shows the trend in half cast factors during the month of June 2016. The half cast factor figures were again below the targeted 50% because smoothwall blasting was not yet implemented.

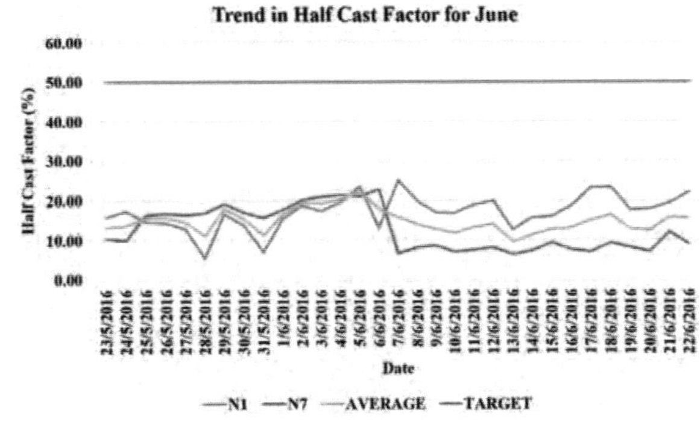

Figure 12. June trend in half cast factor

4.4 Summary of Results

The trial continued to the month of August and significant changes were noted. Average linear advance for the whole trial was 3.07m which was above the targeted advance of 3.0m. The achieved advance was found to be 95.9% of the drilled length of 3.2m. The average powder factor for the whole trial was 1.83kg/t, which is above the target of 1.66kg/t. The difference is attributed to the subsequent decrease in the span as a result of geotechnically poor ground conditions. The targeted half cast factor was not achieved due to bad grounds and the lag in purchasing polypipes for smoothwall blasting. The average half cast factor for the whole trial was 20%.

4.4.1 Blasting profile

Minimum overbreak into the hanging wall, footwall, and sidewall were observed during emulsion charging compared to ANFO loading. Visible barrels were evidence of reduced blast over break. Profiles that were not smooth were observed in extremely poor grounds, thus regions with a tunnelling index less than 0.4. There was reduced throw from emulsion explosives compared to ANFO which actually lessened LHD lashing and scraping time.

4.4.2 Fragmentation

Bulk emulsion proved that 97% of the blasted ore can pass through the 400mm×400mm grizzly apertures without difficulties compared to 94% from ANFO charging.

4.5 Study findings

The study done was to a greater extent conclusive to the following findings:

1. *Advance* improved from an average of 2.80m using ANFO to an average of 3.0m using emulsion explosive. This increased the blasted tonnage from 105t to 113t per 6m end.
2. The *average time taken to charge* an end also decreased from an average of 23 minutes using ANFO to an average of 13 minutes, saving an average of 30 minutes within the mining cycle per team, since 3 (6m gulleys) ends are charged per shift.
3. The increase in explosive quantity utilized per end increased the powder factor to 1.83kg/t despite the increase in advances from 1.66kg/t.
4. The *scaling time* of bad hangings dropped remarkably. The time decreased from an average of 40 minutes to 11 minutes. This was accounted due to a reduction in the generation of bad hangings brought about by the use of emulsion.
5. The *half cast factor* had an average of 20% which was below target. Observations made show that the board profiles were fairly smooth as compared to those produced with ANFO. Table 2 compares emulsion explosives with ANFO. Table 3 presents cost per blasted tonnage for a 6m gulley.

4.6 Comparison of explosives

Table 2. Emulsion vs ANFO

Parameter	**Emulsion**	**ANFO**
Operating cost per blasted tonnage	High	Low
Capital cost	High	Low-ANFO loader is cheaper
Advance	Average 3.0m	Average 2.8m
Powder factor	High	Low
Productivity	Improved productivity	Decreased productivity
Safety	Improved as a result of less bad hangings	Generates a lot of bad hangings
Charging time	Less charging time 13mins per end	More time averages 23mins per end
Water resistance	Excellent water resistance	Dissloves easily
Overbreak	Less overbreak hence minimum PGM dilution	More overbreak resulting in PGM dilution
Throw	Reduced throw hence less LHD loading time	Greater throw hence more scrapping time

Table 3. Cost per blasted tonnage for a 6m gulley

Consumable	Units required	Unit cost ($)	Total cost of units used ($)
ANFO (25kg bag)	7	19.98	139.86
EZ stopper (4.2m)	30	1.43	42.9
Dual Dets (8.4m)	9	2.08	18.72
Megamite cartridges (200mm x38mm)	48	0.45	21.6
Safe start	1	4.19	4.19
Trunkline	1	1.7	1.7
Twisted cable	1	7.32	7.32
Emulsion	208	0.89	185.12
ANFO Total cost (6m round)			**236.29**

The expected tonnage from a 6m gulley after charging with ANFO is 105.84 tonnes; therefore the explosive cost per blasted tonnage is $2.23 per tonne. The expected tonnage from a 6m gulley after charging with emulsion is 113.4 tonnes, which corresponds to an explosive cost per blasted tonnage of $2.48 per tonne. The study was aimed at improving safety through reduction of FOG influenced accidents. The implementation of bulk emulsion offers additional improvements in some areas that the mine has been facing such as minimum advance, poor fragmentation and high powder factor. Emulsion explosives pose high operating costs and capital cost as compared to ANFO, however, the performance results attained through this trial show that the use of emulsion as the main column charge is a worthy sacrifice which will yield benefits over a certain period of time. The high OPEX and CAPEX cannot be compared to the savings that the mine would have realized in achieving the goal of zero harm through avoiding or minimizing FOG related accidents.

5 CONCLUSIONS

The results gathered and analyzed showed that, technically, emulsion explosives are beneficial but high operational and capital costs down-weigh them. The authors recommend the mine to take up emulsion in solution to the problem which prompted this research since this explosive promotes safety at higher productivity in terms of tonnage output. A cost benefit analysis clearly pinpoint to the implementation of emulsion as a result of optimized KPIs. By making the effort to implement cautious blasting practices in tunnel development, the amount of overbreak is limited, which improves rock mass conditions and support integrity with the spin-off of reduced support and remediation costs, thereby better project feasibility. The following recommendations arise from the study:

- At a later stage of the trial, a change of the drilling pattern to one with a reduced burden between the lightly charged perimeter holes and the next line of production holes should be implemented to further improve the state of remaining wall.
- Overtime, there is also need that the trials be carried out in areas of good ground, so that specific results for these areas can be obtained for analysis to strengthen the argument to implement change in explosive to the whole mine.
- The mine should consider using the emulsion explosive in conjunction with a more powerful and small diameter primer like pentolite booster instead of megamite 38 cartridges so as to improve on wall smoothness and get even better results out of emulsion explosives.
- Mechanical scalers are also critical for safety of employees during barring down process.

REFERENCES

Bohanek, V., Dobrilovic, M. and Skrlec, V., 2013. Influence of the Initiation Energy on the Velocity of Detonation on ANFO explosives. *Central European Journal of Energetic Materials,* vol.10, no.4, pp. 555-567.

Budin, M., 2009. *The Benefits of Bulk Emulsion Explosives in UndergroundApplications through String Loading.* [Online] Available at: http://www.infomine.com/library/publications/docs/budin-sme2009.pdf [Accessed 9 March 2016].

Dey, K. and Murthy, V. M., 2010. Investigations on impact of Blasting in tunnels. *International Journal of Geotechnical Earthquake Engineering,* vol. 1, no. 2, pp. 59-62.

Hustrulid, W. A. and Iverson, S. R., 2010. Evaluation of Kiruna mine drifting data using the NIOSH design approach. In: Sanchidrián, ed. *Rock Fragmentation by Blasting.* London: Taylor and Francis Group, pp. 498-504.

Lee, J., Holmberg, R. and Persson, P., 1993. *Rock Blasting and Explosives.* Boca Raton, CRC.

Maranda, A., 2011. ANFO Detonation Parameters. *Central European Journal of Energetic Materials,* vol 8, no. 4, pp. 280-291.

Mather, W., 1997. *Bulk Explosives.* [Online] Available at: https://miningandblasting.files.wordpress.com [Accessed 13 March 2016].

McKown, A., 1984. Design and evaluation of perimeter controlled blasting in fractured and weathered rock. In: *Explosives and blasting techniques*, pp. 261-267.

Oberthür, T., Melcher, F., Buchholz, P. & Locmelis, M., 2012. The oxidized ores of the main sulphide zone, great dyke, zimbabwe: turning resources into minable reserves. [Online] Available at: http://www.saimm.co.za/Conferences/Pt2012/647-672_Oberthur.pdf [Accessed 18 May 2016].

Prendergast, M. D., 1989. The Wedza-Mimosa platinum deposit, Great Dyke, Zimbabwe. In: *Magmatic Sulphides-the Zimbabwe volume.* 1 ed. London: Institution of Mining and Metallurgy, pp. 43-69.

Sellers, E. J., 2011. Controlled blasting for enhanced safety. *The Journal of The Southern African Institute of Mining and Metallurgy,* vol. 111, no.1, pp. 11-16.

Singh, S. P., 1992. Mining industry and blast damage. *Journal of Mines, Metals and Fuels,* vol. 1, no. 12, pp. 465-468.

Szendrei, T., Sellers, E. J. and Cunningham, C., 2006. *European Federation of Explosive Engineers Conference.* Vienna, EFEE.

Advanced DInSAR, PSI-techniques and SAR Methodologies Applied in the Ground Motion Monitoring in Mining Areas.

F. Sánchez, J. Raventós
TRE-Altamira S.L. Barcelona, Spain

ABSTRACT InSAR (acronym for interferometry of Synthetic Aperture Radar) uses the electromagnetic wave signal emitted by radars shipped in satellites orbiting in quasi polar orbits to measure with milimetric accuracy displacements of the earth's surface or objects on it. The first SAR sensors were shipped in the early 1990s providing the possibility to process data from the past. Together with the schedule of new satellite acquisitions it allows investigations on site stability before the star of the project and also during mining operations.

The following paper describes the use of the technology, in real case applications, in mining sites, where open pit slope stability, waste dam, underground excavation settlement monitoring and other infrastructure monitoring is needed.

1 INTRODUCTION

This article is focused on Differential synthetic aperture radar interferometry DInSAR methodology, PSI technique SqueeSAR and other SAR applications such as correlation of amplitude images, RMT, applied in the detection of ground deformation. The paper presents the main characteristics of the methodologies applied by TRE ALTAMIRA for the detection of ground motion in a wide range of magnitudes, from millimeters to meters and includes examples of real case applications in mining environments.

The use of DInSAR and SAR methodologies for detecting ground deformation has increased in the last decade in the mining sector for two main reasons. Firstly , these methodologies provide a large amount of measurement data that are easily comparable with other classical geodesic measurements on going in the mining sector. Secondly, these methodologies provide the advantage of covering large spatial areas at low cost, a regular acquisition of measures over time and the availability of large archives of historical data to perform retrospective studies.

2 TECHNICAL BASIS OF DINSAR

Differential Aperture Synthetic radar interferometry (DInSAR) is a remote sensing technique able to measure surface deformation on the ground using complex satellite radar image datasets. The main information used by DInSAR analysis corresponds to the interferometric phase obtained by the computation of the phase difference between a pair of SAR images, the interferogram. The phase information coming from an interferogram is also related to the topography of the area observed and the deformation occurred in the time between the two images used.

In the case of having a Digital Elevation Model (DEM) of the area of interest the topographic component can be simulated and removed from the phase signal giving as a result the DInSAR phase mainly related to the ground deformation, that is, the motion. However in an in-depth analysis of the DInSAR phase it can be seen that there are other phase components to take into account when performing advanced interferometry:

-Phase noise: SAR sensors on satellites scan the earth surface in a 2D regular pattern. Only the pixels with a low interferometric phase noise will be taken into account to obtain motion measurements and can be good quality measurement points. All the points with a high noise ratio will be removed from the processing applying interferometric coherence and amplitude dispersion criteria.

-Atmospheric contribution: As well as the GPS phase signal, DInSAR phase is also affected by signal distortions due to the propagation of the signal through the atmosphere. Advance DInSAR techniques are able to estimate and extract the signal due to atmospheric artifacts for each SAR image, Ferretti et al., (2001) and Lanari et al., (2004).

-Topographic error: Phase component related to topographic differences between the DEM used and the radar readings can introduce errors in the deformation measurements. The error magnitude depends on the quality of the DEM used and the perpendicular baseline between the two images of the interferometric pairs took as the dataset.

Most of the DInSAR techniques base their deformation estimation in the unwrapping of the interferometric phase. In SAR images the phase information of each pixel is known modulus 2π (wrapped phase). The estimation of the deformation consists in unwrapping the phase information coming from the interferograms to get the real value of the phase. The unwrapping computation is the most critical step in the processing chain and consists in the estimation of the phase ambiguity. In particular, just with one interferogram, (phase differences between two SAR images) in order to get a correct estimation of the phase ambiguity, the phase gradient between close pixels has to be less than π. Pi corresponds to a deformation in the line-of-sight of the satellite (LOS) of lambda/4. Lambda corresponds to the wavelength of the radar signal that is, for example: lambda=5.66 cm for C-band sensors or lambda=3.1 cm for X-band sensors. These values suppose a clear limit in the detection of the spatial maximum gradient but it depends on the spatial distribution of the real pattern. To address this limitation temporal span between images can be reduced.

On the other hand for slow motion phenomena the main scope is the minimum detection limit. In these cases multiple SAR images during long temporal periods can be acquired in order to get a redundant dataset of DInSAR observations. The fact of adding redundancy results in the advantage of canceling noise and atmospheric artifacts from the phase signal. Nowadays some advanced DInSAR techniques to retrieve ground motion information from redundant interferometric datasets can be found in the literature. These include *Permanent Scatterers* (Ferretti et al., 2000) and the Small Baseline Subset (Berardino et al., 2002; Lanari et al., 2004). Other interesting approaches can be found in Werner et al., 2003, Mora et al., 2003 and Hooper et al., 2004. In the best case velocity precision may be below 1 mm/year as described in Colesanti et al., 2003. Most of these techniques deal in PSI techniques, explained in the following point 3.

Since the first description of DInSAR technique using SEASAT SAR data by Gabriel et al., 1989, DInSAR techniques were applied in a wide range of applications as landslides (Tamburini et al., 2013), seismology (Massonet et al., 1993), volcanology (Amelung et al., 2000), glaciology (Goldstein et al., 1993), terrain subsidence (Galloway et al., 1998), etc…
Further details on the applications of advance DInSAR techniques can be found in

Rosen et al., (2000), Hanssen (2001) and Ferretti (2014).

In the mining sector, knowledge regarding the location, dimensions and magnitudes of ground motion in mining areas has important implications in the mine operations. Ground motion data provides valuable information in a wide range of mine tasks for example: in the mine business plan development, health and safety management (critical infrastructures such as tailing dams located upstream of communities), planning during mine closure, geodesic measurements regarding ground stability and control and in the design of the ongoing mine operation.

3 PSI TECHNIQUES AND SQUEESAR™ MILIMETRIC MOTION ESTIMATION

Advanced DInSAR techniques applied to a set of input interferograms (redundancy) are the ones known as advanced PSI techniques. PSI techniques are related to the detection of millimetric and centimetric displacements on the ground surface caused by underground activity or low-motion instability (Adam et al., 2009). The PSI family of algorithms can identify high quality reflectance points (high coherence) in the radar images. Model fitting methodologies are applied to these points to derive the precise height, the mean displacement rate and the temporal evolution of the displacement (time series). During this process atmospheric artifacts are compensated. The basis of the PSI techniques consist of the separation of each cmponent: motion, topographic error, atmospheric artefacts and signal noise from the input interferogram phases. In a step by step procedure, using a stack of input interferograms coming from a minimum image dataset of more than 20 images, atmospheric artefacts are estimated and compensated by using low pass filters as well as deformation and topographic contribution values are extracted through a high resolution analysis.

During years, InSAR analysis was performed on the basis of analyzing the position and control of very coherent radar reflectors called Permanent Scatters (PS) that can be detected in the whole dataset. The PS-InSAR application (Ferretti et al., 2000) was able to retrieve milimetric accuracies in motion estimation with very good results especially in urban areas by removing atmospheric noise. The main limitation of the application was the low point density in non-urban areas or areas without infrastructures. In order to achieve better results especially in non-urban areas, bare ground, mines, hydrocarbon reservoirs, or landslide areas, also reflectors known as Distributed Scatterers (DS) were took into account. The DS measurement point corresponds to areas showing similar SAR signals. The size of the area is dependent on the pixel size and the number of adjacent pixels showing the same SAR signal.

Advanced processing techniques that use both PS and spatial DS significantly increase the density of measurement points in non-urban settings. SqueeSAR (Ferretti et al., 2001) is one of these advanced techniques as it preserves the precision of PS-InSAR while expanding the possibilities of investigation and control to many non-urbanized areas, including mountainous regions. The SqueeSAR algorithm also produces improvements in the quality of the displacement time series. The homogeneous areas that produce DS points normally comprise several pixels. The single time series attributed to each DS is estimated by averaging the time series of all pixels within the DS, effectively reducing noise in the data. The number of PS and DS points typically increases over time as more images are captured, although this is partially offset by active mine operations that tend to alter the landscape or surface and therefore decrease the number of stable measurement targets (Macdonald et al., 2016).

4 RAPID MOTION TRACKING TECHNIQUE FOR DECIMETRIC AND METRIC MOTION ESTIMATION

Rapid Motion Tracking (RMT) is a recently developed technique used for tracking rapid movement rates that exceed those observable with SAR interferometry (Macdonald et al., 2016). While SAR interferometry measures motion using the phase component of a radar signal, metric measurement technique is based on the identification of textures and spatial patterns in the amplitude of SAR images. The amplitude values correspond to the backscattering power or echo of the radar signal on the ground and show patterns and textures. Tracking these patterns between two SAR amplitude images by using image correlation techniques, it is possible to detect the magnitude both in range and azimuth directions.

Rapid motion tracking analysis is performed in the azimuth-range plane (flight direction and line-of-sight plane respectively) where real movement is projected, see figure 1. Two kind of information is delivered in terms of metric motion detection. One corresponds to the motion detected in Range direction and the other one in azimuth direction.

As this technique regards movements starting at 400 millimetres per year and into the range of tens of metres per year, the precision is in the order of approximately 100-200 millimetres instead of the millimetre precision attainable with SqueeSAR (Macdonald et al., 2016).

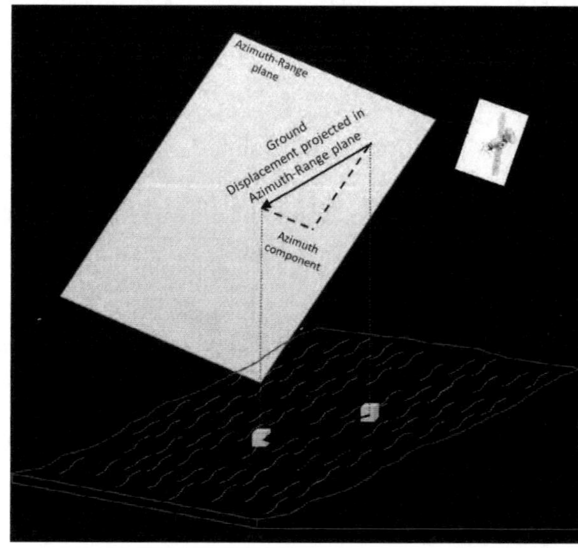

Figura 1: 3D scheme on how metric motion is detected. Real ground motion displacement is measured on range and azimuth directions and then their component are projected on the range-azimtuh directions.

5 FREQUENT UPDATE BULLETIN – EARLY DETECTION MAP

Early detection mapping product is focused in the idea to bring reliable information prior the delivery of the whole SqueeSAR study (which requires a minimum set of input SAR images and a certain acquisition time as well). The basic unit of information to perform and early detection map is one interferogram which can be computed just with two images. The interferogram, as explained before contain the phase differences between two images taken in two different dates. The information of the phase, which is wrapped, are related to the deformation occurred in the period of time analyzed. Interferometric phase has to be unwrapped in order to get the real value of the phase and retrieve the deformation magnitude. This technique is only capable to extract information between closest pixels with a phase gradient of less than pi which corresponds to a deformation value, in the line-of-sight of the satellite, of lambda/4.

Join with the deformation unwrapped from the interferogram, the product also include a visibility map showing the areas where detection was not possible. Visibility map include Amplitude Change Detection, that corresponds to those areas where surface changes occur during the period of study, the Decorrelation Noise where no proper backscatter occur, for example water or snow and No Visibility Areas which correspond to areas where there is no direct visibility between the sensor and the terrain due to the interaction between satellite oblique incidence and the terrain relief. Also a deformation gradient zonation is included in the early detection Map, figure 2. This gradient correspond to such areas where, despite the accumulated magnitude of motion during the period of study, the area presents a high gradient respect to the surrounding areas in a certain threshold in mm/m, magnitude versus distance.

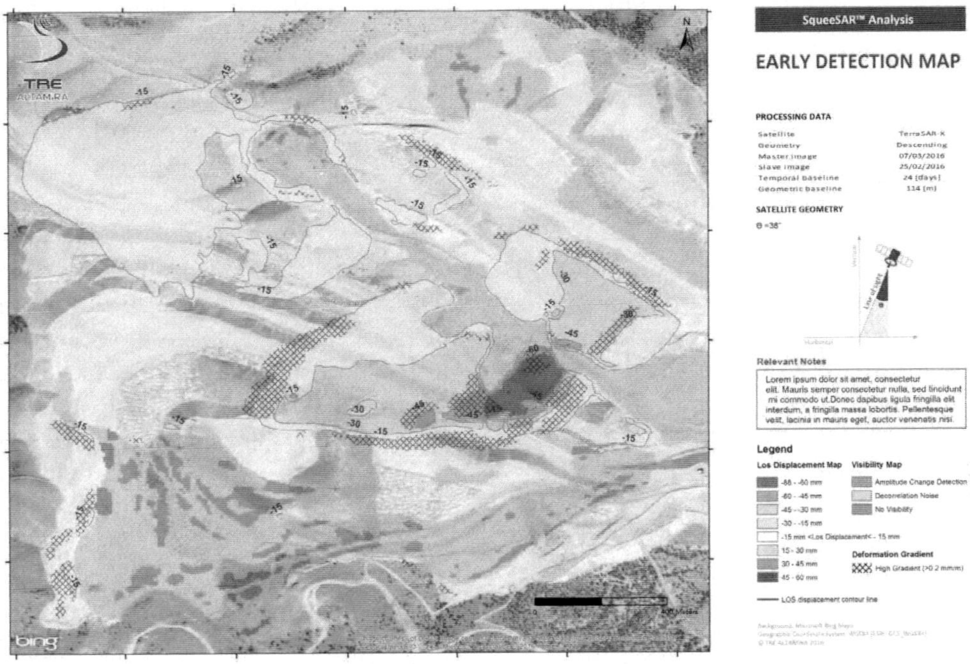

Figura 2: Example of an early detection map over a mining environment.

6 CASE STUDIES

6.1 Súria-Sallent Underground Mining Spain

Surface deformation phenomena such subsidence is frequent in underground mine exploitations triggered by the mining activity. DInSAR processing methodologies comprise wide useful satellite based remote sensing solutions to detect and measure surface deformation covering large areas with millimetre precision and at very high resolution, providing high density of measurement points. This information is very useful to compute the extension and the gradients of the motion areas not only over the mine but also above the mine helping in setting the extent of the movements and their possible affectation to the surrounding infrastructures.

In addition, DInSAR measurements can be combined with other data (geology, mine layouts, GPS, ground pegs) to produce more advanced interpretation of the measurements. In that way advanced products can be derived allowing the establishment of relationships between mine parameters and ground deformations. In this particular work, InSAR measurements were combined with mine maps and GPS data to measure the angel of draw of the excavations.

In a second step of the study, DInSAR data were compared with a dense network of GPS. Once both surveying techniques are in agreement they are combined together to retrieve improved 3D motion maps in terms of accuracy and density. This allows for more advanced analysis, thus creating products that are more useful for mine operators.

6.1.1 Area of interest

The Catalonian Potassic basin, northeast Spain, was exploited during years since the beginning of XX century. The main formation comprises evaporitic rocks formed 40 million years ago and later tectonized during alpine orogeny. Tectonics confers the final emplacement of the salts, mainly located in extruded diapirs in narrow and faulted anticlines and subhoritzontal salt beds in wide and subflat synclines.

The geological structure and the location of potassic salt (the main economic interest) determined the mining method. As in the case of Súria and Sallent mines, due to the subhoritzontal position of the salt beds (low angle synclines), underground room and pillars mining method was applied.

6.1.2 Methodology available data

For this project 18 radar images stack from ALOS satellite were used, covering a temporal period from January 2007 to March 2011. Archive SAR acquisitions were used for the processing, with acquisition every 46 days, in consecutive acquisitions. The SAR sensor of this satellite works in L-band (low frequency and wide wavelength) that is the most appropriated for vegetated areas due to its penetration capacity into slight forest and bush. With this dataset motion maps can be derived at 30 m resolution with precision in the order of 8mm and with densities of about 750 PS/KM2.

In addition, two periods of two years GPS data over both areas of study, 2009-2011 in Súria and 2008-2010 in Sallent, were also available. GPS network was very dense with near 900 points of measurement which allowed for a robust comparison. As observed, this dataset overlaps partially the SAR stack and only two GPS campaigns were available. Therefore, the measurements and the comparison were constrained to the mean deformation rate rather than the time series.

6.1.3 Obtained results

The resulting product is a surface deformation map which represents the motion of the ground in mean annual velocity (mm/year) in the mining area of Súria and Sallent. Displacement measurements are taken in line-of-sight geometry. Also, accumulated displacement for each acquisition date is provided.

In the result, figure 2 three main deformation areas can be seen. These areas correspond to underground mine sites of Súria and Sallent. Rooms and pillars mining method grid, trigger large deformation areas on the surface. Maximum deformation rates in the period analyzed are located in the inner parts of the deformation bowls. Moreover, the fact of left pillars of support also is responsible of gradual and linear subsidence patterns.

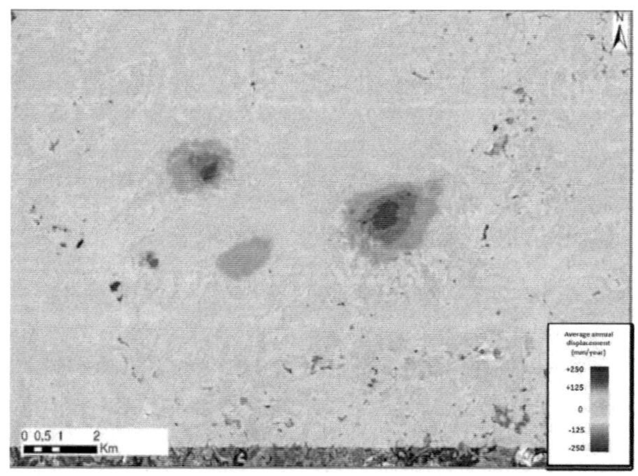

Figura 3: Mean anual velocity map obtained of processing ALOS data over Suria- Sallent area of interest.

In order to make the comparison between GPS data and satellite data, the mean annual velocity for GPS, considering two years of GPS campaign, was calculated analytically. Also a composition of x, y, z GPS deformation data into LOS (line-of-sight) geometry was done. Taking into account: satellite incident angle (inc=36°) and track heading angle (head=346°) to define the vector to re-project GPS in LOS vector, Equation 1. This allows the cross validation analysis between mean velocities (mm/year) of both datasets, and in the same geometry (LOS).

$$LOS = A(xyz) * mov(xyz) \quad A=[-\cos(inc1), sign(head1)*\sin(inc1)*\cos(head1) \quad sign(head1)*\sin(inc1)*\sin(head1)]$$

Equation 1: Analytical transformation from x, y, z GPS to LOS geometry.

A comparison is carried out between GPS points and their nearest SPN points. The parameter selected to compare is the mean annual velocity. Figure 4 show a good correlation between both datasets. As a quality control of the adjustment between data, an analytical result of such comparison is computed using a statistical function RMS (root mean squares) to get the correlation value between GPS and SPN. Values obtained are 22mm/year for Sallent area and 33 mm/year for Súria. Taking into account that GPS precision is about ±20 mm in height (centimetre vertical precision of GPS) correlation adjustment is into acceptable value range.

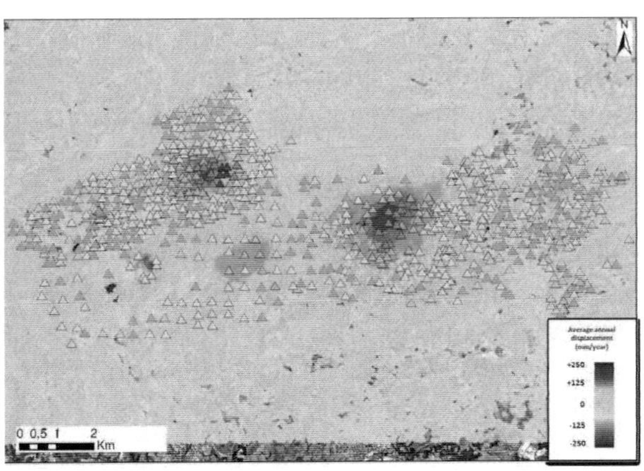

Figura 4: Data obtained from GPS and InSAR. Same annual mean velocity is used (mm/year)

In addition a time series comparison in Sallent area is presented, Figure 5. Temporal interval selected corresponds to the closer disposable dates of SPN to GPS dates in order to match the same period of time. These comparative time series show how SPN describes the evolution of movement along time with displacement data for each acquisition date and how both techniques show the accumulated displacement for that point in the whole period.

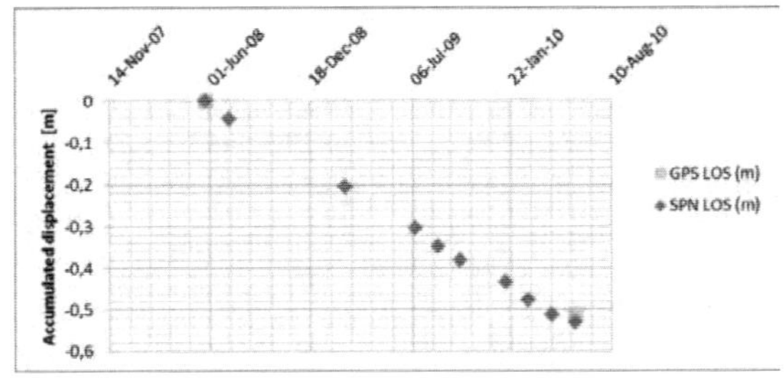

Figure 5: Time series comparison between GPS and InSAR

6.2 Slope Stability Monitoring in an Open Pit Coal Mine Spain

Surface deformation in both underground and open pit mining environment offers a unique case study to apply InSAR and SAR methodologies. In this work SAR image processing techniques as SqueeSAR and RMT were successfully applied to fit monitoring tasks in an active mine environment. From millimetric to centimetric and metric motion magnitudes coming from the open pit excavation, slope dynamics and also produced by the underground mining, located at the south of the area of interest, were analyzed in a monitoring basis of 6 months delivery period to avoid winter snow cover on the area.

Also corner reflectors, artificial measurement points, were installed in the area of interest to assure measurement points in areas with poor point density and also get winter measurement.

6.2.1 Area of interest

The coal mine analyzed is situated in north-eastern Spain (Figure 1). It is a large-scale open-pit consisting of alternating lutites, sandstone, and seams of coal, distributed evenly over the four slopes that make up the open pit. The pit slopes are large, with a total height from between 100 m to 350 m. The mine's benches are 10 m in height and the berms have a width of 5 to 6 m. Although coal is extracted by open-pit mining operations, underground operations are also performed in the southern slope of the mine.

The coal deposit consists of very thick, highly irregular vertical and sub-vertical seams. Of all of these, the Pastora Formation is the one of greatest financial interest, as it is the formation that provides all of the coal production for the current mining centers. Operations currently focus on the western edge of the coalfield. The Pastora and Competidora seams are the thickest seams in the coalfield.

The Pastora seam (Santa Lucia group) runs for 1600 m in a primarily north-south direction, with varying dips, with values close to 30°. Its thickness varies from 7 to 40 m, and within the seam there are bands and wedges of varying thicknesses, shaped like lenses, and varying hardness and composition, ranging from soft slate to sandstone.

The area where mining operations are performed on the Competidora Seam runs some 1680 m in a majorly north-south direction with dips ranging from 60° to 80°. ts thickness varies from 2 to 25 m, with 14 m being the most usual thickness.

6.2.2 Methodology Available data.

During the last four years, a stack of TerraSAR-X images has been acquired over the mine area. These images have been programmed in the Stripmap mode of the satellite (descending orbit), which means an approximate spatial resolution of 3 metres on ground.

Displacements affecting mine slopes can range from some millimeters to several meters. These different behaviors must be studied using different technologies based on SAR data. A first classification of events can be done taking into account the displacement intensity, which implies different radar techniques for their correct measurement:

Millimetric and centimetric movements: This kind of displacement is usually caused by underground activity or low-motion slope instability. SqueeSAR algorithm was used for this purpose. Millimetric precision is obtained and a maximum threshold of ground velocity can be measured depending on the number of SAR images and the temporal period of the study. Typically, this maximum velocity ranges from 15 to 30 mm/year in case of X-band satellites, like TerraSAR-X.

Metric movements: Due to the limitations of PSI techniques regarding the maximum measurable velocity, TRE-Altamira applies their Rapid Motion Tracking (RMT) algorithm to detect metric displacements for the mining industry. These very strong and sometimes fast movements can be found in mine slopes and represent a major problem for the mine safety, mainly in infrastructure and working areas.

As presented in this paper, combination of both methodologies allows obtaining a complete displacement map of the area of interest, no matter the movement magnitude.

Furthermore, two different types of measurement points have been used on the monitoring of this mine. These points can be classified as natural and artificial reflectors:

Natural reflectors: These are the measurable points of the surface under study. They must not change their surface characteristics during the temporal period of monitoring. These natural reflectors can be buildings, infrastructures, bare soils, low-vegetation areas, etc. Not all the points in the image can be classified as natural reflectors, and therefore only a part of the image pixels can be measured by means of radar interferometry.

Artificial reflectors:Trihedral structures that can be installed on surface, in order to ensure measurable points where natural reflector density is very low. Artificial reflectors do not need any additional maintenance, and power supply is not required. Figure 6 shows an example of natural points and artificial reflectors with their time series.

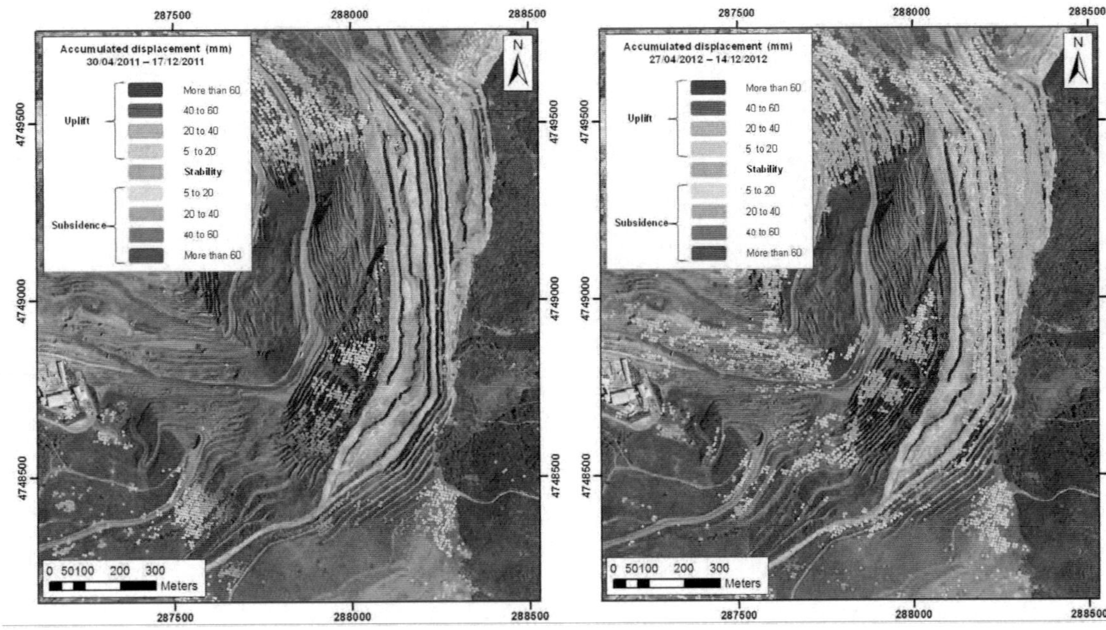

Figure 6: Slope ground accumulated displacement map in the open pit area for two different periods, 2011 (left) and 2012 (right).

6.2.3 Obtained results.

Figure 6 shows displacement maps for two different years 2011 and 2012. Differences on ground deformation can be observed, mainly in the Southern part of the slope, where velocity decreases during 2012. Another important issue is the distribution of measurement points, depending on surface changes in the open pit. For example, lack of points is more present during 2011, when a higher surface activity has been reported.

On the other hand, a large area without detected points is clearly visible in the center of the slope. This is due to extreme displacements, more than 5 meters, presented in the next section. These displacements cannot be measured by means of interferometric phase due to aliasing, in this area and RMT processing was performed, see figure 7.

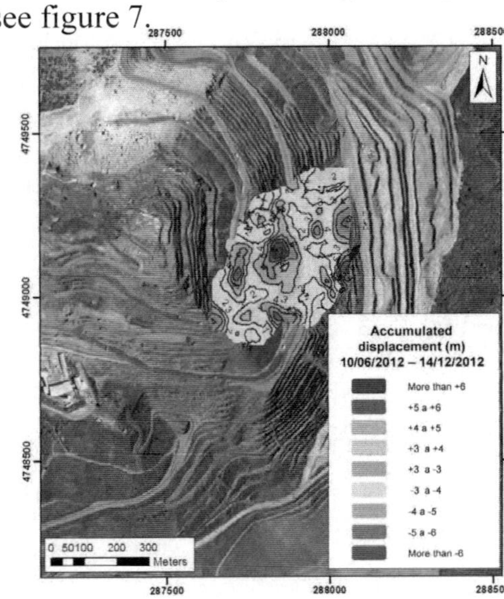

Figure 7 : Central slope ground displacement map for motion up to 3 meters in six months.

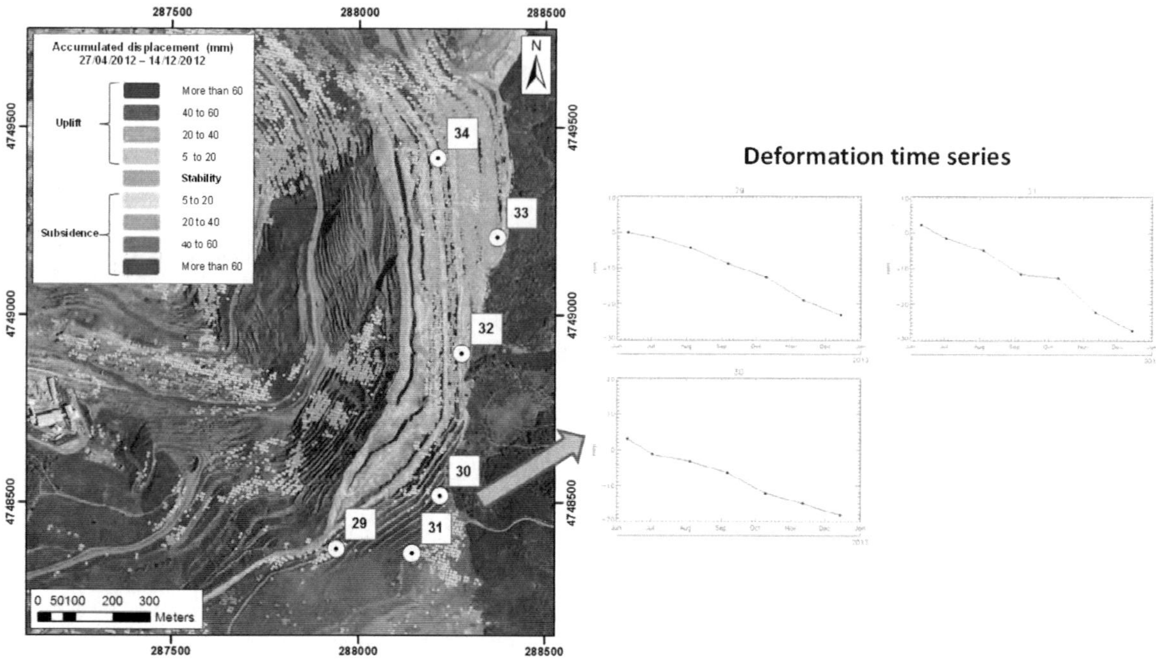

Figure 8: Ground displacement time series (mm) for artificial corner reflectors installed in the eastern slope area.

Ground deformation maps of Figure 6 correspond to natural reflectors measurements. Nevertheless, additional information can be retrieved using artificial reflectors, as it is shown in Figure 8. In this case, six aluminum trihedral reflectors have been installed in the studied slope, covering it from North to South. The three northern reflectors are located in a stable zone, as the natural reflector map shows. Nevertheless, the other three are installed on a deformation area, and their time series confirm the same movement pattern obtained from natural reflectors. This is an interesting study that confirms the functionality of both methodologies, natural and artificial reflectors. The advantage of artificial points is that only 3 or 4 SAR images are necessary for a first processing, while a larger dataset (typically more than 15 acquisitions) is recommended for natural reflectors.

7 CONCLUSION

The paper describes the use of InSAR technology in monitoring different types of mines and the benefits obtained through the combination of different products. SqueeSAR achieves millimetric accuracy by processing a stack of images of the area acquired over time. The bulletins exploit the classic InSAR technique by comparing two images over the site covering a period that typically ranges from 6 to 12 days with an accuracy between 1.5 to 3cm and is useful to detect significant fast displacements over the mine. Lastly the Rapid Motion Tracking technique is applied in order to detect decimetric and centimetric displacements.

The combination of these three techniques results in an integrated service that provides a general picture of the stability of the site allowing an improved risk management of mining operations.

REFERENCES

Adam, N., Parizzi, A. and Crosetto, M. (2009) Practical Persistent Scatterer Processing Validation in the Course of the Terrafirma Project. Journal of Applied Geophysics, vol. 69, pp. 59-65.

Amelung, F., Jonson, S., Zebker, H.A., Segall, P. (2000). "Widespread uplift and 'trapdoor' faulting on Galápagos volcanoes observed with radar interferometry". Nature, 407, 993-996.

Berardino, P., Fornaro, G., Lanari, R., Sansosti, E. (2002). "A new algorithm for surface deformation monitoring based on small baseline differential SAR interferograms". IEEE Transactions on Geoscience and Remote Sensing, 40(11), 2375-2383.

Colesanti, C., A. Ferretti, C. Prati, C. and F. Rocca. 2003, Monitoring landslides and tectonic motions with the Permanent Scatterers Technique. Engineering Geology. 68: 3–14.

Ferretti, A., Prati, C., and Rocca, F. 2000, Nonlinear subsidence rate estimations using permanent scatteres in differential sar interferometry. IEEE Transactions on Geoscience and Remote Sensing, 38(5):2202–2212

Ferretti, A., C. Prati, and F. Rocca. 2001, Permanent scatterers in SAR interferometry. IEEE Trans. Geosci. Remote Sens. 39(1): 8–20.

Ferreti, A. (2014). "Satellite InSAR Data. Reservoir Monitoring from Space". EAGE Publications.

Gabriel, A.K., Goldstein, R.M., Zebker, H.A. (1989). "Mapping small elevation changes over large areas: differential radar interferometry". J. Geophys. Res., 94 (B7), 9183-9191.

Galloway, D.L., Hudnut, K.W., Ingebritsen, S.E., Phillips, S.P., Peltzer, G., Rogez, F., Rosen, P.A. (1998). "Detection of aquifer system compaction and land subsidence using interferometric synthetic aperture radar, Antelope Valley, Mojave Desert, California". Water Resources Research, 34 (10), 2573-2585.

Goldstein, R.M., Englehardt, H., Kamb, B., Frolich, R.M. (1993). "Satellite radar interferometry for monitoring ice sheet motion: application to an Antarctic ice stream". Science, 262, 1525-1530.

Hanssen, R. (2001). "Radar interferometry". Kluwer Academic Publishers, Dordrecht, The Netherlands.

Hooper, A., Zebker, H., Segall, P., Kampes, B. (2004). "A new method for measuring deformation on volcanoes and other natural terrains using InSAR Persistent Scatterers". Geophysical Research Letters, 31, L23611, doi:10.1029/2004GL021737.

Lanari, R., Mora, O., Manunta, M., Mallorquí, J.J., Berardino, P., Sansosti, E. (2004). "A small-baseline approach for investigating deformations on full-resolution differential SAR interferograms". IEEE Transactions on Geosciences and Remote Sensing, 42(7), 1377-1386.

Macdonald, BC., Sharon, R., Muir, S., Iannacone, JP., Falorni, G., Michaud, JS. (2016) InSAR derived synoptic ground surface deformation map of the Porgera Mine, Papua New Guinea. Proceedings of the first Asia Pacific Slope Stability in Mining Conference APSSIM 6-8 September 2016, Brisbane Queensland, Australia.

Massonnet, D., M. Rossi, C. Carmona, F. Adragna, G. Peltzer, K. Feigl, and T. Rabaute. 1993, The displacement field of the Landers earthquake mapped by radar interferometry. Nature. 364: 138–142.

Mora, O., Mallorquí, J.J., Broquetas, A. (2003). "Linear and nonlinear terrain deformation maps from a reduced set of interferometric SAR images". IEEE Transactions on Geosciences and Remote Sensing, 41(10), 2243 –2253.

Rosen, P.A., Hensley, S., Joughin, I.R., Li, F.K., Madsen, S.N., Rodríguez, E., Goldstein, R.M. (2000). "Synthetic Aperture Radar Interferometry". Proc. of the IEEE, 88 (3), 333-382.

Tamburini, A., S. Del Conte, G. Larini, L. Lopardo, C. Malaguti and P. Vescovi. 2013, Application of SqueeSAR™ to the characterization of deep seated gravitational slope deformation: the Berceto case study (Parma, Italy). In Landslide Science and Practice, eds. C. Margottini, P. Canuti, K. Sassa, Springer, 437-443.

Werner, C., Wegmüller, U., Strozzi, T., Wiesmann, A. (2003). "Interferometric point target analysis for deformation mapping". Proceedings of IGARSS 2003, 4362-4364.

Laboratory Measurement of Time-Dependent Deformation Properties of Muddy Siltstone

O. Hamza
University of Derby, Derby, UK

R. Stace
University of Nottingham, Nottingham, UK

ABSTRACT This paper investigates the time-dependent deformation of argillaceous rocks, drawing principally on laboratory experiments carried out on muddy siltstone recovered from an open pit mine located in the midlands, UK. A series of creep tests were conducted on both intact and fractured rock samples to cover the two ultimate structural conditions of rock mass. Based on the creep test results, the relationship between axial strain and time under different axial and deviatoric stresses was investigated. The creep data of both intact and fractured rock samples was successfully fitted to Burgers model to represent creep behaviours of pre and post failure. The study provides improved representation of the time-dependent deformation properties of rock mass, which is essential to enhance geomechanical modelling of long-term stability of abandoned mines and for the application of underground disposal of radioactive waste.

Keywords: Time-dependent, Fractured rock, Argillaceous rocks, Burgers model

1 INTRODUCTION

1.1 Time-Dependent Observation in Underground Mine

Rock mass response may be time-*in*dependent or time-dependent and the material (whether it is intact or fractured) may behave elastically or may yield according to the confining pressure and the applied stress level (Kaiser and Morgenstein, 1981).

An understanding of the time-dependent behaviour has been considered essential for further development in the field of underground mine design and long-term strata control (e.g. Singh, 1975), and more recently it is becoming important for the application of underground disposal of radioactive waste (Liu et al., 2015).

In coal mines, it is generally accepted that after the initial excavation of an entry, most coal-measure rocks deform over time, even when no activity is taking place nearby.

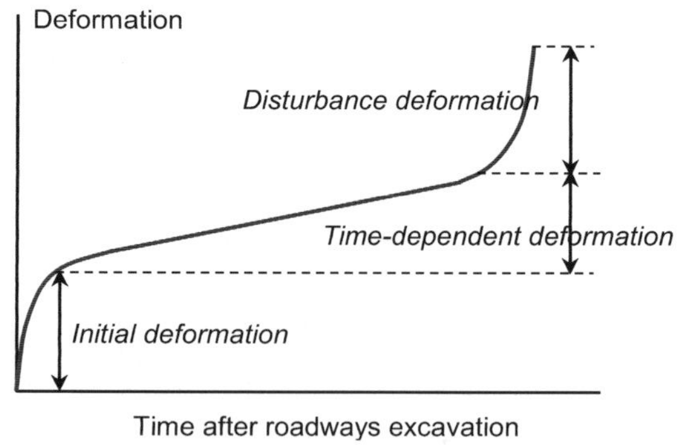

Figure 1. General components of deformation of underground roadways.

This deformation is most evident in the form of roof sag, floor heave, pillar dilation, or shearing along bedding planes and joints. Figure 1 presents graphically the deformation of underground roadways, which (generally) consists of three components. After the initial rapid deformation induced by the roadway excavation, a time-dependent deformation takes place. This rheological deformation might be followed by 'disturbance deformation', which is caused by stress changes due to mining activities in adjacent areas (e.g. Lu, 1984, cited by Wang et al., 2000). These physical events observed in underground excavation have been attributed to time-dependent characteristics, which can also be evident in a small rock sample tested in laboratories as discussed in next section.

1.2 Laboratory Measurement of Time-Dependent Behaviour

Laboratory experiments use three main testing methods to investigate time-dependent behaviour; these are: (i) creep, (ii) relaxation and (iii) loading tests at different stress or strain rates. Creep in particular can be simply observed in much laboratory testing when 'relatively high' stress applied constantly on a cylindrical specimen for a period of time. Figure 3 shows idealised creep curve which exhibits three out of four principle phases of deformations as listed below:

a) Instantaneous elastic strain due to instantaneous load.
b) Primary or transient creep with rapid strain increments, but at a decelerating strain rate.
c) Secondary creep at low, or near constant strain rate.
d) Possible tertiary creep accelerating strain increment to failure, which might take much longer time than in the lab studies (e.g. Jeremic, 1987).

If the specimen is unloaded during the primary creep stage, the deformation can be recovered. However, permanent deformation is resulted if the secondary creep is dominated.

To represent such time-dependent behaviour of rocks in mathematical form, mechanical, phenomenological or rheological models are commonly developed by fitting the experimental data (e.g. Munson, 1997; Günther et al., 2015) on the basis of empirically observed internal variables (e.g. Challamel et al., 2005) or the

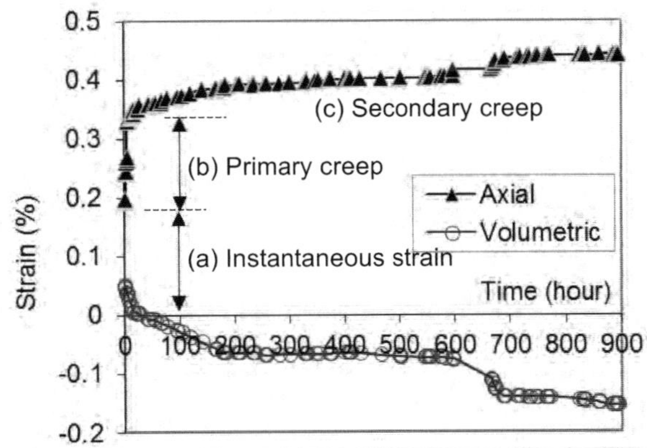

Figure 3. Axial and volumetric strain curves for Lea Hall Sandstone under 35.8 MPa uniaxial stress showing three stages of creep (after Hobbs, 1970).

superposition of several viscous and elastic elements such as the Burgers body model (Cristescu, 2012).

Extensive creep testing has been conducted on coal-measure rocks around the mid of last century. In early stages (prior to 1964), the majority of the time-strain experiments have been carried out in uniaxial compression, at room temperature and, for periods of less than a week (e.g. Phillips 1930, 1931, 1954, Pomeroy, 1956 cited in Valsangkar, and Gokhale, 1972). These studies have been mostly carried out on intact rock samples. However, the time-dependent behaviour of fractured rock represents an important ultimate structural condition of the rock mass. Larson and Wade, 2000, indicated that creep in brittle coal measure rock occurs, likely, because of (i) microcracking along weak bedding planes and (ii) weakening of asperities of joints. Therefore creep along bedding plan and pre-existed joints is also important. This has been investigated in a number of studies as explained below.

Höwing and Kutter, 1985, conducted direct-shear tests to investigate the effects of joint filler on creep behaviour. In this study, creep velocity during the primary phase has followed a power law, up to the point where creep progressed from the primary to the tertiary phase.

Clean joints have been investigated in uniaxial and triaxial compression tests on intact and jointed rock specimens of

Westerly Granite and Navaho Sandstone (Wawersik, 1974). Creep of the rock was observed, particularly when water was present. The creep behaviour of jointed rock has the same character as that of the intact material, but with a larger amount of strain.

Larson and Wade, 2000, investigated friction and creep characteristics of weak planes in mudstone by conducting direct-shear creep tests; non-linear rheological model (Boukharov et al., 1995) was used to fit their experimental data.

More recently, a microcrack-based damage constitutive law, established at the elemental scale, has been used to directly link the time-dependent degradation of elastic stiffness and the induced anisotropy to microcrack propagation (Lu et al., 2014). Although this model is an attractive solution to reproduce the macroscopic creep behaviour on the basis of the microscopic kinetics of microcrack growth, however the model requires suitable calibration in order to obtain reliable results. Therefore phenomenological and empirical approaches might still be a good alternative way to characterise rheological behaviour of both intact and fractured rock, i.e. representing the two ultimate structural conditions of rock mass.

Although extensive research has been conducted on creep of intact as well as jointed rocks, very limited studies exist which investigate both conditions for the same type of rock. In a previous laboratory work, conducted by Hamza et al., 2005, it has been found that creep strain rate of broken rock samples can be significantly greater than that of the intact rock samples (depending on the confinement), indicating that the ability of rock to relax or creep considerably increases in fractured rocks. Despite this difference, a similar phenomenological model has been successfully used (Hamza et al., 2005) to represent the effect of strain rate on strength and stiffness of both intact and fractured silty mudstone individually. It follows that other time-dependent behaviours such as creep of both intact and fractured rock can be modelled using a similar phenomenological approach such as the Burgers creep model. This paper investigates the creep properties of both intact and fractured rock using muddy siltstone. The study can assist in understanding the rheological properties of argillaceous rocks particularly under pre and post failure conditions, which are essential to enhance geomechanical modelling of long-term stability of mines.

2 CREEP TESTING ON MUDDY SILTSTONE

2.1 Materials and Apparatus

The tested rock material (muddy siltstone) was originally obtained from an open pit mine located at Arkwright in the midlands of the UK. The rock consisted of laminated muddy siltstone (also described as silty mudstone in places) which is grey in colour with darker bands. The rock has a dry density of 2.48 ± 0.01 g/cm^3 and specific gravity of approximately 2.75.

To reduce any possible variability in the mechanical properties, the samples (shown in Figure 3) were cored from carefully selected similar blocks, which were mostly dominated by a similar angle of lamination (i.e. 0°-7° to the horizon of specimen flat surface). After coring, the rock, samples were left to dry out for 24 hours at 100°C, and then coated with a very thin layer of flexible varnish to eliminate the effect of humidity so that the unweathered condition of the material would be maintained over the storing and testing periods.

Figure 3. Intact samples of the muddy siltstone.

The tests were conducted at room temperature (about 23°C) on cylindrical dry samples of 74mm in height and 36mm in diameter, which were prepared in accordance with the procedure outlined by ISRM (1981).

A total of 14 samples of the muddy siltstone were successfully tested and reported in this paper. Before conducting the creep experiments, six samples were initially tested to obtain the uniaxial/ triaxial strength and Young`s modulus of the intact and fractured rock at different confining pressures. The remaining eight rock samples were used for the creep tests, where a constant axial stress σ_1 was applied and maintained by a servo-controlled rig machine (shown in Figure 4). The equipment can be used to carry out conventional compression tests and rheological tests such as uniaxial creep. However the required triaxial loading condition was achieved by confining the rock sample using a Hoek cell connected to a manual pump, which can provide constant confining pressure throughout the test. The axial displacement was measured by a high-precision displacement gauge, and the recorded deformation was corrected for the effect of steel Platens used for the application of axial stress.

The fractured rock samples were produced from the intact rock samples after bringing them to failure in conventional triaxial compression tests. A similar pattern of fractures were noticed in which samples developed a major shear plane at about 45° from horizontal. The failed (fractured) specimens were then used to carry out multi-stage creep testing with the prescribed confining pressure σ_3 (ranging between 1 and 6 MPa). Larger values of confining pressure would have allowed better investigation of triaxial creep behaviour, however the chosen pressure range was based on the best performance of the equipment used in the experimental programme.

2.2 Secondary Creep Strain Rate

Secondary or Steady State (SS) creep strain rate (see Figure 2) is probably the most important parameter as far as the time-dependent deformation is concerned. Several research studies (e.g. Hobbs, 1970; Munson 1979; Yang et al., 1999; Cristescu and Hunsche, 1998) conducted over a number of rocks and hard soils have shown that steady-state creep strains are almost independent of the loading history and are a function of the current stress state only. It is important for many engineering applications to know when a steady state (SS) is reached and what magnitude it can be at a certain level of stress.

This was investigated at an early stage of our experimental programme to determine the likely time scale required to establish the steady state creep for the muddy siltstone used in this study.

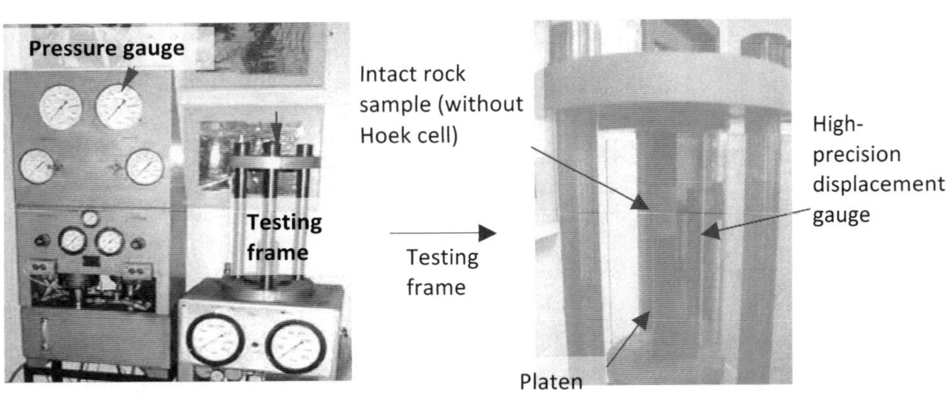

Figure 4. Testing Rig with servo-controlled hydraulic pressure cell system used for conducting the creep tests.

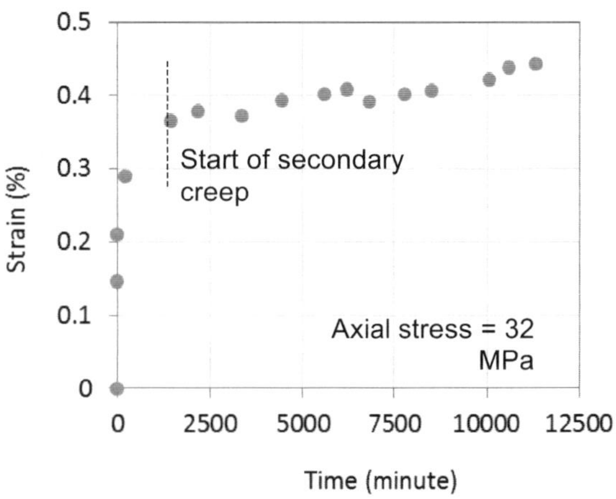

Figure 5. Creep test showing the period required to obtain secondary creep in intact muddy siltstone sample.

Figure 5 shows a typical creep curve conducted at constant axial stress σ_1= 32 MPa (≈57% of UCS); as expected the secondary creep did not start before some time i.e. approximately 1300 minutes (almost 1 day). Therefore, in planning for our experiments, the rock samples were given at least double this time (about 2 days) to creep for each stress increment. This procedure allowed the steady state strain rate (SS) to be obtained, which is of great importance for modelling and characterising time-dependent behaviour.

2.3 Experimental Results

2.3.1 *Uniaxial/ triaxial strength and stiffness*

Prior to conducting any creep experiment, six samples were initially tested to obtain the uniaxial/ triaxial strength and stiffness of the intact and fractured rock. The unconfined compression strength (UCS) and Young's modulus, (E) of the intact samples were found to be approximately 56MPa and 11GPa respectively. However it was not possible to conduct similar (unconfined) testing on any fractured rock sample without confinement. The results of the uniaxial compression tests are summarised in Table 1 and 2, where average Young`s modulus (E) and average peak strength (σ_p) are presented for each confining pressure.

2.3.2 *Creep experiments on intact rock samples*

Figure 6 shows results of multi-stage creep tests conducted on four intact rock samples at four different confining pressure σ_3 (0, 2, 4, and 6 MPa). For each sample, a constant confining pressure was applied, while the axial stress was increased in three stages: σ_1= 31.2, 39.1, and 46.8MPa; these values are approximately equivalent to 40%, 50%, and 60% of the average axial peak strength σ_p of the intact rock at confining pressure σ_3 = 2 MPa. Summary of stresses applied on the intact rock samples is presented in Table 3.

Table 1. Results of uniaxial/triaxial compression tests on intact rock samples

σ_3 (MPa)	0	2	4	6
E (GPa)	11.1	12.6	12.8	14.5
σ_p (MPa)	57.2	78.7	95.4	104.0

Table 2. Results of uniaxial/ triaxial compression tests on fractured rock samples

σ_3 (MPa)	1*	2	4	6
E (GPa)	2.8	3.7	4.2	5.8
σ_p (MPa)	11.7	23.4	37.4	46.0

* Sample is not stable under σ_3 less than 1MPa

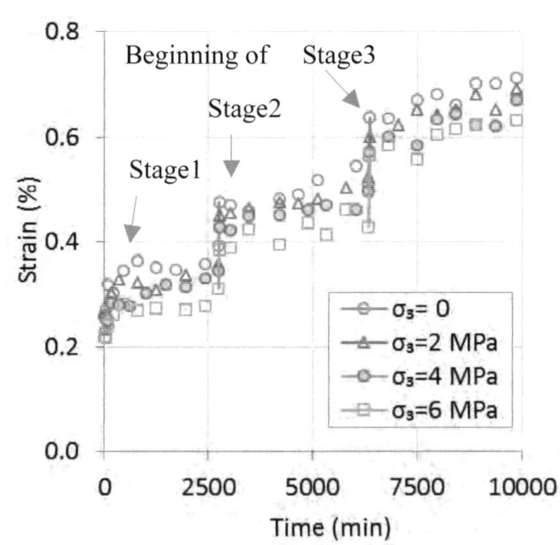

Figure 6. Creep strain obtained from four *intact* rock samples of muddy siltstone.

Table 3. Stresses used for creep experiments on intact rock samples

Test ID		Test 1	Test 2	Test 3	Test 4
σ_3 (MPa)		0	2	4	6
$q=\sigma_1-\sigma_3$	Stage1	31.1	29.3	27.2	25.2
	Stage2	39.1	37.1	35.2	33.1
	Stage3	46.8	44.7	42.8	40.6

2.3.3 *Creep experiments on fractured rock samples*

Table 4 presents stresses applied on the fractured rock samples, where typical confining pressures σ_3 (0, 2, 4, and 6 MPa) were used. However smaller values of axial stress were applied because the fractured rock samples are much weaker than the intact samples (according to the uniaxial/ triaxial properties of both rocks).

Four creep tests on were conducted, in each test, the axial stress was increased in three stages: σ_1= 10.35, 12.8, and 15.10 MPa; these values are approximately equivalent to 40%, 50%, and 60% of the uniaxial/ triaxial peak strength at confining pressure of 2MPa, respectively. The creep results of the fractured rock samples are shown in Figure 7.

Table 4. Stresses used for creep experiments on fractured rock samples

Test ID		Test 1	Test 2	Test 3	Test 4
σ_3 (MPa)		1*	2	4	6
$q=\sigma_1-\sigma_3$	Stage1	9.35	7.39	5.36	3.37
	Stage2	11.8	9.7	7.6	5.7
	Stage3	14.10	12.04	10.08	8.04

* Sample is not stable under σ_3 less than 1 MPa

3 DISCUSSION

3.1 General Observation on Creep Behaviour

The most noticeable finding to emerge from the experimental data is that all stain curves (Figure 6 and 7) showed an initial instantaneous strain followed by two phases of time-dependent strain including transient creep phase (particularly for the first loading stage) and a steady creep phase. Accelerated or tertiary creep was not considered as part of this study because the applied stresses were moderately less than the compressive strength of the rock.

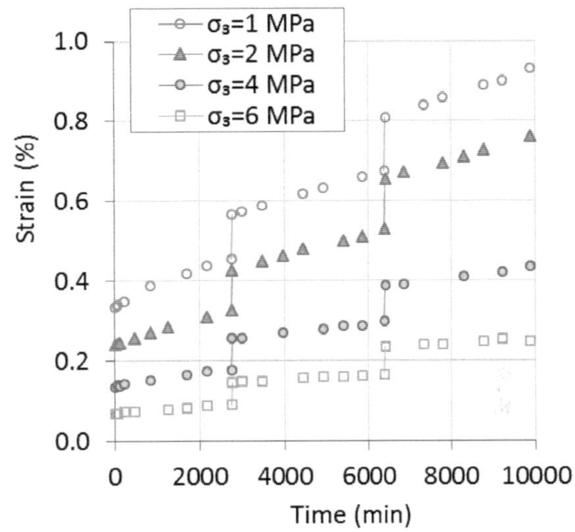

Figure 7. Creep strain obtained from four *fractured* rock samples of muddy siltstone.

The results also indicate that the instantaneous and creep strain are proportional to the deviatoric stress. This is clearly shown in fractured rock samples where larger deviatoric stress resulted in increased creep strain and strain rate.

3.2 The Effect of Confinement on Creep

The experiment conducted on intact rock samples shows that rock exhibit larger creep strain at larger deviatoric stress at almost similar rate regardless of confining pressure. Unlike the results from fractured rock samples where these showed that the creep strain is not only proportional to deviatoric stress but also to the level of confining pressure, where the temptation for creep in the fractured rock decreased with the increase of confinement. This was not clearly shown in the intact rock samples, possibly because of the low level of confining pressure (less than 6 MPa) in comparison with the high strength of intact rock (UCS=56 MPa).

To investigate the effect of confinement on creep behaviour of both intact and fractured rock samples, a further attempt of analysis was carried out, in which the recorded creep strains at 2 days are plotted against deviatoric stress, as shown in Figure 8. From this figure, it is possible to assess the tendency of creep in relation to deviatoric stress for all rock samples. The intact rock samples shows an average value of creep strain of 0.02% per MPa.

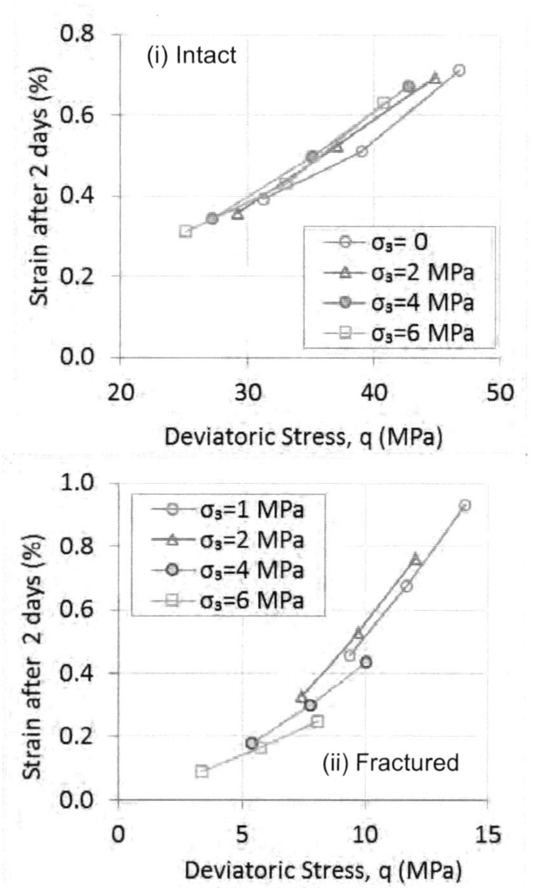

Figure 7. Creep strain recorded after approximately 2 days from the start of each loading stage. (i) Intact rock samples, (ii) Fractured rock samples.

On contrary the fractured rock samples showed large variation: from 0.09 to 0.03% of strain per MPa for confining pressure ranging from 1 to 6 MPa, respectively. The average steady state (SS) creep rate was approximately 0.023% per day for the intact rock, and between 0.05% to 0.01% per day for the fractured rock (corresponding to the increase of confining pressure).

3.3 Analysis of Creep Model

The experimental data obtained were analysed and fitted to the best mathematical expressions or *models*. These models with representative material properties are highly required to predict the geotechnical performance and also for the design applications.

Many constitutive laws (models) considering the effect of time have been developed using (i) Experimental/empirical fitting, (ii) Mechanical models and (iii) Phenomenological models based on physical process theories.

Burgers model is one of the most successful constitutive laws that have been used to model instantaneous and time-dependent deformation. Burgers model possesses instantaneous deformation properties, the primary creep, and steady creep. The constitutive equation of this model can be expressed as:

$$\varepsilon = \sigma\left[\frac{1}{E_m} + \frac{t}{\eta_m} + \frac{1}{E_k}\left(1 - e^{-\frac{E_k t}{\eta_k}}\right)\right] \quad (1)$$

where ε = strain, σ = stress, t = time, and the rest are the model's parameters. In general, the parameters required for Burgers model are:

- Kelvin Parameters: Kelvin modulus (E_k) and Kelvin viscosity (η_k);
- Maxwell parameters: Maxwell modulus (E_m) and Maxwell viscosity (η_m).

In the current study, these parameters were determined through a non-linear, least squares regression analysis using Matlab software package (Mathworks, 2016). The values of parameters obtained for the creep tests are presented in Table 5. The model was found to be suitable to describe the experimental data with good accuracy (R^2 = 0.87 - 0.93).

From Table 5, the fractured rock has generally smaller parameters, thus allowing larger magnitude of creep deformation in comparison with the intact rock. These factors may explain the relatively good correlation between rock strength and its tendency for creep as previously reported by several authors (Parsons and Hedley, 1966; Hobbs, 1970).

The confining pressure has almost similar effect on creep parameters where their numerical values generally increase with the increase of confinement (up to confining pressure of 6MPa). Further studies, which investigate these variables for, will need to be undertaken.

Table 5. Creep parameters of the intact and fractured rock samples under different confining pressures

σ_3 (MPa)	E_m (GPa)	η_m (GPa.d)	E_k (GPa)	η_k (GPa.d)
Intact				
0	11.2	150.3	50.1	50.4
2	11.9	162.4	52.3	52.2
4	12.5	171.5	49.8	53.1
6	13.4	169.8	58.4	52.9
Fractured				
1	2.8	60.5	15.2	23.3
2	3.1	72.4	16.0	23.1
4	4.0	94.8	19.6	28.9
6	4.5	101.4	30.2	30.6

4 CONCLUSION

To understand the long-term stability of underground opening it is important to describe time-dependent behaviour for the entire strain range of rock, i.e. up to and beyond the yield point, where rock is expected to be fractured.

The creep behaviour muddy siltstone has been investigated in this study to determine and compare the time-dependent properties of intact and fractured rock samples.

Within the tested range of stresses used in this study, the confining pressure did not induce any significant changes in the creep properties of intact samples. On the contrary, the creep curves showed that the fractured samples experienced larger variation in stain influenced by the variation of confining pressure. A possible explanation for this might be that confinement plays important role in closing up the rock joints and stiffening up the rock specimen.

It is expected that in fractured rock samples the time-dependent process is largely influenced by the structure and degree of fracturing of the rock materials. However, in this study, no systematic investigation was carried out to assess the level of fragmentation of the fractured samples. The experiments would have been also benefited from measuring the volume and lateral strain.

The experimental data describing the creep of the intact and fractured muddy siltstone under various deviatoric stresses was successfully fitted to Burgers model, adopting elasto-viscoplastic properties. The application of this model to other type of argillaceous rock, with wider range of confining pressures, can be part of a future investigation. Moreover, applying the results to a case study would validate this approach and improve the geomechanical modelling of long-term stability of mines.

ACKNOWLEDGEMENT

All experiments were carried out at the geotechnical engineering laboratory of the University of Nottingham, and funded by the Research Programme of the Research Fund for Coal and Steel (RFCS).

REFERENCES

Boukharov, G.N., Chanda, M.W. and Boukharov, N.G., 1995, June. The three processes of brittle crystalline rock creep. *In International journal of rock mechanics and mining sciences & geomechanics abstracts* (Vol. 32, No. 4, pp. 325-335). Pergamon.

Brown, E.T., 1981. *Rock characterization, testing and monitoring*: ISRM suggested methods.

Challamel, N., Lanos, C. and Casandjian, C., 2005. Creep damage modelling for quasi-brittle materials. *European Journal of Mechanics-A/Solids*, 24(4), pp.593-613.

Cristescu, N. and Hunsche, U., 1998. *Time effects in rock mechanics*. New York: Wiley.

Cristescu, N., 2012. *Rock rheology* (Vol. 7). Springer Science & Business Media.

Günther, R.M., Salzer, K., Popp, T. and Lüdeling, C., 2015. Steady-State Creep of Rock Salt: Improved Approaches for Lab Determination and Modelling. *Rock Mechanics and Rock Engineering*, 48(6), pp.2603-2613.

Hobbs, D.W., 1970, March. The behavior oof broken rock under triaxial compression. In *International Journal of Rock Mechanics and Mining Sciences & Geomechanics Abstracts* (Vol. 7, No. 2, pp. 125-148). Pergamon.

Höwing, K. and Kutter, H., 1985, September. Time-dependent shear deformation of filled rock joints. In A keynote lecture. Fundamentals of rock joints: *Proceedings of the international symposium on fundamentals of rock joints*, björkliden (pp. 15-20).

Jeremic, M.L., 1987. *Ground mechanics in hard rock mining*. A.A. Balkema, Rotterdam.

Kaiser, P.K. and Morgenstein, N.R., 1981, April. Time-dependent deformation of small tunnels—I. Experimental facilities. *International Journal of Rock Mechanics and Mining Sciences & Geomechanics Abstracts*, Vol. 18, No. 2, pp. 129-140. Pergamon.

Larson, M.K. and Wade, R.G., 2000. Creep along weak planes in roof and how it affects stability. *Society for Mining, Metallurgy, and Exploration, Inc*, Vol 310.

Liu, Z.B., Xie, S.Y., Shao, J.F. and Conil, N., 2015. Effects of deviatoric stress and structural anisotropy on compressive creep behavior of a clayey rock. *Applied Clay Science*, 114, pp.491-496.

Lu, Y., Elsworth, D. and Wang, L., 2014. A dual-scale approach to model time-dependent deformation, creep and fracturing of brittle rocks. *Computers and Geotechnics*, 60, pp.61-76.

MathWorks, 2016. MATLAB Documentation.

Munson, D.E. and Dawson, P.R., 1979. *Constitutive model for the low temperature creep of salt* (with application to WIPP) (No. SAND-79-1853). Sandia Labs., Albuquerque, NM (USA).

Munson, D.E., 1997. Constitutive model of creep in rock salt applied to underground room closure. *International Journal of Rock Mechanics and Mining Sciences*, 34(2), pp.233-247.

Parsons, R.C. and Hedley, D.G.F., 1966, November. The analysis of the viscous property of rocks for classification. *In International Journal of Rock Mechanics and Mining Sciences & Geomechanics Abstracts* (Vol. 3, No. 4, pp. 325-335). Pergamon.

Phillips D. W. 1930. The nature and physical properties of some coal-measure strata. *Trans. Instn. Min. Engrs.*, V80, P212-42.

Phillips D. W. 1931. Further investigation of the physical properties of coal-measure rocks and experimental work on the development of fractures. *Trans. Instn. Min. Engrs.*, V82, P432-50.

Phillips D. W. 1954. *The use of stereographic Projection in structural Geology*, London, Arnold.

Pomeroy, C. D. 1956. Creep in Coal at room temperature. *Nature*. V178, P279-80.

Quansheng, L., Hua, Z. and Tao, L., 2004. Study on stability of deep rock roadways in coal mines and their support measures. *Chinese Journal of Rock Mechanics and Engineering*, 23(21), pp.3732-3737.

Singh, D.P., 1975, September. A study of creep of rocks. *International Journal of Rock Mechanics and Mining Sciences & Geomechanics Abstracts*, Vol. 12, No. 9, pp. 271-276. Pergamon.

Valsangkar, A.J. and Gokhale, K.V.G.K., 1972. Stress-strain relationship for empirical equations of creep in rocks. *Engineering Geology*, 6(1), pp.49-53.

Wang, C., Wang, Y. and Lu, S., 2000. Deformational behaviour of roadways in soft rocks in underground coal mines and principles for stability control. *International Journal of Rock Mechanics and Mining Sciences*, 37(6), pp.937-946.

Wawersik, W.R., 1974. Time-dependent behaviour of rock in compression. *Advances in Rock Mechanics Proc. 3rd Congr. Int. Soc. Rock Mechanics*. Colorado:[sn], pp.357-363.

Yang, C., Daemen, J.J.K. and Yin, J.H., 1999. Experimental investigation of creep behavior of salt rock. *International Journal of Rock Mechanics and Mining Sciences*, 36(2), pp.233-242.

Bir Linyit Açık Ocağının Şevlerinin 2 Boyutlu Sonlu Elemanlar Geri Analizi: Bir Örnek Çalışma

2D Back Analysis of Slopes of A Lignite Open Pit Using Finite Element Method (FEM): A Case Study

Z. Sertabipoğlu, Ü Özer
İstanbul Üniversitesi, Maden Mühendisliği Bölümü, İstanbul

H. Tunçdemir
İTÜ Maden Mühendisliği Bölümü, İstanbul

ÖZET Açık ocak maden işletmelerinde cevher üretimi, dekapaj kazısı, nakliyat, atık yönetimi vs. gibi tüm madencilik faaliyetleri şevler oluşturularak yapılmaktadır. Bu tür işletmelerde olası bir şev kayması, yakın alandaki yapılara, insanlara ve ekipmanlara tehlike yaratabilir ve maden üretiminin aksamasına ve üretim maliyetinin artmasına neden olabilir.

Bu çalışmada, Orhaneli açık ocağında kalıcı doğu şevlerinde gelişen şev yenilmelerinin analiz edilmesi ve kayan kütlenin hareketi devam ederken, kütle hareketinin yönü ve yenilme modelinin belirlenmesi amaçlanmıştır. Bu kapsamda, çalışma alanındaki şevin kayma ve kazı öncesi ve sonrasına ait toplam beş kesit hazırlanmış ve Sonlu Elemanlar Yöntemi (SEY) kullanılarak yenilme modelinin İki Boyutlu (2B) Geri Analizi yapılmıştır.

Analizlerin sonucunda, incelenen şevde en son kazı sonrasında kaymanın devam etmeyeceği ancak şev önündeki üretim basamaklarında (ocak içindeki) kömür üretimi için kazı devam ettiği sürece duyarsızlığın artacağı ön görülmüştür.

ABSTRACT All mining activities such as, extraction of mineral, overburden excavation, transport, waste management are carried out by creating slopes in open pit mines. In these type of mines, a possible slope failure may harm to structures, people and equipment which are close to the slope and also may lead to interruption of the production and increment in the cost of mining.

Aim of this study is to analyze instabilities that have developed in the permanent east slopes of the Orhaneli open cast. In addition to this, while mass movement is continuing, it is targeted for determining the direction and failing model of mass movement. In this context, totally five sections, were prepared to understand the geometry of the instabilities and Two-Dimensional (2D) Back Analyses of the failing model have been carried out by using Finite Elements Method (FEM).

As a result of the analysis, it was found that the slope failure examined in this study will not continue at the time after the last excavation. However, it was predicted that instability of benches in front of the slope (inside the pit) would increase as long as the excavation continues for coal production.

1 INTRODUCTION

After a slope fails by sliding, a stability analysis needs to be performed with respect to understanding the failure mechanism for latter deformation activities. If the slope has failed, the factor of safety is considered to be less than (1.0) at the time of failure. For an appropriate method of analysis, it is possible to develop a model of the slope at the time that it failed (Duncan and Wright, 2005). Each failure process in soil and rock slopes can be reasonably monitored as a large scale in situ shear test. With understanding of the failure mechanism and gathering the required data for the failed mass, its geotechnical properties may be specified by a specific method which is named as "Back Analysis (BA)" (Sharifzadeh et al., 2010).

BA can be performed for a determination of the shear strength parameters (cohesion (c), internal friction angle (ϕ) etc.) at the time of failure. It is generally considered that shear strength parameters obtained by back analysis of slope failures ensure more reliability than those obtained by laboratory or in situ testing which is influenced by a scale effect. Therefore, this type of analysis is an effective approach to provide an insight into the underlying failure mechanism and improve the understanding regarding the factors controlling the stability of slopes. This analysis is the most powerful method that can be used to determine the shear strength parameters of a failed material (Sharifzadeh et al., 2010; Ng et al., 2014). In order to realize a successful back analysis strategy, it is essential to understand the processes and mechanisms driving instability (Sharifzadeh et al., 2010). In general, the determination of failure mechanism depending on the material involved may be divided into three broad categories (Sonmez et al., 1998):

(a) Methods being suitable for slopes in soils or soil like materials where the strength of the material can be determined from testing small specimens of the material in the laboratory.

(b) Methods being suitable for slopes in hard jointed rocks where slope stability is controlled by the discontinuities in the rock material.

(c) Methods being suitable for closely jointed rock masses where failure can occur along a shear surface similar to those observed in soil slopes, as a result of a combination of macro and micro jointing, and through the rock substance. Determination of the strength of this category of rock mass is extraordinarily difficult since the size of representative specimens is too large for laboratory testing. When slope failure occurs under such mechanism, back analysis methods could be carefully applied to calculate the rock mass strength (Sonmez et al., 1998; Tüdeş and Ceryan, 2011; Sharifzadeh et al., 2010).

One of the key problems of back analysis is to calculate the factor of safety. The limit equilibrium method was adopted in to the aforementioned researches, and it was assumed that the slip surface was a circular one. Since this method is a statically indeterminate problem, assumptions on the inter-slice shear forces are employed to render the problem statically determinate. Although limit equilibrium analysis is widely used and is a simple method to assess the stability of the slopes, it is difficult to assess the accuracy of the limit equilibrium solution and is not interested with stress distribution along the slope, below or above the sliding surface and deformation and stress distribution after the mass break occurs (Zhang et al. 2013; Eberhardt, 2003; Karaman et al. 2013). Therefore, the stress and displacement distributions in the slope can be calculated by using numerical methods. The analysis of the stability of slopes using limit equilibrium methods necessitates determination of the critical slip surface that yields the minimum factor of safety (FS). Numerical methods solve the governing equation of continuum mechanics based on predefined boundary conditions. The factor of safety is not solved explicitly in numerical methods but can be assessed based on the Shear Strength Reduction (SSR) technique (Dawson, et al. 1999).

The aim of this study is to investigate the instabilities that have developed in the permanent east slopes of the Orhaneli Open Cast of Turkish Coal Enterprises. An analysis is performed for finding the direction and failing model of mass movement, while mass movement continuing. In study area, totally five cross sections are taken before and after the slide experienced at three different time intervals, and a 2D BA failing model have been carried out by using Finite Elements Method (FEM) approximation.

2 STUDY AREA DESCRIPTION

Orhaneli Open Pit Mine (OOPM) is one of the branches of Bursa Lignite Establishment (BLİ) on the management of Turkish Coal Enterprises (TKİ). The distance between Orhaneli County and OOPM is 21 km. The altitude of mine is +500m from the sea level. OOPM is an area next to Gümüşpınar, Sağırlar and Çivili Villages in a space with 1 km, 28km and 30 km respectively (Figure 1). Average seam thickness of coal is 7m, 3.35m in Çivili and Sağırlar sectors. Reserves of them are 23 million, 10.5 million and 6 million tons respectively. However, Gümüşpınar sector is only on service now. A large amount of coal, projected with 1.2 million tons, are planned to be extracted per year. 75 % of it are sent to thermoelectric power plant to burn and produce electricity (Tundemir et al., 2013; Akcay et al, 2013, Sertabipoğlu, 2016).

2.1. Geology

Geology of mine has a complex structure. Stratigraphic time scale of Pre Neogene structures consist of metamorphic schists, recrystallized limestone, and serpentinites. Neogene formation scales covers from clastic rocks in the lowest level. Clastic rocks contains conglomerates, sandstones, gravels and clays at the bottom side and marl and tuffs consisting of lignite seams at the top side which are coal reserves of OOPM. Volcano sedimentary and volcanic rocks are settled on them (Tunçdemir et al., 2013; Sertabipoğlu, 2016).

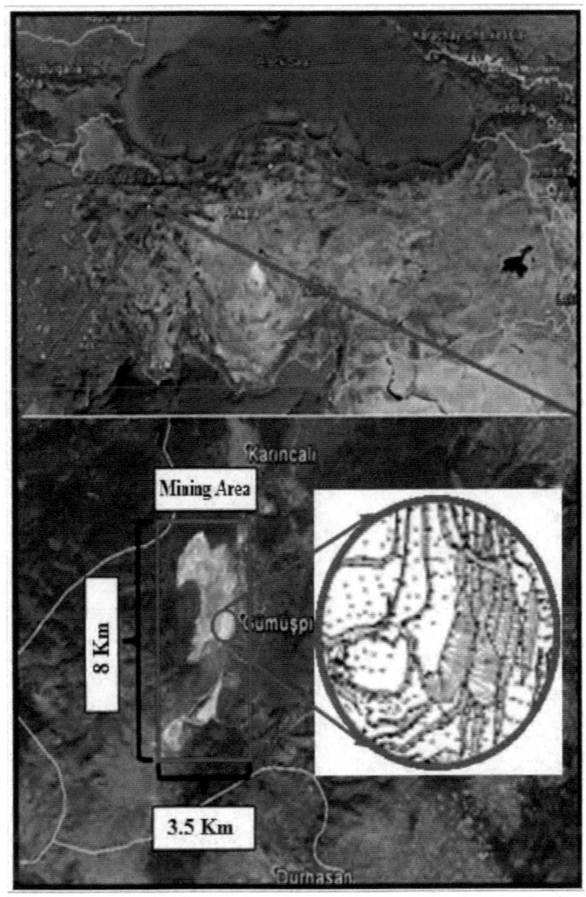

Figure 1 Location map of study area

2.2. History of the study area

The study area (Figure 1) was chosen as the east slopes of the Gümüşpınar sector because three slope failures had experienced on 24th April 2012, 15th January and 2th April 2013. Cross-sections of slope failures determined with regard to failure direction (A-A' section, seen in Figure 2a) can been seen in Figure 2b.

After the third failure (2th April 2013), there was still slope instability hazard despite of naturally decreasing the slope angle in each failure. For this reason, a part of the sliding mass was excavated in September 2013 by the enterprise. However, slope stability could not be completely ensured. Therefore, all sliding mass was excavated down to bedrock level (limestone) and removed from slope, in June 2014 (Figure 2a).

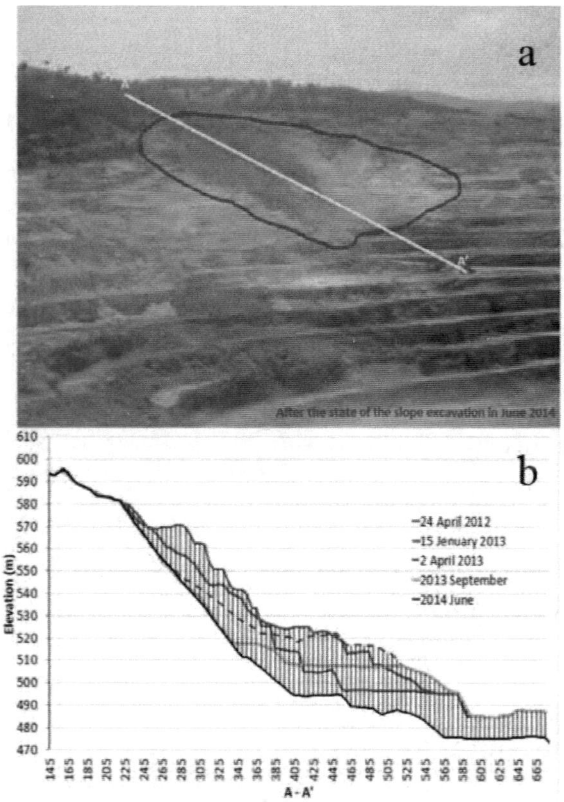

Figure 2 (a) After the state of the slope excavation in June 2014, (b) view of sections of slope failures happened in 2012 and 2013 and post-excavations.

3 BACK ANALYSIS

In order to detect the slope failure mechanism of failures and excavations 2D Finite Element analysis were performed using Phase2 v. 8.0 computer software program (Back Analysis).

In all analyzes, 7 materials (tuffite, marl, sandstone, weak zone, coal, claystone, and limestone and underground water table) were defined for geometric modeling (Figure 3). The material parameter values were obtained from previous experimental studies of study area (Aksoy et al, 2013; Karpuz et al., 2006; Özçelik et al., 2013) in order to create the most adverse conditions in the slope (Table 1). Mohr-Coulomb Failure Criterion (M-C) were used because of the presence of dense heterogeneous zones and discontinuities in the field.

3.1 Analysis of failures

First Failure on 24th April 2012: a pre-failure geometric model was created and 7 materials and underground water table were defined to the model (Figure 3a). Material parameters were used for analysis seen in Table 1. The Strength Reduction Factor (SRF) was found 0.64 (Figure 4a).

Second Failure on 15th January 2013: Due to being disturbed tuffite, marl, and sandstone in failure zone at the time of first failure, shear strength parameter (c and ϕ) values were used for analysis of second failure as the first disturbed material values (Table 2) and assigned to the materials in failure zone of geometric model (Figure 3b). The SRF value for second failure was found 0.69 (Figure 4b).

Third Failure on 2th April 2013: the shear strength parameter values in failure zone at the time of the second failure were taken as the second disturbed material values (Table 2) and assigned to the materials in failure zone of geometric model (Figure 3c). The SRF was found as 0.73 (Figure 4c).

In three analysis, it was found that the mass would continue to slip downward along the surface of limestone rock (base rock) in the direction of the pit bottom. In addition, circular sliding type failings were observed due to closely mixed up media with regard to undefined joint planes or joint sets, which control the form of the failure mode.

3.2 Analysis of excavations

First excavation in September 2013: Since a part of the sliding mass was removed and there was still disturbed material in slope, Shear strength parameter values for third failure (2th April 2013) were used in geometric modelling (Figure 3d) as the third disturbed material values (Table 2). the SRF was found 0.93 (Figure 4d). According to this value, the first excavation carried out in the area does not ensure slope safety.

Second excavation in June 2014: 7 materials and underground water table have been defined to the geometric model showing the latest situation (Figure 3e). Material properties seen in Table 1 were assigned since all disturbed materials have been removed completely by the excavation. The result of analysis shows that there is no slip surface in the slope zone and the failure will not continue (Figure 4e). the SRF value in the production benches (towards into the pit) is 1.14 and very close to the equilibrium value. It is predicted that the instability of benches would increase consequently while the excavation continues for coal production. As a matter of fact, according to information received from the enterprise there has been a further slope failure in these production stages in February 2016.

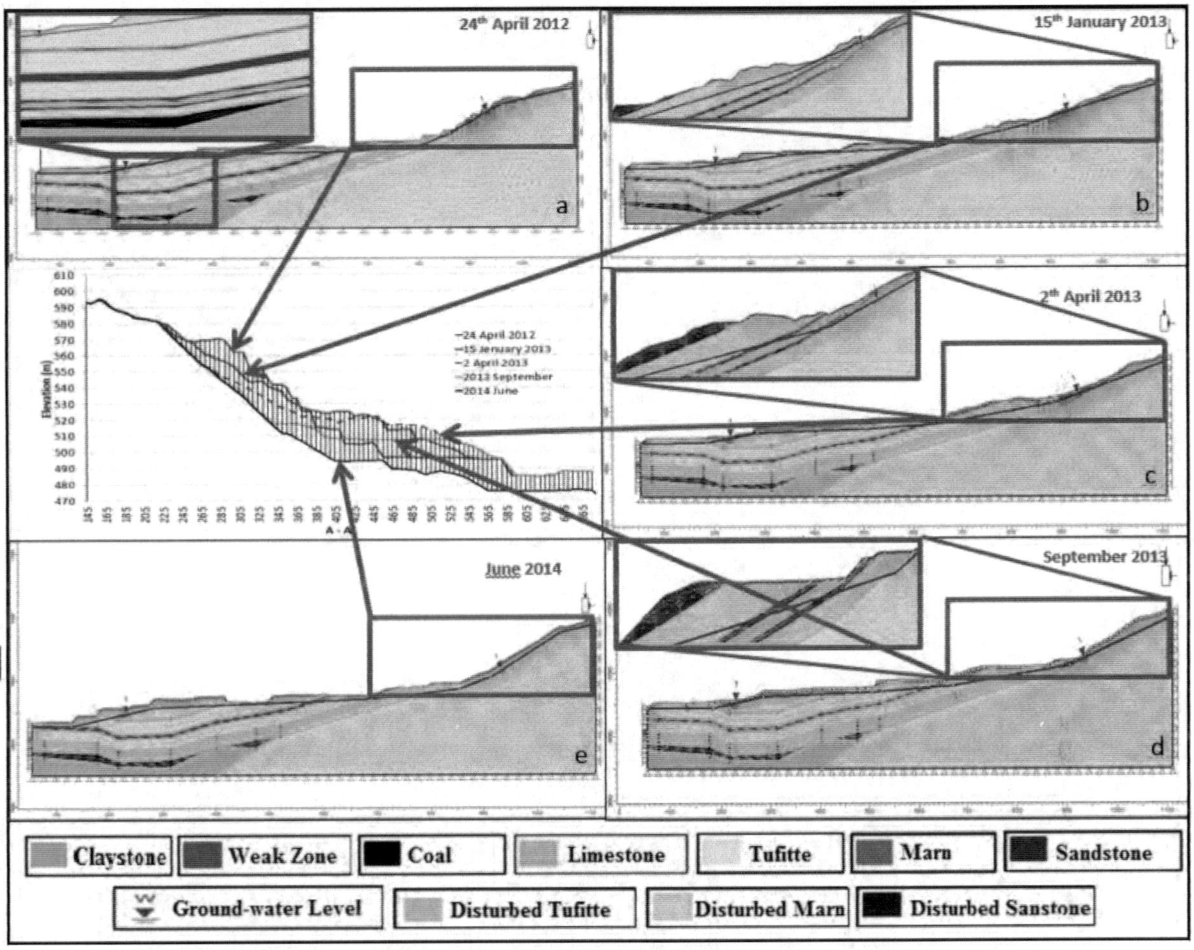

Figure 3 Geometric model of (a) the first failure, (b) the second failure, (c) the third failure, (d) the first excavation and (e) the second excavation

Table 1 Properties of materials

Parameters	Materials						
	Tuffite	Marl	Sandstone	Weak Zone	Coal	Claystone	Limestone
Unit Volume Weight (kN/m^3)	19.00	16.41	18.46	17.65	12.46	20.0	27.37
Friction Angle (ϕ) (°)	10.0	25.0	26.0	11.0	20.7	15.0	38.0
Cohesion (c) (kPa)	18.0	13.0	56.0	5.0	325.0	180.0	500.0
Compressive Strength (kPa)	2200	9000	9800	2200	3800	11000	45100
Tensile Strength (kPa)	220	900	980	220	380	1100	4510
Elastic Modulus (kPa)	10000	400000	335000	50000	500000	394000	2000000
Poisson ratio	0.32	0.30	0.33	0.35	0.20	0.35	0.28

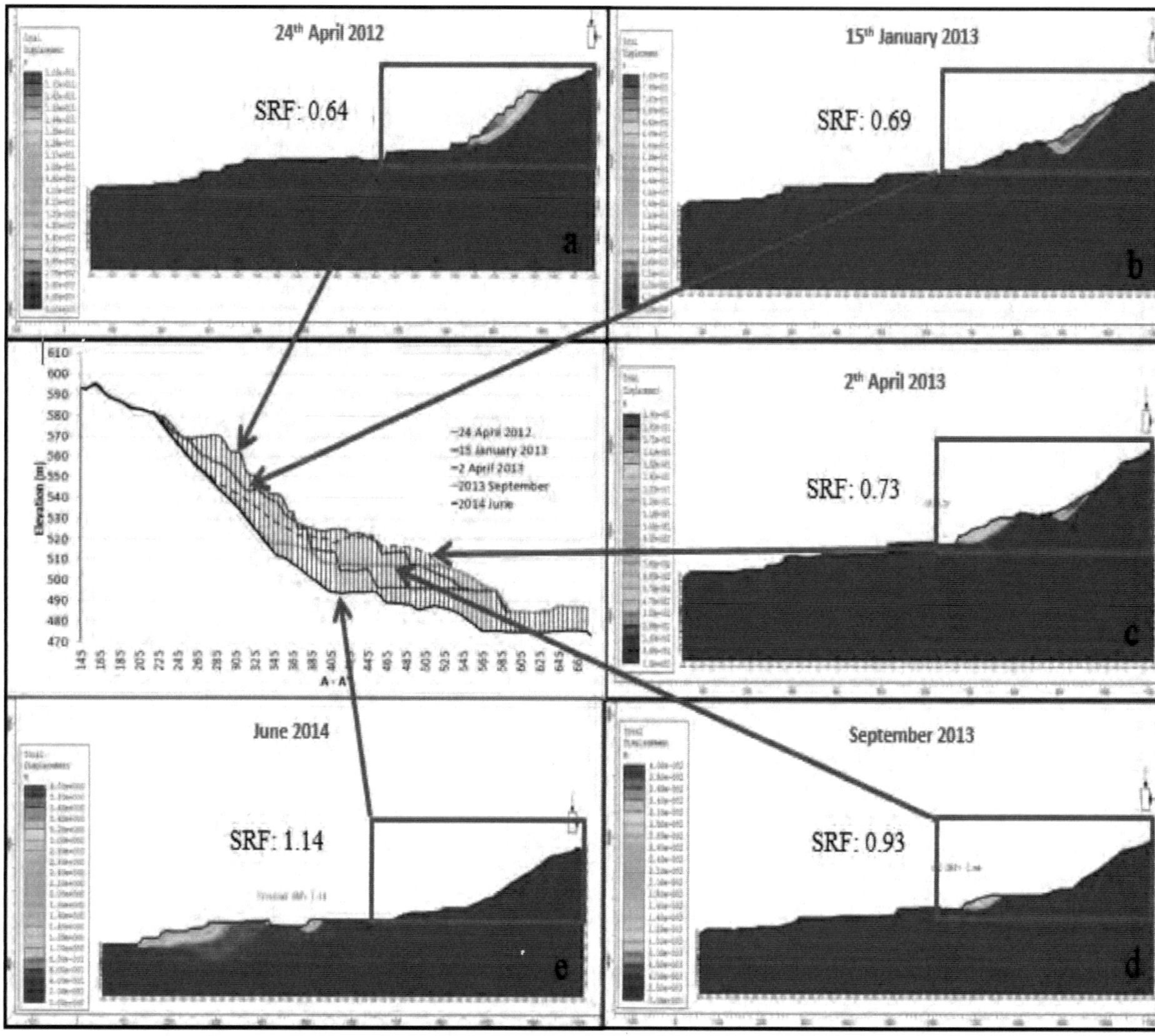

(c) the third failure, (d) the first excavation and (e) the second excavation

Table 2 Shear strength parameters and SRF values of the failures and excavations

	24th April 2012 1st failure		15th January 2013 2nd failure		2th April 2013 3rd failure		September 2013 1st excavation		June 2014 2nd excavation	
	ϕ*	c**	ϕ	c	ϕ	c	ϕ	c	ϕ	c
Tuffite	10	18	10	18	10	18	10	18	10	18
Marl	25	13	18	13	18	13	18	13	18	13
Sandstone	26	56	26	56	26	56	26	56	26	56
			1st disturbed		2nd disturbed		3rd disturbed			
Disturbed Tuffite			7	14	4.71	9.35	4.24	8.42		
Disturbed Marl			20	10	13.72	6.71	12.40	6.04		
Disturbed Sandstone			21	46	14.44	30.87	13.06	27.81		
Weak-zone	11	5	11	5	11	5	11	5	11	5
Coal	20.7	325	20.7	325	20.7	325	20.7	325	20.7	325
Claystone	15	180	15	180	15	180	15	180	15	180
Limestone	38	500	38	500	38	500	38	500	38	500
SRF***	0.64		0.69		0.73		0.93		1.14	

*ϕ: Friction Angle (°) **c: Cohesion (c) (kPa) ***SRF: Strength Reduction Factor

4 CONCLUSIONS

Purpose of the current study is to analyze the instabilities that have developed in the permanent east slopes of the Orhaneli Open Pit Mine (OOPM) between 2012 and 2015 years. The analysis is performed for finding the direction and failing model of mass movement, while mass movement continuing. In this context, totally five cross sections are taken before and after the slide experienced at three different time intervals, and a Two-Dimensional (2D) Back Analyses (BA) failing model have been carried out by using Finite Elements Method (FEM) approximation.

In analysis of failures; it was found that the mass would continue to slip downward along the surface of limestone rock (base rock) in the direction of the pit bottom, towards to the front of the benches despite the natural decrease in overall slope angle and the increase in SRF value for each failure. In addition, circular sliding type failings were observed due to studied rock slope in closely jointed media has no clearly defined joint planes or joint sets, which control the form of the failure mode.

First excavation analysis; It was understood that the excavation carried out in the area does not ensure slope safety.

In second excavation analysis: it was found the slope failure examined in this study will not continue at the time after the last excavation. However, it was predicted that instability of benches in front of the slope (inside the pit) would increase as long as the excavation continues for coal production. As a matter of fact, according to information received from the enterprise, there has been a further slope failure in these production stages in February 2016.

Consequently, experiments conducted on the field and in the laboratories represent only a limited area in mine sight, whereas back analysis provides a more accurate way to obtain the shear strength information of that region. In summary, an accurate and economic package of measures from the engineering point of view can be prepared by the way of projecting the measures to be taken for the designed geometric conditions and the shear strength range calculated by Back Analysis method.

REFERENCES

Akcay O., Akyilmaz O., Tuncdemir H., Guclu E., Yilmazturk S., 2013. Monitoring and Evaluation of The Continuous Surface Changes on Orhaneli Open-Pit Mine, 5. TerraSAR-X / 4. TanDEM-X Science Team Meeting, DLR – Oberpfaffenhofen, 4 syf. CDROM.

Aksoy, O., Onargan T., Küçük, K., Pamukçu Ç., Özacar, V., 2013. Gümüşpınar kömür sahasının optimum değerlendirilme koşullarının belirlenmesi, üretim faaliyetleri halen devam eden basamaklardaki şev stabilitesi problemlerinin çözümüne yönelik çalışmaların yapılması ve işletme projesinin revize edilmesine yönelik proje raporu, 9 Eylül Üniversitesi Maden Mühendisliği Bölümü, s. 141.

Dawson, E. M., Roth, W.H., Drescher, A. 1999. Slope stability analysis by strength reduction. Géotechnique, 49(6), pp. 835-840.

Duncan, J.M., Wright, S.G., 2005. *Soil strength and Slope Stability*, John Wiley&Sons, Inc., New York, NY, USA, pp.309.

Eberhardt, E. 2003. *Rock Slope Stability Analysis - Utilization of Advanced Numerical Techniques*, Vancouver, Canada: Earth and Ocean Sciences, University of British Columbia, pp.41.

Karpuz, C., Koçyiğit A., Tutluoğlu, L., Düzgün, Ş., Koçal, A., Erdem, E., Alkılıçgil, Ç., 2006. Türkiye Kömür İşletmeleri Kurumu TKİ, Orhaneli İşletmesi Açık Ocakları Panolarında şev tasarımı ve Dragline çalışma sisteminin belirlenmesi projesi, ODTU Maden Müh. Bölümü, s.73.

Karaman, K., Ercikdi B., Kesimal, A., 2013. The assessment of slope stability and rock excavatability in a limestone quarry, *Earth Sciences Research Journal*, http://www.revistas.unal.edu.co/index.php/esrj/article/view/35804/44934.

Ng, S.M., Ismail, M. A. M., Abustan, I., 2014. Back analysis of slope failure using Finite Element with Point Estimate Method (FEM-PEM), *Journal of Civil Engineering Research*, 4(3A), pp. 31-35.

Özçelik, Y., Dirik, C.K., Özsayın, E., Yılmazkaya, E., 2013. TKİ-GLİ Müessesesine bağlı Bursa Linyitleri İşletme Müdürlüğüne Ait Gümüşpınar, Sağırlar Ve Çivili sektörlerindeki kömür sahalarının optimum değerlendirilme koşullarının belirlenmesi ve üretim faaliyetleri halen devam eden Gümüşpınar sektöründeki şev stabilitesi problemlerinin çözümüne yönelik çalışmaların yapılması, işletme projesinin revize edilmesi,

Sağırlar ve Çivili sektörlerindeki yer altı rezervlerinin işletilmesine yönelik avan projelerin hazırlanması projesi nihai raporu, Hacettepe Üniversitesi Jeoloji Bölümü, s.104.

Sharifzadeh, M., Sharifi M., Delbari, S. M., 2010. Back analysis of an excavated slope failure in highly fractured rock mass: the case study of Kargar slope failure (Iran), *Environ Earth Sci*, 60, pp.183–192, DOI 10.1007/s12665-009-0178-2.

Sertabipoğlu, Z., 2016, Monitoring of open pit slope movements by using InSAR and slope stability analysis, Istanbul University, *PhD thesis*, p. 227.

Sonmez, H., Ulusay, R. & Gokceoglu, C.,1998. A practical procedure for the back analysis of slope failures in closely jointed rock masses, *Int. J. Rock Mech. Min. Sci.*, vol. 35(2), pp. 219-233.

Tunçdemir, H., Güçlü, E., Akçay, Ö., Akyılmaz, O., Yılmaztürk, S., Motagh, M., 2013. A Methodology for deformation control with InSAR: Orhaneli Open Pit Mine Case Study, *23rd International Mining Congress and Exhibition of Turkey*, Turkey, Antalya, pp. 211-217.

Tüdeş, Ş., Ceryan, N. 2011. A comparative study on the estimation of shear strength of rock masses using rock SSPC system and Hoek-Brown Criterion, *Gazi University Journal of Science*, 24(4), pp. 855-865.

Zhang, K., Cao, P., Bao, R., 2013. Rigorous back analysis of shear strength parameters of landslide slip, *Trans. Nonferrous Met. Soc. China*, 23, pp.1459–1464.

Comparison of Mechanical Properties of Thin Spray-on Liners under Tension and Compression

Püskürtme İnce Kaplamaların Mekanik Özelliklerinin Basma ve Çekme Altında Karşılaştırılması

D. Güner, H. Ozturk

Middle East Technical University, Mining Engineering Department, Ankara, Turkey

ABSTRACT

Thin spray-on liners (TSL's) are polymer based rock support systems used especially in deep underground openings. Although they have higher early strength, elongation and waterproofing performance than the conventional shotcrete, the usage is still limited. Support mechanism of TSL is a research topic that is still investigated. In this study, the mechanical behaviour of two different TSLs were compared by experimental results and numerical analysis. For this purpose, tensile and deformability tests of TSLs were performed according to the related test standards. As a result of laboratory studies, ultimate and yield strength, elastic modulus and tensile modulus, and modulus of resilience values were found. In addition, laboratory tests were simulated in finite element analysis software to find the strain energy density variation during loading. Characterization of TSL products by experiments and numerical analysis is of major importance to understand the support mechanism of TSLs.
Keywords: Thin spray-on liner, Deformability test, Tensile testing, Yield strength, Modulus of resilience.

ÖZET

Püskürtme ince kaplamalar (PİK), özellikle derin yeraltı açıklıklarında kullanılan polimer bazlı kaya tahkimat sistemleridir. Püskürtme betona nazaran yüksek erken dayanım, uzama ve su geçirmezlik performansına sahip olmalarına rağmen kullanımı halen yaygın değildir. PİK'lerin tahkimat mekanizması halen araştırılmaktadır. Bu çalışmada iki farklı PİK ürününün mekanik davranışları deneysel sonuçlar ve sayısal analiz ile karşılaştırılmıştır. Bu amaçla, çekme ve deformasyon deneyleri ilgili test standartlarına uygun olarak gerçekleştirilmiştir. Laboratuvar çalışmaları sonucunda nihai ve akma dayanımları, elastisite ve çekme modülleri ve esneme modülü değerleri bulunmuştur. Ayrıca laboratuvar deneyleri sonlu eleman analiz yazılımı ile yükleme sırasında gerinim enerji değişiminin tayin edilmesi amacıyla simüle edilmiştir. PİK ürünlerinin deneysel ve sayısal analiz yöntemleri ile karakterize edilmesi PİKlerin tahkimat mekanizmasının anlaşılması açısından önem arz etmektedir.
Anahtar Kelimeler: Püskürtme ince kaplama, Deformasyon deneyi, Çekme deneyi, Akma dayanımı, Esneme modülü.

1 INTRODUCTION

Ground support is a challenging issue almost in all rock engineering related fields. Different active and passive support systems are used for rock reinforcement and rock support purposes under varying ground conditions. Shotcrete is the most widely used rock support element in underground projects. As technology is advancing, the mining industry demands improvement in mechanical properties of shotcrete. Although some shotcrete additives, such as accelerator, hardener, and steel-polymer fibers are engineered to meet the demand of industry, researchers started to develop new technology material with the support of private companies. At the end of 1980's, Canadian Mining Industry Research Organization improved the first thin spray-on liner (TSL), a polyurethane-based product. This product has a very fast curing rate, higher adhesion, tensile strength, elongation capability, good waterproofing performance, and is easily applicable with discardable rebound amount in comparison with shotcrete.

General definition of TSL, as accepted by the mining society, is "generally cement, latex, polymer-based and also reactive or non-reactive, multi-component materials applied to the rock surface sprayed by nozzle, in a layer of generally 6 mm or less (3-5mm) thickness material" (Hadjigeorgiou, 2003). Figure 1 shows the application of the TSL.

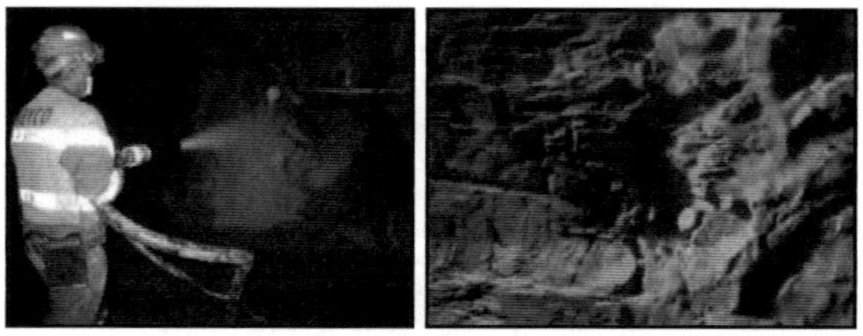

Figure 1. Application of TSL Product (BASF, 2013).

In literature, many different advantages of TSLs were listed. However due to the lack of understanding of its support mechanism, TSLs are now only being used in around 150 mines and underground openings worldwide. Companies and researchers are trying to introduce the support mechanisms of TSLs to the industry by performing different analytical, numerical, and experimental studies.

Currently 17 TSL suppliers are available on the market. Laboratory tests have significant importance to understand the support capacity and to compare existing TSL products. EFNARC (2008) specified the minimum mechanical performance requirements of a TSL product, which could create the difference between TSLs as improved mechanical properties might aid to strengthen their market share.

10 different laboratory test setups were used in the literature. Those tests are;

- tensile strength,
- uniaxial compressive strength,
- punch,
- bonding, adhesion strength,
- double sided shear strength,
- plate pull,
- linear block support,
- gap shear load
- coated core compression

Researchers and the manufacturers still compare ultimate tensile strength, elongation capacities, and bond strength of TSLs. According to the EFNARC (2008); 7-day cured TSL should have at least 2 MPa tensile strength, 20 MPa tensile modulus and 10% elongation at break values.

In addition to the laboratory studies, numerical studies are commonly performed in the last decade in order to simulate TSL laboratory testing and field conditions. These studies are summarized in Table 1.

Table 1. Conducted Numerical TSL Studies in the Literature.

Authors	Model purpose	Numerical method
Tannant and Wang (2003)	Tensile, Block Punch tests and tunnel model	Discrete element
Malan and Napier (2008)	Fractured underground excavation (square)	Boundary element
Richardson et al. (2009)	Bending, Double sided shear test	Finite difference
Dirige and Archibald (2009)	Underground excavation (horseshoe)	Finite difference
Nater and Mena-Cabrera (2010)	Underground excavation (square)	Distinct element
Lee et al. (2015)	Linear block support	Finite element
Guner and Ozturk (2016)	Tensile testing	Discrete element

1.1 Scope of the Study

In the basis of this study, 7-day cured 2 different TSL products were tested under ambient temperature in accordance with ASTM D638 and ASTM D695 test standards. Ultimate and yield strength, elastic modulus and tensile modulus, and modulus of resilience parameters were investigated. Moreover, since the energy absorbance capacity is also another significant parameter for TSL comparison, the strain energy density variation under tensile and compressive loads were determined by simulating the laboratory test by numerical methods.

Tensile tests are commonly performed TSL testing methods by researchers and manufacturers as field studies show that TSLs commonly failed under the influence of tensile forces.

In the literature and practice, researchers generally consider ultimate tensile strength values as a design parameter. However, TSLs are viscoelastic materials and have time dependent material behaviours.

Guner and Ozturk, (2017), performed creep tests on TSLs. In this study, it is observed that, if the applied constant load exceeds the elastic limit of the liner, failure takes place within an hour. In field applications, loading conditions are irreversible which means that there is no possible way of unloading of TSL. Therefore, instead of using ultimate tensile strength as a design parameter, yield strength should be utilized as a more accurate input parameter.

Resilience is the capability of a material to absorb energy when it is deformed elastically. This parameter represents the energy per unit volume that a material can absorb without yielding. The unit of the modulus of resilience is in J/m^3. This parameter is very useful in material comparison, especially in polymer engineering, as it takes both yield strength, and modulus of elasticity into account. If the material behaves linear elastic and brittle, this parameter is equal to strain energy.

2 LABORATORY STUDIES

In order to perform laboratory studies, 2 different TSL products manufactured in Europe, were selected. Laboratory studies were conducted at the Middle East Technical University (METU), Rock Mechanics Laboratory, Ankara.

2.1 Tested Products

Product A has two components; the liquid-component is a stabilised resin latex and the powder-component is a hydraulically curing powder based on special cement. Components are mixed with 2:1 liquid-powder ratio by weight. The setting time of the product is about 30 minutes.

Product B is a one-component polymer based powder for spray application on rock or coal faces for surface support and protection against different conditions. One of the most beneficial sides of the product is its setting time. Laboratory tests proved that the setting of the mixture takes only few minutes.

2.2 Sample Preparation

Sample preparation is one of the most labor intensive parts of this experimental study. In order to obtain homogenous test specimens, all test specimens were prepared in a single mixing batch. Dogbone shaped tensile test specimens were prepared by molding and die cutting technique. Cylindrical compression test specimens with 25 mm diameter and 55 mm height were prepared by coring technique.

Mixing (A), molding (B), and die cutting (C) steps for tensile test specimen preparation are presented in Figure 2.

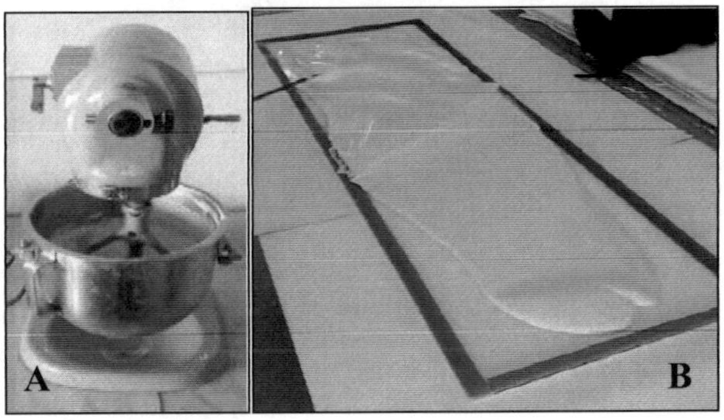

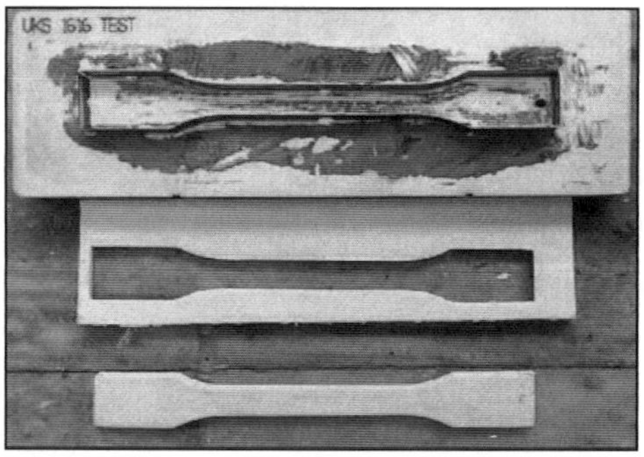

Figure 2. Tensile testing sample preperation steps.

In tensile testing, specimen thickness is directly related to the field application thickness of TSL. As the general application thickness is about 3-5 mm, in this study, 4 mm sample thickness was selected. Specimen dimensions are presented in Figure 3.

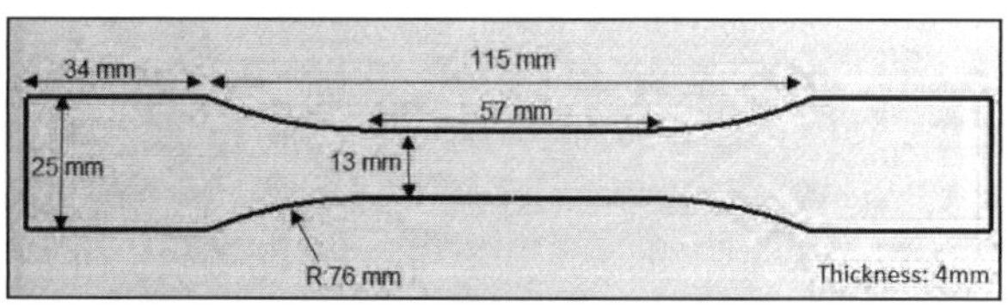

Figure 3. Tensile Test Specimen Dimensions

2.3 Testing Setup

During the laboratory studies, 20 valid test result were obtained. Tensile tests were conducted under slow strain option (6 mm/ min). During the tests, load, axial displacement, and time were recorded continuously. Tensile test set-up is presented in Figure 4.

The deformability tests were performed with a displacement controlled MTS 815 testing machine. Circumferential and axial displacement measurement instruments have a 10 mm capacity with ±0.001mm sensitivity. In the testing part, 10 mm capacity is not enough to measure upper limits. However, yield stress, elastic moduli, and Poisson's ratio were calculated in elastic loading part of the tests. Constant displacement rate, 1.5 mm/min, was applied during tests. Deformability test set-up is presented in Figure 5.

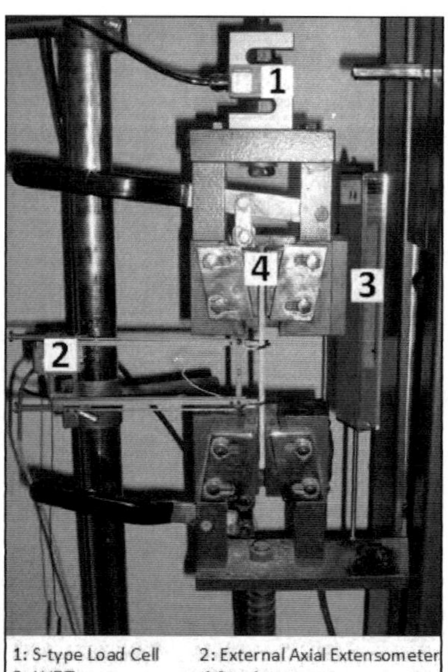

Figure 4. Tensile Test Set-up

Figure 5. Deformability Test Set-up

3 TEST RESULTS

3.1 Tensile Test Results

7-day cured sample test results are presented in this part. As a result of tensile tests, obtained stress strain curves and numerical model simulation results are presented in Figure 6.

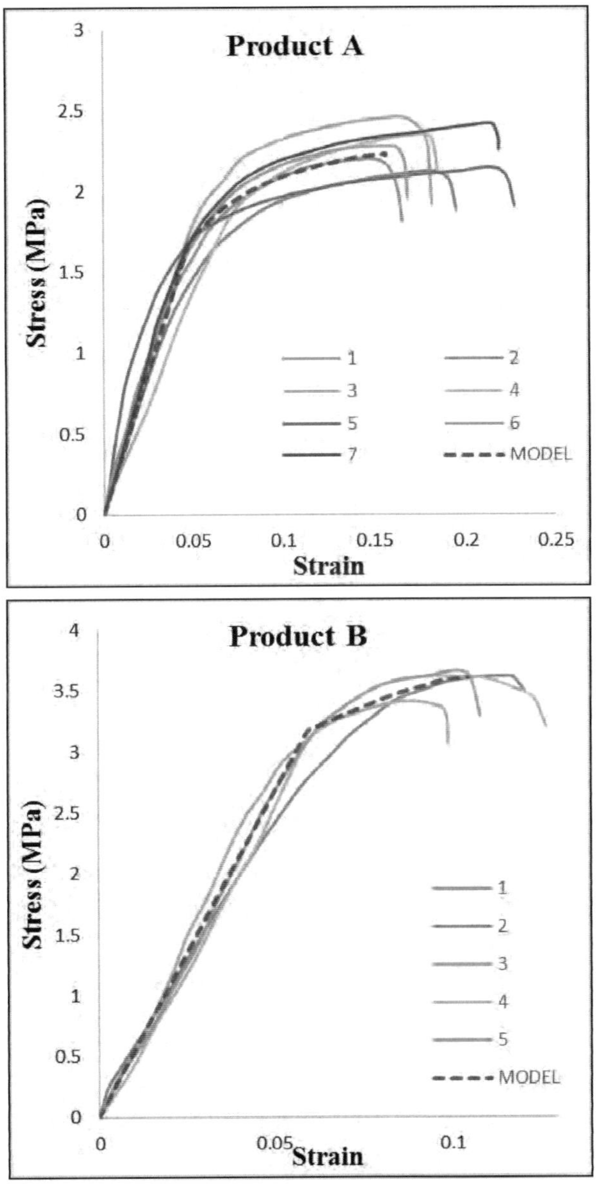

Figure 6. Stress-Strain Curves Obtained From Tensile Testing and Numerical Simulations

As it can be seen from the above figure, Product B is stronger and stiffer. On the other hand, elongation capability of this product is almost half of Product A.

Calculated tensile material properties with average and standard deviation values are presented in Table 2.

Table 2. Tensile Test Results

Product A				
Sample No.	σt (MPa)	Yield Strength (MPa)	Modulus of Resilience (kJ/m³)	Et (MPa)
1	2.20	1.74	47.66	31.76
2	2.19	1.64	37.97	35.42
3	2.45	1.96	50.06	38.37
4	2.39	1.82	48.30	34.29
5	2.19	1.64	41.75	32.21
6	2.28	1.82	46.54	35.59
7	2.15	1.92	45.15	40.82
Avg.	**2.26**	**1.79**	**45.35**	**35.49**
Std. Dev.	**0.11**	**0.13**	**4.19**	**3.24**

Product B				
Sample No.	σt (MPa)	Yield Strength (MPa)	Modulus of resilience (kJ/m³)	Et (MPa)
1	3.53	3.17	83.66	60.06
2	3.63	3.13	108.49	45.15
3	3.40	2.90	68.70	61.21
4	3.63	3.05	74.08	62.79
5	3.63	3.22	94.57	54.82
Avg.	**3.56**	**3.09**	**85.90**	**56.81**
Std. Dev.	**0.10**	**0.13**	**16.02**	**7.17**

As stated in the introduction part, yield strength parameter is more important than ultimate strength values. During support design studies, engineers should consider yield strength value. If the liner exceeds yield strength value in field application, since unloading is not possible, the failure will occur in a short period of time.

Modulus of resilience values give us the maximum energy absorbance capacity in the elastic region of the liner. This research suggests that the modulus of resilience parameter is a key value to compare the support capacity of liners. When we compare resilience values, Product B has higher energy absorbance capacity than Product A.

In addition to laboratory studies, numerical model of the simple tensile test was performed in ABAQUS (2017) software in order to observe the strain energy variation. Figure 7 shows the generated mesh and the axial stress distribution of the sample (in MPa) at the end of the simulation for Product B.

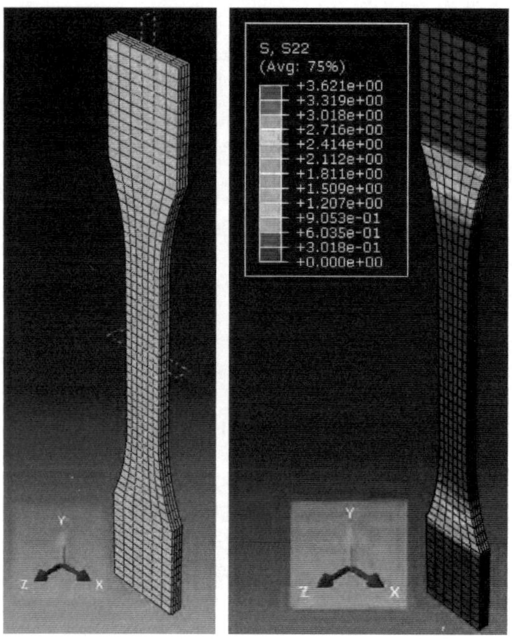

Figure 7. Mesh Geometry and Stress Distribution of the Tensile Model for Product B.

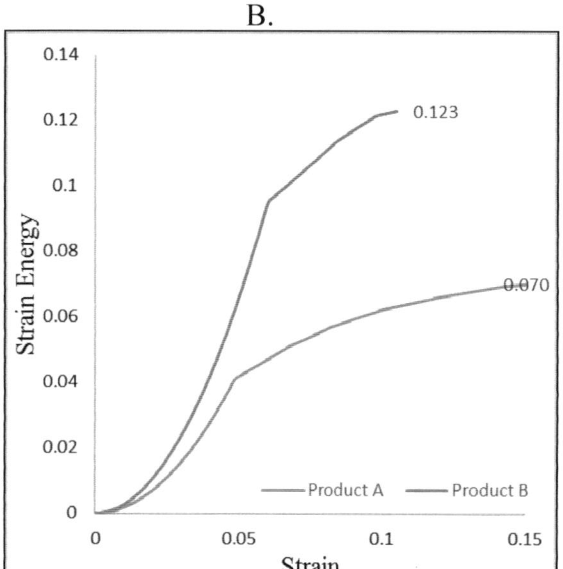

Figure 8. Strain Enery Variations during Tensile Test

As it can be seen from Figure 8, energy absorbance capacity increases exponentially in the elastic part while it increases nearly linear as the strain value increases in the post-peak part.

Product B has higher energy absorbance capacity, however, when plastic strain amounts were compared; Product A has a plastic strain about 8.5% and it is only 3.5 % for Product B.

As a result of tensile testing, it was seen that both products conform to the minimum mechanical requirements suggested by EFNARC (2008). Moreover, Product B has better mechanical properties than Product A. Notable differences were observed in yield strength and modulus of resilience values. Note that, all test conditions (mixing method-time, temperature and humidity, load rate) were kept constant for both products.

Main objective of the numerical model is to find a quantitative parameter, which would enable to compare TSL products. In future studies, interaction between the liner and the rock might be investigated by numerical modelling.

3.2 Deformability Test Results

At the end of the deformability tests, stress-strain and lateral-axial strain curves were plotted. Since both products behave relatively ductile, a distinctive failure state was not observed during the tests. The tests ended after observing a linear elastic path in the stress-strain curve. Note that constant displacement rate was applied (1.5 mm/min) during the tests. Obtained stress strain curves and numerical model simulation results are presented in Figure 9.

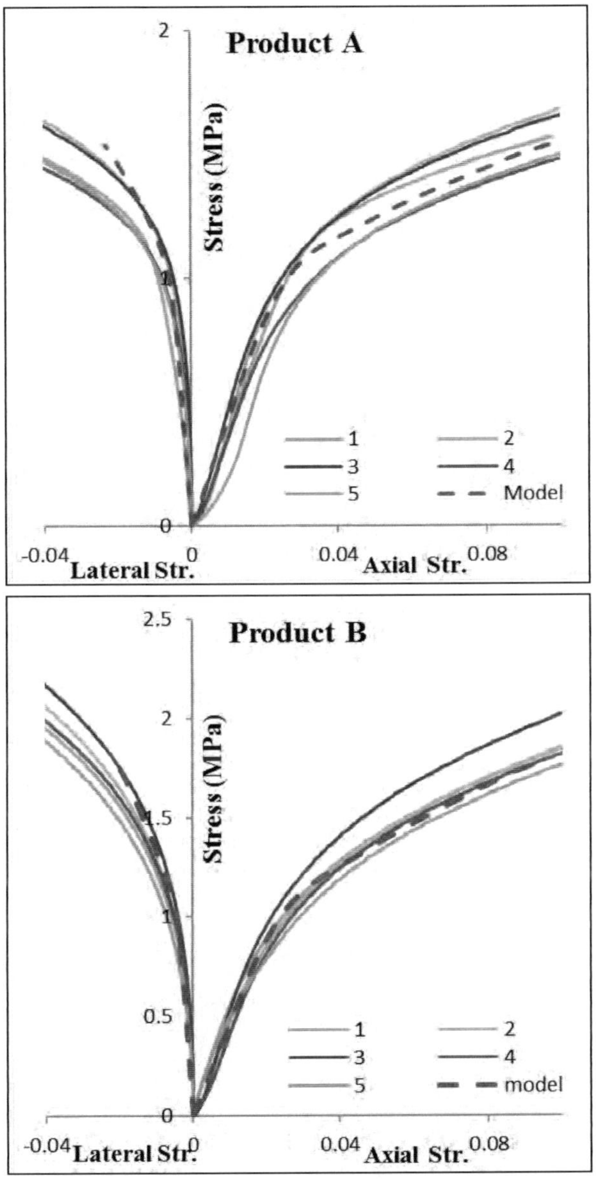

Figure 9. Stress-Strain Curves Obtained from Deformability Test and Numerical Simulations

According to Figure 9, both products have very similar responses to compression loading. They have a bi-linear stress-axial strain behaviour. As a result of the interpretation of the given curves, compressive material properties with average and standard deviation values were calculated (Table 3.).

Table 3. Deformability Test Results

Product A				
Sample No.	**E (MPa)**	**Yield Strength (MPa)**	**Modulus of resilience (kJ/m3)**	**Poisson's Ratio**
1	40.33	1.01	12.73	0.44
2	55.79	0.87	6.72	0.46
3	46.62	0.81	7.11	0.45
4	39.39	0.78	7.67	0.43
5	46.80	0.68	4.99	0.42
Avg.	**45.79**	**0.83**	**7.8444**	**0.44**
Std. Dev.	**6.57**	**0.12**	**2.9078**	**0.02**

Product B				
Sample No.	**E (MPa)**	**Yield Strength (MPa)**	**Modulus of resilience (kJ/m^3)**	**Poisson's Ratio**
1	43.57	0.60	4.15	0.41
2	46.91	0.67	4.76	0.36
3	55.34	0.86	6.61	0.37
4	47.02	0.73	5.61	0.36
5	36.88	0.67	6.18	0.37
Avg.	**45.94**	**0.71**	**5.46**	**0.37**
Std. Dev.	**6.67**	**0.09**	**1.01**	**0.02**

As a result of deformability studies, an average modulus of elasticity value of 45 MPa was found for both products. Also yield strength values were found very similar and showed small variation between 0.7 to 0.9 MPa.

4 CONCLUSIONS

This study presents the tensile and compressive mechanical properties of two different TSLs. A new TSL selection parameter, modulus of resilience, which takes into account yield strength, brittleness of the material is presented for the first time in literature. Moreover, the importance of the yield strength parameter is discussed. During support selection, it is very critical to estimate the required energy absorbance capacity of the support so liner or support material that can withstand the corresponding energy requirement can be used. Therefore, modulus of resilience values might be a significant parameter to compare TSL products. Tested products behave differently under tensile and axial loading.

Main conclusions of this study are as follows;

- Strain energy and modulus of resilience are important parameters for all surface support materials.
- During the selection of the suitable TSL, first of all specified standards should be conformed and then the energy absorbance capacity should be investigated.
- During tensile studies, significant difference is observed between two products, product B has better mechanical performance under tensile loading.
- Numerical models can simulate TSL laboratory testing. For complex studies where laboratory tests can not be performed, numerical models can be an alternative to estimate the work done by the liner.
- Both materials have very similar responses under compression.

REFERENCES

ASTM Standard D-638-10, 2010, Standard Test Method for Tensile Properties of Plastics-1, ASTM International, West Conshohocken, PA, USA DOI: 10.1520/D0638-10

ASTM Standard D-695-10, 2010, Standard Test Method for Compressive Properties of Rigid Plastics, ASTM International, West Conshohocken, PA, USA

BASF, Turkey web page 23 Dec. 2013

Chulho, L., Chang, S., Kicheol, L., Dongwook, K. 2015. Numerical study on contact behavior of TSL (Thin Spray-on Liner). Journal of Korean Tunnelling and Underground Space Association. 2015. Nov, 17(6): 665-674

Dassault Systèmes Simulia Corp., 2016. Abaqus v. 2016. Providence, RI: Dassault Systèmes Simulia Corp.

Dirige, A., Archibald, J. 2009, Numerical Modeling Simulations of Spray-on Liners Support Potential in Highly Stressed and Rockburst Prone Rock Conditions". Proceedings of the 3rd CANUS Rock Mechanics Symposium, Toronto.

EFNARC, 2008. "Specification and Guidelines on Thin Spray-on Liners for Mining and Tunneling". ENC 250TSL v7.2 25-07-08.

Guner, D, and Ozturk,H, 2017. Isothermal Creep Behaviour Investigation Of Thin Spray-On Liners. ARMA, American Rock Mechanics Association, San Francisco.

Guner, D. & Ozturk, H. (2016) Experimental and Numerical Analysis of the Effects of Curing Time on Tensile Mechanical Properties of Thin Spray-on Liners Rock Mech Rock Eng (2016) 49: 3205. doi:10.1007/s00603-016-0997-x

Hadjigeorgiou, J, 2003. Proceedings of the Third International Seminar on Surface Support Liners: Thin Spray-On Liners, Shotcrete, and Mesh Quebec, Canada.

Malan, D.F., Napier, J.A.L. 2008. Numerical Modelling of Tunnel Liner and Fracture Interaction. 6th. International Symposium on Ground Support in Mining and Civil Engineering Construction, SA.

Nater, P., Mena-Cabrera, A. 2010. Thin Sprayed Liners - an Approach With Numerical Models. International Society for Rock Mechanics. Eurock 2010.

Richardson, J., Mitra, R., Saydam, S., 2009. Investigation of Thin Spray-On Liners using Numerical Modeling". ARMA, American Rock Mechanics Association, Asheville.

Tannant, D.D, Wang, C, 2003, "Thin Tunnel Liners Modelled With Particle Flow Code". Engineering Computations, Vol.21, pp. 318-341.

Usability of Thin Spray-on Liners (TSL) for Akarsen Underground Mine in Murgul

Murgul Akarşen Yeraltı İşletmesi için Püskürtülen İnce Kaplama (PİK) Kullanılabilirliği

E. Kömürlü, A. Kesimal

Karadeniz Technical University Mining Engineering Department, Trabzon

ABSTRACT In this study, thin spray-on liners usability instead of shotcrete applied in galleries of Akarsen copper mine in Artvin city of Turkey was investigated carrying out a series of field and laboratory scale experimental studies. According to the results obtained from this study, two liquid component pure polyurea type TSL with 5 mm thickness was found to be usable in the mine. In the topic of TSL design, shotcrete and TSL load bearing capacities have been compared and emprical study results on TSL applications in literature were considered.

ÖZET Bu çalışmada Artvin ili Murgul ilçesi içerisinde bulunan Akarşen yeraltı madeninde kullanılmakta olan püskürtme beton yerine PİK (Püskürtülen ince kaplama) uygulanabilirliği bir dizi saha ve laboratuvar ölçekli deneysel çalışmalar ile incelenmiştir. Elde edilen verilere göre, işletmede 5 mm kalınlığında saf poliüre türü çift bileşenli sıvı fazda uygulanmakta olan PİK malzemesinin kullanılabilir olduğuna karar verilmiştir. PİK tasarımı için uygulamadaki püskürtme beton ile tahkimat performanslarının kıyaslaması yapılmış ve kaya kütlesi özelliklerinin belirlenmesi yolu ile literatürdeki PİK kullanımı üzerine ampirik çalışma sonuçları değerlendirilmiştir.

1 INTRODUCTION

The support reactions of Thin Spray-on Liners (TSLs) have been investigated since 1990s (Pridchard et al., 1998; Archibald and DeGagne, 2001; Archibald, 2004; Öztürk, 2011). The TSL type underground support which can have the property of much earlier curing for supplying support pressure in comparison with the shotcrete liners was firstly applied in Canadian mines. In first applications, polyurethane type polymeric TSLs were used. Then, polyurea type and polyure/polyurethane based hybrid materials were started to be in use. In the late 1990s, latex type, methacrylate type polymeric TSLs, different cement based and cement/polymer based hybrid products were also being used in mining industry. Totally, 6 different producers were supplying TSLs for the mining industry in the late 1990s. As a result of increasing popularity of TSLs in different countries like Canada, Australia, South Africa, USA, Chile, USA, Sweden, the TSL producer number has passed 20, nowadays (Boeg-Jensen, 2013).

TSL usage is advantageous because of its practical application, time-saving fast curing reactions, rapid increase in strength values after being sprayed-on and proper adhesion to rock surfaces. Load bearing capacities of TSLs depend on both TSL material strength and their adhesion to the rock surfaces. As illustrated in Figure 1, failure in the use of TSLs can result from reaching TSL material strength or the adhesion of interface of TSL and rock. Therefore, rock surface has a significant role on the load bearing capacity supplied by the adhesion, in addition to TSL material strength values (Öztürk and Tannant, 2010; Tannant, 2001; Komurlu and Kesimal, 2012).

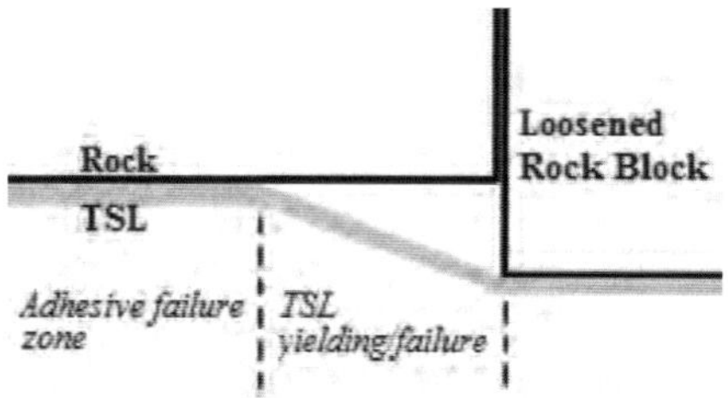

Figure 1. A block being moved and support mechanism of TSL (Komurlu ve Kesimal, 2013)

Instabilities in weak zones having a very short time to be stable can be prevented using fast curing TSLs which can supply early support pressures. Another advantage of TSL using is its effective isolation performance of preventing air ingress to rock masses. Polymeric TSLs are also effective and practical for setting continuous water resisting layers called spraying membranes in rock engineering. In Figure 2, a spraying membrane application is seen as an example of polymer based TSL membrane setting in tunnelling. Polymeric TSL materials have much higher energy absorption capacities than those of the shotcrete liners. Having high dynamic energy resistiviy is a significant advantage of polymeric TSL materials in terms of performance of support materials against different type of dynamic loads such in earthquakes, rock blasting, rock bursting (Yi and Kaiser, 1994; Li and Doucet, 2012; Li et al., 2014; Stacey, 2016). Some other advantages of TSLs are written in Table 1.

TSLs are generally sprayed with a thickness like some millimeters. Because its support reactions lie within those of shotcrete and steel wire meshes, the main aim of TSL use is preventing blocks to fall down from tunnel wall instead of carrying rock masses with high volume. TSLs can be used instead of shotcrete liners with low thickness values like 4 - 5 cm, and cannot be used in supporting rock masses with poor quality (Ferreira and Piroddi, 2011; Stacey, 2001).

In thin shotcrete liners, there may be very few aggregates through their thicknesses. Depending on the binder material, content and contact interface properties, load bearing capacities of thin shotcrete liners with low aggregate number like two or three in the thickness would be quite low that causes a very low load bearing capacity. In contrast, homogeneous material properties of TSLs provide advantage in support lining applications.

Figure 2. TSL application for the aim of water resisting (Wallis, 2008)

Within this study, polyurea type TSL usability instead of shotcrete applied in Akarsen underground copper mine in Murgul district of Artvin city in Turkey was investigated. In the mine, fiber reinforced shotcrete with 5-7 cm thickness is applied. Especially for roof shotcrete applications, up to 30% rebound is faced in the operations. As a result of the rebound problem with high percentages, a big amount of material is lost and working conditions get worse in the mine.

The polyurea type elastomer material was tested in this study. To assess whether TSL is usable in the mine, support performances of shotcrete mix of the mine and polyurea type TSL material were tested together. In addition, TSL usability according to some emprical approachs was investigated via determination of rock mass properties.

A guide of TSL usage details according to the RMR (Rock Mass Rating) value is given in Tables 2 and 3 (Espley et al., 1999). It should be noted herein that the usage details in the tables are suggested in case of using thermoset polymer type TSL materials like polyurethane and polyurea. Other TSL materials like some cement based ones should not be used considering the tables suggested by Espley et al. (1999).

Table 1. TSL and Shotcrete comparison

TSL (Polyurea) properties	Shotcrete properties
Low spraying material amount	High spraying material amount
No rebound problem (Rebound is smaller than 5%)	Material loss due to the rebound problem (Rebound can reach 30% for the roof)
Rapid curing (Thermoset polymer based ones can start to supply support in several minutes).	Need for some hours of curing for start of load bearing.
No need for use of extra chemical additives	Depending on the shotcrete content, different chemical additives like accelerator, plasticizer, retarder are needed.
Good adhesion resulting from getting the shape of wall roughness	Having voids at the contact to wall because of the aggregate content.
High chemical resistivity and having no change in its mechanical properties as a result of contact with underground water.	Weathering due to underground water. Especially, acidic underground water can cause a very fast decrease in mechanical properties of shotcrete.
A good isolation for rock masses can be supplied with a thin liner.	To supply a good isolation, thick liners like those with a thickness more than 7 cm are needed.
Having good resistivity against the dynamic loads, good energy absorption capacity. Therefore, it is advantageous to use near the blasting areas.	Having poor resistivity against the dynamic loads because of brittle material charecteristics.
It is environmental friendly because of no need for use of aggregate and concrete plant in the vicinity of the mine	Need for use of aggregate and concrete plant in the vicinity of the mine
It has tensile strength of 20 MPa - 25 MPa within one hour of curing (For pure polyurea).	It has no significant tensile strength within one hour of curing, around 3 MPa tensile strength within 28 days of curing.
Homogeneous material property in liner thickness	Inomogeneous material property in the thickness. Thin liners can have low bearing capacity depending on cement content and its adhesion to aggregate
TSLs can better the ventilation performance due to the low resistance of air stream in comparison with those of the shotcrete liners.	High resistance of air stream
TSLs can better the lighting performance	No light reflecting property

Tablo 2. TSL using guide (Espley et al. 1999)

Description	RMR	TSL thickness (mm)	Bolt pattern[1] (length x span)
Development drifts (wall)	45-65	2-3	1,8 mx1,5m[2]
	>65	2-3	1,8 mx1,5m[3]
Development drifts (roof)	45-65	3-4	1,8 mx1,1m[4]
	>65	3-4	1,8 mx1,3m[4]
Production heading (wall)	45-65	2-3	1,8 mx1,5m[5]
	>65	2-3	Boltless, spot bolts[5]
Production heading (roof)	45-65	3-4	1,8 - 2,4 mx1,1 m[4]
	>65	3-4	

Table 3. Notes for Table 2 (Espley et al. 1999)

Number	Note
1	Mechanical bolting
2	Bolting after every two rounds of advance
3	Indefinitely delayed with bi-annual audits
4	Install before or immediately after liner
5	Installation can be delayed

2 EXPERIMENTAL STUDY

Experimental studies carried out within this study was explained under relevant subtitles.

2.1. Rock Mass Classification

RMR values of different formations of the mine were evaluated to determine whether the rock mass properties are convenient for the use of TSL. Core samples with a total length more than 1200 meters from the southern mineralisation region which is the remaining last part for being mined in Akarsen (Figure 3). Discontinuity properties (roughness, spacing, weathering, infilling, aperture etc.) were carefully investigated to be rated. "wet" condition was considered for ground water rating. As seen in Figures 4 and 5, core samples were prepared to have length to diameter ratio of 2 for determination of uniaxial compressive strength (UCS) of rock materials. Loading rate was chosen to be 1.4 kN/sec in UCS tests. The RQD (Rock Quality Designation) value which is an important parameter for RMR determination was calculated with length measurements on the cores. In Tables 4-6, RMR values are given as multiplied by blasting and weakness adjustment factors. To designate the adjustment factors, "controlled blasting" condition was considered. For weak ore formations with lower UCS values than 2 MPa the "soft ore zone" condition were considered to choose weakness plane adjustment factor. Otherwise, "no weakness planes" condition were considered.

Figure 3. Some core samples

Figure 4. Core specimen cutting

Figure 5. UCS testing

2.2. Adhesion Tests

To comparatively test the adhesion to rock surface performances of polyurea type TSL and concrete, different rock materials from the Akarsen mine were used in direct tension test applied on rock and liner contact interfaces. As seen in Figure 6, rock core specimens were wrapped using platic sheets to pour polyurea components mix or cement paste. The liquid mix of polyurea components (isocyanate and amine) and water/cement mix were poured to have 3 mm thickness. As soon as possible after pouring the liner materials, another rock core specimen with flat surface were put in the sheet to make second adhesion interface. To hold specimens for applying tensile load on adhesion interface, steel wires inserted through eyebolts then clamped by steel collars to be fixed on the core specimens (Figure 6). The loading rate was selected to be 0.1 kN/s in adhesion tests of both polyurea and cement specimens. For the aim of considering the best adhesion performance of concrete, cement mortar without aggregate was made with the water/cement ratio of 0.4 by weight and tested in this study. For having the best polymerisation performance, the polyurea components need to be contacted at 60 C^{o}. Therefore, the components were heated to 75 C^{o} in a stove before mixing and pouring. In addition to the rock core specimens with moisture content of air, some ore specimens were waited in water tank to make wet specimens for investigating the adhesion performance on the wet surfaces.

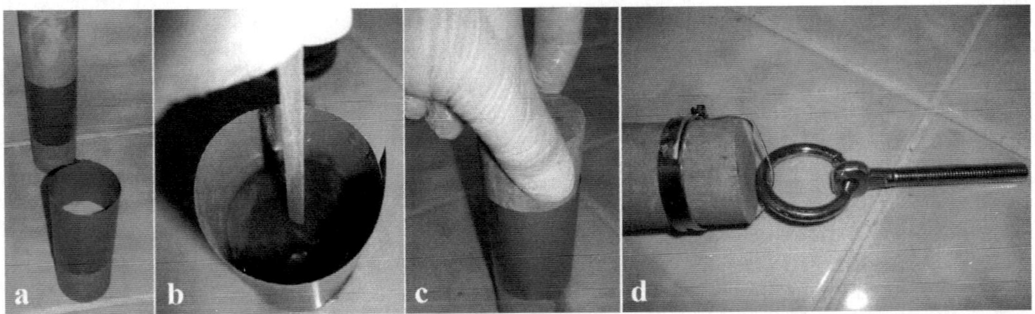

Figure 6. Preparation of TSL adhesion test specimens: a) Plastic sheet wrapped specimen to pour liner materials, b) A fresh TSL components poured specimen, c) Second core specimen putting in plastic sheets after pouring mix of TSL components, d) Specimen holding with steel wire inserted eyebolts and clamps

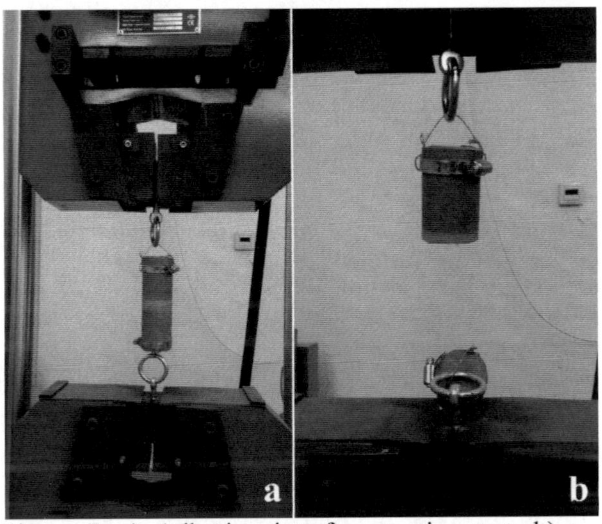

Figure 7. a) Adhesion interface tension test, b) a failed cement adhesion specimen

2.3. Slab Loading Tests

As a typical test for evaluating quality of shotcrete liners, three dimensional bending test applied on the slab specimens with dimensions of 60cmx60cm was carried out to compare load bearing capacities of TSL and shotcrete liners. To investigate the thickness effect on load bearing capacity of shotcrete liners, three different slab thicknesses of 4 cm, 7 cm and 10 cm were tested in this study. The loading rate was chosen to be 0.3 kN/s in slab loading test. Shotcrete slabs were prepared using a shotcrete robot. A mix including 450 kg/ m^3 CEM1 type ordinary Portland cement, 1620 kg/m^3 aggregates with maximum particle size of 12 mm, 225 kg/m^3 water, 12 kg/m^3 plasticizer, 12 kg/ m^3 accelerator and 3 kg/m^3 polypropylene type synthetic fiber additive was sprayed on 60 cm x 60 cm slab moulds with different thicknesses of 4 cm, 7 cm and 10 cm. Double slab specimens were prepared for each of the shotcrete thicknesses. Shotcrete slabs were made during mining operations in the Akarsen mine. In the slab load tests, 28 days cured shotcrete slabs were tested.

In the same mine, TSL slabs were also prepared during pilot applications carried out using a professional polyurea spraying machine. As same with the concrete slabs, double TSL slab specimens were sprayed for each thicknesses of 4 mm, 7 mm and 10 mm. Some photos from the slab loading tests are given in Figures 8-11. Shotcrete slabs were put on the frame of the slab loading test equipment to be loaded, directly. On the other hand, TSL specimens were clamped using 12 cramping frames with rough surface steel clips for holding slabs before loading (Figure 10).

Figure 8. Shotcrete slab loading test

Figure 9. TSL slabs with different thicknesses

Figure 10. TSL slab loading test

Figure 11. A tested TSL slab loading specimen

2.4. Block Bearing (Block Pushing) Test

Load bearing capacities of TSLs applied on triple blocks were tested with deformation controlled tests. In this experimental study, effect of different liner materials with different thicknesses on load bearing capacities was comparatively assessed. As seen in Figure 12, concrete liner and TSL specimens load bearing capacities were tested fixing right and left side blocks using cramp frames, and loading the mid block in the vertical direction. Triple cramping frames were used to fix right and left side blocks. In this experiment, each of the blocks have a size of 200 mm x 100 mm x 30 mm. Tripple blocks were encircled with plastic sheets to make moulds for pouring the liner materials. The shotcrete mix used in the mine was poured on the tripple block system, with a thickness of 4 cm. To compare its load bearing capacity with that of the concrete liner with 4 cm thickness, pure polyurea material with 4 mm thickness were coated on the tripple blocks.

The loading rate in block bearing test was chosen to be 0.2 kN/sec. Both concrete and TSL specimens were cured for 2 days before testing. For preparation of the TSL specimens, polyurea was poured using gun of the spraying machine, with a low pressure.

Figure 12. a) Polyurea coating, b and c) fixing concrete and polymer coated blocks, d) cracked concrete liner, e) Separation of concrete adhesion surface, f) a mid block carried by polyurea adhesion surface, g) totally separated polyurea adhesion surface

2.5. Field Study

A pilot TSL application study was performed in galleries of the Akarsen copper mine. Totally, 150 m^2 TSL with thickness of 5 mm was applied to investigate whether its application is fast and can eliminate the problems of shotcrete. A professional machine with 20 kg/min polyurea spraying capacity was used in the field study. In addition to spraying pure polyurea type TSL, rock bolts were inserted to test TSL usability with rock bolting. Some pictures from the site study are given in Figures 13 and 14.

Professional spraying machines mix the polyurea components (isocyanate and amine) with same amounts by weight, at the 60 C^o temprature for proper polymerisation reactions. In this pilot application, duration for TSL coating after each blasting advance, total material consumption for a unit liner thickness applied underground, existence of material loss resulting from the rebound problem, various difficulties during TSL operations, work safety details of TSL applications were investigated.

Figure 13. TSL spraying application (a and b), Rock bolt hole drilling in TSL applied galleries (c and d), Split set type friction bolt insertion (e, f, g, h and i)

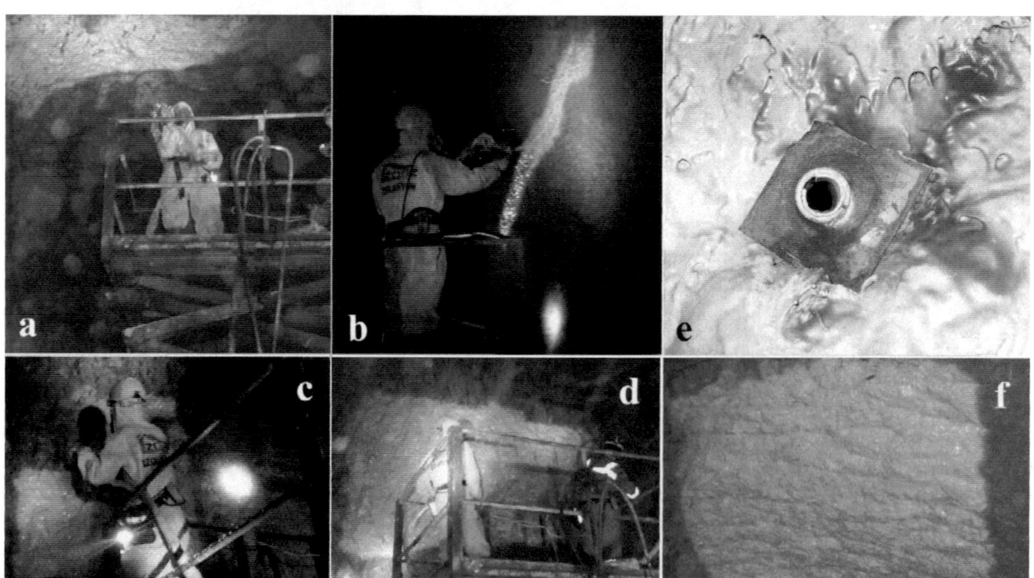

Figure 14. TSL spraying in the mine (a, b, c, d), An inserted rock bolt after TSL application (e), A TSL coated face (f)

3 RESULTS

Results obtained from the rock mass classification studies are given in Table 4. Shotcrete and TSL slab loading test results are respectively given in Table 5 and Table 6. As a result of vertical displacement limit of the loading piston which is 8 cm, polyurea type TSL material slab loading tests were stopped without having a failure. Therefore, the sign of ">" is used for values given in Table 6. As it is seen in Figure 11, TSL slabs were found to keep their shape after being loaded and deform elastically under the loads given in Table 6.

Adhesion test results for cement paste and polyurea liner specimens are given in Tables 7 and 8. The TSL and fresh concrete materials with the curing time of 2 days were tested in block bearing capacity tests considering the advance rate, liner spraying time details and stress influence area approaches. According to the results of block bearing test, TSL with the thickness of 4 mm has higher load bearing capacity than that of the shotcete liner with 4 cm thickness (Table 9). In load bearing capacity tests of blocks with TSL, all specimens had adhesive failure at the TSL contact to rock surface. Therefore, tests were stopped as the migration at the interface propogates to the cramp frame part, without having a failure in the TSL body. On the other hand, failure in the blocks with concrete liners was seen to be in the liner material.

In the use of polyurea type TSL product, material loss due to the rebound problem was found to be negligible. There was no problem of drilling bolt holes on the liner and the TSL product was found to be properly applicable with rock bolting. Rock bolt use with TSL photos can be seen in Figures 13 and 14.

Since 1990s, work safety in TSL applications has been studied as an important topic to improve the application quaility in terms of safety and eliminating the risks. Although there is no toxic gas emission during polymerisation of the TSL material used in this study, full face mask must be used in the vicinity (within 30 meters) of TSL application area in the first hour which is a time for completing major part of the polymerisation. This suggestion should be considered in case of having a good mine ventilation. Otherwise, a longer time should be passed for removing gasses realised due to the polymerisation reactions. The safety precautions written above were suggested to the mine as a result of consultations by some professional polyurea applicators.

Table 4. Geotechnical data of formations from various drillings (UCS: Uniaxial Compressive Strength, ÇD=very low, broken by hand)

Formation name	RQD (%)	UCS (MPa)	RMR_{89}
F2 Mudstone	64	4.32	49
F2 Siltstone	84	2.15	51
F2 Transition zone	55	-	-
F2 Hematitic formation	22	4.12	38
F2 Ore (Bornite)	60	3.37	44
F2 Chalcopyrite ore	80	4.15	45
F2 Ore with hematite	45	1.92	34
F3 Mudstone	79	9.93	52
F3 Mudstone-siltstone tran.	100	9.25	56
F3 Siltstone	91	10.92	55
F3 Limestone	89	12.66	53
F3 Hematitic formation	48	1.61	43
F3 Chalcopyrite ore	82	3.86	44
F3 Dacitic tuff with ore	51	1.87	37
F10 Mudstone	47	5.63	45
F10 Hematitic formation	90	2.58	50
F10 Transition zone	57	-	-
F10 Dacite	70	11.75	58
F10 Ore	94	34.17	61
F11 Mudstone	83	2.23	52
F11 Chalcopyrite ore	100	-	-
F11 tuffite ore	50	ÇD	36
F11 Siltstone	63	7.80	50
F18 Trans. zone with tuff	74	3.65	42
F18 Mudstone	65	6.57	48
F18 Limestone	54	38.15	52
F18 Siltstone	78	20.41	58
F18 Trans. zone-Siltstone	-	10.16	-
F18 Hematitic trans. zone	43	15.02	47
F18 Hematitic formation	55	4.12	45
F18 Ore with zinc	60	7.65	47
F18 Chalcopyritic ore	56	4.14	41
F18 Tuffite ore	0	-	-
F19 Mudstone	63	12.81	44
F19 Transition zone	52	-	-
F19 Limestone	72	2.85	47
F19 Transition zone	10	ÇD	25
F19 Dacite	13	18.46	40
F19 Chalcopyritic ore	94	3.11	48
F22 Tuffite ore	63	ÇD	42
F22 Dacite	100	13.03	64
F22 Transition zone	68	3.54	45
F22 Hematitic formation	53	3.69	43
F22 Ore	86	1.98	40

Polyurea starts to solidify immediately after being sprayed and becomes ready to be drilled for rock bolt instertion right after completing its spraying. Therefore, it is available to start rock bolting without waiting curing reactions of TSL material. However, full face mask should be used by all the people being in the TSL application area within an hour after the liner spraying.

There is no gas emission risk of polymerised polyurea which is a popular isolation material used in different applications needing high chemical durability such in drinking water storage tanks/pools. In sulphuric mines like Akarsen mine where is the case study area of this study, shotcrete liners are weathered because of the acidic ground water condition. In terms of eliminating the chemical weathering problem of the shotcrete, polyurea use instead of shotcrete is considerable. Polyurea is becoming popular for application on sewage walls because of its resistance to chemical degradation from sulphuric acid and hydrogen sulfide gas.

Table 5. Shotcrete slab load test results (PB: Shotcrete, 40: 40 mm thickness, 70: 70 mm thickness, 100: 100 mm thickness)

Numune adı	F_{max} (kN)
PB-40	13.25
PB-40	14.12
PB-70	26.99
PB-70	23.24
PB-100	43.18
PB-100	45.57

Table 6. TSL slab loading test results (4: 4 mm, 7: 7 mm, 10: 10 mm thickness)

Numune adı	F_{max} (kN)
TSL-4	>11.53
TSL-4	>12.38
TSL-7	>17.94
TSL-7	>18.26
TSL-10	>25.01
TSL-10	>23.53

Tablo 7. Cement paste adhesion test results (S.D.: Standard deviation)

Adhesion surface	Interface tensile strength (MPa)	Specimen number	S.D. (MPa)
Siltstone	1.12	3	0.09
Marn	1.73	3	0.11
Limestone	1.47	3	0.12
Tüffite ore	>0.55	2	0.07
Ore (F10)	2.10	3	0.08
Wet Ore F10	1.54	3	0.10

Table 8. TSL adhesion test results

Adhesion surface	Interface tensile strength (MPa)	Specimen number	S.D. (MPa)
Siltstone	1.03	3	0.09
Marn	1.29	3	0.08
Limestone	2.31	3	0.10
Tüffite ore	>0.61	3	0.09
Ore (F10)	2.35	3	0.13
Wet Ore F10	0.74	3	0.05

Table 9. Block bearing test results

Specimen name	F_{max} (kN)
TSL-4	3.60
TSL-4	3.14
TSL-4	3.42
PB-40	2.43
PB-40	2.55
PB-40	2.67

4 COST COMPARISION

Cost for applying shotcrete on 50 m^2 wall area which occurs as a result of 3 meters advance in one step blasting is 635 Euro (2100 TL according to October exchange rate). Shotcrete design used in the mine includes plastic fibre and chemial additives (plasticizer and accelerator). Polyurea cost for TSL application for the same area is 925 Euro for 5 mm thickness of the liner. Polyurea material cost is 3.7 Euro per a kilogram. In this case, TSL is looking to be expensive. However, advantageous point for the use of TSL is rapid polymerisation and no need for waiting curing reactions to start new blasting operation work or rock bolt drilling. In contrast, one shift after shotcrete spraying is needed to pass for start rock bolt hole drilling in the mine. In gallery excavation time, use of TSL is expected to decrease time consumption for 25%. Additionally, repairing costs for cracked shotcrete due to the blasting operations can be eliminated using TSL. Because of the decrease in the operation time, TSL use is seen to be economical for applying 4 mm thickness. Because, it is not practical to keep TSL thickness constant at 4 mm, 5 mm thickness can be applied to minimize the decrease of thickness to a level less than 4 mm. In this case, use of TSL is seen to be more expensive in comparison with shotcrete used in the mine. With a detailed market investigation, TSL applications can be done more economically than stated herein.

5 DISCUSSIONS AND CONCLUSION

According to the results of this study, rock mass properties of the Akarsen mine is not convenient to use a thin polyurea liner like those with 2-3 mm thicknesses. For a safe application, 5 mm thickness of polyurea type TSL was found to be needed except using in the weak ore and transition zones with low RMR values which is not acceptable to be enough for the use of TSL. Therefore, it is suggested that shotcrete liner thickness should be more than 7-8 cm depending on the poor quality rock mass location in mine. In addition to the liner support and rock bolts, steel sets are suggested to be in use for some weak zones with RMR values like 20-30. In Table 10, support suggestions according to the RMR value can be found. As confirmed by the Table 10, 5 cm thickness of shotcrete which is applied in the mine is not sufficient for the weak rock masses needed to use steel arches. As parralel, some instability problems have been faced in the weak ore zones when 5 cm shotcrete is used without applying steel sets. In the weak rock masses, rock bolt performances also gets worse making a reason to prevent applicability of TSL or thin shotcrete liners (Komurlu and Kesimal, 2011; Hemphill, 2013; Kolymbas, 2005).

Although the shotcrete thickness is 5 cm in the mine except of some weak formations, the concrete liner thickness can be many times decreased to 4 cm resulting from inhomogeneities in thickness due to personel reasons, rebound problem etc. Slab load test and block bearing capacity tests results confirmed that TSL with the thickness of 4 mm is applicable as an alternative of shotcrete with the thickness of 4 cm. Another encouraging point is using 28 days cured shotcrete slabs in experimental studies to compare with the TSL samples. When the necessity to carry loads induced from new excavations by short term shotcrete liners like those with several days curing time, TSL with 4 mm thickness is also assessed to be usable instead of shotcrete with 5 cm thickness. Significantly higher load bearing capacities of TSL specimens with 4 mm thickness than those of the shotcrete specimens with 4 cm thickness in the block bearing capacity tests also confirms this assessment.

In parallel, different support guides suggested by Espley et al. (1999) and Bieniawski (1989) which are respectively given in Table 2 and Table 10, TSL with the thickness of 4 mm is usable instead of 5 cm thickness shotcrete. According to Bieniawski (1989), rock bolting space is suggested to be 1.0-1.5 meters and shotcrete thickness is suggested to be 5 cm at the side walls and can increase up to 10 cm at the roof when the RMR value is between 40 and 60. In case of higher RMR values than 45, 4 mm thickness of TSL and rock bolts having 1.1 m spacing are usable according to Espley et al. (1999). It should be noted herein that the Table 2 is suggested for the use of thermoset polymeric TSL materials as same in this study. As an outcome of this study, 4 mm TSL thickness is not suggested to use for the rock masses with lower RMR values than 50, which need for use of shotcrete liners with up to 10 cm thicknesses in crown. This modification for the approach suggested by Espley et al. (1999) is seen in Table 11.

When it is considered that RMR values of the rock masses of the mine were generally evaluated to be around 40 or higher, TSL thickness is inferred to need for being higher than 4 mm and/or the rock bolt pattern details are found to need for being different than those suggested in the approach of Espley et al. (1999). From this point of view, 1 m spacing in the rock bolt applications of the Akarsen mine is assessed to be adequate to continue in the next operations. In the mine part with RMR values 40-50, TSL thickness is suggested to be increased up to 5-6 mm.

To determine those values, different emprical approaches in literature, experiences in the Akarsen mine, results obtained from this experimental study applied on both shotcrete and TSL specimens were considered together. Significantly higher block bearing capacities of TSLs with 4 mm thickness than those obtained from the shotcrete specimens with 4 cm thickness and slab load tests results were some key points to suggest TSL use in the mine.

However, weak rock masses like some tuffites and low strength ore zones in the mine were not assessed to be convenient for TSL applications as a result of low RMR values which are not enough for the use of thin shotcrete liners like those having 5 cm thickness. It should be noted herein that TSLs are usable as alternative for only thin shotcrete liners (Espley et al., 1996; Kömürlü and Kesimal, 2013; Yilmaz, 2010).

Table 10. Support suggestions according to RMR values (Bieniawski, 1989)

RMR value	Excavation	Support		
		Rock bolts, fully grouted, d: 20 mm	Shotcrete	Steel sets
81-100	Full face, 3 m advance	Generally no support is required except spot bolting		
61-80	Full face, 1-1.5 meter advance, complete support maximum 20 m from face	Locally, bolts in crown 3 m long, spaced 2.5 m with occasional wire mesh	5 cm in crown where required	None
41-60	Top heading and bench 1.5-3 m advance in top heading. Commence support after each blasting, complete support maximum 10 m from face	Systematical rock bolts, 4 meters long, spaced 1.5-2 meters with wire mesh in crown	5-10 cm in crown and 3-5 cm in sides	None
21-40	Top heading and bench 1.0-1.5 m advance in top heading. Install support concurrently with excavation, complete support maximum 10 m from face	Systematical rock bolts, 4-5 meters long, spaced 1-1.5 meters with wire mesh	10-15 cm in crown and 10 cm in sides	Light to medium ribs spaced 1.5 m where required
<20	Multi drifts, 0.5-1.5 m advance in top heading. Install support concurrently with excavation, shotcrete should be applied as soon as possible after blasting	Systematical rock bolts, 5-6 meters long, spaced 1-1.5 meters with wire mesh, bolt invert (use floor bolts)	15-20 cm in crown and 15 cm in sides and 5 cm on face	Medium to heavy ribs spaced 0.75 m with forepoling if required

Because its support reactions lie within those of shotcrete and steel wire meshes, the main aim of TSL use is preventing blocks to fall down from tunnel wall instead of carrying rock masses with high volume. Therefore, TSLs are used with rock bolts depending on rock mass quality. Although its stiffness is not alternative for shotcrete stiffness values, load bearing capacities of high strength polymeric TSLs like pure polyurea type ones can be similar to those of thin shotcrete liners (Tannant, 2001; Archibald and Lausch, 1999; Yilmaz et al., 2003; Ozturk, 2011). Because TSLs are only used instead of shotcrete liners with low thickness values like 4-5 cm, it cannot be used in supporting rock masses with poor quality. Low stiffness values should be also considered because of making loosening and dead load occurrence in poor quality rock masses as another reason to prevent TSL usage instead of thick shotcrete liners. TSLs are alternative of thin shotcrete liner which only be able to carry small/medium blocks on wall and have not a significant stiffness to prevent a mass loosening in high volume (Archibald et al., 1996; Kömürlü, 2016; Song et al., 2013). On the other hand, TSL usage is advantageous because of its fast curing reactions and rapid increase in load bearing capacity after being sprayed-on. Having no need for waiting curing reactions of polyurea type TSL to insert rock bolts can supply a bettering in stiffness values of support system (Yılmaz et al., 2003; Jjuuka and Kalumba, 2014). This rapid polymerisation property is one of the most important advantages of pure polyurea type TSL products against cement based or cement/polymer based ones (Boeg-Jensen, 2013; Kömürlü ve Kesimal, 2015).

Although it is more expensive, TSL is advantageous in terms of working conditions, because of its color bettering the lighting performance in mines and having no rebound problem which causes to accumulate material in the galleries to make difficult the working conditions underground. Additionally, no need for concrete plant in the vicinity of mine makes the TSL environment friendly and its cost to decrease against the shotcrete.

In this study, RMR values were determined considering the crack water condition as "wet". However, in case of seeing the water condition to be dripping or flowing, it should be considered that RMR value decreases by from 3 to 7. In this case, decrease in the RMR value should be paid attention to decrease especially in water flowing condition, in decision of the usability of TSL.

In case of excessive deformation of TSLs some tearing problems can be seen at the edges of the rock bolt plates. To solve that problem decreasing induced stresses at the edges of the plates, circular bolt plates can be used instead of square ones (Kömürlü vd. 2014; Li., 2010).

According to the results obtained from this study, polyurea type TSL with 5 mm thickness was found usable as an alternative of 5 cm shotcrete.

In this study, some modifications for the approach suggested by Espley et al. (1999) are presented. In the modifications, rock bolt length is suggested to select in accordance with the underground opening cross-section area. In Table 2, typical mine cross-sections like 4 m x 4 m are considered for rock bolt length selection. However, rock bolt length is related with different opening cross-sections in the modified TSL usage guide given in Table 11. In addition, distance between the face and the last rock bolt is added as a funtion of the rock mass rating value in the modified guide, as seen in Table 13. Because there is not a clear and exact distance between the face and support, it is possible to have personal differences in the application of guide by Espley et al. (1999). Because there is no opening cross-section area details in Table 2, RMR values adjusted in accordance with the opening cross-section area is suggested to use as seen in Eq. (1). Details for Eq. (1) is given in Table 12.

From the results of both slab loading tests and block bearing tests, pure polyurea type TSL with 5 mm thickness is assessed to be usable instead of shotcrete liners with thickness of 5 cm, in the Akarsen mine. The big part of the mine has higher RMR values than 40. According to the modified TSL usage guide given in Table 11, TSL with 5 mm thickness and 3 meters length rock bolts spaced 1 m can be used in 5 m x 4 m opening in a big part of the mine with RMR values varying from 40 to 50. In case of RMR values varying from 50 to 65, 4 mm TSL and 2.3 meters length rock bolts spaced 1.1 m can be used in the mine (bolt spacing can increase up to 1.5 meters for sidewalls). For the zones with lower RMR values than 40, TSL use is not suggested for the mine.

In Table 11, the "boltless and spot bolting" suggestion for higher RMR values than 65 in Table 2 was divided into two groups of 65-80 and 80-100 as parallel to suggestions given in Table 10. The "boltless or spot bolting" is only suggested for higher RMR values than 80 according to Table 10 suggested by Bienawski (1989).

Depending on different parameters like modulus of elasticity, ratio of in situ horizontal stress to in situ vertical stress, geometry of the opening cross-section, advance rate, stress change due to new excavations has an influence distance that typically change from 2-3 times of equivalent diameter from the face (Kavvadas, 2005; Aydan ve Genis, 2010; Komurlu vd., 2015). In accordance with this information, distance of last rock bolt from face was suggested to keep more limited with decreasing in the RMR value as seen in Table 13.

$$RMR^* = RMR89 - C \qquad (1)$$

Table 11. Modified guide for TSL usage with rock bolting**

Description	RMR*	TSL thickness (mm)	Bolt pattern[1] (length x span)
Development drifts (wall)	40-50	4-5	0,60L x 1,5m[2]
	50-65	2-3	0,45L x 1,5m[3]
	65-80	2-3	0,45L x 1,5m[4]
	>80	2-3	Without bolting
Development drifts (roof)	40-50	5-6	0,60L x 1,1m[2]
	50-65	3-4	0,45L x 1,1m[3]
	65-80	3-4	0,45L x 1,3m[4]
	>80	2-3	Boltless, spot bolts[5]
Production heading (wall)	40-50	4-5	0,60L x 1,3 m[2]
	50-65	2-3	0,45L x 1,5m[3]
	65-80	2-3	0,45L x 2,0m[4]
	>80	2-3	Boltless, spot bolts[5]
Production heading (roof)	40-50	5-6	0,60L x 1,0 m[2]
	50-65	3-4	0,45L x 1,1 m[3]
	65-80	3-4	0,45L x 1,5m[4]
	>80	2-3	Spot bolts[5]

* RMR values adjusted in accordance with cross-sectioan area should be used.
** Rock bolts should be made sure to have a higher anchorage capacity than 15 kN per one meter length.

Table 12. Regulation rate (C) for cross-section area

Cross-section Area	<20m^2	20m^2-30m^2	30m^2-40m^2	>40m^2
C	0	5	10	Not valid

Table 13. Notes for Table 11

No.	Note
1	Mechanical bolting
2	Maximum distance from last bolt to face: 0.60 R
3	Maximum distance from last bolt to face: 1.00 R
4	Maximum distance from last bolt to face: 1.40 R
5	Bolting can be delayed

In tables, L is H or W depending whichever is higher (1,4>H/W>0,70), R is H or W depending whichever is lower. Here, H is height of the opening cross-section and W is width of the opening cross-section.

TSLs are not suggested to use in rock masses with smaller RMR values than 40. On the other hand, especially in rock masses with RMR values between 40 and 50, rock bolting performance should be considered and made sure for usability of TSLs, because some rock materials with low strength values and poor slake durability can be significantly weathered due to the drilling and affect the anchorage work of bolts.

REFERENCES

AKDE, 2014. Technical Data Sheet of Whitechem Polyurea 1044, Arnavutköy, İstanbul

Archibald, J.F., Mercer, R.A., Lausch, P., 1995. Innovative ground support using rapid deployment of spray-on liners, Mining Engineering, 47, 856-860

Archibald, J.F., DeGagne, D.O., 2001. "Spray-on Lining Support in Canadian Underground Mining - A Reseach Summary," Journal of the Canadian Institute of Mining, Metallurgy and Petroleum (CIM Bulletin) 94, 1050, 49-56, 2001.

Archibald, J.F. , "Chapter 4 - Surface Support in Mining. ," (p. 73-87) in Publication of the Australian Centre for Geomechanics, 2004, ISBN 0-9756756-0-5

Archibald JF (2004) Canadian laboratory and field testing. In: Potvin Y, Stacey TR, Hadjigeorgiou J (ed) Surface support in mining australian centre for geomechanics, pp 135–139

Aydan, Ö., Geniş, M. (2010) Yeraltı açıklıklarında kaya patlaması olgusu ve önlemleri üzerine değerlendirme (in Turkish). Turkish Journal of Rock Mechanics, 17, 1-62

Bieniawski, Z.T. 1989. Engineering rock mass classifications. New York: Wiley

Boeg-Jensen, P., 2013. Managing the Engineering, Health and Safety Aspects of Thin Spray- On Liner Application. Thesis, Master of Science in Engineering Technology Natural Resources Engineering, Lulea Technical University of Technology, Lulea, Sweden.

Espley, J.S., O'Donnell, J.D., Thibodeau, D., Paradis-Sokoloski, P., 1996. Investigation into the replacement of conventional support with spray-on liners. CIM Bulletin, June 1996, 135-143.

Espley, J.S., Tannant, D.D., Baiden, G. and Kaiser, P.K. 1999. "Design criteria for thin sprayon membrane support for underground hardrock mining". Canadian Ins. Of Mining and Metallurgy Annual Meeting, Calgary

Ferreira, P.H., Piroddi, A., 2011. The Application of GRP and Thin spray on liner support products in a typical block cave mining method to enhance safety and productivity. 6th Southern African Base Metals Conference, Phalaborwa, South Africa

Hemphill, G.B., 2013. Practical Tunnel Construction, John Wiley & Sons, USA

ISRM (2007), "The blue book - the complete ISRM suggested methods for rock characterisation, testing and monitoring: 1974-2006" (Ulusay, R., Hudson, J.A. in eds), ISRM & Turkish Group of ISRM, Ankara

Jjuuko, S., Kalumba, D., 2014. A Review of Application and Benefits of Thin SprayOn Liners for Underground Rock Support in South African Mines, Proceedings of the 8th South African Young Geotechnical Engineers Conference, 17, 18 & 19 September 2014, Spier Conference Centre, Stellenbosch, Western Cape, 1-10

Kavvadas, M., Numerical Analysis in the Design of Urban Tunnels, 11thInternational Conference of IACMAG, keynote lecture, Torino, 2005.

Kolymbas, D. 2005. Tunnelling and Tunnel Mechanics, Springer, Germany

Komurlu, E., 2016. Usability of Various Engineering Polymers as Rock Bolt Materials, PhD. Thesis, Karadeniz Technical University, Trabzon, Turkey

Komurlu, E., Kesimal, A., 2011. Effect of Polymer Fiber on the Shotcrete Tunnel Support. RockMec'2011 Xth Regional Rock Mechanics Symposium. Ankara/Turkey, pp. 47-55.

Komurlu, E., Kesimal, A., 2012. Using Sprayed Polymer as Tunnel Support. 7th Asian Rock Mechanics Symposium, pp. 1486-1499, Seoul, South Korea

Komurlu, E., Kesimal, A., 2013. Tunnelling and Support Materials from Past to Present. The Journal of The Chamber of Mining Engineers of Turkey, Vol. 52, pp. 33-47

Komurlu, E., Kesimal, A., 2015. Improved Performance of Rock Bolts using Sprayed Polyurea Coating. Rock Mechanics and Rock Engineering, Vol. 48, pp. 2179-2182

Komurlu, E., Kesimal, A., Colak, U., 2014. Polyurea type Thin Spray-on Liner Coating to Prevent Rock Bolt Corrosion. 8th Asian Rock Mechanics Symposium, pp. 1389-1397, Sapporo, Japan.

Komurlu E, Kesimal A, Hasanpour, R., 2015. In situ horizontal stress effect on plastic zone around circular underground openings excavated in elastic zones. Geomechanics and Engineering, Vol. 8, pp. 783-799.

Komurlu, E., Kesimal, A., Durmus Demir, A., 2016. A Numerical Study for Determining Ideal Size and Geometry of Dog bone shaped direct tensile strength test specimens. European Rock Mechanics Symposium 2016 (EUROCK 2016), Cappadocia, Turkey, pp. 325-330

Li, C.C., 2010. Field Observations of Rock Bolts in High Stress Rock Masses, Rock Mechanics and Rock Engineering, 43, 491-496

Li, C.C., Doucet, C., 2012. Performance of D-Bolts Under Dynamic Loading, Rock Mechanics and Rock Engineering, 45, 193-204

Li, C.C., Stjern, G., Myrvang, A., 2014. A review on the performance of conventional and energy-absorbing rockbolts, Journal of Rock Mechanics and Geotechnical Engineering, 6, 315-327

Öztürk, H., 2011, Püskürtülen ince kaplamaların elastik malzeme özellikleri, Madencilik, 50: 41-45

Öztürk, H., Tannant, D.D., 2010, Thin spray-on liner adhesive strength test method and effect of liner thickness on adhesion. International Journal of Rock Mechanics and Mining Science, 47: 808-815.

Pritchard, C., Swan, G., Henderson, A., 1998. Tekflex as a Sprayon Screen Replacement in an Underground Hard Rock Mine. 18th Conference On Ground Control in Mining. pp 90-97.

Song, K., Cho, G., Chang, S., Lee, I., 2013. Beam–spring structural analysis for the design of a tunnel pre-reinforcement support system. International Journal of Rock Mechanics and Rock Engineering, 59, 139-150

Stacey TR. Review of membrane support mechanisms, loading mechanisms, desired membrane performance and appropriate test methods. J South African Institute of Mining and Metallurgy, 101, 2001, 343 – 351

Stacey, T.R., 2016. Addressing the Consequences of Dynamic Rock Failure in Underground Excavations. Rock Mechanics and Rock Engineering, 49(10), 4091–4101

Tannant, D.D. 2001. Thin spray-on liners for underground rock support testing and design issues. International Conference on Surface Support Liners: Membranes, Shotcrete and Mesh. Perth: Australian Centre for Geomechanics.

Wallis, S. 2008. Spray-on waterproofing - Finding real application, http://tunneltalk.com/Spray-on-waterproofingFinding-real-application.php

Yi, X., Kaiser, P.K., 1994. Impact testing for rockbolt design in rockburst conditions, International Journal of Rock Mechanics and Mining Sciences & Geomechanics Abstracts, 31, 671-685

Yilmaz, H., Saydam, S., Toper, A.Z., 2003. Emerging Support Concept: ThinSpray-onLiners, 18th International MiningCongressandExhibitionot Turkey-IMCET2003, Antalya, Turkey, 65-72

Yilmaz, H., 2010. Tensile strength testing of thin spray-on liner products (TSLs) and shotcrete. The Journal of The Southern African Institute of Mining and Metallurgy, 559-569

A Numerical Study on Support Performance of Thin Spray-on Liners (TSLs)

E. Komurlu
Karadeniz Technical University Mining Engineering Department, Trabzon, Turkey

A. Durmus Demir
Karadeniz Technical University Civil Engineering Department, Trabzon, Turkey

ABSTRACT In this study, load bearing capacities and support reactions of Thin Spray-on Liners (TSLs) with different thicknesses, deformation modules and adhesion to rock surface values were analysed using ANSYS program, a finite element method software. According to the results obtained from this study, failure mechanism was found to be prone to changing from initiation at the contact to rock surface to occurrence at the TSL material as the liner thickness decreases. Load bearing capacity of the TSLs was found to increase with an increase in modulus of elasticity and adhesion values. Stress concentration in the liner was found to increase with an increase in modulus of elasticity of TSL material. Therefore, the failure in stiff liners is possible to be in the case of TSL material yielding or failing. Adhesion values were determinative for load bearing capacities of the thick liners. As another effect, the stiffness value of TSLs was found to be increased in case of having high adhesion. Totally, failure mechanisms and interactions of rock and the support for 8 different TSL models were determined in the numerical analyses by modelling a roof with a block which is able to be loaded and move vertically.

1 INTRODUCTION

As a support type applied in the mine galleries, thin spray-on liners (TSLs) have been used since 1990s. Because its support reactions lie within those of shotcrete and steel wire meshes, the main aim of TSL use is preventing blocks to fall down from tunnel wall instead of carrying rock masses with high volume. Therefore, TSLs are used with rock bolts depending on rock mass quality. Although its stiffness is not alternative for shotcrete stiffness values, load bearing capacities of high strength polymeric TSLs can be similar to those of thin shotcrete liners (Tannant, 2001; Archibald and Lausch, 1999; Yilmaz et al., 2003; Ozturk, 2011). Because TSLs are only used instead of shotcrete liners with low thickness values like 4 - 5 cm, it cannot be used in supporting rock masses with poor quality (Komurlu and Kesimal, 2013; Boeg-Jensen, 2013; Archibald and DeGagne, 2001; Archibald, 2001). There are different studies on suggesting TSL usage details with rock bolts. One of them is given in Tables 1 and 2 (Espley et al., 1999). Another guide of TSL usage details including some modifications for the Tables 1 and 2 are given in Tables 3-5 which are suggested in accordance with some studies carried out in an undergroud copper mine in Turkey (Komurlu and Kesimal, 2016). It should be noted herein that the usage details in the tables are suggested in case of using thermoset polymer type TSL materials like polyurethane and polyurea whose typical strength and deformability parameters are given in Table 6. Other TSL materials like some cement based ones should not be used considering the tables suggested by Espley et al. (1999) and modifications by Komurlu and Kesimal (2016).

The approaches suggested by Espley et al. (1999) and Komurlu and Kesimal (2016) are usable for typical underground mine openings with rectangular cross-sections. In tables, L is H or W depending whichever is higher (1,4>H/W>0,70), R is H or W depending whichever is lower. Here, H is height of the opening cross-section and W is width of the opening cross-section. In the guide suggested in Table 3, regulation rates for opening cross-section area should be considered as given in Eq. (1).

$$RMR^{*}=RMR_{89}-C \quad (1)$$

Table 1. TSL using guide (Espley et al. 1999)

Description	RMR	TSL thick-ness (mm)	Bolt pattern[1] (length x span)
Development drifts (wall)	45-65	2-3	1,8 mx1,5m[2]
	>65	2-3	1,8 mx1,5m[3]
Development drifts (roof)	45-65	3-4	1,8 mx1,1m[4]
	>65	3-4	1,8 mx1,3m[4]
Production heading (wall)	45-65	2-3	1,8 mx1,5m[5]
	>65	2-3	Boltless, spot bolts[5]
Production heading (roof)	45-65	3-4	1,8 - 2,4 mx1,1 m[4]
	>65	3-4	

Table 2. Notes for Table 1 (Espley et al. 1999)

Number	Note
1	Mechanical bolting
2	Bolting after every two rounds of advance
3	Indefinitely delayed with bi-annual audits
4	Install before or immediately after liner
5	Installation can be delayed

Table 3. Modified guide for TSL usage (Komurlu and Kesimal, 2016)

Description	RMR*	TSL thickness (mm)	Bolt pattern[1] (length x span)
Develop-ment drifts (wall)	40-50	4-5	0,60L x 1,5m[2]
	50-65	2-3	0,45L x 1,5m[3]
	65-80	2-3	0,45L x 1,5m[4]
	>80	2-3	Without bolting
Develop-ment drifts (roof)	40-50	5-6	0,60L x 1,1m[2]
	50-65	3-4	0,45L x 1,1m[3]
	65-80	3-4	0,45L x 1,3m[4]
	>80	2-3	Boltless, spot bolts[5]
Production heading (wall)	40-50	4-5	0,60L x 1,3 m[2]
	50-65	2-3	0,45L x 1,5m[3]
	65-80	2-3	0,45L x 2,0m[4]
	>80	2-3	Boltless, spot bolts[5]
Production heading (roof)	40-50	5-6	0,60L x 1,0 m[2]
	50-65	3-4	0,45L x 1,1 m[3]
	65-80	3-4	0,45L x 1,5m[4]
	>80	2-3	Spot bolts[5]

Table 4. Regulation rate (C) for cross-section area

Cross-section Area	<20m^2	20m^2-30m^2	30m^2-40m^2	>40m^2
C	0	5	10	Not valid

Table 5. Notes for Table 3 (Komurlu and Kesimal, 2016)

No.	Note
1	Mechanical bolting
2	Maximum distance from last bolt to face: 0.60 R
3	Maximum distance from last bolt to face: 1.00 R
4	Maximum distance from last bolt to face: 1.40 R
5	Bolting can be delayed

TSLs are not suggested to use in rock masses with smaller RMR values than 40. On the other hand, especially in rock masses with RMR values between 40 and 50, rock bolting performance should be considered and made sure for usability of TSLs, because some rock materials with low strength values and poor slake durability can be significantly weathered due to the drilling and affect the anchorage work of bolts.

Load bearing capacities of TSLs depend on both TSL material strength and their adhesion to the rock surfaces. As illustrated in Figure 1, failure in the use of TSLs can result from reaching TSL material strength or the adhesion of interface of TSL and rock. Therefore, rock surface has a very significant role on the load bearing capacity supplied by the adhesion, in addition to TSL material strength values (Ozturk, 2012; Ozturk, 2011; Li et al., 2017; Drige and Archibald, 2009; Saydam and Docrat, 2007).

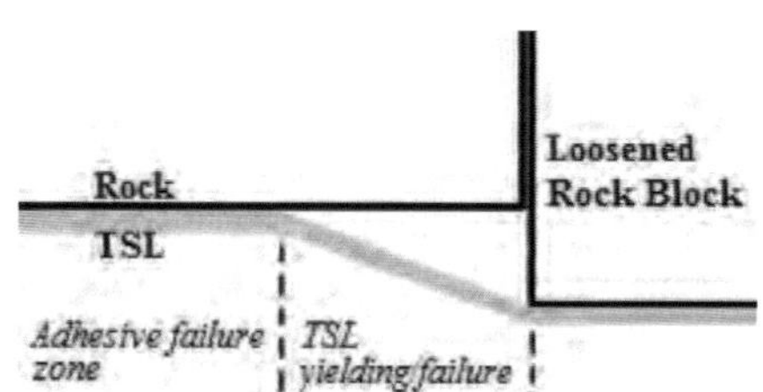

Figure 1. TSL failure mechanisms

Strength of the rocks can influence adhesion of the liners. Weak rocks often results in poor adhesion as the rock itself is the weakest part. Therefore, failure at the rock mass can be faced as TSL is loaded in some specific cases of weak rock conditions (Yilmaz, 2013; Ozturk and Tannant, 2010; Ozturk and Tannant, 2011).

In this study, numerical models with different TSL thicknesses, modulus of elasticity values and adhesion to rock surface values were analysed to investigate variations in load bearing capacities, support stiffness and failure mechanisms.

2 NUMERICAL STUDY

2.1 Methods

To investigate block carrying capacities of TSLs, triple block system which has right and left side fixed blocks and a middle block being able to move vertically was modelled in numerical analyses (Figure 2). In this numerical study, finite element analyses were carried out using the ANSYS software. Downside of all blocks is contacted by TSL models which have different adhesion to rock surface values. As the middle block is loaded in vertical direction, TSL models started to carry load owing to its adhesion to the right and left side blocks.

TSL materials with different modulus of elasticity values were analysed to investigate its effect on support performance. Additionally, different thickness values of TSLs were modelled to investigate the thickness effect on the support reactions of the liner. Details of material properties and variations in different TSL models are given in Table 6 and Table 7. The TSL material properties were selected to be representative for polyurea based TSL materials (AKDE, 2014; BASF, 2009; Komurlu and Kesimal 2015).

In the triple block model, dimension of left and right side fixed blocks are 60cmx60cmx40cm and the middle one size is 40cmx40cmx40cm. The mesh size of the middle block was constant and 2.5 cm x 2.5 cm x 2.5 cm, but right and left side blocks have narrowing meshes from outside of the model to the side of middle block because of the need for sensitive adhesion analyses close to the middle block being moved vertically. Mesh sizes in TSL was chosen as same with those of the rock model in horizontal and 2 mm along its thickness as seen in Figure 3 and Figure 4. To determine enough small mesh sizes, preliminary analyses on models with different meshes were carried out.

For displacement-controlled loading, loads were divided into multiple sub-steps until the maximum load was achieved. Stress distributions and failure mechanisms for all models were investigated. Eight-node solid elements were used for three-dimensional modeling of rock and TSL materials. The modulus of elasticity in models was assumed same for the tension and compression. The modelled rock and TSL materials were defined as linear elastic material until the strength value of the materials are reached. The rock material is modelled to fail as the strength value is reached. On the other hand, TSL material model reaching its elastic deformation limit starts to have yielding deformation under the maximum load level until a high strain level of 25%. Rock and TSL contact was also modelled to be deformed with the linear elastic behaviour and have a very limited deformation without debonding capacity of 0.1 mm. As seen in Figure 5, debonding initiates at the deformation level of 0.1 mm and is completed at 0.2 mm.

Eight different models with low (L) and high (H) modulus of elasticity (E), adhesion (A) and thickness (T) values were analysed and named as LTLALE, LTHALE, LTLAHE, LTHAHE, HTLALE, HTHALE, HTHAHE, HTLAHE in this study. It should not be missed to note that friction between the triple blocks was taken to be zero in the analyses.

Table 6. Mechanical properties of Rock and TSL models

Parameter	Values for Rock	Values for TSL
σ_c (MPa)	80	30
σ_t (MPa)	8	20
c (MPa)	10	15
v	0.25	0.35
E (GPa)	50	Given in Table 7

Table 7. Details about variations in TSL models

Parameters	Values
HE (High Elastic Modulus)	1000 MPa
LE (Low Elastic Modulus)	250 MPa
HT (High Thickness)	8 mm
LT (Low Thickness)	4 mm
HA (High Adhesion)	2 MPa
LA (Low Adhesion)	1 MPa

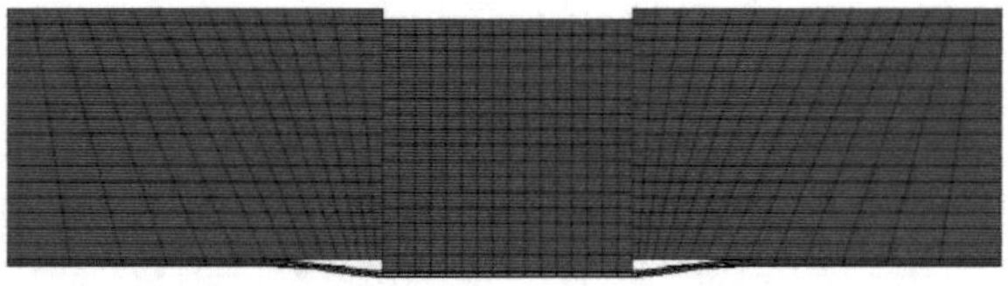

Figure 2. A model after reaching its load bearing capacity

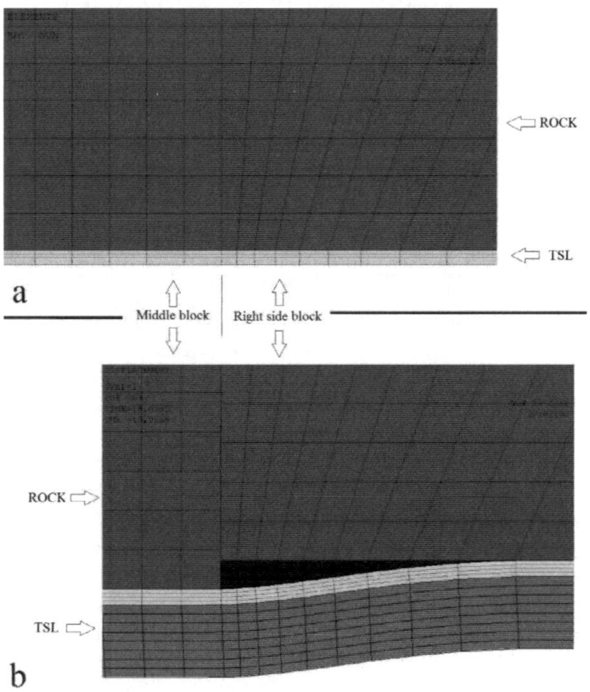

Figure 3. a) Middle block before loading, b) a failed model

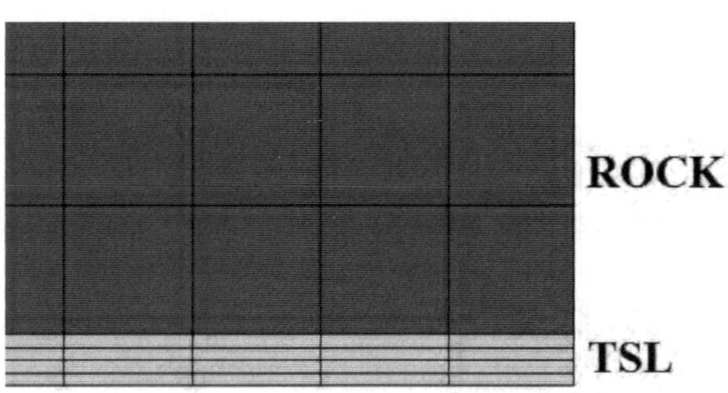

Figure 4. A close view to TSL meshes

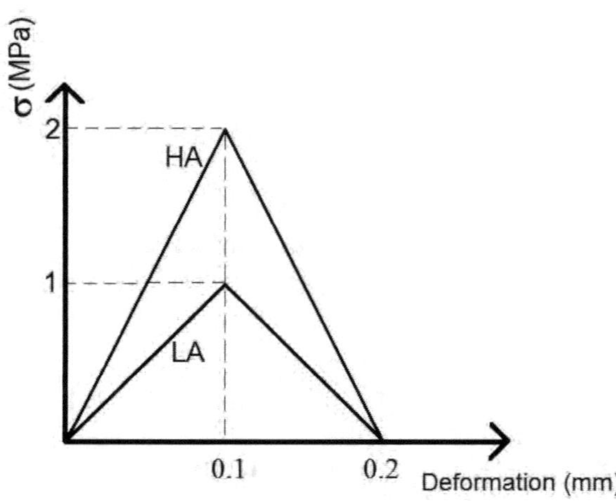

Figure 5. Debonding depending on load and interface deformation

2.2 Results

Results obtained from the numerical analyses are summarized in Tables 8-11. Failure mechanisms of the models were seen to be varied for different models.

In all models, a pre-debonding was observed as the middle block is loaded. Debonding proceeded owing to the displacement of middle block. In some models, TSL materials reached its strength values and started to yield. On the other hand, failure of some other models initiates without reaching tensile strength values of the TSL material, as a result of excessively high debonding during loading the middle block. To stop analyses without having TSL material yielding, 80 mm vertical displacement of the middle block was designated as the limitation. Two different models as examples for TSL yielding and failure due to the debonding at the contact are given in Figure 6 and Figure 7, respectively. Contact status of debonded TSL models is shown in Figure 8.

It should be noted herein that the load levels in Table 9 are given as the loads applied on the middle block, which are the load bearing capacities of triple block system with the TSL coating. In Table 9, F_{pre} is load for start of debonding at the contact (pre-debonding), Δ_{pre} is displacement of the middle block during being acted the pre-debonding load, F_{max} is load bearing capacity of the model after pre-debonding, Δ_{max} is displacement of the middle block in vertical at the step that F_{max} is reached, σ_{max} is maximum tensile stress in TSL, ε_{max} is maximum strain of TSL.

For the models with high modulus of elasticity and low adhesion values, load bearing capacity after pre-debonding (F_{max}) was lower than F_{pre} values as given in Table 9. In all models, an immediate decrease in load on the middle block was measured after reaching the pre-debonding load (F_{pre}). The load level measured directly after the pre-debonding (F_i) values are given in Table 11. Vertical displacements at the load level of F_i (Δ_i) were measured to vary from 1.4 mm to 2.6 mm for all the models. After this displacement, the analyses were continued until the displacement values given in Table 9.

The maximum TSL strain was seen to be at the edges of the middle block in all the models. After pre-debonding load, TSL material yielding stress was seen to be reached in some models as stated in Table 8. For instance, a yielded TSL in LTLALE model with maximising strain level of 9.3% is shown Figure 9. According to the maximum tensile stresses reached in TSL materials, yielded TSL models can be also seen from Table 10. The middle block's vertical displacement at the time for start of yielding which means reaching maximum stress in TSL material can be found in Table 9.

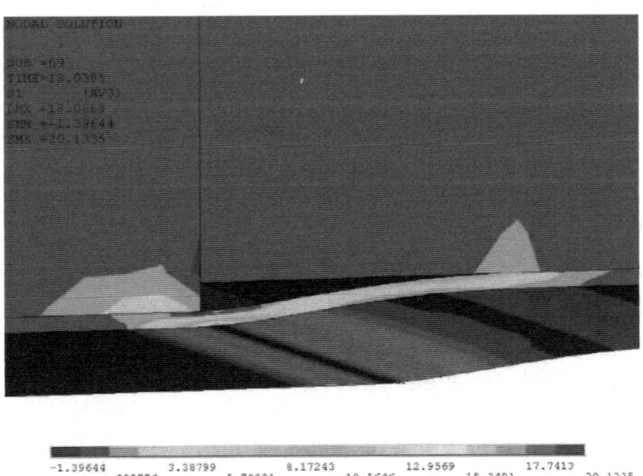

Figure 6. Stresses in TSL model (stress unit is MPa)

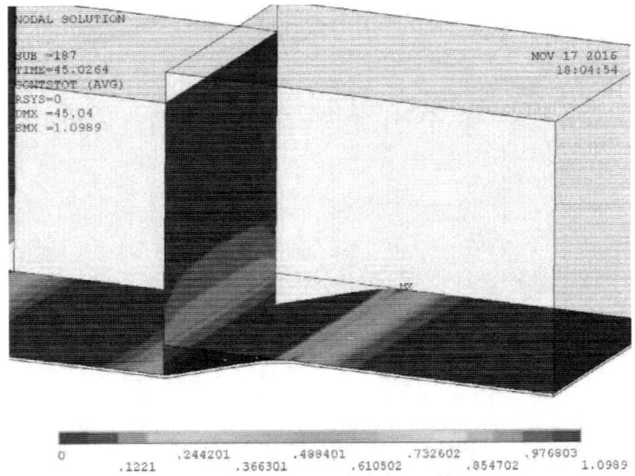

Figure 7. Contact stresses and debonding

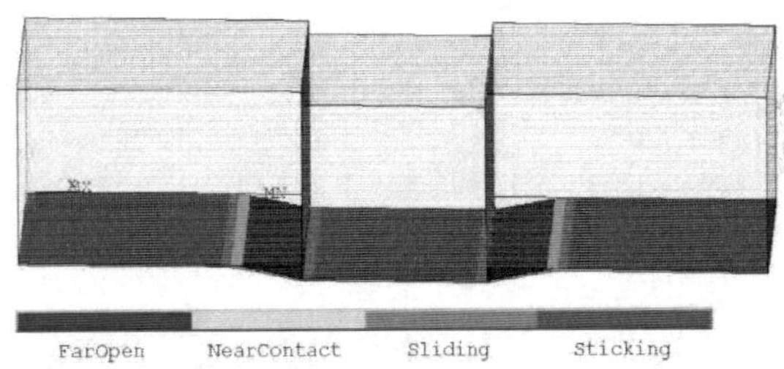

Figure 8. Contact status

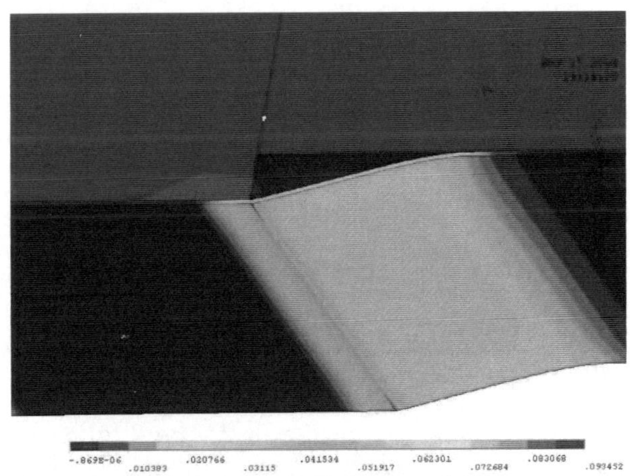
Figure 9. Strain in TSL (LTLALE model)

Table 8. Failure mechanisms of different models

Model	Failure mechanism
LTLALE	Contact debonding
LTHALE	TSL yielding
LTLAHE	TSL yielding
LTHAHE	TSL yielding
HTLALE	Contact debonding
HTHALE	Contact debonding
HTHAHE	TSL yielding
HTLAHE	Contact debonding

Table 9. Load and mid block displacement values from different models

Model	F_{pre} (kN)	F_{max} (kN)	Δ_{max} (mm)
LTLALE	4.72	9.91	80
LTHALE	6.75	17.74	68
LTLAHE	12.76	11.95	40
LTHAHE	15.40	20.85	72
HTLALE	6.84	10.98	80
HTHALE	8.86	21.89	80
HTHAHE	21.60	23.46	18
HTLAHE	18.67	11.80	80

Table 10. The maximum tensile stress induced in TSLs

Model	σ_{max} (MPa)
LTLALE	15.93
LTHALE	20.48
LTLAHE	20.36
LTHAHE	20.03
HTLALE	11.02
HTHALE	16.94
HTHAHE	20.13
HTLAHE	19.24

Table 11. F_i values

Model	F_i (kN)
LTLALE	1.12
LTHALE	1.93
LTLAHE	1.98
LTHAHE	2.90
HTLALE	1.81
HTHALE	3.06
HTHAHE	3.91
HTLAHE	2.41

3 DISCUSSION AND CONCLUSION

In comparison with models for low adhesion values, both load bearing capacity and stiffness values were seen to be higher in models with high adhesion values. Stresses in TSL were found to increase with an increase in the adhesion value. As the failure mechanism variations between those of LTLALE and LTHALE, HTLAHE and HTHAHE models confirmed, failure mechanism is possible to be in the case of yielding in TSL material instead of adhesive failure, because of an increase in adhesion value, depending on the TSL material strength and thickness (Ozturk and Tannant, 2010; Jjuuko and Kalumba, 2014). Under a same load applying on the middle block, a decrease of the stresses in TSL material was obtained as liner thickness increases.

As a result of decreasing stresses in both TSL material and stresses at the contact interface of TSL and rock, load bearing capacity of models are improved owing to an increase in TSL thickness (Jjuuko and Kalumba, 2015; Mason et al., 2015; Boeg-Jensen, 2013). That thickness effect was also confirmed by the results obtained from this study. As it is an expected issue resulting from the increase in thickness, the failure was seen to start at the contact in models with the high thickness (HT), except the HT model with high adhesion and high modulus of elasticity. Therefore, it is verified that the contact part becomes more critical for load bearing performance of thick TSLs (Komurlu et al., 2016; Espley et al., 2001; Yilmaz et al., 2003). Because of the decreasing stresses at the contact due to the increase in TSL thickness, an increase in adhesive load bearing capacity can be clearly observed comparing the F_{pre} and F_{max} values of LTLALE and HTLALE models. As another issue, a rise in the thickness has an effect making the liner support stiffer.

With the increase of modulus of elasticity, load bearing capacity and stiffness values were found to increase. As another effect of the increase in modulus of elasticity, possibility of TSL yielding instead of adhesive type failure increases because of increasing stresses in the TSL material (Wang and Wang, 2011; Mason et al., 2015; Kolymbas, 2005). This effect can be seen in failure mechanism differences obtained from LTLALE and LTLAHE models, and also HTHALE and HTHAHE models.

Although high modulus of elasticity values are preferable for deformation limitations, increase of modulus of elasticity without an increase of strength value causes to decrease in energy absorption capacity of the liners (Stacey, 2016; Stacey, 2001; Barla, 2016; Dirige and Archibald, 2009). In the case of using rigid liner materials, dynamic load bearing capacity of the liners notably decreases. On the other hand, rigid liners are advantageous against static loads as a result of preventing load increase resulting from the excessive deformations such in the joints of blocks which are potential to fall down (Komurlu and Kesimal, 2012; Carranza-Torres and Fairhust, 2000).

In short, thickness, modulus of elasticity and adhesion values of TSLs were found to have significant roles on stress distributions, load bearing capacity and the failure mechanisms of TSLs. To have a proper load bearing performance of TSLs, these parameters should be considered together to assess whether an enough support pressure can be supplied, or not.

REFERENCES

AKDE, 2014. Technical data sheet (TDS) of Whitechem Polyurea 1044 product, Arnavutköy, İstanbul

Archibald, JF, 2001. Canadian Laboratory and Field Testing. *Proceedings of the International Seminar on Surface Support Liners*, Perth: Austral Centre Geomech., August 2001 pp. 1-35.

Archibald, JF, Lausch P, 1999. Thin spray-on linings for rock failure stabilization. *Proceedings 37th U.S. Rock Mechanics Symposium - Rock Mechanics for Industry* (Balkema), Vail, Colorado, pp. 617-624.

Archibald, JF., DeGagne, DO, 2001. Spray-on Lining Support in Canadian Underground Mining - A Reseach Summary. *Journal of the Canadian Institute of Mining, Metallurgy and Petroleum* (CIM Bulletin) 94, 1050, 49-56.

Barla, G, 2016. Full-face excavation of large tunnels in difficult conditions, *Journal of Rock Mechanics and Geotechnical Engineering*, 8, 294-303.

BASF, 2009. Solutions for tunnelling and mining injections (Broschure of injection products), Zurich

Boeg-Jensen, P, 2013. *Managing the Engineering, Health and Safety Aspects of Thin Spray- On Liner Application*. Thesis, Master of Science in Engineering Technology Natural Resources Engineering, Lulea Technical University of Technology, Lulea, Sweden.

Carranza-Torres, C, Fairhust, C, 2000. Application of the convergence-confinement method of tunnel design to rock masses that satisfy the hoek-brown failure criterion, *Tunn. Undergr. Technol.*, 15(2), 187-213.

Drige, APE, Archibaldİ JF, 2009. Numerical modeling simulations of spray-on liners support potential in highly stressed and rockburst prone rock conditions. *Proceedings of the 3rd CANUS Rock Mechanics Symposium*, Toronto, May 2009.

Espley, JS, Tannant, DD, Baiden G, Kaiser PK, 1999. Design criteria for thin sprayon membrane support for underground hardrock mining. *Canadian Ins. Of Mining and Metallurgy Annual Meeting*, Calgary, published on CD-ROM, 7 p.

Espley, JS, Gustas, R, Heilig, J, Moreau, LH, 2001. Thin spray-onliner research and field trial sat INCO .Surface SupportLiners: Membranes. Shotcrete and Mesh. Australian Centre for Geomechanics, Australia.

Jjuuko, S, Kalumba, D, 2014. A Review of Application and Benefits of Thin SprayOn Liners for Underground Rock Support in South African Mines. *Proceedings of the 8th South African Young Geotechnical Engineers Conference*, 17-19 September 2014, Spier Conference Centre, Stellenbosch, Western Cape

Jjuuko, S, Kalumba, D, 2015. Shear-bond strength of thin spray-on liners as a surface support mechanism in underground rock, *The 13th International Congress of Rock Mechanics ISRM Congress 2015*, Montreal, Canada,

Kolymbas, D, 2005. *Tunnelling and Tunnel Mechanics*, Springer-Verlag, Heidelberg, Germany.

Komurlu, E, Kesimal, A, 2012. Using Sprayed Polymer as Tunnel Support. *7th Asian Rock Mechanics Symposium*, pp. 1486-1499, Seoul, South Korea.

Komurlu, E, Kesimal, A, 2013. Tunnelling and Support Materials from Past to Present. *The Journal of The Chamber of Mining Engineers of Turkey*, 52, 33-47.

Komurlu E, Kesimal A, 2015. Improved Performance of Rock Bolts using Sprayed Polyurea Coating. *Rock Mechanics and Rock Engineering*, 48, 2179-2182.

Komurlu, E, Kesimal, A, Demir, S, 2016. Spraying Membrane Layer Effect on Load Bearing Performance of Concrete Liners. *European Rock Mechanics Symposium 2016* (EUROCK 2016), Cappadocia, Turkey, pp. 715-719

Komurlu, E, Kesimal, A, 2016. *Use of Thin Spray-on Liners in Akarsen Undergroung Mine of ETI Bakir Inc.*, Karadeniz Technical University Circulating Capital Project Report, November 2016, Trabzon, Turkey, 27 p.

Li, Z, Nocelli, B, Saydam, S, 2017. Effect of rock strength and surface roughness on adhesion strength of thin spray-on liners. Int J Rock Mech Min Sci., 91, 195-202.

Öztürk, H, 2011. Püskürtülen ince kaplamaların elastik malzeme özellikleri, *Madencilik*, 50: 41-45.

Öztürk, H, Tannant, DD, 2010, Thin spray-on liner adhesive strength test method and effect of liner thickness on adhesion. *International Journal of Rock Mechanics and Mining Science*, 47: 808-815.

Ozturk, H, Tannant DD, 2011. Influence of rock properties and environmental conditions on thin spray-on liner adhesive bond. *Int J Rock Mech Min Sci*. 48(7), 1196–1198.

Ozturk, H, 2012. Fracture mechanics interpretation of thin spray-on liner adhesion tests,

International Journal of Adhesion & Adhesives, 34, 17-23.

Ozturk, H, Tannant, DD, 2011. Influence of rock properties and environmental conditions on thin spray-on liner adhesive bond. *Int J Rock Mech Min Sci.*, 48(7), 1196–1198.

Mason, DP, Fowkes, ND, Yemata, RM, Nnakenyi, CA, 2015. The use of spray on liners for wall stabilization in mines. Online reference: *www.wits.ac.za/media/migration/files/cs-38933-fix/migrated-pdf/pdfs-2/misgliners2015a.pdf*

Saydam, S, Docrat, YS, 2007. Evaluating the adhesion strength of different sealants on Kimberlite. *Proceedings of the 11th Congress of International Society of Rock Mechanics*, Lisbon, Portugal. p. 585-588.

Stacey, TR, 2001. Review of membrane support mechanisms, loading mechanisms, desired membrane performance and appropriate test methods. *J South African Institute of Mining and Metallurgy*, 101, 343 – 351.

Stacey, T.R., 2016. Addressing the Consequences of Dynamic Rock Failure in Underground Excavations, *Rock Mechanics and Rock Engineering*, 49(10), 4091–4101.

Tannant, DD, 2001. Thin Spray-on Liners for Underground Rock Support, *17th International Mining Congress and Exhibition of Turkey(IMCET 2001)*, pp. 57-73, Ankara, Türkiye

Wang, MB, Wang, G, 2011. A stress-displacement solution for a pressure tunnel with impermeable liner in elastic porous media, *Latin American Journal of Solids and Structures*, 9(1), 95-110.

Yilmaz, H, Saydam, S, Toper AZ, 2003. Emerging Support Concept: ThinSpray-onLiners, *18th International Mining Congress and Exhibition of Turkey-IMCET2003*, Antalya, Turkey, 65-72.

Yilmaz, H, 2013. Comparison of tensile-bond strength of thin spray-on liners, *Int J Min Reclam Environ.*, 27, 56–71.

Practice of the Face Perpendicular Preconditioning Technique at a South African Deep Level Gold Mine with a View to Enhance Safety and Productivity

E.G Nephiphidi, T. Zvarivadza
School of Mining Engineering, University of the Witwatersrand, Johannesburg, SA

ABSTRACT The South African gold mining industry is currently faced with challenges of extracting ore at deep levels. Stress accumulates significantly in the rockmass as the depth of mining increases. Seismicity and rockburst events occur as a result of high stresses induced by mining. Correctly implemented preconditioning can ameliorate rockbursts in stope faces. Rockburst related accidents at the studied mine motivated research to study the practice of the face perpendicular preconditioning technique used at the mine, with a view to improve safety and productivity. A preconditioning checklist was developed as an element of identifying gaps between the code of practice and cultural practice. Information to understand the challenges that the workforce experience while doing preconditioning was also gathered. In light of that, subjects that needed optimisation were identified. High gold content was found in areas closed due to seismic events, straining the need to open them up to enhance mine productivity. The research conducted noted that these areas can be opened up by making use of preconditioning and hydraulic fracturing. Various ways on how the practice of the technique can be optimized are discussed in the paper.

Keywords: Preconditioning, Safety, Productivity, Seismicity, Gold Mining

1 INTRODUCTION

According to McMahon (1988), gold mining began in the Republic of South Africa on a large scale in 1886 with the initial development of outcrop in the Witwatersrand area. Rock burst activities were found to be a serious issue that claimed lives of many in the mining industry (Windsor and Thompson, 1993). The studied deep level gold mine is currently mining at depths of 2000m to 3500m from surface. According to Toper et al (2003), mining a reef that extends 2000m further below the earth's surface involves inducing extremely high stresses on the rockmass within the vicinity of an excavation. Stresses are disrupted when an opening is excavated on a rock hence resulting in a new set of stresses being formed or induced around an opening. Violent release of the strain energy can be described as a seismic event which may result in a rockburst (Toper et al., 2003). Seismic event occurrences depend on the magnitude of the released energy; if it is great, extensive damage may result and sometimes fatalities. Continual occurrence of rock burst and seismic events in South African deep level mines led to the discovery of the preconditioning technique. Preconditioning is a rock engineering technique which is used to ameliorate rock bursts activities (Toper et al., 2003). It was first introduced in the early 1950s to make better the rockburst conditions in deep level mines. Preconditioning ensures the redistribution of stresses away from the working face, which reduces the chances of face bursting. The approach used to precondition at the mine was by drilling long holes (3m) on the face and the fractures ahead of the face are created through blasting activities prior to mining production. Preconditioning holes are drilled to increase the depth of fracturing. Having known the unfolding process of the preconditioning technique, rock bursts at the mine remained a problem.

1.1 Objectives of the Study

The research project aimed at reviewing the practice of the face perpendicular preconditioning technique looking into the safety and productivity of the mine. The objectives of the study were as follows:

- Investigating the significance of preconditioning in deep gold mines;
- Identifying the gaps between the code of conducts and the cultural practice; and
- Understanding how preconditioning can be used to open up areas that have been closed down due to seismicity.

1.2 Importance of the Study

Mine workers connecting from one section to another as their daily activity need to work within panels which are safe. One means of achieving this is effective preconditioning. Ineffective preconditioning exposes employees to high levels of fall of ground (FOG) occurrences. Ineffective preconditioning also leads to reduced production rates. The face perpendicular preconditioning technique has been found effective at the mine; this was hypothetically stated based on the perceived improvement in trends of FOG accidents at the mine from the year 2013 to 2015 as seen in Figure 1.

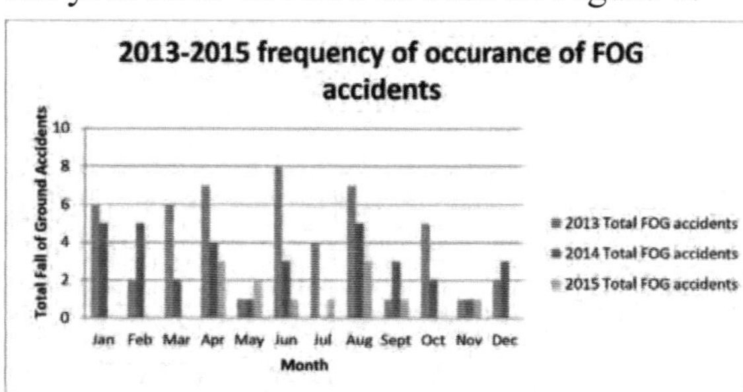

Figure 1. Frequency of FOG accidents occurrence

Preconditioning is an important procedure that must be done to ensure the safety of individuals working on the face. Preconditioning holes were drilled but were not effective as increased instances of falls of ground were noted. The study was conducted on 5 panels of the mine and different observations were noted in different days. Preconditioning ensures safe working areas and if it is done incorrectly, it exposes workers to a risky and unsafe environment.

1.1.1 Practical Significance of the project

It was found essential for the project to be carried out to find ways on how to improve the technique. Improving or optimizing the technique will help reduce the number of accidents experienced at the mine. FOG and seismicity related accidents are the most common accidents at the mine which contributes a huge proportion to the overall accidents that occur at the mine. Accidents do not only affect the victims, but also affect the mines' productivity as working operations are ceased during investigations when fatalities occur. Preconditioning acts as a rock bursts mechanism; correct application of the technique would result in a lot of lives being saved, safer working areas being created, increased advance rates; which will not only benefit the mine but also the mine workers themselves through bonuses which will then increase their take home salary. The increased mines production rates also results in increased total revenue which leads to economic growth. Safety is the main priority in the mining industry; optimization of the face perpendicular preconditioning technique will not only increase the mine's production but will result in much safer working areas.

2 BRIEF ON PRECONDITIONING

2.1 Feasibility Study on Preconditioning

A feasibility study was first conducted by the East Rand Proprietary Mines (ERPM) in the early 1950s by means of field trials to access the feasibility of preconditioning as a technique that can be used to reduce the incidence of rock bursts (Toper, et al., 2000). These field trials that were performed resulted in encouraging outcome that then gave them a go ahead with the technique. Toper et al (2000) came up with the following results after the trials:

- Incidents of rock bursts activities reduced by 36%

- Severe rock bursts reduced in numbers by 73%
- Almost 0% was recorded for a huge drop down percentage for on-shift frequency

Mining Review Africa (2016) stated that FOG incidents have decreased from 25 in 2014 to 22 in 2015 in South African gold mining industries. This shows a continual improvement on safety in the gold mining sectors.

In the 1980s, the Chamber of Mines Research Organization (COMRO) introduced a curriculum to re-investigate the effectiveness of preconditioning as a safe mining method. The re-investigation involved conducting field trials in various mines. The results seen from the trials were encouraging as well which then gave mining companies a go ahead with the technique for safe mining and increased productivity (Toper, et al., 2000).

2.2 Hydraulic Fracturing Process

Hydraulic fracturing process is well known for the measurements of stress magnitudes (Stacey and Wesseloo, 2002). Hydraulic fracturing was previously used mainly for borehole simulations by increasing the water supply and also for gas ejections from the ground through a process of pumping in high pressured water with sand to hold the fissures open in order to pump out the gases from the ground (Granderg, 2013). Determination of stresses using the process of hydraulic fracturing process was first introduced by Fairhurst (1971). Hydraulic fracturing involves the pressurization of the length of a borehole and measurements of the pressure required to fracture the rock or opening existing fractures.

2.2.1 Hydraulic fracturing applied in pre-existing fractures

The conventional method of hydraulic fracturing involves reopening existing fractures, these fractures orientations and pressures that balance the normal stresses acting on fractures needs to be measured (Stacey and Wesseloo, 2002). Packers that are installed are used for such measurements. The in-situ stresses are determined by identifying at least 6 non parallel pre-existing fractures, after hydraulic tests are carried on them six components of in-situ stresses can then be identified (Stacey and Wesseloo, 2002). This method can replace having to use strain gauges to calculate or determine stresses on the rocks.

2.3 Optimization Techniques

Optimisation of the preconditioning blasts will be described based on the research performed by Toper et al (2000) where seismic events, drilling rates, safety implication, rock mass fracturing and hanging wall profiles are looked at.

2.3.1 Seismic events

Seismic data was recorded by the Portable Seismic System (PSS) in potentially seismic areas during the preconditioning optimization work (Toper, et al., 2000). Two seismic events occurred almost simultaneously in two different panels; one panel was not preconditioned whilst the other one was preconditioned. Severe damages were seen on the un-preconditioned panel where working operations had to be stopped for more than a week. Figure 2 shows an unpreconditioned area with severe damage while a preconditioned area shown by Figure 3 displays minor damage.

Figure 2. Unpreconditioned area (Toper, et al., 2000)

Figure 3. Preconditioned area (Toper, et al., 2000)

2.4 Drilling Rates

Drilling of preconditioning holes takes longer (Toper, et al., 2000). The approximate time to drill one 3 m long hole was found to be 12 minutes. When drilling a panel that is 18 m long, six preconditioning holes have to be drilled at a spacing of 3 m. When the hole length increases, the timing increases as well. It was found that this is not always the case because the increase rate was not uniform in some cases but it was due to the increased stresses encountered further ahead of the face which makes drilling difficult. Improved drilling rates were achieved by increasing the amount of explosives. A time study of the precondition holes' drilling rates was conducted and the results are illustrated in Table 1.

Table 1. Comparison of drilling rates of preconditioning holes (Toper, et al., 2000).

Scenario	Minimum	Average	Maximum	Average (meters/minutes)
• Preconditioning (2.4 m; 36 mm)	10'0"	11'38"	12'59"	0.19
• Preconditioning (3.2 m; 36 mm)	10'30"	12'37"	14'16"	0.24
• Preconditioning (3.8 m; 36 mm)	12'59"	14'31"	15'55"	0.25
• Preconditioning (3.2 m; 40 mm)	8'36"	10'45"	12'35"	0.28

Drilling preconditioning holes is relevant and plays a significant role to the drilling rates of production holes. This is illustrated in Table 2.

Table 2: Comparison of face drilling rates of production holes for adjacent preconditioned and un-preconditioned panels (Toper, 2000).

Scenario	Minimum	Average	Maximum	Average (meters/minutes)
• Unpreconditioned	4'34"	5'06"	5'5"	0.21
• Preconditioning (2.4;36mm)	3'56"	4'48"	5'50"	0.23
• Preconditioning (3.2;36mm)	3'00"	3'57"	5'10"	0.28
• Preconditioning (3.8m;36mm)	2'30"	3'06"	3'55"	0.36
• Preconditioning (3.2m;40mm)	1'56"	3'14"	4'31"	0.34

2.4.3 Safety implications

It was found that 72% of all rock related accidents in gold mines were seismic related whilst only 28% of the accidents were gravity related (Koldas, 2001). The study will also focus on fall of ground incidents initiated by both seismic and gravity. The main objective of the preconditioning technique is to reduce the stresses from the face to avoid face bursts, thereby promoting safe workings. Toper et al (2000) reported that since the implementation of the technique 21 months later; no face bursts related incidents were reported. However, slight injuries were associated with the incorrect application and ineffective application of the technique.

Table 3 shows the effectiveness of the preconditioning technique on the mining production rates; it displays the increased gold grades in preconditioned areas than those unpreconditioned. Preconditioning ensures increased advance rates and safe working areas after blasts have been taken.

Table 3. Safety Record during the preconditioning experiment (Toper, et al., 2000).

	Safety record (centares mined/injury)		
	Up-dip cmg/t	Diagonal cmg/t	Combined cmg/t
• Preconditioning	2553	961	1430
• Non-Preconditioning	735	377	562
• Combined	937	528	741

3 STUDY APPROACH

Several tools were utilised in undertaking the study. A preconditioning checklist, containing different preconditioning factors to be assessed was drawn. A preconditioning checklist was used as a practical means of identifying gaps between what is being practiced underground and what is presented on the code of conducts. This was done to be able to identify areas that needed optimization. A preconditioning checklist assisted in giving an overview of how the face perpendicular preconditioning technique can be used to improve safety and enhance productivity.

Accident data from the mine was also analysed in line with preconditioning contributions. Seismicity records were also assessed. Information to understand the challenges that the workforce experiences while doing preconditioning was also gathered.

4 RESEARCH FINDINGS

4.1 Preconditioning Checklist

The studied mine has 52 panels that are currently in operation; only 5 panels which cover approximately 10% of the mine were visited. All panels at the mine uses square packs, roof bolts and in-stope bolting as means of support.

Factors identified on the checklist in Table 4 were supported by observations made underground. Panel dimensions of areas visited are represented in Table 5.

Table 4. Preconditioning Checklist

	Working areas				
Preconditioning Checklist	43/41 South Breast	45/65 South WRS	43/39 North Breast	42/39 South Breast	41/42 South Up-dip
Is there an overhanging face?	No	Yes	No	Yes	No
Are there any unfractured slabs or rock on the face?	Yes	Yes	No	Yes	Yes
Is the hanging wall in good condition?	Slightly	No	Yes	Slightly	Slightly
Were misfires identified from previous blast?	No misfires	No misfires	Yes	No misfires	No misfires
Number of preconditioning sockets? Were the sockets corresponding to the face length?	6 holes on a 20 m panel. Not corresponding	4 holes on a 15 m panel. Not corresponding	6 holes on a 15 m panel. Sockets were corresponding to the panel length	5 holes on a 20 m panel. Not corresponding	6 holes on a 15 m panel. Sockets were corresponding to the panel length
Length of the preconditioning sockets	1.2 m	1 m	1.1 m	0.5 m	0.9 m

Table 5: Panel Dimensons

Working Areas	Stopping Width (m)	Panel Length (m)
43/41 south breast	1.5	15
45/65 south wrs	1.3	16
43/39 north breast	1.6	18
42/39 south breast	1.7	15
41/42 south up-dip	1.5	20

4.2 Analyses of Gaps Found Between the Code of Conducts and Normal Practice Underground

From Table 4, it can be seen that based on the observations made, 60% of the areas visited had overhanging faces while only 40% of the areas had smooth blasted faces. Five areas were visited and only 4 had unfractured slabs of rocks on the face. Bad hanging walls are usually as a result of overcharging holes or incorrect drilling. From the five areas visited; three were in a slightly bad condition, one was irregular with signs of fall of ground and one had a relatively smooth hanging wall. The stoping widths of the areas visited ranged from 1.5-1.8m and no significant change from previous blasts stoping widths were measured. This gives an indication that preconditioning was applied. About 60% of the areas visited, preconditioning holes were not drilled correctly because the holes that were drilled were not corresponding to the face length. Correspondence of the face length to the preconditioning sockets was seen in two out of five working areas. A standard preconditioning socket has to be +1.4 m long considering an advance of 0.9m with a preconditioning hole length of 2.4 m. It is important to observe preconditioning sockets on the face after a blast has been taken because they are used as elements to check if preconditioning was done. Figure 4 shows a 0.7 m preconditioning socket which was found to be 100% off from the standard preconditioning socket; this affects the maximum effectiveness of the technique at the mine.

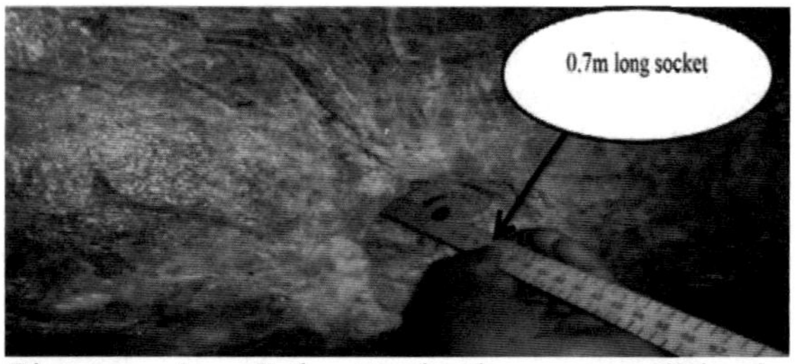

Figure 4. Sub-standard preconditioning socket

4.3 Preconditioning Sockets

Sockets are shot holes that remain after having been charged with explosives and blasted. Sockets are usually suspected of containing explosives; hence it is important to properly wash them after a blast. Precondition sockets are important in stope faces where preconditioning is done. Figure 5 shows a preconditioning cycle representing the length of sockets and the days in which those sockets were created. The mines' advance per blast is 0.9m; preconditioning ensures improved advance, hence it increases it by 0.3m. As a result, the total improved advance will be 1.2m with every blast provided the technique is applied correctly.

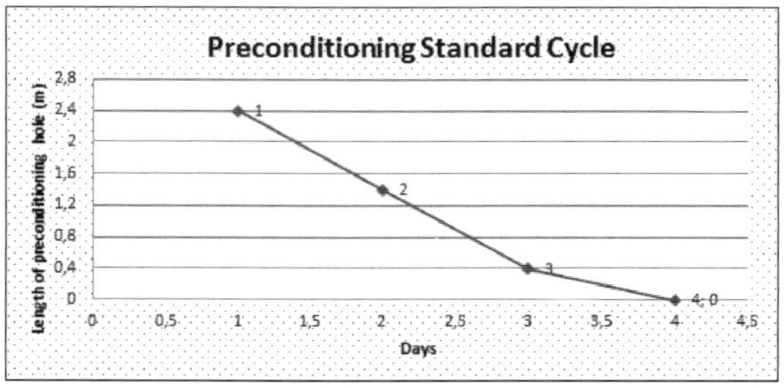

Figure 5. Preconditioning Standard Cycle South

Table 5 shows preconditioning socket lengths after the first blast of the preconditioning cycle in different working areas. Working area 43/41 South Breast was examined for a full preconditioning cycle and all sockets were measured. Figure 6 shows the results of the measured lengths of sockets in their respective days, with day 1 showing the 1st day of the preconditioning cycle. Precondition is said to bring benefits, however in this working area, not much benefits were observed. In day 1, a 1.9 m preconditioning hole was drilled and it gave a socket of 1.2 m with an advance of 0.7 m. This comes as a loss to the mine because time and material are wasted during drilling of preconditioning holes and charging them but get no benefits or positive results from the work done.

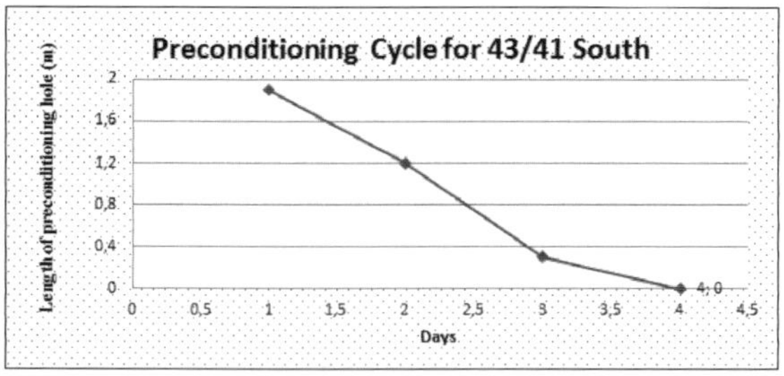

Figure 6. Preconditioning cycle for 43/41

4.4 Challenges Experienced During Operation

Drilling of a preconditioning hole takes twice the time taken to drill a production hole, sometimes it can be longer. Time is the main factor in the mining industry as panels have to be blasted every day and this compromises the safety of workers.

A 2.4m drill steel is used to drill preconditioning holes. When square packs are closer to the face it becomes a challenge to use a 2.4m drill steel due to the limited drilling space which result in workers using a 1.2 or 1.5m drill steels for preconditioning holes. The use of 1.5m drill still for preconditioning holes renders the preconditioning ineffective.

Since preconditioning holes are longer than usual production holes, adequate air pressure is required to drill to ensure fast drilling of the holes

3.5 Solutions to the Problems Encountered

It has been observed that drilling of preconditioning holes takes time. The mine uses a drill steel of 36mm in diameter which takes on average between 10-12 minutes to drill a length of 2.4m, which is a standard length of a preconditioning hole at the mine. It can be advised that the mine change the drill steel to a 40 mm diameter. If a 40 mm drill steel is used, a 3.2m hole can be drilled in a time frame of between 8-10 minutes which improves the drilling time as compared to when a 36mm drill steel is in use (Toper , et al., 2003). When preconditioning is applied correctly, the advance rates increases and this has an advantage of improved production rate. If the required advance is achieved, then enough space between the packs installed from a previous blast and the face will be left for the use of a 2.4m drill steel.

When holes are overcharged with explosives and little tamping is used in drilled holes, the production rate is reduced. This is due to the production of finer fragments as a result of excessive explosives. Ore will be lost during the cleaning process. Stoping costs can be reduced when enough amounts of both the explosives and tamping are used in one hole.

An improved hanging wall is achieved if standard holes drilled on the face are correctly charged and tamped; this ensure safety in the stope as signs of fall of grounds will not be visible and this motivates individuals working knowing that work is being performed in a safe area.

3.6 Seismicity Influence on Safety and Production

A study on areas that have been closed down at the mine due to seismicity was done. This was to assess if they can be opened up for production through the application of preconditioning, and ensuring that mine workers are working in a safe environment. Gold contents found in those areas were compared to those found in areas currently operating at the mine. A comparison of gold contents is represented in Table 6.

Table 6. Current working areas and closed down areas due to seismicity and their gold grades

Working Areas (Closed seismic affected areas)	Gold content found (cmg/t)	Working Areas (Currently in operation)	Gold content found (cmg/t)
VCR41-45 Bottom North	3050	**VCR 40-46 North**	1410
VCR41-64 Top South up dip	4300	**VCR41-48 South**	3875
VCR41-68 North Breast	1525	**VCR43-69A WRS**	3699
VCR43-38 North Up dip	3050	**VC42-40 01 North**	6489
VCR40-45 South Up dip	575	**VCR43-67 South**	1979
VCR41-43 South Brest	4600	**VCR 45 62A 02AN**	1860

Some areas that have been closed down due to seismicity at the mine contained high gold contents than some that are currently operating. This warrants re-evaluation of the areas to see if they can be opened up for production.

3.6.1 Optimization of preconditioning in seismicity affected areas

Hydraulic fracturing was not performed for this study. All recommendations are based on the literature review conducted. Hydraulic fracturing process in this context will be used to eject gases in pre-existing fractures and also determining their stresses by means of suppling high pressured water in the fractures. Sand fillings can be pumped in with water in order to hold the fractures open to inject the gases from previously worked areas i.e. seismic affected areas. The hydraulic fracturing process can be used to measure the stresses existing in the fractures; this will be a good way of knowing if it is then safe to enter the area or not. When the gases have been ejected from areas that were closed due to seismicity and the stresses in the existing fractures have been determined, conclusions can be made on the results if the area is safe for the first preconditioning blast to be taken. If the first blast conducted comes out successfully, it will then give indication that the area would be safe and would have been opened up thereafter for production purposes.

3.7 Key Issues Determining Level of Success of the Technique

It is of paramount importance for the production drilling crew to understand the technique and concepts behind preconditioning. If the general crew understands what the technique is done for, they will be inclined to make sure that it is done correctly.

Continual checks on the procedure weekly or after a preconditioning cycle is essential. On the third day of each preconditioning cycle, all 3 preconditioning sockets should be seen on the face with different dimensions; these are represented on Figure 5. This is another way of checking if the process is applied correctly.

Express support is needed from the production personnel. Production personnel are subjected to severe production pressures which sometimes results in work being short cut by not applying preconditioning. This strain the need for production personnel to always remain disciplined.

Appropriate equipment and material for the technique should always be made available. This includes: drill steels, charging and tamping materials, charging sticks to push back the explosives, high pressured air should be supplied in drilling machines to maximize on the drilling time etcetera.

Production crews need to adjust to the fact that preconditioning entails extra work. Bonus money is received if more than required production is produced. However, through the correct application of the process, advances are achieved always, which will then imply that the workforce will be receiving bonus money.

3.8 Practical Findings to Maximize the Effectiveness of the Preconditioning Technique

When the face-perpendicular preconditioning technique is applied in new areas, it is advisable to maximize on the blast parameters in order to maximize the effectiveness of the preconditioning process. Blast parameters are essential components that formulate the basic understanding that is required for designing an effective preconditioning technique, they include: charge mass, charge type, hole spacing, diameter of drill steel, position of charge and how it is done and the initiation process.

To maximize the effectiveness of the technique, the mine should consider changing to a 3.2 m long drill steel to drill the preconditioning holes as it has shown best advance rates based on the literature conducted by Toper et al (2000).

Preconditioning hole spacing should be kept at 3 m. Constant hole spacing avoids lock-ups of fractures ahead of the working face which reduces stress concentrations leading to reduced chances of face bursting. Spacing between the packs close to the face gives indications of how the preconditioning holes should be spaced as they follow the same standard approach.

It is a known fact that the process of applying preconditioning takes longer, but if it is being applied correctly with every blast it becomes easier to drill the next advance as the stope face would be fractured by the previous preconditioning blast.

5 CONCLUSIONS

Preconditioning is used to ameliorate rockburst events in the mining industry. Increased productivity is ensured through the correct application of the technique. Since the implementation of the face perpendicular preconditioning technique at the mine, a continual improvement in safety and production has been observed. It has been effective in enabling safer mining in seismically hazardous areas. Preconditioning has saved a lot of lives at the mine through reduction of seismic events and rock bursts occurrences in stope panels. Preconditioning has assisted greatly in decreasing the stoping costs because of the minimal support structure required on the stope face and increased advance rates. Accident statistics due to FOG's has shown a gradual improvement in safety at the mine. Optimization of the technique would help achieve even better results than what has been established.

Safety benefits that result because of the correct application of the preconditioning process includes:

- Minimized face bursts hazards
- Reduction in damages by seismicity
- The hanging wall conditions are improved, which will then require less support as it is made stable and the stoping width is controlled and
- Face conditions are improved resulting in a straight break and clean face.

Correct application of the technique also results in improved production and the benefits include the following observations made by the mine.

- The face advance is increased by more than 40%
- Fragment sizes are reduced by more than 30%
- The drilling rate is increased by more than 25%
- Improved hanging wall conditions result in improved stoping width, which controls the blasted width, thereby producing less dilution
- Stoping costs are reduced by more than 15% because minimal support structure is required and
- Workers self-confidence is increased or rejuvenated because the area has been made safe.

REFERENCES

Granderg, A., (2013). What is Hydraulic Fracturing? Available at: https://www.propublica.org/special/hydraulic-fracturing-national. Accessed [01 April 2016].

Karwoski, J. and William, L., (1979). Rock Preconditioning to prevent Rock Bursts- Report on a field Demonstration. *United States Department of The Interior Bureau of Mines*. Vol.3, No.27, pp. 49-70.

Koldas, K., (2001). Rock-related accidents, investigations and inquiries in South African Mines. *International Mining Congress and Exhibition of Turkey-IMCET*. Vol.8, N.o.4, pp. 105-115.

Mining Review Africa, (2016). Harmony Gold employee killed in fall of ground incident at Masimong mine. Available at: http://www.miningreview.com/news/harmony-gold-employee-killed-in-fall-of-ground-incident-at-masimong-mine/. [Accessed 13 July 2016].

Stacey, T. and Wesseloo, R., (2002). Application of indirect stress measurement techniques (non-strain gauge based technology) to quantify stress environments in mines. Available at: www.mhsc.org.za/sites/default/files/GAP%20858%20REPORT.pdf. [Accessed 01 April 2016].

Toper, A. Z.,Kabongo, K.K., Stewart. R.D., and Daehanke. A. (2000). The mechanism, optimization and effects of preconditioning. *The Journal of The South African Institute of Mining and Metallurgy*. Vol. 100, No.1, pp. 7-16.

Toper, A. Z., Janse van Rensburg, A.L., Mile. A.M., Grodner, M. W., and Noble, B.R. (2003). The effect of blasting on the rockmass for designing the most effective preconditioning blasts in deep-level gold mines. 1st ed. Johannesburg: University of the Witwatersrand.

Windsor, C. and Thompson, A., (1993). Rock Reinforcement Practice: Advanced Course, Rock Technology Pty. Ltd., Australia.

The Application of Multivariate Regression Analysis to Predict the Performance of Diamond Wire Saw

R.Mikaeil
Department of Mining and Metallurgical Engineering, Urmia University of Technology, Urmia, Iran.
S.S.Haghshenas
Young Researchers and Elite Club, Rasht Branch, Islamic Azad University, Rasht, Iran.
M.Ataei
Faculty of Mining Engineering, Petroleum & Geophysics, Shahrood University of Technology, Shahrood, Iran.
S.S.Haghshenas
Department of Civil Engineering, Islamic Azad University, Astara Branch, Astara, Iran.
A.S.Haghshenas
School of Industrial and Information Engineering, Politecnico di Milano, Milano, Italy.

ABSTRACT In this research, the performance prediction of diamond wire saw in carbonate rock sawing was investigated on 14 different carbonate rocks in some famous quarries located in Iran. For laboratory tests, rock samples were collected from the studied quarries. Some important mechanical and physical properties of stone such as elasticity modulus, equivalent quartz content and uniaxial compressive strength were determined in the laboratory. The relationship between the performance of diamond wire saw and rock properties were investigated using simple and multivariate regression analysis. Two models with maximum correlation coefficient were proposed for predicting the performance of diamond wire saw. The results showed that the performance of diamond wire saw can reliably be predicted using the proposed models in carbonate rock sawing by only physical and mechanical properties testing.

Keywords: Carbonate Rock, Diamond Wire Saw Performance, Multiple Regression Analysis

1 INTRODUCTION

Nowadays, diamond wire sawing method has been extensively used in dimension stone quarries. Briefly, in this method, firstly, vertical and horizontal holes are drilled through the stone block. Then, the diamond wire is threaded through these holes. The tension and rotation force required for cutting is provided by the movement of the diamond wire cutting machine away from the cut surface on the rail. Water is supplied during the diamond wire cutting operation both as a coolant and as a means of removing waste particles. The diamond wire is simply a steel cable on which small beads

bonded with diamond abrasive are mounted at regular intervals with spacing material placed between the beads. Schematic presentation of a diamond wire-cutting layout and cross-section of diamond bead are illustrated in Figure 1 (Özçelik 1999, Özçelik and Kulaksız, 2000).

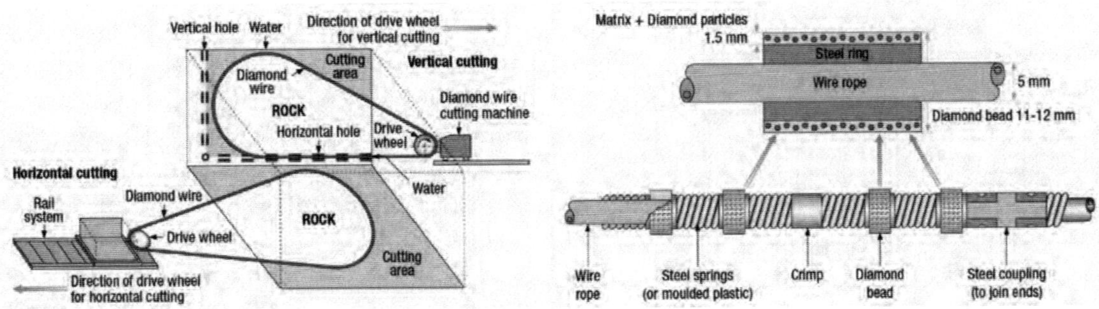

Fig1. Schematic presentation of a diamond wire-cutting layout and cross-section of diamond bead (Ozcelik, 2005a)

The cost of sawing operation is controlled mainly by some of factors. Factors that affect sawing efficiency in the diamond wire sawing method are given in Table 1 (Özçelik, 1999).

Table (1). Factors affecting cutting efficiency in the diamond wire saw method

Non-controlled parameters related to rock characteristics	**Partially-controlled or controlled parameters**	
	Properties of cutting tools and equipment	**working conditions**
Hardness	Machine power	Technical personnel
Strength	Machine position on either vertical or horizontal cutting	Vibration of machine
Water content	Number of beads per meter	
Degree of alteration	Cutting angle between wire and horizontal level	
Discontinuities	Amount of cutting area with respect to angle variation	
Mineralogical properties	Wire speed	
textural characteristics	Amount of water used	
	Bead structure	
	Dimensions of block	

There have been some studies on the sawing performance of diamond wire saw and circular diamond saw. Some researchers have attempted to investigate the effect of type of diamond saw, the saw operating parameters and the properties of the sawn rock on diamond wire and circular diamond saw performance. These studies have been documented in literature (Biasco 1993, Bortolussi et al. 1994, Bortolussi et al. 1990, Cook and Smith 1993, Hallez 1992, Özçelik and Kulaksız, S., 2000, Özçelik et al. 2002, Özçelik, Y and Bayram F., 2004, Özçelik 2005a, Özçelik 2005b, Özçelik 2007, Mikaeil et al. 2008, Mikaeil et al. 2011, Ataei et al. 2011, Ataei et al. 2012, Ghaysari et al. 2012, Mikaeil et al. 2013,Sadegheslam et al 2013, Mikaeil et al. 2014, Mikaeil et al. 2015, Almasi, et al. 2015, Aryafar and Mikaeil 2016, Mikaeil et al. 2016).

In this paper, the effects of the some major rock properties on the performance of the diamond wire saw were examined and relations between the sawing rates are described.

2 FIELD AND LABORATORY STUDIES

During the field studies, 14 marble quarries in Isfahan, Kordestan and Yazd provinces of Iran were visited and the sawing performances of diamond wire saws were measured on their different carbonate rocks. Figure 2 shows the location map of studied quarries (Mikaeil et al. 2008).

Fig 2. Location map of studied mines – Iran (Mikaeil et al. 2008)

2.1 Wire Saw Machine Properties

In Iranian dimension stone quarries, usually similar cutting machines are used. Therefore, in the studied quarries many technical feature of cutting machines were nearly similar. In this study, the characteristics of cutting machine were considered to be constant and were not used in model. The characteristics of cutting machine and operational parameters that were usually used in Iranian dimension stone quarries are listed in Table 2 (Mikaeil et al. 2008).

Table 2. Operational parameters and properties of used cutting machine (Mikaeil et al. 2008)

Parameter	Description
Main motor power (KW)	45
Length of wire (m)	65-80
Linear speed (m/s)	30-35
Rotator diameter (cm)	60
Beads per meter	33-36
Beads type	special for soft rocks

2.2 Rock Properties and Wire Saw Performance

To assess the rock sawability and prediction of production rate, some important parameters such as uniaxial compressive strength (UCS) and elasticity modulus (E). As the most important mechanical properties of rocks and equivalent quartz content (Qc) as the most important abrasivity property was selected and used in the regression analyses simultaneously. For laboratory tests, some rock blocks were collected from the studied quarries. An attempt was made to collect rock samples that were big enough to obtain all of the test specimens of each rock type from the same piece. Each block sample was inspected for macroscopic defects so that it would provide test specimens free from fractures, partings or alteration zones. Then, test samples were prepared from these block samples and standard tests have been completed to measure the above-mentioned parameters following the suggested procedures by the ISRM standards (International Society for Rock Mechanics, 1981). The performance of wire saw and measured UCS, E and Qc of rocks are listed in Table 3. The production rate of wire saw (Pr) was selected as a performance of wire saw (Mikaeil et al. 2008).

Table 3. The performance of wire saw and laboratory tests (Mikaeil et al. 2008)

Mine name / location	Pr (m^2/h)	Qc (%)	E (GPa)	UCS (MPa)
Godarsorkh/ Esfahan	1.28	0.4	5.45	96
Hamzadeh/ Esfahan	6	0.2	2	93.8
Joshghan qali/ Esfahan	5	0.15	2.5	97.5
Laybid/ Esfahan	7	0.15	2.5	87.1
Kandab/ Esfahan	3	0.63	2.3	94.2
Jajarmi/ Esfahan	1.5	0.45	5.2	96.9
Senesar/Kordestan	2.4	0.45	3	96.5
Shanoreh/Kordestan	10	0.075	2.3	59.0
Vihaje/Kordestan	10	0.05	2	59.15
Poshtbaskoli/Kordestan	6	0.25	3.7	64.0
Crysal/kordestan	3.6	0.45	4	68.0
Sorti/Yazd	3.7	0.512	3.7	81.7
Baiatneyriz/Yazd	3.7	0.38	4.5	95.4
Angorakdehbid/Yazd	2.5	0.22	5.2	84.6

3 SIMPLE REGRESSION ANALYSIS

In this step first in order to understand the general relationship between output, i.e. production rate, and input parameters, i.e. equivalent Quartz content (Qc), Elasticity modulus (E), Uniaxial compressive strength (UCS), different simple regression analysis's have been tried between them (Table 4).

Table 4. Different regression models fitted between output and input variables

Model	Linear	Logarithmic	Inverse	Quadratic	Cubic	Power	Exponential
R^2 of Production vs. QC	0.32	0.94	0.86	0.54	0.75	0.91	0.43
R^2 of Production vs. E	0.48	0.52	0.88	0.83	0.86	0.61	0.57
R^2 of Production vs. UCS	0.62	0.72	0.84	0.86	0.85	0.81	0.73

At last the best fitted regression model was chosen (Figure 3). The important note which should be considered is that all these regression analysis's have not got constants in their equations, because statistical analysis's showed that by considering the constant in these equations their R square would be reduced considerably.

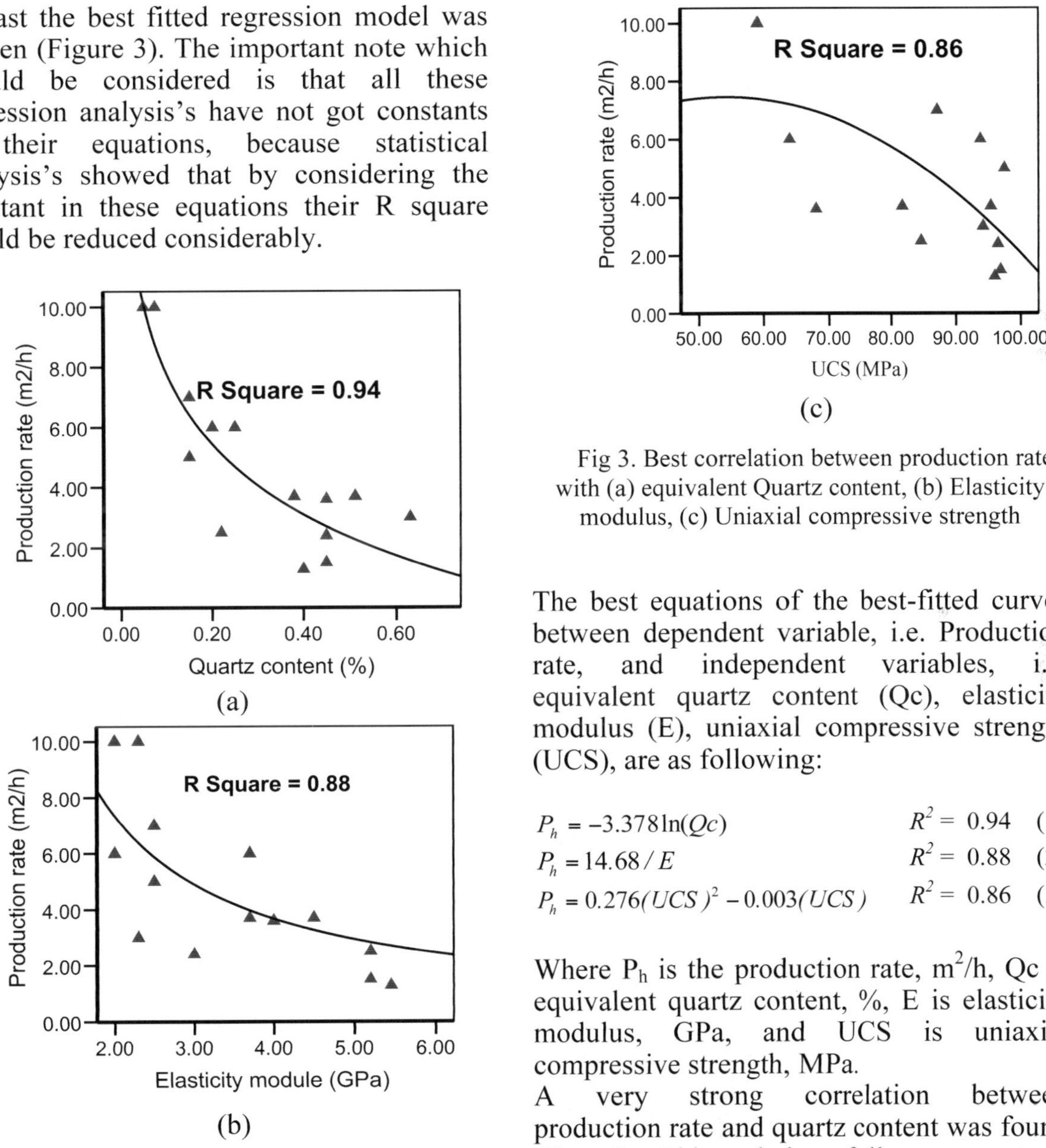

Fig 3. Best correlation between production rate with (a) equivalent Quartz content, (b) Elasticity modulus, (c) Uniaxial compressive strength

The best equations of the best-fitted curves between dependent variable, i.e. Production rate, and independent variables, i.e. equivalent quartz content (Qc), elasticity modulus (E), uniaxial compressive strength (UCS), are as following:

$$P_h = -3.378\ln(Qc) \qquad R^2 = 0.94 \quad (1)$$
$$P_h = 14.68 / E \qquad R^2 = 0.88 \quad (2)$$
$$P_h = 0.276(UCS)^2 - 0.003(UCS) \qquad R^2 = 0.86 \quad (3)$$

Where P_h is the production rate, m^2/h, Qc is equivalent quartz content, %, E is elasticity modulus, GPa, and UCS is uniaxial compressive strength, MPa.

A very strong correlation between production rate and quartz content was found (Eq. 1). This relation follows a natural logarithmic function. Production rate decreases with increasing equivalent quartz content. There is a strong correlation between UCS and E (Eqs. 2 and 3).

Similarly, Production rate decreases with increasing these parameters.

4 MULTIVARIATE REGRESSION ANALYSIS (MVRA)

In section 1, it has been clarified that the output has not direct relationship with inputs, thus instead of using Multivariate regression analysis (MVRA) between output and inputs, this analysis is employed between output and three variables Qc, E and UCS which are defined as below:

$$X = -3.378\ln(Qc) \quad (4)$$

$$Y = 14.68/E \quad (5)$$

$$Z = 0.276(UCS)^2 - 0.003(UCS) \quad (6)$$

And MVRA analysis is used between the output, i.e. production rate, and these three parameters in two ways. Two models included constant, and without constant was produced.

Model 1:

$$P_h = -1.70\ln Qc + 9.29/E - 0.000552(UCS)^2 + 2.45 \quad (7)$$

$R^2 = 0.93$

Model 2:

$$P_h = -2.11\ln Qc + 10.52/E - 0.000276(UCS)^2 \quad (8)$$

$R^2 = 0.98$

There are significant correlations between production rate and quartz content, uniaxial compressive strength and elasticity modulus. The production rate decreases with increasing these parameters.

5 VALIDATION OF MODELS

Validation of the model was carried out by considering the correlation coefficient and the plot of observed production versus estimated production. The correlation coefficients (R^2) of the models are higher than 0.9. These values are good, but it does not necessarily identify the valid models. The significance of R^2 value can be determined by compares between experimental production rates and predicted production rates. The error for each models are given in Table 5.

Table (5). Comparison between experimental production rates with predicted production rates by model 1 and 2

Actual Pr (m^2/h)	Predicted Production Rate (m^2/h)			
	Model 1	Error	Model 2	Error
1.28	0.62	0.66	1.32	0.04
6	4.97	1.03	6.24	0.24
5	4.14	0.86	5.60	0.60
7	5.20	1.8	6.13	0.87
3	2.37	0.63	3.10	0.10
1.5	0.41	1.09	1.12	0.38
2.4	1.76	0.64	2.62	0.22
10	8.97	1.03	9.10	0.90
10	10.25	0.25	10.64	0.64
6	5.05	0.95	4.65	1.35
3.6	3.57	0.03	3.04	0.56
3.7	2.41	1.29	2.42	1.28
3.7	1.13	2.57	1.87	1.83
2.5	2.85	0.35	3.25	0.75

As it can be seen in this Table, the error values are very low, suggesting that the models are valid. To investigate the accuracy of the two models, the estimated production values for each data were plotted against the observed production values and are shown in Figures.4, 5 and 6.

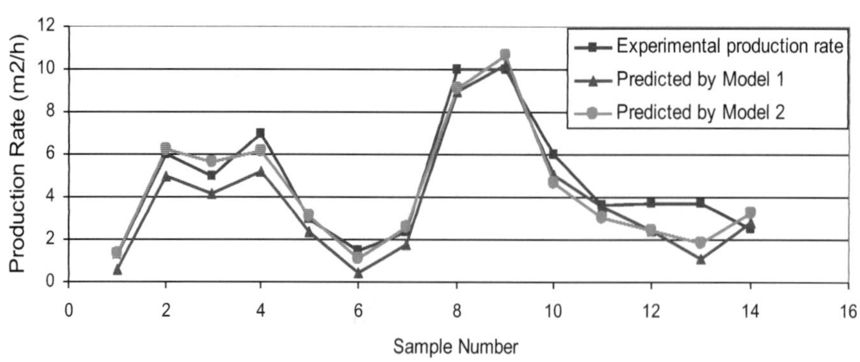

Fig 4. Comparison of experimental and predicted production rate for each sample

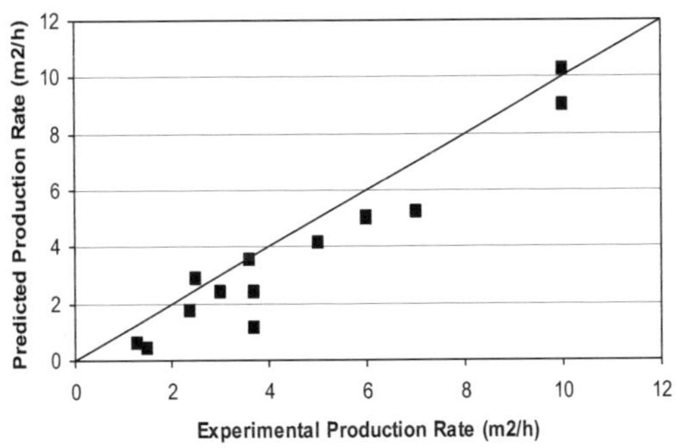

Fig 5. Correlation between experimental production rates with predicted production rates by model 1.

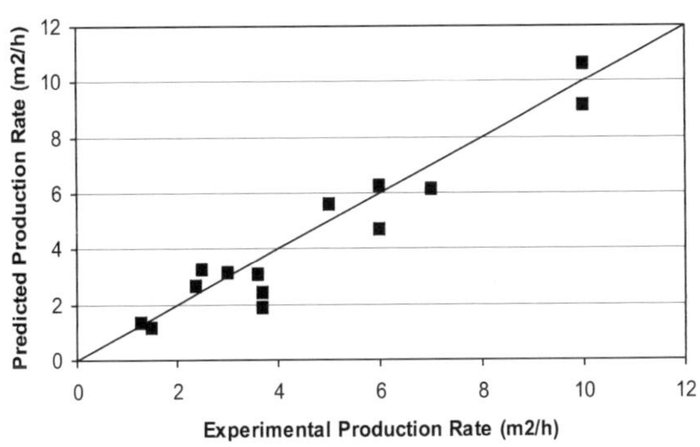

Fig 6. Correlation between experimental production rates with predicted production rates by model 2.

The error in the estimated value is represented by the distance that each data point plots from the 1:1 diagonal line. A point lying on the line indicates an exact estimation. In the plots for models, the points are scattered uniformly about the diagonal line, suggesting that the models are good. It was concluded that the sawing speed of carbonate rocks using diamond wire saws can reliably be estimated using the developed models.

6 CONCLUSION

The diamond wire saws are one of the most important machines used in the ornamental stone extraction. Performance prediction of these saws is important in the cost estimation and the planning of the quarries. A correct estimation of sawability helps to make the stone sawing more efficient. In this paper relationship between production rate and rock properties evaluated using multiple curvilinear regression analysis and estimation models were developed. Equivalent quartz content, elasticity modulus and uniaxial compressive strength are suggested for the estimation of the sawability of carbonate rocks. Two analysis types were employed to predict the production rate. First, a simple regression method was used to understand the relations between the dependent, i.e. production rate, and independent variables, multivariate regression analysis (MVRA) was next used; the results of MVRA were satisfactory and reliable. Afterward, two different models were proposed for predicting the production rate of diamond wire saw, first a model included constant, second without any constant. The results show that production rate of diamond wire saws has a strong relationship with second model but weak relationship with first model. It was concluded that the sawing rate of carbonate rocks using diamond wire saws can reliably be estimated using the developed model.

REFERENCES

Almasi, S.N., Bagherpour, R. Mikaeil, R. Khademian, A. 2015. Influence of Cutting Wire Tension on Travertine Cutting Rate, 24th international mining congress and exhibition of turky, pp 1096-1102.

Aryafar A. and Mikaeil R., Estimation of the Ampere Consumption of Dimension Stone Sawing Machine Using the Artificial Neural Networks, Int. J. Min. & Geo-Eng. Vol.50, No.1, June 2016, pp.121-130.

Ataei M., Mikaiel R., F. Sereshki, N. Ghaysari. Predicting the production rate of diamond wire saw using statistical analysis. Arabian Journal of Geosciences. 2011, 5, 1289-1295.

Ataei M., Mikaeil R., Hoseinie S. H., Hosseini S. M. Fuzzy analytical hierarchy process approach for ranking the sawability of carbonate rock. International Journal of Rock Mechanics & Mining Sciences 50 (2012) 83–93.

Biasco, G., 1993. Diamond Wire for Quarrying Hard Rocks, Industrial Diamond Review, No. 5, pp.252-255.

Bortolussi, A., Ciccu, R., Manca, P.P., Massacci, G., 1994. Computer simulation of diamond-wire cutting of hard and abrasive rock, 103, pp.55-128.

Bortolussi, A., Ciccu, R., Manca, P.P., Massacci, G., 1990. Simulation and Optimization of Rock Cutting with Diamond Wire, International Symposium APCOM, Berlin, pp. 163-176.

Cook, M., Smith, B.A., 1993, A Study of Diamond Wire Bead Rotation, Industrial Diamond Review, No.2, pp. 79-81.

Ghaysari N., Ataei M., Sereshki F., Mikaiel R. Prediction of Performance of Diamond Wire Saw with Respect to Texture Characteristics of Rock, Arch. Min. Sci., Vol. 57 (2012), No 4, p. 887–900.

Hallez, C.P., 1992. Sawing Marble with Diamond Wire, Diamonds in Industry, pp. 24-27.

International Society for Rock Mechanics, 1981. Rock characterisation, testing and monitoring: ISRM suggested methods. Oxford: Pergamon.

Mikaeil, R., Ataei, M., Hoseinie, S.H., 2008. Predicting the production rate of diamond wire saws in carbonate rock cutting, Industrial Diamond Review, No.3, pp.28-34.

Mikaeil R., Yousefi R., Ataei M., Abasian Farani R. Development of a New Classification System for Assessing of Carbonate Rock Sawability, Archives of Mining Sciences, Vol. 56 (2011a), No 1, p. 57–68.

Mikaeil R., Ataei M., Yousefi R., Application of a fuzzy analytical hierarchy process to the prediction of vibration during rock sawing. Mining Science and Technology, 21 (2011b) 611–619.

Mikaiel R., Ataei M., Yousefi R. Correlation of production rate of ornamental stone with rock brittleness indexes. Arabian Journal of Geosciences. 2011c, 6, 115-121.

Mikaeil R., Yousefi R., Ataei M., 2011, "Sawability Ranking of Carbonate Rock Using Fuzzy Analytical Hierarchy Process and TOPSIS Approaches". Scientia Iranica, Transactions B: Mechanical Engineering 18 (2011d) 1106–1115.

Mikaeil R., Ataei M., Yousefi R., "Evaluating the Power Consumption in Carbonate Rock Sawing Process by Using FDAHP and TOPSIS Techniques", Efficient Decision Support Systems: Practice and Challenges – From Current to Future / Book 2", 2011e, ISBN 978-953-307-441-2., pp 478.

Mikaeil R., Ozcelik Y., Ataei M., Yousefi R., 2011f. Correlation of Specific Ampere Draw with Rock Brittleness Indexes in Rock Sawing Process. Arch. Min. Sci., Vol. 56, No 4, p. 741–752.

Mikaeil R., Ozcelik Y., Ataei M., Yousefi R. Ranking the sawability of ornamental stone using Fuzzy Delphi and multi-criteria decision-making techniques. International Journal of Rock Mechanics & Mining Sciences 2013; 58, 118–126.

Mikaeil R., Ataei M., Ghadernejad S., Sadegheslam G. Predicting the Relationship between System Vibration with Rock Brittleness Indexes in Rock Sawing Process. Archives of Mining Sciences, (2014), 59-1, p. 139-153.

Mikaeil R.; Kamran M.; Sadegheslam G.; Ataei M. Ranking the Sawability of Dimension Stone by Using PROMETHEE, Journal of Mining and Environment. 2015,Vol 6, No. 2, 263-271.

Mikaeil, R., Haghshenas, S. S., Haghshenas, S. S., & Ataei, M. (2016). Performance prediction of circular saw machine using imperialist competitive algorithm and fuzzy clustering technique. Neural Computing and Applications, 1-10.

Mikaeil R., Ozcelik Y., Ataei M., Shafiee Haghshenas S. Application of Harmony Search Algorithm to Evaluate the Performance of Diamond Wire Saw. Journal of Mining and Environment, 2016b.

Özçelik, Y., 1999. Investigation of the working conditions of diamond wire cutting machines in marble industry, Ph.D. Thesis, Hacettepe University, Ankara, 242 p.

Özçelik, Y., Kulaksız, S., 2000. Investigation of the relationship between cutting angles and wear on beads in diamond wire cutting method, 9th Mine Planning and Equipment Selection Symposium, Panagiotou & Michalakopoulos (Ed.), pp. 661-666.

Özçelik, Y., Kulaksız, S., Çetin, M.C., 2002. Assessment of the Wear of Diamond Beads in the Cutting of Different Rock Types by the Ridge Regression, Journal of Materials Processing Technology, 127 (3), pp. 392-400.

Özçelik, Y., Bayram F., 2004. Optical investigations of bead wear in diamond wire cutting, Industrial Diamond Review, Vol. 64, pp 60-65.

Özçelik, Y., 2005a. Optimum working conditions of diamond wire cutting machines in the marble industry, Industrial Diamond Review, No.1, pp. 58-64.

Özçelik, Y., 2005b. Effect of Mineralogical and Petrographical Properties of Marble on Cutting by Diamond Wire, CIM Bulletin, Vol.98, No. 1085, pp 1-6.

Özçelik, Y., 2007. The effect of marble textural characteristics on the sawing efficiency of diamond segmented frame saws, Industrial Diamond Review, No. 1, pp. 65-70.

Sadegheslam, G. Mikaeil, R. Rooki, R. Ghadernejad, S. Ataei, M. 2013. Predicting the production rate of diamond wire saw using multiple nonlinear regression analysis, Geosystem engineering.

Predicting the Physico-Mechanical Properties of Konya *Pyroclastic Rocks from the Needle Penetration Index*

S. Kahraman, B. Güneş, T. K. Gün
Hacettepe University, Mining Engineering Department, Ankara, Tırkey

İ. İnce
Geological Engineering Department, Selçık University, Konya, Tırkey

ABSTRACT Well-prepared core samples suggested by standards for the laboratory tests cannot be prepared usually particularly from soft and clay-bearing rocks. For this reason, the needle penetration test was developed for the estimation of properties of soft rocks. This study investigates the predictability of physico-mechanical properties of pyroclastic rocks from the needle penetration index. The needle penetration, the uniaxial compressive, density, and porosity tests were carried out on the pyroclastic rocks from Konya region. The needle penetration index values were correlated to the physico-mechanical properties. Good correlations were obtained between the needle penetration index and both compressive strength and density. However, the correlation between the needle penetration index and porosity is weak. It was concluded that the compressive strength and density of Konya pyroclastics could be predicted from the needle penetration index. The correlation between the needle penetration index and porosity should be further investigated for Konya pyroclastics.

1 INTRODUCTION

The intact rock strength is widely used in mining engineering projects. Smooth core samples suggested by standards for the laboratory tests cannot be prepared mostly particularly from soft and clay-bearing rocks. Indirect test methods such as point load and block punch index tests have been developed to estimate the mechanical properties of rocks. However, the specimen preparation from soft and clay-bearing rocks is still difficult for these indirect tests. For this reason, the needle penetration test which is a simple and non-destructive index test was developed.

The needle penetration test is an easy index test which is applicable both in the field and laboratory and does not require any special sample preparation (Ulusay et al., 2014). This test has been used for the estimation of physical and mechanical properties of weak or soft rocks. Although some researchers have suggested some correlations between the physico-mechanical rock properties and the needle penetration index (NPI) for weak rocks, the studies are limited on this area. In this study, the predictability of physico-mechanical properties of Konya pyroclastic rocks from the NPI was investigated.

2 PREVIOUS STUDIES

The needle penetration test was originally developed for the estimation of the uniaxial compressive strength (UCS) of soft rocks. Although several researchers (Okada et al. 1985; Takahashi et al. 1988; Yamaguchi et al. 1997; Uchida et al. 2004; Aydan et al. 2006; Aydan et al. 2008; Erguler and Ulusay 2007; Park et al. 2011; Aydan 2012; Ulusay and Erguler 2012) have used the needle penetration index (NPI) to estimate the UCS of rocks, some other researchers have also investigated the relations between the NPI and other rock properties such as tensile strength, Young's modulus and P- and S-wave velocities, cohesion, and friction angle (Aydan 2012; Aydan and Ulusay 2013; Aydan et al. 2014).

Ulusay and Erguler (2012) evaluated a database consisting of a total of 725 sample from the previous studies and additional new tests and suggested the following equation:

$$UCS = 0.4NPI^{0.929} \quad (1)$$

where UCS is the uniaxial compressive strength (MPa) and NPI is the needle penetration index (N/mm).

Aydan (2012) derived the following relations for different rock types:

$$E = 0.05NPI \quad (2)$$

$$BTS = 0.02NPI \quad (3)$$

$$V_P = 0.33 + 0.3NPI^{0.5} \quad (4)$$

$$w = 70 - 15.2NPI \quad (5)$$

$$UCS = 0.2NPI \quad (6)$$

where E is the elastic modulus (GPa), BTS is the Brazilian tensile strength (MPa), V_P is the P-wave velocity (km/s), w is the water content (%), UCS is the uniaxial compressive strength (MPa), and NPI is the needle penetration index (N/mm).

Aydan and Ulusay (2013) derived the following relations for Derinkuyu tuff:

$$UCS = 0.3NPI \quad (7)$$

$$E = 0.025NPI \quad (8)$$

Recently, Kahraman et al. (2016) investigated the predictability of physico-mechanical properties of Erciyes pyroclastic rocks from the NPI. They developed the following relations:

$$UCS = 0.16NPI \quad (9)$$

$$BTS = 0.009NPI \quad (10)$$

$$\rho = 1.04NPI^{0.17} \quad (11)$$

$$n = -11.2\ln NPI + 93.84 \quad (12)$$

where ρ is density (g/cm^3), n is porosity (%), and NPI is the needle penetration index (N/mm).

3 SAMPLING

Block samples of pyroclastic rocks were collected from the four different locations (Ardıçlı, Kızılören, Gökyurt) of Konya region. Each block sample was inspected for macroscopic defects so that it would provide test specimens free from fractures, partings or alteration zones.

4 EXPERIMENTAL STUDIES

4.1 Physico-mechanical tests

UCS, density and porosity tests were carried out on the 72 core samples (Figure 1) according to ISRM suggested methods.

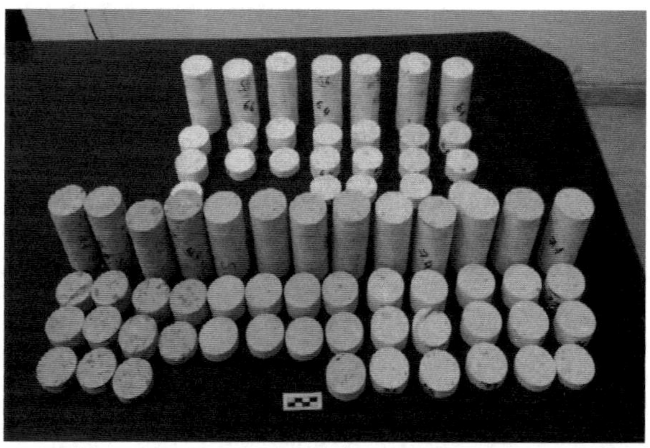

Figure 1. Core samples used in the tests.

4.2 The needle penetration test

The needle penetrometer (Figure 2) is developed by MARUTO Co. in Japan to estimate the UCS of soft rocks. It is a lightweight portable device with a weight of about 700g. The needle is a hardened steel, 0.84mm-diameter rod terminated by a conical tip. The device mainly consists of presser, chuck, penetration scale (0–10 mm, 1 mm graduation), load scale (0–100 N, 10 N graduation), load indicator ring, cap (removable; spare penetration needles

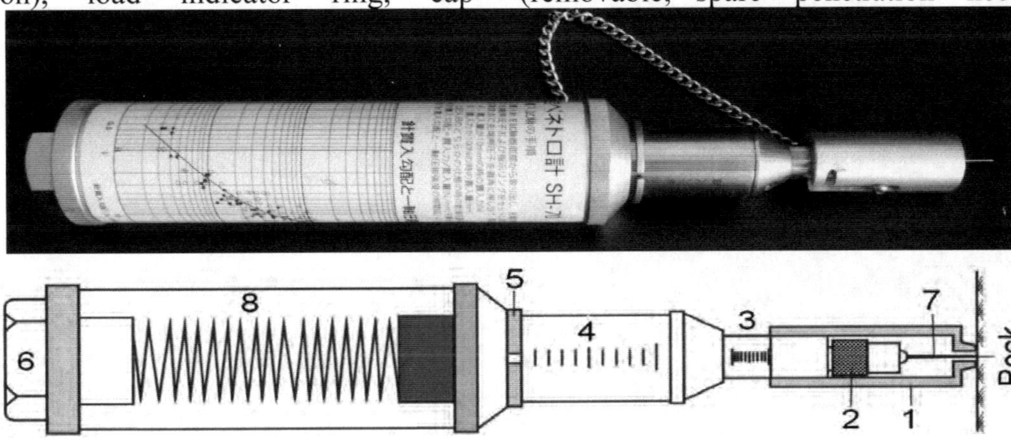

Figure 2. Needle penetrometer and its parts: 1- presser, 2- chuck, 3- penetration scale, 4- load scale, 5- load indicating ring, 6- cap, 7- penetration needle and 8- spring (Ulusay et al., 2014).

contained in the grip), penetration needle and spring mounted in the penetrometer grip (Ulusay et al. 2014)..

The needle of the penetrometer is pushed into the rock until 100 N is reached and the penetration depth is measured from the position of the presser on the penetration scale. The maximum penetration depth (10 mm) may be attained for softer and saturated rocks before the maximum penetration force is reached. In this case, the test stops at this depth, the penetration load is read from the load scale and the needle penetration index (NPI) is calculated from the following equations (Ulusay et al., 2014):

For F = 100 N and D ≤ 10 mm,

$$NPI = 100/D \qquad (13)$$

For D = 10 mm and F ≤ 100 N,

$$NPI = F/10 \qquad (14)$$

where NPI is the needle penetration index (N/mm), F is applied load (N), and D is the depth of penetration (mm).

The needle penetration tests were carried out at least three times on each sample (Figure 3). The NPI values were calculated from the Eqs. (13 and 14) and the results were averaged. Summary of the results of the physico-mechanical and the NPI tests are given in Table 1.

5 EVALUATION OF THE RESULTS

The test results were evaluated using regression analysis. The NPI values were correlated to the physico-mechanical properties of the tested rocks. A good correlation was found between the NPI and the UCS as shown in Figure. 4. The equation of the line is

$$UCS = 0.15NPI \qquad r = 0.80 \qquad (15)$$

A good correlation was also found between the NPI and density as demonstrated in Figure 5. The relation follows a power function. The equation of the curve is

$$\rho = 0.42NPI^{0.32} \qquad r = 0.77 \qquad (16)$$

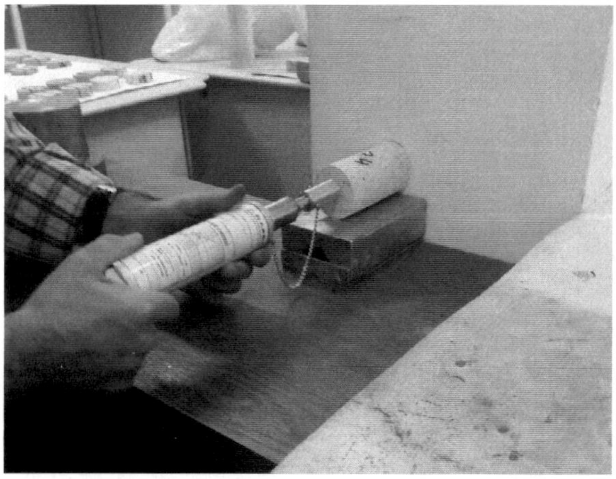

Figure 3. The needle penetration test.

Table 1. Sampling locations and the summary of test results.

Sample Location	Uniaxial compressive strength (MPa)		Density (g/cm^3)		Porosity (%)		Needle penetration index (N/mm)	
	Min.	Max.	Min.	Max.	Min.	Max.	Min.	Max.
Ardıçlı-1	5.8	16.3	1.72	1.98	18.2	28.8	46.0	117.6
Ardıçlı-2	9.3	36.9	1.98	2.18	13.2	24.1	50.0	200.0
Kızılören	11.2	12.3	1.19	1.32	43.1	48.7	33.9	47.6
Gökyurt	6.9	8.5	1.52	1.72	40.1	44.9	54.1	88.9

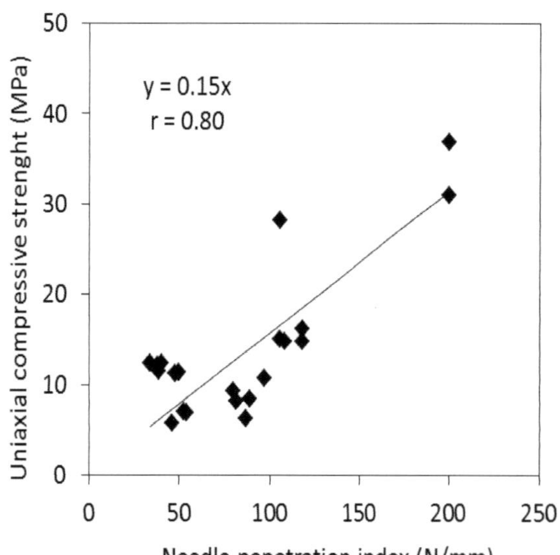

Figure 4. Correlation between the UCS and the NPI.

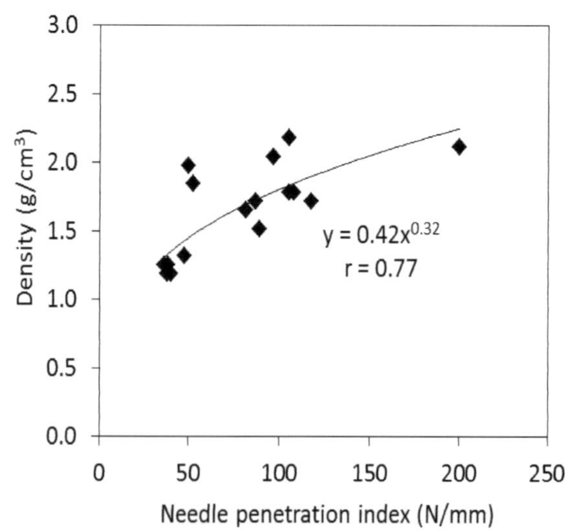

Figure 5. Correlation between density and the NPI.

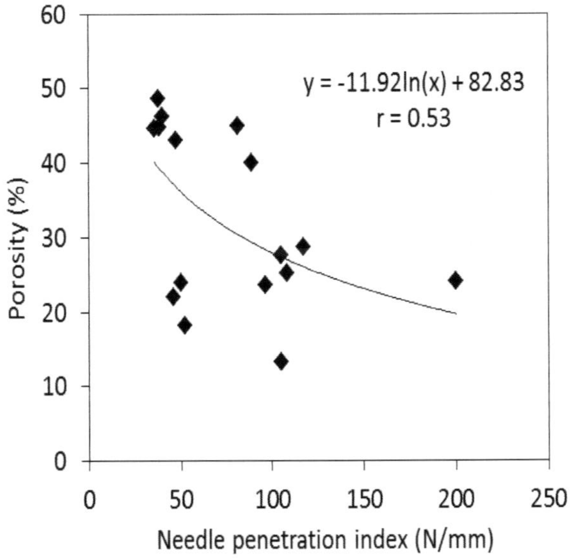

Figure 6. Correlation between porosity and the NPI.

Although good correlations were derived between the NPI and both UCS and density, a weak correlation was found between the NPI and porosity as shown in Figure 6. The equation of the curve is

$$n = -11.92 \ln NPI + 82.83 \quad r = 0.53 \quad (17)$$

Eqs. (15 and 17) are almost the same as the Eqs. (9 and 12) which have been drived by Kahraman et al. (2016) for Erciyes pyroclastic rocks. However, Eq. (16) is a bit different from the Eq. (11).

The significance of r-values of the derived equations can be determined by the t-test, assuming that both variables are normally distributed and the observations are chosen randomly. The distribution of the data was checked by histogram analysis and it was seen that the data generally didn't show normal distribution. For this reason, t-test was not performed.

The significance of regressions was determined by the analysis of variance. In this test, a 95 % level of confidence was chosen. If the computed F-value is greater than tabulated F-value, the null hypothesis is rejected that there is a real relation between the dependent and independent variables. The computed F-values are greater than the tabulated F-values for all derived equations as shown in Table 2. Therefore, it is concluded that the derived equation are valid according to the F-test. Although Eq. (17) is significant, the correlation coefficient of this equation is weak. Therefore, it can be said that the reliability of this equation should be studied in the future.

Table 2. F -test results.

Equation no	F-tabulated	F-test
15	4.07	45.21
16	4.17	53.73
17	4.17	16.18

6 CONCLUSIONS

Seventy two samples of Konya pyroclastic rocks were tested and the relation between the NPI, and UCS, density and porosity was investigated using regression analysis. Good correlations were established between the NPI and both the UCS and density. However, a good correlation was not found between the NPI and porosity.

Concluding remark is that the UCS and density of Konya pyroclastics can be predicted from the NPI. It can also be said that the correlation between NPI and porosity should be further investigated. Further research is also necessary to check whether the results are valid for all types of pyroclastic rocks.

REFERENCES

Aydan, O., 2012. The inference of physico-mechanical properties of soft rocks and the evaluation of the effect of water content and weathering on their mechanical properties from needle penetration tests. *Symposium of ARMA*, Chicago, Paper No. ARMA12- 639 (on CD).

Aydan, O. and Ulusay, R., 2013. Geomechanical evaluation of Derinkuyu antique underground city and its implications in geoengineering. *Rock Mech. Rock Eng.*, 46, 731–754.

Aydan, O., Sato. A., and Yagi, M., 2014. The inference of geo-mechanical properties of soft rocks and their degradation from needle penetration tests. *Rock Mech. Rock Eng*., 47, 1867–1890.

Aydan, O., Seiki, T., Ito, T., Ulusay, R. and Yuzer, E., 2006. A comparative study on engineering properties of tuffs from Cappadocia of Turkey and Oya of Japan. *Proceedings of the Symposium on Modern Applications of Engineering Geology*, Denizli-Turkey, 425–433.

Aydan, O., Watanabe. S. and Tokashiki, N., 2008. The inference of mechanical properties of rocks from penetration tests. *Proceedings of the 5th Asian Rock Mechanics Symposium (ARMS5)*, Tehran-Iran, vol 1, 213–220.

Erguler, Z.A. and Ulusay, R., 2007. Estimation of uniaxial compressive strength of clay-bearing weak rocks using needle penetration response. *Proceedings of 11th Congress on Int. Soc. Rock Mechanics*, Lisbon, vol 1, 265–268.

Kahraman, S., Yumsak, S.M. and Koyuncu, Y.A., 2016. The predictability of physico-mechanical properties of pyroclastic rocks from the needle penetration index. *50th US Rock Mechanics/ Geomechanics Symposium,* Houston, USA, 26-29 June 2016, Paper No. ARMA16-809 (on CD).

Okada, S., Izumiya, Y., Iizuka, Y. and Horiuchi, S., 1985. The estimation of soft rock strength around a tunnel by needle penetration test. *J. Jpn. Soc. Soil Mech. Found. Eng.,* 33, 2, 35–38 (in Japanese).

Park, Y., Obara, Y. and Kan, S.S., 2011. Estimation of uniaxial compressive strength of weak rocks using needle penetrometer. *Proceedings of 12th ISRM International Congress on Rock Mechanics*, Beijing, 795–798.

Takahashi, K., Noto, K. and Yokokawa, I., 1988. Strength characteristics of Kobe formation in Akashi Strata (No. 1). *Proceedings of 10th Japan National Conf. on Geotech. Eng*., 1231–1232 (in Japanese).

Uchida, N., Etoh, Y., Ono, H. and Miura, N., 2004. Strength evaluation of deep mixing soil–cement by needle penetration test. *J. Jpn. Soc. Soi.l Mech. Found. Eng.,* 52, 7, 23–25 (in Japanese).

Ulusay, R. and Erguler, Z.A., 2012. Needle penetration test: evaluation of its performance and possible uses in predicting strength of weak and soft rocks. *Eng. Geol.,* 149–150, 47–56.

Ulusay, R., Aydan, O., Erguler, Z.A., Ngan-Tillard, D.J.M., Seiki, T., Verwaal, W., Sasaki, Y. and Sato, A., 2014. ISRM Suggested Method for the Needle Penetration Test. *Rock Mech. Rock Eng.,* 47, 1073–1085.

Yamaguchi, Y., Ogawa, N., Kawasaki, M. and Nakamura, A., 1997. Evaluation of seepage failure response potential of dam foundation with simplified tests. *J. Jpn. Soc. Eng. Geol*., 38, 3, 130–144.

Oluk Derinleştirme ve Rölyef Kesme Yöntemlerinin Keski Aralığı Açısından Karşılaştırılması

A Rational Comparison of Groove Deepening to Relief Cutting in Terms of Tool Spacing with Drag Tools

O. Z. Hekimoglu
Mugla Sitki Kocman University, Department of Mining Engineering, Kotekli, Mugla

ÖZET Bu bildiride kayaç kömür ve doğaltaş kazısında kullanılan makinelerin keski dizilim tasarımında kullanılan oluk derinleştirme ve rölyef kesme yöntemlerinin rasyonel bir karşılaştırılması anlatılmaktadır. Karşılaştırma, keskilerin gerçek kesme hareketlerinin çeşitli kesme aralığı değerlerindeki benzetişimi'nin laboratuvarda tam ölçekli doğrusal kesme deneyleri yapılarak gerçekleştirilmiştir. Rölyef kesmede keski kuvvetleri ve özgül enerji değerlerinin daha küçük olduğu gözlenmiştir. Bu durum, keskilerin süpürdüğü alan içerisinde tanımlanan 'etkin kesme alanı' değerinin daha küçük olmasından kaynaklandığı anlaşılmıştır. Daha önce düz kaya yüzeyi üzerinde yapılan deneylerde elde edilen en az özgül enerji değeri bu benzetişim deneylerinde görülmemiştir. Deney sonuçları ayrıntılı olarak sunulmuş ve tartışılmıştır.

ABSTRACT This paper mentions a rational comparison between relief cutting and groove deepening for tool lacing design of partial-face cutting machines employed for mechanical excavations of rocks, coal and stones. The comparison was carried out by simulating actual cutting actions of picks at different spacing to depth ratios respectively, through full-scale laboratory linear cutting experiments with drag tools. Relief cutting exhibited lower tool forces and lower specific energy. This merit was attributed to lower 'effective area' defined within cross sections swept by tools. Furthermore, with simulated trials there found to be no marked minimum in specific energy, as opposed to previous cutting experiments on flat rock surface. The results were presented and discussed in details.

1 INTRODUCTION

In mechanical excavation, roadheaders, continuous miners and coal shearers, are widely employed for underground mining and tunnelling, while chain saw machines are employed for production of natural stones in quarry mining. Point attack type cutter picks, among all drag tools, are predominantly employed by roadheaders and drum shearers for excavations of rocks and coal. Radial tools are no longer common in underground mining and tunnelling today, whereas the ridged-type radial picks are extensively employed for chain saw machines in quarry mining.

Arrangement of cutter picks on a given cutting unit of above-mentioned machines, known as tool lacing, is one of the important factors affecting machine performance. Tool spacing, circumferential spacing, number of tracking cutters, angle of wrap and tilt angle are the fundamental parameters in tool lacing. Among them, tool spacing and tracking cutters are considered here in this paper. The lateral distance between the immediate neighbouring tools along the axis perpendicular to the direction of cutting is known as 'line spacing (S_L)', while that between the tools of the same sequence is termed 'cut spacing (S_C)'. A pick is said to perform a 'relief cutting' when it hits the midway between the tools of a preceding sequence, whereas the cutting mode is called 'groove deepening' if the picks in a given sequence exactly follow the same lines or the tracks of the tools of preceding sequence (Fig. 1). With groove deepening, the line spacing equals cut spacing, while with relief cutting, the cut spacing is always a multiple of line spacing.

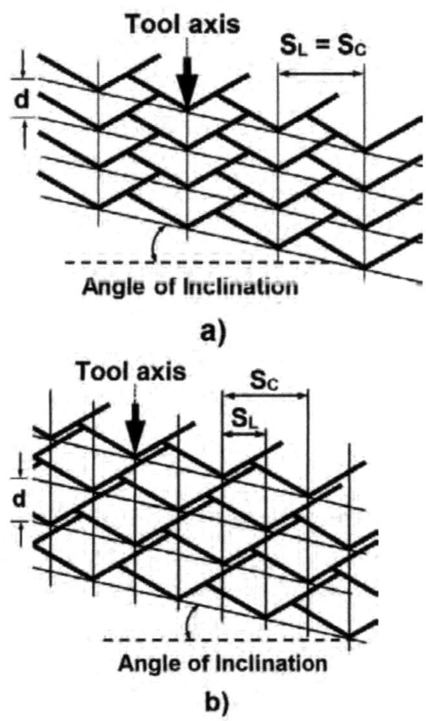

Figure 1. Definition of tool spacing with respect to (a) groove deepening and (b) relief cutting.

Laboratory linear cutting trials on flat rock surface showed that a minimum specific energy occurs when the line spacings are set to a certain value at a given depth of cut which is called optimum spacing to depth ratio (s/d). The limit of interaction between adjacent grooves was reported to occur at '2 tan θ', where θ is half breakout angle of the groove which generally varies between 52^0 and 70^0 depending upon rock properties (Roxborough 1973, Roxborough and Rispin1973, Roxborough and Phillips 1975).

Early studies emphasized that the groove deepening should be avoided, since it is inefficient (Evans and Pomeroy 1966, Roxborough and Phillips 1975). These findings were, however, based upon the results of laboratory trials carried out on a flat rock surface and their results are not necessarily applicable to the case of tools on an actual cutting head or a drum which cut into a surface formed by preceding cutting sequences. Laboratory trials simulating actual cutting action of picks in relief cutting mode with point attack tools showed that specific energy decreased with spacing to depth of cut ratio (s/d) rapidly at first and then more slowly, while tool forces increased continually (Hurt and Evans 1981). (Fig. 2). Lower specific energy continues beyond the point of interaction

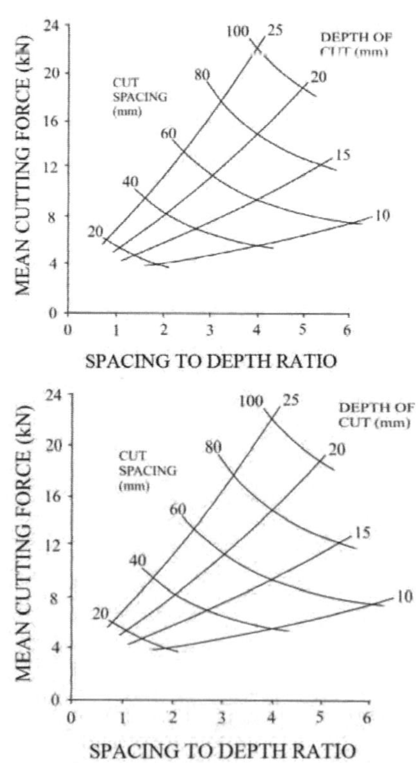

Figure 2. Effect of cut depth and spacing on mean cutting force and specific energy (Hurt and Evans 1981).

between the adjacent grooves, at the expense of high tool forces, in contrast to the previous laboratory investigations on flat rock surface. A compromise was, therefore, noted to be set between optimum cutting efficiency and acceptable level of tool forces, thus, the tool life, rather than cutting efficiency, was suggested to be the primary selection criterion. A laboratory linear cutting investigations on simulating a cycloidal action of a drum shearer at a fixed tool spacing for groove deepening and relief cutting respectively, showed that, with groove deepening, the tool forces were initially high and then gradually attained the same value of relief cutting at higher cutting depths (Fig. 3) (Hekimoglu et al. 2003). Underground investigations on coal shearers demonstrated significant increase in overall machine performance, when groove deepening lacing was modified to relief cutting (Hekimoglu et al. 2003). The cutting rate was reported to increase while tool wear was minimized when the tool lacing of a chain saw machine which was designed in groove deepening was partially modified to relief cutting, and the fundamental procedures for tool lacing of roadheaders cutting head were noted to be very similar to that of chain saw machines (Hekimoglu 1984). Chain saw machines operating in practice were reported to adopt groove deepening concept, and relief cutting was suggested for these machines (Copur 2010 8). The profile shape of grooves deepened by successive cuts of picks arranged in groove deepening mode was investigated through laboratory linear cutting experiments (Sun and Li 2014). The cutting was reported to be successful and breakout angles were consistent with optimum s/d value. However, pick damage was observed and, breakout angles were drastically decreased after the first pass, at an excessive tool spacing.

Despite the advantages of relief cutting, the groove deepening mode has, however, been still persistently employed in practice with many rock, coal and stone cutting machines, probably due to a lack of conclusive investigations in this field. The author of this paper has found no extensive investigations in published literature, on simulating the practical actions of these two cutting modes for a rational comparison. Such an investigation is believed to fill an important gap in this subject.

This paper describes a rational comparison between groove deepening and relief cutting through full-scale laboratory simulations trials with actual drag tools. The results are presented, and discussed in details.

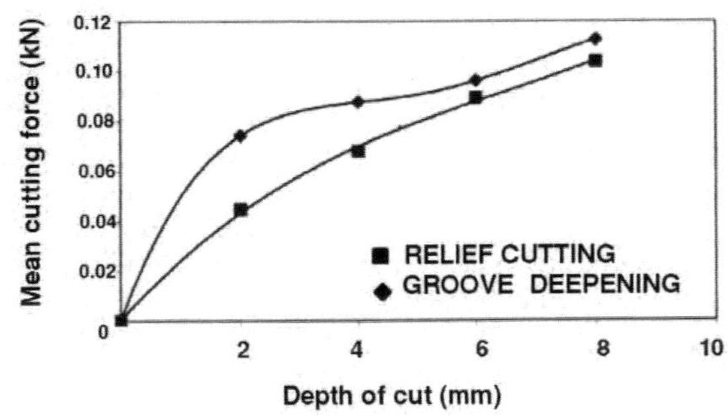

Figure 3. Variation of mean cutting force with depth of cut at constant spacing (Hekimoglu et al. 2003).

2. LABORATORY TRIALS

2.1 Description of laboratory full-scale linear cutting experiments

Simulating the actual cutting action of picks depends upon the type of tool lacing, since various design combinations are available in practice. Tools having their axes parallel to each other, like vane picks found on the cylindrical surface of a cutting drum with zero degree tilt angle, were considered for simplicity. The perimeter of a cutting sequence formed by tools of this type, normally describes a slope rather than a horizontal line during actual cutting, due to transverse motion of tools with helical arrangement as seen in Figure 1. A horizontal perimeter was, however, preferred for the sake of simplicity. The cutting action of a shearer drum of fixed web depth, fitted with two vanes (two spirals) and operating at 12 mm advance per revolution, was simulated at different tool spacing. A fine-grained homogeneous and isotropic sandstone obtained from Springwell, England, was used for the cutting experiments, since the scope of the investigations was to compare the variables under the same rock and operating conditions. Some mechanical and physical properties of the rock sample are presented in Table 1.

Table 1. Some mechanical and physical properties of the rock sample.

Uniaxial compressive strength	43.24±1.51 MPa
Indirect tensile strength	2.99 ± 0.22 MPa
Dynamic elastic moduli	1.79×10^4 MN/m^2
Shore Hardness	36.70 ± 6.29
Schmidt Hammer Rebound Number	52.03 ± 1.07
Cone Indenter Hardness	1.98 ± 0.41
Bulk density	2.21 gram/ cm^3
Quartz content	63 %

Laboratory full scale linear cutting trials were conducted by a 0.66 m instrumented shaping machine capable of producing in line thrust force of 50 kN, at the University of Newcastle Upon Tyne in England. Radial picks having -5° rake angle and 10° back clearance angle with 50 mm gauge (tool reach), designed for cutting medium strength rocks, and slender-type point attack picks, having 75° cone angle and 55° angle of attack were used. Slender type picks were preferred due to limitation of machine which could not withstand to high forces that are generated by heavy duty picks. Utmost care was paid to use sharp picks for each single cut throughout the cutting trials. The trials were carried out both in groove deepening and relief cutting respectively.

Initially, the experiments were carried out on flat surface of the rock. The simulated cuts were instrumented, when consistent groove profiles were obtained. This procedure is illustrated in Figure 4 as an example for groove deepening trials, and similar procedure is also the case with relief cutting. Debris collected for each cut was weighed and corresponding cutting and normal force components were recorded, and the specific energy was calculated.

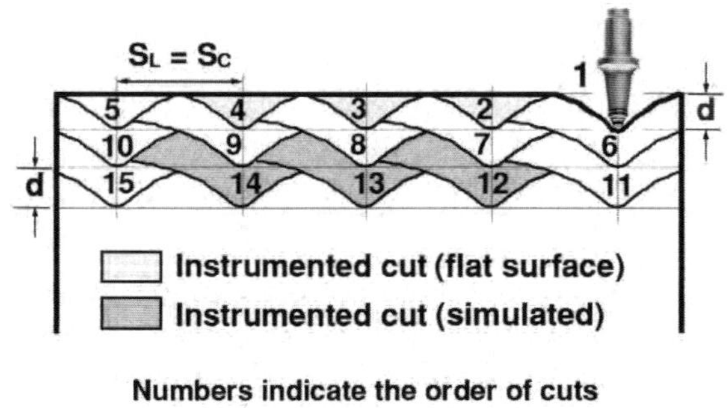

Figure 4. Depiction of laboratory simulation trials for groove deepening.

2.2 Results of laboratory linear cutting trials

It is important to note that the trials for groove deepening with radial picks could only be carried out up to s/d value of 6, since, the shaping machine was stalled at s/d of 8 (Fig. 5). Following experiments planned for groove deepening with point attack picks, therefore, had to be abandoned to avoid a potential damage to the shaping machine. The trials for relief cutting were, however, successfully resumed for the all s/d values planned for both pick types.

Figure 5. Cut failure at s/d of 8 during trials for groove deepening.

Variations in mean tool forces, yield and specific energy with s/d ratios are presented for radial tools and for point attack tools, respectively in Figure 5 and 6. The cutting force and normal force components for both modes of cutting increased towards higher values of s/d. The groove deepening, however, exhibited higher values than did relief cutting for the all s/d values except 2.

Magnitude of forces with point attack tools was higher than that radial picks. Ratio of

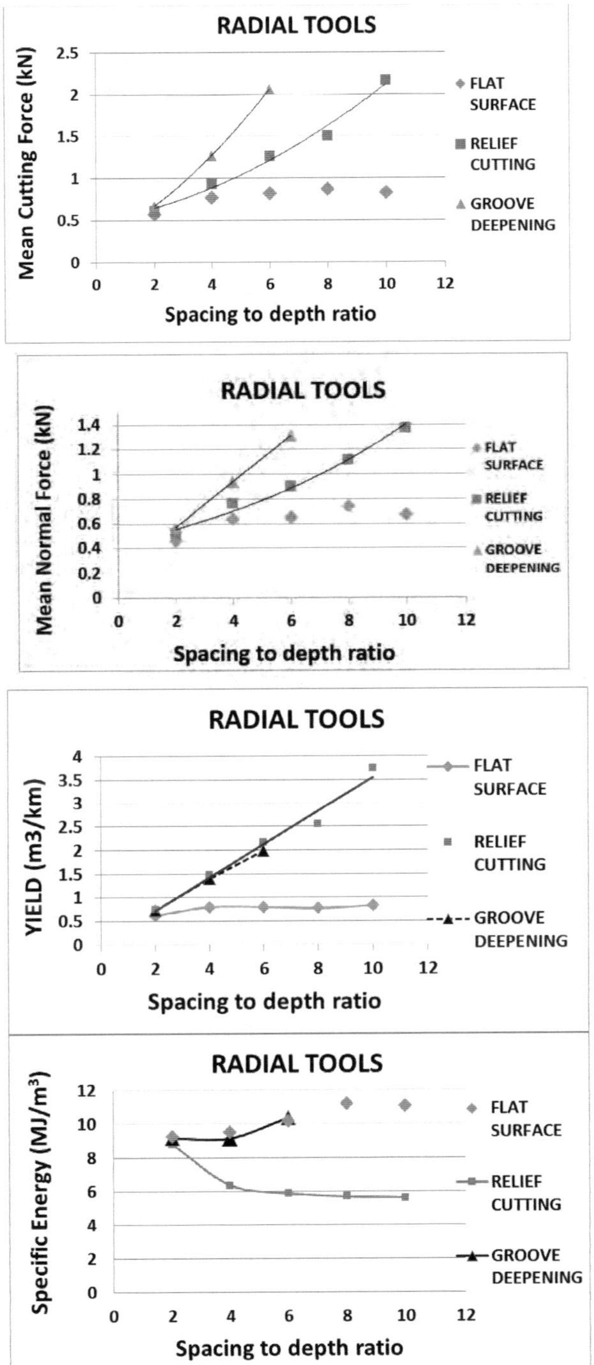

Figure 6. Variation of mean tool forces, yield and specific energy with s/d ratios with radial tools.

normal force to cutting force was less than 1 for radial tools, whilst being vice versa for point attack tools. The values for yield shown in the figures, indicate that in both cutting modes the same cross sectional areas

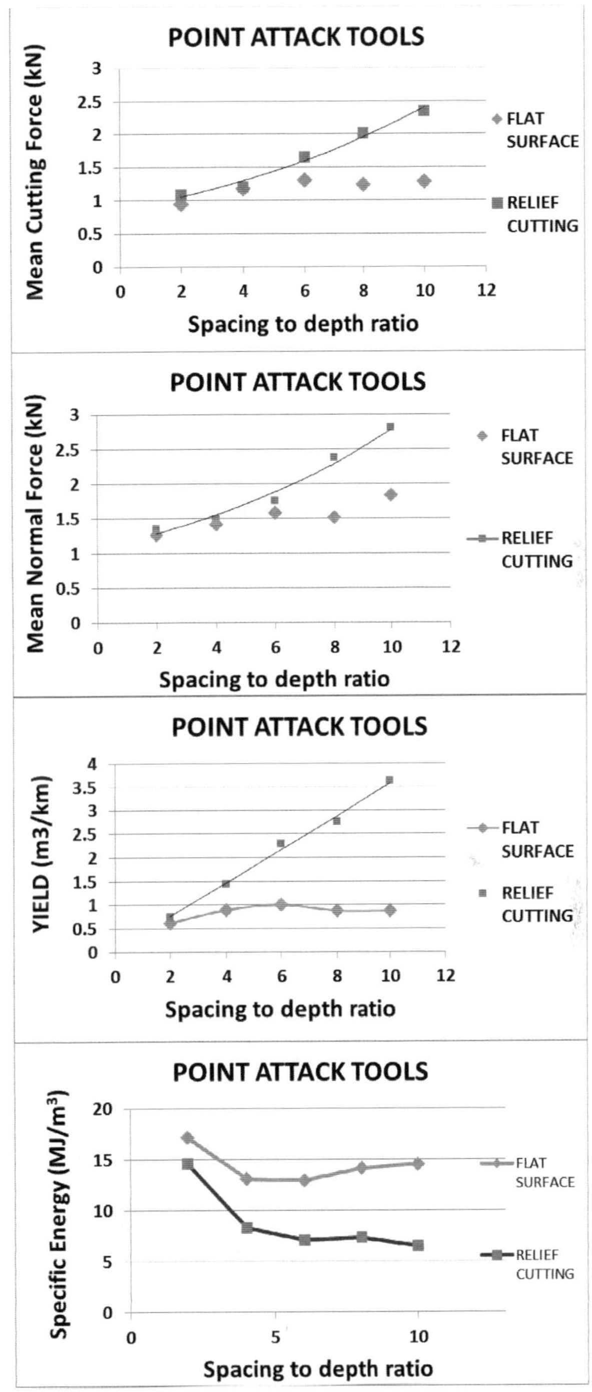

Figure 7. Variation of mean tool forces, yield and specific energy with s/d ratios with point attack tools.

were physically removed with tool forces of different magnitudes at all s/d values but 2. With relief cutting the specific energy decreased rapidly up to s/d of 4 and then more slowly after this value, whereas it was higher with groove deepening than relief cutting for all measured s/d ratios except 2, and ceased to exist after s/d value of 6. The results of trials on flat rock surface indicated that minimum specific energy tends to occur between s/d values of 4 and 5.

The average breakout angle for unrelieved cutting on flat rock surface was measured to be around 66° for radial picks, while being approximately 67.5 ^{o} for point attack picks. The breakout angles were observed to vary with s/d values of each cutting mode in such a way that being asymmetrical at s/d of 2, and reasonably symmetrical at s/d of 4, and beyond this value they were not consistent. The highest breakout angle observed for groove deepening was found to be always lower than that of unrelieved cut on flat rock surface, whereas the opposite was true for relief cutting mode.

3 DISCUSSION

3.1 On the difference between tool forces

With relief cutting lower tool forces were exhibited than groove deepening especially after s/d value of 2, despite the fact that the same cross sectional areas were physically removed in both cutting modes. This results is important to explain the superiority of relief cutting. Before proceeding to the discussion, it is, however, worthwhile to consider the findings reported by Hurt and Evans (1981). They noted that when cutting with point attack picks, the rock surface disintegrates into a series of scallops which have an approximate 'V' cross-section, and the apex angle of 'V' (half of which is also known as `breakout angle) showed a roughly constant value in experiments where depth of cut and tool angle were varied. The failure surfaces were assumed to be planes at right angles to the plane of the diagram making angles θ with the vertical, and thus a V-shaped section of rock is removed as given in Figure 8. The breakage mechanism observed with groove deepening and relief cutting during the laboratory simulation trials described in this paper may be explained better on the bases of V-shaped cross-section of grooves.

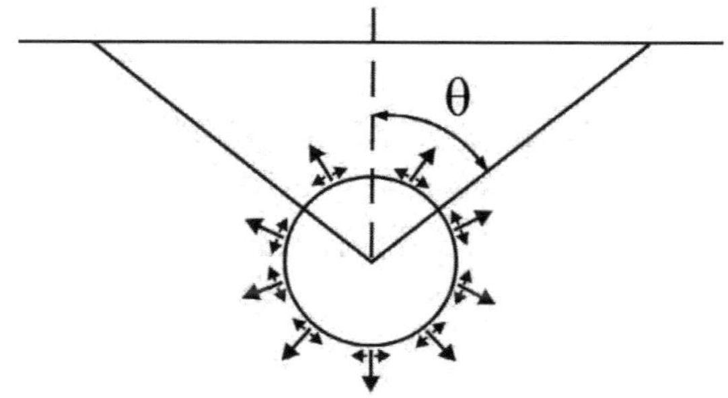

Figure 8. Illustrations of assumptions for the theoretical calculation of breakout angle (Hurt and Evans, 1981).

It is to note that at s/d value of 4 there is no significant difference between the respective values of tool forces measured for flat surface cutting and relief cutting with radial picks; in fact they are almost equal with the case of point attack picks, as in Figure 6 and 7. Furthermore, the corresponding yield values for flat surface cutting were also found to constitute nearly 55 to 60 % of those of relief cutting, indicating that at s/d of 4, the cross-sectional area swept in relief cutting is about 40 to 45% more than that of flat surface cutting. Hence, the magnitude of tool force required for sweeping the area of triangle ABC in flat surface cutting is very similar or almost equal to that required to remove the total area of polygon ACDB in relief cutting, as shown in Figure 9. Therefore, with relief cutting, the extra cross sectional area of triangle ABC at the top portion of the polygon can be discarded, since it does not impose any confinement on the cutting action of the pick. Otherwise the magnitude of tool forces with relief cutting should have been about 40 to 45% more than those of flat surface cutting. The effective cutting action of the tool, then, seems to take place within an area depicted by a triangle which has an apex angle twice the breakout angle of the rock, and with a base line defined by a point where the confined side of apex angle intersects with free surface, or the longer side instead, if there is no tool confinement. The part remaining within this BCD triangle may be considered as an "effective area", for tool force calculations. The methodology for construction of triangles representing effective areas may be better explained, if profile views of the grooves cut by adjacent tools are taken into account, in accordance with observations during the simulation trials, as follows.

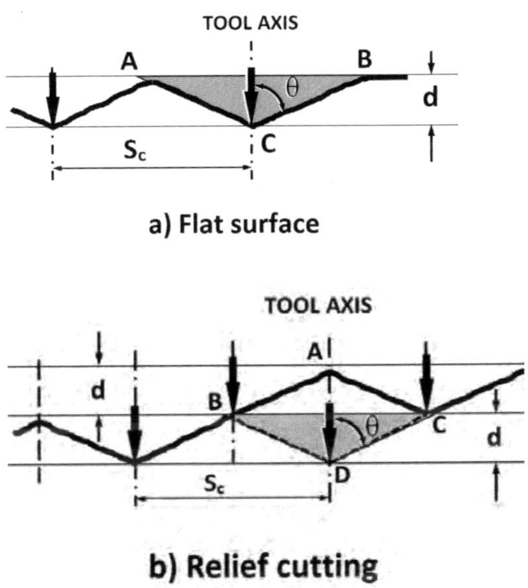

Figure 9. The profile view of breakouts between adjacent grooves at s/d value of 4, as observed during the laboratory trials; θ is half breakout angle, d is depth of cut and S_C is cut spacing.

The state of breakouts at s/d of 2 is illustrated in the form of conceptual drawings in Figure 10. The bold line denotes the profile view of grooves already cut by the neighbouring tools. The pick hits the rock at point C, and tends to break it by forming a 'V-shaped' groove with an apex angle twice normal breakout angle of rock obtained from unrelieved grooves on flat rock surface. The point at which the longer side of this angle or any side which lies in the confined section of cut, intersects with free rock surface (the point B in the figure) determines the location of base line of the isosceles triangle ABC. The last point (the point E) where the base line intersects with an uncut rock surface while moving from B up to A defines the length of upper boundary of the effective area which is the length of line EB in the figure. Any part of rock lying below this line, whether previously cut or uncut, is counted in the effective area (the shaded area in the figure), while those areas which are above this line is totally discarded. During the trials, the longer side of the apex angle was seen to terminate easily through point B for both cutting modes. This is because the tool is able to exert its full potential without any confinement, due to very close tool spacing.

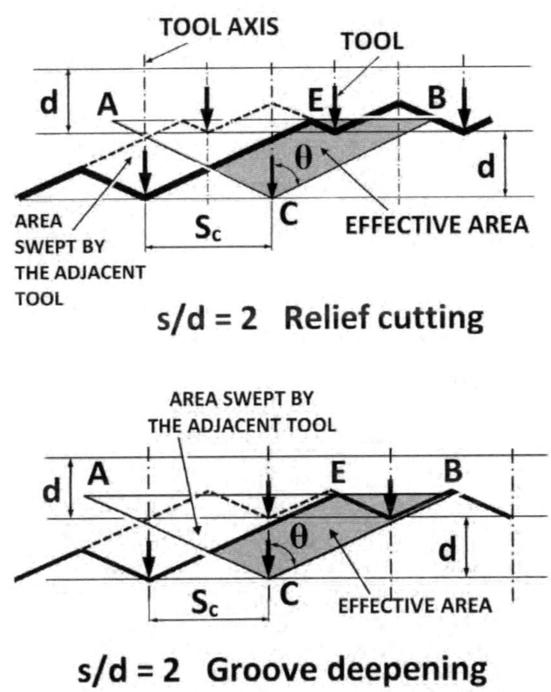

Figure 10. Depiction of breakouts at s/d of 2.

At s/d of 4, with relief cutting, the tool hits the rock at point D, and the right side of the apex angle tends to terminate at point E through point C, in an effort to create its complete breakout angle (Figure 11). During the trials, the fracture was, however, seen to terminate at the nearest free surface located at point B, meaning that the pick is unable to create its complete breakout angle. A question may, then, arise in such that why the fracture did not terminate at point E as it did so with groove deepening at s/d of 2?

This may be attributed to the fact that the tool is unable to exert its full potential at s/d

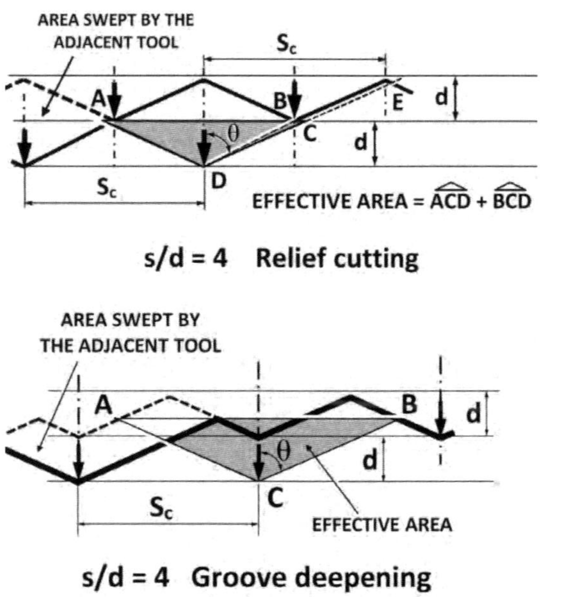

Figure 11. Depiction of breakouts at s/d of 4.

of 4, due to greater pick spacing which is twice that of s/d of 2. The tool was, therefore, too confined to induce the intended fracture at the target point of E. The above-mentioned tool confinement has to be taken into account with the effective area. It may be defined by an additional areal value of a triangle (BCD) added to that of the main triangle (ACD) to obtain the total effective areal value, as shown in Figure11. With groove deepening, the rock fracture was observed to terminate at Point B which defines the location for the base line of the isosceles triangle. It is to note that the uncut rock above the base line was discarded, since it is out of the effective area, as explained in Figure 9. The effective area (the shaded area in the figure) calculated for groove deepening was higher than that of relief cutting, verifying the higher tool forces with groove deepening as measured at this s/d.

The state of profiles of grooves at s/d of 6 is shown for both cutting modes, respectively in Figure 12. The groove surfaces were observed to be no longer smooth, while being more pronounced with groove deepening which also exhibited greater effective areas, compared to previous s/d values, probably due to higher tool spacing. Much more irregular grooves particularly with groove deepening were observed at s/d of 8, as shown in Figure 5 and 13. This was probably the reason why the tools were failed to accomplish cutting at this stage of the simulation trials with groove deepening.

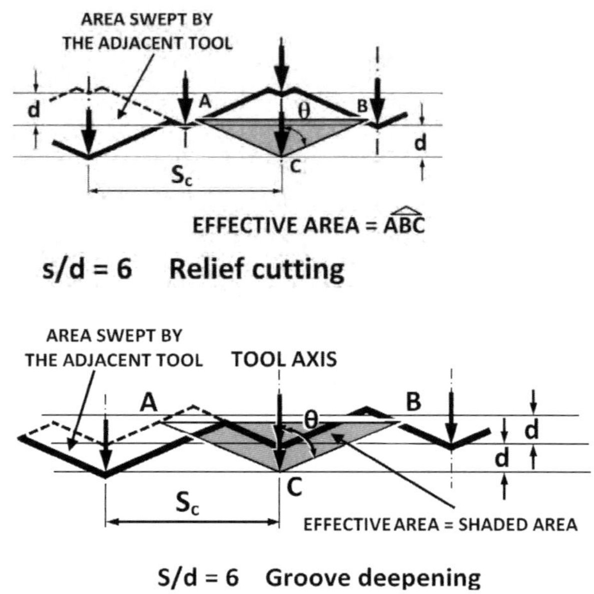

Figure 12. Depiction of breakouts at s/d of 6.

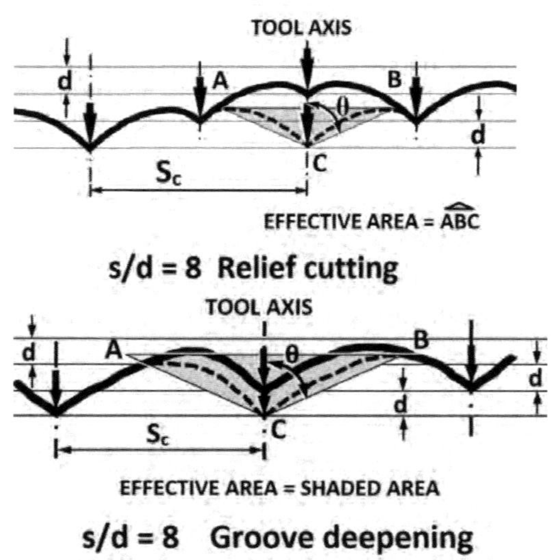

Figure 13. Depiction of breakouts at s/d of 8.

The method described above may be validated by plotting all measured tool force values against the all calculated areal values for both relief cutting and groove deepening together. It is seen that there is a good correlation between tool forces and their corresponding effective areas for both modes of cutting with both radial and point attack

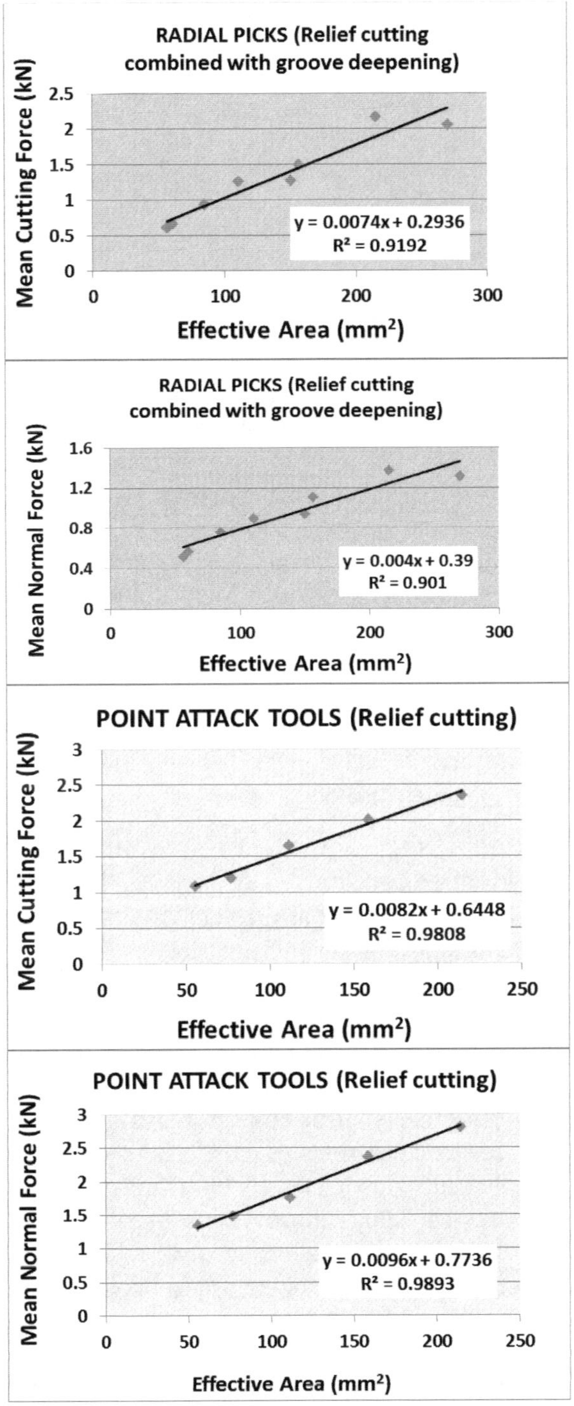

Figure 13. Correlation of mean tool forces with effective areas for both tool types.

tools (Figure 13). It is also important to emphasize that for the sake of simplicity and reliability, the breakout angle adopted for this method is that obtained from unrelieved cuts on a flat rock surface. This was because it was difficult to obtain a consistent value from relieved cutting.

3.2 On the optimum tool spacing

The advantages demonstrated with relief cutting over groove deepening may be explained by considering the state of rock fracture in relation to the spatial position of tools during cutting. With relief cutting, the line spacing can be easily set in fractions of cut spacing, e.g. half of cut spacing as with this study, whereas this is absolutely impossible with groove deepening, since the picks have to follow exactly the path of tools located in the preceding sequence. As a result of this, the point on a free surface where the rock fracture terminates emerges to be much closer to the pick tip in relief cutting than that in groove deepening. The effective areas with groove deepening are, therefore, greater, particularly after the s/d value where interaction between adjacent grooves takes place. It is probably for this reason why tools in groove deepening ceased to cut after the s/d value of 6, while the cutting successfully resumed in relief cutting with all s/d values investigated during the laboratory trials. This merit makes the relief cutting more advantageous when cutting in cycloidal motion, and also helps to explain the trend of force variations presented in Figure 3. The depths begin with zero, and ultimately terminate at maximum value of 8 mm at fixed tool spacing, due to the characteristics of cycloidal cutting. In reality, each value of cutting depths shown along 'x axis' of the graph also represent their corresponding s/d ratios in a reverse order, e.g. at 6 mm depth the corresponding s/d is 3, while at 2 mm depth it is 9. The higher force values occurring at shallow depths or at higher s/d values with groove deepening is attributed to the generation of

Table 3. Analyses of a sample drum for torque, torque fluctuations and specific energy.

s/d	Mean Torque (kNm)	Torque Fluctuations (kNm)	Volume of Cut Material per 'Adv/rev' (10^{-3} x m^3)	Specific Energy (MJ/m^3)
2	2.509	±0.087	1.0368	15.20
4	1.387	±0.103	1.0368	8.40
6	1.263	±0.140	1.0368	7.65
8	1.162	±0.174	1.0368	7.04

much greater tool forces at higher s/d ratios as demonstrated during the laboratory simulation trials described here in this study.

The trials also demonstrated that with relief cutting, specific energy initially showed sharp decrease up to s/d between 4 and 5, and kept decreasing slowly, after this point. This may infer that s/d values higher than that what is known "optimum s/d" in flat surface cutting may also be considered for a tool lacing. This idea should, however, be compatible with other parameters, such as drum balance and torque. An imaginary drum having two starts with fixed dimensions was considered to evaluate this idea, in terms of specific energy, torque and torque fluctuations which were standard deviations of mean torque values. Only vane tools were assumed to exist on the whole drum surface, to avoid complicated corner cutting situation. The calculations were carried out at 90° cut sector for each s/d respectively at a constant advance per revolution, and the results are presented in Table 3.

As seen, the decrease in both specific energy and drum torque is not significant after s/d value of 4, while torque fluctuations rise considerably. The drum balance is known to be inversely affected, as the total number of tools decreases. It is important to note that the fluctuations in torque are related to the availability of full depth of sump throughout the course of cutting process. In practice, the drums and cutting heads do not always operate in full depth of sump, owing to changing face and operating conditions. Under such circumstances, the total number of active picks 'in cut' will reduce, and level of fluctuations in both torque and reaction forces will consequently increase. Fluctuations in torque and reaction forces always have to be avoided, since they inflict heavy damages on major machine components. Hence, the s/d values higher than optimum s/d value which is between 4 and 5 for these experiments are not suitable, since a small decrease in both specific energy and torque after this value cannot be afforded at the expense of high vibrations.

4 CONCLUSIONS

Within the limit of results and experimental conditions, as well as the limit of rock type and parameters investigated, following conclusions can be drawn:

1) The top portion of the cross section of a groove did not impose any significant effect on the magnitude of tool forces. The rock fracture effectively took place mainly in the lower portion of the whole cross section described by a triangle which has an apex angle twice the breakout angle of unrelieved groove. The areal quantity of this triangle considered together with tool confinement which was termed 'effective area' was employed to calculate the tool forces. The calculated effective areas were found to be in good correlation with the measured tool forces. The effective areas calculated for groove deepening were generally greater than those of

relief cutting, accounting for both higher tool forces and specific energy observed with this cutting mode.

2) With simulated trials, specific energy decreased drastically at first and then slowly after s/d of 4. it, however, ceased to continue with groove deepening after s/d ratio of 6, due to greater effective areas, while being continuous with relief cutting at all s/d ratios investigated. There was no marked minimum specific energy value as emphasized with experiments on flat rock surface. However, tool spacing at which interaction between adjacent grooves takes place, i.e. at '2 d tan θ, was recommended to avoid detrimental effects of fluctuations in torque and reaction forces.

ACKNOWLEDGEMENTS

The author wishes to thank Dr. R. J. Fowell (formerly at the University of Newcastle Upon Tyne, now at the University of Leeds) for his invaluable support, and Turkish Coal Enterprises (TKI) for financial support. The views expressed here are the author's own and not necessarily those of TKI.

REFERENCES

Copur, H., 2010Linear stone cutting tests with chisel tools for identification of cutting principles and predicting performance of chain saw machines. *Int J Rock Mech Min Sci.;* 47(1):104–120.

Evans, I. and Pomeroy, C.D, 1966, *The Strength, Fracture and Workability of Coal,* Oxford, Pergamon Press.

Hekimoglu, O.Z., Ayhan, M. and Tiryaki, B., 2003, Laboratory and in-situ investigations of tracking cutters for computer- aided design of shearer drums, *CIM Bulletin,* Sept. Vol. 96 (1073), pp 72-75.

Hekimoglu, O.Z., 2014. Studies on increasing the performance of chain saw machines for mechanical excavation of marbles and natural stones, *International Journal of Rock Mechanics & Mining Sciences,* 72, 230–241.

Hurt, K.G. and Evans, I., 1981, Point attack tools: An Evaluation of functions and use for rock cutting, *The Mining Engineer*, March. pp 673-675.

Roxborough, F.F., 1973. Cutting rock with picks, *Mining Engineer,* June pp. 445-455.

Roxborough, F.F. and Rispin, A. 1973. The mechanical cutting characteristics of the lower chalk, *Tunnels and Tunnelling*, 5:45–67. 1973.

Roxborough, F.F. and Phillips, H.R., 1975, The mechanical properties and cutting characteristics of the Bunter sandstone. *Report by Department of Mining Engineering, University of Newcastle upon Tyne for Transport and Road Research Laboratory, Department of the Environment,* England.

Sun, Y. and Li, X.S., 2014, Ineffective Rock Breaking and its Impacts on Pick Failures, *In: Proceedings of the 31st International Symposium on Automation and Robotics in Construction and Mining (ISARC 2014)* Sydney, Australia. pp. 754-761.

A Statistical Approach for Predicting Surface Settlement Due To Tunneling (Case Study: Tehran Metro Line 7)

Saeed Jamali
Department Of Mining Engineering, Isfahan University Of Technology, Isfahan, Iran

AliReza Yarahmadi Bafghi, Javad Gholamnejad
Department Of Mining Engineering, Yazd University, Yazd, Iran

Amir Khademian*
Department Of Mining Engineering, Isfahan University Of Technology, Isfahan, Iran

ABSTRACT Excavation of underground structures, regardless of its depth, disturbs the soil and rock masses. As a result, the soil loses its initial balance, and move to a new balance condition. In this process, the settlement of ground surface inevitably happens. Generally, researcher knowledge about field properties is associated with uncertainties and thus deterministic approaches encounter some challenges. The probabilistic techniques enable modeling uncertainties by analyzing their dispersion effect on the global behavior of the structure. In this research, a specific statistical settlement design problem considered in order to investigate the maximum surface settlement probability distribution. A three-dimensional finite difference model is used traditionally, but instead of deterministic data, statistical parameters are inputted in the numerical analysis. The geomechanical parameters, including cohesion, friction angle and Young's modulus, are reproduced in separate categories based on their statistical distribution functions and were used in a repetitive procedure. The results of probabilistic analyses show that the maximum settlement has a normal distribution with a mean of 18.11mm and a standard deviation of 1.902. It is concluded that the maximum settlement in the case study area is likely 99% less than 23.86 mm. So, the maximum settlement is admissible and the damage is negligible.

Keywords: Settlement Due to Tunneling, Finite Difference Method, Probabilistic Modeling, Tehran Metro

1 INTRODUCTION

The growth of many cities has resulted in the need for increased infrastructure. As urban space becomes more limited, subsurface structures such as tunnels are becoming more efficient in providing the required infrastructure. In 1863 the first underground railway line was opened in London. Since then many cities worldwide have implemented underground transport systems. Excavation of underground structures, regardless of its depth, disturbs the soil and rock masses. As a result, the soil loses its initial balance, and move to a new balance condition. In this process, the settlement of ground surface inevitably happens. Surface structures damage is the most important effects of settlement. Thus, in recent urban tunneling designs, greater attention is being focused on controlling settlement and ground surface destructive effects (Franzius 2004).

The adverse consequences of shallow tunnels excavation have been discussed by many geotechnical researchers and tunneling engineers since last decades. Studies have shown that different factors like type and strength characteristics of overburden, depth of the tunnel, in situ stress conditions, method of excavation and the supporting system influence the amount of the surface settlement (Do, et al. 2014; Do and Dias 2017).

The first comprehensive studies on settlement due tunneling was Peck's investigation. He stated that the transverse settlement trough can be described by a Gaussian error function which has been widely accepted since then. This mathematical description defined by the maximum settlement and the trough width parameter represent the mean and the standard deviation of the original Gaussian, respectively (Maraš-Dragojević 2012).

Delafuente has presented a model based on the finite element analysis which provide us with a rapid and easy assessment in longitudinal settlement trough (De la Fuente and Oteo 1996). Sagaseta, based on the performed experiments in Madrid and Karakas subway, presented and used some - finite element models for predicting settlements successfully (Oteo and Sagaseta 1996). Wang et al., using the corrected models of Gous based on FLAC software and the finite difference method, presented the diagram of the surface settlement due to tunnel excavations in loose soil and then formulated the settlement induced by excavation of two other parallel tunnels (Wang, et al. 2000). Melis et al., using the finite difference method and FLAC3D software, tried to predict the profile of the settlement due to shield tunneling of Madrid subway (Melis, et al. 2002). They studied the reaction models of elastic and Mohr-Coulomb to predict the settlement in different distances. Carter and Small, through finite element analysis and Tunnel3D and ABAQUS software, studied the effect of excavation of tunnel on support system in a neighboring tunnel in Austria (Liu, et al. 2008). They concluded that the effect of excavation of a tunnel on the support system of its neighbor tunnel strongly depends on the position of two tunnels relative to each other. Huang and Li, using finite element 3D, presented a simple analytical resolution for studying the effect of excavation of subway tunnel on the amount of settlement at the columns of a bridge (Huang, et al. 2009). Darabi studied the settlement in line 3 of Tehran subway and compared the results of numerical method with those artificial neural networks that were very near to each other (Darabi, et al. 2012). Mirhabibi er al. studied the effect of surface settlement on neighbor buildings in line 1 of Shiraz subway (Mirhabibi and Soroush 2012). They investigated the effect of different parameters like depth of tunnel, the distance between the centers of tunnel, weight and width of buildings. Elmanan and Elarabi discussed and compared results of empirical method derived by Peck, 3D and 2D Finite Element Method for surface settlement of the Second Heinenoord Tunnel. They calculated that 3D FEM can predict ground deformations very well (Elmanan and Elarabi 2012). Camós, et al. has presented a model for probabilistic prediction of settlement due to tunneling based on Gaussian profile model (Camós, et al. 2016). Strokova has used finite element method for investigating the effect of the coefficient of earth pressure at rest (K0) on surface subsidence caused by tunneling (Strokova 2015).

Geotechnical models are usually associated with considerable amounts of uncertainty. In previous research work, determination of uncertainty of various methods and models introduced for settlement prediction is mainly neglected. Generally, researcher knowledge about field properties is associated with uncertainties and thus deterministic approaches encounter some challenges. The probabilistic techniques enable modeling uncertainties by analyzing their dispersion effect on the global behavior of the structure. In this research, the uncertainty of model used for settlement estimation is characterized through a statistical approach. To avoid the input parameter uncertainty effects, a three-dimensional finite difference model is used, but instead of traditional deterministic data, statistical parameters are inputted in the numerical analysis in order to investigate the maximum surface settlement probability distribution.

2 METHODOLOGY

In this research work, numerical model is combined with statistical method for estimating surface settlement due to tunneling. For considering the effect of uncertainty of input data, the geomechanical parameters of soils, including cohesion, friction angle and Young's modulus, are reproduced based on their statistical distribution functions. Because, the shotcrete and segments are manmade and so are homogenous, their properties are imputed as Deterministic parameters. Then, the produced data are categorized randomly in separate sets and are used as input data of numerical model in a repetitive procedure. A wide range of engineering and scientific disciplines use simulation methods based on randomized input, often called Monte Carlo methods.

2.1 Monte Carlo simulation

The Monte Carlo method (MCM) is a simulation method and also a sampling technique. In each simulation a particular set of values of the random variables generated in accordance with prescribed probability distributions is used. By repeating the process, a sample of solutions is obtained. A sample from a Monte Carlo simulation is similar to a sample of experimental observations, except that it is a computer-generated experiment. Therefore, the results of Monte Carlo simulations may be treated statistically. The Monte Carlo method can be applied to predict or study the performance and response of a large system and to analyze complex probability problems, that is, its major usefulness and advantage over analytical methods would be most apparent in problems where analytical models are mathematically intractable.

The Monte Carlo method is particularly effective when the process is strongly nonlinear or involves many uncertain inputs, which may be distributed differently. To perform such a study, the analyst generates a random value for each uncertain variable and performs the calculations necessary to yield a solution for that set of values. This gives one sample of the process. The trials are repeated many times, giving many samples of the process. Once a large number of runs have been completed, it is possible to study the output statistically (Baecher and Christian 2005).

2.2 Finite difference method

The finite difference approximations for derivatives are one of the simplest and of the oldest methods to solve differential equations. The advent of finite difference techniques in numerical applications began in the early 1950s and their development was stimulated by the emergence of computers that offered a convenient framework for dealing with complex problems of science and technology. Theoretical results have been obtained during the last five decades regarding the accuracy, stability and convergence of the finite difference method for partial differential equations. The principle of finite difference methods is close to the numerical schemes used to solve ordinary differential equations. It consists in approximating the differential operator by replacing the derivatives in the equation using differential quotients. In FDM, Materials are represented by elements, or zones, which form a grid that is adjusted by the user to fit the shape of the object to be modelled, and every derivative in the set of governing equations is replaced directly by an algebraic expression written in terms of the field variables (e.g., stress or displacement) at discrete points in space; these variables are undefined within elements. This program simulates the behavior of structures built of soil, rock or other materials that may undergo plastic flow when their yield limits are reached (Griffiths and Smith 2006).

3 DETERMINING PROBABILITY DISTRIBUTION OF MAXIMUM SURFACE SETTLEMENT (CASE STUDY: TEHRAN METRO-LINE 7)

Tehran Metro Line 7 covers an approximate length of 26.38 km in a South-East/North direction with 23 underground stations. It is excavated using Earth Balanced Pressure - Tunnel Boring Machine (EPB-TBM) in two sections: East-West section and South-North section. In this research, 45 meter of South-North section that is located under Navab Highway parallel to the Tohid tunnel (between Q7 and R7 stations) is selected for investigating surface settlement. The numerical model as shown in figure 1 is generated based by FLAC3D software. It is an explicit finite difference code that has been widely used in the geotechnical analysis field and tunneling modeling (Liu, et al. 2012; Xie, et al. 2016). The model is confined by appropriate boundaries along the x, y and z, whereas the upper surface in z direction is free to move, as it corresponds to the effective ground surface and other side in x and y directions are fixed to prevention of any movement. Also, bottom boundary in z direction is fixed too. To model the weight of over-ground neighboring structures uniform pressures have been applied on the upper surface.

The mechanical behavior of the soils is adopted to be elastic-plastic according to the Mohr-Coulomb yield criterion. The input data including deterministic properties of soils, the shotcrete and segments elastic properties are given in Table1 and 2, respectively.

In the following, the amount of surface settlement induced by excavation of line 7 of Tehran metro is estimated by using mentioned procedure in section 2 (Methodology). First, a proper amount of input data are reproduced for each soils' parameters based on their statistical distribution function. The statistical properties of soils are presented in Table 3. Next, the reproduced data are randomly put together to create a set of input data for finite difference model. Finally, the numerical model is solved for each data set.

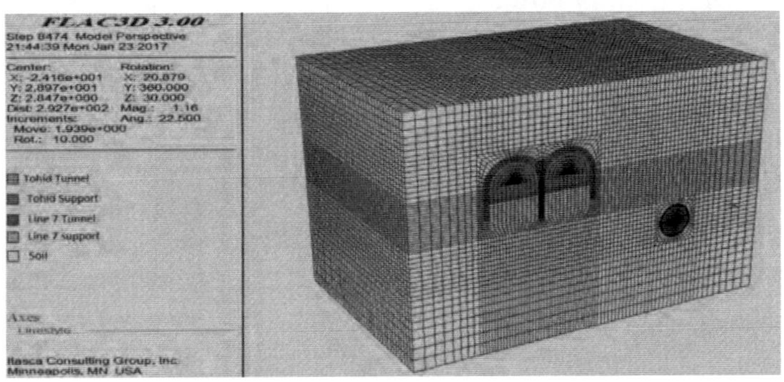

Figure 1. Finite difference model of tunnel line 7

Table 1. Deterministic properties of soils

Property	Depth(m)			
	0-15	15-30	30-45	45-60
Unified Soil Classification System symbol	GM , GC	GM, GC, SM, SC	GM, GC , SM, SC	GM, GC, SM, SC
Density (kg/m^3)	1750	1800	1900	1950
Passion ratio	0.35	0.32	0.3	0.3

Table 2. Strength properties of Support system

property	Shotcrete	Segment
Elasticity modulus (GPa)	9	30
Passion ratio	0.2	0.25
Compressive strength (MPa)	16	30

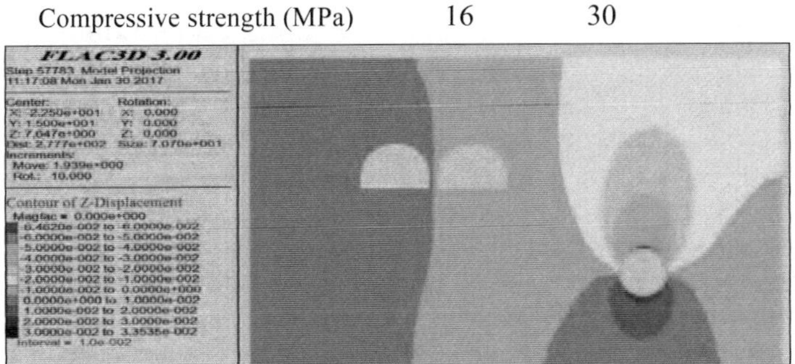

Density (kg/m^3)	2000	2800

Figure 2. Vertical displacement contours around the tunnel

Table 3. Statistical properties of soils

Property	Distribution function type	Depth(m)							
		0-15		15-30		30-45		45-60	
		mean	SD	mean	SD	mean	SD	mean	SD
Elastic modulus(MPa)	Lognormal	107.5	11.25	155	12.5	215	17.5	290	20
Cohesion (KPa)	Normal	12.5	1.25	20	2.5	30	2.5	37.5	1.25
Friction angle	Normal	33	0.5	35	0.5	37	0.5	39	0.5

4 RESULTS AND DISCUSSION

The vertical displacement contours as a result of one of the tunnel excavation simulation is shown in Figure 2.

In figure 3, the transverse section of surface settlement is shown as a result of one of the tunnel excavation modeling. The maximum amount of settlement and the trough width parameter in this model were almost 17mm and 16m, respectively.

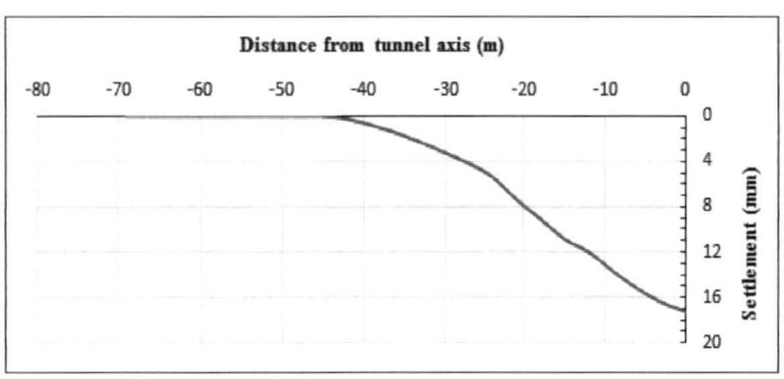

Figure 3. the transverse section of surface settlement

After frequent solving of numerical model, the maximum surface settlement at each run is collected as the model output, and the histogram of the maximum surface settlement is obtained as presented in figure 4.

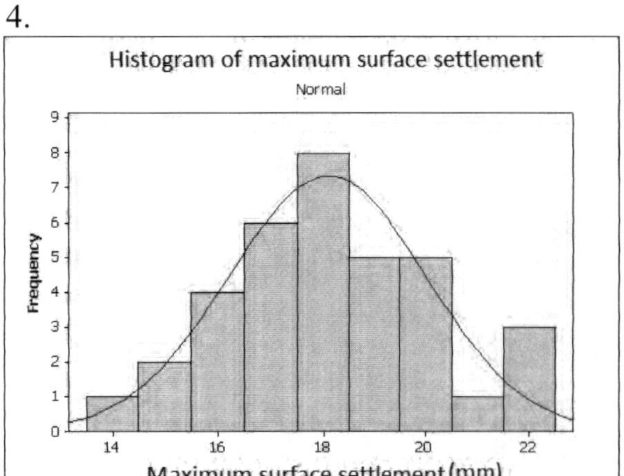

Figure 4. the histogram of the maximum surface settlement

The Anderson-Darling Normality test on results show that the distribution function of the maximum surface settlement is in the form of normal distribution with a mean of 18.11mm and a standard deviation of 1.902 (Figure 5). In other words, the maximum settlement in the case study area was likely 99% less than 23.86 mm. So, the maximum settlement is admissible, and the damage is negligible.

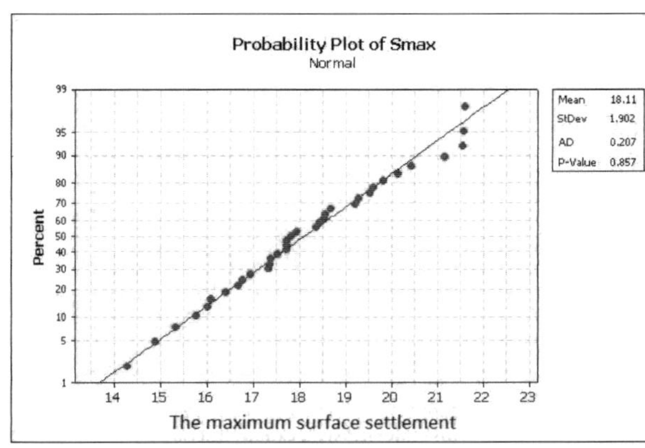

Figure 5. The Anderson-Darling Normality Test of the maximum surface settlement

The results obtained in this paper showed that the maximum surface settlement follows Normal distribution. The statistical approach developed for settlement estimation can model the input parameter uncertainty effects rather than traditional deterministic approaches.

5 CONCLUSION

In this research, by combining finite difference method and Monte Carlo simulation, a statistical approach is developed for settlement estimation. To avoid the input parameter uncertainty effects, statistical parameters are inputted in the numerical analysis instead of traditional deterministic data. The obtained results showed that the maximum surface settlement was in the form of normal distribution with a mean of 18.11mm and a standard deviation of 1.902. In other words, the maximum settlement in the case study area was likely 99% less than 23.86 mm. So, the maximum settlement is admissible and the damage is negligible. The results confirmed that presented probabilistic approach can be used to get possibility of the occurrence of phenomena in geosciences fields.

REFERENCES

Baecher, G. B., and Christian, J. T., 2005. *Reliability and statistics in geotechnical engineering*, John Wiley & Sons.

Camós, C., Špačková, O., Straub, D., and Molins, C. *Probabilistic approach to assessing and monitoring settlements caused by tunneling*. 2016, *Tunnelling and Underground Space Technology*, 51, pp. 313-325.

Darabi, A., Ahangari, K., Noorzad, A., and Arab, A. *Subsidence estimation utilizing various approaches– A case study: Tehran No. 3 subway line*. 2012, *Tunnelling and Underground Space Technology*, 31, pp. 117-127.

De la Fuente, P., and Oteo, C., Theoretical research on the subsidence originated by the underground construction in urban areas. presented at the *Proceedings of the Danube International Symposium, Romania*, 1996.

Do, N.-A., Dias, D., Oreste, P., and Djeran-Maigre, I. *2D tunnel numerical investigation: the influence of the simplified excavation method on tunnel behaviour*. 2014, *Geotechnical and Geological Engineering*, 32(1), pp. 43-58.

Do, N. A., and Dias, D. *A comparison of 2D and 3D numerical simulations of tunnelling in soft soils*. 2017, *Environmental Earth Sciences*, 76(3), pp. 102.

Elmanan, A. A., and Elarabi, H. *Analysis of surface settlement due to tunnelling in soft ground by using empirical and numerical methods*. 2012.

Franzius, J. N. (2004). "Behaviour of buildings due to tunnel induced subsidence." University of London.

Griffiths, D. V., and Smith, I. M., 2006. *Numerical methods for engineers*, CRC press.

Huang, M., Zhang, C., and Li, Z. *A simplified analysis method for the influence of tunneling on grouped piles*. 2009, *Tunnelling and Underground Space Technology*, 24(4), pp. 410-422.

Liu, H. Y., Small, J. C., and Carter, J. P. *Full 3D modelling for effects of tunnelling on existing support systems in the Sydney region*. 2008, *Tunnelling and Underground Space Technology*, 23(4), pp. 399-420.

Liu, J., Qi, T., and Wu, Z. *Analysis of ground movement due to metro station driven with enlarging shield tunnels under building and its parameter sensitivity analysis*. 2012, *Tunnelling and Underground Space Technology*, 28, pp. 287-296.

Maraš-Dragojević, S. *Analysis of ground settlement caused by tunnel construction*. 2012, *Građevinar*, 64(07), pp. 573-581.

Melis, M., Medina, L., and Rodríguez, J. M. *Prediction and analysis of subsidence induced by shield tunnelling in the Madrid Metro extension*. 2002, *Canadian Geotechnical Journal*, 39(6), pp. 1273-1287.

Mirhabibi, A., and Soroush, A. *Effects of surface buildings on twin tunnelling-induced ground settlements*. 2012, *Tunnelling and Underground Space Technology*, 29, pp. 40-51.

Oteo, C., and Sagaseta, C. *Some spanish experiences on measurement and evaluation of ground displacements around urban tunnels*. 1996, *Geotechnical Aspects of Underground Construction in Soft Ground (RJ Mair and RN Taylor, eds.), London*, pp. 731-736.

Strokova, L., Modeling of tunneling-induced ground surface movement. presented at the *IOP Conference Series: Earth and Environmental Science*, 2015, IOP Publishing, pp. 012030.

Wang, Z., Sampaco, K., Fischer, G., Kucker, M., Godlewski, P., and Robinson, R., Models For Predicting Surface Settlements Due To Soft Ground Tunneling. presented at the *North American Tunneling 2000*.

Xie, X., Yang, Y., and Ji, M. *Analysis of ground surface settlement induced by the construction of a large-diameter shield-driven tunnel in Shanghai, China*. 2016, *Tunnelling and Underground Space Technology*, 51, pp. 120-132.

Blasting for Optimum Leaching Performance: Electronic Blast Engineering of Desired Fragmentation

W. R. Adamson
(Davey Bickford)

H. Parra
(Davey Bickford Chile)

ABSTRACT Optimization of gold mining processes can be related to a number of factors; probably none more critical than the size distribution of the blasted ore that is delivered to the extraction process plant. However there are additional blasting engineering outcomes that can further enhance the amenability of delivered ore to efficient extraction of gold. These outcomes relate to stress/strain conditioning of the fragmented ore.

The extraction of gold from oxide orebodies through leaching processes, either Run Of Mine (ROM) or leaching of crushed ore, is an economic option for the operation of low to middle grade deposits throughout the world, however the efficiency of the leaching process (percentage gold recovery and leaching/percolation rate) can be influenced significantly by the characteristics of the size distribution and hardness of broken ore.

Occasionally extraction plant managers can inform the Drill and Blast engineers of the desirable size range for heap feed material. The challenge is then transferred to this critical operation (D&B) to deliver the appropriate broken material, applying both explosive energy and timing sequences to achieve the target result.

The use of precise and flexible initiation allows the design engineer to create a timing scenario that can minimize both the oversize fraction and managed the percentage of fines that appear in the leaching heap.

The Global Technical Solutions ("GTS") team of Davey Bickford has developed on-site, applied blast engineering projects that demonstrate the feasibility of tailoring blast fragmentation results for optimised leaching recoveries (and reduced energy consumption in those cases where ore is pre-crushed prior to dump loading), based on plant engineering requests for maximum delivered size fractions. Measurement techniques inluding VoD, photographic fragmentation analysis, QA/QC audits and Point Load Testing of rock hardness were used to verify the successful delivery of the requested results. This paper reports on the principal techniques applied and the results obtained at a ROM Heap Leach gold mine in Latin America.

1 INTRODUCTION

In order to guarantee their viability in an increasingly challenging marketplace, the majority if not all of currently operating mining initiatives are involved in a constant search for means of optimization of their processes. The final objective translates to a reduction in the global cost of extraction of mineral value through a progressive increase in full-process productivity.

Davey Bickford Global Technical Solutions ("GTS") has worked with an open pit gold producing project located in North America and is actively searching for means of optimizing their operations, maximizing gold production at a high level of productivity. In support of this mission the GTS team has engaged with Drill and Blast staff at the mine to deliver an engineering solution that will tailor fragmentation to a set of characteristics that favour more efficient Run of Mine ("ROM") leaching performance.

This document details the preliminary results obtained in four separate similar sized blasts in one mineralogical ore type; two of these blasts were initiated using standard pyrotechnic detonators and traditional delay selections, while the other two events were fired using the DaveyTronic® electronic initiation system and modified initiation sequences and delay selection. In addition to fragmentation and rock condition measurements, the project also contemplates microscopic sample analysis and column leaching tests for all blasts. Nonetheless this test work is still underway.

2 METHODOLOGY:

In order to draw valid and accurate comparisons between specific results obtained for pyrotechnic and electronic initiation, the first task in the development of the study was the selection of a lithological domain that, as well as being representative of the orebody, contained an average gold grade that is classified as medium (0.3g/t – 1.0g/t Au) by the mine management.

This lithological domain was identified as Hydrothermally Altered Rhyolite (RAH). Mineralogical and geotechnical characteristics were made available to GTS team members.

That target domain defined, four typical sized ore production blasts were charged and fired, split as indicated previously between pyrotechnic and electronic initiation technologies. For three of the four blasts a full QA/QC audit was conducted for the three dimensional location of all hole collars, hole depths and explosive charge profiles. For operational reasons the second electronic blast was not similarly evaluated prior to blasting.

Within each blast event studied, a smaller polygon was defined from within which four (4) truckloads of blasted ore were obtained for further detailed and quantitative analysis. The location of each polygon was selected in order to eliminate any influence of edge (free face) effects and guarantee that the ore material was exclusively medium grade. The extracted material was discharged in four separate piles in a previously selected analysis area of the pit, identifying each source blast and initiation methodology including system type and initiation sequence used. The material from each of the four samples piles was then subjected to a detailed fragmentation analysis using photographic techniques and subsequently screened; classified into three (3) fractions as described in Table 1.

Table 1: Screened Mineral Fraction

Fraction	Inferior Size Limit (in)	Superior Size Limit (in)
Fine	0	1.25
Intermediate	1.25	4
Coarse	4	-

In each of the Fine and Intermediate fractions randomly selected, approximately equi-sized samples were taken (minimum of 40 samples in each case) and reserved for Point Load Index testing and microscopy analysis to determine whether a reduction in rock strength and respective increase in micro-crack presence were present. At the time of writing this Paper only the PLI test work has been completed, with both microscopy and column leach testing yet to be completed by the client.

Finally a specially designed vibration measurement process was executed in the same lithological domain as that studied in this project. The results from this work will allow the estimation of the stress levels induced by different drill and blasting design configurations which, it is intended, can be related to the expected degree of micro-fracturing and rock fabric conditioning achieved. The test work also allowed the quantitative estimation of the P-wave velocity ("V_p") for the lithological domain studied.

3 DESIGN AND IDENTIFICATION OF TEST BLASTS:

The project considered four blasts located in the South-West sector of the pit; two detonated with pyrotechnic initiation and two fired using electronic initiation. The pyrotechnic delay blasts are designated as "Baseline Blasts"; baseline against which the electronic initiation events are compared, these being defined as "Modified Blasts". Table 2 describes the design parameters associated with each test.

Table 2: Detailed Design Parameters

Parameter	1st Baseline Blast	2nd Baseline Blast	1st Modified Blast	2nd Modified Blast
ID	V_{1P}	V_{2P}	V_{1E}	V_{2E}
Date of Blast	29 March 2016	19 May 2016	25 May 2016	21 July 2016
Pattern (BxS) (m)	5.75 x 6.75	5.75 x 6.75	5.75 x 6.75	5.75 x 6.75
Hole diam. (in)	8.5	8.5	8.5	8.5
Hole length (m)	9	9	9	9
Explosive Charge*	ANFO/Emulsion	ANFO/Emulsion	ANFO/Emulsion	ANFO/Emulsion
Density of Charge	0.82/1.2	0.82/1.2	0.82/1.2	0.82/1.2
Charge Length**	4.6/4.3	4.6/4.3	4.6/4.3	4.6/4.3
Booster	Pentex CD 150g	Pentex CD 150g	Pentex CD 150g	Pentex CD 150g

Note * **Large percentage of holes were found to contain water, requiring dewatering and lining. When water presence was so dynamic as to prevent effective lining, package emulsion or a 50/50 Heavy ANFO blend was used.**

Note ** Competent rock was charged with 4.6m while less competent rock was charged with 4.3m

The principal quality of the electronic initiation concept is associated with the precision, accuracy and flexibility of the timing assigned to each initiation sequence. In this way there is safety and certainty in the hole detonation sequence, faithful to the proposed blast design sequence. The flexibility allows the application of a large variety of blast design configurations according to the different requirements and objective results, whether these be increased energy levels between blast holes, increased displacement between rows and/or the definition of preferential directions of initiation wave train focusing, amongst other variables.

On the other hand, non-electric (pyrotechnic) systems display significant variability in the timing precision, accuracy and flexibility; and any expectations regarding likely blast outcomes must include or take into account the impact of timing errors with a coefficient of variation that is often measured in double digits. These limitations in performance limit the available options for more appropriate initiation sequences and delay selections when confronting specific needs for a given blast. Table 3 describes the delay sequences applied for each test.

Table 3: Delay and Initiation Sequence Selection

Timing Design Factor	Base Case (pyrotechnic)	Modified Case (electronic)
Inter-charge timing (ms)	9	5
Inter-row timing (ms)	65	100
Down-hole timing (ms)	600	N/A

Each of the selected blasts included a designated sample area from which the sample material was extracted. As previously mentioned, each sample polygon was chosen in accordance with the following requirements.

- Area specified by mine staff as intermediate grade material (0.3g/t – 1.0g/t Au)
- Area located within the body (core) of the total blast, well distanced from any likely edge effects
- Common lithological domain confirmed by Mine Geology Department as RAH

4 IMPLEMENTATION QUALITY

The quality of the implementation of any drill and blast design including explosive loading is fundamentally important when seeking to evaluate and eventually model the blast results' quality. A deficient implementation with significant deviations will likely produce negative results in terms of fragmentation, dilution, noise (air over-pressure), vibration and potential damage. Equally critical is the possibility that the blasting engineer loses faith in the results apparently delivered by a design that was, in fact, not really implemented.

At the beginning of this project an evaluation of the design implementation quality typically achieved at the host mine was conducted; consistent and significant deviations were encountered and reported to the client. Given the impact that implementation quality has on blast outcomes, this type of analysis was deemed to be essential for the analysis of fragmentation and other blast outcomes. Data were requested from mining engineering staff, however only three of the four blasts were evaluated. The results of this process are presented as follows

4.1 Drilling Accuracy – Hole Collars

This particular evaluation considers the accuracy and precision of the location of the drill holes. An arbitrary definition of acceptable deviation from design was set at ±0.3m. This accuracy criterion was then applied to all the blast-hole survey data available for each blast. Table 4 lists the number of blast holes evaluated per blast and the proportion of each data set that met the accuracy criterion established.

Table 4: Blast Hole Locations

Blast ID	No. of Holes Evaluated	Acceptable %
V_{1P}	179	26.0%
V_{2P}	524	16.6%
V_{1E}	489	20.0%
V_{2E}	No information available	No information available

Clearly, for the analyzed blasts, the quality of drilling measured in the three blasts where data were obtained, is low. The least inaccurate blast was found to be the first pyrotechnic blast, V_{1P}. This level of compliance is comparable with the cases measured at the beginning of the study.

Subsequently the same data were analyzed to determine the implication for deviation in terms of burden and spacing. In this aspect of drilling accuracy, the critical factor relates to the relationship between the distribution of explosive energy and the consistency of inter-charge firing times with respect to real burden dimensions. Figure 1 and Tables 5 and 6 list the state of quality in this sense.

Table 5: Summary of deviation statistics for drilled Burden

Blast ID	Burden % Acceptable	Burden Average Value (m)	Burden Standard Deviation (m)
V_{1P}	62.5%	5.83	0.40
V_{2P}	53.3%	5.62	0.55
V_{1E}	49.7%	5.72	0.59
V_{2E}	No information	No information	No information

Table 6: Summary of deviation statistics for drilled Spacing

Blast ID	Spacing % Acceptable	Spacing Average Value (m)	Spacing Standard Deviation (m)
V_{1P}	59.6%	6.78	0.42
V_{2P}	43.9%	6.7	0.68
V_{1E}	36.3%	6.74	0.73
V_{2E}	No information	No information	No information

Tables 5 and 6 suggest a relatively acceptable level of accuracy in terms of mean values. Nonetheless in both burden and spacing the standard deviation exceeds the acceptable levels of error suggesting that the incidence of irregular levels of energy available throughout the blast is likely.

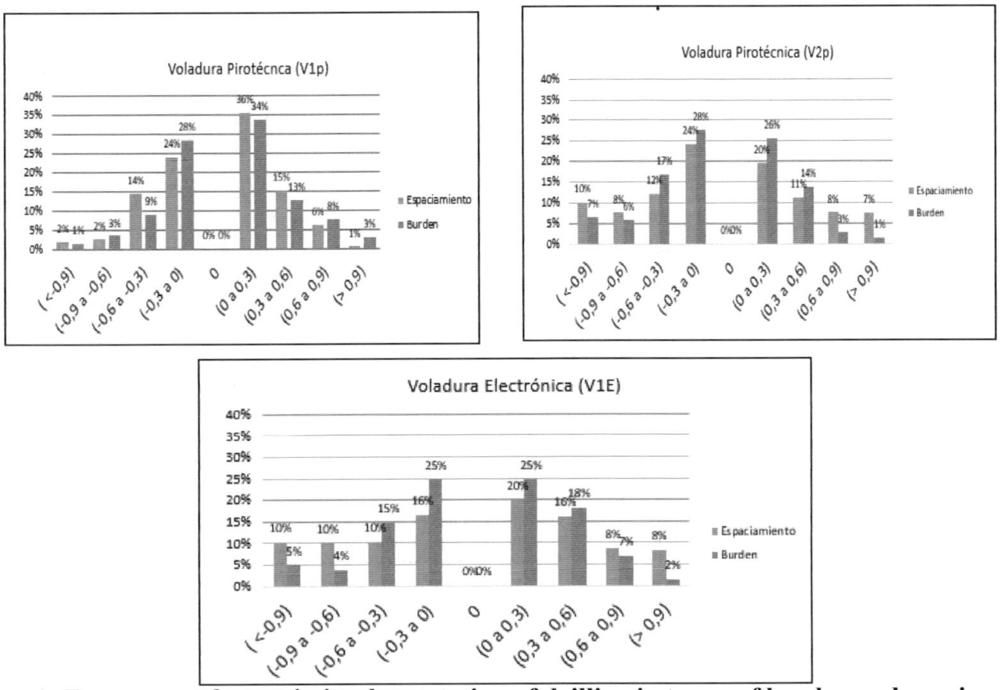

Figure 1: Frequency of error in implementation of drilling in terms of burden and spacing

4.2 Drilling Accuracy – Blast Hole Lengths

As a means of quantifying compliance with drill hole depth design, Global Technical Solutions engineers used an error criterion of ±0.5m, equivalent to approximately 5% of the design blast-hole lengths at the mine.

Figure 2 illustrates the quality of blast hole drilling compliance for the three blasts where QA/QC measurements were made. In general the majority of the measured holes are compliant (with values of 66%, 72% and 69% applying respectively to V_{1P}, V_{2P} and V_{1E} respectively).

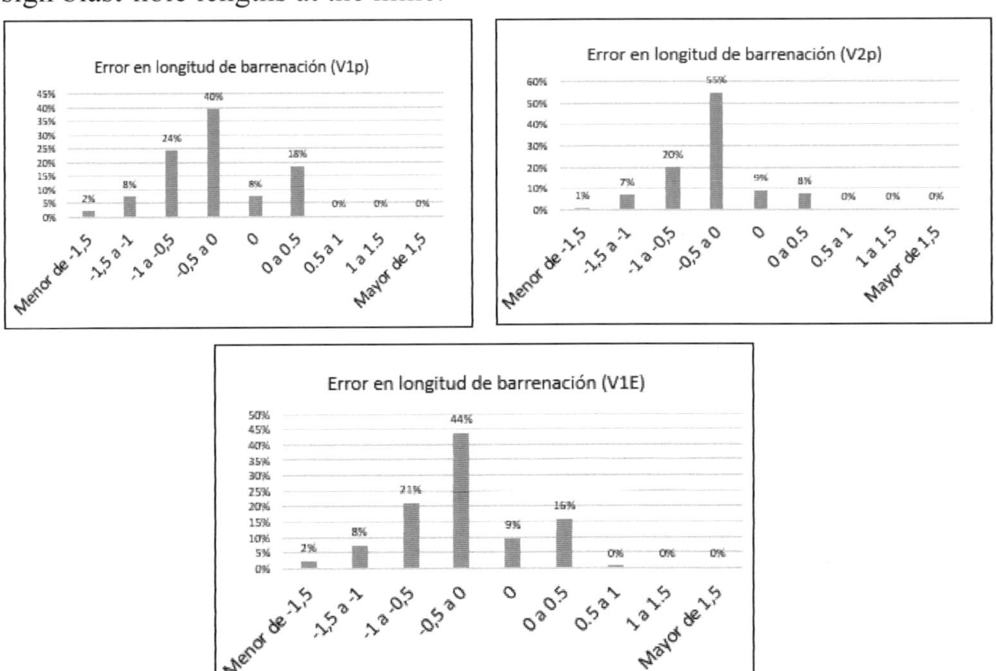

Figure 2: Histogram analysis of hole depth implementation error (Note: Depth Error = Real Depth – Design Depth)

There is however a slight bias in the analyses towards a shorter average blast-hole length than that specified in the design. Such a tendency will produce a potentially hard floor in the pit.

4.3 Explosive Column Length

The last implementation quality variable analyzed was the degree of compliance between design and actual explosive charge profiles. This datum was calculated in each case from the subtraction of measured stemming deck length from the full, real blast-hole depth. Once again, using an arbitrary quality of standard of ±0.5m, these data sets were evaluated for each blast where the QA/QC audit process was applied. Table 7 summarizes the findings of this exercise.

Table 7: Percentage of Acceptable Blast Hole Charge Lengths

Blast ID	Acceptable Charge Lengths
V_{1P}	96.9%
V_{2P}	99.0%
V_{1E}	97.7%
V_{2E}	No information available

These results suggest a high level of quality control in the explosive charging process for all blasts, independent of the initiation system technology applied.

5. FRAGMENTATION

Taking the material removed from the four central polygons, one for each blast, the fragmentation quality was analyzed using scaled image photography digitalization techniques.

Figure 3: Images of extracted samples of blasted rock, heaped for fragmentation analysis

Considering the common characteristics of the rock mass in which trial blasting was done, as well as the use of identical blast designs for each type of initiation system test, it was decided to combine the four data sets into two; pyrotechnic and electronic for the purposes of quantifying the quality of fragmentation delivered. This quantitative classification is summarized in Table 8. The size distributions for each case have been described in a cumulative form in Figure 4.

Table 8: Combined Fragmentation Distributions; Pyrotechnic vs. Electronic

Percentage Passing	Particle Size Pyrotechnic (in)	Particle Size Electronic (in)	Reduction in Size Electronic - Pyrotechnic
P_{10}	1.93	1.65	14.5%
P_{20}	2.64	2.19	17.0%
P_{30}	3.37	2.71	19.6%
P_{40}	4.21	3.27	22.3%
P_{50}	5.26	3.92	25.5%
P_{60}	6.66	4.72	29.1%
P_{70}	8.60	5.77	32.9%
P_{80}	11.32	7.33	35.2%
P_{90}	15.39	10.05	34.7%
P_{100}	33.66	27.74	17.6%
No.of Images Analyzed	76	63	

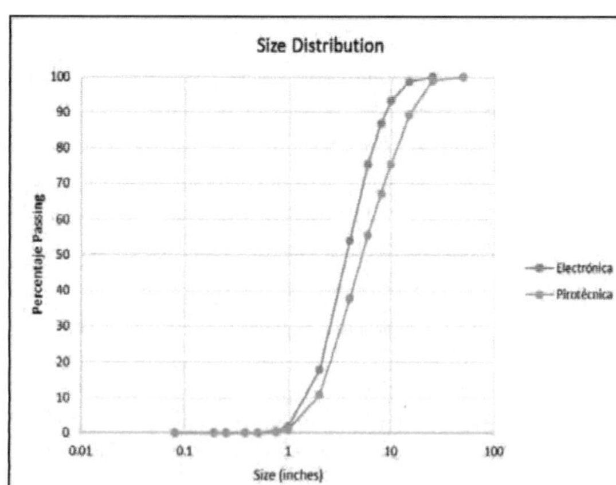

Figure 4: Size distribution curves obtained for Pyrotechnic and Electronic Initiation

These curves summarize the fragmentation quality achieved following consolidation of the data sets into two classes; pyrotechnic and electronic. The results obtained by electronic blasting show significantly improved results with a clear shift to the finer end of the spectrum.

It is notable that the difference between the two fragmentation tendencies increases as the size fraction becomes coarser, suggesting that, in this rock mass, the electronic timing regime employed is more effective in reducing the size of coarser fragments. This could change with alterations in rock mass, explosive type and timing assignation between charges. Certainly the positive impact of electronic timing is well marked, reaching as high as a 35.2% reduction in the 80th percentile size fraction, a size description that has been described as important for leaching performance by a number of plant managers.

Both size distribution curves were also analyzed for uniformity, calculated using Equation 1 as shown below.

$$I_U = (P_{80} - P_{20}) / (P_{50}) \qquad \text{Equation 1}$$

This Uniformity Index is an indicator of the heterogeneity of the fragmentation described in Figure 4. In the case of the present project the values obtained for pyrotechnic and electronic blasting were 1.65 and 1.31 respectively, suggesting a better trend towards homogeneity in blasted rock fragmented sizes in electronically fired events. This in turn supports the Value Proposition that combining electronic blast initiation with the appropriate level of skilled blast engineering will produce long term productivity benefits in mining and gold extraction at mines such as that discussed in this paper.

Following the photographic evaluation of the fragmentation distributions for the four blasted ore samples, just described, each blasted rock pile was screened into three size classifications as described in Table 1; that is into Fine, Intermediate and Coarse size ranges.

Each of these separate piles of screened rock was surveyed accurately for volume (m^3) such that the percentage volume in each size range could be graphed for the consolidated samples for both pyrotechnic and electronic initiated blasts.

In total a little over 318m^3 of electronically blasted rock was compared with a sample of 337m^3 of pyrotechnically blasted rock. How these volumes were distributed between the fine, intermediate and coarse fractions is described in Figure 5, below, recalling that the fine fraction is described as between 0 and 1.25 inches, intermediate is defined as 1.25 – 4in. and coarse material is greater than 4in. average size.

The graph also shows the results of the screening classification of the evaluated material following the consolidation of the blasts for each initiation system. A simple and quick comparison between the results demonstrates that the electronic initiated blasts produce a more regular distribution of fragments across the size fraction ranges; lower production of coarse and increased production of fines when compared with pyrotechnic results.

There is a noteworthy agreement between the Uniformity Index results quoted previously and the percent distribution of volumes of rock across the three screened ranges. Thus a greater degree of volumetric regularity (uniformity) across the size ranges of the electronically blasted material is seen.

This clear set of results points to the fact that electronic blasting has delivered muckpiles that are easier to load and haul, from a mining perspective. Additionally, according to the plant engineering staff, this condition favors increased leaching performance. If the trends described here can be maintained in time then there is a confident expectation that gold recovery in the ROM heap leaching performance will increase.

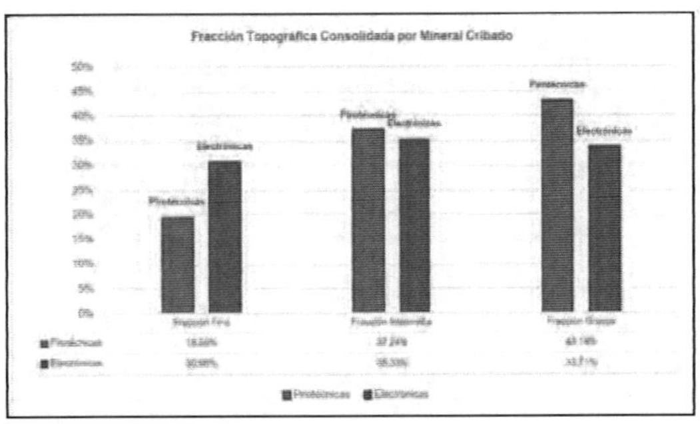

Figure 5: Classification of blasted ore according to screened size intervals

6. GENERATION OF MICRO-FRACTURES:

The potential of each type of initiation system to deliver blast outcomes with an elevated presence of micro-fracturing in the rock matrix was analyzed by subjecting a representative randomly selected sample of rock fragments to Point Load Test evaluation. Furthermore the results were subjected to statistical inference analysis to ensure that observed differences were statistically significant.

6.1 Point Load Testing ("PLT")

The simple purpose of this test is to estimate the resistance of a rock sample to an axially applied stress load. For the effects of the current project the objective was to infer the existence of changes in micro-fracture content by observing variations in Point Load Index strength, for samples of rock broken by electronic and pyrotechnic blasting. It was expected that the electronic blasting rock samples would present a weaker response to the point load due to an increase in micro-fracture content. The tested samples were taken from the screened and classified muck piles, extracted from the sample extraction polygons within the body of each of the four blasts. Figure 6 shows the PLT equipment in use as well as a selection of typical samples.

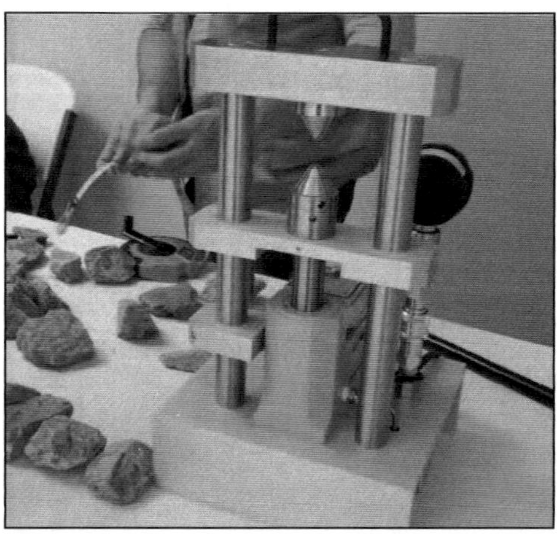

Figure 6: PLT Equipment

The PLT results were used to estimate Uniaxial Compressive Strength ("UCS") for each sample and the generated statistical tendencies interpreted in a loss of rock strength for each case, assuming the incidence of a form of "strain-softening" due to the generation of fresh micro-fracturing. On this basis, the results of the pyrotechnic and electronic blasted ore were compared and contrasted.

This test work was carried out for samples taken from the fine and intermediate classified samples for each blasted muckpile; Tables 9 and 10 report the results obtained from this measurement series for the fine and intermediate size fractions.

Table 9: Statistical Analysis of PLT – Fine Fraction

Initiation System	No. of Samples	Mean Resistance (MPa)	Variation
Pyrotechnic	47	69.37	1428.79
Electronic	39	49.60	647.31

Table 10: Statistical Analysis of PLT – Intermediate Fraction

Initiation System	No. of Samples	Mean Resistance (MPa)	Variation
Pyrotechnic	50	53.08	1463.49
Electronic	40	36.30	528.24

These results indicate that the average strength of the rock sample matrix is significantly lower when blasted using electronic initiation than when relying on the performance of the pyrotechnic technology; approximately 30% weaker in terms of the PLT-estimated UCS values.

It has also been demonstrated that the level of variability in the PLT strength results is much lower for the electronic blasting samples than for the pyrotechnic equivalents (54% less for the fine fraction and 63% less for the intermediate fraction). This is very probably related to the much more consistent timing performance of electronic detonators as opposed to the pyrotechnic equivalent.

These trends are confirmed visually in Figures 7 and 8, which are "box and whisker" representations of the differing strength properties between pyrotechnic and electronic initiation for fine and intermediate size fractions respectively.

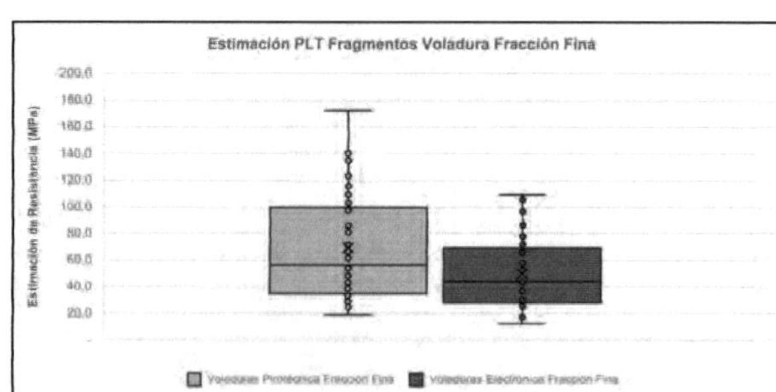

Figure 7: Box and Whisker representation of PLT data for fine fraction

These analyses strongly suggest that the blasts initiated with the electronic system create fragments of rock with lower insitu strength, probably due to an enhanced degree of micro-fractures or micro-cracking, than those fragments generated using pyrotechnic technology and delay times.

It appears that the greater level of temporal distribution of explosive energy throughout the rock mass, due to the finer timing control afforded by electronic initiation system allows such a result to be engineered into the blast outcome.

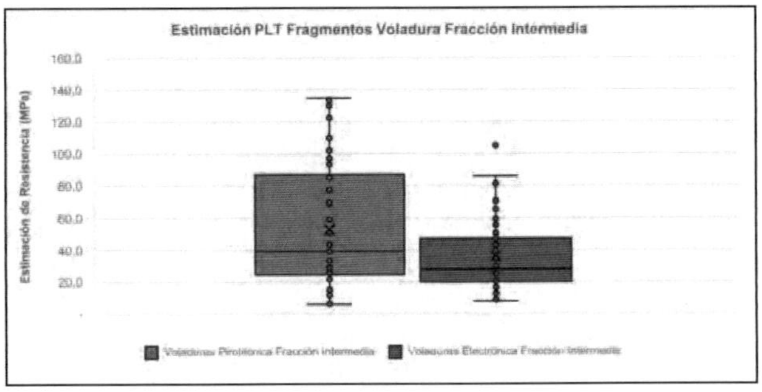

Figure 8: Box and Whisker representation of PLT data for intermediate fraction

6.2 Statistical Inference Analysis

Given the importance of the conclusions being drawn from the analyses previously described, it is appropriate to challenge the statistical significance of the interpretations herein described.

Therefore an analysis of variance (ANOVA) was employed to separate the sources of the variation in results of the two initiation system test series. The objective of the analysis is to compare the diverse values of rock strength measurement to determine if the initiation system cases differ significantly.

The results of the ANOVA analysis allow the conclusion that in the case of both size fractions studied (fine and intermediate) one of the two initiation systems statistically differs significantly from the other. Comparisons of variance of the means between the groups and within the groups indicates that the rock strength values reported for each of the two initiation systems are in fact separate populations with different characteristics.

Finally, upon evaluation of the average values of the rock strength estimates and considering the results of this ANOVA process, it is deemed reasonable to propose that the reduction in the strength of those rock fragments coming from blasts initiated by DaveyTronic® is real and can be attributed to the creation of micro-fractures due to the stress intensity generated in the rock mass by precisely initiated explosive charges.

7. VIBRATION MEASUREMENT AND MODELLING

In order to broaden the characterization of the relevant lithological rock mass domain and increase the scope of potential future work, the GTS team designed and carried out a vibration measurement exercise that allowed the adjustment (calibration) of a near field vibration model and a quantitative estimate of V_p for the subject rock mass. These two achievements allow subsequent analyses to predict likely vibration intensities from given explosive charge designs and estimate an inter-charge delay period that will deliver enhanced fragmentation performance for any sector of the mine.

The measurement of the vibration pulse produced by the detonation of a known explosive charge permits the evaluation and understanding of the behavior of that pulse as it moves through a rock mass of known characteristic properties.

In the near field, high intensity vibration induced at short distances from the explosive charge produces high induced stress/strain levels that induce fracturing (fragmentation and conditioning) of the rock mass. Based on these considerations a near field vibration model was calibrated in order to allow the confident prediction of vibration intensity in the RAH lithological domain at the mine, for any configuration of explosive charge. In the future it will be useful to equate this vibration intensity with probability of micro-fracture stimulation.

The calibrated near field vibration model has yet to be applied in a more detailed manner, however the V_p measurement data were used to determine a first estimate of inter-charge delay programming time. An analysis process based on the application of a Lagrange diagram was employed to estimate the 5ms delay applied during this project. More detail regarding this process will be provided in a future report.

It should be understood that the 5ms delay time is only applicable with any validity in the RAH lithological domain; any change of domain would require re-calibration of the vibration work described here.

8. FINAL COMMENTS AND DISCUSSION

Following the completion of the analyses described in this report it can be concluded that electronic initiation systems display a technically superior functionality that in turn allows the generation of improved conditions for increasing the recovery of gold through ROM heap leaching. The results obtained suggest that the electronic initiation system delivers a more controlled and consistent liberation of explosive energy, optimizing the useful application of that energy to create a finer and more homogeneous distribution of rock fragments that have lower strength properties than those generated by the pyrotechnic system.

The lower variability in rock fragment characteristics is due to the virtually inexistent timing scatter for the detonation of the electronic detonators, leading to a more predictable and controllable interaction between adjacently detonating charges. Pyrotechnic delays with their high level of inherent timing scatter do not allow the same efficiency in energy utilization.

Fragmentation analysis comparing electronic blasting vs. pyrotechnic processes shows that an average reduction of 35.2% is achieved in the size of the P80 fraction (11.32" – 7.33") in favour of the use of electronic initiation technology.

The distributions of sizes of blasted rock fragments are also demonstrably more uniform for the electronically initiated blasts, evidenced by the calculation of the Uniformity Index for each distribution as well as a volumetric measurement of the screened and separated size fractions. The electronic blasted rock showed an U.I. value of 1.31 vs. the 1.65 for the pyrotechnic blast induced fragmentation. The three screened size fractions for the electronically blasted rock were distributed in volumes of approximately 33% each, supporting the case for more uniform fragmentation being delivered in that manner.

In general terms, a more uniform distribution of rock fragments will promote more efficient downstream mining processes. It is also generally understood that coarser sized ore fragments will lead to reduced levels of exposition of gold ore to the leaching solvent for equivalent sized samples of leachable ore. A more even distribution of smaller fragments will promote better leaching recovery for the same leaching process conditions (geometry, heap dimensions, solvent deployment methodology, etc.).

The fully quantified demonstration of these benefits will require additional experimentation, particularly the completion of the microscopic examination of rock samples to confirm micro-fracturing behavior and column leach testing to measure recovery performance improvement.

However, rigorous Point Load Index evaluation of a large number of rock samples from both electronic and pyrotechnic blasts demonstrated that the former system produced fragments that were 28% and 31% weaker/softer for fine and intermediate size fractions respectively. An increased presence of micro-fractures is the most probable cause for this observed trend.

A statistical analysis of variance for the previous data sets indicated conclusively that the different sample populations are indeed separate and the differences in behavior between them are significant.

A specially designed and executed vibration measurement program was conducted to calibrate a near field vibration model and obtain an estimate of V_p that was used to select the inter-charge delay time of 5ms.

In conclusion, despite the existence of demonstrable errors in drilling accuracy (leading to the production of some experimental variability) it has been demonstrated that well engineered, well executed, and electronically initiated blasting offers reliable means to improve the mining and gold extraction efficiency in ROM heap leach operations.

As a closing remark, it should be considered that it is possible to sufficiently blast condition rock fragments such that they display the type of reduced mechanical resistance that has been described in this paper. Parra et al described a number of detailed studies in which a number of industry accepted tests of rock toughness, including the Bond Work Index, the JKMRC aXb term, and others have displayed clear and unambiguous decreases in magnitude – closely associated with the generation of new micro-fracturing in the rock matrix. Several of such studies have also related this degradation on rock matrix strength to significant savings and efficiencies in the energy consumption and throughput in comminution circuits around the world. However, it has also been reported that such decrease in rock strength and the assumed increase in micro-fracture content *do not always seem to trigger an additional increase in leaching performance.* Parra (2013) and Parra et al (2015) proposed a reasonable explanation for this initially seeming dichotomy. There are three types of micro-cracks that can be present in blast conditioned rock fragments; intra-granular (where cracking is restricted to within individual grains), inter-granular (where cracking occurs along grain boundaries) and trans-granular (where the cracking induced actually traverses multiple grains). The level of cracking that occurs may well be associated with the probability of enhancing contact between leaching solvent solutions and mineral grains (gold, copper, etc.) within the matrix of the rock. This level of cracking is probably also closely linked with the intensity of the stress conditioning induced by the blasting event – itself a function of explosive energy and electronic timing sequences (McKenzie and Adamson, 2011).

The authors of this current paper await the final results of the microscopy and column leaching studies being undertaken by the host mining operation.

9. REFERENCES

McKenzie, C.K. and Adamson, W.R., 2010, *Explorando el paradigma de selección de tiempos electrónicos ultra-cortos*, IX Jornadas de Tronadura, Santa Cruz, Asociación Chilena de Ingenieros Explosivistas

McKenzie, C.K. and Adamson, W.R., 2011, *Exploring optimized delay timing for fragmentation*, Explo 11 Conference, AustIMM, Melbourne, 2011, pp 97-104

Parra, H., 2013, *Blast induced fragmentation conditioning and its impact on impact breakage and leaching performance*, PhD, University of Queensland, St. Lucia

Parra, H., Onederra, I. and Michaux, S., 2014, *Effect of blast-induced fragment conditioning on impact breakage strength,* Mining Technology, vol. 123, 2014

Parra, H., Onederra, I., Michaux, S., Kuhar, L, McFarlane, A. and Chapman, N., 2015, *A study of the impact of blast induced conditioning on leach performance*, Minerals Engineering, 74 (2015), pp 1-12

Blasting Optimization in a Phosphate Rock Mine in Brazil

A. C. Silva[1], L. T. B. Mendonça[1], P. A. A. Martins[2], R. A. G. Silva[2], G. Birro[2], H. C. S. Nadler[2]

1 Federal University of Goiás, Catalão, Brazil

2 CopeBrás, Catalão, Brazil

ABSTRACT The cost reduction in mines is always welcome and a permanent goal all around the world. In search of better fragmentation in blasting with the use of explosives and an optimization in the subsequent stages of comminution, tests in different mine fronts in a phosphate rock mine (Chapadão Mine, located in the municipality of Ouvidor, state of Goiás, Brazil) were carried out. The tests were realized modifying operational parameters of the blast layout plan such as stemming, blast layout and explosive type. After basting digital images analysis using Split Desktop 3.1 were conducted to evaluate the fragmentation of the rocky massifs and its size distribution. It is important to highlight that the comparisons in this study were realized focusing the reduction of the particle size distribution over the P80 for each blasting.

1 INTRODUCTION

This paper presents a project developed at Chapadão mine, in the municipality of Ouvidor-GO Brazil, coordinates 18°08'S latitude and 47°48'W longitude, the figure 1 presents the map with the location of the municipality. The Copebras Industries Ltd one companies of the CMOC international's portfolio. The Copebras phosphate mine has been in operation since 1976 and has a target for the production of 1,200 tph of crushed ore. Currently in operation, there are two processing plants in operation in Ouvidor and two chemical plants. In specifically, the two chemical plants are located in the city of Catalão-GO and in Cubatão-SP. Moreover, the Copebras main products are GTSP, SSP, TSP fertilizers and dicalcium phosphate, one of the components utilized in animal feed. Beyond these products, by-products such as barium sulphate (Barite) and Niobium are obtained over the process.

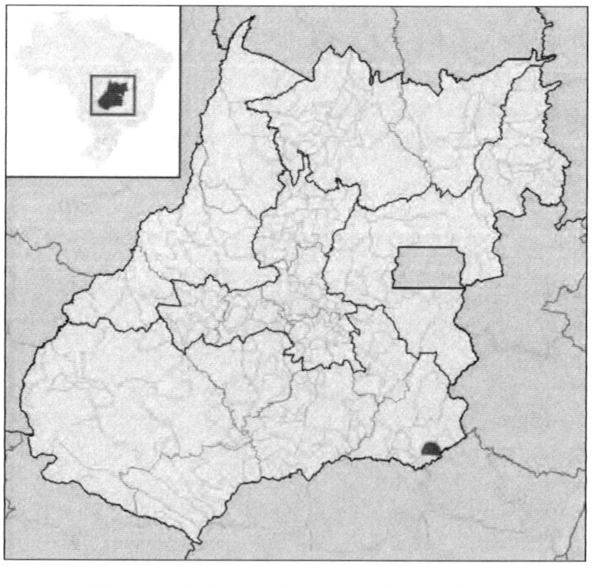

Figure 1. Location Ouvidor-GO.

The Copebras Phosphates Mine is located not in the Catalão I alkaline complex, but inserted in the tectonic context of the Igneous Province of Alto Paranaíba. This mineral province consists of a shallow alkaline ultramafic magma chamber, with polyphase evolution, with predominance of bebedourites and phlogopitites in the borders and phoscorite and nelsonite veins in the centre, stock works of carbonatites and picrites cutting all lithologies, exemplified in figure 2. The phosphate ore from the Chapadão mine results from a supergenic residual concentration on ultramafic alkaline phoglopitized rocks, intruded by rocks from the series phoscorites and carbonatites (GIBSON et al., 1995).

Figure 2. Stock work of dolomitic (whitish) carbonatite dikes in phoscorite (grey) showing the incidence of dikes and veins of different compositions.

Three main different series of rocks occur at the Copebras mine: the phoscorite series is represented by phoscorites, nelsonites, apatitites and magnetitites; the carbonatite series is represented by dolomite carbonatite and calcite carbonatite; the bebedourite series is represented by metasomatic phoglopitites, derived from metasomatism of primary ultramafic rocks and by late-stage conduitbreccias. (PALMIERI et al., 2011). Due to the large amount of The NW portion of the dome, mainly in the Phosphate from Copebras and in the two niobium Mining, it is suggested to adopt the name of carbonatite phoscorite I complex, as it is used in Kovdor, on the Kola island (KRASNOVA et al., 2004).The table 1 shows the lithotypes according to their tenacity rate.

The material from the crushing is used in the formation of two piles homogenization, one of 64,000 tons and another 80,000 tons of maximum capacity, the composition of these piles ranges from 20 to 50% of detonated material. According to Clark (1980), the use of explosives is common in mining activities that precede the mining stage, is a fast and economical method as they produce a high-energy release in fractions of seconds and can contemplate bulk large to be disaggregated at one time.

Table 1. Lithotypes tenacity scale.

Tenacity (+) ↓	Friable
	Hardest Semi-Compact Apatite
	Hardest Compact Apatite
	Hardest Semi-Compact Silexite
	Hardest Compact Silexite

The rock blasting is a crucial operation as it has a direct influence on downstream process. As the first step of rock fragmentation, where the in situ rock is reduced resulting in best conditions for the next operations that most commonly are: loading, hauling, ore crushing, milling and mineral processing (STIEHR, 2011).

Due to the high relevance of blasting, the expected results for the phosphate rock at Chapadão mine are defined to achieve a P80 value lower than 400 mm, which means that 80% of the blasted fragments passing in 400 mm, the grizzle dimension of the primary crusher system. In order to achieve this result blast designs must be accurately defined and followed.

Aiming to achieve the target, the blasting team has been assessing in detail the benefits of changing different parameters of the blast plan considering the geological characteristics of each blast domain and measuring the fragmentation using the software Split Desktop 3.1.

2 METHODOLOGY

To ensure a structured and analytical approach, the technical team have defined which historical data should be used and compared with the field tests following as the main premise the changing in only one parameter per time. The data were organized in a standard database elaborated by the blast team comprehending 20 variables, with 10 blast plan parameters and three blast Key Performance Indicators (KPI's). The main performance indicators are the P80 and the blast unit cost (US$/t).

The principal target of the Project was a 20% reduction in the rock blasting unit cost. For this purpose, 56 testing blasts were realized between 13/06/2016 and 23/11/2016. Some of these tests were peers compared fixing all variables and changing only one parameter per time comprehending three operational steps as showed on the flowchart in figure 3. Each step was composed by two tests that included two distinct blasts totalizing the quantity of six tests and nine blasts.

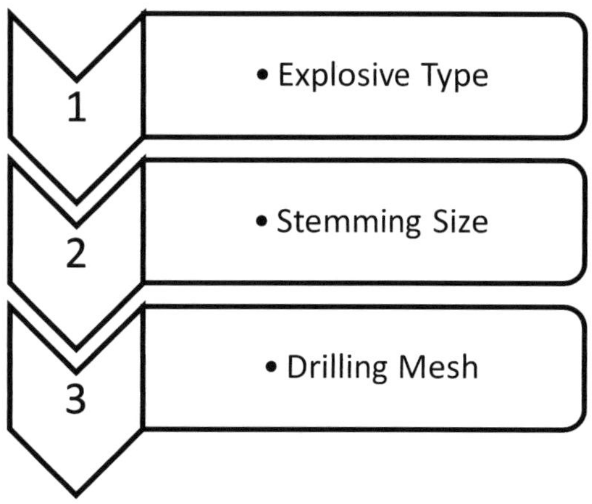

Figure 3. Test Steps.

Table 2 shows the fixed parameters. The tie-in sequence defined was a closed diagonal or closed "V" as is showed in figure 4 and did not suffer any change during the tests period. However other parameters were changed according to it respective test, but only changing one parameter per time fixing it for post analysis.

The changed parameter will be always be indicated in the last line of its respective tables.

Table 2. Fixed Parameters for tests

Fixed parameter		Value
Hole Diameter (m)		0.1016
Bench Inclination Angle (°)		0
Bench Height (m)		5.00
Surface Delay Time (ms)		17
Depth Delay Time		500
Booster (g)		225
Stemming Material		Rock Dust
Explosive Density (g/cm³)	ANFO	0.80
	Emulsion	1.15

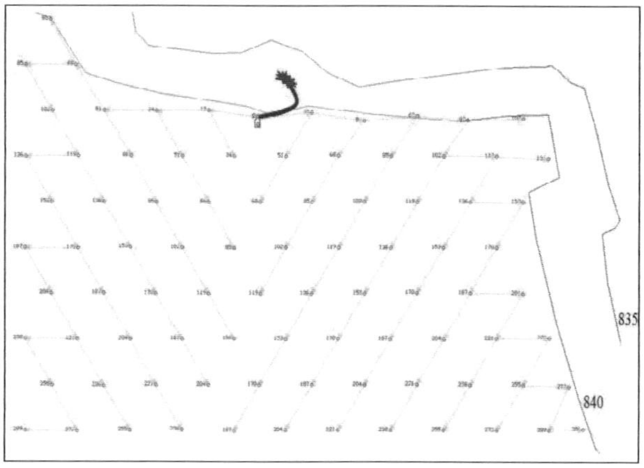

Figure 4. Example of tie-in sequence.

2.1 Explosive Type

The first step of the tests was planned exclusively by a comparison of the granular explosive ANFO 3500 and a pumped emulsion with a detonation velocity (VOD) about 50% higher than the previous one. Table 3 present tests A and B and their respective results (blasts 1, 2, 3 and 4).

2.2 Stemming Size

The second step was compounded by the assessment of the blasting results using different sizes for stemming.

The goal of this kind of evaluation is to ensure that the energy and gases from the explosives detonation are confined, avoiding energy losses in the hole collar region, what would affect negatively the rock fragmentation.

Table 3 – Tests A and B

	Test A		Test B	
Number Blast	1	2	3	4
Date Blast	10/06	11/07	06/10	24/10
Number Holes	248	143	129	143
Drilling Mesh (m²)	5.20	5.20	8.75	8.75
Sub Drilling	0.50	0.50	0.30	0.30
Litothype	Hardest Compact Silexite	Hardest Compact Silexite	Hardest Semi-Compact Silexite	Hardest Semi-Compact Silexite
Stemming Size (m)	1.40	1.40	2.00	2.00
Loading Rate (g/t)	453.54	651.97	215.05	309.14
Explosive Type	ANFO	Emulsion	ANFO	Emulsion

Table 4 – Tests C and D

	Test C		Test D	
Number Blast	5	6	7	8
Date Blast	21/07	27/06	01/08	03/08
Number Holes	132	235	253	245
Drilling Mesh (m²)	5.20	5.20	8.75	8.75
Sub Drilling	0.50	0.50	0.5	0.5
Litothype	Hardest Compact Apatite	Hardest Compact Apatite	Hardest Compact Silexite	Hardest Compact Silexite
Stemming Size (m)	ANFO	ANFO	Emulsion	Emulsion
Loading Rate (g/t)	377.95	453.54	344.40	387.46
Explosive Type	2.00	1.40	1.80	1.40

2.3 Drilling Pattern

For the last step of the Project, a drilling pattern expansion was defined resulting in a powder factor reduction. The expansion presented very satisfactory results achieving the expressive number of 40% compared with the average pattern used in the first period of 2016.

On table 5, the tests "E" and "F" are presented and its respective blasts. The presented blasts 2, 5 and 8 were already showed on tables 2 and 3; however, the focus on this step is the comparison of a different variable.

Table 5- Tests E and F

	Test E		Test F	
Number Blast	2	8	5	9
Date Blast	11/07	03/08	21/07	04/08
Number Holes	143	245	132	170
Sub Drilling	0.50	0.50	0.50	0.50
Litothype	Hardest Compact Silexite	Hardest Compact Silexite	Hardest Compact Silexite	Hardest Compact Silexite
Stemming Size (m)	1.40	1.40	2.00	2.00
Explosive Type	Emulsion	Emulsion	ANFO	ANFO
Loading Rate (g/t)	651.97	387.46	377.95	224.61
Drilling Mesh (m²)	5.20	8.75	5.20	8.75

2.4 Digital Images Analyses

After the three steps, the natural sequence of the assessment was peers comparisons among blasts taking in to account the same variations on blast design parameters.

As the focus, the team recorded the total explosives costs by blasted tonnage, the P80 results, which quantified the rock fragmentation differences and by the end, the crushing plant productivity was inferred using the amount of ore directing passing through the grizzle.

The ideal target defined for P80 was 400 mm. This value was defined accordingly the dimensions of the grizzle opening that feeds the primary crusher at the crushing plant. To ensure that most of the blasted fragments achieved the expected result, photo analysis were token at the muck piles (blasted ore) using a digital camera and the images treated in the Split Desktop software.

The muck pile pictures were acquired immediately after the blasts and imported to the software respecting the constraints such as lack of luminosity, correct angle and scale. The two scale reference spherical objects used have 25.40 cm of diameter and were positioned accordingly the guidelines and after acquisition, the images were edited and delineated. The quantity of images acquired for each blast variated with the dimensions and other characteristics of the blasted ore; however, the range was limited by 4 to 6 pictures that best represented the muck pile following a zigzag pattern through the blasted muck pile, as is showed in figure 5.

Figure 5. Correct strategy to get images from the blasted rock.

After carefully follow the previous steps, the taken images were imported, edited and delineated in the Split Desktop software, an example is showed in figure 6, where is presented a comparison among the delineated and non-delineated image.

The process of photo analysis is ended when the previously described edition process is made for all selected pictures and the set of all results is condensed to build the general particle size distribution (PSD) curve that represents the analysed blast. By this curve, the amount (percentage) of passing particles for a determined size was indicated in a millimetre scale, configured in the software.

(a)

(b)

Figure 6. (a) Blasted muck pile and (b) Delineation of blasted fragments.

3 RESULTS AND DISCUSSION

The results originated from the blasts were assessed in terms of cost variations, P80 results from the PSD and real powder factor.

A pre-estimation of the powder factor for each blast is presented on tables 2, 3 and 4, however this variable can be changed for operational reasons, that is why were considered only actuals powder factor avoiding the impacts of in field variations. The vertical line in the position of 400 mm highlights the defined target.

3.1 Tests A and B

Table 6 summarizes the results for tests A and B, which compares fragmentation and costs results after the change of explosive used in the column.

In this test, there were a significant powder factor reduction where around 48% (334.30 g/t) less explosives were consumed on blast 1. It can be explained looking exclusively to the type of explosive, as in blast 2 was used emulsion that has a 30% higher density compared with ANFO (ammonium nitrate and fuel oil) and the characteristic of fill all spaces in the hole. Figure 7 shows the respective PSD curves from tests 1 and 2.

Table 6. Tests A and B results

	Test A		Test B	
Number Blast	1	2	3	4
Loading Rate (g/t)	359.84	694.14	217.17	369.74
P 80 (mm)	383.16	290.63	254.54	359.74
< 400 mm	82%	90%	95%	85%
Cost Variation ($/t)	-	+ 45%	-	+ 40%

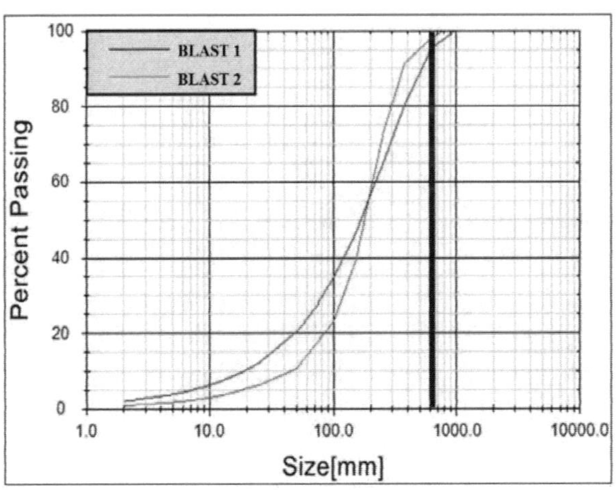

Figure 7. PSD curves for blasts 1 and 2.

Regarding fragmentation, the P80 from blast 2 was more satisfactory, resulting in a fragmentation around 24% (92.53 mm) inferior than blast 1. The litothype of this test was compact silexite, which represents a high tenacity when compared with the others litothypes presented in the mine. However, the high detonation velocity of the emulsion (7,000 m/s) can achieve best results against ANFO, but is important to mention that both results are below the proposed P80 target of 400 mm.

There is an important consequence related with the quantity of material that passes directly through the grizzle without the second break by the hydraulic hammer. This process can result in delays and increase the trucks cycle time (waiting time) and for some moments, the oversized blocks are left to receive a secondary blast in the pit. Therefore, it is possible to conclude that 82% of the fragments on blast 1 passed directly in the grizzle while on blast 2 the number was 90% that is, both blasts had an excellent productivity because the major volume of its fragments did not ask for a second breakage.

As the last assessment, the total blasts costs were compared and observed that blast 1, despites its small advantage in terms of fragmentation and productivity, presented a 45% higher cost than blast 2. This was considered an extremely elevated variation indicating that the utilization of emulsion is economically unfeasible up to now.

On figure 8, the PSD curves for blasts 3 and 4 are showed, both regarding test B.

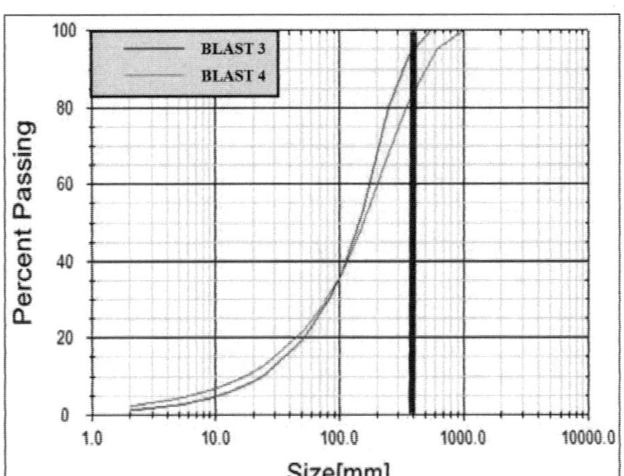

Figure 8. PSD curves for blasts 3 and 4.

Analysing the results from test B it is possible to conclude that the powder factor on blast 4 was around 41% (152.57 g/t) higher than on blast 3, which is justified by the same conditions mentioned on test A.

The fragmentation on blast 3 showed a better result, 39% (105.22 mm) lower than the fragmentation on blast 4. On this case, the geological litothype presents a lower resistance, indicating that ANFO is more efficient for blasting this kind of rock as suggested by (Jimeno et al., 2003).

The productivity on blasts 3 and 4 were respectively around 95% and 85% of passing, highlighting the very significant result originated from blast 3 where only 5% of the fragments did not passes directly through the grizzle.

The total blasts costs variation were 40% as blast 4 presented a higher cost.

Test B presented the best results with ANFO utilization in all analysed criteria, which means a better fragmentation, productivity and lower cost.

Comparison between both tests A and B presented significant advantages for the use of ANFO, mainly in terms of costs, where was achieved up to 40% of reduction. Regarding fragmentation, all blasts achieved the proposed target.

3.2 Tests C and D

Table 7 present the results of tests C and D and its respective performance indicator, including the P80 showed in the PSD curves in figures 9 and 10.

Table 7. Tests C and D results

	Test C		Test D	
Number Blast	5	6	7	8
Loading Rate (g/t)	694.14	359.84	217.17	369.74
P 80 (mm)	219.41	351.33	236.59	357.13
< 400 mm	100%	80%	95%	85%
Cost Variation ($/t)	-	+ 19%	-	+ 13%

The test C was composed by blasts 5 and 6, both in the litothype apatitic compact with different stemming sizes. On blast 5, the stemming size was 30% (0.60 m) higher than blast 6 and consequently the powder factor from blast 6 was 17% (75.59 g/t) higher than blast 5.

Even with a higher powder factor, blast 6 did not achieve a better rock fragmentation probably because the excessive energy ejected the stemming material affecting the efficiency of keeping the explosives and gases confined in the hole during the necessary time to act in the rock breakage. (Cavadas, 2012).

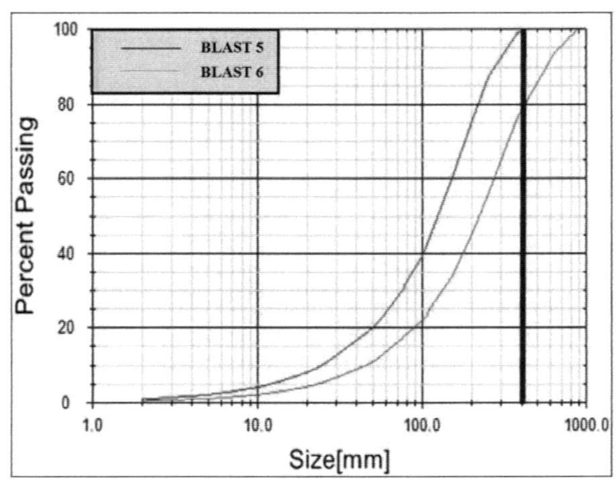

Figure 9. PSD curves for blasts 5 and 6.

The fragmentation on blast 5 presented advantageous, as it was 48% (131.92 mm), lower than blast 6. The percentage of passing fragments (lower than 400 mm) on blast 5 represents the total efficiency that is 100% of material passing in the grizzle while on blast 6 only 80% of the material passed directly through the grizzle.

Regarding the blast total costs, blast 6 was 19% higher than blast 5, which is justified by the higher charge column length when used the stemming size of 1.40 m. The increasing in the stemming size showed to be extremely satisfactory for both productivity, due to the better fragmentation, and blast cost reduction.

The test D was composed by blasts 7 and 8 and its PSD curves are presented in figure 10.

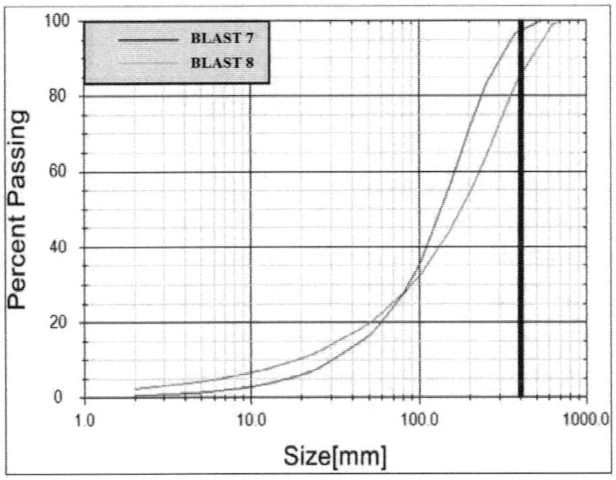

Figure 10. PSD curves for blasts 7 and 8.

On blasts 7 and 8, there were also variations in the stemming size, however on that case the variation was 23% (0.40 m) with blast 7 performed with a higher length.

Similarly test C, the higher stemming length blast presented advantageous for all performance indicators.

The fragmentation on blast 7 showed a P80 34% (117.54 mm) lower than blast 8. Considering the quantity of passing fragments, blast 7 also had better results as it presented 95% compared with 85% from blast 8.

Once again, the increasing in the stemming size has improved the productivity in the crushing plant reducing the breakage by hydraulic hammer and it costs was 13% lower than blast 8.

The gains obtained in both tests clarify the great relevance of stemming size as an individual parameter for blast efficiency, that is, its correct definition is intrinsically connected with good fragmentations results that were observed on blasts 5 and 7, where a 30% more efficient comminution was achieved when compared with its peers.

3.3 Tests E and F

The last step comprehended the results from blasts E and F presented on table 8.

Table 8 – Results from tests E and F.

	Test E		Test F	
Number Blast	8	2	9	5
Loading Rate (g/t)	383.09	694.14	230.86	361.04
P 80 (mm)	357.13	290.63	208.21	219.41
< 400 mm	85%	90%	100%	100%
Cost Variation ($/t)	-	+ 42%	-	+ 32%

The blast 2 presented a better fragmentation result, around 19% (66.50 mm) lower than blast 8, and the percentage of passing fragments was 90% on blast 2 and 85% on blast 8, as can be observed on figure 11. However, the total cost of blast 2 was 42% higher than the costs on blast 8.

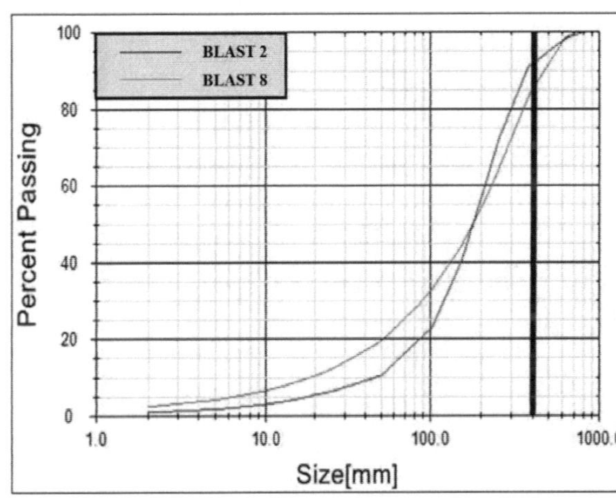

Figure 11. PSD Curves for blasts 2 and 8.

The fragmentation to the smaller drilling grid was more efficient, but the total explosives and accessories costs by blasted tonnage were very higher, which could not justify its implementation because the P80 results for both blasts achieved the expected productivity with similar crushing productivity.

The PSD curves for blasts 5 and 9 from test F are showed in figure 12.

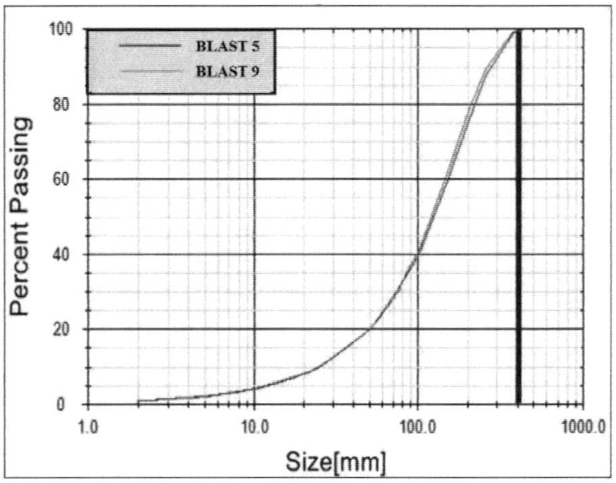

Figure 12. Fragmentation curves for blasts 5 and 9.

Once again, the results of a 40% of increased drilling grid on test F were assessed. The higher grid blast presented a fragmentation result around 8% (11.20 mm) lower than the blast 5. Both blasts showed the maximum expected efficiency that is 100% of the material passing directly in the grizzle and improving the crushing productivity.

Compared with blast 9, blast 5 resulted in 32% higher costs. However, the expansion of the drilling grid was very advantageous for this geological litothype, as it presented better fragmentation result and blasting lower costs.

The comparison between tests E and F could prove that a cost reduction is obtained by expanding the drilling grid and still keep a desired fragmentation. A special attention is given to blast 9 which showed a lower fragmentation when compared with blast 5, contradicting the definition of (Cavadas, 2012) who defends that a drilling grid reduction is totally related with the necessity of achieve better fragmentations.

There were designed comparatives tests for different rock blasting which were based on changes of specific parameters of the blast design. However, they only indicates some directions for future improvements in the blast process.

The changes in some variables were implemented according to the gains achieved and demonstrated in the tests and can be better expressed by comparing the scenarios before and after the optimization.

The graphic presented in figure 13 shows in percentage the gains achieved with fragmentation considering the grizzle feed size of 400 mm and the total costs reduction achieved in the blasting operation.

By adopting the new blast design parameters it was possible to reduce the fragmentation by 31%, increase the crushing productivity by 12% and reduce blast costs by 42%, surpassing the previous expectation defined at the beginning of the project.

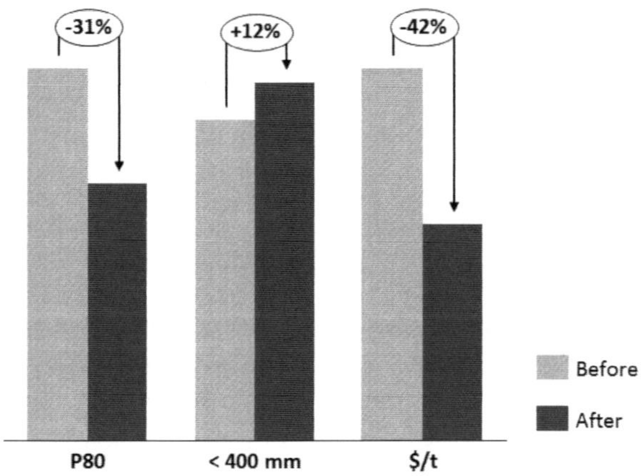

Figure 13. Before and after blasting optimization analysis

4 CONCLUSIONS

The definition of blast design parameters is essential to achieve the desired results in the rock blast process. By changing only one parameter, the blast result can be very different and it was confirmed as expected analysing how the P80 results are different after just one change. However, some parameters have more influence than others do.

It was also noticed that some changes in the blast design could improve the quality and fragmentation compliance and still obtain significant cost reduction for explosives and accessories.

The application of emulsion showed good fragmentation results for hardest rocks as the compact silexite, but the elevated emulsion cost when compared with ANFO and the low demand for the detonation of this litothype (less than 20%) are considered barrier for its implementation. The explosive ANFO presented best fragmentation results for most of the rocks, and it is justified by the lower cohesive force and tenacity of the rock massifs, that is, the percentage of passing fragments and consequently the percentage of material that directly passes through the grizzle that feeds the crushing plant.

However, the comparison between which types of explosive to be used for rock blasts on Chapadão Mine was favourable to ANFO utilization, as it resulted in a balance between costs and fragmentation.

The proposed change of increasing the stemming size presented expressive fragmentation gains and is viewed as very simple operational strategy for reducing the explosive consumption by hole and at the same time improves the explosive energy confinement. The operation adopted the stemming size length of 2.00 m.

By the end, it was noticed that the drilling grid expansion in the tests of the third step had as a result the fragmentation lower than 400 mm besides to reduce significantly the explosive and accessories consumption. The methodology applied was proved very efficient for aleatory changes based on field experiences and systematic assessment using the presented tools.

ACKNOWLEDGEMENTS

The authors thank financial support from the Brazilian agencies CNPq, CAPES, FAPEG and FUNAPE. In addition, we like to thank Copebras for the allowing this research and Federal University of Goiás.

REFERENCES

KRASNOVA, N.I., BALAGANSKAYA, E.G., GARCIA, D., 2004. Kovdor – classic phoscorites and carbonatites. In: Wall, F. e ZAITSEV, A.N. (eds), Phoscorites and carbonatites from mantle to mine: the key example of the Kola alkaline province. *Mineralogical Society*. pp. 99-132.

GIBSON, S.A., THOMPSON, R.N., LEONARDOS, O.H., DICKIN, A.P., MITCHELL, J.G., 1995. The Late Cretaceous impact of the Trindade mantle plume - evidence from large-volume, mafic, potassic magmatism in SE Brazil., 1995. *Journal Petrol* 36, pp. 189-229.

CLARK, G.B., Industrial High Explsoives: Composition and Claculations for Engineers., 1980. *Colorado School of Mines,*Golden, v.76, pp. 11-16.

JIMENO, C.L., JIMENO., E.L., BERMÚDEZ, P.G. Manual de Perfuracion y Voladura de Rocas, 2003. *Instituto Tecnológico Geominero de Espanã,* Madri, 88 p.

STIEHR, J.F. Blaster's Handbook 18th., 2011. *International Society of Explosives Engineers,* Cleveland,. pp. 76-77.

CAVADAS, P.M.M., Optimização do Desmonte numa Mina a céu aberto com aplicação de Air Decks, 2012. *Faculdade de Engenharia Universidade do Porto,* Porto, pp. 70-83.

ORICA. Technical Data Sheet, 2010. *Orica Mining Service*, Melbourne. pp 1-2.

SPLIT. Manual de uso do software. 2011. *Split Engineering, LLC.* v.2, pp. 30-88.

Preconditioning Effect of Blasting and Its Role in Mineral Extraction Chain

A. Khademian*, R. Bagherpour, S. N. Almasi, S. Jamali

Department of Mining Engineering, Isfahan University of Technology, Isfahan, Iran

ABSTRACT Operations chain of mineral supply in a mine starts with blasting in most cases. There are two important outcomes in operational blasting. The first is the size distribution of fragments which is referred as the seen result and the second outcome is related to the inner damage exerted on the fragments which is known as the unseen effect of blasting. In fact, blasting can reduce the rock fragments strength by generating microcracks inside them. This reduction in strength prepares the materials further for the downstream mining processes. This phenomenon is known as the preconditioning effect of blasting. The aim of this study is to show the influence of this phenomenon on different mining processes, including crushing, grinding and leaching processes. For this purpose, the impact of preconditioning on fragments strength and their microcrack content have been analyzed. Then, the role of preconditioning in crushing and grinding process have been demonstrated. At the end, the potential advantage of this phenomenon in leaching operation performance has been discussed. Results show that, involvement on this phenomenon in optimization approaches could bring many economic benefits for the mining complex.

Keywords: Blasting, Preconditioning, Crushing, Grinding, Microcrack, Leaching

1 INTRODUCTION

Operations chain of mineral supply in a mine starts with blasting in most cases. On the other hands, the objective of operations sequences of blasting, crushing and grinding is to bring a certain degree of liberation to the minerals, while blasting is the most efficient size reduction activity in terms of energy consumption compared to other two activities. Blasting can achieve an energy efficiency between 15% and 30%, in contrast to grinding which is approximately 2% efficient (Eloranta 1997). From the environmental and economic point of view, a policy of using energy where it is most efficient is essential. So, it is preferred to shift the size reduction activities to the blasting stage as much as possible. Therefore, blasting optimization is a prerequisite for optimization of the whole mining complex.

Blasting operation is mainly carried out for breaking rock masses into small fragments. There are two important outcomes in operational blasting. The first is the rock fragmentation and the size distribution of fragments which is referred as the *seen* result of blasting and it is usually evaluated qualitatively by visual inspection. The second outcome is related to the inner damage exerted on the fragments which is known as the *unseen* effect of blasting (Workman and Eloranta 2003). In fact, blasting can reduce the rock fragments strength by generating microcracks inside them. This reduction in strength, facilitates the crushing and grinding of the materials and in other words prepares the materials further for the comminution processes. This phenomenon is known as the conditioning or preconditioning effect of blasting. Fragment conditioning by blasting is defined in literature as the alteration of physical properties of the rock after blasting due to the extent of pre-existent cracks and the creation of new fractures (Parra et al. 2015).

Over the years, only the seen result of blasting has been attributed to the subsequent mining processes. Preconditioning effect of blasting came to researcher's attention when several observations showed some changes in grindability of minerals as a result of alteration of blasting parameters. There have been many studies on this phenomenon during last three decades. These studies sometimes succeeded to isolate the impact of microcracks and sometimes they studied the coupled impacts of seen and unseen results of blasting. JKMRC institute, Queen's University, Lafarge institute and NTNU are the pioneers in this field (Khademian and Bagherpour 2017). All of these researches showed that, blast induced fragment conditioning influences many downstream processes in a mining complex. This study aims to characterize the exact effect of blasting parameters such as spacing, stemming, delay timing, on the degree of fragments conditioning and their influence on different subsequent mining and processing operations. This work is accomplished in the following sections.

2 INFLUENCE ON MICROCRACKS NETWORK

Upon detonation of explosive in blasthole, two types of loading are applied on rock body. First, a shock wave loading, that travels outward from the blast hole, followed by a longer duration gas pressurization loading (Kutter and Fairhurst 1971). Olsson et al. (2002) noted that gas loading had no significant influence on the generation of Microfractures and the shock wave is the primary responsible for microfracturing. Shock waves can be classified into two types of P compressive wave and S shear wave. At low levels of stress, The P compressive wave (primary wave) opens existing cracks and propagates them. As the stress level exceeds the dynamic tensile strength of intact rock, two sets of new microcracks will be generated. One set is created by tangential tensile strength at the front of compressive wave and the other set is due to tensile tail of the compressive wave. Tangential stresses generate radial cracks, while the tensile tail generates some concentric cracks around the blasthole (Figure 1) (Thum and Hettwer 1978). The primary P wave is followed by a slower S shear wave, which is developed by blastholes deformation. The S-wave can enlarge existing microcracks formed by p-wave or generate new microcracks (Nielsen and Malvik 1999). The total surface of microcracks in rock mass may be 10 to 100 times greater than all rock fragments surfaces resulting from blasting (Revnivtsev 1988).

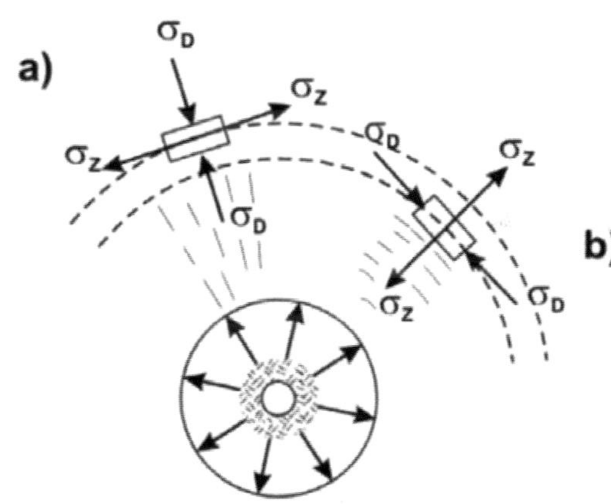

Figure 1. (a) Development of radial microcracks by tangential tensile stress at the front of P-wave, (b) concentric cracks generated by tensile tail at the back of the wave (Nielsen and Malvik 1999).

Although all of the preconditioning effect are consequences of microcracks network alteration, there are not detailed studies that focus on the change of microcrack network parameters of rock fragments. Most of researches in this area are accomplished by Nielsen. Nielsen and Malvik (1999) focused their work on the induced changes in the microcrack content of fragments. By changing the number of detonating cords used as the main charge, they investigated the effect of higher powder factor on the microcrack content of two kinds of taconite and one ilmenite samples. For these purpose, they measured the number of each type of microcracks by fluorescent microscopic examination. Results showed that the increase in blasting applied energy leads to increasing the density of all types of microcracks. The results on their study are summarized in Table 1.

Table 1. Quantitative assessment of the crack density in thin sections of taconite (Nielsen and Malvik 1999)

Sample	Small crack	Boundary cracks	Grain cracks
Taconite 1	98	97	45
Taconite 2	146	130	72
US taconite 1	55	48	15
US taconite 2	73	116	23
US taconite 3	84	103	35
Ilmenite ore 1	40	9	31
Ilmenite ore 2	74	15	48
Ilmenite ore 3	97	18	80

HAMDI et al. (2002) adopted a different approach to investigate the microcrack content after a blast. They separated the crack porosity from pore porosity by proposing a new method based on continuity index (CI). Sampling from three production blasts, they compared the continuity index and crack porosity of different samples according to their distance to blasthole. Results showed that the crack porosity is highly correlated to the proximity to the blasthole, while it has a significant negative correlation with continuity index (Figure 2). Muñoz et al. (2007) also showed that microfractures type and quantity has a direct relationship with proximity to blasthole in granite. Their findings showed that boundary microcracks are the dominant microfracture produced by production blasts, while intragranular and intergranular microcracks are produced around the pre-splitting blastholes.

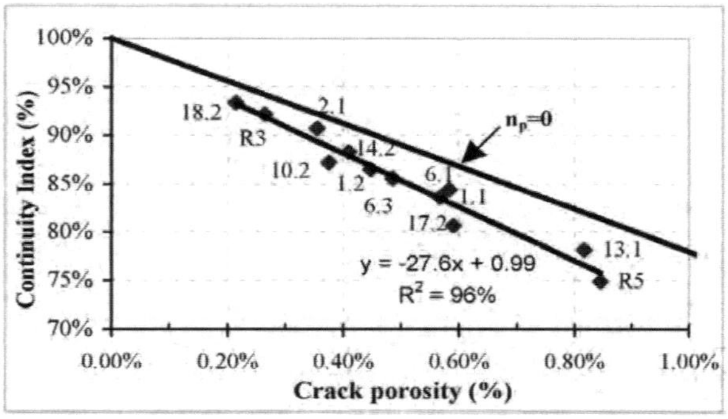

Figure 2. Correlation between crack porosity and continuity index (HAMDI et al. 2002).

3 INFLUENCE ON FRAGMENTS STRENGTH

As mentioned before, application of explosive energy generates some new microcracks inside the rock body and propagates the existing ones. This implies that some degrees of internal damage occur to the rock and the residual strength of resulting rock fragments is not the same as intact rock. In mining operations, it is desirable to reduce the strength of rock fragments as much as possible because it enhances the mineral handling in various mining stages. This work may be accomplished by changing the blasting operation's parameters (such as powder factor, stemming, etc.) in the desired direction. For this purpose, it is required to characterize how different parameters of blasting affect the strength of rock fragments.

Kim and McCarter (1998) studied the effect of blasting shock waves on some strength characteristics of rocks by using a explosively driven split Hopkinson pressure bar. Their results indicated that p-wave velocity of rocks is significantly influenced by blasting while their UCS is less affected. Also, using experimental blastings on 12 granodiorite blocks, Katsabanis et al. (2003a) evaluated the damage level of rock in various distances from the blastholes by measuring the P-wave velocity. Results showed that the damage level lay between 0.35-0.5 close to the blast-holes and then decreased radially outward. Point load tests also confirmed these results. Besides, this work showed that damage level is also affected by the delay timing. In addition, the impact of changes in the applied energy of blasting on mechanical characteristics of rock fragments has been investigated and it turned out that shifting the applied energy, decreases the elastic and strength characteristics of the fragments (Katsabanis et al. 2003b). Michaux and Djordjevic (2005) conducted some experimental blasting using three types of explosive, ANFO, Riogel and PETN. They examined the strength of rock fragments by point load index experiment (Table 2) and concluded that increasing the applied energy in blasting, leads to a significant reduction in point load index of granite rock fragments. Point load index of different samples are illustrated in Figure 3. In one of the latest studies, Catalan and Onederra (2016) investigated the influence of blast holes pattern on the degree of preconditioning in an underground mine and concluded that "*ultra blast preconditioning*" generates the highest degrees of damages to the extracted rocks.

Table 2. Point load index results (Michaux and Djordjevic 2005)

Sample	Average	Std dev
Unstressed	6.54	2.39
Black Powder 1	6.43	1.54
Black Powder 2	5.72	1.93
Riogel 1	5.48	1.89
Riogel 2	4.59	2.16
PETN 1	3.52	1.37
PETN 2	3.59	2.04

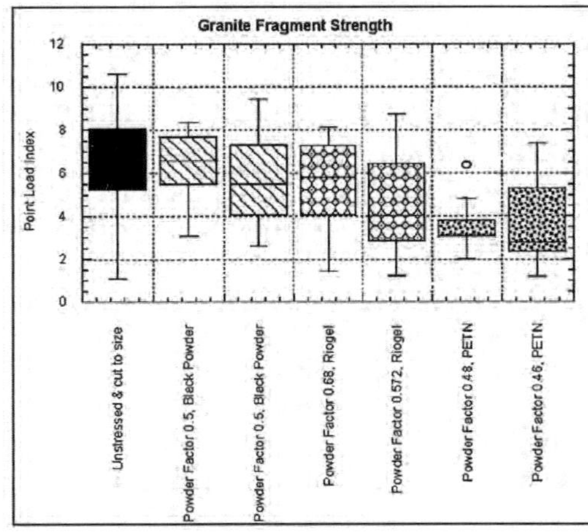

Figure 3. point load index of rock fragments

4 INFLUENCE ON MINERAL PROCESSING

It has been proven that the preconditioning effects of blasting carry through to mineral processing stages too. These effect are especially noticed in two sections of comminution and separation (leaching). The following sections discuss how the blasting may enhance the mineral processing performance in the two sections.

4.1 Influence on Comminution Process

As mentioned previously, blasting operation generates fractures in the rock fragments which leads to a decrease in their strength and thus decreases their resistance against impact and abrasion factors which are the two key elements in comminution processes. In this way, for a unit amount of minerals, the required comminution energy for achieving a desired size distribution will decrease. In other words, from operational point of view, this phenomenon increases the throughput of comminution circuits in unit time. This phenomenon occurs by altering the microcrack content of rock fragments. crushing mainly takes place by propagation of macrofractures inside the ore, while microcracks play the main role in grinding stages. Therefore, it is generally considered that grinding process is more affected than crushing by this phenomenon. However, the degree to which this happens strongly depends on the survivability of microcracks during the whole mining processes, from blasting to grinding.

There is evidence that Bond work index is significantly reduced by heavier blasting. Nielsen and Kristiansen (1996) assessed the impact of changing of the blasting parameters on the work index and mill throughput. By conducting experimental blasting on taconite, gabbro, quartz diorite and monzonite rock samples, they found out that the work index of fragments is highly influenced by any changes in blasting parameters. For example, shifting the powder factor to its double amount, decreased the taconite work index from 6.7 to 3.9 kWh/ton, while the it is 14.4 kWh/ton for the unblasted ore (Figure 4). Subsequent studies showed that increasing the applied energy in blasting may enhance the mill throughputs too, however this effects vanishes gradually for materials with p80 under 200 to 300 micrometers (Nielsen and Lownds 1997; Nielsen and Malvik 1999). Based on these results, Workman and Eloranta (2003) showed that increasing the powder factor to 0.45 kg/ton in taconite blasting, decreases the required work input and total costs by 30% and 26% respectively.

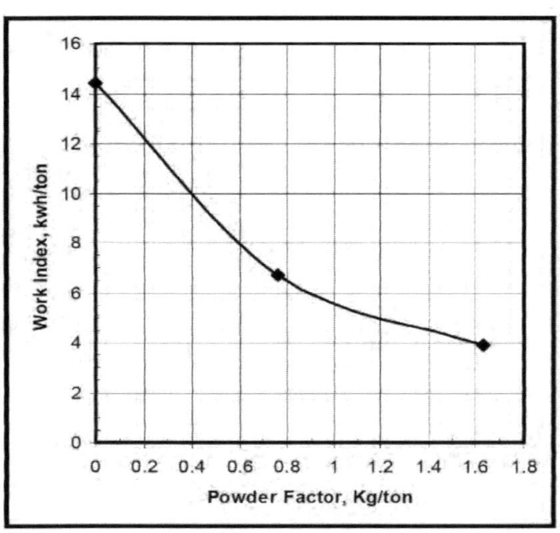

Figure 4. Relationship between powder factor and Bond work index (Workman and Eloranta 2003)

Also, Michaux and Djordjevic (2005) studied the influence of increasing the applied energy in blasting on the comminution circuit. Simulating a SAG mill, they investigated the performance of four batches of rock fragments which were blasted with different energies while they had the same size distribution. In this way, all sample had the same feed size. The results showed that increasing the blasting applied energy can enhance the mill throughput up to 21%. Detailed results of this study are summarized in Table 3 and Figure 5.

Table 3. Explosive energy applied and SAG mill performance (Michaux and Djordjevic 2005)

Explosive	Energy applied in blast (kJ/kg)	Throughput (tph)
Black Powder	1.593	1465
Riogel	1.995	1616
PETN	2.704	1873

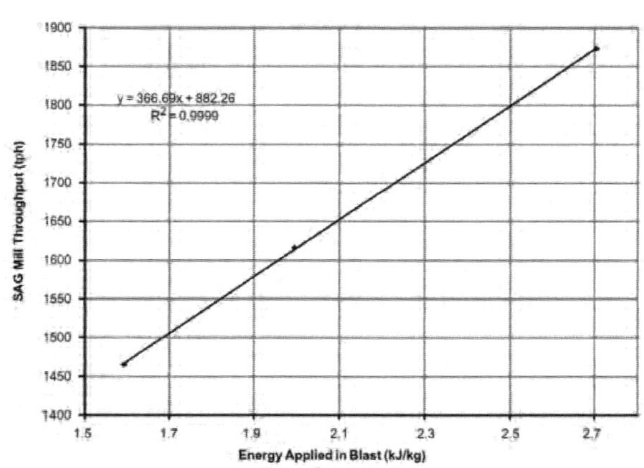

Figure 5. SAG mill throughput and explosive energy applied (Michaux and Djordjevic 2005).

4.2 Influence on Leaching Process

Recent studies in the field of fragment conditioning are focused on the impact of preconditioning effect on mineral recovery in the leaching process. Generating the blasting induced microcracks and porosities inside the rock fragments, may increase the chance of exposing the ore to extracting solvents in leaching processes. Fribla (2006) investigated the effect of increasing the powder factor on microcracks generation and then its subsequent effect on leaching process recovery of 3 different types of oxidized copper ore. Results showed that a higher powder factor in all cases leads to higher percent copper recovery. Fribla concludes that the higher recovery is achieved by increasing the contact surfaces between the leaching solution and the porous solid, which is an outcome of higher microfracturing. For example,

Figure 6 shows the kinetics of leaching process for B-1 and B-2 sample, while B-1 and B-2 are the same copper ore blasted with 2 and 4 grams of PETN respectively.

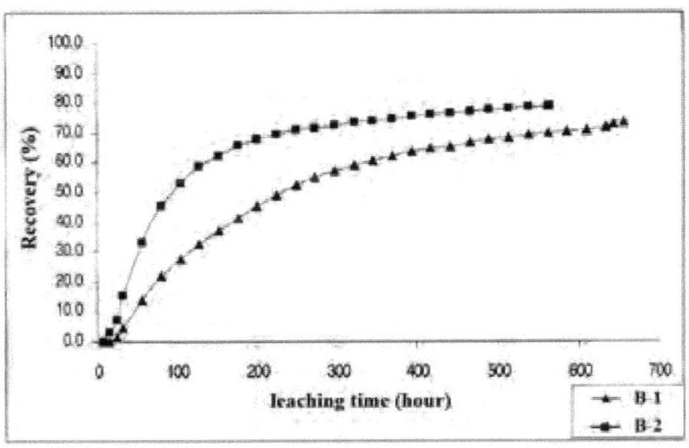

Figure 6. Kinetics of leaching process for B-1 and B-2 samples (Fribla 2006).

In another study, Parra et al. (2015) studied the effect of blasting induced preconditioning on the Cu recovery of a leachable synthetic material. Implementing a zoning approach, they divided the whole blasting media into several conditioning zones. The results of leaching tests showed that the ore recovery in low preconditioning zones is almost the same as unblasted materials, while it rises significantly in blasthole surroundings area as much as 39%.

5 CONCLUSION

The purpose of this study was to characterize the preconditioning effect of blasting on different mining stages and processes. For this purpose, all important finding in this field has been discussed in different sections and the following results are concluded:

- Blast-induced pre-conditioning is a real phenomenon, which can bring many benefits to mining operation and mineral processing.
- Blasting operation can influence the microcrack content of rock fragments. Heavier blasting leads to higher density on all types of microcracks in rock materials.
- Application of explosive energy reduces the residual strength of the resulting fragments. The internal damage is highly correlated with the level of applied energy in blasting.
- An increase in explosive energy applied corresponds to a decrease in Bond work index of rock materials and an increase in mill throughput.
- Blasting operation enhances the recovery of ore in leaching process, which is a result of increased exposing of ore to extracting solvent. Applying higher powder factor in blast leads to higher recovery.
- Preconditioning effect of blasting should be noticed and involved in all optimization approaches in all mining complexes.

REFERENCES

Catalan, A., and Onederra, I. *Modelling of preconditioning by blasting in block and panel caving*. 2016, *Mining Technology*, pp. 1-18.

Eloranta, J., The efficiency of blasting versus crushing and grinding. presented at the *Proceedings of the 23rd Conference of Explosives and Blasting Technique. Las Vegas, Nevada. February*, pp. 1-7.

Fribla, H. M. (2006). "Investigación de la relación entre el microfracturamiento producido por el explosivo y la disolución del metal en una mena oxidada de cobre." Minas.

Hamdi, E., du Mouza, J., and Alain, F. J., Influence of Rock Mass Structure on Blast Efficiency. presented at the *Proceedings of the Seventh International Symposium on Rock Fragmentation by Blasting*.

Katsabanis, P., Kunzel, G., Pelley, C., and Kelebek, S., Damage development in small blocks. presented at the *Proceedings of the annual conference on explosives and blasting technique, ISEE*; 1999, pp. 363-374.

Katsabanis, P., Gregersen, S., Pelley, C., and Kelebek, S., Small scale study of damage due to blasting and implications on crushing and grinding. presented at the *Proceedings of the annual conference on explosives and blasting technique, ISEE*; 1999, pp. 353-362.

Khademian, A., and Bagherpour, R. *Environmentally sustainable mining through proper selection of explosives in blasting operation.* 2017, *Environmental Earth Sciences*, 76(4), pp. 166.

Kim, S. D., and McCarter, K. M. *Quantitative Assessment of Extrinsic Damage in Rock Materials.* 1998, *Rock Mechanics and Rock Engineering*, 31(1), pp. 43-62.

Kutter, H., and Fairhurst, C., On the fracture process in blasting. presented at the *International Journal of Rock Mechanics and Mining Sciences & Geomechanics Abstracts*, Elsevier, pp. 181IN1189-1188IN10202.

Michaux, S., and Djordjevic, N. *Influence of explosive energy on the strength of the rock fragments and SAG mill throughput.* 2005, *Minerals engineering*, 18(4), pp. 439-448.

Muñoz, C., Guerra, N., and Mancilla, P. *Caracterizacion del daño inducido por voladuras en la matriz de macizos rocosos.* 2007, *VII Jornadas de Tronadura, Asociacion Chilena de Ingenieros Explosivistas, Puerto Varas.*

Nielsen, K., and Kristiansen, J., Blasting-crushing-grinding: Optimisation of an integrated comminution system. presented at the *Proc 5th Int Symp on Rock Fragmentation by Blasting, B Mohanty ed*, pp. 269-277.

Nielsen, K., and Lownds, C. M. *Enhancement of taconite crushing and grinding through primary blasting.* 1997, *International Journal of Rock Mechanics and Mining Sciences*, 34(3), pp. 226. e221-226. e214.

Nielsen, K., and Malvik, T. *Grindability enhancement by blast-induced microcracks.* 1999, *Powder Technology*, 105(1–3), pp. 52-56.

Olsson, M., Nie, S., Bergqvist, I., and Ouchterlony, F. *What causes cracks in rock blasting?* 2002, *Fragblast*, 6(2), pp. 221-233.

Parra, H., Onederra, I., Michaux, S., Kuhar, L., McFarlane, A., and Chapman, N. *A study of the impact of blast induced conditioning on leaching performance.* 2015, *Minerals Engineering*, 74, pp. 1-12.

Revnivtsev, V., We really need revolution in comminution. presented at the *Proceedings of XVI International Minerals Processing Congress*, pp. 93-114.

Thum, W., and Hettwer, A., 1978. *Sprengtechnik im Steinbruch- und Baubetrieb*, Bauverlag, Wiesbaden.

Workman, L., and Eloranta, J., The effects of blasting on crushingand grinding efficiency and energy consumption. presented at the *29th 'Explosives and blasting technique conference,*, International Society of Explosives Engineers.

Air Flow Pattern Analysis of Line Brattice Ventilation System Using CFD

T. Feroze[1] and B. Genc[2]

University of the Witwatersrand, Johannesburg, South Africa

ABSTRACT Line brattice ventilation system in underground coal mines is used to control the quantity, movement, and direction of air inside the development headings. The quality of the ventilation i.e. the quantity of air and the air flow pattern provided by the line brattice ventilation system varies with the variation in the associated system drivers. It is important to monitor the quality of ventilation to avoid build-up of methane and accidents. However, it is possible to measure the quantity of ventilation using sensors, but it is not possible to study the complete air flow patterns inside a heading experimentally. The study of the air flow pattern is required to identify potential regions of low air flow and recirculation for monitoring of methane accumulation using correctly positioned sensors. A validated Computational Fluid Dynamics model has been used to analyse the air flow pattern inside empty headings of different dimensions, ventilated using various settings of the line brattice and three Last Through Road air velocities. The study has identified regions of low air flow (regions of potential methane build-up) which should be monitored continuously to avoid build of methane to dangerous levels.

1 INTRODUCTION

The ventilation of the development headings in underground coal mines is one the most important area of ventilation, Hartman et al. (2012). It is considered to be the most difficult to achieve, Bise (2013) because of the complexity of the airflows. Insufficient and improper ventilation of the development heading can often cause quick build-up of methane and increase the probability of accidents. Therefore, it is vital to obtain thorough knowledge about the effect of the heading ventilation system upon airflow patterns at the face. The ventilation of these development headings is carried out using auxiliary ventilation systems which generally consists of Line brattice (LB), ducted fans, and Jet fans.

LB ventilation system is the cheapest and quickest method to provide ventilation to the development headings. It is manufactured of plastic sheeting with or without fabric reinforcement, Hartman et al. (2012). It is hung from the roof of the heading to direct the air moving in the Last Through Road (LTR) into the heading. The quality of the ventilation provided by an auxiliary ventilation system is generally quantified in terms of the quantity delivered, movement, and the direction of the air inside the heading. The quality of LB ventilation system is affected by the variation in the system variables like heading dimension, LB settings and the velocity of the air in the LTR.

The quantity of air delivered by the LB can be measured at the exit of the LB and at the face of the heading using different types of anemometers. Tracer gas and smoke tube has also been used for this purpose,

McPherson (2003) and Thimons and Kissel (1976). Conventionally, smoke tube and tracer gas has been used in the mining industry to study airflow pattern, Luxner (1969), and leakage through the seals at the stoppings, Singh et al. (2004). Tracer gas has also been used to measure the airflow pattern along the longwall faces as well Krog et al. (2014). Ndenguma et al. (2014) used smoke to investigate the air flow patterns in a scaled down model of a mining section. The study of the air flow patterns inside the heading is however, much complex to be studied comprehensively through experiments. Since each auxiliary ventilation system is dependent on a number of system drivers, it becomes an even bigger challenge to carryout large number of experiments.

Computational Fluid Dynamics (CFD) has been used in the mining industry as an alternate solution to study the various aspects of ventilation. CFD requires only a few experiments to validate the numerical model. A number of studies using CFD have been conducted for underground coal mines, to study the airflows around the face area. Wala et al. studied the face ventilation in the presence of a continuous miner, Wala et al. (2003) and (2007). Wang et al. (2011) studied the air flow and dust-isolation at a fully mechanized working face. Zhang et al. (2011) studied the air flow at the face of the heading to find the suitability of a push pull ventilation system. Ranjan et al. (2014) used two dimensional CFD model to examine the air flow pattern in a bord and pillar mine.

The current study using a validated CFD model in ANSYS Fluent software package is looking at the air flow patterns inside empty headings ventilated using LB. This is done by varying some of the system drivers associated with the LB ventilation system, that is, LB settings (LB length in the LTR, LB angle in the LTR, LB length inside the heading, LB distance from the wall of the heading), heading dimensions (heading height and depth) and LTR velocity.

2 METHODOLOGY

The effect of the system drivers associated with LB ventilation system on the ventilation of empty heading as given below and shown in Figure 1 were studied. The cases generated by all the possible combination of these variables were studied to identify the effect of each variable. The same cases were used to study the general air flow pattern inside the empty heading, to recognize the distinct features of these air flow patterns.

- Distance of the LB from the wall in the heading (b) = 0.5 and 1m
- Length of the LB in the LTR (c) = 3 and 6m
- Length of the LB in the Heading (d) = 5, 7.5, 10, and 15m
- Velocity of air in the LTR (v) = 1, 1.5, and 2m/s
- Heading Depth (HD) = 10 and 20m
- Heading Height (HH) = 3 and 4m
- Heading Width (HW) = 6.6m
- Angle of LB in the LTR (θLB) = 0, 7.5, and 15.□

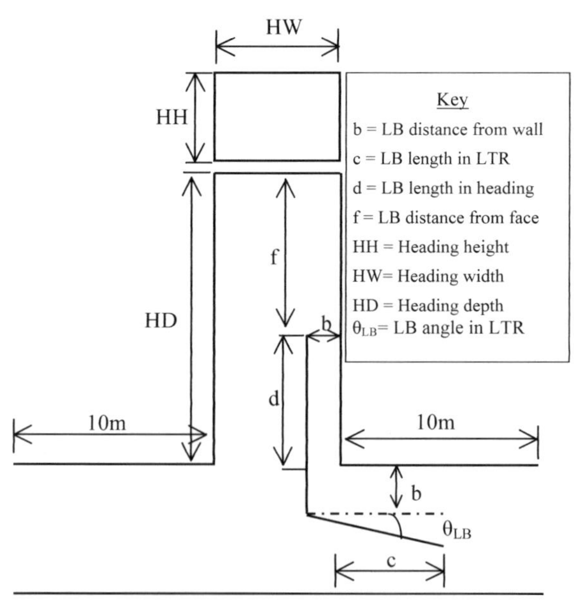

Figure 1 Parameters Varied for the Study

3 CONSIDERATIONS FOR NUMERICAL MODELING

3.1 Modeling and Meshing

ANSYS Fluent Workbench was used for the modeling and meshing of the problem domain. A constant length of 10m was used for all the cases on the down and upstream side of the heading as shown in Figure 1. A hexahedral mesh was used with a mesh size of 0.04m for all the cases. The mesh size was chosen after carrying out a mesh independence test.

3.2 Boundary Conditions

Velocity inlet and outflow boundary conditions were used at the inlet and outlet of the LTR. Wall boundary condition was used for all the walls of the heading and LTR (ANSYS Fluent Theory Guide, 2015).

3.3 Turbulence Model

k-e realizable turbulence model with enhanced wall treatment was used in ANSYS Fluent. The solution was calculated using a second order scheme. A convergence criterion of the order of 10-5 was used. The turbulence model used was selected after conducting the various validation studies (Feroze and Phillips, 2015, and Feroze and Genc 2016).

4 RESULTS AND DISCUSSION

The air flow in the heading for all the cases followed a similar trend as far as the distinct air flow and low flow rate/recirculation regions are concerned. Low velocity regions were formed at the corners, centre of the heading and behind the angled LB. Seven such most prominent regions were formed as shown in Figure 2. These regions can become potential zones of methane accumulation in the LB ventilation system. The extent of these regions changed with the variation in the settings of the LB, i.e. the distance from the wall, the length of the LB, and the angle of the LB. The variation of the size of these regions with the change in these settings is briefly discussed here. It is important to highlight here that streamlines along with velocity vectors have been used in all the Figures given in this section to clearly show the general trends of the airflow.

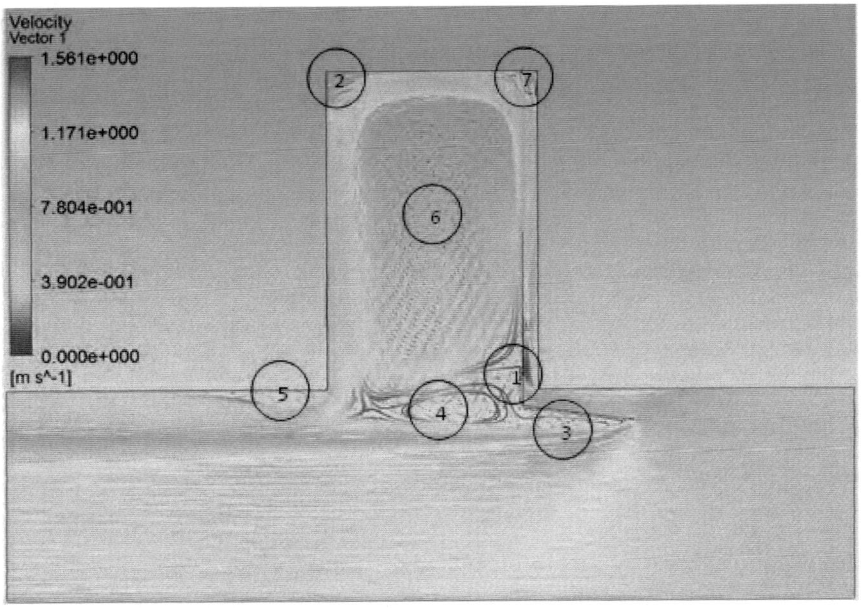

Figure 2 Typical Low Air Flow and Recirculation Zones of LB Ventilation System

4.1 Region 1

When the air turns into the heading in the LB-Wall channel then due to centrifugal force and flow separation it turns towards the LB. This results in higher air flow rate close to the LB than the wall of the heading as shown in Figure 3. Since the air closer to the LB (Figure 2) turns back without reaching the face. Therefore, when higher quantity of air (high velocity) is present close to the LB, higher percentage of air leaving the LB turns back without reaching the face of the heading. This can be avoided by using the smallest possible LB to face distance. Small recirculation regions can also be seen in Figure 3, but are not dangerous as far as the accumulation of methane is concerned.

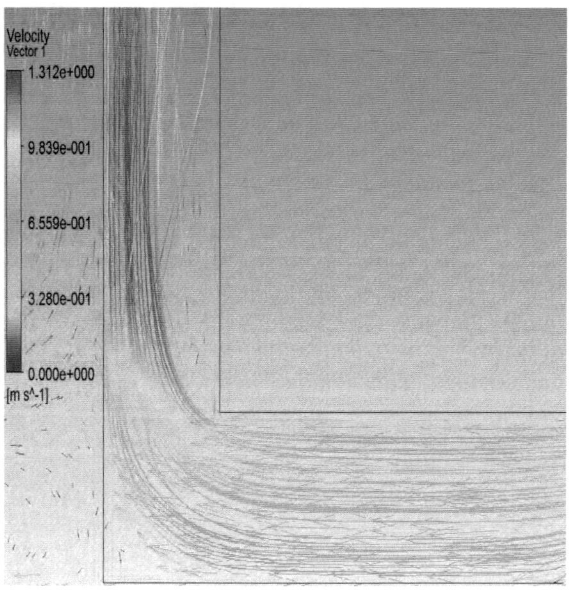

Figure 3 Typical air flow zone 1

4.2 Region 2

This region was formed due to the sharp corner at the left face of the heading and remained almost the same for all cases. This region for different cases can be seen in Figures 4 through 8. This can be a potential zone for accumulation of methane emitted from the face of the heading.

4.3 Region 3

A cavity was created behind the LB when it was used with an angle. The size of the region increased with the increase in the size of the LB inlet, due to the increase in angle of the LB as shown in Figures 6 through 8. This zone may not be dangerous since it is not close to face of the heading.

4.4 Region 4

The air after moving along the face of the heading turns, and moves parallel and close to the side wall (left) of the heading to join the main stream of air in the LTR. At the entrance of the heading, this air divides into three parts, the air close to the wall (which joins the air in the LTR), the air farthest from the wall (which turned counter clock wise to create recirculation), and the layer of air at the middle of these two streams. This region of recirculation is created by this middle stream of air, which fails to move straight or turn completely into the heading and slows down creating this low flow region. This zone joins zone 3 creating a bigger zone in the cases where the LB was used with an angle as shown in Figures 6 through 8. Methane may not accumulate here since it is away from the face of the heading.

4.5 Region 5

The air after moving along the face of the heading turns down to join the air moving downstream in the LTR. The airflow of the main air stream in the LTR is very less behind the LB for all the cases and is not sufficient to turn the returning air downstream. The returning air therefore, continues its downward movement until it meets the high velocity main stream, resulting in the formation of this zone (the presence of the sharp corner also contributed to this separation). When the LB was used with 1m distance from the face or used with an angle the region behind the LB got bigger, the main air stream in the LTR thus got shorter (in width) increasing the distance for the return air to join with it. Therefore, this zone became bigger and bigger with the increase in the entrance length of the LB as shown in Figures 4 through 8.

4.6 Region 6

The most prominent feature of the LB heading ventilation system is the presence of a low flow region (region 6) in the centre of the heading. The air after ventilating the face turns in the opposite direction to join the main stream in the LTR. It travels along the left wall, and a very little portion of this air farthest from the wall turns counter clock wise and moved up. This air flow was very low but prominent in all the cases as shown in Figures 5 through 9. The size of this zone varies with the LB-wall distance and LB-face distance as shown in Figures 5 through 9. The slow moving air in this region may be very dangerous and allow layering of tenacious gases in this region. This region should always be monitored for the presence of methane in headings ventilated using LB ventilation system.

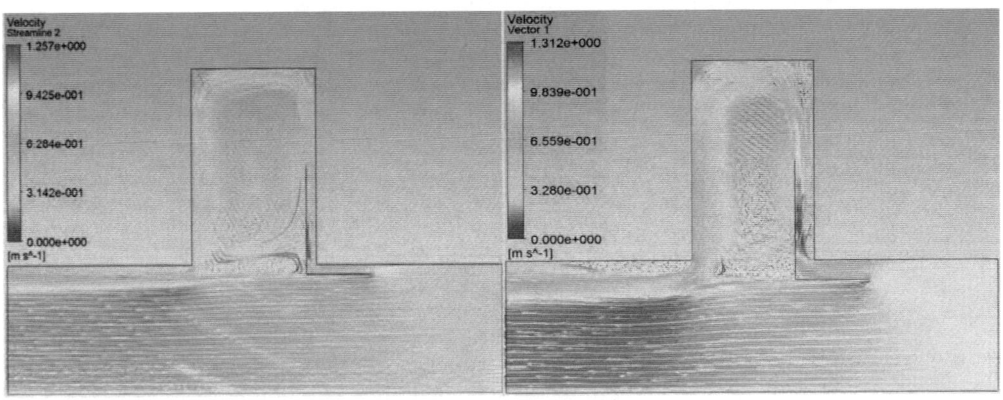

Figure 4 10m long headings using LB settings of; 0° angle, 5m length of LB in heading, 3m length in LTR, 0.5m and 1m LB to wall distance (left to right)

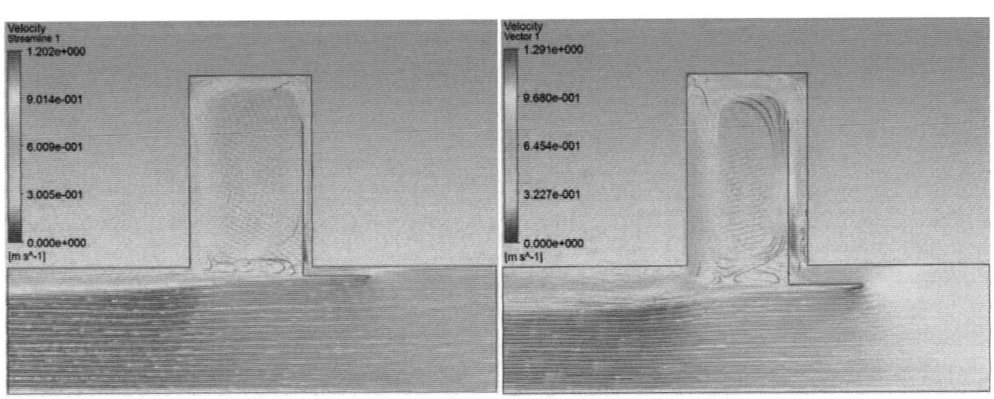

Figure 5 10m long headings using LB settings of; 0° angle, 7.5m length of LB in heading, 3m length in LTR,0.5m and 1m LB to wall distance (left to right)

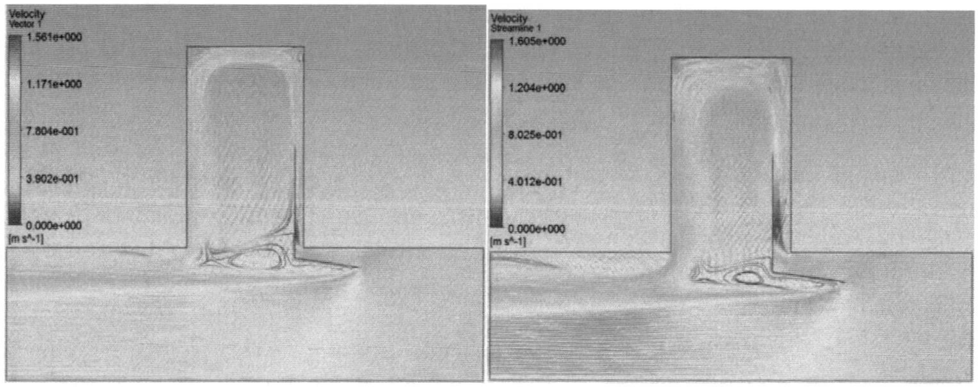

Figure 6 10m long headings using LB settings of; 7.5° angle, 5m length of LB in heading, 3m length in LTR, 0.5m and 1m LB to wall distance (left to right)

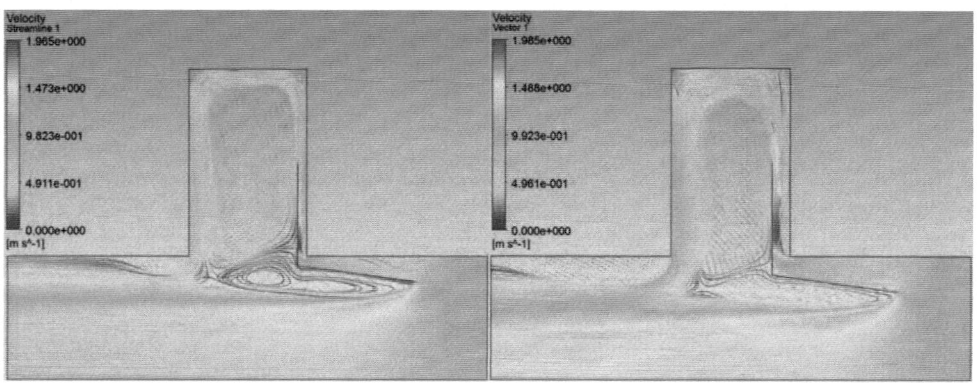

Figure 7 10m long headings using LB settings of; 7.5° angle, 5m length of LB in heading, 6m length in LTR, 0.5m and 1m LB to wall distance (left to right)

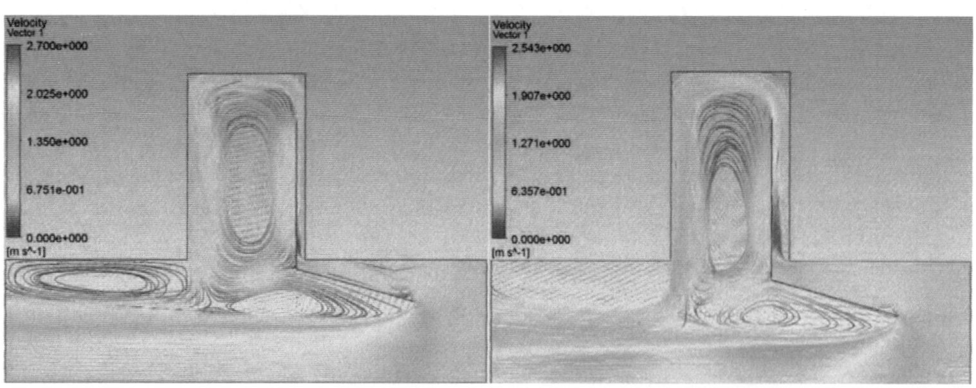

Figure 8 10m long headings using LB settings of; 15° angle, 7.5m length of LB in heading, 6m length in LTR, 0.5m and 1m LB to wall distance (left to right)

4.7 Region 7

Another recirculation region, although visible from the top view and appears similar to Region 2, but, is actually different. The difference could be seen from the side view of the heading by plotting the air flow velocity vectors on the vertical planes drawn at different distances from the right wall. The air exiting the LB hit the face, turns left on reaching the face, sweeps the face and moves (returns) parallel to the side wall of the heading, towards the LTR. However, a part of this air, close to the right wall, creates a recirculation zones as shown in Figure 9 (figure only shows a part of the plane after the exit of the LB). The air at the bottom of the heading rises and forms a counter clockwise recirculation zone close to the face, and the air at the top turns downward and forms a clockwise recirculation zone. The extent (length and width) of this region varies with the distance of the LB from the right wall and the distance of the LB exit from the face of the heading.

The more the distance of the LB from the wall of the heading and the distance of the LB exit from the face of the heading the longer this zone became as shown in Figure 9.

The zone became smaller and disappeared, away from the right wall of the heading (stayed approximately up till the extent of the LB), where all the air turned left, and swept the face of the heading as shown in Figure 10. Therefore, shorter the distance of the LB from the wall, and the shorter the distance of the LB exit from the face, the shorter is the length and width of this region. This is a potential zone for the accumulation of methane and should always be monitored.

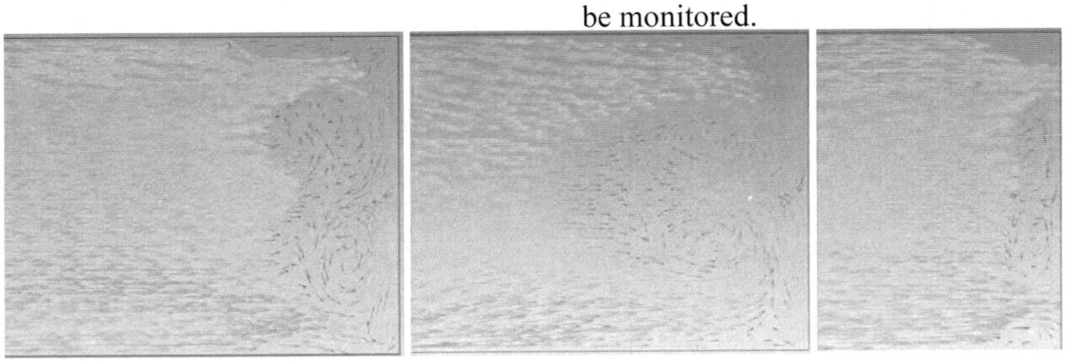

a) 5m long LB with LB to wall distance of 0.5m **b) 5m long LB with LB to wall distance of 1m** **c) 7.5m long LB with LB to wall distance of 0.5m**

Figure 9 Air flow vectors on a vertical plane drawn at a distance of 0.1m from the right wall for 10m long heading

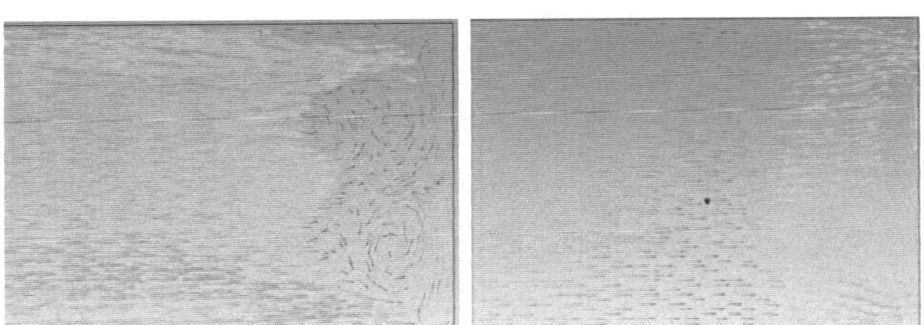

Figure 10 Air flow vectors on a vertical plane c the right wall for 10m long heading using a 5m

5 CONCLUSION

The results of a comprehensive study related to effects of the system variables associated with the LB ventilation system on the quality of ventilation were presented. It was shown that CFD is suitable to carry out the analysis of the air flow patterns of the LB ventilation system. The results showed the typical air flow pattern of the LB ventilation system and identified critical recirculation zones which can cause methane accumulation. The results may help the mining industry in the installation of the monitoring devices at the potentially dangerous areas of the heading ventilation using LB to ensure safer mining.

6 ACKNOWLEDGEMENT

The work presented in this paper is part of a Ph.D. research study in the School of Mining Engineering at the University of the Witwatersrand. The authors would like to acknowledge the Wits Mining Institute (WMI), University of the Witwatersrand, for making the Digital Mine facility available for the research, and the financial assistance required to purchase the high performance PC and the CFD software.

REFERENCES

Hartman, H. L., Mutmansky, J. M., Ramani, R. V., and Wang, Y. J, 2012. Mine ventilation and air conditioning. Third Edition. Wiley-Interscience Publication John Wiley and Sons, Inc.

Bise, C. J, 2013. Modern american coal mining: Methods and applications, Published by Society of Mining, Metallurgy and Exploration, Inc.

Mcpherson, M. J, 1993. Subsurface ventilation and environmental engineering. New York: Champan & Hall.

Feroze, T., Phillips, H. W, 2015. An initial investigation of room and pillar ventilation using CFD to investigate the effects of Last Through Road velocity. 24th International Mining Congress and Exhibition of Turkey (IMCET 15). Antalya, Turkey, (pp. 970-977).

ANSYS (R) Academic Research 2015. Release 15.0, Help System, ANSYS Fluent Theory Guide, ANSYS, Inc.

Feroze,T., and Genc, B, 2016. Ventilation of underground coal mines - A computational fluid dynamics study. Mine Ventilation Society of South Africa (MVSSA) Conference, Johannesburg, South Africa, (pp. 89-95)

Thimons, E.D,. and Kissell, F.N, 1974. Trace gas as an aid in mine ventilation analysis, Bureau of Mines Report of Investigation/1974.

Luxner, J. V, 1969. Face ventilation in underground bituminous coal mines. Airflow and methane distribution patterns in immediate face area - line brattice, Report of investigations 7223, US Department of Interior, Bureau of Mines.

Singh, A.K., Ahmad, I., Sahay, N., Varma, N.K., and Singh, V.K. 2004. Air leakage through underground ventilation stoppings and in situ assessment of air leakage characteristics of remote filled cement concrete plug by tracer gas technique. Journal of The South African Institute of Mining and Metallurgy, Volume 104, Issue 2, (pp.101-106).

Krog, R.B., Schatzel, S.J., and Dougherty, N.D, (2014). Methane emissions and airflow patterns along longwall faces and through bleeder ventilation systems. International Journal of Mining and Mineral Engineering (IJMME), Volume 54, (pp. 328-349).

Ndenguma, D.D., Dirker, J., and Burger, N.D.L, 2014. A computational fluid dynamics model for investigating air flow patterns in underground coal mine sections. Journal of the Southern African Institute of Mining and Metallurgy, Volume 114, Issue 6, (pp. 419-426).

Wala, A.M., Jacob, J., Huang, P.G., and Brown, J.T, 2003. A new approach to evaluate mine face ventilation. Journal of Society for Mining, Metallurgy and Exploration (SME), Volume 55, Issue 3, (pp.25-30).

Wala, A.M. Vytla, S., Taylor, C.D., and Huang, G, 2007. Mine face ventilation: A comparison of CFD results against benchmark experiments for the CFD code validation. Journal of Society for Mining, Metallurgy and Exploration (SME), Volume 59, Issue 10, (pp. 49-55).

Wang, P. F., Feng, T., Ronghua, L, 2011. Numerical simulation of dust distribution at a fully mechanized face under the isolation effect of air curtain. Journal of Mining Science and Technology, Volume 21, Issue 1, (pp. 65–69).

Zhang, X., Zhang, Y., and Tien, J. C, 2011. The Efficiency Study of the Push-Pull Ventilation System in Underground Mines. 11th Underground Coal Operators' Conference. University of Wollongong & Australasian Institute of Mining and Metallurgy, NSW, (pp. 225-230).

Ranjan, M., Kumar, K., and Ghosh, S.Kr, (2013). Mine ventilation in a bord and pillar mine using CFD. International Journal of Emerging Technology and Advanced Engineering (ICERTSD), Volume 3, Special Issue 3, (pp.389-393).

Critical Structures Assessment by Solving a Complex Ventilation Network

N. Ianc, D. Cioclea, C. Lupu, I. Gherghe, A. Matei, F. Radoi, C. Boanta

National Institute for Research and Development in Mine Safety and Protection to Explosion, INSEMEX Petrosani, Romania

ABSTRACT The primary protection to ensure the optimal work conditions in underground, is represented by the ventilation. The ventilation system consists of one or more main ventilation stations and the grafted ventilation network on the whole complex of related mine workings. Air circulation through a network of related underground mining works is a complex process that is part of an interdisciplinary field that requires knowledge of mathematics, physics, aerodynamics, geotechnical, mining technique, thermotechnics etc. Ventilation structures that induce a high degree of instability at the network level ventilation, are called critical ventilation structures and must be given special attention from those responsible for network management ventilation. In order to identify the critical ventilation structures at an underground mine, are used specialized programs such as 3D CANVENT and high end IT equipments for modeling, solving, optimization and simulation of the networks ventilation.

1 INTRODUCTION

The main measure for explosion protection is the achieving a proper ventilation at the underground mine workings.

During the course of the phenomenon of explosion due to dynamic wave energy important mechanical effects are produced both in the affected mining works and in the ventilation constructions.

An explosion directly affects the ventilation network by changing the functional parameters of the main fans. This causes a different post event natural distribution of the air flows at the branch level (Cioclea et al, 2013).

To achieve the required air flow in each branch, it is essential to solve the network ventilation in order to identify the optimal solutions at branch level. In practical terms, ensuring the necessary flow to working places in mining adjacent construction is performed using ventilation structures (Teodorescu et al, 1980).

Construction of ventilation structures is used for isolation, separation respectively directing the airflow ventilation circuits and sub-circuits. Both theoretically and practically any ventilation structure introduces in the network additional strength, which considerably increases the total resistance value of the mine (Patterson, 1992). In these circumstances, active fan of the main ventilation station must develop a depression proportional to the total flow required at the mine (Băltărețu and Teodorescu, 1971).

Under a complex network of ventilation, ventilation structures may occur that induces at the network level a degree of instability obliging the active fan to move the operating point from the stable to the unstable previous area - or the subsequent are of the point of maximum depression (Crăciunescu et al, 1993).

Ventilation structures that induce a high degree of instability at the network level ventilation, are called critical ventilation structures and must be given special attention by those responsible for network management ventilation (Gherghe, 2004). In order to identify the critical ventilation structures at an underground mine, are used specialized programs such as 3D CANVENT and high end IT equipments for modeling, solving, optimization and simulation of the networks ventilation (Cioclea, 2011).

CANVENT 3D software is a Windows application and has been designed to support the planning, design and analysis of mine ventilation systems.

As previously disclosed in other research works, the program CANVENT 3D uses the iterative procedure Hardy-Cross and by a number of iterations, the network ventilation is solved and balanced. After entering all mining works, ventilation network can be simulated (solved) and should be done only when the branches, nodes and the fans are all introduced.

After the program, has solved and balanced the ventilation network, any access and each image will automatically be updated by adding new current data.

After clearing and balancing the network can be visualized simulation results in tables or shown on network diagram Network Ventilation Diagram and the output data generated during the simulation process can be directly edited in 2D / 3D, either as input or output data.

For both input and output data can be specified an order of editing for nodes, branches and fans.

To identify the critical ventilation structures is used the simulation process successive at the level of every ventilation structure.

In a simulation by applying successive gradient strength is observed that there is no influence on the operating mode for the active fan in the sense of not changing the operating point in this case the ventilation structures is considered normal.

Further, it is considered the resistance R, as the strength of the ventilation structure obtained from the measurements, respectively, after the optimization of the ventilation network, apply a gradient resistance ± X, thereby obtaining a new value of the resistance Ry, Ry = R ± Rx.

After the successive simulation and resistance gradient is applied, it is found that there is little influence on the operating mode, for the active fan, so as to modify the operating point in the range specified for the stable regime, the ventilation structure is considered harmless, however regular measures for its control it will be taken.

After the successive simulation and resistance gradient is applied, it is found that there is a major change on the operating mode for the fan active, meaning the movement of operating point outside the range specified for the stable regime, in the range specified to the unstable regime but subsequent to the point of the maximum depression, then the ventilation structure is considered dangerous and measures should be taken to its permanent monitoring.

After the successive simulation and resistance gradient is applied, it is found that there is a radical change on the operating mode for the active fan, meaning the movement of the operating point to the unstable regime, but prior to the maximum depression point that includes pumping, the ventilation structure is considered critical and will be taken measures to eliminate its criticality.

The danger generated by critical ventilation structures is given by the fact that following the subjective modifications of the relating aerodynamic parameters, induce an extremely high degree of instability at the ventilation network level, receptively shocks on dynamic motor unit-fan, which can lead untimely removal from service of the fan.

2 CASE STUDY

For solving the ventilation network of the Uricani Mine, in the case of eliminating the critical character of the ventilation structures is used the program 3D CANVENT.

Also for this purpose is used the ventilation network of Uricani Mine modeled solved and optimized in the previous stage of the research.

To eliminate the criticality of ventilation structures of the Uricani Mine were applied 2 methods, namely:

- Method of reducing the total resistance of the critical ventilation structure;
- Method of dispersion the total resistance of the critical ventilation structure.

2.1 Applying the Method of Reducing the Total Resistance of the Critical Ventilation Structure

For this purpose, was the basic ventilation network as follows:

- They were introduced five additional nodes 350; 351; 352; 353; 354;
- Were removed two existing branches 352-333; 254-255;
- They were introduced 10 new ramifications 324-350; 350-325; 332-352; 352-333; 332-351; 351-352; 254-354; 354-255; 254-353; 353 -354;
- Were removed two ventilation buildings on ramifications 352-333; 254-255;
- They were introduced five equivalent ventilation structures on the ramifications 324-350; 332-352; 332-351; 254-354; 254-353.

Network modeled resolved, optimized and simulated is shown in Fig. 1.

Compared with the situation of optimized ventilation network in the previous stage, there were obtained the following results:

- On the evacuation circuit of the polluted air related to the V S bloc, airflow at the 500-level increased from 34.83 m^3 / s to 35.39 m^3 / s (branch 207-208)
- On the evacuation circuit of the polluted air related to the III N bloc (branch 302-308) presented an increase from 13.03 m^3 / s to 15.03 m^3 / s
- At the level of the mine (ramifications 343-345 259-348 respectively) flows showed an increase from 86.34 m^3 / s to 94.40 m^3 / s on the ventilation shafts E, respectively W.
- At the main ventilation stations (345-346 respectively ramifications 348-349) flows showed an increase from 88.65 m^3 / s to 96.57 m^3 / s on the ventilation shafts E, respectively W.

From the analysis of the functional parameters of active fans before and after applying the method to a single critical ventilation structure are the following:

- Applying the method to reduce the overall resistance of the critical ventilation structure by creating parallel links to the branches level on which are located relevant critical ventilation structures, had a major impact on the active fan located at the main West ventilation station. Thus, the total resistance of the network of ventilation dropped from 0.755 NS^2 / m^8 to 0.745 NS^2 / m^8, the negative pressure developed by the fan dropped from 1469.86 to 1465.31 Pa, and the flow of circulated air rose from 44 13 m^3 / s to 44.35 m^3 / s.
- Applying the method of reducing the total resistance of the critical ventilation structure by creating parallel links to the branches on which located relevant critical ventilation structures, had a major impact on active fan located at the main station of Eastern main ventilation station. Thus, the total resistance of the network of ventilation dropped from 0.891 NS2 / m^8 to 0.523 NS2 / m^8, negative pressure developed by the fan dropped from 1766.45 to 1425.25 Pa, and the flow of circulated air rose from 44 52 m^3 / s to 52.22 m^3 / s.

2.2 Applying the Method of Dispersion the Total Resistance of the Critical Ventilation Structure

For this the basic ventilation network was modified as follows:

- Were eliminated three ventilation buildings on the ramifications 324-325; 332-335; 254-255;
- Were introduced six equivalent ventilation structures on the ramifications 323-324; 320-324; 333-334; 333-335; 255-256; 255-258.

The ventilation network model, solved, optimized and simulated is shown in Fig. 2.

Compared with the situation of optimized ventilation network in the previous stage, there were obtained the following results:

- On the evacuation circuit of the polluted air related to the V S bloc, airflow at the 500-level decreased from 34.83 m^3 / s to 34.37 m^3 / s (branch 207-208)
- On the evacuation circuit of the polluted air related to the III N bloc (branch 302-308) the flow presented an increase from 13.03 m^3 / s to 15.45 m^3 / s
- At the level of the mine (ramifications 343-345 259-348 respectively) flows showed an increase from 86.34 m^3 / s to 97.19 m^3 / s on the ventilation shafts E, respectively W.

- At the main ventilation stations (345-346 respectively ramifications 348-349) flows showed an increase from 88.65 m^3 / s to 99.32 m^3 / s on the ventilation shafts E, respectively W.

From the analysis of the functional parameters of active fans before and after applying the method to a single critical ventilation structure are the following:

- Applying the total resistance dispersion method of critical ventilation structure by distributing the equivalent resistances on the parallel branches located upstream or downstream of the ramifications on which were placed relevant critical ventilation structure, had a major impact on the active fan disposed at East main station ventilation. Thus, the total resistance of the network of ventilation dropped from 0.891 NS^2 / m^8 to 0.536 NS^2 / m^8, the negative pressure developed by the fan dropped from 1766.45 to 1442.81 Pa, and the flow of circulated air rose from 44 52 m^3 / s to 51.89 m^3 / s.
- Applying the total resistance dispersion method of critical ventilation structure by distributing the equivalent resistances on the parallel branches located upstream or downstream of the ramifications on which were placed relevant critical ventilation structures, had a major impact on the active fan disposed at West main ventilation. Thus, the total resistance of the network of ventilation dropped from 0.755 NS^2 / m^8 to 0.606 NS^2 / m^8, the negative pressure developed by the fan dropped from 1469.86 to 1364.17 Pa, and the flow of circulated air rose from 44 52 m^3 / s to 47.43 m^3 / s.

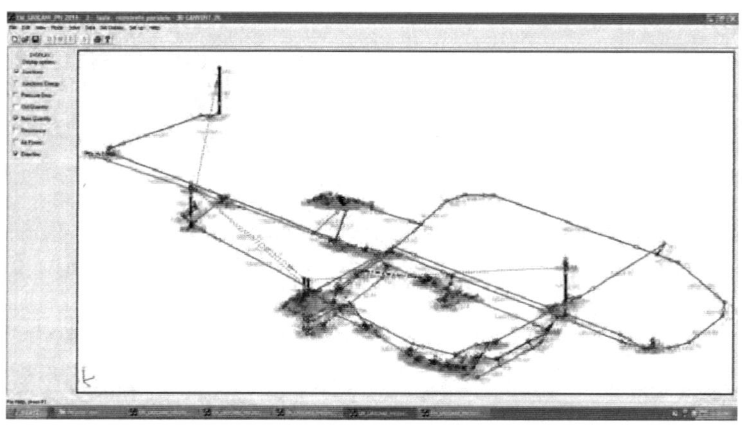

Fig. 1 Simulation of the parallel resistances

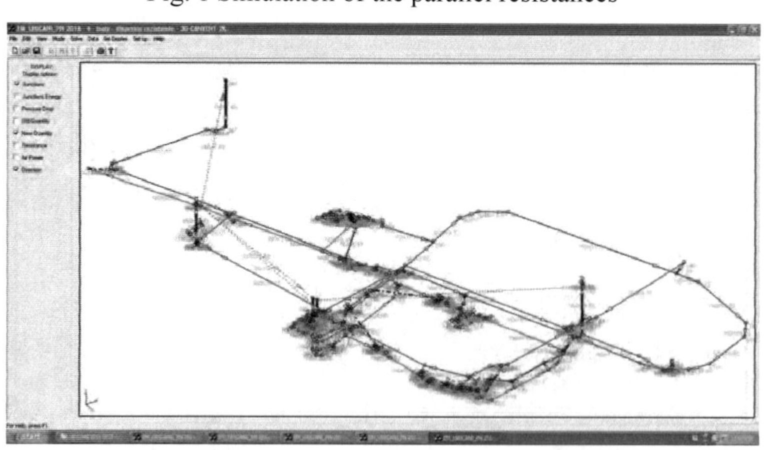

Fig. 2 Simulation of the total resistence disspersion

3 CONCLUSIONS

Ventilation structures that induce a high degree of instability at the network level ventilation, are called critical ventilation structures and must be given special attention from those responsible for network management ventilation.

To identify the critical ventilation structures is used the simulation process successive at the level of every ventilation structure.

To eliminate the criticality of ventilation structures of the Uricani Mine were applied 2 methods, namely:

- Method of reducing the total resistance of the critical ventilation structure;

Method of dispersion the total resistance of the critical ventilation structure.

REFERENCES

Băltărețu, R., Teodorescu, C., 1971, *Ventilation and work protection in the underground*, Editura Didactică şi Pedagogică, Bucureşti.

Cioclea, D., 2010-2011, *The diminishing of explosion hazard of coal mine from Jiu Valley by computerized handling of ventilation systems*, Insemex Study, Petroşani.

Cioclea, D, Lupu, C., Gherghe, I., 2013, *Guide for the dimension of industrial ventilation instalations*, Insemex Ed, Petroşani.

Crăciunescu, B., Mustață, T., Marinică, I., 1993, *Frame method for exploitation of coal in Jiu Valley Basin*, ICPM Study, Petrosani.

Gherghe, I., 2004, *Rationalising the ventilation network of Jiu Valley Mines due to their closure*, Insemex Study, Petrosani.

Patterson A. M., 1992, *The Mine Ventilation Practitioner's* DATA BOOK, M.V.S. of South Africa.

Teodorescu, C., Gontean, Z., Neag, I, 1980, Mining ventilation, Technical Ed. Bucureşti

3D Canvent software, manual***, Technical documentation of Uricani Mine.

A New Approach to Underground Stope Layout Optimization

Y. A. Sari, M. Kumral
McGill University, Montreal, CANADA

ABSTRACT As opposed to variety of approaches and tools for open pit mine layout optimization, there are more opportunities for enhancement for computerized underground mine planning. A new, ranking-based stope layout optimization approach is introduced in this paper. After modeling the economic value of the deposit, a score is calculated for each block that is the sum of the economic value of the block and its surrounding blocks. First, the sublevels are determined by comparing the total sum of scores of each level and ranking them in decreasing order; taking into account the minimum and maximum level heights. An adjustable number of combinations are attempted before making the final decision. If the number of combinations is increased, whole space can be searched. Then, each level is examined separately. A similar ranking approach is used for the stope selection at each level. The approach has been shown to create feasible sublevel stopes.

1 INTRODUCTION

As resources are rapidly exhausting near the surface of the earth, the mining industry is forced to intensify their underground extraction processes. This increases the importance of developing computerized optimization tools for underground mining methods such as sublevel stoping. Computerized stope layout optimization is a relatively less studied subject compared to open-pit planning applications, leaving more opportunities for enhancement for researchers.

Current approaches for stope layout optimization are mainly categorized into rigorous algorithms, approaches using geologic models and heuristic algorithms (Ataee-pour, 2005).

Dynamic programming (Jalali & Ataee-pour, 2004), mixed integer programming (MIP) (Ovanic & Young, 1995) and maximum flow (Bai et al., 2013) approaches are among rigorous algorithms to solve the stope layout problem. Although these approaches work well on simple, small deposits, they have their disadvantages as the deposit becomes more complicated: dynamic programming is optimal only in two dimensions, MIP is only optimal row by row and impractical on too large deposits, maximum flow approach does not work well with inclined orebodies.

Approaches using geologic models do not use the economic model of the orebody; they use the grades of the geological model directly (Cheimanoff et al., 1989; Deraisme & de Fouquet, 1984). These approaches focus on geometrically forming stopes by only extracting the ore and avoiding the non-valuable sections.

Heuristic algorithms consist of floating stope algorithm (Alford, 1996; Alford et al., 2007), maximum value neighborhood algorithm (Ataee-pour, 1997; Ataee-pour, 2004) and preference based profit maximization approach (Sens & Topal, 2008). Floating stope algorithm is analogous to moving cones method for the open-pit layout optimization. All possible stopes with minimum stope sizes are generated and all stopes that are above the cut-off grade are recorded. Some stopes from the generated list are selected. Problems arise when stopes overlap. Maximum value neighborhood algorithm aims to locate the best neighborhoods of blocks. It takes into account of the minimum feasible stope size and identifies the best neighborhoods by comparing maximum net values. Although reportedly it solves the problem of overlapping stopes, the results change according to the starting point. Preference based profit maximization approach is similar to floating stope algorithm as it enumerates possible stopes and saves the positive ones. It deals with the overlapping stopes problem by eliminating stopes containing common blocks. The selection preference is established according to a user defined criteria. The weakness of existing heuristic algorithms is that they do not search the whole parameter space, only providing a guideline for the planner. Additionally, they do not consider varying but stable stope level heights. Consistent level height is accomplished through setting the level height to minimum stope height.

2 METHODOLOGY

The proposed approach breaks down the stope layout design process into three steps: (1) preparing the model and calculating scores, (2) designing the sublevels and (3) designing the stopes at each level as demonstrated in Figure 1. In this section, the process is explained in detail.

2.1 Preparing The Model

As all the extracted is processed, applying a cut-off to blocks and converting to economic model is not necessary at this point. If a stope is extracted, all blocks in the stope will have an economic value proportional to their grades. All non-extracted blocks will have the economic value of zero. Hence, instead of converting to economic model, the block model with grades is converted to a heuristic score model. The score model emphasizes the strategical value of a block as well as its own value emerging from its grade. The score of a block is calculated by summing the grade of the block and the neighboring blocks, discounted by a factor defined by the user. The scoring helps to locate the high-grade ore concentrated regions in the deposit, providing a guidance to following design steps.

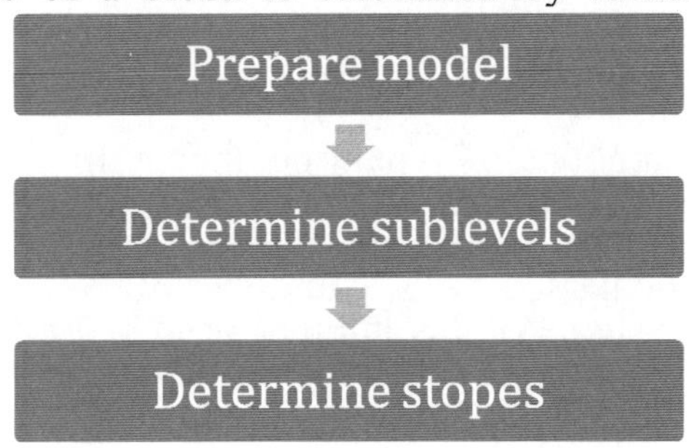

Figure 1: Summary of the approach

2.2 Sublevel Design

Sublevels are an important part of the development process of the mine. As previously mentioned, the majority of the current approaches take the minimum stope height as the sublevel height. To provide a more practical design, the proposed approach asks for a minimum and maximum stope height and decides on the sublevel heights considering the orebody characteristics.

Practically, sublevels are selected where the orebody is richest to allow the extraction near these points. In our approach, we follow the same intuition and prioritize the rich areas. The procedure is as follows:

- All candidate sublevels are calculated by summing the scores of the blocks in every sublevel.
- Among the candidate sublevels, the levels that violate the stope height constraints are eliminated to speed up the search process.
- The candidate sublevels are ranked according to their scores.

- A candidate combination is created as such: The sublevel at the top of the list is picked and each of the next sublevels are attempted to add to the combination if stope height constraints are not violated.
- The score of the candidate combination is calculated by adding up the scores of each sublevel in the combination. The combination and the outcome are stored.
- The top candidate sublevel in the list is removed and the combination process is repeated.

The amount of repetition in the final step depends on the user. If the number of repetitions is equal to the number of candidate sublevels, each possibility will be exhausted. As this number is increased, the search will be longer and slower, but the outcome will be closer to optimal.

In the end, among the candidate combination list, the one with the highest score is selected and registered as sublevels.

2.3 Stope Layout Design

As the sublevels are determined with the above step, each area between the sublevels is considered separately for stope layout planning. As with the sublevel design step, stope layout also uses the heuristic score model to identify the ore concentrated regions in the slabs. For each area in between two sublevels the following procedure is followed:

- If the height between sublevels are less than the maximum allowed stope height, it is considered as the stope height. Otherwise, the maximum allowed stope height is measured from the sublevel and determined as the stope height.
- All blocks in the determined area are listed and sorted in decreasing order.
- Analogous to the sublevel design algorithm, the first block in the list is selected and a stope is formed where the block is centered. The width and length of the stope is optimized by trying every combination and deciding on the stope dimensions where the economic value of the stope is maximized.
- Following the list of blocks, each block is attempted to add to the stope combination and added if the stopes do not overlap. If an overlap occurs, the algorithm moves to the next block in the list.
- When the list is exhausted the first block on the list is removed. The process is repeated by taking the next block on the top of the list, forming a stope as the primary stope in the combination.

As it is the case with the sublevel design, this process is repeated for a user specified number of times. Likewise, in the end, the combination that generates the highest economic value is selected. When this algorithm is applied to each area between sublevels and the resultant stopes constitute the final stope layout.

In this method, there are three main adjustable parameters. The first parameter plays role on determining the scores of the blocks. As this parameter grows, the number of neighboring blocks to be considered increases. It is logical that the block score should depend on the other blocks in a probable stope. However, is this parameter is set too large, unrelated blocks will also be considered.

Second parameter is the number of repetitions of the final step of the sublevel design stage of the method. This parameter will yield better results if it is set high enough. Nevertheless, the overall run time will increase as this parameter is increased. Moreover, because the of the manner the model is constructed, the most likely combination is tried first and with a high probability, the most economical combination will be selected in the first several trials. In short, it is not necessary to set this parameter as high as possible.

Third parameter is the number of repetitions of the final step of the stope layout stage of the method. Because this stage is analogous to the sublevel design stage, the parameter is also analogous and likewise, it is not needed to be maximized but it should be set high enough such that at least a few combinations will be compared. The most practical way to set the parameters is to experiment with them and tune them to the minimum where the improvement in the economic value is ceased.

3 CONCLUSION

In this paper, a novel heuristic approach is proposed for stope layout design. The heuristic is based on the practical engineering applications.

However, combined with the computational tools, the parameter space is explored more efficiently, potentially leading into better designs and higher economical profits.

The presented approach has a high potential of generating near-optimal results because if the search parameters are selected well, it explores the promising design possibilities very effectively. If the repetition parameter is set high, although the algorithm will take longer to execute, almost all feasible designs will be generated and compared to each other, guaranteeing a near-optimal result. The fact that the heuristic is inspired by real practices increases the confidence that even with smaller repetition parameters the result will be at least comparable to the plan of an engineer.

In future work, the algorithm will be tested on a real case study and the resultant economic value will be compared to the plan by a mining engineer. The potential of the heuristic approach will be tested through increasing the number of repetitions of final steps of sublevel design and stope layout design algorithms and observing the change in the economic value of the plan. In addition, stope sequencing will also be adapted to this algorithm.

ACKNOWLEGMENTS

The authors thank to Natural Sciences and Engineering Research Council of Canada (NSERC) for the support of this research (#488262).

REFERENCES

Alford, C., 1996. Optimisation in underground mine design. *International Journal of Rock Mechanics and Mining Sciences and Geomechanics Abstracts*, 5, pp.220.

Alford, C., Brazil, M., & Lee, D. H., 2007. *Optimisation in underground mining*. Springer, New York, 624 p.

Ataee-pour, M., 1997. A new heuristic algorithm to optimise stope boundaries. *Proceedings of the Second Regional APCOM Symposium on Computer Applications and Operations Research in the Mineral*, 501.

Ataee-pour, M., 2004. Optimisation of stope limits using a heuristic approach. *Mining Technology,* 113(2), pp.123-128. doi:10.1179/037178404225004959

Ataee-pour, M., 2005. A Critical Survey of the Existing Stope Layout Optimization Techniques. *Journal of Mining Science,* 41(5), pp.447-466. doi:10.1007/s10913-006-0008-9

Bai, X., Marcotte, D., & Simon, R., 2013. Underground stope optimization with network flow method. *Computers & Geosciences,* 52, pp.361-371.

Cheimanoff, N., Deliac, E., & Mallet, J., 1989. GEOCAD: an alternative CAD and artificial intelligence tool that helps moving from geological resources to mineable reserves. *21st International Symposium on the Application of Computers and Operations Research in the Mineral Industry*, pp.471-478.

Deraisme, J., & de Fouquet, C., 1984. Recent and future developments of "downstream" geostatistics,". *Verly, G., and others, eds., Geostatistics for natural resources characterization: Dortrecht, Reidel*, pp.979-999.

Jalali, S., & Ataee-pour, M., 2004. A 2D Dynamic Programming Algorithm to Optimise Stope Boundaries. *Proceedings of the 13th International Symposium on Mine Planning and Equipment Selection–MPES,* 4, pp.45-52.

Ovanic, J., & Young, D., 1995. Economic optimisation of stope geometry using separable programming with special branch and bound techniques. *Third Canadian conference on computer applications in the mineral industry*, pp.129-135.

Sens, J., & Topal, E., 2008. A new algorithm for stope boundary optimisation. *Proceedings of the The AusIMM New Leaders' Conference.*

Proposal of a Waste Material In-Pit Dumping Solution Using a Strategic Mine Planning Approach ,to Improve Waste Dump Storage Capacity in a Confined Mining Property.(Case Study of the Mashamba East Open Pit Mine in the Democratic Republic of the Congo)

P.Mukonki
University of Lubumbashi,Lubumbashi,Democratic Republic of the Congo)

A.Muhota
University of Lubumbashi,Lubumbashi,Democratic Republic of the Congo)

ABSTRACT: Mashamba pit is located South-East 10Km from Kolwezi, one of the richest copper areas of the Democratic Republic of the Congo. Optimization scenarios have provided a Life of Mine (LOM) of 6 years using datamine NPV scheduler software. Ultimate pit design has been executed using Minesight software. Mashamba pit is located in a confined area, when considering property boundaries and mine dependencies (ore and waste dump locations).The North Waste dump for the life of mine is designed to contain 70.4Million tonnes of waste while the South west Dump is designed to contain 76.3 Million tonnes of waste. The concern is to have the north waste dump, becoming an obstacle for the pit to be extended northward, when considering economic opportunities dues to the rise of the commodity price. This paper outlines the need of delaying the North West dump implementation in order to avail mining space for future North pit expansion.
Keywords: Trategic Mine Planning, Open Pit, Waste Dump Storage Capacity, Mine Optimization, Mine Planning, Mine Scheduling

1 INTRODUCTION

1.1 Site Location

Mashamba open pit is located in the city of Kolwezi, and it is a satellite pit of a sedimentary deposit called "Lambeau Geologique de Kolwezi" located in the major copperbelt inside the Democratic Republic of the Congo. The main geological structures are summarized in the figure 1 below.

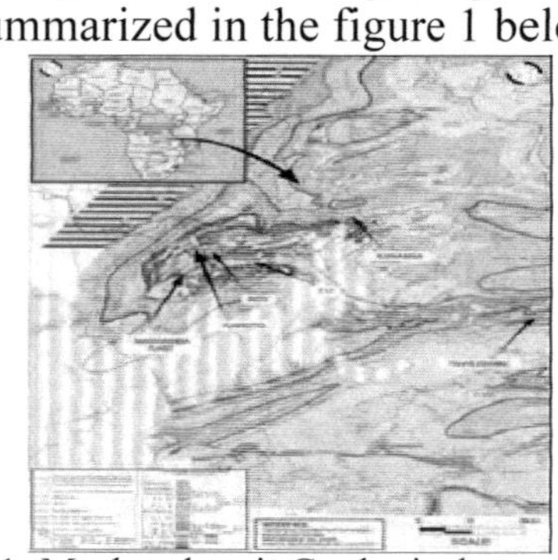

Figure1: Mashamba pit Geological structures

Mashamba East open pit mine has been initially optimized and provided a Life of Mine (LOM) of 6 years at 3.2Mt of ore per year as a target to achieve. The mine has been planned and provided 4 major pushbacks. Mashamba East pit has a potential to produce more ore but the mine is restricted by tight property boundary that doesn't allow flexibility in terms of waste dump storage capacity. Currently the plan is to have a north waste dump (70Mt) of dumping capacity and a south waste dump with 76.4Mt of waste dumping capacity.

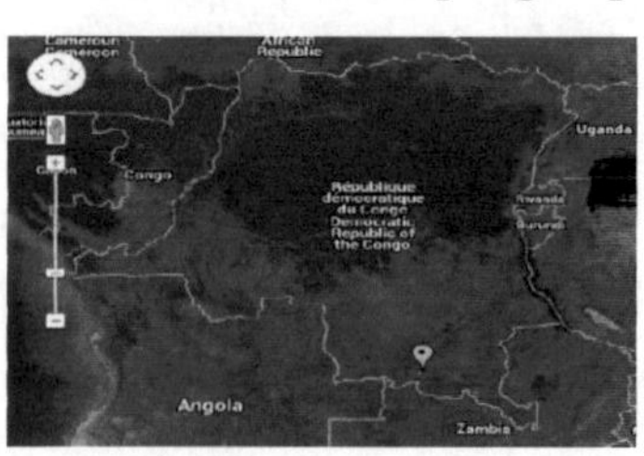

Figure2: Kolwezi city location in the Congo

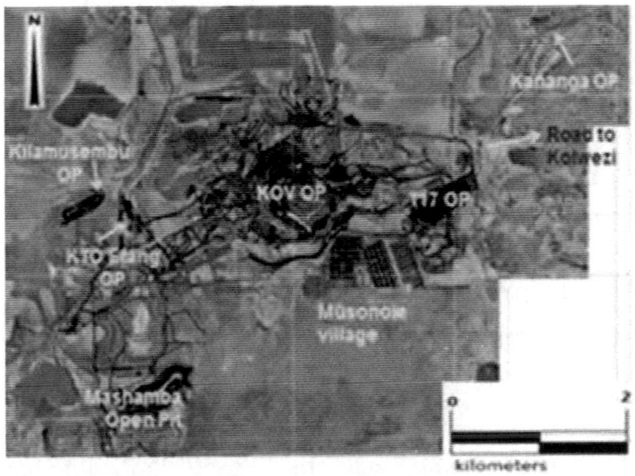

Figure3: Mashamba pit Location in the Katanga Mining limited property

1.2 Property Ownership

Kamoto Copper Company SA (KCC) owns the Material Assets since 2006, including the mining and exploitation rights related to the Mining Assets. Katanga Mining Limited (KML) holds a 75% stake in KCC. La Generale des Carrieres & des Mines (GCM) and La Société Immobilière du Congo (SIMCO), state-owned mining companies in the DRC, own the other 25% of KCC.

2 METHODOLOGY

The methodology used in this approach is based on the concept of dumping waste material inside the pit itself, while it is mined.

The strategy here is to present a mining sequence based on an accelerated mining schedule, different from the initial long term mine plan. The difference here will be to increase the annual ore mining rate by 25% from 3.2Mt/year to 4.0Mt/year with the immediate consequence to shorten the mine life by one year (from 6years to 5 years). The reason for this is to quickly provide in-pit space to allow in-pit dumping to take place as soon as possible to match the end of the construction of the south waste dump and the start of the north waste dump. The aim on this approach is to dump in the pit, the maximum waste material (70.4Mt) to avoid building the North Dump too early as shown on the figure 4.

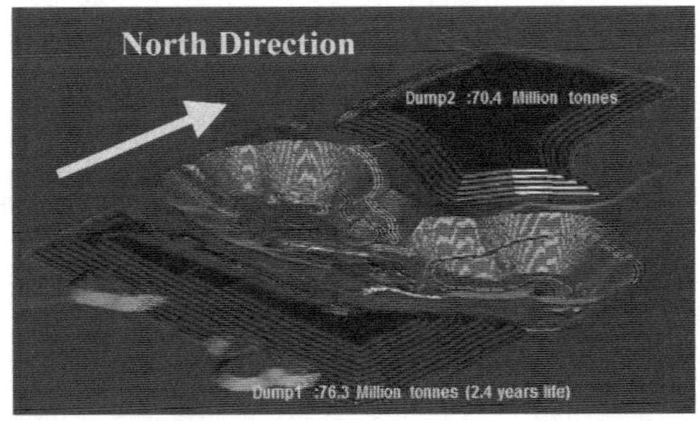

Figure4: Mashamba pit Waste dumps locations

Figure5: Mashamba pit with the potential north Pit expansion.

Existing ramps and pit access will be used to implement the in-pit dumping strategy.

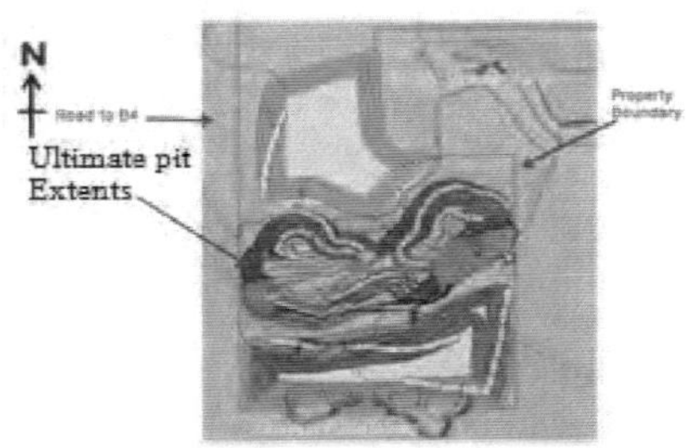

Figure6: Mashamba ultimate pit configuration

2.1 Mashamba Pit Mining Sequence

The mining sequence is representing graphically how the Mashamba pit has been initially developed on a long term approach for 6 years as shown on the figure 7.

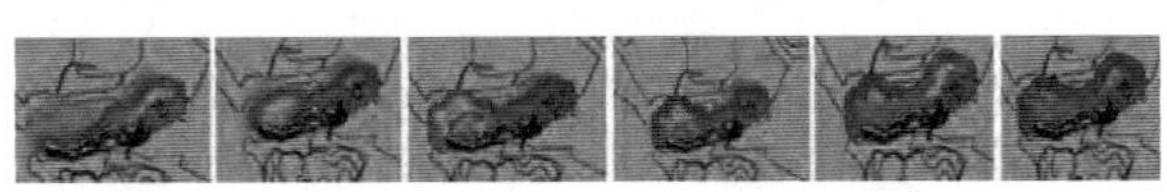

Yearl year2 year3 year4 year5 year6

Figure 7: Mashamba previously proposed schedule

The table 1 shows the ore and waste quantities associated with the graphical sequence of the figure 7.

Table1.Mashamba east pit 6 years mining Schedule.

Year	Rock Million tonnes	ORE Million tonnes	Waste Million Tonnes
YEAR 1	40,3	3,25	37.05
YEAR 2	37,5	3,25	34.25
YEAR 3	29,2	3,25	25.95
YEAR 4	25,9	3,25	22.65
YEAR 5	26,8	3,25	23.55
YEAR 6	1,8	1,03	0.77
Total	162	17,2	144.8

2.2 Mashamba Pit Modified Sequence

The long term mining schedule has been modified as mentioned precedently, to accelerate the development of in-pit available space for waste dumping. The new schedule has been run through the NPV scheduler software and has provided the following graphical sequence and the following table:

Year1 year2 year3 year4 year5

Figure 8: Mashamba modified (accelerated) schedule

Table2.Mashamba east pit 5 years Accelerated mining schedule.

Year	Rock Million tonnes	ORE Million tonnes	Waste Million Tonnes
YEAR 1	41.5	4.00	37.05
YEAR 2	47.0	4.00	43.0
YEAR 3	27.5	4.00	23.5
YEAR 4	42.5	4.00	38.5
YEAR 5	30.7	1.30	29.4
Total	162	17,2	144.7

2.3 Mashamba In- Pit Dumping Sequence

In-pit dumping available capacity

When considering how Mashamba open pit is mined in terms of ore and waste stripping, in regards of the table2, and by comparison with the South waste dump capacity shown on figure 4, it can be easily seen that 88.5 Million tonnes of dumping capacity is available for a start ,when combining year1 and year2, total material mined. In terms of dumping strategy it is advisable to start dumping on the south waste dump first, to its maximum capacity, (76.3Mt) before to start dumping inside the pit.

Pit development versus in-pit dumping

In terms of pit development, the sequence on figure 8 shows that the eastern sides of Mashamba pit is completely free from year3 to start the in-pit dumping strategy.

3 DUMP DESIGNS PARAMETERS

Geotechnical recommendations suggested keeping an overall dump slope of 23.3 degrees (note that the natural angle of repose is 37 degrees for the slope angle of waste material).

A summary of waste dump designs parameters is shown on figure 9.

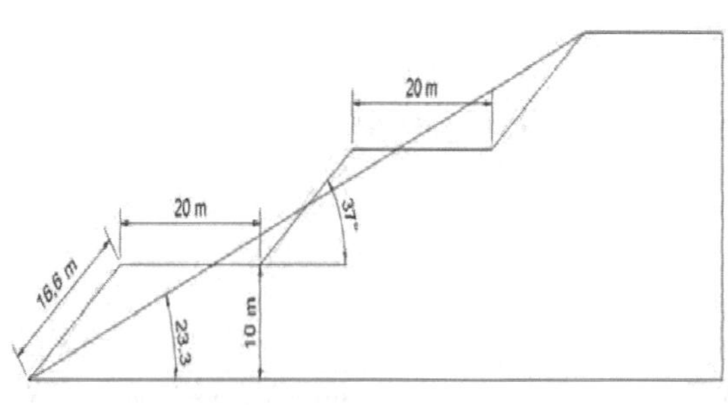

Figure 9: Mashamba waste dump design Parameters

4 DUMPING GRAPHICAL SEQUENCE

The Graphical dumping sequence will be based on a 10m lift height to mimic the dump design parameter and for the in-pit dumping; the height of the bench (10m) will be also used as the dumping height.

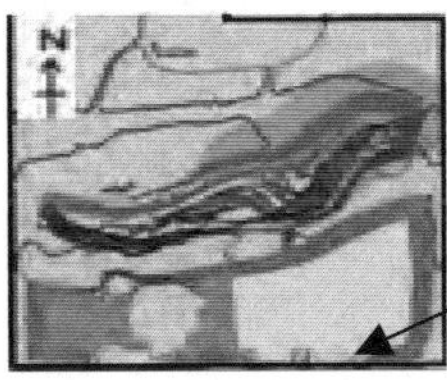

Figure 10: Year1 (in red, the mining sequence, in blue, the waste material dumping sequence)

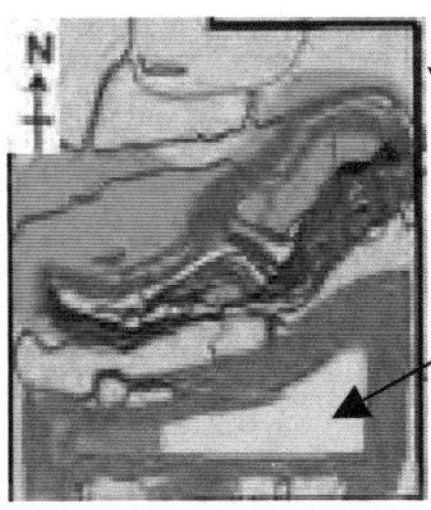

Figure 11: Year2 (in red, the mining sequence, in blue, the waste material dumping sequence)

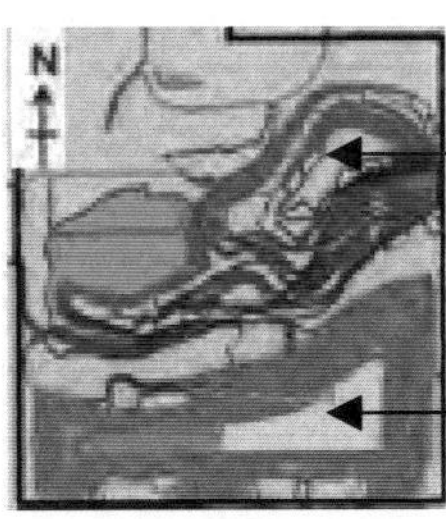

Figure 12: Year3 (in red, the mining sequence, in blue, the waste material dumping sequence)

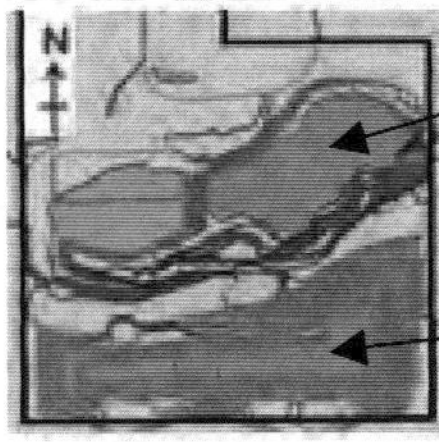

Figure 13: Year3 (beginning of the in-pit Dumping sequence)

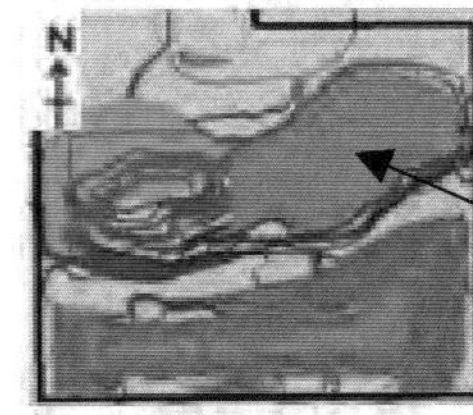

Figure 14: Year4 waste material dumped.

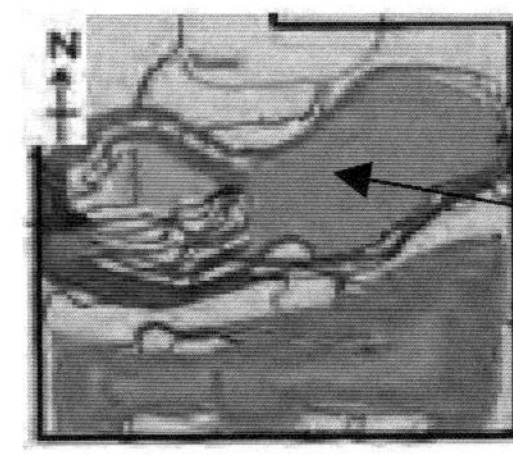

Figure 15: Year5 waste material dumped.

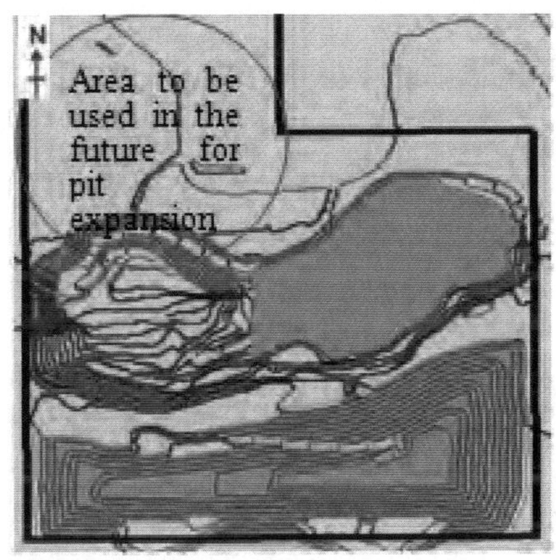

Figure 16: Final property configuration.

5 RESULTS AND DISCUSSION

The study that has been conducted here was based on the assumptions that in the future, the copper price may rise to a level where, the north waste dump could have become an obstacle to expand the pit to get to the ore. This study was also based on the fact that the property limits will stay the same and then, by the way, become a concern for the mine planning engineer, when considering the alternative of long distance haulage to other potential waste dump if the pit has to be expanded to north. Multiple scenarios have revealed to us that a 25% increase in the ore target (from 3.2Mt to 4.0Mt) will accelerate the pit development and can lead to avail substantial space to allow the in-pit dumping strategy to take place at year3 of the new 5 years mining schedule, so that instead of constructing the north waste dump, the in-pit dumping can directly be implemented.

It has been determined that, instead of a 6years schedule as presented and proposed precedently, it is required to accelerate mining the pit by implementing the 5years schedule as proposed in this study. This may need to have additional mining equipment, (situation to be analyzed by mine planning engineers and issue to be addressed to the top management for decision making).

6 CONCLUSIONS

The study, as presented in this paper, was conducted in the spirit of looking at an alternative solution in terms of the future of the mine and the future of the commodity price. The main idea was to suggest a technical approach of thinking, in case the space becomes a concern (because it will likely become a concern).

In the paper, the approach wasn't focused only at the mine planning side of the long term plan, but also on how the available space could be managed as good as possible, to still have an alternative in case the mining circumstances varies.

When looking at the final Mashamba East Pit configuration topography (figure16), it can be seen that there is also an option of balancing the in-pit dumping strategy in such a way that the filled waste material stays at the very bottom of the pit to postpone eventual rehandling material costs. Deeper analysis regarding on how to technically proceed on the in-pit dumping will be conducted if opportunity arises in the future.

REFERENCES

Hustrulid, W. and Kuchta, (3rd edn), 1998. Open pit Mine planning and design Vol1-Fundamentals, A.A.Balkema, and Rotterdam-Brookfield.

Ian, C.R, 1998. Mining Economics and strategy, Society for Mining and Exploration, Littleton, Colorado.

Elahi, Z.E., Kakaie. and Yousefi, A. 2011. A new algorithm for optimum open pit design: Floating cone method III, journal of mining and environment, Vol N°2, 2011, (pp.118-125).

Dagedelen, K.2013.Open pit optimization-strategies for improving economics of mining projects through Mine planning.17th International Mining Congress and exhibition of Turkey-IMCET 2001.

Whittle G. and Burks, S.2010.Simulatneous mining and mineral processing enterprise optimization for the platinum industry. The 4th International platinum Conference, platinum in transition "Boom or Burst". The South African Institute of Mining and Metallurgy, 2010.

Optimization of Underground Haul Roads Using an Evolutionary Algorithm

A.G. Yardimci, C. Karpuz
Middle East Technical University, Ankara, Turkey

ABSTRACT Underground mine haul road carries the main traffic of mining activities. Composed of straight and spiral sections with a fixed maximum gradient and minimum turning radius constraints, haulage road aims to connect the underground mine portal to the production levels with a navigable path. Most widely, haulage path is designed by skilled mine design experts. Considering haul road design controls the initial capital investment and operating cost parameters, an optimum design that minimizes the path length is obviously desirable. Designing an optimum path satisfying the kinematical constraints of a mine car for the sake of navigability is a complicated topological problem, especially in complex underground networks. This paper presents a genetic algorithm implementation for predicting the near optimum navigable haul road. The path predicted by the algorithm is compared with a manually designed underground mine haul road and improvement in terms of length are described.

Keywords: Underground Mining, Haul Road Path Optimization, Genetic Algorithm

1 INTRODUCTION

In an underground mine, main haul road connects the surface portal, to the crosscuts driven through the ore production levels. Cost of main haul road construction is classified under the title of development stage capital expenditures. In addition, path length influences the long-term mine economy by operating costs (fuel consumption). Optimum design for an underground main haul road should satisfy the navigability constraints with the shortest path connecting the production levels to the surface portal.

Mining has always been an economically significant industry (Brazil et.al., 2003). Optimization is a hot topic among the mining society in terms of cost minimization and profit maximization. Compared to the extensive research on open pits, there are limited attempts for underground mines. Brazil et.al. (2014) declares that the complex topology of underground mines contributes to the limited interest of researchers. As an early attempt, Lee (1989) investigated underground network optimization.

This study presents a novel method to predict an optimum path for an underground main haul road using an evolutionary algorithm. Objective is to determine the shortest navigable path visiting all the crosscut entries and the surface portal. Motion of an underground mining equipment is simulated by a proven mathematical approach that models the shortest route of a visual car on a curved path. Genetic algorithm, implemented in Matlab environment, is used to search for the optimum path in an iterative manner. However, in each iteration, which is called as generation in genetic algorithm, the model learns from the past experience and improves the fitness value.

A simple underground main haul road layout contains two distinct types of sections:

- straight paths with descending gradient (decline)
- helical paths (similar to a curve travelling around a cylindrical surface and joining the upper elevations to the lower elevations by a fixed gradient).

The proposed algorithm searches for an optimum valid path composed of the appropriate section types. A valid path does not violate the navigability constraints that are minimum turning radius and maximum gradient.

3D path planning with multiple constraints is a complex computational problem. As the number of variables increase, the global optimum becomes harder to reach because of the large solution set. Complex underground mines with hundreds of production levels can be either solved in long computation times or the problem becomes insoluble. In a previous study, authors (Yardimci & Karpuz, 2016) recommend a heuristic algorithm, which adds some extra constraints to reduce the search space. However, it is based on expert opinion and the result approximates to the near global optimum. A new idea presented in this study recommends genetic algorithm that learns from the past experience as it evolves. In this approach, the initial solution provided into the algorithm directly affects the quality of prediction. Although, the genetic algorithm is theoretically capable of determining the global optimum, it is not guaranteed. The solution will be most likely a near global optimum. However, the computational performance considerably increases. As the near global optimum does not show any remarkable variation from the global optimum, genetic algorithm is a smart alternative for predicting an optimum underground mine haul road path.

Motion of an underground mining equipment is modelled as a Dubins car. According to the principle proposed by the famous mathematician Lester Dubins (1957), the shortest path for a virtual car controlled by the heading angle is one of the six alternative curved paths. Later, the overall path is discretized and control variables are optimized by the genetic algorithm.

A Dubins path has three motion primitives, which are denoted by the initials of ‘Straight’, ‘Left’ and ‘Right’: {S, L, R}. These primitives represent a constant action over a given section. The six alternative curved paths are denoted by a sequence of three symbols that is called as a “word”. The following six ‘words’ are a set of candidates for the shortest path connecting two nodes: {RSL, LSR, RSR, LSL, RLR, LRL}.

Because ‘R’ and ’L’ primitives represent the turns, they can also be denoted by the ‘C’, which is the initial of ‘Curve’. By this way, the Dubins paths can be shortly represented by two base words and they are {CCC, CSC}.

Dubins path has long been used to predict the optimum route of an aircraft. Similar to the underground mining vehicles, turning radius and gradient constraints determine the physical limits of aircrafts. Although, the kinematical model can be represented by the fully coupled equations of motion, Dubins car simulating the underground equipment with a simpler model offers some advantages.

Originally, Dubins path was presented in two dimensions. Nevertheless, an underground mine ramp is a three dimensional element. Obviously, the principle requires to be modified.

Benchmarking is a crucial topic in path planning. Lack of a validation problem or an analytical solution makes it hard to check the quality of the proposed path planning algorithm. For this purpose, a simple underground mine scheme is generated. The proposed algorithm is compared to the easily predictable shortest path.

Main haul road of an underground iron mine in the Mideastern Turkey is investigated as a complex verification problem. Manual design is compared with the path predicted by the proposed genetic algorithm. Haul road length is remarkably improved.

2 PROBLEM DESCRIPTION

The problem is determining the shortest navigable path that connects a set of main nodes regarding the kinematical constraints. Main nodes are surface portal and crosscut entries.

Commonly, a safety buffer is required to leave between the orebody and the main haul road in order to decrease the effects of production blasting on the permanent opening. The algorithm should make predictions inside a region of interest excluding this buffer region.

Figure 1 presents a simple layout for an idealized underground mine.

Assumptions, givens and outputs of the algorithm are listed below.

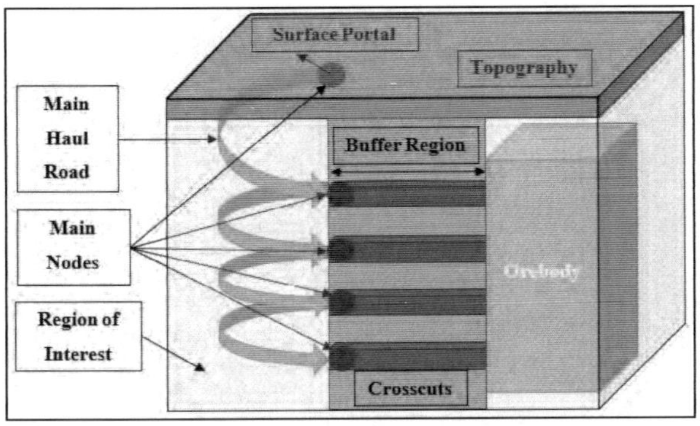

Figure 1 Layout of an idealized underground mine

Assumptions:

- The algorithm works for a single orebody.
- Visiting sequence of main nodes is defined.
- Either elevations of the two succeeding main nodes are the same or the latter one is lower. In other words, the ramp declines or travels on the same elevation but cannot climb up.
- Each of the two main nodes are connected by one of the three types of paths that are listed below:
  - A straight path.
  - A 'Curve-Straight-Curve' (CSC) path.
  - A 'Curve-Curve-Curve' (CCC) path.

Givens:

- Main node coordinates.
- Region of interest
- Buffer Region
- Kinematical constraints:
  - Minimum turning radius (m)
  - Maximum gradient (%)

Outputs:

- x, y, z coordinates of the equally spaced nodes on the shortest valid path.

2.1 Objective Function

Objective function is the mathematical model of that minimize the haul road path. The function has three cost factors, which are the path length cost, the missed node cost and the gradient cost. If the algorithm predicts a solution out of the feasible boundaries for any of these three factors, then the cost adds to the overall path cost. This overall cost is called as the fitness value in genetic algorithm terminology.

Given:

$$\textit{Main node coordinates} = (x_i, y_i, z_i) \quad \text{for } i = 1 \ldots n$$

$$tr_{\min} = \text{minimum turning radius, m}$$

$$\psi = \text{maximum gradient, \%}$$

Objective:

$$\text{Minimize} \left\{ w_1 \times LC + w_2 \times MNC + w_3 \times GC \right\}$$

Where:

$$LC = \text{Cost of path length}$$

$$MNC = \text{Cost of missed nodes}$$

$$GC = \text{Cost of gradient}$$

Subject to:

$$0 < w_i < 1 \qquad \text{for } i = 1 \ldots 3$$

$$\sum_{1}^{n} w_n = 1 \qquad \text{for } i = 1 \ldots 3$$

The objective function has three weighting factors denoted by w_1, w_2, and w_3. These factors determine the effect of the cost factors on the overall cost. Sum of these factors are one.

3 PATH PLANNING USING AN EVOLUTIONARY ALGORITHM

Each section, constricted by two of the main nodes, are divided into three sub-sections either representing a Dubins 'word' or a straight path. In a Dubins path, the first section is a turning path, the middle section can be either a turning path or a straight path and the third section is a turning path. There are five control variables in each section, that will be optimized by the genetic algorithm. Number of dummy nodes in each section depict the motion of a vehicle in unit simulation time. Another variable is the turn control, which determines whether the middle section is a turning path or a straight path. Finally, gradient between the two main nodes is controlled by a gradient variable. Chromosome structure of the genetic algorithm can be seen in Figure 2.

Step # primary section	Step # secondary section	Turn control secondary section	Step # third section	Gradient control

Figure 2 Chromosome structure

3.1 Kinematical Model of an Underground Mining Equipment

Kinematical model simulates the motion of an underground mining equipment throughout the simulation. A realistic model can be as complex to consider the body forces and moments driven by the gravity, propulsion and aerodynamic forces. Non-linear fully coupled ordinary differential equations of motion are proven to be valid for a vehicle moving along three axes with six degrees of freedom (Ergezer and Leblebicioglu, 2013). This method lacks of closed form solutions, which makes it hard to solve the equations. Alternatively, numerical methods provide more steady state solutions. Dubins car is a simple kinematical model to represent the motion of an underground mining equipment. Figure 3 represents an illustration of a 2D Dubins car controlled by the heading angle and turning radius.

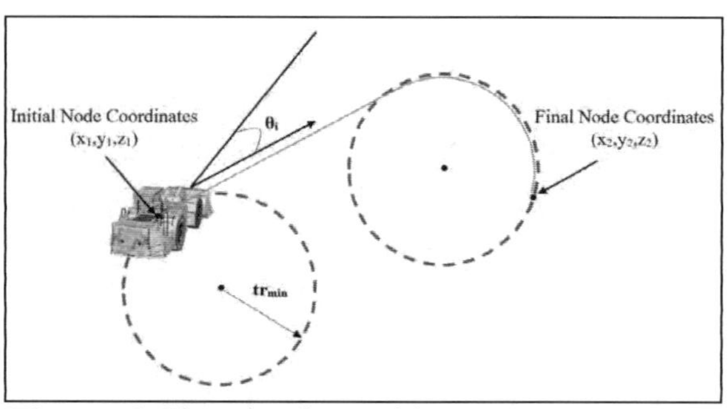

Figure 3 Sketch of a Dubins car

A Dubins car is a bounded speed and no reversing planar vehicle with constriction to move along paths of bounded curvature (Savla et.al., 2008). In this study, motion of an underground mining equipment is modelled as a Dubins vehicle and the governing equations can be seen below.

$$x_i = v_i \cos(\theta_i) \qquad i = 1 \ldots n \quad (1)$$

$$y_i = v_i \sin(\theta_i) \qquad i = 1 \ldots n \quad (2)$$

$$z_i = -v_i \tan(\theta_i) \qquad i = 1 \ldots n \quad (3)$$

$$\theta_i = \frac{u_i v_i}{tr_{\min}} \qquad i = 1 \ldots n \quad (4)$$

Where;

i = node number
n = maximum number of nodes
x_i = x coordinate of the ith node
y_i = y coordinate of the ith node
z_i = z coordinate of the ith node
θ_i = heading angle of the ith node
ψ_i = *max. gradient of the ith node*
u_i = turn control
v_i = velocity
r_{min} = minimum turning radius

3.2 Discretization of Travel Path

As a Dubins path has three distinct sub-sections of either curved or straight, the algorithm divides the section between each two main nodes into three equally spacing sub-sections. In each sub-section, velocity of the Dubins car is fixed; therefore, distance travelled in unit time is also constant. The heading angle changes at a fixed rate in curved sub-sections and constant in straight sub-sections. Sub-sections are composed of equally spacing dummy nodes. Maximum number of dummy nodes to search in a sub-section is provided as an input.

3.3 Initial Population Generation

Genetic algorithm starts optimization from an initial candidate solution named as seed path and this initial population controls the latter generations. Ergezer and Leblebicioglu (2013) states that a good initial population satisfying the physical constraints is required for a better near optimal solution.

This study makes use of a heuristic path planning algorithm presented in a previous study (Yardimci & Karpuz, 2016). Opposite to the advantages of starting from a good initial population, optimization may also trap inside local optimums as a result of a poor initial population.

3.4 Genetic Algorithm Operators

Classical genetic algorithm operators are used to improve the fitness values.

3.4.1 Crossover

Crossover operator produces a child from two parent chromosomes that contains the genetic heritage of both of the parents. Selection procedure enriches the population with better individuals by cloning the good strings. Afterwards, crossover operator performs to obtain better off springs. Cross over has three steps. First of all, two individual strings are randomly selected. Later, a random position is selected to define the genes to be kept and the ones to be migrated to the other string. Finally, two parent individuals are swapped from the randomly selected positions.

Figure 4 presents a sample cross over operation. Fitness value of any chromosome, defines the probability of being a parent of a crossover operation. As the fitness value increase, chromosomes have a higher probability of selection.

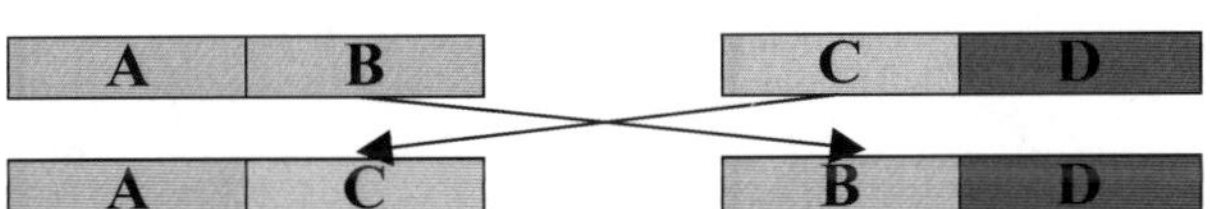

Figure 4 A sample view of a cross over operation

3.4.2 Mutation

Crossover operation is followed by the mutation. Mutation avoids the algorithm from trapping in a local optimum by randomly disturbing the chromosomes.

While cross over operator highlights the strong genes in the current population, mutation operator explores the new ones. In addition, mutation keeps the genetic diversity in the population. A sample mutation operation can be seen in Figure 5.

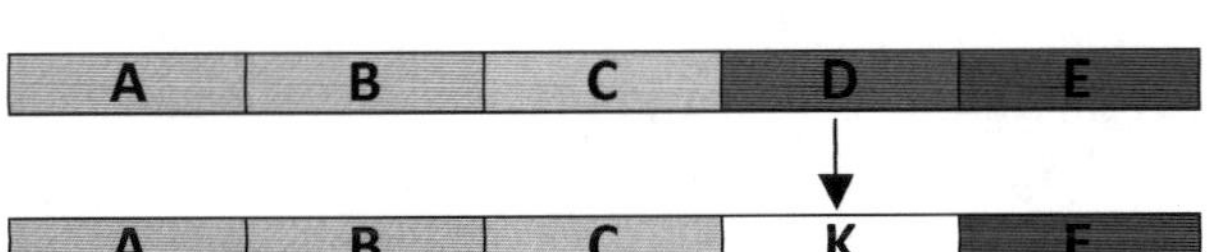

Figure 5 A sample view of a mutation operation

3.5 Algorithmic Prediction of Haul Road Path

The algorithm is implemented in Matlab Global Optimization toolbox with Genetic Algorithm solver. A graphical user interface is developed for data input. Figure 6, outlines the flowchart of the algorithm.

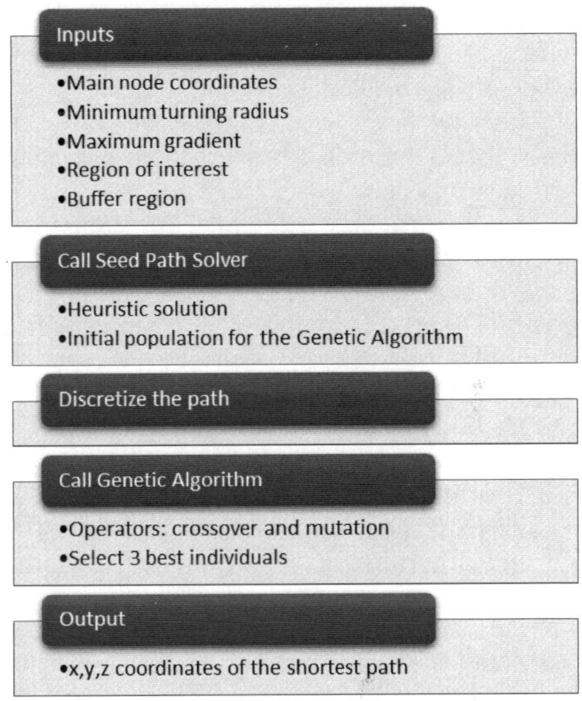

Figure 6 Flowchart of the genetic algorithm

The algorithm discretizes between the main nodes. Next, the heuristic algorithm offers a seed path for the genetic algorithm. Using the seed path, a population of 200 individuals is generated. Next, the algorithm calculates the optimum curved path between each successive main nodes and compute the objective value for each path. The best three individuals are kept as parents of the next generation. Later, classical mutation and crossover operators are applied to obtain off springs with better objective values. The algorithm removes 90% of the total population if there is no longer decrease in the fitness value. A new population is generated and combined with the proceeding individuals. If the objective values decrease stops at the same levels the algorithm terminates, if not, the same procedure is applied until the steady state is reached. By this way, trapping on the local optimum solutions is avoided.

4 CASE STUDY

4.1 Research Area

The case history used in this study belongs to an underground iron mine located in the Mideastern Turkey. In the region, there are other surface and underground mines.

The ore body was formed by flow of hydrothermal fluids from Precambrian aged primer iron deposits.

The major ore minerals are hematite, goethite, and siderite. Altered siderite and iron minerals transform into limonite and goethite predominated by atmospheric conditions, where surface reaction mechanisms are active. Tectonism controls the ore body - country rocks interrelation.

4.2 Results and Discussion

The main haul road path predicted by the genetic algorithm is 50 m shorter than the conventionally designed path. In Figure 7, the red string presents the manually designed path and the blue string shows the path predicted by the genetic algorithm.

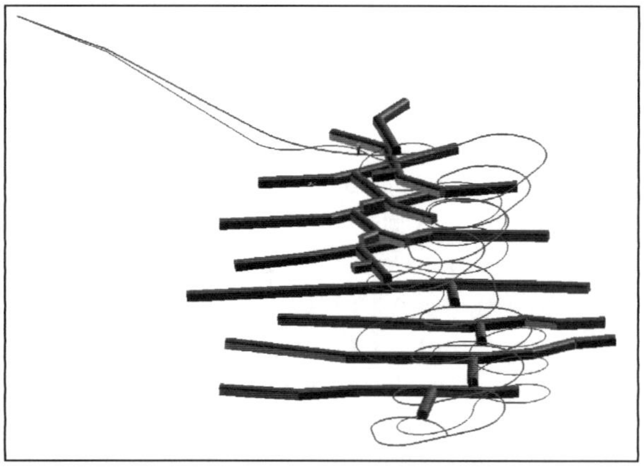

Figure 7 Comparison of the conventional (red string) and algorithmic (blue string) designs of the haul road path

Including excavation and supporting costs, average cost of advancing in a tunnel is around 2000 $/m. Thus, improvement in the haul road length saves 100,000 $. In addition to the reduced cost of capital expenditures in the development stage, operating cost is also decreased by less fuel consumption. Table 1 presents the improvement in path length compared to the manual design.

Table 1 Summary of the path planning by genetic algorithm

		Manual Design	Optimization by Genetic Algorithm
Kinematic Constraints	Minimum Turning Radius (m)	17.5	
	Maximum Gradient (%)	12.5	
Results	**Path Length (m)**	1497	1447
	Improvement (%)	3.3	

A standard workstation (i7 processor, 32 GB RAM) takes around 20 minutes to compute a near global optimum path by the proposed genetic algorithm. Compared to the heuristic algorithm, there is an appealing improvement in the computation time.

5 CONCLUSION

Underground mine design influences the early stage development costs and the operating costs throughout the mine life. Compared to the successive attempts in open pit mining, there is a huge gap in underground mine optimization. Main haul road is a permanent underground space that carries the vast of traffic. Operating cost is closely related with the length of this path.

Conventionally, skilled experts carry out manual mine designs depending on their experience. The shortest path of an underground mine main haul road can be easily estimated for a small scale, simple mine layout. However, it is not such easy to predict the optimum route for a large scale and complex underground mine.

This study introduces a genetic algorithm to predict the shortest path for underground mine main haul road. The proposed algorithm is validated by a simple underground mine layout specifically prepared as a benchmark problem. Next, the main haul road of an underground iron mine located in Mideastern Turkey is investigated by the proposed algorithm. Near global optimum solution offers a considerable improvement in the path length.

Although the global optimum is not guaranteed, the near global optimum offers a significant improvement. In addition, the evolutionary algorithm increases the computational performance.

The future research will be focused on including geological characteristics and rock mass quality as a constraint. Additionally, forbidden regions like faults, joints or aquifers will be avoided.

REFERENCES

Brazil, M., Grossman, P. A., Rubinstein, J. H., & Thomas, D. A. (2014, March-April). Improving Underground Mine Access Layouts Using Software Tools. *Interfaces, 44*(2), 195-203.

Brazil, M., Thomas, D. A., Weng, J. F., Rubinstein, .. H., & Lee, D. H. (2005). Cost Optimisation for Underground Mining Networks. *Optimization and Engineering, 6*(2), 241-256.

Dubins, L. E. (1957). On Curves of Minimal Length with a Constraint on Average Curvature, and with Prescribed Initial and Terminal Positions and Tangents. *American Journal of Mathematics*, 497-516.

Ergezer, H., & Leblebicioglu, K. (2013, January). Path Planning for UAVs for Maximum Information Collection. *IEEE Transactions on Aerospace and Electronic Systems, 49*(1), 502-520.

Lee, D. H. (1989). Industrial case studies of Steiner trees. *NATO.* Denmark.

Savla, K., Frazzoli, E., & Bullo, F. (2008, July). Traveling Salesperson Problems for the Dubins Vehicle. *IEEE Transactions on Automatic Control, 53*(6), 1378-1391.

Yardimci, A. G., & Karpuz, C. (2016). Shortest Path Estimation Considering Kinematical Constraints of Main Haulage Roads in Underground Mines: A Heuristic Algorithm. *6th International Conference on Computer Applications in the Minerals Industries, CAMI2016.* Istanbul.

Mine Planning and Optimization Techniques Used in Surface Mining – Position Paper

J. Kwiri and B. Genc
University of the Witwatersrand, Johannesburg, South Africa

ABSTRACT Planning and optimisation in surface mines make use of a number of techniques to solve specific problems at different stages of the mining value chain. Each technique has its own capabilities and limitations and as such, some problems remained unresolved within the planning and optimization area. The study showed that optimization techniques are applied to four main areas, which are cut-off grade, production scheduling, mining limits and the mine value chain. Most common techniques used include genetic algorithm, particle swarm optimization, Lerchs-Grossmann, and enterprise optimization. Despite the advances in the application of these techniques, challenges remain. These challenges include the difficulty in forecasting costs, commodity prices and exchange rates. Increasing demands from stakeholders exacerbate these challenges. Minimizing these challenges, require the use of two or more techniques to solve the problem and to involve all stakeholders in the development of the optimization techniques.

Key Words: Mine Planning, Cut-Off Grade, Production Scheduling, Optimization

1 INTRODUCTION

The success of existing mining operations relies on efficient mine planning and efficient management of available resources. Large surface mines with an enormous footprint brings complexity to this management. In this study, the authors seek to discuss the planning and optimization techniques applied to surface mining operations to improve their productivity and profitability. In addition, most surface mines are getting to their mature stage with remaining resources having lower grades, which demand highly efficient operations and higher commodity prices to justify their extraction.

On the other hand, commodity prices continue to drop with the fluctuations becoming highly unpredictable. The drop in commodity prices is against the predictions made by most companies and as a result, mining companies are operating at the extreme end of the operating cost curve where there is significant level of restructuring. Operating costs continue to rise due to rising labour costs, energy and fuel costs. Furthermore, productivity (where productivity relates to equipment operating efficiency, man-shift efficiency, and capital productivity) is low.

These challenges require new strategies to be applied within the mining sector. These strategies include the application of efficient mine planning and optimization techniques. This paper discusses planning and optimization techniques that are commonly used in surface mines to improve productivity.

2 PLANNING AND OPTIMIZATION OF SURFACE MINES

The stages in the mining value chain consists several stages, which include discovering, establishing, exploiting, beneficiating, selling and rehabilitating (Anon, 2016a). A detailed representation of the mining value chain is shown in Figure 1 (Anon, 2010) showing the stages and areas within each stage where optimization techniques can be applied. Planning and optimization process applied to each stage enables efficient utilization by re-engineering the mining business process to get maximum profits through the exploitation of optimization techniques (Anon, 2010).

2.1 Planning and Optimization Process

Planning and optimization process that adds value includes long term and short planning models. Such process seeks to simultaneously optimize the stages in the mining value chain in addition to correct and practical implementation of the set methodologies. A mine design criteria document that sets out all the assumptions used at each stage of the mining value chain is created at the beginning of a planning process (Anon, 2016a).

Once the creation of the document is finalized, optimum pit limits and production schedules at desired cut-off grades are designed. The generation of pit limits is aided by the use of software packages for example Gemcom Whittle software. Anon (2016a) identifies the following steps in the process of optimization;

1. Recognising and stating the problem that needs to be optimized;
2. Setting the objective for optimizing;
3. Formulating appropriate models;
4. Validation of the models using algorithms;
5. Generating solutions;
6. Interpreting the solutions and selecting the best alternative; and
7. Implementation of the selected solution

The generation of each solution is done several times with different scenarios. It is after such a number of simulations that the best solution is selected, that maximises the profitability or one that satisfies the objective function. Traditionally, optimization processes operated in isolation and their effectiveness was not beneficial to the overall business profitability. This has prompted for a variety of techniques that have been developed to reduce the ineffectiveness of such systems that operate individually.

2.1 Types of Planning and Optimization Process Techniques

Irrespective of the type of optimization technique, the characteristics are common to all types. These characteristics are, the presence of an objective function, varying degrees of constraints and the search space. The search space is significantly large and as such, the presence of constraints limit the search space to the most feasible ones. Optimization techniques can be classified into stochastic techniques and deterministic techniques (Hickman, n.d).

2.1.1 Stochastic

The nature of these techniques is to use some form of randomness when searching for feasible solutions (Hickman, n.d). The input to these techniques take into account the variabilities of each input parameter. Their outputs do not give one optimum solution but a range of solutions with some form of distribution to give the user a feel of variability for the solution (Hickman, n.d). The use of stochastic techniques works best when using evolutionary algorithms.

2.1.1 Deterministic

Hickman (n.d) defines deterministic optimization techniques as those techniques that require a definitive starting point from which a sequence of possible solutions are generated. The number of sequence points is defined by the technique that is being used and the technique terminates each sequence of points once a solution has been reached or when the number of iterations has been exhausted (Hickman, n.d). Linear programing, integer programing and mixed integer linear programming use this type of technique.

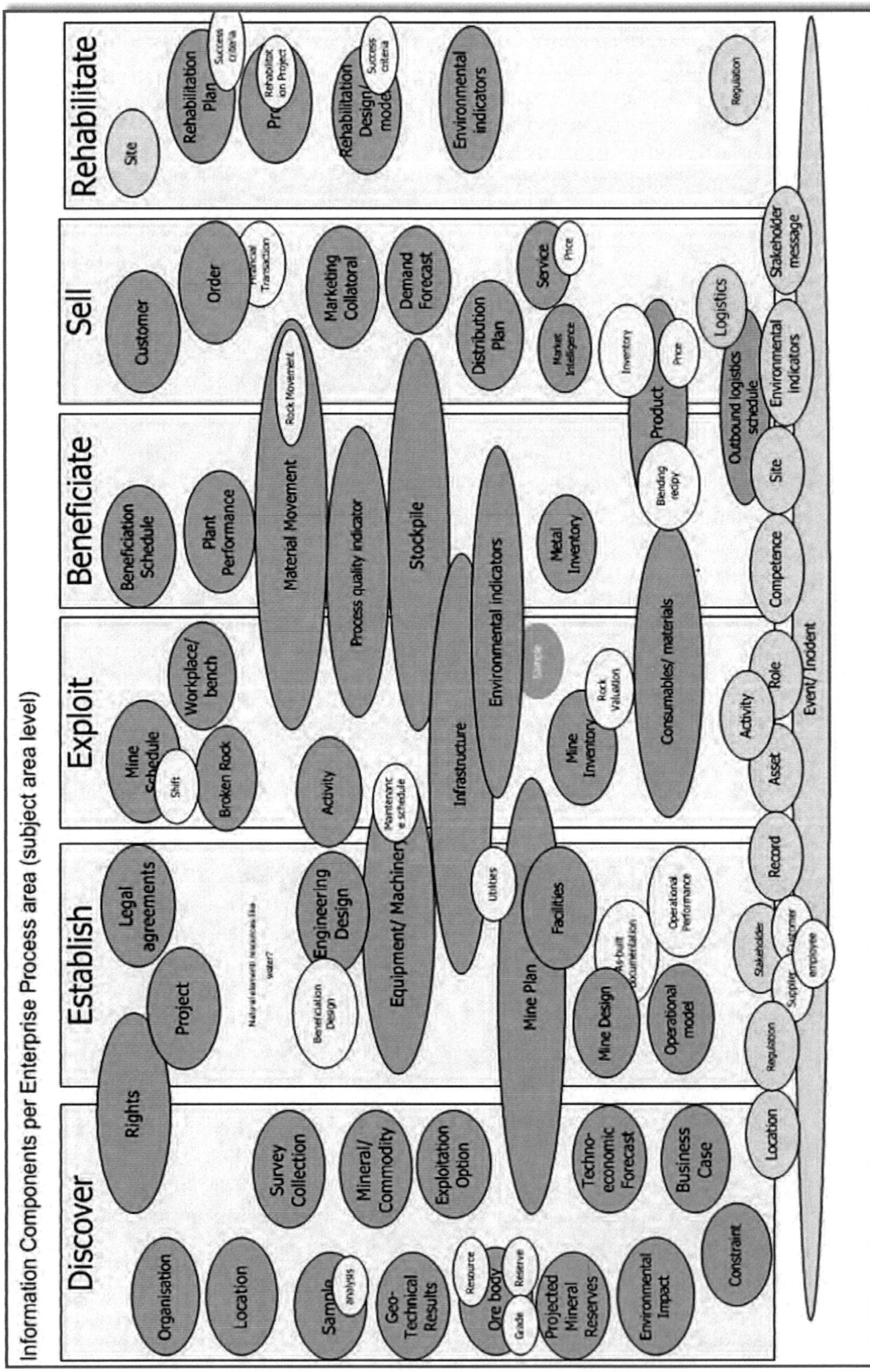

Figure 1 The mining value chain (Anon, 2010)

2.2 Planning and optimization techniques

Figure 1 showed the generic mining value chain for a surface mining operation. Each stage in the value chain has its own optimization techniques. Cumulatively, for the whole value chain, the techniques can be in the order of hundreds. The list below, though not exhaustive, shows a list of optimization techniques that have been applied to the optimization of different stages of the mining value chain.

- Integer programing
- Tabu search
- Multi criteria decision analysis
- Enterprise optimization
- Split shell optimization
- Artificial neural networks
- Network flow analysis
- Quadratic programming
- Combinatorial optimization
- Mixed integer programming
- Floating cone optimization
- Maximum value neighbourhood
- Simulated annealing
- Genetic algorithm
- Particle swarm optimization
- Ant colony optimization
- Linear programming
- Mixed integer programming
- Non-linear programming
- Geometric programming
- Graphing theory
- Korobov

The above mentioned techniques have been applied in four (4) main areas of optimization in in surface mines. These areas are cut-off grade, production scheduling, mining limits and the mining value chain. Each optimization area, has several optimization techniques that are used. The most common technique under each area of optimization is explained in sections that follow.

3 CUT-OFF GRADE OPTIMIZATION

The concentration of minerals in the earth's crust is critical to mining as it defines the cut-off grade (Lane, 1997). Cut-off grade is defined as the grade that distinguishes ore and waste (Dagdelen, 2001). The determination of cut-off grade uses the economics of breakeven analysis. This cut-off grade maximize the net present value and is a function of mining, economic and processing parameters. The economic parameters are influenced by the market and as a result the effect of market changes has a significant effect on the cut-off grade (Anon, 2016b). The choice of cut-off grade is therefore important as it defines the amount of reserves that are exploitable and hence the profitability and life of the mine (Dagdelen, 2001).

A number of techniques have been used to optimize the cut-off grade and to determine how the cut-off grade changes with changes in market conditions. These techniques include the control theory (Yingliang, 1998), genetic algorithm (Cetin and Dowd, 2002), equivalent grade factors (Ataei and Osanloo, 2003a), Lane's cut-off grade theory and marginal grade index system (Xi, 1985), combination of genetic algorithm and grid search (Ataei and Osanloo, 2003b) and heuristics (Craig *et al,* 2014). These techniques have been applied to a variety of commodities and different mining regions. The variability in these techniques is due to the varying nature of the mining optimization problems presented to each individual mine. The nature of the problems result in a real world problem being forced to fit the techniques compromising the quality of the solution obtained thereafter (Craig *et al,* 2014). Most of these techniques do not usually take into account blending considerations rendering them ineffective (Rendu, 2008).

The need to optimize mines with a number of commodities, stockpiles and multiprocessing streams is not solved by the mentioned techniques (Dagdelen and Kawahata, 2007). A metaheuristic technique is used to address such challenges. This technique works for multi metal deposits (Abrand *et al*, 2014; Craig *et al*, 2014). Although the technique is relatively new, it has proved to give better results even when used for a single metal deposit (Craig *et al*, 2014). The technique however, requires extensive knowledge of the limitations and interpretation of results.

Metaheuristic technique is composed of genetic algorithm and other optimization methods. The self-adaptiveness of genetic algorithm allows it to be blended with other techniques easily (Craig *et al*, 2014). The technique consists of three optimization levels which are the master level and two slave levels (Craig *et al*, 2014). The master level is responsible for managing cut-off grade variations and the extraction sequence permutations (Craig *et al*, 2014). Additionally, the slave level linear program (Program that is responsible for the optimum flow of material for the different processes and stockpiles) is managed by the master level (Craig *et al*, 2014). Metaheuristic technique requires a definition of the ultimate pit or mining limits, capacities for mining and processing, forecasts on operating costs and commodity prices and a distribution for the grade in the orebody (Abrand *et al*, 2014). The components of the technique are shown in Figure 2 with each function for individual technique.

The technique generates an initial population by producing an extraction sequence subject to economic and mining constraints (Craig *et al*, 2014). Each member of the population is created its own grade profile that is multi process, a stockpile cut-off grade profile and stockpile availability profile. Once the profiles have been created, fitness value (which in this case is the net present value) is calculated and the members of the population are ranked according to fitness values (Craig *et al*, 2014). The slave level (linear programming optimization technique) is called for by the master level during fitness value calculation. Iterative procedures are done to produce offspring population that repeats the process thereby producing a fitness value of better quality (Craig *et al*, 2014). The ability of genetic algorithm to be used in combination with other optimization techniques produces an adaptable cut-off grade optimization technique. Being based on nature, the suitability of the technique to varying complexities of real world problems gives it a competitive advantage over other techniques (Craig *et al*, 2014) and this technique is heavily used by Maptek software packages.

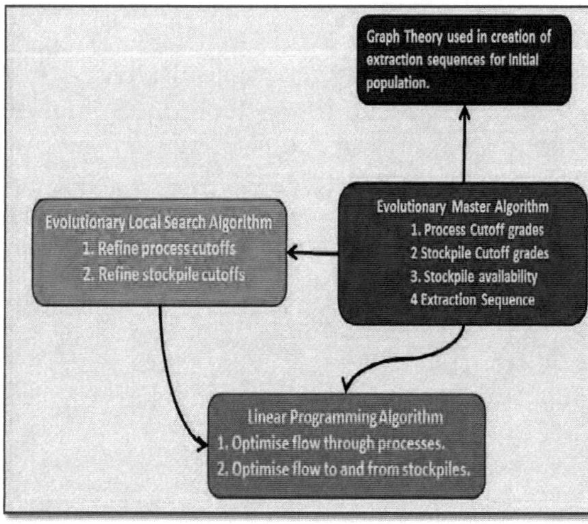

Figure 2 Metaheuristic optimization for cut-off grades (Craig *et al*, 2014)

4 PRODUCTION SCHEDULING

Production scheduling in surface mines is defined as the sequence in which blocks of the ore orebody are extracted and moved in order to maximize the net present value subject to mining, economic and processing constraints (Caccetta and Hill, 1987). The definition of block models gives a starting point for the production scheduling optimization task. Block models divide the orebody into small blocks (Khan and Niemann-Delius, 2014) which are dependent on pit slopes, orebody dip, grade distribution and the type of equipment being used (Caccetta and Hill, 1987). Each block is assigned certain attributes using either Kriging or inverse distance interpolation techniques. These attributes include density, grade and tonnages. Using the cut-off grade optimization techniques, the blocks are then divided into ore and waste (Khan and Niemann-Delius, 2014).

The size of the blocks created can be in the order of millions thus requiring movement over a long period of time. This movement of blocks demands adequate scheduling considering the physical and operational constraints (Khan and Niemann-Delius, 2014). Production scheduling thus requires efficient optimization techniques in order to realise the benefits. Several optimization techniques have been applied. These techniques include mixed integer programming (Gershon, 1983), branch and cut (Caccetta and Hill, 2003), dynamic programming (Onur and Dowd, 1993), fundamental tree algorithm (Ramazan, 2006) and Lagrangian parameterization (Dagdelen and Johnson, 1986).

The limitations of these techniques lies with computational costs when applied to a complex real world problem. Although the techniques have been applied to a variety of commodities, their use and the solutions they give are still questionable on quality. Easier and less computational cost techniques have increased usage among practitioners in optimization of production scheduling (Khan and Niemann-Delius, 2014). These techniques include genetic algorithm (Denby and Schofield, 1994), simulated annealing (Kumral and Dowd, 2002) and ant colony optimization (Niemann-Delius and Sattarvard, 2013). The techniques highlighted before have been used extensively. However, particle swarm optimization is used more often by planning and optimization practitioners.

Particle swarm optimization (PSO) technique is stochastic (Anon, 2016a) and is based on the social interactions of group of living animals or creatures for example birds, fish and herds of animals (Khan and Niemann-Delius, 2014). In each group, there are individual animals or creatures. These individuals in the concept of PSO represent a potential solution to the production scheduling optimization problem. The application of PSO in production scheduling starts by generating a list of blocks to be mined (Khan and Niemann-Delius, 2014). The generated blocks have precedence relationship that makes the production scheduling problem "an optimum depth determination problem" (Angeline, 1998). The technique selects a block at random and the block is assigned the period it was selected. Once a block has been selected and assigned a period, the population of blocks from which the selection was done is updated (Khan and Niemann-Delius, 2014). The process is repeated until there are no more blocks to choose from or when the number of periods to assign each block are finished for that particular solution (Khan and Niemann-Delius, 2014).

The solutions obtained are encoded by determining the maximum depth to be mined for each column of blocks and the depth variable is updated for each iteration done (Angeline, 1998). Back transformation is applied to the depth variable in order to obtain the period assigned to each block or period that can be assigned to that particular block. Constraints are included by normalising the solution and applying a penalty method for each violation of the constraint capacities (Michalewicz and Schoenauer, 1996).

5 OPTIMIZATION OF MINING LIMITS

The definition of what is economically mineable is defined by the mining limits. These limits (size and shape) of an open pit are determined by applying design and production constraints (Hustrulid *et al*, 2013) These limits determine blocks of ground that should be mined and blocks of ground that should remain (Dagdelen, 2001). The determination of which blocks to move is based on assumed mining costs, processing costs and commodity prices. This enables each block to be given an economic value which will further determine if the cost to remove waste above it, is less than the block value (Dagdelen, 2001). The basis of this analysis is the economics of breakeven, where the undiscounted profit for each block can just cover the undiscounted mining cost for the same block (Dagdelen, 2001).

Techniques used for the optimization of mining limit include Lerchs-Grossmann (Lerchs and Grossmann, 1965), floating cone method III (Elahizeyni *et al*, 2012), split shell (Gallagher and Kear, 2001), graph theory (Abbaspour, 2011), dynamic programming (Wight, 1989), and artificial neural network (Tolwinski and Underwood, 1996). From the techniques mentioned, the Lerchs-Grossmann is widely used and it is from this technique that most optimization has been done. The Lerchs-Grossmann (LG) technique is used to determine the pit that is economic and mineable. The first conception of the technique was by Lerchs and Grossmann in 1965. LG technique uses the concept of dynamic programming in the development of the ultimate mining limit. The constraints to the development of the mining limit are the geometry of the pit, which is the slope of the pit and is determined geotechnically (Lerchs and Grossmann, 1965). The technique assumes that every point within the mining boundaries has got a known cost and mineral value or mineral content. LG technique seeks to maximize the volume of ore extracted (Anon, n.d).

The approach divides the pit into a two dimensional (2D) pit, making parallel vertical sections (Lerchs and Grossmann, 1965). The contour of a section is then determined by three lines representing pit bottom and two walls at a certain angle. The extraction costs for the specific section are determined from limiting the pit to three lines. Iterations are done to determine sets of line configurations that produces the best results (Anonymous, n.d). The angle for the two walls is assumed to constant throughout the optimization process (Ataei and Osanloo, 2000).

The technique is simple in two dimensional form and more complex when considered in three dimensional form. Once the orebody has been divided into parallel vertical sections, blocks are created. Each block is assigned a block economic value (BEV) (Anon, 2016a). The summation of BEVs for blocks in a single column gives block column values (M_{ij}) (where the row = i and the column = j) (Lerchs and Grossmann, 1965). M_{ij} value shows the amount of profit which can be possibly made under given assumptions when that block is mined. After calculating the M_{ij} values, pit values (P_{ij}) are then calculated using the formula;

$P_{ij} = M_{ij} + \max (P_{i+k,\, j-1})$ ....(1)

where the values of k can be = [-1, 0, 1]

The maximum column contributions can be seen from pit values if the block is to be part of the contour. The tracing of the ultimate pit is done using arrows drawn during the calculation of the pit values (Anon, 2016a). The values that are included in the ultimate mining limits are the ones with positive values for P_{ij} (Lerchs and Grossmann, 1965). Negative pit values are not included in the ultimate pit. This ultimate pit is one which has the maximum cumulative value (Hustrulid *et al*, 2013). Unlike other optimization techniques for mining limits, Lerchs-Grossmann technique defines the pit limits optimally, maximizing the undiscounted profit. Other techniques in particular heuristic techniques, give sub-optimum results (Dagdelen, 2001).

6 THE MINING VALUE CHAIN

The techniques described so far have been applied to specific areas within the stages of the mining value chain. Their approach therefore, isolates and optimize one area which results in giving best results for that particular stage and or process. Optimization of the mining value chain uses a different approach which incorporates all the stages of the mining value chain. This approach, from a systems engineering perspective gives better results as the effect of optimizing one area on other areas is looked at. There are, however complexities when having an integrated approach. One of the challenges is having to deal with a number of departments which may not see the benefit of integrated optimization.

The techniques that have been applied in optimizing the mining value chain are the mineral resource management system (Badenhorst *et al*, 2003), theory of constraints (Mathu, 2014), reconciliation and stockpiling (Macfarlane, 2013) and enterprise optimization (Whittle and Burks, 2010). The use of these techniques, which are relatively new considering the age of mining business, have received great appreciation across the planning and optimization field. It is the ability of the technique to incorporate all stage optimization to "global" optimization that enable them to give better results. Their use, however requires extensive discipline across all departments and the technique that have gathered more appreciation is the enterprise optimization developed by Whittle and others (Whittle and Burks, 2010).

6.1 Enterprise Optimization Applied to the Mining Value Chain

According to Whittle and Burks (2010), enterprise optimization is a technique that addresses all areas of the mining value chain for single and multicommodity mines, for mines with multiple pits and where there are issues of stockpiling and or blending considerations. Enterprise optimization technique divides the value chain into ten steps as shown in Figure 3.

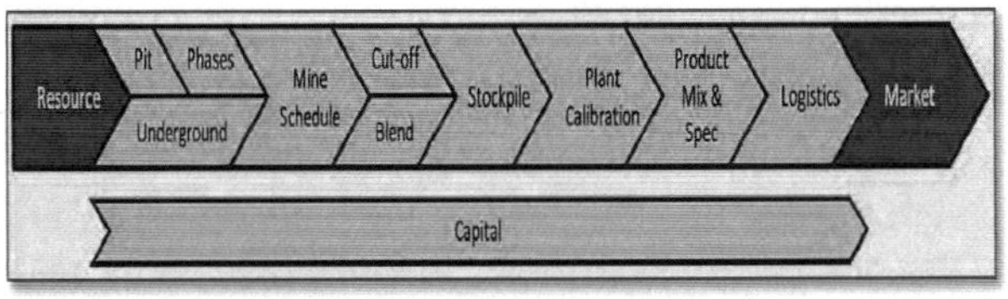

Figure 3 Enterprise optimization applied to the mining value chain

The mining sector is a price taker. Any stage or process between the resource and the market can be optimized as it is under the control of planning and optimization practitioners. These stages are simultaneously optimized thereby removing the competition for capital resources and minimizing the limitations of techniques that optimize in isolation. The process uses a search algorithm and some linear programming in order to obtain effective simultaneous optimization (Whittle and Burks, 2010). The steps in the optimization process are;

6.1.1 Pit Design

Lerchs-Grossmann technique or other conventional optimization methods are used to determine the pit shells and ultimately the final pit layout (Whittle and Burks, 2010).

6.1.2 Phases

Subject to the constraint that an outer phase cannot be mined before an inner phase when increasing in depth, phases in a pit can be mined one after the other or at the same time (Whittle, 2010).

6.1.3 Scheduling

Higher grades drive the optimization process, making the average grade high at the beginning of extraction period. A variety of scheduling techniques can be used to solve the scheduling problem (Whittle, 2010).

6.1.4 Cut-off grade

Cut-off grade techniques are applied to demonstrate that a high net present value is obtained by bringing high grades in the early years and reducing the life of mine. The cut-off grade used is dynamic, compared to the single cut-off grade, traditionally applied (Whittle and Burks, 2010).

6.1.5 Stockpile

Marginal material is stockpiled and processed later when commodity prices rise or when there is plant spare capacity. Costs related to rehandling and changing ore characteristics due to material oxidation should be taken into account (Whittle, 2010).

6.1.6 Simultaneous Optimization

The first five steps are optimized concurrently with the use of Prober software using the pit shells obtained in the first step. The management of first five steps is in most cases done within the mining department (Whittle and Burks, 2010).

6.1.7 Process Calibration

The processing of the ore results in changing recoveries and throughputs. This step tries to increase the residence time, thus improving the recovery and throughput (Whittle, 2010).

6.1.8 Product Specification

In order to get the required product, ore to the plant may be blended. Although difficult to achieve, blending provides an optimization opportunity (Whittle and Burks, 2010). Blending can be based on geometallurgical properties or on the basis of grade or product size (Anon, 2016b).

6.1.9 Logistics

The product produced may be sold in small quantities or in bulk and can be shipped using different modes of transport (Whittle and Burks, 2010). This product can be bought locally or internationally creating differences in prices and capacities to be transported. The Optimiser software, incorporates all these challenges (Whittle and Burks, 2010).

6.1.10 Capital

The nine steps described, optimize the processes using already available equipment, thus looking at all possible techniques to exploit bottlenecks before any capital expenditure (Whittle and Burks, 2010). Once the bottlenecks are fully exploited, considerations are done on using capital for expansion projects.

The use of enterprise optimization though relatively new, eliminates optimization techniques that work in isolation, which in most cases are counterproductive to overall business profitability.

7 CHALLENGES IN PLANNING AND OPTIMIZATION OF SURFACE MINES

Mining is based on a depleting resource and the prices of commodities and costs are forecasted using different techniques. These two are the major challenges in planning and optimization processes. Commodity prices have become very volatile and extremely unpredictable. The depleting nature of the resource and the existence of unknown geological structures exacerbate the problems to planning and optimization. The techniques described herein attempt to develop ways of fully exploiting the resources. Each optimization area has got its own challenges and the techniques used also have their own limitations.

Environmental considerations are becoming extremely important in modern mining. There is an increase in stakeholder requirements and this increase is coupled with uncertainty of the requirements. Inclusion of these factors and obtaining a balance is very difficult, and failure to include them only leads to decreased profits. Additionally, the techniques available assumes a number of parameters which in their nature may be dynamic. This prompts planning and optimization practitioners to force real world problems to fit into the developed techniques. The solution obtained thereafter does not represent the actual real world solution.

Although heuristic techniques have tried to combine other techniques in order to solve different problems, there exist a difficulty in creating one technique that can address all the challenges. The use of particle swarm optimization requires the application of other techniques as it converges prematurely giving sub-optimum solutions (Khan and Niemann-Delius, 2014). Techniques that are not heuristic create formulations that are too large and as such they are costly to solve and difficult to formulate requiring a significant amount of time to solve (Caccetta and Hill, 1987).

Smit and Lane (2010) identified additional challenges to the field of planning and optimization. These challenges include;

- The difficulty in connecting labour profiles and production profiles presenting complications in developing the operating cost models;
- The continuous changes in mining layouts which shows that there is no sticking to planned layouts. The newly implemented layouts are not optimized which affects optimization in the specific area and the whole value chain;
- The relaying of planned objectives across all levels of work has a disconnection which results in planners optimizing against the strategic vision of the company as they locally try to optimize within their departments;
- The time allocated to conduct optimization process is very little considering the number of alternatives and scenarios that must be run in order to get an optimized solution; and
- The response to continuously changing business environment is very poor especially to changing exchange rates, commodity prices and costs. This suggests inefficiency of optimized plans.

The failure of most mines to meet expected targets shows that there is a gap in planning and optimization. Lane (1997) also identifies that failure to optimize cut-off grade is a result of not using different cut-off grades, which are dynamic in nature.

8 CONCLUSION

The paper focused on discussing planning and optimization techniques that are used in surface mining operations and highlighting the challenges that still exist in this area. Four main areas were found to be the focus of optimization and these areas are cut-off grade, production scheduling, mining limits and the mining value chain. Each area of optimization has a number of planning and optimization techniques. The common techniques are genetic algorithm for cut-off grade, particle swarm optimization for production scheduling, Lerchs-Grossmann for mining limits and enterprise optimization for the mining value chain. Despite extensive use of these techniques, challenges still exist to planning and optimization. Some of these challenges include the difficulty in accurately forecasting costs, commodity prices and exchange rates. These challenges are exacerbated by the recent increase in demands from stakeholders and environmental demands which must be included in the optimization process. However, the use of planning and optimization techniques remain crucial to the profitability of mining operations. Their use therefore needs modifications or the development of new techniques.

REFERENCES

Abbaspour, H, 2011, *Optimum Ultimate Pit Limit (UPL) Design by Graph Theory*, (Online), Available at; https://www.researchgate.net/publication/301295358 Accessed on 12 September 2016

Abrand, J, Fa-ben, L, Chun-ni, B, and Qing-hua, G, 2014, *The optimization and application of cut-off grades of multiple metal open pit mines based on equivalent grade*, Metallurgical and Mining industry, Number 6. Page 83 – 91

Angeline, P, 1998, *Evolutionary optimization versus particle swarm optimization: philosophy and performance differences*, in Evolutionary Programming VII, V. W. Porto, N. Saravanan, D. Waagen, and A. E. Eiben, Computer Science, Volume 1447 page 601–610

Anon, 2010, *Exploration and Mining Business Reference Model*, (Online), Available at; opengroup.co.za/sites/default/files/EM%20Model%20graphic%20v%201.00_1.pdf Accessed on 12 September 2016

Anon, 2016a, *Applied Operations Research in Mineral Resource Management Course lecture notes,* University of the Witwatersrand, Johannesburg, South Africa, June 2016

Anon, 2016b, *Planning and Optimisation of Surface Mines Course lecture notes,* University of the Witwatersrand, Johannesburg, South Africa, August 2016

Anon, n.d, *Optimized pit design*, (Online), Available at; http://files.carlsonsw.com/mirror/manuals/Carlson_2017/source/Geology/Block_Model/Optimized_Pit_Design/Optimized_Pit_Design.html Accessed on 11 September 2016

Ataei, M. and Osanloo, M, 2000, *Using 2d Lerchs and Grossmann algorithm to design final pit limits of Sungun copper deposit of Iran*, International Journal of Engineering, Volume 13, Number 4, page 81 – 89

Ataei, M. and Osanloo, M, 2003a, *Using equivalent grade factors to find the optimum cut-off grades of multiple metal deposits,* (Online), Available at; https://www.researchgate.net/publication/223494840_Using_equivalent_grade_factors_to_find_the_optimum_cut-off_grades_of_multiple_metal_deposits Accessed on 7 September 2016

Ataei, M. and Osanloo, M, 2003b, *Using a Combination of Genetic Algorithm and Grid Search Method to Determine Optimum Cut-off Grades of Multiple Metal Deposits*, International Journal of Surface Mining, Reclamation and Environment, Volume 18, Number 1, page 60 – 78

Badenhorst,S.N, Diedericks, A.W, Fouche, P.A.P, 2003 *A systems approach in implementing minerals resource management at a base metal mine (Rosh Pinah Zinc Corporation, Namibia),* The Journal of the South African Institute of Mining and Metallurgy, January 2003, page 11- 22

Caccetta, L and Hill, P.S, 1987, *Optimization techniques for open pit mine scheduling*, (Online), Available at; www.mssanz.org.au/MODSIM99/Vol%203/Hill.pdf Accessed on 7 September 2016

Cetin, E. and Dowd, P.A, 2002, *The Use of Genetic Algorithms for Multiple Cut-off Grade Optimisation*, Application of computers and operations research in the mineral industry: Proceedings of the 30th International Symposium, page 769 – 780

Craig, S, Deb, K and Myburgh, C, 2014, *Applying Modern Heuristics to Maximising NPV through Cut-off Grade Optimisation*, (Online), Available at;

www.maptek.com/pdf/insight/AUSIMM_Applying_Modern_Heuristics.pdf Accessed on 12 September 2016

Dagdelen, K, 2001 *Open pit Optimization – Strategies for improving economics of mining projects through mine planning* 17th International Mining Congress and Exhibition of Turkey, page 117 -121

Dagdelen, K and Kawahata, K, 2007, *Value creation through strategic mine planning and cut-off grade optimization* (Online), Available at; https://www.researchgate.net/publication/285891275_Value_creation_through_strategic_mine_planning_and_cutoff-grade_optimization Accessed on 12 September 2016

Dagdelen, K and Johnson, T.B, 1986, *Optimum open pit mine production scheduling by Lagrangian parameterization*, Proceedings of the 19th International Symposium Application of Computers and Operations Research in the Mineral Industry (APCOM '86), page 127–142

Denby, B and Schofield, D, 1994, *Open-pit design and scheduling by use of genetic algorithms*, Transactions of the Institution of Mining and Metallurgy, Section A: Mining Technology, Volume 103, page A21–A26

Elahizeyni, E, Kakaie, R, and Yousefi, A, 2012, *A new algorithm for optimum open pit design: Floating cone method III*, Journal of Mining and Environment, Volume 2, Number 2, page 118 – 125

Gallagher, M.S and Kear, R.M, 2001, *Split shell open pit design concept applied at De Beers Venetia Mine South Africa using the Whittle and Gemcom software*, The Journal of the South African Institute of Mining and Metallurgy, Volume 101, Number 8, page 401- 410

Gershon, M.E, 1983, *Mine scheduling optimization with mixed integer programming*, Mining Engineering, Volume 35, Number 4, page 351–354

Hickman D, n.d *Particle swarm optimization for energy minimization of molecular systems,* (Online), Available at; files.matlabsite.com/docs/thesis/th930929039.pdf Accessed on 7 September 2016

Hustrulid W, Kuchta M, and Martin R, 2013 *Open Pit Mine Planning and Design Volume 1 - Funtamentals,* 3rd Edition, CRC Press, Boca-Raton, Florida, page 409,450

Lane K.F, 1997 *The Economic Definition of Ore – Cut-off grades in Theory and Practice,* 3rd Edition, COMET Strategy Pty Ltd Queensland, Australia, page 4 - 5

Lerchs, H and Grossmann, I.F, 1965, *Optimum design of open pit mines*, Canadian Institute of Mining Transactions, Volume 68, page 17–24

Khan, A, Niemann-Delius, C, 2014 *Production Scheduling of open pit mines using particle swarm optimization algorithm*, Advances in Operations research, Volume 2014, article 208502, page 1-9

Kumral, A and Dowd, P.A, 2005, *A simulated annealing approach to mine production scheduling*, Journal of the Operational Research Society, Volume 56, Number 8, page 922–930

MacFarlane, A.S, 2013, *Reconciliation along the mining value chain,* The Journal of the South African Institute of Mining and Metallurgy, Volume 113, Number 8, page 11- 22

Mathu, K, 2014, *Applying the Theory of Constraints in the South African Coal Supply Chain*, Mediterranean Journal of Social Sciences, Volume 5, Number 9, page 131 – 141

Michalewicz, Z and Schoenauer, M, 1996, *Evolutionary algorithms for constrained parameter optimization problems*, Evolutionary Computation, Volume 4,Number 1, page1–32

Niemann-Delius, C and Sattarvand, J, 2013, *A new metaheuristic algorithm for long-term open-pit production planning*, Archives of Mining Sciences, Volume 58, Number 1, page 107–118

Onur, A.H and Dowd, P.A, 1993, *Open-pit optimization. Part 2. Production scheduling and inclusion of roadways*, Transactions of the Institution of Mining and Metallurgy Section A: Mining Technology, Volume 102, page 105–113

Ramazan, S, 2006, *The new fundamental tree algorithm for production scheduling of open pit mines*, European Journal of Operational Research, Volume 177, Number 2, page 1153–1166

Rendu J, 2008 *Introduction to Cut-off Grade Estimation,* 1st Edition, Society for Mining, Metallurgy and Exploration, Littleton, Colorado page 16

Smit, A and Lane, G, 2010, *Mine optimization and its application using the Anglo Platinum Mine Optimization Tool,* The 4th International Platinum Conference, Platinum in transition 'Boom or Bust, The Southern African Institute of Mining and Metallurgy, page 317 – 328

Tolwinski, B and Underwood, R, 1996, *A scheduling algorithm for open pit mines*, Journal of Mathematics Applied to Business and Industry, Volume 7, page 247 – 270 (Online), Available at; http://imaman.oxfordjournals.org/content/7/3/247.full.pdf Accessed on 8 September 2016

Whittle, G, 2010, *Enterprise Optimization*, (Online), Available at; www.whittleconsulting.com.au/documents/articles/Enterprise%20Optimisation.pdf Accessed on 12 September 2016

Whittle, G and Burks, S, 2010, *Simultaneous mining and mineral processing enterprise optimization for the platinum industry*, The 4th International Platinum Conference, Platinum in transition 'Boom or Bust, The Southern African Institute of Mining and Metallurgy, page 329 – 338

Wright, E. A, 1989, *Dynamic Programming in Open Pit Mining Sequence Planning: A Case Study*, Proceedings of the 21st Application of Computers and Operations Research in the Mineral Industry Symposium, page 415 – 422

Xi, X, 1985, *The marginal grade index system and lane's cut-off grade theory*, The Journal of Quantitative and Technical Economics. 1985, Volume 4, page 29 – 34

Yingliang, X, 1998, *Optimization of cut-off grade in open pit based on control theory*, Transitional Nonferrous Metallurgical Society of China, Volume 8, Number 2, page 353 – 356

Image Processing for Characterization of Iron Ore Pellets Using Geostatistical Instruments

T. Y. Yunsel, O. Sivrikaya
Adana Science and Technology University, Adana

ABSTRACT Approximately in the last 20 years, the computational methods have advanced very fast and affected a vast range of discipline. Also, new dimensions in the applications have appeared as parallel to development of computing technology. One of the study methods requiring exhaustive data processing in the earth related science is geostatistics.

Iron ore pellets are produced by using magnetite or hematite concentrate together with a binder and water addition. The raw pellets are sintered to produce the sufficiently strong pellets. Some parameters such as type of the binder, addition amount of the binder and sintering conditions affect the quality of the product pellets. The physical quality of the product pellets can be determined by physical pellet testing methods. Physical quality of pellets can be considered as function of sintering conditions.

In this study, sintering properties of magnetite pellets produced with bentonite binder are examined by geostatistical image processing method. Polished samples of the pellets were prepared and micro images were taken by a stereomicroscope. The micro images are processed by ISATIS software and converted to digitised images. Models are analysed using geostatistical instruments like kriging and image filtering.

The obtained data are used to examine the sintering properties of the pellet matrix. The differences in the pellet matrix (core, inner shell and outer shell) may be used to understand the sintering properties of the pellets. The results may be correlated with the physical quality parameter of the pellets.

1 INTRODUCTION

High grade iron ores can be directly utilized, after crushing and sizing, in reduction facilities. However low grade iron ores require upgrading after grinding generally through concentration. Sufficiently upgraded concentrate are agglomerated to be charged to reduction facilities.

Iron ore pelletizing is a world-wide applied agglomeration method for iron ore concentrates to be used in iron-steel industry Meyer, 1980; Ball et al. 1973). Iron ore pellets are produced by using magnetite or hematite concentrate together with a binder and water addition.

This process requires a lot of stages; pre-treatment of the ores, addition of binders and additives, balling of the mixtures, green ball production, thermal treatment (sintering) of the green balls. The green pellets are sintered to produce the sufficiently strong pellets.

The characters of the concentrate at the hand, types of the binders and additives and sintering conditions affects the quality of the product pellets. Physical, chemical and metallurgical properties determine the quality of the product pellets. The quality of pellets are determined applying of related pellet standards such as ASTM or ISO standards. Many researches were conducted to determine optimum iron ore pellet production in literature.

The sintering conditions are accepted as one of the important parameters affecting the quality of products pellets especially compressive strength and porosity. It is well known that the sintering conditions affects the bonding of the ore grains found in pellet matrix. Therefore it is thought that physical qualities of the pellets may be correlated with their physical visual structures.

In this study cross-sectional physical visual structure of the iron ore pellets were analysed by image processing method. This method is based on kriging which is basically used in the geosciences in two-three dimension (Yünsel, 2012; Ersoy, Yünsel and Cetin, 2004). The study basically consists of moulding the iron ore pellets in a resin, splitting iron ore pellets to half, polishing of the samples, digitising taken images and numerical & graphical analyses respectively.

2 AIM OF THE STUDY

The aim of the study was to show possible characterization of the sintering properties of pellets using digital imaging. Therefore an iron ore pellet produced with a binder addition and sintered at determined temperature was selected. The sintering properties of the pellets produced with different binder addition and sintered at different temperatures can be compared.

Sintering phenomenon which is one of the important pellet properties gives the durability and strength to the pellets is investigated through image processing techniques. For this reason geostatistical software was used on the images of the sintered pellets produced with bentonite binder addition and sintered at a certain temperature and duration.

In the present study the magnetite concentrate pellets produced using laboratory-scale equipment, were used as pellets samples for the image examination. Polished samples were prepared using the polishing equipment shown in **Figure 1.** After the preparation of the polished samples the photos were taken using the stereomicroscope equipped with a 3MP camera (**Figure 2**).

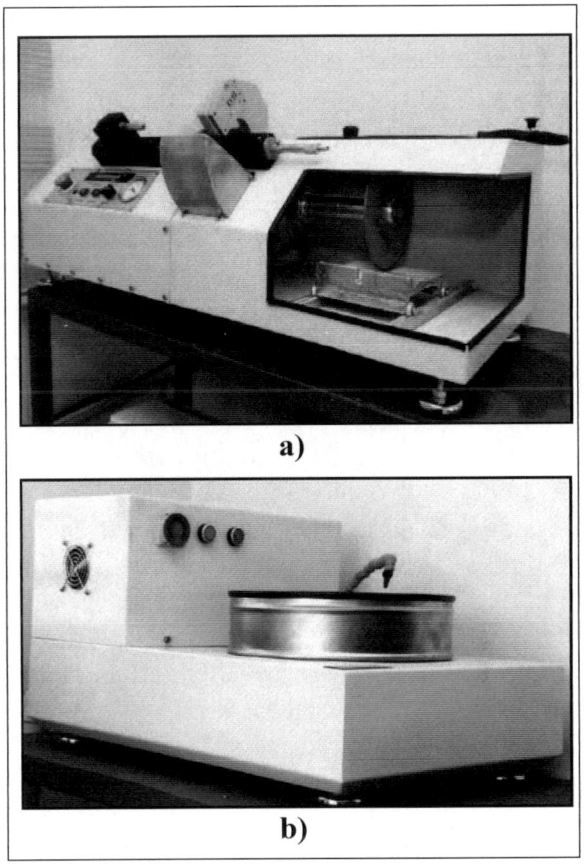

a)

b)

Figure 1. Polished sample preparation equipment: **a)** Disc cutter-trimmer tool **b)** Polishing tool.

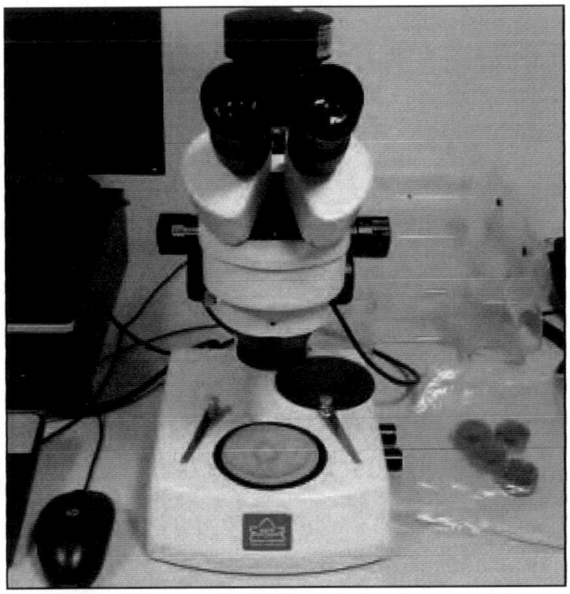

Figure 2. Stereomicroscope for getting sample photos.

Using these polished pellet samples, the different phases (core, inner shell or mantle and shell) in the sintering matrix were tried to be identified by image processing techniques.

It is thought that, the sintering parameters affecting the strength of pellets can be associated by graphically and statistically in this way. Image processing based pellet analyses were studied previously by some researchers (Nellrosa and Thurley, 2011).

The photos taken by stereomicroscope's camera were converted to digitized models by importing photo into software media. Basically ISATIS software (Geovariances, 2006) is used for image analysis. The imported and digitized images were processed and analysed using kriging techniques and structure of pellets were modelled.

3 MATERIAL AND METHODS

The pellets used in the image processing were produced by laboratory-scale pelletizing as explained elsewhere (Sivrikaya and Arol, 2011; 2012; 2013a; 2013b; Sivrikaya et.al. 2013). Some of the pellets were selected to examine their cross-sectional images through image processing. Pellets used in the analyses is given in Figure 3.

Raw moulded pellet samples were selected randomly and prepared for taking photo. In the polished samples preparation the same steps likewise regular rock samples preparation were followed.

During polishing process, polishing performance was always monitored to get best image from the samples because of polishing quality affects the photo quality.

After preparation of polished samples, a few photos of samples were taken to distinguish and model the internal structural difference. Because it is expected that the different phases may be associated with bonding of the grains and the pellet strength which is tried to observe using microscopic image analysis of polished pellet samples.

Figure 3. Green and dry pellets and sintering of pellets in the box-furnace.

4 EXAMINATION OF PELLETS

The analysed sample was produced using 0.66% bentonite binder. The sample was sintered at 1000°C for 30min after dried for 24hr in a laboratory scale oven. After that the sample is splitted into 2 equal pieces to investigate the transformations and differences in the sintering matrix. One of these samples was selected for polishing and subjected to regular polished sample preparation process steps for microscope analysis. A successful polished sample, directly affects the image quality to be digitized. Thus samples were polished as carefully as possible. A few photos of sample were taken and the clearest one was selected for digitizing. The selected image of sample processed for analysis is presented in **Figure 4.**

Figure 4. Stereomicroscope image of the pellet sample produced using bentonite binder.

The obtained photo image is digitized as importing into analysing software so that it fits the grid design presented in **Table 1.** Grid design parameters are leaved completely up to user's choice. Large number of the grid size defining image resolution, affects the time consumption for calculations. On the contrary, small number of grids reduces the image resolution which makes difficult to distinguish the inner structure parts of sample. Thus optimum grid design parameters should be chosen according to parameters like computer specifications, raw image quality etc. An imported and digitized sample image can be seen in **Figure 5.** Raw image colour scale is a random scale and these scale is adjusted according to grid design of imported image. In this image, a number is assigned to every grid node in accordance with the colour value of raw image at assigned node.

Table 1. Importing grid design of raw image

	X	Y
Origin	0	0
Mesh	1	1
Nodes number	696	672

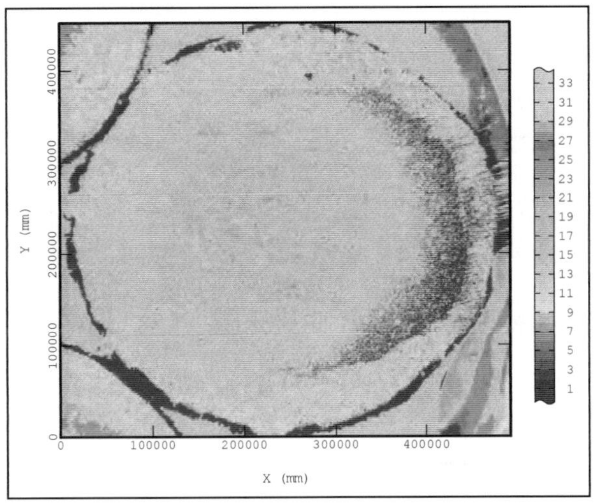

Figure 5. A digitized image of imported raw image of selected pellet sample.

After importing and digitising the raw image, first step is trimming the locations outside of the investigation area to take into account the only sample surface for further calculations. A cropped image is shown in **Figure 6.** As it can be seen in **Figure 3, 4, 5,** an imported image may not contain only the studied area. Thus limitations need to be done by either auxiliary software or this software's itself by creating polygons around study area.

A detailed image processing is required prior to a judgment of structure and modelling image data to get more detail making easy to distinguish the structures. To get a better image, a series of process including image filtering is applied to digitised image. Image filtering is very essential part of smoothing and revealing more detail of raw images. Kriging is applied to smooth the raw data similar to application in geosciences in 2-dimension.

Figure 7 presents the variogram modelling to be applied for image filtering. Variogram analysis is carried out on digitised image data and numeric variogram parameters is given in Table 2.

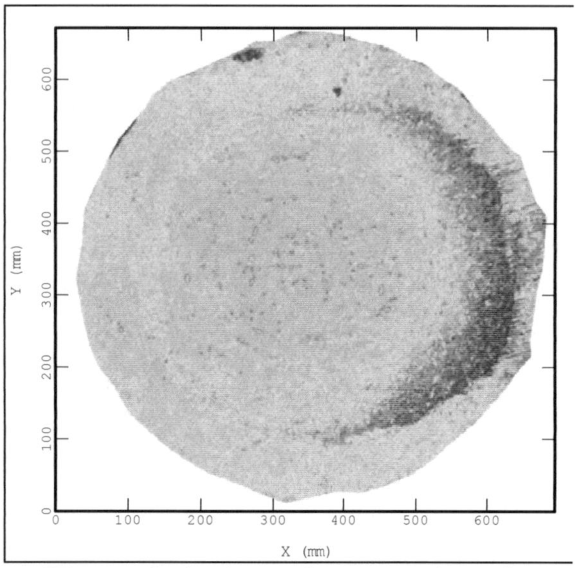

Figure 6. A trimmed presentation of imported and digitized raw sample image.

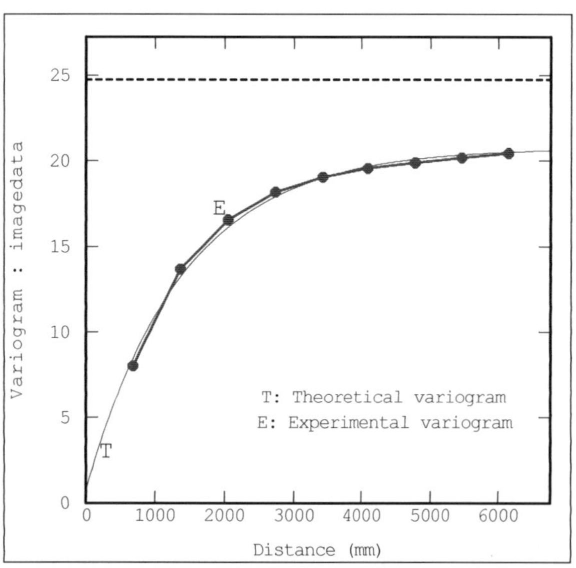

Figure 7. Variogram modelling of image.

Figure 8 presents the histogram of the processed image digits distribution. Digits distribution represents the frequency individual digits which is assigned in colour scale. Every colour is assigned to a number representing the structure in the image.

Table 2. Theoretical variogram modelling parameters.

	Structure
Variogram type	Spherical
Nugget effect	0.75
Structural effect	20
Range	420

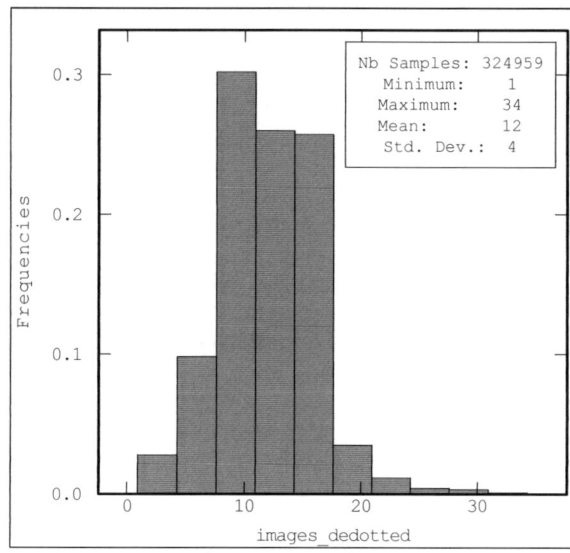

Figure 8. Histogram of processed image digits.

In the light of this knowledge, the structure in the image can be separated to main regions by adjusting the colour scale intervals. For this purpose image is splited into 3 study regions: Red, Blue and Yellow.

To get a clear discrimination for these three structure, a special colour scale was created to reveal the location of each part determined above. Distributions of the determined structures are presented in **Figure 9.** According to **Figure 9** the plotted structure borders and their ratios are calculated easily as it can be seen in **Figure 9-10** and **Table 3.** A very important point in this stage is creating a scale and determining the scale interval of the colour scale which determines transition of the structures. This colour scale is adjusted manually by user as parallel to original raw image.

The yellow colours in **Figure 9** shows the residues of binder and the contact points of the binders to the magnetite grains. This part showing binder residues are said to be concentrated at the centre area of the pellet. The ratio of this part to whole section of pellet was calculated to be 0.74%. The blue and red colours were determined to be the sintering matrix of the pellet. The blue colour indicating the normal sintering part was calculated to be 72.76%. The red colour is said to be the better sintered part in the sintering matrix of the pellets. Interestingly, the change of the sintering is more at the right side of the cross-sectional area of the pellet. It is thought that the reason of that sintering difference is more probably due to the position of the pellet during the furnace. In addition, it can be seen that a different structure in circular shape at the centre of the cross-sectional area of the pellet. Actually the red colour is used to reveal this kind of segment. When this image of cross-sectional area of the pellets compared with the raw photo of the pellets, it can be said that there is a circular core-shaped sintering region at the centre. This finding imply that the sintering develops from the outer shell of the pellets towards to the inner core and the centre of the pellet due to the thermal treatment which is effective on the sintering of the pellets. As a conclusion, the better sintering occurs at the outer shell of the pellet due to the more effective thermal effect and therefore the strengthening mechanism occurs from the outer shell of the pellets towards to the inner core and the centre of the pellet. The red coloured part indicating the better sintering part was calculated to be 26.10% of the whole cross-sectional area of the pellet. Furthermore, the porosity which is one the important pellet properties was tried to be determined. However the porosity cannot be determined and estimated at this magnification. Also the pellet sintered at this temperature has the presence of micro pores rather than macro pores. As a result pores in the pellet matrix could not be determined.

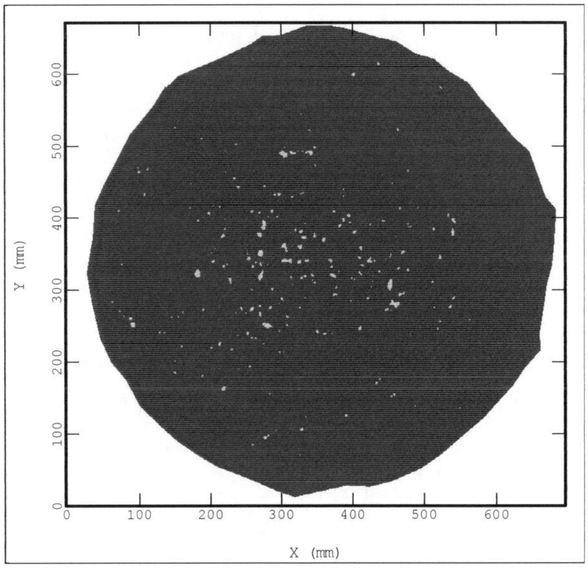

Figure 9. Visual presentation of phase distribution in iron pellet surface.

Table 3. Rational presentation of phase distribution in iron pellet surface.

Region	Sample number	Percentage
Red	86138	26.51%
Blue	236432	72.76%
Yellow	2389	0.74%

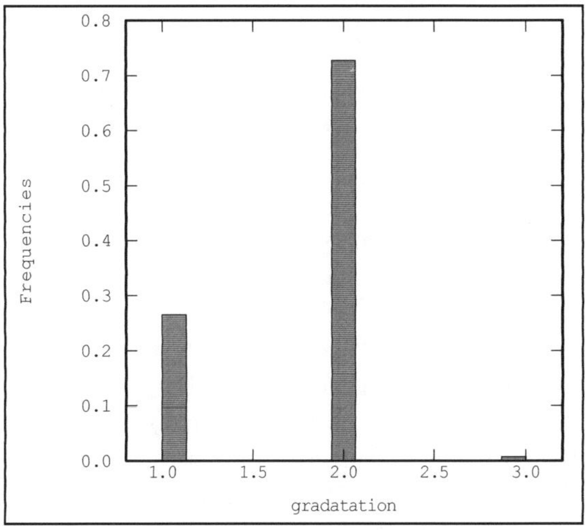

Figure 10. Histogram of the regional distribution of regions.

5 RESULTS

In order to examine and characterise the cross-sectional visual structure of iron ore pellet sample through application of image processing technique by geostatistical methods the following steps were followed.

- A pellet sample is selected randomly and moulded in a casting die with a resin.
- This process followed by polished sample preparation which includes cutting the sample through and polishing one face to prepare getting high quality micro-photos.
- The taken photos digitised as importing into dedicated grid designed by user.
- Image processing techniques (kriging) is used to smooth and reveal more details of the digitised image.

Results obtained from examination may be summarised as follows:

- The results showed that cross-sectional structure can be characterised using digital imaging.
- As a result of pellet image examination the different sintering phases of the pellet sample is exhibited by image and their ratios were showed as histogram and percentages.
- Cross-sectional structure of an iron ore pellet reflects the physical characteristics of pellet and may be used to characterise the success of pelletizing production process.
- Examinations show that the sintering occurs from the outer shell of the pellets towards to the inner core and the centre of the pellet due to the thermal treatment which is effective on the sintering of the pellets. As a conclusion the better sintering occurs at the outer shell of the pellet due to the more effective thermal effect and therefore the strengthening mechanism occurs from the outer shell of the pellets towards to the inner core and the centre of the pellet.
- Success of the digital imaging procedure is highly depend on the quality of images and it is thought that additional image analyses should be carried out with more magnified images to get more detail such as porosity.
- Digital imaging method may be subjected to continuous quality control systems in mining applications such as coal and other industrial mineral feed.
- Also this technique may be developed in the direction of pattern recognition to characterise the internal structure shape of varied micro sample.
- In this study a 3MP camera was used. The success of the application may be improved using high resolution cameras.

ACKNOWLEDGEMENT

The authors extend their thanks to Unit of Scientific Research Projects Coordination, Adana Science and Technology University mainly supporting this study.

REFERENCES

Ball D. F., Dartnell J., Davison J., Grieve A., Wild R., Agglomeration of iron ores, American Elsevier Publishing company, Inc., New York, 1973.

Geovariances, 2006, ISATIS Software, Centre de Geostatistique Fountainblue France.

Meyer K., Pelletizing of iron ores, Springer-Verlag Berlin Heidelberg New York Verlag Stahleisen GmbH, Düsseldorf, 1980.

Ersoy, A., Yünsel, T.Y., Cetin, M., 2004, Characterization of Land Contaminated By Past Heavy Metal Mining Using Geostatistical Methods, Archives of Environmental Contamination and Toxicology, 46, 162-175

Nellros F., Thurley M.J., 2011 Automated image analysis of iron-ore pellet structure using optical microscopy, Minerals Engineering, Volume 24, Issue 14, pp. 1525–1531.

Sivrikaya O. and Arol A.İ., Pelletization of magnetite ore with colemanite added organic binders, Powder Technology, Vol. 210, Issue. 1, pp. 23-28, 2011.

Sivrikaya O, Arol A. I, 2012, The bonding/strengthening mechanism of colemanite added organic binders in iron ore pelletization, International Journal of Mineral Processing, 110-111, 90-100

Sivrikaya O, Arol A. I, 2013a, Method to improve preheated and fired strengths of haematite pellets using boron compounds with organic binders, Ironmaking & Steelmaking,, 40-1, 1-8(8)

Sivrikaya O, Arol A. I, 2013b, An investigation of the relationship between compressive strength and dust generation potential of magnetite pellets, nternational Journal of Mineral Processing, 123, 158-164

Sivrikaya O., Arol A. I., Eisele T & Kawatra S. K., 2013, The effect of calcined colemanite addition on the mechanical strength of magnetite pellets produced with organic binders, Mineral Processing and Extractive Metallurgy Review, 34-4, 210-222

Yünsel, T.Y, 2012, A Practical Application of Geostatistical Methods to Quality and Mineral Reserve Modelling of Cement Raw Materials, The Southern African Institute of Mining and Metallurgy, 112, 239-249

Numerical Simulation of Iron Ore Pellets Drying in the Industrial Pelletizing Bed

H. Nankali, H. Amani, S. S. Razavi, E. K. Alamdari
Amir Kabir University, Tehran ,Iran

ABSTRACT The drying of green iron oxide pellets accounts for 25% of the consumed energy for iron pellet production, hence any improvement in this process could lead to lowering the energy demands and process costs. In this study drying of a single iron ore is simulated using commercial Computational Fluid Dynamics (CFD) package and User Defined Functions (UDF). A solid-film method is developed to overcome software limitations and to model simultaneous heat and moisture transfer and kinetics of drying. Four stages of drying is observed in this model: 1-Surface evaporation at the beginning of process, 2-Surface evaporation along with internal evaporation below boiling point and emersion of a dry front, 3-Internal evaporation and shrinking of asymmetric wet core, 4-Evaporation of residual internal moisture at boiling point. Asymmetric wet core is formed in the second stage due to variation of heat and mass transfer coefficients on pellet's surface springing from surrounding fluid flow and changes into an acentric sphere at the final stages of drying. Fluid flow, heat and mass transfer equations are solved numerically and the results will be compared to experimental data.
Keywords: Drying , Iron oxide pellet, CFD, Numerical simulation

1 INTRODUCTION

Pelletizing is an agglomeration process in which iron ore fines are processed into a round shaped strong material which is used as a raw material for iron-steel making industry. Pellets are produced in rotary pelletizing discs from iron ore concentrate and additives like water, bentonite, etc. These green pellets are soft and must meet certain mechanical, chemical and physical properties before being fed into blast furnace or direct iron reduction process (Meyer 1980, Sadrnezhad et al 2008, Barati 2008).

Amongst the induration systems available “traveling straight grate” developed by Outokumpo is the most used worldwide. Figure 1 shows a schematic of Induration process. The process can be divided into four stages: drying, preheating, firing and cooling. Here the drying process is investigated which accounts for approximately 25% of energy consumption in this process. As seen in figure 1 drying stage is divided into two zones, Up-Draft Drying and Down-Draft Drying, this is done to ensure that the pellets at the bottom layer don't get deformed by weight force of top layers and also the pressure applied by the hot gas. It also helps to prevent re-condensation of the moisture in bed. It is of great importance to control the drying rate so that the moisture content of pellets is removed before entering the preheating zone on one hand, and to limit the vapor pressure inside the pellets on the other hand to prevent spalling (Meyer 1980, Sadrnezhad et al. 2008,Tsukerman et al. 2007, Barati 2008).

There are various models available for simulating the drying process, Küçükada et al. 1994, Sadrnezhad et al. 2008 and Barati 2008 used shrinking core model to describe drying process, but the most comprehensive was the work of Ljung et al. that used CFD to model drying of iron pellets. Following the precious work of Ljung et al. 2011 in this paper a more realistic method is developed for predicting drying rate at the final stage of drying where spalling may occur.

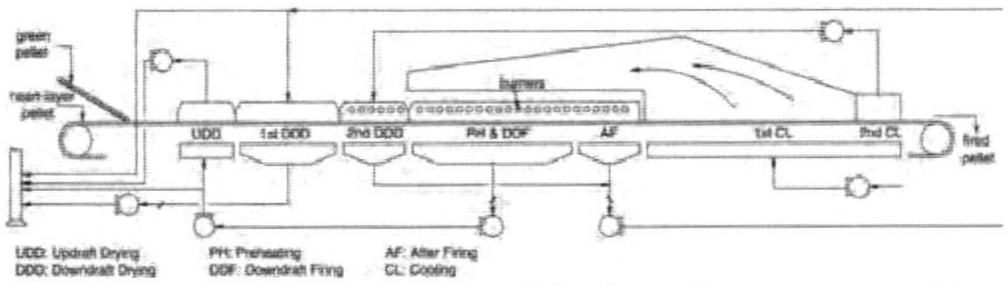

Figure 1. Schematic of induration process(Barati 2008).

2 METHODS

2.1 Governing Equations

The conservation equations of mass, momentum and energy which govern the flow of fluid in the domain are as follows (Ansys Fluent theory guide):

$$\frac{\partial \rho}{\partial t} + \nabla . \left(\rho \vec{v}\right) = 0 \qquad (1)$$

$$\frac{\partial}{\partial t}\left(\rho \vec{v}\right) + \nabla . \left(\rho \vec{v}\vec{v}\right) = -\nabla p + \nabla . (\bar{\bar{\tau}}) + \rho \vec{g} + \vec{F} \qquad (2)$$

$$\frac{\partial}{\partial t}(\rho E) + \nabla . \left(\vec{v}(\rho E + p)\right) = = \nabla . (k_{eff} \nabla T - \sum_j h_j \vec{J}_j + (\bar{\bar{\tau}}_{eff} . \vec{v})) + S_h \qquad (3)$$

Where ρ, t and $\vec{v}$ are density of fluid, time and velocity vector respectively. p is static pressure, $\bar{\bar{\tau}}$ is stress tensor, $\rho \vec{g}$ and $\vec{F}$ are the gravitational body force and external body forces. E, k_{eff}, T, h_j, j_j and S_h are internal energy, effective conductivity, temperature, sensible enthalpy of species j, diffusion flux of species j and other source terms.

2.2 Gas Properties

Following properties of air are calculated dynamically in each cell using UDFs. All temperatures are in Kelvin.(Tsukerman et al 2007)

Density (kg/m^3):

$$\rho_g = \frac{219.38(1+X)}{(X+0.622)T} \qquad (4)$$

Where X is water mass fraction in air on a dry basis.

Viscosity (kg/ms):

$$\mu = 10^{-7}(4.02 + 7.46\times10^{-1}\,T - 5.72\times10^{-4}\,T^2 + 2.99\times10^{-7}\,T^3 - 6.25\times10^{-11}\,T^4) \qquad (5)$$

Thermal conductivity (J/m s K):

$$k = \frac{2.11\times10^{-3}T^{1.5}}{T+123.6} \qquad (6)$$

Heat capacity (J/kg K):

$$Cp_{air} = \frac{8314}{29}\left(3.36 + 0.58\times10^{-3}T - \frac{0.02\times10^5}{T^2}\right) \qquad (7)$$

2.3 Drying

Drying process can be represented in four stages, 1-Surface evaporation at the beginning of process, 2-Surface evaporation along with internal evaporation below boiling point and emersion of a dry front, 3-Internal evaporation and shrinking of the wet core, 4-Evaporation of residual internal moisture at boiling point therefore two mechanisms are considered for drying process, evaporation below boiling temperature and boiling of water at boiling point which will be described in details.

2.3.1 *Surface and internal evaporation*

Below the boiling temperature of water, the mass flux of water is driven by heat transfer from surrounding gas and is limited by relative humidy of the gas. The mass flux of water, $\dot{m}$, is calculated from (Incropera et al. 2007)

$$\dot{m} = h_m \frac{m_w}{R}\left(\frac{p_{v,sat(Ts)}}{T_s} - \frac{p_{v,sat(T\infty)}}{T\infty} RH\right) \qquad (8)$$

Where R is universal gas constant and h_m is mass transfer coefficient which can be calculated from heat and mass transfer analogy using following equation (Incropera et al. 2007)

$$\frac{h}{h_m} = \frac{k}{D_v . Le^{1/3}} \qquad (9)$$

Where k is thermal conductivity, D_v is vapor-air diffusion coefficient and Le is Lewis number.

Relative humidity is defined as

$$RH = \frac{P_v}{P_{vsat}} \qquad (10)$$

Where $p_{v,sat}$ is saturation vapor pressure and is calculated using Antoine's equation in (Pa)

$$P_{sat} = 133.3\exp\left(A - \left(\frac{B}{C+T}\right)\right) \qquad (11)$$

With constants A=18.3036, B=3816.44 and C=-46.13.

Vapor diffusion coefficient at the surface is calculated from (Tsukerman et al. 2007)

$$D_v = 2.58\times10^{-5}\left(\frac{T}{300}\right)^{1.5} \qquad (12)$$

and is corrected inside the pellet using Burgmann's equation

$$D_{eff} = \varepsilon^{1.5} D_v \qquad (13)$$

Where ε is porosity of the pellet.
Evaporation energy source term, φ, is calculated as

$$\varphi = -\dot{m}h_v \quad (14)$$

And h_v is calculated from (Tsukerman et al. 2007)

$$h_v = 3.16\times10^6 - 2495.45T + 0.29T^2 \quad (15)$$

2.3.2 *Boiling*

When pellet's temperature reaches the boiling point of water we are at the final stage of drying. Ljung et al. assumed a mass transfer coefficient large enough to keep the temperature steady at the boiling temperature, but in this model the amount of heat that is absorbed in each control volume in every time step is calculated using UDFs and is used boil the water inside the control volume. following equation is used to calculate mass flux of water:

$$\dot{m} = -\frac{\rho C_p (T - 373.15)}{h_v \Delta t} \quad (16)$$

3 MODELING

This CFD model is developed using Commercial code Ansys Fluent 17 and User Defined Functions (UDFs) to give detailed information about the drying process of a single iron ore pellet and to investigate the influence of pellet's shape on the drying rate.

3.1 Geometry

The simulation is run in a horizontal cylindrical furnace with a radius of 5cm and length of 15cm in which the hot air enters the domain from a velocity inlet at left and exits at right from a pressure outlet. Single pellet is placed on the axis of furnace at 5cm from the velocity inlet. Here an axisymmetric boundary condition is used and half section of the domain is modeled which makes the simulation very efficient in comparison to 3D models figure 2.

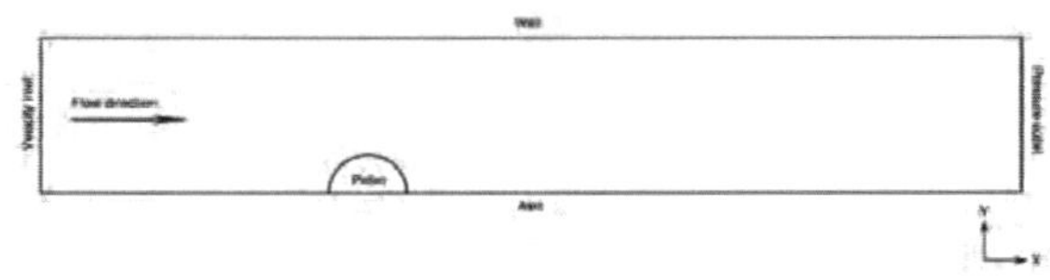

Figure 2. Geometry.

Assumptions

Due to the limitations of Fluent in prediction of correct heat transfer coefficient on the surface of a porous media (Partopour et al. 2016) in this simulation a solid surface with zero thickness is assumed around the pellet which leads to right boundary layer formation and correct heat transfer coefficient calculation.

Following assumptions are also made for this simulation:

- Velocity of air in the pellet is neglected.
- No bond water is present in the pellet.
- Heat transfer inside the pellet is only by conduction.
- Bulk flow of water vapor in boiling temperature is modeled using Fick law.
- Effect of water vapor on the surrounding fluid flow is neglected.
- Heat loss from furnace walls is neglected.

Since diffusion of species through the walls is impossible in Fluent, two films are assumed at each side of the pellet-gas interface and the transport of species is simulated through these films using UDFs.

3.2 Boundary and Initial Conditions

Velocity of inlet gas is set to 1.5 m/s and the temperature is assumed to be 433 K. Although it can be easily investigated with current model, the relative humidity of inlet gas for this simulation is assumed to be 0%. The initial temperature of pellets is set to 298 K. The initial moisture content of the pellets is set to 7.5 wt% which corresponds to a porosity of 30% and dry weight of pellets a-e is 3.6 g. The properties of pellet are set according to Table 1. The minor and major axes of the pellets are set as Table 2.

Table 1. Material properties of iron ore

Property	Value	
Density	5270	kg/m^3
Thermal Conductivity	5.27	W/m.K
Specific Heat	586	J/kg.K

Table 2. Major and minor axes of pellets

Case	Major axis(cm)	Minor axis(cm)
a	0.600	0.600
b	0.632	0.540
c	0.667	0.486
d	0.703	0.437
e	0.741	0.394

3.3 Simulation Technique and Numerical Accuracy

Simulation is carried out on a 2.2 GHz dual core personal pc thanks to high efficiency of axisymmetric model and the code. Numerical solutions have different sources of errors. To minimize the errors affecting this numerical solution a grid independency study has been made on 3 different grids choosing average wall heat transfer coefficient as a key factor.
All equations were solved using Second Order Upwind discretization scheme from a First Order Upwind initial guess using a blending factor of 0.75 and the solution for all equations has reached convergence in every time step with residual convergence criteria set according to table 2.
The turbulence model used for this simulation is k-ω SST due to its good ability to predict low Reynolds regions as well as high Reynolds regions. A y+=1 condition is met for pellet-gas interface mesh as recommended.

4 RESULTS AND DISCUSSION

The velocity flied is simulated around the pellets and the boundary layer separation is seen at the angel of 80 degree from the front stagnation point for case a in Figure 3 which is in agreement with the literature data. The convective heat transfer is calculated by Fluent and for obtaining mass transfer coefficient the heat transfer coefficient is calculated using UDFs and the results are presented in Figure 4. As expected the heat transfer coefficient in front side of the pellet is highest and its lowest value is on the side of turbulence wake. The average wall heat transfer coefficient for cases a-e is calculated 47.95, 48.46, 49.35, 50.15 and 50.89 $W/m^2.K$. Since upstream side surface of the pellets are increased from case a-e the above results were expected.

As temperature of the pellet rises it reaches the wet-bulb temperature which can be calculated using a heat balance between convective heat transfer from gas and evaporative heat loss from liquid as long as heat and mass transfer analogy holds true. Here the wet-bulb temperature from simulation is calculated to be 309.6 that is in very good agreement with 309.7 calculated from heat balance equation.

Four stages of drying can be seen in this model, before reaching wet-bulb temperature the rate of capillary diffusion of water to the surface is high enough to keep the surface moist and most of the evaporation occurs at the surface , since heat transfer at the surface vary by location (Fig 4), at some points the rate of water diffusion is gets lower than evaporation rate at the surface so a dry shell is formed in upstream side but other parts of the surface are still wet, this is where surface evaporation and internal evaporation co-exist (Fig 5). As drying process continues the dry shell covers the surface and a wet core is formed inside the pellet that indicates 3rd stage of drying where internal evaporation is the main mechanism of drying, the shape and location of wet core is presented in Figure 6. The wet core is asymmetric because of heat transfer distribution at the surface of the pellets.

When temperature inside pellet reaches the boiling temperature of water, in cases that the inlet gas temperature is higher than boiling temperature, the mechanism of drying changes and in this stage bulk flow of vapor may occur if the rate of heat transfer to the pellet is high enough. In a separate case study 3 different inlet velocities where simulated with an inlet gas temperature of 423 K to see the models ability to predict the drying rate at the final stage, the results presented in Figures 7, 8 indicate that the higher the rate of heat transfer to the pellet, the higher the chance of spalling because the drying rate increase in the final stage gets more intense which is in agreement with experimental data.

Total drying time for cases a-e is

calculated from simulations 326, 316, 310, 298.5 and 280.5 seconds respectively (Fig 9). As seen in the results by increasing inlet velocity drying time is decreased due to higher heat transfer rate and also as can be seen in Figures 7, 8 the amount of water inside the pellet at the point of boiling is also higher which may lead to an increase in internal pressure and spalling. The results obtained from different shapes of pellets shown in Figure 9 shows a decrease in time and remaining water at boiling point at the same time which is favorable.

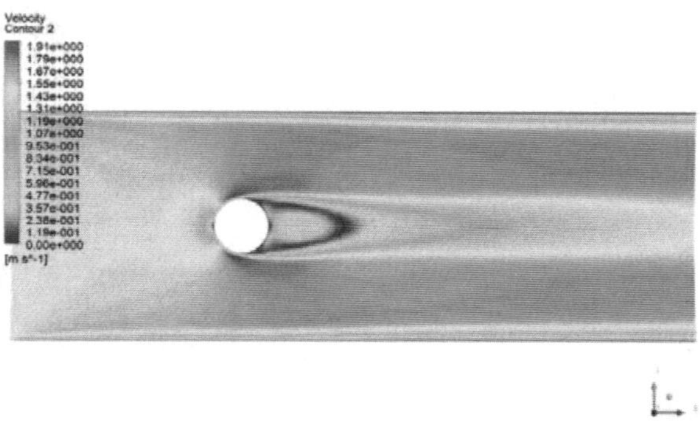

Figure 3. Velocity filed around spherical pellet.

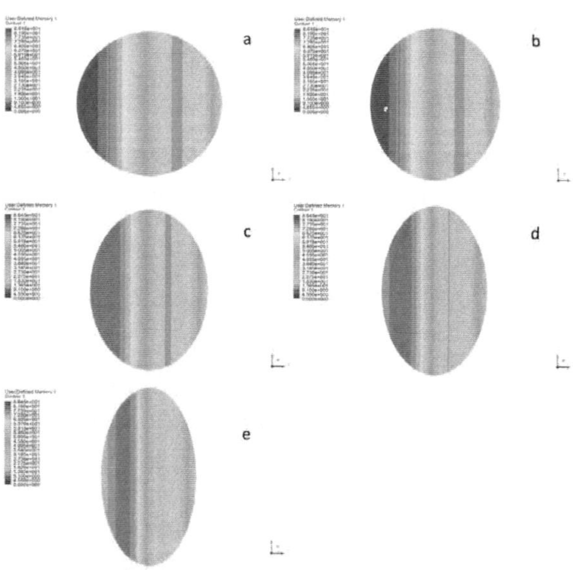

Figure 4. Contours of heat transfer coefficient at the surface of pellets.

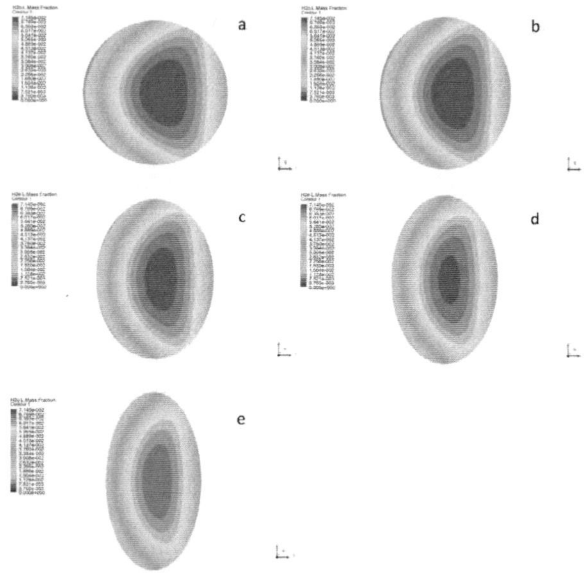

Figure 5. Contour of mass fraction of water in pellets at t= 100s

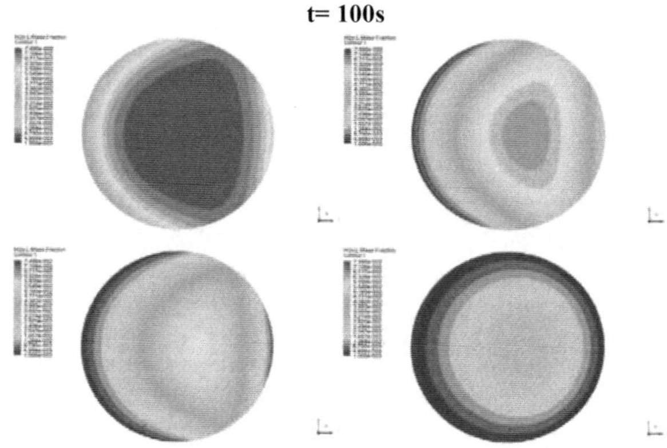

Figure 6. Contour of water mass fraction at t= 43,143,193,293 s.

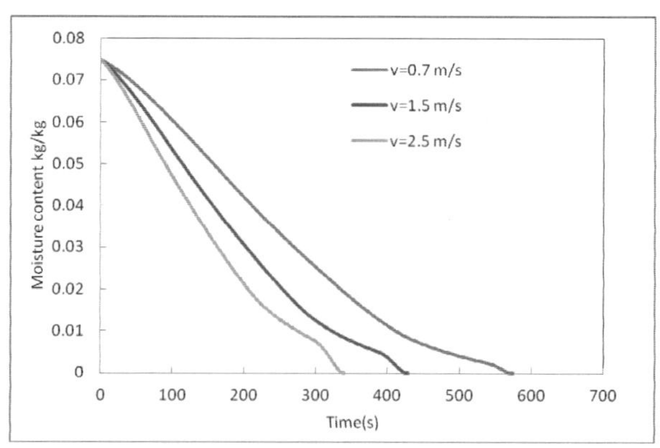

Figure 7. Variation of moisture content with time for different inlet velocities.

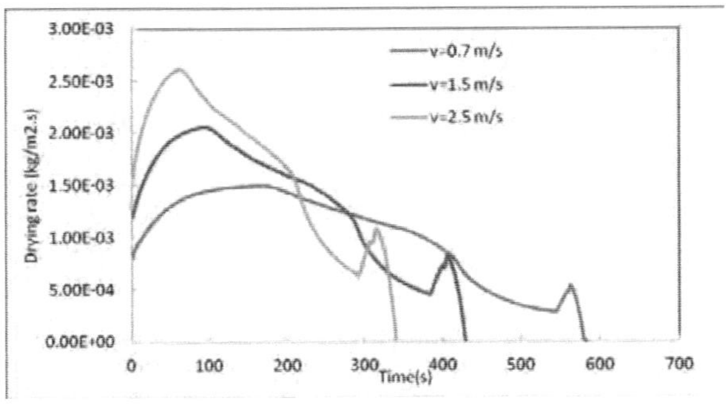

Figure 8. Variation of drying rate with time for different inlet velocities.

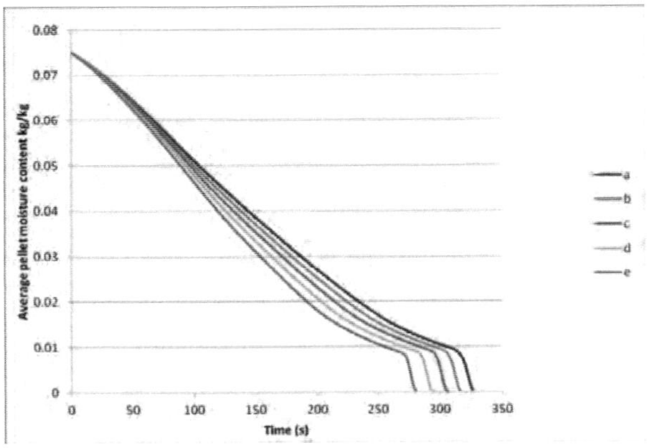

Figure 9. Average moisture content over time for different pellet shapes.

5 CONCLUSION

The numerical simulation results show that drying process is affected by velocity field and as simulation results suggest four stages of drying were observed. It was shown that deformation of green pellets in the bottom layers of bed as a result of gravity forces and gas pressure can reduce drying time and the amount of water evaporated at boiling point which is favorable. It was also demonstrated that doubling the inlet gas velocity can reduce drying time by 20% in the investigated range.

This model gives insight to details of drying process and can be used to develop larger scale models to simulate and predict the behavior of pellets in travelling grate and for optimization of the drying process to consume less energy and produce higher quality products.

REFERENCES

Ansys, Fluent Theory guide ,Ansys Fluent Ver. 17.

Barati, M., 2008. Dynamic simulation of pellet induration process in straight-grate system, *Int. J. Miner. Proces*, 89, pp. 30-39.

Incropera , F.P., Dewitt, D.P., Bergman, T.L., Lavine, A.S., (6) , 2007. *Fundamentals of Heat and Mass Transfer*, John Wiley & Sons, Inc., Hoboken.

Küçükada, K., Thibault, J., Hodouin, D., Paquet, G., Caron, S., 1994, Modelling of a Pilot Scale Iron Ore Pellet Induration Furnace, *Canadian Metallurgical Quarterly*, 33, pp. 1-12.

Ljung, A.L., Staffan Lundström, T. , Daniel Marjavaara, B., Tano, K. , 2011. Convective drying of an individual iron ore pellet – Analysis with CFD, *International Journal of Heat and Mass Transfer*, 54, pp. 3882-3890.

Meyer, K., 1980. *Pelletizing of Iron Ore*, Springer-Verlag, New York.

Partopour, B., Dixon, A.G., 2016, Computationally efficient incorporation of microkinetics into resolved-particle CFD simulations of fixed-bed reactors, *Computers & Chemical Engineering*, 88, pp. 126-134.

Sadrnezhad, S.K., Ferdowsi, A., Payab, H., 2008, Mathematical model for straight grate iron ore pellet induration process of industrial scale, *Comput. Mater. Sci.* ,44, pp. 296-302.

Sivrikaya, O., Arol, A. I., 2011, Pelletization of magnetite ore with colemanite added organic binders, *Powder Technology*, 210, pp. 23-28.

Sivrikaya, O., Arol, A. I., 2013, An investigation of the relationship between compressive strength and dust generation potential of magnetite pellets , *International Journal of Mineral Processing*, 123, pp. 158-164.

Sivrikaya, O., Arol, A. I., Eisele, Kawatra, T., Komar, S., 2013, The Effect of Calcined Colemanite Addition on the Mechanical Strength of Magnetite Pellets Produced with Organic Binders, *Mineral Processing and Extractive Metallurgy Review*, 34, pp. 210-222.

Tsukerman, T., Duchesne, C., Hodouin, D., 2007, On the drying rates of individual iron oxide pellets, *Int. J. Miner. Process*, 83, pp. 99–115.

Residual-based evaluation for spatially-varying processes

B. Tutmez
Department of Mining Engineering, Inonu University, Malatya, Turkey

ABSTRACT Evaluation of variability in a mining site is one of the main motivations of geosciences uncertainty analysis. In parameter estimation, both maximum-likelihood-based methods and ordinary least squares (OLS) correspond constant variance, which is the variability of a measurement is the same in any case of the values of the explanatory variables connected with it. When the assumption of constant variance is not satisfied, a data transformation and use of generalized least squares regression (GLS) is required. GLS approach considers the inequality of variance in the measurements. In the present study, GLS fitting is performed by residual maximum likelihood (REML) method in which each observation is represented additively in terms of fixed and random effects. By using the Bauxite Mine data, the performances of the REML-based GLS model are also compared with conventional Geographically Weighted Regression (GWR) Model performances and the results are discussed.

1 INTRODUCTION

Evaluating reserve parameters such as grade or thickness from a spatial data analysis framework has increasing importance in mineral exploration. Spatial data include both attribute and areal information. The use of conventional techniques in resource appraisals is generally limited because of natural variability the relationships resourced from the location. When using the non-spatial statistical techniques, it is important that the samples are independent of one another. If sample locations were designed due to some geological knowledge of the deposit or assumption the results may contain bias (Moon et al. 2006)

Some parametric and non-parametric regression methods were expressed for examining relationships in non-spatial data (Loader 1999, Fox et al. 2016). The Geographically Weighted Regression (GWR) has been presented as a regression-based effective algorithm for the analysis of spatially varying relationships. (Fotheringham et al. 2002). Recently this algorithm has been and discussed in different works related with ore deposits appraisal. (Tutmez et al. 2011; Zeng et al. 2016).

Conventional geostatistical methods also provide the application of probabilistic tools to regionalized variables. The evaluations performed by the geostatistical methods are tailored to the intrinsic structure of the variable and not only to the sampling quantity or geometric pattern (Cressie 2015). At the difference of traditional statistics, whatever the complexity and irregularity of the real phenomenon, geostatistics search to reveal a structure of spatial correlation.

Variability analysis has critical importance in mineral exploration. In spatial data analysis, a data transformation and use of generalized least squares regression (GLS) can be desired when the assumption of constant variance is not satisfied. GLS approach considers the inequality of variance in the measurements. As discussed in Tutmez and Dag (2013), spatially correlated errors can be modelled by a regression-based kriging procedure and an alternative, residual maximum likelihood method (Stacey et al. 2006) has been addressed as an alternative solution. GLS fitting provides some possibilities for using the residuals produced by the linear model and by this way the residual maximum likelihood (REML) can be implemented. In this study, by using the Bauxite Mine data, the results of the REML-based GLS model are compared with the OLS and the GWR models. By this way, the importance of residual analysis is revealed using the performances of the algorithms.

2 METHODOLOGY

2.1 A reminder on spatial correlation

In parameter estimation, both maximum-likelihood-based methods and OLS correspond constant variance, which is the variability of a measurement is the same in any case of the values of the explanatory variables connected with it. However, the main motivation of a spatial analysis is to identify the nature of relationships that exist between variables. For this purpose and to eliminate the weakness in conventional regression forms, the GWR model has recently been developed by allowing local variations in rates of change. In this structure, the coefficients in the model rather than being global estimates are specific to a location.

The use of GLS aims to handle the problem where we would expect neighbouring values of the response variable to be correlated. The great superiority of this approach is that the errors are allowed to be correlated and to have unequal variances.

2.2 Spatial analysis by regression

Conventional regression is used to estimate the quantitative functional relationships between response variable and one or more predictor variables from the measured data. A common feature of this analysis is that it is applied globally. In addition, the spatial relationships cannot be considered. However, it is often desirable to examine the relationship at a more local scale.

The GWR model was proposed for local estimations (Fotheringham et al. 2002). This technique calculates the coefficients of the regression equation for each plot, in a window that moves across a spatial field. In this model, the observations are weighted in reference to their distance from the kernel centre:

$$\hat{\beta}(u_i, v_i) = [\boldsymbol{X}^T \boldsymbol{W}(u_i, v_i)\boldsymbol{X}]^{-1}\boldsymbol{X}^T \boldsymbol{W}(u_i, v_i)y \quad (1)$$

In eq. (1) $\hat{\beta}$ represents an estimate of β, and $\mathbf{W}(u_i, v_i)$ is an n by n matrix whose off-diagonal elements are zero and diagonal elements are geographical weights of each of the n observed data for regression point *I*. Despite different weighting structures, the Gaussian function is preferred.

2.3 Residual maximum likelihood (REML)

To analyze the non-independence of the residuals due to spatial correlation, GLS is one of the alternative methods. In the OLS linear regression, the residuals ε are independently and identically distributed with the same variance σ^2:

$$y = \boldsymbol{X}\beta + \varepsilon, \ \varepsilon \sim \mathcal{N}(0, \sigma^2 \boldsymbol{I}). \quad (2)$$

On the other hand, the errors are taken into account themselves a random variable η which has a covariance form:

$$y = \boldsymbol{X}\beta + \eta, \quad \eta \sim \mathcal{N}(0, \boldsymbol{V}). \quad (3)$$

In eq. (3), V indicates a positive-definite variance-covariance matrix of the errors (Lark and Cullis 2004). Thus, rather than a single residual variance σ^2, to contain the spatial dependence implicit in the variance-covariance matrix **V** the likelihood of parameters is extended. The log-likelihood structure can be expressed as follows:

$$l(\beta, \Theta \backslash y) = c - \frac{1}{2} log|\boldsymbol{V}| - \frac{1}{2}(y - \boldsymbol{X}\beta)^T \boldsymbol{V}^{-1}(y - \boldsymbol{X}\beta) \quad (4)$$

In eq. (4), c is a constant and $\mathbf{V}$ is established using the distances between the measurements and the variance parameters Θ. Based on second order stationary, Θ covers the covariance parameters $\theta = [\sigma^2, s, a]$, i.e., the total sill, nugget proportion, and range (Rossiter 2012).

In REML method, each observation is represented additively in terms of fixed and random effects. The algorithm first estimates Θ without having to known β. After that, covariance/variogram function **C**, $C_{ij} = C(x_i - x_j)$, and $\boldsymbol{V}$ is calculated. Thus, the fixed-effects regression parameters are estimated as follows:

$$\hat{\beta}_{GLS} = (\boldsymbol{X}^T \boldsymbol{V}^{-1} \boldsymbol{X})^{-1} \boldsymbol{X}^T \boldsymbol{V}^{-1} y\ . \quad (5)$$

3 CASE STUDY

3.1 Field and data set

The application area, Antalya – Akseki Bauxite region corresponds to approximately half of the bauxite reserves of Turkey. In a recent work, Dag and Mert (2008) evaluated the thickness variability in this bauxite reserve. In our present study, the information for 40 boreholes is considered. The data set includes the coordinates as well as chemical compositions (%) such as SiO_2, TiO_2 and Al_2O_3. To illustrate the importance of residual analysis on the solution, the model considers spatial variables (x and y) and TiO_2 (%) as the independent variables, and Al_2O_3 concentration as the target variable, respectively. Figs.1 and 2 show the relationships between the coordinates and chemical compositions, respectively.

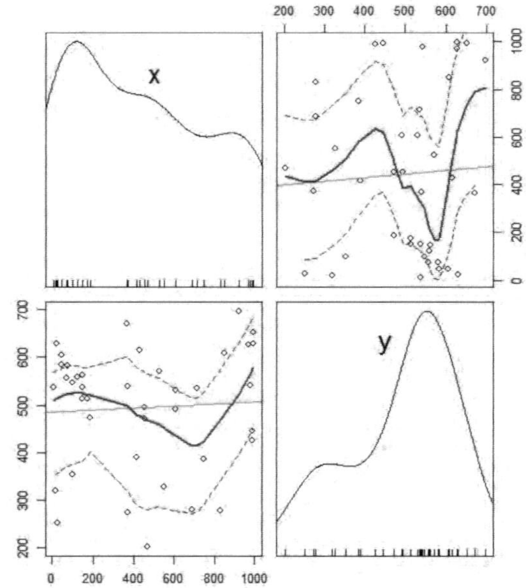

Figure 1. Variability in locations.

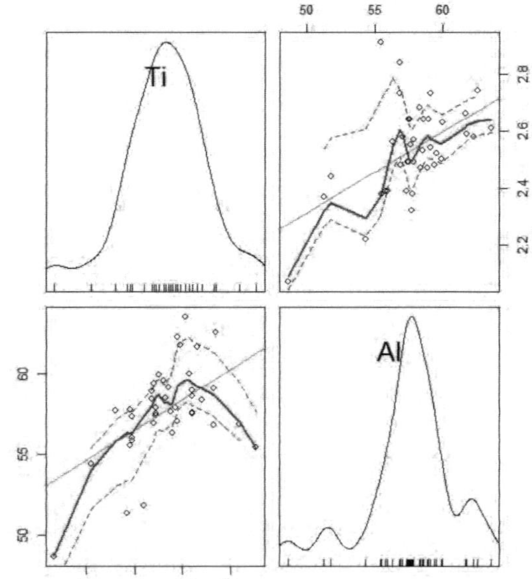

Figure 2. Variability in chemical variables.

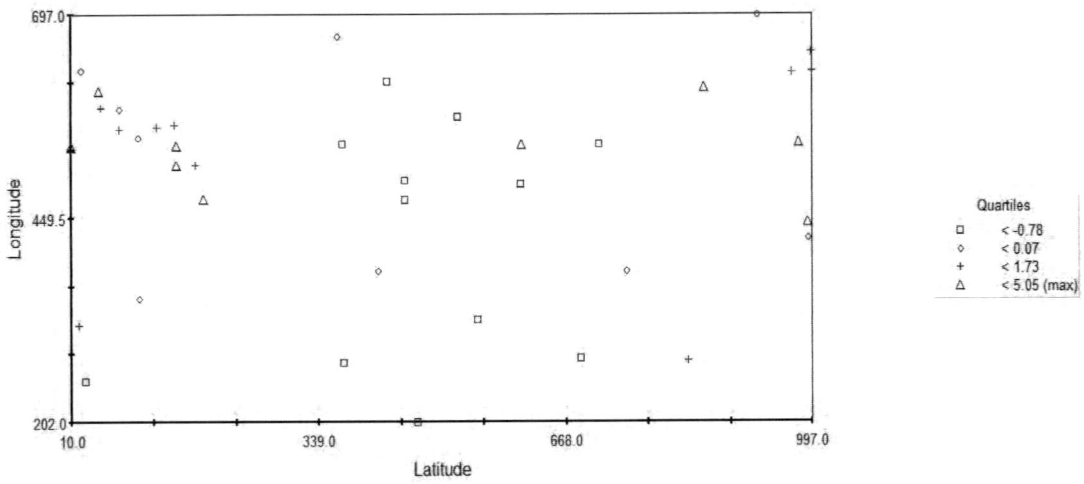

Figure 3. Scatter plot of OLS residuals.

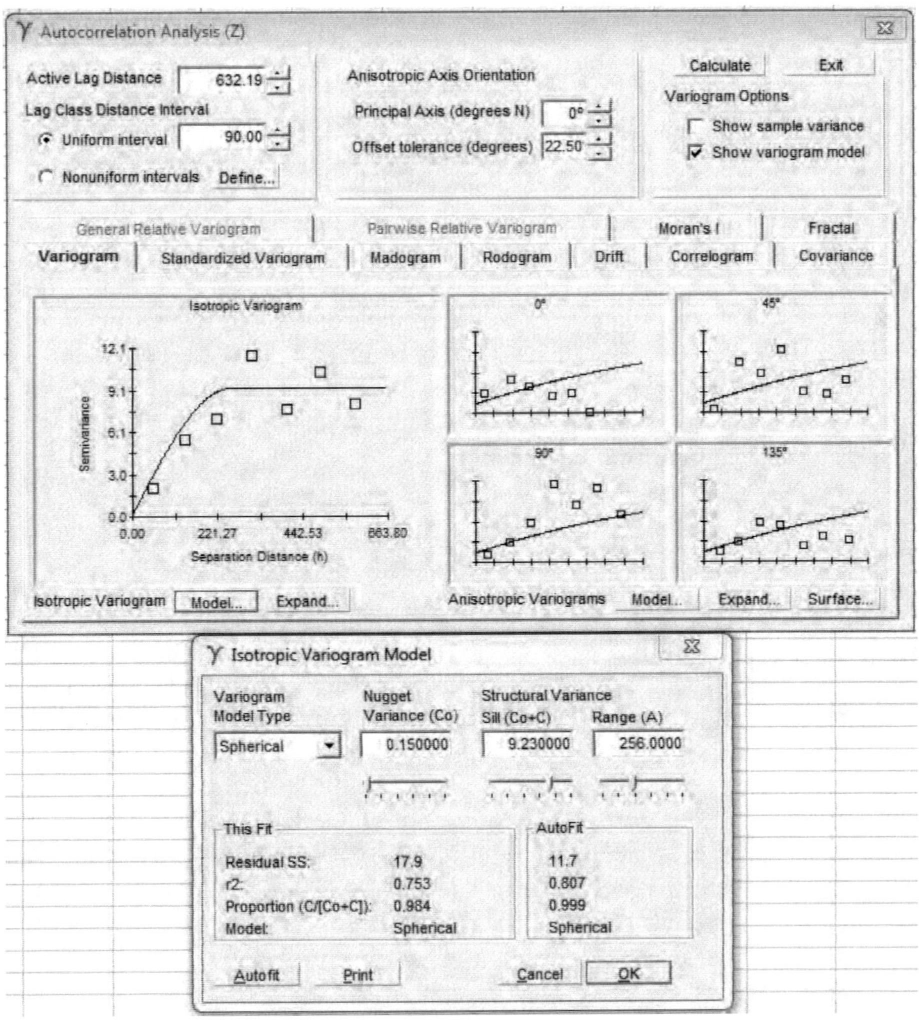

Figure 4. Determination of spatial variability of residuals.

3.2 Results and discussion

The main motivation of the paper, REML analysis has been performed using the residuals produced by linear least squares model. First the spatial correlations among the coordinates and residuals were illustrated (Fig. 3) and then the fitted variogram model parameters have been obtained for using in future GLS analysis (Fig. 4).

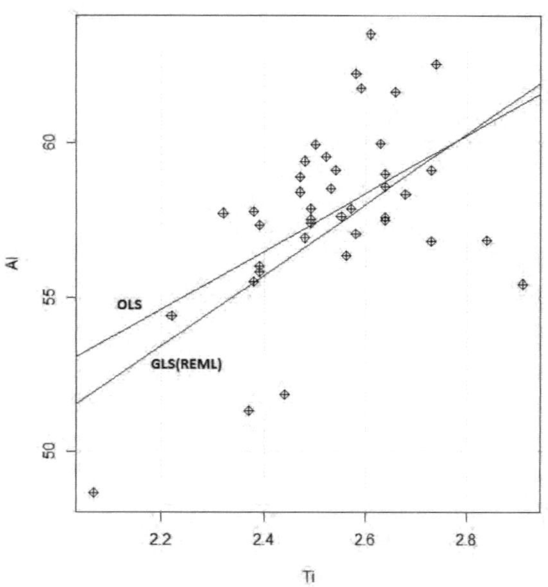

Figure 5. Regression lines of methods.

To compute the GLS relation using residuals, the maximum likelihood algorithm given in section 2.3 has been imp implemented. The results provided by OLS and REML models are illustrated in Fig. 5.

In addition to the results of the OLS model, the proposed model has been compared with the GWR model. By comparing the Akaike Information Criterion (AIC) for three competing models, we can decide which is closest to reality (Burnham and Anderson, 1998). The simplistic form of AIC can be expressed as follows:

$$AIC = 2n \log_e \hat{\sigma} + n \log_e 2\pi + n + tr(\boldsymbol{S}) \quad (6)$$

Table 1. Model performances.

MODEL	AIC
OLS	190.65
GWR	187.43
REML	181.73

where *tr(**S**)* is trace of S matrix and $\widehat{\hat{\sigma}^2}$ is estimated standard deviation, $\hat{\sigma}^2 = RSS/n$, respectively.

Table 1 summarizes the model capacities. The REML model produced lower AIC value and thus, residual maximum likelihood estimation has the best performance.

Fig. 6 illustrates the final residuals produced by the models. The REML model has high fluctuations. However, the GWR model has lower standard deviation (smoothing) and this output corresponds a restrained model structure. However, from a general look, the residuals produced by all models have similar flows and this can be accounted by the similar calculation perspectives (regression) and also the natural variability of the site.

4 CONCLUSIONS

Use of error based-GLS analysis is practiced in a Bauxite mine site. The GLS fitting has been conducted by the residual maximum likelihood (REML) method and performances of this novel algorithm have been compared with the traditional least squares (OLS) and Geographically Weighted Regression (GWR) Models. The results showed that residual based-spatial analysis and geographically weighted solution can provide explanatory successful results as well as high performance measures.

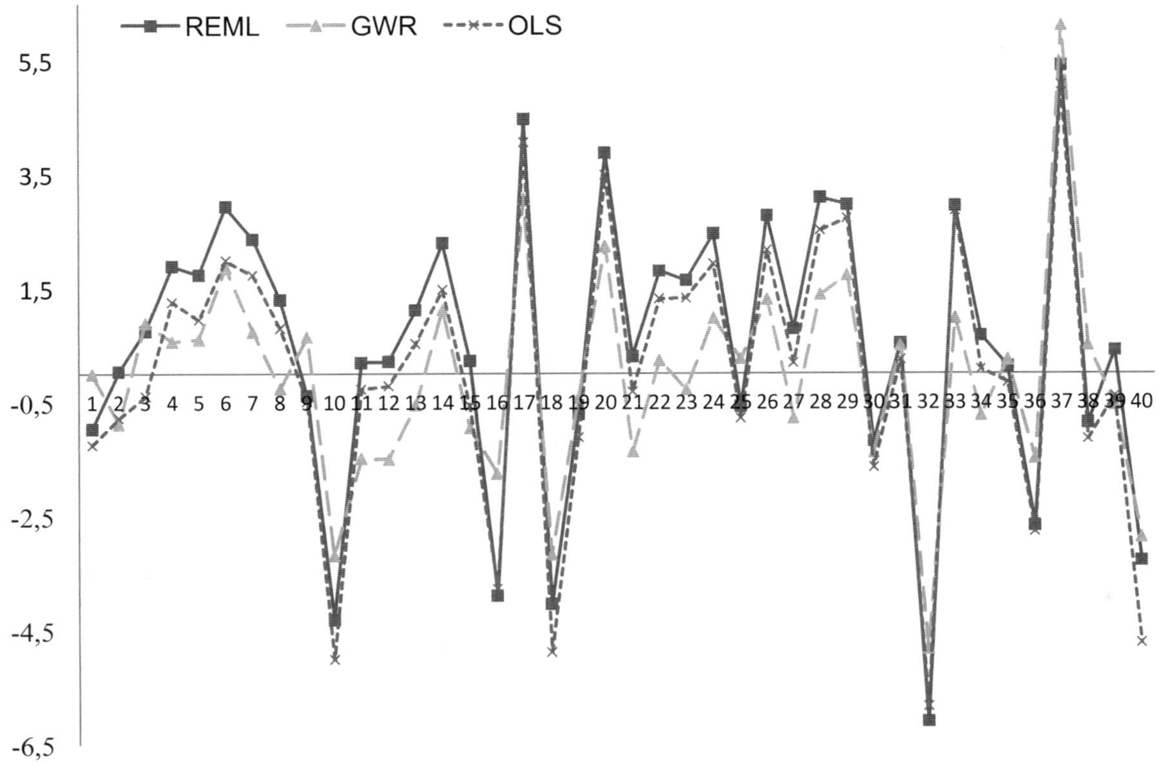

Figure 6. Deviations of model errors.

REFERENCES

Burnham, K.P., Anderson, D.R., 1998. *Model selection and inference: a practical information-theoretic approach*, Springer, New York.

Cressie, N., 2015. *Statistics for spatial data*, revised edition, Wiley.

Dag, A., Mert, B.A., 2008. Evaluating thickness of bauxite deposit using indicator geostatistics and fuzzy estimation, *Resource Geology*, 58(2), 188-195.

Fotheringham, A.S., Brunsdon,C., Charlton, M. 2002. *Geographically weighted regression*, John Wiley & Sons, Great Britain.

Fox, J.T., Kim, K.i., Yang, C., 2016. A simple nonparametric approach to estimating the distribution of random coefficients in structural models, *Journal of Econometrics*, 195, 236-254.

Lark, R.M., Cullis, B.R., 2004. Model based analysis using REML for inference from systematically sampled data on soil. *European Journal of Soil Science*, (4), 799-813.

Loader, C. 1999, *Local regression and likelihood*, Springer, New York.

Moon, C.J., Whateley, M.E.G., Evans, A.M. 2006. *Introduction to mineral exploration*, Blackwell, India.

Rossiter, D.G., 2012. *Introduction to the R project for statistical computing for use at ITC*. Enschede (NL), 4.0 edition.

Stacey, K.F., Lark, R.M., Whitmore, A.P., Milne, A.E., 2006. Using a process model and regression kriging to improve predictions of nitrous oxide emissions from soil, *Geoderma*, 135, 107-17.

Tutmez, B., Kaymak, U., Tercan, A.E., 2011. Local spatial regression models: a comparative analysis on soil contamination, *Stoch Environ Res Risk Assess*, 26, 1013-1020.

Tutmez, B., Dag, A., 2013. Mapping water chemical variables with spatially correlated errors, *Environmental Ecological Statistics*, 20, 19-35.

Zeng, C., Yang, L., Zhu, A-X, Rossiter, D.G. .., 2016. Mapping soil organic matter concentration at different scales using a mixed geographically weighted method, *Geoderma*, 281, 68-82.

Data Warehousing and Big Data Applications in Mining Industry
Madencilik Sektöründe Veri Ambarı ve Büyük Veri Uygulamaları

M. Erkayaoğlu
Orta Doğu Teknik Üniversitesi, Maden Mühendisliği Bölümü, Ankara

ÖZET Madencilik sektöründe artan maliyetler ve üretim baskısı nedeniyle maden yönetimi ve planlaması veriye daha bağımlı hale gelmektedir. Madencilikte kullanılan ekipmanlarda veri toplamayı ve veri paylaşmayı sağlayan teknolojilerin bulunması sonucunda operasyonel veri miktarı artmaktadır. Üretim, bakım-onarım, maliyet gibi farklı veri kaynakları tarafından üretilen verinin karar verme aşamasında ve yönetimde etkin şekilde kullanılması için verinin güvenilir şekilde toplanması, temizlenmesi, farklı verilerle entegre edilmesi ve raporlanması gerekmektedir. Veri ambarı, geçmiş veya gerçek zamanlı verinin sorgu, analiz ve raporlanması amacıyla geliştirilen bir teknoloji olup modern madenlerde hem maden işletme hem de cevher hazırlama tesislerinde kullanılmaktadır. Oluşan veri miktarının, hızının ve karmaşıklığının sürekli olarak artması, madencilik sektöründe büyük veri uygulamalarının da kullanılması için uygun ortamı oluşturmaktadır. Teknolojik yatırımın yanı sıra verinin etkin şekilde kullanılması için gerekli etmenlerden bir diğeri yönetim bakış açısının değiştirilmesi ve işletme kültürünü veriye dayanan kararlar alacak biçimde yeniden şekillendirmektir. İşletmenin temel performans kriterleri tanımlanarak yönetilmesi ve çalışanların performans değerlendirmelerinin veriye dayalı şekilde yapılması için entegre edilmiş madencilik verisine ihtiyaç duyulmaktadır. Bu bildiri, madencilikte var olan temel veri tiplerini tanıtmakta ve modern madenlerde veri ambarı ve büyük veri uygulamalarına ilişkin teknolojileri tanıtmaktadır.

Anahtar Kelimeler: Veri Ambarı, Büyük Veri, OLAP, Nesnelerin İnternet'i

ABSTRACT Due to the increased costs and production pressure in the mining sector, mine management and planning becomes more dependent on data. The amount of operational data increases as a result of data gathering and data sharing technologies being available on mining equipment. Reliable collection, cleaning, integration and reporting of data is essential to ensure that the data generated by different data sources such as production, maintenance, and cost are used effectively in decision-making and management. Data warehousing is a technology developed for the query, analysis, and reporting of historical or real-time data and is utilized in modern mines both in mining operations and mineral processing plants. The continuous increase in the amount, speed, and complexity of generated data creates the suitable environment for big data applications in the mining industry. Besides technological investment, another factor required for the effective use of data is to change the management perspective and to reshape business culture with the aim of decision-making based on data. Integrated data related to mining is required to manage the mine based on key performance criteria definitions and to conduct performance evaluations of employees based on data. This paper introduces the basic data types available in mining and introduces the technology related to data warehousing and big data applications in modern mines.

Keywords: Data warehousing, Big Data, OLAP, Internet of Things (IoT)

1 INTRODUCTION

Technological improvement in software and hardware solutions in the mining industry changed the perspective of mine management. Increased operation costs, lower grades, and tight production schedules lead the mining industry to become more data dependent. Every stage of operation in mining has the potential to be equipped with a technology that generates data for decision making and advanced analysis. Equipment used for overburden removal, drilling, blasting, haulage, crushing, grinding, and other production stages of surface mining can be tracked for productivity, maintenance, and safety purposes. Similarly, underground mines that utilize drills, rock cutting machines, support systems, and other machinery have the potential to be managed in a more data-oriented perspective by generating, collecting, and analyzing data. The equipment intensive nature of mining leads the industry to vast amount of data that is generated by mobile equipment and stationary plant facilities. These different types of data sources create complexity in data management, analysis, and reporting. New technologies such as Big Data gained the attention of various industries including mining that will face the problem of data that exceeds the amount, speed, and complexity limits of conventional solutions. The introduction of Internet of Things (IoT) devices will enable more efficient and safe working conditions by generating more data for closer tracking of production. Management perspective and decision making culture of mine managers is another key aspect of utilizing data on a daily basis. Mining engineers that are trained to analyze data and evaluate performance by using quantitative measures will lead the industry to utilize integrated data collected from different parts of the production. The infrastructure that has been used to collect and analyze data in mining will have to be updated to handle the high frequency and high volume of various data types generated by a modern mining operation.

2 TYPES OF DATA IN MINING INDUSTRY

Mining industry utilizes various mobile equipment and stationary machines for development, production, materials handling, and processing purposes. The main data sources of the mining industry can be grouped into mining and mineral processing for both surface and underground operations. These processes generate relational and process type of data that are handled by different technologies. Unit operations of mining such as rock breakage by drilling and blasting and materials handling by loading, haulage, hoisting are performed by mobile equipment that should be tracked for machine health and productivity purposes. GPS based systems are commonly used in fleet management systems that monitor movement of mobile equipment and generate relational data. The mineral processing part of mining uses stationary machines that are equipped with sensors. These sensors provide process type of data that has a different structure compared to relational data and should be integrated with other data sources for continuous improvement. Besides these conventional types of data that are originated from mining and mineral processing equipment, there is another type of data that was introduced to the industry together with big data technology. Unstructured data, first defined by Buneman et al. (1996), consists of various files such as images, videos, email messages, webpages, etc. that do not fit into a predefined data structure today. It is crucial to be familiar to the types of data in mining prior to any business intelligence (BI) application to aid management. Luhn (1958) defined the basic expectations of a business intelligence system as providing information based on user demand.

2.1 Relational Data

Mobile equipment used in surface and underground mines generate relational data about productivity and machine health. This type of data can be stored in tables and fits data models that is used for data storage and data processing. Machines equipped with GPS provide location data and allows engineers to analyze production by using quantitative measures. As an example, definitions such as availability and utilization can be converted into more dynamic measures by using real-time data. Operators are playing an important role for productivity, machine health, and safety of mobile mining equipment. The relational data of mobile equipment can either be manually entered by the operator or automatically recorded by fleet management systems as seen in Figure 1.

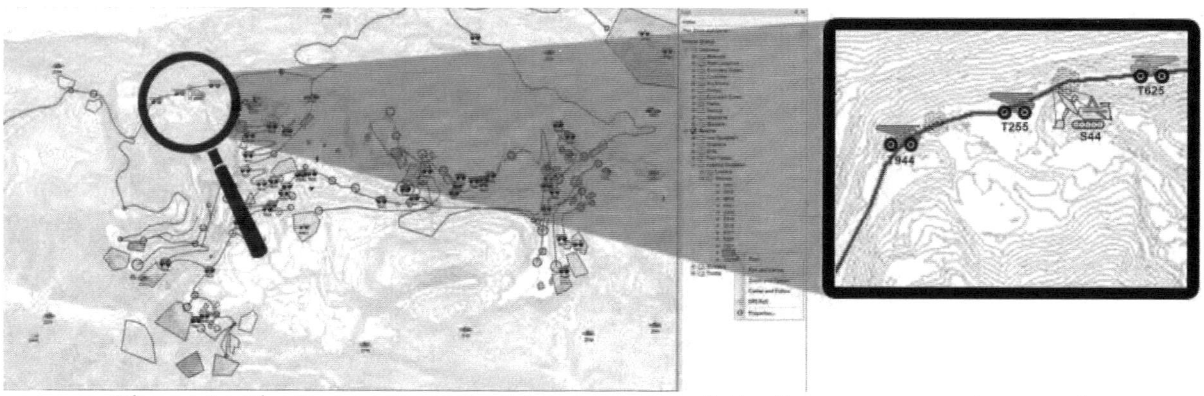

Figure 1 Fleet management system overview (http://www.wencomine.com/)

Underground mines can be considered as more challenging for connectivity and data collection. Similar to mobile equipment used in surface mining, underground mining equipment can also be tracked by systems that record data from on board systems and sensors. The major advantage of this type of data is the concept of defining relationships between tables where the data is stored. Engineers can browse this integrated data by using different dimensions such as location, time, equipment brand, and others to analyze production. This way, all processes related to a certain part of production can be used for reporting and other performance measurement tools. Drill monitoring systems, as seen in Figure 2, are another major system that provide relational data.

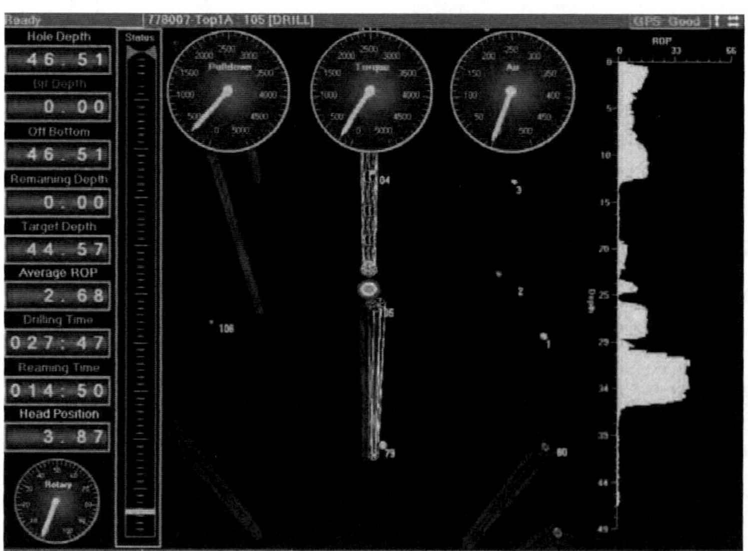

Figure 2 Drill monitoring system overview (http://www.tbirdpac.com)

The system mainly consists of transducers connected to the drilling machine that record measures such as torque, rotary speed, depth, etc. Together with the location data collected by GPS antennas on the drilling machine, data is sent via Wi-Fi to a central data server. Similarly, jumbos equipped with guidance systems aid operators to follow daily production plans in underground mines.

2.2 Process Data

Equipment and machines in mineral processing plants are commonly interconnected to be controlled and supervised by operators from a centralized system. Supervisory Control and Data Acquisition (SCADA) systems, as seen in Figure 3, and Distributed Control Systems (DCS) are main components of smart ore tracking in mining industry.

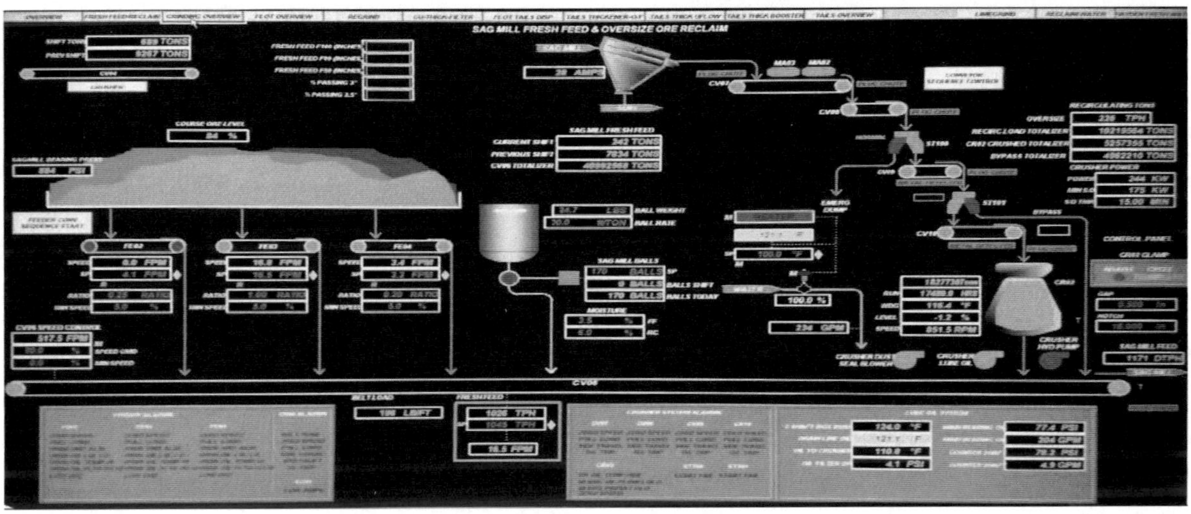

Figure 3 SCADA system overview

These systems provide an overview of mineral processing equipment by displaying operating conditions and these can be extracted for further analysis. The data generated by equipment used for crushing and grinding, classification, flotation, and other processes consists commonly of a value and a timestamp. The main difference of this type of data from relational data is its high frequency and small size. Sensor based systems on equipment provide data for the tags defined in historians. Historians are designed to extract analog data in a very efficient way without disrupting the functioning of the equipment. Special data compression and exception algorithms are used to ensure that data is recorded in a reliable way. As the sensors are commonly used for equipment health purposes, it is also possible to create predefined alarms and provide valuable information to the technicians about engine oil pressure or cooling water temperature.

Another example of process type of data can be generated by rock cutting equipment in underground mines. Shearers and continuous miners used in longwall mining can be equipped with data acquisition systems such as Programmable Logic Controllers (PLC) that provide tag-based data.

Once data is read from on board sensors and sent to a data server, it is possible to connect to this server to extract and integrate data with other sources as seen in Figure 4.

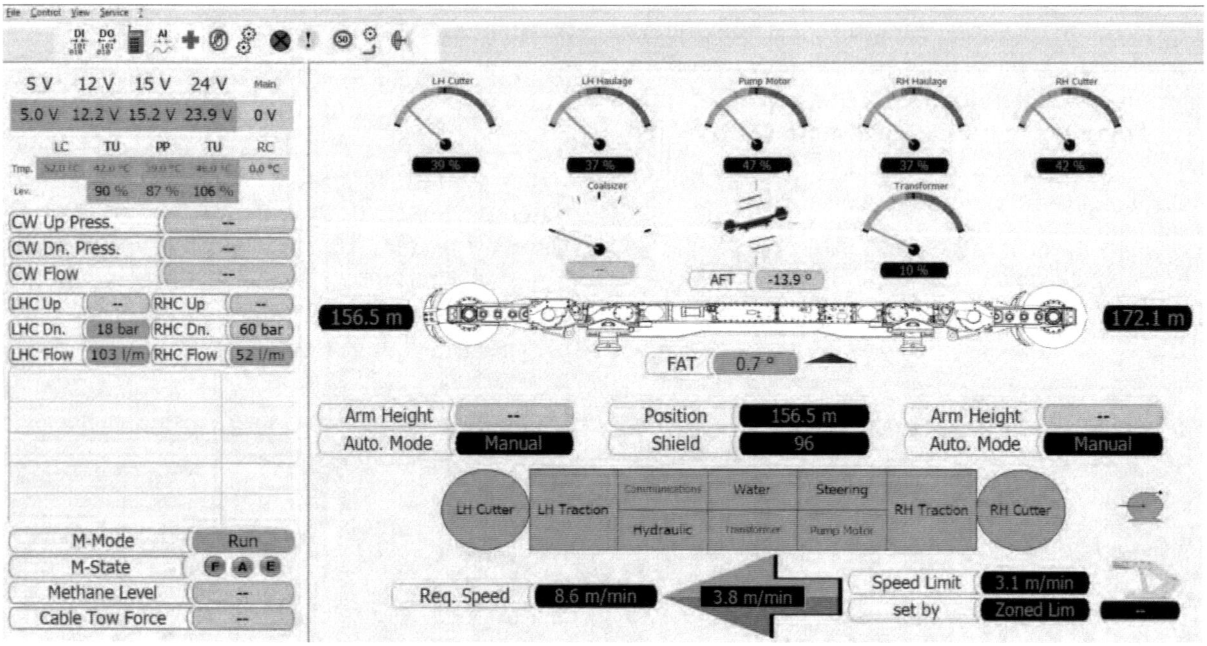

Figure 4 Condition monitoring of continuous miner (http://www.cat.com/)

2.3 Unstructured Data

It is discussed that 80-85% of business related information originates from unstructured data which is majorly text based. Recent development in technologies related to big data enabled to gain knowledge from unstructured data such as social media feeds, images, and other data. These data sources would require comprehensive data manipulation to fit a predefined data model and create an additional layer of complexity for integration.

Mining industry has been and will always be in the center of the discussion about environmental protection, sustainability, and energy. Information that can be extracted from comments and articles that are published in websites or social media about the mining industry are invaluable. Any mining project can face opposition from the society about its environmental impact especially in an era of technology where the power of social media tools is significant. Information about the opinion of the society can be extracted by sentiment analysis and other text mining methods. This type of data is unstructured by its nature such as Facebook like counts, Twitter feeds, and other social media tools. Programming the existing data layer for this type of data to fit predefined data models is not a feasible option as these data sources are continuously changing their data infrastructure. The speed and volume of data that can be used for getting the opinion of the society on a specific mining project exceeds the limits of the processing power of conventional hardware used for this purpose. MapReduce, introduced by Dean and Ghemawat (2004), is a technology that distributes the computation that can be a sorting and filtering process between multiple processors on servers. Another implementation that will become popular in the mining industry and will generate unstructured data is IoT.

3 DATA WAREHOUSING IN MINING INDUSTRY

Data warehousing can be defined as a collection of databases in an integrated way where the time perspective can be historical or real-time (Inmon, 2005). Its benefits in management and decision-making have been proven in various industries. Mining is a highly complex production field where technology is utilized heavily on equipment. Mobile equipment that is part of the fleet management system provides relational data whereas plant equipment generates non-relational process data. Integration of these unique data sources for business intelligence on a real-time basis is a challenge. Various researchers investigated the potential of data warehousing in the mining industry by implementing different tools that are based on operational data (Benito and Dessureault 2008, Kahraman and Dessureault 2011, MacDonald 2013).

One of the most commonly used tools that are built on data warehouses are Online Analytical Processing cubes (OLAP) as seen in Figure 5.

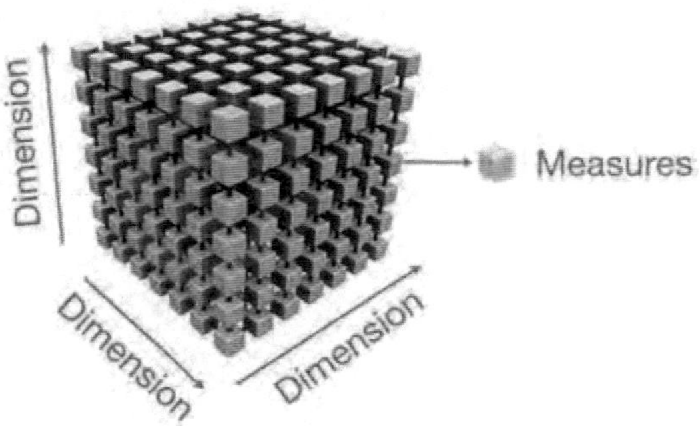

Figure 5 OLAP cube structure

OLAP cubes create a flexible environment where measures can be sliced and diced by using dimensions such as location, time, or any other existing property. Danckert and Masterman (2011) presented centralized data used in orebody knowledge. OLAP and Extract-Transform-Load (ETL), commonly referred in data warehousing studies, were used to access detailed information in block models. Existing block models and newly developed ones used for reserve estimation and mine planning are stored in a centralized data repository.

Such kind of centralized data is utilized in an operation room for integrated planning and maximizing system efficiency. Centralizing data and managing it from a control room is different from integrating different source systems.

4 BIG DATA APPLICATIONS IN MINING INDUSTRY

The continuous increase in the amount, speed, and complexity of generated data creates the suitable environment for big data applications in the mining industry. Big data is commonly identified by its unique characteristics. The amount of data that has a direct impact on data storage and processing infrastructure is defined as volume. Cost of IT hardware continues to decrease although the amount of data that is seen as potential information sources increases more rapidly. Cloud based solutions offer cost affordable storage and computing power for big data implementations (AWS, Azure, etc.). Another characteristic of big data is the variety, which specifies the level of integration as different types of data will result in a more difficult structure. Velocity of the data is another concern of industries that aim to implement big data as especially social media feeds could easily generate large scale real-time data that can not be handled by conventional methods. Mining industry will encounter challenges in data handling and integration when IoT devices become more commonly utilized.

4.1 Internet of Things (IoT)

IoT is a concept that covers all devices that can be interconnected via Internet without any computer. Mining industry is a field where both machinery and personnel can be equipped with smart devices, sensors, and other tools that can enhance communication and provide vital information. One of the first implementations of IoT based devices in mining industry is in the field of safety. Location and proximity sensors that were embedded in proprietary hardware were not cost efficient whereas IoT devices are comparably low-priced. Sensors that track gas concentrations that can be contaminant, poisonous, or hazardous are able to trigger alarms and automatically start on-demand ventilation systems. Wearable devices equipped with GPS that can also monitor basic health parameters of miners will become more popular, especially in underground mines, generating vast amount of data ready to be integrated with other data sources. There is research conducted by Delsing et al. (2015) on converting consumables such as rock bolts into smart tools communicating with other bolts in tunnels and providing information to track stability. Similar to people tracking, ore tracking is another field that will benefit from cost affordable hardware in quality control and mine-to-mill applications. The data provided by these devices will enable the optimization of operational stages, aid predictive maintenance, and enhance safety on site.

4.2 Cloud Based Data Solutions

The improvement of cloud based systems provided various industries the ability to utilize data storage and computation capacity without investing in hardware. Researchers such as Singh and Kaur (2016), started to investigate different options for big data analytics on cloud based systems. Simmhan et al. (2013) introduced a cloud-based platform for integration, storage, analysis, and visualization of dynamic data. The main advantage of these platforms is that the company can choose which licensed software will be installed on the virtual machine and the technical specifications such as data storage capacity and technology, number of processors, amount of memory, and others. Manekar and Pradeepini (2015) stated main benefits of cloud based platforms as affordable investment, transfer of risks, and scalable infrastructure. These indicate that analytics that could not be handled with conventional IT infrastructure can now easily be implemented by mining companies that choose to finance their data warehouse on the cloud by making monthly payments to utilize a high-end computer infrastructure. Internet connectivity and network infrastructure are the essential parts of a data oriented mining operation. However, the main challenge in mining industry is not the platform that is used for data management or analysis, it is the management mindset that does not consider data as a valuable asset.

5 CONCLUSIONS

Mining industry is utilizing various types of machinery and equipment that generate data. Different types of data need to be integrated on a semantic layer to be analyzed and used in mine management. This requires investment in technology that provides data collection capability, IT infrastructure that is required for data gathering and analysis, and most importantly qualified mining engineers that have the skill set to utilize data. Besides technological investment, the most crucial factor required for the effective use of data is to change the management perspective and to reshape business culture with the aim of decision-making based on data. Integrated data related to mining is required to manage the mine based on key performance criteria definitions and to conduct performance evaluations of employees based on data. This paper introduces the basic data types available in mining and introduces the technology related to data warehousing and big data applications in modern mines. It is inevitable that mining industry will be challenged by the continuously increasing amount of different data types. Mining engineers should get more familiar with tools that are used to analyze data and make data a center piece of their daily working routine.

6 REFERENCES

Benito, R., Dessureault, S., 2008, Estimation of Incremental Haulage Costs by Mining Historical Data and their Influence in the Final Pit Limit Definition, *Mining Engineering,* Vol. 44, Issue 9.

Buneman, P., Davidson, S., Hillebrand, G., Suciu, D., 1996, A query language and optimization techniques for unstructured data, *SIGMOD '96 International Conference on Management of Data,* pp. 505-516.

Danckert, G., Masterman, J., 2011, Identify, Convert and Sustain - Resource Development in the Pilbara - the Rio Tinto Way, *Iron Ore Conference*, pp. 589-595.

Dean, J., Ghemawat, S., 2004, MapReduce: Simplified Data Processing on Large Clusters, *6th symposium on operating systems design and implementation.*

Delsing, J., Eliasson, J., Pereira, P., Gebart, J., 2015 The IoT Rockbolt, *5th International Conference on the Internet of Things (IoT).*

Inmon, W.H., 2005, Building the Data Warehouse. Indianapolis, Ind: *Wiley*.

Kahraman, M.M., Dessureault, S., 2011, Development of a Real-Time Adherence to Mine Plan Tools as Part of an Integrated Remote Mine Control Centre, *35th APCOM Symposium*, pp. 819-825.

Luhn, H.P., 1958, A business intelligence system, *IBM Journal of Research and Development archive*, Vol. 2, Issue 4, pp.314-319.

MacDonald, J., 2013, Managing a Mining Operation Requires Actionable Production Intelligence, *Mining Engineering*, Vol.55, Issue 6.

Manekar, A., Pradeepini, G., 2015, Cloud Based Big Data Analytics A Review, *International Conference on Computational Intelligence and Communication Networks*.

Singh, R., Kaur, P., 2016, Analyzing performance of Apache Tez and MapReduce with hadoop multinode cluster on Amazon cloud, *Journal of Big Data, 3:19*.

Simmhan, Y., Aman, S., Kumbhare, A., Liu, R., Stevens, S., Zhou, Q., Prasanna, V., 2013, Cloud-Based Software Platform for Big Data Analytics in Smart Grids, *Computing in Science & Engineering*.

Developing a Novel Method for Selecting More Efficient Blasting Pattern in Sungun Copper Mine

F. Asadi Ouriad
Faculty of Engineering, Tarbiat Modares University, Tehran, Iran

R. Bagherpour
Department of Mining Engineering, Isfahan University of Technology, Isfahan, Iran

M. Yari
Department of Mining Engineering, Isfahan University of Technology, Isfahan, Iran

M. Khoshouei
Department of Mining Engineering, Isfahan University of Technology, Isfahan, Iran

ABSTRACT
Blasting is one of the most significant parts of extraction in mining projects. Choosing an unfitting blasting is the basis of negative actions such as inappropriate fragmentation, back break, fly rock, etc. In detail, the selection of the most proper and efficient blasting pattern between all examined patterns can be concerned as a Multi Criteria Decision Making (MCDM) problem.
In this investigation, primary, executed patterns were assessed using Data Envelopment Analysis (DEA). In the second step, SAW Method is implemented for evaluating the more efficient patterns and distinguishing the most suitable pattern. In conclusion, most efficient and suitable pattern is selected for Sungun Copper Mine.

Keywords: Blasting, MADM, DEA, copper mine, SAW

1. INTRODUCTION

Blasting operation is one of important sections in the mining developments. Improper blasting pattern could consequence numerous practical, economic and security difficulties (Hudaverdi 2012; Kecojevic and Radomsky 2005; Monjezi and Rezaei 2011). In the mining operations, the major goal of blasting process is to attain a appropriate rock fragmentation (Chakraborty et al. 2004; Crum and Crum 1990; Latham et al. 2006; Morin and Ficarazzo 2006; Ozkahraman 2006; Shim et al. 2009). In contrast, the explosive energy is not completely consumed for rock breaking and just 20–30% of this is almost consumed for the breakage and the rest part is exhausted to undesirable phenomena like ground vibration, air blast, fly rock, back break, etc. (Singh and Singh 2005). Similarly, ecologists are progressively worried about mining actions; henceforth, there should be abundant determination to control unwelcome blast-induced ecological difficulties.

Convenient methods could only offer a guesstimate to the result for forthcoming faultless outcome of blasting in view of practical, ecological and security factors and the final appropriate plan can be recognized by means of a trial-error procedure (Inanloo Arabi Shad and Ahangari 2012). Existing investigational approaches of blasting pattern designation are not truthful sufficiently, since they are location-based (Inanloo Arabi Shad and Ahangari 2012; Yari et al. 2014b). For confirming a planned preliminary pattern, an examination of the gained outcomes would cause to revision of the design factors (Jimeno 1995). Thus, this method is inefficient and executes additional costs on the process.

MADM is a procedure for assessing alternatives, ranking and assorting the finest accessible alternatives, This method is also used in other mining interests is acceptable conclusion has been achieved.(Yari et al. 2016a; Yari et al. 2016b)

Assumed the being of diverse limitations and numerous alternatives, it is problematic to choose the most appropriate pattern. Consequently, it is essential to implement a system to improve the design. The designated pattern should be practical from both practical and economical aspects. Safety and environment are the additional significant subjects to be concerned in the blasting process (Hudaverdi 2012; Kecojevic and Radomsky 2005).

The chief purpose of blasting operation is gaining an appropriate fragmentation of rock, while reducing the negative influences (Monjezi and Rezaei 2011). In the earlier researches, fragmentation is considered as the most vital goal of the blasting (Ghasemi et al. 2012b; Kulatilake et al. 2010; Michaux and Djordjevic 2005; Morin and Ficarazzo 2006; Sanchidrián et al. 2006). Up to now, numerous investigations concerned about blasting process organization. For instance, academics represented some accurate surveys observing fly rock and offered causative opinions (Amini et al. 2011; Bajpayee et al. 2003; Bajpayee et al. 2002; Bajpayee et al. 2004a; Bajpayee et al. 2004b; Ghasemi et al. 2012a; Ghasemi et al. 2012b; Kecojevic and Radomsky 2005; Little and Blair 2010; Monjezi et al. 2012; Monjezi et al. 2011b; Ning 1999; Rehak et al. 2001; Rezaei et al. 2011; Stojadinović et al. 2011; Tota et al. 2001).

Anywhere that back break has remained as blasting problem in a new bench, the lesser quantity of back break is considered as the blasting pattern assessment parameter (Gate et al. 2005; Khandelwal and Monjezi 2012; Monjezi et al. 2011a; Monjezi et al. 2012; Monjezi and Dehghani 2008; Monjezi et al. 2010b). Furthermore, several activities have been carried out to reduce ground vibration (Ak et al. 2009; Ak and Konuk 2008; Bakhshandeh Amnieh et al. 2012; Dehghani and Ataee-Pour 2011; Erarslan et al. 2008; Guosheng et al. 2011; Hudaverdi 2012; Iphar et al. 2008; Monjezi et al. 2010a; Monjezi et al. 2011c; Shuran and Shujin 2011). The important subject of these investigations is to identify impacting criteria in the blasting process improvement.

DEA is a non-parametric technique implemented for assessing the comparative efficiency of decision-making units (DMUs) rendering various inputs and outputs (Cooper et al. 2007). It could also be carried out to produce native weights of alternatives from pair-comparison matrixes in the analytic hierarchy process (AHP) (Ramanathan 2006). This technique has been implemented in diverse arenas of engineering (Athanassopoulos et al. 1999; Hermans et al. 2009; Kao and Liu 2009).

So as to attain a general assessment of blasting patterns, some facets (criteria) such as fragmentation, back break, fly rock and blasting expenses should be measured (Jimeno 1995). Therefore, because of the existence of several blasting properties (consequences), it is complicated to choose the most applicable pattern(Yari et al. 2015a; Yari et al. 2013). Then, novel mathematical-based approaches such as Simple Additive Weighting (SAW), a division of multi attribute decision making (MADM), could be implemented. So, in cases with numerous alternatives, it is better to boundary the exploration space by neglecting incompetent alternatives and bearing in mind efficient ones(Yari et al. 2014a; Yari et al. 2014c; Yari et al. 2015b). The research could be achieved implementing approaches such as DEA (Li et al. 2007). In this research, the most efficient applied blast patterns of mine were selected by DEA method. After this stage, the most appropriate blasting pattern is selected implementing SAW technique for Sungun Copper Mine.

2. DATA ENVELOPMENT ANALYSIS

Data Envelopment Analysis (DEA) is offered by Charnes et al. in 1978 to complete investigates directed by Farrel (1957). This method is used for assessing comparable DMUs efficiency by concerning numerous inputs and outputs (Sowlati et al. 2005). DEA and MADM collective method could be applied for assessing service processes by means of a standing system (Cooper et al. 2005). Commonly, DEA methods can be classified into two chief divisions, i.e. input-orientated and output-orientated. Input oriented models contain the models in which input values can be frequently reduced without changing the generated outputs values, while in another cluster, the output values can be proportionally amplified during saving the inputs values unaffected (Allen and Thanassoulis 2004; Bal et al. 2010).

The efficiency value is a proportion of the biased sum of productions to the biased sum of inputs. The comparative efficiency (wo) of specific DMUs is gained by resolving the next fractional programming, w_o = 1. It means that DMU is efficient whereas w_o <1 indicates the inefficiency of the DMU.

$$w_0 = Max \frac{\sum_{r=1}^{s} u_r y_{ro}}{\sum_{i=1}^{m} v_i x_{io}}$$

$$\frac{\sum_{r=1}^{s} u_r y_{rj}}{\sum_{i=1}^{m} v_i x_{ij}} \leq 1 \quad j = 1,2,...,n \tag{1}$$

$$u_r \geq 0, \quad r = 1,2,...,s$$

$$v_i \geq 0, \quad i = 1,2,...,m$$

Where:
j is the DMU index, j=1, ..., n,
r is the output index, r=1,..., s,
i is the input index, i = 1,...,m,
yrj is the value of the r-th output for the j-th DMU,
xij is the value of the i-th input for the j-th DMU,
ur is the weight given to the r-th output; and vi is the weight given to the i-th input.

The program (1) could be transformed into a linear programming (2) by making the biased sum of the inputs to 1. This method, that is the first appropriate kind of DEA techniques, is stated by Charnes, Cooper and Rhodes (CCR) (Al-Refaie and Al-Tahat 2011). Answer of the program can be create with constant return to the scale (CRS).

$$w_0 = Max \sum_{r=1}^{s} u_r y_{r0}$$

$$\sum_{i=1}^{m} v_i x_{i0} = 1$$

$$\sum_{r=1}^{s} u_r y_{rj} - \sum_{i=1}^{m} v_i x_{ij} \leq 0 \qquad i = 1,2,...,n \tag{2}$$

$$u_r \geq 0, \quad r = 1,2,...,s$$

$$v_i \geq 0, \quad i = 1,2,...,m$$

Another kind of DEA methods presented by Banker, Charnes and Cooper (BCC). Dissimilar to CCR model, in the BCC method, the answer comes with variable return to scale (VRS) (Chu et al. 2008). The BCC model could be as exposed in equation (3) (Sheikhalishahi et al. 2013)

$$w_0 = Max \sum_{r=1}^{s} u_r y_{r0} + c_0$$

$$\sum_{i=1}^{m} v_i x_{i0} = 1 \tag{3}$$

$$\sum_{r=1}^{s} u_r y_{rj} - \sum_{i=1}^{m} v_i x_{ij} + c_0 \leq 0 \quad j = 1,2,...,n$$

$$u_r \geq 0, \quad r = 1,2,...,s$$

$$v_i \geq 0, \quad i = 1,2,...,m$$

Where Co indicates returns to scale (RS) and is free in sign.

At what time which there is more than one efficient DMU, a balancing concept has to be used to identify the most efficient alternative.

In a blasting process complaint in which the entire environment is rock with asymmetrical crack and joint systems, it is incredible to anticipate whether all DMUs are working at an ideal scale. In blasting patterns act, it is unbearable to precisely compute the outputs growth or reduction rate. Consequently, this problem is resolved by implementing a BCC technique that is one of VRS models.

3. SIMPLE ADDITIVE WEIGHTING (SAW)

SAW method (Jaberidost, Olfat, & dkk, 2015) is recognized as a weighted sum model. The important idea of saw technique is seeking a weighted sum of act ranking of each alternative on all attribute. SAW model require the normalization process matrix decree (x) to a scale are comparable with all rating the alternatives. A formula to normalized follows:

$$R_{ij} = \begin{cases} \dfrac{x_{ij}}{\underset{i}{Max}\, x_{ij}} & \text{J is beneficial attribute} \\ \dfrac{\underset{i}{Min}\, x_{ij}}{x_{ij}} & \text{J is non beneficial attribute} \end{cases} \qquad (4)$$

r_{ij} is rating performance normalization of alternatives

A_i in attribute C_j; i = 1.2... m and j = 1.2, ..., n.

Preference value for each alternative assumed as:

$$V_{[i]} = \sum_{[n=1]}^{[n]} w_{[J]} r_{[ij]} \qquad (6)$$

Greater V_i value indicates that alternative A_i is more suitable.

4. SUNGUN COPPER MINE

The Sungun copper mine is the main open-pit copper mine in Iran which is located in North-West of Iran, East Azerbaijan province. This mine is in the prime phases of extracting. Geological location of Sungun copper mine is shown in figure 1.

Fig1 Geological location of Sungun

5. DISCUSSION

For choosing the most inexpensive, well-organized and suitable blasting pattern in view of practical aspects implementing DEA-SAW technique, in first stage all alternatives should be defined. all patterns should be assessed through 5 attributes (Yari et al. 2015b):

- Powder factor (PF): It must be concerned that the cost of explosive should be diminished.
- Specific drilling (SD): specific drilling growth improves fragmentation.
- Fly rock (FR): The lesser value of this index indicates the appropriate results.
- Back break (BB): Back break is the maximum distance of crack spread at the back of new bench.
- Fragmentation (K): the goal of suitable blasting pattern is reduction of fragmentation up to optimum size.

For omitting inefficient patterns, output-orientated BCC model of DEA is implemented for 50 patterns. Consequently, 27 patterns have been designated as more efficient patterns using BCC method (Table 1). Relative comparison of different attributes is mentioned in Table 2 with details. Weights of attributes are presented in Figure 2.

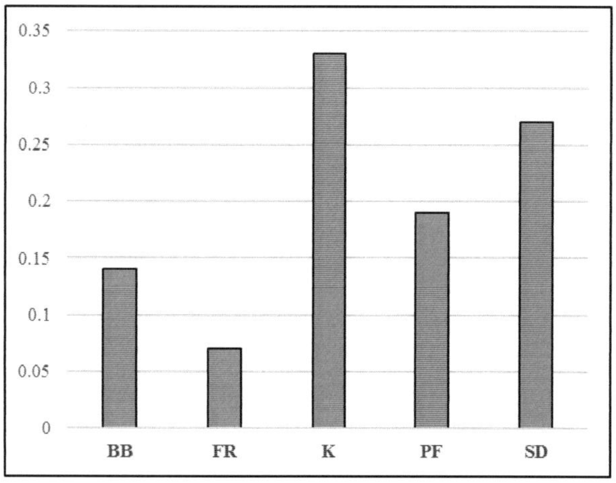

Figure 2. Attributes weights

For solving MADM, it is required to arrangement decision matrix. This matrix contains 6 columns and 28 rows. 27 efficient blasting patterns in Sungun Copper Mine have been considered with the 5 indices and assessed using SAW method. The Final results of DEA-SAW model for assessing 50 blasting patterns is presented in Table 3.

Table 1. 27 Efficient patterns

	Hole diameter	Hole Length	Spacing	Burden	Stemming
	inch	m	m	M	m
A1	5.5	12.1	4.5	3.5	3.8
A 2	6	11.5	4.5	3.5	3
A 3	5.5	12.5	4.5	3.5	3.6
A 4	6	12.3	4.5	3.5	3.6
A 5	5	13	5	4	3.8
A 6	6	11.8	4.5	3.5	3.8
A 7	5.5	12	4	3	3.2
A 8	6	12.8	5	4	4.1
A 9	5	13.5	5.5	5	4.5
A 10	5.5	11.5	4	3	3.2
A 11	5.5	11.5	4.5	3.5	3.6
A 12	5	13.5	5	4	4.1
A 13	6	13.2	5	4	3.5
A 14	5	11	4.5	3.5	3.8
A 15	5.5	13	4.5	3.5	4.1
A 16	5.5	12	4.5	3.5	3.8
A 17	5.5	13	5	4	3
A 18	5	13.2	5.5	4.5	3.8
A 19	6	12	5	4	4.1
A 20	5.5	12.5	5	4	4.3
A 21	5	13.2	5	4	4
A 22	5	11	3.5	3	3
A 23	5	12.8	4.5	3.5	4.1
A 24	5	11.5	4	3	3
A 25	6	12.9	5	4	4.1
A 26	5.5	12.5	4.5	4	4
A 27	5	11.8	4	3	3.2

Table 2. Matrix of values, mean and standard deviation

Pattern	PF	SD	FR	BB	K	Pattern	PF	SD	FR	BB	K
A1	0.36	0.05	72	2.5	31	A16	0.4	0.05	75	3	30
A2	0.34	0.05	75	2	31.5	A17	0.59	0.06	80	5	24.7
A3	0.42	0.05	76	3	30	A18	0.59	0.07	80	5	24.6
A4	0.43	0.05	76	3	31	A19	0.59	0.07	82	5.5	24.9
A5	0.4	0.05	75	3	32	A20	0.52	0.06	79	5	26.3
A6	0.41	0.05	76	3	29	A21	0.54	0.06	79	5	25.7
A7	0.38	0.05	75	2	30.1	A22	0.34	0.05	73	2	31
A8	0.59	0.07	80	5	24.7	A23	0.52	0.06	78	5	26.7
A9	0.85	0.09	85	9	20.8	A24	0.34	0.05	73	2	31.6
A10	0.37	0.05	74	2	30.2	A25	0.52	0.06	79	5	26.8
A11	0.4	0.05	76	3	31	A26	0.46	0.06	75	4	28.2
A12	0.59	0.06	78	5.5	24.7	A27	0.37	0.05	74	2	30
A13	0.59	0.07	81	5.5	24.6	Average	0.478	0.0578	77.11	3.907	27.99
A14	0.4	0.05	76	3	30	STEV	0.12	0.0101	3.13	1.687	3.057
A15	0.59	0.07	80	5.5	24.7						

Table 3. Final results of DEA-SAW model

Rank	Alternative	Score	Rank	Alternative	Score	Rank	Alternative	Score
1	A22	0.891999029	10	A14	0.827982903	19	A12	0.734445028
2	A24	0.887835357	11	A11	0.820673345	20	A23	0.73282721
3	A2	0.887225701	12	A3	0.820427739	21	A25	0.731302962
4	A27	0.883519209	13	A5	0.814441415	22	A18	0.701644928
5	A10	0.882018572	14	A4	0.809604151	23	A8	0.700526155
6	A7	0.877615084	15	A26	0.749442148	24	A13	0.696581005
7	A1	0.857454158	16	A17	0.737828774	25	A15	0.696008289
8	A6	0.831926851	17	A21	0.737639504	26	A19	0.692718904
9	A16	0.828603685	18	A20	0.736125254	27	A9	0.644801342

6. CONCLUSION

Blasting process in open-cast mines is one of the most vital processes concerning practical, economic and security aspects. Blasting plays a significant role in all future phases.

MADM approaches are valuable systems for assessing blasting patterns because it is problematic to make decision about the most appropriate blasting pattern. This difficulty is because of the diversity of the functioned patterns and the variety of attributes which meaningfully effect the assessment of blasting patterns. Among different techniques, DEA-SAW method is one of the most appropriate means for assessing blasting patterns in Sungun copper mine. Through this method, pattern 22 was selected as the most proper blasting pattern.

References

Ak H, Iphar M, Yavuz M, Konuk A (2009) Evaluation of ground vibration effect of blasting operations in a magnesite mine Soil Dynamics and Earthquake Engineering 29:669-676

Ak H, Konuk A (2008) The effect of discontinuity frequency on ground vibrations produced from bench blasting: A case study Soil Dynamics and Earthquake Engineering 28:686-694

Al-Refaie A, Al-Tahat MD (2011) Solving the multi-response problem in Taguchi method by benevolent formulation in DEA Journal of intelligent Manufacturing 22:505-521

Allen R, Thanassoulis E (2004) Improving envelopment in data envelopment analysis European Journal of Operational Research 154:363-379

Amini H, Gholami R, Monjezi M, Torabi SR, Zadhesh J (2011) Evaluation of flyrock phenomenon due to blasting operation by support vector machine Neural Computing & Applications:1-9

Athanassopoulos AD, Lambroukos N, Seiford L (1999) Data envelopment scenario analysis for setting targets to electricity generating plants European Journal of Operational Research 115:413-428

Bajpayee T, Bhatt SK, Rehak TR, Engineer G, Mowrey GL, Ingram DK (2003) Fatal accidents due to flyrock and lack of blast area security and working practices in mining Journal of mines, metals and fuels 51:344-349

Bajpayee T, Rehak T, Mowrey G, Ingram D A Summary of Fatal Accidents Due to Flyrock and Lack of Blast Area Security in Surface Mining, 1989 to 1999. In: PROCEEDINGS OF THE ANNUAL CONFERENCE ON EXPLOSIVES AND BLASTING TECHNIQUE, 2002. ISEE; 1999, pp 105-118

Bajpayee T, Rehak T, Mowrey G, Ingram D (2004a) Blasting injuries in surface mining with emphasis on flyrock and blast area security Journal of Safety Research 35:47-57

Bajpayee T, Verakis H, Lobb T An Analysis and Prevention of Flyrock Accidents in Surface Blasting Operations. In: PROCEEDINGS OF THE ANNUAL CONFERENCE ON EXPLOSIVES AND BLASTING TECHNIQUE, 2004b. ISEE; 1999, pp 401-410

Bakhshandeh Amnieh H, Siamaki A, Soltani S (2012) Design of blasting pattern in proportion to the peak particle velocity (PPV): Artificial neural networks approach Safety Science 50:1913-1916

Bal H, Örkcü HH, Çelebioğlu S (2010) Improving the discrimination power and weights dispersion in the data envelopment analysis Computers & Operations Research 37:99-107

Chakraborty A, Raina A, Ramulu M, Choudhury P, Haldar A, Sahu P, Bandopadhyay C (2004) Parametric study to develop guidelines for blast fragmentation improvement in jointed and massive formations Engineering geology 73:105-116

Chu M-T, Shyu JZ, Khosla R (2008) Measuring the relative performance for leading fabless firms by using data envelopment analysis Journal of Intelligent Manufacturing 19:257-272

Cooper WW, Seiford LM, Tone K (2005) Introduction to data envelopment analysis and its uses: with DEA-solver software and references

Cooper WW, Seiford LM, Tone K (2007) Data envelopment analysis: a comprehensive text with models, applications, references and DEA-solver software. Springer,

Crum S, Crum S (1990) Fractal concepts applied to bench-blast fragmentation Proc Rock Mechanics Contributions and Challenges 919

Dehghani H, Ataee-Pour M (2011) Development of a model to predict peak particle velocity in a blasting operation International Journal of Rock Mechanics and Mining Sciences 48:51-58

Erarslan K, Uysal Ö, Arpaz E, Cebi MA (2008) Barrier holes and trench application to reduce blast induced vibration in Seyitomer coal mine Environmental Geology 54:1325-1331

Gate W, Ortiz B, Florez R Analysis of rockfall and blasting backbreak problems. In: Paper ARMA/USRMS, proceedings of the American rock mechanics conference, 2005. pp 671-680

Ghasemi E, Amini H, Ataei M, Khalokakaei R (2012a) Application of artificial intelligence techniques for predicting the flyrock distance caused by blasting operation Arabian Journal of Geosciences:1-10

Ghasemi E, Sari M, Ataei M (2012b) Development of an empirical model for predicting the effects of controllable blasting parameters on flyrock distance in surface mines International Journal of Rock Mechanics and Mining Sciences 52:163-170

Guosheng Z, Jiang L, Kui Z (2011) Structural safety criteria for blasting vibration based on wavelet packet energy spectra Mining Science and Technology (China) 21:35-40

Hermans E, Brijs T, Wets G, Vanhoof K (2009) Benchmarking road safety: lessons to learn from a data envelopment analysis Accident Analysis & Prevention 41:174-182

Hudaverdi T (2012) Application of multivariate analysis for prediction of blast-induced ground vibrations Soil Dynamics and Earthquake Engineering 43:300-308

Inanloo Arabi Shad H, Ahangari K (2012) An empirical relation to calculate the proper burden in blast design of open pit mines based on modification of the Konya relation International Journal of Rock Mechanics and Mining Sciences 56:121-126

Iphar M, Yavuz M, Ak H (2008) Prediction of ground vibrations resulting from the blasting operations in an open-pit mine by adaptive neuro-fuzzy inference system Environmental Geology 56:97-107

Jimeno C (1995) Rock drilling and blasting. AA Balkema, Rotterdam, Brookfield,

Kao C, Liu S-T (2009) Stochastic data envelopment analysis in measuring the efficiency of Taiwan commercial banks European Journal of Operational Research 196:312-322

Kecojevic V, Radomsky M (2005) Flyrock phenomena and area security in blasting-related accidents Safety science 43:739-750

Khandelwal M, Monjezi M (2012) Prediction of Backbreak in Open-Pit Blasting Operations Using the Machine Learning Method Rock Mechanics and Rock Engineering:1-8

Kulatilake P, Qiong W, Hudaverdi T, Kuzu C (2010) Mean particle size prediction in rock blast fragmentation using neural networks Engineering Geology 114:298-311

Latham J-P, Van Meulen J, Dupray S (2006) Prediction of fragmentation and yield curves with reference to armourstone production Engineering geology 87:60-74

Li S, Jahanshahloo GR, Khodabakhshi M (2007) A super-efficiency model for ranking efficient units in data envelopment analysis Applied Mathematics and Computation 184:638-648

Little T, Blair D (2010) Mechanistic Monte Carlo models for analysis of flyrock risk Rock fragmentation by blasting:641-647

Michaux S, Djordjevic N (2005) Influence of explosive energy on the strength of the rock fragments and SAG mill throughput Minerals engineering 18:439-448

Monjezi M, Ahmadi M, Sheikhan M, Bahrami A, Salimi A (2010a) Predicting blast-induced ground vibration using various types of neural networks Soil Dynamics and Earthquake Engineering 30:1233-1236

Monjezi M, Amini Khoshalan H, Yazdian Varjani A (2011a) Optimization of Open pit Blast Parameters using Genetic Algorithm International Journal of Rock Mechanics and Mining Sciences 48:864-869

Monjezi M, Amini Khoshalan H, Yazdian Varjani A (2012) Prediction of flyrock and backbreak in open pit blasting operation: a neuro-genetic approach Arabian Journal of Geosciences 5:441

Monjezi M, Bahrami A, Varjani AY, Sayadi AR (2011b) Prediction and controlling of flyrock in blasting operation using artificial neural network Arabian Journal of Geosciences 4:421-425

Monjezi M, Dehghani H (2008) Evaluation of effect of blasting pattern parameters on back break using neural networks International Journal of Rock Mechanics and Mining Sciences 45:1446-1453

Monjezi M, Ghafurikalajahi M, Bahrami A (2011c) Prediction of blast-induced ground vibration using artificial neural networks Tunnelling and Underground Space Technology 26:46-50

Monjezi M, Rezaei M (2011) Developing a new fuzzy model to predict burden from rock geomechanical properties Expert Systems with Applications 38:9266-9273

Monjezi M, Rezaei M, Yazdian A (2010b) Prediction of backbreak in open-pit blasting using fuzzy set theory Expert Systems with Applications 37:2637-2643

Morin MA, Ficarazzo F (2006) Monte Carlo simulation as a tool to predict blasting fragmentation based on the Kuz–Ram model Computers & geosciences 32:352-359

Ning K (1999) Prevention Measures for Controlling Flyrock in Engineering Blasting [J]. BLASTING,

Ozkahraman H (2006) Fragmentation assessment and design of blast pattern at Goltas Limestone Quarry, Turkey International journal of rock mechanics and mining sciences 43:628-633

Ramanathan R (2006) Data envelopment analysis for weight derivation and aggregation in the analytic hierarchy process Computers & Operations Research 33:1289-1307

Rehak T, Bajpayee T, Mowrey G, Ingram D Flyrock issues in blasting. In: PROCEEDINGS OF THE ANNUAL CONFERENCE ON EXPLOSIVES AND BLASTING TECHNIQUE, 2001. ISEE; 1999, pp 165-176

Rezaei M, Monjezi M, Yazdian Varjani A (2011) Development of a fuzzy model to predict flyrock in surface mining Safety Science 49:298-305

Sanchidrián J, Segarra P, López L (2006) A practical procedure for the measurement of fragmentation by blasting by image analysis Rock mechanics and rock engineering 39:359-382

Sheikhalishahi M, Ebrahimipour V, Farahani MH (2013) An integrated GA-DEA algorithm for determining the most effective maintenance policy for ak-out-of-n problem Journal of Intelligent Manufacturing:1-8

Shim H-J, Ryu D-W, Chung S-K, Synn J-H, Song J-J (2009) Optimized blasting design for large-scale quarrying based on a 3-D spatial distribution of rock factor International Journal of Rock Mechanics and Mining Sciences 46:326-332

Shuran L, Shujin L (2011) Applying BP Neural Network Model to Forecast Peak Velocity of Blasting Ground Vibration Procedia Engineering 26:257-263

Singh T, Singh V (2005) An intelligent approach to prediction and control ground vibration in mines Geotechnical & Geological Engineering 23:249-262

Sowlati T, Paradi JC, Suld C (2005) Information systems project prioritization using data envelopment analysis Mathematical and Computer Modelling 41:1279-1298

Stojadinović S, Pantović R, Žikić M (2011) Prediction of flyrock trajectories for forensic applications using ballistic flight equations International Journal of Rock Mechanics and Mining Sciences 48:1086-1094

Tota EW, Mudge K, Branson JW, Georgiou PN, Gavrilovic M, Watson JD (2001) Method and apparatus for flyrock control in small charge blasting. Google Patents,

Yari M, Bagherpour R, Almasi N (2016a) An Approach to the Evaluation and Classification of Dimensional Stone Quarries with an Emphasis on Safety Parameters Rudarsko-geološko-naftni zbornik 31:15-26

Yari M, Bagherpour R, Almasi N (2016b) NAČIN PROCJENE I KLASIFIKACIJE KAMENOLOMA OBLIKOVANOGA KAMENA S NAGLASKOM NA SIGURNOSNE PARAMETRE Rudarsko-geološko-naftni zbornik 31:15-26

Yari M, Bagherpour R, Jamali S, Asadi F (2015a) Selection of Most Proper Blasting Pattern in Mines Using Linear Assignment Method: Sungun

Copper Mine/Wybór Najodpowiedniejszego Schematu Prowadzenia Prac Strzałowych W Kopalni Miedzi Sungun Z Użyciem Metody Przyporządkowania Liniowego Archives of Mining Sciences 60:375-386

Yari M, Monjezi M, Bagherpour R (2013) Selecting the most suitable blasting pattern using AHP-TOPSIS method: Sungun copper mine Journal of Mining Science 49:967-975

Yari M, Monjezi M, Bagherpour R (2014a) ISTRAŽIVANJE OPERACIJA MINIRANJA KORISTEĆI METODU ODLUČIVANJA Rudarsko-geološko-naftni zbornik 29:69-79

Yari M, Monjezi M, Bagherpour R (2014b) A novel investigation in blasting operation management using decision making methods Rudarsko-Geolosko-Naftni Zbornik 29:69

Yari M, Monjezi M, Bagherpour R, Jamali S (2014c) Developing a mathematical assessment model for blasting patterns management: Sungun copper mine Journal of Central South University 21:4344-4351

Yari M, Monjezi M, Bagherpour R, Sayadi A (2015b) Blasting operation management using mathematical methods. In: Engineering Geology for Society and Territory-Volume 1. Springer International Publishing, pp 483-493

Incorporation of Tonnage Uncertainty in to Long-term Production Scheduling Model by Using Stochastic Integer Programming

E. Moosavi
Department of Mining Engineering, South Tehran Branch, Islamic Azad University, Tehran, Iran

ABSTRACT Tonnage uncertainty is a critical factor in strategic mine planning and optimization of long term sequencing. This paper presented an optimization model that is shown to effectively incorporation tonnage uncertainty in to long-term production scheduling model by using a chance constrained binary integer programming model for open pit mine. The probability distribution function of tonnage in each ore block is used as a stochastic input to the optimization model. It is assumed that ore blocks tonnage to be a random variable with the known expected value and variance. In order to solve this model containing probabilistic parameters, chance constrained programming in used and linear model with uncertain parameters is converted to a nonlinear model with deterministic parameters. Then the nonlinear objective function is approximated by a linear one. The tonnage uncertainty cannot be eliminated; therefore, the best solution is to quantify uncertainties, reduce these uncertainties as far as investment permits, and finally manage the associated risk during the scheduling procedure. The latter can be achieved by explicit incorporation of uncertainty into the pit design and production scheduling process.

1 INTRODUCTION

The open pit mine production scheduling problem is the process of defining the sequence of mining blocks and simulating their removal whilst accumulation results in each time period to meet a predetermined target.

Mine scheduling often includes consideration of grade, tonnage and impurities in each period and it is these elements that make the scheduling problem more complex than the mining of ore to required amount. Input data for open pit mine scheduling include the following (Westcoft, 1991).

- A block model or mining database, which represents the economic boundaries of the orebody.
- A target amount of material to be mined per period, including grade and tonnage of ore and waste.
- A definition of time period for the mine plan (day, weeks, months or years).

In fact, mine scheduling plays an important role at all stages in the life of an open pit mine that major element of mine scheduling is the optimization of long- term production scheduling. Geological uncertainty is seen as the major contributor to not meeting project expectations. The major source of geological risk is uncertainty in grade and tonnage. Optimization in mine scheduling has been accepted as a set of techniques that introduce mathematical models.

The effects of geological uncertainty on approaches to optimizing open pit mine design have been shown in recent studies. Ravenscroft (1992) discusses risk analysis in mine production scheduling, where the use of stochastically simulated orebodies showed the impact of grade uncertainty on production scheduling and concluded that conventional mathematical programming models cannot accommodate quantified risk. Dowd (1997) proposed a framework for risk integration in surface mining projects. Dowd also considered some other random variables like commodity price, mining costs, processing costs, etc. Dimitrakopoulos et al. (2002) showed the substantial conceptual and economic differences of risk- based frameworks. The need for optimization methods that can integrate uncertainty raises the need for efficient simulation methods, as discussed in Godoy (2003) and Dimitrakopoulos and Luo (2004). Ramazan and Dimitrakopoulos (2004) developed efficient Mixed Integer Programming (MIP) formulations to generate feasible mining patterns of optimized probabilistic production schedules. Pursuing this line of thought, Ramazan and Dimitrakopoulos (2004) presented scheduling models and tested on how to generate MIP formulations using fewer binary variables. It also presented alternative approaches to MIP modeling for efficiency in solving the formulations with different mine data sets.

Although all traditional methods fail to consider the risk of not meeting production targets caused by the uncertainty in the optimization process. Dimitrakopoulos and Ramazan (2004) showed a new, risk-based production scheduling formulation for complex, multielement deposits. The formulation is based on expected block grades and probabilities of grades being above required cutoffs. Godoy and Dimitrakopoulos (2003) developed an algorithm which addresses the generation of optimal condition under uncertainty. At first, they generated production scheduling on each simulated ore body and then, they combined the mining sequences to produce a single schedule that minimizes the chance of deviating from production target; however, it still does not directly and explicity account for grade uncertainty. Therefore, Gholamnejad et al. (2006) developed a stochastic programming based model. In this model, grade uncertainty is incorporated explicity in the mathematical programming model for long- term production scheduling by applying chance constrained programming approach. This model generates the schedules that maximizes NPV of the total revenue and simultaneously, minimizes the risk of the schedules which is origined from ore grade uncertainty and also maintains a predetermined reliability with respect to satisfying probabilistic constraints. To deal with this draw back, Ramazan and Dimitrakopoulos (2007) presented an efficient new stochastic integer programming mathematical model that can consider multiple simulated orebody models to optimize long- term production scheduling. This model constructed as maximizing NPV of the mining operation, with a managed risk of not meeting production targets in terms of ore tones, grade and quality.

This paper presents an optimization model that is shown to effectively incorporation tonnage uncertainty in to long-term production scheduling model by using a chance constrained binary integer programming model for open pit mine. In the case of production scheduling, the objectives are to maximizes the expectation of net present value (NPV) and minimizes the variance function simultaneously.

2 MIP FORMULATIONS FOR LONG-TERM OPEN PIT PRODUCTION SCHEDULING

In long- term production scheduling of open pit mines, MIP models are usually constructed to maximize the overall net present value of the mining project. The general MIP form of open pit production scheduling is presented as follows, with some variations in the slope and reserve constraints. In this section describe binary integer programming model of long-term production scheduling problem.

2.1 Definition of Symbols

The variable and constant factors for the mathematical formulation of long-term production scheduling model are defined below:

T: is the total number of scheduling periods.

t: is the scheduling periods index, t=1, 2,…,T.

n: is the block identification number, n=1, 2,…,N.

N: is the total number of blocks to be scheduled.

r: is the discount rate in each period.

BEV^t_n: is the block economic value to be generated by mining block n in period t.

$$BEV^t_n = \begin{Bmatrix} \tilde{O}_n \cdot \left[(P^t - SP^t) \cdot G_n \cdot R - P^t_c - M^t_{co} \right] - \\ (M^t_{cw} \cdot W_n) \end{Bmatrix} \quad (1)$$

O_n: is the block tonnage in block n.

$\tilde{O}_n$: is block tonnage which is a random variable (n=1,2,…,N).

$E(\tilde{O}_n)$: is expected value of $\tilde{O}$n.

$Var(\tilde{O}_n)$: is variance of $\tilde{O}_n$.

$Cov(\tilde{O}_n, \tilde{O}_m)$: is covariance between $\tilde{O}_n$ and $\tilde{O}_m$.

W_n: is the waste tonnage in block n.

G_n: is the average grade of block n.

P^t: is unit selling price of the metal in period t.

SP^t: is unit selling cost of the metal in period t.

R: is total metal recovery.

P_c^t: is unit processing cost of the ore in period t.

M_{co}^t: is unit mining cost of the mineralized material in period t.

M_{cw}^t: is unit mining cost of the waste/overburden material in period t.

NPV^t_n: is the net present value to be generated by mining block n in period t:

$$NPV^t_n = \frac{BEV^t_n}{(1+r)} \quad (2)$$

G^t_{max}: is the upper bound average grade of material sent to the mill in period t.

G^t_{min}: is the lower bound average grade of material sent to the mill in period t.

PC^t_{max}: are the upper bound total tons of ore processed in period t.

PC^t_{min}: are the lower bound total tons of ore processed in period t.

MC^t_{max}: is the upper bound total amount of material (waste and ore) to be mined in period t.

MC^t_{min}: is the lower bound total amount of material (waste and ore) to be mined in period t.

Y: is the total number of blocks overlaying block k.

K: is the index of a block considered for extraction in period t.

y: is the counter for the m overlaying blocks.

2.1.1 The objective function

The objective function of the MIP model is constructed as the maximization of a profit function. It is evaluated computing the net present value of the planning.

$$\textbf{\textit{Maximiz}}\ z = \sum_{t=1}^{T} \sum_{n=1}^{N} NPV^t_n \cdot b^t_n \quad (3)$$

2.1.1.1 The model constraints

The proposed scheduling optimization model in Eq. (3) contains a series of constraints. They include as well as the more traditional of constraints of grade blending mill requirements, mill capacity, upper and lower bounds for ore quality parameters, mining capacity and wall slope. This type of constraints can be expressed by:

Grade blending constraints:

Upper bound constraints: The average grade of the material sent to the mill to be less than or equal to a certain value, G^t_{max}.

$$\sum_{n=1}^{N} \left(G_n - G^t_{\max} \right) \cdot O_n \cdot b^t_n \le 0 \quad for \quad t = 1,2,\dots,T \quad (4)$$

Lower bound constraints: The average grade of the material sent to the mill to be greater or equal to a certain value, G^t_{min}.

$$\sum_{n=1}^{N}\left(G_n - G^t_{\min}\right)\cdot O_n \cdot b^t_n \geq 0 \qquad for \qquad t = 1,2,...,T \quad (5)$$

Reserve constraints.
A block cannot be mined more than once. The following formulations are written for each block.

$$\sum_{t=1}^{T} b^t_n = 1 \qquad for \quad n = 1,2,...,N \qquad (6)$$

Processing capacity constraints.
These upper and lower bounds are necessary to ensure a smooth feed of ore.

Upper bound constraints: Maximum production capacity of the plant that cannot be more than the processing capacity (PC^t_{max}) in any period, t.

$$\sum_{n=1}^{N}\left(O_n \cdot b^t_n\right) \leq PC^t{}_{\max} \qquad for \quad t = 1,2,..., T \qquad (7)$$

Lower bound constraints: Minimum production capacity of the plant that cannot be less than a certain amount (PC^t_{min}) in any period, t.

$$\sum_{n=1}^{N}\left(O_n \cdot b^t_n\right) \geq PC^t{}_{\min} \qquad for \quad t = 1,2,...,T \qquad (8)$$

Mining capacity constraints.
Mining capacity constraints represent the actual available equipment capacity (MC^t_{max}) for each period, t.

$$\sum_{n=1}^{N}\left(O_n + W_n\right)\cdot b^t_n \leq MC^t{}_{\max} \qquad for \quad t = 1,2,...,T$$

To force the mentioned model to produce balance waste production throughout the periods, a lower bound(MC^t_{min}) may need to be implemented as follows:

$$\sum_{n=1}^{N}\left(O_n + W_n\right)\cdot b^t_n \geq MC^t{}_{\min} \qquad for \quad t = 1,2,...,T \quad (10)$$

Slope constraints.
Slope constraints represent the all overlying blocks that must be mined before mining a given block have to be determined. For example, if y is the index for the block that have to be removed before being able to remove block k in period, t. the sequencing formulation is written for each y as follows:

Using one constraint for each block per period:

$$Y \cdot b^t_k - \sum_{y}^{Y}\sum_{r=1}^{t} b^r_y \leq 0$$
$$for\, t = 1,2,...,T \qquad and \quad k = 1,2,...,N \qquad (11)$$

The sequencing formulation can also be written as in Eq. (12). The two types of formulations are discussed in Ramazan and Dimitrakopoulos (2004) in terms of solution times.

Using Y constraints for each block per period:

$$b^t_k - \sum_{r=1}^{t} b^r_y \leq 0 \qquad (12)$$
$$for\, t = 1,2,...,T \qquad and \quad k = 1,2,...,N$$

3 FORMULATION OF THE LONG-TERM PRODUCTION SCHEDULING UNDER TONNAGE UNCERTAINTY

Stochastic integer programming is defined as an extension of mixed integer programming (MIP) with uncertainty in one or more of the related coefficients (Escudero, 1993). This tends to increase problem size and complexity when compared with scheduling formulation based on MIP (2001). Birge and Louveaux (1997) discussed different approaches in stochastic integer programming formulations. The problem of stochastic (or better, chance- constrained) programming is here defined as follows, select certain random variable as function of random variables with known distributions in such a manner as (Charnes & Cooper, 1959):

a) To maximize a functional of both classes of random variables subject to
b) Constraints on these variable which must be maintained at prescribed levels of probability.

As mentioned before, ore blocks tonnage ($\tilde{O}n$) are not known with certainty at the beginning of each planning period. We only have statistical information about the random tonnages; therefore, objective function contains random parameters. The main difficulty of such models is due to optimal decision that have to be taken prior to the observation of random parameters.

In this model, to specify the effects of tonnage uncertainty on the mining sequence over the life of a mine. In this section the chance constrained programming approach is exploited to handle block tonnage uncertainty for the proposed binary integer model; then the chance constrained binary integer programming model of long- term production scheduling problem is reduced to its deterministic equivalent. The model contains an objective function and a set of constraints as follows:

3.1 The new objective function

As is clear from equation (3) due to stochastic nature of NPV^t_n, Z is also a random variable with the expected value and variance of:

$$E(Z) = \frac{1}{(1+r)^t}\cdot\left\{\left[\begin{array}{l} E(\tilde{O}_n)\cdot\left[\left(P^t - SP^t\right)\cdot R\cdot G_n - P^t_c - M^t_{co}\right] \\ -\left(M^t_{cw}\cdot W_n\right)\end{array}\right]\cdot b^t_n\right\} \quad (13)$$

$$Var(Z) = \frac{1}{(1+r)^{2t}}\cdot\left\{\begin{array}{l}\sum_{t=1}^{T}\sum_{n=1}^{N}\left[\left[\left(P^t - SP^t\right)\cdot R\cdot G_n - P^t_c - M^t_{co}\right]\cdot b^t_n\right]^2\cdot Var(\tilde{O}_n) \\ +\sum_{t=1}^{T}\sum_{n=1}^{N}\sum_{m=1}^{N}\left[\left(P^t - SP^t\right)\cdot R\cdot G_n - P^t_c - M^t_{co}\right]\cdot\left[\left(P^t - SP^t\right)\cdot R\cdot G_m - P^t_c - M^t_{co}\right]\cdot \\ b^t_n\cdot b^t_m\cdot Cov(\tilde{O}_n,\tilde{O}_m)\end{array}\right\} \quad (14)$$

We setup our objective to maximization of expected value of net present value and minimization of standard deviation simultaneously; as a result, the objective function of open pit long- term production scheduling can be written as:

$$Maximize \quad \bar{Z} = k_1\cdot E(Z) - k_2\cdot\sqrt{Var(Z)} \quad (15)$$

Where k_1 and k_2 are nonnegative coefficients and reflect the relative importance of maximization of expected value and minimization of standard deviation of net present value. If $k_1 = k_2$ then these two objectives are of the same importance for the decision maker. Thus the objective function can be re-written as follows:

$$Maximize \quad \bar{Z} = k_1\cdot\sum_{t=1}^{T}\frac{1}{(1+r)^t}\cdot\left\{\sum_{n=1}^{N}\left[\left(\begin{array}{l}\left[\left(P^t - SP^t\right)\cdot R\cdot G_n - P^t_c - M^t_{co}\right]\cdot \\ E(\tilde{O}_n) - \left(M^t_{cw}\cdot W_n\right)\end{array}\right)\cdot b^t_n\right]\right\}$$

$$-k_2\cdot\left[\frac{1}{(1+r)^{2t}}\left\{\sum_{t=1}^{T}\sum_{n=1}^{N}\left[\left(P^t - SP^t\right)\cdot R\cdot G_n - P^t_c - M^t_{co}\right]^2\cdot b^t_n\cdot Var(\tilde{O}_n) + \sum_{t=1}^{T}\sum_{n=1}^{N}\sum_{\substack{m=1\\ n\neq m}}^{N}\begin{array}{l}\left[\left(P^t - SP^t\right)\cdot R\cdot G_n - P^t_c - M^t_{co}\right]\times \\ \left[\left(P^t - SP^t\right)\cdot R\cdot G_m - P^t_c - M^t_{co}\right]\cdot b^t_n\cdot b^t_m\cdot Cov(\tilde{O}_n,\tilde{O}_m)\end{array}\right\}\right]^{1/2} \quad (16)$$

Let define:

$$A^t_n = \frac{1}{(1+r)^t}\left\{\left[\left(P^t - SP^t\right)\cdot R\cdot G_n - P^t_c - M^t_{co}\right]\cdot E(\tilde{O}_n) - \left(M^t_{cw}\cdot W_n\right)\right\} \quad (17)$$

$$B^t_n = \frac{1}{(1+r)^{2t}}\left[\left(P^t - SP^t\right)\cdot R\cdot G_n - P^t_c - M^t_{co}\right]^2\cdot Var(\tilde{O}_n) \quad (18)$$

$$C^t_{nm} = \frac{1}{(1+r)^{2t}}\left[\left(P^t - SP^t\right)\cdot R\cdot G_n - P^t_c - M^t_{co}\right] \times\left[\left(P^t - SP^t\right)\cdot R\cdot G_m - P^t_c - M^t_{co}\right]\cdot Cov(\tilde{O}_n,\tilde{O}_m) \quad (19)$$

Thus the objective function can be simplified as:

$$Maximize \quad \bar{Z} = k_1\cdot\sum_{t=1}^{T}\sum_{n=1}^{N}A^t_n\cdot b^t_n - k_2\cdot\sqrt{\sum_{t=1}^{T}\sum_{n=1}^{N}\left[B^t_n\cdot b^t_n + \sum_{m=1}^{N}C^t_{nm}\cdot b^t_n\cdot b^t_m\right]} \quad (20)$$

3.2 Linearization of the nonlinear model for long- term production scheduling

A major difficulty in using chance constrained programming is the need for a nonlinear algorithm. In this section linear approximation method is used for solving the nonlinearity of the problem; therefore, the objective will be linearized the functions and constraints so that linear programming algorithms can be used.

Suppose that x_i and x_j are two dependent random variables. The definition of the correlation between variable x_i and x_j is:

$$\rho_{ij} = \frac{Cov(x_i,x_j)}{\sqrt{Var(x_i)} \times \sqrt{Var(x_j)}} \quad (21)$$

$$Since \quad -1 \le \rho \le 1$$

Then:

$$-1 \le \frac{Cov(x_i,x_j)}{\sqrt{Var(x_i)} \times \sqrt{Var(x_j)}} \le 1 \quad (22)$$

Therefore:

(23)

$$-\sqrt{Var(x_i)} \times \sqrt{Var(x_j)} \le Cov(x_i,x_j) \le \sqrt{Var(x_i)} \times \sqrt{Var(x_j)}$$

This means that the maximum amount of the covariance of two variables would be the product of standard deviation of each of them. Hence replacing Cov(x_i,x_j) with $\sqrt{Var(x_i)} \times \sqrt{Var(x_j)}$ in equation (16) and considering $\sqrt{x_1^2 + x_2^2 + ... + x_n^2} \le x_1 + x_2 + ... + x_n$ (where $x_1, x_2, ..., x_n$ are positive variables) the linear equivalent of the objective function can be written as:

$$Maximize \quad \overline{Z} = \quad (24)$$

$$k_1 \cdot \sum_{t=1}^{T} \frac{1}{(1+r)^t} \left\{ \sum_{n=1}^{N} \left[\left(\begin{matrix} [(P^t - SP^t) \cdot R \cdot G_n - P_c^t - M_{co}^t] \cdot \\ E(\tilde{O}_n) - (M_{cw}^t \cdot W_n) \end{matrix} \right) \cdot b_n^t \right] \right\} -$$

$$k_2 \cdot \frac{1}{(1+r)^t} \left\{ \sum_{t=1}^{T} \left(\sum_{n=1}^{N} [(P^t - SP^t) \cdot R \cdot G_n - P_c^t - M_{co}^t] \cdot b_n^t \cdot \sqrt{Var(\tilde{O}_n)} \right) \right\}$$

Let define:

$$D_n^t = \frac{1}{(1+r)^t} \left\{ \begin{matrix} [(P^t - SP^t) \cdot R \cdot G_n - P_c^t - M_{co}^t] \cdot \\ E(\tilde{O}_n) - (M_{cw}^t \cdot W_n) \end{matrix} \right\} \quad (25)$$

$$E_n^t = \frac{1}{(1+r)^t} [(P^t - SP^t) \cdot R \cdot G_n - P_c^t - M_{co}^t] \cdot \sqrt{Var(\tilde{O}_n)} \quad (26)$$

Thus the objective function can be simplified as:

$$Maximize \quad (27)$$

$$\overline{Z} = k_1 \cdot \sum_{t=1}^{T} \sum_{n=1}^{N} D_n^t \cdot b_n^t - k_2 \cdot \sum_{t=1}^{T} \sum_{n=1}^{N} E_n^t \cdot b_n^t$$

The proposed scheduling optimization model in Eq. (27) contains a series of constraints that same written constraints in Eqs. (4) through (12).

4 CONCLUSION

In this paper we have demonstrated an optimization model that is shown to effectively incorporation tonnage uncertainty in to long- term production scheduling model by using a chance-constrained binary integer programming in open pit mines. At first, traditional linear integer programming model for long- term production scheduling is set up as base model. Then tonnage uncertainty is incorporated at the mentioned base model. Stochastic objective function could not to be handled directly in the optimization process; consequently, using chance constraints, a deterministic from of the model was obtained.

This process leads to converting stochastic linear equations to deterministic nonlinear equivalents. In this model objective function is nonlinear. Because of difficulties in solving large scale nonlinear models, a linearization process was applied and nonlinear functions are approximated by linear ones. The resultant linear model can be solved by using popular linear zero- one programming algorithms. The objective function in this model is constructed as maximizes the expectation of NPV and minimizes the variance function simultaneously. So the NPV to be generated from actual mining can be nearer to optimal. In fact, the NPV to generated from this model is the achievable NPV.

Comparing the results from the proposed model with those from the deterministic one shows that in the proposed model, the total number of blocks with higher confidence in the first periods of exploitation is greater than that of the deterministic model while all the production constraints are satisfied. This can result in decreasing the risk of not meeting production targets during the first period of the mining process.

REFERENCES

Westcoft, P.C. 1991. Mine scheduling and optimization, *Mining Industry Optimization Conference*, Sydney, Melbourne: Australasian Institute of Mining and Metallurgy, pp. 55-58.

Rovenscroft, P.J, 1992. Risk analysis for mine planning by conditional simulation, *Trans Instn Min. Metall*, Sec.A: Min. Industry, 101, May-August, pp. A82-A88.

Dowd, P.A., 1997. Risk in minerals projects: analysis, perception and management, *Trans. Instn Min. Metal*, Sec.A: Min. Industry, 106, pp. A9-A18.

Dimitrakopoulos, R., Farrelly, C.T., Godoy, M., 2002. Moving forward from traditional optimisation: grade uncertainty and risk effects in open-pit design, *Transactions of the Institutions of Mining and Metallurgy*, Section A: Mining Technology, 111, pp. 82-88.

Godoy, M.C., 2003. The efficient management of geological risk in long- term production scheduling of open pit mines, *PhD thesis*, University of Queensland, Brisbane, pp. 256.

Dimitrakopoulos, R, Luo, X, 2004. Generalized sequential Gaussian simulation on group size v and screen-effect approximations for large field simulations, *Mathematical Geology*, 36(5): pp. 567-591.

Ramazan, S, Dimitrakopoulos, R, 2004. Traditional and new MIP models for production planning with in-situ grade variability, *International Journal of Surface Mining, Reclamation and Environment*, Vol 18, No. 2, pp. 85-98.

Ramazan, S, Dimitrakopoulos, R, 2004. Recent application of operations research in open pit mining, *SME Transactions*, Vol 316, pp. 73-78.

Dimitrakopoulos, R., Ramazan, S, 2004. Uncertainty based production scheduling in open pit mining, *SME Transactions*, Vol 316, pp. 106-112.

Godoy, M, Dimitrakopoulos, R, 2003. Managing risk and waste mining in long-term production planning of open pit mine, *SME Annual meeting & exhibit*, pp. 24-26, Cincinnati Ohio.

Gholamnejad, J, Osanloo, M, Karimi, B, 2006. A chance-constrained programming approach for open pit long-term production scheduling in stochastic environments, *The Journal of the South African Institute of Mining and Metallurgy*, Vol. 106, pp. 105-114.

Ramazan, S, Dimitrakopoulos, R, 2007. Stochastic optimization of long term production scheduling for open pit mines with a new integer programming formulation, *Orebody Modelling and Strategic Mine Planning*, The Australasian Institute of Mining and Metallurgy, Spectrum Series, vol. 14, 2nd Edition, pp.385-392.

Escudero, L.F, 1993. Production planning via scenario modelling, *Annals of Operation Research*, 43, pp. 311-335.

Ramazan, S, 2001. Open pit mine scheduling based on fundamental tree algorithm, *PhD thesis*, Colorado School of Mines, Golden.

Brige, S.R, Louveaux, F, 1997. Introduction to stochastic programming, *Operation Research*, pp. 421.

Charnes, A, Cooper, W.W, 1959. Chance-constrained programming, *Management Science*, Vol. 6, pp. 73-79.

Classification of Draglines Failure Types Using Multilayer Perceptron and Radial Basis Function

Amir Taghizadeh Vahed
PhD Student, Mining Engineering Department, Middle East Technical University, Ankara, Turkey

Nuray Demirel
Assoc. Prof., Mining Engineering Department, Middle East Technical University, Ankara, Turkey

ABSTRACT The availability and utilization of draglines, which are massive and expensive machines used for overburden stripping, are a paramount concern for coal producers. Dragline's breakdown and prognostic of its failure types have a significant impact on a maximizing mine's production rate as well as minimizing maintenance and overall operating costs. In this sense, Machine Learning approaches have utilized in order to tackle the optimization issues.

The main objective of this research study is prediction of a classification of the failure types for a walking dragline, operating in Tunçbilek coal mine in Turkey, using Multilayer Perceptron and Radial Basis Function approaches. It also aims to compare and assess the utilization and performance of two different algorithms in machine maintenance data classification. The research methodology essentially entails five stages: (i) acquisition of data and its preprocessing including whitening smoothing and noise elimination; (ii) coding and generalizing function approximation; (iii) tuning the function; (iv) implementation of multi-layer perceptron and radial basis function algorithms; (v) evaluation of results and comparison of two different machine learning algorithms for dragline reliability.

The main novelty of this study is the utilization of machine learning approaches for dragline maintenance optimization for the first time.

1 INTRODUCTION

The field of reliability is not new and its application area is various. History of reliability backs to II World War, which was used by the German army for rockets. The reliability subject is used in various types of categories. According to Dhillon *et al.* (1992), the reliability-related studies can be classified as reliability optimization, reliability growth modeling, nuclear system reliability, human reliability, and machine reliability. The application of reliability has become widespread and penetrated to most of the industries. Accordingly, the reliability methods are widely applied in the mining industry. The breakdown, as an element of reliability, takes essential place in this subject, which results in increased maintenance costs and production losses. Schwartz (1990) claimed that production lost could be 50,000 British pounds per pit at some collieries owing to equipment break-down.

In order to achieve higher efficiency of equipment, which causes the study of reliability, equipment design investigation has taken place over the years. Correspondingly, fully computerized machines and equipment, as for instance, are used for representing a major role of reliability in the mining subject. Moreover, reliability methods, thanks to the utilization of modern monitoring techniques and application of expert systems, could diagnose efficiency and eventually prevent formerly any failures occurs. Hence, reliability assists to higher availability and thus, better performance of equipment in mining.

Coal mining is a pioneer in maintenance and reliability applications. According to Dhillon *et al.* (1992), coal mines are the witness of a phenomenal growth of reliability usages. The author mentioned that high volume of research activities are based on the technology of mining in order to the reliability issue which has applied in coal mines. Reliability of coal mine is employed in i) transportation and material handling equipment, ii) drilling and excavation equipment, iii) mine monitoring and control equipment, iv) mine electrical equipment, and iv) auxiliary support equipment.

Various arrays of algorithms have been developed in order to deal with equipment maintenance which was mentioned by Altshuler (1969). In the recent studies (Ali et al. (2015), Mosallam et al. (2014), Zhang et al. (2013), Benkedjouh et al. (2013), Le Son et al. (2013), Hu et el. (2012), Rodriguez et al. (2013)) machine learning approaches have been utilized in order to prognostic action and reliability. Rubinshtein (1975) stated that the reliability of mining equipment is based on the stochastic character because of random variation in the conditions of operations. This paper proposes the utilization of "energy of destruction" idea including the application of Least Squares in order to achieve better accomplishments for the expected value of reliability. In order to apply Least Squares (LS) methods, Artificial Neural Network (ANN) which uses LS as a core engine, has been utilized. Multi-Layer perceptron (MLP) and Redial Basis Function (RBF), as methods of ANN, have been implemented for the classification of break-down types owing to data acquisition from Tunçbilek coal mine in Turkey. As a final desired result, it would be possible to increase the required period for sequential break-downs by predicting an accurate break-down type; subsequently, the final aim of maximizing production as well as minimizing the operating costs will be gained.

In order to apply MLP and RBF, a dataset has been created based on acquisitioned data from a dragline which has been working in the coal mine for 23 years. The dataset encompasses two vectors which are input vector and target vector. The input vector consists two sequential times which two maintenances have taken place for the dragline. The target vector, moreover, is types of break-downs as shown in Table 2. In case of accessing trimmed dataset, pre-processing procedure has been carried out which includes: i) data cleaning (i.e. noise cleaning), ii) data integration, iii) data reduction or selection, and iv) data transformation. Sample of dataset has been shown in Table 1.

Table 1. Sample dataset

Mean Time to Failure (Hour)	ID of Failure Type
1.6597	1
3.5625	1
14.41	2
5.5625	6
0.5938	1

The paper is organized as: section 2 presents application of multi-layer perceptron in the mentioned data-set. Section 3 represents the utilization of radial basis function in the data set, and final section contains conclusion.

2 APPLICATION OF MULTI-LAYER PERCEPTRON

Based on the dataset, which contains data with discrete variables the classification approach can be carried out. One of the famous methods which are utilized for classification is Multi-layer Perceptron (MLP) algorithm. In this study, MLP has been applied for classification of break-down type in order to time duration between any sequential maintenance. Figure 1 depicts the distribution of break-down type and time to failures.

In Table 2, the break-downs types, which has been converted from linguistic format to numeric one (ID), has shown as below.

Table 2. Types of break-downs

Break-down type	ID
Mechanical Failures	1
Electrical Failures	2
Energy Related Failures	3
Major Revision	4
Maintenance	5
Power Cut	6

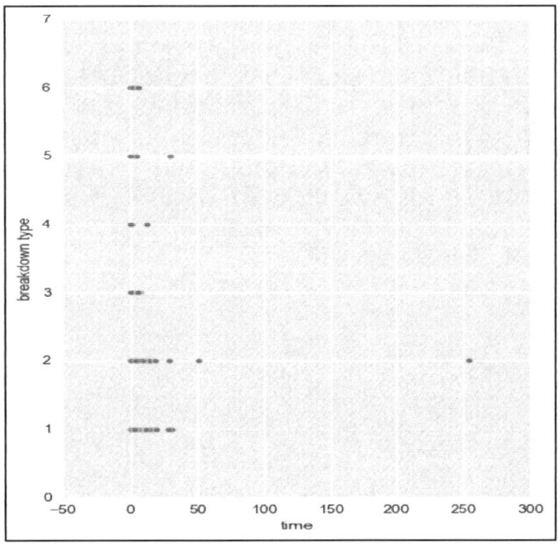

Figure 1. Distribution of data

In this study, applied MLP encompass four layers of neurons (i.e. the shematic view of MLP network has been shown in Figure 2). In the first layer, which is called input layer, a time period between two sequential break-downs vector has been entered. As far as, MLP is a supervised method, the final layer which is labeled taregt layer contains a break-down types vector. Besides, among these two layers, hidden layers are placed. Gathered data (i.e. the observations of the dragline) from input layer goes to the first hidden layer and thanks to the activation function given in Equation 1 (Jun *et al.*, 1995) data goes forward and in the second hidden layer will be take place.

$f(t) = (1)/(1+\exp(-\alpha t))$ Equation (1)

f(t) is an activation function (i.e. called Sigmoid function) and α is the slope of the function. t is input values which are driven from the input vector. Correspondingly, data in the first hidden layer has been assigned by Equation 2 (Rosenblatt, 1958):

$$a_m = \sum_{l=1}^{L} W_{lm} X_l \quad \text{Equation (2)}$$

a_m is the calculated value for hidden unit; W is, indeed, the weights for emelent, and the X is element of the input vector.

Each neuron's output even linear or non-linear in order their activation function are estimated by below-mentioned equation (Rosenblatt, 1958):

$$s_m = f(a_m) \quad \text{Equation (3)}$$

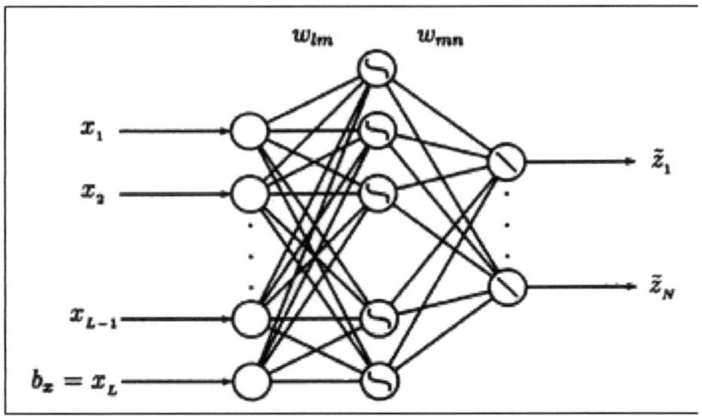

Figure 2. Schematic view of multi-layer perceptron network

The primary task of MLP is an estimation of appropriate weights for a model, so an optimization problem is faced. Regards to the calculated weights, the errors (i.e. the error is the difference between the actual values of outputs and predicted values for inputs) must be minimized. Various types of methods for weight calculation are utilized. In this study, Gradient Descent (GD) method (i.e. the gradient method has been invented by Bernard Widrow and Ted Hoff in 1960) has been applied. Update of weights for step *(k+1)* which done by GD is shown as below mentioned equations (Widrow *et al.*, 1960):

$$w(k+1) = w(k) + \Delta w(k) \quad \text{Equation (4)}$$

and

$$\Delta w(k) = -\eta \Delta_w E(k) + \omega \Delta w(k-1) \quad \text{Equation (5)}$$

η is learning rate ; and, ω is momentum.

The model has been coded and generalized for outfit data which is called the test data, and test data portion is 30 percent of all data (i.e. the number of implemented data is equal with 1303). The results of prediction by MLP have been shown in Figure 3.

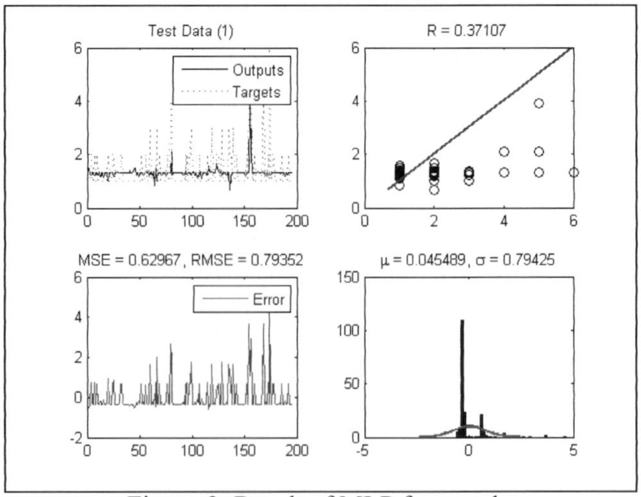

Figure 3. Result of MLP for test data

As shown in Figure 3, the regression for the test dataset is 37 percent. The Root Mean Square Error (RMSE), indeed, is 0.79. The approximated function has been generalized for dataset and shows MLP could not cover an appropriate regression rate. The low regression rate and high error caused due to non-linear inherent of data. Thus, a structure which uses method for handling nonlinearity can be advantageous. Radial Basis Function as a method which maps nonlinearity of features to a linear space will be suitable.

3 APPLICATION OF RADIAL BASIS FUNCTION NETWORK

The second considered algorithm is Radial Basis Function (RBF) which is mostly used for the classification. The advantage of RBF is mapping the non-linear observation to a kernel which shaped the feature to the linear one. Thus, application of RBF is required lower CPU consumption as well as time. RBF's schematic network has been illustrated in Figure 3.

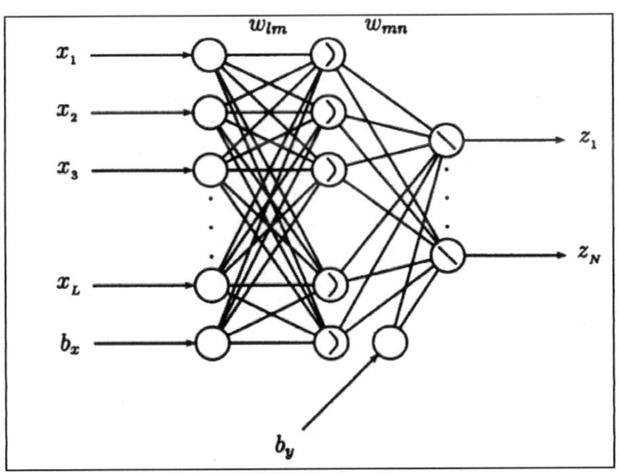

Figure 4. Schematic view of radial base function network

Based on Figure 4, elements of the input vector send to kernels and the classification is calculated based on the distance of the input's element from the Centers. The aim of RBF is minimizing the total distance from Centers. The distance function is calculated by the equation below:

$$S_m = f(||x - C_m||) \qquad \text{Equation (6)}$$

x is elements of the input vector and C are selected centers for the valuation of distances. In this study, Gaussian function has been utilized for estimation of distances and shown in Equation 7 (Jun *et al.*, 1995).

$$f(t) = exp(r^2/\sigma^2) \qquad \text{Equation (7)}$$

r is the radius which represents a distance from a center and σ is spread. The spread is a tuning feature of RBF which has a major role. In this study, the spread is 0.1 which calculated by the trail-error procedure. Regards to all mentioned calculation, the predicted output vector is estimated by Equation 8:

$$\hat{y} = p(\textstyle\sum_{m=1}^{L} w_{mn} s_m) \qquad \text{Equation (8)}$$

$\hat{y}$ is the predicted output vector; moreover, p is the probability of generalized model. S is a parameter which is calculated by equation 6.

For achieving the weight vector, the vectored format of Equation 8 is utilized which is represented as below:

$$y = Sw \qquad \text{Equation (9)}$$

The final aim of RBF application is justifying the weight vector, so the weight vector is evaluated by Equation 10:

$$w = (S^T S)^{-1} S^T y \qquad \text{Equation (10)}$$

The model has been shaped by coding and implementing some hypothesis which has been assigned by a tail-error procedure (i.e. Trail-error is the only method for tuning parameter in ANN approach). 1303 elements of input, as well as output, are assigned to the model and the result of the training shown in the figure below:

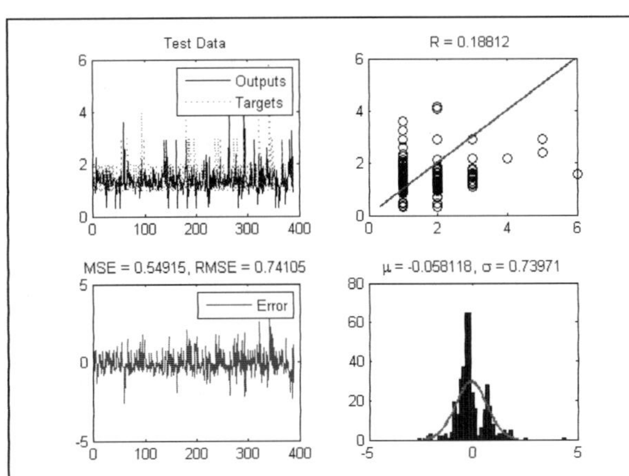

Figure 5. Result of RBF for test data

Based on the results (Figure 5), regression rate has not proved in contrast to MLP algorithm, yet RMSE has decreased, so the modeled function works slightly better.

4 CONCLUSION

Draglines reliability plays a crucial role in the production rate of a mine as well as its operating costs, particularly in coal mines. An accurate prediction of break-down types for draglines deals with the achievement of the desired objective function (i.e. in this study, the objective function is maximizing the production and minimizing the operating cost). In previous studies, authors applied statistical approaches in order to describe the data and fit a model based on data's distribution; indeed, the statistical methods utilize for small data sets. However, recent datasets contain ample data with lots of features, and they represent the high dimensionality. In this sense, the application of machine learning methods is more effective.

In this study, MLP and RBF, as algorithms of machine learning, have been applied for creating approximation functions in order to predict the dragline's failure-types. 1303 data are utilized which were gathered from a dragline in Tunçbilek coal mine. The data are related to a dragline break-down types and the time means which two sequential maintenances have taken place. The approximated functions have been created and regards to evaluating the accuracy of the generated functions RMSE parameter has been calculated. The results of RMSE of MLP, as well as RBF, are given in Table 3.

Table 3. Results for MLP and RBF

RMSE	MLP	RBF
All Data	0.70	0.70
Training	0.64	0.69
Test	**0.79**	**0.74**

RMSE of RBF is lower than MLPs, which shows RBF has higher performance in contrast to MLP. This outcome has been caused due to RBF maps the non-linearity feature of data to the linear dominant, so the RBF is a better resemblance of the data. The results show the MLP and RBF could model the dataset which can be utilized for draglines' failure-types prediction. In the case of improving the regression rate, the application of algorithms which consider more features is recommended for future studies.

REFERENCES

Altshuler, V. M. 1969. A Method of Constructing a Mathematical Model to Study the Reliability of Mine Transportation Systems, Soviet Mining Science, 5, (pp.72-76).

Ben Ali, J., Fnaiech, N., Saidi, L., Chebel-Morello, B., & Fnaiech, F. 2015. Application of empirical mode decomposition and artificial neural network for automatic bearing fault diagnosis based on vibration signals. *Applied Acoustics*, *89*, 16–27.

Benkedjouh, T., Medjaher, K., Zerhouni, N., & Rechak, S. 2013. Remaining useful life estimation based on nonlinear feature reduction and support vector regression. *Engineering Applications of Artificial Intelligence*, *26*(7), (pp.1751–1760).

Dhillon, B. S., & Anude, O. C. 1992. Mining Equipment Reliability: A review. *Microelectronics Reliability*, 32, 8, (pp.1137–1156).

Hu, C., Youn, B. D., Wang, P., & Taek Yoon, J. 2012. Ensemble of data-driven prognostic algorithms for robust prediction of remaining useful life. *Reliability Engineering and System Safety*, *103*, (pp.120–135).

Jun, H., & Claudio. M. 1995. The Influence of the Sigmoid Function Parameters on the Speed of Back propagation Learning. *Natural to Artificial Computatuin*. (pp.195-201).

Le Son, K., Fouladirad, M., Barros, A., Levrat, E., amd Iung, B. 2013. Remaining useful life estimation based on stochastic deterioration models: A comparative study. *Reliability Engineering & System Safety*, *112*, (pp.165–175).

Mosallam, A., Medjaher, K., & Zerhouni, N. 2014. Data-driven prognostic method based on Bayesian approaches for direct remaining useful life prediction. *Journal of Intelligent Manufacturing*, (pp.1037–1048).

Rodriguez, J. A., Hamzaoui, Y. El, Hernandez, J. A., Garcia, J. C., Flores, J. E., & Tejeda, A. L. 2013. The Use of Artificial Neural Network (ANN) for Modeling the Useful Life of the Failure Assessment in Blades of Steam Turbines. *Engineering Failure Analysis*, *35*, (pp. 562–575).

Rosenblatt, F. 1958. The Perceptron a Probabilistic Model for Information Storage and Organization in the Brain. *Cornell Aeronautical Laboratory, Psychological Review*, 65, 6, (pp.386–408).

Rubinshtein, B. S. 1975. Some Aspects of the Reliability of Mining Equipment. *Journal of Mining Science*, 11, 4, (pp.349–353).

Schwartz, M., Ed. 1990. Lubricants and Condition Monitoring. *Colliery Guardian*, 238, (pp.7-12).

Zhang, Z., Wang, Y., & Wang, K. (2013). Fault diagnosis and prognosis using wavelet packet decomposition, Fourier transform and artificial neural network. *Journal of Intelligent Manufacturing*,*24*(6),(pp.1213–1227). https://doi.org/10.1007/s10845-012-0657-2

Simulating the Haulage System Of Abadeh Fireclay Mine by Using Arena Software

Dr. M.H. Basiri*
Assistant Professor, Department of Mining & Materials Engineering, Tarbiat Modares University, Tehran, Iran.

S. Alamdari
Department of Mining & Materials Engineering, Tarbiat Modares University, Tehran, Iran.

ABSTRACT In open pit mines, loading and transportation costs devote more than half of the overall operational costs. The suitable allocation of machinery in haulage systems can reduce mining expenses as well as leads to increase efficiency. In this regards, simulation provides a superior insight to the mining systems.

In this study, the haulage system of Abadeh Fireclay mine, located in the Fars Province of Iran, has been studied by the use of simulation techniques and Arena software. Arena is an object-oriented technique. Design the model was carried out in a cycle-oriented manner. In this research, after modeling the current system of the mine, increasing rate of production has been studied according to the performance of the active machineries. The results indicated that the efficiency of loading machines was low and the unit costs were increased due to the shortage of the number of trucks.

Keywords: Abadeh Fireclay Mine, Haulage system, Simulation, Arena software.

1 INTRODUCTION

In open pit mining, a haulage system is used to transport material from the production area to the desired location. The haulage system depends on the mining method employed. Operations of trucks and shovels constitutes 50% to 60% of the overall operations' cost in open-pit mines (Ercelebi & Bascetin, 2009). Therefore, efficient utilization of the haulage system could result in savings and reducing the overall mining cost. The selection of effective loading and hauling machines becomes an important factor to optimize the haulage system.

Various methods and techniques are created to identify, improve and make a decision about systems performance to solve problems. Classic methods, operations research techniques and artificial intelligence methods can be considered examples of these methods. Simulation is one of the most admired operations research tools to understand and improve the systems performance due to its ability to deal with complex models, flexibility, power, and ease of use (Alamdari, 2014).

This paper presents a simulation study of haulage system of Abadeh Fireclay mine, located in the Fars Province of Iran, by using the Arena software.

2 SIMULATION IN MINING OPERATIONS

Simulation is the imitation of the real world processes or systems over time. Examples of Systems that can be simulated include transportation logistics, business processes, mining operations, emergency response systems etc. Simulations can apply a number of rules and procedures, thereby increasing the understanding of the interaction between variables and their importance in the system performance and suggesting possible system modifications (Banks, 2009).

Using simulation technique, lots of complicated problems can be solved with fewer simplifications. Discrete event simulation has proved itself to be an effective tool for complex process analysis.

The use of computer simulation in mining operations was first reported by Rist at 1961, where a model was built to determine the optimum number of trains for a haulage level (Sturgul, 1999). Since then several studies has shown a wide applicability of simulation studies on various operations in both underground and open pit mines in the world. Also, the use of simulation has progressed to several mining aspects such as queuing theory, scheduling, decision-making, location models, etc.

Other examples include development of a simulation model to study the train transportation in underground coal mines in Germany (Wilke, 1970), and development of a computer program that assists in analysis shovel-truck operations in the mines. Ataeepour and Baafi developed a statistical simulation model to analysis the operating performance of the truck - shovel system, by using ARENA software (Ataeepour & Baafi, 1999). In other study, Askari-Nasab et al presented a Mixed Integer Linear Programming (MILP) model to determine the optimum number of shovels and trucks required to meet the plan's goals and also analyzed the operations of trucks and shovels in an open pit mine using simulation software (Askari-Nasab, Torkamani, Badiozamani, & Tabesh, 2012). Salama et al studied the use of discrete event simulation for underground haulage equipment selection and compared two different haulage units with the aim of improving production (Salama, Greberg, & Schunnesson, 2013).

ARENA is a powerful simulation tool, developed by Rockwell Software Company on the basis of SIMAN/Cinema (Rockwell Automation, 2012).

Stages of the research had been done based on the simulation process steps. Simulation activity usually consists of a number of stages as follows: 1-definition of purpose, 2- data acquisition, 3 model construction, 4- model verification, 5- model validation, 6- model experimentation. 7- analysis of results, 8-documentation, and 9- implementation (Banks, 2009).

After definition of the issue, collection of required data symmetrical to the goal must be done. This stage is the most time consuming one. The model construction is done next, through which the logic flow of the real system must be represented carefully by the model components. After that, verification and validation of the model will be accomplished. Then, the constructed simulation model will be executed and results will be obtained and analyzed. For achieving the optimum solution, the model is run under various condition and parameters of the system. Specifications of the model that provided the best results are selected as the optimum solution. Everything done during the simulation process is documented and a recommendation is made for implementation of the model.

3 THE SIMULATION CASE STUDY

This paper presents the case study of an open pit mine, Abadeh Fireclay Mine, which is the biggest Iranian fireclay mine located 250 km north of Shiraz city-Iran (Figure 1**Hata! Başvuru kaynağı bulunamadı.**).

Based on reserve estimation, currently more than 300 million tons of minerals are in this mine.

Figure 1: Abadeh Fireclay mine location

Due to geotechnical structures, drilling and blasting operations are not applied in production operation in this mine.

In Abadeh fireclay mine, 3 trucks with the capacity of 30 tons, 3 loaders, 3 excavators and 24 trucks with the capacity of 20 and 25 tons, operate mining extraction in one shift per day.

To simulate the system, time data was collected by using direct timing study of activities during the shift. The collected data comprised; trucks and loaders cycle times, economic data, machines available time, ore production and etc.

Timing include loading time, trucks travelling time in different routes, dumping time, machines available time, and etc. Table 1 presents the existing equipment in the haulage system.

Table 1: Equipment date of haulage system

Loading point	Extraction type	Loading machine	Truck capacity (ton)
ES1	Ore	Excavator*	25
ES2	Waste	Loader*	25
ES3	Waste	Loader	25
ES4	Ore	Excavator	20
ES5	Ore	Loader	30
ES6	Waste	Excavator	20

* Excavators capacity is 1.8 m^3, Loaders capacity is 4.5 m^3

4 MODELING HAULAGE SYSTEM

In this study, for modeling the haulage system, the ARENA simulation software was applied. It is based on discrete event simulation principles and uses a full graphical user interface to set up the model. It utilizes statistical distribution functions to model variations in process times.

To simulate the haulage system of the mine, the initial study was carried out on the target system. Modeling and input data analysis was conducted after collecting the necessary data and designing the conceptual model of system. Then simulation of model was implemented in ARENA software and after verification and validation of the model, the simulation results was analyzed.

Schematic layout of Abadeh fireclay mine haulage system with six loading points and two dumping stations in the ARENA environment is shown in Figure2.

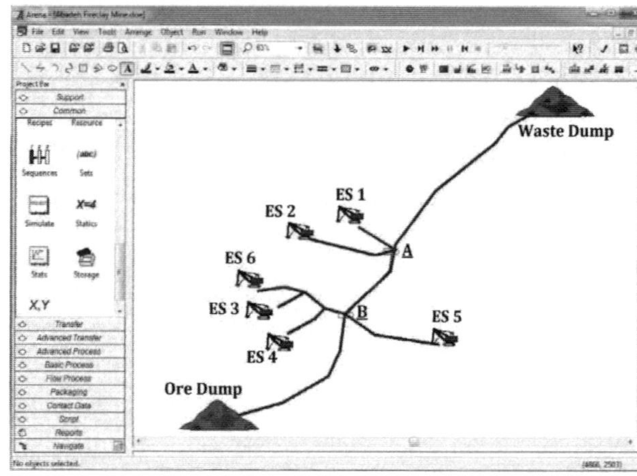

Figure2: Layout of haulage system in ARENA

In this research, Arena's Input Analyzer software was used to fitting and selecting the suitable distribution of timing data. For example, loading time of the 30-tons dump truck through 31 recorded data has normal distribution NORM (3.07, 0.255) in terms of minutes (**Figure3**). By investigate Chi-Sequare and Kolmogorov-Smirnov (K-S) tests, and because the P-value is greater than the alpha level (0.05), thus the fitted distribution is not refutable.

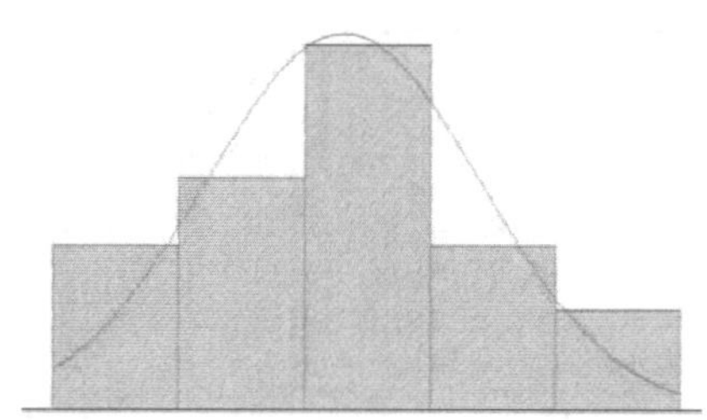

Figure3: Fitted Distribution of Loading Time for 30-tons Truck

For example, the summary of timing for 30-tons truck is shown in **Table 2**. The timing data distribution of other machines were obtained and used in the simulation model.

Table 2: 30-tons truck Timing and distribution

Variable type	Variable distribution
Travel from ES5 to B (full)	NORM(2.03, 0.266)
Travel from B to Ore Dump (full)	3.24 + 1.02 * BETA(2.08, 2.38)
Travel from Ore Dump to B(empty)	NORM(3.36, 0.19)
Travel from B to ES5 (empty)	NORM(1.9, 0.152)
Load Time	NORM(3.07, 0.255)
Dump Time	NORM(1.03, 0.137)

It is assumed that the amount of extraction in all periods is identical and average production is required, the system has been modeling for a period of one month.

In this study, Naylor and Finger method was used to validation the model (Naylor & Finger, 1967): Step 1: Build a model that has high face validity, Step 2: Validate model assumptions, Step 3: Survey simulation output and compare with real system. In this study, the monthly production of ore and waste had been selected as a benchmark to compare the model with the real system and simulation outputs validation (Table 3**Table 3**).

Table 3: Mine production

Extractive Type	Production (ton)	
	Real system	Simulation Outputs
Ore	185000	180400
Waste	195000	186860

According to these results, there were no significant differences at 95% confidence level between outputs of the simulation model and real system in the period of system study, therefore, the simulation model was validated.

Table 4.

The loaders waiting time was high, thus the system efficiency was decreased and the production costs were increased. Increasing the number of trucks leads to increase the loading machines performance and the production's loading points.

5 ANALYSIS THE HAULAGE SYSTEM PERFORMANCE

Loading machines efficiency, monthly production, production costs, and the possibility of increasing production with current machines was investigated in current haulage system of Abadeh Fireclay mine. The system has been evaluated by providing comparison graph in term of delivered tonnages in a month. The optimum number of truck was selected by economic analysis of results to achieve maximum production with minimum cost as well as considered the efficiency of loading and hauling machines. For example, Figure 4 presents the performance of ES5 loading point.

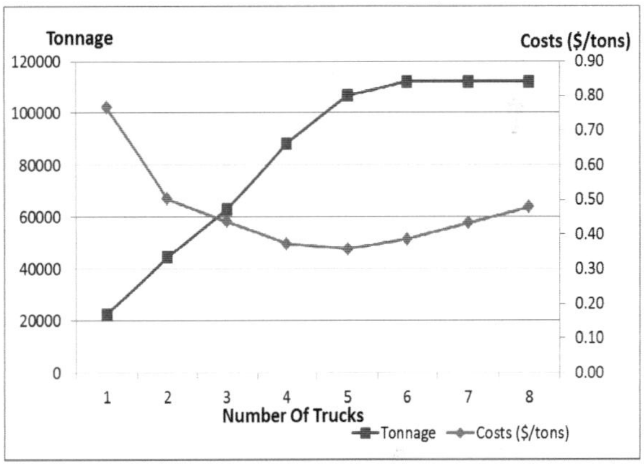

Figure 4: Comparison graph of ES5 loading point

This procedure was conducted for all loading points and the result was evaluated. According to Figure 4, **with 5 trucks, the monthly productions of ES5 point is 106,680 tons with a least costs of 0.36 $/ton. The summarized of the simulation results of current system performance is shown in**

The ability of increasing rate of production has been studied according to the current machinery. According to the results by allocating 52 trucks (5 trucks with the capacity of 30 tons, 29 trucks with the capacity of 25 tons and 18 trucks with the capacity of 20 tons), monthly production increase to 626700 ton and production costs will be decrease till 0.41 dollar per ton.

Table 4: Simulation results of current haulage system

Loading Point	current system				optimum system			
	Trucks Number	Production (ton)	Production Costs($/ton)	Loader utilization%	Trucks Number	Production (ton)	Production Costs($/ton)	Loader utilization%
ES1	4	48900	0.50	68	6	69225	0.39	97
ES2	5	77150	0.84	58	10	131895	0.43	99
ES3	5	60650	0.64	42	13	140100	0.50	98
ES4	5	68320	0.38	70	8	97480	0.36	98
ES5	3	63180	0.44	59	5	106680	0.36	95
ES6	5	49060	0.50	57	10	81320	0.37	99
Total	**27**	**367260**	**0.56**		**52**	**626700**	**0.41**	

6 CONCLUSION

This paper presented a simulation technique to evaluate the haulage system in Abadeh Fireclay open-pit mine. The simulation models were designed using ARENA software.

According to the simulation results, due to the shortage of trucks, the efficiency of loading machines in this mine was low and the unit production cost was increased. The number of trucks can be increased to 52 devices with current machines. This means that production can be increased up to 70 percent and the production costs have the potential to be reduced 27 percent. Future studies in this area should consider failures of equipment to generate more accurate assessment of the system performance.

REFERENCES

Alamdari, S., 2014. *Optimization based on simulation of the production system in Abadeh fireclay mine.* (M.Sc. thesis), Tarbiat Modares University, Iran.

Askari-Nasab, H., Torkamani, E., Badiozamani, M. m., & Tabesh, M., 2012. *Alignment of Short-Term and Operational Plans Using Discrete Event Simulation.* Paper presented at the Proceedings of Society for Mining, Metallurgy & Exploration (SME).

Ataeepour, M., & Baafi, E., 1999. ARENA simulation model for truck-shovel operation in despatching and non-despatching modes. *International Journal of Surface Mining, Reclamation and Environment, 13*(3), 125-129.

Banks, J., 2009. *Discrete Event System Simulation, 4th Edition*: Pearson Education India.

Ercelebi, S., & Bascetin, A., 2009. Optimization of shovel-truck system for surface mining. *Journal of The Southern African Institute of Mining and Metallurgy, 109*, 433-439.

Naylor, T. H., & Finger, J. M., 1967. Verification of computer simulation models. *Management Science, 14*(2), B-92-B-101.

Rockwell Automation., 2012. Arena Simulation Software Ver. 14.0.

Salama, A., Greberg, J., & Schunnesson, H., 2013. The use of discrete event simulation for underground haulage mining equipment selection. *Accepted for publication in the International Journal of Mining and Mineral Engineering.*

Sturgul, J., 1999. Discrete mine system simulation in the United States. *International Journal of Surface Mining, Reclamation and Environment, 13*(2), 37-41.

Wilke, F., 1970. *Simulation studies of computer controlled traffic in underground large coal mines.* Paper presented at the Decision making in the Mineral Industry, 9th International Symposium Can IMM, Special.

Short-Term Production Planning for Multi-Metal Open-Pit Mines with Equipment Size Constraints

Y. A. Sari, N. Germain, M. Kumral
McGill University, Montreal, CANADA

ABSTRACT After long-term production plans are completed and the operation starts, more information about deposit is gathered from the blast-hole samples. This more specific information is then used to plan the mining production in short range basis. The daily process of discriminating and diverting the mined parcels to ore or waste destinations is called short-term production planning. Although classifying the parcels according to the cut-off grade serves as a valuable guide, the produced plan is not practically viable because the equipment size does not allow to treat each parcel separately. Usually, consecutive parcels have to be mined and diverted together. This connection leads to complex relationships between consecutive parcels. An optimization model based on an exact method is constructed for the short-term production planning of a multi-metal mine that accounts for the equipment size constraints. Unlike meta-heuristic based approach, it guarantees optimality. The performance of the model is demonstrated through a case-study.

1 INTRODUCTION

Mining operations are planned at three levels: long-term (strategic), medium-term (tactical) and short-term (operational) operations. Long-range planning focuses on maximizing the Net Present Value (NPV) of a project and determining the overall mining sequence. Mid-range planning is concerned with managing the objectives of the long-range plan while taking into consideration more complex and detailed operational parameters. Short-range planning manages the day-to-day performance of the mining operation to meet the targets established by the medium-term plan.

During short-range planning, the mine engineer must draw dig-limits from the block model to be used as operation guidelines for grade control purposes. Dig-limits form the interface between the ore and the waste and need to have mineable shapes. The grade information arrives from blast holes. Using this information, planning units known as "selective mining units (SMU)" are formed. SMUs are analogous to mining block units used in surface mine planning, but blocks are formed using drill-hole data which is much more sparse than blast holes; forcing the blocks to be larger than SMUs. Ore and waste decisions are made for each SMU. However, the decision of an SMU cannot be made independently of other SMUs. Due to the large dimensions of dig-line equipment, consecutive SMUs have to be mined together and be sent to the same destination. The exact number of consecutive SMUs depends on the equipment being used. This complex relationship among SMUs complicates the formulation of a model.

Lacking computational tools, dig-limits have traditionally been drawn by hand. However, with the advancing software technology, there has been attempts to develop algorithms that will automatically calculate the dig-limits, given an SMU model of a bench. Current approaches consist of heuristic, meta-heuristic and clustering techniques. Allard et al. (1994) brought the idea of connectivity index such that the value of an SMU to be considered together with its grade and its location. An analogous approach to floating cone method has been developed that floats a circle in a bench (Richmond & Beasley, 2004). If average grades of the SMUs within the circle are above the cut-off grade, all SMUs in the circle are flagged as ore and waste otherwise. If it is flagged as ore, the circle is extended until the average grade falls below the cut-off grade. A local search heuristic for short-term bench planning has been developed by Richmond (2004) to minimize ore loss and mining dilution that satisfies the dig-limit constraints. Another greedy heuristic approach was developed by Wilde and Deutsch (2015) where an engineer can input their manual short-term plan and profit is attempted to increase iteratively by re-arranging the ore-waste boundaries in the plan.

The meta-heuristic approaches mainly create an initial solution where ore-waste boundaries were drawn manually or randomly, then attempted to optimize the solution by changing the plan at every iteration and update the plan according to an acceptance criterion. There has been approaches that used simulated annealing (Isaaks et al., 2014; Norrena & Deutsch, 2000, 2002) and meta-heuristics (Ruiseco & Kumral, 2016; Ruiseco et al., 2016). A unique, clustering approach was developed by Tabesh and Askari-Nasab (2013) where they assigned similarity indices for SMUs and applied hierarchical clustering, although the shape of clusters could not be controlled, thus violating the dig-limit constraints.

Efforts having done by researchers given above indicate that software support is essentially based on non-optimal methods for open pit bench ore-waste boundary determination. Current study summarizes an optimal method and enhances it to optimize a multi-metal mine with irregularities.

2 METHODOLOGY

A mixed integer linear programming (MILP) approach was developed by Sari and Kumral (2017) to solve the ore-waste boundary problem in mining benches. The algorithm can be outlined as follows:

- The algorithm reads the parameter files to retrieve parameters such as the bench grades, length, and width.
- A frame is created starting at each SMU that has the dimensions of the dig-line equipment.
- A constraint defines a valid frame to be a frame where all SMUs are sent to the same destination.
- A constraint ensures that each SMU is assigned to at least one frame that intersects with that SMU.
- The objective function is to selected best set of frames such that the economic value of the bench will be maximized.

For strip mining operations, it might be desirable to have irregular benches when dykes of unsuitable material are encountered. An example to this can be observed in **Hata! Başvuru kaynağı bulunamadı.**. In the cases where there are irregularities in a bench, data pre-processing needs to be conducted to identify which frames are inbounds and out-of-bounds with respect to the bench. A PowerShell script has been written to perform the pre-processing step. The details of the script are the following:

- Find the maximum and minimum SMU indices of the bench in X and Y directions.
- Create a rectangular array based on these maxima and minima and populate the array with the SMUs
- Assign a value of 1 to non-empty SMUs and a value of 0 otherwise

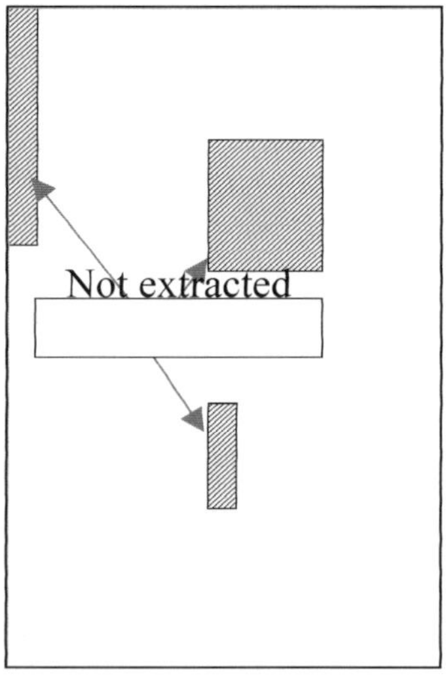

Figure 1: A bench sector with irregularities

The preprocessing step is followed by two small modifications in the MILP model:

- A frame is valid where all SMUs are sent to the same destination and does not contain a non-extracted SMU.
- If an SMU is marked with a value of 1 at the above pre-processing step, the SMU is not assigned to a frame.

With these modifications, the non-extracted SMUs are effectively ignored by the MILP model while for the rest of the SMUs, the procedure works as with the regular benches.

3 CASE STUDY

In order to test the approach, a bench sector was created with the dimensions 25x15 SMUs in X and Y directions. Three non-extracted areas (shown as shaded areas) are added as given in **Hata! Başvuru kaynağı bulunamadı.**. Other parameters are given in **Hata! Başvuru kaynağı bulunamadı.**.

Table 1. Case study parameters

Parameter	Value
Ore prices	40 and 2 \$/g
Block mass	100 tons
Mining cost	15 \$/ton
Processing cost	15 \$/ton
Dig-line equipment width	4 SMUs

To visualize the character of the orebody, a cut-off is applied to the bench sector and illustrated in **Hata! Başvuru kaynağı bulunamadı.**. In that figure, SMUs with the value of 1 are above the cut-off whereas with the value of 0 are below or equal to the cut-off grade. Although this illustration would be a valuable guide to a mining engineering while drawing the ore-waste boundaries, it is not directly applicable as a plan.

The pre-processing step was applied to the bench sector to flag the SMUs. The outcome was solved by the above given MILP model using CPLEX. The resulting ore-waste boundaries are visualized in **Hata! Başvuru kaynağı bulunamadı.**. As can be observed, the dig-line equipment width constraint of 4 SMUs is not violated. Another observation to be made is ore SMUs are concentrated in the areas where most SMUs are above the cut-off grade. However, where the concentration of below cut-off SMUs are high, even though there may be some above cut-off SMUs, they are not extracted. This is logical since processing an SMU is another 15 \$/ton. If the total revenue of all SMUs in a 4x4 diameter from processing does not supply for the mining and the processing cost, the area is not extracted.

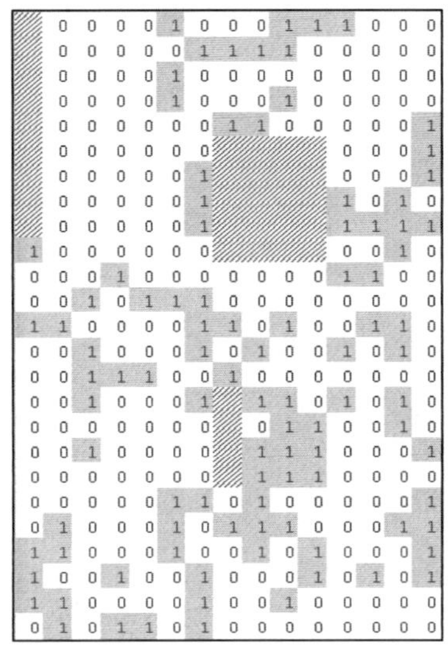

Figure 2. Bench sector after having applied the cut-off grade

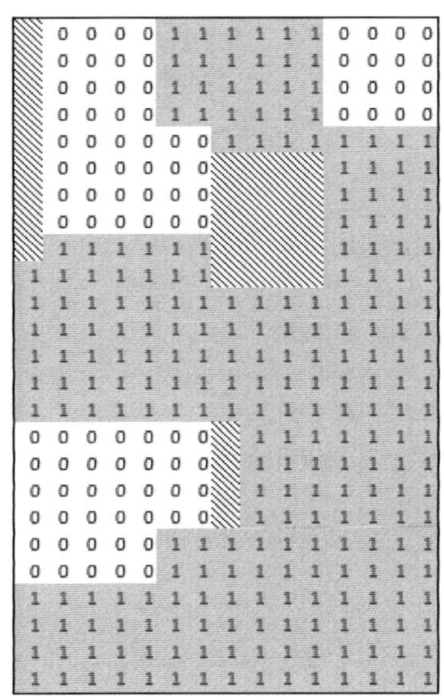

Figure 1. The resulting ore-waste boundaries

4 CONCLUSION

Drawing ore-waste boundaries in a bench is a part of short-term open pit mine planning. It is an optimization problem that is traditionally done manually by mining engineers. In this paper, we present a computerized ore-waste discrimination technique that will optimize the overall profit from an irregular mine bench. This is accomplished through first pre-processing the data for non-extracted areas and second, solving the problem using a MILP model and solver. The technique is demonstrated through a case study. It was observed that the equipment constraints were not violated and the ore-destined areas are in agreement with the cut-off grade map of the bench sector. High grade concentrated regions were flagged as ore whereas the poor regions were flagged as waste.

ACKNOWLEGMENTS

Financial support from NSERC (#435661) is gratefully acknowledged by the authors.

REFERENCES

Allard, D., Armstrong, M., & Kleingeld, W. (1994). The need for a connectivity index in mining geostatistics *Geostatistics for the Next Century* (pp. 293-302): Springer.

Isaaks, E., Treloar, I., & Elenbaas, T., 2014. Optimum Dig Lines for Open Pit Grade Control. *9th International Mining Geology Conference*, pp.425-432.

Norrena, K., & Deutsch, C., 2000. Automatic Determination of Dig Limits Subject to Geostatistical, Economical and Equipment Constraints. *Center for Computational Geostatistics (CCG), University of Alberta, Edmonton, Alberta, Canada.*

Norrena, K., & Deutsch, C., 2002. Optimal determination of dig limits for improved grade control. *APCOM 2002: 30 th International Symposium on the Application of Computers and Operations Research in the Mineral Industry*, pp.329-339.

Richmond, A., 2004. Integrating multiple simulations and mining dilution in open pit optimisation algorithms. *Orebody Modelling and Strategic Mine Planning Conference.*

Richmond, A., & Beasley, J., 2004. Financially efficient dig-line delineation incorporating equipment constraints and grade uncertainty. *International Journal of Surface Mining,* 18(2), pp.99-121.

Ruiseco, J. R., & Kumral, M., 2016. A Practical Approach to Mine Equipment Sizing in Relation to Dig-Limit Optimization in Complex Orebodies: Multi-Rock Type, Multi-Process, and Multi-Metal Case. *Natural Resources Research*, pp.1-13.

Ruiseco, J. R., Williams, J., & Kumral, M., 2016. Optimizing Ore–Waste Dig-Limits as Part of Operational Mine Planning Through Genetic Algorithms. *Natural Resources Research*, pp.1-13.

Sari, Y., & Kumral, M., 2017. Dig limit optimization through mixed integer linear programming in open pit mines. *Journal of Operational Research Society,* accepted.

Tabesh, M., & Askari-Nasab, H., 2013. Automatic creation of mining polygons using hierarchical clustering techniques. *Journal of Mining Science,* 49(3), pp.426-440.

Wilde, B., & Deutsch, C. V., 2015. A Short Note Comparing Feasibility Grade Control with Dig Limit Grade Control. *Retrieved May,* 10, pp.2015.

Field Performance of a New Generation 35 m^3 Range Electric Rope Shovel-(2800XPB) - A Case Study

M. Özdoğan, H. Özdoğan

İdeal Makine Danışmanlık Ltd. Şti., Ankara

ABSTRACT Performance of a 35 m^3 electric rope shovel is given and discussed from a bituminous coal mine in Kuzbass Coal basin, Siberia, based on daily monitoring reports (A four days period in May 2015). Parameters like cycle times, cycle time segments, dig times, total number of cycles, total tonnes moved, total bank measure m^3 moved, payloads, hourly production rate etc. are given and depicted, interpreted and discussed.

Keywords: Electric rope shovel, Cycle time, Cycle time segments, Dipper payloads, Production rate, Daily production, Process rate

ÖZET Sibirya, Kuzbass kömür havzasında bir taşkömürü işletmesinde çalışan 35 m^3 kapasiteli bir elektrikli halatlı yerkazı makinasının başarımı, günlük izleme raporlarına dayanılarak incelenmiştir (Mayıs 2015 tarihinde dört günlük bir süre). İş döngüsü, iş döngüsü evreleri, kazı süreleri, toplam döngü sayısı, toplam kaldırılan örtü miktarı, m^3 ve ton olarak, kepçe yararlı yükleri, saatlik kazı oranı vb verilmiş, yorumlanmış, tartışılmış ve grafik olarak gösterilmiştir.
Anahtar Kelimeler: Elektrikli halatlı yerkazı makinesi, İş döngüsü, İş döngüsü evreleri, Kepçe yükü, Üretim hızı, Günlük üretim, İşleme hızı

1 INTRODUCTION

Digging and loading is the heart of any earthmoving operation. The entire system's productivity is governed by the digging and loading equipment's performance. Primary digging and loading tools are electric rope shovels (ERS), hydraulic face shovels (HFS), hydraulic face bachoes (HFB) and wheel loaders (WL). These digging and loading equipment are equally good on condition that they fit into conditions and characteristics the mine site. Specific needs of the mine site and operating bench conditions are primary selection considerations of digging and loading equipment. The overall goal is to maximize profit with a digging and loading system that meets targeted hourly production at the lowest cost-per-tonne of material moved.

Electrical infrastructure, longevity of the operation, annual production target, face material, material fragmentation, penetration characteristics, material angle of repose, face height and angle, bench width, maneuvering space, multiple loading faces, floor conditions, job set-up, blasting restrictions are the site considerations are taken into account in selection of the primary digging and loading equipment (Anon. a, 2016).

Mine sites equipped with an electrical infrastructure have the option to choose electric rope shovels and/or electric powered hydraulic shovels for their loading tool needs. Using electric machines lowers the operation's overall cost and reduces emissions. They do require trail cables connected to substations, which limits their mobility. For non-electrified mine sites, hybrid rope shovels, diesel-powered hydraulic shovels and mechanical and hybrid wheel loaders remain the primary loading equipment options, (Anon c, 2016). Diesel-powered machines offer a distinct advantage in mobility and flexibility, despite a higher operating cost. However, with the introduction of our off-grid power solution, non-electrified mines now have the option to use electric rope shovels. Greenfield mines in the planning process must weigh the long-term cost implications between investing in an electrical infrastructure for electric-powered loading tools and using diesel-powered equipment only.

The expected longevity of the mine site also plays a role in loading equipment selection. Electric rope shovels are designed to have an operating life of 120,000 hours, making them a viable option for mines with estimated lives that meet or exceed a timeframe of 20 years. For mine sites with shorter estimated lives, hydraulic shovels and wheel loaders may be more viable options. Hydraulic shovels and wheel loaders are designed to have an operating life of approximately 50,000 to 60,000 hours without exchange of major steel structures, dependent upon application and proper maintenance. If it is a captive coal mine to an electric power generating plant, as a rule of thumb electric rope shovels and/or electric hydraulic shovels to be favoured due to the consumption of the home generated electricity at prime cost by the equipment, (Binns, 2015).

2 OVERVIEW

The electric rope shovel monitored have a rated payload of 60,5 tonnes and a nominal dipper capacity of 35 m^3 (Anon d., 2016) operating at a bituminous coal mine in Kuzbass Coal Basin. Four days are examined in May 2015 for average cycle times, cycle time segments, process rates and dipper payloads, Table 1. Loose unit weight of the overburden in the dipper is 1,78 tonnes per m^3. Bank measure unit weight is 2,23 tonnes per m^3. The ERS operated on two shifts a day basis. Day shift starts at 08AM and ends at 08PM, and night shift is from 08PM to 08 AM.

Cumulative operating hours on the 2800 ERS are as follows: Equipment running hours reading is 23,100 h., the crowd motor reading is 19, 951h, hoist motor reading is 19,966 h, swing motor reading is 21, 163h and propel motor reading is 1,349 h. (Anon b., 2015)

The shovel is matched with 180 tonnes and 230 tonnes Belaz trucks. The shovel fills 180 tonnes trucks in (3) three passes and fills 230 tonnes trucks in (4) four passes. The original equipment manufacturers recommend 3 to 4 passes are best match in filling the off-highway trucks. One may think that why ideal dipper and truck match would be such that one or two dipper passes fill the truck body. Yes, it would be best to fill the truck in one or two passes, but it would usually be detrimental to the truck frame and body. That is why it is not recommended by the OEM, (Binns, 2015).

As the number of passes increase, production decreases while costs increase; the best number of dippers is around three passes where the production tonnes per hour curve and cost curve USD per tonne intersect. According to OEM P&H, the optimum number is three passes.

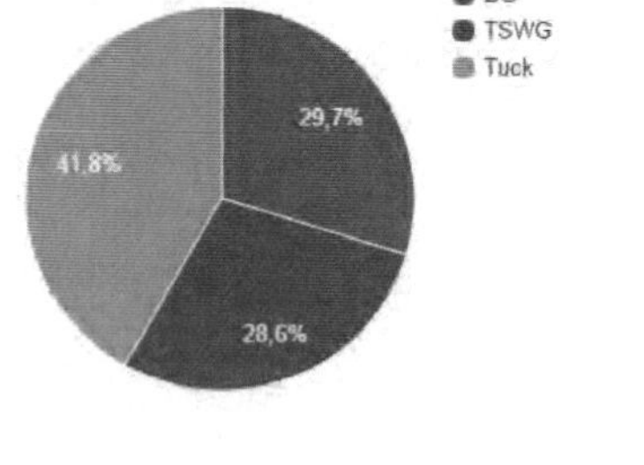

Figure 1a. Percentages of cycle time components of the electrical rope shovel

The main measure of ERS field performance is the cycle time. ERS CT comprises of dig (DG), swing to dump (SWG to DMP), dump (DMP) and swing back to face (SWG to F), normally. However, P&H PrevailSystem programmers preferred recording the phases as Dig (DG), Total Swing (TSWG) and Tuck time (TCK) segments: Swing includes dipper swing to truck and swing back to face, whereas tuck includes crowding and retracting of dipper for positioning to dump over the truck body, lowering the dipper to truck tray prior to dumping the load and opening the dipper door and unloading time of the dipper.

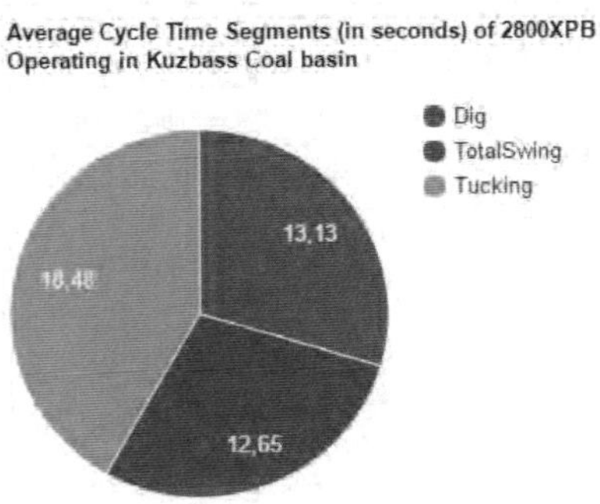

Figure 1b. Durations of cycle time components of the 35 m^3 shovel

There are two studies made on cycle times and cycle time segments of electric rope shovels operating at Turkish surface coal mines. Ceylanoğlu (1991) made a survey on 12 m^3 capacity 2100BL electric rope shovels in Yatağan area. The researcher explored five cases and reported the CT components in four segments. Dig (DG), Swing with loaded dipper (SWG-L), Dump (DMP), Swing with empty dipper (SWG-E). Here are the results he reported: Dig time, s (DG): 9,09±0,79; swing time with loaded dipper, s (SWG-L): 5,73±0,84; dumping time, s (DMP): 4,1±0,59; swing time with empty dipper, s(SWG-E): 8,27±1,04. Average total cycle time (CT) is 26,94±2,58 seconds.

Hindistan (1997) explored 919 cycles in 22 cases on 15 m^3 2300XP and M191M electric rope shovels operating in Tuncbilek, Soma and Çan coal basins for CTs and CT segments. In his research he reported CT components in three segments. Dig (DG), swing to truck (SWG-L) and swing back to face (SWG-E). Here is his facts and figures: Dig time, s (DG): 7,88±1,38; swing to truck, s (SWG-L): 9,95±1,39; swing back to face, s (SWG-E): 10,39±0,89 and the total cycle time (CT) is 28,21±3,29 seconds.

Table 1. Dig time over cycle time and total swing time over cycle time ratios

ERS Model	Dipper, m^3	Total Swing Time, s	TSWG/ CT, %	Dig Time, s	DG/ CT, %	CT @ 90° Swing Angle, s
2100BL	12	14,00	0,52	9,09	0,34	26,94
2300XP / M191M	15	20,34	0,72	7,88	0,28	28,21
2800XPB#1	35	12,65	0,29	13,13	0,30	44,28
2800XPB#2	35	12,62	0,28	6,62	0,30	44,65
4100C	45	14,5	0,30	15,3	0,31	49,00

Swing phase time is affected by swing arc (swing angle), and the radius of the swing gear (roller circle) between the under carriage and upper frame, the number and power of swing motors, and swing gear box reduction ratio, and by the skill and experience of operator. For example, a shovel and truck layout with a 70° swing arc has a shorter swing time than the one having 90°. It is interesting to note that new generation electric rope shovels have shorter swing times with respect to total cycle times, eventhough, they have bigger roller circle to swing. This might be due to the fact than the new generation ERSs usually have a third swing motor and gear box to have faster swing times. The old generation ERSs had two swing motors traditionally, See Table 1.

3 CYCLE TIME DETAILS

The P&H PrevailsSystem® classifies the cycle time segments as Digging (DG), Total Swinging (TSWG) and Tucking (TCK). Tucking time includes counter breaking of swinging motion, crowding and lowering (positioning of dipper on the truck tray) and opening dipper door and emptying the rock material into the truck body, See Figures 1a and b, and Table 2.

Focus on the analysis of the components of the cycle time may reveal the bottlenecks of the total cycle time. Dig time is a function of bench geometry (bench height), rock characteristics and fragmentation, equipment design features, operator's skill and experience, rake angle etc. Swing is a function of roller circle diameter, number of swing motors and reduction ratio of gearboxes, matching truck size, operator skill and experience. Dumping phase is a function of dipper design, payload, truck spotting, stickiness of material, dumping height and skill and experience of operator. Dig phase is one of the most important segments of the CT which governs the dipper fill factor and payload; dig segment constitutes about 1/3rd the total cycle time, See Figure 1a and b, and Table 2.

Table 2. Cycle time segments and process rates of the shovel in Kuzbass Basin

Parameter	Day 1	Day 2	Day 3	Day 4	OverallAverage
Utilisation, U %	85	86	88	89	**87±1,58**
Cycle Time, s	47,3	44,6	42,9	42,3	**44,28±1,94**
Dig Time, s	14,4	13,0	12,7	12,4	**13,13±0,77**
Swing Time, s	14,3	11,8	11,6	12,9	**12,65±1,07**
Tuck Time, s	18,6	19,8	18,6	16,9	**18,48±1,03**
ProcessRate, %	91	96	100	102	**97±5**

Please note that rated cycle time (CT_r) is taken as 43 seconds, (CT_f) stands for the field cycle time of the equipment. Process Rate is CT_r/CT_f . CT @ 90° Swing Angle, s

The monitored data is recorded by P&H PrevailSystem® (Anon b, 2015).

CT and CT phases are functions of equipment parameters (design features of equipment and dipper, and age of equipment), loading equipment and hauler match parameters, operator parameters, bench geometry and rock parameters, diggability (loadability) and rock blasting and fragmentation parameters, mainly.

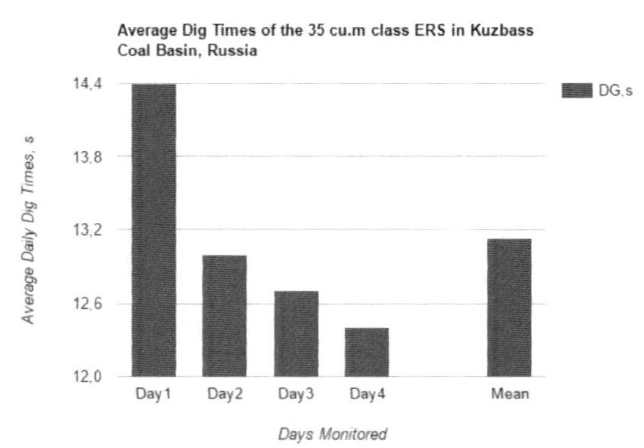

Figure 2. Average daily dig times of the electric rope shovel

A good indicator of productivity is the average time it takes to fill the dipper. A proper loading equipment productivity study measures the four-part loading cycle: Dig

(DG), swing to dump (SWG to DMP), dump (DMP), swing back to face (SWG to F). The total cycle time is the sum of segments 1 through 4. Dipper fill (DG) times are averaged and will vary cycle to cycle as the digging face changes. Handling oversize pieces of material or digging the toe of a bench will increase individual dig times. It may also be affected by the operator's skill and experience level, See Figures 2 and 3.

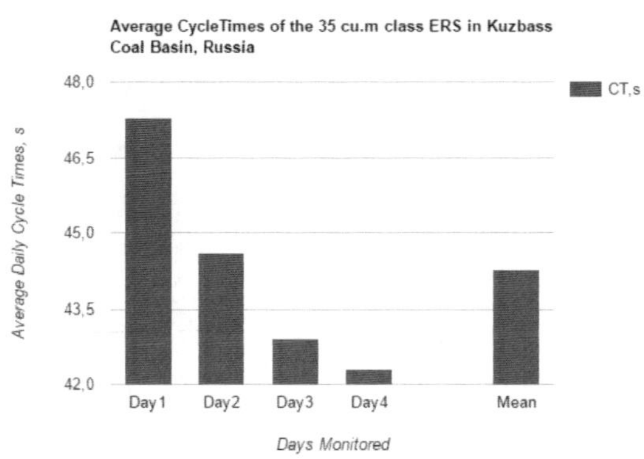

Figure 3. Average daily cycle times of the shovel

There seems to be a positive relation between the average dig times and the total cycle times of the shovel as expected as long as the swing angle and the truck shovel layout do not change, swing and dumping factors do not vary much since they are the design features of the equipment, See Figures 2 and 3. Therefore, dig time duration is the main factor influencing the total cycle time of the equipment.

The cycle time phases of earthmoving equipment that have a slewing upper frame are entirely different than that of wheel loaders. As a rule of thumb, equipment with slewing upper frame have more favourable cycle times than that of similar sized mining wheel loaders; however, there may be exceptions to this rule, depending on the size and diameter of the roller circle and swing gear, number and HP of the swing motors.

4 PRODUCTIVITY DETAILS

4.1 Production Rate

Production rate is the production per unit time; unit time usually is hour and may be expressed in terms of m^3 and/or tonne.

Table 3. Daily productivity data of the shovel operating in Kuzbass region

Parameter	Day 1	Day 2	Day 3	Day 4	OverallAverage
U %	85	86	88	89	**87±1,58**
Runtime, h	20,3	20,6	21,1	21,5	**20,88±0,46**
NonProdTime, h	3,5	3,0	2,9	2,5	**2,98±0,36**
Production, t	70,210	84,483	90,732	94,446	**84,968±9234**
Production, m^3	39,174	47,137	50,624	52,696	**47,408±5152**
No.of cycles,count	1190	1433	1539	1602	**1441±157**
ProductionRate,t/h	3,459	4,101	4,300	4,393	**4063±365**
ProdRate, m^3/h	1,930	2,288	2,399	2,451	**2267±203**
DipperPayload, t	59,00	58,96	58,96	58,96	**58,97±0,00**
CT @ 90° Swing Angle, s. The monitored data is recorded by P&H PrevailSystem® (Anon b, 2015).					

The electric rope shovel operates two shifts a day; day shift 08AM to 08PM, night shift 08PM to 08 AM. Shifts are 12 hours ones.

4.2 Hourly Production Rate

A four-day average of hourly production rate is 2267±203 m^3/h and in terms of tonnes the hourly production rate is 4063±365 tonnes/h, See Figure 4 and 5.

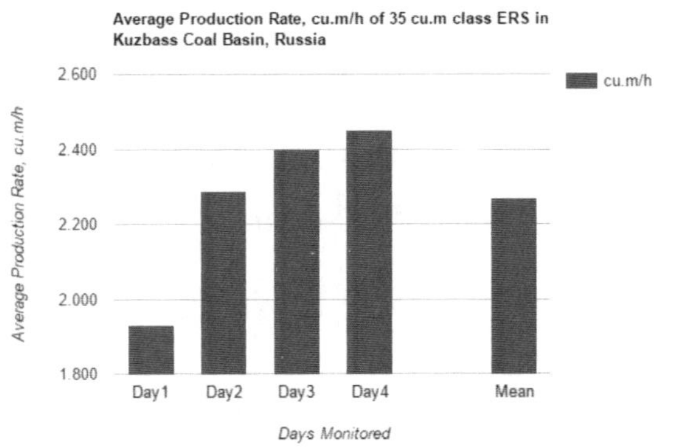

Figure 3. Average hourly production rates of the shovel

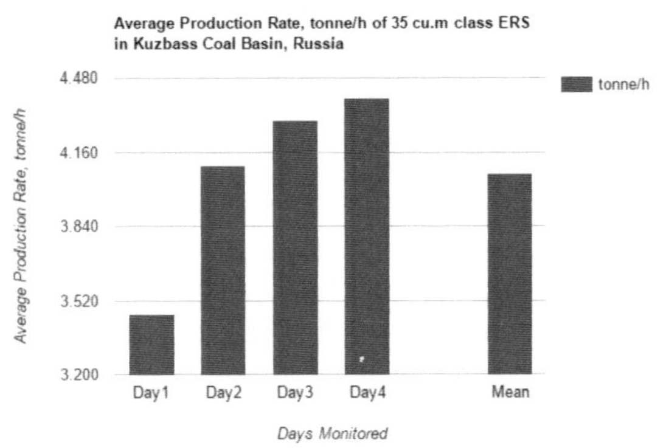

Figure 4. Average hourly production rates of the shovel

4.3 Daily Production Rate

The shovel is programmed to operate 24 hours a day on two 12 hour shifts basis. As an average of four days, productive running time average is 20,88±0,46 hours, and non-productive time average is 2,98±0,36, See Table 3.

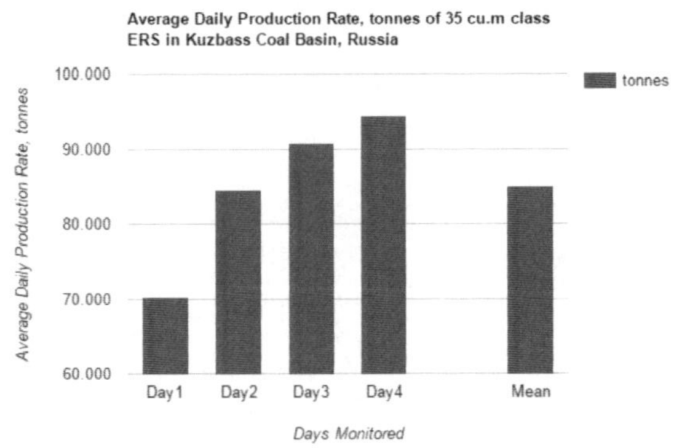

Figure 5. Average daily productions of the shovel in tonnes

Daily production rates are illustrated in Figures 5 and 6.

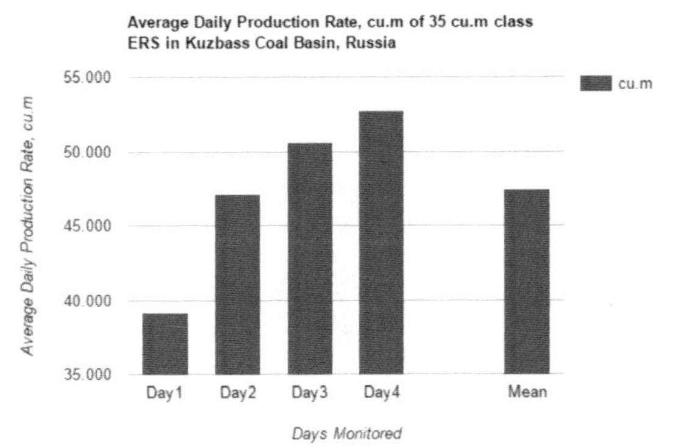

Figure 6. Average daily productions of the shovel in m^3

4.4 Daily Process Rate (Ratio)

Process rate for an electric mining shovel is defined as the ratio of the rated cycle time (CT_r over CT_f) to field cycle time, See Figure 7. The closer the field CT to the rated CT, the higher process rate is; and consequently, the higher the production rate is.

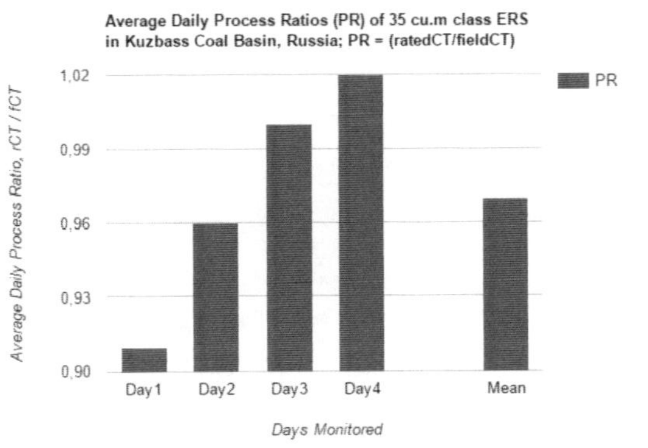

Figure 7. Average daily process rates of the shovel

4.5 Daily Dipper Payloads

The average deviation of the dipper payload (2%) is in the limits of ±5 percent deviation allowed off the target payload. Dipper payload is an important productivity indicator like cycle time of the electric rope shovel. Therefore, it has to be monitored whether it is in the vicinity of the target payload recommended for the loading tool or not.

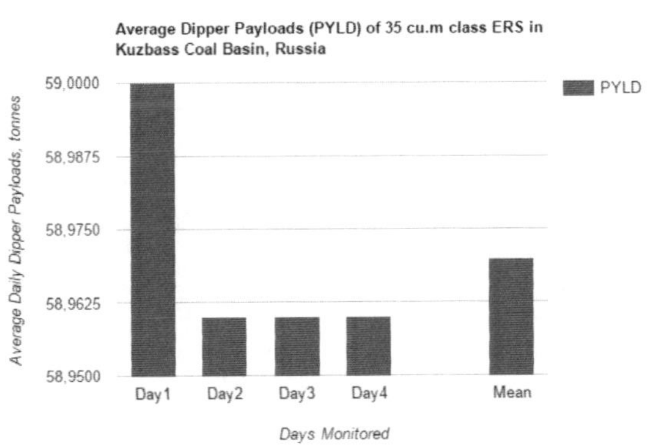

Figure 8. Average daily dipper payloads of the shovel in tonnes

5 CONCLUDING REMARKS

Electric rope shovels are in the heart of the earthmoving operations as digging and loading equipment. Higher performance electric rope shovels secure lowest cost per tonne operations as long as bench floor, bench geometry, rock material fragmentation are favourable and having good operators, properly matching truck fleet, good pit planning and administration. Providing that good and steady supply of quality parts and consumables, proper maintenance and repair programs exist. The average utilisation (U%) of the shovel monitored is 87±1,58 and average Process Rate (PR%) 97±5.

The facts and figures cited in the text are to be only considered as guidelines, because of the fact that there are so many factors (minesite, equipment, human, environmental) affecting the cycle times and cycle time phases, and dipper payloads.

Cycle time is the major performance indicator of both the operator and the equipment. Focusing on the analysis of the components of the cycle time is of supreme importance in improving, shortening, the cycle time. Another major productivity indicator is the dipper payload. The cycle time secured may be good, shorter, however, if the bucket is not filled to the designated target load, it has no meaning, at all.

The electric rope shovel with a rated 60,5 tonnes dipper payload and a dipper volume of 35 m^3 range explored had an average cycle time of 44,28±1,94 seconds. Average cycle time comprises of dig time 13,13±0,77 seconds, total swing time 12,65±1,07 seconds, and tuck time 18,48±1,03 seconds. According to the period of observation time recorded, average cycle time ratios of the segments are as follows: 29,7 % of the CT is digging time, 28,6 % of the CT is total swing time and 41,8 % of the CT is tucking time.

The new generation electric rope shovels seem to have shorter total swing times compared to the old generation ones. In the new generation rope shovels the ratio of the dig times and total swing times over the total cycle times are about the one third of the cycle times based on the cases explored.

Good blasting and fragmentation techniques play an important role in the productivity and life expectancy of the loading equipment, whether it is an electric rope shovel, a hydraulic shovel, or a wheel loader. In general, the improvements in loading equipment productivity offsets the higher costs associated with drilling and blasting.

REFERENCES

Anon.a,2016; http://www.witraktor.com/sites/default/files/media/Finland/Koneet/Kaivoskoneet/PDFs/valintaopas_kaivoslastauslaitteet.pdf

Anon. b, 2015; P&H PrevailSystem® Reports, U.K.

Anon. c, 2016; http://www.joyglobal.com/surface-mining/hybrid-excavators

Anon. d, 2016; http://www.joyglobal.com/product-details/2800xpc

Binns, D., 2015 : Personnel Communication, *Plymounth, U.K.*

Ceylanoğlu, A., 1991; "Performance Monitoring of Electrical Power Shovels for Diggability Assessment in Surface Coal Mines", Ph.D. Thesis, METU, Ankara, July 1991 (Unpublished Ph.D. Dissertation).

Hindistan, M. A., 1997; "Development of a Computer Based monitoring System and Its Usage for Power Shovels' Monitoring", Ph.D. Thesis, METU, Ankara, November 1991 (Unpublished Ph.D. Dissertation).

Ranking and Assessment of Tunneling Projects Risks Using Fuzzy MCDM (Case Study: Toyserkan Doolayi Tunnel)

S.S.Haghshenas
Young Researchers and Elite Club, Rasht Branch, Islamic Azad University, Rasht, Iran.

Y. Ozcelik
Hacettepe University, Department of Mining Engineering, Ankara, Turkey.

S.S.Haghshenas
Department of Civil Engineering, Islamic Azad University, Astara Branch, Astara, Iran.

R.Mikaeil
Department of Mining and Metallurgical Engineering, Urmia University of Technology, Urmia, Iran.

P.S. Moghadam
Department of Civil Engineering, Islamic Azad University, Bandaranzali Branch, Bandaranzali, Iran.

ABSTRACT In tunneling projects, risk assessment has a fundamental role as one of the important indices in achieving the predetermined purposes of the project. Assessing the tunneling projects' risk is always faced with vague and uncertain problems. Therefore, in assessing the projects' risk, the use of ranking methods based on studies and experience of experts is very effective. Thus, this research addresses the study and ranking of the tunneling project's risks based on a fuzzy multi-criteria decision-making method. In this research, after conducting studies for the project, 11 risks and 3 criteria were selected for investigation and ranking of the project's risk which are analyzed using the Fuzzy Analytical Hierarchy Process (FAHP). The present study was conducted on Toyserkan Doolayi Tunnel in the south of Hamadan Province, Iran. According to the results obtained from this research, the swelling of rock's risk has the highest rank and terrorist attacks' risk has the lowest rank from among the existing risks in the research.
Keywords: Tunneling Projects, Risk Assessment, Fuzzy MCDM, FAHP

1 INTRODUCTION

Study and assessing the risk of projects dealing with deep excavations is one of the most significant criteria affecting the accurate implementation and development of such constructional infrastructures. The mechanized tunneling projects are one of the most important constructional infrastructures in which the ranking and assessment of risk are used as one of most important and influential parameters in designing and implementation of such projects. In fact, the purpose of ranking risks is addressing the most important ones considering the limitation of resources. Given the significance of this subject, different studies were conducted to study and rank the tunneling risk. In a study conducted by Linkov et al (2006), the tunneling projects' risk were assessed and matched using multi-criteria decision making methods. Results obtained from this research showed a proper match between risk assessment and multi-criteria decision making methods. Ranking of geological risks in Golab tunnel in the center of Iran was conducted by Nezarat et al (2015). In their researches, they ranked the tunnel's geological risks using the Fuzzy Analytical Hierarchy Process (FAHP) and a team of experts. The geological hazards were studied by Mikaeil et al (2016) using meta-heuristic algorithms. Finally, they validated the results with the real observations collected from the project. This research was conducted on Ardabil-Mianeh Railway Tunnel and its validation results showed the proper use of meta-heuristic algorithms in assessing the tunneling projects' risk. The risk assessment and management in Ghomrud Tunnel was addressed by Haghshenas et al (2016) using the fuzzy clustering technique. They assessed the risks existing in the implementation of this project using soft computing and geological units. Other various researches were conducted on the tunneling projects' risk using different methods such as Fault Tree, Even Tree and Monte Carlo Simulation (Hyun et al., 2015, Vílchez et al., 2011, Amigun et al., 2011).

In this research, the risk assessment was done for Toyserkan Doolayi Tunnel project in the south of Hamadan Province, Iran. For this reason, with the cooperation of a team of experts and data collection based on the conducted researches and experiments using FAHP method, the ranking of existing risks is done based on the following criteria: Repeat chance (C1), occurrence possibility (C2), and efficacy (3). Finally, using FAHP method, a high level of precision and skill in data analysis can be obtained.

2 METHODOLOGY

In recent years, with the increasing growth of complex and uncertain problems, the use of soft computing with a high ability in solving these kinds of problems has increased significantly. Methods such as meta-heuristic algorithm, fuzzy logic and neural networks also use the soft computing and have a wide application in different areas of science and industries (Rad et al, 2012; Rad et al, 2014; Mikaeil et al, 2016a; Mikaeil et al, 2016b; Armaghani et al, 2016).

Among soft computing methods, the theory of fuzzy sets has a special place in terms of solving imprecise problems and modeling uncertainties. The use of this theory in decision making problems is one of the most widely use and most effective applications of this theory. Fuzzy analytical hierarchy process (FAHP) method is one of the most fundamental branches of this theory in decision making problems in the area of risk management, construction and prioritization of mining projects and water resources management.

Chang's FAHP method which has a high ability in solving decision making problems has been used in many researches. The process of using this technique with m options and n criteria in 7 steps for fuzzy triangular numbers is as follows (Chand,. 1996).

2.1 The First Step: Hierarchical Graph Drawing

In the first step, the hierarchical graph is drawn at three levels based on the number of criteria and options under study and the desired objective.

2.2 The Second Step: Defining Fuzzy Numbers for Pairwise Comparisons

In order to conduct pairwise comparisons, fuzzy numbers are determined. These numbers can be defined as fuzzy triangular numbers or fuzzy trapezoidal numbers.

2.3 The Third Step: Forming Pairwise Comparison (A) Matrix with Fuzzy Numbers

Matrix of pairwise comparison (A) will be as follows:

$$\tilde{A} = \begin{pmatrix} 1 & \tilde{a}_{12} & \cdots & \tilde{a}_{1n} \\ \tilde{a}_{21} & 1 & \cdots & \tilde{a}_{2n} \\ \vdots & \vdots & \ddots & \vdots \\ \tilde{a}_{n1} & \tilde{a}_{n2} & \cdots & 1 \end{pmatrix} \quad (1)$$

Fuzzy numbers of the matrix are as follows:

$$a_{ij} = \begin{cases} 1 & i=j \\ 1,3,5,7,9 \text{ or } 1^{-3},3^{-1},5^{-1},7^{-1},9^{-1} & i \neq j \end{cases}$$

2.4 The Forth Step: Calculation of Si for Each Row of Pairwise Comparison Matrix

Si is a fuzzy triangular number which is obtained from equation (2).

$$S_i = \sum_{j=1}^{m} M_{gi}^{j} \otimes [\sum_{i=1}^{n}\sum_{j=1}^{m} M_{gi}^{j}]^{-1} \quad (2)$$

Furthermore, if i, j and M_{gi}^{j} indicate line number, column number and triangular fuzzy numbers of pair comparison matrices, therefore, we have the quantities of $\sum_{j=1}^{m} M_{gi}^{j}$ ، $\sum_{i=1}^{n}\sum_{j=1}^{m} M_{gi}^{j}$ ، $\left[\sum_{i=1}^{n}\sum_{j=1}^{m} M_{gi}^{j}\right]^{-1}$ calculated according to the following equations (3) , (4) and (5) (Chand,. 1996).

$$\sum_{j-1}^{m} M_{gi}^{j} = \left(\sum_{j=1}^{m} l_j, \sum_{j=1}^{m} m_j, \sum_{j=1}^{m} u_j\right) \quad (3)$$

$$\sum_{i=1}^{n}\sum_{j=1}^{m} M_{gi}^{j} = \left(\sum_{i=1}^{n} l_i, \sum_{i=1}^{n} m_i, \sum_{i=1}^{n} u_i\right) \quad (4)$$

$$\left[\sum_{i=1}^{n}\sum_{j=1}^{m} M_{gi}^{j}\right]^{-1} = \left(\frac{1}{\sum_{i=1}^{n} u_i}, \frac{1}{\sum_{i=1}^{m} m_i}, \frac{1}{\sum_{i=1}^{n} l_i}\right) \quad (5)$$

2.5 The Fifth Step: The Value of Magnitude in Relation to Each Other

In this stage, if $S_1 = (l_1, m_1, u_1)$ and $S_2 = (l_2, m_2, u_2)$ are two fuzzy numbers, determination of the value of magnitude in relation to each other is based on equation (6). The correlation is shown in Figure (1) (Chand,. 1996)

$$V(S_2 \geq S_1) = hgt(S_1 \quad S_2) = \mu s_{2(d)} \begin{cases} 1 & \text{if } m_2 \geq m_2 \\ 0 & \text{if } l_1 \geq u_2 \\ \dfrac{l_1 - u_2}{(m_2 - u_2) - (m_1 - l_1)} & \text{otherwise} \end{cases}$$

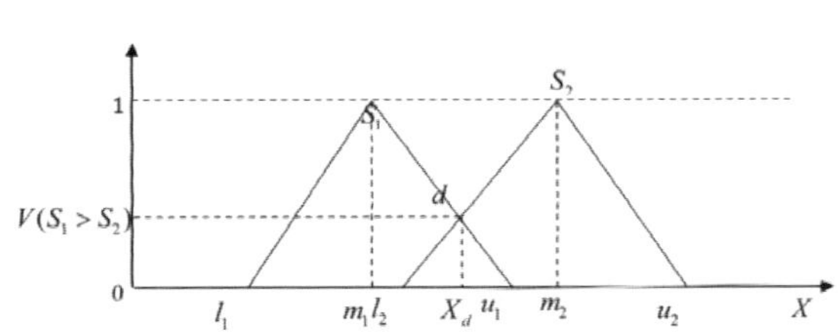

Fig 1. The largeness degree of two triangle fuzzy numbers S_1, S_2 in relation to each other

Therefore, the order of magnitude of triangular fuzzy digit is found according to the (K) of the triangular fuzzy digit through equation (7).

$$\begin{aligned} &V(S \geq S_1, S_2, \ldots, S_k) \\ &= V\left[(S \geq S_1) and (S \geq S_2) and \ldots and (S \geq S_k)\right] \\ &= \text{MinV}(S \geq S_i), \quad i=1,2,3 \end{aligned} \quad (7)$$

2.6 The Sixth Step: Weight Calculation of Each Criterion and Alternative within Pair Matrices

In this step, the weight of each criterion and alternative in their pairwise comparison matrices and non-normalized weight vector are obtained through equations (8) and (9).

$$d'(A_i) = MinV(S_i \geq S_k) \qquad \text{K-1,2,...,n , } k \neq i \quad (8)$$

$$w'(d'(A_i), d'(A_2), \ldots, d'(A_n))^T \quad (9)$$

2.7 The Seventh Step: Calculating and Determining the Final Weight Vector

Eventually, the final weight vector is calculated through normalizing the weight vector and from equation (10).

$$W = (d(A_1), d(A_2), \ldots, d(A_n))^T \quad (10)$$

3(6)TOYSERKAN DOOLAYI TUNNEL

Toyserkan Doolayi Tunnel is one of the most strategic tunneling projects in the west of Iran which is located at kilometer 5 of south Toyserkan with the purpose of reducing the distance of the inter-city transportation route and removal of Doolayi Ghat. The length of this tunnel is approximately 777 m and its cross-sectional area is 72 square meters. This tunnel is located in the route of one of the most intense and active structural zones in Iran (Sanandaj-Sirjan zone). Different types of sedimentary, igneous and metamorphic rocks relating to different geological periods are available in this region. The hornfelsic unit forms the most rocks in the route of tunnel. Basic and granitic masses in this zone along with schists and amphibolites are located in this region. One of the most important characteristics of this zone is its complex tectonic which had affected it during different geological periods. Different landslides occurred due to special petrological conditions, clay deposits in the route of tunnel and high groundwater level (Ghasemi et al.2004; Jalilvant and Haghshenas, 2013; Jalilvand et al., 2014).

Figure (2) shows the location of tunnel.

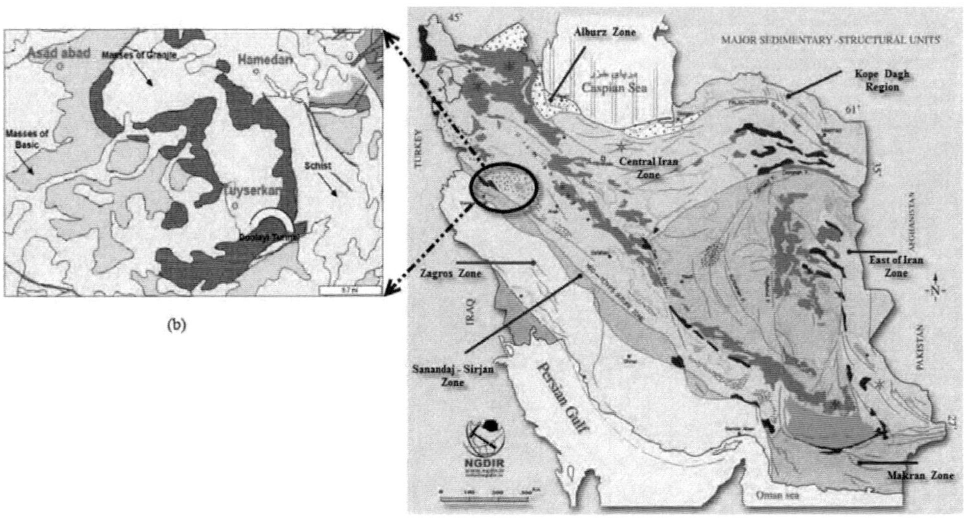

(b) (a)

Fig.2. (a) Geological Map of Iran (1:1000.000 Scales) and (b) Geological Map of Tunnel's Zone (National Geoscience Database of Iran)

4 MODELLING AND ANALYSIS

In the modeling, for assessing and ranking the risks existing in Toyserkan Doolayi Tunnel, first a team is formed with skillful experts. Then, among more than 20 common risks in tunneling projects, based on the conditions of Toyserkan Doolayi Tunnel, 11 project's risks are provided according to Table (1). In addition, three criteria are determined to assess these risks, including occurrence possibility (C1), repeat chance and efficacy (C2), respectively (C3). Figure (3) shows the hierarchical graph for ranking the project's risks.

Table (1). A set of serious risks of Toyserkan Doolayi Tunnel

Risk Sorts	Title of Risk	Number of Risk
Accidents and Failure Risks	Machineries Failure	R_1
Technical Risks	Design Mistakes	R_2
Natural Hazards Risk	Earthquake, etc.	R_3
Geological Risks	Squeezing	R_4
	Instability of Wall	R_5
	Water in Flow	R_6
	Face Tunnel Instability	R_7
	Swelling of Rock	R_8
Project Estimations (PE)	Construction Delay	R_9
	Price Changes	R_{10}
Unpredictable Risk	Terrorist Attacks	R_{11}

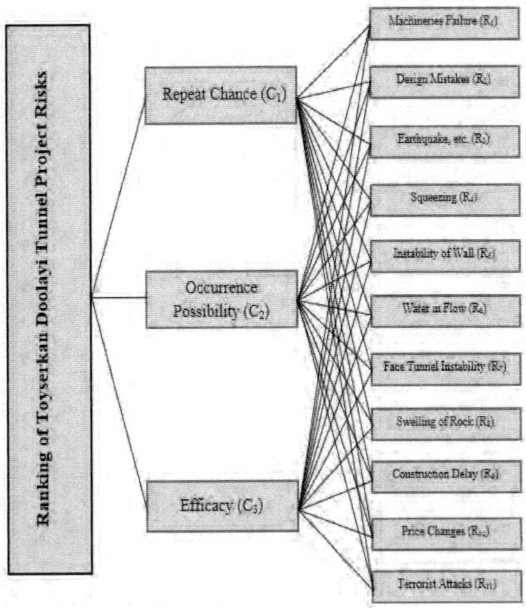

Fig.3. Hierarchical structure of Toyserkan Doolayi Tunnel project risks

In addition, the fuzzy numbers of this study are shown in Table (2) as triangular types.

Table (2). Linguistic variables to rank

Fuzzy number	Fuzzy number
$1 = (1,1,1)$	$1^{-1} = (1,1,1)$
$2 = (1,2,4)$	$2^{-1} = (\frac{1}{4},\frac{1}{2},1)$
$3 = (1,3,5)$	$3^{-1} = (\frac{1}{5},\frac{1}{3},1)$
$4 = (2,4,6)$	$4^{-1} = (\frac{1}{6},\frac{1}{4},\frac{1}{2})$
$5 = (3,5,7)$	$5^{-1} = (\frac{1}{7},\frac{1}{5},\frac{1}{3})$
$6 = (4,6,8)$	$6^{-1} = (\frac{1}{8},\frac{1}{6},\frac{1}{4})$
$7 = (5,7,9)$	$7^{-1} = (\frac{1}{9},\frac{1}{7},\frac{1}{5})$
$8 = (6,8,10)$	$8^{-1} = (\frac{1}{10},\frac{1}{8},\frac{1}{6})$
$9 = (7,9,11)$	$9^{-1} = (\frac{1}{11},\frac{1}{9},\frac{1}{7})$

According to the opinions of experts, pairwise comparisons matrices for three criteria including repeat chance (C1), occurrence possibility (C2) and efficacy (C3), respectively are formed based on table (3) and the normalized weights of each one are calculated based on equations (1)-(10) as follows and shown in Table (4).

Table (3). Fuzzy pair-wise comparison matrix of Criteria

	C1			C2			C3			$\sum_{j=1}^{m} M_{gi}^{j}$		
C1	1	1	1	0.167	0.25	0.5	0.2	0.333	1	1.37	1.58	2.5
C2	2	4	6	1	1	1	1	2	4	4	7	11
C3	1	3	5	0.25	0.5	1	1	1	1	2.25	4.5	7

$$\sum_{i=1}^{n}\sum_{j=1}^{m} M_{gi}^{j} = (7.62, 13.08, 20.5)$$

$$\left[\sum_{i=1}^{n}\sum_{j=1}^{m} M_{gi}^{j}\right]^{-1} = (0.049, 0.076, 0.131)$$

$$S_1 = (1.37, 1.58, 2.5) \otimes (0.049, 0.076, 0.131)$$
$$= (0.049, 0.076, 0.131)$$
$$S_2 = (4, 7, 11) \otimes (0.049, 0.076, 0.131)$$
$$= (0.196, 0.532, 1.441)$$
$$S_3 = (2.25, 4.5, 7) \otimes (0.049, 0.076, 0.131)$$
$$= (0.11, 0.342, 2.686)$$
$$S_1 \geq S_2 = 0.241 \; , \; S_1 \geq S_3 = 0.494$$
$$S_2 \geq S_1 = 1 \quad , \; S_2 \geq S_3 = 1$$
$$S_3 \geq S_1 = 1 \quad , \; S_3 \geq S_2 = 0.929$$

Table (4). Normal Weight Matrix of Each Criterion

	C1	C2	C3
Not normal weight	0.241	1	0.929
Normal weight	0.111	0.461	0.428

Then, the pairwise comparisons matrix for existing risks is formed in relation to each criterion and the significance of the project's risks is calculated in relation to each criterion based on equations (1)-(10). Figures (4)-(6) show the normalized weights of each option in relation to each criterion.

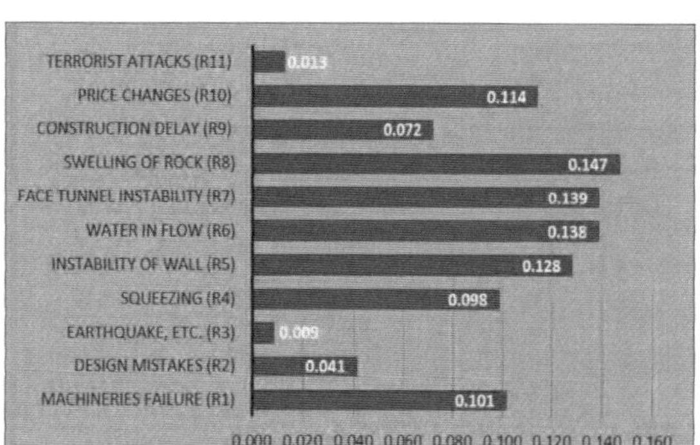

Fig.4. Normalized Weight of Risks in Relation to the Criterion of Repeat chance

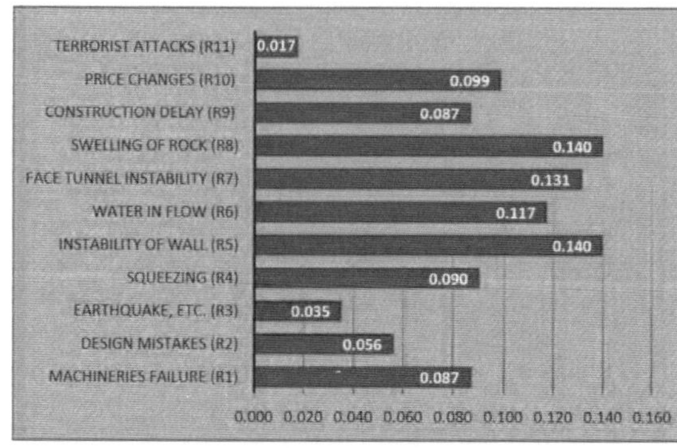

Fig.5. Normalized Weight of Risks in Relation to the Criterion of Occurrence Possibility

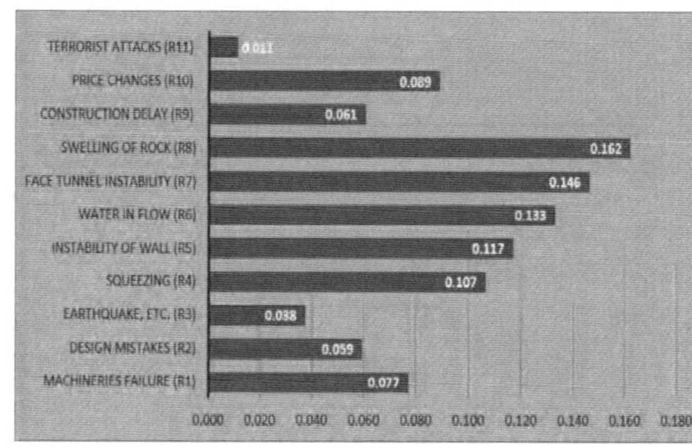

Fig.6. Normalized Weight of Risks in Relation to the Criterion of Efficacy

In the final step, table (5) is formed according to the results obtained from the normalized weights of criteria and options and the ranking of all risks is done. Figure (6) shows the final ranking of Toyserkan Doolayi Tunnel's risks. Accordingly, the swelling of rock's risk (R8) with score 0.153 has obtained the highest significance. The obtained results show that the highest score relates to risks with geological risks' sources, which according to the geology of the region, they properly match to the ranking in this study. The important point here is to conduct more field studies and experiments before the implementation and design phases.

Table (5). Results and Final Score of Each of the Options

	Repeat Chance (C_1)	Occurrence Possibility (C_2)	Efficacy (C3)	Final score
	0.111	**0.461**	**0.428**	
Machineries Failure (R1)	0.087	0.101	0.077	0.089
Design Mistakes (R2)	0.056	0.041	0.059	0.05
Earthquake, etc. (R3)	0.035	0.009	0.038	0.024
Squeezing (R4)	0.09	0.098	0.107	0.101
Instability of Wall (R5)	0.14	0.128	0.117	0.125
Water in Flow (R6)	0.117	0.138	0.133	0.134
Face Tunnel Instability (R7)	0.131	0.139	0.146	0.141
Swelling of Rock (R8)	0.14	0.147	0.162	0.153
Construction Delay (R9)	0.087	0.072	0.061	0.069
Price Changes (R10)	0.099	0.114	0.089	0.102
Terrorist Attacks (R11)	0.017	0.013	0.011	0.013

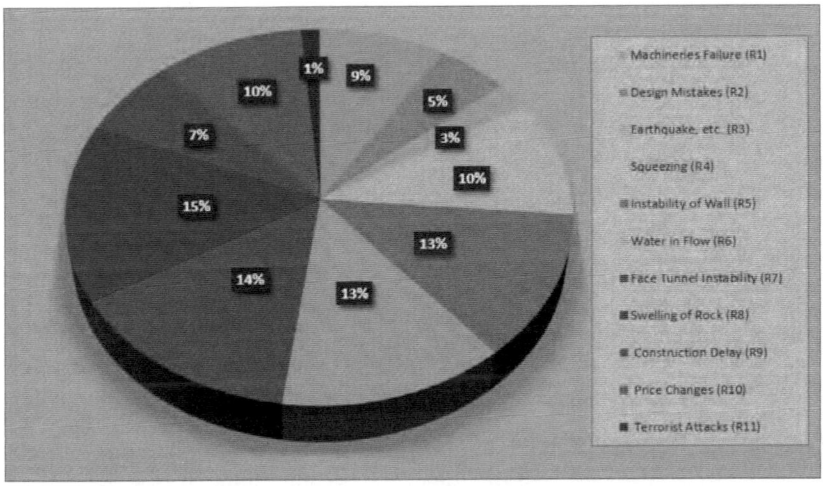

Fig.7. Ranking of Project Risks in Toyserkan Doolayi Tunnel with FAHP

5 CONCLUSION

Ranking of tunneling projects risks is one of the most important sections in prediction and planning of design section, risk management, project management in tunneling projects. Connecting two specialized area of risk and tunneling is significantly important. Furthermore, tunneling projects are traditionally classified as one of the most risky construction projects in terms of the possibility to solve unknown and less predictable issues.

Therefore, in this study, FAHP method is used which is a powerful method in the

Decision-making problems. In this study, considering 3 criteria and 11 risks for Toyserkan Doolayi Tunnel project and after conducting analyses, it was determined that the swelling of rock with score 0.153 has the highest risk and terrorist attacks (R11) with score 0.013 has the lowest one in this ranking. Based on the prioritization of risks in this study, the highest scores belong to geological risks. Therefore, it suggested to the project's executives that further studies are conducted on the geology of the region in the design phase and before the implementation phase. In addition, for further studies, it is suggested that a sensitivity analysis is conducted in the area of tunnel's risk and its ranking using other decision making methods, including fuzzy Vikor in order to better determine the high ability of soft computing techniques in the area of risk.

REFERENCES

Amigun, B., Petrie, D. and Görgens, J., 2011. Economic risk assessment of advanced process technologies for bioethanol production in South Africa: Monte Carlo analysis. *Renewable Energy*, *36*(11), pp.3178-3186.

Armaghani, D.J., Hasanipanah, M. and Mohamad, E.T., 2016. A combination of the ICA-ANN model to predict air-overpressure resulting from blasting. *Engineering with Computers*, *32*(1), pp.155-171.

Chang, D.Y., 1996. Applications of the extent analysis method on fuzzy AHP. *European journal of operational research*, *95*(3), pp.649-655.

Ghasemi I, Bahlol B, Toti F., 2004 the relationship between microscopic structure and mechanical behavior of the Alvand aureole rocks and its effect on duplex stability Tuyserkan tunnel. *The 5th International Conference of Engineering Geology and Environment. Tehran, Iran: Tarbiat Modarres University*. pp. 895–904.

Haghshenas, S.S., Haghshenas, S.S., Barmal, M. and Farzan, N., 2016. Utilization of Soft Computing for Risk Assessment of a Tunneling Project Using Geological Units. *Civil Engineering Journal*, *2*(7), pp.358-364.

Hyun, K.C., Min, S., Choi, H., Park, J. and Lee, I.M., 2015. Risk analysis using fault-tree analysis (FTA) and analytic hierarchy process (AHP) applicable to shield TBM tunnels. *Tunnelling and Underground Space Technology*, *49*, pp.121-129.

Jalilvand, P., Haghshenas, S.S., Haghshenas, S.S., Javan, M.H., Astara, I., Rasht, I. and Arak, I., 2014, May. Evaluation of Dynamic Resistance of the Toyserkan Doolayi Tunnel by Rock Bolt and Reinforced Shotcrete Composite System. In *Tunneling and Underground Construction* (pp. 376-384). ASCE.

Jalilvand P, Haghshenas S S., 2013. The study stability of Toyserkan Doolayi Tunnel using reinforces shotcrete and rock bolt under static condition. The 23rd International Mining Congress and Exhibition of Turkey. pp. 1299 – 1305.

Linkov, I., Satterstrom, F.K., Kiker, G., Batchelor, C., Bridges, T. and Ferguson, E., 2006. From comparative risk assessment to multi-criteria decision analysis and adaptive management: Recent developments and applications. Environment International, 32(8), pp.1072-1093.

Mikaeil, R., Haghshenas, S.S., Haghshenas, S.S. and Ataei, M., 2016. Performance prediction of circular saw machine using imperialist competitive algorithm and fuzzy clustering technique. *Neural Computing and Applications*, pp.1-10.

Mikaeil, R., Haghshenas, S.S., Shirvand, Y., Hasanluy, M.V. and Roshanaei, V., 2016. Risk Assessment of Geological Hazards in a Tunneling Project Using Harmony Search Algorithm (Case Study: Ardabil-Mianeh Railway Tunnel). *Civil Engineering Journal*, *2*(10), pp.546-554.

Mikaeil, R., Ozcelik, Y., Ataei, M. and Shaffiee Haghshenas, S., 2016. Application of harmony search algorithm to evaluate performance of diamond wire saw. *Journal of Mining and Environment*.

National Geoscience Database of Iran. General geology and structural units of Iran, geological map of Iran (1:1000.000 scales). http://www.ngdir.ir/.

Linkov, Igor, F. K. Satterstrom, Gc Kiker, Cd Batchelor, Td Bridges, and Ed Ferguson. "From comparative risk assessment to multi-criteria decision analysis and adaptive management: Recent developments and applications." *Environment International* 32, no. 8 (2006): 1072-1093.

Rad, M.Y., Haghshenas, S.S., Kanafi, P.R. and Haghshenas, S.S., 2012. Analysis of Protection of Body Slope in the Rockfill Reservoir Dams on the Basis of Fuzzy Logic. In *IJCCI* (pp. 367-373).

Rad, M.Y., Haghshenas, S.S. and Haghshenas, S.S., 2014. Mechanostratigraphy of cretaceous rocks by fuzzy logic in East Arak, Iran. In *The 4th International Workshop on Computer Science and Engineering-Summer, WCSE*.

Vílchez, J.A., Espejo, V. and Casal, J., 2011. Generic event trees and probabilities for the release of different types of hazardous materials. *Journal of Loss Prevention in the Process Industries*, *24*(3), pp.281-287.

Investigating a Novel Model for Risk Assessment in Decorative Stones Quarrying

F. Asadi Ouriad
Faculty of Engineering, Tarbiat Modares University, Tehran, Iran

R. Bagherpour
Department of Mining Engineering, Isfahan University of Technology, Isfahan, Iran

M. Yari
Department of Mining Engineering, Isfahan University of Technology, Isfahan, Iran

M. Khoshouei
Department of Mining Engineering, Isfahan University of Technology, Isfahan, Iran

ABSTRACT Quarrying projects are from hazardous fields of engineering which needs further attention of risk assessment. Constructed buildings with dimensional stones produced by quarrying, have industrialized in Iran. Currently, Iran, after China and Italy, is classified in third place in world production and consequently, risk assessment is one of the most vital steps of quarry organization. Risk Breakdown Structure (RBS) is one of the main phases of risk assessment. For this purpose, comprehensive structure of decorative stones quarrying is shaped and separated into 17 chief levels and 128 sublevels. The difficulty of recognising diverse strictures, made it essential to use Multi Attribute Decision Making (MADM) means for ranking related hazards. For real-world application, the foremost dangers of Ghasre Dasht marble mine are assessed by means of 'Technique for Order of Preference by Similarity to Ideal Solution' (TOPSIS) system bearing in mind 10 main factors under Fuzzy environment. In conclusion, after examining 21 experts' views and categorization of key hazards, the most threatening risks of dimensional stone quarrying are offered.

1. INTRODUCTION

Currently, investigations in the plans are powerfully influenced by hazards and any choice in projects need risk organization. Considering Project Management Body Of Knowledge (PMBOK), risk is recognised as uncertainty and consequently its' outcome can unfavourably cause to injuries and safety reduction (Guide 2001). Additional definition of risk is an unwanted event or condition which its' existence consequences positive or negative impressions on the plan (Pinto and Ireland 1998).

The hazard organisation method contains recognizing, analyzing and replying to plan hazards in order to maximize the effect of positive actions and minimizing the possibility and impression of negative fates. According to PMBOK, the risk management process is separated into six key stages (Institute 2008):

- Scheduling Risk Management,
- Recognising Risks,
- Qualitative Risk Analysis,
- Quantitative Risk Analysis,
- Development risk replies,
- Risk control by monitoring,

Quarrying is one of the most dangerous schemes in designation and operation phases, and needs substantial consideration to risk control .Documentation of risk producing issues, consciousness about effect of risks and ranking of them, are essential stages of risk management.

In 1997, Steffen recognized risk causes in open pit mine scheduling. Based on his investigation risk in these mines are separated into two chief classes: technical hazards and management hazards. He similarly recommended three breakdown approaches for risk organization in mining processes (Steffen 1997). Risk assessment actions in the assortment and plan of equipment for the mining processes are debated by Joy in 2004. For achieving this goal, the probability of risk occurrence and maximum impact of risk on production facilities are divided into five classes) Joy 2004(.

Arends et al. offered a method to forecast the risk management expenses in the underground mining plans (Arends et al. 2004). Risks of underground coal mining have been measured by Duzgun and Einstein in Turkey pits. In this investigation, based on the occurrence of roof collapse during one year, risks have been assessed using two chief mechanisms: possibility of occurrence and roof collapse consequence (Duzgun and Einstein 2004). Heuberger estimated the mining schemes in view of geological and economic uncertainty. He applied "@Risk" software for risk calculation in a gold deposit based on the DCF process (Heuberger 2005).

Soons et al. offered a qualitative method for risk assessment of fire in the underground projects (Soons et al. 2006). For assessing the risks of underground constructions, Reilly et al. stated the time and cost aspects as probability function. For this goal, correlation coefficients and independency of risks are considered and two core factors of risk assessment (probability and consequence) have been implemented (Reilly and Brown 2004; Reilly and Parker 2007). Evans and Brereton concluded that one of the drawbacks of common risk assessment methods is that more consideration is essential to avoid negative results. They highlighted that positive and negative effects must be measured synthetically and therefore established a fresh technique for sustainable development supplies (Evans et al. 2007). Fuentes considered probable risks in the copper mining in Chile and evaluated risks using Monte Carlo model (Fuentes et al. 2008). Dehghani and Ataee-pour definited economic and engineering ideas of uncertainties in source survey in quarry plan. They similarly offered recommendations to eradicate or decrease risks to an satisfactory level (Dehghani and Ataee-pour 2012). Sayadi, et al, have assessed risks of mines economically, and have examined the risks of the Songun copper mine, using Monte Carlo simulation (Sayadi Anari et al. 2007).

All cited studies show that a comprehensive method have not covered all existing risks of quarrying, and commonly, definite risks are measured with respect to situation studies. Correspondingly, it is noteworthy that in most of risk assessment investigates, mostly two directories of "consequence" and "probability" were considered as key risk valuation parameters. Additionally, some academics believe that these measures are not reliable (Chapman and Ward 1996; Sayadi Anari et al. 2007). The main disadvantages of these approaches is that low possible risks with high consequence and high probable risks with low consequence are supposed to be equivalent (Pipattanapiwong 2004).

For ordering a significant set of alternatives based on their comparative position rendering to experts' opinions, Multi Attribute Decision Making (MADM) method can be advantageous (Pomerol and Barba-Romero 2000; Shirland et al. 2003).

In this research, all affecting parameters of risks are defined using 21 experts' belief as Risk Breakdown Structure (RBS) in dimensional stone quarries in 17 main levels and 128 sublevels. In the next phase, weights of valuation issues are assessed using Fuzzy-AHP. Lastly, using TOPSIS, all determined risks concerning significant factors are measured and the most threating hazards are determined.

2. RISK BREAKDOWN STRUCTURE

Hillson in 2002 defined the Risk Breakdown Structure (RBS) as a structure that is used for effective and operative risk evaluation. This description of RBS normally is alike to Work Breakdown Structure (WBS). RBS is a pyramid structure of possible hazards which can support directors to control additional hazards of the plan. In the sublevel of the RBS, comprehensive risk parameters are accessible (Iranmanesh et al., 2007).

An inclusive RBS can be valuable for determining the risks of a project, but does not include all hazards of all similar projects. Hence, a suitable RBS is organized for each plan based on its detailed features. The hazards related to mining schemes commonly splits into internal and external hazards. Internal hazards are because of geological circumstances and external hazards are instigated by external situations such as economic and social limitations (Hillson, 2003).

Fuentes categorized hazards of the mining process as: the geological hazards, geotechnical hazards, plan dangers, working hazards, environmental hazards, economic risks, macroeconomic risks, political hazards and contract risks (Fuentes et al., 2008). Serious hazards of mining process in Mongolia have been articulated by Chinbat and Takakuwa as (Chinbat and Takakuwa, 2009): owners' economic difficulties, inappropriate management, practical risks, government bureaucracy documentation, Incorrect assessment of material, employees negligence, rail transport postponements, scarcity of experts (skilled employee), transfer interruption of technologies, government checkers' limitations, fluctuations in rules, fuel lack, unforeseen environmental events, deficient investment, organization/human confrontation, fates through manufacturing processes.(Yari et al. 2016c)

All stated researches indicate that the organizations of hazard issues in quarries have not sufficient comprehensiveness. Concerning the diversity of perils which can happen in quarrying operation, without any methodical technique for recognizing and managing risks, makes quarrying plans extra risky projects. RBS presents an operative mean for documentation and organization of hazards (MacDermott et al., 1996, Sayadi Anari et al., 2007). The current research offers an inclusive structure of risks for dimensional stone quarries. Figure 1 and Figure 2 show Risk Breakdown Structures for internal and external causes of hazard in dimensional stone quarries.

3. ASSESSING ATTRIBUTES FOR RISK EVALUATION

Because of the conventional methods' disadvantages, for risk evaluation and grading (by means of two factors: probability and consequence), in this study, after complete examination, 10 impressive attributes were recognized for risks evaluation (Table I).

4. MULTI ATTRIBUTE DECISION MAKING (MADM)

MADM is a procedure for assessing alternatives, ranking and assorting the finest accessible alternatives(Yari et al. 2013). In MADM problems, there are some alternatives that should be graded (Reza et al. 2014; Yari et al. 2016a; Yari et al. 2016b). Each problem has numerous attributes that would specify alternatives to describe problems precisely (Hwang and Yoon 1981). The attributes in a decision matrix are dissimilar. Occasionally, attributes have positive or negative characteristic. Consequently, appropriate alternatives will offer the top state for diverse attributes (Hwang and Yoon 1981).

4-1- Fuzzy-AHP Method

In decision making problems each attribute is assumed a specific weight and is defined by means of these "weights". There are diverse approaches for calculating these weights. In this research, Fuzzy-AHP system is implemented for controlling of proportional importance of attributes. Proportional importance is gained from pairs comparison matrices. In these problems, just assignment of weights for all attributes are the key purpose (Saaty and Vargas 2001).

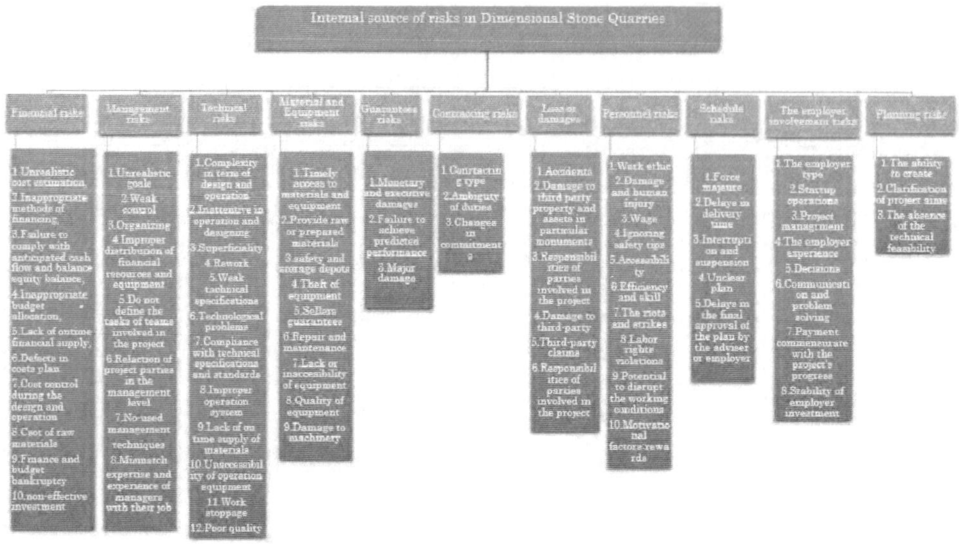

Figure 1. Risk Breakdown Structure for internal risks in dimensional stone quarries

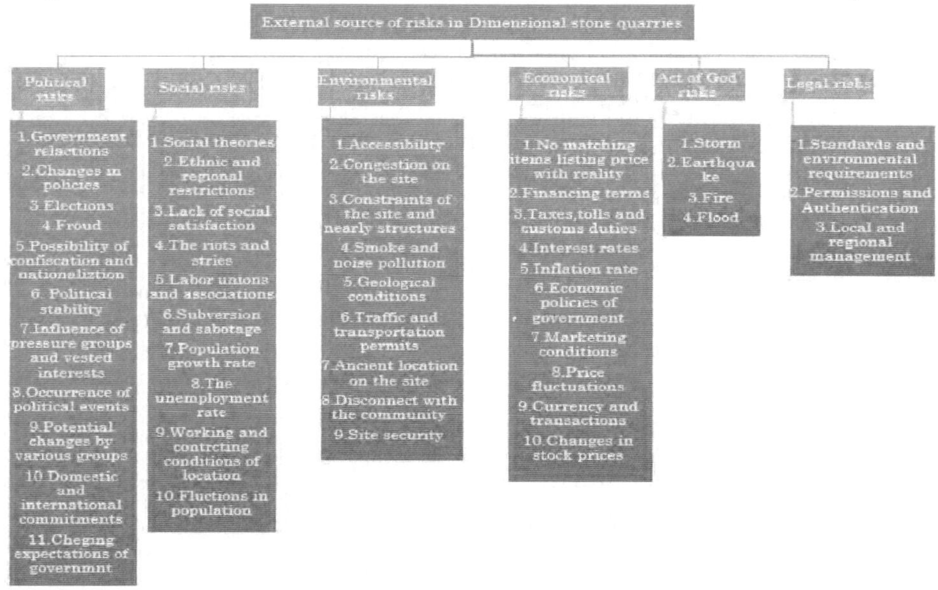

Figure 2. Risk Breakdown Structure for external risks in dimensional stone quarries

Impressive attributes were recognized for risks evaluation (Table I).

Table I. Important attributes in risk assessment process for dimensional stone quarries

Attribute	Description
Event probability	Expert opinion about the Expected rate of Event risk
Impact on project time	Negative effect of risk on the project time
Impact on project cost	Negative effect of risk on the project cost
Impact on project quality	Negative effect of risk on the project quality
Impact on project performance	Negative effect of risk on the project performance
Manageability	The ability to manage and respond to risk
Continually repeating	Repetition rate of risk facing
Exposure	The ability to predict risk occurrence
Proximity	Proximity of time of the risk occurring
Confidence level	confidence level of estimated risk values

The Fuzzy-AHP method could be regarded as one progressive technique which established based on classic AHP. This procedure, contains the next phases (Özdağoğlu and Özdağoğlu 2007):

Step 1. Breakdown the multipart structure into a hierarchical form.

Step 2. Arrangement a pair comparisons matrices:

$$\begin{bmatrix} 1 & a_{12} & a_{13} & \cdots & a_{1n} \\ 1/a_{12} & 1 & \cdots & \cdots & \cdots \\ 1/a_{13} & \cdots & 1 & \cdots & \cdots \\ \cdots & \cdots & \cdots & \cdots & \cdots \\ 1/a_{1n} & 1/a_{2n} & \cdots & \cdots & 1 \end{bmatrix} \quad [1]$$

Where:

$$a_{ij} = \begin{cases} 1 & i = j \\ 1,3,5,7,9 \text{ or } 1^{-1},3^{-1},5^{-1},7^{-1},9^{-1} & i \neq j \end{cases}$$

[2]

$$S_i = \sum_{j=1}^{m} M_{g_i}^{j} \otimes \left[\sum_{i=1}^{n}\sum_{j=1}^{m} M_{g_i}^{j}\right]^{-1}$$

of S_i for pa

$$S_i = \sum_{j=1}^{m} M_{g_i}^{j} \otimes \left[\sum_{i=1}^{n}\sum_{j=1}^{m} M_{g_i}^{j}\right]^{-1} \quad [2]$$

Where $M_{g_i}^{j}$ are Triangular Fuzzy Numbers (TFNs). To get $\sum M^{j}$ achieve "fuzzy adding [3] ... process ... for a spe...

$$\sum_{j=1}^{m} M_{g_i}^{j} = \left(\sum_{j=1}^{m} l_j, \sum_{j=1}^{m} m_j, \sum_{j=1}^{m} u_j\right)$$

$$\sum_{j=1}^{m} M_{g_i}^{j} = \left(\sum_{j=1}^{m} l_j, \sum_{j=1}^{m} m_j, \sum_{j=1}^{m} u_j\right) \quad [3]$$

Where l is the minor boundary value, m is the most talented value and u is the higher boundary value .to find $\sum_{i=1}^{n}\sum_{j=1}^{m} M_{g_i}^{j}$ complete the"fuzzy adding process" of

$M_{g_i}^{j} (j = 1, 2, , m)$:

$$\sum_{i=1}^{n}\sum_{j=1}^{m} M_{g_i}^{j} = \left(\sum_{i=1}^{n} l_i, \sum_{i=1}^{n} m_i, \sum_{i=1}^{n} u_i\right) \quad [4]$$

Step 4. The grade of comparison of $M_2 = (l_2, m_2, u_2) \geq M_1 = (l_1, m_1, u_1)$ is determined (Zhu et al. 1999).

$$V(M_2 \geq M_1) = \begin{cases} 1 & \text{if } m_2 \geq m_1 \\ 0 & \text{if } l_1 \geq u_2 \\ (l_1 - u_2)/(m_2 - u_2) - (m_1 - l_1) & \text{otherwise} \end{cases} \quad [5]$$

The grade of comparison for a convex fuzzy number to be larger than k convex fuzzy numbers $M_i (i = 1, 2, \ldots, k)$ can be definite.

Step 5. Calculating weights of attributes in pair comparisons matrix:

$$d'(A_i) = \min V(S_i \geq S_k),\ k = 1, 2, \quad , n \quad , k \neq i \quad [6]$$

Then the weight vector is specified by:

$W = (d(A_1), d(A_2), \quad , d(A_n))^T$

Step 6. Calculating normalized weight vectors:

$$W = (d(A_1), d(A_2), \quad , d(A_n))^T \quad [7]$$

Where W is non-fuzzy number. To assess the dangers, experts first select the associated linguistic variable, then convert them into triangular fuzzy numbers (Chang, 1996) as Table II.

4-2- TOPSIS Method

Technique for Order Preference Similarity to Ideal Solution (TOPSIS) model was proposed by Hwang &Yoon in 1981. This model has broad applications and is one of the best Multi Attribute Decision Making models. In this method, m alterative evaluates by n attribute and each problem is a geometrical system including m point in an n-dimensional space. In this technique, the most suitable alternative with positive ideal solution (the best possible solution) has the shortest distance, and with negative ideal solution (the worst possible state) has the maximum distance. It is assumed that value of each index is uniformly increasing or decreasing. TOPSIS includes six stages for solving decision making problem:

1. Converting decision making matrix to a normalized matrix using Eq. (6):

$$n_{ij} = a_{ij} / \sqrt{\sum_{i=1}^{m} a_{ij}^2} \quad [8]$$

2. Creating weighted normalized matrix by assuming W vector which is obtained from opinions of the experts:

ND, W_{n*n} = weighted normalized matrix, (comparative priority of the attributes is on the main diameter relative to each other).

3. Determining positive ideal solution and negative ideal solution.
4. Calculating distance from v+ and v- Distance between i-th alterative and ideal is using Euclidean method.
5. Calculating relative proximity of VI to ideal solution. This relative proximity is defined.
6. Ranking of alternatives: Based on the descending order of CL_i^+, one can rank the available alternatives of the assumed problem.

5.ILLUSTRATION

Ghasre Dasht Marble Mine is a dimensional stone quarry which is situated in north east of Fars Province, Iran. Stone resource of this deposit contains Marble stone with remarkable tonnage of Marls. Figure 3 demonstrates the geological location of Ghasre Dasht quarry.

For gathering experts' views two part questionnaires are organized: The first part is designed for defining the weights of 10 attributes implementing Fuzzy-AHP and other part is a decision matrix to assess the hazards using the TOPSIS method. In this study, ideas of 21 experts are gathered as linguistic variable for weight of attributes and hazard scores relative to each attribute. In the following phase, all weights of attributes are intended based on experts' ideas. Fuzzy weights of attributes are accessible in Table III and final weights of them are indicated in Figure 4.

After assessing weight of attributes and implementing all mentioned stages of TOPSIS, the ranking process of hazards is directed. The hierarchical assembly of the problem is exposed in Figure 5. The offered outcomes in Table IV presented that economical risk is the greatest threatening risk for Ghasre Dasht quarry.

6.CONCLUSION

Quarrying as one of the dangerous fields of production needs extra deliberation of risk evaluation. Risk Breakdown Structure (RBS) as one of main phases of hazard valuation is shaped for decorative stone quarrying and separated into 17 main levels and 128 sublevels. In the following, the chief hazards of Ghasre Dasht marble quarry is assessed using TOPIS bearing in mind 10 main factors. In conclusion, subsequently examining 21 experts' views and categorisation of key hazards, economic risk is determined as most threatening danger of quarrying.

Table II. Fuzzy number of linguistic variable

Linguistic	Very High	High	Medium High	Medium	Medium Low	Low	Very Low
Fuzzy	(5,7,9)	(3,5,7)	(1,3,5)	(1,1,3)	(1/5,1/3,1)	(1/7,1/5,1/3)	(1/9,1/7,1/5)

Figures 3. The geological location and general view of Ghasre Dasht mine

Table III. Fuzzy weights of attributes

Attribute	Fuzzy weight			Attribute	Fuzzy weight		
Event probability	0.069	0.107	0.136	Manageability	0.018	0.099	0.136
Impact on project time	0.137	0.167	0.185	Continually repeating	0.071	0.099	0.118
Impact on project cost	0.026	0.114	0.183	Exposure	0.061	0.085	0.120
Impact on project quality	0.039	0.111	0.168	Proximity	0.071	0.068	0.162
Impact on project performance	0.031	0.079	0.124	Confidence level	0.032	0.071	0.176

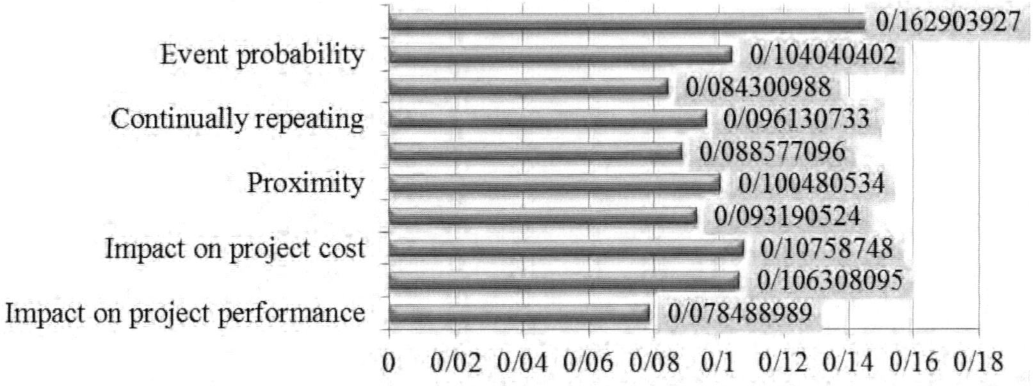

Figure 4. The final weights of attributes

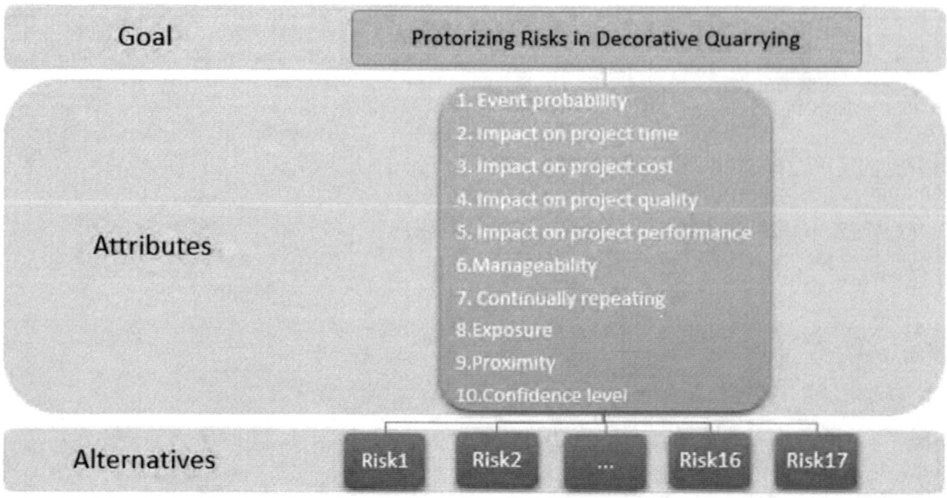

Figure 5. The hierarchical structure of problem

Table IV. Ranking of risks using Fuzzy-VIKOR (most threatening risk is first)

Risk	Rank	Risk	Rank	Risk	Rank
Act of God	7	**Contracting risk**	10	**Material and Equipment risk**	6
Economical risk	1	**Financial risk**	4	**The employer involvement risk**	8
Environmental risk	17	**Guarantees risk**	14	**Planning risk**	5
Legal risk	12	**Personnel risk**	9	**Schedule risk**	2
Political risk	11	**Loss or damages risk**	16	**Technical risk**	13
Social risk	15	**Management risk**	3		

REFERENCES

Arends G et al. (2004) Risk Budget management in progressing underground works: International Society for Trenchless Technology (ISTT) and International Tunnelling Association (ITA) Joint Working Group Report Tunn Undergr Sp Tech 19:29-33

Chapman C, Ward S (1996) Project risk management: processes, techniques and insights. John Wiley,

Dehghani H, Ataee-pour M (2012) Determination of the effect of operating cost uncertainty on mining project evaluation Resour Policy 37:109-117

Duzgun H, Einstein H (2004) Assessment and management of roof fall risks in underground coal mines Safety Sci 42:23-41

Evans R, Brereton D, Joy J (2007) Risk assessment as a tool to explore sustainable development issues: lessons from the Australian coal industry Int J Risk Assess Manag 7:607-619

Fuentes JE, George T, Whittaker J (2008) Measuring and mitigating risk in mining operations. marzo,

Guide A PROJECT MANAGEMENT BODY OF KNOWLEDGE (PMBOK® GUIDE). In: Project Management Institute, 2001.

Heuberger R (2005) Risk analysis in the mining industry J south afr inst min metall 105:75-80

Hwang CL, Yoon K (1981) Multiple attribute decision making: methods and applications: a state-of-the-art survey vol 13. Springer-Verlag New York,

Institute PM A Guide to the Project Management Body of Knowledge: PMBOK® Guide. In, 2008. Project Management Institute,

Joy J (2004) Occupational safety risk management in Australian mining Occup Med-C 54:311-315

Özdağoğlu A, Özdağoğlu G (2007) Comparison of AHP and fuzzy AHP for the multi-criteria decision making processes with linguistic evaluations Istanb Ticaret Üniversitesi Fen derg 6:65-85

Pinto JK, Ireland L (1998) Project management handbook. Jossey-Bass Publishers,

Pipattanapiwong J (2004) Development of multi-party risk and uncertainty management process for an infrastructure project. Kochi University

Pomerol J-C, Barba-Romero S (2000) Multicriterion decision in management: principles and practice vol 25. Springer,

Reilly J, Brown J (2004) Management and control of cost and risk for tunneling and infrastructure projects Tunn Undergr Sp Tech 19:330

Reilly J, Parker H Benefits and Life-Cycle Costs of Underground Projects. In: Proceedings, AITES-ITA World Tunnel Congress, Prague, 2007. pp 679-684

Reza Y, Raheb B, Reza K, Hossein MN, Mojtaba Y (2014) Development of 2D computer program to determine geometry of rock mass blocks International Journal 2:008

Saaty TL, Vargas LG (2001) Models, methods, concepts & applications of the analytic hierarchy process vol 34. Springer,

Sayadi Anari AR, Monjezi M, Heidari M, Vahidi M (2007) ECONOMIC EVALUATION AND RISK ANALYSIS OF SUNGUN COPPER MINE IRANIAN JOURNAL OF MINING ENGINEERING (IRJME)

Shirland LE, Jesse RR, Thompson RL, Iacovou CL (2003) Determining attribute weights using mathematical programming Omega 31:423-437

Soons C, Bosch J, Arends G, van Gelder P (2006) Framework of a quantitative risk analysis for the fire safety in metro systems Tunn Undergr Sp Tech 21:281-281

Steffen O (1997) Planning of open pit mines on a risk basis J south afr inst min metall 97:47-56

Yari M, Bagherpour R, Almasi N (2016a) An Approach to the Evaluation and Classification of Dimensional Stone Quarries with an Emphasis on Safety Parameters Rudarsko-geološko-naftni zbornik 31:15-26

Yari M, Bagherpour R, Almasi N (2016b) NAČIN PROCJENE I KLASIFIKACIJE KAMENOLOMA OBLIKOVANOGA KAMENA S NAGLASKOM NA SIGURNOSNE PARAMETRE Rudarsko-geološko-naftni zbornik 31:15-26

Yari M, Bagherpour R, Jamali S, Shamsi R (2016c) Development of a novel flyrock distance prediction model using BPNN for providing blasting operation safety Neural Computing and Applications 27:699-706

Yari M, Monjezi M, Bagherpour R (2013) Selecting the most suitable blasting pattern using AHP-TOPSIS method: Sungun copper mine Journal of Mining Science 49:967-975

Zhu K-J, Jing Y, Chang D-Y (1999) A discussion on extent analysis method and applications of fuzzy AHP Eur J Oper Res 116:450-456

Measuring People and the Impact on The Mining Value Chain. Student Assignment-Based Observations.

C. Birch

University of Witwatersrand, Johannesburg, South Africa

ABSTRACT How people are measured, motivated and managed and if it is beneficial to the business has been a question asked many times in the mining industry. The background to this research paper is the MSc (Mineral Resource Management) at University of Witwatersrand student assignments where this has been a topic for 2014, 2015 and 2016. The focus of the students' research is whether the Key Performance Indicators (KPIs) which people wish to achieve drive, value through the mining value chain or are actually destroying value. This question was posed to the students to research on their operations where they are employed. The assignments have been analysed to identify current trends and effectiveness of people measurement systems on various African mining operations. Furthermore, the effectiveness of this type of assignment based research is considered where students each gather information at their own mining operations and the overall pattern is then subsequently analysed to get an overall feeling for the current situation regarding the core question. The impacts of these measuring systems on the overall value through the mining value chain were considered and recommendations made to improve the key performance measuring systems. The advantage of using student assignments for gathering information is that numerous countries and commodities can be compared quickly and with no real costs to the university.

Keywords: Mineral Resource Management; Key Performance Indicators (KPIs); Theory of Constraints; Student Assignments; Value Drivers.

1 INTRODUCTION

In the competitive world of mining, it is vital that everyone performs or value will be lost. Mining managers introduce performance targets and set minimum acceptable goals. For each division in the mine, different key performance indicators (KPIs) are used for this purpose. It is observed, however, that when personnel focus on their own KPIs, it can be to the detriment of the overall mining value chain. How performance management systems are implemented, is critical to the success of mining organisations (Seotlela & Miruka, 2014). These performance management systems should be reviewed on a regular basis to ensure their continued effectiveness. Implementing changes to performance management systems often fail, however, if management neglects to start with change management, communication and training of employees regarding the entire process (Seotlela & Miruka, 2014). If performance management systems are seen to favour certain employees over others, dissatisfaction and conflict can occur. A recent study on a South African underground coal mine identified how crews working on the different seams were given different KPIs to ensure comparable payments for equal effort. The crews on the seam with the higher production target are dissatisfied as they feel the bonus system is unfair (Kemp, 2016). This shows that there was insufficient communication with the affected crews when the bonus system was implemented. The research showed that the performance management system was actually fair and suggested only minor changes be implemented. The coal production from the lower seam was found to be lower than the upper seam due to more challenging mining conditions justifying the different performance targets (Kemp, 2016).

Neingo did a detailed study of the productivity drivers on the Anglo Platinum Mines (Bothopele, Kroondaal and Mototolo). She identified that labour availability and utilisation are becoming increasingly important especially considering the unionised nature of the mining industry. The performances measures that were identified can be split into labour-based and financial-based measures. The labour-based measures are square meters per total employee and refined platinum ounce per total operating employee (Niengo & Cawood, 2014). The financial-based measures are cash-on-mine costs per tonne milled, as well as the cash operating costs per refined or equivalent refined ounce of platinum. She noted that productivity is well understood at the higher levels of management and their remuneration includes performance-based rewards for these measures. On the contrary, the remuneration mix at the lower levels in the organisation does not include these measures and is independent of these (Neingo, 2014).

Various studies have been conducted into incentive schemes across different countries. These have identified that there are differences in these forms of remuneration between Europe and the USA, as well as between large and smaller firms (Bryson, et al., 2012). The share of the overall remuneration that incentive schemes contribute, is found to be between 10-15% in some European countries and rises to as much as 40% in Scandinavian countries, as well as the USA (Bryson, et al., 2012). Traditionally in mining companies in South Africa, performance incentive schemes contribute highly to the overall remuneration for the employees directly in production departments, whilst in services and support departments this contributes far less. This study attempts to look at the various KPIs used for these types of remuneration incentive schemes across various commodities and across various African countries.

The two week long Mineral Resource Management Module forms part of a 50/50 MSc degree in Mining Engineering at the University of the Witwatersrand. This course is intended to enable the student to understand the role of a Mineral Resource Manager and develop a Mineral Resource Management system on their own operations. On concluding the module, the student should have a better understanding of how Mineral Resource Management is developing into Mineral Asset Management. This is due to the changing environment of the mining industry. The student will then be better positioned to link all the elements of the mining value chain with the focus on the overall return on invested capital rather than traditional measures of value. Unfortunately, it is found that on individual mines, this type of management is still in the development phase and many mines are still performance-managing their operations with the focus on measures like square metres mined, tonnage milled and final product produced (Neingo, 2014).

The background to this research paper is the student assignments where performance-management systems and their respective KPIs has been a topic for 2014, 2015 and 2016. The question posed to the students to research on the operations where they are employed is as follows:

"People are motivated by different things and rewarded in different ways for achieving set goals. On the mines, the bonus schemes are typically used to reward people for achieving the targets.

Investigate the various performance-management schemes on YOUR operation (or an operation you are currently involved with) and determine if they actually achieve what they intend to achieve and if they support the flow of quality ore to the plant. Is there conflict arising between different departments as a result of how people are measured and what they are focused on? Is there an impact of the behaviour on the value of the operation (Net Present Value)?"

The assignments have been analysed to identify current trends and effectiveness of people-measurement systems on various Southern African mining operation. The focus of the students' research is whether the Key Performance Indicators (KPIs) which people wish to achieve, drive value through the mining value chain or are actually destroying value. Furthermore, the effectiveness of this type of assignment-based research is considered where students each gather information at their own mining operations and the overall pattern is then subsequently analysed to get an overall feeling for the current situation regarding the core question.

A total of 26 assignments were analysed representing students from South Africa, Zimbabwe, Namibia, Botswana, Zambia and Ghana. The commodities are gold, platinum, diamonds, coal, copper, zinc and cement. Only the assignments which received a passing grade were included in the assessment. The students focused their research primarily on the production departments on the mines rather than their own departments (predominantly mine surveyors, geologists and mine planners in the classes) as this is the key value driving activity on the operations. Most students successfully identified how the personnel were measured and rewarded from the semi-skilled, skilled and management levels. The impacts of these measuring systems on the overall value through the mining value chain were considered and recommendations made to improve the key performance-measuring systems. The advantage of using student assignments for gathering information is that numerous countries and commodities can be compared quickly and with no real costs to the university.

1 CLASS DEMONSTRATION

Games can be very beneficial in demonstrating complex concepts (Özkan & Ulutas, 2011). To demonstrate the concept of KPIs and performance management, the Dice Game is used. Goldratt's Dice Game is a well-known demonstrator for the Theory of Constraints (TOC) (Theory of Constraints Institute, 2016). The game basically simulates a system with a product moving through from station to station. In the Mineral Resource Management class, the dice game is placed into the mining context. The product can be any ore mined and sold, and typically six stations are used (although five may be used depending on the number of students in the class). These stations are Mining, Tramming, Hoisting, Crushing, Milling and Refining and represent a mining value chain (Figure 1).

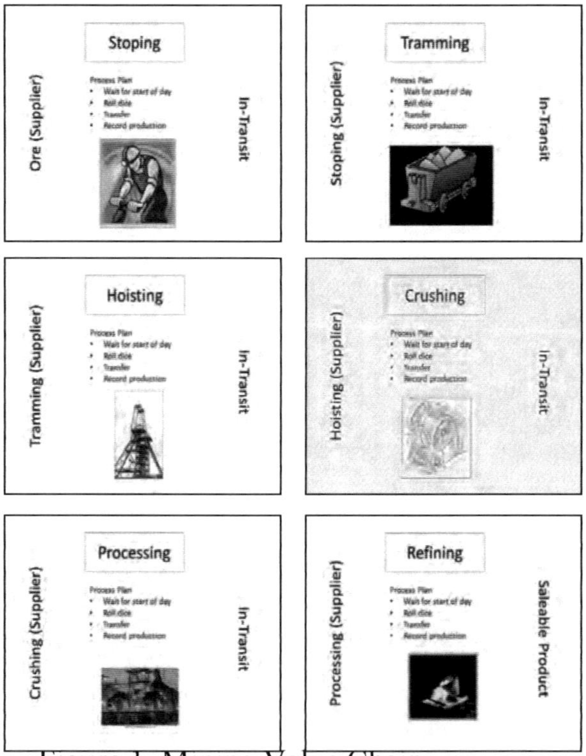

Figure 1. Mining Value Chain

The students sit at the stations with three counters (match-sticks) between each of the stations. To the left of the Stoping Station are full match boxes, and two the right of the Refining Station are empty match boxes. Once the students are all in their places (the number of students in the class governs the number of games running simultaneously), they are given the instructions as well as their "call". In the mining industry, the "call" is the required production to receive the "bonus". Crews who do not achieve their monthly call, do not earn a production bonus. Those who exceed their call earn the bonus, which is often calculated to increase almost exponentially for crews exceeding the call by significant margins. For the dice game, the call is the average result of 20 throws of a six-sided dice (which is 70). The students record their own throws to determine if they achieve their call or not. However, they can only pass the number of matches they have to their left, and they cannot have more than 6 matches between stations at any time (indicating a buffer capacity).

Because the system starts "fully loaded", the student at the Refining Station throws his/her dice first. The number of matches on the dice indicates the station's daily production. The student can then move that number of matches. If it is less than 3 for the first round, the full amount is transferred. If it is more than 3, then only 3 matches are passed because that's all that is available in the proceeding buffer. As the game progresses and the buffers fill up, so more matches can be passed.

The game becomes very animated and the students quickly identify who is exceeding the call and those falling below the targets. This can be enhanced by appointing a "manager" who identifies those falling behind and blaming them for the mine falling down on the overall production.

After the 20 rounds have passed, the number of matches which are to the right of the Refining Station are tallied. This represents the total production from the mine for the month (20 production shifts). After many observations running this game (usually three games per class, two to three classes per year), groups seldom produce more than 54 units. At this point, the students who achieved more than 70 for their cumulative throws are rewarded with a chocolate. This represents their "bonus" for exceeding "call". The post-game discussion is then focused on why the overall system fails to meet the monthly call where typically over half of the students were now eating their chocolates. Typically the comments include recommendations that increasing the buffer capacity (set at 6 matches) would improve the overall production. There are numerous simulations available online and the effect of removing the buffer capacity is demonstrated using one of these (Ganesha, 2016). Removing the buffer capacity has actually no measurable difference in the overall production from the system. This is the first surprise for the students. In reality, increasing buffer capacities on mines is an incredibly expensive exercise requiring extensive capital investment. This capital expense is shown by these simulations not to solve production problems from mines.

The system is then reloaded. All the buffers are filled with the three matches and the constraint of six matches is back into place. The difference with the second run of the game is that personal performance is limited and controlled. If a student throws 1-3 on the dice, they can move three matches. If they throw 4-6 on the dice, they can move four matches. The game proceeds and the first observation is the calm atmosphere in the classroom. The students are not required to record their individual throws and so the focus is on the throughput through the entire system. By removing the downside (one and two matches) at the expense of the upside (five or six matches), variability is reduced and the entire system performs better. Typically, the game now results in throughputs of 60-63 matches. Still slightly below the call of seventy matches but significantly better than the earlier 46-51 matches typically achieved. There are no individual rewards and so the only measure to be rewarded is the overall performance.

At this point, the students who did not receive chocolates for the previous game are given a chocolate too. It is noted that increasing buffer size has very little overall benefit. Reducing variability between stations has a far greater positive impact. Individual stations may be smaller because there is no benefit in overproduction if the following stations cannot process the production. The observations of people behaviour are as significant as the primary learning lesson which is the increased production from the mine. The second game is far less animated than the first game and the room is calm. The students are not focused on their individual throws as these are not being recorded. The only number of concern is the overall production.

Taking these lessons into the real world can aid managers in developing key performance indicators. Individual performance should never be rewarded if the overall business fails to meet the target. The optimum production will always be slightly lower than the average from the individual stations, but by lowering variability the gap can be far narrower. Removing variability can result in significant savings as each station can be made smaller because excess production does not benefit overall production. Spending capital to increase the sizes of buffers between stations is also fruitless expenditure.

2 STUDENT FINDINGS

The findings from the gold, platinum, coal, diamonds, as well as the other commodities (nickel, copper and cement) are summarised and presented separately.

2.1 Gold Mines

The performance management schemes on nine gold mines from South Africa, Ghana and Namibia have been investigated by the students. On Navachab Gold Mine in Namibia, semi-skilled employees are paid incentive schemes based on volume, tonnes, cost and gold produced. Supervisory and management staff are rewarded on their merit appraisal results and management employees are also included in a share allocation. Conflict has been identified with employees complaining that their performance management schemes include parameters over which they have no direct control. The mine in Ghana (Obuasi Mine) was under-performing at the time of the investigation. The mine had recently introduced a new performance management scheme where the primary KPI was gold ounces per total employee costed. There had been a significant improvement and the student was positive that the change in performance management was contributing to the improvement.

The rest of the gold mines investigated are in South Africa. The common methods for incentivising people are primarily based on square metres mined (for narrow tabular gold deposits) and tonnes milled. The criticism with these two KPIs is that they can result in conflict. When the square metres mined are the dominant KPI, tonnes are often left in back areas and not delivered to the processing plant. When tonnage is the key KPI, stoping widths increase causing higher dilution and lower grades being delivered to the processing plant. All the performance-management schemes have various penalties in place to counter the resultant behaviour, ensuring sufficient quality ore is delivered to the processing plant. However, many of these penalties are difficult to implement and semi-skilled workers are not always informed as to how these penalties are established. This can result in industrial action and conflict with the departments (survey, sampling, ventilation, geology and safety) who implement the penalties.

Some of the mines have included additional incentive schemes based on the overall company performance (financial) to allow the employees to share in the profits. The purpose of this is to try challenge all employees to focus on the financial performance of the company including minimising waste. Profits for gold mining companies are primarily determined by the commodity price and thus the incentive scheme becomes more luck than a true indicator of the employees' efforts.

2.2 Platinum Mines

Students from the three largest South African platinum producers submitted assignments (Anglo-American Platinum, Impala Platinum and Lonmin).

At Lebowa Platinum (Anglo-American Platinum), the incentive scheme is balanced between tonnes milled, 4E (platinum, palladium, rhodium and gold) produced, costs and safety. At Tumela Mine, the semi-skilled employees are measured on efficiencies (m or m^2 per man) with penalties for poor safety and sweepings. The skilled employees are measured on a suite of KPIs which include those for the semi-skilled employees, as well as additional KPIs including compliance to the health and safety management system (audit results), stoping widths and costs. Minor differences are noted across the Anglo-American Platinum Group due to their specific mining conditions, but all consider multiple KPIs with emphasis on production, safety and efficiency.

At Impala Platinum, the employee incentive scheme is heavily weighted with safety issues. The quality side of the incentive scheme is in the form of penalties, and as with the gold mines, it is noted that these are difficult to police and implement. The KPIs used at Lonmin are similar to Impala, with the addition of a "people" parameter. This is related to employees staying away from work without permission (AWOP) with very strict penalties. Since the scheme was implemented in 2012, a marked improvement in mining volumes was noted, but not a significant improvement in the final product.

2.3 Coal Mines

Only two coal mines were reviewed by students. One underground (bord and pillar) mine in Zimbabwe and a truck & shovel operation in South Africa. The main focus of the employee incentive schemes for both of these operations is coal produced and safety. Students noted that, at both of these operations, the coal quality is not a KPI. This was felt to be problematic because the value of the coal is directly influenced by the quality. Poor discipline during mining can severely impact on the mines' revenues.

2.4 Diamond Mines

Two students reviewed Debswana in Botswana. The mines in the group pay incentive schemes based on the overall performance of the operation, rather than individual performance. While the one student felt this was problematic, the other commented that at Jwaneng Mine there has never been any industrial action. This student stated that by not identifying and rewarding individual performance, the sense of teamwork is enhanced.

2.5 Base metals and other commodities

A Zinc mine in Namibia, copper mine in Zambia and a cement quarry company with operations in South Africa, Zimbabwe, Mozambique, the Democratic Republic of the Congo, Rwanda and Ethiopia were reviewed. At the zinc mine, there is no individual incentive scheme. The incentive scheme considers the individual employees' salary and the incentive payment is a percentage of this. This is felt to be problematic because the lowest paid workers will always receive a lower bonus even if they put in more effort than a higher paid employee.

On the copper mine, the student reviewed the incentive scheme in the metallurgical plant rather than the mine itself. It was noted that the incentive scheme encourages throughput and quality. Costs are not considered a KPI and the student identified that it could lead to a waste culture.

At the cement quarry, numerous incentive schemes are in place. These include a scorecard bonus (individual team performance including retaining skills, safety and health, customer service, quality, corporate governance, operational efficiency, production and energy consumption), share scheme, inventive idea scheme (Cheaper Better Faster), best employee and absenteeism prevention. The student feels with all these different schemes and the amount of KPIs being focused on, employees are richly rewarded for their efforts above and beyond just production. It was also noted that there is an overall feeling of wellness and satisfaction amongst the employees.

3 CONCLUSIONS

Students from various countries and representing various commodities were tasked with an assignment to review the incentive scheme KPIs in place on their mines. This occurred over three years (2014, 2015 and 2016). The main commodities covered are gold, platinum, coal, diamonds, base metals and cement quarries. Countries include metal mines in South Africa, Namibia, Zimbabwe, Ghana and Zambia whilst the quarries are in South Africa, Zimbabwe, Mozambique, the Democratic Republic of the Congo, Rwanda and Ethiopia. This approach to research gives a very cost effective manner to cover various commodities and various countries. The problems are that some students did poor quality research and these results were not included in the study. Two students even failed to identify which mine or even commodity their research was taking place in. The question posed in the assignment was very broad and as a result the students approached the study in very different ways. Some students focused on the production departments only whilst other students considered a far broader base including support and services, as well as the management structures.

The study shows that a simple set of KPIs focused primarily on volume parameters can result in conflict with quality ore delivered to the plant. Including various quality-based parameters resulting in penalties, can cause conflict when these are implemented. The ideal incentive scheme KPIs appear to be those which encourage the entire team to work towards a common goal.

REFERENCES

Bryson, A. et al., 2012. CEP Paying For Performance: Incentive Pay Schemes and Employees' Financial Participation, London: The London School of Economics and Political Science.

Ganesha, 2016. Goldratt's Theory of Constraints: Online Simulation. [Online] Available at: http://www.ganesha.org/leading/toc.html [Accessed September 2016].

Kemp, J. J., 2016. An Investigation into the Production Bonus System of Continious Miner Sections in an Underground Coal Mine, Johannesburg: Unpublished research report, University Witwatersrand.

Neingo, P. N., 2014. Drivers, Measures and Rewards for Productivity and Financial Impacts/Benefits in the Mining Value Chain at Bothopele, Kroondal and Mototolo Mines: An Investigation of Impacts from Short Term Planning to Mill., Johannesburg: MSc, University of the Witwatersrand..

Niengo, P. N. & Cawood, F. T., 2014. Correlation of productivity trends with market factors at three selected platinum mines. Johannesburg, The Southern African Institute of Mining and Metallurgy.

Özkan, F. & Ulutas, B., 2011. Evaluating the Effect of Teaching Strategies and Learning Styles to Students' Success. Elazig, Turkey, Fırat University.

Seotlela, R. P. J. & Miruka, O., 2014. Implementation Challenges of Performance Management Systems in the South African Mining Industry. Mediterranean Journal of Social Sciences, 5(7), pp. 177-187.

Theory of Constraints Institute, 2016. What is the Theory of Constraints. [Online] Available at: http://www.tocinstitute.org/theory-of-constraints.html [Accessed September 2016].

Economic Impact of Bauxite Beneficiation in West Kalimantan, Indonesia: Input-Output Analysis

F. Firmansyah[1,2], C. Drebenstedt[1]

TU Bergakademie, Freiberg, Germany[1]

Ministry of Energy and Mineral Resources, Republic of Indonesia[2]

ABSTRACT Government of Indonesia implemented raw mineral material export ban policy effectively in early 2014. All of mining license permit and contract of work holders were forbidden to export their raw mineral material and obliged to process it in Indonesia. Bauxite is also part of this policy and West Kalimantan province has the largest production and reserves of bauxite. Therefore, we need to investigate the economic impact of this policy in West Kalimantan, Indonesia. The methodology contains the projection of West Kalimantan Input-Output Table for the year 2010 to 2030 by including alumina refineries as a new classification, the linkages and multipliers analysis, and output, income and employment impact analysis with three scenarios. Our findings confirm the necessity of bauxite beneficiation. The final result of this research could help policy and decision makers especially in Indonesia to choose the optimum option in utilizing mineral resources for the welfare of the people.

Keywords: Bauxite, Beneficiation, Input-Output Analysis, Linkages, Multiplier

1 INTRODUCTION

Beside coal, nickel, copper and gold, bauxite is one of the back-bone mining commodities for supporting Indonesian economy as an export commodity. Indonesian bauxite ores production grew fast in the recent year especially in 2013 before the implementation of the raw mineral material export ban policy (see Fig. 1). Unfortunately, before this implementation, all of bauxite ore productions are exported because there are no alumina refineries or other bauxite processing facilities in Indonesia. Meanwhile, Indonesia's bauxite resources and reserves are not very abundant compare to other bauxite ores exporter country (see Fig. 2). Therefore, to get optimum benefit of bauxite, it is necessary to develop bauxite beneficiation by encouraging bauxite mining permit holders to process bauxite domestically.

Bauxite resources and reserves in Indonesia are accumulated mainly in three provinces, namely West Kalimantan, Central Kalimantan and Riau Islands. West Kalimantan has 82% of the total bauxite resources and 68% of total bauxite reserves (Center for Geological Resources, Indonesia, 2014). For that reason, West Kalimantan is chosen as the locus of this paper.

Bauxite beneficiation can be done in several ways. Bauxite can be processed as calcined product, special products, chemical products, cement products and alumina production (Hill & Sehnke, 2006). Almost 85% of mined bauxite is converted into aluminium metal while 10% goes to nonmetal uses in various forms of alumina and the remaining 5% goes to non-metallurgical-grade bauxite application. The market for non-metallurgical-grade bauxite is mature and only consumes five million tons of these products. On the other side, the market for alumina production still grows significantly. For example, in 2015 the world aluminium production which needs alumina grew from 50 million tons to 58 million tons (US Geological Survey, 2016). This amount of aluminium production approximately needs about 116 million tons of alumina. Based on the aforementioned demand, it is suggested to build alumina refinery in order to beneficiate bauxites.

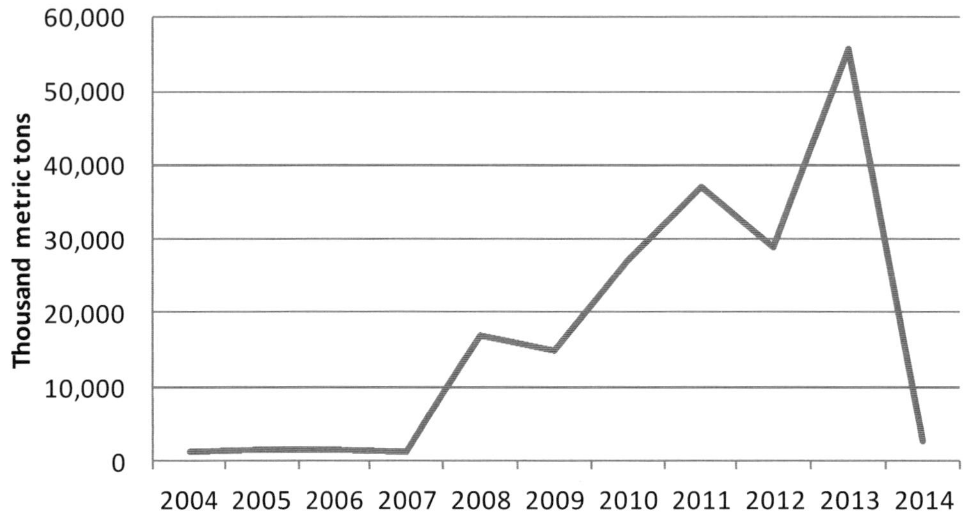

Figure 1. Indonesia's bauxite ores export and production 2004-2014 (US Geological Survey, 2016)

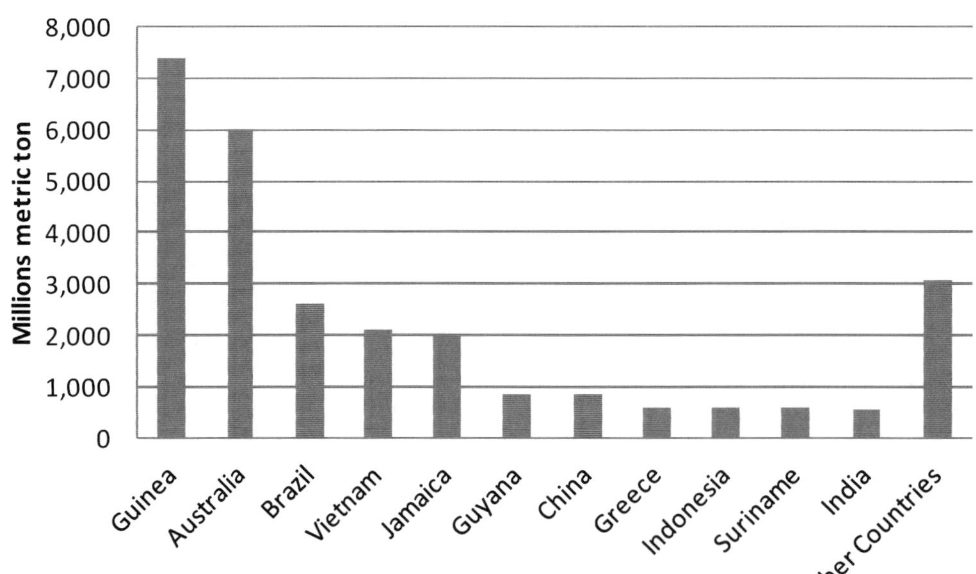

Figure 2. World bauxite reserves (US Geological Survey, 2014) and (Center for Geological Resources, Indonesia, 2014)

2 RESEARCH METHOD

The aim of this paper is to analyze the impact of the bauxite beneficiation development to the West Kalimantan economy using input-output analysis. The analysis is based on West Kalimantan Input-Output Table for 2010, 54x54 sectors, domestic transactions on the basis of producer prices (Badan Perencanaan Pembangunan Daerah Provinsi Kalimantan Barat, 2012). Moreover, this table is aggregated and added with alumina refineries as a new sector into 10 x 10 sectors. Thus, this table is projected until the year 2030 by using the assumptions that the economic growth is 4.5%. The output of alumina refineries is the result of processing 11.2 million metric tons of bauxite production or equal with 4 million metric tons alumina production. Meanwhile, the inputs of alumina refineries sector are based on the summary of the feasibility study report of alumina refinery from Ministry of Energy and Mineral Resources of Indonesia.

2.1 Input-Output Analysis

The relationship between the entries in Input-Output Table and GDP are as follows:

$$\sum_{j=1}^{n} x_{ij} + F_i = X_i + M_i \quad (1)$$

$$\sum_{i=1}^{n} x_{ij} + V_j = X_j \quad (2)$$

And $X_i = X_j$ then the formula can be written as follows:

$$\sum_{j=1}^{n} x_{ij} + F_i - M_i = \sum_{i=1}^{n} x_{ij} + V_j$$

$$or \quad F_i - M_i = V_j \quad (3)$$

The proportion of intermediate input from i sector to the total input of j sector is referred as coefficient of intermediate input (Miller & Blair, 2009) and obtained by the formula:

$$a_{ij} = \frac{x_{ij}}{X_j} \quad or \quad x_{ij} = a_{ij}.X_j \quad (4)$$

By substituting equation (4) into equation (1) with import (Mi) is excluded, the equation becomes as follows:

$$\begin{aligned} a_{11}X_1 + a_{12}X_1 + \ldots + a_{1n}X_1 + F_1 &= X_1 \quad (5)\\ a_{21}X_2 + a_{22}X_2 + \ldots + a_{2n}X_2 + F_2 &= X_2\\ \ldots + \ldots + \ldots + \ldots + \ldots &= \ldots\\ a_{n1}X_1 + a_{n2}X_2 + \ldots + a_{nn}X_1 + F_n &= X_n \end{aligned}$$

If the composition of the equation (4) is simplified into matrix form, it is obtained:

$$\begin{aligned} \mathbf{AX} + \mathbf{F} &= \mathbf{X}\\ \mathbf{X} - \mathbf{AX} &= \mathbf{F}\\ (\mathbf{I} - \mathbf{A})\mathbf{X} &= \mathbf{F} \quad (6) \end{aligned}$$

The output can be calculated as the effect of final demand induction, as follows:

$$\mathbf{X} = (\mathbf{I} - \mathbf{A})^{-1}\mathbf{F} \quad (7)$$

where:
I = the identity matrix of size n × n
F = final demand matrix of size n x 1
A = matrix of input coefficients / technical of n x n

Matrix (I - Ad) -1 is a matrix multiplier which is suitable for measuring changes in domestic output, due to the change in domestic final demand. This matrix is also known as Leontief inverse matrix (Bulmer, 1982).

2.1.1 Linkage analysis

Linkages analysis is used to calculate the output impact of the fact that basically the industrial sectors in the economy affect each other (Nazara, 2005). There are two types of linkages, namely backward linkage and forward linkage. The calculation of forward linkage can be done in two ways, namely by using Leontief inverse matrix and Ghosian inverse matrix.

Forward and backward linkages indexes using Leontif inverse matrix are calculated as follows:

$$\boldsymbol{\beta}_i = \frac{(1/n)\sum_j b_{ij}}{(1/n^2)\sum_i\sum_j b_{ij}} \quad (8)$$

$$\boldsymbol{\alpha}_j = \frac{(1/n)\sum_i b_{ij}}{(1/n^2)\sum_i\sum_j b_{ij}} \quad (9)$$

Forward linkage/supply multiplier using Ghosian inverse matrix is calculated as follows:

$$\boldsymbol{SuM}_j^I = \sum_{j=1}^{n} \vec{b}_{ij} \quad (10)$$

where:
b : Leontief inverse matrix;
$\vec{b}$: Ghosian inverse matrix

2.1.2 Linkage analysis

Multiplier analysis is used to describe what happens to the endogenous variables, which are sectors output, if there is a change in exogenous variables, such as final demand, in the economy (Nazara, 2005). In this paper, we will calculate the Type I and II multiplier. Type I will capture the direct and indirect effect. Type II will also capture the induced effect as the result of the introduction the household consumption as endogenous variable. The multiplier analysis is calculated as follows:

$$\boldsymbol{OM}_j^I = \sum_{j=1}^{n} b_{ij} \quad (9)$$

$$\boldsymbol{IM}_j^I = \frac{\sum_{j=1}^{n} b_{ij} I_j^T}{I_j^T} \quad (10)$$

$$\boldsymbol{VM}_j^I = \frac{\sum_{j=1}^{n} v_j b_{ij}}{v_j} \quad (11)$$

$$\boldsymbol{EM}_j^I = \sum_{j=1}^{n} \frac{l_j b_{ij}}{l_j} \quad (12)$$

$$\boldsymbol{SM}_j^I = \frac{\frac{\sum_{j=1}^{n} v_j b_{ij}}{v_j}}{\sum_j^n k_j b_{ij}} \quad (13)$$

where:
b : Leontief inverse matrix;
OM : Output multiplier;
IM : Income multiplier;
EM : Employment multiplier;
VM : Value added multiplier;
SM : Surplus multiplier;
l : Coefficient of salary;
v : Coefficient of value added;
k : Coefficient of capital

2.1.3 Impact analysis

Impact analysis is used to compare the economic benefits of some condition of export control scenarios, namely: (a) all bauxites are exported, (b) bauxites partially are processed domestically and (c) all bauxites are processed domestically. The impact analysis consist of output, income and employment impact. The scenarios are based on the assumptions as follows: the additon of production are 2.8 million metric tons of bauxite ores per year, selling price of bauxite ore is US$ 30/metric tons, selling price of smelter-grade alumina (SGA) is US$ 425/tons.

3 RESULT AND DISCUSSION

Table 1 shows the result of linkage analysis among sectors in the West Kalimantan economy the year 2030. The industry sector has the highest backward linkage index, while mining sector has the lowest. The sectors with the yield of backward linkage above the average (> 1) will give enormous effect through investment on the economy. Moreover, agriculture sector has highest forward linkage index, while alumina refinery has the lowest. The sectors with the yield of forward linkage above the average (> 1) will be more stimulated by the economy than other sectors. Alumina refinery's backward linkage index ranking is 2 and its forward linkage index ranking is 10. This small forward linkage index of alumina refinery sector can be explained by the lack of its downstream sector. The output of this sector is not used by other sectors in West Kalimantan and it is exported to other province in Indonesia or other country. In addition, linkage analysis using supply-driven or also known as supply/input multiplier shows alumina refinery sector ranking is 10. We can notice that this sector has important position in West Kalimantan economy. Based on the range of backward and forward linkage index's ranking, we can classify alumina refinery sector as intermediate primary production because of its large backward and small forward linkage indexes (Lixon, Thomassin, & Hamaide, 2008). Figure 3 shows the sectoral linkage pattern of alumina refinery sector.

Table 1. Key Sectors Analysis: Forward & Backward Linkage

	Sector	Backward Linkage	Rank	Forward Linkage Demand-driven	Rank	Forward Linkage Supply-driven	Rank
1	Agriculture	0.9396	6	1.3239	1	0.9955	5
2	Mining	0.8159	10	1.0841	4	1.0113	4
3	Industry	1.3269	1	0.9269	7	1.2228	3
4	Electricity and Water	0.8535	9	0.7668	9	0.8854	7
5	Construction	1.0685	3	0.8398	8	0.8167	8
6	Trades	1.0080	4	1.2161	3	0.9561	6
7	Transportation	0.9444	5	1.2216	2	1.3494	1
8	Finance	0.8567	8	0.9309	6	1.2245	2
9	Services	0.8655	7	0.9434	5	0.8125	9
10	Alumina Refinery	1.3209	2	0.7466	10	0.7259	10

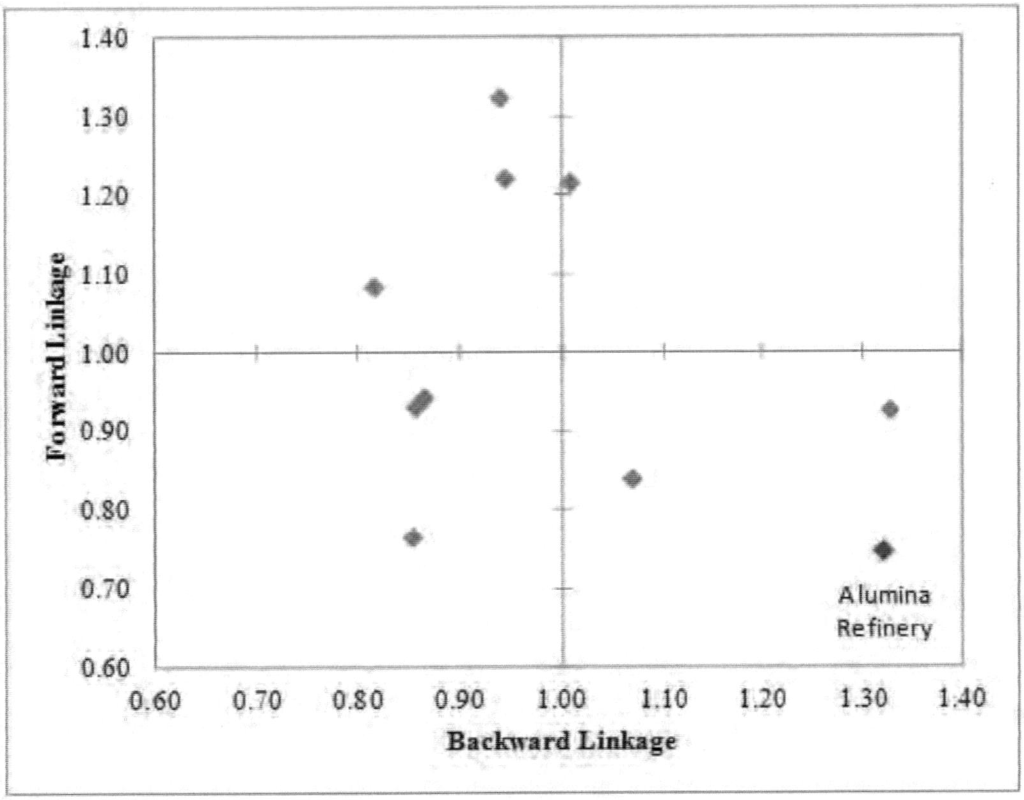

Figure 3. Sectoral linkage pattern of alumina refinery sector

Table 2 describes the total output multiplier among sectors in the West Kalimantan economy. The industry sector is the highest in Type I output multiplier, while in Type II the alumina refinery sector is the highest. The lowest in Type I output multiplier is mining sector but in Type II is electricity and water sector. We can notice here that the introduction of household consumption as an endogenous factor in multiplier analysis changes the rankings. In general the alumina refinery sector gives relatively higher output multiplier than other sectors.

Table 3 describes the total income multiplier among sectors in the West Kalimantan economy. The alumina refinery sector has the highest Type I and II income multiplier, while mining sector has the lowest.

Table 4 describes the total employment multiplier among sectors in the West Kalimantan economy. The alumina refinery sector has the highest Type I employment multiplier, while mining sector has the highest in Type II. Meanwhile the electricity and water sector has the lowest Type I and II employment multiplier.

Table 5 describes the total value added multiplier among sectors in the West Kalimantan economy. The alumina refinery sector has the highest Type I and II value added multiplier. Meanwhile the mining sector and electricity and water sector have the lowest Type I and II value added multiplier consecutively.

Table 6 describes the total surplus multiplier among sectors in the West Kalimantan economy. The alumina refinery sector has the highest Type I and II surplus multiplier, while the electricity and water sector has the lowest.

Table 7 describes the sector distribution of the output, final demand and gross regional domestic product (GDRP) in the West Kalimantan economy. The services sector has the highest contribution to the output, final demand and gross regional domestic product while the electricity and water sector has the lowest. The alumina refinery sector has relatively high contribution as it gives 6.99% to the output, 9.3% to the final demand and 8.42% to the GRDP.

Table 8 shows the result of impact analysis based on three scenarios. Scenario 3 gives the highest total output, income, and employment, while scenario 1 gives the lowest. Meanwhile, scenario 2 gives almost three times higher than scenario 1 and more than half of scenario 3. This result confirmed the advantage of bauxite beneficiaton development through alumina refinery sector.

Table 2. Total Output Multiplier: 2030 West Kalimantan IO Table 10 sectors

	Sector	Initial	First Round	Indust Sup	Con-sump-tion	Total	Type I	Type II
1	Agriculture	1	0.1955	0.0630	0.6209	1.8794	1.2585	1.8794
2	Mining	1	0.0749	0.0179	1.0993	2.1922	1.0928	2.1922
3	Industry	1	0.6052	0.1720	0.5532	2.3305	1.7773	2.3305
4	Electricity and Water	1	0.1104	0.0329	0.4538	1.5971	1.1432	1.5971
5	Construction	1	0.3088	0.1224	1.0050	2.4363	1.4312	2.4363
6	Trades	1	0.2589	0.0912	0.5400	1.8901	1.3501	1.8901
7	Transportation	1	0.2007	0.0643	0.5742	1.8392	1.2650	1.8392
8	Finance	1	0.1220	0.0255	0.6479	1.7955	1.1476	1.7955
9	Services	1	0.1170	0.0423	0.9941	2.1535	1.1593	2.1535
10	Alumina Refinery	1	0.6766	0.0927	0.8034	2.5727	1.7693	2.5727

Table 3. Total Income Multipliers: 2030 West Kalimantan IO Table 10 sectors

Sector	Initial	First Round	Indust Sup	Con-sump-tion	Total	Type I	Type II
1 Agriculture	0.2437	0.0469	0.0143	0.1347	0.4395	1.2509	1.8034
2 Mining	0.5112	0.0245	0.0042	0.2384	0.7782	1.0561	1.5225
3 Industry	0.0908	0.1408	0.0400	0.1200	0.3916	2.9908	4.3117
4 Electricity and Water	0.1914	0.0241	0.0074	0.0984	0.3213	1.1646	1.6790
5 Construction	0.4081	0.0577	0.0277	0.2180	0.7115	1.2094	1.7435
6 Trades	0.1870	0.0573	0.0209	0.1171	0.3823	1.4181	2.0444
7 Transportation	0.2230	0.0443	0.0147	0.1245	0.4065	1.2646	1.8231
8 Finance	0.2805	0.0316	0.0060	0.1405	0.4587	1.1342	1.6351
9 Services	0.4489	0.0300	0.0093	0.2156	0.7038	1.0876	1.5679
10 Alumina Refinery	0.0516	0.3182	0.0247	0.1742	0.5688	7.6518	11.031

Table 4. Total Employment Multipliers: 2030 West Kalimantan IO Table 10 sectors

Sector	Initial	First Round	Indust Sup	Con-sump-tion	Total	Type I	Type II
1 Agriculture	0.0320	0.0036	0.0012	0.0134	0.0503	1.1504	1.57
2 Mining	0.0006	0.0012	0.0003	0.0237	0.0259	3.5170	42.74
3 Industry	0.0281	0.0157	0.0032	0.0119	0.0591	1.6745	2.10
4 Electricity and Water	0.1075	0.0012	0.0006	0.0098	0.1191	1.0164	1.11
5 Construction	0.0036	0.0048	0.0025	0.0217	0.0326	2.9897	8.94
6 Trades	0.0142	0.0048	0.0018	0.0117	0.0324	1.4655	2.29
7 Transportation	0.0023	0.0033	0.0013	0.0124	0.0192	2.9973	8.44
8 Finance	0.0332	0.0027	0.0005	0.0140	0.0504	1.0972	1.52
9 Services	0.0114	0.0011	0.0008	0.0215	0.0347	1.1638	3.05
10 Alumina Refinery	0.0020	0.0048	0.0014	0.0173	0.0255	4.1619	13.01

Table 5. Total Value Added Multipliers: 2030 West Kalimantan IO Table 10 sectors

Sector	Initial	First Round	Indust Sup	Con-sump-tion	Total	Type I	Type II
1 Agriculture	0.7450	0.1337	0.0436	0.4093	1.3316	1.2380	1.7874
2 Mining	0.8808	0.0579	0.0127	0.7247	1.6761	1.0802	1.9030
3 Industry	0.3792	0.4399	0.1177	0.3647	1.3015	2.4705	3.4323
4 Electricity and Water	0.8650	0.0757	0.0225	0.2992	1.2624	1.1135	1.4594
5 Construction	0.6632	0.2057	0.0837	0.6626	1.6151	1.4363	2.4354
6 Trades	0.7064	0.1708	0.0635	0.3560	1.2967	1.3316	1.8356
7 Transportation	0.6588	0.1375	0.0451	0.3785	1.2200	1.2772	1.8517
8 Finance	0.8659	0.0961	0.0183	0.4272	1.4075	1.1321	1.6254
9 Services	0.8470	0.0752	0.0292	0.6554	1.6069	1.1233	1.8971
10 Alumina Refinery	0.1621	0.5781	0.0656	0.5296	1.3355	4.9726	8.2409

Table 6. Total Surplus Multipliers: 2030 West Kalimantan IO Table 10 sectors

	Sector	Initial	First Round	Indust Sup	Con-sump-tion	Total	Type I	Type II
1	Agriculture	0.4629	0.0720	0.0239	0.2186	0.7774	1.2072	1.679
2	Mining	0.3083	0.0263	0.0068	0.3870	0.7285	1.1076	2.363
3	Industry	0.2273	0.2607	0.0638	0.1947	0.7465	2.4281	3.285
4	Electricity and Water	0.6293	0.0390	0.0122	0.1598	0.8402	1.0812	1.335
5	Construction	0.1796	0.1109	0.0462	0.3538	0.6905	1.8745	3.844
6	Trades	0.3850	0.0916	0.0351	0.1901	0.7019	1.3292	1.823
7	Transportation	0.3303	0.0728	0.0249	0.2021	0.6301	1.2956	1.908
8	Finance	0.4988	0.0521	0.0099	0.2281	0.7889	1.1241	1.581
9	Services	0.1897	0.0342	0.0161	0.3500	0.5900	1.2652	3.110
10	Alumina Refinery	0.0453	0.1863	0.0324	0.2828	0.5468	5.8303	12.077

Table 7. Sector Output Distributions

	Sector	Total Output (mil. US$)	Per-cent (%)	Final Demand (mil. US$)	Per-cent (%)	GRDP (mil. US$)	Per-cent (%)
1	Agriculture	4568.20	18.78	3416.00	18.69	3144.06	18.57
2	Mining	2000.09	8.22	1234.00	6.75	1145.08	6.76
3	Industry	1069.50	4.40	509.00	2.78	492.00	2.91
4	Electricity and Water	251.80	1.04	206.00	1.13	199.97	1.18
5	Construction	3086.85	12.69	2779.00	15.20	2693.02	15.90
6	Trades	4867.14	20.01	3653.00	19.98	3484.00	20.57
7	Transportation	1975.08	8.12	740.00	4.05	462.80	2.73
8	Finance	683.03	2.81	377.00	2.06	369.00	2.18
9	Services	4122.50	16.95	3666.00	20.05	3517.96	20.77
10	Alumina Refinery	1700.00	6.99	1700.00	9.30	1426.03	8.42
Total		24324.19	100.00	6044.19	100.00	18280.00	100.00

Table 8. Final Demand Impacts

	Scenario 1			Scenario 2			Scenario 3		
Sector	Output (mil. US$)	In-come (mil. US$)	Em-ploy-ment	Output (mil. US$)	In-come (mil. US$)	Em-ploy-ment	Output (mil. US$)	In-come (mil. US$)	Em-ploy-ment
1	23.73	5.78	6845	57.69	14.07	16644	91.66	22.35	26443
2	86.97	44.44	474	135.55	69.26	738	184.12	94.09	1003
3	21.18	1.92	5364	53.07	4.81	13438	84.95	7.71	21512
4	0.93	0.18	897	5.07	0.97	4900	9.22	1.76	8903
5	1.89	0.77	62	8.09	3.30	265	14.29	5.83	469
6	22.74	4.25	2900	56.66	10.59	7225	90.57	16.93	11550
7	12.63	2.82	259	32.50	7.25	666	52.37	11.68	1074
8	6.65	1.87	1985	15.18	4.27	4535	23.72	6.67	7085
9	7.40	3.32	756	62.50	28.06	6385	117.61	52.79	12015
10	0	0	0	212.50	11.00	3750	425.00	22.00	7500
Total	184.12	65.35	19542	638.82	153.58	58547	1093.5	241.80	97553

4 CONCLUSION

This paper has applied input-output analysis for examining the economic impact of bauxite beneficiation development in West Kalimantan economy. Bauxite beneficiation development was performed by adding alumina refinery as a new classification sector. The alumina refinery sector's backward linkage's rank is high relative to other sectors especially mining sector. Meanwhile, both its forward linkage's rank is very low relative to other sectors indicating the lack of downstream sector linkage. Based on the composition of backward and forward linkage, this sector can be classified as intermediate primary production.

Multipliers of alumina refinery sector generally show higher average values than other sectors especially income and employment multiplier. These values indicate that this sector has the potential to create new output that capable of driving the West Kalimantan economy.

Contribution of alumina refinery to the West Kalimantan economy relatively high with 6.99% to the output, 9.3% to the final demand and 8.42% to the GRDP.

Impact analysis show that the scenario 3 (all of bauxite ores are processed) gives the highest total output, while scenario 1 (all of bauxite ores are exported) gives the lowest This result confirms the necessity of bauxite beneficiaton development in West Kalimantan.

The results from this paper give valuable information and could help in choosing the optimum option in utilizing mineral resources for the welfare of the people in Indonesia. To further our research, we intend to optimize the sectors in the input-output table to get the optimum GDP using linear programming.

REFERENCES

Badan Perencanaan Pembangunan Daerah Provinsi Kalimantan Barat, 2012. *2010 West Kalimantan Input Output Table*, Badan Perencanaan Pembangunan Daerah Provinsi Kalimantan Barat, Pontianak.

Bulmer, V. T., 1982. *I-O Analysis in Developing Countrie,*. John Wiley & Sons Ltd, London.

Center for Geological Resources, Indonesia., 2014. *Executive Summary Pemutakhiran Data dan Neraca Sumber Daya Mineral Status 2014*. Center for Geological Resources, Geology Agency, MEMR of Indonesia.

Hill, V. G., & Sehnke, E. D., 2006. Bauxites. in *Industrial Minerals & Rocks: Commodities, Markets, and Uses,* Littleton: Society for Mining, Metallurgy and Exploration, Inc., (pp. 227-261).

Lixon, B., Thomassin, P. J., & Hamaide, B., 2008. Industrial output restriction and the Kyoto protocol: An input–output approach with application to Canada. *Ecological Economics*, (pp. 249-258).

Miller, R. E., & Blair, P. D., (2nd Ed.), 2009. *Input–Output Analysis: Foundation and Extension*., Cambridge University Press, New York.

Nazara, S., (2nd Ed.), 2005. *Analisis Input Output,* Lembaga Penerbit FE-UI, Jakarta.

US Geological Survey., 2014. *Mineral Commodity Summary*, US Geological Survey.

US Geological Survey., 2016. *Mineral Commodity Summary*, US Geological Survey.

Development of a Strategy to Deal with the Business Risk of Productivity Improvement: A Perspective of the Platinum Mining Sector in Zimbabwe

J. Kwiri, T. Zvarivadza
University of the Witwatersrand, Johannesburg, South Africa

ABSTRACT The Platinum Group Metal mining sector is a significant contributor to Zimbabwe's economy. However, the sector is based on non-renewable resource. Despite the need to fully exploit the resources in order to gain maximum returns, there exists, business risk that when not addressed, the resources are put to waste. These risks include productivity improvement, to which this study addresses. Such a business risk requires the development of a strategy. Currently, the platinum mining industry in Zimbabwe, experiences low productivity in a market of low commodity prices which lead to failure of fully exploiting the resource for maximum benefits. To improve the low productivity, and benefit from expected commodity price upswing, a conservative strategy has been developed which entails an integrated approach to productivity improvement where growth is the main objective with innovative culture. This is coupled with diversification and forward integration with joint ventures with the Zimbabwean government.

Keywords: Strategy, Productivity, Platinum Mining, Business Risk

1 INTRODUCTION

Zimbabwe has a number of minerals, of which forty are mineable mineral resources (KPMG Zimbabwe, 2015). The mineralisation is dominant within the Great Dyke and the greenstone belts. It is within the Great Dyke that the world's second largest platinum group minerals (PGMs) deposits occur. The main commodities currently being exploited in Zimbabwe are platinum group metals, coal, diamonds, chrome and gold. Gold and PGMs are the largest contributors to the country's GDP. The resources of PGMs are estimated to be 2.8 billion tonnes and three major players exist which are Mimosa, Zimplats and Unki (KPMG Zimbabwe, 2015). The current PGM production has been heavily affected by the closure of Bimha Mine, under Zimplats and the continual declining commodity prices (KPMG Zimbabwe, 2015), leading to a drop in platinum metal production of 500 kilograms for the period from 2014 to 2015. The government has been encouraging mining companies to beneficiate the PGMs. The development of more PGM deposits has been limited by "under investments in exploration and production" (Hawkins, 2009). The PGM industry also suffers from a "severe and rapidly worsening skills shortage, exacerbated by the precipitous decline in the country's capacity to regenerate skills" (Hawkins, 2009).

Extended periods of low commodity prices result in "unprecedented impacts on earnings, balance sheets and investor perceptions" are affecting the mining industry (Earnest and Young, 2016). The high risk averse capital markets have forced the most mining companies to focus on short term (Earnest and Young, 2016). The threat to this growth is exacerbated by the low productivity, which is amongst the identified risks by Earnest, and Young (2016). Productivity in this context is defined as the overall business productivity covering all resources which are equipment, people, the orebody and the system. The current business environment and the unpredictable commodity prices have prompted for a need to re-evaluate current operating strategies to remain competitive. It is imperative that sustainable and robust productivity improvement measures be implemented if the industry has to remain competitive.

2 EXTERNAL AND INTERNAL ENVIRONMENT ANALYSIS

The environment within which it operates (School of Mining Engineering, 2016) influences the PGM sector. This environment can be external or internal. The rate of change of the environment affects the way the PGM mining sector is shaped. In order to remain competitive, the PGM sector has to be in step with the changes to the environment (School of Mining Engineering, 2016). The process of strategy formulation extensively identifies all elements of the two environments so as to identify the opportunities and strengths that exist for exploitation to gain competitive advantage.

2.1 External Environment Analysis

The external environment is where the PGM mining sector draws its inputs and sells its products to (School of Mining Engineering, 2016). This external environment is divided into three sub-environments which are; operating environment, industry environment and remote environment. The industry environment is adequately descried by Porter's five forces and the remote environment by the (Political, Economic, Social and Technological) PEST analysis. The operating environment looks at labour, suppliers, competitors, customers and creditors.

2.1.1 Operating environment analysis

The PGM mining sector in Zimbabwe has a number of competitors across the globe. This includes PGM mining sectors largely in South Africa and Russia. The production contribution of this sector in Zimbabwe is such that it is currently uneconomic to fully beneficiate the ore to the same standard as the South African competitors. The high skill exodus and the inability of the education sector to regenerate skills make the skill shortage a binding constraint in the long term for the PGM mining sector (Hawkins, 2009).

Although there is surtax waiving for capital goods, there is little trust towards the stability of the regulations (Hawkins, 2009). The PGM mining sector exports all its products to Rustenburg plant in South Africa for further processing, with some companies exporting matte and others exporting concentrates. This, in essence, makes the Rustenburg plant the customer for the PGM mining sector in Zimbabwe. All the existing companies within the PGM mining sector are foreign owned and as such credit facilities locally have little effect to the sector. This, however, presents a challenge to the global competition for credit as the sector has to look for credit through parent companies.

2.1.2 Industry environment analysis

The industry environment is adequately described by Porter's Five Forces which are barriers to entry, supplier power, buyer's power, substitute availability and competitive rivalry (School of Mining Engineering, 2016).

2.1.2.1 Threat of new entry

The current economic conditions globally make it difficult to acquire the huge amounts of capital required to start business in this sector. The existing regulation seeks to boost the country's investment portfolio and as such, it is attractive to new entries. The unavailability of local processing plant forces the new entries to have large volumes of production to realise profits as the product has to be exported for further processing. This makes the threat to new entry weak.

2.1.2.2 Supplier power

Locally, the number of suppliers of PGMs is low compared to the same sector in South Africa. Currently in Zimbabwe, there is no threat to any forward integration by the existing suppliers, as the current suppliers need more players to justify forward integration for example having a processing plant. Thus the supplier power can be considered to be weak.

2.1.2.3 Buyer power

The existing buyer is the Rustenburg Processing Plant. Arrangements have to be made such that the quality of ore suits the plant requirements. The buyer can therefore choose from a number of suppliers, although with the recent declining in production the buyer has limited choice. The buyer is sensitive to price volatility of PGMs and the decision to receive ore for processing is dependent on the price on the market. The buyer power can therefore be taken to be moderately strong.

2.1.2.4 The threat of substitution

The current substitutes to the main use of PGMs fail to give the same convenience as they are still expensive to produce on large scale. Considerable research to substitute the use of platinum in automotive industry is being done. Once a substitute has been found, there is a large risk of the PGM mining sector being obsolete. The threat of substitution is taken to be very strong should a breakthrough happen.

2.1.2.5 Competitive rivalry

There is a small number of competitors for the local PGM sector, although globally there are a number of significant players. The PGM sector is large and matured globally and is still growing locally. The risk of losing money should a company decide to leave the industry is large as the markets are not lucrative enough. The threat of being acquired by a competitor is very high as some competitors seek to grow their portfolio and other competitors in different sectors seek to diversify their portfolios.

2.1.3 Remote environment analysis

The area analyzed by the remote environment include political, economic, social and technology. (School of Mining Engineering, 2016).

2.1.3.1 Political

The extent to which a government influences the PGM mining sector is considered under political factors. Within the operating business environment, the following factors affect the external environment of the PGM sector.

- The government passed a legislation on resource nationalization requiring a stake of 51%, which is to be paid through future dividends (KPMG Zimbabwe, 2015).
- The new requirement to present plans on how the ore can be further beneficiated before exportation, else an increase in export tax on raw materials.
- Investment regulation is supportive as the government seeks to increase its foreign direct investments.

2.1.3.2 Economic

These factors are the “determinant of an economy’s performance that directly impact the PGM mining sector” (School of Mining Engineering, 2016). These factors include;

- The inflation rate that is still negative, since the dollarization in 2009 (KPMG Zimbabwe, 2015);
- The growth rate has been lowered from 3.9% to 1.5% which shows a declining growth rate (KPMG Zimbabwe, 2015); and
- The commodity prices still fluctuates and the sector is very sensitive to this volatility.

2.1.3.3 Social

These factors analyse the cultural trends, demographics and how they affect the market for PGMs. The factors include;

- A decreasing purchasing power of automobiles by consumers thus impacting the demand for PGMs in the automotive industry; and
- The increase in awareness of the green trade alliance has increased the awareness of community and the companies need to have well planned and executed environmental rehabilitation plans (World Economic Forum, 2010)

2.1.3.4 Technology

Technological factors include;

- There is little that the sector or the government is spending on research and development;
- Globally fully automated machinery that can be used by the PGM mining sector is being developed which aims to bring down the operating costs and improving the safety records; and
- The technology level within the local PGM mining sector is relatively good as all operations in the sector are highly mechanized.

2.2 Internal Environment Analysis

This analysis focuses on the strengths and weaknesses of the PGM mining sector with the idea to know the capabilities of the sector. The strength, weakness, opportunity and threat (SWOT) analysis is used. The result of the analysis enables the sector to determine the core competencies that can be used for competitive advantage (Ovidijus, 2016).

2.2.1 Strengths

The strength of the PGM sector include;

- The deposits for PGMs are relatively shallow giving easy access;
- The workforce is highly literate, with the lowest level employee being able to communicate in English;
- The Mines within the PGM mining sector are highly mechanised;
- Existing companies are relatively low cost producers (School of Mining Engineering, 2016); and
- Most of the mines in this sector have a positive balance sheet, which enables them to consider expansion plans (KPMG Zimbabwe, 2015).

2.2.2 Weaknesses

The weaknesses of the PGM sector are;

- There is still poor productivity despite the high mechanisation levels (Earnest and Young , 2016);
- The low life expectancy of the workforce is a great challenge in the long term;
- The majority of the human resource is not skilled for mechanised mining;
- The sector has poor presence in the world's market despite having the second largest deposits; and
- There is high employee turnover.

2.2.3 Opportunities

The opportunities for the PGM sector are;

- Global market growth;
- Improved productivity with improved asset utilisation;
- Political and economic stability of the country; and
- New technology that can drive costs down

2.2.4 Threats

The threats for the PGM sector include;

- Takeover of operations by the government;
- Declining safety performance;
- Rising salary levels and cost of living;
- Increase in corporate tax;
- Low growth rate;
- Ageing experienced workforce;
- Fluctuations in exchange rate; and

- Rising competition from other countries.

3 STRATEGY FORMULATION

The conversation model has been used as it provides development of scenarios and tools for effective decision making. The model is split into two sections which are defining the game, and playing the game.

3.1 Defining the Game

The game is defined by describing the context, scope, players, uncertainties, possible scenarios and rules of the game.

3.1.1 Defining the context

According to the Fraser Institute (2016), the mining industry faces challenges in investment capacities and Earnest and Young (2016) among the risks identified, productivity improvement along with access to capital topping the list in PGM mining sector. The recent economic environment has forced out a number of exploration activities within the PGM mining sector (Earnest and Young, 2016). Investors are still not convinced with the projection of commodity price increase as prices are tending to be very unpredictable with probably no hope for a prolonged upswing.

Despite the country's declining investor attractiveness, as presented by the World Economic Forum (2010), the PGM mining sector has managed to increase its expansion projects as initiated by Mimosa and Zimplats. The existing mining and mineral policies play a major role in reducing investor confidence. There is therefore need to improve the mining and mineral policies.

3.1.2 Scope

The PGM mining sector considered is local. The primary products are PGMs. This sector is still yet to be fully exploited should the policies improve to attract investors, making the sector an economic hub sought after. The government has put in place ways in which the industry can expand. These ways include a review of the mining rights, promoting joint ventures, investment in global technologies and improving the country's infrastructure. According to Hawkins (2009), there still exists a challenge on improving "capacity utilisation and production volumes". Should these be improved, investor confidence will be increased.

3.1.3 Players in the game

The players in the game are;

- The mining companies;
- The International Monetary Fund;
- Chinese market;
- Western markets and the European Union (EU);
- Other PGM sectors in South Africa;
- Local community;
- The local government; and

3.1.4 Rules of the game

The rules of the game are;

- The PGM resources are non-renewable;
- PGMs will always be in demand as long no substitute exist;
- The PGM sector is a price taker;
- Labour law and practice to meet international standards;
- Royalties and taxation polices (product is exported raw, with the sector incurring large taxes);
- The PGM sector subscribes to the Chamber of Mines Zimbabwe; and
- Health and safety law compliance is priority.

3.1.5 Key uncertainties

The key uncertainties include;

- Power supply;
- Exploration prospects;
- Commodity prices;
- Productivity improvement initiatives;
- Access to capital and policy certainty;
- Viability of growth potential;
- Availability of skilled labour;
- Investor interest and political stability; and

- Security of property rights ownership.

Key uncertainties that have highest impact and highest uncertainty are productivity and commodity price. These uncertainties are used to develop scenarios using the scenario game board shown in Figure 1. From Figure 1, the regions represented are described below.

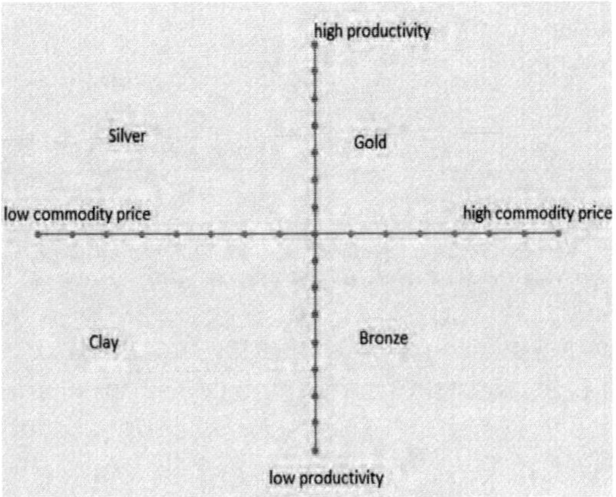

Figure 1 Scenario game board

3.1.5.1 Gold

The PGM sector has not yet realised this scenario. The down swing in economic growth and declining commodity prices probe for a positive upswing of both the growth rate and the commodity price. This upswing will be driven by another "scramble for Africa" as developed markets seek to secure raw material, increasing the demand. Unknown to this scenario is the period to which it is going to be high growth and high commodity prices as forecasting is tending to be very unpredictable. The PGM mining sector needs to re-evaluate its existing strategies in order to benefit from this scenario.

3.1.5.2 Silver

This scenario describes the current situation, where commodity prices are low. However, the PGM sector realises that there seems to be no upswing in prices, and as such, the sector put in measures to improve its efficiencies within the business value chain.

3.1.5.3 Bronze

The declining commodity prices start to show improvement. The threat of new substitutes eventually decay as research shows that there is no feasible substitute. The "scramble for Africa" aids to the increase in commodity prices. The PGM sector however, does not realise the need to improve productivity as the high commodity prices cover the costs.

3.1.5.4 Clay

This scenario describes the current situation, where commodity prices are low and the productivity is low. The increasing demand for substitutes eventually materialise and the demand for PGMs declines resulting in low commodity prices. The persistent downswing of commodity prices forces the sector to stop capital investments and productivity continues to be a major risk.

3.1.6 Indicators

The PGM mining sector in Zimbabwe is in the Clay scenario where the productivity and commodity prices are low. In order to move from clay to gold, the sector needs to develop a competitive strategy. This development also results in indicators for improved business environment. These indicators are used by investors as a measure to see what the sector needs to do should the sector want to be attractive. The indicators for improved productivity are;

- High production volumes with decreased costs; and
- High efficiencies on resource (human, orebody, equipment) utilization.
- The indicators for commodity prices are;
- Country's economic and global economic upturn where there is increased manufacturing hence high demand for commodities; and
- Decreasing global interest rates. This decline, according to Frankel (2014) is related to an increase in commodity demand.

3.2 Playing the Game

Playing the game consists SWOT analysis, developing options, making decisions and developing measurable outcomes (School of Mining Engineering, 2016).

3.2.1 SWOT analysis

SWOT analysis is developed from internal analysis of the PGM mining sector.

3.2.2 Options

The PGM mining sector seeks to achieve the following objectives;

- To have the highest productivity;
- To attract highly skilled personnel;
- To offer employment opportunities; and
- To aid in improving the country's growth rate

The options available to deal with the productivity improvement risk are;

1. The PGM sector chooses to do nothing;
2. The sector put measures to improve productivity;
3. The current players decide to exit the mining business;
4. The sector collaborates with Original Equipment Manufacturers to develop new technology; and
5. Sector to invest in exploration and development activities.

4 STRATEGY SELECTION

The Space Matrix (Anon, n.d) is used to analyse the internal and external characteristics of the PGM mining sector in order to define the nature of the strategy to be selected. The internal dimensions that the matrix uses are financial strength and competitive advantage and the external dimensions are industry strength and environmental stability (Anon, n.d).

4.1 Internal Dimension

4.1.1 Financial strength

The factors considered under financial strength and their respective scores are in Table 1. The scores range from zero (0) to six (+6) with +6 being the best and zero (0) being the worst.

Table 1 Financial strength factors and scores

Financial Strength	
Factor	**Score**
Return on Investment	5
Liquidity	4
Capital	3
Risk Level	1
Arithmetic Mean	3.3

4.1.2 Competitive advantage

The factors considered under competitive advantage and their respective scores are shown in Table 2. The scores range from negative six (-6) to zero (0) with zero (0) being the best and negative six (-6) being the worst.

Table 2 Competitive advantage factors and scores

Competitive Advantage	
Factor	**Score**
Market share	-4
Product quality	-4
Product Life Cycle	-3
Arithmetic Mean	-3.7

4.2 External Dimension

4.2.1 Industry strength

The factors considered under industry strength and their respective scores are shown in Table 3. The scores range from zero (0) to six (+6).

Table 3 Industry strength factors and scores

Industry Strength	
Factor	**Score**

Growth Potential	4
Profit Potential	3
Financial Stability	1
Technological knowhow	1
Arithmetic Mean	2.25

4.2.2 Environmental stability

The factors considered under environmental stability and their respective scores are shown in Table 4. The scores range from negative six (-6) to zero (0).

Table 4 Environmental stability factors and scores

Environmental Stability	
Factor	**Score**
Technological Change	-3
Rate of Inflation	-1
Demand Variability	-1
Arithmetic Mean	-1.7

The arithmetic mean for competitive advantage (CA) and industry strength (IS) are added and the mean is calculated. The result is plotted along the x- axis. Resultant mean = -0.725. The arithmetic mean for financial strength (FS) and environmental stability (ES) are added and the mean is calculated. The result is plotted along the y-axis. Resultant mean =1.6. The two arithmetic mean are plotted on the Space matrix in Figure 2 to identify the appropriate nature of the strategy.

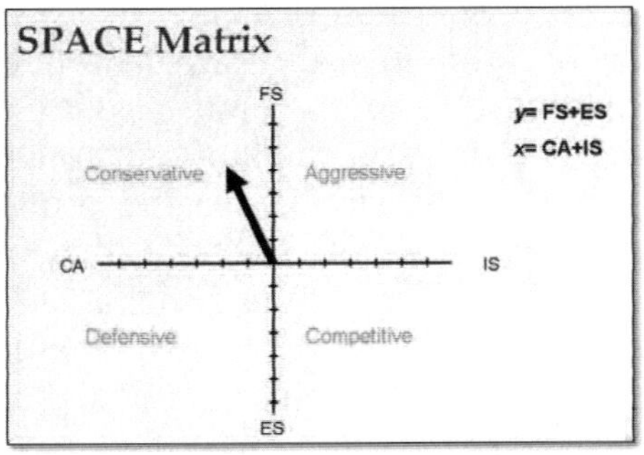

Figure 2 Space matrix (Anon, n.d)

A strategy of conservative nature is required. The nature of this strategy entails the following;

• The strategy seeks to penetrate the market through growth. This means that the volumes produced by PGM sector are to be increased for the sector to be competitive in the market.

• The role and uses of the products of the PGM mining sector suffers a big threat to substitution. As such, the PGM sector under this strategy requires investment into research and development to find other uses of the products.

• The strategy requires that, for the PGM mining sector to gain considerable benefit from its resources, the level of beneficiation be increased, so that the product sold would have increased value.

4.3 Developed Strategies

Two strategies that have a conservative approach are;

1. Slow and steady productivity improvement, which involves;
   - Using optimization techniques to identify bottlenecks and exploiting them;
   - Improving plant recoveries and efficiencies on equipment and human resources; and
   - Using the returns on exploration and development.
2. Growth, expansion and diversification with joint ventures with the government, which involves;
   - Being actively involved in the making of policy that increases attractiveness to investors;
   - Seeking capital to invest in expansion projects;
   - Investing in research and development for more uses of platinum thus maintaining market presence;
   - Being involved in skills regeneration and being attractive enough to reduce skill exodus;
   - Reduction of human operated equipment, by considering automation; and
   - Diversifying into chrome and gold mining as they are in the same proximity with PGM deposits.

4.4 Selecting the Appropriate Strategy

The election of the best strategy between the two strategies uses the Quantitative Strategic Planning Matrix (QSPM) (Forest et al, 2009). The SWOT analysis factors are given weights and the attractiveness of each factor is determined. The determination of weighting and attractiveness for external factors and internal factors is slightly different. Internal factors are evaluated on three categories, which are the weighting (importance), rating and the score (Ovidijus, 2016). The weighting shows the extent to which a factor is important to the PGM mining sector as some factors are of more value than the others, if the industry is to be successful. A value of 0.01 for weighting means the factor is not important but a factor of 1 means that factor is very important (Forest at al, 2009). The rating has a score that varies from one (1) to four (4) and the rating indicates if the factor is a major, given a value of four (4) or the factor is a minor (given a value of one (1) (Ovidijus, 2016).

The score is the product of the weighting and the rating. External factors are also determined from three categories, which are weighting (importance), probability and score (Forest at al, 2009). The weighting factor shows to what extent will the external factor impact the business with 0.01 having no impact and one (1) having very high impact. The probability of that factor happening has values ranging from one (1) with low probability to four (4) with high probability (Ovidijus, 2016). The score is obtained by multiplying the probability and the weighting. The probability in external factors or the rating in internal factors is also called the attractiveness score (AS) and the score is called total attractiveness score (TAS). The total score for all factors is calculated and the strategy with the highest score is chosen. Table 5 shows all the SWOT factors and the resultant calculations, which is strategy two.

5 STRATEGY IMPLEMENTATION

The development and selection of appropriate strategy is followed by a guided and disciplined approach to implementation. Successful implementation of selected strategy requires action plans that are developed from an integrated team that involves all levels of work and from all stakeholders. Such action plans may include;

- Improving the productivity by fully "sweating" existing assets;
- Maintaining a positive balance sheet
- Using raised capital on expansion projects and development;
- Establishing or revamping existing skill generation centres with joint ventures with original equipment manufacturers; and
- Investing in research and development giving the PGM mining sector a competitive advantage.

6 CONCLUSIONS

The PGM mining sector in Zimbabwe faces a number of business risks which, if not considered and acted upon, will result in failed sectors like the local gold and asbestos sectors. Productivity improvement is the highest business risk with the significant impact to the success of the sector during commodity price unpredictability. Four scenarios have been developed that enabled the development of two strategies. These four scenarios are gold, silver, bronze and clay. The strategies developed from these scenarios are slow and steady productivity improvement and growth, expansion and diversification with joint ventures with the government. Using the quantitative strategic planning matrix, the strategy is the growth, expansion and diversification with joint ventures with the government. The implementation of this strategy requires integration of all stakeholders and a closer look at the business value chain to improve the efficiencies and exploit fully any existing constraints.

Table 5 SWOT factors and calculations

Quantitative Strategic Planning Matrix			Strategy 1		Strategy 2	
		Weight	AS	TAS	AS	TAS
Key Internal Factors						
Strength	The deposits for PGMs are relatively shallow giving easy access;	0,10	3	0,30	3	0,30
	The workforce is highly literate, with the lowest level employee being able to communicate in English;	0,20	4	0,80	4	0,80
	The Mines within the PGM mining sector are highly mechanised;	0,15	3	0,45	3	0,45
	There is a strong innovative culture especially with the new generation of the human resource; and	0,10	2	0,20	4	0,40
	Most of the mines in this sector have a positive balance sheet, which enables them to consider expansion plans.	0,05	3	0,15	4	0,20
Weakness	There is still poor productivity despite the high mechanisation levels;	0,05	2	0,10	3	0,15
	Locally, the access to capital is still challenging;	0,10	3	0,30	4	0,40
	The experienced human resource is ageing;	0,10	3	0,30	4	0,40
	The sector has poor presence in the world's market despite having the second largest deposits;	0,05	2	0,10	3	0,15
	There is high employee turnover.	0,10	3	0,30	4	0,40
	Total	1,00				
Key External Factors						
Opportunities	Market and economic growth globally;	0,15	4	0,60	3	0,45
	Improved productivity with improved asset utilisation;	0,15	4	0,60	4	0,60
	Political and economic stability of the country;	0,10	3	0,30	3	0,30
	New technology that can drive costs down;	0,10	2	0,20	2	0,20
	Government incentive for capital projects and leniency on importing capital goods; and	0,05	2	0,10	4	0,20
Threats	Takeover of operations by the government;	0,10	3	0,30	4	0,40
	Security of property rights and increased corporate tax;	0,10	3	0,30	3	0,30
	Rising salary levels and cost of living;	0,05	3	0,15	4	0,20
	Low growth rate and fluctuations in exchange rate;	0,10	3	0,30	4	0,40
	Increased research into substitution of PGMs especially in the automotive industry.	0,10	3	0,30	3	0,30
	Total	1,00				
Total Sum of Total Attractiveness Score				6,15		7,00

REFERENCES

Anon, n.d Análisis estratégico interno-externo (Online). Available at: https://www.goconqr.com/en/p/4113757-matriz-space-slide_sets Accessed on 22 August 2016.

Earnest and Young (2016) Business risks facing the mining and metals 2015 – 21016 (Online). Available at: http://www.ey.com/Publication/vwLUAssets/EY-business-risks-in-mining-and-metals-2015-2016/$FILE/EY-business-risks-in-mining-and-metals-2015-2016.pdf Accessed on 22 August 2016.

Forest, R.D, Fred, R.D and Meredith, E.D (2009) The Quantitative Strategic Planning Matrix (QSPM) Applied To A Retail Computer Store The Coastal Business Journal, Volume 8, Number 1, pg. 42-52.

Frankel, J (2014) Monetary influences on commodity prices (Online). Available at: https://www.hks.harvard.edu/fs/jfrankel/CP.htm Accessed on 22 August 2016.

Fraser Institute (2016) Fraser Institute Annual Survey of Mining Companies 2015 (Online). Available at: https://www.fraserinstitute.org/sites/default/files/survey-of-mining-companies-2015.pdf Accessed on 22 August 2016.

Hawkins, T (2009) The Mining Sector in Zimbabwe and its potential contribution to recovery United Nations Development Programme Comprehensive Economic Recovery in Zimbabwe Working Paper series (Online). Available at: http://cryptome.org/kimberly/kimberly-003.pdf Accessed on 22 August 2016.

KPMG Zimbabwe (2015) Zimbabwe Country Mining Guide (Online). Available at: https://www.kpmg.com/ZW/en/IssuesAndInsights/articlespublications/Documents/Zimbabwe%20Country%20Mining%20Guide.pdf Accessed on 22 August 2016.

NEDGROUP Securities (2009) Special Expose: Mining in Zimbabwe (Online). Available at: https://www.proactiveinvestors.com/upload/SponsorFile/File/2/mining_in_zimbabwe051009_2.pdf Accessed on 22 August 2016.

Ovidijus, J (2016) SWOT Analysis (Online). Available at: https://www.strategicmanagementinsight.com/tools/swot-analysis-how-to-do-it.html Accessed on 22 August 2016.

School of Mining Engineering, 2016. Strategic Planning in Mining Course lecture notes, University of the Witwatersrand, Johannesburg, South Africa.

United States of America Department of State (2015) Zimbabwe Investment Climate Statement (Online). Available at: http://www.state.gov/documents/organization/242012.pdf Accessed on 22 August 2016

World Economic Forum (2010) Mining and Metals Scenarios to 2030 (Online) Available at: http://www3.weforum.org/docs/WEF_Scenario_MM2030_2010.pdf Accessed on 22 August 2016.

The Effect of Plaster Stemming For Large Hole Diameter Stripping Blasting: A Case Study

Alçı Sıkılamanın Geniş Çaplı Bir Dekapaj Patlatmasında Etkisi

H. Cevizci
Sıleyman Demirel University, Mining Engineering Department, Isparta

ÖZET Bu çalışmada, bir geniş çaplı dekapaj patlatmasında alçı sıkılama uygulamasının etkisi incelenmiştir. Daha önce, 3,5, 4, ve 5 inç çaplı deliklerde, kil ve kireçtaşı ocaklarında alçı sıkılama başarı ile uygulanmıştır. Fakat 9"çaplı deliklerde ilk uygulamadır.
Alçı sıkılama daha iyi yığın ötelemesi ve birim hacim için daha düşük maliyet sağlamıştır. Patatma kaynaklı titreşim ve hava şoku değerleri, daha iyi enerji kullanımından dolayı, biraz artmıştır. Ancak ölçülen değerler izin verilen limitlerin altındadır. Maksimum parçacık hızı 6,83 mm/s'den 13,6 mm/s'ye çıkmıştır.

Anahtar Kelimeler: Alçı sıkılama, dekapaj patlatması, sıkılama, parçalanma, titreşim.

ABSTRACT In this study, the plaster stemming application for blasting at a large hole diameter lignite stripping blasting is studied. Recently, plaster stemming application was successfully carried out at some limestone and clay quarries which hole diameter was 3.5 inches, 4 inches and 5 inches. But, it is first time to applying large hole diameter (9 inches).
Plaster stemming provides better displacement and more economic per unit volume. Blast induced vibration and air shock values increased slightly due to more blast energy being available for rock breakage. However, measured values were stayed under the permitted limit for blast damage criteria. Using plaster stemming peak particle velocity was increased from 6.83 mm/s to 13.6 mm/s.

Key Words: Plaster stemming, stripping blasting, stemming, fragmentation, vibration.

1 INTRODUCTION

Open pit mining is increasing day to day all part of world. Of course, capacity of excavation and stripping blasting is increasing too.

The stemming of blasthole collars in surface mines with an inert material redirects blasting energy to the rock more efficiently, thus the energy is utilized more effectively in breaking rock (Cevizci, 2013). In this procedure, high efficiency of blockage is important since the blast gases should not be allowed to escape due to loose stemming material. More efficient stemming with better confinement therefore increases the generation of fines. In addition, better rock breakage can be obtained. On the other hand, scatter distance is increased, giving rise to a looser muck pile that can be more easily loaded and transported (Ozkahraman, 2006).

In this study, a new stemming material was investigated with the aim of increasing the blast energy directed to the rock. For this purpose, quick-setting moulding plaster was used as a stemming material. Apart from the work of Cevizci (2012), there is no previous work found in the literature citing the usage of this material. Blasting tests were carried out in quarries using both the suggested new stemming method and classical stemming material, and performance measurements carried out by image analysis of fragmented rock piles.

Drill cuttings are the most common stemming material used in open pits and quarries, since they are most readily available at blast sites and are cheap. However, dry drill cuttings eject very easily from blastholes without offering much resistance to the explosion. Thus, a great percentage of blast energy is wasted and lost to the atmosphere. Cevizci (2012) carried out blasting tests by changing the blasting parameters of open pit blasts and obtained better results with the plaster stemming method in two different limestone quarries and one clay quarry. These blasting trials were carried out on same benches and under same rock conditions. The new method employs a plaster prepared as a thick paste, which hardens in less than 25–30 minutes after application. The hardened plaster creates a very strong plug, therefore the stemming column length can be reduced and the explosive column length increased. This increased explosive column results in better rock breakage than similar holes stemmed with dry drill cuttings. Also, this increased utilization of hole length reduces specific drilling costs due to increased burden and spacing distances. Blasthole drilling constitutes a major cost in blasting operations. Another advantage of the new method is better fragmentation, with more induced cracking in the rock mass.

In one series of blast tests, blasting costs per unit volume of rock were reduced to 16 % by increasing burden and spacing distances (Cevizci, 2012). In addition, better fragmentation was obtained by using the plaster stemming method. Blast trials showed that plaster stemming produced finer material. In the same blast tests, +30 cm size fragments reduced to 5.4 % of the total, compared to 37.7 % in the conventional method of drill cuttings stemming. With this method of stemming, vibration and air shock values increased slightly due to more blast energy being available for rock breakage, but these increased values were small and under the permitted limit for blast damage criteria. Plaster stemming method provides better confinement and so higher pressure. As this pressure is used for better breakage of rock, vibration values slightly increases.

At Bastas limestone quarry, a 25 cm of drill cuttings were placed between the explosive and the plaster paste (Cevizci, 2012). Length of plaster column was 45 cm. The top 55 cm of the drill hole was filled with drill cuttings. With plaster stemming round, total length of stemming was 1.25 m and no fly rock was generated at the top level bench, similar to the drill cuttings stemming round test trial. In addition, the scattering of the muck pile with the plaster stemming round was similar to scattering of the drill cuttings stemming round.

2 METHOD

The study was carried out at Tuncbilek open pit of Garp Linyitleri Isletmesi Coorperation. The quarry is located near of Tuncbilek town of Kutahya Region. A summary of the properties of material at the blast site, blast patterns, measurements of blast tests, and features of the bench faces is shown in Table 1.

Table 1. Summary of properties of material at Tuncbilek open pit lignite stripping blasting quarry blast site, blast patterns and measurements of blast tests

Blast Tests	Dip direction/ angle of dip /angle of blast direction relative to dip direction of discontinuity	Stem. length (m)	Average burden (m)	Average spacing (m)	Bench height (m)	+ 50 cm size fraction (%)	Surface /column charge/ bottom delays (ms)	Specific charge (kg/m^3)	Specific drillling (m/m^3)
Drill cuttings stemming round	185/25/90	4	7.23	8.04	14	15.6	25/450/ 475	0.373	0.022
Plaster stemming round	185/25/90	6	8.13	10.25	14	25.9	25/450/ 475	0.332	0.015

Both fast-setting moulding plaster and drill cuttings were used as stemming material at different lengths in similar blastholes on the same quarry bench. A thick milky moulding plaster was prepared by mixing ten units of plaster powder and seven units of water in a barrel, and charged into the blastholes as shown in Figure 1. This wet paste hardens in 25–30 minutes. The design of the stemming, using the tests at Tuncbilek lignite stripping blasting is shown in Figure 2.

Figure 1. Application of plaster paste to hole collar at Tuncbilek open pit

Wet plaster should not be placed in contact with ANFO, which is water-sensitive, thus 100 cm of drill cuttings were placed between the explosive and the plaster paste. The length of plaster column was 40 cm. The top 260 cm of the drill hole was filled with drill cuttings instead of leaving empty. Weight of drill cuttings can contribute the stemming system in terms of better plug. This had the advantage of protecting the hole from loose stones dropping in.

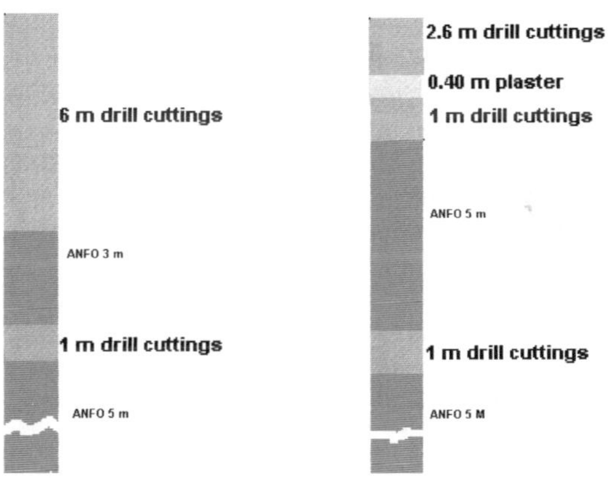

(a)Drill cutt. St. (b) Stemming with plaster
Figure 2. Blast hole stemming at Tuncbilek open pit lignite stripping blasting

3 BLAST TRIALS AND RESULTS AT TUNCBILEK LIGNITE STRIPPING BLASTING

Two blast rounds sixty meters apart were carried out with the plaster stemming and with the classic conventional drill cuttings stemming, each consisting of one row of six holes on the same quarry bench. Drill cuttings stemming carried out first round. Second round was carried out by using plaster stemming. Both two rounds all 229 mm (9 inches) diameter holes were drilled. Compared to blasting with drill cuttings stemming, burden and spacing distances were approximately 20 % larger at plaster stemming method.

Drill cutting stemming method is generally used globally in open pits and quarries. Therefore, the stemming blast holes with drill cutting procedure was a standard procedure and details of this procedure is not given in the paper instead the new and more efficient plaster stemming method is more emphasized. For comparison of two stemming procedure, test blasts were carried out in the same location. Therefore, the rock structure and strength were similar. In both rounds ANFO, with one primer containing of 500 g dynamite was used. A quick hardening moulding plaster was used for plaster stemming.

3.1. Charging holes in the drill cuttings stemming test at Tuncbilek lignite stripping blasting

Each blast hole was charged with 250 kg ANFO and detonation initiated with three primers (Dinex 100G dynamite) 1 kg in weight. Two primers 1 kg in weight were used as primer at the bottom charge of holes. Totally, 150 kg of ANFO was used at the bottom charge of holes. One primer 1 kg in weight was used as primer for the column charge of holes. Totally, 100 kg of ANFO was used at the column charge of holes. The firing was started with one electric cap. At the surface 25 ms, 450 ms for column charge and at hole bottom 475 ms Nonel millisecond caps were used.

3.2. Charging holes in the plaster stemming test at Tuncbilek lignite stripping blasting

Each blast hole was charged with 300 kg ANFO and detonation initiated with three primers 1 kg in weight. Two primers 1 kg in weight were used as primer at the bottom charge of holes. Totally, 150 kg of ANFO was used at the bottom charge of holes. One primer 1 kg in weight was used as primer at the column charge of holes. Totally, 150 kg of ANFO was used at the column charge of holes. The firing was started with one electric cap. At the surface 25 ms, 450 ms for column charge and 475 ms for hole bottom Nonel millisecond caps were used.

Drill cutting stemming method is generally used globally in open pits and quarries. Therefore, the stemming blast holes with drill cutting procedure was a standard procedure and details of this procedure is not given in the paper instead the new and more efficient plaster stemming method is more emphasized. For comparison of two steming procedure, test blasts were carried out in the same quarry bench. Therefore, the rock structure and strength were similar. A quick hardening moulding plaster was used for plaster stemming.

4 RESULTS

4.1. The evaluation of blast trials

The blasting results of two stemming method rounds were compared. Rock pile fragmentation was evaluated using Split Desktop image-analysis software and verified standard ''compare photo'' method (Ozkahraman, 2006). The rock piles from the blasting tests at Tuncbilek lignite stripping blasting are shown in Figure 3 and Figure 4. The cumulative percentage of retained size at the Tuncbilek lignite stripping blasting tests is given in Table 2.

Figure 3. Rock pile of blast round with drill cuttings stemming at Tuncbilek open pit lignite

The first round was carried out using drill cuttings stemming with six holes. The length of the stemming was 6 m. The second round was carried out by plaster stemming with six holes. The length of the stemming was 4 m. Each blasthole was filled with 150 kg ANFO initiated with two primers with 1 kg in weight for bottom charge in the case of the drill cuttings stemming method. Each blast hole was filled with 100 kg ANFO initiated with one primer with 1 kg in weight for column charge in the case of the drill cuttings stemming method. Each blast hole was filled with 150 kg ANFO initiated with two primers with 1 kg in weight for bottom charge in the case of the plaster stemming method. For the plaster stemming method, the quantity of ANFO was 150 kg per blast hole and initiated with one primer with 1 kg in weight for column charge. The total length of ANFO column in the plaster stemming method was 2 m greater than for the conventional method of drill cuttings stemming.

Figure 4. Rock pile of blast round with plaster stemming at Tuncbilek open pit lignite stripping blasting

Table 2. Comparison of cumulative percentage of retained size (oversize) from blast trials with plaster stemming and drill cuttings stemming

Fragment size (cm)	Drill cuttings stemming (%)	Plaster stemming (%)
100	2.4	6.7
70	7.4	15.1
50	15.6	25.9
40	22.6	33.9
30	32.8	44.4
20	47.5	58.2
15	57.3	66.7
10	69.0	76.3
5	83.0	87.4

The blasted area was 291 m^2 for the drill cuttings stemming trial, and blasted volume was 4071 m^3 in situ. The specific charge was found to be 0.373 kg/m^3 and the specific drilling was 0.022 m/m^3. The blasted area was 417 m^2 for the plaster stemming blast trial and yielding the blasted volume was 5836 m^3 in situ. The specific charge was 0.332 kg/m^3 and the specific drilling was 0.015 m/m^3.

The total length of holes for the plaster stemming trial at Tuncbilek lignite stripping blasting quarry was 27.22 m less than for the drill cuttings stemming round. Because at plaster stemming method 5836 m^3 in situ and at drill cuttings stemming method 4098 m^3 in situ rock was obtained with same drill hole (90 m). This resulted in 30 % less drilling per unit volume rock. The cost saving for only drilling calculated was $ 326.64 (27.22 m X12 $/m). At this site, specific drilling and specific charge decreased because a larger burden and spacing were applied with the plaster stemming method. In order to fragment the same volume of rock as for the plaster stemming round, an additional hole length of 27.22 m should be drilled for the drill cuttings stemming round. Consequently, the profit per unit volume (m^3) was $ 0.158 and total profit by using plaster stemming was $ 922 (Table 3). Therefore, the plaster stemming trial was found to be economic as well as giving near fragmentation. For instance, the +50 cm size fraction increased from 15.6 % to 25.9%. But, this did not create a big problem for loading and hauling. Because, excavation capacity of machines of the open pit is high level. Therefore, near fragmentation is obtained with $ 922 profit.

Table 3. Comparison of blasting cost of plaster stemming versus drill cuttings stemming

Cost item	Drill cuttings stemming ($)	Plaster stemming($)
ANFO	1761	2113
Nonel caps	94.5	94.5
Initiating el. cap	0.5	0.5
Dynamite	51	51
Plaster and labor	-	20
Drilling cost	1080	1080
Total Cost	2987	3359
Fragmented rock(m^3)	4071	5836
Unit cost ($/m^3)	0.734	0.576
Specificc.(kg/m^3)	0.373	0.332
Specificdrill.(m/m^3)	0.022	0.015

4.2. Environmental effects of blast trials

Vibration and air noise levels for both stemming methods were measured with an Instantel Minimate Blaster placed 262 m away from the blastholes.

Peak particle velocity (PPV) known to be a function of site conditions (i.e. geological conditions) and scaled distance is calculated from following equation (Devine et.al., 1966),

$$SD = D/W^{(1/2)} \quad [1]$$

Where D is the distance from the blast face to the vibration monitoring point and W is weight of explosive per delay for surface blasting.

In the plaster stemming method, 152 kg explosive per delay was used compared to 152 kg in the drill cuttings stemming. This increase in explosive charge caused an increase in PPV value from 6.83 mm/s to 13.6 mm/s (Figure 5 and Figure 6), whereas it should be same.

The vibration and air shock values measured for both the drill cuttings stemming and the plaster stemming at the Tuncbilek lignite stripping blasting quarry test trials were under the safety limits specified in the limit criterion. At the top level bench, small quantities of fly rock were generated, but this did not constitute a major problem. In addition, the plaster stemming round resulted in a slightly more scattered muck pile owing to more blast energy directed to the rock, but this did not create a big problem either. Using plaster stemming method, displacement of rock pile is better than drill cuttings stemming round. Loading of the rock pile was easier due to the looser particles.

5 CONCLUSION

Blasting experiments carried out in Tuncbilek lignite stripping blasting quarry proved that the plaster stemming results in better confinement in blast holes. Additionally, the new method offers a more profitable solution. The cost of drilling for one meter of hole length is approximately $12. On the other hand, lower cost was obtained for unit volume rock (0.158 $ per one m^3 rock) in spite of near fragmentation. Totally, 922 $ profit was obtained for one round using plaster stemming.

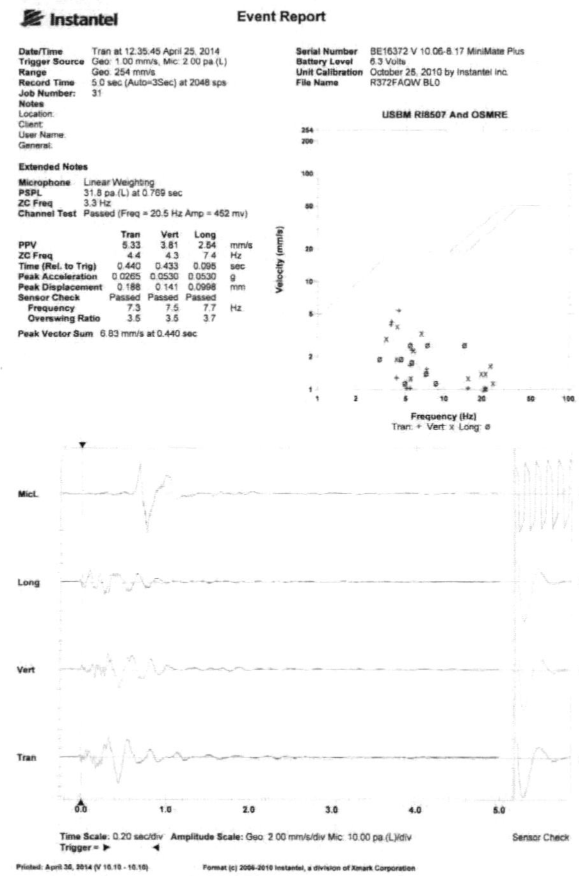

Figure 5. Measured vibration and noise levels in the drill cuttings stemming test at Tuncbilek lignite

This increase in explosive charge caused an increase in PPV value from 6.83 mm/s to 13.6 mm/s, whereas it should be same. This means that we can more efficiently use blast energy for breaking rock. Plaster stemming method provides better confinement and so higher pressure. As this pressure is used for better breakage of rock, vibration values slightly increases.

With plaster stemming, more of the hole length is better utilized by increasing the loaded length of hole, resulting in better breakage at the hole collar. The increased length of loading in the hard cap rock region improves the cap rock breakage, thus reducing the creation of oversized boulders and increasing both efficiency and profit (Cevizci and Ozkahraman, 2012). It was observed that a plaster stemming column 0.4 m and total 4 m in length provided a more robust sealing than 6 m of drill cuttings used in the classical method.

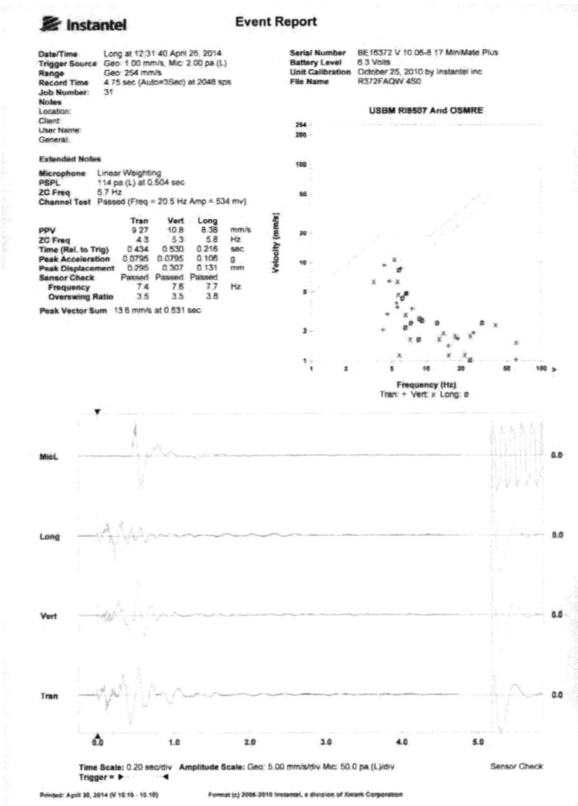

Figure 6. Measured vibration and noise levels in the plaster stemming test at Tuncbilek lignite

The energy utilization rate increased with the plaster stemming method by creating effective confinement with better clogging. In addition, increased vibration levels, fly rock and air shock levels are all indicators of the increase in energy usage. Nevertheless, these increased values were small and under the permitted limit for blast damage criteria.

Recently, this stemming method was found acceptence by some quarries. It was reported from these quarries that new method meets their needs for aggregate production with minimum cost.

ACKNOWLEDGEMENT

Author thanks to Garp Linyitleri Isletmesi Coorperation, Mehmet Mert and Busra Demir for their support, encouragement and help.

REFERENCES

Cevizci, H. A newly developed plaster stemming method for blasting. The J. of The South African Inst. of Min. and Metall., December. p. 1071- 1078, 2012.

Cevizci,H. A new stemming application for blasting: a case study, Rem: Rev. Esc. Minas vol.66 no.4 pp.513-519 Ouro Preto Oct./Dec. 2013. http://dx.doi.org/10.1590/S0370-44672013000400017.

Cevizci, H., Özkahraman, H.T. The effect of blast hole stemming length to rockpile fragmentation at limestone quarries. International Journal of Rock Mechanics and Mining Sciences, v. 53, p. 32-35, 2012.

Devine JF, Beck RH, Meyer AVC, Duvall WI, Effect of charge weight on vibration levels from quarry blasting Bureau of Mines. 1966. Washington.

Özkahraman, H.T. Fragmentation assessment and design of blast pattern at Goltas limestone quarry, Turkey. International Journal of Rock Mechanics& Mining Sciences 43, p.628-633. 2006.

Mermer ve Pirit Atıklarının Toprak Rehabilitasyonunda Kullanılması

Use of Marble and Pyritic Wastes in Soil Remediaton

A. İ. Arol
Middle East Technical University, Mining Engineering Department, 06800 Ankara, Tırkey
G. Tozsin
Atatürk University, Mining Engineering Department 25400 Oltu, Erzurum, Turkey

ÖZET Madencilik genellikle büyük ölçekli bir işlemdir ve dünyanın her köşesinde büyük miktarda atık üretilmektedir. Bu şekilde üretilen atıklar sadece malzeme ve para kaybı değil aynı zamanda düzgün bir şekilde ele alınmazsa ciddi çevre sorunlarına neden olur. Son yıllarda, madencilik atıklarının azaltılması ve/veya alternatif hammaddeler olarak onlardan fayda sağlanması için çalışmalar yürütülmektedir. Farklı uygulamalarda bakır ve mermer atıkları olmak üzere iki farklı maden atığını kullanmak için çalışmalar yapılmıştır. Bakır konsantrasyon tesisindeki pirit atıklarının, alkali ve kireçli toprakları ıslah etmek için kullanılabileceği ve mermer atıkların, asitli toprakların pH'sını artırmak için tarım kireci yerine geçebileceği araştırılmıştır. Bu çalışmada, yukarıda bahsedilen araştırmaların sonuçları özetlenmiştir.

ABSTRACT Mining is usually a big scale operation and large amounts of wastes are produced in every corner of the world. The wastes so produced are not only a loss in material and money but also cause serious environmental problems if not properly handled. In recent years, there have been efforts to reduce the mining wastes and/or benefit from them as alternative raw materials. Studies were carried out to utilize two different mining wastes, namely copper and marble wastes, in different application areas. It was found that pyrite wastes from copper concentration plant can be utilized to remediate the alkaline and calcareous soils, and marble wastes can replace agricultural lime to increase the pH of acidic soils. In this paper, results of these studies have been summarized.

1 INTRODUCTION

Mining has always been in the center of industrialization on which the modern economies have grown. No mineral commodities can be produced or used without any wastes. The wastes generated by the mining industry at large can be classified in different ways. Based on the state of the matter, the wastes could be gaseous, liquid and/or solid. They could be hazardous or non-hazardous based on their hazard potential. The nature of wastes could be acidic, basic or neutral. Wastes could come from underground or surface mines, mineral processing plants or other plants where minerals are used. They could be fragmented by blasting, crushing or grinding. Therefore, their sizes could range from a few meters down to a few micrometers, or even nanometers. Classification could be done on the basis of different physical properties, such as hardness, shape, surface area, water absorption, cation exchange capacity, color, etc. Old abandoned mines could also be classified as a waste in terms of land.

Quantity of wastes can vary for different mining operations (Blight, 2011). For example, more wastes than mined ore are generated in precious and base metal mining; 80-90% of natural stones quarried, 40-60% of coal processed, excluding the overburden, may be discarded as waste. Iron ore, industrial minerals mining and processing may also generate substantial amount of wastes.

Historically, wastes generated in mineral industries have been mostly used as fill material or, stored in dams or waste dumps. Some wastes can be reused as a source of valuable minerals when the economy is favorable. In recent years, sustainability in management of mining waste has gained importance. Sustainability is sometimes expressed as "zero waste" or 3R's, namely reduce, re-use and recycle, mining waste management hierarchy (Lottermoser, 2011). A slightly different hierarchy, reduce, reprocess, down cycle and dispose, has been proposed (Lebre and Corder, 2015; Lebre et al., 2016). In whichever way this structure is defined, re-use or down cycle of the wastes generated will be a part of the mining practice for sustainability.

2 UTILIZATION OF PYRITIC TAILINGS IN ALKALI AND CALCAREOUS SOILS

Alkaline soils, characterized by high pH (>8.5) and exchangeable sodium percentage (ESP>15%), occupy large parts of arid and semi-arid areas of the world (Richards, 1954). Alkaline soil contains sufficient exchangeable sodium to affect the plant growth adversely (Suarez et al., 1984; Gupta and Abrol, 1990). Therefore, the reclamation of alkaline soils has great importance for environmental and agronomical purposes. Gypsum ($CaSO4 \cdot 2H2O$) has been known to be an amelioration agent for alkaline soil for more than 100 years; however, it has been used only rarely because of the high cost involved in the exploitation, transportation and crushing (Wang et al., 2008).

In the reclamation of alkaline soils, pyrite waste products from mine tailings might be an alternative treatment agent. Tailings from sulfide mines usually contain large amount of pyrite (FeS2), which is the main source of acid rock drainage and a major concern to the environment. Pyrite waste must be disposed off under strictly controlled conditions (Pulles et al., 1996; Jovanovic, 1998; Annandale et al., 1999, Shimada et al., 2012; Novo et al., 2013). It has a great potential for restroring the productivity of calcareous sodic lands and improving the fertility of the soil. Oxidation of pyrite leads to the production of sulfuric acid, which reacts with native CaCO3 to produce soluble Ca to replace adsorbed Na from the exchange complex of sodic soil and brings down to pH and ESP (Fig. 1) (Tiwari et al., 1988; Qadir et al., 1996; Qadir and Oster, 2004). The exchange of Ca with Na in the soil exchange complex results in flocculation of soil particles and the restoration of porous structure and high permeability of soil (Chun et al., 2001).

Calcareous soils, characterized by the poor bioavailability of plant nutrients, a high base status, and pH between 7.5 and 8.5 depending on the quality and quantity of carbonate minerals present, cover more than 30 percent of land surface of the earth (Chen and Barak 1982 ; Marschner 1995 ; Ylivainio 2010). In areas with calcareous soils, plant growth is usually retarded and/or plant quality is adversely affected as a result of high pH and unavailability of essential micronutrients: iron (Fe), copper (Cu), zinc (Zn), and manganese (Mn) (Jacobs 2008). Pyrite can be also used as a source of micronutrients needed for a plant growth, particularly Fe, Cu, Zn and Mn. Therefore, it can be used as a fertilizer in calcareous sodic soils where these elements are deficient.

Several experiments have been conducted to study the relative performance of pyrites and gypsum in reclaiming alkaline and calcareous soils. Tozsin et al. (2014, 2015a, 2015b) was conducted studies to determine the effectiveness of waste pyrite from a copper concentration plant on alkaline and calcareous soil amelioration as compared with gypsum and to evaluate its effects on soil structural improvements and plant growth.

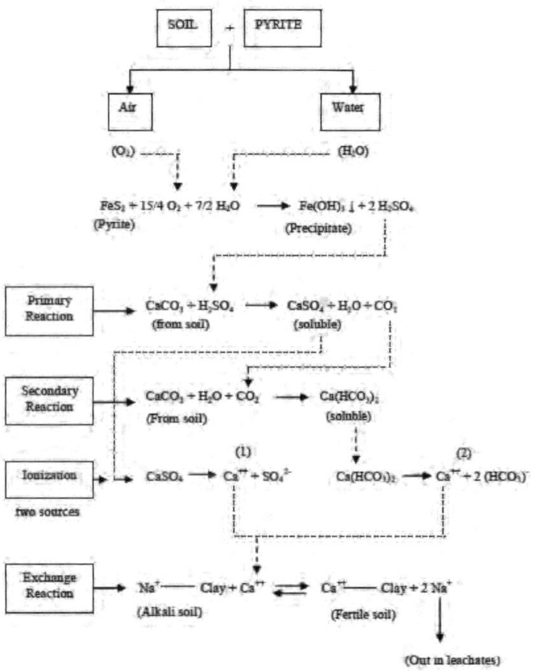

Fig. 1. Sequence of chemical reactions of pyrite in soils in the reclamation of alkali soils (Somani, 1994).

Tozsin et al. (2014, 2015a) conducted the column-leaching tests to determine the effectiveness of application of waste pyrite obtained from the tailings stream of Kure Copper Concentration Plant, Turkey on characteristics of alkaline soil obtained from Saraykoy, Ankara, Turkey as compared with gypsum application. Polyvinyl chloride (PVC) columns were hand packed with air-dried and homogenized sodic soils (Fig. 2a, 2b). The columns were open at the top and closed at the bottom, except for an outlet for collection of the leachates.

Fig. 2a. Column test set-up

The results of these studies indicated that the pH and ESP of the alkaline soil significantly decreased ($p<0.01$) from 9.27 to 7.77 and from 83.33% to 11.77%, respectively with waste pyrite application. Tozsin et al. (2014, 2015a) suggested that waste pyrite could be good ameliorating agent and an alternative to gypsum in reclamation of alkaline soils.

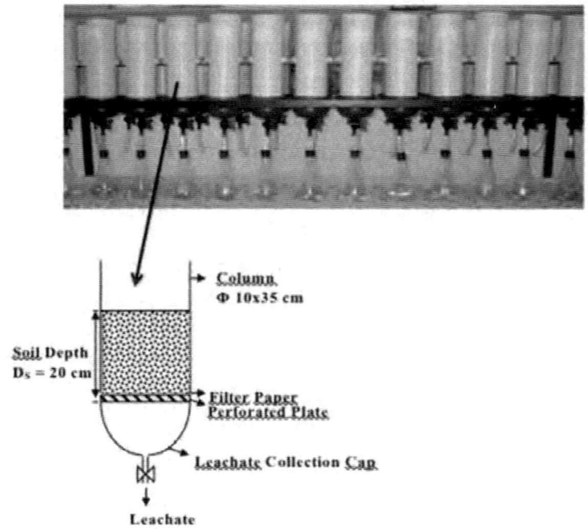

Fig. 2b. Schematic representation of the column leaching test

In other study, Tozsin et al. (2015b) investigated the effectiveness of waste pyrite applications on calcareous soils deficient in plant micronutrients. For this purpose, greenhouse pot experiments were conducted (Fig. 3). Soil was placed into plastic pots and wheat seeds were sown in each pot. The maximum and minimum temperatures were 27 and 15 °C, respectively. One month after the seed sowing, plants were harvested.

Fig. 3. The effects of waste pyrite treatments on wheat growth.

The results showed that while the applications of waste pyrite to the calcareous soils improved the levels of Fe, Cu, Zn and Mn and increased the dry matter yield of wheat, it did not cause pollution or toxicological problems in the soil. These studies indicated that waste pyrite is a good ameliorating agent and an alternative to gypsum in reclamation of alkaline soils and remediation additive for calcareous soils deficient in plant micronutrients.

3 UTILIZATION OF MARBLE WASTES IN ACIDIC SOIL REMEDIATION

Copper flotation typically produces tailings containing residual sulfides. Acid mine drainage (AMD) is produced when sulfide-bearing materials, primarily pyrite associated with mining/processing wastes or overburden, are exposed to oxygen and water (Akcil and Koldas, 2006; Ercikdi et al., 2009). AMD waters are usually characterized by strong acidity (pH < 3) and high concentrations of sulfate and toxic metals (Bhatti et al., 2012; Manafi et al., 2012). The decrease in pH values caused by sulfuric acid formation increases the solubility of heavy metals in AMD waters. Consequently, the mobilities of inorganic contaminants such as Cd, Cr, Cu, Ni, Pb, and Zn are enhanced, posing a serious threat to the environment (Ciccu et al., 2003; Cihangir et al., 2012; Jha et al., 2012).

Various strategies for acid neutralization and metal retention are employed for the remediation of sites contaminated with AMD (Komnitsas et al., 2004). The most commonly used techniques are the application of alkaline substances into the AMD water or directly into sulfide-rich wastes. Once in aqueous solution, the alkaline substances dissolve and help to neutralize acidic solutions and immobilize metals via the precipitation of the metals as hydroxides, carbonates, and sulfides (Hale et al., 2012; Sánchez-Andrea et al., 2014).

Liming is a common method to increase the pH of acid soils. The liming material reacts with carbon dioxide and water in the soil to yield HCO3-, which removes the H+ from the solution, raising the soil pH (Safari and Bidhendi, 2007; Muthukrishnan and Oleske, 2008). Naturally occurring minerals such as lime (predominantly CaCO3) are commonly used to reclaim acid soils (Wang et al., 2011). Marble is a crystalline metamorphic limestone, basically containing calcite (CaCO3) and maybe dolomite (CaMg(CO3)2) (Segadães et al., 2005). As a result of the marble production activities, marble-manufacturing industry produces high amount of waste. Marble wastes are generated by quarries (marble quarry waste-MQW) and processing plants (marble cutting waste-MCW). The proportion of marble discharged as waste during block production at the quarries is equal to 40-60% of the overall production volume. The waste generation rate at marble processing plants is around 30-35% and this varies according to shape and kind of blocks being cut (Celik and Sabah, 2008). The large amount of marble wastes can cause environmental problem and economic loss if it is not used. However, this type of waste has a good potential to be used as an amendment for the neutralization of acid soils (Xenidis et al., 2002; Karasahin and Terzi, 2007; Aruntas et al., 2010; Bilgin et al., 2012).

MQW and MCW can be utilized as a neutralizing material for copper flotation tailings (CFT). The neutralization of acidity by MQW and MCW occurs via the interaction of these wastes with sulfuric acid (AMD) and the subsequent formation of gypsum (Eq. (1)):

$$CaCO3 + H+ + SO42- + 2H2\ O \rightarrow CaSO4.\ 2H2O + HCO3 \quad (1)$$

The removal of aqueous iron by MQW and MCW proceeds via Eq. (2):

$$Fe3+ + 3HCO3 \longrightarrow Fe(OH)3 + 3CO2 \quad (2)$$

The application of a CaCO3-containing material such as MQW and MCW as a remediation agent for CFT increases the pH value, resulting in metal precipitation and a reduction in trace element mobility (Hale et al., 2012).

Tozsin (2016) conducted a study to investigate the feasibility of using MCW taken from marble cutting, sawing, and polishing factories in Afyon, Turkey as an acid-neutralizing agent to inhibit AMD and immobilize heavy metals from copper flotation tailings (sulfide- bearing wastes) obtained from Murgul Copper Mines in Artvin, Turkey. Continuous-stirring shake-flask tests were conducted for 40 days, and the pH value, sulfate content, and dissolved metal content of the leachate were analyzed every 10 days to determine the effectiveness of MCW as an acid neutralizer. MCW was demonstrated to be an effective amendment to reduce the dissolved metal and sulfate concentrations in CFT drainage. The application of MCW to CFT removed over 80 wt% of heavy metals (Cd, Cr, Cu, Ni, Pb, and Zn) and reduced the sulfate and iron concentrations from 4558 to 838 mg/L and from 536 to 0.01 mg/L, respectively, after 40 days.

In other study, Tozsin et al. (2014a; 2015) investigated the effectiveness of marble waste applications on neutralization of soil acidity. MQW and MCW taken from Afyon, Turkey were applied to an acid soil obtained from Kelali garden, Inece village, Bulancak town, Giresun, Turkey at different rates and their effectiveness on neutralization was evaluated by a laboratory incubation test. The MQW and MCW doses were thoroughly mixed with air-dried soil and incubated at 25 °C and sub-sampled after 15, 30, 45, 60 and 75 days, respectively. The soil pH increased from 4.71 to 6.36 and 6.84 by applications of MCW and MQW, respectively at the end of a 75-day incubation period. Results of this study indicated that MQW and MCW, significantly increased soil reaction under all conditions. However, the increasing rate in soil pH was dependent on incubation period and application rate. Tozsin et al. (2014a; 2015) suggested that MQW and MCW could be used as soil amendments for the neutralization of acid soils and thus the negative impact of marble wastes on the environment could be reduced.

The Black Sea Region climate together with the geological formation of the area has resulted in the formation of the acidic soil. Soil acidity greatly affects the availability of plant nutrients. Therefore, whenever needed, agricultural lime is used to improve the fertility. Tozsin et al. (2014b) also conducted a study if the agricultural lime can be replaced by MQW and/or MCW in hazelnut fields.

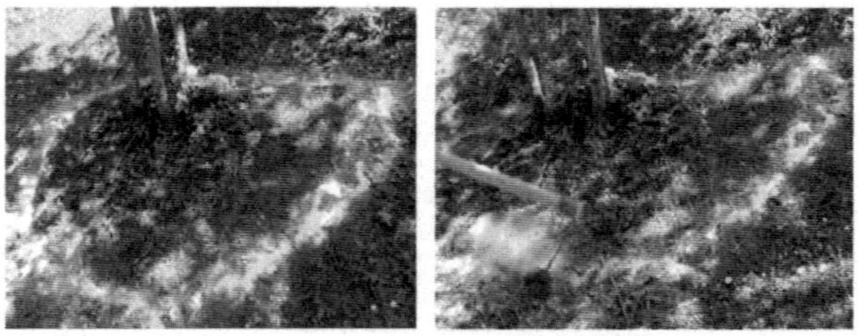

Fig. 4. Application of agricultural lime and marble wastes to the hazelnut trees

The study was conducted in the Kelali garden, Inece village, Bulancak town, Giresun, which is located at the Eastern Black Sea Region of Turkey. Field tests in a selected orchard were conducted for one year to determine the effect of MQW and MCW on the soil pH and Tombul hazelnut yield. Marble wastes at different rates were applied to an acid soil (Fig. 4). The results showed that marble wastes had a significant effect on the neutralization of the soil as well as on the hazelnut yield. The soil pH was increased from 4.71 to 5.88 upon marble waste application at rates equal to agricultural lime requirement. Hazelnut yield increased from 1120.3 kg/ha on the field with no marble waste treatment to 1605.5 kg/ha with marble wastes. This study indicates that MQW and MCW could be used in the hazelnut fields for the neutralization of acidic soil to increase the yield.

REFERENCES

Akcil, A., Koldas, S., 2006. Acid mine drainage (AMD): causes, treatment and case studies. Journal of Cleaner Production, 14 (12-13), 1039-1188.

Annandale, J.D., Jovanovic, N.Z., Benade, N., Tanner, P.D., 1999. Modelling the long-term effect of irrigation with gysiferous water on soil and water resources, Agriculture, Ecosystems and Environment, 76, 109–119.

Aruntas, H.Y., Guru, M., Dayi, M., Tekin, I., 2010. Utilization of waste marble dust as an additive in cement production. Materials and Design, 31 (8), 4039-4042.

Bhatti, T.M., Bigham, J.M., Vuorinen, A., Tuovinen, O.H., 2012. Chemical and bacterial leaching of metals from black schist sulfide minerals in shake flasks. International Journal of Mineral Processing, 110 (111), 25-29.

Bilgin, N., Yeprem, H.A., Arslan, S., Bilgin, A., Gunay, E., Marsoglu, M., 2012. Use of Waste Marble Powder in Brick Industry. Construction and Building Materials, 29, 449-457.

Celik, M.Y., Sabah, E., 2008. Geological and technical characterization of Iscehisar (Afyon-Turkey) marble deposits and the impact of marble waste on environmental pollution. Journal of Environmental Management, 87, 106-116.

Chen,Y., and P. Barak. 1982. Iron nutrition of plants in calcareous soils. Advances in Agronomy, 35, 217–40.

Chun, S., Nishiyama, M., Matsumoto, S., 2001. Sodic soil reclaimed with by-product from flue gas desulfurization: corn production and soil quality. Environmental Pollution, 114, 453–459.

Ciccu, R., Ghiani, M., Serci, A., Fadda, S., Peretti, R., Zucca, A., 2003. Heavy metal immobilization in the mining-contaminated soils using various industrial wastes. Minerals Enginnering, 16 (3), 187-192.

Cihangir, F., Ercikdi, B., Kesimal, A., Turan, A., Deveci, H., 2012. Utilisation of alkali-activated blast furnace slag in paste backfill of high-sulphide mill tailings: Effect of binder type and dosage, Minerals Engineering, 30, 33-43.

Ercikdi, B., Cihangir, F., Kesimal, A., Deveci, H., Alp, I., 2009. Utilization of industrial waste products as pozzolanic material in cemented paste backfill of high sulphide mill tailings. Journal of Hazardous Materials, 168 (2-3), 848-856.

Gupta, R.K., Abrol, I.P., 1990. Salt-affected soils: Their reclamation and management for crop production. Advances in Soil Science, 11, 223-288.

Hale, B., Evans, L., Lambert, R., 2012. Effects of cement or lime on Cd, Co, Cu, Ni, Pb, Sb and Zn mobility in field contaminated and aged soils. Journal of Hazardous Materials, 199-200, 119-127.

Jacobs, L. 2008. Micronutrients needed by crops. In MWEA Biosolids Conference. Bay City, Michigan.

Jha, R.K.T., Satur, J., Hiroyoshi, N., Ito, M., Tsunekava, M., 2012. Suppression of pyrite oxidation by carrier microencapsulation using silicon and catechol. Mineral Processing and Extractive Metallurgy Review, 33 (2), 89-98.

Jovanovic, N.Z., Barnard, R.O., Rethman, N.F.G., Annandale, J.G., 1998. Crops can be irrigated with lime-treated acid mine drainage, Water SA, 24, 113–122.

Karasahin, M., Terzi, S., 2007. Evaluation of marble dust in the mixture of asphaltic concrete. Construction and Building Materials, 21, 616-620.

Komnitsas, K., Bartzas, G., Paspaliaris, I., 2004. Efficiency of limestone and red mud barriers: laboratory column studies. Minerals Engineering, 17 (2), 183-194.

Lèbre, É. and Corder, G., 2015, Integrating industrial ecology thinking into the management of mining waste, Resources, 4, 765–786.

Lèbre, É., Corder, G. D. and Golev, A., 2016, Sustainable practices in the management of mining waste: A focus on the mineral resource, Minerals Engineering, In Press, Corrected Proof, Available online 18 December 2016.

Lottermoser, B.G., 2011. Recycling reuse and rehabilitation of mine wastes. Elements 7, 405–410.

Manafi, Z., Abdollahi, H., Tuovinen, O.H., 2012. Shake flask and column bioleaching of a pyritic porphyry copper sulphide ore. International Journal of Mineral Processing, 119, 16-20.

Marschner, H. 1995. Mineral nutrition of higher plants, 2nd ed., 347–64.London: Academic Press.

Muthukrishnan, S., Oleske, M., 2008. Effects of lime amendment on the pH of engineered soil mix for the purposes of bioretention. Journal of Irrigation and Drainage Engineering, 134, 675-679.

Novo, L.A.B., Covelo, E.F., Gonzalez, L., 2013. Phytoremediation of amended copper mine tailings with Brassica juncea. International Journal of Mining, Reclamation and Environment, 27, 215-226.

Pulles, W., Heath, R., Howard, M., 1996. A Manual to assess and manage the impact of gold mining operations on the surface water environment (Pretoria, South Africa), Water Resources Commission, Report No. TT 79/96.

Qadir, M., Qureshi, R.H., Ahmad, N., 1996. Reclamation of a saline-sodic soil by gypsum and Leptochloa fusca. Geoderma, 74 (3-4), 207–217.

Qadir, M., Oster, J.D., 2004. Crop and irrigation management strategies for saline-sodic soils and waters aimed at environmentally sustainable agriculture. Science of The Total Environment, 323, 1–19.

Richards, L.A., 1954. Diagnosis and improvement of saline and alkali soils; Agricultural Handbook 60, United States Department of Agriculture, Washington, DC.

Safari, E., Bidhendi, G.N., 2007. Removal of manganese and zinc from Kahrizak landfill leachate using daily cover soil and lime. Waste Management, 27, 1551- 1556.

Sánchez-Andrea, I., Sanz, J.L., Bijmans, M.F.M., Stams, A.J.M., 2014. Sulfate reduction at low pH to remediate acid mine drainage. Journal of Hazardous Materials, 269, 98-109.

Segadães, A.M., Carvalho, M.A., Acchar, W., 2005. Using marble and granite rejects to enhance the processing of clay products. Applied Clay Science, 30, 42-52.

Shimada, H., Kusuma, G.J., Hiroto, K., Sasaoka, T., Matsui, K., Gautama, R.S., Sulistianto, B., 2012. Development of a new covering strategy in Indonesian coal mines to control acid mine drainage generation: A laboratory-scale result. International Journal of Mining, Reclamation and Environment, 26, 74-89.

Suarez, D.L., Rhoades, J.D., Lavado, R., Grieve, C.M., 1984. Effect of pH on saturated hydraulic conductivity and soil dispersion. Soil Science Society of America Journal, 48, 50-55.

Tiwari, K.N., Kumar, A., Carter, M.R., Gupta, U.C., 1988. Evaluation of sedimentary iron pyrites as an ameliorant for a saline-sodic soil in Uttar Pradesh, India. Arid Soil Research and Rehabilitation, 3, 361–368.

Tozsin, G., Arol, A.I., Cayci, G., 2014. Evaluation of pyritic tailings from a copper concentration plant for calcareous sodic soil reclamation. Physicochemical Problems of Mineral Processing, 50(2), 693-704.

Tozsin, G., Arol, A.I., Oztas, T., Kalkan, E., 2014a. Using marble wastes as a soil amendment for acidic soil neutralization. Journal of Environmental Management, 133, 374-377.

Tozsin, G., Oztas, T., Arol, A.I., Kalkan, E., Duyar, O., 2014b. The effects of marble wastes on soil properties and hazelnut yield. Journal of Cleaner Production, 81, 146-149.

Tozsin, G., Oztas, T., Arol, A.I., Kalkan, E., 2015. Changes in the chemical compositon of an acidic soil treated with marble quarry and marble cutting wastes. Chemosphere, 138, 664-667.

Tozsin, G., Arol, A.I., Cayci, G., 2015a. Use of waste pyrite as an alternative to gypsum for alkaline soil amelioration. International Journal of Mining, Reclamation and Environment, 29(3), 169-177.

Tozsin, G., Arol, A.I., 2015b. Pyritic tailings as a source of plant micronutrients in calcareous soils. Communications in Soil Science and Plant Analysis, 46, 1473-1481.

Tozsin, G., 2016. Inhibition of acid mine drainage and immobilization of heavy metals from copper flotation tailings using a marble cutting waste. International Journal of Minerals, Metallurgy and Materials, 23 (1), 1-6.

Wang, S.J., Chen, C.H., Xu, X.C., Li, Y.J., 2008. Amelioration of alkali soil using flue gas desulfurization byproducts: Productivity and environmental quality, Environmental Pollution, 151, 200–204.

Wang, N., Xu, R.K., Li, J.Y., 2011. Amelioration of an acid ultisol by agricultural byproducts. Land Degradation and Development, 22, 513-518.

Ylivainio, K. 2010. Effects of iron (III) chelates on the solubility of heavy metals in calcareous soils. Environmental Pollution, 158, 3194–200.

Xenidis, A., Mylona, E., Paspaliaris, I., 2002. Potential use of lignite fly ash for the control of acid generation from sulphidic wastes. Waste Management, 22, 631-641.

Manufacture and Characterization of Ferroniobium Alloy Briquettes

A. C. Silva[1], K. S. Macedo[1], E. M. S. Silva[1], M. R. Barros[1], D. F. Lopes[1], D. Y. Marinho[1], V. O. Morato[1], L. T. B. Mendonça[1], D. L. Florêncio[1], L.F. Silva[2]

1 Federal University of Goiás, Catalão, Brazil

2 NioBrás, Catalão, Brazil

ABSTRACT Brazil is the largest producer of Niobium alloys, with a production of 73,668 tons of Nb_2O_5 concentrate in 2013, corresponding to a 92.81% of the world production. The Brazilian Nb ores grades vary from 0.51 to 2.71%. Ferroniobium is an important iron niobium alloy, with niobium content of 60-70%. It is the main source for niobium alloying of HSLA steel and covers more than 80% of the worldwide niobium production. In 2013, Brazil produced 46,555 tons of ferroniobium, being around 90% of it for export. In the final stage of the Fe-Nb production, the alloy needs to be crushed in a jaw crusher, which generate fin particles (around 26% of the feed below 10 mm in NioBrás process). The selling price of the fine particles have a significant loss in the market, since the fine particles cannot be used directly in steel working. Given this scenario, a process to agglomerate the ferroniobium fine particles into briquettes with size and chemical qualities similar to original products was developed. During the briquetting tests organic (zinc stearate) and inorganic (iron powder and water) binders were used. The briquettes were characterized both physically and chemically. For the physical characterization, the briquettes were tested for mechanical resistance, drumming, compressive strength, water resistance, thermal shock and Brinell hardness. The test contained addition of 3% of zinc stearate, 2% iron powder and 1% water reached a niobium content above 62% and the levels of contaminants below the standard ferroniobium specifications. The tested briquettes have chemical and physical properties compatible with sales specifications.

1 INTRODUCTION

Niobium is a trace element present in Earth crust with average grades between 0.3 and 1% of Nb_2O_5. However, some Brazilian deposits can reach exceptional grades around 3% due to supergene enrichment in weathered zones (MENDES, 2005). Niobium is a metallic, soft, ductile solid and has a high melting point (2,469 °C) one of the highest of the periodical table. It is resistant to corrosion, mainly due to the formation of a thin film in the oxide surface, called passivation layer. When combined Niobium exhibit different oxidation numbers, being +5 the most common. At room temperature Niobium, do not react with hydrogen, air, water or acids, except for hydrofluoric and its mix with nitric acid (SOUSA et al., 2013).

Industrial applications of Niobium include, but are not limited to, high-strength low-alloy steel (HSLA), stainless steel and super alloys. The Niobium consumption in micro alloyed structural steels reaches above 80% of its demand (DAMASCENO, 2006).

Brazil holds the Niobium bigger deposits in the world, distributed between the states of Minas Gerais, Amazonas, Goiás and Rondônia, with a total reserve around 10.8 Mt of pyrochlor exploitable reserves. Brazil is also the bigger producer with an annual production around 88.8 kt of Nb_2O_5 concentrate and 51.7 kt of Ferro Niobium Alloy (Fe-Nb), with Nb content ranging from 60 to 70%. Canada holds the second position in the world marketing with approximately a reserve of 200 kt and annual production around 5 kt (DNPM, 2015). Figure 1 shows

Brazilian's Niobium production between the years 2000 and 2014. It is possible to see a development in the Niobium marketing until 2010, when the steel marketing started to face a severe economic crisis. An improvement in the marketing was noticed after 2012 and 2014 end up being, together with 2009, the years with higher production in the period.

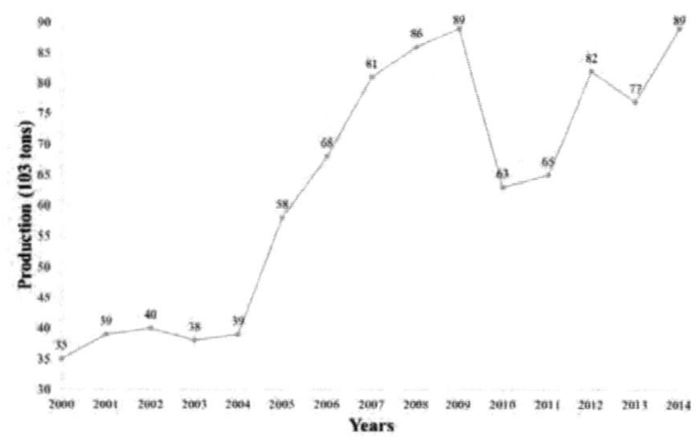

Figure 1. Brazilian's Niobium production (2000-2014).

Brazilian's most exported Niobium product is Fe-Nb, responding for over 90% of the Niobium exportations. According to the Brazilian official Mineral Agency (DNPM, 2015) in 2014, the total exportation of this product was 71 kt, corresponding to US$ 1.7 billion.

Industrial activities such as mining and metallurgy can produce many types of residues, including fine particles. Several different initiatives have been adopted to minimize the production of fine particles and/or to recover then. These particles can be generated in process such as ore blasting, transport, hauling, screening, comminution and even handling. The recovery of fine particles can lead to an increase in the production or in the production of by-products. Taking that in consideration many efforts have been done in order to reduce the emissions of fine particles and to try to give then a better destination, avoiding potential environmental and health issues.

NioBrás is company from the Chinese group CMOC and is one of the two companies that produce Fe-Nb in Brazil (around 35% of the Brazilian production). Situated in the state of Goiás the company production is destined to exportation. In NioBrás, after the aluminothermy, the final stage of the Fe-Nb production is the comminution of the freshly produced alloy, normally using a jaw crusher (see figure 2). This stage often generate particles under 10 mm, considered fine particles and not able to be used direct in the steel making process because only of their size. Other operations that generate fine particles are Fe-Nb screening and handling (especially the commutation between convey belts), reaching a production around 95 tons per month (equivalent to 26% of the Fe-Nb production) of fines. Fe-Nb fine particles are divided in three different classes according to their granulometry and generation process: under 2 mm (type A), between 1 and 5 mm (type B) and between 3 and 15 mm (type C). Figure 3 shows samples of the three types of fines.

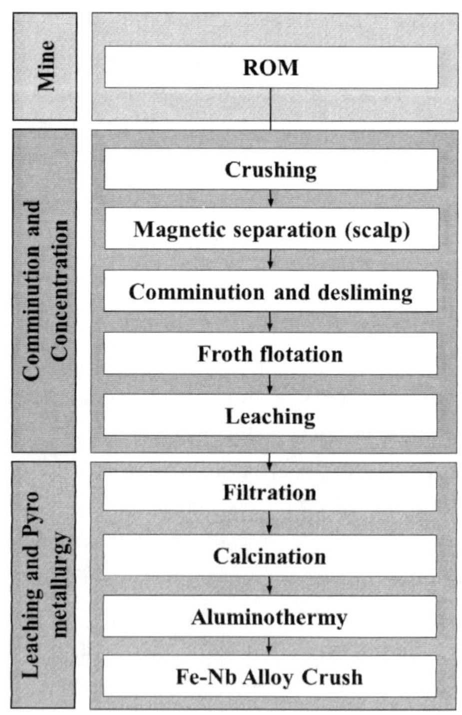

Figure 2: Fe-Nb alloy production flowsheet.

Figure 3: Fe-Nb fine particles: (a) under 2 mm or type A, (b) between 1 and 5 mm or type B and (c) between 3 and 15 mm or type C.

Seeking a more efficient production concept regarding the increase of Fe-Nb incorporation in the metal bath during the steel making process, the agglomeration of the Fe-Nb fine particles was studied, once this by-product could be very economic attractive to both supplier and consumer.

The agglomeration process aim in produce material with size adequate to the future application and the correct choose between the available methods depends on a very careful analyses. According to Carvalho & Brinck (2010), this involves not only laboratory work in order to produce the agglomerate, but also its physical and chemical characterization. Among the agglomeration techniques, the most important ones in mineral processing are pelleting, sintering and briquetting.

Moraes & Kawatra (2011) tested how binders successfully used in the pelletizing of Brazilian hematitic iron ores, using disk pelletizer, would perform in the processing of an American magnetite concentrate by balling drums. Although the results were very good, the operational cost of the pelleting was relatively high when compared with the other two methods.

Telles, Espinosa & Tenório (2013) developed a sintering process to agglomerate electric arc furnaces dust in order to recycle this material in the steelmaking. In their process, the dust had to be first agglomerated as micro-pellets and then sintered.

Lucena *et al* (2008) were able to produce a low-cost charcoal wastes briquetted to be used in blast furnaces and for energy generation using a piston extruder. On the other hand, Silva *et al* (2016) were able to agglomerate limestone fines using a batch briquetting process. In both case the operational cost where small.

2 METHODOLOGY

Taking in consideration the fact that the Fe-Nb fine particles has very small moisture content, high melting point (1,900 ºC) and is very abrasive the briquetting was choose as the most feasible technique to agglomerate the fine particles.

2.1 Fe-Nb Briquette Manufacture

Samples of the Fe-Nb fines from the alloy comminution, screening and handling were sent to two different companies in Brazil to be agglomerated called Höganäs and Renova.

The differences, beside the manufacture costs and geometry, were the briquetting technology used by the two companies. Höganäs used a piston extruder and Renova used a high-pressure roll press. Each company produced two different briquettes composed by a blend of the fine particles type composition (see Table 1). The briquettes were labelled as H (from Höganäs) and R (from Renova) followed by the percentage of the fines.

Table 1. Briquettes composition regarding the fine particles type.

Briquettes	A	B	C
R100-0-0	100	0	0
R0-100-0	0	100	0
H50-40-10	50	40	10
H60-40-0	60	40	0

Three binding agents were used two organic and one inorganic. The organic binders were zinc stearate ($Zn(C_{18}H_{35}O_2)_2$) and zinc palmitate ($Zn(C_{16}H_{31}O_2)_2$) in a proportion of 70 and 30% respectively. Iron powder with very low impurities grade was used as inorganic binder. Table 2 shows the proportion of water and binders used in the briquettes manufacture. The homogenization of the particles was performed in a mixing tank, where the fine particles were added first and then water and the binders.

Table 2. Binders and water composition percentage (in mass) in the briquettes.

Organic binder (%)	Iron powder (%)	Water (L)
3.0	2.0	1.0

During the briquettes manufacture samples of the material where collected: before, during and after the briquetting. The samples where than homogenised and sent to X-ray diffraction with Rhodium anode and Nondispersive Infrared spectroscopic (NDIR).

Figure 4 shows one photograph of the four types of briquettes. It is possible to notice the differences in geometry (Höganäs' briquettes were cylinders and Renova's were soap like shape) and surface quality in the manufactured briquettes.

(a) (b) (c) (d)

Figure 4. Fe-Nb briquettes: (a) R100-0-0; (b) R0-100-0; (c) H50-40-10 e (d) H60-40-0.

2.2 Fe-Nb briquettes characterization

The manufactured briquettes where tested in the Mineral Processing Lab of the Federal University of Goiás (LaMPPMin). All tests were performed in triplicate.

2.2.1 Shatter Test

Shatter test (or Impact Resistance Test or Drop Test) determines the briquette strength to withstand repeated drops, simulating impacts that occur naturally during handling and transportation. Before heat treatment (curing), the impact strength of the briquettes can be determined by dropping it from a height of 0.3 m direct onto a steel plate with 10 mm thickness. The procedure must be repeated until the briquette loses 5% of its initial mass (CARVALHO & BRINCK, 2010).

According to the same authors if the briquette had been subjected to heat treatment to increase its mechanical strength, then the height of the drop should be changed to 1.5 m. In both cases (cured or not) the drop height can be changed to 0.6, 0.9, 1.2 and 1.5 m, for example. For uncured briquettes, 3 drops from 0.3 m are considered a reasonable value, while for cured ones the number of drops shift to 10.

Eighteen briquettes from each type were submitted to shatter tests from 0.3, 0.5, 0.75, 1.0, 1.25 and 1.5 m height.

2.2.2 Uniaxial compression and tension

An important test to measure the maximum load that a briquette can support before its rupture is the uniaxial compression test. A compressive load was applied direct to the briquette using a hydraulic press until the briquette rupture, as described by the Brazilian Association of Technical Standards (ABNT) NBR 12767/92.

To evaluate the tension resistance the method created by Professor Lobo Carneiro was used. This method calculates the tensile strength through a diametric compression test.

Altogether twenty-six briquettes where used in both tests. Briquettes manufactured by Renova could not be tested using Lobo Carneiro test because of their geometry.

2.2.3 Water absorption

According to Cunha et al. (2006) in this test, briquettes must be complete submerged in water for 24 hours. After this time, the briquettes must be removed from water and put to rest in room temperature for 10 minutes in order to remove the water layer adsorbed in their surface and then weighted. Then they must be dried in a drying oven at 80 °C for 2 hours. In order to return the briquettes to room temperature they must be put to rest again for 20 minutes and weighted again. The amount of water that is absorbed by the briquette is important when the briquettes can be stored in open areas (CARVALHO & BRINCK, 2010).

Altogether forty-eight Fe-Nb briquettes where tested.

2.2.4 Thermal shock

Thermal shock occurs when a thermal gradient causes different parts of an object to expand by different amounts. This differential expansion can be understood in terms of stress or of strain, equivalently. At some point, this stress can exceed the strength of the material, causing a crack to form.

The shock temperature can be defined as the maximum temperature that the briquette resist without lose more than 10% of its mass, without collapse or explode, during a fast temperature elevation.

Briquettes were weighted and placed in a steel cylinder with holes in the lid and placed in a muffle oven with temperatures ranging from 300 to 550 °C and 20 minutes of residence time. After this time, the briquettes were removed from the oven and let rest in room temperature before a new weighing. The maximum temperature reached was the temperature when the briquette start to combust (defined as ignition temperature). Sixty briquettes where submitted to thermal shock test.

2.2.5 Brinell hardness test

According to Callister and Rethwisch (2013), in the Brinell test, a hard, spherical indenter is forced into the surface of the material to be tested. The diameter of the used tungsten carbide indenter was 10.00 mm (0.394"). Standard loads range between 500 and 3000 kg in 500-kg increments; during a test, the load is maintained constant for a specified time (between 10 and 30 s). Harder materials require greater applied loads. The Brinell hardness number, HB, is a function of both the magnitude of the load and the diameter of the resulting indentation. Twenty-five briquettes where used measure the Brinell hardness.

2.2.6 Tumbler drum test

The tumbler drum test is derived from the R-556 of the MICUM standard for coke of the International Organization for Standardization (ISO). In this test, 1.5 kg of briquettes are placed in a steel test drum with 29 cm long and 30.5 cm diameter fitted with four steel angles with 6.5 cm wider, fixed lengthwise inside the drum. The drum is rotated at 25 rpm for eight minutes (200 revolutions in total).

The effect of tumbling on the briquette is tested as follows. The briquette is first sieved by hand to remove material that passes a 31.75 mm (1.25") screen. This is to prevent the finer particles cushioning the effect of rumbling on the coarse material during the test. After tumbling, the material is sieve-analysed to measure the size reduction that has occurred. These results are combined with the sieve analysis of the fine material excluded from the tumbler test to give a total evaluation of the briquette quality.

3 RESULTS AND DISCUSSION

3.1 Characterization of the Fe-Nb Fines

Figure 5 shows the granulometric analyse of the Fe-Nb fine particles before the agglomeration. It is possible to notice that particles type A (under 2 mm) and type B (between 1 and 5 mm) are well classified, fitting to their label. On the other hand, particles type C (between 3 and 15 mm) where actually smaller than expected. In fact, particles type C were smaller than 7 mm.

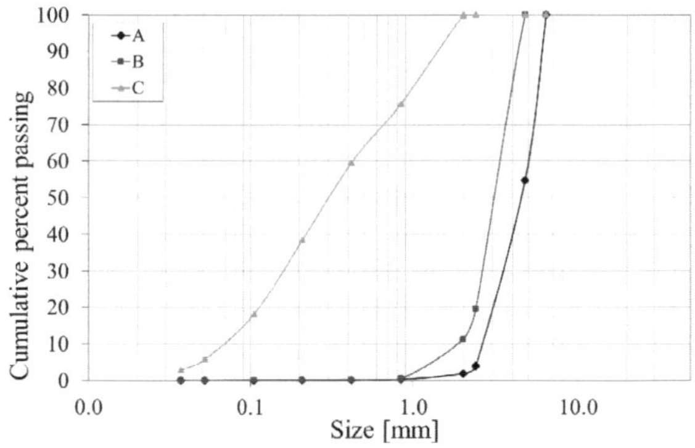

Figure 5. Granulometric analyse of the Fe-Nb fine particles before the agglomeration. (a) under 2 mm or type A, (b) between 1 and 5 mm or type B and (c) between 3 and 15 mm or type C.

Table 3 shows the X-ray diffraction results of the Fe-Nb fine particles characterization compared with the standard composition of this alloy as expected by the steel making market. Iron and niobium content must be at least 87% of the alloy composition. Due to the mineral association at NioBrás mine site, the major contaminants are P, Ti and Mn. It is possible to notice that all three types of fine particles have composition adequate to be used in steel making.

Table 3. Standard Fe-Nb alloy composition and X-ray diffraction results of the fine particles.

Type	Element (%)										
	Nb	P	Fe	Ti	Mn	Si	Ta	Pb	S	Al	C
Standard	62- 67	< 0.20	25-30	< 1.00	< 0.50	< 3.00	< 0.50	< 0.20	< 0.15	< 1.00	< 0.15
A	64.56	0.15	29.18	0.33	0.30	2.95	0.31	0.19	0.07	0.40	0.14
B	65.17	0.17	28.65	0.25	0.30	2.85	0.31	0.19	0.09	0.31	0.14
C	65.25	0.15	27.97	0.24	0.35	2.68	0.29	0.18	0.07	0.23	0.13

3.2 Fe-Nb Briquettes Characterization

3.2.1 Shatter Test

Table 4 shows photographs of the briquettes before and after (main body and fragments) shatter test for the four different briquettes manufactured.

Table 4. Fe-Nb briquettes before and after (main body and fragments) shatter test.

Briquette	Before	After	
		Main body	Fragments
R100-0-0	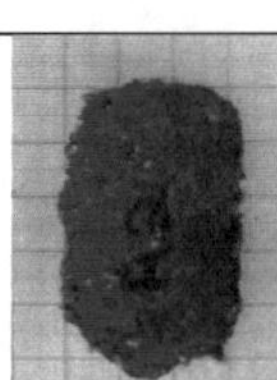		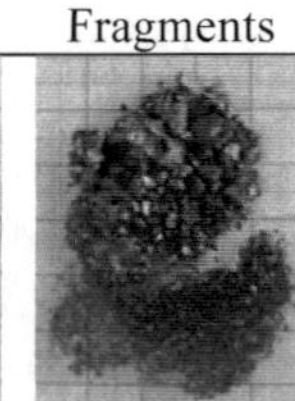

R0-100-0

H60-40-0

H50-40-10

Figure 6 summarizes the average resistance of the briquettes regarding the number of drops at different heights supported in the shatter test.

Comparing the photographs before and after the test (Table 4) and the Figure 5 is possible to notice that cylindrical briquettes manufactured by Höganäs had a high tendency in complete fragment themselves after a few drops, even from the smallest heights. The higher resistance was observed for the briquette R-0-100-0 manufactured by Renova. Although this briquette was able to resist more drops than the others at any tested heights, for height higher than 1.0 m all type of briquettes showed a similar behaviour losing 5% or more of their initial mass in less than 10 drops.

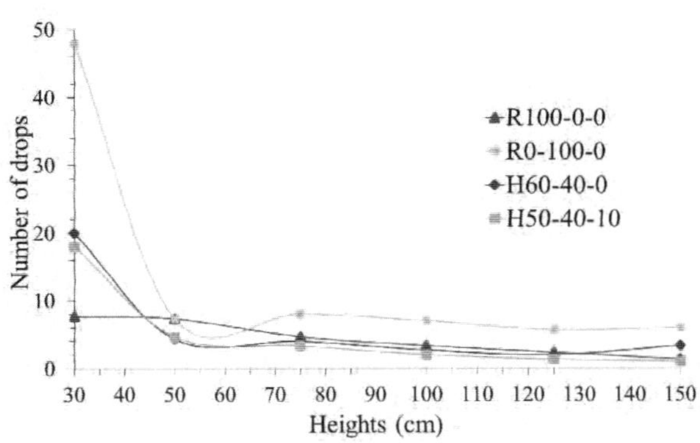

Figure 6. Average shatter test results.

3.2.2 *Uniaxial compression and tension*

Figure 7 shows the average results found in the uniaxial compression test for the four types of briquettes and the results for the uniaxial tension test for the cylindrical briquettes.

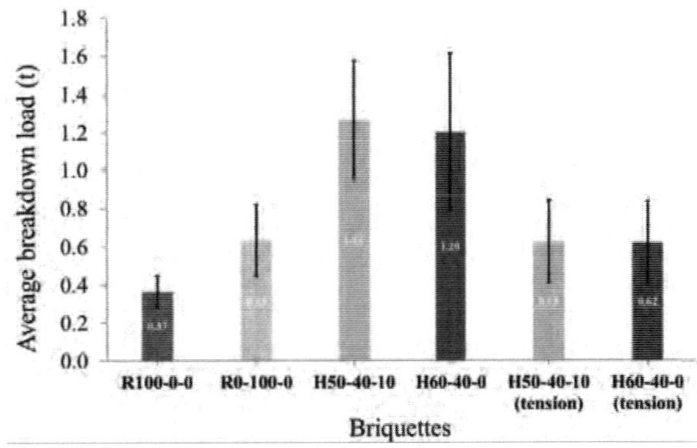

Figure 7. Average uniaxial breakdown load (compression and tension) results.

The cylindrical briquettes, manufactured by Höganäs, obtained the best results in the uniaxial compression test. This fact could be related with their geometry once the cylindrical shape provided a better contact surface. The soap shaped briquettes has a poor contact surface with the hydraulic press, normally being just one contact point and not a surface.

Regarding the tension, the cylindrical briquettes showed a resistance around 50% of the compression resistance. Compared with other materials this is a high value, for example concrete tension resistance is around 10% of the compression resistance.

3.2.3 Water absorption

Table 5 shows three different stages of the water absorption test (before immersion in water, after and after the briquette had been dried in drying oven). In one hand, it was not possible to see any major visual changes on them, but in the other hand, the briquettes supported the immersion, even 48 hours, without any physical degradation or decomposition, remaining almost intact.

Table 5. Fe-Nb briquettes before and after water adsorption test (briquettes immerged for 48 hours).

Briquette	Before	After Immersion	After Drying
R100-0-0			
R0-100-0	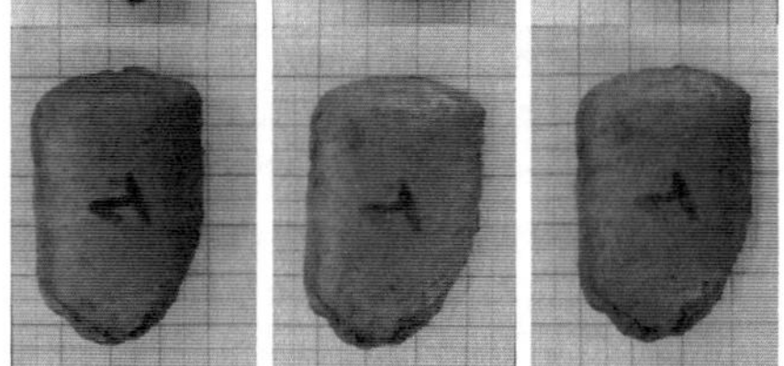		
H60-40-0			
H50-40-10			

Figure 8 shows the water absorption test results. It is possible to notice that all briquettes had their mass enhanced after the test at least by 1% after 48 hours of immersion in water. Although the Renova briquettes had the high mass gain during the test, the briquettes behaviour was quiet unusual, since a plateau was expected describing a water saturation level.

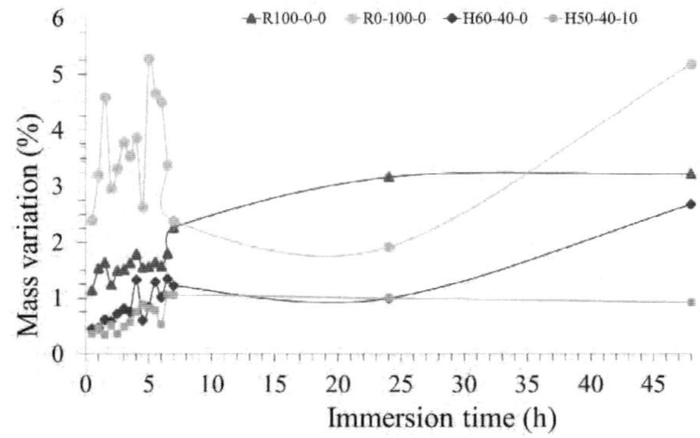

Figure 8. Average water absorption test results.

This behaviour was noticed only by one type of briquette manufactured by Höganäs (H50-40-10). The other cylindrical type showed a peak in mass variation for 48 hours immersion. Explanations for this behaviour could be the hydration of the binder agents, such as the iron powder, or chemical reactions (Fe-Nb alloy or residues of the aluminothermy) process with water.

3.2.4 Thermal shock

The first result from this test was unexpected. The briquettes start combusting after a certain temperature (see Figure 9). Since the Fe-Nb alloy was produced through aluminothermy, it was expected that all combustible material were complete burnt in that process. That was not the case. Every test performed at temperatures above 600 °C ignited the briquettes. The flame produced remained burnt for almost 12 hours after the briquettes ignition, no matter with one of the four types tested.

Table 6 shows photographs of the briquettes before and after (main body and fragments) thermal shock test for the four different briquettes manufactured.

Figure 10 shows the average fragments percentage generated at different temperatures during the tests. No clear pattern could be observed between the briquettes behaviour regarding the fragments generation during the test. It is possible to highlight that the briquette R100-0-0 supported 100 °C more than the briquette R0-100-0 and H50-40-10 before they start combusting.

Figure 9. Briquettes combusting inside the steel cylinders.

Table 6. Fe-Nb briquettes before and after (main body and fragments) thermal shock test.

Briquette	Before	After	
		Immersion	Drying
R100-0-0			
R0-100-0			
H60-40-0			
H50-40-10			

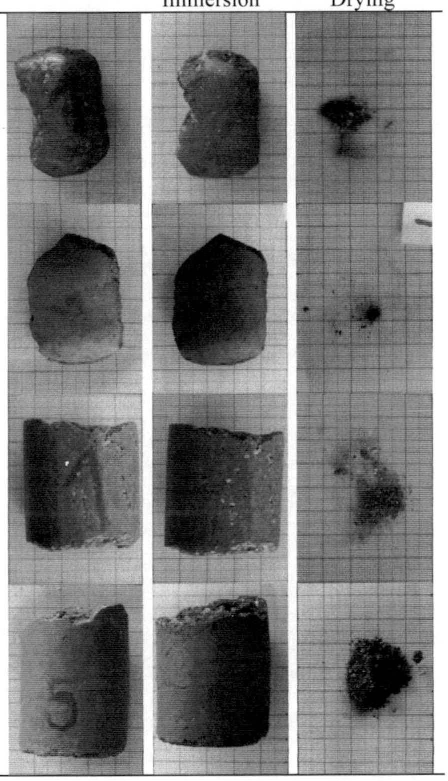

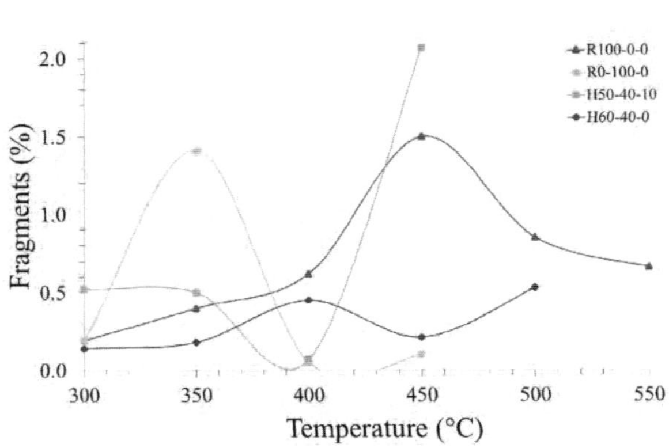

Figure 10. Average briquettes fragments percentage generated at different temperatures.

3.2.5 Brinell hardness test

Figure 11 shows the average results found in the Brinell test. As expected, the found results were quite low for all briquettes. Since they are agglomerated materials, formed mainly the mechanical forces their resistance to indentation must be low and, in this case, compatible with a polymeric material or minerals with Mohs hardness bellow gypsum.

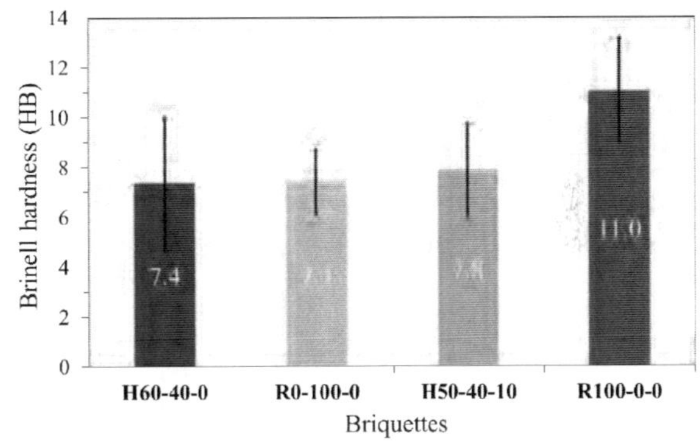

Figure 11. Briquettes average Brinell hardness.

3.2.6 Tumbler drum test

Figure 12 shows the average granulometric analyse of the briquettes after the tumbler test. Considering the material passing in sieve with aperture 3/4" (or 19.0 mm) only the briquette R100-0-0 had a considerable resistance, producing 34% of its mass bellow this aperture. This result is consistent with the Brinell hardness result, since the R100-0-0 was the briquette hardener one and the tumbler test evaluate the shear resistance of the material. The other briquettes had results above 80% of passing (98.5% for R0-100-0, 89.2% for H60-40-0 and 81.9% for H50-40-10).

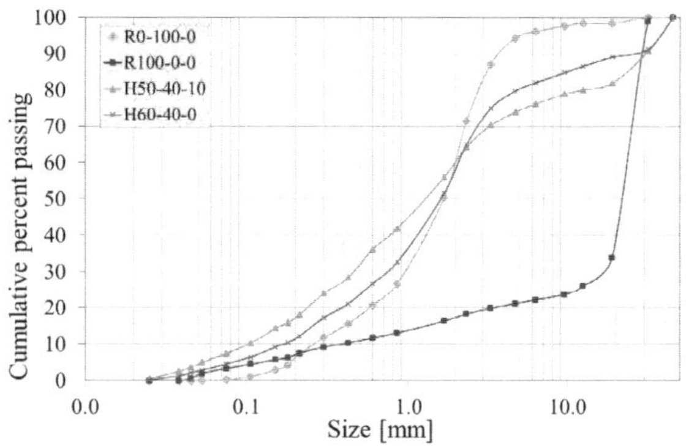

Figure 12. Granulometric analyse of the briquettes after the tumbler test.

4 CONCLUSIONS

Regarding the shatter test, all tested briquettes succeeded in the test considering the literature parameters. However, the found results were heterogeneous and a trend could not be stablished. Briquettes manufactured by Höganäs (cylinders) showed results slightly more homogeneous than the ones manufactured by Renova.

It was possible to notice in the uniaxial compression test that briquettes manufactured by Höganäs had better results, however showing a standard deviation higher than Renova's briquettes. This fact may be due to the geometry of them, once Höganäs' briquettes have a better stress distribution because of their cylindrical shape. The same briquettes showed a higher than expected uniaxial tension resistance.

All briquettes were able to resist the water absorption test retaining their shapes. For higher immersion time Höganäs' briquettes complete disintegrate, not retaining either shape or mechanical resistance. In some briquettes, immersed for 48 hours, it was possible to notice some oxidation spots on them. Only one briquette (R0-100-0) did not gain mass after the test.

Regarding the thermal shock, all briquettes were successful. Although it was not possible to stablish a trend from the obtained results, the briquettes ignition temperature was stablished bellow 550 °C (450 °C for R0-100-0 and H50-40-10, 475 °C for H60-40-0 and 550 °C for R100-0-0.

Briquettes showed a low Brinell hardness, which was expect one they were only agglomerated through mechanical stress, without the addition of any kind of chemical binder.

At last, regarding the tumbler test all types of briquettes generated fine particles during the test. Although the briquette R100-0-0 showed a higher resistance compared with the others.

Summarizing the found results both fabrication methods can be industrially adopted for briquette production with parameters acceptable by the steel making marketing. All fine particle blends tested were successful regarding the briquette chemical composition and mechanical properties. It was not possible to choose one manufacture method between the two tested because the found results were quite different. In the end, the authors suggest the adoption of the cylindrical briquettes manufactured by Höganäs as the best choice, combining good results in most of the tests and the lower manufacture costs.

ACKNOWLEDGEMENTS

The authors thank financial support from the Brazilian agencies CNPq, CAPES, FAPEG and FUNAPE. In addition, we like to thank NioBrás for the samples donation and chemical analyses and Federal University of Goiás.

REFERENCES

Associação Brasileira de Normas Técnicas (ABNT), 1992. *NBR 12767/92 - Rochas para revestimento - Determinação da resistência à compressão uniaxial*.

CALLISTER, W. D., RETHWISCH, D. G., 9th ed., 2013 *Materials Science and Engineering: An Introduction*. John Wiley & Sons.

CARVALHO, E. A., BRINCK, V. 4th ed., 2010. Briquetagem. Tratamento de Minérios, CETEM/MCT, Rio de Janeiro, 963 p.

CUNHA, A. F., MOL, M. P. G., MARTINS, M. E., ASSIS, P. S., 2006. Caracterização, Beneficiamento e Reciclagem de Carepas Geradas em Processos Siderúrgicos. *REM: Revista Escola de Minas*, Ouro Preto, v. 59, n. 1, pp. 111-116.

DAMASCENO, E. C., 2006. Disponibilidade, suprimento e demanda de minérios para metalurgia. *Série Estudos e Documentos* - CETEM/MCT.

DNPM, 2015. Departamento Nacional de Produção Mineral. Disponível em: <http://www.dnpm.gov.br/conteudo.asp?IDSecao=68&IDPagina=2263>

LUCENA, D. A., MEDEIROS, R. D., FONSECA, U. T., ASSIS, P. S., 2008. Fine charcoal agglomeration and its feasibility for blast furnace usage and for energy generation. *Tecnologia em Metalurgia e Materiais*, São Paulo, v.4, n.4, pp. 1-6.

MENDES, M. W. D., 2005. *Obtenção de pós de Nb a partir da redução alumino térmica com ignição por plasma.* [Master dissertation] – Universidade Federal do Rio Grande do Norte, Natal, 78 p.

MORAES, S. L., KAWATRA, S. K., 2011. Evaluation of the use of combinations of binders in the pelletization of magnetite concentrate by balling drum technology. *Tecnologia em Metalurgia, Materiais e Mineração*, São Paulo, v. 8, n. 3, pp. 168-173.

SILVA, A. C., SILVA, E. M. S., BARROS, M. R., MARINHO, D. Y., 2016. Limestone fines reuse for agriculture through briquetting. *Tecnologia em Metalurgia, Materiais e Mineração,* São Paulo, v. 13, n. 4, p. 365-372.

SOUSA, R. M. F., FERNANDES, L. E., GUERRA, W., 2013. Nióbio. *Química Nova na Escola*, v.35, pp. 68-69.

TELLES, V. B., ESPINOSA, D. C. R., TENÓRIO, J. A. S., 2013. Iron ore sinter production using electric arc furnace dust as raw material. *Tecnologia em Metalurgia, Materiais e Mineração, São Paulo*, v. 10, n. 1, pp. 72-77.

Reclasification of Praid Salt Mine by State of Gas Emission

E. Chiuzan, D. Cioclea, A. Matei, I. Gherghe, C. Boantă, C. Tomescu

National Institute for Research and Development in Mine Safety and Protection to Explosion, INSEMEX Petrosani, Romania

ABSTRACT The purpose of this paper is to determine the rate of methane (explosive gas) and of carbon dioxide (axphysants gas) in the mine workings of Praid salt mine. Currently in Romania there are six salt mines which exploit the salt through underground mine workings: Praid salt mine, Tg. Ocna salt mine, Slanic Prahova salt mine, Ocnele Mari salt mine, Cacica salt mine, and Dej salt mine. Of these six, in three salt mines occurred gas and salt eruptions: Praid salt mine, Slanic Prahova salt mine and Tg. Ocna salt mine. The purpose of the present paper work is to establish the methane release (explosive gas) and the carbon dioxide release regime in mine workings from Praid salt mine in order to verify its classification.

This verification of the Praid salt mine classification has been carried out during October 2012 and there have been taken into account the following elements:

- Geological and technical-mining conditions of the salt deposit related to Praid mining perimeter;
- Results of the quantitative and qualitative measurements from the underground mine workings regarding:
  a) Circulated air flows;
  b) Measured methane and carbon dioxide concentrations;
  c) Absolute flows registered;
  d) Accumulation capacity of gas in the deposit;
  e) Gas concentrations and pressure in the massif.
- Existing evidences and documentations regarding the previous gas occurrences and their manner of manifestation;
- Establishing the way in which the gas is released into the underground atmosphere of the mine workings.

Based on the observations and on the carried out measurements, on the analysis of the studied geological and mining elements, there has been made the classification proposal of the Praid salt mine from the point of view of gas emissions (methane and carbon dioxide).

1 INTRODUCTION

The presence of marsh gas in underground atmosphere of Praid salt mine was highlighted both by geological drillings deposit performed for exploring the deposits and by opening, preparation and exploitation workings [1].

Systematic measures carried out in 1976 by INCD - INSEMEX Petrosani, to determine the category that Praid salt mine falls into in terms of the gas emanation, revealed the existence of marsh gas in the holes that were bored in the salt massif for shooting the coal face and the 657 run to open a new horizon, marsh gas between salt and surrounding waste rock measured concentrations of marsh gas went up to 84% vol.

2. GENERAL REGARDS ON DEPOSIT'S AND MINING PERIMETER'S GEOLOGY

Praid rock salt deposit from the mine is part of the saliferous area on the eastern frame of Transylvania basin at the base of Gurghiu Mountains. The deposit is located in the central part of Sovata - Praid – Corund area and takes the form of a dome flanked by sedimentary or volcanic miopliocene or quaternary formations.

Horizontally the deposit has an approximately circular shape (NE - SW axis and NW-SE axis of 1.2 - 1.4 km). Vertically, based on the structural boreholes performed, it is appreciated that Praid diapir dome of rock salt has a height of 2.6 to 2.8 km, being the deepest and thickest rock salt massif in Romania. The salt diapir is flanked by sedimentary rocks, partially covered by volcanic formations. Its quality, in terms of impurities (found in the form of impregnations including enclaves in size up to several cm, made of clay, marl, sandstone, gypsum, anhydride, but also massive sterile insertions frequently occur) places it behind other deposits in exploitation in the country.

3. OPENING, PREPARATION AND EXPLOITATION OF DEPOSITS

The first underground salt mine dates back to 1762 and is located in the south-west side of Spatele Sarii hill.

The systematic exploitation of salt began in 1787 when exploitation of Iosif mine began, the Paralela mine (1864), Elisabeta mine (1873) and Gh. Doja mine (1947).

The current mine located under the Gh. Doja mine was opened from the surface by an adit located at elevation +497.2 m, which continues with a run that descends to elevation +474.5 m, respectively +427.0 m. Opening the depth of the deposit was performed by a blind shaft from elevation +427.0 m to elevation +208 m and through the auto run between elevation +474.5 m and elevation +170 m (Fig. 1).

In 1991, the preparation and opening of a new sector began, Telegdy mine, accessible through a side split of the main transportation gallery. Extracting salt from the massif in Telegdy mine is performed by small rooms and rectangular pillars exploitation method. The rooms have a width of 16 m, height of 8 m and length up to 20 m. The pillars are 14 m wide, 8 m in height, 14 m long and have a thickness of 8 m.

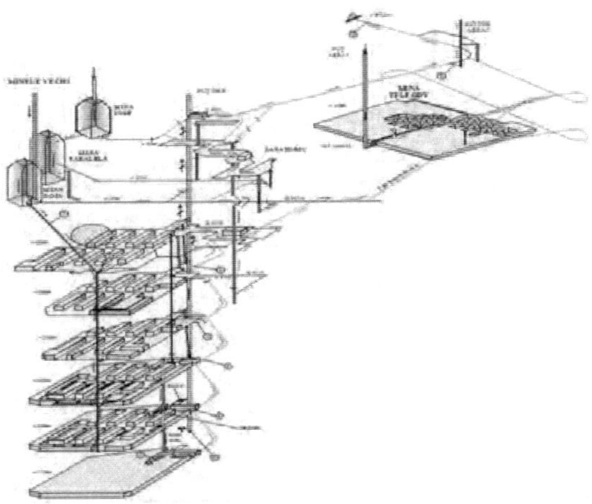

Fig. no. 1 Praid salt mine

Currently production activities are carried out in blocks no. 4, 8, 9 and 17. After exhaustion of the reserves at horizon +228 m salt exploitation will continue through the horizon +208 m, respectively horizon 170 m (horizon under way of being opened). Salt exploitation is performed through the small rooms and rectangular pillars exploitation method. The rooms have a width of 20 m, 12 m in height and between 20 and 275 m in length. Pillars are 20 m wide, 12 m high with a length between 95-100 m.

Between the horizons there is safety floor, having a thickness of 8 m. Exploitation of different floors is performed in descending order.

Beside the extraction of salt, an opened all year long asthmatic asylum is located at horizon 50 (elevation 407.0 m).

4. PRAID MINE VENTILATION SYSTEM

Praid salt mine general ventilation is achieved through an aspirant- forced draught system, the two main ventilation installations being located underground. The air flow direction is descended from the old mines, Iosif, Paralela and Gh. Doja mines, situated above horizon +339, being under the influence of the depression created by the fan. Fresh air is discharged on the ventilation rising up to horizon +230.0 m and through the network of ventilation galleries, at the level of each horizon. Return air is exhausted to surface through the chamber's pathway, auto transportation runs from each horizon, blind extraction shaft, blind shaft ramp horizon +426.0 m, main auto transportation run, coast gallery.

5. MEASUREMENTS PERFORMED IN UNDERGROUND WORKINGS

To establish the scheme of gas emissions, respectively the classification, depending on the absence or presence of methane and carbon dioxide, quantitative and qualitative measurements were carried out, in the main and secondary return air exhaust currents, in the active, inactive and reserve mine workings [2].

In addition to these measurements, specific measurements were also performed to establish the presence of methane during some technological operations such as:

- Punching blowholes in the salt massif;
- Performing the path at the coal face base;
- Measurement of gas concentrations in the coal face, rooms ceiling and in general exhaust air currents of the horizons and mine. Measurements in active mine workings were performed also for a preparation work, where partial ventilation was stopped for 24 hours.

Prospecting for marsh gas in the salt massif around mine workings, was performed through advance boreholes in which were inserted gas collecting probes. These were sealed for 24 hours. Also to this purpose were performed measurements in a horizontal bore, performed toward the salt-sterile limit [4].

6. MEASUREMENT RESULTS

The results of the measurements performed in the underground atmosphere and the salt massif highlighted the following:

• Absence of methane in the mine's primary and secondary return air exhaust currents;

• Presence of methane in the salt massif around mine workings having the following concentrations:

► 3 ppm in the boreholes performed in the preparatory work front – auto transportation run to horizon +188 m;

► 1% vol. in the G2 test drilling, located at horizon +208, in chamber 6400, at more than 24 hours after sealing;

• Presence of carbon dioxide in the atmosphere of mine workings in concentrations up to 0.3% vol. (value measured in the cesspool of the blind pit);

Lack of gas pressure in advance boreholes performed in the salt massif;

• No blower of methane or other manifestations of gases were shown.

Considering the results of measurements and observations performed by INSEMEX Petrosani in Praid salt mine underground workings during 1985, 1986, 1989, 1994, 1996, 1998, 2001, 2003, 2005, 2007, 2010 and 2012, respectively 2014 we made the suggestion of differentiating Praid underground mine workings into "mine areas " and properly classifying them in terms of gas emission as follows:

• absence of methane in technological processes at coal faces, such as: performing advance boreholes and scaling coal faces or after performing shooting activities at coal faces;

• absence of methane in the underground atmosphere if the partial ventilation system is shut for a period of over 24 hours;

• presence of methane in the salt massif and marginal waste rock around underground workings(advance boreholes and pits F-166, inclined F-166, F-166 bis);

• absence of carbon dioxide in the working's atmosphere and its presence in the salt massif;

• lack of gas pressure in advance boreholes performed in the salt massif;

• no blower of methane or other manifestations of gases were shown.

7. CONCLUSIONS AND RECOMANDATIONS

Considering the results of measurements and observations performed by INSEMEX Petrosani in Praid salt mine underground workings combined with previous occurrences of gases we made the recommendation to keep the classification as "1st Category gassy" considering methane emissions and as "1st CATEGORY" considering carbon dioxide emissions [5].

Given the dynamics of methane concentrations in underground workings over the years and the experience gained by NRDI INSEMEX Petrosani in this field, we recommend changing the classification criteria for salt mines by dividing them into clearly delineated mine areas, intended for opening, preparation and exploitation of the salt massif through underground workings.

Salt mines, mining areas or underground workings where the presence of methane hasn't been detected, will be classified as: non gassy.

Salt mines, mining areas or underground workings where the presence of methane has been detected, will be classified as gassy:

- of non gassy conditions, active, inactive and in reserve mine workings found under the depression/pressure created by main ventilation fans;
- and maintaining the classification as "1st Category gassy" for:
  ▪ dead-hole coal faces ventilated through partial ventilation;
  ▪ coal faces where the following activities are performed: cutting path with cutting machinery, drilling holes for blasting, blasting the coal face, loading;
  ▪ chambers where bore holes are performed for analyzing the salt deposit or sterile intercalations and the salt / sterile contact area;
- lack of classification in terms of carbon dioxide emissions since the presence of this gas in the underground atmosphere, hasn't been detected in concentrations greater than 1%.

When opening a new horizon, it will be considered as non gassy only if the analyzing works/drillings haven't revealed the presence of marsh gas and the horizon's classification will be made only after performing the first circuit of the mine ventilation under the general depression of the mine.

---

[1]Note: –clearly defined parts of the salt mine intended for salt extraction, for example: active , inactive, reserve chambers or dead-holes opening or preparatory activities.

REFERENCES

1. X X X - *Technical report presenting Praid salt mine;*
2. E. Chiuzan – *Assessing the gas emanation state - methane and carbon dioxide - to verify the classification category of Praid salt mine, studies* - NRDI INSEMEX 2006-2016;
3. D. Tulica, I.Toth – *Analysis of gas parameters in the salt massif from Praid salt mine*, NRDI INSEMEX study 1986;
4. *Test procedure for measuring gas emissions in underground workings* - NRDI INSEMEX Petrosani
5. *Assessment procedure on approval of documentation concerning the classification of mines in terms of gas emanation* - NRDI INSEMEX Petrosani.

Effects of Geology on Production Rate: Open Pit Coal Mining Case Study

T. M. Malambule and T. Zvarivadza
University of the Witwatersrand, Johannesburg, South Africa

ABSTRACT The study focused on the effects of geology on production rate at an open pit colliery. The aim of the study was to evaluate how geological structures directly affect the production rate at this colliery. These structures include dolerite intrusions, faults, paleo-highs and paleo-lows. The study showed that dolerite intrusions have burnt the coal seam to ash. Faults have displaced the coal seams to unmineable depths. Paleo-highs and Paleo-lows have caused major delays in production due to irregularity of the floor that enormously influenced the production of coal. The abrupt change in the elevation of the coal seam makes the coal difficult to mine. The geological structures on monthly production, from July to December 2015 were evaluated as part of the research. It is recommended that the design of the pit, the access, the haul roads and maintenance must be planned separately and monitored frequently to reduce delays in production.

Key Words: Geology effects, coal mining, production rate, geological structures, open pit

1 INTRODUCTION

Clear understanding of the influence of geological structures on production rate is key to a profitable and sustainable mining operation. Careful planning and designing has to be done around geology to enhance a mine's productivity. This paper presents findings from a study of open pit coal mining in South Africa. The seam of economic interest is No. 2 coal seam whereby bituminous coal (Thermal coal) is extracted. The Middle to Bottom seams are extracted using the drill and blast as well as the truck and shovel excavation methods. The mine is situated in the Karoo Supergroup. The Karoo Supergroup is of economic importance since it hosts all the coal deposits of the subcontinent (Johnson et al., 1996). The project area has paleo-highs, paleo-lows and is cross-cut by geological structures such as faults and igneous Karoo intrusions. This study focuses on the effects of geology on production rate at the Colliery. Particular emphasis is on examining and documenting the geological structures such as intrusions, faults and their impact on the coal seams. The aim of this study is to evaluate how these geological structures directly affect the production rate of the mine.

The quality of coal tends to decrease in areas where there are dominantly dolerite sills and dyke intrusions crosscutting the coal seams. These intrusions burn the coal to ash thereby reducing the coal quality. The faults on the other hand can have the following effects on coal seams: (1) they can displace the coal seams to depths that make the seams impractical to mine and (2) they can also reduce the stability of the mining face thus inducing rock falls. Therefore, it is important to identify the faults in the mining area and determine their orientations and amount of throw (vertical displacement). This information can be factored into mine plans and designs and help to reduce the risk of unexpected seam displacement and rock falls ahead of mining. Failing to identify the igneous intrusions and faults before mining activities commence may compromise production targets as well as workers' safety.

The results from this study are beneficial as these observations can be used for proper mine planning at the Colliery to adequately plan which coal seams to target and which areas to avoid thereby reducing downtime.

The seams in this study are not flat due to undulations within this area. The thickness of the coal seam differs within this region. Towards the basement the coal seam becomes thinner. The seam which is less than 0.5 m is uneconomical to mine.

1.1 Study Background

The project was done at the Colliery, which is an open cast mining operation where bituminous (thermal coal) coal is the commodity being mined. In 2013, mining commenced at the Colliery and the mine was expected to produce 3 million tonnes of coal per annum. The area was mapped as fairly flat with no major geological structures that will disrupt mining activities. However, a year into production the mining personnel noticed that the mine was under producing and the targets were not being met. New geological mapping information revealed several dykes, sills and faults present in the mining area.

1.2 Problem Statement

The main problem at the mine is the complexity of geological structures which causes production delays and which makes the mine unable to reach the planned production. The main problems associated with the topography are the fluctuated paleo-lows and paleo-highs, which affect ramp design. The coal seam has undulating layers which result in slow mining rate and cause delays in production. In some cases the mine has had to close ramps due to very steep areas. Designing a ramp in steep areas is difficult since the sliding angle necessarily needs to be less than 15^0 for the mine equipment to travel safely. The steep gradients make it difficult for the equipment to travel across the pit safely. The coal seams that occur in close proximity to the dolerite dykes and sills become burnt due to the overheating of the igneous intrusions. The burnt coal turns into ash and becomes uneconomical to mine.

1.3 Aims and Objectives of the Research

The study focuses on the effects of geology on the production rate at the Colliery. The study also aims to make recommendations on how the mine can optimize production, with these geological structures being considered, as they occur naturally.

The objectives of the research were to study the following:

- Geological Structures- conduct detailed geological mapping to identify the faults and dykes that crosscut the area, and determine the orientation of the structures so that it is known ahead of mining
- Paleo floor- the base of the Colliery is mostly dominated by low quality coal due to the presence of Dolerite sills
- Effect of geological structures on ramp design- Determine the general topography of the region so that proper ramps can be designed
- Coal losses- the mine has lost coal due to deep weathering and increasing depth

1.4 Justification of the Research

The study is beneficial to the mine since it assists the mine to resolve the current production rate problem. The authors have observed that most of the tonnages at the mine are lost due to geological features. It was necessary to conduct this study in order to come up with new strategies on how the mine can optimize production with the geological structures taken into account. The study assists the mine to maximize its profit. When geology is complex, the operating costs escalate, for example, when mining through the paleo-high, the machine has to consume more fuel and that has a negative impact on costs. The authors noticed that the mine has many breakdowns, which result from the effect of geological structures.

2 STUDY APPROACH

Quantitative research method was used to conduct the studies. Data was collected from the mine to come up with interpretations that will benefit the mine and solve the problem. For the research project to reach the objectives of the study and to come up with valid conclusions, the following parameters were taken into account:

- Amount of Coal losses due to the presence of geological structures were investigated
- Analysis of production rate using data from the mine
- Photographs were taken for all geological structures
- Investigated the types of geological Structures in the mine and the effect thereof
- Investigated the effects of geology on blasting
- Investigated the effects of overburden thickness on the mining costs
- Studied the stratigraphy of the mine location and how the stratigraphy affects production losses
- Investigated how the geometry of the coal seam affect the production in Mining
- Investigated further effect of geological structures on Ramp Design
- Effect of geology on tyres was considered

3 IMPACT OF GEOLOGY ON COAL MINING

Geology has an impact on the coal mining industry since there are various geological structures that may influence mining. These geological structures include igneous intrusions, faults and fluctuating paleo-surface on mine design and production. The mining activities at the Colliery are largely controlled by geological structures within the area.

3.1 Paleo-floor

The basement of the Karoo Supergroup varies in thickness and rock types, due to geological conditions that prevailed during the time of formation. During the period of ice flow, rock particles were deposited at different speeds. Larger particles settled as a result of their higher densities and lighter particles were deposited on top of the larger particles as a result of their lighter densities. This settling of particles of varying sizes created uneven topography. The change in basement geology plays a major role in the nature of the paleo-topography created (Hancox & Götz, 2014). The basement has a greater impact when it comes to the distribution of coal seams in the studied Coalfield.

3.2 Paleo-highs

A paleo-high is experienced when the coal seam has been displaced to highly elevated regions of the mine. The abrupt change in the elevation of the coal seam makes the coal difficult to mine. If the incorrect equipment is used it may become damaged when climbing steeper slopes than it was designed for. The angle of repose may result in challenges during mining. The geology of the area also affects blast design and is particularly important to remember that the design changes due to variable depth of blasthole. In most places, the coal seams thin out against paleo-highs and thicken in paleo-lows. Another risk is that if these paleo-highs are not properly defined from the start, money could be wasted, for example, when removing the waste burden only to find out that the coal has thinned out. It is essential to consider the geological features before mining progresses in order to be aware of the costs for that particular region and to be well informed on how production rates are to be affected.

3.3 Paleo-lows

The geological conditions that accompany coal seams in lower regions make it difficult for mining to occur. In such regions, the rate of mining is slower compared to the rate in flat lying areas. This results in low production rates because of the undulating ground conditions. Instead of mining efficiently, the operator spends most of the time adjusting the excavator in order to make sure that it does not get damaged. When mining through paleo-lows, too much waste needs to be removed so that the coal seam can be exposed, this results in higher operational costs. Blasting coarser material is also one of the major problems since fragments need to be blasted twice until such material can be transported. The blasting pattern also changes due to geology, blast holes have variable depth.

Figure 1 represents the paleo-low, the coal seam is found in the lower positions causing difficulties in mining. There are cases where the coal seam disappears completely when there is a paleo-low due to increased depth. The paleo-low has been crosscut by the plane which is one of the geological features that causes disruptions.

Figure 1. Paleo-low

3.4 Weathering

Weathered sandstone is found in the northern parts of the mining area. In this region, silicates and clay minerals are associated with sandstone, and play a major role in environmental geochemistry of coal. Mechanical weathering breaks down coal through physical processes and it becomes mud, therefore coal reports to waste. Production slows down if the coal is weathered, and there is a loss in profit if such coal is mined without checking its physical and chemical properties. The coal that is weathered affects its beneficiation process (Jha et al., 2014). Deep weathering causes explosive-energy distribution to be disrupted and that results in large boulders.

3.5 Coal Losses

The loss of coal in mining can be caused by a number of factors. These include (i) weakening of the floor of the seam, which makes mining difficult due to the presence of soft and powdery coal (ii) weathering, which results in the contamination of the coal, thus reducing the quality of coal and (iii) the presence of geological structures such as igneous intrusions.

3.6 Effect of Geological Structures on Ramp Design

The way in which the ramp is designed depends on the geology of the mine. If the area of concern is steeply dipping, it means machines (excavators, trucks and vehicles) will consume more fuel when ascending to highly elevated regions, thus increase the operating costs. The maximum angle of ramp is supposed to be 15^0; this enables machinery to travel safely without any disturbances. Mining equipment does not work efficiently and to its full capacity under unstable geological conditions such as paleo-highs and paleo-lows. Construction of the ramp in the areas where there are paleo-lows can cause complications in mining. It is not easy to construct the ramp as there are costs associated with the construction of ramps, and the costs are normally high in areas of paleo-lows, paleo-highs and dolerite dykes. When confronting a paleo-high, the ramp needs to be flattened in order for the coal to be accessible and to ensure that the angle of the ramp does not exceed 15^0. In the areas where there is a dyke, it is important to drill and blast so that it will be easy to construct a ramp. The costs associated with drilling and blasting are high in areas where there is a dyke, and this can negatively affect mining operations.

Figure 2 illustrates a ramp where coal is transported and this allows easy travelling for machinery. The inclination of this ramp is less than 15^0.

Figure 2. Coal transportation ramp

3.7 Normal Fault

Normal faults are characterized by a hanging wall that has moved down relative to the footwall. As a result of this fault, mining operations were delayed because the coal seam was untraceable. Once the production has been delayed it becomes difficult to reach production targets and the profit generated will be lower than expected.

Figure 3 illustrates the coal seam crosscut by a fault. The coal seam has been displaced by a fault. When there is a fault, the mine has to spend more money and time to remove waste in order to access the continuation of the coal seam.

Figure 3. Coal seam crosscut by a fault

4 RESULTS AND ANALYSIS OF RESULTS

The results presented are those obtained from the mine over a period of 6 months from July to December 2015. The mine has monthly targets to reach in order to pay the running costs and make a profit, but it is sometimes difficult to achieve the projected targets. It is particularly important to conduct studies to ensure that all parameters that affect production are frequently monitored. Therefore, it is necessary to do proper mine planning in order to avoid delays in production during mining and profit losses which can result due to extra incurred operational costs. The main areas of interest in this current research are those that were previously affected by geological structures thereby delaying production.

4.1 Waste Production

At the Colliery, removal of waste before extracting coal is rightfully incorporated as part of the initial phases of planning and production. Inclusion of waste production during the planning process is a crucial step as the coal is not accessible if the overburden has not been removed. There are high costs associated with overburden removal; the mine has to allocate machinery only for overburden. The waste production can affect the costs of the mine if blasting is done incorrectly, hence Run-of mine (ROM) coal production will be affected. Waste is normally taken to the mine dumps and there are costs associated with transportation. The removal of waste is one of the most important components of surface mining costs. The variations in overburden thickness can affect the machinery and efficiency of its operations. The mine has to remove softs (free dig), softs (drill and blast), hards and interburden, so that coal can be exposed.

Table 1 summarizes the amount of waste material that is mined each month. The mine has planned material which refers to the amount of waste that needs to be removed before extracting coal. The actual survey refers to the amount of waste that was removed each month. The table summarizes waste production from July 2015 to December 2015. Variance refers to the difference between Actual Survey column and the planned column; refer to Table 1 and Table 2.

Table 1. Amount of waste material bank cubic meter (bcm) mined per month

Month	Material Planned (bcm)	Actual Survey (bcm)	Variance (bcm)
July 2015	506,765	495,423	-11,342
August 2015	516,027	258,246	-257,781
September 2015	593,195	569,345	-23,850
October 2015	561.700	600,123	38,423
November 2015	750,565	687,265	-63,300
December 2015	635,321	518,151	-117,170

From Table 1, it can be seen that in July 2015, the mine planned to remove 506,765 bcm, but the actual survey indicates that only 495,423 bcm was removed. The negative variance of -11,342 bcm is an indication that the mine took more time to remove waste; therefore there were delays when it comes to ROM production coal, causing Net present value (NPV) to be lower. This variation is caused by several parameters such as geology, equipment breakdown, absenteeism, machines availability, rainfall, labour unrest etc. These factors can severely affect production rate at the mine, causing loss of profit. Loading hards such as competent sandstone is challenging than loading softs, therefore the loading cycle increases with the size of fragmentation. Huge boulders also cause delays due to their accumulation on the loading pads of an excavator, the mine has to use dozing machines to clean boulders to avoid damages to the trucks' tyres.

In August 2015, the amount of waste that was supposed to be removed is 516,207 bcm but only 258,246 bcm was removed, which is much lower than anticipated, hence production is affected. It is essential for the mine to ensure that there are enough machines to remove overburden, this will enable the mine to reach production targets.

In September 2015, the mine removed 569,345 bcm instead of 593,195 bcm of material which was planned. This results in a variance of -23,850 bcm. In October 2015, the mine planned to remove 561,700 bcm but it ended up removing 600,123 bcm and there was a variance of 38,423 bcm. In November 2015, the mine removed 687,265 bcm instead of 750,565 bcm, the variance was -63,300 bcm. It becomes difficult to mine coal under a circumstance where the overburden is too large and that has a negative impact on costs. In December 2015, the mine removed 518,151 bcm instead of 635,321 bcm and the variance was -117,170 bcm.

Figure 4 shows the monthly waste production, graph of the planned material to be removed and the actual survey. Generally the amount of waste that is planned to be removed is higher than the actual waste that has been removed. This is due to several delays that are encountered during mining and that has an impact when it comes to the amount of waste, which has been removed.

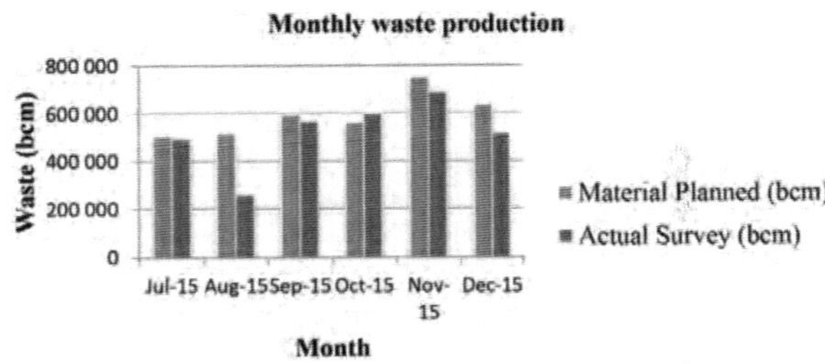

Figure 4. Waste production per month

The results from Table 1 are illustrated graphically in Figure 4, there are variations on the monthly overburden that is supposed to be removed. The month with the lowest waste that was removed is August, only 258,246 bcm was removed. This delay has affected the monthly production target due to large amount of waste that is left on the floor. The mine has failed to remove the planned waste due to several disruptions such as increase in cycle time and breakdown. The planning crew, planned to remove 292,978 bcm (hards) but the actual survey indicates that only 188,086 bcm (hards) were removed. This causes negative variance of -104,892 bcm, resulting in production delays. The mine lost 2212.6 hours of work from 32 machines as a result of breakdowns, hence production was delayed. The high wall in Block 5/8 has caused slow production; operators were not mining at the rate at which they were supposed to mine because the high wall was not taken into consideration during planning.

4.1.1 Cycle time

Authors noticed that most of the delays are caused by the waiting period in the pit. An excavator takes about 5 minutes waiting for the next truck to come to load waste. This is due to the complex design of the pit as a result of geology, access routes are limited. The tipping area is too confined, therefore trucks wait longer period than expected. The road is also not wide enough to allow trucks from two opposing directions to pass each other. Waste material that falls on the road also increases the cycle time, the operator has to reduce the normal travelling speed. Proper monitoring must be done to ensure that there are no spillages on the road that might delay production.

Figure 5 illustrates an excavator in the pit, which is waiting for the truck to come and load waste. This excavator waited for almost 5 minutes without any truck coming. When the lost times are added, the production can be severely affected. It was noticed from the pit that the waiting area is too small only few trucks can be accommodated. This causes trucks to take more time going to the loading area.

Figure 5. Excavator in the pit waiting for truck

4.2 Coal Production

The ROM coal production plays a major role in generating profit for the mine. It is therefore important for the mine to meet the monthly production targets in order to cover its operational costs. Geological structures also affect the production of coal, for example when the coal has been intruded by dolerite intrusions, the quality decreases. The coal becomes completely burnt by these intrusions and it becomes uneconomical to mine it. The different seams where current mining selections take place in the Colliery are Mid seam (MM), Bottom Seam A and Bottom Seam B unit (MAB), Bottom Seam C1 unit (MBC1), Bottom Seam C2 unit (MBC2) and Bottom Seam D unit (MBD).

Table 2 summarizes the amount of coal production that was mined out in each month. The Coal Planned refers to the amount of coal that is predicted to be mined, while actual survey column refers to the amount of coal that was mined out. Variance refers to the difference between Coal planned and actual survey.

Table 2: Coal production per month

Month	Coal Planned (tonnes)	Actual Survey (tonnes)	Variance (tonnes)
July 2015	309,359	273,375	-35,984
August 2015	336,925	139,228	-197,697
September 2015	167,947	136,681	-31,266
October 2015	223,434	204,029	-19,405
November 2015	247,624	209,511	-38,113
December 2015	306,345	306,183	-162

4.2.1 July 2015 coal production

According to Table 2, in July 2015 the mine planned to mine out 309,359 tonnes of coal, but the actual survey indicates that only 273,375 tonnes were mined out. This variation is caused by several parameters such as geology, equipment breakdown, absenteeism and machine availability. The production in July was affected by breakdowns; 5762.8 hours were lost in the same month due to the breakdowns of 50 equipment, this results in lower production rate. In July 2015, there was a considerable gain in MM coal, which added up to 3,065 tonnes. Weathering that was encountered in Block 101/9 (the main block) did not affect a wide area as anticipated. Deep weathering in the same block led to a loss of coal in different layers. In MBAB Coal, there was a loss of 8,636 tonnes and in MBC Coal, there was a loss of 22,387 tonnes.

In block 101/8 and 101/9; 2424 tonnes of MBC2 Coal was left in the floor. After proximate analysis was conducted it was discovered that coal had low volatiles, therefore the coal could not be mined. Coal with low volatiles is uneconomical and it must be blended with high quality coal from the South of the mine to reach the required quality and to optimize production. Deep weathering extended to block 101/12, and coal losses were expected in that block. Block 6/7 has paleo-low which delays production because of the coal seam irregularity. High wall in Block 5/8 makes it difficult to mine, since safety factors need to be taken into consideration when mining this block.

Figure 6 illustrates a Dolerite dyke in Block 101/10, which disrupt production rate. This dolerite dyke is 2.5 m wide and it disrupts the whole block. When dolerite intrudes into geological strata, particularly coal strata, it is known to heat up the coal and may lead to (i) a decrease in the coal quality or (ii) total destruction of the coal seam. Dolerite dykes also pose a threat to mining because they add onto the instability of the mining face and induce poor ground conditions.

Figure 6. Dolerite dyke that cut through the coal seam

4.2.2 August 2015 coal production

In August 2015, the mine only managed to obtain 139,228 tonnes of coal instead of 336,925 tonnes of coal which was predicted. This resulted in a large negative variance of -197,697 tonnes and the production is very small due to large amount of waste that was not removed during that month because of insufficient machines.

4.2.3 September 2015 coal production

The target of 167,947 tonnes of coal to be mined was not reached as planned instead only 136,681 tonnes were mined. The mine spent most of the time removing the overburden and that caused production to be low. The main reason for this slow production is deep weathering in a washout and there is a loss of MBAB coal in Block 101/10. Most of the coal was left on the floor due to low volatile matter; it will be mined at a later stage to blend it with high quality coal from the South. There is deep weathering in Block 102/9 to 102/12, which has also resulted in low production since some of the coal has reported as waste. Another BC2 coal was intentionally left on the floor from Block 101/11 since it has low volatile matter.

The rate of rainfall was a bit higher during this month and that also delayed production, rainfall was 58mm. During rainfall, it becomes difficult to mine and transport coal due to poor visibility and the road is slippery causing machines not to travel well. The other delay in production was caused by breakdowns, whereby 3210.1 hours were lost, 40 machines underwent breakdowns. The floor in Block 6/6 has rolled towards Block 6/7 and that has negatively affected production from this block, it becomes difficult for the operator to mine. The mining has been disrupted in Block 6/7 due to paleo-low; Block 6/6 is higher than Block 6/7. The difference in elevation is 2.72m, which has caused the floor in block 6/6 to slide down to Block 6/7.

4.2.4 October 2015 coal production

The mine planned to mine 223,434 tonnes of coal but they only managed to mine 204, 029 tonnes and the variance is -19,405 tonnes. This delay in production is caused by sharp elevation dip in Block 3/12, which makes mining difficult. The BC2 coal in Block 6/8 and 102/6 remain on the floor due to low volatile matter. There was a gain in MBAB volume since parting 2 (P2) has been mined with the MBAB in Block 6/8 and Block 4/10. The other delays in production were caused by breakdowns, 3476.4 hours were lost. The mine has increased machines, now there are 58 machines instead of 50 machines, 44 machines underwent breakdowns. The rate of rainfall was 18mm and the production was delayed because of rainfall.

4.2.5 November 2015 coal production

The mine did not reach the target only 209,511 tonnes of coal was mined instead of 247,624 tonnes. The rate of rainfall was 116 mm in November and this resulted in production delays. During high rate of rainfall, water ponding on the safety berm need to be drained in order to allow free travelling. If there is water on the safety berm, it becomes difficult for the machines to operate efficiently and that causes slow production. The operation stops a bit during high level of rainfall since there is poor visibility and the road is slippery, that's where the mine loses production due to rain. The breakdowns were also high, only 4384.4 hours were lost from 55 machines. There was coal loss of approximately 1884 tonnes of MM in Block 102/7 because of weathering in this block. There was a gain in MBC volumes because of mining floor coal in Blocks 6/8,101/9 and 102/7. The volatile matter in BC2 coal was low especially for Block 102/8 and Block 102/9.

4.2.6 December 2015 coal production

The target was to mine 306,345 tonnes of coal but the actual survey indicates that only 306,183 tonnes of coal was mined and the variance is -162 tonnes. The coal loss was encountered in MM coal in Block 102/9 and 6/9 because of weathering. The paleo-highs in Block 6/9 and Block 3/12 caused a loss in MBC1 and MBC2 volumes. The BC1 and BC2 packages were mined selectively in Block 102/8 and Block 102/9 in order to optimize quality. The rate of rainfall was 57 mm, it has delayed production causing operation to stop. In December the mine decided to increase the number of machines from 58 to 65. A total of 52 machines underwent breakdowns and this resulted in 5316.7 hours lost due to breakdowns.

Figure 7 shows a graphical representation of the amount of coal produced per month and the predicted amount of coal product. Geological structures also have an influence when it comes to the amount of coal that is produced per month.

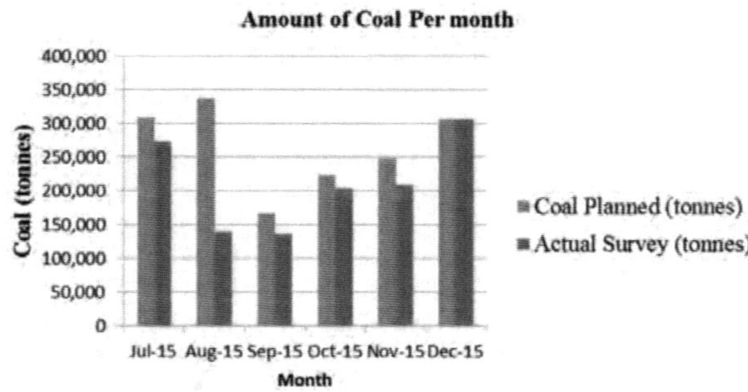

Figure 7. Coal Production Per month

The amount of coal that the mine has planned to achieve varies in each month. The mine rarely reaches the target due to several factors that delay the production rate such as complex geology, cycle time and absenteeism. In August 2015, the production was extremely low; this is due to the large amount of overburden that was not removed. The mine planned to mine out 336,925 tonnes of coal but the actual survey indicates that only 139,228 tonnes of coal were mined. In terms of variance, it was extremely high approximately -257,961 bcm was not removed. The ROM coal production is affected as well as a result; low profit was generated in that month affecting the NPV.

4.3 Rainfall

The Colliery is made up of weathered sandstone, which has a lot of clay material and makes the surface impermeable. The area is generally muddy during rainfall, more water is absorbed until it is saturated and that water accumulates on surface. The topography of the mine causes water to accumulate in lower regions causing difficulties in mining. Problems arise on production when the trucks are supposed to haul material from the loading area into the tipping area due to wet surface. It becomes difficult for the machinery to work effectively on a wet surface; sometimes trucks get trapped, causing delays in production. The cycle time of production is prolonged and the number of loads produced per day decreases. Working under slippery conditions is difficult and pose risk to mining, causing more accidents which might result in downtime.

Figure 8 illustrates the amount of water that has accumulated on the floor as a result of rainfall. The mine has to pump out this water in order to allow mining to take place.

Figure 8. Water on the floor blocking access to the seam

Figure 9 illustrates a graphical representation of rainfall from July 2015 to December 2015. The amount of rainfall has an influence when it comes to production achieved per month. The mining can stop if large amount of rainfall is experienced per month.

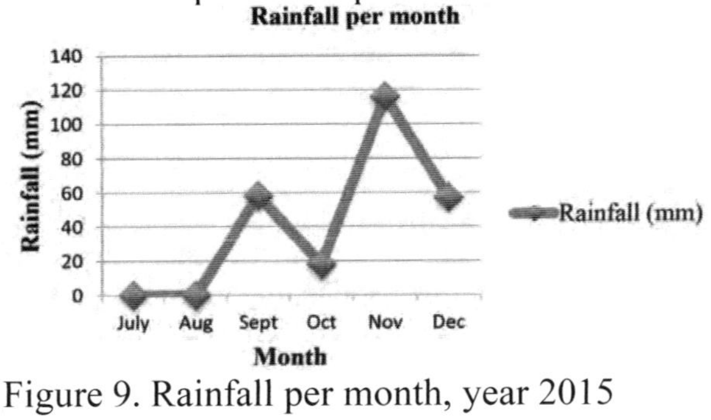

Figure 9. Rainfall per month, year 2015

The level of rainfall decreases the production of the mine. Between July 2015 and August 2015, there was no rainfall hence the ramp did not encounter problems. The production decreased in September 2015 and October 2015 where there was rainfall of approximately 58 mm and 18 mm, respectively. During November 2015, the mine experienced 118 mm of rainfall and the production was disrupted due to downtime. The heavy rainfall causes the floor to become slippery and the efficiency of equipment decreases. The floor of the mine is not flat due to geology, as a result large amount of water accumulates in the workings and coal becomes contaminated. Once the coal is contaminated, more washing needs to take place, hence production is delayed.

4.4 Breakdowns

The large number of breakdowns causes the equipment to be less utilized, therefore production decreases. The decrease in production is caused by less availability of the machine. Most of the breakdowns are encountered when the machine is loading huge boulders from the floor. Huge boulders result from poor blasting; therefore it is necessary to apply secondary blasting in order to reduce the size of the rock. Sometimes these boulders accumulate on the loading pads of an excavator and it gets damaged causing delays in production.

Table 3 illustrates the number of breakdowns at the mine. Total hours lost per month is the summation of all equipment.

Table 3. Breakdowns per month

Year: 2015	Number of breakdowns	Hours Lost	Total Machines
July	47	5762.8	50
August	32	2212.6	50
September	40	3210.1	50
October	44	3476.4	58
November	55	4384.4	58
December	52	5316.7	65

The mine has experienced a large number of breakdowns over a period of 6 months. These breakdowns affect production due to downtime. The floor of the mine is irregular because of complex geology. Equipment gets damaged as the mining progresses and the time is lost while waiting for someone to repair the machine.

Figure 10 illustrates an excavator that has undergone breakdown. The machine stopped for 2 hours, therefore production was delayed. The floor was very poor due to geology. The coal seam was not flat; it was difficult for the operator to adjust the machine.

Figure 10. An excavator that has underwent breakdown

4.5 Overall Production

The waste results indicate that the monthly planned production is not reached at the Colliery. This is due to several delays such as breakdown, machine availability and geology. There is correlation between the amount of waste removed and the amount of coal that is produced. If the mine removes less amount of waste than planned, the ROM coal production is affected; therefore fewer tonnages will be produced in that particular month. It is necessary to ensure that waste is removed as planned in order to avoid delays in ROM coal production.

4.6 Economic Value/Benefits

The results from this study contribute vastly to future mine planning by improving the amount of tonnages mined per month, and the mine will thus have a greater amount of coal to sell to Eskom. The mine should generate more profit with reduced breakdowns and fewer lost hours during breakdowns. It is beneficial to conduct this study in order to reduce operational costs that are associated with the complexity of geology at the Colliery. Geological structures lower production, which affects the profit of the mine, hence operational costs also increases. High production will enhance the revenue of the mine leading to high NPV, hence more investors will be interested in investing into the company. The proper mine plan will enable the mine to reach the planned production in order to cover its operational costs.

4.7 Opportunities for Value Addition

It is crucial to come with mechanisms to lessen the effects of geological structures on the mine's production rate, and therefore enhance mine value. The study proposes the following in this light:

- Monitor plans on the daily basis with both the Surveyor and the Mining Planner
- Multiple access to a specific block will be more efficient than adding more machines so that the planned coal from the pit can go where it is supposed to be without any interferences
- Geological structures in the mine do affect the utilization of the fleet; so it is proposed that the design of the pit, the access, haul roads, maintenance must be planned separately and monitored frequently to reduce the delay in production
- Since the north side of the pit has low quality coal and most of it is left at the stockpiles for a longer period to increase quality, it will be recommended to increase the storage capacity of the stockpiles
- The production can be improved by increasing the availability and utilization of equipment
- The mine should consider adding more dumps in order to avoid delays, such as waiting period

• It is highly recommended to mine one block in the cases where the floor is falling towards another block
• Proper maintenance of the equipment must be done to reduce the number of breakdowns. In so doing, projected production might be reached
• Use sandstone or gravel on haul road to cover clay material as it has a higher infiltration rate than clay; therefore it can be easily compacted. During dry weather conditions the surface can be sprayed with water to suppress the dust.

5 CONCLUSIONS

This study investigated various geological factors that influence the production rate at an open pit Colliery in the Karoo Supergroup. The study considered factors such as the geological structures (faults), igneous intrusions (dykes), paleo-highs and paleo-lows present in the mining area, as well as the amount of rainfall and the efficiency of the equipment used to mine. It was observed that the geological structures have caused major delays in production, which has resulted in lower tonnages of coal produced per month than anticipated. The mined region is dominated by normal faults that displace the coal seams to depths that are considered impractical to mine. As a result of challenging geological mapping, most of these faults were not anticipated during mining. This resulted in mining activities being halted in regions where the coal seams disappeared resulting in reduced production and lower revenue.

In addition to faults, the mining region is also crosscut by dolerite dykes of varying sizes and orientation. In most cases it was observed that the quality of the coal located in close proximity to the dykes differed from the coal located further away from the dykes. The coal located near the dykes had higher ash content and was therefore of lower quality compared to the coal that was located further away from the dykes. This significantly affected the production because the coal that was initially thought of as minable had been 'burnt' by the dykes and had been considered uneconomical to mine. The mine encountered a lot of coal losses as a result of deep weathering and low quality. Therefore adequate and thorough mapping of geological structures is vital so as to ensure proper planning and mining.

The undulating paleo-surface also contributed to the low production achieved by the mine. The paleo-highs have resulted in an unexpected increased elevation of the coal seams and have resulted in the mining activities being slowed down. The coal seam gets thinner towards the paleo-highs and if these paleo-highs are not properly defined, money could be wasted by removing waste to access coal that has thinned out.

Lastly, geological structures related equipment breakdown cause the mine not to function at its optimum rate, thus compromising the set target for production. The equipment breakdown affects the cycle time which in turn negatively affects the production. These findings can be factored into mine plans to ensure that the mine functions to its full potential.

REFERENCES

Hancox, P.J. & Götz, E., 2014. South Africa's coalfields — A 2014 perspective. International Journal of Coal Geology, 132, pp.170–254.
Jha, P.K., Das, T.K. & Soni, A.B., 2014. Effect of Weathering on Coal Quality. [Online] Available at: http://www.ipedr.com/vol75/31_ICQM2014-014.pdf [Accessed 23 July 2016].
Johnson, M.R. et al., 1996. Stratigraphy of the Karoo Supergroup in Southern Africa: an overview. Journal of African Earth Sciences, 23(1), pp.3-15.

Land Use Land Cover Change Analysis for an Abandoned Surface Coal Mine

E. İşleyen, A. Torun, H.Ş. Düzgün

Middle East Technical University, Mining Engineering Department, Ankara

ABSTRACT Monitoring land use land cover changes for abandoned coal mines are critical for assessing the mine closure and reclamation activities. As surface coal mines have large foot prints on the Earth's surface, performing monitoring activities using remote sensing is one of the effective approaches. In this study, the results of land use and land cover (LULC) change analysis for an abandoned coal mine are presented. Research area is located in Ovacık province of Çankırı. Coal production stopped in the mine since 2005 with no reclamation measures taken. LULC analyses are performed with satellite imagery data for 2007 and 2011. Conditions of the mine site in the years 2007 and 2011 are visualized by LULC mapping for both years. Method of maximum likelihood classification is utilized to classify each pixel of image to a specified land use class. Changes between 2007 and 2011 in LULC characteristics influenced by the mining activities are investigated. Considering the differences between values attained to corresponding pixels, changes in LULC characteristics of the abandoned mine after four years are observed.

1 INTRODUCTION

Mine reclamation is an integral part of mining activities, which aims at restoring the mining site to an acceptable state that has been disturbed by the mining operations. The main goal of restoration is increasing quality of land and water regimes for physical, chemical, biological characteristics which has been changed and degraded during the mining process. Reclamation efforts have the intension to achieve to incrementally enhance land use and land cover characteristics to support the region from ecological, social and economical aspects.

Monitoring the land either during or after mine reclamation has been studies in multiple studies. Chitade et.al. (2010), Khan&Javed (2012), Wu et.al (2013), Szostak et.al. (2015) and Baruah et.al. (2016) introduced use of multi-source multi-spectral image analysis along with GIS technology for assessment of LULC at mine sites including coal mines by using mainly parametric image analysis methods. Monitoring a surface coal mine by image classification and analysis by using non-parametric – machine learning- image analysis methods such as support vector machines (SVM) has been conducted by Demirel et.al. (2011) and Petropoulos et.al (2012) and Arican (2015).

Comber et al. (2005) define land cover is as the observed material on the Earth's surface. On the other hand, land use is how the land is utilized for social and economic purposes. In mining industry, especially the surface mines bring about significant inevitable disturbances to the natural state of the land. Land use land cover (LULC) analysis serves for understanding the extent of the disturbance.

Assessing LULC changes sequentially has been valued by researchers in order to monitor land degradation. Significant land use changes are observed in areas affected by mining activities. Performing LULC analysis at pre, post and during mining stages gives a comprehensive information about the imposed disturbances on the land and how the land recovers or responds to the post-mining reclamation measures. Remote sensing technologies offers an effective approach to perform LULC analysis. Information from satellite images plays an important role in identifying the land cover classes. Also, utilization of GIS tools for image processing increases the applicability of this approach. In this study, the recovery of the land without any reclamation measures after years of mining activities is investigated by analyzing the satellite images between 2007 and 2011 using change analysis in LULC.

2 CASE STUDY AREA

The area is located in Ovacık region of Çankırı, in the northern Turkey. Topography of the region consists of small hills with an average elevation of 1300 m. Emil (2010) records the history of mining activities in the study area. The first mining activities started in the region in 1984 by Köseoğlu Mining Company. Proven reserve was estimated as 144,000 tonnes and the company planned to operate the mine for 72 years. Thickness of the coal seam was specified between 0.3-0.8 m with 78-80 degree inclination. Calorific value of the coal was stated as 4500-500 kcal/kg. Both underground and surface mining methods were applied in Ovacık Coal Mine. Köseoğlu Mining operated the mine until 2005, producing 225,000 tonnes of coal (new coal reserves found during operation). After a short time, Uzay Yolu Mining Company bought the production rights. However, the company did not start coal production. Concession agreement was cancelled in 2008 (Emil, 2010). Aerial view of the research area can be seen in Figure 1.

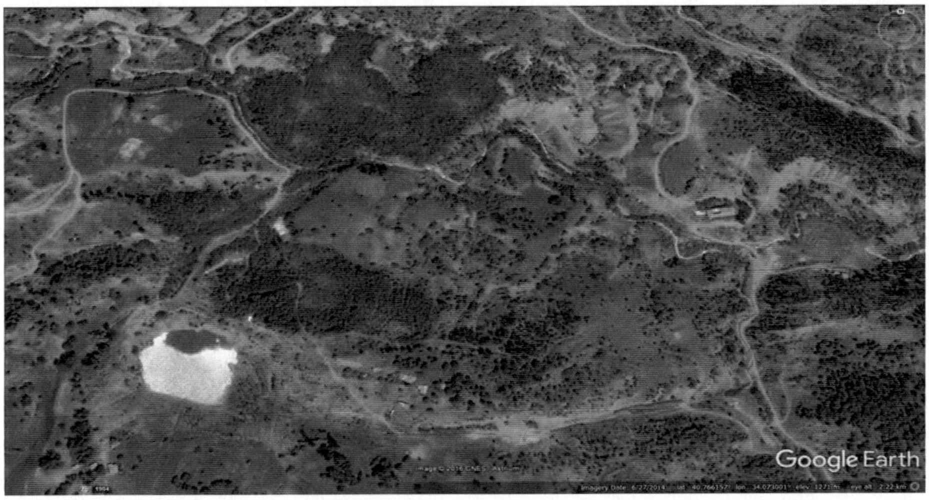

Figure 1: Aerial view of research area (Adapted from Google Earth)

There are several environmental and safety problems in the field, due to lack of mine closure and reclamation measures. Steep slopes of the open pit pose slope stability issues. Also, slopes of the dump sites have been under erosional processes. Acid mine drainage is observed in the field along drainage pathways.

The data set obtained for the study area is listed in Table 1.

Table 1: Data for Ovacık Coal Mine

Data Type	Date
ASTER Imagery	12/09/2007
Worldview-2 Imagery	30/08/2011

3 LULC CLASSIFICATION AND CHANGE ANALYSIS

Classification is the process of allocating objects to classes or categories. Remote sensing data commonly classified using two different approaches; identifying regions with similar characteristics (unsupervised classification) or matching regions to predefined training samples (supervised classification). Supervised classification method is one of the most commonly employed in LULC mapping since it enables users to have some control over the classes (Comber, 2008). This method is also adopted in this study. Mining activities cause land degradation by resulting in unexpected changes in ecosystems and environmental processes and pollution of water, soil and air (Ellis, 2007). Monitoring LULC changes through time is a main prerequisite to make better use of mine reclamation plans and to follow the ongoing effects of mining activities on land. Emil (2010) investigated the LULC change of Ovacık Coal Mine region between 1951 and 2008 to demonstrate the effects of mining on land (Table 2). Mining activities caused the origination of new land cover classes with a significant decrease in the size of the forestland.

The workflow to realize the methodology developed in this study is given in Figure 2. In this study, field works were conducted in the beginning and at the end of the study, for modeling and accuracy assessment, respectively.

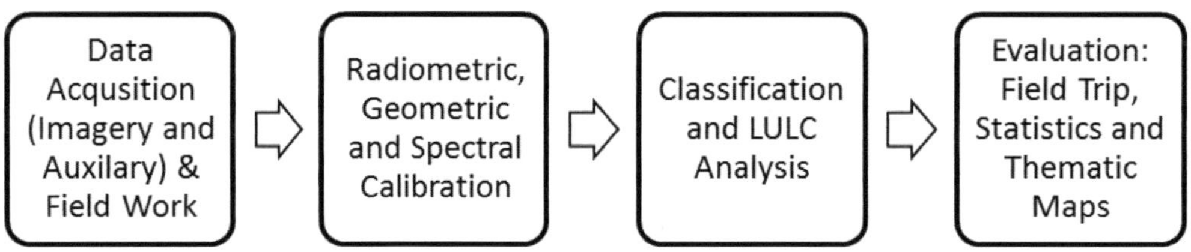

Figure 2: Workflow

Table 2: LULC change matrix between 1951 and 2008 (Emil, 2010)

LULC Change in m^2		2008								
		Agriculture	Bareland	Forest	S. Buildings	Excavation	Dump	Coal Stock	Lake	Total
1951	Agriculture	56059	180	27			3			56269
	Bareland	17688	220530	69482	14468	30634	66202	7594	6404	433002
	Forest	34877	11766	396810	3702	37595	88046	11		572807
	S. Buildings				1173					1776
Total		108624	232476	466319	19943	68229	154251	7605	6404	1063851

Emil (2010) conducted LULC analysis on ASTER imagery taken in 2007 using maximum likelihood classification method. Resultant LULC map of Ovacık Coal Mine is directly used in this study for classification change analysis. In order to create LULC map of the region for 2011, Worldview-2 satellite image is used. Accuracy assessment of the classification is performed. LULC maps are created using the image classification tools in ArcGIS environment.

4 RESULTS

Eight LULC classes are observed in the region, namely agriculture, bareland, forest, sparse buildings, excavation, dump site, coal stock area and lake. LULC map for 2007 (Figure 2) has a noisy appearance caused by the quality of the ASTER imagery and therefore has slightly low accuracy with an overall accuracy of 41%. However, some classes such as forest, agriculture and excavation which covers a large part of the area, shows satisfactory classification results with higher user's and producer's accuracy (Emil, 2010). LULC map for 2011 (Figure 3) has more smooth appearance with an overall accuracy of 70%. Forest, agriculture and bareland classification yield accurate results, reflected by high user's and producer's accuracy. Confusion matrices for both classification results are presented in Table 3 and Table 4.

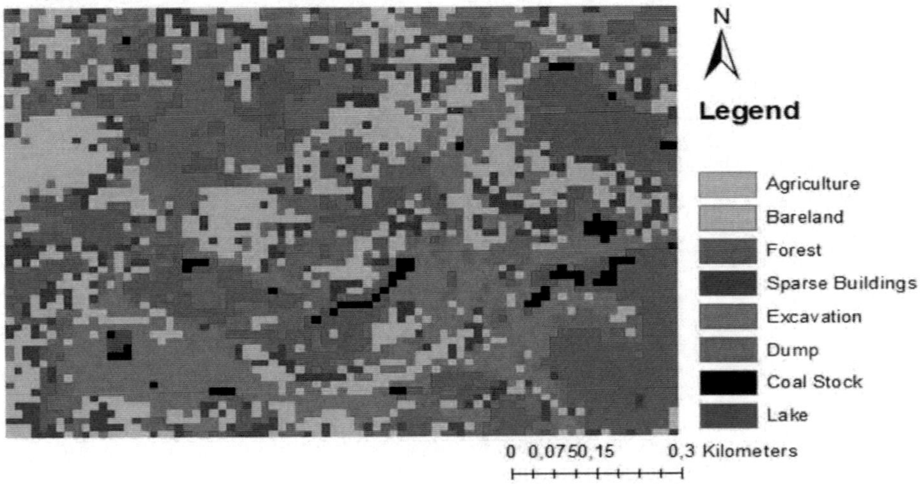

Figure 3: LULC Classification at 2007 (Emil, 2010)

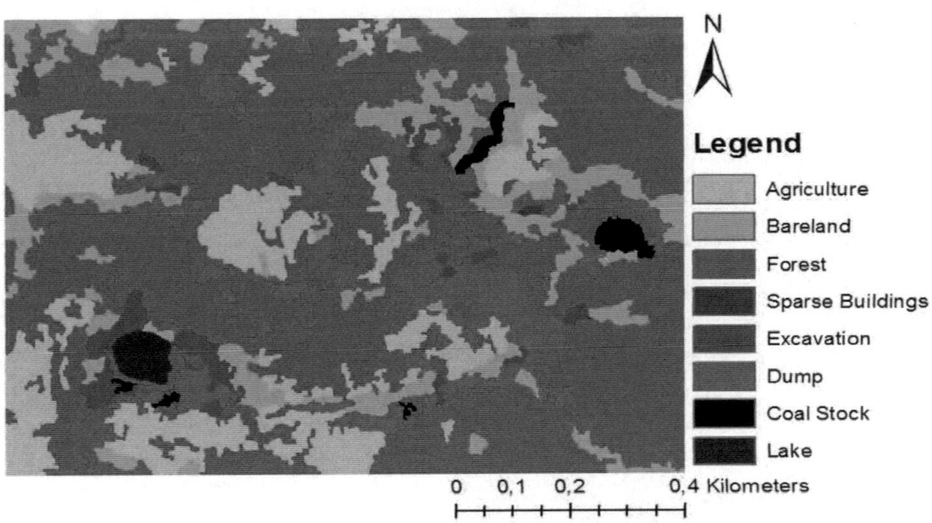

Figure 4: LULC Classification at 2011

Table 3: Confusion matrix of classification results for 2007 (Emil, 2010)

				Ground Truth (pixels)						
Class	**A**	**B**	**F**	**S**	**E**	**D**	**C**	**L**	**Total**	**U.A.**
A	71	17	54	3	1	6	0	0	152	0.47
B	1	55	35	0	6	5	0	0	102	0.54
F	8	11	183	0	0	35	0	0	237	0.77
S	22	24	58	3	0	3	0	0	110	0.03
E	12	83	31	2	51	10	5	1	195	0.26
D	4	44	126	0	0	67	0	0	241	0.28
C	0	2	3	0	0	2	1	0	11	0.09
L	2	0	0	0	0	0	0	0	2	1
Total	118	236	490	8	58	128	6	6	1050	
P.A.	0.60	0.23	0.37	0.38	0.88	0.52	0.17	0.33		
Overall Accuracy = 41%					**Kappa Coefficient=0.2812**					

Table 4: Confusion matrix of classification for 2011

				Ground Truth (pixels)						
Class	**A**	**B**	**F**	**S**	**E**	**D**	**C**	**L**	**Total**	**U.A.**
A	79	5	65	0	6	0	0	0	155	0.51
B	2	86	25	0	10	0	0	0	123	0.70
F	19	25	495	3	10	23	1	0	576	0.86
S	0	0	0	1	0	0	0	0	1	1
E	0	5	7	0	15	43	1	0	146	0.51
D	3	31	53	0	15	43	1	0	146	0.29
C	0	0	0	0	0	0	10	0	10	1
L	0	0	0	0	0	0	0	10	10	1
Total	103	152	645	4	56	68	12	10	1050	
P.A.	0.77	0.56	0.77	0.25	0.26	0.63	0.83	1		
Overall Accuracy = 70%					**Kappa Coefficient=0.523**					

The LULC changes between 2007 and 2011 are evaluated by comparing areal distribution of each class and presented with a change matrix (Table 5). Also, area covered by each class are given as a percentage of the total area.

Table 5: LULC change matrix between 2007 and 2011

LULC Change in m^2		2011									
		Agriculture	Bareland	Forest	S. Buildings	Excavation	Dump	Coal Stock	Lake	Total	Percent (%)
2007	Agriculture	70795	15925	45050	0	1800	11025	225	225	145225	13.8
	Bareland	8075	48475	32875	0	5625	12725	1800	2025	111600	10.6
	Forest	27950	29575	161800	0	3150	14850	900	225	238450	22.7
	S. Building	17775	19350	51075	450	1800	12825	675	0	103500	9.8
	Excavation	31950	30700	49100	0	74400	12425	4725	4950	208700	19.9
	Dump	41825	17325	120975	0	5625	38925	1800	225	226700	21.6
	Coal Stock	900	1125	8325	0	450	3375	1575	0	15750	1.5
	Lake	0	0	0	0	0	225	0	900	1125	0.1
Total		199450	162475	469200	450	92850	106375	11700	8550	1051050	
Percent (%)		19	15.5	44.6	0	8.8	10.1	1.1	0.8		

5 DISCUSSION

Mining induced changes in the LULC was detected by Emil (2010) by comparing the pre- and post-mining stages of the research area. Mining activities caused new LULC classes to emerge and the disturbance on the land was substantial. However, after the mining activities are stopped in 2005, Ovacık Coal Mine was abandoned without any mine closure and reclamation measures. The LULC maps for both years are given in the previous section. Classification of 2007 is prone to some misclassification due to the low resolution of the imagery data, with an overall accuracy of 41% and kappa coefficient of 28%. However, large LULC classes such as forest, bareland and agriculture are classified with enough accuracy for a comparison. Classification of 2011 yielded good results for all classes with overall accuracy of 70% and kappa coefficient of 52%. Investigating the change matrix given in the previous section, reveals that the maximum change is observed on the forest class, due to the forestation works on the dump site. This also explains the decrease in the size of the dump class. Agricultural areas are increased because of the increased agricultural activities of the nearby villages. Observed decrease for sparse building class is due to the misclassified pixels in the LULC map for 2007. Buildings used by mining company are still in place and a few small houses are also built by villagers. In addition to that, excavated areas (e.g. open pit) are still present with a small amount of decrease in size and still possess safety threats for the people around the region.

This study emphasizes the need for mine closure and reclamation measures as the severity of land degradation of the abandoned mine site is still at considerable level. Areas disturbed by mining activities exhibit very slow recovery over the time that is passed.

6 CONCLUSION

The analysis for determining change and trend LULC characteristics within the scope of this study has been realized by analysis and classification of multitemporal multi-spectral satellite imagery. The findings show that there are significant changes in the LULC classes of forestry, vegetation and agricultural as well as bareland and dump area, respectively at the site.

In this study, LULC change between 2007 and 2011 for Ovacık abandoned surface coal mine was investigated. LULC changes prove that the lack of reclamation activities cause the region to have serious environmental and safety problems that will continue to exist for years. Effective mine closure and reclamation plans should be enforced on mining companies to prevent contamination on soil and water sources, safety hazards and to eliminate negative social and economical effects of mining.

REFERENCES

Arican, İ. (2015). *Land Use Land Cover Change Analysis of Afsin Elbistan Coal Basin with Two Different Classification Methods* (Master's thesis supervised by Prof. Dr. S. Duzgun). Middle East Technical University, Ankara, Turkey.

Baruah, J & Baruah, B. K. & Kalita, S. & Choudhury, S. K. (2016). Impact Analysis of Open Cast Coal Mining on Land Use/Land Cover using Remote Sensing and GIS Technique in LedoMargherita Region of Assam, India, *Imperial Journal of Interdisciplinary Research (IJIR)* , Vol-2, Issue-8, 2016 , ISSN: 2454-1362.

Chitade A.Z., & Katyar, S.K. (2010). Impact Analysis of Open Cast Coal Mines on Land Use/ Land Cover Using Remote Sensing and GIS Technique: A Case Study, *International Journal of Engineering Science and Technology*, Vol. 2 (12), 2010, 7171-7176.

Comber, A., Fisher, P., & Wadsworth, R. (2005). What is land cover. *Environment and Planning B: Planning and Design* , 32 (2), 199-209.

Comber, A. (2008). The separation of land use from land cover using data primitives. *Journal of Land Use Science*, 3 (4), 215-229.

Demirel, N & Emil, M.K. & Duzgun, H.S. (2011). Surface coal mine area monitoring using multi-temporal high-resolution satellite imagery, *International Journal of Coal Geology*, 86 (2011) 3–11.

Emil, M. K. (2010). *Land degradation assessment for an abandoned coal mine with geospatial information technologies* (Master's thesis supervised by Prof.Dr.S.Duzgun). Middle East Technical University, Ankara, Turkey.

Ellis, E., & Pontius, R. (2007). Land-use and land-cover change. Retrieved January 19, 2017, from Encyclopedia of Earth. Available at http://editors.eol.org/eoearth/index.php?title=Land-use_and_land-cover_change.

Google Earth v 7.1.7.2066. Ovacık Coal Mine, Çankırı, Turkey. 40.766157° lat, 34.073001° lon, 1727 m elev, 2.22 km eye alt. CNES/Astrium 2014. http://www.earth.google.com (January 19, 2017).

Khan, I & Javed, A (2012). Spatio-Temporal Land Cover Dynamics in Open Cast Coal Mine Area of Singrauli, M.P., India, *Journal of Geographic Information System*, 2012, 4, 521-529.

Petropoulos, G.P & Partsinevelos, P. & Mitraka, Z (2012). Change detection of surface mining activity and reclamation based on a machine learning approach of multi-temporal Landsat TM imagery, *Geocarto International*, 2012, 1–20.

Szostak, M. & Wezyk, P. & Hawrylo, P. & Pietrzykowski, M. (2015). The analysis of spatial and temporal changes of land cover andland use in the reclaimed areas with the application of airborne orthophotomaps and LANDSAT images, *Geodesy And Cartography*, Vol. 64, No 1, 2015, pp. 75-86.

Wu, K. & Ye, X. & Qi, Z. & Zhang, H (2013). Impacts of land use/land cover change and socioeconomic development on regional ecosystem services: The case of fast-growing Hangzhou metropolitan area, China, *Cities 31* (2013) 276–284.

Masif ve Bozunmuş Beyaz Tüflerin Küre Deney Örnekleri üzerinde Suda Dağılmaya Karşı Duraylılık İndeks (SDI) Değerlerinin Kıyaslanması

A Comparison of Slake Durability Index (SDI) Values of Weathering and Massive White Tuffs on Sphere Test Samples

H. Ankara
Eskişehir Osmangazi Üniversitesi, Maden Mühendisliği Bölümü, Eskişehir/Türkiye
S. Yerel Kandemir
Bilecik Şeyh Edebali Üniversitesi, İnşaat Mühendisliği Bölümü, Bilecik/Türkiye
F. Çiçek
Eskişehir Osmangazi Üniversitesi, Maden Mühendisliği Bölümü, Eskişehir/Türkiye

ÖZET Suda dağılmaya karşı Duraylılık İndeksi Testi (Slake Durability Index-SDI) yerbilimleri-mühendislik problemlerinde zayıf ve killi kayaçların parçalanma karakteristiğinin belirlenmesi için yaygın olarak kullanılmaktadır. Bu çalışmanın asıl amacı; bozunmuş ve masif beyaz tüf örneklerinin SDI indeks değişimi küre ve yuvarlak test örnekler üzerinde değerlendirmektir. Bu amaç için beyaz tüf kaya blok örnekleri Derbent-Eskişehir-Türkiye beyaz tüf ocağından alınmıştır. Bozunmanın etkisini ölçmek için küre ve yuvarlak biçimli deney örnekleri üç grup olarak hazırlanmıştır. İndeks değerleri 12 çevrimlik SDI deneyine tabi tutularak üç farklı grup için belirlenmiştir. 12'nci çevrim sonundaki SDI değerleri masif tüf örnekleri için % 72,48, yuvarlak örnekler için % 69,03 ve bozunmuş küre tüf örnekleri için % 64,67 olarak bulunmuştur. Çoklu çevrim sonucu belirlenen SDI değerleri; masif ve bozunmuş tüflerle kıyaslandığı zaman, masif küre deney örneklerinin SDI değerleri bozunmuş ve masif yuvarlak deney örneklerine göre daha yüksek bulunmuştur. Bozunmuş küre deney örneklerinin SDI değeri 7'nci çevrime kadar yüksek çıkmasına rağmen, 7'nci çevrimden sonra SDI değerleri hızlı düşüş göstermiştir. Sonuç olarak, standart iki çevrimlik SDI deneyi çoklu çevrim ile kıyaslandığında, geleneksel iki çevrimlik SDI deneyi bozunmuş beyaz tüflerde kabul edilebilir bir duraylılık göstermemiştir.

ABSTRACT The Slake Durability Index (SDI) test is widely used to determine the disintegration characteristic of the weak, and clay-bearing rocks in geo-engineering problems. The main purpose of this study is to assess the changes of SDI values of weathering and massive white tuffs on sphere and rounded test samples. For this purpose, white tuff block samples were collected from white tuff quarry in Derbent village, Eskişehir, Turkey. Three different sets for assessment of influence of weathering were prepared as the test samples which had sphere and rounded shapes. Index values were determined for the three different sets subjected to the SDI test for 12 cycles. The SDI values at the end of twelve cycles were found to be 72.48 % for massive sphere tuff sample, 69.03 % for rounded sample, and 64.67 % for weathering sphere tuff sample. When the SDI values determined from multiple-cycle tests were compared with massive and weathering tuffs, the SDI value of massive sphere test sample set was found to be higher than weathering and rounded sample test samples. Although the SDI value of weathering sphere test sample set was higher than rounded test sample set up to 7 cycles, the SDI values of weathering sphere test sample after 7 cycles were rapidly decreased. As a result, it is emphasized that two-cycle conventional SDI test did not appear to offer an acceptable indication of the durability of weathering white tuff rocks when compared with the SDI values for multiple-cycle test.

1 INTRODUCTION

Weathered tuffs are a transitional material between weak rock and soil, slope stability and bearing capacity failures as well as flow and slide problems are often encountered in weathered tuffs (Chigira and Yagi 2006; Crosta et al. 2005; Yassaghi and Salari-Rad 2005; Delibalta 2012). Clays form as a result of chemical alteration of the primary minerals in tuffs, it is essential that the weathering states of tuffs are determined according to generally accepted criteria so that their long-term engineering behavior in excavations and foundation pit slopes can be assessed satisfactorily (Kıncal et. al., 2010).

Slake durability testing can be employed while quantifying the weathering state of tuff samples (Topal, 2002). Determining the weathering state using slake durability test results can be improved by adding the geometrical dimensions and shape changes to the test data. This may be achieved by examining the tuff particles after a certain number of cycles in the test and quantifying the particle shapes using the fractal dimension approach (Turk et al. 1987; Vallejo 1994; Kolay and Kayabali 2006). In the majority of the previous studies, the aim of this test is to provide an index that is related to resistance of rocks against degradation when it is subjected to two Standard cycles of wetting and drying (Franklin and Chandra, 1972); Singh et al. (2004) by ISRM (2007) and ASTM (1998). However, some investigators (Ulusay et al. 1995; Gökceoğlu et al. 2000) emphasized that two-cycle slake durability testing did not appear to provide an acceptable indication of the durability of clay-bearing rocks. Ulusay et al. (1995) and Gökçeoğlu et al. (2000) carried out three-cycle, four-cycle and five-cycle tests, respectively. A five-cycle slake durability test was performed on the tuff samples with different weathering grades in order to assess the relationships between Id values and the number of cycles (Kıncal et. al., 2010). A four-cycle SDI test was carried out on white tuff with the saturation of the equal size sphere test samples (Ankara et. al., 2013a).

In this study, changes of SDI values of massive and weathering white tuffs were investigated using multiple-cycle SDI test. Disintegration behaviors of massive and weathering tuffs were compared with the SDI test results.

2 MATERIALS AND METHOD

2.1 Rock Sample

In this study, white tuff block samples were collected from white tuff quarry in Derbent village, Eskişehir, Turkey. Tuff extracted in this place is called as Derbent white tuff. The mineralogical structure of white tuff consists of Quartz, K-Feldspar, biotite and phenomenon crystals of opaque minerals. Besides, various metamorphic rock and white pumice pieces can be observed. Phenomenon crystals, rock pieces and white pumice are buried in the structure of slightly decayed tuff. In tuffs, the existence of volcanic glass pieces is very common (Lünel, 1974; Sözmen, 2000; Daloğlu, 2008). Some physicomechanical properties of the white tuff are weak, very prous, low uniaxial compressive strength, low point load strength, and slake durability index. The results of previous studies on the white tuff were summarized in Table 1. The chemical content of the white tuffs was determined using XRF. Chemical composition consists of SiO_2 69.62 %, Al_2O_3 13.30 %, K_2O 5.09 %, Na_2O 1.93 %, Fe_2O_3 1.51 %, CaO 1.06 %, TiO_2 0.10 %, LO I 7.7 %.

2.2 Sphere Sample Preparation Method

The massive rounded test samples were prepared according to ISRM (2007) and TSE (1990). The preparation of equal-sized spherical test samples consists of three stages. First stage is to cut cubes whose size is accord with final spherical sample diameter from collected rock samples. At this stage, the length of cube edge is calculated as follows; the final sample weight is required to be 50 g by standards. Therefore, the diameter of the spherical sample is calculated from volume and dry

Table 1. Summary of some physico-mechanical properties of the white tuffs

Properties	Ayday and Göktan (1990)	Binal et. al. (1997)	Topal and Sözmen (2003)	Daloğlu* (2008)
Porosity (%)	28.30	33.10	38.82	41.61
UCS (MPa)	8.15	11.00	10.00	11.39
Slake durability–Id_2 (%)	92.20	87.00	91.00	93.80

* Derbent white tuff samples

density of the rock. After the determination of diameter, the edge length of the cube is set to be the diameter plus 2 mm.

Second stage is to cut cubes to form a pre-spherical shape. A pre-spherical shape is performed in two steps. In the first step, the edges of cube are cut. The newly formed corners of the samples after first step are cut in the second step. The final stage is to obtain equal sized spherical samples from pre-spherical shaped samples by means of an instrument modified for this purpose. Pre-sphere samples are placed into sphere preparation machine (Covington brand). Special cutter cups having a diameter of 35 mm were designed and manufactured (Ankara et. al., 2011; Ankara et. al., 2013a, b, c; Ankara et. al., 2015).

Table 2. Slake durability index test results

Indexes	Massive %	Rounded %	Weathering %
Id_1	98.46	96.62	97.36
Id_2	96.33	94.36	95.39
Id_3	94.25	91.88	93.33
Id_4	92.22	89.36	91.09
Id_5	90.00	86.96	88.36
Id_6	87.68	84.00	84.90
Id_7	85.21	81.44	81.47
Id_8	82.77	78.70	77.11
Id_9	80.06	76.37	73.66
Id_{10}	77.49	73.99	70.22
Id_{11}	74.87	71.55	67.25
Id_{12}	72.48	69.03	64.67

2.3 Slake Durability Index Test

In the present investigation, the test was conducted on the white tuff samples in accordance with ISRM and ASTM standards by using a Standard slake durability test apparatus. Two different equal sized spherical sample sets were prepared as massive and weathering white tuff. The weight of sample for the SDI test is about 50 g each. The sample sets are shown in Figure 1.

The slake durability tests were performed on these sample sets according to ISRM test procedure. Tests were repeated for twelve cycles. The condition of samples at the end of twelve cycles is shown in Figure 1. The test results were summarized in Table 2.

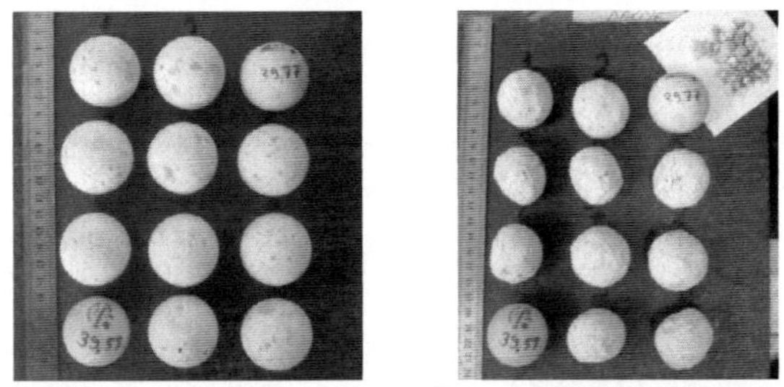

Before test (massive sphere) After test

Before test (massive rounded) After test

Before test (weathering sphere) After test

Figure 1. The condition of samples at the end of twelve cycles.

The results of massive and rounded test samples were compared to the results of the weathering tuff test sample in which spherical Derbent tuff samples were subjected to slake durability index test (Figure 2).

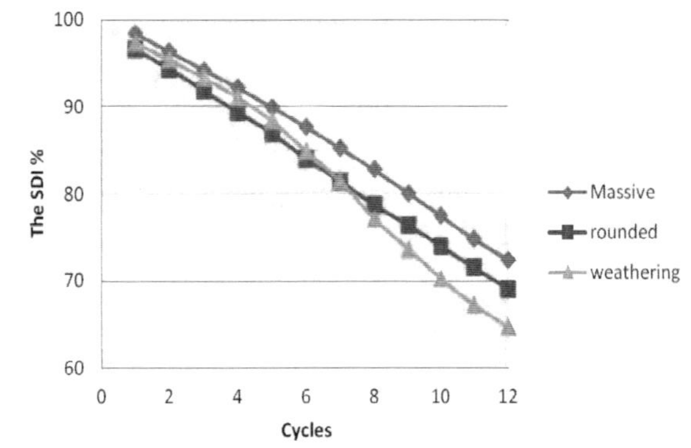

Figure 2. The SDI values of three test samples

As it can be seen this figure, the results of the weathering samples after twelve cycles were lower than the massive sphere and rounded samples . Although the SDI value of weathering sphere test sample set was higher than rounded test sample set up to 7 cycles, the SDI value of weathering sphere test sample after 7 cycles was rapidly decreased.

3 CONCLUSIONS

In this study, Slake Durability Index (SDI) test were carried out on the equal sized sphere and rounded test samples taken from Eskisehir-Derbent quarry on white massive and weathering tuffs. According to standart SDI test, the SDI values after second cycle were found to be % 94.36 for massive rounded sample, and % 95.39 for weathering sphere tuff sample. In addition, the SDI values at the end of twelve cycles were determined to be 72.48 % for massive sphere tuff sample, 69.03 % for massive rounded sample, and 64.67 % for weathering sphere tuff sample. The SDI value of weathering tuff test sample at the end of twelve cycles was lower than massive rounded test samples. However, the SDI value of weathering sphere tuff test sample between 1 and 7 cycles was found higher than rounded test samples, and then it rapidly decreased. As a result, it is emphasized that two-cycle conventional SDI test did not appear to offer an acceptable indication of the durability of weathering white tuff rocks when compared with the SDI values of massive tuff samples for multiple-cycle test.

KAYNAKLAR

Ankara, H., Aksoy, M., Yerel, S. and Keser, Y., 2011. A new sample preparation method for slake durability index test. 4th Balkan Mining Congress, Ljubljana-Slovenia. pp. 571-575.

Ankara, H., Aksoy, M. and Yerel, S., 2013a. Suda dağılmaya karşı duraylılık testi için kayaçlardan eş boyutlu küresel örneklerin hazırlanması, Bilimsel Araştırma Projesi, Eskişehir Osmangazi Üniversitesi, Eskişehir, Türkiye,

Ankara, H., Aksoy, M., Yerel, S., Keser, Y. and Çiçek, F., 2013b. Determination of slake durability index on saturated spherical samples, 5th Balkan Mining Congress, Ohrid-Macedonia, pp. 571-575.

Ankara, H., Aksoy, M., Yerel, S. and Keser, Y., 2013c. The determination of slake durability index with spherical samples. Key Engineering Materials, 548, pp. 247-252.

Ankara, H., Yerel Kandemir, S. and Çiçek, F., 2015. Comparison of slake durability index (SDI) values of sphere and rounded marl samples, World Multidisciplinary Earth Sciences Symposium, WMESS, pp. 93-98.

ASTM, 1998. D4644 Standard test method for slake durability of shales and similar weak rocks. Annual book of ASTM standards. American Society for Testing and Materials.

Ayday, C. ve Göktan, R.M., 1993. Yazılıkaya (MIDAS) anıtı civarında gözlenen kaya blok devrilme ve kayma mekanizmaları, *Türkiye Jeoloji Kurultayı Bülteni*, 8, pp. 155-159.

Binal, A., Kasapoğlu, K. E. ve Gökçeoğlu, C., 1997. Eskişehir-Yazılıkaya çevresinde yüzeylenen volkanosedimanter kayaçların donma-çözülme etkisi altında bazı fiziksel ve mekanik parametrelerin değişimi, *Yerbilimleri Uygulama ve Araştırma Merkezi Bülteni*, 19, pp. 17-40.

Chigira, M. and Yagi, H., 2006. Geological and geomorphological characteristics of landslides triggered by the 2004 Mid Niigata prefecture earthquake in Japan. Eng. Geol., 82, pp. 202–221.

Crosta, G.B., Imposimato, S., Roddeman, D., Chiesa, S. and Moia, F, 2005. Small fast-moving flow-like landslides in volcanic deposits: the 2001 Las Colinas Landslide (El Salvador). Eng. Geol., 79, p. 185–214.

Delibalta, M.S., 2012. Kömür açık işletmelerinde pasa şev stabilitesinin hipoplastik model ile tespiti, Niğde Üniversitesi Mühendislik Bilimleri Dergisi, 1, 1, pp. 50-59.

Daloğlu, G., 2008. Eskişehir-Derbent tüflerinin yapı taşı olarak değerlendirilebilirliği, MSc. Thesis, ESOGÜ Fen Bilimleri Enstitüsü, Eskişehir, Turkey.

Franklin, J.A. and Chandra, R., 1972. The Slake Durability Test, International Journal of Rock Mechanics and Mining Sciences, 9, pp. 325-341.

Gökçeoğlu, C., Ulusay, R. and Sönmez, H., 2000. Factors affecting the durability of selected weak and clay-bearing rocks from Turkey, with particular emphasis on the influence of the number of drying and wetting cycles. Eng. Geol., 57, pp. 215–237.

ISRM, 2007. Ulusay, R. and Hudson, J. A. (Eds.), The complete ISRM suggested methods for rock characterization, testing and monitoring, 1974-2006, suggested methods prepared by the commission o testing methods, International Society for Rock Mechanics, Compilation Arranged by The ISRM Turkish National Group, Ankara, Turkey.

Kıncal, C., Koca, M.Y., Özden, G. and Demirbasa N., 2010. Fractal paremeter approach on weathering grade determination of Çeşme (Izmir, Turkey) Tuffs, Bull Eng. Geol. Environ., 69, pp. 617-629.

Kolay, E., Kayabali, K., 2006. Investigation of the effect of aggregate shape and surface roughness on the slake durability index using the fractal dimension approach, Eng. Geol., 86, 4, pp. 271–284.

Lünel, T., 1974. A Preliminary study of the tertiary volcanic and sedimentary rocks, Gümele, Eskişehir, Türkiye Jeol. Kur. Bült., 18, pp. 31-50.

Singh, T.N., Verma, A.K., Singh, V. and Sahu, A., 2004., Slake durability study of shaly rock and its predictions. Environ. Geol., 47, pp. 246–252.

Sözmen, B., 2000. Investigation of detoration mechanism of Yazılıkaya Tuffs in Midas Monument, MSc. Thesis, METU, Ankara, Turkey.

Topal, T., 2002. Quantification of weathering depths in slightly weathered tuffs. Environ. Geol., 42, 6, pp. 632–641.

Topal, T. and Sözmen, B., 2003. Deterioration mechanisms of tuffs in Midas monument, Engineering Geology, 68, pp. 201-223.

TSE 8543, 1990. Kayaçların şişme ve suya dayanıklılık özelliklerinin tayini, Türk Standartları Enstitüsü, Ankara, p. 7.

Turk, N., Grieg, M.J., Dearman, W.R. and Amin, F., 1987. Characterization of rock joint surfaces by fractal dimension. In: Proceedings of 28th US symposium on rock mechanics, Rotterdam, Balkema, pp. 1223–1236.

Ulusay, R., Arıkan, F., Yoleri, M.F. and Çağlan, D., 1995. Engineering geological characterization of coal mine waste material and an evaluation in the context of back-analysis of spoil pile instabilities in a strip mine SW Turkey. Eng. Geol., 40, pp. 77–101

Vallejo, E.L., 1994 Fractal analysis of the slake durability test. Can. Geotech. J., 31, pp. 1003–1008

Yassaghi, A. and Salari-Rad, H., 2005. Squeezing rock conditions at an igneous contact zone in the Taloun tunnels, Tehran-Shoumal freeway, Iran: a case study. Eng. Geol., 42, pp. 95– 108.

Atomik-Kuvvet Mikroskobu ile Parlatılmış Mermer Yüzey Morfolojisinin İncelenmesi

Examination of Polished Marble Surface by Using the Atomic-Force Microscopy (Afm)

H. Ankara
Eskişehir Osmangazi University, Department of Mining Engineering, Eskişehir, Tırkey

S. Yerel Kandemir
Bilecik Şeyh Edebali University, Department of Civil Engineering, Bilecik, Tırkey

H. Aras
Eskişehir Osmangazi University, Department of Mechanical Engineering, Eskişehir, Tırkey

S. Pat
Eskişehir Osmangazi University, Department of Physics, Eskişehir, Tırkey

ÖZET Bu çalışmada morfolojinin parlatılmış mermer yüzey kalitesi üzerindeki etkisi Atomik-Kuvvet Mikroskobu (AFM) kullanılarak araştırılmıştır. Bu amaç için on iki çeşit mermer örneği için sekiz farklı konumda tarama yapılarak, parlatılmış mermer yüzeylerden yüzey morfolojisinin üç boyutlu görüntüleri elde edilmiştir. Daha sonra yüzey morfolojisi ile yüzey pürüzlülüğü arasındaki ilişki araştırılmıştır. Bütün mermerlerde yüzey pürüzlülüğün artmasıyla sünek akma morfolojik oluşumu mermer yüzey kalitesine etkisi olduğu sonucuna varılmıştır. Parlatılmış yüzeydeki pürüzlülük azalması yarı küresel biçimlerin oluşumu mermer yüzey kalitesini artırdığı belirlenmiştir.

ABSTRACT In this study, the influence of morphology on quality of polished marble surfaces has been investigated using Atomic-Force Microscopy (AFM). For this purpose, the three dimensional images of surface morphology were obtained from the polished marble surfaces by scanning on the eight different positions for twelve different kinds of marble samples. After that, the relationship between the surface roughness and the surface morphology was examined. Consequently, it has been concluded that the occurrences of ductile flow morphology with increasing in polished surface roughness in all marble samples has been appeared to have an effect on marble surface quality. The forms of semi-spherical morphology with decreasing in the polished surface roughness have been determined in the improvement of the marble surface quality.

1 INTRODUCTION

Natural Stone is widely used as decorative materials in terms of beautiful color, pattern, brightness and smooth surface. However the brightness and roughness are the most important quality criterion and are obtained from precision machining processes. Surface grinding and polishing processes are applied in the final stages to obtain glossy and smoothing surface. Grinding and polishing of marble surfaces with traditional abrasives pointed out that a thin scale glossy film might exist on the polished marble surface. This film may play the most important role in determining the brightness of stone surface. The mechanisms concerning the glossiness on machined stone surfaces have been studied widely but are not yet understood (Huang et. al., 2002; Huang and Xu, 2004; Görgülü and Ceylanoğlu, 2008). The relationship between roughness and glossiness is reciprocal. When the surface glossiness starts to increase conspicuously, surface roughness value decrease rapidly (Görgülü and Ceylanoğlu, 2008). The determination of surface quality should not be only depending on glossiness and roughness but also surface morphology (topography) which can be analyzed reaching fine results. The surfaces layered polished materials are of great interest at the nanometer scale and are studied using nanoindentations with atomic force microscopy (Caro et. al., 2001; Fraxedas et. al., 2002). AFM is a powerful technique for determining the three-dimensional surface morphology.

For this purpose, in this study, relationship between roughness and morphology of polished stone surfaces was investigated on twelve different stone types by scanning the AFM to determine marble surface quality.

2 EXPERIMENTAL STUDY

The surface morphologies and the roughnesses were analyzed using an Ambios Q-scope atomic force microscope. The three-Dimensional images and roughnesses of the AFM were obtained using a 10 x10 micrometer scale. The AFM measurements were taken from eight different points at these marble surfaces. These polished surface images are given in Fig. 1.

The images in figure 1 indicate obvious traces of ductile flow on the polished marble surfaces when the rougnesses of marble surfaces are the highest value. The amount and areas of the ductile flows have decreased as the roughness value minimized. The morphologies of polished surface in the minimum roughnesses have converted to semi-spherical appearance instead of ductile flows.

3 SINGLE LINKAGE CLUSTER METHOD

The steps in the Single Linkage Cluster Method are given as follows (Sharma, 1996):

• The number of clusters and the number of observations are determined.

• The Euclidean distances between the pairs of cluster are calculated.

• *N*-dimensional symmetric Euclidean Distance Matrix (DME) is formed from using Euclidean distances.

• The Similarity Matrix (SIME) is computed usingDME. SIME is represented by the similarity values between pairs of cluster.

• Maximum similarity value between all pairs of cluster is selected and this cluster pair is joined to form a newcluster.

• Above steps are repeated $n-1$ times, until all clusters are combined in a single cluster.

• After, clusters are represented graphically in what is called a dendrogram.

Relationship between the roughness in Table 1 and similarity for 12 marble types was evaluated using the cluster analysis.

Table 1. The Roughness of the Polished marble surfaces

Marble Types			Roughness µm	
			Min-Ra	Max-Ra
1	Rosso Levanto	Elazığ Vişne	0.0044	0.0358
2	Perlato Giallo	Akçadağ Kaysısı	0.0064	0.0262
3	Best Cream	Akçadağ Krem	0.0061	0.0214
4	Perlato Rosa	Akçadağ Gülü	0.0066	0.014
5	White Onix	Beyaz Oniks	0.004	0.0454
6	Blue Onix	Mavi Oniks	0.0037	0.0765
7	Honey Onix	Bal Oniks	0.0057	0.0228
8	Premium Gray	Karaman Gri	0.006	0.0199
9	Cream Karaman	Karaman Krem	0.0066	0.0202
10	Gloden Beige	Darende Bej	0,005	0.0955
11	Black Pearl	Siyah İnci	0,0064	0.0393
12	Breccia Adonis	Konglomera	0,0061	0.0884

Dendrograms were constructed using the clusters joined versus the similarity values (Fig. 2). Examining the dendrogram for 12 marble types, it was seen that all cluster combines with minimum 61.85 % similarity (Fig. 2). It could be said that all marble has a high similarity between the roughness and the surface morphologies.

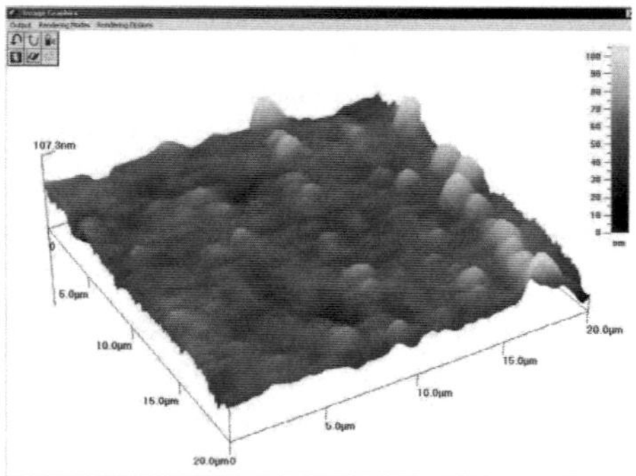

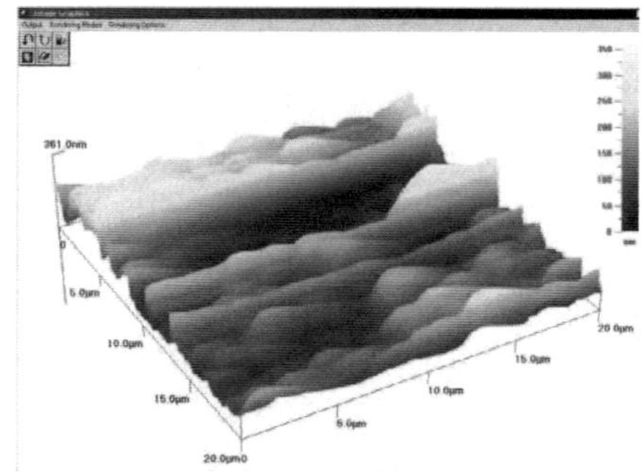

Rosso Levanto

Ra=0.00438 μm

Ra=0.0358 μm

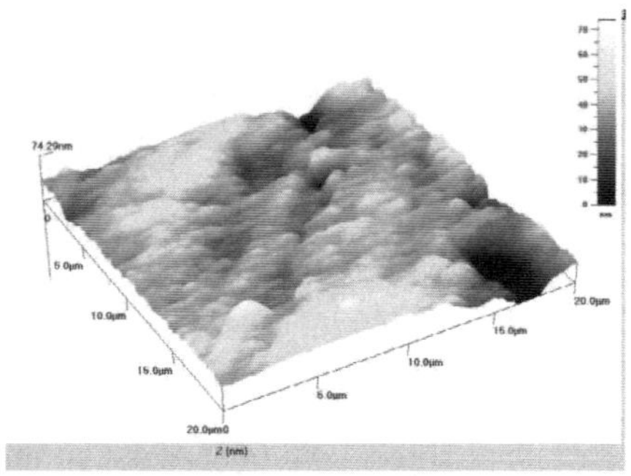

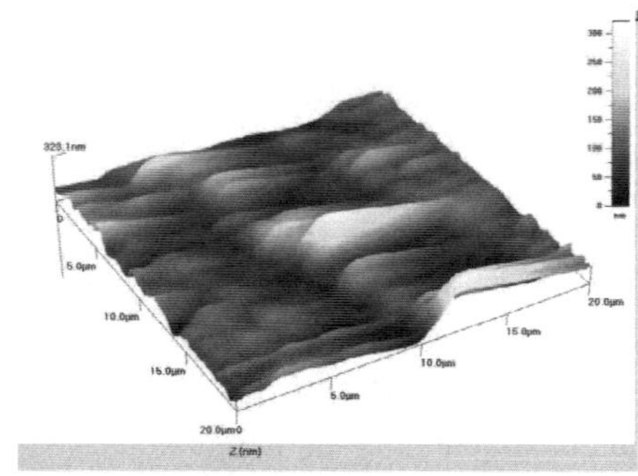

Perlato Giallo

Ra=0.00640 μm

Ra=0.0262 μm

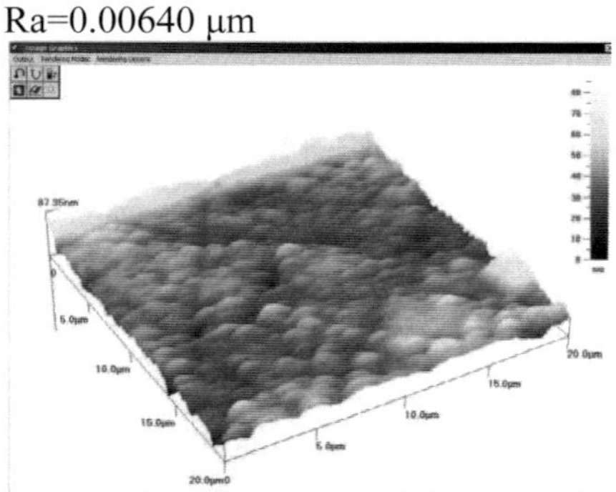

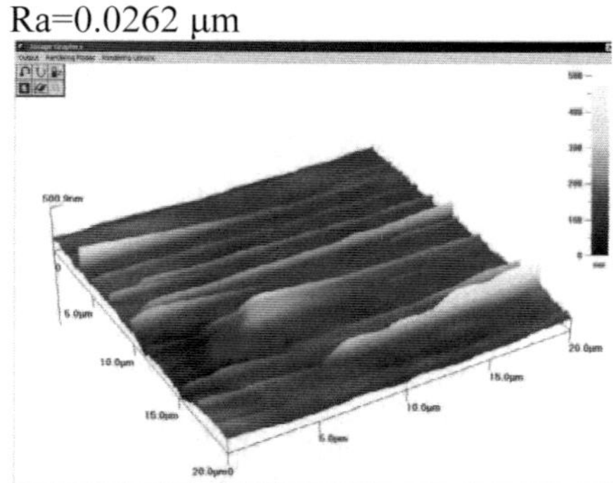

Best Cream

Ra=0.00606 μm

Ra= 0.02140 μm

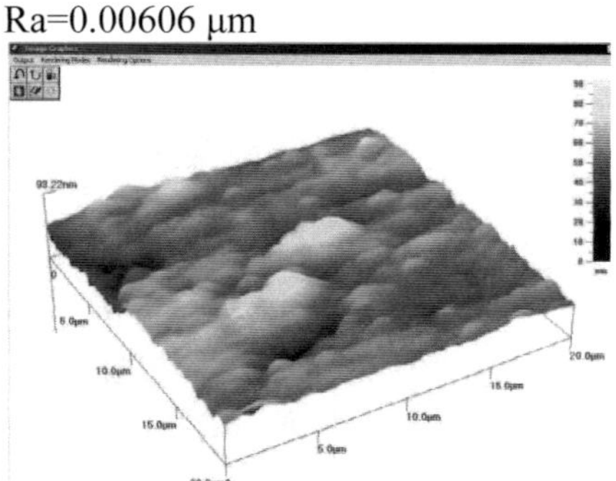

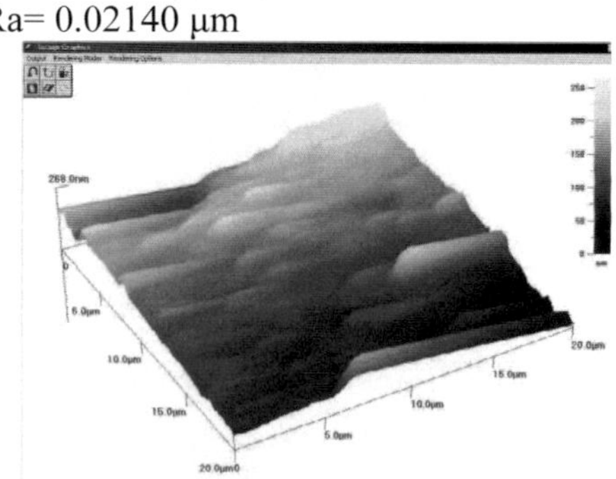

Perlato Rosa

Ra=0.00655 μm

Ra=0.014 μm

Figure 1. The AFM Images of the Polished Marble Surfaces.

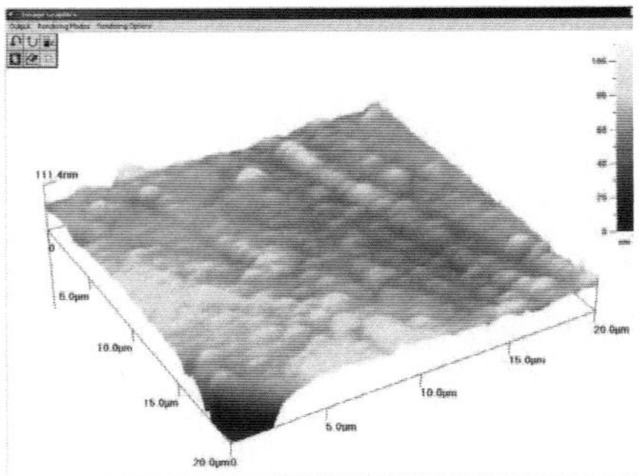
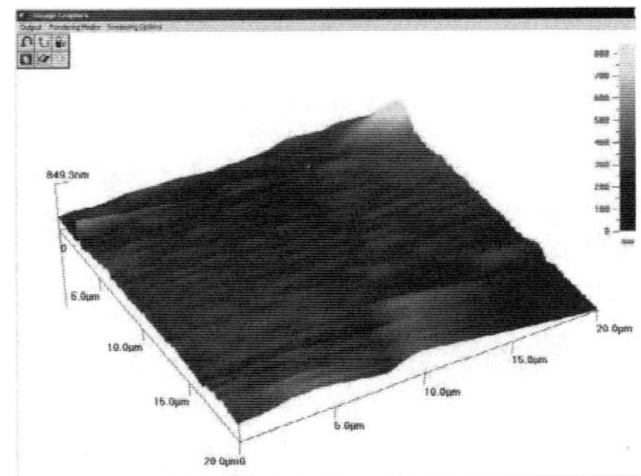

WHITE ONIX

Ra= 0.00399 μm Ra= 0.0454 μm

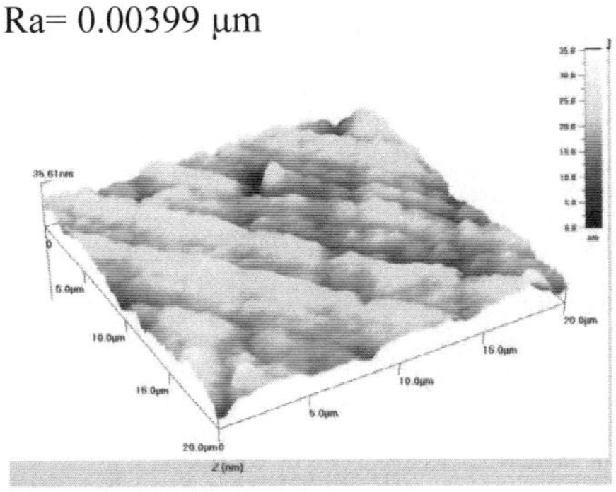
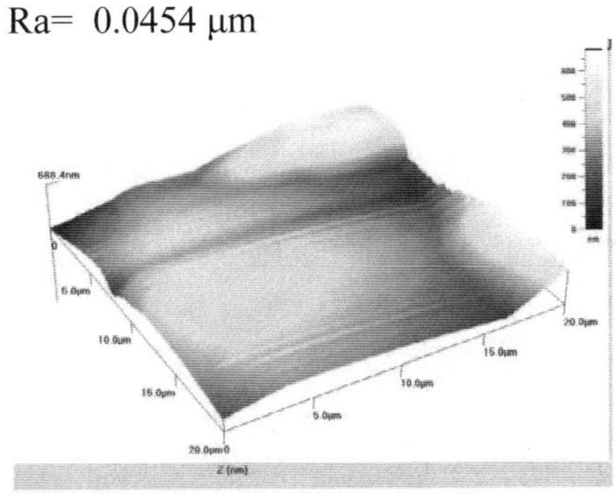

BLUE ONIX

Ra= 0.00368 μm Ra= 0.0765 μm

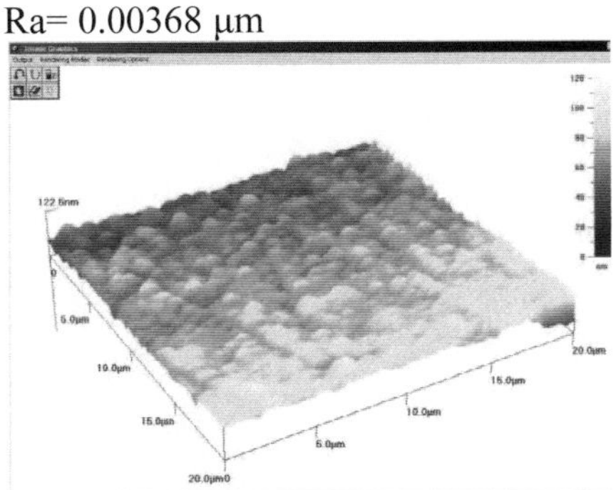
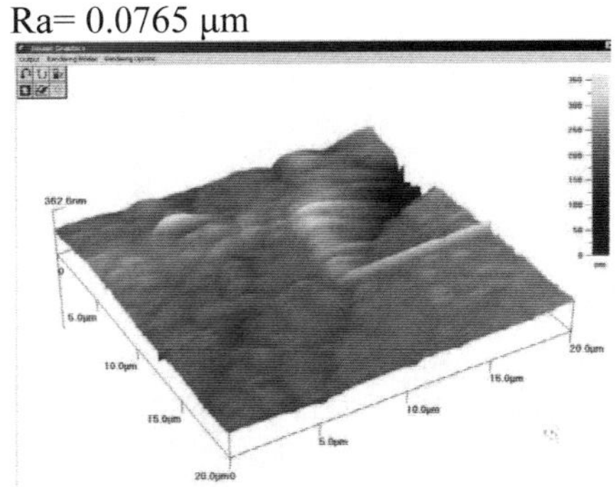

HONEY ONIX

Ra= 0.00567 μm Ra= 0.0228 μm

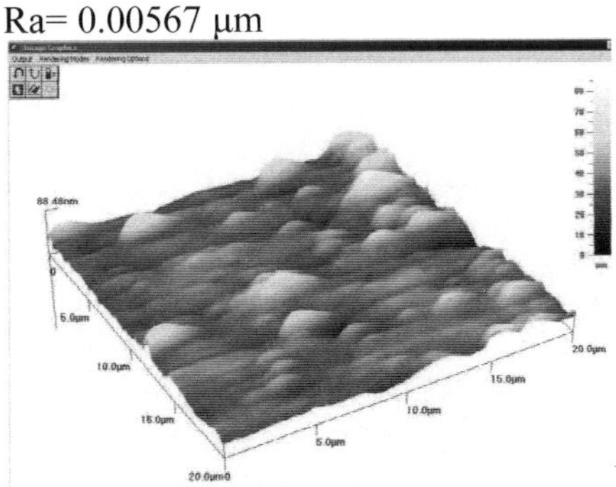
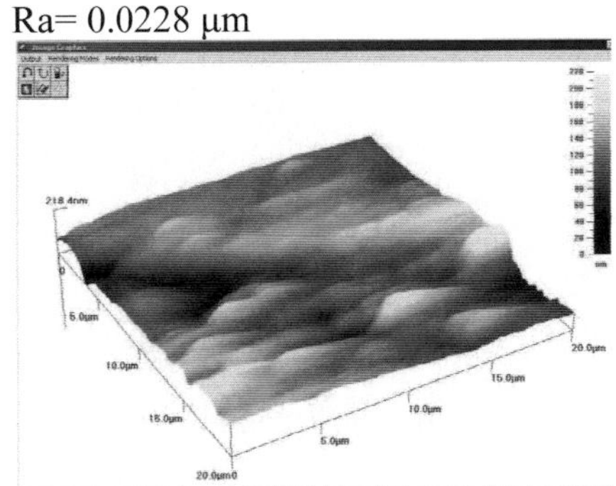

PREMIUM GRAY

Ra= 0.00598 μm Ra= 0.0199 μm

Figure 1(Cont.). The AFM Images of the Polished Marble Surfaces.

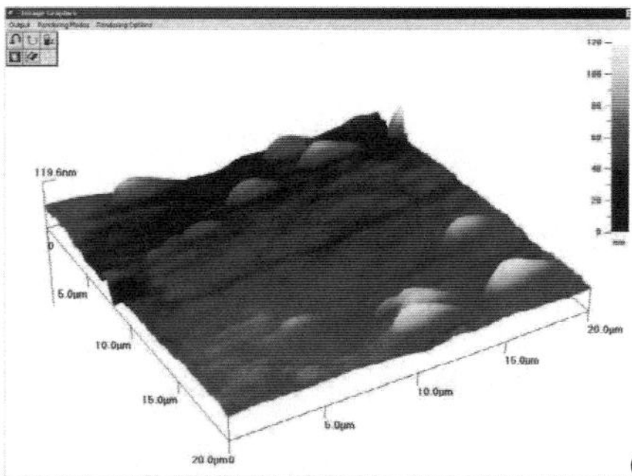
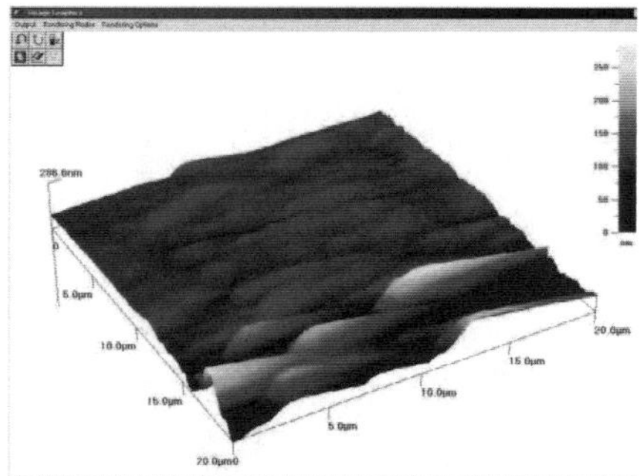

CREAM KARAMAN

Ra= 0.00664 μm

Ra= 0.0202 μm

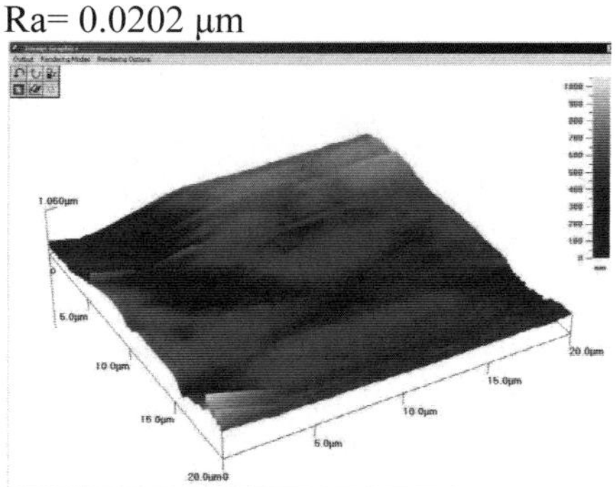

GOLDEN BEIGE

Ra= 0.00497 μm

Ra= 0.0955 μm

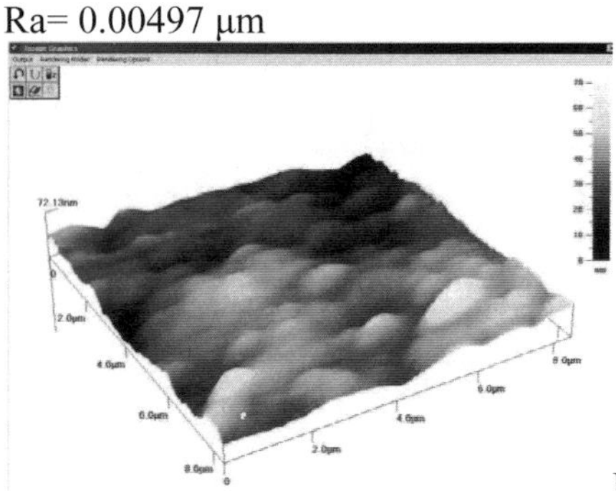
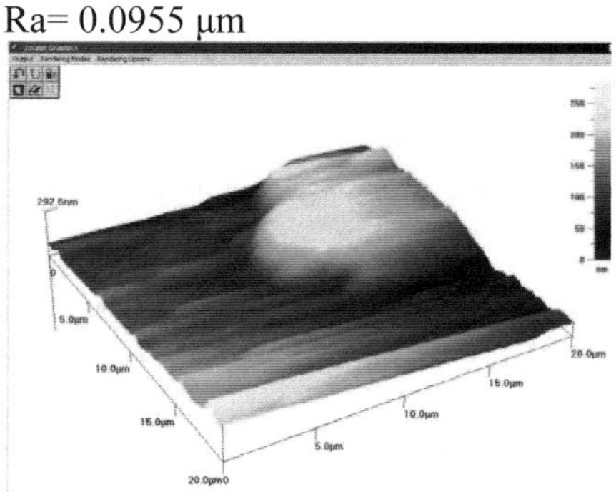

BLACK PEARL

Ra= 0.00635 μm

Ra= 0.0393 μm

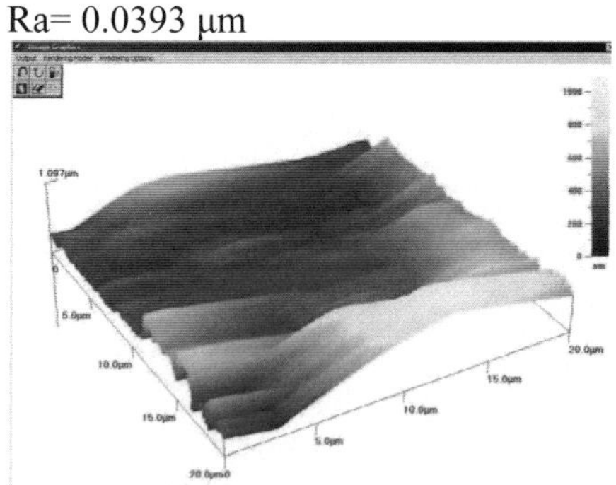

BRECCIA ADONIS

Ra= 0.00610 μm

Ra= 0.0884 μm

Figure 1(Cont.). The AFM Images of the Polished Marble Surfaces.

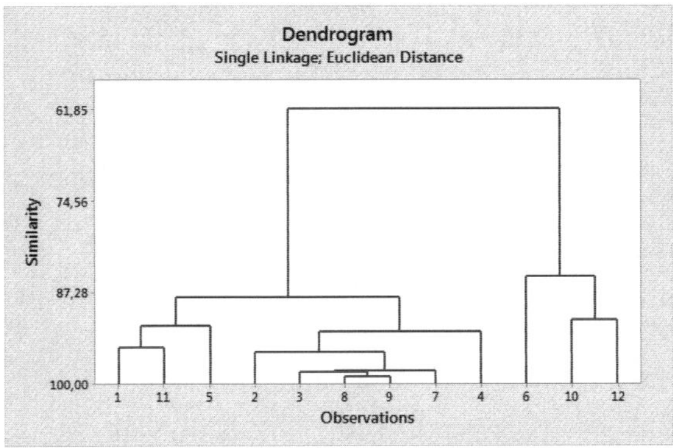

Figure 2. Dendrogram for polished marble surface roughness.

4 RESULTS

It could be said that the occurrences of ductile flow morphology in increasing in roughness in the polished surfaces of all marble samples has been appeared to have an effect on marble surface quality. The occurrences of semi-spherical morphology in decreasing roughness in the polished surface have been determined in the improvement of the marble surface quality. This idea was also supported by cluster analysis.

REFERENCE

Caro, J., Gorostiza, Sanz, P., F. and Fraxedas, J., 2001, Nanomechanical propertiesof surfaces of molecular organic thin films, Synth.Met., 121, pp. 1417–1418.

Fraxedas, J., Garcia-Manyes, S., Gorostiza, P. and Sanz, F., 2002. Nanoindentation: toward the sensing of atomic interactions, Proc. Natl. Acad. Sci. USA 99, pp. 5228–5232.

Görgülü, K. and Ceylanoğlu, A., 2008. Evaluation of continuous grinding tests on some marble and limestone units with silicon carbide and diamond type abrasives, Journal Of Materials Processing Technology, 204, pp. 264–268

Huang, H., Li, Y., Shen, J.Y., Zhu, H.M., Xu, X.P., 2002. Micro-structure detection of a glossy granite surface machined by the grinding process. J. Mater. Process. Technol. 129, 403–407.

Huang, H. and Xu, X. P., 2004. Interfacial interactions between diamond disk and granite during vertical spindle grinding, Wear, 256 , pp. 623–629.

Sharma, S., 1996. Applied Multivariate Techniques. John Wiley & Sons, Inc., New York, pp. 188–193.

Utilization of Brown Fused Alumina as an Alternative Abrasive in Abrasive Waterjet Cutting of Rock

S. Kaya, G. Aydin, I. Karakurt

Karadeniz Technical University, Mining Engineering Department, Trabzon, Turkey

ABSTRACT As an alternative abrasive, utilization of Brown Fused Alumina (BFA) in abrasive waterjet (AWJ) cutting of marble is investigated in this study. The performance of the BFA is compared with the garnet which is widely used as an abrasive in the AWJ cutting applications. Marble samples are cut with the constant operating variables of the AWJ (traverse speed, abrasive mass flow rate, waterjet pressure and standoff distance). Cutting performances of these two abrasives are evaluated based on the cut depth, kerf width, kerf angles and surface roughness. The results showed that the BFA can be effectively used as an alternative to garnet in the AWJ cutting of rock such as marble.

Keywords: Abrasive waterjet, rock cutting, abrasives, brown fused alumina, garnet

1 INTRODUCTION

Abrasive waterjet (AWJ) technology is one of the non-traditional machining processes. The AWJ technology presents many advantages such as precise shape cutting, extended tool life, process automation, complex free-form cutting, no dust, better working conditions, and environment. These features make the technology an environmentally friendly technique over other traditional cutting processes such as circular sawing in natural stone machining and processing applications (Miranda and Quintino 2005; Vundavilli et al. 2012).

In the AWJ cutting (Fig. 1), a high speed stream of abrasive particles and water mixture passing through a narrow neck (nozzle) remove the particles from the material surface. Abrasive particles are employed in order to cut harder materials such as glass, metals and rocks (Karakurt, 2011). Selecting the appropriate abrasive for AWJ cutting applications can significantly affect the cutting performance and profitability of equipment. Garnet is generally used an abrasive material in the AWJ cuttings (90%) (Mort, 1995). Nevertheless, there are some materials in the market that can be evaluated as an abrasive in the AWJ. In this study, as an alternative abrasive, utilization of Brown Fused Alumina (BFA) in the AWJ cutting is investigated.

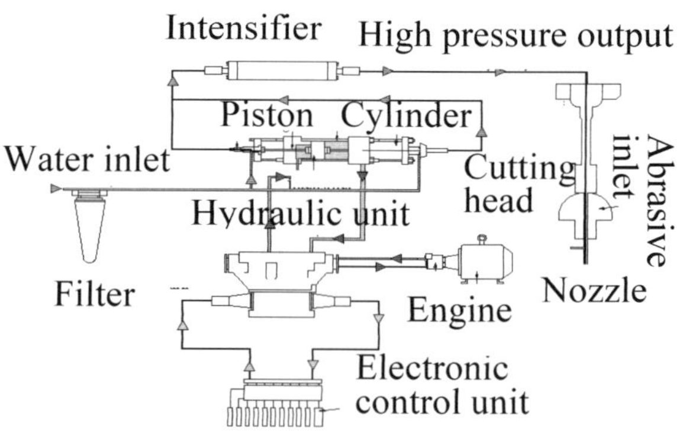

Figure 1. AWJ system (Kartal, 2015)

2 EXPERIMENTAL STUDY

2.1 Material

Marble with the traditional name of Crema Eda (Sivrihisar Beige) was used as the cutting material. It was preferred due to its homogenous structure. This structure enables to constant rock properties and determines the abrasives performance effectively for each cutting. Some properties of the marble are shown in Table 1. Mohs hardness of the marble was approximately 3. Samples are prepared in a thickness of 3 cm, length of 20 cm and width of 10 cm (Fig. 2). The marble chemically consist of 55.5% Cao, 41.75% loss of ignition, 2.20% MgO, 0.24% SiO_2 and 0.21% Fe_2O_3.

Table 1. Some properties of the marble

Standard	Property	Value
ASTM-C170	Compressive Strength (MPa)	88.80
ASTM-C880	Flexural Strength (MPa)	8.85
ASTM-C97	Density (g/cm^3)	2. 68
ASTM-C97	Water absorption (%)	0.19

Chemical composition of the abrasives are presented in Table 2. The BFA and garnet were provided in a narrow range of size (150-300 µm) in order to minimize the effect of size distribution on the cutting performance. Figure 3 shows garnet and BFA particles that examined under microscope (20X). As can be seen, the BFA particles have similar grain shapes (the edges of sharp and slightly round) with garnet. It is also seen that garnet contains different liberated fractions while the BFA seems homogenous.

Table 2. Chemical compositions of abrasives

Content (%)	Garnet	BFA
Al_2O_3	23	96
SiO_2	35	1
Fe_2O_3	33	0.1
MgO	7	0.2
TiO_2	-	2.5
CaO	1	0.2
MnO	1	-

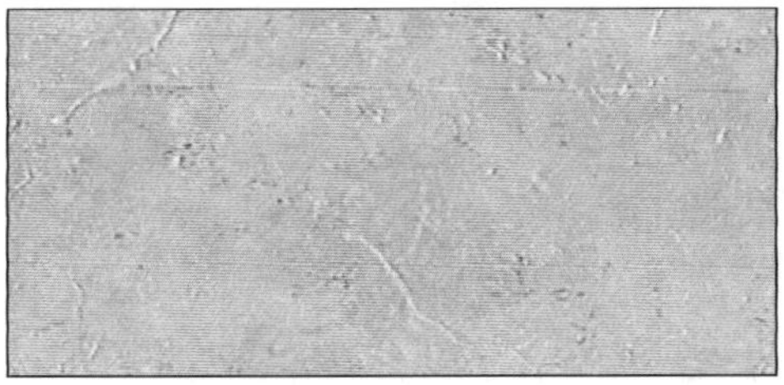

Figure 2. Surface of a marble sample.

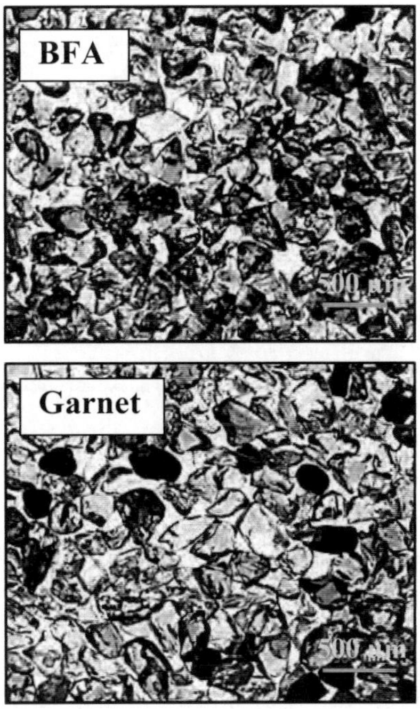

Figure 3. Microscope images of the BFA and garnet

2.2 Experimental Method

The experiments are conducted on an AWJ cutter (Fig. 4). The nozzle diameter is 1.1 mm and its length is 75 mm. The diameter of orifice is 0.33 mm. The marble samples are cut through their lengths and each sample is cut three times in the same condition using each abrasive. Owing to the variability and accuracy of the experimental data, each sample was cut three times in the same conditions.

Figure 4. AWJ used in the study

All the operating variables are maintained at constant values (Table 3). For the performance assessment of the abrasives, cutting depth, cutting width, kerf angle and surface roughness are considered.

Table 3. Levels of cutting parameters

Cutting parameters	Value
Traverse speed (mm/min)	100
Abrasive flow rate (g/min)	350
Standoff distance (mm)	4
Water pressure (MPa)	200
Abrasive size (μm)	150-300

Totally 27 [9 (on the each cutting line) x 3 (repetition)] measurements were taken for cutting depth and width and the average was recorded as the final reading for the cutting depth and width for each abrasive. Kerf angle is also determined as a performance output. Kerf angle is a quantity that is often used to reflect the inclination of the kerf wall from the top surface to the bottom of the kerf (Wang and Guo, 2003). It is an undesirable geometrical feature inherent to AWJ machining. A schematic diagram of the kerf profile is shown in Figure 5.

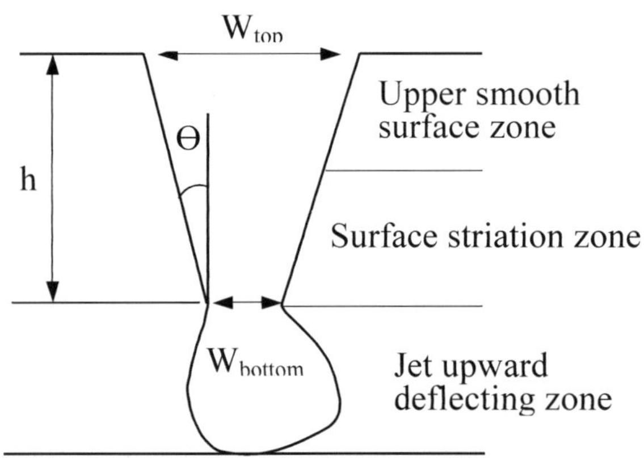

Figure 5. Schematic illustration of kerf profile produced by AWJ (Liu, 2004)

The kerf angles (kerf angle-access and kerf angle-exit) for each cut were determined using Eq. 1. Kerf angle-access and kerf angle-exit were measured for each cut from the marble surfaces where the AWJ cutting were started and finalized, respectively. Three measurements were made for the variable included in the Eq. 1 and the kerf angles for access and exit were calculated. Averages of the measurements were recorded as final kerf angle-access and kerf angle-exit.

$$\Theta = \tan^{-1} [(W_{top} - W_{bottom}) / 2h] \quad (1)$$

Where; Θ is the kerf angle (degree), W_{top} and W_{bottom} are the top and the bottom kerf widths, respectively, and h is the kerf depth (mm).

Another performance output considered in this study is surface roughness. Surface roughness measurements of the cut surfaces of marble samples were made using a stylus-type profilometer, Mitutoyo Surftest SJ-301 (DIN EN ISO 4287). Due to the variability of surface finish data, totally 27 measurements for three cutting were taken at the upper zone (smooth zone) of the cut surface and the average was taken as the final surface roughness.

3 RESULTS AND DISCUSSION

A summary of the experimental results are presented in Table 4.

Table 4. A summary of the experimental results

Content	Garnet	BFA
Cutting width (mm)	2.21	2.13
Cutting depth (mm)	15.62	16.84
Ra (μm)	4.59	4.34
Rq (μm)	5.63	5.38
Rz (μm)	22.52	21.90
Kerf angle-access	5.58°	4.53°
Kerf angle-exit	4.98°	3.90°

As can be seen from Fig. 6, lower cutting width and higher cutting depth were obtained using the BFA. These results indicated that the BFA showed the better cutting performance. With the comparison of garnet, improvement in the cutting width and the cutting depth were calculated as 4% and 8%, respectively for the BFA.

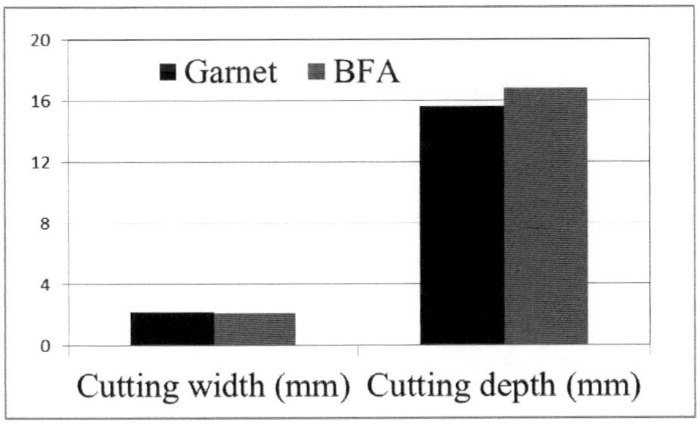

Figure 6. Abrasive performances based on cutting width and cutting depth

Kerf angle is a quantity that is often used to reflect the inclination of the kerf wall from the top surface to the bottom of the kerf. It is a special and undesirable geometrical feature inherent to AWJ machining. Higher kerf angles indicate the poor cutting performances since the inclined surfaces are required additional treatments for smoothing. As can be seen from Fig. 7, the BFA shows better performance over garnet in terms of kerf angles. The BFA produced smaller kerf angles of 19% for access and 22% for exit as compared to garnet.

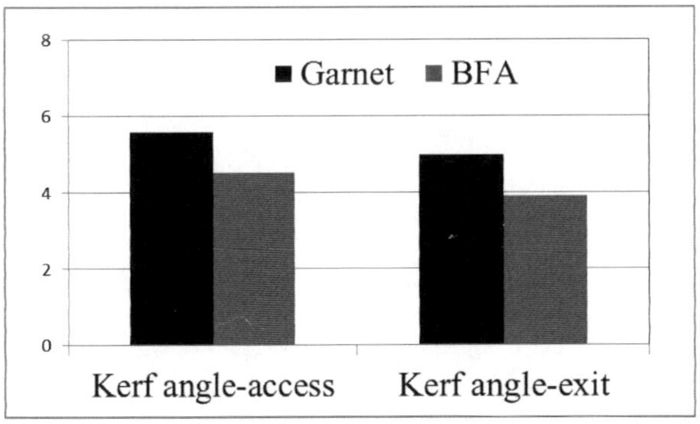

Figure 7. Performance comparison of abrasives based on kerf angles

Surfaces produced by the AWJ, can be divided into two zones along the depth of cut as cutting wear and deformation-wear zone. Uniform surface texture occurs at the top zone of the cut since water jet particles impact the kerf wall at shallow angles. This uniform surface is generally assessed by a surface roughness measure. The performances of two abrasives in terms of the surface roughness are presented in Fig. 8. Similar values of surface roughness were recorded in the cuttings with two abrasives.

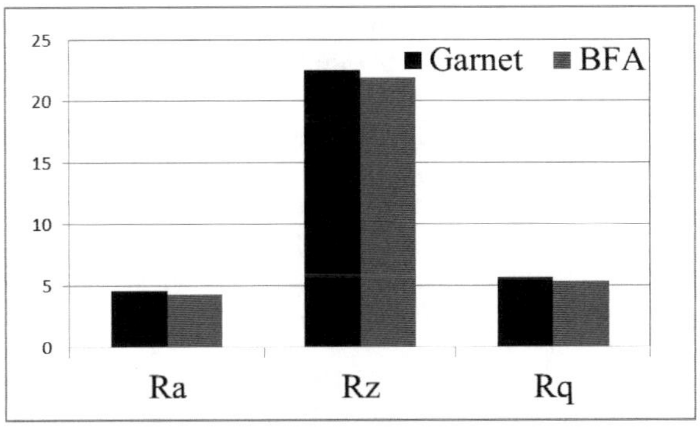

Figure 8. Performance comparison of abrasives based on surface roughness

4 CONCLUSIONS

In this study, as an alternative abrasive to garnet, performance of Brown Fused Alumina (BFA) in the AWJ cutting of marble was investigated. Results indicated that the BFA showed better performance than garnet based on all performance outputs. Therefore, it can be recommended for an alternative abrasive in the AWJ cuttings. For further studies, the cutting performance of the BFA should be also investigated for different levels of operating parameters.

REFERENCES

ASTM, 2015. Standard Test Method for Flexural Strength of Dimension Stone, C880, West Conshohocken, PA.

ASTM, 2015. Standard Test Methods for Absorption and Bulk Specific Gravity of Dimension Stone, C97, West Conshohocken, PA.

ASTM, 2016. Standard Test Method for Compressive Strength of Dimension Stone, C170 West Conshohocken, PA.

DIN EN ISO 4287. Geometrical Product Specifications (GPS) – Surface texture: Profile method – Terms, definitions and surface texture parameters, Germany.

Karakurt, I., 2011. Aşındırıcı Su Jeti (ASJ) ile Kayaç Kesmede Parametrik Değişimlerin ve Malzeme

Dokusal Özelliklerinin Kesme Performansı Üzerindeki Etkilerinin Araştırılması, Doktora Tezi, K.T.Ü., Fen Bilimleri Enstitüsü, Trabzon.

Kartal, F., 2015. AISI 105 Çeliğinin Aşındırıcı Su Jeti ile Tornalanmasında Farklı Parametrelerin Talaş Kaldırma Oranı ve Yüzey Pürüzlülüğü Üzerindeki Etkilerinin İncelenmesi, Doktora Tezi, Karabük Üni., Fen Bilimleri Enstitüsü, Karabük.

Liu, H., 2004. A Study of the Cutting Performance in Abrasive Waterjet Contouring of Alumina Ceramics and Associated Jet Dynamic Characteristics, PhD Thesis, School of Mechanical, Manufacturing and Medical Engineering, Queensland University of Technology.

Miranda, M.R., Quintino, L., 2005. Microstructural study of material removal mechanisms observed in abrasive waterjet cutting of calcareous stones, *Mater Charact,* 54, 4-5, pp. 370–377.

Mort, A, 1995. Results of abrasive water jet market survey, *Proceedings of 8th American Water Jet Conference,* Vol. 1, pp. 259-289, Houston, Texas, USA.

Vundavilli, R.P., Parappagoudar, M.B., Kodali, S. P., Benguluri, S., 2012. Fuzzy logic-based expert system for prediction of depth of cut in abrasive water jet machining process, *Knowl Based Syst,* 27, pp. 456 - 464.

Wang, J. and Guo, D.M., 2003. The Cutting Performance in Multipass Abrasive Waterjet Machining of Industrial Ceramics, *Journal of Material Processing Technology*, 133, pp. 371-377.

Analyzing Land Cover Dynamics Through Spatial Statistics in Relation to the Turkish Mining Sector

Türkiye Madencilik Sektörüne İlişkin Olarak Arazi Örtüsü Dinamiklerinin Mekansal İstatistikler Yoluyla Analizi

A.Ç. Dikmen
Ministry of Environment & Urbanization, Ankara, Turkey

A. Gül
Department of Civil Engineering, Dokuz Eylul University, Izmir, Turkey

ABSTRACT There are needs for considering social, economic and ecological decisions together in planning regional and national development. This requires better understanding of land use dynamics through the use of real spatial data and regular monitoring of changes in a proper planning framework. Similar requirements apply to mining areas where land cover changes are fast. Inspection and regular monitoring of land use changes are more facilitated by means of advancing technologies. With the help of technological decision support mechanisms that include remote sensing tools and geographic information systems, it is easier to have access to accurate, reliable and up-to-date information on land-related processes. CORINE (Coordination of Information on the Environment) project is one of the significant land management projects in the context of GMES (Global Monitoring for Environment and Security) Initiative of the European Union. With CORINE, it became possible to develop land cover maps that include land information in a spatial context. In the presented study, it is aimed to unfold development dynamics through spatial statistics in the context of CORINE data and assess spatial characterizations that relate to the mining sector in Turkey.

ÖZET Bölgesel ve ulusal kalkınma planlarında sosyal, ekonomik ve ekolojik kararların bir arada değerlendirilmesi büyük önem taşımaktadır. Bu yönde bir değerlendirme de, gerçek mekansal verilerin ve uygun bir planlama çerçevesinde yürütülen düzenli izleme programlarının kullanımıyla, arazi kullanımı dinamiklerinin iyi anlaşılmasını gerekli kılmaktadır. Benzer gereklilikler, arazi örtüsü değişikliklerinin hızlı geliştiği madencilik sahaları için de geçerli olmaktadır. Arazi kullanımı değişikliklerinin incelenmesi ve düzenli izlenmesi, gelişen teknoloji nedeniyle oldukça kolay hale gelmiştir. Uzaktan algılama araçları ve coğrafi bilgi sistemlerini de içeren teknolojik karar destek yapılarının yardımıyla, araziyle ilgili süreçler konusunda doğru, güvenilir ve güncel bilgiye erişim mümkün olmuştur. CORINE (Çevresel Bilgilerin Koordinasyonu) projesi, Avrupa Birliği'nin GMES (Çevre ve Güvenlik İçin Küresel İzleme) girişimi kapsamındaki en önemli arazi yönetimi projelerinden birisi olarak öne çıkmaktadır. CORINE ile, mekansal bir çerçevede arazi bilgilerini içeren arazi örtüsü haritalarının hazırlanması mümkün hale gelmiştir. Sunulan çalışmada, CORINE verileri üzerinden mekansal istatistikler yoluyla arazi gelişim dinamiklerinin ve Türkiye'de madencilik sektörüne ilişkin olarak mekansal karakteristiklerin ortaya çıkarılması amaçlanmıştır.

1 INTRODUCTION

Identification of spatial patterns of different land uses and the way they change over time represent a vital source of information for understanding the impacts of sector policies and for planning/orienting new policy frameworks and actions. This sort of identification is also significant for mining and mineral extraction areas which may bring in considerable impacts on socio-economic situation as well as the environment. Indeed, one of the most significant causes of environmental problems that increase in parallel to the industrialization is the improprieties in using the land (Dikmen and Gül 2015). The facts that land cover changes are comparably faster in mining areas and there are needs for considering economic and ecological decisions together in planning the development require better understanding of land use dynamics, monitoring of changes in a proper planning framework and identification for the causes of problems through real spatial data, which would together serve for taking proper land use decisions.

A series of studies focused in the last decade on land use/cover issues and change dynamics in Europe especially after the release of European datasets on land cover information. A recent study analyzed land cover changes observed in Europe between 1990 and 2000 and defined relevant Land Cover Flows (LCFs) as the processes of land cover change (EEA 2006). In this study, 1892 possible one-to-one changes between the 44 different land cover classes included in the CORINE (Co-ordination of Information on the Environment) Land Cover (CLC) dataset were grouped into the so-called flows of land cover and were organized on three levels of information to supply a rapid vision of processes taking place over the land.

Along with such predictive approaches to describe future conditions on land, understanding global and regional dynamics on land use/cover changes through exploratory data analyses still holds key importance in order for assessing relevant policy impacts, redirecting policy tools and/or designing the conceptual bases of future scenarios. Indeed, land use/cover changes and their sector impacts are very dynamic, both in time and in space. The spatial aspects of land use/cover systems play a crucial role in understanding the complex nature of landscape pattern dynamics.

In the presented study, land cover dynamics were analyzed through the use of outputs from the extensions of the CORINE project in Turkey, and a series of detailed information on the spatially explicit changes experienced in relation to the mining sector in Turkey were displayed with tabular and spatial outputs.

2 MATERIALS, METHODOLOGY AND RESULTS

2.1 Primary data sources on land cover information

CLC database is the most fundamental data set employed in the study. CLC2000 data, for which the CLC1990 inventory and its updates are the key reference data sets, provides an inventory of the Earth surface features for managing the environment. The data is mainly for the year 2000, with plus or minus a few years (JRC 2005). CLC data for the reference year 2006 (CLC2006) is a direct continuation of previous CLC mapping campaigns and was produced by integrating the data of land cover changes 2000–2006 with the CLC2000 data.

Figure 1. CORINE map for the year 2006.

2.2 Aggregated figures for unfolding land cover accounts and the rates of change

The most basic information that can be gleaned from recent land cover processes in Turkey is the total amount of change. Surface areas of the four sub-classes (CORINE level 2) under the artificial surfaces theme are given in Table 1 associated with the change figures for the two subsequent monitoring periods. Total coverage areas for the mineral extraction areas (CORINE level 3 sub-class listed under the mine, dump and construction sites) in Turkey in the years 1990, 2000 and 2006 as well as the rates of changes experienced in the two data monitoring periods, 1990-2000 and 1990-2006 are given in Table 2.

Table 1. Areal distribution of Level 2 classes of artificial surfaces in Turkey.

	1.1 Urban fabric (ha)	1.2 Ind., comm., transp. units (ha)	1.3 Mine, dump, const. sites (ha)	1.4 Artif., non-agri. areas (ha)
1990 stock	767.995	86.174	76.557	31.778
2000 stock	902.731	167.203	102.980	47.616
1990-2000 change	17.5 %	94.0 %	34.5 %	49.8%
2006 stock	913.558	182.079	112.700	49.945
2000-2006 change	1.2 %	8.9 %	9.4 %	4.9 %

Table 2. Land cover areas for mineral extraction sites and change rates in the two monitoring periods.

	Years			Periods	
	1990	2000	2006	1990-2000	1990-2006
		Area (ha)		Change (%)	Change (%)
Mineral Extraction Areas	43.997	67.329	74.999	+ 56.59	+ 74.43

In order for unfolding the change dynamics between the years 1990 and 2006 for mineral extraction sites, the shares of these areas inside the generic coverage of all artificial surfaces were computed at the statistical region unit level of NUTS2 (nomenclature of territorial units for statistics). Figure 2 displays the spatial distribution of the computed relative shares. Apparently the sub-regions Ankara, Manisa (covering Manisa, Afyonkarahisar, Kütahya and Uşak provinces), Tekirdağ (covering Tekirdağ, Edirne and Kırklareli provinces), Balıkesir (covering Balıkesir and Çanakkale provinces), Aydın (covering Aydın, Denizli and Muğla provinces), Bursa (covering Bursa, Eskişehir and Bilecik provinces), Adana (covering Adana and Mersin provinces) and Kayseri (covering Kayseri, Sivas and Yozgat provinces) experienced comparably higher increase rates in the period 1990-2006 against the decrease rates associated with some of the sub-regions which included İstanbul, Hatay and Mardin, for example.

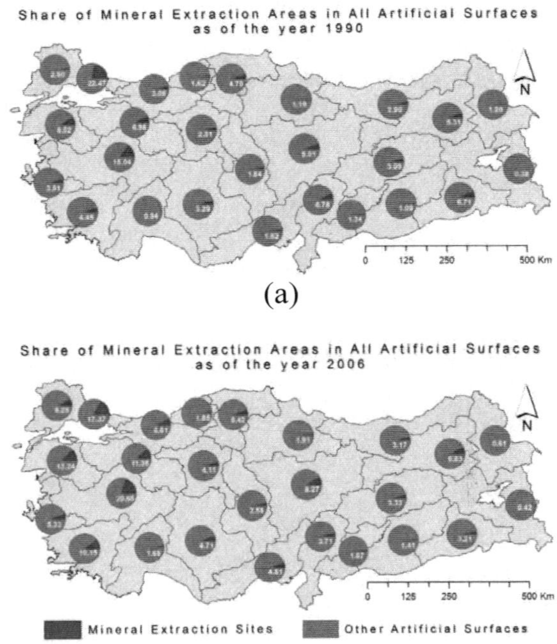

(b)

Figure 2. Mineral extraction areas in the years (a) 1990 and (b) 2006.

In the presented study, the total coverage areas of mineral extraction areas were also computed at the districts level for the years 1990, 2000 and 2006. Figure 3 shows the spatial distribution of the mining areas in Turkey's districts.

Another important source of data employed in the study is the land accounts for Europe and the associated LCFs (EEA 2006). In the context of CLC data, the total amounts of LCFs give insights into the development processes over the land. The top level classification of flows, which distinguishes the changes between broad land cover classes and the changes internally among these classes, contains a total of nine flows. In the study, only those relating to the sprawl of mines and quarrying areas in the previously agricultural, forest and semi-natural land were considered for the targeted spatial assessments. As can be seen in the Figures 4 and 5 which both indicate the land cover flows in the total period 1990-2006 at the expense of the agricultural areas and the forest and semi-natural land, the sprawl of mineral extraction areas existed greatly in southern Marmara region, in Elbistan district, and similarly in some other districts shown in the maps.

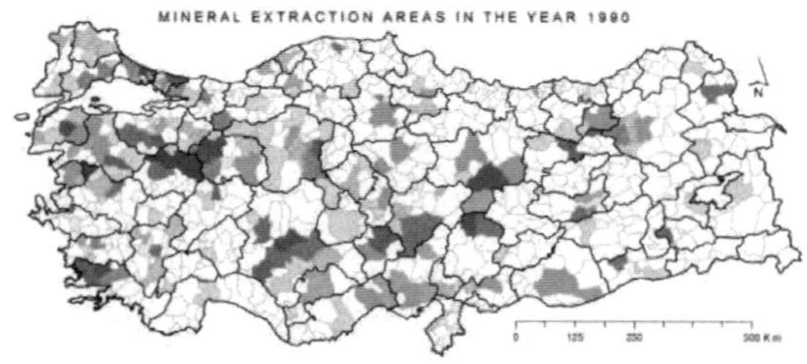

(a)

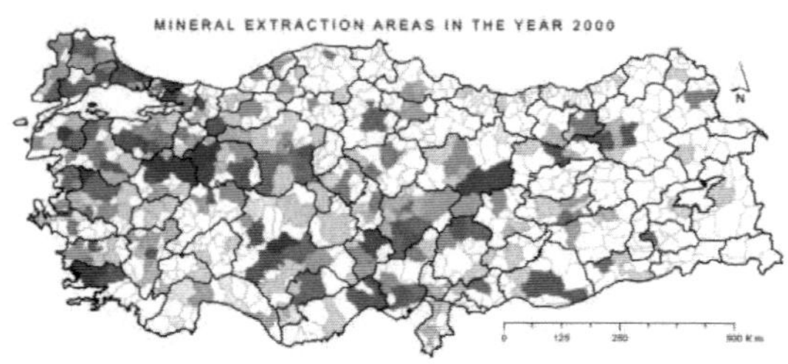

(b)

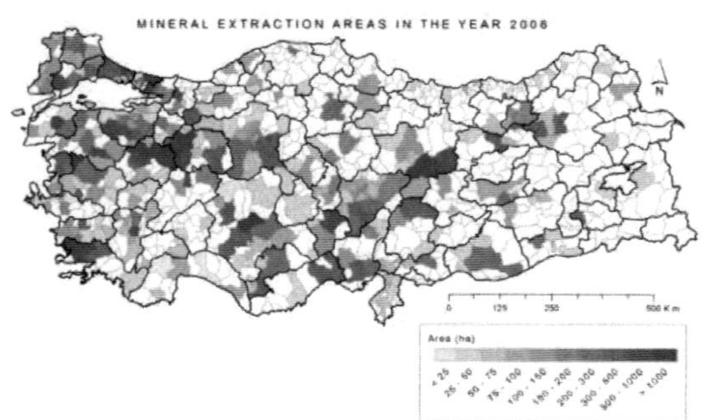

(c)

Figure 3. Mineral extraction areas inside the districts, for the years (a) 1990, (b) 2000 and (c) 2006.

2.3 Basic spatial statistics with potential uses in explaining land cover changes

The summary statistics such as total stock areas for mineral extraction areas and the rates of changes or flows into mines land cover depicted for administrative units provide an overview of the land processes. In order to explain the distributional characteristics of land use/cover patterns and to look for trends in land changes, other types of spatial measures might still be required. The most basic types of descriptors for spatial distributions (so-called centrographic statistics) include mean centre (or weighted mean centre) and standard deviational ellipse. Mean centre is a basic descriptor of spatial distributions and is a point constructed from the average *X* and *Y* values for the input feature centroids. It provides a single point summary measure for the location of distributions. In the presented study, mean centre calculations were used to observe any directional shifts from 1990 to 2006.

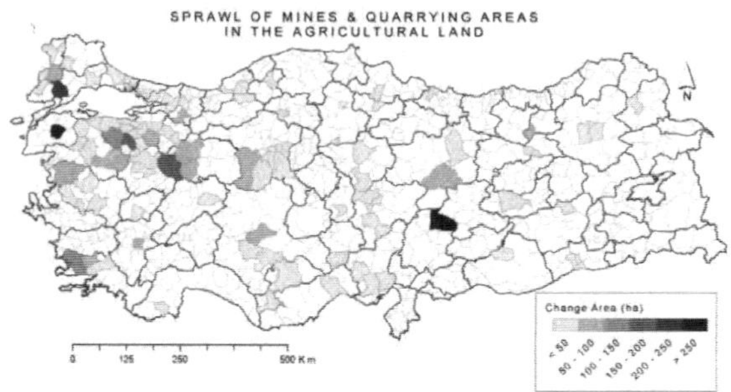

Figure 4. Sprawl of mining areas in the agricultural land.

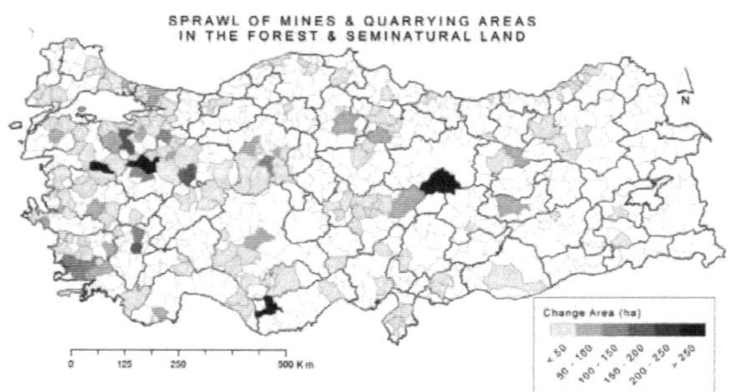

Figure 5. Sprawl of mining areas in the forest/semi-natural land.

Standard deviational ellipse (also called directional distribution), on the other hand, is a good single measure of the dispersion of incidents around the mean centre and captures any directional bias (Ebdon 1988), Cromley 1992). It is a powerful spatial statistic to show the dispersion of land cover types in two dimensions, so it is able to account for irregularities in the landscape. An ellipse measures the shape and orientation of the distribution by summarizing the maximum and minimum variance along the *X*- and *Y*-axes and is able to account for a non-uniform arrangement of events (distribution of a certain land cover type in this case) (Yuill 1971, Kent and Leitner 2007). It shows directionality on the major axis as the direction of maximum spread of the incidents and the minor axis as the direction of minimum spread. In the assessments covered by this study, standard deviational ellipses were used to determine the maximum and minimum spread directions of the mines land cover and to explore how well they follow the geographic distribution of geographic/administrative territories (i.e. if the land cover is evenly distributed or concentrated on some parts). Weighted mean centers shown in Figure 6 indicate that there is an eccentric development of mineral extraction areas due to comparably larger cover areas in the western part. Again in the figure, the orientation of the standard deviational ellipse displays a tilted dispersion around the mean centers in the northwest-southeast alignment. A little shift in the mean centre to the west is also visible from a comparison between the years 1990 and 2006, potentially indicating comparably higher sprawl areas in the west.

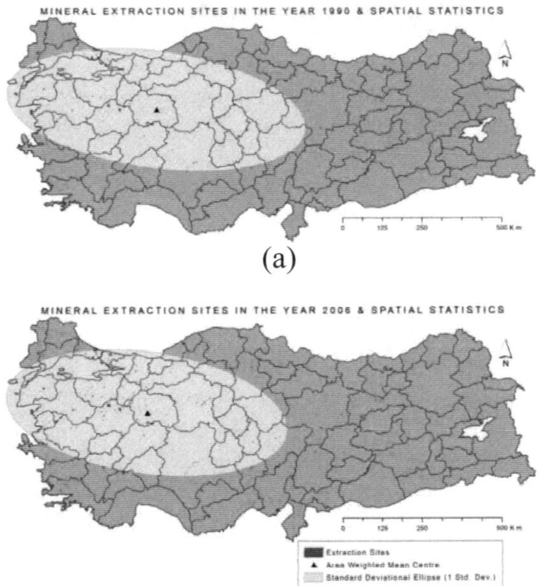

Figure 6. Spatial statistics for mineral extraction areas in (a) 1990 and (b) 2006.

In the study, standard deviational ellipses were also generated at the scale of geographic regions in Turkey, again for the two land cover data monitoring years 1990 and 2006 (Fig. 7). In the output maps, standard deviational ellipses for the region territories were also generated by using the corresponding territorial coverage areas in order to allow more practical evaluations about the existence of non-homogeneous dispersions.

Indeed, it is clear from the figure that the extraction areas in the regions of Central and Southeastern Anatolia have the most homogeneous dispersion in the regional territories while those in Aegean, Mediterranean and Eastern Anatolia regions have more distorted distribution patterns in either direction.

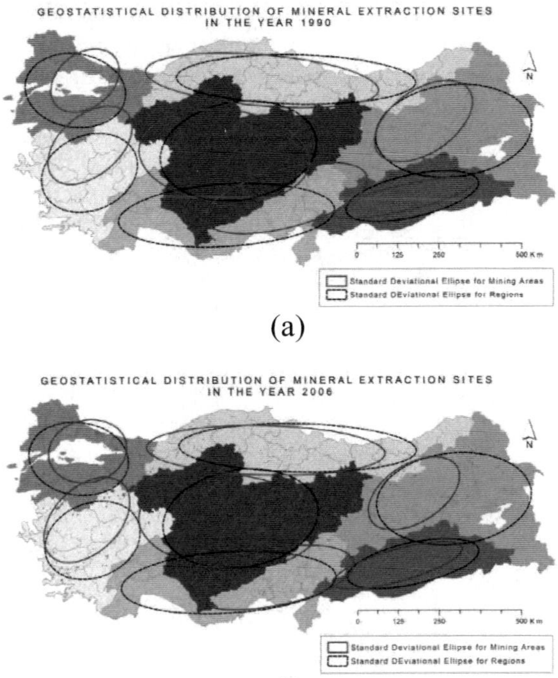

Figure 7. Standard deviational ellipses for the mineral extraction areas grouped at the geographic regions scale.

3 CONCLUSIONS

Apart from its significance as an important driver on global environment politics, land use together with its associated local/global processes of change is an important determinant for regional development and the ways it is shaped through policy implementations. In this regard, any proper assessment on land use needs thorough understanding of local/global quantities of land use/cover changes, similitudes and clustered behaviors of the regions experiencing these changes and the potential mechanisms that drive changes. The development of spatially comprehensive datasets on the land issues and the extensive use of geographic information tools and software applications greatly contribute to these assessments, and thus facilitates the policy implementations which depend on continuous/periodic monitoring of environment and other related economic sectors. Periodic and frequent updates of CORINE or similar land monitoring projects would yield the making of more sound land use decisions, while any precision increase in the land images that are used in developing such databases would contribute to the efficiency increases in the targeted activities.

To this end in the presented study, land cover stocks relating to the mining areas in Turkey were assessed together with the change rates monitored in the two data periods; 1990-2000 and 2000-2006. With the help of the developed datasets on land cover and the software tools that help analyze geographic data/information at the spatial scale, the variable nature of the mining land cover in different administrative units was presented through the use of tabular, graphic and map outputs. These analyses were also supported by a series of geo-spatial statistics for detecting any specific spread or change along any direction, and thus any particular orientation. When the analytical outputs from the study are examined in the context of the Turkish mining sector, it is observed that the areal coverage of mineral extraction areas increased by 56.59% and 74.43% as of the years 2000 and 2006, respectively, with reference to the conditions in 1990.

It is noteworthy that the performed assessments do not consider the factors that originally drive the changes and are only based on statistical analysis of the observed changes. Originally, patterns of land use, land cover change, and land management are shaped by the interaction of economic, environmental, social, political, and technological forces on local to global scales. Thus, for more detailed assessments, the statistical analysis of change rates needs to be integrated with more extensive assessments that focus more on potential drivers of land-related changes, such as socio-economic factors, policy interventions on land use, etc. This can be achieved by performing complementary analyses that search for correlations between the actual land cover changes relating to the mining sector and the changes in their potential drivers, both on time and spatial scales.

REFERENCES

Cromley, RG, 1992. *Digital Cartography*, Prentice Hall: Englewood Cliffs, NJ.

Dikmen, AÇ, Gül, A, 2015. Land Use Dynamics in Turkey and Spatial Changes Experienced in the Mining Sector, *Proceedings of the 5th Mining and Environment Symposium*, 26-27 November 2015, Antalya, Turkey.

Ebdon, D, 1988. *Statistics in Geography* (second edition with corrections), Blackwell: Oxford.

JRC, 2005. *Image2000 and CLC2000 - Products and Methods*, Maria Vanda Nunes de Lima (Ed.), European Commission, Directorate-General Joint Research Centre, Institute for Environment and Sustainability, Land Management Unit, Ispra.

EEA, 2006. *Land accounts for Europe 1990-2000*, EEA Report No 11/2006, European Environment Agency, Copenhagen.

Kent, J, Leitner, M, 2007. Efficacy of standard deviational ellipses in the application of criminal geographic profiling, *J Investig Psych Offender Profil*, 4, 147–165.

Yuill, RS, 1971. The standard deviational ellipse; an updated tool for spatial description, *Geografiska Annaler, Series B, Human Geography*, 53(1), 28-39.

MINERAL PROCESSING / EXTRACTIVE METALLURGY

High Pressure Grinding Rolls: Moving From Cement to Minerals

H. Dündar, N.A. Aydoğan, H. Benzer
Hacettepe University, Dept. of Mining Engineering, Beytepe, 06800 Ankara, Turkey

ABSTRACT The idea of interparticle crushing resulted in invention of the high pressure grinding rolls (HPGR) in 1984. The first installations were in the cement industry and it was mainly used for pre-grinding the ball mill feed. The advantage of this was that the capacity of the downstream ball milling is increased, thus reducing the specific energy consumption of the overall circuit. After having proved its success in energy efficiency, more of the grinding required has been transferred to the HPGR by introducing various circuit configurations in cement such as pre-grinding mode, hybrid-grinding mode, semi-finish grinding mode and finish grinding mode. By mid-nineties, HPGR installations have been started in minerals industry, mainly in iron ore pelletizing and diamond. With the improvement in wear protection, this was followed by applications in hard rock grinding as copper, gold and platinum ore. The HPGR has been introduced to minerals industry as a replacement for tertiary crushers, SAG mills and rod mills. In some applications it appears as a pebble crusher in AG milling. This paper presents the energy and size reduction figures of the HPGRs from various circuit configurations applied in cement and minerals industry.

Keywords: High pressure grinding rolls, cement grinding, hard rock grinding.

1 INTRODUCTION

Comminution of a bulk material between two surfaces under compression has long been studied (Carey, 1934; Schönert et al., 1965; Feige, 1979) and eventually Schönert (1979) showed that a single particle breakage by compressive loadings is more energy efficient than conventional crushing. His research led to the patenting of high pressure comminution process in 1982, followed by the development of high pressure grinding rolls (HPGR).

The first commercial application of the HPGR was in 1985 in a cemet plant and early applications were limited to precrushing the ball mill feed in clinker grinding circuit in order to increase throughput. After having proved its energy saving capacity more of the grinding required has been transferred to HPGR. As a result, various circuit configurations were achieved in cement grinding (Kellerwessel, 1993; 1996). Patzelt (1992) showed that HPGR based circuit configurations in cement were 10 to 50% energy efficient compared to conventional ball mill grinding. Similar results were presented by Aydogan et al. (2006).

Its potential on energy saving and capacity improvement then made the HPGR attractive to minerals industry. However, the roll surface wear was the major issue restricting the acceptance of the HPGR into minerals industry. The progress in roll surface design led to increased wear life, thus increasing the number of applications in minerals industry. By mid-nineties, HPGR installations took place at iron ore pellet feed preparation and kimberlite crushing followed by applications in hard rock grinding as copper, gold and platinum ore. HPGRs have been used in place of tertiary crushers, SAG mills and rod mills in minerals industry. However, HPGRs in minerals industry have remained underutilized compared to those in cement. The main reason for that is the high competency of the ores, which are highly abrasive compared to cement clinker.

This paper discusses the cement and minerals applications of the HPGR in different circuit designs.

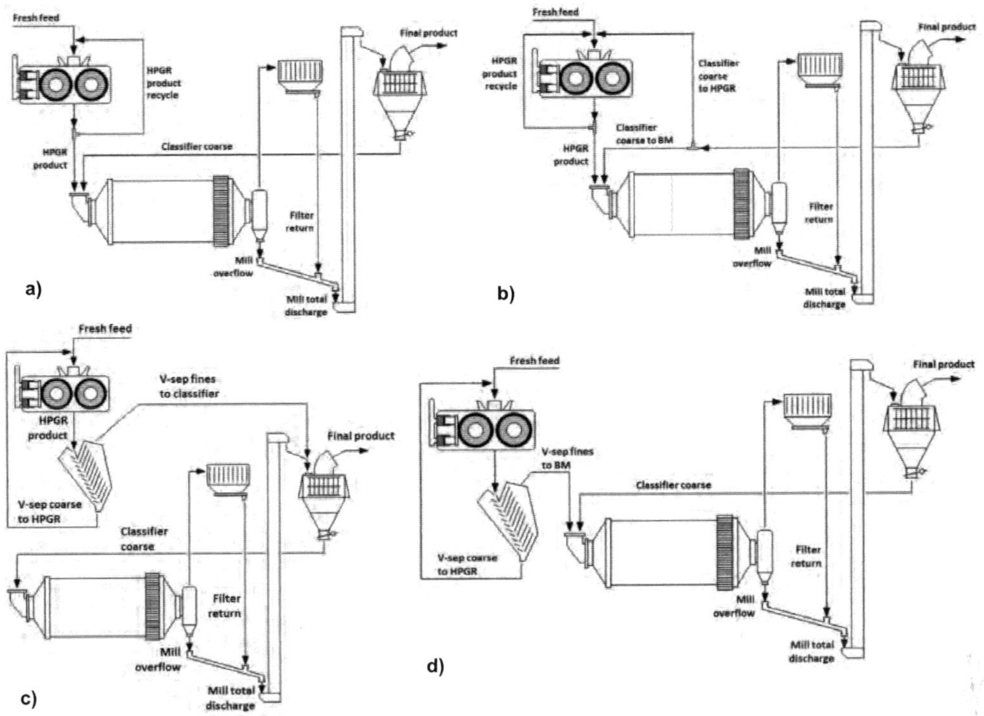

Figure 1. Cement grinding with HPGR-Ball Mill-Air Classifier a) open circuit pregrinding b) hybrid grinding c) semi-finish grinding d) closed circuit pregrinding

2 CEMENT

HPGR is used in a wide range of circuit configurations in cement grinding. After pre-grinding applications ahead of existing ball milling, the HPGR was then highly utilized by introducing various circuit designs (Figure 1). The HPGR incorporated circuits offer reduction in energy consumption of the overall circuit. Besides the HPGR-Ball Mill circuits, the HPGR is also used standalone in cement grinding in closed circuit with an air classifier.

Aydogan et al. (2006) discussed the benefits of the HPGR for different circuit configurations in the cement industry. Table 1 gives the performance figures of the circuits in terms of specific energy consumption and the final product size. From Table 1, the overall specific energy consumption of the circuit decreases as the work transferred to the HPGR is increased with more recycle.

3 MINERALS

The status of the HPGR in hard rock applications is different from that in cement grinding. HPGRs are used in base metals, precious metals, diamond industry and iron ore pellet feed preparation. Unlike cement grinding, the HPGRs in minerals industry find application in crushing rather than grinding. Because of the abrasiveness of the ores, the specific grinding forces of the rolls are set to relatively lower values compared to cement HPGRs. As a result, coarser product is obtained at relatively low power consumption.

The larger diameter rolls allow coarser feeds to be processed and to control the size of the product closing the HPGR with a screen, or edge recycling are commonly used in minerals applications. This section presents the data from gold and platinum mines.

Table 1. Specific energy consumptions of the circuits with final product size (Aydogan et al., 2006)

Circuit configuration	HPGR (kWh/t)	HPGR-Ball mill (kWh/t)	Final product size, P80 (micron)
Open circuit HPGR with no recycle/ closed circuit BM	4.95	34.19	33
Open circuit HPGR with recycle/closed circuit BM	8.93	29.57	68
Hybrid grinding	-	29.85	29
Closed circuit HPGR–closed circuit BM	8.02	21.65	28
Semi-finish grinding	9.80	23.03	31

3.1 HPGR with Edge Recycle in Gold Ore Crushing

An extensive testwork was conducted around the HPGR circuit in a Gold Mine in Ghana. The HPGR was situated after the closed circuit crushing stage and was used to prepare feed material for heap leaching. The circuit was configured in such a way that the edge material from the HPGR discharge was recycled for further crushing while the product from the centre reported to the heap leach section (Figure 2). The installation below the rollers allows separating the edge material and the centre product (Figure 3). The technical specifications of the HPGR are given in Table 2. The main feature of the machine is variable speed drive system. The variable speed drive system makes the rolls' speed adjustable. The throughput of the HPGR can be controlled by adjusting the rolls' speed, depending on the bin level, to maintain the choke feeding.

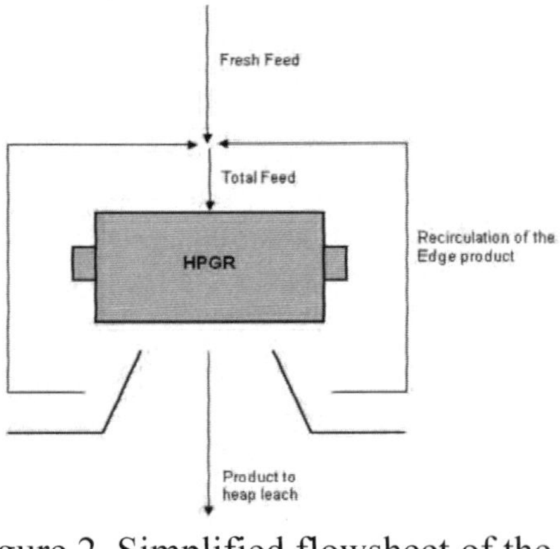

Figure 2. Simplified flowsheet of the HPGR circuit

Figure 3. Installation below the rollers to separate the edge and centre products

During the testwork, sampling around the HPGR circuit is performed for varying conditions. Fresh feed, edge and centre products were sampled by belt cuts after crash stopping the HPGR. The samples were processed to obtain particle size distributions.

Table 2. Technical specifications of the HPGR

Roll diameter	1700 mm
Roll length	850 mm
Drive motors	2x980 kW

Figures 4 to 6 show the relationships between the operating variables. For the same material, the throughput of the HPGR is affected by the roll speed and the roll gap (Figure 4). There is an interaction between the operating pressure and the roll gap, i.e. as the operating pressure increases the roll gap is reduced (Figure 5). The roll speed and the operating pressure determine the power draw; the higher the roll speed and the pressure, the higher is the power draw (Figure 6).

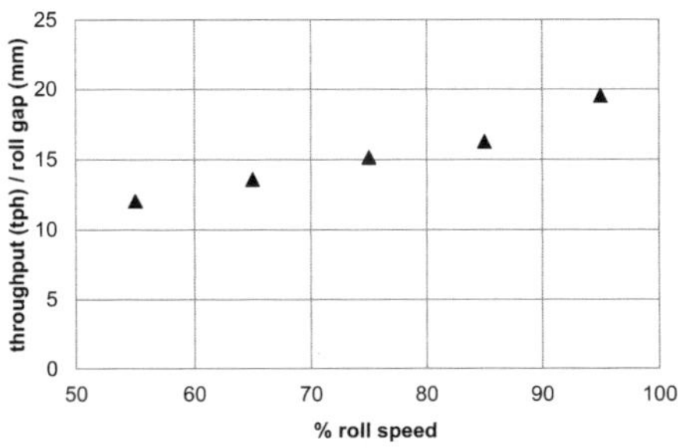

Figure 4. Rolls' speed vs. throughput/gap

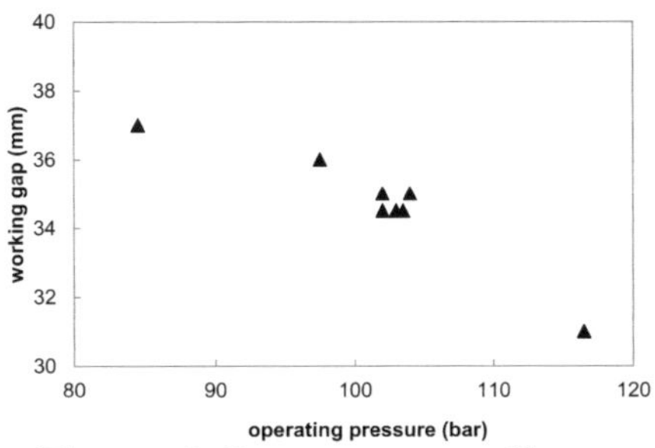

Figure 5. Pressure vs. roll gap

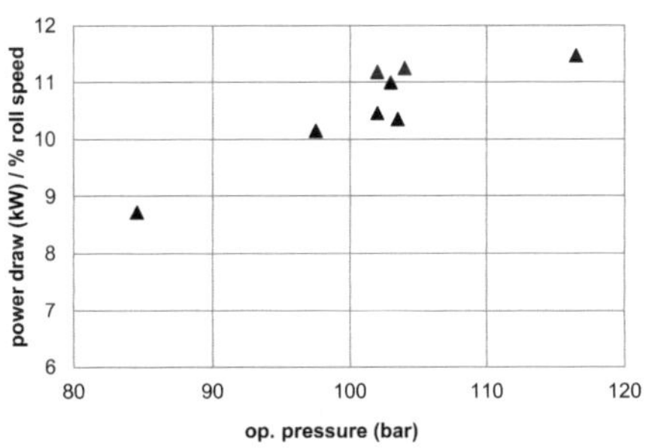

Figure 6. Pressure vs. power draw/rolls'speed

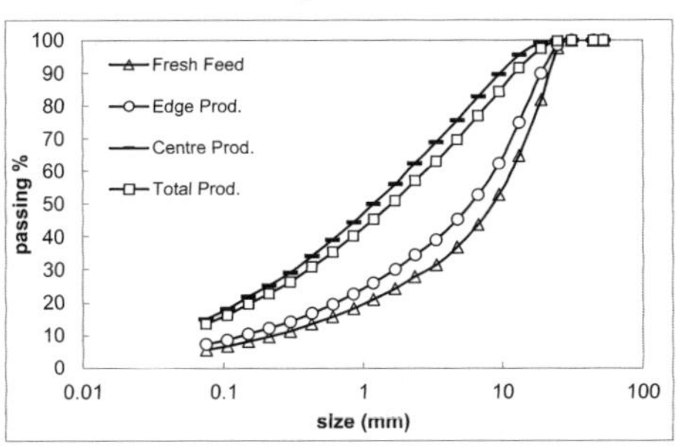

Figure 7. PSDs around the HPGR

The pressure gradient along the roll length is relatively low at the edges. Therefore the product obtained from the edges are coarser than that obtained from centre. From this point of view, circuit configuration with edge recycling was developed to control the product size of the HPGR. Figure 7 gives the PSDs around the HPGR for one the tests conducted. As can be seen from figure, the edge product is coarser than the centre product. The other tests results are similar to the data given in Figure 7, therefore only one of them is presented.

During the testwork the maximum operating pressure was 116 bars and the corresponding unit energy consumption of the HPGR was 2.08 kWh/t.

3.2 Closed Circuit HPGR in Platinum Ore Crushing

Three sampling surveys were performed around the HPGR circuit in a Platinum Mine in South Africa. The HPGR circuit is situated prior to the ball milling circuit for pre-crushing the ore. The HPGR operates in closed circuit with a 6.3mm aperture sized screen. The technical specifications of the HPGR are given in Table 3.

Table 3. Technical specifications of the HPGR

Roll diameter, mm	2200
Roll length, mm	1550
Drive power, kW	2x2800
Roll speed, m/s	2.4

The first sampling campaign was performed when the HPGR was operating under standard conditions in terms of pressure. The second and third ones were performed for lower and higher pressure values respectively.

Figure 8 gives the ratio of the screen OS tonnage to the HPGR discharge tonnage with respect to the operating pressure. As can be seen from figure, the screen oversize ratio is decreasing with the increased pressure as more fines were produced. Then the unit energy consumption of the HPGR circuit goes up to 2.37 kWh/t.

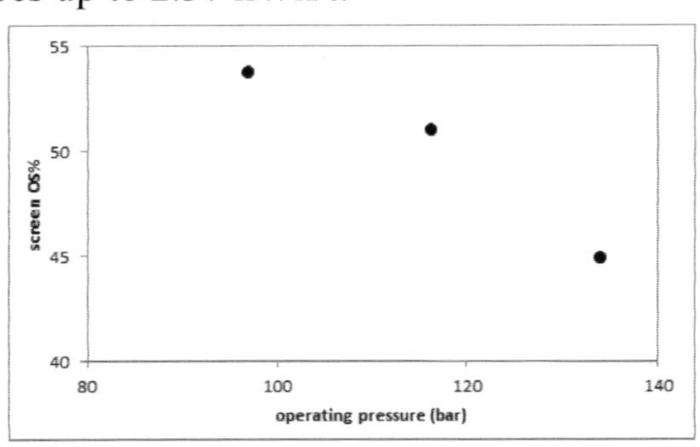

Figure 8. Screen OS ratio to HPGR disch.

4 DISCUSSION

The major issue in hard rock crushing with high compression is the competency of the ores. Therefore the mineral HPGRs are designed with low specific grinding forces to protect the rolls. With the progress in roll surface design the high compression technology moves forward in hard rock applications.

The mineral HPGRs are designed to process coarser feeds with high capacity. Since the roll diameter determines the feed top size, larger rolls are commonly preferred in minerals industry. Depending on the roll diameter and low operating pressures, the HPGRs in mineral run at higher roll gaps, hence higher throughputs. However, as a result of the higher roll gap the product gets coarser. Figure 9 shows a rough comparison of the product size distributions of the HPGR products obtained from a gold mine and a cement plant.

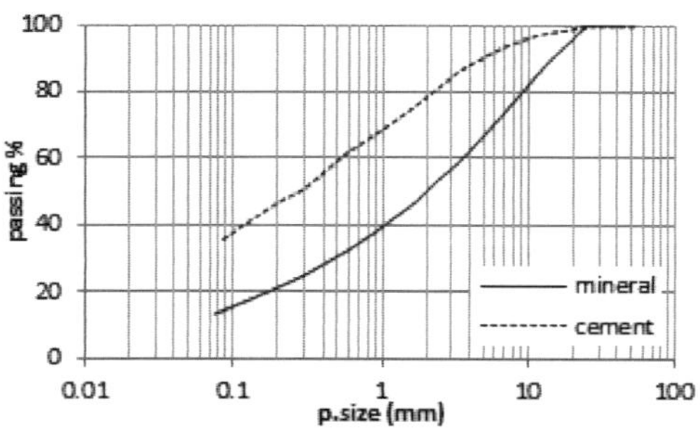

Figure 9. PSDs of the HPGR products from cement and minerals

Regarding the low operating pressure in minerals industry the specific energy consumption of the HPGR is limited to 3-4 kWh/t while it is up to 10 in cement depending on the circuit configuration. Figure 10 gives the specific grinding forces applied in minerals and cement. The data comes from the site surveys carried out for different operating conditions and different circuit configurations. As can be seen from figure the specific grinding forces in cement are considerably higher than that in minerals.

5 CONCLUSION

High pressure grinding rolls are running successfully in cement industry by reducing the energy consumption of the overall circuit. On the other hand, in minerals industry they are widely used for handling high throughputs, as an alternative to tertiary crushing, SAG and rod milling as well as pebble crushing. The roll surface design makes the HPGRs available for minerals industry, but there still needs to be progress in applying higher grinding forces to highly utilize the machine for energy efficient processes.

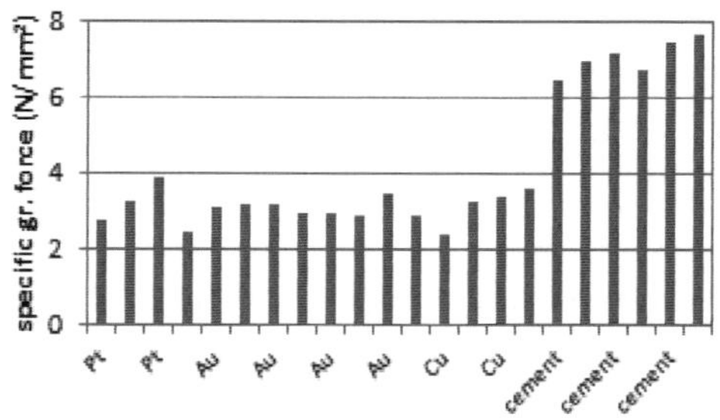

Figure 10. Specific grinding forces applied in cement and minerals

REFERENCES

Aydogan, N.A., Ergun, L., Benzer, H., 2006. High pressure grinding rolls (HPGR) applications in the cement industry. Minerals Engineering 19 (2006), 130–139.

Carey, W.F., 1934, Trans. Inst. Chem. Engrs. 12, 179/85

Feige, F., 1979, Beitrag zum Bewegungs- und Beanspruchungsverhalten einer Kornerschicht im Walzenspalt. Dissertation to the Bergakademie Freiberg, (unpublished).

Kellerwessel, H., 1996. High pressure particle bed comminution, state of the art, application, recent developments. Eng. Min. J., February, 45–52.

Kellerwessel, H., 1993. High pressure particle bed comminution of mineral raw materials. Aufbereitungs-Technik 34 (5), 243–249.

Patzelt, N., 1992. High pressure grinding rolls, a survey of experience. IEEE Cement Industry Technical Conference, Dallas/Texas 10 (14), 180.

Schonert, K., 1982, Method of fine and very fine comminution of materials having brittle behaviour. U.S. Patent 4,357,287.

Schönert, K., 1979. Aspects of the physics of breakage relevant to comminution. In: Fourth Tewksbury Symposium, University of Melbourne, pp. 3.1–3.30.

Schonert, K., v.d. Ohe, W., Rumpf, H., 1965, Technische Feinzerkleinerung mit Einzelkornbeanspruchung zwischen zwei Flächen. Chemie-Ingenieur-Technik 37, S.259/264.

Numerical Study of Parameters Affecting Industrial Screening Performance

A. Aghlmandi Harzanagh, E.C. Orhan, Ş.L. Ergün
Hacettepe University, Department of Mining Engineering, 06800, Ankara, Turkey

ABSTRACT Screening is one of the most widely used unit operations in mineral processing plants. In crushing circuits which is the most frequent area of use of screens, the proper selection and sizing, optimization and their operation as efficient as possible is essential in terms of the performance and profitability of crushing circuit and the whole plant. In this study, the effects of some important design and operating variables on the efficiency of screening and on the product characteristics investigated by means of DEM modelling of industrial screens. Additionally, screening tests were conducted using pilot scale industrial screen. The prediction capability of the simulator was revealed by the comparison of the data obtained using the simulator and the data obtained from the pilot scale tests under the same conditions. As a result, an engineering tool was obtained that could be used in studies such as equipment selection, sizing, design and the evaluation and optimization of screen efficiency.

1 INTRODUCTION

Screening is one of the most common unit operations in mineral processing plants. It is used for various purposes starting from the production at the mine field until making it suitable for use as a definite end-product. Classification, size limitation of crusher feed, dewatering, slime cleaning, solid recovery, washing, etc. are some of the main utilization purposes of screening in mineral processing plants(Mular et al., 2002). Screening is widely applied in ore preparation and many other areas (food, textile, etc.). The design, sizing and separation efficiency of screens has a direct effect on the quantity and properties of the target products, the total power consumption and the efficient operation of the crushers and consequently the profitability of the mineral processing plant. Therefore, a large number of studies have been done in terms of modeling of the screens in the past years. However, a large number of variables affecting the performance and characteristics of the equipment and also the relation between some of these variables, led to the development of empirical or semi-empirical models. This kind of models, need a database from a large number of plant performance data and also require a great deal of experience to be applied correctly. Therefore, the application of modeling techniques for equipment sizing is limited.

As an alternative to experimental models, particle-based simulation approaches such as the discrete element method (DEM) first presented by (Cundall and Strack, 1979) have begun to be used in mineral processing. Initially these applications were implemented in low-grain simulations like ball motion in ball mills, but with the increased capacity of the computers, these applications also began to be applied to multi-grain processes such as screening process. By observing inter-grain interactions in the sieving process, DEM can provide a predictive model for screening process without the need for experimental effort (Elskamp and Kruggel-Emden, 2015).

In the recent years, DEM studies have been carried out for modeling the screening process. Li et al. (2003)'s study was one of the first attempts in 3D modelling of screens. Following this investigation (CHEN and Tong, 2009), (Cleary et al., 2009a), (Dong and Brake, 2009), (CHEN and TONG, 2010), (Zhao et al., 2011), (Delaney et al., 2012) , (Dong and Yu, 2012), (Xiao and Tong, 2012), (Kruggel-Emden and Elskamp, 2014), (Elskamp and Kruggel-Emden, 2015) and Jahani et al. (2015)'s studies focused on modeling of inclined and especially banana screens. Most of these studies considered spherical particles in DEM simulations but there are some examples of employing non-spherical particles (Cleary et al., 2009a) (Cleary et al., 2009b) (Kruggel-Emden and Elskamp, 2014). There is an apparent lack of validation data in most of the mentioned investigations.

Numerical modelling and simulation of screens have benefits in terms of understanding screening process and related phenomena like segregation, investigations on the effects of process variables which lead to process optimization and finally design of efficient screening plants.

In this study, DEM was employed to investigate the effects of some operating and design parameters on screening performance of vibrating pilot scale screen. Variations in surface inclination, vibration amplitude and vibration frequency have been investigated. Finally, the validity of a reference simulation has been investigated using a pilot scale vibrating screen.

2 THE DISCRETE ELEMENT METHOD (DEM)

The discrete element method is a numerical technique used to predict the behavior of collision dominated particle flows. Each particle in the flow is tracked and particle-particle and particle-boundary collisions are simulated. The DEM variant used here is sometimes called a 'soft particle method'. The particles are allowed to overlap and the extent of overlap is used in conjunction with a contact force law to give instantaneous forces from knowledge of the current positions, orientations, velocities and spins of the particles (Cleary and Morrison, 2009). In this study, the Hertz–Mindlin's contact force law was used to simulate the particle motion. It states that the repulsive force resulting from a collision is calculated from the amount of normal overlap, δn, and tangential overlap, δt, (soft-sphere approach) (Just et al., 2013). This granular model uses the following formula for the frictional force between two granular particles, when the distance r between two particles of radii R_i and R_j is less than their contact distance $d = R_i+R_j$. There is no force between the particles when $r > d$:

$$F= (k_n\delta n_{ij}- \gamma_n v n_{ij})+ (k_t\delta t_{ij}-\gamma_t v t_{ij}) \quad (1)$$

The first term is the normal force (F_n) between the two particles and the second term is the tangential force (F_t). The normal force has two terms, a spring force and a damping force. The tangential force also has two terms: a shear force and a damping force. The shear force is a "history" effect that accounts for the tangential displacement (tangential overlap) between the particles for the duration of the time they are in contact.

The quantities in the equation are as follows:

k_n elastic constant for normal contact
δn_{ij} $d - r$ = normal overlap (overlap distance between the two particles)
γ_n viscoelastic damping constant for normal contact
vn_{ij} normal relative velocity (normal component of the relative velocity of the two particles)
k_t elastic constant for tangential contact
δt_{ij} tangential overlap (tangential displacement vector between the two spherical particles which is truncated to satisfy a frictional yield criterion)
γ_t viscoelastic damping constant for tangential contact
vt_{ij} tangential relative velocity (tangential component of the relative velocity of the two particles).

Static friction is obtained by keeping track of the elastic shear displacement throughout the lifetime of the contact. The magnitude of δt_{ij} (the tangential overlap) is truncated as necessary to fulfill a local Coulomb yield criterion: $F_t \leq \mu F_n$, where μ is the grain–grain friction coefficient. Therefore, the contact surfaces are treated as sticking when $F_t < \mu F_n$, and as slipping when the Coulomb yield criterion is satisfied (Chand et al., 2012).

Considering that the shear modulus (G) can be calculated from Young's modulus and Poisson ratio, the Hertz–Mindlin contact model depends on the following material parameters (Just et al., 2013).

Coefficient of restitution, e
Young's modulus, Y
Poisson ratio, ν
Coefficient of static friction, μ_s
Coefficient of rolling friction, μ_r.

The maximum overlap between particles is determined by the stiffness k_n of the spring in the normal direction. Typically, average overlaps of 0.1–0.5% are desirable, requiring spring constants of the order of 104–106 N/m in three dimensions. The normal damping coefficient γ_n is chosen to give the required coefficient of restitution e (defined as the ratio of the post-collisional to a pre-collisional normal component of the relative velocity) (McBride and Cleary, 2009).

For each particle, the translational and rotational motion, which can be described by Newton's second law of motion, are determined by (Jahani et al., 2015):

$$\dot{v} = \frac{\Sigma F}{m} + g \quad (2)$$

$$\dot{\omega} = \frac{\Sigma M}{I} \quad (3)$$

where v is the vector of a particle velocity, F is the contact force acting on a particle, m and g are the mass of a particle and the gravitational acceleration, ω is the vector of angular velocity, and M and I denote the moment caused by the tangential force and the moment of inertia. By knowing the various forces (contact and gravitational forces) acting on particles, the velocity and the trajectory of each individual particle are computed by integration with time. This allows all particles to be registered in a predefined domain and, therefore, the interactions between particles and with boundaries can be precisely calculated using the local particle and boundary properties (Li et al., 2003). In this study, open source LIGGGHTS solver was used to perform simulations.

3 SIMULATION CONDITIONS

In this work, a pilot scale inclined vibrating screen was used as schematically shown in Fig. 1. Geometrical conditions of the screen and reference conditions of simulations are listed in Table 1 and Table 2 respectively.

Table 1. Geometrical conditions of the simulated screen

Length (mm)	Width (mm)	Opening (mm)	Open Area (%)	Material
900	300	10.5	27.44	Polyurethane

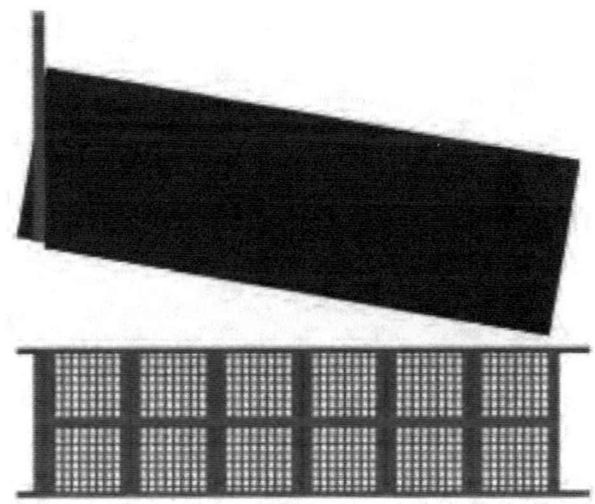

Fig. 1. Schematic illustration of simulated pilot scale screen

Table 2. Reference conditions of simulations

Particle size, d (mm)	26.6 (class8)	20.6 (class7)	15.7(class6)	12.1(class5)	9.5 (class4)	6.7 (class3)	4.7 (class2)	2.8 (class1)
Feed size distribution (%)	10	10	20	8	14	14	14	10
Feed height (mm)	275							
Vibration motion	Linear							
Particle density (kg/m3)	2700							
Young's modulus (N/m2)	5×10^{7}							
Poisons ratio	0.45							
Coefficient of restitution	0.3							
Sliding friction coefficient	0.5							
Rolling friction coefficient	0.01							
Time Step (s)	5×10^{-6}							
Simulation Duration (s)	25							

As seen in Fig. 1, the screen consists of 6 square panels in length and 2 panels in width.

To investigate how different parameters, affect screening performance, 12 different simulations were carried out in different tonnage, screen inclination, vibration amplitude and vibration frequency while keeping other parameters constant. Table 3 shows the parameter values of performed simulations.

Table 3. List of parameter values in performed simulations

Case	1	2	3	4
Screen Inclination (°)	5	10	15	20
Vibration Amplitude (mm)	3	4	5	6
Vibration Frequency (Hz)	15	20	25	30

A simulation begins by discharging the particles from feed-end. The particles that reach the screen surface will either pass the surface or flow along the screen towards discharge-end. It is important to evaluate particles movement data after achieving macroscopic steady state, which inlet flow rate is equal to the sum of oversize and undersize outlet flow rates for each size class. Fig. 2 shows a typical illustration of simulations.

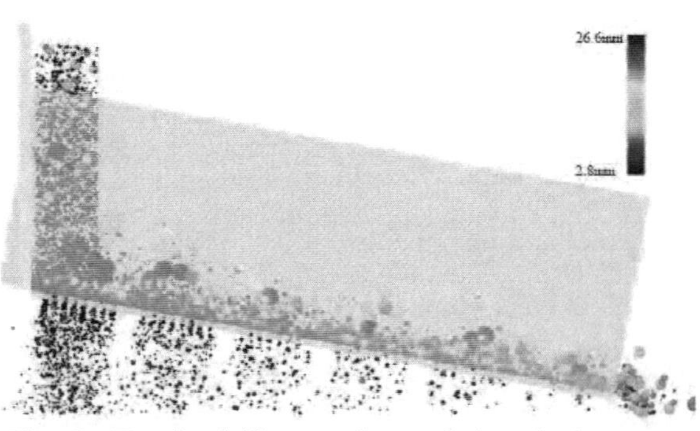

Fig. 2. Typical illustration of simulations

4 RESULTS AND DISCUSSION

After performing simulations according to Table 3, the data obtained is examined. Screening efficiency based on the recovery of undersize materials in overflow stream and partition curve which represents the efficiency of the screening for each size fraction, give a clear knowledge about the effects of different variables on screening performance.

4.1 Effects of Screen Inclination

For examining the effects of screen inclination on screening performance, four simulations at 5°, 10°, 15° and 20° screen inclinations were performed. Table 4 shows the screening efficiency and mass of undersize stream for the simulations and Fig. 3 shows the partition curves.

Table 4. Screening efficiency and undersize percentage for different screen inclinations

Screen Inclination	Screening Efficiency (%)	Mass of Undersize Stream (%)
5°	69.75	35.63
10°	54.96	30.40
15°	45.69	25.65
20°	39.80	21.53

As it is apparent from Table 4, screening efficiency and undersize percentage are decreasing as the screen inclination's increasing. As the screen inclination increases, the probability of collision between particles and screen surface decreases and this reduces performance.

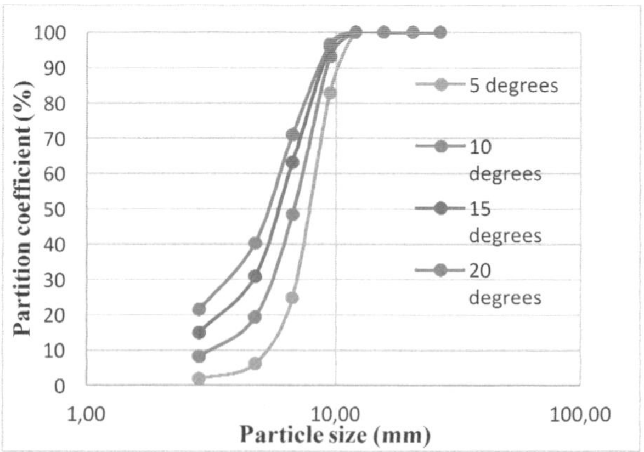

Fig. 3. Partition curves for different screen inclinations

Partition coefficient is defined as the percentage of feed reported to the oversize product. The inclination of the partition curve represents the efficiency of the screening. Partition curve also shows the cut -size of the screen and from Fig. 3 it is apparent that performance and cut -size of the screen are decreasing with screen inclination.

4.2 Effects of Vibration Amplitude

For examining the effects of vibration amplitude on screening performance, four simulations at 3, 4, 5 and 6mm vibration amplitudes were performed. Table 5 shows the screening efficiency and mass of undersize stream obtained from the simulations.

Table 5. Screening efficiency and mass of undersize stream at different vibration amplitudes

Vibration Amplitude(mm)	Screening Efficiency (%)	Mass of Undersize Stream (%)
3	73.85	41.01
4	63.38	35.15
5	54.96	30.40
6	47.24	26.06

It is apparent from Table 5 that vibration amplitude has a significant effect on both screening efficiency and undersize mass flow and increasing in vibration amplitude reduces the screening efficiency and mass of undersize stream. Low amplitudes increase the probability of contact between particle and screen surface so increase the efficiency.

Fig. 4 shows the partition curves at different amplitudes. It proves that the performance and cut-size of screening are much higher in small amplitudes.

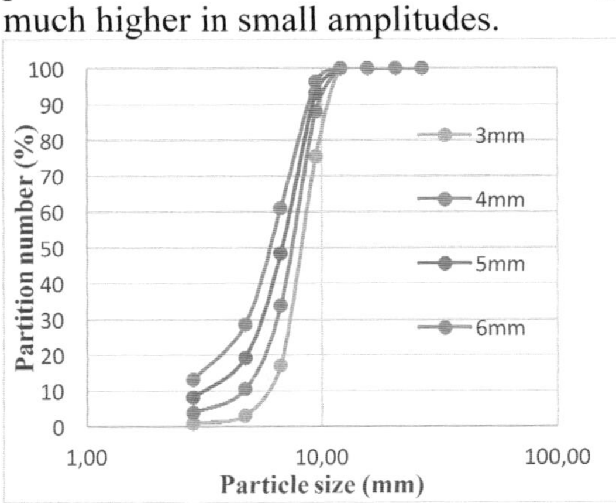

Fig. 4. Partition curves for different vibration amplitudes

4.3 Effects of Vibration Frequency

Simulations at 15, 20, 25 and 30 Hz frequencies is performed for investigating the effects of vibration frequency. Simulation results showed that in lower frequencies screening is more efficient rather than higher frequencies. Table 6 shows the screening efficiency and mass of undersize stream for the simulations and Fig. 5 shows the relevant partition curves.

Table 6. Screening efficiency and undersize stream mass percentage at different vibration frequencies

Vibration Frequency(mm)	Screening Efficiency (%)	Mass of Undersize Stream (%)
15	77.40	42.73
20	63.75	34.26
25	54.96	30.40
30	47.50	26.37

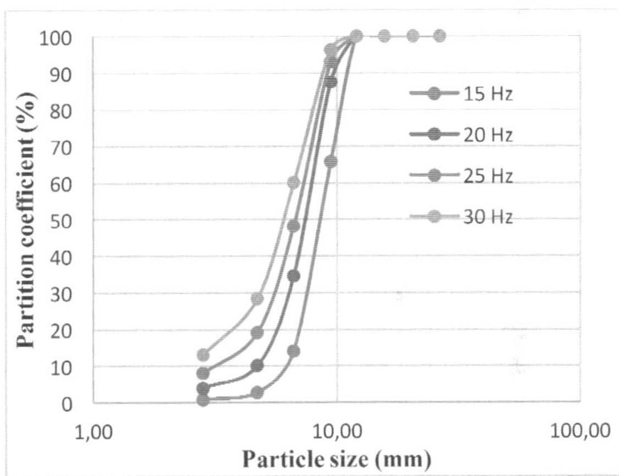

Fig. 5. Partition curves for different vibration frequencies

5 VALIDATION ATTEMPTS

For validation of the data extracted from DEM simulations, a pilot scale vibrating screen with the same dimensions was used. Features like screen surface shape, screen inclination, vibration amplitude, frequency and direction have been measured and same quantities have been imported to the simulation model. In the case of material to be screened, an aggregate sample of -30 mm size was taken and the sample was classified to different size fractions and the same particle size distribution that was imported to DEM model. After importing material properties like density, Young Modulus, Poison's ratio and etc. a simulation has been performed and an experimental test has been performed at the pilot scale screen. Based on the data extracted from experimental test and simulation, Table 7 shows the comparison of separation efficiency and mass of undersize stream of experimental test and simulation. Fig. 6 also shows related partition curves.

Table 7. Comparison of separation efficiency and undersize mass percentage at experimental test and simulation

	Experimental	Simulation
Screening Efficiency (%)	47.55	45.31
Mass of Undersize Stream (%)	27.97	27.96

Despite using spherical particles in simulations which is not realistic, the comparison of the results shows good agreement between experimental and simulation data.

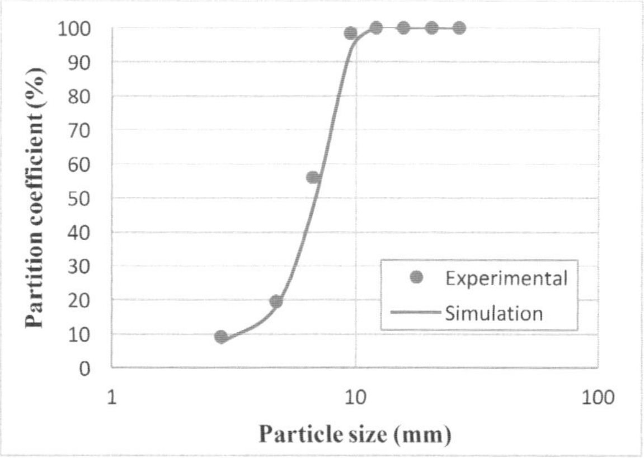

Fig. 6. Comparison between experimental and simulation partition curves.

As seen in Fig. 6 there is a good overlap especially at fine and coarse particle sizes between experimental and simulated partition numbers, but in the case of near-mesh particles (3rd and 4th size classes) there is a small amount of shift towards coarser size range and the partition number is lower for simulated data. This means that simulated test is more efficient in the case of near-mesh particles which probably is due to the use of spherical shapes to represent particles.

6 CONCLUSIONS

Particle flow in typical pilot scale vibrating screen has been numerically simulated by a three-dimensional DEM model. The effects of screen inclination, vibration amplitude and vibration frequency on the performance of screening process have been studied. The following conclusions can be extracted from this study:

- Based on the screen inclination ranges that were tested during this study, screening performance can be improved by reducing screen inclination. A significant increase in screening efficiency, undersize percentage and cut point size has been observed by reduction of screen inclination. Obviously, further increase of the feed flowrate would cause an increase in the bed-depth and a reduction in the overall performance.
- Screening performance is very sensitive to both vibration amplitude and frequency. It is observed that screening performance is relatively high in low vibration amplitude and frequencies and this leads to higher cut point size in the screening operation. This statement would obviously be correct for spherical particles and the amplitude and frequency values would have a lower limit with irregular shaped particles.

Despite nonrealistic spherical shape of used particles in DEM simulation, there is a good agreement between simulated and experiment data extracted. The studies are being continued towards the simulation of irregular shaped particles which would provide more accurate results.

Acknowledgement

The authors gratefully acknowledge the financial support of The Scientific and Technological Research Council of Turkey (TÜBİTAK) to the project entitled "Numerical Modelling of Industrial Screening" (215M368).

REFERENCES

Chand, R., Khaskheli, M.A., Qadir, A., Ge, B., Shi, Q., 2012. Discrete particle simulation of radial segregation in horizontally rotating drum:

Effects of drum-length and non-rotating end-plates. *Phys. A Stat. Mech. its Appl. 391*, (pp.4590–4596).

CHEN, Y., TONG, X., 2010. Modeling screening efficiency with vibrational parameters based on DEM 3D simulation. *Min. Sci. Technol. 20*, (pp.615–620).

CHEN, Y.H., Tong, X., 2009. Application of the DEM to screening process: a 3D simulation. *Min. Sci. Technol. 19*, (pp.493–497).

Cleary, P.W., Morrison, R.D., 2009. Particle methods for modeling in mineral processing. I*nt. J. Comut. Fluid Dyn. 23*, (pp.137–146).

Cleary, P.W., Sinnott, M.D., Morrison, R.D., 2009a. Separation performance of double deck banana screens – Part 1: Flow and separation for different accelerations. *Miner. Eng. 22*, (pp.1218–1229).

Cleary, P.W., Sinnott, M.D., Morrison, R.D., 2009b. Separation performance of double deck banana screens – Part 2: Quantitative predictions. *Miner. Eng. 22*, (pp.1230–1244).

Cundall, P.A., Strack, O.D.L., 1979. A discrete numerical model for granular assemblies. *Géotechnique 29*, (pp.47–65).

Delaney, G.W., Cleary, P.W., Hilden, M., Morrison, R.D., 2012. Testing the validity of the spherical DEM model in simulating real granular screening processes. *Chem. Eng. Sci. 68*, (pp.215–226).

Dong, K.J., Brake, I., 2009. DEM simulation of particle flow on a multi-deck banana screen. *Miner. Eng. 22*, (pp.910–920).

Dong, K.J., Yu, A.B., 2012. Numerical simulation of the particle flow and sieving behavior on sieve bend/low head screen combination. *Miner. Eng. 31*, (pp.2–9).

Elskamp, F., Kruggel-Emden, H., 2015. Review and benchmarking of process models for batch screening based on discrete element simulations. Adv. *Powder Technol. 26*, (pp.679–697).

Jahani, M., Farzanegan, A., Noaparast, M., 2015. Investigation of screening performance of banana screens using LIGGGHTS DEM solver. *Powder Technol. 283*, (pp.32–47).

Just, S., Toschkoff, G., Funke, A., Djuric, D., Scharrer, G., Khinast, J., Knop, K., Kleinebudde, P., 2013. Experimental Analysis of Tablet Properties for Discrete Element Modeling of an Active Coating Process. *AAPS PharmSciTech 14*, (pp.402–411).

Kruggel-Emden, H., Elskamp, F., 2014. Modeling of Screening Processes with the Discrete Element Method Involving Non-Spherical Particles. *Chem. Eng. Technol. 37*, (pp.847–856). 9

Li, J., Webb, C., Pandiella, S.S., Campbell, G.M., 2003. Discrete particle motion on sieves—a numerical study using the DEM simulation. *Powder Technol. 133*, (pp.190–202).

McBride, W., Cleary, P.W., 2009. An investigation and optimization of the "OLDS" elevator using Discrete Element Modeling. *Powder Technol. 193*, (pp.216–234).

Mular, A.L., Halbe, D.N., Barratt, D.J., 2002. *Mineral Processing Plant Design, Practice, and Control*. SME, Vancouver, British Columbia.

Xiao, J., Tong, X., 2012. Particle stratification and penetration of a linear vibrating screen by the discrete element method. *Int. J. Min. Sci. Technol. 22*, (pp.357–362).

Zhao, L., Zhao, Y., Liu, C., Li, J., Dong, H., 2011. Simulation of the screening process on a circularly vibrating screen using 3D-DEM. *Min. Sci. Technol. 21*, (pp.677–680).

Applicability of DIP Methods on Particle Size Distribution of Unsized Aggregates

Boyutlandırılmamış Agregaların Tane Boyut Dağılımlarının Belirlenmesinde Dijital Görüntü İşleme Yöntemlerinin Uygulanabilirliği

İ. Kurşun, M. Terzi, T. D. Tombal
İstanbul Üniversitesi, Maden Mühendisliği Bölümü, İstanbul

U. Çınar
Yeni Yüzyıl Üniversitesi,İş Sağlığı ve Güvenliği Bölümü, İstanbul

ÖZET Patlatmadan cevher hazırlamaya kadar bir dizi ardışık işlem kademesinden oluşan madencilik faaliyetlerinde tane boyut dağılımının hassas bir şekilde belirlenmesi hem verim hem de maliyet açısından büyük bir öneme sahiptir. Sayısal görüntü işleme yöntemleri cevher hazırlama disiplininde zaman tasarrufu sağlaması ve güvenilir veri sağlaması açısından çeşitli uygulama alanları bulmuştur.

Bu çalışmada özel bir şirkete ait kırmataş tesisinden alınan numunelerin tane boyut dağılımları elek analizi ve sayısal görüntü işleme analizi yöntemleri kullanılarak belirlenmiş ve bu yöntemlerin endüstriyel ölçekte uygulanabilirliği değerlendirilmiş ve karşılaştırılmıştır. Bu kapsamda, laboratuar ölçekli görüntü işleme analizi amacıyla bir pilot düzenek kurulmuştur. Numunelerin tane boyut dağılımı ölçümleri bu pilot düzenek kullanılarak sayısal görüntü işleme yöntemi ile ve konvansiyonel elek analizi yöntemi ile gerçekleştirilmiştir. Her iki yöntemle elde edilen sonuçlar, temsili numunelerin güvenilir ve tekrar edilebilir tane boyut dağılımlarının elde edilmesi noktasında birbirleri ile karşılaştırılmıştır.

ABSTRACT Many stages of mining operations which consists sequential processes from blasting to mineral processing, sensitive determination of the particle size distribution is important in terms of efficiency as well as affordability. Digital image processing methods used in mineral processing discipline found different application areas due to providing accurate data in relatively short time. In this study, the particle size distribution analysis of the samples taken from privately owned aggregate processing plants using sieve analysis and digital image processing methods were conducted and accordingly a comparison of these methods in terms of the applicability on industrial scale were realized. In this context, a pilot setup was assembled for the laboratory-scale image processing analysis purposes. Particle size distribution measurements of the samples were conducted by digital image processing method using this pilot setup and conventional sieve analysis methods. The obtained results were compared to each other to achieve reliable and reproducible particle size distribution values of the representative samples.

1 INTRODUCTION

As it is the case in various industrial processes, determination of particle size distributions of various raw materials reduced to fine particles such as glass, cement, paper, plastic, ceramic, which are process input and/or output, is of great importance also in mining. An accurate determination of particle size distribution in many phases of mining which is consisting of a series of operation from blasting to mineral processing is important both in terms of efficiency and also economy (Kursun, 2009).

Various methods are used to determine particle size distributions and shape factors.

Sieve analysis, based on principle of aperture size from which certain particles pass through whilst others cannot, is generally used to determine particle size within the scope of mineral processing. However, taking representative samples from conveyor belt type process flows is in general only possible through stopping the process (Saklara et al., 2000; Bernhardt, 2000; Kippax, 2005; Ulusoy et al, 2006; Ulusoy, 2008; Kursun and Ozdemir, 2009).

To determine particle size distribution of a material composed of particles in various sizes, to find out the relation between particle size and quantities of this sample; in other words to determine in which size and quantities are present in a given material that performed with standard sieves in laboratories is called laboratorial sieving (Wills, 1985; Kesking 1986).

Image processing is a practice which carried out in order for measured or captured electronic (digital) image data to be transformed in accordance with the purpose by use of a computer software. Image processing is more commonly used to process available captured images; that is to say it is used to change, to split or enhance available pictures and graphics (Tovev and Hunslow, 1995; Yılmaz, 2007).

Image processing is a process different from signal processing, which is conversion of data into a format usable by another device following to measurement and evaluation by means of data capture, or a practice related to transfer of data from an electronic media to another electronic media. Images are consisted of different contents captured from image source directed to a specific target. The said images can be exemplified as ultrasound images, electron microscope and computer images. Preliminary stages applied to images reduce noise (image blur, sharpness, poor image quality) on images. To this end, low, medium and high level processes are performed on the images. In low level operations, acuity of input and output images is ensured with filtering. As for medium level operations, segmentation and recognition steps are performed for recognition and classification of the objects on images. High level operations involve analysis of images for recognition of objects on images. By analyzing the images in the computer, the content of the objects in the images are detailed. Image processing would be performed by this detailing phase (Jähne, 2005; Gonzalez and Woods, 2005).

This study is conducted in order to make particle size measurements in crushed stone plants by use of digital image processing techniques. In this context, a lab scale digital image processing set-up was assembled. Systematic imaging for image processing and processing of the captured images was realized using this set-up, and then finally, reliability and accuracy analyses were made on the obtained data and results were discussed.

2 MATERIALS AND METHODS

The unsized aggregate samples used for experimental studies in this project were supplied from a crushed stone plant operating in Cendere Region of Istanbul. Approximate chemical content of the sample is given in Table 1.

Table 1. Approximate chemical content of the sample

Content	Assay Value (%)
CaO	2,00
SiO_2	57,00
Fe_2O_3	12,00
MgO	4,00
Al_2O_3	18,00
SO_3	0,10
P_2O_5	0,20

First, the samples were subjected to sieve analysis and their particle size distribution was determined. The obtained particle size distribution results from sieve analysis were used as a reference for reliability analysis on the data to be obtained from digital image analyses.

The laboratory-scale digital image processing set-up is consisting of CMOS camera with high-resolution lens, mounting parts, telescopic support legs, white background, LED lights, computer and image processing software. Schematic view of the set-up is given in Figure 1.

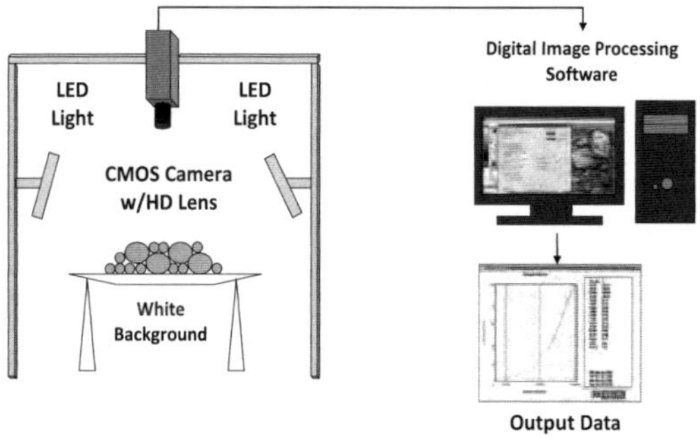

Figure 1. Schematic view of the lab set-up

The sample was homogenous spreading through the displaying surface with white background of the setup to simulate the conditions in a crushing plant to a degree. Additionally, the sample was not arranged in any way to prevent overlapping of particles. Subsequently, the high resolution image of the sample was captured using the pilot setup and Split Desktop Software was used for determination of the PSD of the samples with DIP method.

Principal stages used in image processing by the software are as follows:

- Obtaining digital images automatically or manually,
- Pre-processing of the unacceptable images that have lighting problems,
- Defining the fragments within each DIP algorithm,
- Applying statistical algorithms to two-dimensional fragment areas to define three-dimensional fragment volumes of each image,
- Processing various images together for average distribution,
- Transmitting output data to the screen, hard disc and network control systems (Ozdemir et al, 2005).

Software's working principle is based on the determination of the border and length of the particles in 2D. After this process, the software gives user a chance to correct the border of the particles. (Ozdemir et al. 2005, 2007).

3 RESULTS AND DISCUSSION

The sieve analysis result of the sample is given in Table 2.

Table 2. Sieve analysis results

Size (mm)	Weight (%)
-32.5+16.0	23.27
-16.0+12.0	36.70
-12.0+10.0	17.86
-10.0+8.0	7.02
-8.0+6.0	7.70
-6.0+4.0	6.43
-4.0+0.0	1.02
Total	100.00

Cumulative undersize graph of the sample is given in Figure 2.

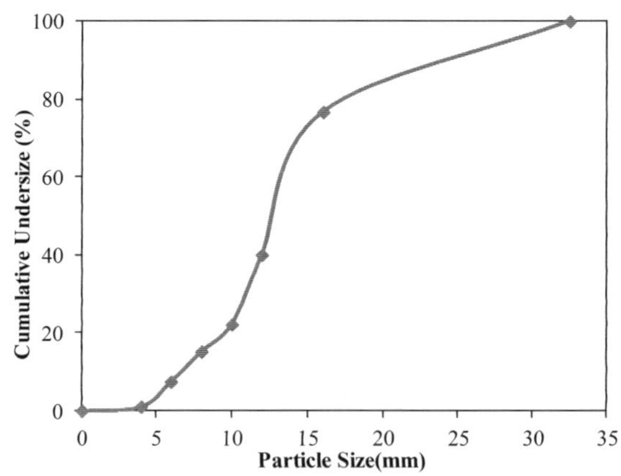

Figure 2. Cumulative undersize graph

According to cumulative undersize graph, d_{20}, d_{50} and d_{80} sizes of the sample were determined as 9.60 mm, 12.72 mm and 17.30 mm, respectively.

The image of the sample captured using the set-up is given in Figure 3. In order to allow processing of these images by software, an object with known dimensions was positioned in the center.

Figure 3. Raw image obtained with the set-up

Different sections of the irregularly distributed aggregate samples were captured using the set-up and these images were processed with the help of Split Desktop software.

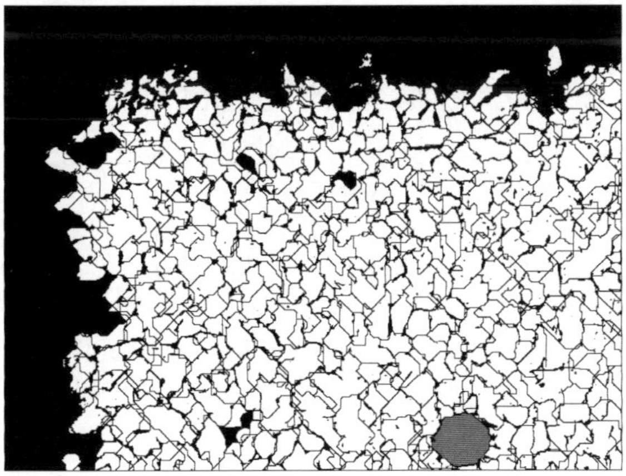

Figure 4. Images processed with Split Desktop software (Section 1)

Images of first section processed with the help of Split Desktop software is given in Figure 4. The particle size distribution graph based on reference object was plotted by processing the particles with identified boundaries and it is given in Figure 5.

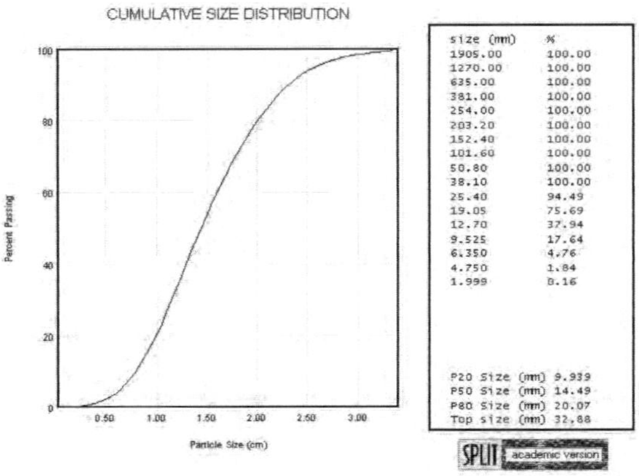

Figure 5. Particle size distribution graph (Section 1) obtained by Split Desktop software.

According to particle size analysis of the Section 1 realized by digital image processing; d_{20}, d_{50} and d_{80} sizes of the sample were determined as 9.94 mm, 14.49 mm and 20.07 mm, respectively.

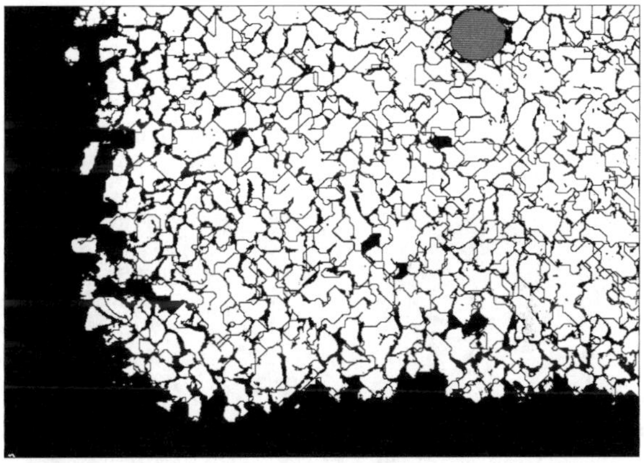

Figure 6. Images processed with Split Desktop software (Section 2)

The processed image of Section 2 and the particle size distribution graph obtained from processing of said image are given in Figure 6 and Figure 7, respectively. d_{20}, d_{50} and d_{80} sizes of the sample were determined as 10.59 mm, 15.73 mm and 22.81 mm, respectively according to particle size analysis by digital image processing of the Section 2.

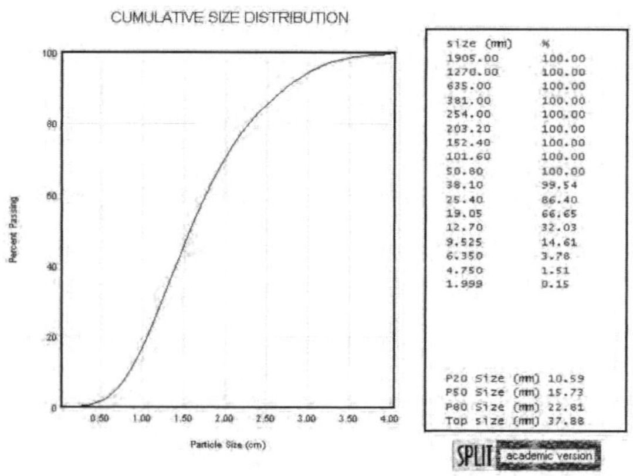

Figure 7. Particle size distribution graph (Section 2) obtained by Split Desktop software.

Images of third section processed with the help of Split Desktop software is given in Figure 8. The particle size distribution graph plotted by Split Desktop software is given in Figure 9.

According to particle size analysis of the Section 3 realized by digital image processing; d_{20}, d_{50} and d_{80} sizes of the sample were determined as 10.38 mm, 15.09 mm and 20.88 mm, respectively.

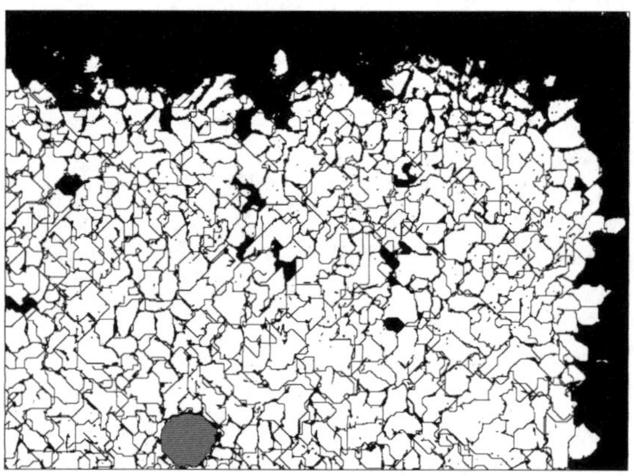

Figure 8. Images processed with Split Desktop software (Section 3)

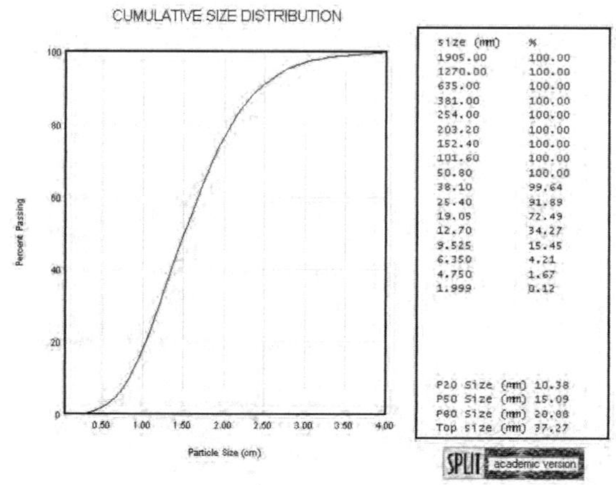

Figure 9. Particle size distribution graph (Section 3) obtained by Split Desktop software.

The processed image and particle size distribution graph of Section 4 are given in Figure 10 and Figure 11, respectively.

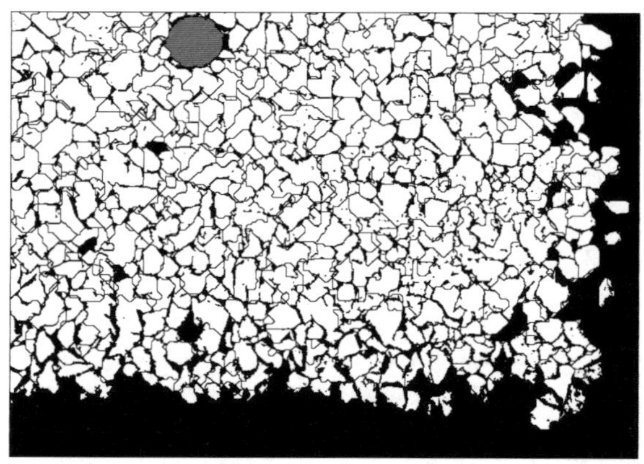

Figure 10. Images processed with Split Desktop software (Section 4)

d_{20}, d_{50} and d_{80} sizes of the sample were determined as 10.07 mm, 14.79 mm and 20.68 mm, respectively according to particle size analysis by digital image processing of the Section 4.

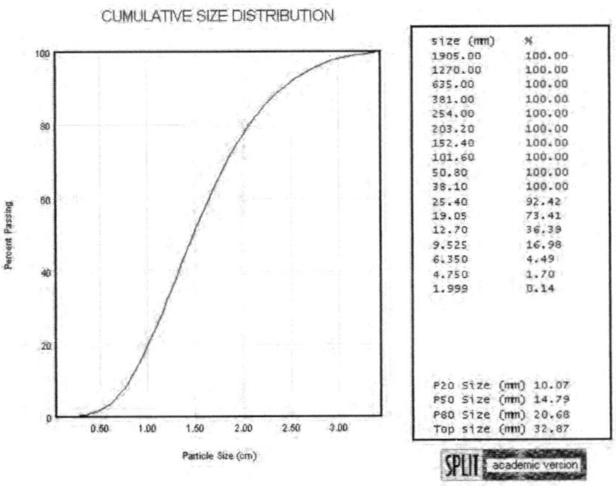

Figure 11. Particle size distribution graph (Section 4) obtained by Split Desktop software.

4 CONCLUSION

In this study, the particle size distribution analysis of the samples taken from privately owned aggregate processing plants using sieve analysis and digital image processing methods were conducted and accordingly a comparison of these methods in terms of the applicability on industrial scale were realized. In this context, a lab scale digital image processing set-up was assembled. Systematic imaging for image processing and processing of the captured images was realized using this set-up.

In laboratory study, maximum error was 31.85% for d_{80} of Section 2 whereas minimum error was 3.54% for d_{20} size of Section 1. Average error for d_{20}, d_{50} and d_{80} sizes were 6.72%, 18.12%, and 22.02%, respectively. Laboratory data were obtained by taking average of 4 sections taken from a single image. Thus, error margin for each section had values close to the average. Final error percentages obtained by taking average of sizes identified processing image sections and comparing them with sieve analysis results are given in Table 3.

When we compared the results of experimental studies with real data; it is observed that real and experimental values for d_{20} and d_{50} sizes were generally determined to be close to each other. However, values of d_{80} size were above real values. The most important reason for this is presumed to be that some particles were detected as a single particle when particle edges were determined and thus larger particle sizes were obtained. Additionally, the resulted deviations between the methods can be attributed to a fact that the DIP method takes only a two dimensional projected image of a particle, therefore elongated particles which were likely to pass a certain sieve size might be considered as oversize in DIP method.

As a conclusion, the obtained results supported that the DIP method can be used, especially with applying a correction coefficient due to varying degrees of error in d_{80} sizes, in PSD analysis of unsized aggregates as an alternative to conventional methods such as the sieve analysis. Further studies are planned for the evaluation of this experimental setup in a crushing plant environment not only for the evaluation of the industrial applicability of the procedure and but also for to investigate the minimization of resulted errors.

Table 3. Results of percentage error

Section	**Sieve Analysis**			**Digital Image Processing**			**Error (%)**		
No	d_{20}	d_{50}	d_{80}	d_{20}	d_{50}	d_{80}	d_{20}	d_{50}	d_{80}
1	9,60	12,72	17,30	9,94	14,49	20,07	3,54	13,92	16,01
2				10,59	15,73	22,81	10,31	23,66	31,85
3				10,38	15,09	20,88	8,13	18,63	20,69
4				10,07	14,79	20,68	10,25	15,03	21,11
						Mean	6,72	18,12	22,02

ACKNOWLEDGEMENTS

This work was supported by the Scientific and Technological Research Council of Turkey (TUBITAK) with project number 214M625.

REFERENCES

Bernhardt, C. 2000. Particle Size Analyses, Classification and Sedimentation Methods. Chapman & Hall.

Gonzalez, R.C., Woods, R.E., (3th Ed.), 2007. *Digital Image Processing*, USA, New Jersey: Prentice Hall.

Jähne, B., (6th Ed.)., 2005. *Digital Image Processing,* Berlin: Springer.

Keskin, Y., 1986. *Kömür Hazırlama Yöntemleri,* Zonguldak: T.T.K. İnsangücü Eğitim Şube Müdürlüğü Yayınları.

Kippax, P., 2005. Measuring Particle Size Using Modern Laser Diffraction Techniques, *Paint & Coating Industry Magazine*. pp. 1-5.

Kursun, I., 2009. Particle Size and Shape Characteristics of Kemerburgaz Quartz Sands Obtained By Sieving, Laser Diffraction, And Digital Image Processing Methods, *Mineral Processing and Extractive Metallurgy Review*, vol.30, pp. 346-360.

Kursun, I. and Ozdemir, K., 2009. An Example of the Application of Digital Image Processing Methods to Mineral Processing: Determination of Particle Size of Akpinar Quartz Sands, *21st International Mining Congress and Exhibition of Turkey*, 6-8 May 2009, Antalya, Turkey, Chamber of Mining Engineers of Turkey, pp. 365-374.

Ozdemir, K., Kahriman, A. and Dogan, T., 2005. Patlatma Sonrası Parça Boyut Dağılımının Ekskavatör Kazı Süresine Olan Etkisinin Araştırılması, *II. İş Makinaları Sempozyumu ve Sergisi*, Istanbul, Turkey. pp. 273-282.

Ozdemir, K., Tuncer, G., Kahriman, A., Ozer, U. and Karadogan, A., 2007. The relation between Excavator Bucket Loading Time and Particle Size Distrubution of Shot Rock, *Proceedings of the 33rd Annual Conference on Explosives and Blasting Technique, Volume I*, Nashville, Tennesse, USA. pp. 303-313.

Saklara, S., Bayraktar, I. and Oner, M., 2000. The Methods Used In Analysis of Fine Particle Size, *Mining Bulletin*, no. 39.

Tovey, N.K. and Hunslow, M.W., 1995. Quantitative micro-porosity and orientation analysis in soils and sediments, *Journal of the Geological Society London*, vol. 152, no. 1, pp. 119-129.

Ulusoy, U., 2008. Application of ANOVA to Image Analysis Result of Particles Produced by Different Milling, *Powder Technology*, vol. 188, pp.133-138.

Ulusoy, U., Yekeler, M., Bicer, C. and Gulsoy, Z., 2006. Combination of the Particle Size Distributions of Some Industrial Minerals Measured by Andreasen Pipette and Sieving Techniques, *Particle and Particle Systems Characterization*, vol. 23, pp.448-456.

Wills, B. A., 1985. Mineral Processing Technology, Oxford: Pergamon Press.

Yılmaz, A., 2007. Real time security application with image processing using camera, MSc. Thesis, Halic University, Graduate School of Science, Istanbul

Konvansiyonel Kapalı Devre İşleminin Farklı Çimento Üretimlerinde Performansının Değerlendirilmesi

Performance Evaluation of a Conventional Closed Circuit Process at Different Cement Productions

Ö. Genç
Muğla Sıtkı Koçman Üniversitesi, Maden Mühendisliği Bölümü, 48000, Kötekli Yerleşkesi, Muğla
A. H. Benzer
Hacettepe Üniversitesi, Maden Mühendisliği Bölümü, 06540, Beytepe, Ankara

ÖZET CEMII tipi Portland çimentosu üretimlerinde, konvansiyonel Polysius® iki-kamaralı bilyalı değirmen ve O-Sepa® yüksek verimli dinamik havalı sınıflandırıcı (FLSmidth) kapalı devre işleminin performansı denge koşullarında incelenmiştir. Devre etrafında iki örnekleme çalışması yürütülmüştür. Devrelerin madde denkliği JKSimMet Cevher Hazırlama Simülatörü kullanılarak yapılmıştır.

Ø4xL12.4m boyutlarındaki bilyalı değirmenin boyut küçültme performansı farklı besleme koşullarında değerlendirilmiştir. Bilyalı değirmenin boyut küçültme performansının CEMII Portland kompoze çimento üretimi tipinde iyileştiği belirlenmiştir. CEMII Portland kompoze çimento üretiminde istenilen son ürün incelik özelliklerini elde edilebilmek için, havalı sınıflandırıcının rotor hızı (%) düşürülmüştür. Devredeki dinamik havalı sınıflandırıcının ayrım verimliliği, verimlilik eğrisi yaklaşımı temel alınarak belirlenmiştir (Munn vd., 2005). Öğütme yardımcısının tipi ve miktarının değiştirilmesine ek olarak havalı sınıflandırıcının optimum koşullarda çalıştırılmasıyla, CEMII Portland kompoze çimento üretiminde devre kapasitesinde yaklaşık olarak %9 oranında bir artış elde edilebilmiştir.

ABSTRACT Performance of a conventional Polysius® two-compartment ball mill and a O-Sepa® high efficiency dynamic air classifier (FLSmidth) closed circuit process was investigated at the steady state conditions at CEM II type Portland cement productions. Two sampling surveys were performed around the circuit. CEM II type Portland calcareous and composite cements were produced in sampling survey-1 and survey-2 respectively. Mass balance of circuits were performed by using the JKSimMet Mineral Processing Simulator.

Size reduction performance of the ball mill with dimensions of Ø4xL12.4m was evaluated in different feed conditions. Size reduction performance of the ball mill was determined to be improved at CEM II Portland composite cement production mode. Air classifier rotor speed (%) was decreased in order to achieve required fineness specifications of the final product in CEM II Portland composite cement production. Separation efficiency of the dynamic air classifier in the circuit was determined on the basis of efficiency curve approach (Munn et.al., 2005). Approximately 9% circuit capacity increase was achieved in CEM II Portland composite cement production by changing the type and amount of grinding aid in addition to operating the air classifier at its optimum conditions.

1 INTRODUCTION

Turkey is one of the major cement producers in the world with a cement production capacity of 77 million tons per year (United States Geological Survey, 2015) and ranked fourth in the world. According to the Turkish Cement Manufacturers' Association statistics (2015), installed cement production capacity of the Turkish Cement Sector is 126.141 (×1000ton).

Cement grinding is an energy intensive process such that, 2% of the total electrical energy produced in the world is consumed in the grinding stage of cement production and approximately 40% of the total electrical energy consumption in a cement plant is required for finish grinding stage only (Norholm,1995). Thus, even a slight reduction in grinding energy consumption will reduce the operating costs in a significant amount. In this respect, knowledge about the operational conditions and performances of the mills and classifiers in grinding circuits will enable the optimization of the grinding process which will consequently reduce the energy consumption of the process.

In this study, grinding and classification performance of a two-compartment ball mill and an air classifier closed circuit process were evaluated when operated at CEM II type Portland calcareous and composite cement types. Operation of the air classifier at its optimum conditions in addition to the optimum level of grinding aid addition with a different brand to the ball mill, resulted an increase in ball mill and air classifier performances at CEM II Portland composite cement production at a higher circuit capacity condition.

2 MATERIALS AND METHODS

2.1. Sampling Surveys

Extensive sampling surveys were carried out to determine particle size distributions around the circuits and inside the mills in CEM II type Portland calcareous and Portland composite cement_productions. Circuit flowsheet with the typical sampling points is given in Figure 1 for the CEM II Portland composite cement production case. CEM II composite cement was obtained by grinding Portland cement clinker, to provide mechanical strength, with mineral additive materials such as gypsum ($CaSO_4{:}2H_2O$) and calcareous (limestone) and natural pozzolana (trass). Limestone and trass are added as setting time regulators to cement. Pozzolana is a siliceous and aluminous material that contains volcanic material such as pumice or volcanic ash. Trass was not fed to the circuit in CEM II Portland calcareous cement production.

Polysius® two-compartment ball mill and a O-Sepa® high efficiency dynamic air classifier were operated in closed circuit to obtain the required cement types. Some of the design specifications for the ball mill and air classifier are tabulated in Table 1 and 2. Simplified flowsheet of the conventional closed circuit process is shown in Figure 1. There is a electrofilter at the discharge of the mill which collects very fine particles (100% -150μm) from the mill by air sweeping. Samples were collected from the streams shown in Figure 1 at the steady state condition of the circuits.

Table 1. Two-compartment ball mill specifications

Parameter	Value
Diameter (m)	4
Compartment-1 length (m)	3.9
Compartment-2 length (m)	8.4
Design capacity (t/h)	85
Mill revolution (rpm)	16
Mill filter fan volumetric flowrate (m3/h)	81430

Table 2. Air classifier specifications

Air classifier brand	O-Sepa N-2500
Rotor height (mm)	1390
Rotor diameter (mm)	2530
Maximum rotor speed (rpm)	221
Motor speed (rpm)	1500

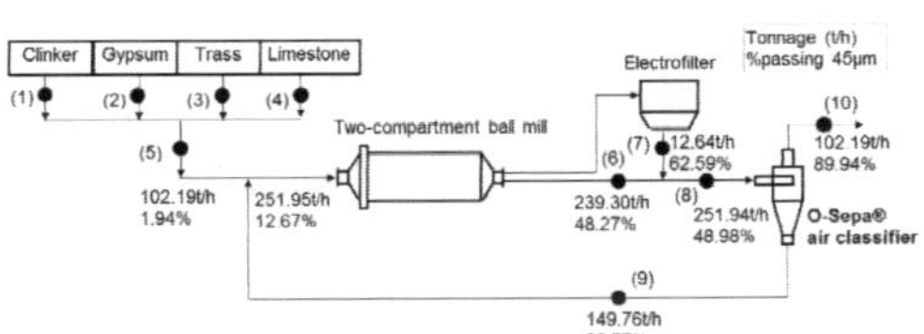

Figure 1. Simplified flowsheet of the sampled circuit in CEM II Portland composite cement production

Sampling points correspond to the following streams in Figure 1:

(1) Clinker feed
(2) Gypsum feed
(3) Trass feed
(4) Limestone feed
(5) Mill fresh feed
(6) Mill overflow
(7) Electrofilter return
(8) Air classifier feed
(9) Air classifier reject
(10) Air classifier fine

The circuits were operated with grinding aid addition in both cement production types. Different brands of grinding aids with different amounts were applied due to the mill feed grindability differences. Grinding aid types and amounts are tabulated in Table 3.

Table 3. Grinding aid addition

Cement production type	Grinding aid type	Addition (g/ton)
CEM II portland calcareous cement (survey-1)	Grace (CBA 1701)	400
CEM II portland composite cement (survey-2)	Chryso (EPCT 200)	1000

2.1.2. Mill inside sampling surveys

After completing circuit sampling, ball mill was crushed-stopped for inside mill sampling in both surveys. Both compartments of the ball mill were sampled by approximately one meter along the long axis of the mill from an approximate depth of 40cm below the mill ball charge level. Mill inside sampling locations are tabulated in Table 4.

Table 4. Mill inside sampling locations in grinding compartments 1 and 2

Compartment-1 axial length (m)	Compartment-2 axial length (m)
0 (inlet)	0 (inlet)
1	1
2	2
3	3
3.9	4
	5
	6
	7
	8.6

2.2. Determination of particle size distributions

Particle size distributions of the samples were determined by dry sieving from the top size which was 50mm down to 150µm. Sub-sieve sample of -150µm material was dry sized down to 1.8µm by using a Sympathec® laser diffractometer. Ro-tap screening and laser sizing results were mathematically combined to obtain the full size distribution from the top size down to 1.8 µm.

2.3. Material characterization

Standard Bond ball mill work index values of the mill feed were determined using a Standard Bond ball mill with an effective diameter of 30.5cm and a length of 30.5cm which was operated without liner. Bond index values were determined according to the Bond (Austin et.al., 1984) using a 90µm Tyler test sieve. Test results are tabulated in Table 5. It was determined that, circuit feed in Portland composite cement production was more harder than that of Portland calcareous cement. So, it was expected that, material in composite cement production would show more resistance to grinding in the ball mill.

Table 5. Standard Bond work index values of circuit feed samples

Sampling surveys	Bond work index (kWh/t)
CEM II Portland calcareous cement (survey-1)	14.12
CEM II Portland composite cement (survey-2)	15.86

3 RESULTS AND DISCUSSION

3.1. Mass Balancing

Mass balance calculations were performed to estimate the statistically adjusted tonnage flow rates and particle size distributions by using the mass balance module of the JKSimMet Mineral Processing Simulator. Calculated (mass balanced) tonnage flowrates and fineness values are given in Figures 2 and 3. Agreement between the experimental and calculated particle size distributions are plotted in Figures 4 to 7. A very good agreement was obtained which indicated a successful sampling operation. So mass balanced values could be used in performance evaluation study.

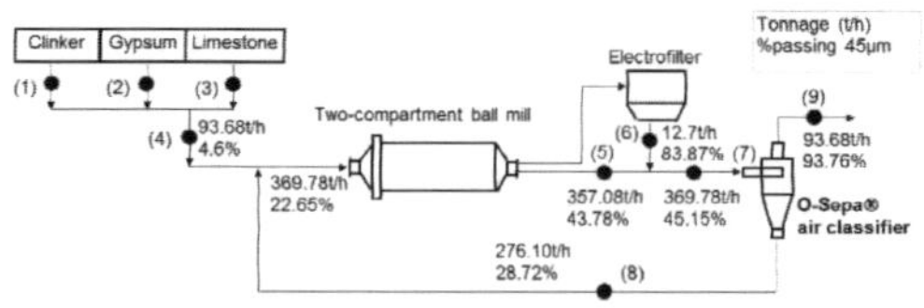

Figure 2. Mass balanced flowsheet of the sampled circuit in CEM II Portland calcareous cement production (survey-1)

Sampling points correspond to the following streams in Figure 2:

(1) Clinker feed
(2) Gypsum feed
(3) Limestone feed
(4) Mill fresh feed
(5) Mill overflow
(6) Electrofilter return
(7) Air classifier feed
(8) Air classifier reject
(9) Air classifier fine

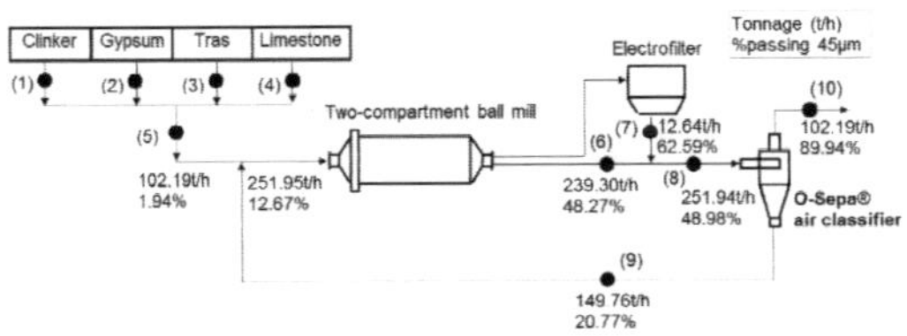

Figure 3. Mass balanced flowsheet of the sampled circuit in CEM II Portland composite cement production (survey-2)

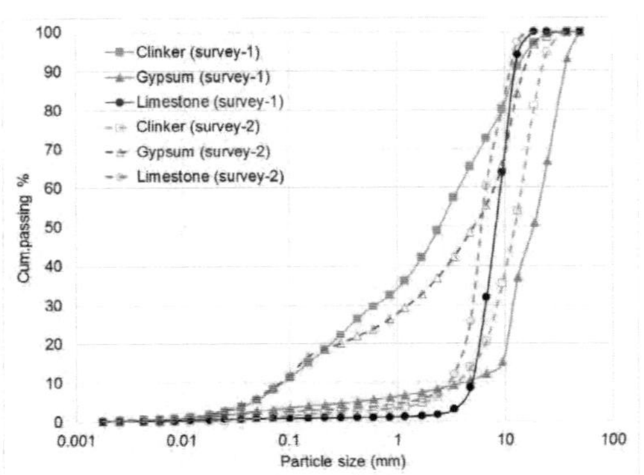

Figure 4. Experimental particle size distributions of circuit fresh feed materials in survey-1 and survey-2

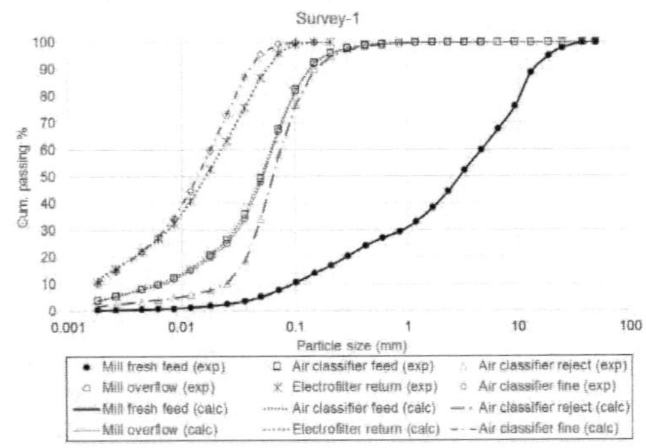

Figure 5. Experimental and mass balanced particle size distributions around the circuit in CEM II Portland calcareous cement production (survey-1)

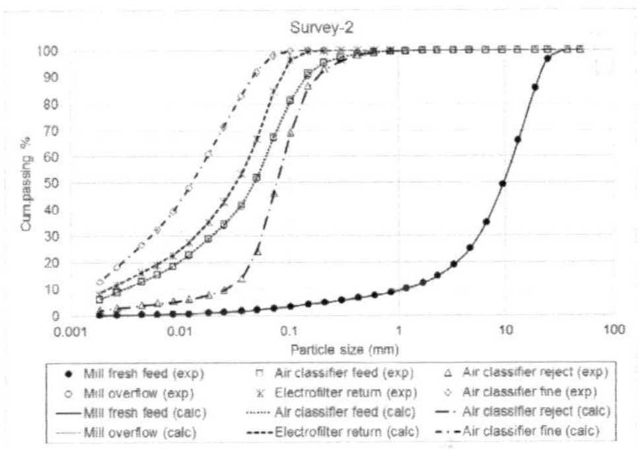

Figure 6. Experimental and mass balanced particle size distributions around the circuit in CEM II Portland composite cement production (survey-2)

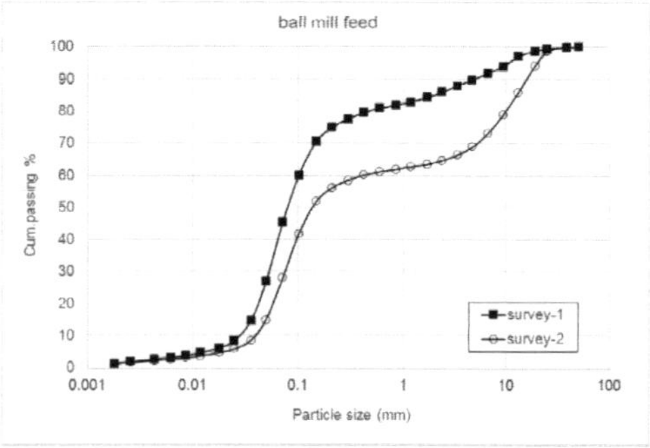

Figure 7. Ball mill feed particle size distributions in survey-1 and survey-2

3.2. Size Reduction Performance of the Two-compartment Ball Mill

Cumulative 90%, 80% and 50% passing sizes which are x90, x80 and x50 sizes of the mill samples respectively are determined to evaluate the size reduction progress along the long axis of the compartments. Results are tabulated in Tables 6 and 7 for the sampling surveys.

Table 6. x90, x80 and x50 sizes for mill inside samples in compartment-1 in survey-1 and survey-2

Axial sampling length (m)	Survey-1 Comp-1 x90 (mm)	Survey-2 Comp-1 x90 (mm)
0 (inlet)	5	11.5
1	9.5	4.0
2	2.5	3.0
3	1.5	1.1
3.9	0.95	1.5
	Comp-1 x80 (mm)	Comp-1 x80 (mm)
0 (inlet)	0.48	2.0
1	0.50	0.95
2	0.27	1.08
3	0.35	0.51
3.9	0.35	0.73
	Comp-1 x50 (mm)	Comp-1 x50 (mm)
0 (inlet)	0.800	0.093
1	0.075	0.094
2	0.075	0.130
3	0.076	0.103
3.9	0.080	0.140

Table 7. x90, x80 and x50 sizes for mill inside samples in compartment-2 in survey-1 and survey-2

Axial sampling length (m)	Survey-1 Comp-2 x90 (mm)	Survey-2 Comp-2 x90 (mm)
0 (inlet)	0.450	0.500
1	0.330	0.360
2	0.270	0.310
3	0.200	0.270
4	0.166	0.200
5	0.150	0.200
6	0.149	0.170
7	0.146	0.160
8.6	0.136	0.150
	Comp-2 x80 (mm)	Comp-2 x80 (mm)
0 (inlet)	0.200	0.250
1	0.170	0.200
2	0.151	0.175
3	0.130	0.150
4	0.096	0.130
5	0.105	0.125
6	0.104	0.105
7	0.100	0.108
8.6	0.099	0.102
	Comp-2 x50 (mm)	Comp-2 x50 (mm)
0 (inlet)	0.076	0.085
1	0.072	0.076
2	0.072	0.076
3	0.065	0.069
4	0.055	0.059

5	0.055	0.056
6	0.055	0.056
7	0.054	0.052
8.6	0.053	0.050

There should be a size reduction progress along the long axis of the mill if the grinding performance was high. Size reduction progress was found to be achieved at coarse size ranges in the first compartment which were indicated by the 90% and 80% cumulative passing size in survey-1 (Table 6). Since coarser grinding media was applied in the first compartment, size reduction performance was higher as compared to that obtained at fine size ranges along the compartment length as coarser balls will break coarser particles more efficiently. Size reduction progress was achieved in the first two meters of the first compartment length as indicated by the 80% cumulative passing size. Size reduction progress could not have been observed at fine size ranges which were indicated by the 50% cumulative passing size in survey-1. Compartment-1 inlet sample was found to be finer as compared to the sample collected from the first meter of the compartment-1 length at coarse size ranges (coarser than 0.425mm) which could be linked to the air classifier reject material accumulation at the mill inlet at the crash-stop condition.

The second compartment of the ball mill was lined by classifying liners and finer grinding media size distribution was applied in the second compartment. As classifying liners were applied, ball size distribution will be classified from coarser to finer towards the end of the compartment length. If classifying liners are working efficiently and a true ball size classification was achieved (Genç, 2008), an efficient grinding process will be performed and grinding performance of mill is expected to be high.

It was observed that, size reduction was achieved in the second compartment at coarse size ranges as indicated by the 90% cumulative passing size in survey-1. However, it became more difficult to grind particles at finer size ranges which was indicated by the 50% cumulative passing size.

A systematic size reduction progress was achieved at coarse size ranges as indicated by the 90% cumulative passing size in survey-2. However, size reduction progress was not systematic at fine size ranges as particle size (i.e. x80 and x50) fineness fluctuated towards the end of the first compartment. For example, x80 and x50 sizes became coarser at the end of the first compartment length. Size reduction progress was found to be systematic in the second compartment of the mill which indicated the higher size reduction performance.

Consequently, grinding performance of the ball mill was determined to be higher, although coarser material was fed to the mill. This could be linked to the type and amount of the grinding aid applied.

Cumulative 80% passing particle size (x80) and cumulative 50% passing particle size (x50) were determined for the total mill feed and mill discharge. Determined particle sizes in addition to the size reduction ratio which was defined as a ratio of F80 and P80. F80 and P80 values were cumulative 80% passing sizes of the mill feed and mill discharge respectively. As it is a fine grinding operation, median size of the distributions were also considered in determining the size reduction performance. F50 and P50 were cumulative 50% passing sizes of the mill feed and the mill discharge respectively. Related particle sizes and the size reduction ratios are tabulated in Table 8.

Table 8. Overall size reduction on the basis of x80 and x50 particle sizes (x: mill feed or discharge stream)

	F80 (mm)	P80 (mm)	F80/P80
Survey-1	0.45	0.1	4.5
Survey-2	10	0.1	100
	F50 (mm)	**P50 (mm)**	**F50/P50**
Survey-1	0.08	0.050	1.6
Survey-2	0.14	0.046	3.0

Higher size reduction values were obtained in survey-2. Size reduction ratio was found to increase at both coarse and fine size ranges. When x50 sizes were considered increase in size reduction was lower as compared to the case when x80 sizes were considered. This because, it is more difficult to achieve size reduction at finer sizes. Consequently, size reduction performance of the mill was increased in survey-2.

3.3. Separation efficiency of the O-Sepa® air classifier

The O-Sepa® air classifier is known to be the world standard in high-efficiency separation. It has many advantages in terms of cost savings such that, reduced specific power consumption, increased grinding efficiency, low maintenance, integral cooling capability (http://www.flsmidth.com).

3.3.1. Air Classifier Separation Efficieny

The tromp (efficiency or partition) curves were plotted to determine the classification performance of the O-Sepa® air classifier in the sampling surveys. The tromp curve indicates the probability of a particle in the air classifier feed that will be returned to the mill. If there is a high probability for coarse particles and lower probability for fine particles, the separation efficiency is higher and better. Partition coefficients were calculated from Equation 1 to plot the efficiency curves.

$$\text{Partition coefficient} = \frac{Uu}{Ff} \times 100 \quad (1)$$

U : Air classifier reject flowrate (t/h)
F : Air classifier feed flowrate (t/h)
u : Percentage of particle size i in air classifier reject stream (%)
f : Percentage of particle size i in air classifier feed stream (%)

O-Sepa® air classifier feed size distributions are compared in Figure 8 for the sampling surveys. Efficiency curves for the air-classifier are given in Figure 9. Performance parameters of the efficiency curves are tabulated in Table 9.

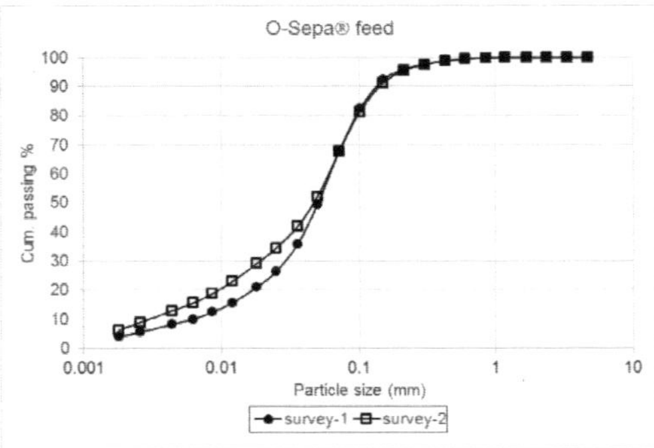

Figure 8. O-Sepa® air classifier feed size distributions in sampling surveys

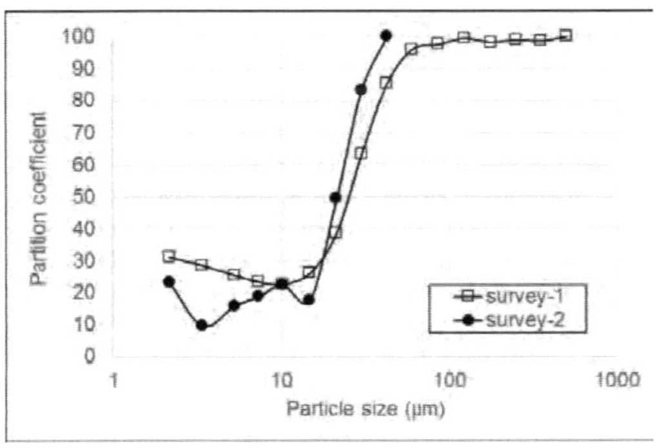

Figure 9. Efficiency curves of the air classifier in survey-1 and survey-2

Table 9. Air classifier performance parameters

Parameters	Survey 1	Survey 2
d50 (µm)	25	21
By-pass (%)	22.29	9.44
Fish-hook (%)	8.87	13.61

Cut size or separation size (d50) defines the size for which 50% of the particles in the feed report to the underflow stream (Svarovsky, 1984). By-pass defines the percentage of feed material reporting to air classifier reject without classification. Fish-hook parameter defines the difference between the maximum percentage of fine material amount that appears in underflow of the air classifier and the by-pass percentage.

The fineness of the product is adjusted by the speed of the rotor of the air classifier. Air classifier feed size distribution in sampling survey-2 got finer as shown in Figure 8. Thus, O-Sepa® rotor speed was decreased from 75% to 67% in order to attain the required final product fineness during the operation. Air classifier air speed was recorded as 65% in both sampling surveys.

Air classifier rotor speed was decreased as the feed size distribution became finer in survey-2. Finer separation size (d50) was obtained with a lower by-pass value which indicated a higher separation performance. Higher by-pass value in survey-1 was due to the increase in air classifier feed rate. 8.87% fish-hook amount indicated the increase in the amount of particles finer than 10.16µm in the underflow stream of the air classifier in survey-1. This amount of particles should have been reported to the overflow stream of the classifier under ideal separation conditions. 13.61% fish-hook amount indicated the increase in the amount of particles finer than 3.38µm in the underflow stream of the air classifier in survey-2. Increase in the fish-hook amount in survey-2 could be linked to the finer feed size distribution as compared to that of in survey-1. Perfect separation can be achieved with zero fish-hook and by-pass amount.

4 CONCLUSION

Grinding performance of the two-compartment ball mill was improved by changing the type and amount of the grinding aid as the mill feed size distribution became coarser and harder. Air classifier separation performance was improved by operating the air classifier at a lower rotor speed. Consequently, approximately 9% circuit capacity increase was achieved in CEM II Portland composite cement production at optimum operational conditions.

Acknowledgements

Authors would like to acknowledge Assist.Prof. Okay Altun from Hacettepe University, Department of Mining Engineering.

References

Austin, L.G., Klimpel, R.R., Luckie, P.T., 1984. Process Engineering of Size Reduction: Ball Milling, AIMEPubl., NY.

Genç, Ö., 2008. An investigation on the effect of design and operational parameters on grinding performance of multi-compartment ball mills used in the cement industry. PhD Thesis, Hacettepe University, Mining Engineering Department, Turkey (in English).

http://www.flsmidth.com

Munn, N., Morrell, S., Morrison, R.D., Kojovic T., 2005., Mineral Comminution Circuits Their Operation and Optimization, JKMRC, Brisbane.

Norholm, A.,1995. Notes on energy conservation, FL Smidth and Co. a/s, Seminar in İstanbul, Turkey.

Svarovsky, L, 1984. Hydrocyclones. Holt, Rinehart & Winson Ltd., Eastbourne.

U.S. Geological Survey, 2015. Mineral Commodity Summaries Turkish Cement Manufacturers' Association Statistics.

Effect of ash reduction on the grindability of some Turkish Brown coals

D. Çuhadaroğlu,
Department of Mining Engineering, Zongıldak Bülent Ecevit University, 67100, Zongıldak, Tırkey

A. A. Sirkeci,
Istanbıl Technical University, Department of Mineral Proccessing, Istanbıl, Tırkey

M. Bilen, S. Kizgut, S. Yılmaz,
Department of Mining Engineering, Zongıldak Bülent Ecevit University, 67100, Zongıldak, Tırkey

C. E. Yılmaz
Oreks Madencilik Ltd. Şti. Çanakkale, Tırkey

ÖZET Kömür organik ve inorganik malzemelerin çökelmesinden oluşan çökelek olarak düşünülebilir. İnorganik malzemeler olarak kil mineralleri, quvars, karbonatlar, sülfitler ve sülfatlar bulunur. Bu çalışmada Türkiye'de bulunan 10 farklı sahadan numuneler alınmıştır. İlk olarak alınan kömür numuneleri yoğunlukla ayrılarak temizlenmiştir. İkincil olarak, orijinal ve temiz kömürlerin kül içerikleri belirlenmiştir. Son olarak, temiz ve tüvenan örnekler için HGI değerleri saptanmıştır. Bu sonuçlar incelendiğinde, tüvenan kömürlerde Gürmin-Merzifon örneğinin en yüksek HGI (68.5) değerine, Saray-Tekirdağ örneğinin en düşük HGI (43.5) değerine sahip olduğu görülmüştür. Temiz Kömürler içinse Ilgın-Konya örneğinin en yüksek HGI (84.4) değerine, Orhaneli-Bursa örneğinin en düşük HGI (40.8) değerine sahip olduğu görülmüştür. HGI ve kül içeriği analiz sonuçları birlikte düşünüldüğünde, kömürdeki kül yapıcılar yüzdürme yöntemiyle azaltıldıkça HGI değerlerinin yükselme eğiliminde olduğu anlaşılmıştır. Daha fazla mineral madde içeriği olan kömürler diğerlerine göre daha zor kırılır ve öğütülebilir.

ABSTRACT Coal can be regarded as an aggregation of organic and inorganic materials. Most of the minerals associated with the inorganic material are mainly clay minerals, quartz, carbonates, sulfides, and sulfates. In this study, brown coal samples were collected from 10 different brown coal pits in Turkey. Firslty, these coal samples were cleaned in terms of density seperation. Secondly, ash contents of original and clean coal samples were determined. Thirdly and finally, HGI (Hard Groove Index) values were determined for each sample of clean and run of mine coals. According to obtained HGI results for the run of mine coal samples; Gürmin-Merzifon brown coal has the highest HGI value as 68.5, while Saray-Tekirdağ brown coal has the lowest HGI value as 43.5, respectively. HGI results for the clean coal samples; Ilgın-Konya brown coal has the highest HGI value as 84.4, while Orhaneli-Bursa brown coal has the lowest HGI value as 40.8, respectively. Evaluating the all results of HGI and ash content analysis, as ash making minerals reduce down with density seperation, HGI values are tend to increase. Coals containing higher mineral contents are more difficult to fracture compared to the ones having less mineral content.
Keywords: Ash reduction, HGI, grindability, Brown coal

1 INTRODUCTION

Ash reduction is gaining importance in terms of lignite utilization. Turkey has huge amount of lignites reserves in variable locations. Lignites has higher values of ash content and this makes them hard to utilize, especially grind. In terms of size reduction, lignites are most of the time harder to grind due to their higher ash contents comparing to hard coal. Khoshjavan et al. (2013) claimed that volatile matter, hydrogen, Btu, nitrogen and fixed carbon have the greatest effect on HGI and moisture, oxygen and ash and total sulphur have the least effect on HGI. Although they claimed as ash or mineral matter content of coal has less effect on HGI, according to Sengupta (2002), it was claimed that high-ash coals are difficult to grind compared to low-ash coal, which means that mineral matter are difficult to grind. In terms of coal grindability many factors are effecting, i.e. not only coal ash content but also coal petrology. According to Trimble and Hower (2003), rank is an important factor in coal grindability from which it can be interpreted as ash content which is strongly related to rank has important role in coal grindability. In the study of Delibalta and Toraman (2012), significant amount of increase in grindability were achieved for the studied Turkish lignites after microwave radiation were exposed. They concluded that expansions after microwave treatment in gangue minerals (such as pyrite) are the probable cause for the improvement in coal grindability.

Sahoo et al. (2011) tried to improve the grinding characteristics of Indian coals by microwave pre-treatment and they found out that microwave treated coal ground much more rapidly thant the untreated coal. According to Xia et al. (2014), under the same grinding time, -200 mesh coals were getting finer when microwave pre-treatment had been increased. Bhattacharya and et al. (1998) have evaluated the sink-float test data to estimate the grindability for two different coals. They found out that Lalmatia non-coking coal showing the grindability to be monotonically decreasing with the increase in specific gravity, with the smallest HGI for the heaviest fraction. Bagdigi coking coals from the study of Bhattacharya et al. (1998) however showed the same trend but HGI was not the smallest in the heaviest fraction. Deniz et al. (2013) studied three Turkish lignites and evaluation of sink-float test data helped them to predict the Bond Grindability Index. Three coal sample studied by Deniz et al. (2013) showed higher specific gravity fractioned coals have higher Bond Grindability Index. This basically shows increase in ash content makes coals harder to grind. According to Bilen et al. (2016) however, variations in HGI could be partially attributed to the differing mineral matter contents of coal seams in the same pit.

In this study, 10 different lignites from various locations in Turkey were studied in terms of their grindability relationship between their ash content before and after sink-float pretreatment.

2 EXPERIMENTAL METHOD

In the experimental studies, samples from 10 locations (Saray, Orhaneli, Dursunbey, Milas-Ekizköy, Yatağan, Çayırhan, Ilgın, Ermenek, Sorgun and Gürmin-Merzifon) in Turkey as can be seen in Figure 1 were examined.

Sink and float analysis was carried out on these samples to have clean coal samples. In order to carry out the density seperation, samples were sieved into 0.5 mm and mostly ash making clay minerals under 0.5 mm particles were removed. Density seperation was realized with ZnCl2 solutions which have the density of 1.60 g/cm3.

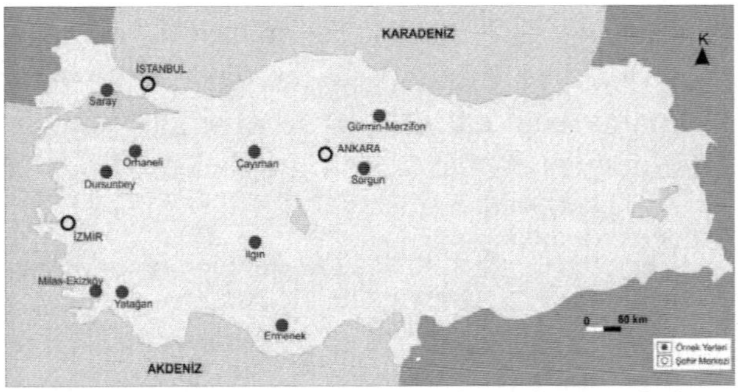

Figure 1. Map showing the locations of the studied coal samples.

Grindability experiments were carried out with HGI grindability testing method. HGI experiments were conducted with HGI testing environment regarding ASTM D 49. In order to carry out these experiments, screens such as 4.75 mm, 2.36 mm, 1.18 mm and 0.60 mm were used and step by step 1000 g of coal sample was prepared in the size group of -1.18 +0.60 mm. 50 g sample was taken out of this size reduced sample and it was ground in the mill (HGI testing equipment) for 20 rpm. Ground sample was sieved in 0.075mm for 20 minutes. At the end of this process, undersize fraction of 0.075 mm was weighted and it was employed in Equation 1, representing W.

$$HGI=13,74+(6,93xW) \quad [1]$$

W = The amount (g) of undersize fraction of 0.075 mm sieve.

Coals regarding to their HGI grindability measurements;

HGI	Grindability Property
<50	Hardly ground coals
50<HGI<70	Neither hardly nor easily ground coals
>70	Easily ground coals

can be classified.

3 RESULTS AND DISCUSSION

In the first step of this study, ash content variation for the run of mine coals and 1,60 gr/cm3 cleaned coals was obtained. Density separated fractions has lower content of ash as seen in Table 1. Sink and float test was used in order to remove the mineral impurities from run of mine coals.

Table 1. Ash content of run of mine coals and 1.60 gr/cm3 cleaned coals.

Samples	Ash content (%) of run of mine coals	Ash content (%) of 1,60 gr/cm^3 cleaned coals
Saray	14.53	11.18
Orhaneli	9.20	10.44
Dursunbey	24.86	16.49
Milas-Ekizköy	13.07	10.44
Yatağan	15.97	13.12
Çayırhan	26.64	17.06
Ilgın	38.77	26.38
Ermenek	26.37	17.85
Sorgun	5.68	4.19
Gürmin-Merzifon	45.02	23.33

Referring to Table 1, Gürmin-Merzifon coal sample was one of the most ash making minerals removed out with sink-float tests. Almost half of the ash making minerals were removed. For Orhaneli and Sorgun coal sample has already low ash content starting from sampling from the seams. It is hard to remove some impurities due to size reduction was not carried out before sink-float test. Depending upon the achievement of ash making minerals removal, utilization of these high ash lignites would be easier.

Table 2 however shows the variation of HGIs before and after sink and float analysis. Referring out of the literature studies, HGI of clean coal should be higher due to removal of ash making minerals. Expected trend was observed in Table 2 with some exceptions. From Table 1 and Table 2, it can easily be understood that ash content of coals more or less related to their HGI grindability.

Table 2. HGI grindability of run of mine coals and 1.60 gr/cm3 cleaned coals.

Samples	Run of mine coals		1.60 gr/cm^3 density cleaned coals	
	HGI	Grindability	HGI	Grindability
Saray	43,54	Hardly Gorund	58,79	Neither hardly nor easily ground
Orhaneli	47,00	Hardly Gorund	40,77	Hardly Gorund
Dursunbey	45,80	Hardly Gorund	44,93	Hardly Gorund
Milas-Ekizköy	47,00	Hardly Gorund	44,23	Hardly Gorund
Yatağan	59,48	Neither hardly nor easily ground	61,56	Neither hardly nor easily ground
Çayırhan	49,78	Hardly Gorund	58,2	Neither hardly nor easily ground
Ilgın	67,79	Neither hardly nor easily ground	84,43	Easily Ground
Ermenek	47,00	Hardly Gorund	44,93	Hardly Gorund
Sorgun	50,47	Neither hardly nor easily ground	56,01	Neither hardly nor easily ground
Gürmin-Merzifon	68,49	Neither hardly nor easily ground	55,32	Neither hardly nor easily ground

Coal samples like Dursunbey, Milas-Ekizköy, Yatağan, Çayırhan and Ermenek did not show noticable change in terms of HGI measurements regarding run of mine and clean coals. For the abovementioned coal samples however there was not a noticeable ash content changing via sink and float, especially Milas-Ekizköy and Yatağan. Saray, Ilgın and Gürmin-Merzifon samples has the widest range in terms of run of mine and cleaned coal HGIs. This may be because of several factors influencing HGI grindability of coal such as volatile matter, hydrogen, Btu, nitrogen and fixed carbon as mentioned by Khoshjavan et al. (2013). This is also mentioned by the authors Bilen et al. (2016) as, variations in HGI could be partially attributed to the differing mineral matter contents of coal seams in the same pit. However in the study of Bilen et al. (2016), original coal samples have higher HGIs than the floated coal samples, and they claimed that lower the mineral matter content in floated fractions lower the values of HGI are. This is basically true for low ash content samples and this was also considered in the study of Trimble and Hower (2003). The high ash content samples are more inconsistent in compositions and a set of low ash samples ensures more uniform sample set, which would end up less fluctuation. Considering only low ash content samples such as Saray, Orhaneli, Milas-Ekizköy, Yatağan and Sorgun; especially Orhaneli and Milas-Ekizköy showed same trend as Bilen et al. (2016). Saray sample although has low ash content, floated coal sample of Saray has higher HGI, which can not be explained with removing ash making materials result in low HGI. This increase in HGI for Saray sample can be explained by other factors such as pyrite as claimed by Delibalta and Toraman. Not only Saray sample but also Yatağan and Sorgun has higher HGI in floated fractions, which is also not really supported in literature. However Yatağan and Sorgun has little difference in terms of HGIs in original and floated fractions. This may be because of the relation between grindability with other factors such as moisture, coal petrography, coal rank and etc.

Mineral matter content in coal samples especially for lignites play an important role in their grindability. However not only mineral matter content can be denoted as promoting parameter in the variation of HGIs, but also for high ash content coals especially lignites moisture, coal rank and coal petrography might be reason of this inapprehensible variation.

4 CONCLUSION

This study focused on the HGI grindability variation of 10 lignite samples from variable locations in Turkey. Sink and float analysis carried out on these coal samples and their ash contents were decreased to some degrees. It was expected that removal of ash making materials would enable these lignites to be utilized better easily in size reduction processes. For this purpose, cleaned and original coal samples were analyzed in terms of their grindability.

Removal of mineral matter content from original coals would make lignite utilization easier. In order to understand this, cleaned and original coal samples were investigated in terms of their HGI grindabilities. Some coal samples showed expected result, which lower the mineral content easier the grindability is. However throughout this whole study, it can be concluded that variation in HGI can be partially attributed to the differing mineral matter contents of coal for original and cleaned coals. This is also supported with literature studies, since grindability of coal is not only a function of mineral matter content. In order to investigate the effect of removal of ash making minerals in terms of utilization of lignites, number of the samples should be increased and influence of other factors such as coal petrography, moisture content and coal rank should be limited and considered. With further studies the effect of mineral matter removal effect on HGI would be better understood.

REFERENCES

Bhattacharya S., Anand V., Banerjee P. (1998) Estimation of grindability from sink-float test data for two different coals, International Journal of Mineral Processing Volume 53, Issues 1–2, pp: 99–106

Bilen M., Kızgut S., Yilmaz S., Baris K. and Cuhadaroglu D. (2016) Grindability of Coal Changing with Burial Depth, International Journal of Coal Preparation and Utilization DOI: 10.1080/19392699.2016.1196199.

Delibalta M. S. and Toraman O.Y. (2012), The Effect of Microwave Energy on Grindability of a Turkish High-Ash Coal, Energy Science and Technology, Vol. 3, No. 2, pp: 46-49 DOI:10.3968/j.est.1923847920120302.293

Deniz V., Umucu Y., Çayırlı S. (2013) Prediction of the Bond Grindability Index from the Sink-float Test Data of Coals, Energy Sources, Part A: Recovery, Utilization, and Environmental Effects, Volume 35- Issue 15 pp:1385-1391

Khoshjavan S., Khoshjavan R., Rezai B. (2013) Evaluation of the effect of coal chemical properties on the Hardgrove Grindability Index (HGI) of coal using artificial neural networks, The Journal of The Southern African Institute of Mining and Metallurgy, Volume 113, pp: 505-510

Sahoo B.K., De S., Meikap B.C. (2011) Improvement of grinding characteristics of Indian coal by microwave pre-treatment Fuel Processing Technology Volume 92, Issue 10, pp: 1920–1928.

Sengupta (2002) An assessment of grindability index of coal, Fuel Processing Technology Volume 76, Issue 1, 20 pp: 1–10

Trimble A.S. and Hower J.C. (2003), Studies of the relationship between coal petrology and grinding properties, International Journal of Coal Geology Volume 54, Issues 3–4, pp: 253–260

Xia W.C., Yang J.G., Zhu B. (2014), The Improvement of Grindability and Floatability of Oxidized Coal by Microwave Pre-treatment Energy Sources, Part A, Volume 36 pp: 23–30 DOI: 10.1080/15567036.2011.653621

Prediction of Recovery of Boron in terms of Moisture and Grade of Run of Mine Ore

M. Bilen, S. Kizgut,
Department of Mining Engineering, Zonguldak Bülent Ecevit University, 67100, Zonguldak, Turkey

E. Çiçikci
Eti Maden İşletmeleri, Bandırma Balıkesir 10219 Turkey

ÖZET Bu çalışmada Eti Maden'e ait Bandırma Bor Tesisinden bir aylık süreyle (Ekim 2015) her gün örnekler alınmıştır. Alınan bu toplamda 31 örneğin nem tayini yapılarak tenörleri saptanmıştır. Tüvenan cevherin nemi ve tenörü tesisin geri kazanımında önemli rol oynamaktadır. Numunelerin alındığı periyot için konsantrenin de tenörü belirlenerek bir ilişki tayin edilmeye çalışılmıştır. Tesisin geri kazanımın hesaplanmasında günlük 500 ton tinkal beslendiği (aylık 15000 ton) ve günlük 200 ton (aylık 6000 ton) ürün alındığı varsayılmıştır. Beslemedeki tinkal neminin %2.77 ile %5.54 arasında değiştiği gözlemlenmiştir. Her ne kadar bu nem içeriği çok değişmese de bunun geri kazanımdaki artış veya azalış ile ilişkili olduğu söylenebilir. Girişteki tinkal tenörünün %22.67 ile %27.75 arasında olduğu bilinmektedir. Bu %5'lik fark da ürünün tenörün de etkili olabilir ve girişteki tenörün ürün tenörünü arttıracağı ve dolayısıyla da geri kazanımı arttıracağı düşünülmüştür. Bu çalışmada beslenen cevherin nemi ve tenörünün geri kazanımla olan ilişkisi incelenmiş ve geri kazanımı formulize eden yeni bir yöntem önerilmiştir. Bu yeni yönteme göre sadece beslemedeki nem ve tenör içeriğiyle geri kazanım tahmin edilebilecektir. Geri kazanım için tesisten alınan örnekler ve datalarla hesaplanan ve tahmin edilen arasında ilişki grafikleştirilmiş ve istatistiksel olarak yüksek korelasyon katsayısına sahip olduğu görülmüştür. Bu önerilen yeni yöntemle besleme cevherinin tenörünün ve neminin etkisi daha iyi anlaşılacaktır.

ABSTRACT In this study, Tincal samples from Bandırma Boron Plant were taken for the period of October 2015. Tincal samples from the feed stream were collected for a one-month period on daily basis. A total number of 31 samples were analyzed in terms of moisture content and B_2O_3 grade. Moisture content and grade of the run of mine ore have an effect on the recovery of the boron processing plant. Concentrate grade was also determined at the same period where feed samples were collected and a correlation was tried to be established. Recovery of the plant was calculated each day as assuming the daily feed amount of Tincal as 500 tonnes (monthly 15000 tonnes) and daily product of 200 tonnes (monthly 6000 tonnes). Feed Tincal moisture contents vary between 2.77% and 5.54%. Although this range of moisture content is not a big deal, it might be related with increase or decrease in recovery percentages. B_2O_3 grade of the feed ore varied between 22.67 % and 27.25%. This approximately 5% range could be meaningfull, an increase in feed grade would also cause an increase in product B_2O_3 grade as expected. In this study, feed moisture and feed grade was correlated with recovery and a new method to predict recovery is proposed. According to this new method, recovery of B_2O_3 would be easily calculated by just employing the moisture content and grade data of run of mine ore. Very high correlation coefficient (R^2=0.99) between measured and predicted recovery was achieved. This proposed method would enable one to better understand the effect of moisture content and tenor of run of mine ore on recovery.

Keywords: Ash reduction, HGI, grindability, Brown coal

1 INTRODUCTION

Known reserves of borate minerals are large particularly in Turkey, South America and the United States, and production from Turkey and United States will continue to dominate the world market (Kistler and Helvacı 1994). Colemanite, ulexite and tincal are the main boron ores reserved in Turkey and these reserves account up to 63 % of estimated total reserves of boron all over the world (Christogerou et al. 2009). Tincal is one of the economical boron minerals and the studied plant was utilizing the Tincal as raw material. In Table 1, some boron minerals and their composition, formulazition, B2O3 and H2O content was presented. Figure 1 represents the borate districts of western Turkey (Kistler and Helvacı 1994).

Table 1. Common Boron minerals and their composition, formulazition, B2O3 and H2O content (Saçak et al. 2015, Roskill 1995, Garrett 1998).

Boron Mineral	Composition	Formulaziton	B_2O_3 %	H_2O %
Tinkal	Na2O.2B2O3.10H2O	Na2B4O7.10H2O	36.5	47.2
Tinkalkonit	Na2O.2B2O3.5H2O	Na2B4O7.5H2O	47.8	30.3
Kernit	Na2O.2B2O3.4H2O	Na2B4O7.4H2O	50.9	26.4
Uleksit	Na2O.2CaO.5B2O3.16H2O	NaCaB5O9.8H2O	43.0	35.6
Kolemanit	2CaO.3B2O3.5H2O	Ca2B6O95H2O	50.9	21.9
Priseit	5CaO.6B2O3.7,5H2O	Ca5B12O23.7,5H2O	50.7	15.4
Borasit	5MgO.MgCl2.7B2O3	Mg3B7O13Cl	62.2	-
Hidroborasit	CaO.MgO.3B2O3.6H2O	CaMgB16O11.6H2O	50.5	26.1
Inyonit	2CaO.2B2O3.13H2O	2CaO.3B2O3.13H2O	37.6	42.1
Aşarit	2MgO.B2O3.H2O	12Mg2O.B2O3.H2O	41.4	10.7
Sassolit	B2O3.3H2O	H3BO3	56.4	43.6

In this study, Tincal samples were studied from Eti Maden (Eti Mine) located in Bandırma, Turkey. Tincal is mostly composed of tincal and in addition to this mineral clay, calcite, and dolomite are more or less is in the form of this ore. Tincal ore randomly consist of calcium, sodium-calcium and magnesium boron compounds.

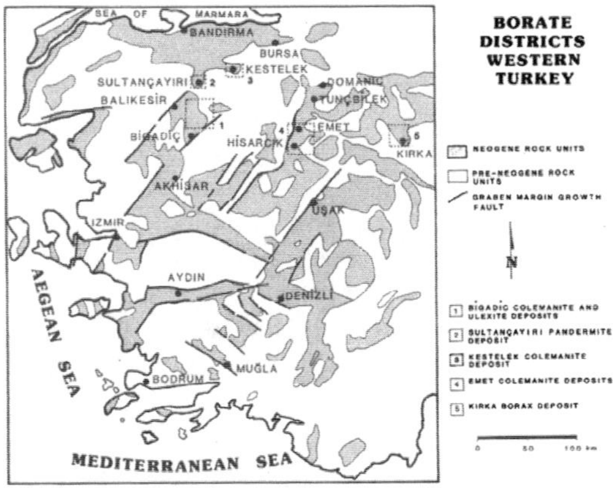

Figure 1. Borate district of western Turkey (Kistler and Helvacı 1994)

Tincal ore is dissolved in water or in main solvent at 93-96 °C. After the process of solution, mixture is drained with the equipments such as decanter, thickener and polish filter. Obtained pure solution is crystallized in vacuum crystallator at 63-66 °C in the form of borax pentahidrat. Pulp of the crystal-solution is separated in centrifuges into main solution and crystals and these crystals are dried and packaged. Main solution on the other hand is sent to solution unit (Saçak et al. 2015). Process flow chart is presented in Figure 2.

Moisture content of raw material is important for most of the processes. In this study recovery of boron changes monthly based within 5 % range and this must be because of either feed conditions or operational constraints. As assuming operational parameters are constant, which is already a good assumption without any maintanence then feed conditions of raw material is mostly the reason of this variance of boron recovery. What is meant by feed conditions is moisture content and grade of run of mine ore. In this study, depending on moisture content and grade of run of mine, new method of prediction the recovery was developed.

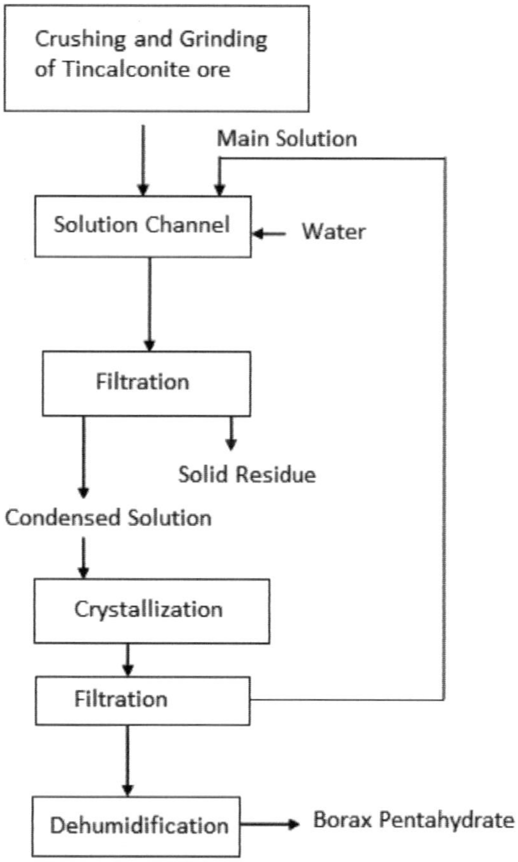

Figure 2. Process flow chart of borax pentahydrate production (Saçak et al. 2015).

2 EXPERIMENTAL METHOD

Moisture experiments were carried out on feed tilcalconite run of mine ore. This analysis was carried regarding ASTM (2011) D3173. This standard is basically for coal moisture content. However the idea of effect of moisture content on the process can be similarized to the fact that moisture in coal affects its utilization.

In order to carry out the moisture content analysis, samples were weighted firstly and then they were dried at 105 °C till their weight get constant. Finally they are weighted again and the ratio of the difference between initial weight and final weight is basically the moisture content of that specific sample.

In the context of this study, XRF and XRD analysis were not carried out, and the grade data was provided from monthly records of the plant. The change of the grade of the run of mine ore for 31 days was reported in October 2015.

3 RESULTS AND DISCUSSION

In this study, the moisture content and grade of the run of mine Tincal ore were determined and depending on these a prediction method is employed. In order to figure out the recovery of the plant, simply the product grade should be known. Moisture content change and grade measured in feed and product every day was presented in Table 2.

Table 2. Moisture Content and Grade of run of mine ore for 31 days.

Day	Moisture Content (%)	B_2O_3 Grade of Feed (%)	B_2O_3 Grade of Product (%)
1	4.75	24.8	48.79
2	5.54	23.91	48.48
3	3.68	26.14	48.78
4	5.48	22.67	48.66
5	3.83	23.93	48.73
6	5.14	24.86	48.79
7	4.88	25.14	48.66
8	4.22	26.67	48.79
9	5.18	24.15	48.89
10	3.41	26.88	48.69
11	2.77	24.27	48.74
12	4.72	22.94	48.67
13	4.29	25.97	48.74
14	4.18	26.09	48.68
15	4.75	25.9	48.78
16	4.07	26.56	48.77
17	4.22	24.75	48.79
18	3.91	26.06	48.86
19	4.14	24.73	49.08
20	3.65	25.65	49
21	4.19	25.69	49.01
22	4.66	24.52	48.84
23	5.01	23.88	48.9
24	4.02	25.44	48.98
25	3.94	25.11	49.03
26	4.64	26.8	48.97
27	4.38	26.37	49.21
28	4.69	26.17	48.92
29	3.6	27.25	49.08
30	4.32	26.66	49.07
31	4.87	25.17	48.99

Recovery of the plant can easily be calculated with some valid assumptions. It should be kept in mind that assumptions of daily feed amount of Tincal as 500 tonnes (monthly 15000 tonnes) and daily product of 200 tonnes (monthly 6000 tonnes) are valid assumptions and they should be evaluated in terms of recovery calculations. The purpose of this study was to develop another prediction method of recovery calculation with employing moisture content and grade of run of mine Tincal ore. In order to achieve this, some data are tabulated as in the following.

Referring to Table 2, moisture content of the run of mine ore changes between 2.77% and 5.54%. B_2O_3 grade of the feed ore varied between 22.67 % and 27.25%. Product B_2O_3 grade varies between 48.48 % and 49.21 %. This is not a wide range of product B_2O_3 grade. However calculated recovery changes between 72.04 % and 84.86 %. By employing moisture content and grade data of run of mine ore, a new method of recovery calculation is proposed as shown in Equation 1.

Recovery=154.09+(5.38*10-3)*MC-(3.03*T) [1]

(MC: moisture content, T: Grade or tenor of the run of mine Tincal ore)

Equation [1] is employed for 31 data presented in Table 2 and recovery data of the plant and predicted recovery percentages are given in Table 3.

Referring to Table 3, the predicted values of recovery is almost same as the real recovery percentages. Comparision of these data was realized on a graph (Figure 3) and R^2 was calculated to ensure about the confidence level of Equation 1.

R^2 was obtained as 0.99 for calculated and predicted recovery percentages. This ensures one to evaluate Eqn.1 in order to have quick estimate about recovery percentage of a stable plant. Not only Eqn.1 helps out with quick estimates but also importance of moisture content and grade of feed ore can be understood regarding to the coefficients of this variables.

Table 3. Recovery of the plant (%) and predicted recovery of the plant (%).

Day	Recovery of the plant (%)	Predicted Recovery of the plant (%)
1	78.69	78.97
2	81.11	81.67
3	74.64	74.91
4	85.86	85.43
5	81.45	81.60
6	78.50	78.79
7	77.42	77.94
8	73.18	73.30
9	80.98	80.94
10	72.46	72.66
11	80.33	80.57
12	84.86	84.61
13	75.07	75.42
14	74.63	75.06
15	75.34	75.64
16	73.45	73.64
17	78.85	79.12
18	75.00	75.15
19	79.39	79.18
20	76.41	76.39
21	76.31	76.27
22	79.67	79.82
23	81.91	81.76
24	77.01	77.03
25	78.10	78.03
26	73.09	72.91
27	74.65	74.21
28	74.77	74.82
29	72.04	71.54
30	73.62	73.33
31	77.85	77.85

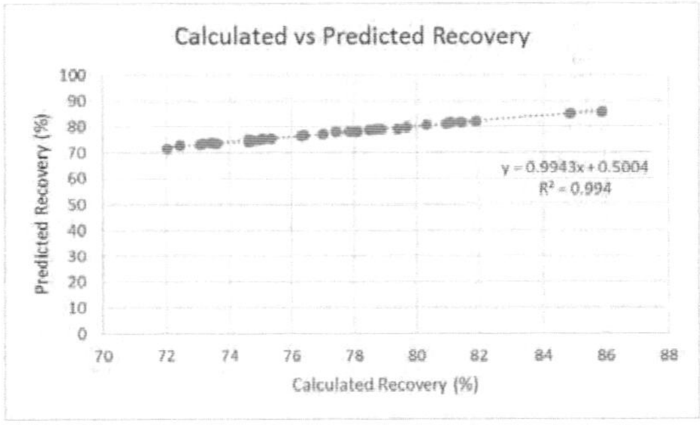

Figure 3. Calculated vs predicted recoveries (%) of the plant.

Referring back to the Eqn.1, it can be interpreted as the moisture content does not play as important role as feed ore grade in the recovery percentage.

4 CONCLUSION

In this study, a new prediction method of recovery percentage in a boron plant was studied by evaluating feed ore conditions i.e. the moisture content and feed ore grade. Boron production is one of the most in Turkey comparing other countries and Turkey has up to 63 % of total boron reserves. Process of borax production is not so complicated though some factors were not considered in the previous studies in terms of the recovery. Recovery is as definition, ratio desired component content in the feed over produced amount of this desired component in the products. However in every energy utilizing plant, moisture content of the feed plays important role in recovery. And higher the feed content of the desired material, higher the recovery should be achieved.

A new method and equation was executed to predict the recovery with the inlet conditions of feed ore. This formulization basically helps plant operators for quick estimates of what recovery would be at the end. Not only this new proposed prediction method would be helpful in terms of quick estimates but also one can better understand the effect of feed ore moisture content and grade on the recovery.

Future studies should focus more on improvements on the recovery of desired minerals. In order to have better process in boron plants, overall system parameters should be checked with more validated assumptions..

REFERENCES

ASTM (2011) D3173 Standard Test Method for Moisture in the Analysis Sample of Coal and Coke, ASTM International, West Conshohocken, PA, https://doi.org/10.1520/D3173-11.

Christogerou A., Kavas T., Pontikes Y., Koyas S., Tabak, Y., Angelopoulos G.N. (2009) Use of boron wastes in the production of heavy clay ceramics Ceramics International Volume 35, Issue 1, Pages 447–452

Garrett, D.E. (1998.) Borates, Handbook of Boron Deposits, Processing, Properties and Use, Academic Press

Kistler, R.B. and Helvaci, C. (1994) Boron and borates. In: D.D. Carr (Editor), Industrial Minerals and Rocks, Society of Mining, Metallurgy and Exploration, Inc., 6th ed., pp. 171-186.

Roskill Reports on Metals and Minerals (1995), The Economics of Boron, Roskill Information Services Ltd, London

Saçak D., Şara O. N., Kocakerim M. M., Yartaşı A., Dilmaç Ö. F. (2015) Sülfürik Asitle Kolemanitten Borik Asit Üretimi Ve Tinkalden Boraks Pentahidrat Üretim Proseslerinde Kütle Ve Enerji Denkliği Çalişmalarinin Yapilmasi, Project Report 18.11.2013-18.05.2015

Investigation of Effective Parameter in Copper Solvent Extraction Using Chemorex CP-150

F. Kiani Broujeni
M.Sc. in Mining Engineering, University of Tehran, Kian Mine Pars Co.

M. H. Golpayegani
M.Sc. in Mineral Processing, University of Kashan, Kian Mine Pars Co.

H. Rafeie
M.Sc. in Mineral Processing, Islamic Azad University, Science and Research Branch, Kian Mine Pars Co.

A. Mirzaeian
B.Sc. in Mining Engineering, University of Tehran, Kian Mine Pars Co.

ABSTRACT In this paper, copper solvent extraction (SX) of pregnant leach solution (PLS) obtained from heap leaching stage of ChahMousa's oxidized ore (in Semnan Province, Iran) using chemorex CP-150 as extractant and kerosene as diluent have been studied. Optimized value of extractant to diluent ratio and pH of aqueous phase in this study have been investigated. Also, isotherm curve of extraction and Mc. Cabe-Thiele diagram for optimized experimental pH of 3.17 and common industrial pH of 2 have been drawn, separately. Moreover, value of sulfuric acid 180 g/L usage and interfering ions concentration including Fe^{3+} and Mn^{2+} in strip solution have been considered in stripping stage. Based on results, optimized values of extractant to diluent ratio and pH are 5% and 3.17, respectively. According to Mc. Cabe-Thiele's diagram results, at aqueous to organic ratio of 1, theoretical stages of SX at pH rates of 3.17 and 2 are one and two stages, respectively. Also sulfuric acid to PLS ratio is 20% in stripping stage. Furthermore, it has been observed that interfering ions concentrations are in allowed range.

Key words: solvent extraction; copper; pregnant leach solution; Chemorex Cp-150

1 INTRODUCTION

Hydrometallurgical processes to obtain copper metal are becoming increasingly important due to their relative simplicity and environmental compatibility, as compared to traditional thermal processes. Heap leaching has the advantages of low investment and operating costs. Nowadays, heap leaching is a vital technique within the mining industry for treating low-grade oxidized ores [Arias, 2003]. Heap leaching operating costs are generally lower per ton of ore processed than the operating costs associated with an agitated tank leach circuit, mainly because of reduced equipment and manpower requirements. In addition, heap leaching is typically conducted at a coarser size fraction than agitated tank leaching, thus reducing the power requirement and reagent consumption [Kime, et al., 2016, Banza, et al., 2002, Sole, et al., 2005, Reddy, and Priya, 2005].

Also, it is difficult to reach pure copper without impurities from PLS using classical methods such as cementation by scrap iron. However, copper in pure form can be obtained by using solvent extraction method. Considering the industrial and economic importance of copper, there is a great need in separation and recovery of copper using cost effective commercial extractants. Many hydroxyoximes compounds such as LIX 87QN, LIX 54, LIX 973, LIX 65, LIX 64, LIX 70, LIX 36 and Chemorex CP-150 reagents used for the extraction, separation and recovery of copper from PLS [Reddy, and Priya, 2005]. Among hydroxyoximes compounds reagents, Chemorex CP-150 used commercially for copper ions extraction of aqueous solutions [Soeezi, et al., 2015].

The ChahMousa's Oxidized ore is a copper mine in Semnan Province in Iran. It contains various amounts of copper that can be as low as 0.8 %. This low-grade ore is treated by heap leaching process following by solvent extraction. The counter current operation of ChahMousa's ore, includes two-stage solvent extraction by chemorex CP-150 and kerosene as extractant and diluent respectively. Also, the purification process

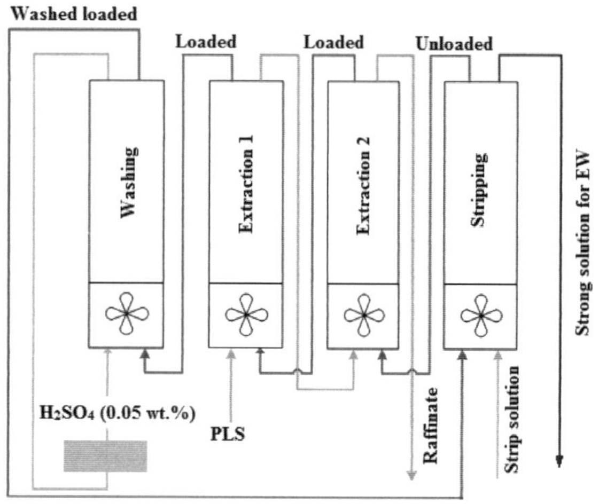

contains one stripping stage and one washing stage as shown in Figure 1. According to the observations, it seems that the current production capacity is by far less than designed capacity of the plant. It is in fact due to changes in copper grade values during mining period which can be seen in PLS concentration and system instability.

Figure 1. ChahMousa's oxidized ore SX operation flowsheet.

In this investigation, effective parameters such as extractant to diluent ratio in SX and stripping stage and pH of aqueous phase have been investigated in order to optimize the process. Moreover, Isotherm curve and Mc. Cabe-Thiele diagram for optimized experimental pH of 3.17 and common industrial pH of 2 have been drawn, separately in order to determine the number of theoretical stages of extraction.

1.1 MATERIALS AND METHODS

1.2 MATERIALS AND INSTRUMENTS

In this investigation, pregnant leach solution (PLS) obtained from heap leaching stage of ChahMousa's oxidized ore (in Semnan Province, Iran) has been used as aqueous phase. Also, organic phase has been made of Chemorex CP-150 as extractant and Kerosene as diluent. Specification of mentioned PLS has been shown in Table 1. In order to mix two phases in extraction and stripping stage, a mechanical stirrer has been used. A digital pH meter has been used to adjust the pH. A decanter funnel has been used to separate aqueous and organic phases. Sulfuric acid 180 g/L has been used in stripping stage. Concentration of Cu, Mn and Fe has been measured through atomic absorption spectroscopy (AAS) using uniqam939 model.

Table 1: Specification of ChahMousa's Oxidized ore's pregnant leach solution.

Concentration of Cu (g/l)	Concentration of Fe (g/l)	Concentration of Mn (g/l)	pH
1.68	0.589	2.546	3.17

1.3 SOLVENT EXTRACTION PROCEDURE

Solvent extraction of copper was performed using Chemorex CP-150 (Chemorex, China) as the extractant. The pH was regulated using H_2SO_4 (20 wt. %) and NaOH (10 wt. %). The mixtures were shaken until equilibrium was reached (5 min) using mechanical agitator. The loaded organic phase was mixed with stripping solution (water acidified to 180 g/L H_2SO_4) and shaken until equilibrium was reached (5 min). The phases were allowed to separate, followed by filtration on filter papers and the aqueous phases were saved for analysis. The aqueous phases were analyzed for Cu, Fe and Mn contents by atomic absorption spectrophotometry (AAS). Concentrations of copper and impurity in organic phases were calculated from the difference between their initial concentrations in PLS and their concentrations in the raffinates at fixed organic-to-aqueous (O/A) phase ratios. The experiments were carried out at room temperature (+ 25°C).

1.4 RESULT AND DISCUSSION

1.5 EFFECT OF PH

One the vital parameters in solvent extraction processes is pH of the pregnant leach solution., Therefore, an optimized pH value, i.e., maximum copper extraction and minimum Fe^{3+} and Mn^{2+} extraction, has to be investigated. pH rates of 1.5, 1.75, 2, 2.25, 2.5 and 3.17 have been investigated in order to find the effect of pH rate on solvent extraction process. Results and conditions of the mentioned experiments have been shown in Table 2. Also copper, iron and manganese extraction rate at different pH rates has been shown in Figure 2. At pH rate of 3.17, diagram mutation toward slow increase in copper extraction shows the key role of pH increasing. With pH rate increasing, iron extraction experienced a mild decrease and manganese extraction was not affected. According to the results of experiments, optimized pH is 3.17. According to the following equation, concentration of H^{+} ions decreased with increasing pH rate and as a result, copper extraction increased.

Table 2. Effect of pH on solvent extraction of Cu^{2+}, Mn^{2+} and Fe^{3+} from PLS.

Experiment number	PLS		Raffinate			Cu Recovery (%)	Fe Recovery (%)	Mn Recovery (%)
	Weight (gr)	pH	Fe (g/l)	Cu (g/l)	Mn (g/l)			
1	560	1.5	0.441	0.711	2.464	57.68	25.13	3.22
2	560	1.75	0.445	0.637	2.458	62.08	24.45	3.47
3	560	2	0.446	0.621	2.458	63.04	24.28	3.47
4	560	2.25	0.448	0.605	2.457	63.99	23.94	3.51
5	560	2.5	0.511	0.485	2.452	71.13	21.39	3.68
6	560	3.17	0.515	0.291	2.445	82.38	12.56	3.97

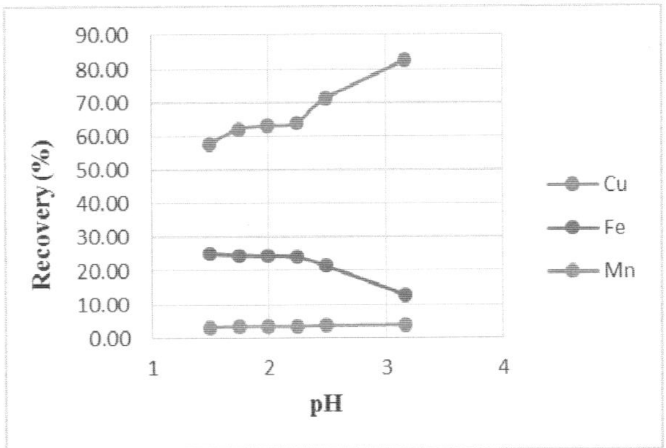

Figure 2. Effect of pH on extraction rate.

1.6 Extractant to diluent ratio

In order to optimize extractant to diluent ratio, organic solutions with Chemorex CP-150 to Kerosene ratio of 2.5:97.5, 4:96, 5:95, 7.5:92.5, 10:90 and 15:85 has been made. pH and aqueous to organic ratio in these experiments were 3.17 and 1:1, respectively. Copper ion extraction rate from aqueous phase at various Chemorex CP-150 concentrations has been shown in Figure 3. According to results, maximum loading capacity factor has been reached in 5:95 extractant to diluent ratio for 1.68 g/L of copper in PLS.

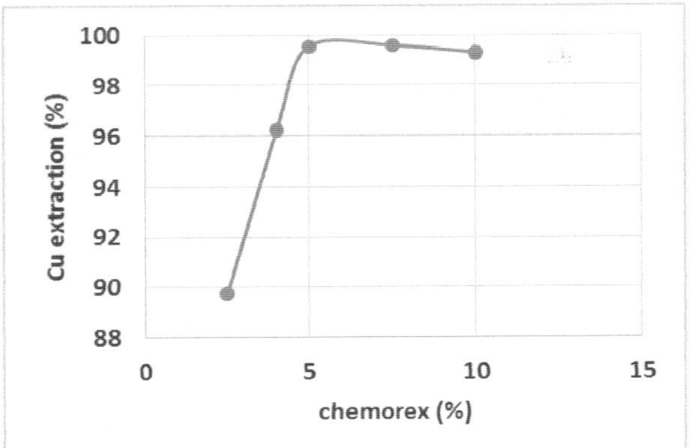

Figure 3. Effect of Chemorex CP-150's concentrations on Cu extraction.

1.7 Determination of theoretical steps of extraction

In a solvent extraction process, theoretical steps of extraction can be estimated using Distribution Isotherm and Mc. Cabe Tail diagram.

Table 3. Experiments designed to draw Mc. Cabe Tail diagram.

Experiment number	A/O	Aqueous phase volume (ml)	Organic phase volume (ml)	Stripping solution volume (ml)	Stripping Solution to PLS ratio
1	0.2	250	50	10	4
2	0.5	250	125	25	10
3	0.8	250	200	40	16
4	1	250	250	50	20
5	1.5	250	375	75	30
6	2	250	500	100	40

Table 4. Extraction rate of Cu^{2+}, Fe^{3+} and Mn^{2+} for various aqueous to organic ratios.

Experiment number	A/O	Raffinate			Cu Recovery (%)	Fe Recovery (%)	Mn Recovery (%)
		Cu (g/l)	Fe (g/l)	Mn (g/l)			
1	0.2	1.09	0.672	2.541	35.11	0.66	0.20
2	0.5	0.3	0.675	2.514	82.14	0.85	1.26
3	0.8	0.05	0.659	2.474	97.02	1.4	2.83
4	1	0.04	0.693	2.445	97.73	2.3	3.97
5	1.5	0.015	0.698	2.400	99.1	5.7	5.73
6	2	0.01	0.697	2.380	99.58	5.6	6.52

In order to obtain Distribution Isotherm, various organic to aqueous ratio are mixed in a constant temperature. After reaching equilibrium, two phases (aqueous and organic) are separated and metal concentration are measured in them. Copper concentration in aqueous and organic phases are plotted [Barid and Hanson, 1983].

Distribution Isotherm and an operating line are used in order to draw and analysis of Mc. Cabe Tail diagram. As the mentioned operating line is a straight line, it can be drawn by any 2 points of the line. Also, it can be drawn by a point and the line slope. Slope of this line is aqueous to organic ratio and the one point which are needed to draw the line (X_0) is the concentration of copper at the starting time of process. With having these all, Y_1 will be reached by the following equation.

$$Y_1 = A/O(x_0 - x_n) + y_{n+1} \quad (1)$$

Furthermore, in order to draw Mc. Cabe Tail diagram, experiments have been done as shown in Table 3 by considering extractant to diluent ratio and optimized experimental pH of 5:95 and 2, respectively.

Extraction rate of Cu, Fe and Mn for various aqueous to organic ratios has been shown in Table 4. According to the results, copper concentration in raffinate decreased by the increase in organic to aqueous ratio and this is mainly because of organic extractant value increase. Distribution isotherm according to results has been shown in Figure 4. Also, Mc. Cabe Tail diagram based on distribution isotherm has been drawn as shown in Figure 5. According to the Equation 1, by considering that organic to aqueous ratio is 1 and the hypothesis that Copper ions transfer to organic phase completely, X_n will be 0. Moreover, copper concentration in organic phase (y_{n+1}) has been assumed 0. As a result, according to equation 1, y_{n+1} will be reached and operating line can be drawn with having the mentioned one point and line slope.

As seen in Figure 5, two operating lines with aqueous to organic ratio of 1 and 1.5 has been drawn in Mc. Cabe Tail diagram. According to the diagram, one extraction stage is adequate for A:O=1 and below, however, there are two extraction stages at ChahMousa's plant with A:O=1 that by considering mentioned result, which is unnecessary. It seems necessary for the plant to work at accurate A:O ratio in order to prevent extractant dissipation and also using both solvent extraction's mixer-settlers in the plant. Also, the process needs two extraction stages at aqueous to organic ratio of 1.5 according to the diagram. Moreover, extractant usage diminished and solvent extraction process is optimized at this A:O ratio.

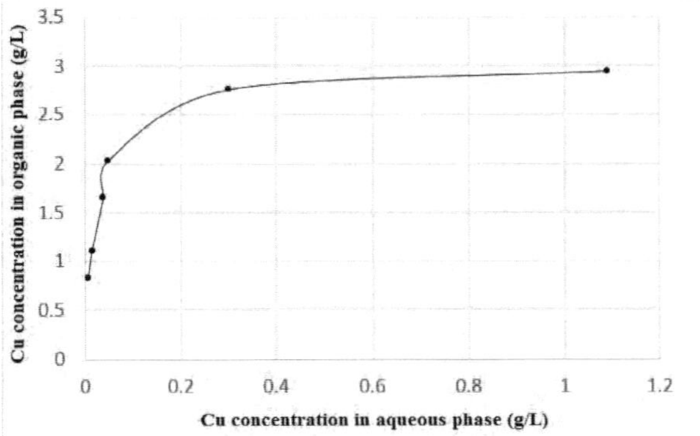

Figure 4. Distribution Isotherm for Cu extraction.

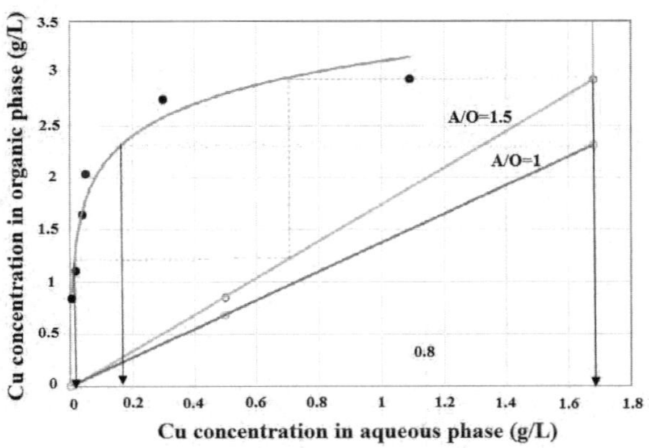

Figure 5. Mc. Cabe Tail diagram at pH=3.17.

In addition to optimized experimental pH (3.17), Mc. Cabe Tail diagram also has been drawn for the industrial pH of 2 (Figure 6). In this pH, there is less possibility for impurities migration from heap leaching stage to the solvent extraction stage. Also, interfacial crud production which cause in more separation time and less plant production, capacity decreased. Presence of interfering ions such as Fe^{3+}, Al^{3+} etc. is one of the reason of crud increasing which transformed to hydroxide polymer forms as a result of interfering ions hydration and hydrolysis by increasing of pH. These hydroxide polymer forms caused to formation and increasing of crud in combination with organic phase [LIU, Jian-she and QIU, 2002]. Polymerization phenomenon of $Fe(OH)_2$ has been shown in Figure 7a. On the other hand, fine silica particles in aqueous phase have Silanol group at their surfaces mainly because of hydration and hydrolysis which can form hydrogen bonds between their particles and also with organic phase and make stable polymerization (Figure 7b) [Zhuravlev, 2000]. Mc. Cabe Tail diagram for pH of 2 has been shown in Figure 6. As seen in this figure, two extraction stage is needed for this pH rate and at the A:O ratio of 1.

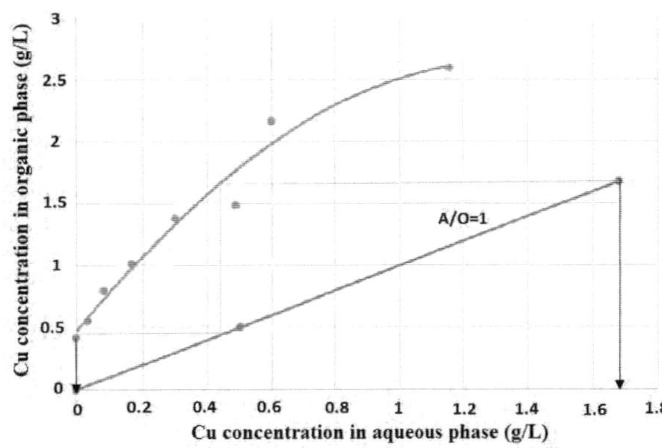

Figure 6. Mc. Cabe Tail diagram at pH=2.

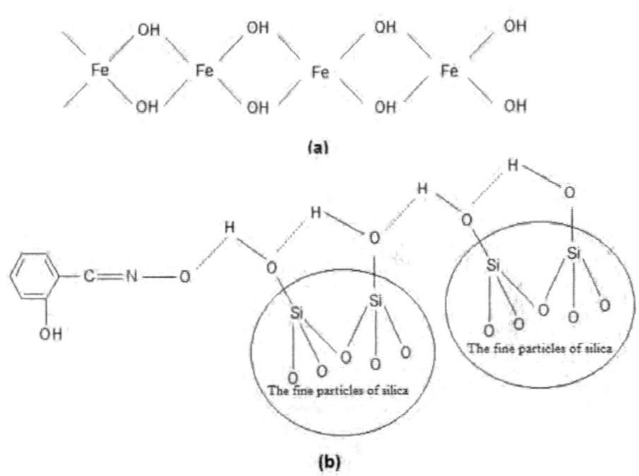

Figure 7. Polymerization of particles in Crud: (a) Polymerization of iron hydroxyl (b) Hydrogen bonds between silica particles and with organic.

1.8 Determination of organic to aqueous ratio at stripping stage

According to Table 8, stripping solution to PLS ratios which have been tested are 0.04, 0.1, 0.15, 0.2 and 0.4. Also, Cu, Fe and Mn concentrations in stripping solution at pH rates of 2 and 3.17 has been shown in Tables 5 and 6 respectively. According to Tables 5 and 6, it is clear that both stripping stages and whole process recovery at stripping solution to PLS ratios of 0.2 and 0.4 (which ChahMousa's plant runs in this ratio currently) are nearly the same. As a result, it is observable that stripping solution to PLS ratio of 0.2 is adequate for the stripping stage.

Table 5. Results of stripping stage at pH=3.17.

Experiment number	A/O in SX stage	Stripping solution to PLS ratio	Cu (g/l)	Mn (g/l)	Fe (g/l)	Cu recovery in stripping stage (%)	Total recovery (%)
1	0.2	1.5	10.58	0.238	0.114	73	25.83
2	0.5	1.75	9.98	0.193	0.126	72.29	59.4
3	0.8	2	8.78	0.062	0.088	86.18	83.61
4	1	2.25	7.22	0.103	0.073	87.92	85.95
5	1.5	2.5	4.94	0.051	0.090	88.99	88.21
6	2	3.17	3.81	0.045	0.084	91.09	90.71

Table 6. Results of stripping stage at pH=2.

Experiment number	A/O in SX stage	Stripping solution to PLS ratio	Cu (g/l)	Mn (g/l)	Fe (g/l)	Cu recovery in stripping stage (%)	Total recovery (%)
1	0.2	1.5	10.58	0.238	0.114	73	25.83
2	0.5	1.75	9.98	0.193	0.126	72.29	59.4
3	0.8	2	8.78	0.062	0.088	86.18	83.61
4	1	2.25	7.22	0.103	0.073	87.92	85.95
5	1.5	2.5	4.94	0.051	0.090	88.99	88.21
6	2	3.17	3.81	0.045	0.084	91.09	90.71

As the stripping solution is the electro winning stage's feed, therefore, concentrations of impurities such as Fe and Mn have to be at standard range. Standard range of concentrations of impurities has been shown in Table 7. By comparing amount of impurities from Tables 5 and 6 with standard range, it seems that their concentrations are in allowed range at stripping stage.

Table 7. Allowed range of Mn^{2+} and Fe^{3+} in the copper welectrowinning process [Barid and Hanson; 1983].

Mn	Fe
100-500 (ppm)	1-3 (g/l)

2 CONCLUSION

Copper solvent extraction from pregnant leach solution (containing 1.68 g/L Cu) of ChahMousa's oxidized ore using Chemorex CP-150 as extractant and kerosene as diluent has been investigated. Optimized extractant to diluent ratio and optimized pH value of aqueous phase have been studied which were 5:95 and 3.17, respectively. Distribution Isotherm and Mc. Cabe Tail diagram have been drawn for the optimized experimental pH of 3.17 and normal industrial pH of 2. According to the Mc. Cabe Tail diagram, at aqueous to organic ratio of 1 (A:O=1) theoretical extraction stages at optimized experimental pH and normal industrial pH are 1 and 2 respectively. According to the results, it is clear that Copper extraction increased by the increase in pH value. Also, it has been proven that at pH 3.17 and with PLS copper concentration of 1.68g/L, aqueous to organic ratio has to be 1.5 for two extraction stages according to the Mc. Cabe Tail diagram. Moreover, stripping solution to low grade PLS at stripping stage has been studied and has been calculated as 20% toward PLS. Furthermore, Mn and Fe ion concentrations have been measured and investigated at strip solution and it has been proved that impurity concentrations are in allowed range.

REFERENCES

Arias, J.A., 2003. Heap leaching copper ore using sodium nitrate. Google Patents.

Kime, M-B., Kanowa, E.K., Bafubiandi, M., and Diyambi, S.N., 2016. Hydrometallurgy Conference 2016: Sustainable Hydrometallurgical Extraction of Metals, 309-320.

Banza, A.N., Gock, E. and Kongolo, K. 2002. Base metals recovery from copper smelter slag by oxidising leaching and solvent extraction. *Hydrometallurgy*, 67, 63–69.

Sole, K.C., Feather, A.M. and Cole, P.M. 2005. Solvent extraction in southern Africa: An update of some recent hydrometallurgical developments. *Hydrometallurgy*, 78, 52–78.

Reddy, B.R., Priya, D.N., 2005. Process development for the separation of copper (II), nickel (II) and zinc (II) from sulphate solutions by solvent extraction using LIX 84 I. *Separation and Purification Technology*, 45,163-167.

Soeezi, A., Rahimi, E and Mohaghegh, N., 2015. Investigating the Effect of Concentration and stirring time on Copper Extraction Process from Pregnant Liquid Solution (PLS), Case study: Sungun Copper Mine. Int'l Journal of Research in Chemical, Metallurgical and Civil Engg, 2, 33-35.

Barid, Lo.T., Hanson M.E; 1983. Handbook of solvent extraction. GSA.

LIU, Jian-she., QIU, C., 2002. Mechanism of crud formation in copper solvent extraction". Journal of Central South University of Technology 9, 169-172.

Zhuravlev, L.T., 2000. "The surface chemistry of amorphous silica. Zhuravlev model" journal of Colloids and Surfaces 173, 1-38.

Solvent Extraction of Tin by D_2EHPA

N. Shirvani, E. K. Alamdari

Department of Mining and Metallurgical Engineering, Amirkabir University of Technology, Tehran, Iran

ABSTRACT In this study, by solvent extraction method and using Di-(2-ethylhexyl) phosphoric acid as the organic extractant, tin was extracted from the acidic solution of HBF_4. The extractant D_2EHPA is a kind of organic acid phosphate. It extracts metals by replacing metal ions in a solution with hydrogen (i.e. cation exchange). Fluoroboric acid is the aqueous solvent used in this experiment. In each stage the extraction (rate) from the aqueous phase at different concentration of acid, various pH and three different temperatures 0°C, 15 °C and 20 °C was studied. The final concentration of the acidic solution after extraction was obtained by titration with a solution of 0.01M EDTA. In general the extraction of tin by D_2EHPA in a solution containing HBF_4 is high. Therefore it may be possible to use this method for selective extractions in solutions containing tin to separate tin from other metals. It depends on the kind of the other metals present in the solution and the ability of D_2EHPA for extraction of them in the (desired) pH range. The tin extraction in these experiments was between 84.02% to 94.29% at 0°C, 90.87% to 96.57% at 15°C and 66.90% to 90.86% at 20°C.

Keywords: Solvent Extraction, Tin, D2EHPA, Fluoroboric Acid, EDTA, Titration.

1 INTRODUCTION

Solvent extraction is a common type of industrial hydrometallurgical operation to recover and refine tin. Extraction by an organic liquid or solvent extraction is a process in which an aqueous solution is contacted by an insoluble liquid, therefore some components of the aqueous solution are selectively separated into the organic phase (Gaskell, n.d.) (Keshavarz Alamdari, et al., 2002).

In this type of extraction, extractant molecules in reaction with metal ions in aqueous phase, form an organic-metal complex and it remains stable in the organic phase if there is equilibrium. Equilibrium conditions depend on some factors like temperature, equilibrium acidity or pH, the concentration of extractant and concentration of metal ions. Hence solvent extraction is an equilibrium process containing two insoluble phase, aqueous and organic phases in equilibrium (Keshavarz Alamdari & Sadrnezhad, n.d.).

Different extractants are used in solvent extraction like acidic extractants, amines and chelators. Cyanex 301, D_2EHPA and PC_88A are acidic extractants; amines contain Alamine 302_1, tetra alkyl amine, Di-iso-two-decylamine, Alamine 336, Alamine 304_1 and three ethylamine. Lix 63 and Lix84_1 are chelators.

In this study, the effect of pH for tin extraction from acidic solutions by D_2EHPA was studied at 0°C, 15°C and 20°C to determine the optimum conditions for extraction. There is no data about stable chemical potential range for Tin(II) in acidic solution of HBF_4 medium. Hence pH range for the experiment is estimated by HSC. Tin(II) is not stable in alkaline environment and will be converted to Tin (IV) or oxidized or deposit. Generally after a while there will be some deposit in the solution. So in this study the experiments were done in acidic medium, for short times and immediately after preparing the aqueous solution.

2 EXPERIMENT

In this experiment Tin (II) oxide, organic solvent D_2EHPA, Kerosene, fluoroboric acid, ammonia, EDTA, sodium acetate, Eriochrome black T and distilled water were used. For all experiments at 0°C, 15°C and 20°C aqueous and organic phases have been ready at 20°C. In this case, organic phase contains 10% D_2EHPA; 25 mL of D_2EHPA was poured in a 250 mL volumetric flask by a 50 mL pipette and it was filled up to 250 mL by the diluent kerosene. Aqueous solution of $Sn(BF_4)$ that contained 5.2 g/L Sn was prepared from tin(II) oxide, fluoroboric acid, boric acid and distilled water. Then 20 ml of each solution was poured into an Erlenmeyer flask of 100 mL volume. The first series of tests were carried out under these conditions. But, for next tests HBF_4 was gradually added to the test solutions containing 20 mL acidic solution and 20 mL organic solution to examine the influence of pH on the extraction of tin.

At 0°C and 15°C prepared samples were put in an incubator (type of IN17-D and model of A.M 27) for 30 minutes to reach the intended temperature. The incubator shaker was operated for agitation of aqueous and organic phases at a speed of 160 rpm for 30 minutes. The experiment at 20°C was carried out similarly at the ambient temperature, but, the incubator was not used. After mixing, the resultant solution was transferred to a separation funnel for standing 5 minutes to achieve phase separation. pH was measured for aqueous solution and the aqueous phase at the tested pH was used to determine its residual tin concentration by titration.

Aqueous solution was titrated by EDTA with the concentration of 0.01M. The amount of EDTA needed to prepare the solution with the concentration of 0.1M was put in an oven at 80°C for 24 hours to get out its moisture. Then it was achieved the concentration of 0.1M by distilled water. After that the solution was diluted up to ten times to reach the solution with the concentration of 0.01M. To determine the residual concentration of tin in the aqueous phase by titration after extraction, 2 ml of each sample was poured in a beaker and it was diluted to 12 ml by distilled water.

In order to prevent precipitation of tin, sodium acetate was added to the solution to form a soluble complex. Eriochrome black T was used as an indicator to determine the color change of the solution to purple. Then using some ammonia (approximately 0.1 ml) changed pH range of solution to alkaline which is an appropriate to keep the complex stable. Also the purple color was changed to blue. EDTA was poured in a burette to achieve the used volume by that. Therefore by knowing the concentration and volume of EDTA and volume of aqueous phase, the concentration of the aqueous solution was achieved.

3 RESULTS AND DISCUSSIONS

The results of measurement of pH and titration of the aqueous solution with the concentration of 5.2g/L after mixing with the organic phase at 0°C, 15°C and 20°C for 30 minutes and then separated of it, is presented respectively in tables 1, 2 and 3. As it stands, changes in extraction does not have a certain and exact relation with pH. It supposed to be because of tin (II) oxidation.

Table 1 effect of pH on extraction of tin at 0°C

pH	Extraction percent
1.586	88.56
1.591	90.87
1.616	89.73
1.686	94.29
1.761	93.15
1.777	85.16
1.778	84.02

Table 2 effect of pH on extraction of tin at 15°C

pH	Extraction percent
1.259	95.43
1.72	92.01
1.744	94.29
1.758	96.56
1.759	96.57
1.802	92.01
1.806	90.87

Table 3 effect of pH on extraction of tin at 20°C

pH	Extraction percent
1.259	95.43
1.72	92.01
1.744	94.29
1.758	96.56
1.759	96.57
1.802	92.01
1.806	90.87

The diagram in Figure 1 shows the effect of pH on extraction percent of tin at 0°C. Also Figures 2 and 3 show the data obtained at 15°C and 20°C respectively.

4 CONCLUSION

Generally tin extraction by acidic extractor, D_2EHPA is high in acidic environment of HBF_4, so it make it possible to use this method to separate tin from other metals presenting in a solution, which named selective extraction. It depends on D_2EHPA ability to extract other metals dissolved in the solution and pH range for extracting them. The extraction of tin was between 84.02% and 94.29% at 0°C, between 90.87 and 96.57 at 15°C and between 66.90 and 90.86 at 20°C.

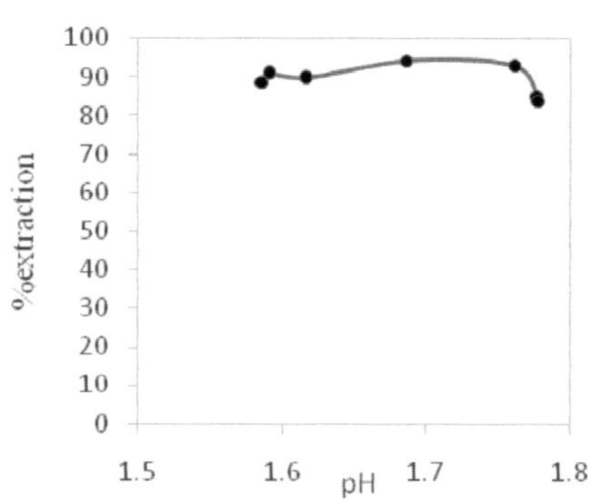

Fig 1 effect of pH on extraction of tin solution with the solution of 5.2g/L at 0°C

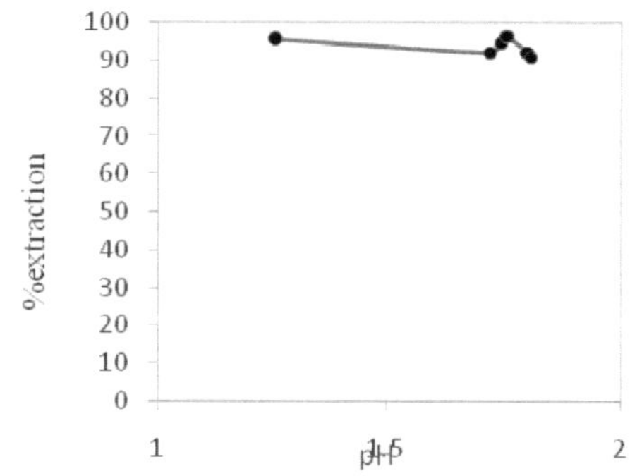

Fig 2 effect of pH on extraction of tin solution with the solution of 5.2g/L at 15°C

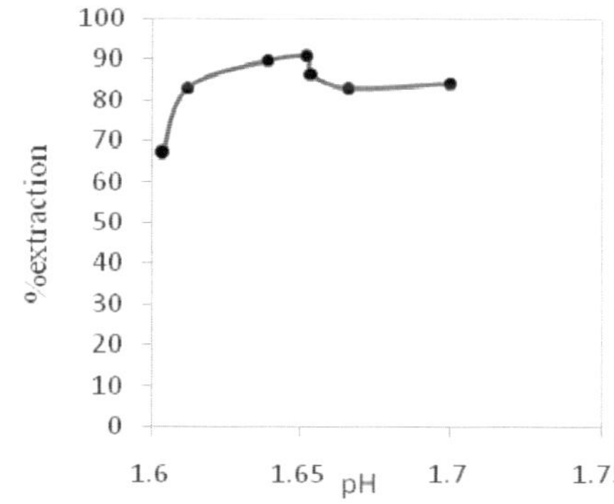

Fig 3 effect of pH on extraction of tin solution with the solution of 5.2g/L at 20°C

5 REFERENCES

Gaskell, D. R., n.d. *Introduction To The Thermodynamics of Materials.* s.l.:s.n.

Keshavarz Alamdari, E. et al., 2002. *effectof TBP as a Modifier for Extraction of Zinc and Cadmium with a mixture of D2EHPA and MEHPA.* Cape Town, South Africa, s.n.

Keshavarz Alamdari, E. & Sadrnezhad, S. K., n.d. *The Thermodynamics of Solvent Extraction of Molybdenum by Organic Solvent TBP.* s.l.:s.n.

Recovery of Lead from the Residue of Zinc Leaching

H.Soltani, A.V.Kamal, E.K.Alamdari
Department of mining and metallurgical engineering, Amirkabir University of Technology, Tehran, IRAN

ABSTRACT In the present study, lead recovery from the filter cake of the zinc leaching using acetate ion is discussed. The residue of zinc leaching contained 7.6% zinc and 3.4% lead. The lead and zinc, were in the form of sulfate in the residue. In order to dissolve lead sulfate, various systems containing acetate ions were studied. The best result in the combination of acetic acid, ammonia and barium acetate, in 40-120 minutes, L / S = 10 and 2g of barium acetate per 10 grams of soil was obtained.

Keywords: Leaching, Lead, Zinc Leaching Filter Cake, Hydrometallurgy, Recovery.

1 INTRODUCTION

Lead and zinc are the most important non-ferrous metals after copper and aluminum (Ruşen, A., A.S. Sunkar, and Y.A. Topkaya, 2008). Today, due to lower grade resources available, the industries are looking for ways to recover these metals from industrial and mining waste. In industries using raw soil to produce zinc, the grade of lead is higher in the residues and by accumulation of these residues along these industries, a good source of lead mining is achieved. Also, due to the toxic element lead in the waste, the accumulation and maintenance of that, serious damage to the environment and groundwater contamination risks also arise. Thus, separation and extraction of lead from leach residue of zinc, in addition to significant cost savings, contribute to the health of the environment.

The lead contained in the waste from the production of zinc, because generally this process is done with the help of sulfuric acid, is in the form of lead sulfate. Traditional processes used for the leaching of lead sulfate are based on the pyrometallurgy that in addition to high costs, have also more pollution. The solution to overcome these problems is to utilize hydrometallurgical processes (SHIRCHINNAMJIL N., C.Y., FANG Z H., 2008). This process involves leaching and solvent extraction and solubility information can be useful in selecting an appropriate leaching system (SHIRCHINNAMJIL N., 2008).

The leaching of lead and lead retrieval considering the type of lead compound, are done in different ways. One of the lead sulfate leaching methods that are generally in the Zn leaching waste, is the use of common salt (NaCl) (Ruşen, A, 2011; Bahram, B. and M. Javad, 2009; Farahmand, F., et al., 2009). Also several preceding studies on the leaching of lead sulfate by means of sodium sulfide were done and as a result of their response, the lead sulfide was separated in the form of precipitation. This method is also used in order to precipitate the lead in the form of sulfide from the lead chloride (Raghavan, R., P. Mohanan, and S. Swarnkar, 2000). Also the dissolution of lead sulfate with sodium hydroxide is examined. One of the advantages of sodium hydroxide compared to the acidic type solution agent, is the maximum solubility of lead and zinc and the minimum available impurities, which reduces operational costs after the dissolution. But one of the major problems of alkali dissolution is the formation of gelatinous deposits and the reduction of the capacity of filtering operations (Nagib, S. and K. Inoue, 2000). Carbonate recovery of lead sulfate is also taken into consideration.

In this method the lead sulfate reacts with the help of a carbonated agent such as sodium carbonate and the lead carbonate precipitate is formed (Havuz, T., B. Dönmez, and C. Çelik, 2010). According to the works done by Naghib et al., the dissolution of heavy metals, hydrochloric acid, sulfuric acid, acetic acid and sodium hydroxide were investigated. The studied material was the residue of the burning unit of industrial wastes containing heavy metals such as lead, which became clear that increasing the concentration of acetic acid results in increasing the amount of the lead extraction (Nagib, S., 2000). In this study, the dissolution of lead on the basis of hydrometallurgical methods and with the help of acetic acid was examined and the impact of factors such as temperature, time, concentration of acetic acid and the process promoters was investigated.

2 MATERIALS AND EXPERIMENTAL PROCEDURE

2.1 Materials

In this study, zinc filter cake or residue of the leaching process of zinc from the mineral ore, was used. In order to determine the composition of the residue, a gram of the control soil, which was completely milled, was boiled in 20 cc of nitric acid (32.5 %) with 60 cc of hydrochloric acid (18.5 %) at high temperatures with some distilled water and then sifted. The analysis of the resulting solution is shown in table 1.

Table 1- Analisys of the control soil

Zn (mg/lit)	Pb (mg/lit)	% Zn	% Pb
76.1	34.5	7.61	3.45

Also the zinc filter cake, due to the agglomeration, was tested for wet sieving to determine the soil grading and from each fractions of grading, an analysis was taken that the obtaining results are shown in table 2.

2.2 Experimental

Due to the abundance of sulfate ions inside the system, the effect of neutralizing factors of this ion in the leaching of lead was investigated. First the soil with 7M acetic acid was under leaching process, then with a similar condition barium acetate, ammonium acetate, acetic acid combined with ammonia and acetic acid, ammonia and barium acetate were studied. Considering the presence of sulfate ion, the lead is present in the form of $PbSO_4$ in the system (table 3). So it is anticipated that the reactions of this complex in the mentioned leaching systems are as given in formulas 1 through 5.

(1) $PbSO_4 + 2\ CH_3COOH = H_2SO_4 + Pb(CH_3COO)_2$

(2) $PbSO_4 + Ba(CH_3COO)_2 = BaSO_4 + Pb(CH_3COO)_2$

(3) $PbSO_4 + 2\ CH_3COONH_4 = (NH_4)_2SO_4 + Pb(CH_3COO)_2$

(4) $PbSO_4 + 2\ CH_3COOH + 2\ NH_3 = (NH_4)_2SO_4 + Pb(CH_3COO)2$

(5) $2\ PbSO_4 + 2\ CH_3COOH + 2\ NH_3 + Ba(CH_3COO)_2 = (NH_4)_2SO_4 + 2\ Pb(CH_3COO)_2 + BaSO_4$

Although the leaching was done for one time, the soil had a high amount of zinc in itself, so with a leaching by sulfuric acid at PH=1, L/S=20 and temperature of 90 °C, for about 85% of the zinc was removed initially, then the leaching process was done in a system containing acetate ion to determine the impact of zinc ion omission.

Table 2- Results of the wet Sieve testing of the zinc filter cake.

Sieve No.	The remaining weight on the Sieve (kg)	Wt% on any Sieve	The cumulative weight	The cumulative percentage	%Zn	%Pb
<400	25.96	65.79	25.96	65.79	7.33	3.64
400	1.44	3.65	27.40	69.44	3.21	4.08
325	7.76	19.67	35.16	89.10	5.04	8.80
170	1.76	4.46	36.92	93.56	4.61	5.14
100	1.36	3.45	38.28	97.01	5.18	3.66
50	0.56	1.42	38.84	98.43	0.89	0.14
35	0.62	1.57	39.46	100.00	0.65	0.17

Table 3- Various leaching systems to remove the sulfate ion and the lead leach.

	S (g)	L (ml)	NH_3(25%) (ml)	Acetic (ml)	Time (min)	T (°C)	BaAc (g)	NH_4Ac (g)
7M	50	500	0	160	120	90	-	-
BaAc	50	500	0	160	120	90	20	-
NH_4Ac	50	500	0	160	120	90	-	20
NH_3+HAc	50	500	50	50	120	90	-	-
NH_3+HAc+BaAc	50	500	50	50	120	90	2	-

Other factors affecting the leaching such as temperature, time and solid to liquid ratio were also examined. At first, the highest stable temperature of the system 90°C for the leaching was tested and then the same test was conducted at ambient temperature of 30°C to determine the effect of temperature on the process (table 4). To study the effect of time, the tests were started with small period of times and continued toward longer period of times with small steps; So that by increasing the time, the steps became longer (tables 5 and 6). Also in different ratios of liquid to solid, the tests were repeated and the effect of this parameter on the percentage of leaching was investigated (table 7).

Ultimately, the impact of the amount of barium acetate on the leaching process of the zinc filter cake in the presence of specific acetic acid and ammonia was studied (table 8).

Table 4- Tests for the investigation of the effect of the temperature on the recovery process of the lead from the zinc filter cake

	S (g)	L (ml)	NH_3(25%) (ml)	Acetic (ml)	Time (min)	T (°C)
NH_3+HAc	10	200	10	10	180	90
NH_3+HAc	10	200	10	10	180	30

Table 5- Tests for the investigation of the effect of the time on the recovery process of the lead from the zinc filter cake in the short periods of time.

	S (g)	L (ml)	NH_3(25%) (ml)	Acetic (ml)	Time (min)	T (°C)
NH_3+HAc	40	400	80	60	5	90
NH_3+HAc	40	400	80	60	10	90
NH_3+HAc	40	400	80	60	15	90
NH_3+HAc	40	400	80	60	20	90
NH_3+HAc	40	400	80	60	30	90
NH_3+HAc	40	400	80	60	40	90

Table 6- Tests for the investigation of the effect of the time on the recovery process of the lead from the zinc filter cake in the long periods of time.

	S (g)	L (ml)	NH_3(25%) (ml)	Acetic (ml)	Time (min)	T (°C)
NH_3-HAc	10	200	10	10	80	90
NH_3-HAc	10	200	10	10	120	90
NH_3-HAc	10	200	10	10	300	90

Table 7- Tests for the investigation of the effect of the solid to liquid ratio on the recovery process of the lead from the zinc filter cake.

	S (g)	L (ml)	NH_3(25%) (ml)	Acetic (ml)	Time (min)	T (°C)
NH_3-HAc	10	50	10	10	90	90
NH_3-HAc	10	100	10	10	90	90
NH_3-HAc	10	200	10	10	90	90
NH_3-HAc	10	300	10	10	90	90

Table 8- Tests for the investigation of the effect of the amount of the barium acetate on the recovery process of the lead from the zinc filter cake with the presence of specified portions of the acetic acid and ammonia.

	S (g)	L (ml)	NH_3(25%) (ml)	Acetic (ml)	Time (min)	T (°C)	BaAc (g)
NH_3-HAc-BaAc	10	100	20	15	90	93	0
NH_3-HAc-BaAc	10	100	20	15	90	93	0.5
NH_3-HAc-BaAc	10	100	20	15	90	93	1
NH_3-HAc-BaAc	10	100	20	15	90	93	2
NH_3-HAc-BaAc	10	100	20	15	90	93	5

The processes were done inside a beaker and on a magnetic stirrer that was equipped with a heater; the beaker was blocked with a watch glass and a thermometer was used to control the temperature of the container for the whole time. The solid-liquid separation at the end of the process with the help of a vaccum pump and within Büchner funnel connected to a vaccum beaker was done, so that the separation of the solid from the solution was done at least time (less than 20 seconds). The analysis of the solutions by the atomic absorbtion spectroscopy (AAS) was performed.

3 RESULTS AND DISCUSSION

In conclusion and the comparison of the results, the criteria is based on the recovery percentage of the lead. However, the zinc recovery percentage has been reported for more information.

The obtaining results from various leaching systems mentioned in table 3, is shown in table 9.

Table 9- Obtaining results from various leaching systems to remove the sulfate ion and the lead leach.

	Recovery Zn%	Recovery Pb%
7M	45.69	0.07
BaAc	59.67	0.75
NH_4Ac	57.35	3.22
NH_3+HAc	73.42	32.90
NH_3+HAc+BaAc	66.71	34.20

As it is illustrated in figure 1, the best system for the leaching of the lead was the combination of acetic acid, ammonia and the barium acetate. Also, comparing all five leaching systems for lead, it is recognizable that combination of acetic acid, ammonia compound and barium acetate, acetic acid and ammonia, ammonium acetate with 7M acetic acid, barium acetate with 7M acetic acid and finally 7M acetic acid, were effective in the leaching of the lead, respectively. The present of the acetic acid cause the dissolution of lead. At the other hand ammonia cause deposition of sulfate ions, then with this method the lead could be separated. Also barium acetate can release the acetate ions in the solution, and improved the recovery of the lead.

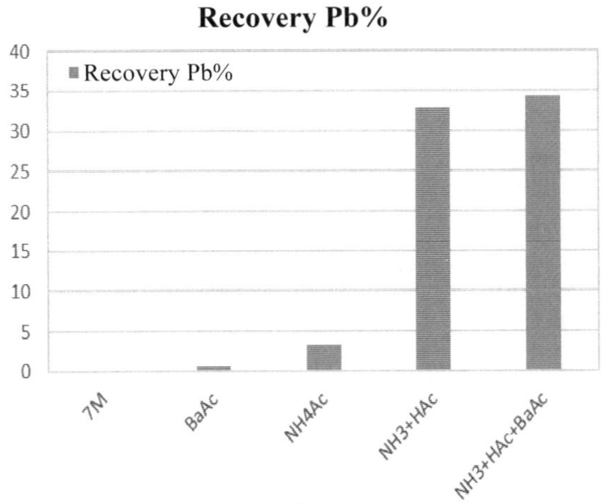

Figure 1- Obtaining results from various leaching systems to remove the sulfate ion and the lead leach.

The results of the investigations about the effect of a leaching process with sulfuric acid to reduce the zinc content in the system, are given in table 10.

As it stands, this step had little effect on the amount of the lead recovery and the cost imposed for a stage of leaching with sulfuric acid, this step is not recommended.

Table 10- The effect of leaching for one time with the sulfuric acid to decrease the zinc content in the system on the recovery percentage.

	Recovery Zn%	Recovery Pb%
Without Pre-Leaching	71.09	45.36
With Pre-Leaching	9.69	50.30

The obtaining results of the effect of temperature on the recovery percentage of the lead are given in table 11. Based on these results, it is predicted that the temperature has little or no effect on the leaching process of the lead. However, more rigorous testing is needed to verify this claim.

Table 11- The effect of the temperature on the recovery percentage of the lead.

	T (°C)	Recovery Zn%	Recovery Pb%
NH_3+HAc	90	73.42	32.90
NH_3+HAc	30	73.53	32.89

The effect of time on the recovery of the lead is given in tables 12 and 13.

Table 12- Tests for the investigation of the effect of the time on the recovery process of the lead from the zinc filter cake in the short periods of time.

	Time (min)	Recovery Zn%	Recovery Pb%
NH_3+HAc	5	28.15	28.76
NH_3+HAc	10	28.58	27.30
NH_3+HAc	15	28.57	26.46
NH_3+HAc	20	30.74	24.39
NH_3+HAc	30	30.63	32.05
NH_3+HAc	40	33.22	35.96

Table 13- The effect of the time on the recovery process of the lead from the zinc filter cake in the long periods of time.

	Time (min)	Recovery Zn%	Recovery Pb%
NH_3-HAc	80	71.97	33.04
NH_3-HAc	120	71.64	32.90
NH_3-HAc	300	70.04	24.64

As it is shown in figure 2, by increasing the time from 5 minutes to 20 minutes, decrease in the recovery percentage occurred that it would have various reasons. But the important thing is the increase of the recovery percentage after 20 minutes. Also, in another test the time period of more than 80 minutes was investigated and the results showed a decrease in the recovery percentage for the period of time exceeding from 80 minutes. So the best range of time for the process is predicted between 40 and 120 minutes. It seems after 60 minutes some adverse reactions will be started and causes decrease of the lead recovery, so one hour is the optimum time for lead recovery in this case.

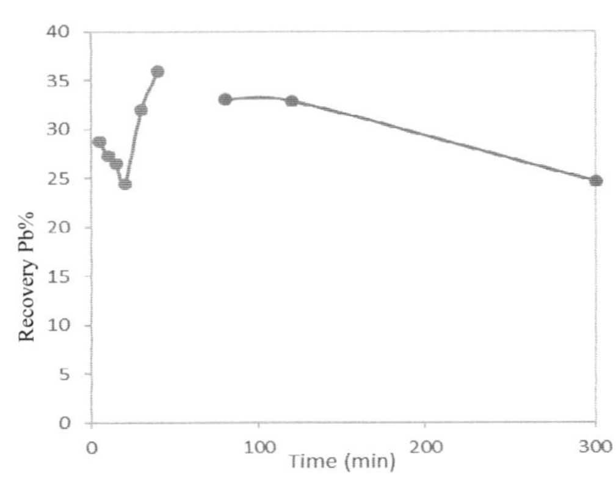

Figure 2- The effect of the time on the recovery process of the lead from the zinc filter cake

Then in the next step, the effect of liquid to solid ratio (L/S) on the recovery percentage of the lead was studied. The results of these experiments are shown in table 14. As illustrated in figure 3, the best result was attained in L/S=10 and by increasing this ratio, a decrease occurred again.

It seems increasing in the measure of L/S, causes increasing of the concentration of different ions in the solution and also causes antagonism effect on the system.

And at the other hand decreasing in L/S even less than 10 make problems for solution of soil, and generally it has negative effect on lead recovery percentage.

Table 14- The effect of the solid to liquid ratio on the recovery process of the lead from the zinc filter cake.

	S (g)	L (ml)	Recovery Zn%	Recovery Pb%
NH_3-HAc	10	50	62.30	38.91
NH_3-HAc	10	100	66.80	42.90
NH_3-HAc	10	200	65.03	32.46
NH_3-HAc	10	300	70.03	27.54

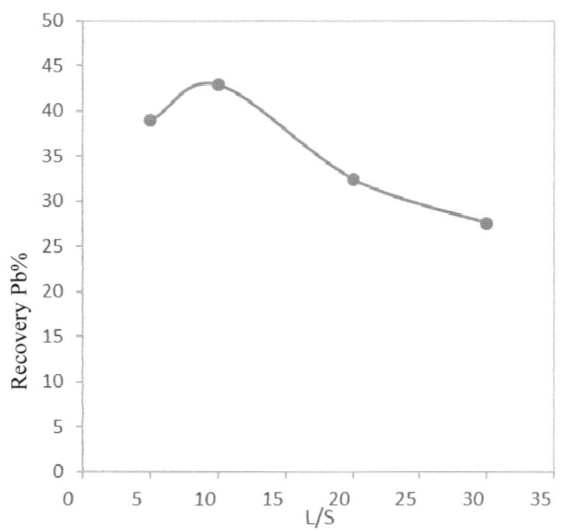

Figure 3- The effect of the solid to liquid ratio on the recovery process of the lead from the zinc filter cake.

The effect of adding barium acetate to the leaching system is shown is table 15. The effect of increasing the amount of barium acetate is shown in figure 4. With increasing the amount of barium acetate up to 2g, the recovery percentage of the lead was increased relatively and with increasing this amount up to 5g, the percentage of the lead leaching was decreased. So the maximum amount of the recovery was obtained for 2g of barium acetate.

Increasing in measure of barium acetate make the recovery better because barium ions can help the removal of sulphate ion, but increasing of these ions more than specific amount, increasing the concentration and causes reaction to occur in reverse direction. It means lead in this situation will be deposited.

Table 15- The effect of the amount of the barium acetate on the recovery process of the lead from the zinc filter cake with the presence of specified portions of the acetic acid and ammonia

	BaAc (g)	Recovery Zn%	Recovery Pb%
NH_3-HAc-BaAc	0	63.96	33.48
NH_3-HAc-BaAc	0.5	63.65	34.78
NH_3-HAc-BaAc	1	66.71	34.20
NH_3-HAc-BaAc	2	63.59	40.20
NH_3-HAc-BaAc	5	59.82	19.13

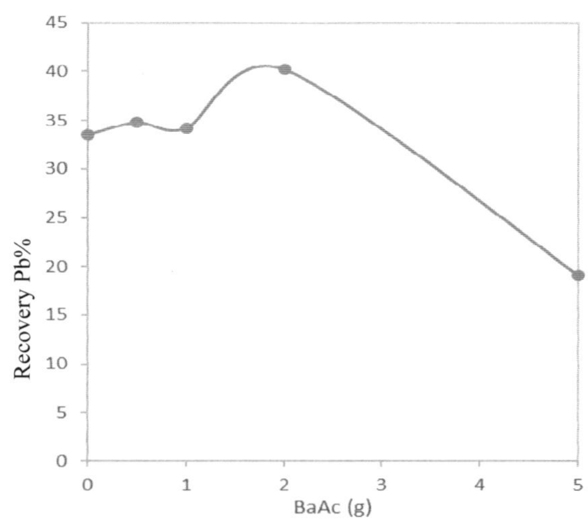

Figure 4- The effect of the amount of the barium acetate on the recovery process of the lead from the zinc filter cake with the presence of specified portions of the acetic acid and ammonia.

4 CONCLUSION

Due to the presence of sulfate ions in the system, it is required to use a leaching solution that not only dissolves the lead but also separates the sulfate from the lead ion. The best leaching solution is a combination of acetic acid, ammonia and barium acetate.

Leach process of the zinc with the sulfuric acid before the leaching of the lead and decreasing the amount of zinc, has a positive but little effect on the recovery percentage of the lead.

It is predicted that the temperature of the leaching process for the lead has no effect or little effect on the process. However, it is recommended to perform more rigorous tests and to examine temperatures between 30°C and 90°C and less than 30°C to verify the claim is approved.

The best range for the time period is predicted between 40 and 120 minutes and 90 minutes is recommended in order to ease the process.

The best ratio of liquid to solid (L/S) for the recovery of the lead was L/S=10 and the maximum recovery percentage is attained by the addition of 2g of barium acetate for each 10g of the soil.

REFERENCES

Bahram, B. and M. Javad, Chloride leaching of lead and silver from refractory zinc plant residue. Research Journal of Chemistry and Environment, Vol, 2011. 15: p. 2.

Farahmand, F., et al., Brine leaching of lead-bearing zinc plant residues: Process optimization using orthogonal array design methodology. Hydrometallurgy, 2009. 95(3): p. 316-324.

Havuz, T., B. Dönmez, and C. Çelik, Optimization of removal of lead from bearing-lead anode slime. Journal of Industrial and Engineering Chemistry, 2010. 16(3): p. 355-358.

Nagib, S. and K. Inoue, Recovery of lead and zinc from fly ash generated from municipal incineration plants by means of acid and/or alkaline leaching. Hydrometallurgy, 2000. 56(3): p. 269-292.

Raghavan, R., P. Mohanan, and S. Swarnkar, Hydrometallurgical processing of lead-bearing materials for the recovery of lead and silver as lead concentrate and lead metal. Hydrometallurgy, 2000. 58(2): p. 103-116.

Ruşen, A., A.S. Sunkar, and Y.A. Topkaya, Zinc and lead extraction from Çinkur leach residues by using hydrometallurgical method. Hydrometallurgy, 2008. 93(1-2): p. 45-50.

SHIRCHINNAMJIL N., C.Y., FANG Z H., Leaching of silver from Boorchi Ag-Pb ore in Mongolia with acidic thiourea solution. Journal of Process Engineering, 2008. 4: p. 725-730.

Mathematical Modeling and Simulation of Solid-Liquid Mixing In a Conventional Mixer

Z. Bagheri, E. Keshavarz Alamdari

Department of Mining and Metallurgical Engineering, Amirkabir University of Technology, Tehran, Iran

H.Khosravi

Department of Mechanical Engineering, Amirkabir University of Technology, Tehran, Iran

S.Parvizi

Department of Mechanical engineering, Shahid Rajaee Teacher Training University, Tehran, Iran

ABSTRACT Solid-liquid mixing is one of the operations that useful in various sectors of industry. In order to increase the efficiency of the mixing, process should be designed in order to preventing accumulation of solid particles or non-uniformity. In all studied cases, at first time, solid was in the bottom of tank but in real case solid cannot be in bottom and it pour with certain mass flow from upper surface of tank. In this study has been tried to simulate this system. The effect of impeller speed on mixing performance was also investigated by using Eulerain-Eulerian model and standard k-ε turbulence model.

Keywords: Mixing, Solid-Liquid, CFD, Numerical simulation

1 INTRODUCTION

Solid-liquid mixing, which is one of the most important processes used in various industries. Irregularity of solid particles can lead to increased energy consumption For multiphase flow modeling there are two models: the Euler-Euler and Euler-Lagrange, In simulating solid–liquid flows, the Eulerian–Lagrangian model would be a more realistic approach because it simulates the solid phase as a discrete phase and so allows particle tracking(Liu et al 2013). This approach requires a significant amount of computational time and huge memory space. In Eulerian–Eulerian approach both liquid and solid are treated as continua, interpenetrating and interacting with each other in the computational domain (Hosseini et al 2010). There are several model to simulate turbulence, among them the standard *k*-ε model is most widely used model because it is robust, economical, and rapid. Besides, it gives stable calculations and reasonable results for many flow domains (Hosseini et al 2010). In this paper the effect of velocity on suspension of solid is investigated. For validation, experimental results of Guida are used (Fig 1.a).

2 METHODS

2.1 Governing Equations

Based on the conservation principles of mass and momentum, the continuity and momentum equations in the stirred tank are given below(Qi et al 2013):

Continuity equation

$$\frac{\partial(\rho_q \varphi_q)}{\partial t} + \frac{\partial \rho_q u_{iq} \varphi_q}{\partial x_i} = 0 \qquad 1$$

Momentum equation

$$\frac{\partial(p_q u_{iq} \varphi_q)}{\partial t} + \frac{\partial(p_q u_{iq} u_{jq} \varphi_i)}{\partial x_i} = -\varphi_q \frac{\partial \rho}{\partial x_i} + \frac{\partial}{\partial x_i}\{\emptyset_q \mu_{qeff}\left(\frac{\partial u_{iq}}{\partial x_j} + \frac{\partial u_{jq}}{\partial x_i} - \frac{2}{3}\frac{\partial u_{kq}}{\partial x_k}\delta_{ij}\right)\} \qquad 2$$

2.2 Turbulence Model

The two-equation turbulence model consists

of the differential transport Eq. (3) for the turbulent kinetic energy, k, and Eq. (4) for its dissipation rate, ε, thus:

$$\frac{\partial(\rho_q\varphi_q\varepsilon)}{\partial t}+\frac{\partial(\rho_q\varphi_q u_{iq}\varepsilon)}{\partial x_i}=\frac{\partial}{\partial x_i}\left[\varphi_q\left[\mu_q+\frac{\mu_{qeff}}{\sigma_\varepsilon}\right]\frac{\partial\varepsilon}{\partial x_i}\right]+\rho_q\varphi_q\frac{\varepsilon}{k}(C_1P_k-C_2\varepsilon) \quad 3$$

$$\frac{\partial(\rho_q\varphi_q k)}{\partial t}+\frac{\partial\rho_q\varphi_q u_{iq}k}{\partial x_i}=\frac{\partial}{\partial x_i}\left[\varphi_q\left[\mu_q+\frac{\mu_{qeff}}{\sigma_\varepsilon}\right]\frac{\partial k}{\partial x_i}\right]+\rho_q\varphi_q(P_k-\varepsilon) \quad 4$$

where C1, C2, k, and ε are parameters in the standard k–ε model and the following values are selected: C1 = 1.42, C2 = 1.9, k = 1.0, and ε = 1.314 (Liu et al 2013).

2.3 Stirred Tank Configuration and Grid Topology

A flat bottom cylindrical tank with a diameter T = 0.205m, equal to the liquid height H, and four baffles at a width of 0.1T equally spaced along the tank wall are used in this simulation. PBT impeller with four blades with a diameter of 0.68 meters was used for agitation. Impeller clearance is equal to 0.041. Characteristics of Liquid phase and solid phase Glass bead are listed in the Table 1:

MRF approach was used to simulate the rotating impeller. That's why the system is composed of two volumes: The outer and inner volume. For more accurate calculations mesh of inner volume was smaller. The total number of mesh in the computational domain is 745000 (Fig 1.b).

3 RESULTS AND DISCUSSION

For illustrate the effect of velocity on particle suspension, there are 6 CFD solid concentration contours (400,600 and 800 rpm) in the figure 2 , solid with mass flow = 0.0000785 m^3/s pour into tank for 10 sec. the counters are shown in ten second with 400 rpm the maximum solid sedimentation is

Table 1 Physical Properties

	density	Particle diameter (m)	Volume fraction (vol%)
Liquid	1000	…..	….
solid	2500	0.0001	10

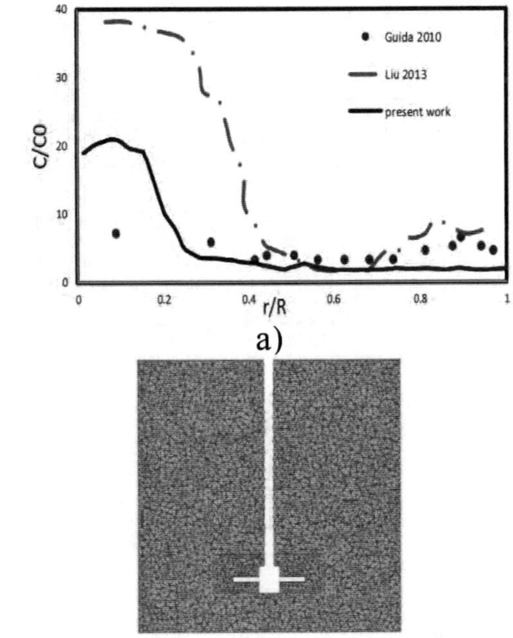

Fig 1. (a) Guida experimental and CFD results, (b) mesh grid

happened so it is not appropriate. In 25sec at 600 and 800 rpm, there are little solid sedimentation but in 600 rpm cloud height is less than 800 rpm. So the best velocity for having excellent suspension is 800 rpm. In figure 3, there are three vector of solid velocity that show increase of movement of solid particle in all of height in 800 Rpm.

4 CONCLUSION

Solid-liquid are simulated while solid pour from upper surface. The simulation are conducted by Euler-Euler model with a standard K-ε turbulence. Results show the suspension of solid increased with an increase in impeller speed.

REFERENCES

1)*Qi, N., Zhang, H., Zhang, K., Xu, G., & Yang, Y. (2013). CFD simulation of particle suspension in a stirred tank. Particuology, 11(3), 317-326.*

2) *Liu, L., & Barigou, M. (2013). Numerical modelling of velocity field and phase distribution in dense monodisperse solid–liquid suspensions under different regimes of agitation: CFD and PEPT*

experiments. Chemical Engineering Science, 101, 837-850.

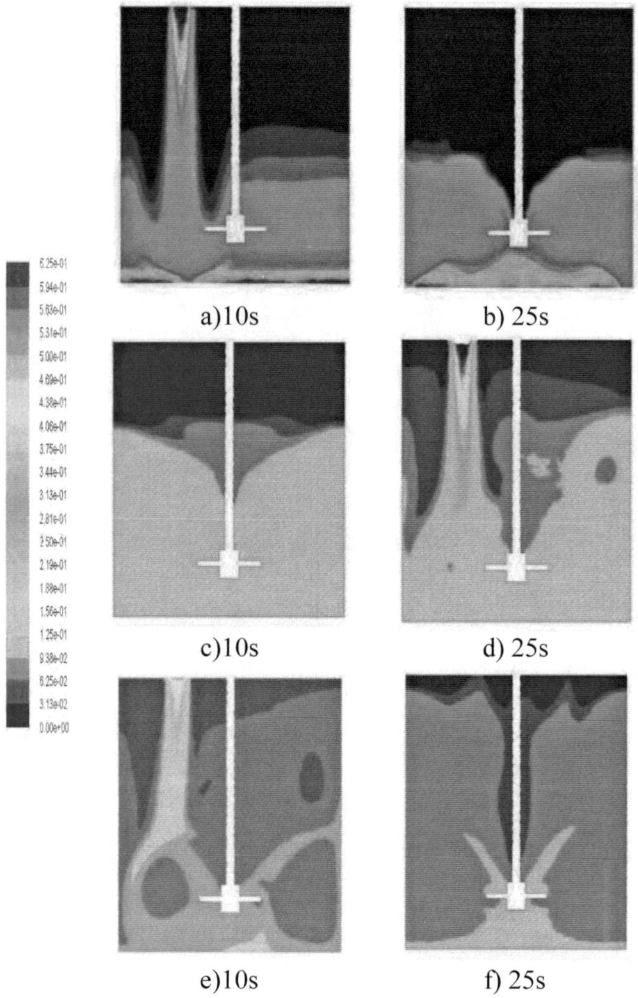

Fig 2. a) 400 rpm 10s b)400 rpm 25s
c) 600 rpm 10s d) 600 rpm 25s
e)800rpm 10s f) 800rpm 25s

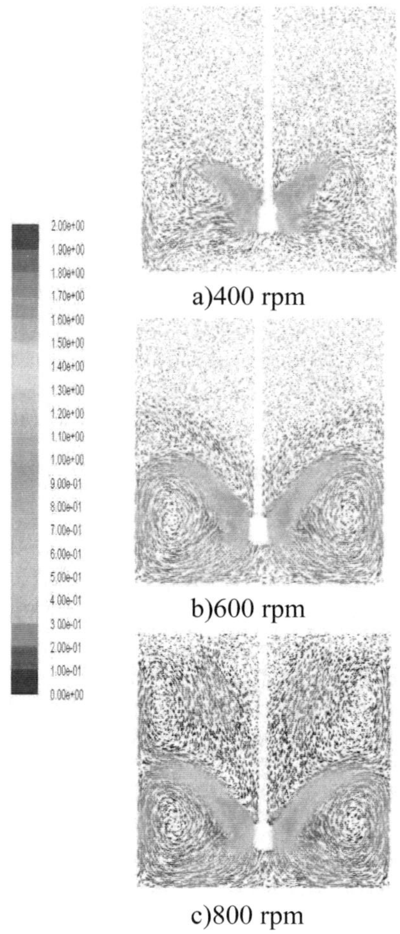

Fig 3. Solid velocity vector profiles in the solid–liquid phase stirred tank at 25 sec

3) *Guida, A., Nienow, A. W., & Barigou, M. (2010). PEPT measurements of solid–liquid flow field and spatial phase distribution in concentrated monodisperse stirred suspensions. Chemical Engineering Science, 65(6), 1905-1914*

4) *S. Hosseini and D. Patel, 'Study of Solid– Liquid Mixing in Agitated Tanks through Computational Fluid Dynamics Modeling', Ind. ..., pp. 4426–4435, 2010*

A Preliminary Study on Nitric Acid Pre-treatment of Refractory Gold/Silver Ores

O. Celep, E.Y. Yazici, H. Deveci

Hydromet B&PM Group, Mineral&Coal Process Div., Mining Eng. Dept., Karadeniz Technical University, 61080, Trabzon, Turkey

ABSTRACT Nitric acid is a strong oxidant capable of rapidly oxidizing sulphide minerals. This study was undertaken to evaluate the nitric acid leaching as a potential pre-treatment process prior to cyanidation of refractory gold/silver ores. The nitric acid pre-treatment experiments for refractory silver ore and gold/silver concentrate were performed for 2.5 hours at 5M HNO_3 and 80°C. Before and after nitric acid pretreatment, cyanide leaching tests (1.5 g/L NaCN, 0.3 L/min air) were carried out over a period of 24 h. Direct cyanidation of both refractory silver ore and gold/silver concentrate resulted in low extractions of only ≤71% over 24 h. After nitric acid pretreatment, the extraction of gold and silver from both materials substantially improved with almost complete leaching of gold and silver from the concentrate and extensive leaching of silver from the ore. This improvement by nitric acid pre-treatment can be attributed to the decomposition of sulfide minerals that encapsulate gold and silver. These findings suggest that nitric acid leaching can be suitably exploited for pre-treatment of refractory gold and silver ores prior to cyanide leaching.

1.INTRODUCTION

Gold and silver are commonly occluded within the host mineral matrix in refractory ores. For this reason, gold and silver extractions can be remarkably low in cyanide leaching (Bhappu, 1990; Marsden and House, 2006; Celep et al., 2009). Many studies appear to focus on refractory gold ores rather than silver ores for elimination of refractoriness. Extraction of silver from its ores or gold ores is often lower than gold (Zhou et al., 2009; Zhou, 2010). This is often concomitant with silver mineralogy, which is more complex than that of gold, complicating the extraction of silver. Roasting, pressure oxidation and bio-oxidation are commercially available pretreatment processes for refractory ores to improve cyanide leaching of gold or silver (La Brooy, 1994). These pretreatment processes are based essentially on the decomposition of sulphide minerals (e.g. pyrite, arsenopyrite) to ensure the exposure of occluded gold/silver to cyanide solution. Gold and silver can be extracted in dilute solutions of cyanide in the presence of air/oxygen. Gold is oxidised by oxygen to form the Au(I) cyanide complex $[Au(CN)_2^-]$ in alkaline cyanide solution. The overall dissolution of gold is appropriately described by the following reaction equation as proposed by Elsner (Marsden and House, 2006):

$$4Au + 8CN^- + O_2 + 2H_2O \leftrightarrow 4Au(CN)_2^- + 4OH^- \quad (1)$$

Balaz (2000) reported very limited silver extractions (often ≤10%) from cyanide leaching of refractory silver sulphides such as proustite, pyrargyrite, tennantite and tetrahedrite. However, dissolution of native silver and acanthite (Ag_2S) readily occurs according to Eq. 2 and 3 respectively (Fleming, 1992, Zhang et al., 1997, Senanayake, 2006; Luna and Lapidus, 2000) through the formation of a stable silver–cyanide complex (i.e. $Ag(CN)_2^-$, logK=20.5).

$$4Ag^0 + 8CN^- + O_2 + 2H_2O \rightarrow 4Ag(CN)_2^- + 4OH^- \quad (2)$$

$$Ag_2S + 4CN^- + O_2 + 0.5H_2O \rightarrow 2Ag(CN)_2^- + 0.5S_2O_3^{2-} + OH^- \quad (3)$$

Nitric acid has been studied as a strong oxidant for the treatment of refractory pyritic and arsenopyritic ores and concentrates since the early 1980s (Marsden and House, 2006). Nitrogen species-catalyzed process (NSC) is tested for the treatment of refractory gold-bearing materials (La Brooy, 1994). Nitric acid has the potential to selectively dissolve metals such as Cu, Ni, Pb, Ag, etc. over metals such as Au and Sn (Bas et al., 2014). Despite its comparatively high cost, nitric acid can be readily regenerated from its reaction products (NO and NO_2) (Habashi, 1999; Li, 2009). Nitric acid leaching of metals, pyrite and arsenopyrite occurs as follows:

$$3M^0 + 8HNO_3 \rightarrow 3M(NO_3)_2 + 4H_2O + 2NO_{(g)} \quad \text{(M: Cu, Pb, Zn, Ag etc.)} \quad (4)$$

$$2FeS_2 + 10HNO_3 \leftrightarrow 2Fe^{3+} + 2H^+ + 4SO_4^{2-} + 10NO + 4H_2O \quad (5)$$

$$3FeAsS + 12HNO_3 \leftrightarrow 3FeAsO_4 + 4H^+ + 2SO_4^{2-} + 4H_2O + S + 12NO \quad (6)$$

This study was undertaken to evaluate the nitric acid leaching of refractory gold and silver concentrate and ore as a potential pre-treatment process ahead of cyanidation.

2.MATERIAL AND METHOD

In this study, samples of an arsenical silver ore and a gold/silver concentrate were used. The ore sample were prepared for the experiments by crushing in a jaw crusher (-4 mm) and grinding in a rod mill at 50% pulp density. The concentrate sample was used as received (d_{80}: 80 μm). The chemical composition of both samples (Table 1) was determined using ICP-ES (Inductively Coupled Plasma-Emission Spectroscopy) after four acid digestion, fire assay fusion and X-ray fluoressans (XRF). Gold and silver contents were determined by cupelation method with a gravimetric finish. The X–ray diffraction analysis was carried out using a Rigaku X-ray diffractometer (D/Max-IIIC).

The silver ore was found to contain 63.7% SiO_2, 11.7% Al_2O_3, 1.31% As, 127 g/t Ag. The gold/silver concentrate has also chemical composition of 32.64% SiO_2, 2.57% Al_2O_3, 26.55% Fe_2O_3, 110 g/t Ag and 62 g/t Au.

The ground ore sample was riffled using a Quantachrome Mini-riffler to prepare representative sub-samples. A number of polished resin mounts were prepared and used for MLA using an FEI Quanta 600F scanning electron microscope (SEM).

The nitric acid pre-treatment tests were performed for a leaching period of 2.5 hours at 5M HNO_3 and 80°C. The cyanide leaching tests (1.5 g/L NaCN, 25% w/w, 0.3 L/min air, 24 h) were carried out in a glass reactor, which was agitated at 700 rpm. The concentration of NaCN in solution was followed by silver nitrate titration using p–dimethylamino–benzal–rhodanine (0.02% w/w in acetone) as the indicator. If required, cyanide (as NaCN) was added to keep the level of titratable NaCN at the 1.5 g/L for the duration of leaching. The consumption of cyanide was determined based on the addition of cyanide. pH was controlled at ≅10.5 by using 1 M NaOH.

Table 1. Chemical and mineralogical compositions of the ore samples used in the study

Compound	Content (%)		Element	Content (g/t)		XRD Results	
	Au/Ag Ore	Ag Ore		Au/Ag Ore	Ag Ore	Au/Ag Ore	Ag Ore
SiO_2	32.64	63.7	Au	62	-	Quartz	Barite
Al_2O_3	2.57	11.7	Ag	110	127	Pyrite	Quartz
$BaSO_4$	-	5.5	As	-	1.31%		Clays
Fe_2O_3	26.55	5.19	Cu	3037	-		Mica
CaO	3.49	0.73	Zn	6893	7182		
MgO	0.46	0.69	Pb	2670	7252		
Na_2O	0.03	0.2	Sr	42.9	491		
K_2O	0.67	4.14	Sb	62.7	699		
TiO_2	0.13	0.47	Ni	159	245		
P_2O_5	0.05	0.16	V	26	86		
Tot. S	17.9	0.99	Cd	110.5	103		
Tot. C	0.9	0.21	Ga	12.9	13.1		
LOI	19.7	7					

3. RESULTS AND DISCUSSION

3.1. Mineralogical characterization

The mineralogical analysis of the silver ore indicated that it is composed of mainly quartz, barite, mica and the group of clay minerals. Silver (127 g/t) was observed to be present mainly as acanthite. Free grains of acanthite were 57% (by weight) in abundance. The remainder appears to occur as closely associated with and/or encapsulated in other mineral phases. This may well lead to its limited exposure to cyanide solutions with the resultant low extractions for silver (Fig. 1). The ore was also found to contain mainly Pb-Fe and Fe-Pb arsenates and sulphate-arsenates, and lesser extent antimony minerals represented by Sb oxide, Sb-Fe-Pb oxide and Sb-Pb-Fe arsenate. The concentrate sample was determined to contain sulphides; mainly pyrite (32.65 wt%) and, to a less extent, sphalerite (5.37 wt%), galena (2.12 wt %) and anglesite (1.13 wt%) as well as traces of arsenopyrite, chalcopyrite, covellite and bournonite.

Figure 1. An acanthite grain in a particle that is composed of quartz, K-feldspar, Pb-Fe illite and muscovite within silver ore (left); Electrum grains in a particle that is composed of an aggregate of fine grains of quartz, K-feldspar, goethite and clay within gold/silver ore (right)

3.2. Effect of nitric acid pre-treatment on metals extraction

Figure 2 illustrates the extent of dissolution of some metals in the nitric acid treatment stage. Extractions of As and Fe from silver ore were limited to be ≤55%. Leaching of silver was insignificant (1.3%). Extensive leaching of metals from the concentrate was observed in the nitric acid treatment. Dissolution of As and Cu was almost complete, accompanied with a Fe extraction of 80%. These findings suggest that interfering phases (e.g. As and Cu minerals) with cyanide leaching can be eliminated by nitric acid treatment. A significant dissolution of Ag (43%) was also observed to occur in this test. The leaching residues from nitric acid treatment of both ore and concentrate were separated from the solution and submitted to cyanide leaching for gold and silver extraction.

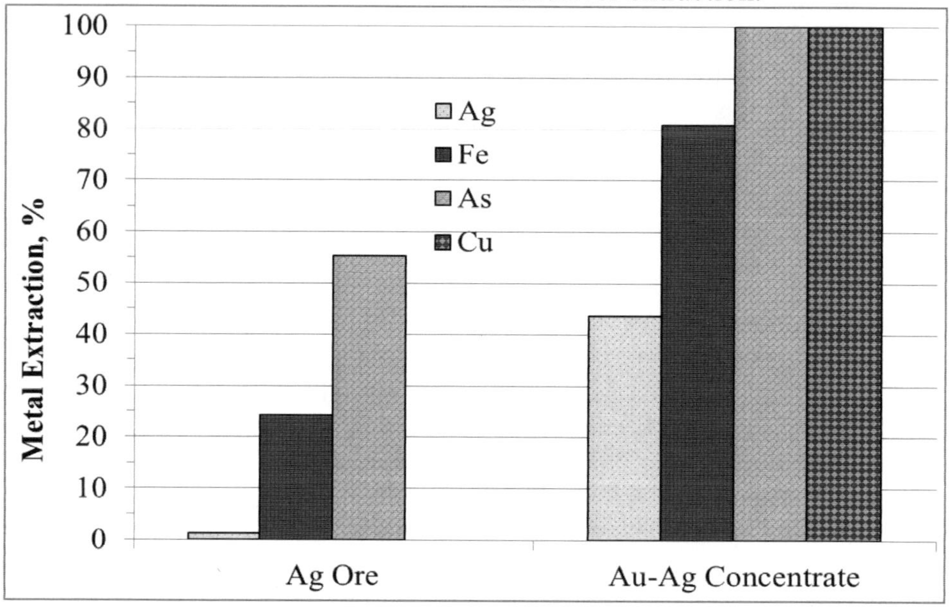

Figure 2. Extractions of some metals during nitric acid pre-treatment from the silver ore and the concentrate (Conditions: 5 M HNO_3 and 80°C).

3.3. Cyanide Leaching of Gold and silver After Pretreatment

Figure 3 illustrates the effect of nitric acid pre-treatment on the subsequent cyanide leaching of gold and silver from the ore and concentrate over 24 h leaching period. In the direct leaching of the ore (d_{80}: 50 μm) without pretreatment, silver extraction was determined to be 71%. Nitric acid pretreatment led to a 20% improvement in the extraction of silver, which was determined to be 91% over 24 h of cyanide leaching. It is pertinent to note that only 2% of silver was leached during nitric acid treatment.

The concentrate (d_{80}: 80 μm) as received (i.e. no nitric acid pretreatment) was more refractory in character under the cyanide leaching conditions tested. In this regard, direct cyanide leaching resulted in limited gold and silver extractions of~46%. After nitric acid pretreatment, cyanide leaching of gold and silver were substantially improved with their almost complete extractions (Fig. 4). This improvement as a result of nitric acid pre-treatment may well be attributed to the decomposition of sulfide minerals leading to the exposure of the encapsulated gold and silver as well as the removal of interfering soluble metals such as copper in the concentrate. Furthermore, nitric acid pretreatment was found to significantly reduce cyanide consumption (by up to 54%) from 4.1 kg/t to 1.8 kg/t for the concentrate.

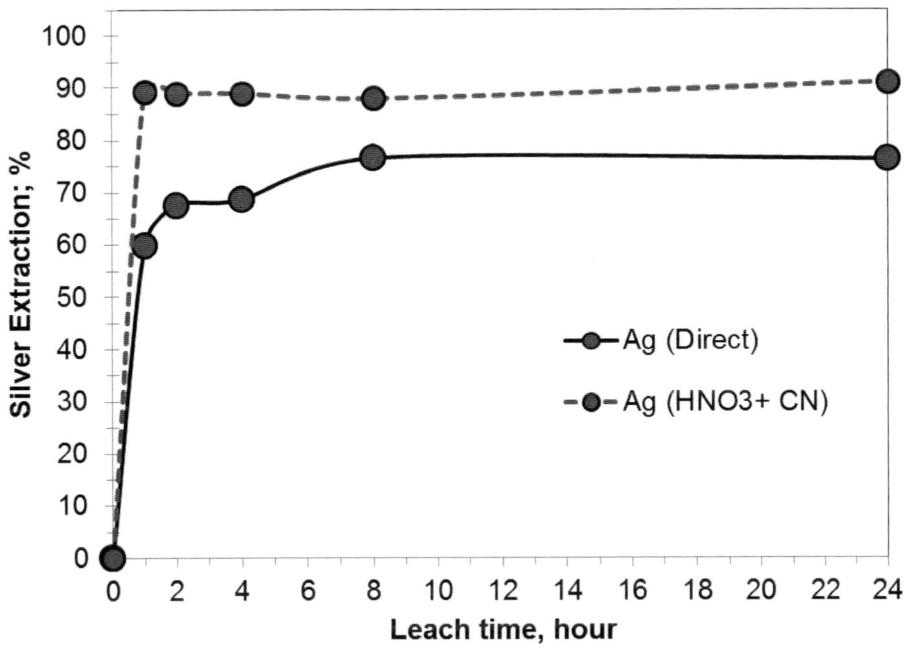

Figure 3. Effect of nitric acid pre-treatment on the extraction of silver from the ground ore (d_{80}: 50 μm) in cyanide leaching (Cyanidation conditions: 1.5 g/L NaCN, pH 10.5, 25±3°C, flow rate of air: 1.8 L/min., 26%w/w).

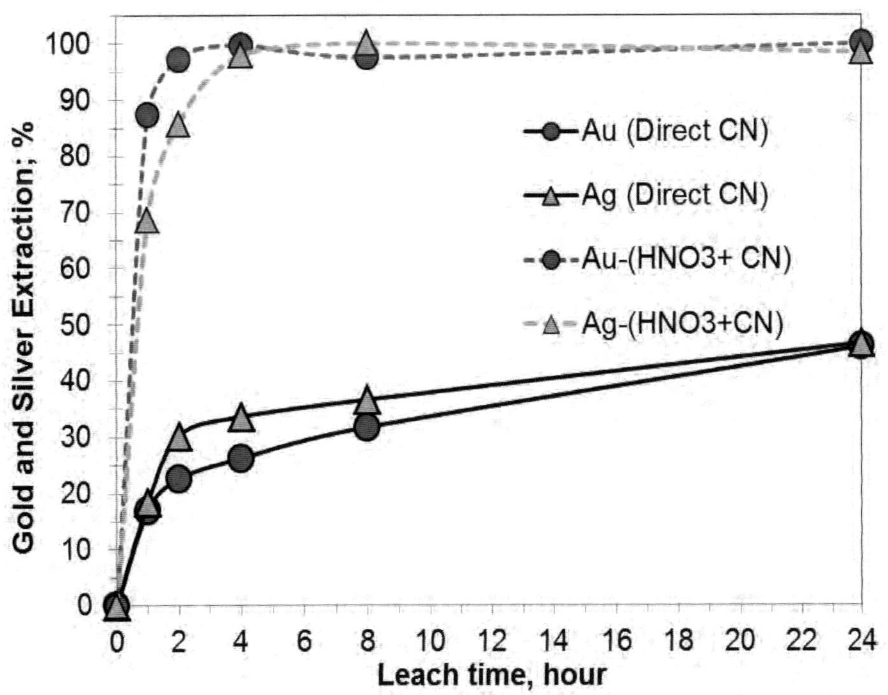

Figure 4. Effect of nitric acid pre-treatment on the extraction of gold and silver from the as-received concentrate (d_{80}: 80 μm) in cyanide leaching (Cyanidation conditions: 1.5 g/L NaCN, pH 10.5, 25±3°C, flow rate of air: 1.8 L/min., 26%w/w).

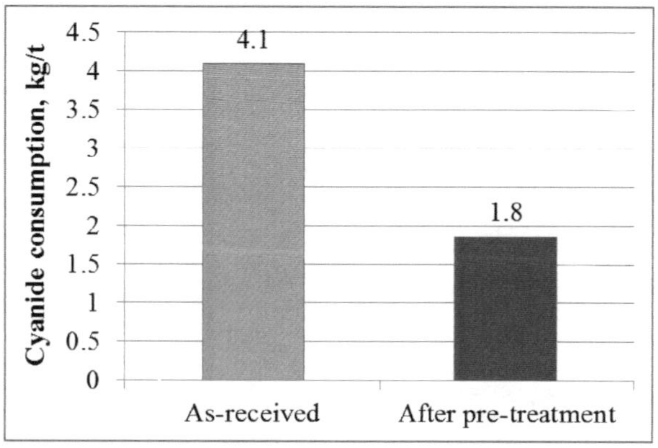

Figure 5. Effect of pre-treatment on consumption of cyanide over 24 h leaching period.

4. CONCLUSION

This study demonstrates the beneficial effect of oxidative chemical pretreatment via nitric acid leaching on the extraction of gold and silver prior to cyanidation of refractory gold/silver ore/concentrate. Direct cyanidation tests confirmed refractoriness of the ore and concentrate samples with low gold and silver extractions. Nitric acid pretreatment appeared to significantly improve the extraction of silver by 20%, from the silver ore and extractions of gold and silver by 55% from the gold and silver concentrate. These improvements can be attributed to the decomposition of refractory silver phases in the ore and the sulfide minerals such as pyrite that enclose gold and silver. Nitric acid treatment has also shown to remove potential interfering metals such as with cyanide leaching. These findings show that a chemical oxidation method such as nitric acid can be suitably used for pre-treatment of refractory gold and silver ores prior to cyanide leaching.

ACKNOWLEDGEMENTS

The authors would like to express their sincere thanks and appreciation to the Research Foundation of Karadeniz Technical University (Project No: 9200) and TUBITAK (Project No: 213M539) for the financial support, and to Yıldızlar Holding and to TUPRAG Efemçukuru Gold Mine and the late Dr. Alp ASLAN for kindly providing the samples.

REFERENCES

Baláž, P. (2000): Extractive Metallurgy of Activated Minerals-Elsevier, Amsterdam, p.278.

Bas, A.D., Deveci, H., Yazici, E. (2014): Treatment of manufacturing scrap TV boards by nitric acid leaching- Separation and Purification Technology, 130:151-159.

Bhappu, R.B. (1990): Hydrometallurgical processing of precious metal ores- Mineral Processing and Extractive Metallurgy Review, 6: 67-80.

Celep, O., Alp, I., Deveci, H., Vicil, M. (2009): Characterization of refractory behaviour of a complex gold/silver ore by diagnostic leaching-Transactions of Nonferrous Metals Society of China, 19: 707-713.

Fleming, C.A. (1992): Hydrometallurgy of precious metals recovery-Hydrometallurgy, 30: 127-162.

Habashi, F. (1999): Textbook of Hydrometallurgy, second ed., Metallurgie Extractive, Quebec, Canada.

La Brooy, S.R., Linge, H.G., Walker, G.S. (1994): Review of gold extraction from ores, Mineral Engineering, 7,10: 1213-1241.

Li, D.X. (2009): Developments on the pretreatment of refractory gold minerals by nitric acid- World Gold Conference, The Southern African Institute of Mining and Metallurgy,145-150.

Luna, R.M. and Lapidus, G.T. (2000): Cyanidation of silver sulphide- Hydrometallurgy, 36: 171–188.

Marsden, J.O. & House, C.L. (2006): The Chemistry of Gold Extraction, 651, SME, Colorado.

Senanayake, G. (2006): The cyanidation of silver metal: Review of kinetics and reaction mechanism- Hydrometallurgy, 81: 75-85.

Zhang, Y., Fang, Z., Muhammed, M. (1997): On the solution chemistry of cyanidation of gold and silver bearing sulphide ores. A critical evaluation of thermodynamic calculations-Hydrometallurgy, 46: 251–269.

Zhou, J. (2010): Process mineralogy of silver ores and applications in flowsheet design and plant optimization- In Proceedings of the 42nd Annual Meeting of Canadian Mineral Processors, 143–161, Ottawa, Ontorio, Canada.

Zhou, J., Mermillod-Blondin, R., Cousin, P. (2009): Deportment of gold and silver in the Pinos Altos composite and leach residues: Implications for recovery improvement-In Proceedings of the World Gold Conference, The Southern African Institute of Mining and Metallurgy,75–83.

Use of Fly Ash Aggregates in Production of Light-Weight Concrete
Uçucu Küllerinden Elde Edilen Agregaların Hafif Beton Üretiminde Kullanılabilirliği

C. Bursa, M. Tanrıverdi, T. Çiçek
Dokız Eylül University, Department of Mining Engineering, Izmir, Turkey

ABSTRACT In this study, it has been planned to produce light-weight aggregates and light-weight concrete from fly ash of Seyitomer Thermal Power Plant. After the specifiying of the physical and chemical properties of fly ash, various types of pellets were produced from lime and gypsum which are mixed of five different rates. The most durable pellets were found to consist of fly ash (88%), lime (6%) and gypsum (6%) mixture. Three types of light-weight concrete have been produced by mixing different rates of Portland cement (16%, 20%, 24%) with the light-weight aggregates. The concretes have been subjected to several mechanical tests after certain durations of time (1, 3, 7, 14 and 28 days). Compressive strength and flexural tensile strength have been investigated in mechanical tests. The volumetric unit weights of the concrete samples remained below 2 ton/m^3. According to the mechanical tests, the light-weight concrete with 20 % cement addition has been found to be much more durable.

ÖZET Bu çalışma kapsamında, Seyitömer Termik Santrali uçucu küllerinden hafif agrega ve hafif beton üretilmesi planlanmıştır. Uçucu küllerin fiziksel ve kimyasal özellikleri belirlenip, beş farklı oranda kireç ve alçı karışımları denenerek çeşitli peletler (hafif agregalar) üretilmiştir. Üretilen hafif agregalar üzerinde deneysel çalışmalar yapılarak, oluşan fazlar araştırılarak, en dayanıklı hafif agregaların % 88 uçucu kül, % 6 kireç ve % 6 alçı karışımı ile oluştuğu saptanmıştır. Elde edilen bu hafif agregalar ile değişik oranlarda portland çimentosu karışımı (% 16, % 20, % 24) kullanılarak 3 farklı hafif beton üretilmiştir. Üretilen hafif betonların farklı yaşlarda (1, 3, 7, 14 ve 28) mekanik özellikleri incelenmiştir. Mekanik özelliklerden basınç dayanımı ve eğilme dayanımı araştırılmıştır. Numunelerin birim hacim ağırlık değerleri tüm numunelerde 2 ton/m^3 'ün altında olduğu gözlemlenmiştir. Mekanik deneylere göre % 20 çimento katkılı hafif betonun daha dayanıklı olduğu tespit edilmiştir.

1. INTRODUCTION

Fly ash, which is a waste from thermal power plants, pollutes the soil, air and water, and therefore causes great harm to the environment due to the storage problems. Today the amount of fly ash produced worldwide is about 650 million tons a year. However, despite this huge output, very little part of fly ash is used in the construction industry. The fact that a large part of fly ash remains without being used in many countries shows the importantance of the utilization of fly ash (Koçkal, 2007).

The common name for the materials such as sand, gravel, and crushed stone in concrete use is aggregate. Aggregates which occupies a volume of nearly 60-75% in the concrete is an important component. Aggregates are divided into two types namely; fine and coarse aggregates according to their grain sizes. The most expensive material used in concrete manufacture is cement. On the other hand, aggregate is a cheap material. Thus, using as much aggregate as possible for a desired quality reduces the cost. The ones whose specific gravity is less than 2,4 t/m^3 are called lightweight aggregate. Expanded clay, pumice, and perlite are in this category. The concretes manufactured with these aggregates are called light-weight concrete. Fly ashes can be used in cement production and as concrete agent. When fired until 1100-1200 °C, round grain aggregate can be obtained. These grains are used in light-weight concrete manufacture (Turkish Ready Mixed Concrete Association [THBB], 2016).

Studies concerning the use of fly ashes in aggregate production have been of great importance especially in recent years. In the studies using fly ash instead of a part of fine aggregate for the concrete mixture, positive results have been obtained. There are many studies regarding the light aggregates obtained from fly ashes. In these studies, the fly ash produced as artificial coarse lightweight aggregate by sintering was used in light-weight concrete production and its effects on concrete characteristics was studied. On the other hand, it is seen that the coarse light-weight aggregate produced by mixing high amount of fly ash and a little portland cement is used in light coarse production and load-bearing concrete is obtained. In another study, fly ashes manufactured as artificial aggregate with clay was successfully used with building (debris) waste. Additionally, in another study using fly ashes as reused aggregate in concrete after a series of processes, similar results were obtained in the concretes including natural aggregate and waste aggregate. It was reported that in the construction of a nuclear power plant in England, 1.300 tons of sintered fly ash were used in light-weight concrete as light-weight aggregate (Aruntaş, 2006).

The main of this study is to produce sound light-weight aggregates and light-weight concrete with fly ash obtained from Seyitömer Thermal Power Plant. In the scope of this aim, several pellets were produced by trying lime and gypsum in three different proportions and theses aggregates were utilized in production of light-weight concrete.

2.MATERIAL AND TEST PROCEDURE

The fly ash sample used in this study was taken from Seyitömer Thermal Power Plant's electro particle filtration equipment. The slaked lime with a very fine grain size was supplied from Torbalı Kimtaş A.Ş. lime factory in order to be used in light-weight aggregate production. In the production of lightweight concrete, TIP 1 42.5 Portland cement which was obtained from Çimentaş Cement Factory in Izmir was used.

In the scope of this study;

- Physical and chemical properties of the fly ash were determined.
- Several pellets were obtained by trying fly ash and gypsum mixtures in three different portions.
- After the characterization and endurance tests of light- weight aggregates produced, the aggregates enabling the optimum conditions was chosen.
- Of the selected optimum light-weight aggregates, nearly 100 kg were produced.
- Through these lightweight aggregates, 3 different types of light-weight concretes were obtained by using the mixture of portland cement and fly ash additives with different amounts.
- Compressive strength were determined by using cubical samples with a size of 70x70x70 mm.
- Flexural strength was determined by using prisms with a size of 50x50x300 mm

Considering the production of fly ashes from Seyitömer Thermal Power Plant, it was researched whether aggregates at sufficient mechanical endurance can be obtained by producing aggregates in curing cabinet prepared in different proportions of lime and gypsum. After determining the conditions of optimum light-weight aggregates, the light-weight aggregates produced according to these conditions were used to obtain light-weight concrete. Light-weight concretes prepared by adding fly ash and cement with different amounts were exposed to some mechanical tests.

3.EXPERIMENTS

3.1 Characterization of Materials

Sieve, chemical and mineralogical analyses were made on fly ash. $SiO_2+Al_2O_3+Fe_2O_3$ amount is averagely 88.71% and due to being over 70.00 % according to ASTM C 618 and CaO being less than 10 %, the sample is in the F class (low-lime) fly ash class. 3 % limit value desired for SO_3 determined at TS EN 450 standard applies for the SO_3 value obtained through the analysis (1.49 %). Ignition loss complies with all the standards with a value of 2.26.

X-Ray diffraction analysis of ash and aggregate samples was made through the X-ray Diffractometer device Rigaku MiniFlex 2 model in the X-Ray laboratory of our department. Quartz (SiO_2), Haematite (Fe_2O_3), Anorthite ($CaAl_2Si_2O_8$), Mullite ($Al_4Si_2O_{10}$) and Magnesioferrite ($MgFe^{+++2}O_4$) were determined to be the main phases in Seyitömer fly ash.

3.2 Production of light-weight aggregate

Seyitömer fly ash, after its characterization studies, is mixed as shown in Table 1. Pelletizing processes were made in the pelletizing disc with trademark of Evreka Apparatenbau GmbH (Fig 1). The pelletizing conditions are shown in Table 2. Firstly, the determination of moisture and strength tests for the aggregates then the curing cabinet tests were made. It was found that around 23 % moisture was sufficient for good balling.

Table 1. The mixing ratio of lightweight aggregates.

Mixture No	Quantity of fly ash (%)	Quantity of cement (%)	Quantity of gypsum(%)
1	88	12	-
2	88	10	2
3	88	8	4
4	88	6	6
5	88	4	8

Table 2. Conditions of pelletizing.

Pan slope	30°
Scraper slope	25°
Speed	60 rev./ min.

Figure 1. Pelletizer disc and light-weight aggregates.

Since aggregates produced should be resistant against falling (drop) from about a half meter height drop tests were performed. Drop test results of the 88 % fly ash and 12 % lime moist light-weight aggregates are given in Figure 2.

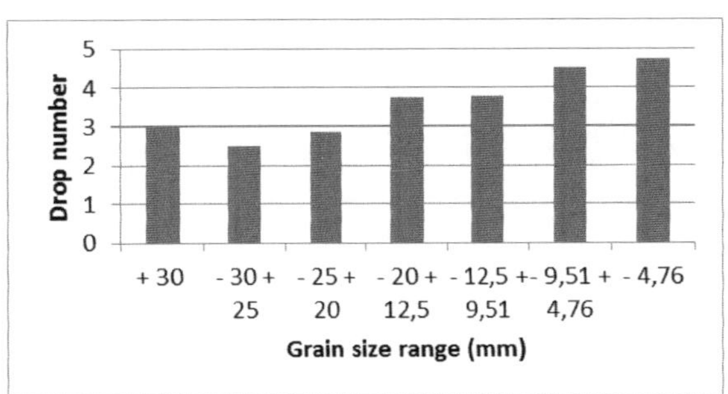

Figure 2. Results of drop test.

Endurance tests of wet aggregates were made through Chatillon & Sons trademark press machine. Crushing strength test results of light-weight aggregates consisting of 88 % fly ash and 12 % lime are given in Figure 3.

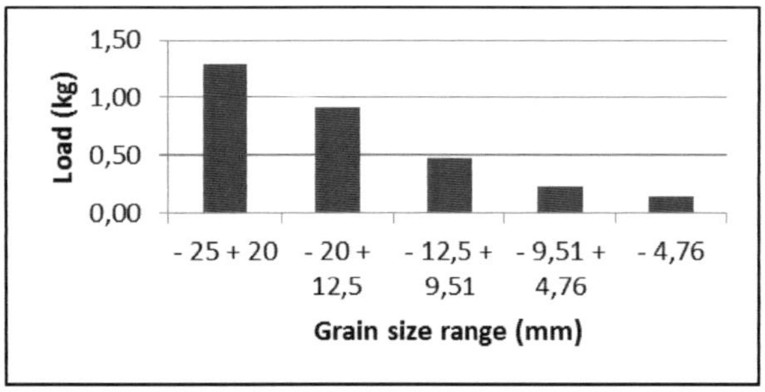

Figure 3. Results of wet strength test.

Pellets made up of 5 different mixtures were put into the curing cabinet model TK 252 with Nuve trademark operating at relatively 95 % moisture and 50 °C temperature.

Endurance tests of 1, 3, 7, and 14 days for light-weight aggregates in the curing cabinet were performed with the hydrolic press device of 5 tons.

The optimum mixture giving the highest endurance in pressure endurance tests were determined from the chart in Figure 4.

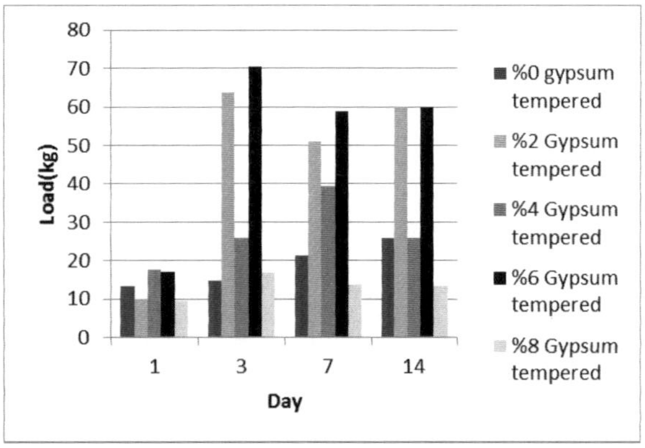

Figure 4. Comparing the results of compressive strength test with lightweight aggregates

To test the water absorption ability, samples are kept in water for twenty-four hours. Later, the water absorption rate was calculated. The sample was determined to have 42.16 % water absorption rate.

XRD analysis of 5 different pellets produced were made. Ettringite and Tobermorite (Oyelit) formations are expected to be seen in the XRD results of the pellets of high endurance. As the XRD results are compared, many differences were observed. Ettringite and oyelite formations were seen in the sample with 6 % gypsum content at most. Gypsum and lime completely reacted and formed other mineral phases in the sample with 6 % gypsum and 6 % lime content.

In order to find the real density of the pellets produced under optimum process conditions, measurement were made using Micromeritics AccuPyc II 1340 device. The real density of the most enduring light-weight aggregates were determined to be at 2.2089 g/cm^3. Using the results of water absorption tests, after finding the volume of the pores in the pellet, unit bulk density of the lightweight aggregates were calculated as 1.1436 g/cm^3. This value is between 800 kg/m^3 and 2000 kg/m^3 and meet the light-weight aggregate conditions.

SEM analysis was performed by using Superprobe 733 model device with the trademark Joel. In order to examine the samples, firstly, samples appropriate for the device were prepared. The sample should be conductive in order to make examination with SEM device. Thus, the samples were made C coating. Later on, observations were made at 10 kV power and with 1000x, 2000x, 4000x, and 6000x magnifications in the SEM device. Ettringite and tobermorite formations were observed in the SEM image.

In Figure 5, SEM images of light-weight aggregates obtained from the mixture without gypsum and the mixture with % 6 gypsum are compared. Entringite and tobermorite formations (acicular crystals) were observed in the SEM image of the light-weight aggregate including 6 % Gypsum.

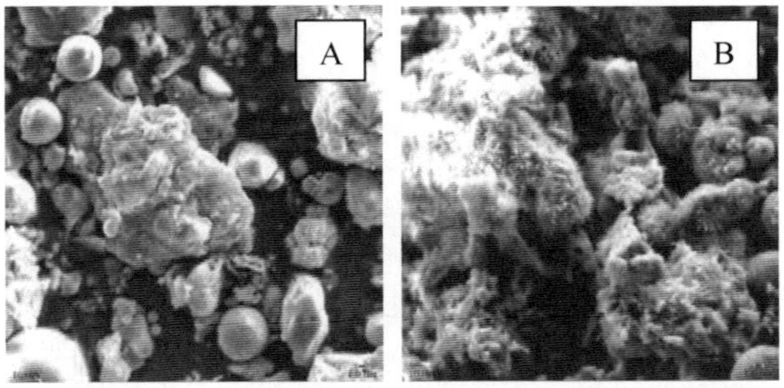

Figure 5. A) SEM images of lightweight aggregates obtained from the mixture without gypsum in 2000x magnification, B) SEM images of lightweight aggregates obtained from the mixture with % 6 gypsum in 2000x magnification.

3.3 Production of light-weight concrete

After determining optimum production parameters for the pellets, light-weight concrete tests were performed. Screen analysis of the pellets used in the production of light-weight concrete is given in Table 3.

Table 3. Sieve analysis of the pellets.

Grain size (mm)	Wt (%)
-12,5+9,51	31,11
-9,51+4,76	42,94
-4,76+3,35	19,98
-3,35+2	3,16
-2	2,81
Total	100

The sizes and the number of concrete specimens for the mechanical strength tests are given in Table 4.

Table 4. The sizes and the number of samples

Test types	The sample sizes	The sample number
Compressive Strength Tests	70*70*70 mm cube	45
Flexural Strength Tests	50*50*300 mm prism	45

3.3.1 Compressive strength tests

In this experimental study, 3 different amounts of fly ash and cement mixture were obtained with the produced lightweight aggregates (Fig 6).

Figure 6. The mixture of fly ash, cement and lightweight aggregate.

The contents of the concrete are shown in Table 5.
The concretes obtained were prepared by using mallet with the process of placing the concrete in the molds (Figure 7).

Table 5. The amounts of samples.

The mixture number	1 Wt (%)	2 Wt(%)	3 Wt (%)
Lightweight Aggregate	55	55	55
Portland Cement	16	20	24
Seyitömer Fly Ash	29	25	21

Figure 7. Light-weight concrete mixture to fill the mold.

Pressure endurance tests of the hardened concrete samples at the ages of 3, 7, 14, and 28 were made according to the standards of TS EN 12390-3. Pressure endurance test was made with a hydrolic press (Fig 8).

Figure 8. Putting the samples into the press machine.

In the pressure endurance tests, the mixture indicating the highest endurance was determined from the graph in Figure 9. As seen from Figure 9, the compressive strength on the 28th day are higher than those on the other days. It was also determined that number 2 mixture was more durable than the other mixtures.

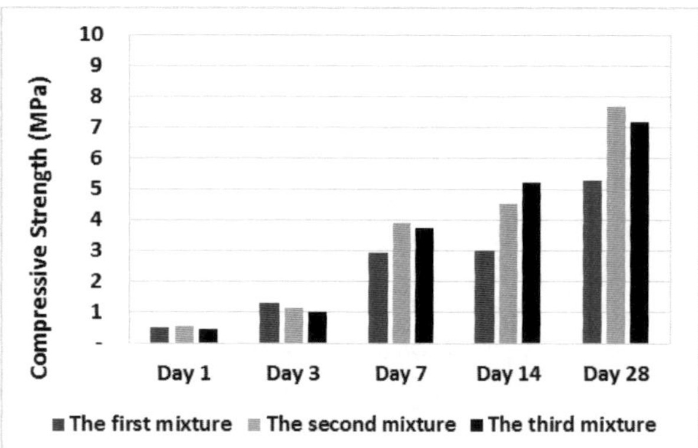

Figure 9. Comparing the compressive strength test results of light-weight concretes.

3.3.2 Flexural strength tests

The specimens for the flexural strength tests were obtained by using prism-shaped molds. (Fig 10).

Figure 10. Light-weight concrete mixture to fill the mold.

The samples were exposed to flexural strength tests after removing the concretes at the daily ages of 1, 3, 7, 14 and 28 from their molds carefully. Flexural strength test were carried out by using a hydrolic press machine with a capacity of 5 tonnes (Fig 11).

Figure 11. Putting the samples into the press machine.

The mixture indicating the highest flexural strength was determined from the graph in Figure 12. As it is seen from Figure 11, the results of the flexural strength on the 28th day are higher than those on the other days. It was also determined that number 2 mixture was more durable than the other mixtures.

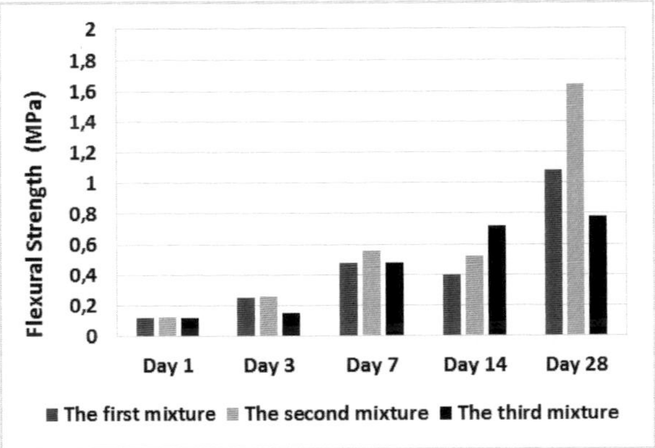

Figure 12. Comparing the flexural strength test results of light-weight concretes.

Fracture surface of lightweight concrete samples after flexure strength test is shown in Figure 13.

Figure 13. Fracture surface of light-weight concrete samples after flexure strength test.

4. CONCLUSIONS

The type of fly ash of Seyitömer power plant falls into the F class (low lime) fly ash according to ASTM C, whereas according to TS EN 197-1 it can be classified as V class.

The hardening mechanism of the fly ash/lime/gypsum pellets under moist conditions based mainly on formation of ettringite and tobermorite.

According to the results of endurance tests, aggregates produced with 88 % fly ash, 6 % gypsum, 6 % lime mixture were determined to be the most enduring light-weight aggregates with 65 kgf in average.

Light-weight aggregates showed a water absorption rate of about 42 %, due to their high amount of open pores. However there is no limit value of condition for water absorption of lightweight aggregates.

It was determined that the real density of the light-weight aggregates was 2.2089 g/cm^3. Unit volume weight of the aggregates was determined as 1.1436 g/cm^3.

It was determined that the mean wet unit volume weight of the light-weight concrete was 1.47 g/cm^3 while the mean dry unit volume weight was 1.25 g/cm^3. Therefore, this result meets the condition of light-weight concrete according to the standards of TS EN 206-1.

According to compressive and flexural strength tests, it was determined that the concrete mixture showing the highest endurance was light-weight concrete produced with 20 % cement addition.

The comparison of the produced light-weight concretes with 20 % cement addition and the features of the light-weight concretes according to the TS EN 206-1 are shown in the below given table. According to the standards of TS EN 206-1, light-weight concrete produced in this study belong to lightweight concrete category D1,4 and LC 8/9.

	Unit volume weight	**Compressive strength**
Produced light-weight concrete	1250 kg/m^3	7,8 MPa
Standards of light-weight concrete	TS EN 206-1	TS EN 206-1
Standard values	>800 kg/m^3 and <2000 kg/m^3	Min 9 MPa
Category	D 1,4	LC 8/9

In order to increase the formation of ettringite and tobermorite, and hence the mechanical strength of the fly ash aggregates high temperature curing can be recommended.

REFERENCES

ACI Committee 213R-03 (2003). Guide for structural lightweight aggregate concrete, American Concrete Institute, ACI.

Ağar, E. & Taşdemir, Y. (2007). Silindirle sıkıştırılabilen beton yollar, Türkiye hazır beton birliği, Retrieved May 10, 2015 from, http://www.thbb.org/Article.aspx?ID=4.

Ağar, E. (2007). Seyitömer termik santral uçucu küllerinin yapı sektöründe kullanılabilirliğinin araştırılması, Msc Thesis, Dokuz Eylül University, İzmir.

Aruntaş, H.Y. (2006). Uçucu küllerin inşaat sektöründe kullanım potansiyeli, Gazi Üniversitesi, Mühendislik Mimarlık Fakültesi Dergisi, 21, 1, 193-203.

ASTM C 618 (1994). Specification for fly ash and raw or calcined natural pozzolan for use as a mineral admixture in portland cement concrete, Annual Book of ASTM Standards.

Çiçek, T. & Tanrıverdi, M. (2007). Lime based steam autoclaved fly ash bricks, Construction and Building Materials, 21, 6, 1295-1300.

Çiçek, T. & Tanrıverdi, M. (2014). Buharlı otoklavda uçucu kül ve kireçten tuğla üretiminde optimum proses şartlarının pilot çapta belirlenmesi ve endüstriyel boyutta uygulamasının araştırılması, Tübitak Projesi, Program kodu 1001, Proje No. 111M694.

Koçkal, N. U. & Özturan, T. (2007). Sinterleme sıcaklığının uçucu kül hafif agregaların özelliklerine etkisi, 7. Ulusal Beton Kongresi, Maya Basın Yayını, 133-143.

TS EN 206-1 (2002). Beton- Bölüm 1: Özellik, performans, imalat ve uygunluğu, Türk Standartları Enstitüsü, Ankara.

Türkiye Hazır Beton Birliği (2016). Hafif agrega, 20 Ekim 2016, http://www.thbb.org/teknik-bilgiler/agrega/.

Influence of Ball and Powder Filling Ratios (J_b and f_c) on the Dry Grinding Process of Calcite

Kalsitin Kuru Öğütme Prosesinde Bilye ve Malzeme Doluluk Oranlarının (J_b ve f_c) Etkisi

M. Uçurum

Industrial Engineering Department, Bayburt University, 69000, Bayburt, Turkey

ABSTRACT Turkey has important quantity of industrial minerals and calcite is one of them and is used as filler in the industries such as plastics, rubber, and paint, to gain a variety of features to products. In order to use calcite as filler, some specific properties such as fine product size are required. In the world, conventional ball mill technology is used to obtain calcite with micronized on the industrial scale. In the mill, ball filling ratio (J_b) and powder filling ratio (f_c) directly affect product size, energy consumption, and grinding costs. In this study, influence of important operating parameters such as ball and calcite filling ratios (J_b and f_c) on the grindability of calcite ore was examined to obtain a fine product (<100 μm) in the conventional ball mill. The batch test results show that the parameters play an important role in the fine grinding process of calcite.

ÖZET Türkiye önemli miktarda endüstriyel minerallere sahip olup bunlardan birisi de plastik, kauçuk, ve boya gibi endüstriyel ürünlere çeşitli özellikler kazandırmak amacıyla kullanılan kalsittir. Kalsitin dolgu minerali olarak kullanılması için mikronize boyutlara öğütülmesi gibi bazı spesifik özelliklere sahip olması gerekmektedir. Dünyada endüstriyel boyutta mikronize kalsit üretimi için konvansiyonel bilyeli değirmen teknolojileri kullanılmaktadır. Konvansiyonel bilyeli değirmenlerde bilye doluluk oranı (J_b) ve malzeme doluluk oranı (f_c) ürün boyutunu, enerji tüketimini ve öğütme maliyetini direk olarak etkilemektedir. Bu çalışmada, önemli öğütme parametrelerinden olan bilye ve malzeme doluluk oranlarının konvansiyonel bilyeli değirmende kalsitin ince öğütülmesine (<100 μm) olan etkileri araştırılmıştır. Kesikli deney sonuçları kalsitin ince öğütülmesinde söz konusu parametrelerin önemli rol oynadığını göstermiştir.

1 INTRODUCTION

Ball mills are the most common and versatile type of tumbling mill. They are remarkable in that they can operate over a very wide range of conditions and geometries. Ball mills may be used for primary, secondary, tertiary and regrind applications. In the 1950s and 1960s ball mills dominated primary grinding, however they were supplanted in the 1970s by AG and SAG mill–which also replaced two stages of crushing. Nowadays ball mills still dominate secondary grinding applications, but are facing increasingly in competition with closed circuit AG/SAG mill circuits and stirred mills (Napier-Munn, et. al., 1996). At product sizes finer than 80% passing 75 μm, the efficiency of ball mill grinding rapidly decreases. The practical limit to ball mill product fineness is considered to be 40–45 μm (Gao and Weller, 1994). Although ball mills can still be found in the regrind circuits in a number of mineral processing plants, their energy consumption is high, and size reduction efficiency is low. The need to grind the ore finer to liberate the minerals and the dramatic increase in the cost of energy in the mid-1970s resulted in more efficient comminution processes being sought (Shi et. al., 2009).

The energy formed in tumbling ball mills is limited by the number of revolutions at which the grinding media are centrifuged in the mill wall, therefore, the energy intensity is relatively small and it is necessary for a much long grinding time for the production of fine particles (Jankovic and Valery, 2004).

Industrial grinding machines used in the mineral processing industries are mostly of the tumbling mill type. These mills exist in various forms, e.g., ball, rod, pebble autogenous, and semi-autogenous. Balls and rods are the traditionally used media shapes in tumbling mills (İpek, 2007). Ball mills are typically operated close to their maximum power draw. At the maximum power draw, it is assumed that the ball charge is well mixed and that the void spaces between the balls are filled with particles. In reality, however, particles can influence the ball charge in various ways, causing the maximum power draw to shift depending on the nature of the influence. The ball load contributes to the bulk of the charge mass; consequently, a change in the location of its center of gravity significantly affects the power drawn by the mill. It is therefore worthwhile to study the behavior of the ball charge and the influence of particles on it. From such a study, one can infer the conditions within the charge that lead to maximum power draw and optimal throughput. Ball mills are typically operated close to their maximum power draw. At the maximum power draw, it is assumed that the ball charge is well mixed and that void spaces between the balls are filled with particles. In reality, however, particles can influence the ball charge in various ways, causing the maximum power draw to shift depending on the nature of the influence. The ball load contributes to the bulk of the charge mass; consequently, a change in the location of its center of gravity significantly affects the power drawn by the mill. It is therefore worthwhile to study the behavior of the ball charge and the influence of particles on it. From such a study, one can infer the conditions within the charge that lead to maximum power draw and optimal throughput (Kiangi and Michael, 2008).

The energy input to a mill increases with the ball charge, and reaches a maximum at a charge volume of approximately 50% (Wills, and Munn, 2006).

Ball mills are relatively simple grinding devices from an operational perspective. The typical operational philosophy is to put as much grinding medium in a rotating tube as it can take and then introduce a feed material. If the feed material is softer than the grinding media, it will be reduced in size. If, after some time, the grinding medium is worn, then it is simply replaced (Fortsch, 2006).

In this study, the influence of important dry grinding parameters such as ball and powder filling ratios (J_b and f_c) on the production of fine calcite particles was investigated by using conventional ball mill.

2 MATERIAL-METHOD

The sample used in the experiments was obtained from Niğde calcite basin (Mertas¸ Co.) in Turkey. According to chemical analysis, the sample contains 54.25% CaO (Table 1). The sample was crushed to a nominal top size of -2 mm by using jaw and roller crusher as closed circuit with 2 mm screen. The particle size distribution of the crushed sample was determined by wet screening, and the obtained results are given in Figure 1. It can be seen that the d_{50} size is about 0.6 mm particle size.

Table 1. Some properties of calcite sample (www.mertasmikronize.com.tr)

Property	%
SiO_2	0.08
Cao	54.25
MgO	0.43
Fe_2O_3	0.01
Al_2O_3	0.02
CO_2	44.14
LOI.	43.11
Hardness	3 mosh
Density	2.7 g/cm3
Whiteness (%)	98.50
Brightness (%)	97.00
pH value	8.7
Moisture content	0.30
Oil absorbsion (%)	16.0
Dop oil absorbsion (%)	21.0

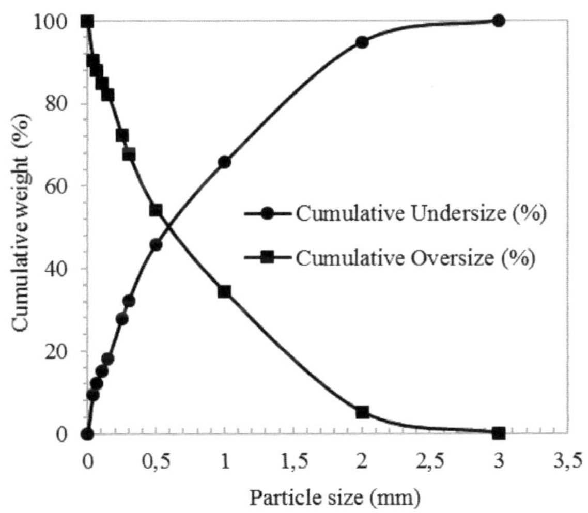

Figure 1. Particle size distribution of the crushed sample

The grinding machine was a laboratory-scale ball mill made of stainless steel with the dimensions 200×200 mm. The grinding media were stainless steel balls with a density of 8.000 kg/m^3 and with four different diameters (i.e., 1, 2, 3, and 4 cm). Powder density was found to be 2.7 g/cm^3. The grinding experiment was performed as a batch process; after each test, all media and ground samples were removed from the mill and the media separated from the products by dry sieving.

To investigate the effect of the experimental grinding parameters, including ball filling ratio (J_b), and powder filling ratio (f_c), on the micronized calcite product size, a series of experiments were conducted.

The critical speed (N_c), ball filling ratio (J_b), and powder filling ratio (f_c) were calculated using Equations 1, 2, and 3, respectively:

$$N_c = [42.3 / (D - d)^{0.5}] \text{ rpm } (D = \text{mill diam. and } d = \text{ball diam.; in meters}) \quad (1)$$

$$J_b = [(\text{mass of balls / ball density}) / \text{mill volume})] \times (1 / 0.6) \quad (2)$$

$$f_c = [(\text{mass of powder / powder density}) / \text{mill volume})] \times (1 / 0.6) \quad (3)$$

It is well known that the main function of time in the grinding process is to produce ultra-fine sized particles and controllable steepness factor (SF). The main objective of second section is to express the grinding SF in relation to the fragmentation mechanism, which could be used for process control. The granulometry or particle size distribution of an extender is characterized in general by the particle size distribution curve. The most important points are the ''top cut'' (d95), which gives the diameter of extender particle so that at least 95% of the particles by weight are finer, i.e., in principle the coarsest particle present, and the ''median particle size'' (d_{50}), which represents the size of particle so that 50% of the particles by weight are coarser or finer (Werner 2000). These terms are of particular importance when considering fine extenders. An additional property is the ''steepness'' of a particle size distribution curve, defined by the SF (Strauch 1998). This is the quotient of the d_{50} and the d_{20} values corresponding to the size of particle such that 20% by weight of those present are finer. The SF defines certain properties in the finished paint. The SF can be calculated from the particle size distribution curve of the relevant extender using Equation (4). A curve with SF greater than 2 is described as ''broad'' and that with SF less than 2 as ''narrow'' or ''steep.'' In other words, a fine extender with a broad particle size distribution curve contains a wider range of different size particles, e.g., from under 0.1 to 3 mm. An extender with a narrow particle size distribution curve contains particles within a smaller size range. Ideally, such an extender would be monodisperse or iso-diametric, i.e., all particles would be of the same size. In the following text, an extender with a very high proportion of particles within a very narrow particle size range will be described as CaCO3 ID (for iso-diametric) (Werner 2000).

$$SF=d_{50}/d_{20} \quad (4)$$

3 RESULTS

A series of grinding tests were carried out to determine the effect of ball filling ratio (J_b) on fine grinding of calcite and J_b values was varied between 0.20 and 0.45. The variables of grinding were kept constant as given in the Table 2. The results showed that Jb=0.45 ball filling ratio given the best results. As it can be seen from the results that fine calcite product containing d_{50}=45 µm and d_{80}=135 µm (Figure 2 and 3). At low Jb=0.20 ball filling ratio, d_{50} and d_{80} values was increased significantly as 115 and 300, respectively.

In the second phase of experimental studies grinding tests were carried out at the optimum conditions obtained from fine grinding studies investigating the effect of powder filling ratio (f_c) on the particle size and steepness factor (SF). The powder filling ratio was varied between 0.10 and 0.40 while other conditions kept constant (Table 3). The best results obtained at f_c=0.10 value. As it can be seen from the results that fine product containing d_{50}=12 µm, d_{80}=40 µm (Figure 4, 5) and SF=1.20 (Table 4). It was observed that the particle size and SF was increased at high f_c values.

Table 2. Experimental conditions of ball filling ratio

Condition	Variable
Operational speed (% of N_c)	80
Ball filling ratio (J_b)	0.20, 0.25, 0.30, 0.42, 0.45
Ball size distribution (1-2-3-4 cm)	30-30-20-20
Powder filling ratio (f_c)	0.20
Grinding aid ratio (% of powder)	0.25
Grinding time (min.)	15.0

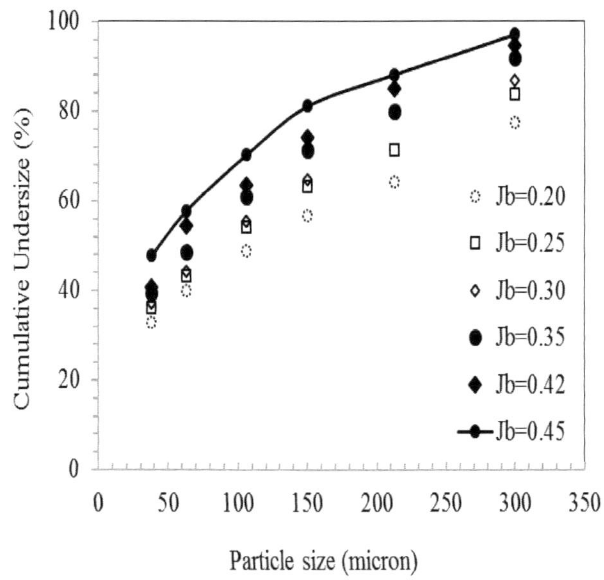

Figure 2. Particle size distribution of ball filling ratios

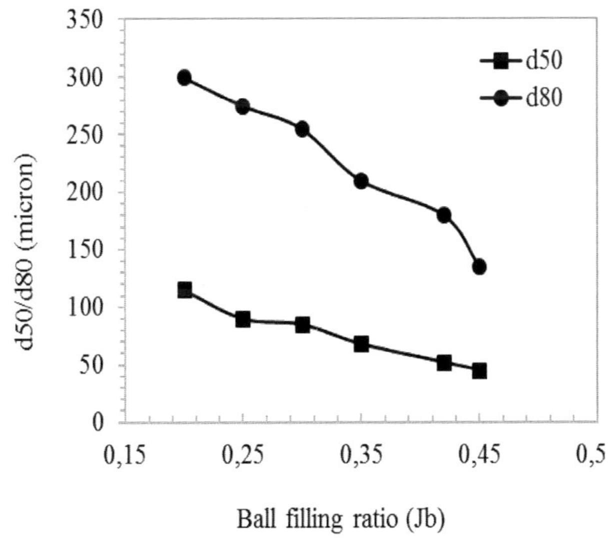

Figure 3. d50/d80 values of ball filling ratios

Table 3. Experimental conditions of talc filling ratio

Condition	Variable
Operational speed (% of N_c)	80
Ball filling ratio (J_b)	0.45
Ball size distribution (1-2-3-4 cm)	30-30-20-20
Calcite filling ratio (f_c)	0.1, 0.15, 0.20, 0.30, 0.40
Grinding aid ratio (% of powder)	0.25
Grinding time (min.)	15.0

SF values were calculated for the fc=0.1, 0,15, 0,20, 0,30 and 0.40 and feed calcite sample. Results are given in Table 4. The SF values decrease linearly from 7.0 down to 1.20 when the fc values decrease from 0.40 to 0.10 min.

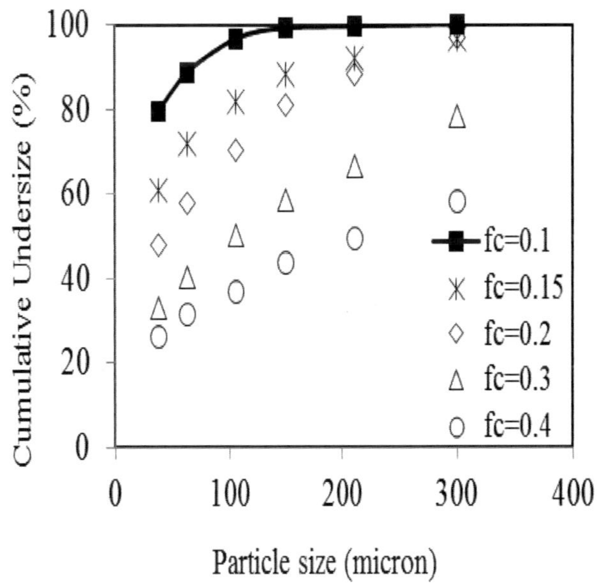

Figure 4. Particle size distribution of powder filling ratios

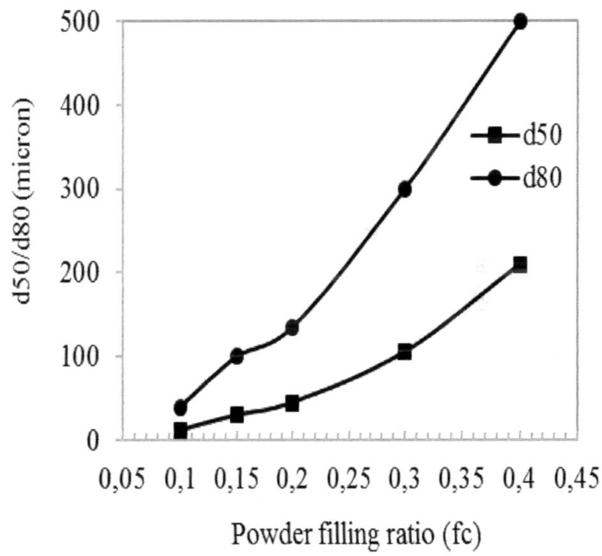

Figure 5. d50/d80 values of powder filling ratios

Table 4. SF values of fc values

Powder filling ratio (fc)	SF
0.10	1.2
0.15	2.0
0.20	3.4
0.30	4.0
0.40	7.0
Feed	7.5

The variation of mill filling can also influence power consumption. More energy is needed when there is an increase of the mass inside the mill. On the other hand, the variation of the center of mass as the percentage of mill filling changes also plays a major role. For greater mill fillings the center of mass is dislocated towards the mill center, reducing energy consumption (Francioli, 2015). It has been found that mill charge is one of the important operating parameters of the mill, which has to be measured and controlled (Kolacz, 1993). For a given balls, it is undesirable both to underfill and overfill the mill powder. At lower powder filling much of the energy of the tumbling balls is taken up in the ball to ball contact given low rate of breakage. The mill power is about the same whether the powder filling is low or normal, so a low rate of breakage gives a low energy of grinding. As the amount of powder is increased, the collision spaces between the balls and filled and higher rates of breakage are obtained. When all the effective spaces in which collision between tumbling balls are occurring are filled with powder, the rates of breakage reach a maximum. Further addition of powder increases the mill charge but does not give increased breakage because the collision zone is already saturated. A plateau of almost constant breakage rates is obtained. Eventually, overfilling lead to damping of the collision by powder crushioning and agglomeration, the powder bed expands to give poor collision effect, and the breakage rates decrease (Kolacz, 1995).

In the grinding tests, fine calcite product was achieved by using high ball filling ratio and lower powder filling ratio, therefore, this situation constitutes a question mark on the economic fine grinding of calcite ore.

Acknowledgements

The author would like to thank Ömer Güleç and Murat Cingitaş.

4 CONCLUSION

A laboratory scale ball mill was used for grinding of calcite ore. The influence of ball mill operating parameters such as ball filling ratio and calcite filling ratio, was systematically examined. The best results, obtained at optimum conditions in the ball mill experiments, are as follows: 0.45 for ball filling ratio (J_b) and 0.10 for powder filling ratio (f_c). The final product size (fineness) and SF were used in the evaluation of the test results. The SF of the calcite powder product at 0.10 powder filling ratio was very narrow (1.20), with average particle size of 12 μm, while more than 80% of particles were less than 40 μm.

REFERENCES

A Brief Introduction to Neural Networks. Available at http:// ww.mertasmikronize.com.tr/index.php/kalsit/ kalsit-nedir 380 (accessed on May 29 st., 2014).

Fortsch, D.S., 2006, Ball charge loading – impact on specific power consumption and capacity, IEEE-IAS Cement Industry Committee (http://ieeexplore.ieee.org/stamp/stamp.jsp?tp=& arnumber=1635707)

Francioli, D.M., 2015, Effect of operatıonal varıables on ball mıllıng, Universidade Federal do Rio de Janeira, Escola Poltecnica.

Gao, M.W., Weller, K.R., 1994, A comparison of tumbling mills and stirred ball mills for wet grinding. In: The Fifth Mill Operators' Conference, AusIMM, Roxby Downs, Australia, 16–20 October.

Ipek, H., 2007, Effect of grinding media shapes on breakage parameters, Part. Part. Syst. Charact. 24 229–235.

Jankovic, W. 2004,Valery, Fine and ultra fine grinding-the facts and myths, Proceedings of the 6th Annual I.I.R. Crushing and Grinding Conference, Perth

Kiangi K.K., Michael, H.M., 2008, Particle filling and size effects on the ball load behaviour and power in a dry pilot mill: Experimental study, Powder Technology 187 79–87.

Kolacz, J. 1993, Instrumentation and control to a high pressure roller crusher and air-swept ball mill circuit processing. Mineral Processing Conferences, Lulea, Sweden, February, 123-133.

Kolacz, J., 1995, Optimization of fine grinding systems employing air swept ball mill, Mineral Processing Recent Advances and Future Trends, Editors: SP Mehrotra, R. Shekhar, pp. 205215.

Napier-Munn, T.J., Morrell, S., Morrison, R.D. Kojovic, T., 1996, Mineral comminution circuits: their operation and optimization. ISBN 0 646 28861 x.Julius Kruttschnitt Mineral Research Centre.

Shi, F., Morrison, R., Cervellin, A., Burns, F., Musa, F., 2009, Comparison of energy efficiency between ball mills and stirred mills in coarse grinding, Minerals Engineering 22 673–680.

Strauch, D., P. Belger, H. Hofer, and M. Merz.. 1988, Pltiss-Staufer AG. U.S. Patent, 4,767,464.

Werner, R. 2000. Effect of extenders with narrow and broad particle size distributions on the properties of coatings. Journal of Coatings 440 Technology 72 (903):71–76.

Wills, B. A., and T. N. Munn. 2006, Mineral Processing Technology, Elsevier Science & Technology Books. ISBN: 0750644508.

Effect of Chemical Dosage on Color Parameters of Hydrophobic Calcium Carbonate Obtained in Aqueous Solution
Yaş Ortamda Elde Edilen Hidrofobik Kalsiyum Karbonatın Renk Parametreleri Üzerinde Kimyasal Dozajın Etkisi

M. Uçurum
Industrial Engineering Department, Bayburt University, 69000, Bayburt, Turkey

ABSTRACT Ultra-fine ground calcium carbonate is widely utilized as a performance mineral in rubber, plastics and paper industries. Both micronized calcium carbonate and a chemical (sodium oleate) treated calcium carbonate products, obtained in the aqueous solution, were used in the experiments. The color property is an important factor that affects the acceptability of the industrial mineral products. The current study describes the relationship between micronized calcium carbonate and coated calcium carbonate on the basis color parameters (L^*, a^*, b^*), whiteness index (*WI*), saturation index (*SI*), hue angle (*H*) and total color difference (ΔE). The results showed that the processing variables (chemical dosage) did not have a significant effect on the color parameters of hydrophobic calcium carbonate products.

ÖZET Çok ince öğütülmüş kalsiyum karbonat kauçuk, plastik ve kâğıt endüstrisinde performans arttırmak amacıyla geniş bir şekilde kullanılmaktadır. Deneysel çalışmalarda, mikronize kalsiyum karbonat ve bir kimyasal (sodyum oleat) kullanılarak yaş ortamda elde edilmiş kaplı kalsiyum karbonat kullanılmıştır. Endüstriyel mineraller ürünlerinin kabulünde renk özellikleri önemli bir yere sahiptir. Bu çalışmada; mikronize ve kaplı kalsiyum karbonatın renk parametreleri (L^*, a^*, b^*), parlaklık (R_{457}), beyazlık indeksi (*WI*), doygunluk indeksi (*SI*), hue açısı (*H*) ve toplam renk farkı (ΔE) çalışılmıştır. Çalışmaların sonunda, kimyasal dozajının kaplı kalsiyum karbonatın renk değerlerinde bir değişime yol açmadığı tespit edilmiştir.

1 INTRODUCTION

Calcite is the most widely used white pigment and filler because of its brightness and high $CaCO_3$ percentage, high whiteness degree, very low ratio of silica and iron in impurities, in which it is surpassed only by a few other materials. In general, calcium carbonate is used in various areas of plastics, textiles, rubbers, adhesives, paint sand waste water treatment. On the other hand, calcium carbonate is also an important product of the chemical industry. Therefore, many studies have already been performed on the crystallization kinetics and processes of calcium carbonate (Sayan, 2005). Finely ground calcium carbonate has been widely applied as filler material. However, it is essential to make the hydrophilic calcium carbonate particles compatible with the matrix, such as plastics, rubber and adhesives (Ding et. al., 2007). Surface modification of fillers to give improved properties to a polymer composite is a topic that has received enormous attention over the last 30 years. Improvements in mechanical properties, dispersion of the filler (which leads to improved properties), improved rheology and higher filler loading have all been reported to accrue by rendering the surface more hydrophobic and hence compatible with the polymer or by enabling the filler to bond covalently, through hydrogen or ionic bonds to the polymer; or by changing the physical nature of the interface so that energy absorption can occur (Roger and Hancock, 2003).

The aspect and color of the calcite surface is the first quality parameter evaluated by consumers and is critical in the acceptance of the product. Industrial applications require specific properties and characteristics. Among the most valuable characteristics is the color, a function of the parent rock and its alteration (Murray, 2002). Color is an easily perceived physical characteristic of materials, and wields great importance for manufacturers (Soriano et. al., 1998). In polymineral natural samples with complex crystallochemistry, the study of color is more complicated than in minerals of high purity (or even synthetic ones) where diffuse reflectance spectroscopy techniques are employed (Burns, 1993). The intrinsic reflectance factor (SO brightness, R_{457}) measured with a reflect meter having the characteristics described in ISO 2469, equipped with a filter or corresponding function having an effective wavelength of 457 nm and a width at half-height of 44 nm, and adjusted so that the UV-content of the illumination incident upon the test piece corresponds to that of the CIE illuminant C (Scan-CM 59:01, 2001). Color science seeks to link the fundamental properties of light and matter to our perception of color and our ability to capture and generate color. As such, color science has long been important to a wide range of manufacturing industries (Hunt, 1998). Numerous systems exist for the systematization-objective measurement of colors. The most important are: Commission International de l' Eclairage (CIE), Deutche Institut fur Normung-German Standard Institute (DIN), Natural Color System (NCS). Ostwald's and Munssel's systems. CIE and DIN chromatic systems are scientifically recognized as the basis for numerical evaluation of colors and the calculation of differences between them. Modern colorimetry is based on the Grassmann's laws of additive mixing (Zarubica, 2005). An organization called Commission Internationaledel'Eclairage (CIE) determined the standard values that are used worldwide to measure color. The values used by CIE are called L*, a*, b* and the color measurement method is called CIELAB. Symbol L* represents the difference between light (where L* = 100) and dark (where L* = 0). a* represents the difference between green (−a*) and red (+a*), and b* represents the difference between yellow (+b*) and blue (−b*). Using this system any color that corresponds to a place on the CIELAB coordinate system was shown in Figure 1 (Sharafudeen, 2012).

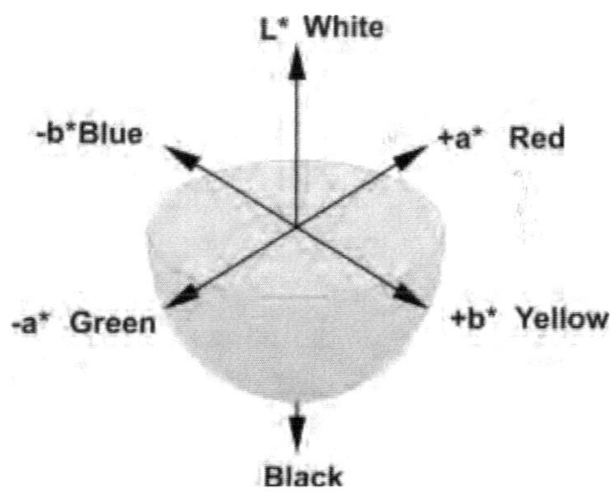

Figure 1. The CIELAB color space system (Camelo and Gómez, 2004)

The hue angle is traditionally measured starting at the direction corresponding to pure red. The simplest way to derive an expression for this angle is to project the vector (1; 0; 0) corresponding to red in the RGB space and an arbitrary vector c onto a plane perpendicular to the achromatic axis, and to calculate the angle between them. For the derivation of an expression for the saturation of an arbitrary color c, we begin by looking at the triangle which contains all the points with the same hue as c, as shown in Figure 2. The intersection of this triangle and the iso-brightness surfaces are lines parallel to the line between c and its brightness value on the achromatic axis L(c) = [L (c); L (c); L (c)]. Traditionally, the saturation is calculated as the length of the vector from L(c) to c divided by the length of the extension of this vector to the surface of the RGB cube. Moreover, it is clear that this definition of the saturation depends intimately on the form of the brightness function chosen (i.e. on the slopes of the iso-brightness lines) (Hanbury, 2002).

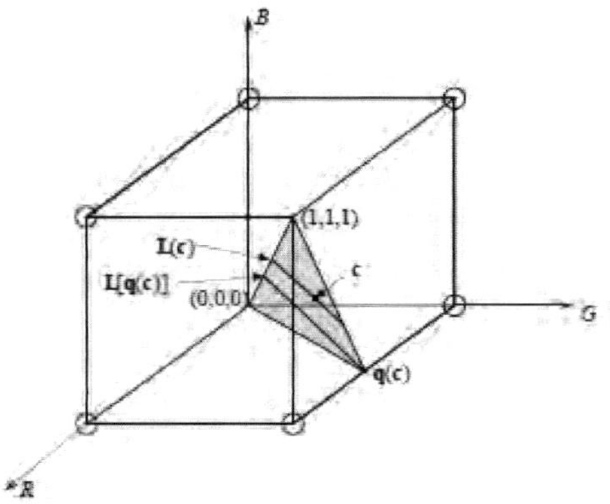

Figure 2. The triangle which contains all the points with the same hue as c. The circled corners mark the edges of the cube containing the points furthest away from the achromatic axis (Hanbury, 2002).

A whiteness index (WI) has been described based on the distance of a color value from a nominal white point, represented in CIELAB color space as L*=100, a*=0 and b*=0. In spectral terms a white material is one whose reflectance across the visible wave length range is constant and high (i.e. close to 100% or reflectance factor of 1). Varying shades of gray to black have a constant reflectance with the perfect black having a reflectance of 0% (Saricoban and Yilmaz, 2010). The CIELAB, or CIE (1976) L*a*b* values have a perceptual meaning: L* is the lightness which relates to the physical intensity of a color, whilst a* and b* are coordinates on the red–green and yellow–blue color axes respectively. This scheme is designed such that a constant difference in color, ΔE, defined by the Euclidean distance. It should give a constant 'perceived' color difference-regardless of the location in the color space. The smallest perceivable difference for two colored patches contacting one another is approximately 0.5–1.0 ΔE units (Joiner et. al., 2008)

Arithmetic mean or simply the mean of a variable is defined as the sum of the observations divided by the number of observations. If the variable x assumes n values X_1, X_2 ... X_n then the mean, x, is given by:

$$X_{ort.}=(X_1+X_2+X_3+\ldots\ldots\ldots X_n)/n \qquad (1)$$

Equation 1 is for the ungrouped or raw data. Karl Pearson introduced the concept of standard deviation in 1893. It is the most important measure of dispersion and is widely used in many statistical formulae. Standard deviation is also called Root-Mean Square Deviation. The reason is that it is the square–root of the mean of the squared deviation from the arithmetic mean. It provides accurate results. Square of standard deviation is called Variance. It is defined as the positive square-root of the arithmetic mean of the Square of the deviations of the given observation from their arithmetic mean. The standard deviation (SD), given Equation 2, is denoted by the Greek letter (σ). The coefficient of variation (Equation 3) is obtained by dividing the standard deviation by the mean and multiply it by 100.

$$\text{Standard Deviation } (\sigma)=[\Sigma X^2/n]^{1/2} \qquad (2)$$

$$\text{Coefficient of variation (CV)} = [(\sigma/X_{ort}]*100 \qquad (3)$$

If we want to compare the variability of two or more series, we can use CV. The series or groups of data for which the CV is greater indicate that the group is more variable, less stable, less uniform, less consistent or less homogeneous. If the CV. is less, it indicates that the group is less variable, more stable, more uniform, more consistent or more homogeneous (Homco, 2004).

The color is an important factor of calcium carbonate products especially after a chemical processing. The colors of the products are affected by a variety of parameters such as chemical dosages. The objective of this study was to follow the changes of basis color parameters (L*, a*, b*), brightness (R_{457}) whiteness index, saturation index, hue-angle and total color difference on the surface of hydrophobic calcium carbonate products which were obtained under different sodium oleate (SO) to calcium carbonate mass ratio (0.005, 0.010, 0.015, 0.020 and 0.030).

2 MATERIAL

There are two samples in the study, one is calcium carbonate powder (CC), and the other is chemical treated calcium carbonate (CTCs) which is hydrophobic products. The CC sample is taken from Niğde/Turkey and its main properties are given in Table 1. The modifier reagent were used such as sodium oleate (SDO) obtained from Sigma. Sodium oleate (SDO) has very different color parameter values from calcium carbonate powder, especially in terms of b*. The characteristics of the chemical reagent used in the study are listed in Table 2. In the experiments, distilled water was used.

2.1 Method

In the surface treatment, calcium carbonate powder+ chemical + water were prepared in a glass vessel at room temperature and then the mixture was mixed by a mechanical stirrer at 2000 rpm throughout the duration of the experiments. The duration of each experiment was set to be 15 minutes. Samples were removed from the reaction vessel and filtered through a filter. Chemical treated calcite products, hydrophobic calcium carbonate, were washed with distilled water and then they were dried at 50 °C in the oven.

Table 1. Main properties of calcium carbonate (CC)

Purity (%)	Brightness	Density (kg/m^3)	Size (d_{50}) (µm)	Specific surface area (m^2/cm^3)	Color Parameters			
					R_{457}	L*	a*	b*
>98.00	96.43	2.70	2.99	3.13	95.00	98.60	0.06	1.05

Table 2. Properties of chemical reagent

Denomination	Molecular Formula	Molecular Weight (g/mol)	Color Parameters			
			R_{457}	L*	a*	b*
Sodium Oleate (SDO)	$C_{18}H_{33}NaO_2$	304.44	79.45	96.45	-0.70	9.55

The dosage of SDO was chosen different sodium oleate (SO) to calcium carbonate (CC) mass ratio (0.005, 0.010, 0.015, 0.020 and 0.030) and pulp density used in all the experiments was 15%. Four tests were carried out for each mass ratio. In all calculations and display, the average of the four values was used.

Floating test was performed as first characterization study. It is described by Sheng et.all., (2006) as follow. Floating test is to measure the ratio of floated product to the overall weight of sample after mixing in water and stirring vigorously. The ratio is called active ratio (AR). The higher the active ratio, the better the modification effect is. The test is formulated as follows;

$$AR(\%) = \frac{Mp}{Mp + Mt} * 100 \qquad (4)$$

where AR (%) is the active ratio, M_f the mass of the floated product and M_s the mass of the sink product (Sekulic et.al., 2009).

Color changes of original calcium carbonate (CC) and CTCs were measured with colorimetric CIE L*a*b* method using a Datacolor Elrepho450 xs. Whiteness index (WI), saturation index (SI), hue angle *(H)* and total color difference *(ΔE)* were calculated using measured Equation 5,6,7 and 9, respectively by using *L*, a** and *b** values (Maskan, 2001; Homco et. al., 2004).

$$WI=100-[(100-L^*)^2+a^{*2}+b^{*2}]^{0,5} \qquad (5)$$

$$SI=[a^{*2}+b^{*2}]^{0,5} \qquad (6)$$

$$H=\arctan(b^*/a^*) \qquad (7)$$

$$\Delta E=[(L_0-L^*)^2+(a_0-a^*)^2+(b_0-b^*)^2]^{0,5} \qquad (8)$$

where subscript "o" refers to the color reading of control calcium carbonate sample used as the reference and a larger *ΔE* indicates greater color change from the reference calcium carbonate sample.

3 RESULTS

In the present study, the floating test was followed to evaluate the effect of surface modification and active ratio was determined for each obtained CTCs. The average active ratio of the products was shown in Figure 3. As can be seen, from Figure 3, that the average active ratio increases gradually depending on SO to CC mass ratio. At lower mass ratio (0.005), the active ratio was 85.50%. The value was achieved almost 100% AR with 0,010 and after. Since is reported in literature that a greater active ratio implies better surface modification effectsn (Ding et. al., 2007; Wu and Lu, 2003). The results exhibit that with the applied solution method, high hydrophobicity of the calcium carbonate surface was reached with sodium oleate modifier. Sayan (2005) obtained that Na-Oleate chemically adsorbs on the $CaCO_3$ surface, which interact with oleate ions, giving the surface Ca-Oleate product. Ca-Oleate is water insoluble and remains on the $CaCO_3$ surface. This result is a good agreement with the result obtained by Kawashima (1986). This chemical relation can be written according to Eq.9 (Sayan 2005).

$$CaCO_3+2C_{17}H_{33}COONa \longrightarrow (C_{17}H_{33}COO)_2Ca+Na_2CO_3 \qquad (9)$$

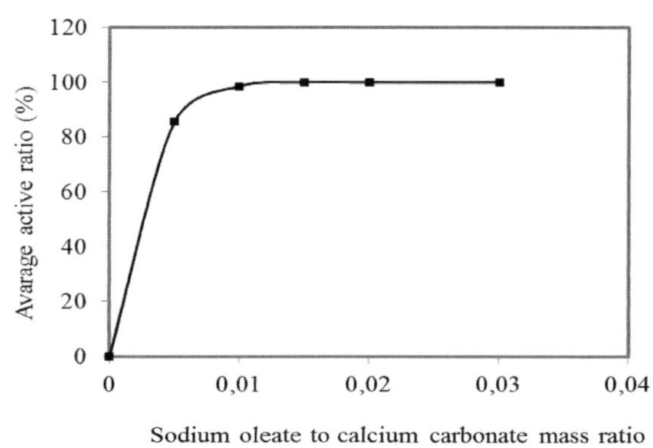

Figure 3. The effect of sodium oleate to calcium carbonate mass ratio on the active ratio of CTCs

As can be seen in Table 3, the Arithmetic mean, Standard deviation and Coefficient of variation of R_{457}, L*, a* and b* color parameters of CTCs products are noted. The values make it clear that there has not been a considerable change in Standard deviation and coefficient variation of R_{457}, L* and b* values of the products belonging to each mass ratios. However, regarding the value of a*, it can be said that the statistical parameters vary in a wide range. The reason for this can be explained as the value a* having very small measurement results and its variation range being high in itself.

Table 3 shows variation of brightness (R_{457}) value as a function of SDO to CC mass ratio between 0.005 and 0.03. The results observed are generally close to each other and almost not dependent of mass ratio. As it can be seen from the results that CTCs products have almost the same values with CC in terms of brightness. The lightness (*L**) value of the calcium carbonate sample (CC) was 98.60; as expected, after chemical process with SDO changed the *L** values, which was due to chemical adsorption between calcium carbonate and chemical. Especially, the chemical is known to influence the color properties of calcium carbonate products. *L** values of CTCs shows that the L* of CTCs products decrease in a very small value with increasing chemical dosage, apparently going through a maximum at 0.03 mass ratio. However, as a result, chemical dosage does not appear to cause large changes of the L* value of the calcium carbonate. Because the value of calcium carbonate (CC) is 98.60 and the parameter was decreased up to 98.56 for CTCs.

Table 3 indicates that an increase in SDO dosage decreased the redness (*a**) of CTCs up to 0.03 mass ratio from 0.06 to 0.02. On the other hand, products yellowness (*b**), very important parameters of hydrophobic calcium carbonate, was affected by chemicals dosage. As can be seen in the table, yellowness (*b**) increased as the SDO dosages increased. Therefore, chemical dosage had a partial effect on the yellowness of the CTCs. Thus, as SDO to CC mass ratio increased from 0.005 to 0.03, the *b** increased from 1.05 to 1.23.

Experimental results are plotted in Figure 4 which indicates a slightly decrease in WI with increasing sodium oleate to calcium carbonate mass ratio between 0.005 and 0.03 from 98.22 to 98.10. As seen in Figure 5, the saturation index (SI) increased with increasing chemical amount for CTCs products. The highest SI of CTCs was achieved with 0.03 mass ratio.

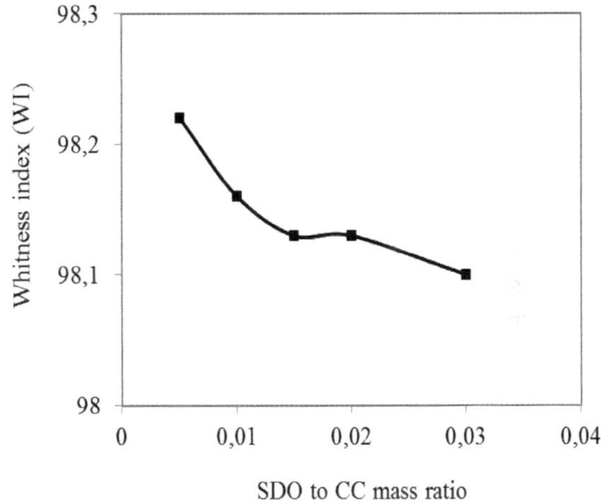

Figure 4. Whiteness index (WI) of experiments

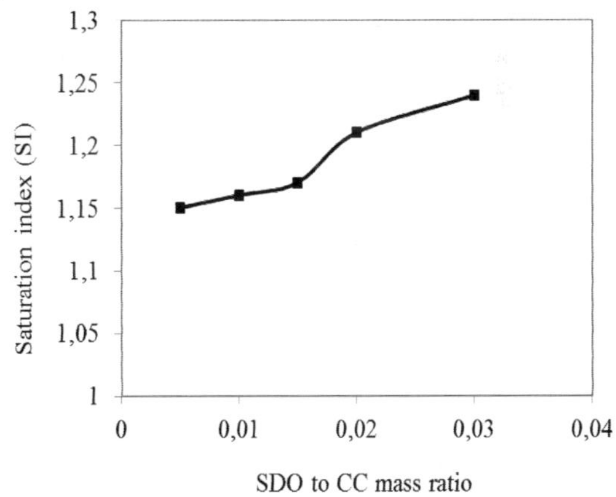

Figure 5. Saturation index (SI) of experiments

Table 3. Four experimental results of each sodium oleate to calcium carbonate mass ratio

0.005 SDO to CC mass ratio	R_{457}	L*	a*	b*	0.010 SDO to CC mass ratio	R_{457}	L*	a*	b*
1	95.09	98.67	0.05	1.11	1	94.66	98.52	0.01	1.16
2	94.79	98.56	0.06	1.13	2	94.92	98.60	0.04	1.11
3	94.97	98.61	0.04	1.09	3	94.87	98.60	0.05	1.15
4	94.89	98.58	0.04	1.10	4	94.68	98.55	0.04	1.16
Mean	94.94	98.60	0.05	1.15	Mean	94.78	98.57	0.04	1.15
SD	0.123	0.05	0.01	0.017	SD	0.112	0.039	0.017	0.024
CV (%)	0.130	0.05	20.00	1.50	CV (%)	0.118	0.04	42.50	2.09

0.015 SDO to CC mass ratio	R_{457}	L*	a*	b*	0.020 SDO to CC mass ratio	R_{457}	L*	a*	b*
1	94.70	98.54	0.04	1.17	1	94.79	98.60	0.02	1.21
2	94.80	98.57	0.03	1.15	2	94.67	98.54	0.02	1.19
3	94.61	98.51	0.05	1.18	3	94.72	98.57	0.03	1.21
4	94.69	98.54	0.04	1.18	4	94.71	98.55	0.03	1.18
Mean	94.70	98.54	0.04	1.17	Mean	94.72	98.57	0.02	1.20
SD	0.05	0.025	0.08	0.014	SD	0.05	0.025	0.08	0.013
CV (%)	0.05	0.025	200.00	1.20	CV (%)	0.05	0.025	400.00	1.07

0.030 SDO to CC mass ratio	R_{457}	L*	a*	b*
1	94.64	98.55	0.01	1.24
2	94.69	98.56	0.02	1.22
3	94.73	98.58	0.01	1.22
4	94.73	98.59	0.00	1.24
Mean	94.70	98.56	0.02	1.23
SD	0.04	0.022	0.01	0.01
CV (%)	0.04	0.022	50.00	0.81

Figure 6 shows the impact of chemical addition on hue angle (H). H values increased with for the chemical additions. Total color difference ΔE, which is a combination of parameters L^*, a^* and b^* values, is a colorimetric parameter extensively used in the modified calcium carbonate products to characterize the variation of colors depending on processing conditions. An increase in ΔE was observed (Figure 7) with SDO addition from 0.01 to 0.02 mass ratio. The smallest perceivable difference for two colored patches contacting one another is approximately 0.5–1.0 ΔE units (Joiner et. al., 2008). Therefore, between products cannot be indicated that a very large color change.

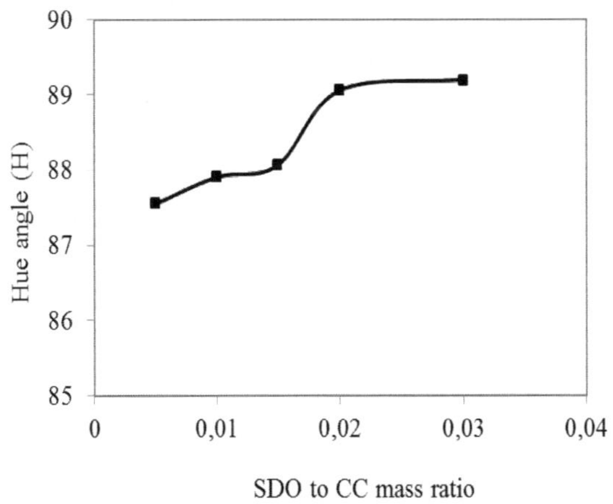

Figure 6. Hue angle (H) of experiments

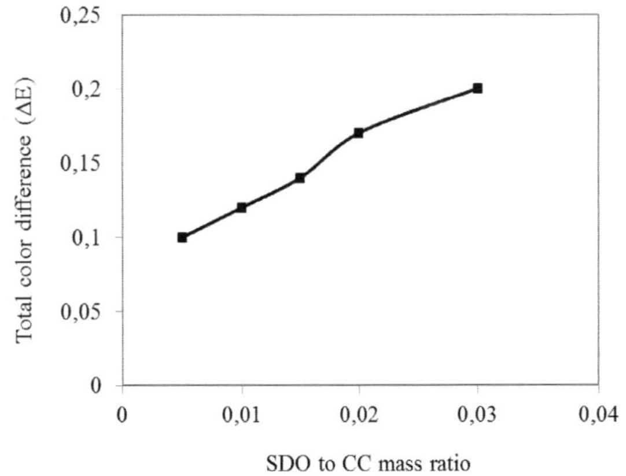

Figure 7. Total color difference (ΔE) of experiments

4 CONCLUSION

Experimental results showed that hydrophobic calcium carbonates can be obtained by aqueous solution process with sodium oleate. The basic color parameters (L*, a*, b*), brightness (R_{457}), whiteness index (WI), saturation index (SI), hue angle (*H*) and total color difference (ΔE) were slightly affected by the chemical (SDO) to calcium carbonate mass ratio. SDO addition decreased the lightness and whiteness of hydrophobic calcium carbonate in a minor proportion. Yellowness (*b**), a very important parameter for hydrophobic calcium carbonate, was observed to be generally at low levels and almost independent of chemical dosage. The application of this range is based on keeping the chemical dosage at SDO, respectively, while changing the parameter from 1.05 to 1.23 at suitable increments. On the other hand, total color change of the CTCs was influenced very little by chemical addition; even increased by the addition of SDO. As a result, the processing variables (chemical dosage) did not have a significant effect on the color parameters such as L*, a*, b*, brightness (R_{457}), whiteness index, saturation index, hue-angle and total color difference on the surface of hydrophobic calcium carbonate products.

REFERENCES

Burns RG., 1993, Mineralogical applications of crystal field theory, Cambridge Topics in Minerals Physics and Chemistry, 2nd ed. Cambridge Univ. Press, Cambridge, UK.

Camelo AFL, Gómez PA., 2004, Comparison of color indexes for tomato ripening, *Horticultura Brasileira, Brasília*, v.22, n.3

Ding, H., Lu, S, Deng, Y., Du, C.-X., 2007, Mechano-activated surface modification of calcium carbonate in wet stirred mill and its properties, *Trans. Nonferrous Met.*, 318, 1100–1104.

Ding H, Lu S, Deng Y, Du CX., 2007, Mechano-activated surface modification of calcium carbonate in wet stirred mill and its properties, *Trans. Nonferrous Met.* 318 Soc. China 17.

Hanbury A., 2002, The taming of the hue, saturation and brightness colour space**,** The Seventh Computer Vision Winter Workshop was held on February 4 – 7, Austria.

Hunt RWG., 1998, *Measuring colour*, 3rd ed. Fountain Press.

Homco RCL, Ryan KJ, Wicklund SE, Nicolalde CL, Lin S, Mckeith FK. et al., 2004, Effects of modified corn gluten meal on quality characteristics of a model emulsified meat product, *Meat Science,* 67.

Joiner A, Hopkinson I, Deng Y, Westlan S., 2008, Review of tooth colour and whiteness, *Journal of Dentistry* ,365.

Jothikumar J., 2005, Statistics Higher Secondary–First Year. *Tamilnadu Textbook Corporation College Road* , Chennai- 600 006.

Kawashima Y, T. Handa, H. Takeuchi, Y. Li, H. Takenaka, and S. Y. Lin, Spherical agglomeration of calcium carbonate dispersed in aqueous medium containing sodium oleate *Powder Technology* 46, 61 (1986)

Maskan M., 2001., Kinetics of colour change of kiwifruits during hot air and microwave drying, *Journal of Food Engineering*, 48.

Murray H., 2002, Industrial clays case study, Report of the Mining, Minerals and Sustainable Development Project, Vol. 64. *International Institute for Environment and Development and World Business Council for Sustainable Development,* UK.

Roger N.R, Hancock M., 2003, General principles guiding selection and use of particulate materials. *Particulate-filled polymer composites*, 2nd Edition. Rapra Technology Limited, Shawbury, Shrewsbury, Shropshire, SY4 4NR, United Kingdom.

Saricoban C and Yilmaz MT., 2010, Modelling the effects of processing factors on the changes in colour parameters of cooked meatballs using response surface methodology, *World Applied Sciences Journal* 9 (1): 14-22.

Sayan P., 2005, Effect of sodium oleate on the agglomeration of calcium carbonate, *Cryst. Res. Technol*. 40, No. 3.

Sekulic Z, Mihajlovic S, Dakovic A, Kragovic M, Stanic T, 2009, Modification of calcite with stearic acid using the solution method, *7th Industrial Minerals Symposium and Exhibition*, İzmir, 218–224.

SCAN-CM 59:01, 2001, *Preparation of tablets for the measurement of ISO brightness.*

Sharafudeen R., 2012, The manufacturing process parameters affecting color and brightness of TiO_2 pigmen, Sharafudeen *International Journal of Industrial Chemistry*.3:26

Sheng Y, Zhou B, Wang C, Zhao X, Deng Y, Wang Z, 2006, In situ preparation of hydrophobic CaCO3 in the presence of sodium oleate, *pplied Surface Science* 253, 1983–1987.

Soriano M., Melgosa M, Sanchez-Maranon M, Delgado G, Gamiz E, Delgado R., 1998, Whiteness of talcum powders as a quality index for pharmaceutical uses, *Color Res. Appl*. 15.

Wu, W., Lu S.C, 2003, Mechano-chemical surface modification of calcium carbonate particles by polymer grafting, *Powder Technology*, 137, 41–48.

Zarubica AR, Miljković MN, Purenović MM, Tomić VB. 2005, Colour parameters. whiteness indices and physical features of marking paints for horizontal signalization, Facta Universitatis. Series: *Physics. Chemistry and Technology* Vol. 3, No 2.

Influence of Chemical Impurities of Rocks on Cutting Efficiency and Wire Specific Consumption in Travertine Quarries

S. N. Almasi, R. Bagherpour, A. Khademian
Department of Mining Engineering, Isfahan university of Technology, Isfahan 8415683111, Iran

ABSTRACT There are various parameters that can affect the cutting rate and production in dimension stone quarries. Natural impurities are one of the most important affecting parameters in limestones. Targh travertine quarry can be divided into three geological zone: calcareous, calcareous with iron oxide and calcareous with iron oxide and manganese oxide. The exact composition of minerals and impurities of the quarry have been specified by an extensive sampling from rock chips produced by percussive drilling in the quarry. Then, the required data of 127 cuttings with three dimensions of 48, 64 and 80 m2 in different working faces have been recorded, including cutting area, cutting time and wire length. After analysis regarding determination of cutting rate and specific wire consumption, the correlation of these two parameters with impurities content has been analyzed in each individual zone. Results showed that there is a negative correlation between cutting rate and impurities content. It seems that silica and magnesium are the most effective impurities in the calcareous zone. Also, by increasing the cutting area in all three zones, the cutting rate has been increased and the wire specific consumption has been decreased.

1 INTRODUCTION

Diamond wire cutting machines are indispensable machines used in several stages ranging from block production in natural stone quarries to the final product. Today these machines are widely used in dimension stone quarries, and have evolved continuously in line with the consumers' demands so far and have opened a new era in natural stone quarrying. Generally, the diamond wire cutting method, as shown in Fig. 1, is used in dimension stone mining and decorative cutting tests, cutting and lifting basic construction structures, railway bridges, old concrete chimneys and dams. One of the most important approaches regarding to the dimension stone quarries is the operation productivity, which had drawn attentions of many researchers and experts.

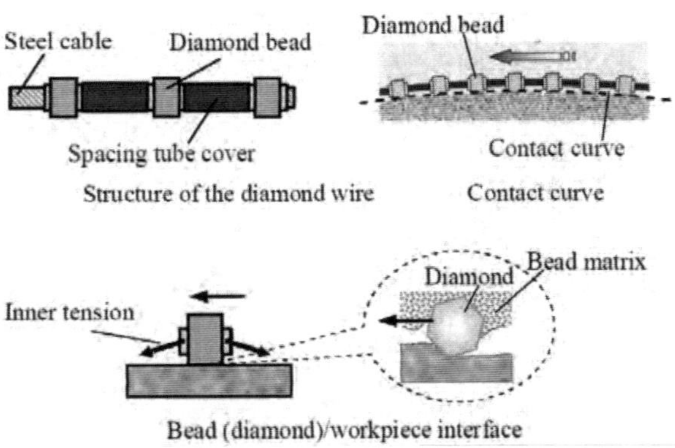

Figure 1. Diamond wire saw and its details (Guoqin et al. 2015)

This factor may be evaluated in multiple domains including rock characteristics, geomechanical parameters, management parameters, equipment, expendable materials and manpower performance. Rock cutting ability has a close relationship with rock characteristics and the performance of cutting machine and cutting tools. Generally, affecting factors in stone cutting process can be classified into three major class, including: a) stone characteristics b) cutting characteristics and c) management characteristics. Cutting characteristics and management characteristics are among controllable parameters, while stone characteristics are categorized into uncontrollable parameters (Mikaeil, 2011).

All of these parameters are summarized in Table 1 (gheysari et al. 2012). In recent years, there has been many efforts to establish a reasonable relationship between these affecting parameters and the cutting ability of rocks and machinery performance. Most of previous researches are focused on rock sawing with cutting disks, while this study evaluates the performance of rock cutting with diamond wire saw method. Yavuz et al. 2011 investigated the influence of texture and mineralogical properties of silica stones on their cutting efficiency and concluded that textural properties of rock have the highest impact on the production costs on finished slabs. Aydin et al. 2013 completed the former research by establishment of significant relationships between technical parameters and coarseness of slab face and showed that the coarseness of the sawed face is absolutely affected by their mineral composition and texture. In another research, Kahraman et al 2005 investigated different affecting parameters on diamonds grits wear and concluded that in addition to operational parameters, physical and structural characteristics of rock are involved in cutting efficiency. Other researchers believed that rock brittleness index is can affect the cutting rate and operation efficiency significantly. Also, there are some literature in the field of diamond wire saw cutting, which are focused on establishment of some relationships between cutting performance and geomechanical, physical and chemical properties of rocks. Some of these works are discussed below. Ozcelik et al (2004) found out that there is a meaningful relationship between texture coefficient of carbonate rocks and their unit wear rate and cutting rate, so that increasing the texture coefficient lead to decrease in unit wear of diamond grits and the cutting rate.

Table 1. Effective parameters of diamond wire saw exploitation performance (Almasi 2017)

Uncontrollable parameters	Controllable parameters	
Rock characteristics	Cutting characteristics	Management characteristics
▪ Physical characteristics	▪ Tools characteristics	Manpower skills
Density	Machine power	Cutting machine vibration
Porosity	Machine required voltage	Environmental conditions
Texture	Pulley diameter	
Equivalent quartz content	Beads type (sintered or electroplated)	
Water absorption coefficient	Beads matrix structure	
Water content	Size, type and density of diamond grits	
▪ Mechanical characteristics	▪ Operational characteristics	
Strength parameters	Peripheral speed of wire	
Toughness	Pull-back force	
abrasiveness	Cutting wire entrance angel	
Brittleness	Block dimensions	
elasticity	Wire length	
▪ Structural characteristics	Cutting angel (vertical, horizontal)	
Discontinuities	Number of beads per unit length	

In another study, Ozcelik et al (2004) carried out optical investigations for visual identification of wear on various diamond beads. For this purpose, they investigated 6 kinds of used cutting wire with specified lifetime and 1 new wire by using microscopic images. Based on their results, they made some recommendations for the diamond beads production. Cai et al. (2007) conducted a research on the effects of cutting wire entrance angel and concluded that increasing the cutting angel will increase the cutting rate and this effect intensifies in higher speeds. Besides, they found that for a constant cutting area, any increment in cutting speed leads to a decline in cutting wire lifetime. In other hand, they believe that by reducing the cutting speed with the aim of increasing the lifetime of the cutting wire, the energy consumption will increase. Also, Ozcelik 2007 investigated the effect of bedding plane and anisotropy of rocks of the cutting machine efficiency. Ataei et al. (2012) used a statistical analysis to develop a prediction model for production rate of diamond wire saw method. It should be mentioned that the investigation of rock cutting rate, diamond bead we are rate, productivity of performance are continued by researchers in recently years and the good results have obtained (Yarahmadi et al. a, b, 2014, Almasi et al. 2015, Bagherpour et al. 2015, Khademian 2015, Yarahmadi et al. a, b, 2015, Yilmazkaya et al. 2015, Yari et al. 2016, Almasi et al. 2017)

Mineral composition and rock forming elements of a rock is undoubtedly one of the influencing parameters in cutting efficiency and beads wear rate. Reviewing the above mentioned literature shows that there is no notable research regarding to the effect chemical properties of rock on cutting process. This study intends to investigate the influence of different chemical components and impurities of rocks on the cutting efficiency in real operational scale (in quarry).

2 GEOLOGICAL AND GEOGRAPHICAL SITUATION OF THE QUARRY

Experimental rock cuttings of this study have been carried out in Targh quarry, which is one of the largest operating travertine quarry in Iran. The exploitation of rock blocks is carried out by diamond wire saw method in this quarry. This travertine quarry is located in Isfahan province near Natanz city that has geographical coordinate of 39 S 572979 3692785 in (WGS 84) UTM system. It lies near Natanz road, after Yahia Abad and Targh villages (Fig. 2).

Figure 2. The location of Targh quarry in Iran map.

Travertine rocks in this area have a wide variety of colors, which is a sign for different chemical impurities. The majority of iron impurities in travertine rocks are deposited by exposing the iron ion-carrying groundwater to atmosphere (Glover and Robertson 2003). The intense hardness of impurities has destructive effects on beads structure and cutting efficiency.

2.1 Zoning of Deposit

Targh travertine deposit can be classified into 3 zones based on rock color and its chemical elements and impurities. These zones are summarized below:

Zone A: this zone is consisted of light-colored and cream-colored rocks, which are the highest-grade rocks of the deposit.

Zone B: this zone is consisted of some chocolate-colored and copper-colored rocks.

Zone C: this zone is consisted of some grey and silver color rocks, which is due to presence of magnesium compounds.

The most important chemical elements and impurities of each zone are summarized in Table 2.

Table 2. Average percentage of impurities in different zones

zone	MgO	SiO_2	Fe_2O_3	MnO	Other	LOI
Zone A	.158	.078	.075	.000	.084	43.163
Zone B	.284	.267	.221	.000	.143	43.423
Zone C	.384	1.042	.079	.118	.604	43.043

2.2 Impurities Variation Scheme in Zones

Figure 3 and 4 show the variation scheme of different impurities in all three zones.

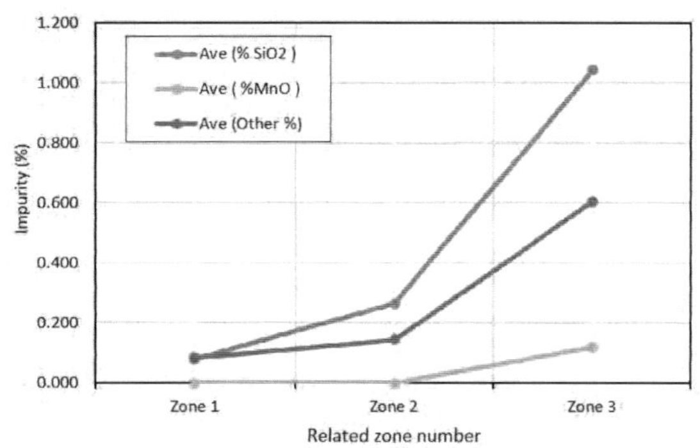

Figure 3. Different chemical compound concentration in different zones

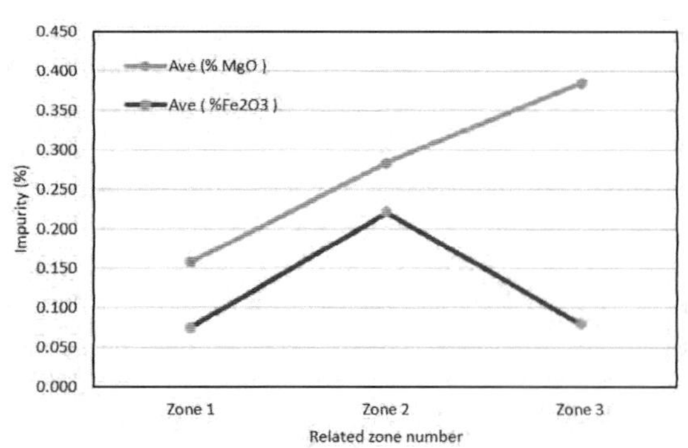

Figure 4. Concentration of iron and magnesium ions in different zones

Zone A has the highest content of lime with hardness of 3 in Mohs scale, while the percentage of impurities of this zone is the lowest one among other zones. Therefore, it is expected to have the best conditions of cutting process in this zone. In zone B, the percentage of impurities rises while the majority of this rise belongs to dolomite, silica and iron oxides, which all have higher hardness compared to limestone. Therefore, all of these components induce a negative impact on cutting performance in zone B.

3 CUTTING WIRE EFFICIENCY IN DIFFERENT ZONES

In order to investigate the relationship between chemical impurities and cutting performance, some experimental cutting operations have been performed in Targh travertine quarry in real operational scale. In order to eliminate the geologically induced errors and enhancing the reliability of obtained data, more than 35 cuttings with different areas of 48, 64 and 80 m2 has been performed in each zone. The total cuttings number in zone A reaches to 50 cuttings. The required cutting related data, including cutting area, wire length and cutting time, have been measured and registered during all cuttings. All of these data are reported in Table 3.

Cutting efficiency (C_e) of the diamond wire is defined as the total rock surface cut by the diamond wire in unit time. The specific consumption (S_c) of diamond wire is defined as the required length of the wire for cutting the unit surface area of the rock. These two parameter could be calculated by Equations 1 and 2.

$$C_e = (n \times A)/T \quad (1)$$

$$S_c = L/(n \times A) \quad (2)$$

Where C_e is the cutting efficiency in terms of m^2/s and T is the spent time for cutting n faces and A is the area of each face. S_c is the wire specific consumption in terms of m/m^2, n is the number of faces cut by a single wire in its lifetime, L is the length of the wire and

A is the area of each face. The calculated values of C_e and S_c for each zone is summarized in Table 3.

Table 3. Cutting efficiency and specific consumption in different zones

Zone	Number of blocks	Machine Amperage	Wire length (m)	Cutting area (m^2)	Total area (m^2)	Cutting time (h)	Specific consumption (m/m^2)	Cutting efficiency (m^2/h)
Zone A	19	55	36	48	912	137	0.039	6.657
	18	55	40	64	1152	148	0.035	7.784
	16	60	45	80	1280	161	0.035	7.950
Zone B	13	45	36	48	624	102	0.058	6.118
	13	45	40	64	832	136	0.048	6.118
	12	50	45	80	960	133	0.047	7.218
Zone C	13	50	36	48	624	120.5	0.058	5.178
	12	50	40	64	768	136.5	0.052	5.626
	11	55	45	80	880	136.5	0.051	6.007

4 DISCUSSION

According to Table 3, independent of cutting dimension, cutting efficiency decreases from zone A to zone C, while the specific consumption of cutting wire increases. The average values of Ce and S_c for all zone have been illustrated in Figure 5 and 6.

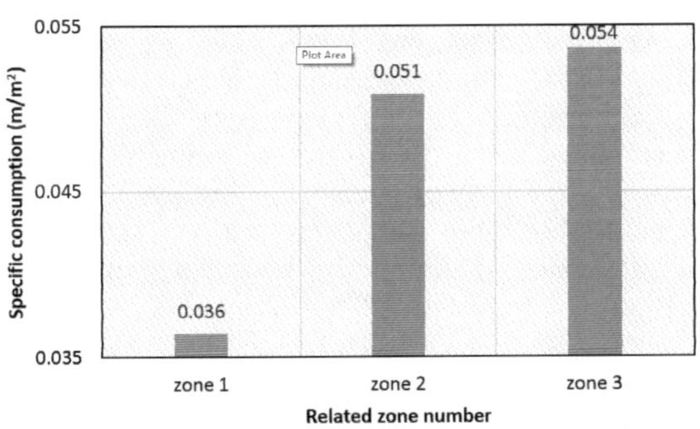

Figure 5. The average specific consumption of wire in different zones

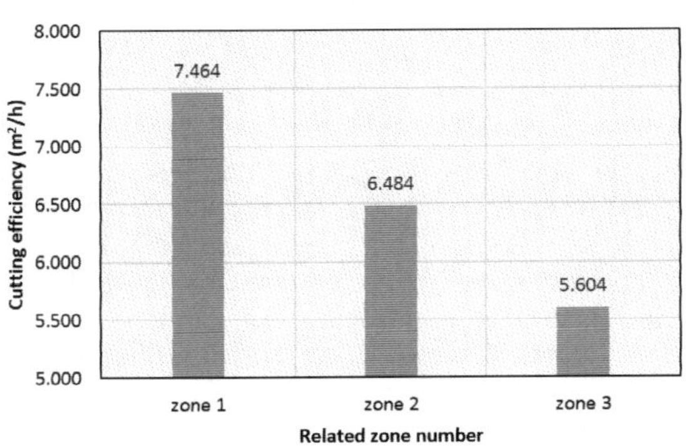

Figure 6. The average cutting efficiency in different zones

The result indicate that by increasing the percentage of impurities between all zones, the cutting efficiency have declined and the specific consumption of cutting wire has increased. This trend is evident for all zones. On the other hand, the number of faces cut by an individual cutting wire decreases from zone A to zone C, which is an indication for hardness of rocks in zone B and zone C. the results of this study have another outcome, which is the increasing of cutting efficiency by increasing the dimension of the cutting.

5 CONCLUSION

Mineral composition and chemical impurities are among the uncontrollable parameters in diamond wire cutting method in dimension stone quarries. Different inherent chemical compound in rocks induce some changes in cutting rate and therefore affect the production efficiency in quarries. Chemical studies in three zones (A, B and C) of a travertine quarries showed that the chemical impurities rise from zone A to zone C. Then the results of some experimental diamond wire cuttings showed that the cutting efficiency decreased from zone A to zone C and the wire specific consumption increased in the same domain. This trends showed that the percentage of impurities have a negative correlation with cutting efficiency and a direct positive correlation with specific consumption of cutting wire. On the other hand, the number of faces cut by an individual cutting wire decreases from zone A to zone C, which is an indication for hardness of rocks in zone B and zone C. therefore, implementing proper measures can decrease the production costs and optimize the working conditions. These proper measures could be selecting the proper operating parameters including, cutting wire speed, pull-back force, proper face dimension, etc.

ACKNOWLEDGEMENT

Authors would like express their gratitude to Ahrare-e-Sepahan company for their priceworthy cooperation in this research.

REFERENCES

Almasi S. N, Bagherpour R, Mikaeil R, and Ozcelik Y. 2017. Developing a new rock classification based on the abrasiveness, hardness, and toughness of rocks and PA for the prediction of hard dimension stone sawability in quarrying, *Geosystem Engineering*, (On line) http://dx.doi.org/10.1080/12269328.2017.1278727.

Almasi S.N, Bagherpour R, Mikaeil R, and Khademian A. 2015. Influence of Cutting Wire Tension on Travertine Cutting Rate, 24th International Mining Congress and Exhibition of Turkey-IMCET'15 Antalya, Turkey, pp. 1096-1102.

Ataei M, Mikaiel R, Sereshki F, and Ghaysari N. 2012. Predicting the production rate of diamond wire saw using statistical analysis. Arabian Journal of Geosciences 5(6):1289-1295.

Aydin G, Karakurt I, and Aydiner K. 2013. Investigation of the surface roughness of rocks sawn by diamond sawblades. International Journal of Rock Mechanics and Mining Sciences 61:171-182.

Bagherpour R, Khademian A, Almasi S.N, and Aalaei M. 2014. Optimum cutting wire assembly in dimension stone quarries, Journal of Mining and Metallurgy, 50 A (1): 1 – 8

Cai O, Careddu N, Mereu M, and Mulas I. 2007. The influence of operating parameters on the total productivity of diamond wire in cutting granite. IDR Industrial diamond review(3).

Ghaysari N, Ataei M, Sereshki F, and Mikaiel R. 2012. Prediction of Performance of Diamond Wire Saw with Respect to Texture Characteristics of Rock/Prognozowanie Wydajności Pracy Strunowej Piły Diamentowej W Odniesieniu Do Charakterystyki Tekstury Skał. Archives of Mining Sciences 57(4):887-900.

Glover C, and Robertson AH. 2003. Origin of tufa (cool-water carbonate) and related terraces in the Antalya area, SW Turkey. Geological journal 38(3-4):329-358.

Guoqin H, and Xipeng X. 2013. Sawing Performance Comparison of Brazed and Sintered Diamond Wires, Chinese journal of mechanical engineering, 26 (1): 1-7.

Kahraman S, Fener M, and Gunaydin O. 2005. A brittleness index to estimate the sawability of carbonate rocks. Impact of Human Activity on the Geological Environment EUROCK 2005: Proceedings of the International Symposium EUROCK 2005, 18-20 May 2005, Brno, Czech Republic: CRC Press. p 233.

Khademian A, Bagherpour R, Almasi S.N, and Aalaei M. 2015. Optimum Distance Between Cutting Machine And Working Face In Travertine Exploitation With Diamond Wire Cutting Method, 24th International Mining Congress and Exhibition of Turkey-IMCET'15 Antalya, Turkey, pp. 1103-1110.

Mikaeil R, Ataei M, and Yousefi R. 2013. Correlation of production rate of ornamental stone with rock brittleness indexes. Arabian Journal of Geosciences 6(1):115-121.

Ozcelik Y. 1999. Investigation of the working conditions of diamond wire cutting machines in marble industry: Ph. D. thesis, Hacettepe University, Ankara, Turkey, 242 pp.(In Turkish).

Özçelik Y. 2007. The effect of marble textural characteristics on the sawing efficiency of diamond segmented frame saws. IDR Industrial diamond review(2).

Ozcelik Y, Polat E, Bayram F, and Ay A. 2004. Investigation of the effects of textural properties on marble cutting with diamond wire. International Journal of Rock Mechanics and Mining Sciences 41:228-234.

Yarahmadi R, Bagherpour R, and Khademian A. 2014. Safety risk assessment of Iran's dimension stone quarries (Exploited by diamond wire cutting method). Safety Science 63: 146–150.

Yarahmadi R, Bagherpour R, Khademian A, Mirzaie H, and Kakaie R. 2015. Developing a Matlab code for determine geometry of rock mass blocks and its applications in mining and rock mechanic engineering. Journal of Mining and Metallurgy 51A (1): 41 – 49.

Yarahmadi R, Bagherpour R, Kakaie R, Mirzaie N, Hossein B, and Yari M. 2014. Development of 2D computer program to determine geometry of rock mass blocks. International Journal of Mining Science and Technology 24: 191–194.

Yarahmadi R, Bagherpour R, Luis M. O. S, and Taherian S. G. 2015. How to determine the appropriate methods to identify the geometry of in situ rock blocks in dimension stones Environmental Earth Science, DOI 10.1007/s12665-015-4672-4.

Yari M, Bagherpour R, and Almasi N. 2015. An Approach to the Evaluation and Classification of Dimensional Stone Quarries with an Emphasis on Safety Parameters, the Mining-Geology-Petroleum Engineering Bulletin, pp. 15-26, DOI: 10.1177/rgn.2016.3.2.

Yilmazkaya E, and Ozcelik Y. 2016. The Effects of Operational Parameters on a Mono-wire Cutting System: Efficiency in Marble Processing, Rock Mechanic and Rock Engineering, 49:523–539.

Yavuz H, Ozkahraman T, and Demirdag S. 2011. Polishing experiments on surface quality of building stone tiles. Construction and Building Materials 25(4):1707-1711.

Surname, Initial(s), (ed.), year.. Title of paper, *Congress, Symposium, etc. name* (serial number if needed), volume, page interval (e.g., pp.46-52).

Mechano-Activated Surface Modification by Stirred Milling with Metallic Soaps and Fatty Acid

O.Y.Toraman
Ömer Halisdemir University, Faculty of Engineering, Mining Engineering Department, 51245 Nigde, Turkey
Ömer Halisdemir University, Industrial Raw Materials & Building Materials Application & Research Center, 51245 Nigde, Turkey
H.Kose, B.Bitirmis
Nigtas R&D Center, Nigtas Limited Company, Nigde, Turkey

ABSTRACT This paper investigates the surface modification of calcite ($CaCO_3$) from the Niğde region of Turkey with sodium oleate (SO), sodium stearate (SDS) and stearic acid (SA) complex as modification agents, which are incorporated into the calcite with dry ultra-fine grinding in a laboratory stirred mill (750 ml). The effect of surface modification is evaluated by a floating test, which is characterized by the active ratio, colour properties and oil absorption. The results indicate that the hydrophilic surface of calcite becomes hydrophobic after the incorporation of SO, SDS and SA by dry stirred milling.

1 INTRODUCTION

Calcium carbonate ($CaCO_3$) containing calcite is most widely used filler in the plastics, rubber, paper, paint and ink industry (Solomon and Hawthorne, 1983). In particular, its reaction with fatty and other organic acids, especially stearic acid, has been used for many years to improve their compatibility with, and dispersion in, polymers. The coated fillers are much more hydrophobic than the uncoated ones, which reduces water pick-up, and have also been shown to have an effect on polymer morphology and hence modifying the properties of polymers (Hancock and Rothon, 2003).

Conventional surface modification technologies, which consist of the heating mix modification and packing modification methods, have been applied widely in the treatment of fillers and pigments. However, conventional techniques have some drawbacks, such as weak stirring capabilities, low materials mixing degree, uneven dispersion between agent and materials, and, in particular, a lack of reaction foundation. Moreover, the modification effect is weak when using conventional techniques and the product quality is unstable. Thus, the improvement of mechanical performance of the product is not significant (Ding et al., 2007).

Mechano-activated surface modification is a modification method that could provide better results than the conventional methods by utilizing mechano-activated effects during ultrafine grinding. Mechano-activated effects are physical, and mechanical changes occur in the near surface region where the solids come into contact with each other under mechanical forces (Frank et al., 2005; Mohamed and Wakeel, 2005). On the other hand, dry particle coating is a relatively new and alternative approach to wet coating methods and it has drawn attention of many researcher using different equipments such as high shear mixer, vibrating mill, planetary ball mill, and jet mill (Jeong et al., 2009; Lefebvre et al., 2011; Sato et al., 2012; Hait and Chen, 2014).

In this study, it was investigated the modification of calcite with sodium oleate (SO), sodium stearate (SDS) and stearic acid (SA) complex, which is incorporated into the calcite with dry ultra-fine grinding in a stirred mill. Moreover, the modified $CaCO_3$ before and after grinding+surface modification was characterized by the active ratio and oil absorption method.

2 EXPERIMENTAL

2.1 Samples and Reagents

Calcite powder used was supplied by the Nigtas Limited Company, Nigde, Turkey. A chemical analysis of the sample is reported in Table 1. The particle population density of feed sample is given in Fig. 1. The mean size of particles in the sample was 23.48 μm, and the specific surface area was 0.941 m^2/g. Chemical reagents (SO, SDS and SA) which were used in the surface modification, were obtained from Sigma-Aldrich. Distilled water was used in the experiments.

Table 1. Chemical composition of the sample (wt%).

$CaCO_3$	$MgCO_3$	Fe_2O_3	SiO_2	Al_2O_3	Total
99.5	0.2	0.01	0.01	0.02	99.74

Physical properties of grinding media and chemical reagents used in the experimental studies are shown in Table 2 and Table 3, respectively.

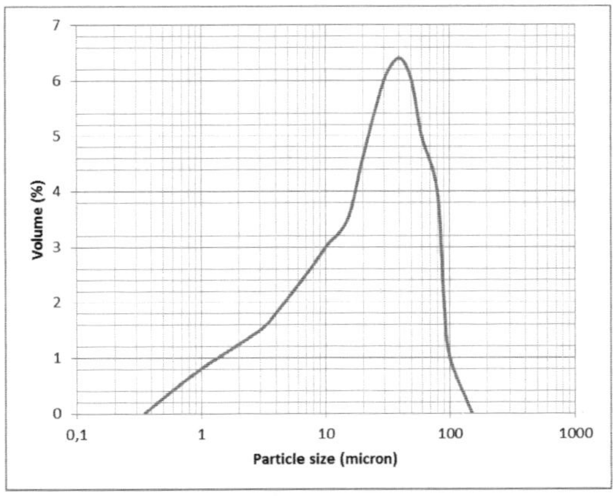

Fig 1. Particle population density of feed sample.

Table 2. Physical properties of grinding media.

Composition	Specific gravity (kg/m^3)	Hardness
Al_2O_3(95%)+SiO_2 (5%)	3600	H_v 1200 kg/mm^2

Table 3. Chemical reagents used in experimental studies.

Reagent	Code	Density (g/cm^3)	Molecular formula	Molecular Weight (g/mol)
Metallic soap				
Sodium oleat	SO	0.90	$C_{18}H_{33}NaO_2$	304.45
Sodium stearate	SDS	1.02	$C_{18}H_{35}NaO_2$	306.47
Fatty acid				
Stearic acid	SA	0.94	$C_{18}H_{36}O_2$	284.48

2.2 Test and Evaluation Method

Laboratory scale modification tests were conducted in a stirred media mill, Union Process, USA, and its volume 0.75 L. The grinding media consists of Al_2O_3 beads with diameter of 2-3 mm. After the modification process, the obtained products were denoted as NK1, NK2, NK3, NK4 and NK5. The effect of the surface modification was evaluated by the floating test, which measures the ratio of the floated product to the overall weight of the sample, after they are mixed in water and stirred vigorously.

$$AR(\%)=M_p/(M_p+M_t)*100 \quad (1)$$

where AR is the active ratio (%), M_p the mass of the floated product and M_t the mass of the non-floated product (Sekulic et al., 2009). A greater active ratio implies better surface modification effects (Ding et al., 2007; Wua and Lu, 2003; Sheng, 2004). The samples were characterized with the Mastersizer Hydro 2000 laser diffraction particle size analyzer (Malvern, UK). Analyses of the colour properties of the samples with a Datacolor Elrepho spectrophotometer were performed.

An organization called Commission Internationale del'Eclairage (CIE) determined the standard values that are used worldwide to measure color. The values used by CIE are called L*, a*, and b*, and the color measurement method is called CIELAB. Symbol L* (Lightness) represents the difference between light ("pure white") (where L*=100) and dark ("black") (where L*= 0); a* (Redness-Greenness) represents the difference between green (−a*) and red (+a*); and b* (Yellowness-Blueness) represents the difference between yellow (+b*) and blue (−b*) (Sharafudeen, 2012). The colour measurement method is called CIELAB. The CIELAB values are calculated from the red green and blue filters of the colorimeters and are particularly suited to describing near white samples according to the following equations (Christidis et al., 2004):

$$L^*=116(Y/Y^n)^{1/3}-16 \quad (2)$$
$$a^*=200[((X/X^n)^{1/3}-(Y/Y^n)^{1/3}] \quad (3)$$
$$b^*=200[((Z/Z^n)^{1/3}-(Y/Y^n)^{1/3}] \quad (4)$$

where X, Y and Z are the tristimulus values for the samples arising from the colourimetric system and X^n, Y^n and Z^n are those of a surface colour chosen as the nominal white stimulus. Using this system and colour that correspons to a place on the Cylindrical CIELAB color space system was shown in Fig. 1.

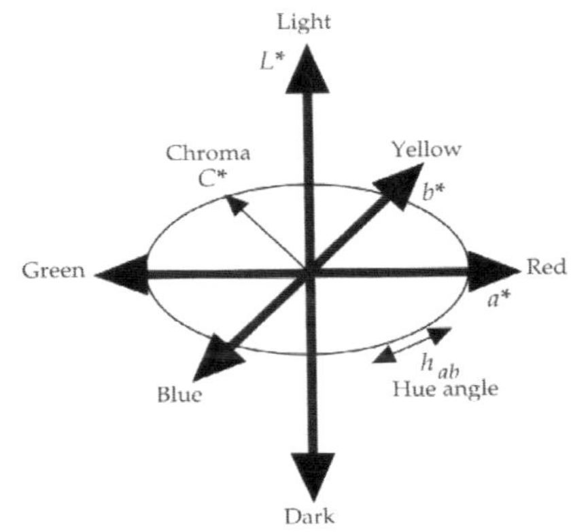

Figure 1. Cylindrical CIELAB colour space (Ochs, 2000)

Summary of experimental conditions is also shown in Table 4.

Table 4. Summary of experimental conditions.

Item	*Experimental condition*
Bead filling ratio	0.70
Sample filling ratio	0.05
Rotation speed of stirrer	600 rpm
Grinding time	10 min
Internal volume of grinding pot	750 ml
Temperature	Room temperature
Material of grinding media	Alumina
Grinding media size	2-3 mm
Powder sample	
Specific gravity	2.7 g/cm^3
Size (average)	23.48 μm
SSA	0.941 m^2/g

3. RESULTS AND DISCUSSION

3.1 Active ratio of modified $CaCO_3$

It is desirable that coated calcite products used in the industry have an active rate value close to 100%. Active ratios of feed and coated products can be seen from Table 5.

Table 5. Active ratios of feed and coated products.

Modifier dosage	*Active ratio, %* $\geq$
Feed	-
1% SO	>98.4
1% SA	>98.7
0.5% SO+0.5 SA	>99.9
1% SDS	>99.1
0.5% SDS+0.5 SA	>99.2

Uncoated calcite and coated products (NK1, NK2, NK3, NK4 and NK5) are shown in Fig. 1.

Figure 2. Uncoated calcite (left) and coated products (NK1, NK2, NK3, NK4 and NK5) (right)

3.2 Particle size distribution of modified $CaCO_3$

Table 6 shows the summary of particle size distributions of feed and coated products.

Table 6. Particle size distributions of feed and coated products.

Modifier dosage	d_{50} *μm*	d_{97} *μm*	-2 *μm*	d_{100} *μm*
Feed	23.48	89.4 3	8.38	138. 04
1% SO	3.03	10.8 8	38.5	19.9 5
1% SA	3.12	11.9 6	38.2 3	22.9 0
0.5% SO+0.5SA	3.12	11.8 7	37.9 4	22.9 0
1% SDS	2.96	11.2 5	39.3 4	22.9 0
0.5% SDS+0.5 SA	2.64	9.82	43.0 2	19.9 5

3.3 Oil absorption of modified $CaCO_3$

The decrease in the dop oil absorption rate is regarded as an indication of the coating quality. Depending on the coating quality of the coated calcite products, the amount of dop oil absorption is reduced by 50% or more. It can be seen from Table 7, the dop oil absoption of $CaCO_3$ decreased with modifier reagent from 25.10% to 11.89% with 0.5% SO+0.5 SA.

Table 7. Dop oil absorption ratios of feed and coated products.

Modifier reagent dosage	*DOP oil absorption, g/100g* ≤
Feed	29.83
1% SO	15.97
1% SA	16.02
0.5% SO+0.5 SA	11.89
1% SDS	19.86
0.5% SDS+0.5 SA	19.92

3.4 Colour properties of modified $CaCO_3$

Colour parameters (Ry, L*,a*,b*) of feed sample and products coated are shown in Table 8. Whiteness is an important specification of micronized mineral filler products, and it is important for marketing purposes that whiteness be high (≥95%). Notably, the lightness *(L*)* values of the ground calcite products slightly increased from 97.56 to 98.1 with 0.5% SO+0.5 SA complex.

Table 8. Colour parameters (Brightnes Ry, L*, a*, b*) of feed and coated products.

Modifier reagent dosage	*Ry*	*L**	*a**	*b**
Feed	93.83	97.56	0.17	1.50
1% SO	95.97	98.42	0.01	0.94
1% SA	95.29	98.15	0.02	0.98
0.5% SO+0.5 SA	95.28	98.15	0.03	0.98
1% SDS	95.04	98.05	-0.01	0.97
0.5% SDS+0.5 SA	94.71	97.92	-0.02	0.83

4 CONCLUSIONS

In this study, mechano-activated surface modification with SO, SDS and SA as modification reagents incorporated with dry ultra-fine grinding in a stirred mill was investigated. The results showed that modified calcite powder with an active rate of 99.9%, a mean particle size of 3.12 µm, oil absorption ratio of 11.89%, Ry of 95.28% can be obtained with mechano-activated surface modification method under 1% modifier reagent dosage (0.5% SO+0.5%SA) with 1.8 m/s stirring velocity, 10 min modification time, 70% ball filling volume ratio and 5% sample and ball mass ratio. The results obtained from the experiments indicate that dry mechano-activated surface modification with metallic soaps and fatty acid is an effective method for the surface modification of calcite using a stirred media mill.

Acknowledgment

The authors would like to thank Ömer Halisdemir University Industrial Raw Materials & Building Materials Application & Research Center and the Nigtas limited company, Nigde, Turkey, for their support in this work.

Nomenclature

- d_{50} Average particle size, µm
- d_{97} 90% cumulative weight passing size of the ground product, µm
- d_{100} 100% cumulative weight passing size of the ground product, µm
- SSA Specific surface area, m^2/g
- SO Sodiuum oleate
- SA Stearic acid
- SDS Sodium stearate
- Ry Brightness of calcite
- L* Lightness of calcite
- a* Redness-greenness of calcite
- b* Yellowness-blueness of calcite

References

Christidis, G.E., Sakellariou, N., Repouskou, E., Markopoulos, Th., 2004. Influence of organic matter and iron oxides on the colour properties of a micritic limestone from Kefalonia, Bulletin of the Geological Society of Greece vol. XXXVI, 72-79.

Ding, H., Lu, S., Deng, Y., Du, C.X., 2007. Mechano-activated surface modification of calcium carbonate in wet stirred mill and its properties, *Trans. Nonferrous Met. Soc. China,* 17, 1100-1104.

Frank, S., Stefan, M., Jorg, S., Wolfgang, P., 2005. The influence of suspension properties on the grinding behavior of aluminum particles in the submicron site in stirred media mills, *Powder Technol.,* 156, 103-110.

Hait, S.K., Chen, Y., 2014. Optimization of milling parameters on the synthesis of stearic acid coated $CaCO_3$ nanoparticles, *J.Coat.Technol. Res.,* 11(2), 273-282.

Hancock, M., Rothon R.N. (Eds.), 2003. Particulate-filled Polymer Composites, Rapra Technology Limited, United Kingdom, p. 57.

Jeong, S.B., Yang, Y.C., Chae, Y.B., Kim, B.G., 2009. Characteristics of the treated ground calcium carbonate powder with stearic acid using the dry process coating system, *Materials Transactions*, 50(2), 409-414.

Lefebvre, G., Galet, L., Chamayou, A., 2011. Dry coating of talc with fumed silica: Influence of the silica concentration on the wettability and dispersibility of the composite particles, *Powder Technology*, 372-377.

Mohamed, I., Wakeel, A., 2005. Effect of mechanical treatment on the mineralogical constitutes of Abu-Tartour phosphate ore Egypt, *Int. J. Miner. Process.,* 75, 1012-1015.

Ochs, M. J., 2000. Development of a Linear Version of the CIECAM97s Color Appearance Model (https://www.cis.rit.edu/research/thesis/bs/2000/ochs/thesis.htm)

Sato, A., Serris, E., Grosserau, P., Thomas, G., Chamayou, A., Galet, L., Baron, M., 2012. Effect of operating conditions on dry particle coating in a high shear mixer, *Powder Technology,* 229, 97-103.

Sekulic, Z., Mihajlovic, S., Dakovic, A., Kragovic, M., Stanic, T., 2009. Modification of calcite with stearic acid using the solution method, *7th Industrial Minerals Symposium and Exhibition,* 218-224, İzmir.

Sheng, Y., Zhou, B., Zhao, J., Tao, N., Yu, K., Tian, Y., Wang, Z., 2004. Influence of octadecyl dihydrogen phosphate on the formation of active super-fine calcium carbonate, *J. Colloid Interface Sci.,* 272, 326-329.

Solomon, D.H., Hawthorne, D.G., 1983. Chemistry of Pigment and Fillers, Willey & Sons, NY, p.108.

Wua, W., Lu, S.C., 2003. Mechano-chemical surface modification of calcium carbonate particles by polymer grafting, *Powder Technol.,* 137, 41-48.

Sharafudeen, R., 2012. The manufacturing process parameters affecting color and brightness of TiO_2 pigment, International Journal of Industrial Chemistry (IJIC), 3(27), 1-7.

POSTERS

Gaziantep CYS Taşocağında Patlatma Kaynaklı Yer Sarsıntısı Sorununun Elektronik Ateşleme Sistemleri Kullanılarak Azaltılması

Reducing of Blast Induced Vibration Problem at Gaziantep CYS Quarry by Using Electronic Initiation Systems

S. Kara
Davey Bickford, France
Ö. Düzgün, Ö.G.İlik
Kapeks Explosives, Ankara, Turkey
Y. İnan
CYS Construction, Gaziantep, Turkey
M. Aygün, C. Doğan
Ideal Explosives, Adana, Turkey

ABSTRACT In the second decade of the commercial use of electronic detonators, a new concept of blasting was introduced in the quarry industry using the latest generation of initiation “electronic system”. Electronic initiation provides an efficient means to deliver the optimum timing to decrease the impact of the blasting operation on the neighborhood without decreasing the blasting quality and production. The increase of the population in Turkey has resulted in the expansion of urbanization and the need for crushed rock for construction. Rapid answers for this requirement includes either expanding existing quarries or the opening of new ones. During the expansion of quarries, it is inevitable that production impacts will be challenged by neighboring districts, particularly due to blast induced vibration. In order to reduce the complaints caused by vibration and to maintain production capacity without blast restrictions, the quarries will suffer negative impacts on their production cycle. As a response to this scenario example, the CYS quarry in Gaziantep decided to try electronic detonators and applied trials. The purpose of this study is to increase number of blast holes per blast and to reduce the ground vibration levels measured at Karpuzkaya Village which is the nearest site to the quarry by using electronic detonators.

In this paper, the applied blasting designs used during the trials, application of electronic detonators and the results obtained with recorded data are presented.

1 INTRODUCTION

Despite the considerable improvements in mechanical rock excavation, drilling and blasting is still the most economical method available for loosening and fragmenting the insitu rock mass in quarries and mines. But the blasts induce ground vibrations and generate a negative impact on nearby structures: building slops roads, dams, etc. and also on the productivity of the quarry operation itself. The approach of the production areas to sensitive structures or residences typically leads to an increase in vibration and air blast level. Often the charge weight per delay time must then be reduced to keep the blast vibrations within the compliance limits, which inevitably increases production costs.

This negative impact can be measured and controlled or totally eliminated using actual technologies. Many companies propose and use advanced vibration management services to measure, analyze and predict the blast induced vibrations with a proprietary model for the management and control of vibration at sensitive locations.

This service can be used for the blast design and optimization for the maximum allowable vibration level in sensitive locations. The need for this service is increasing as urbanization encroaches on quarries, restricting their ability to operate or the quarries expansion encroaches on the cities, inducing population complaints.

There has been engineering and technological approach for the Gaziantep CYS quarry to reduce the vibration impact after complaints from the neighboring village. This paper will demonstrate that by introducing the change of only the initiation system from a non-electric initiation to an electronic initiation the impact on the neighborhood was reduced. The use of a predictive model will be the second step and will be demonstrated in the future studies.

2 INITIATION SYSTEMS

The first initiation system was invented by Davey Bickford in 1831 called a *Safety Fuse* used to initiate a blast and give the blaster a short, known time to reach in a safe location. 1906 was the invention of the *Detonating cord* and in 1920 *Electrical detonator* was invented respectively. This first mentioned detonator helps to reduce the Unit Charge but still remains sensitive to the behavior. In the eighties, the second generation of detonator called *Non electric* detonators was introduced. This detonator became more useful due to its lower sensitivity to the behavior but the only way to check the blast connection was still a visual checking. In the nineties, new technologies allowed the introduction in the market of the *Electronic* detonator which presents more safety and accurate timing compare with the two other technologies (Fig.1).

With all of these new technologies many mining and quarries companies still use the electric detonator for their price despite the various negative impacts on safety and production.

Fortunately, not all companies are doing this and now more and more companies start to understand to use the new technologies to improve the safety, productivity and operation cost.

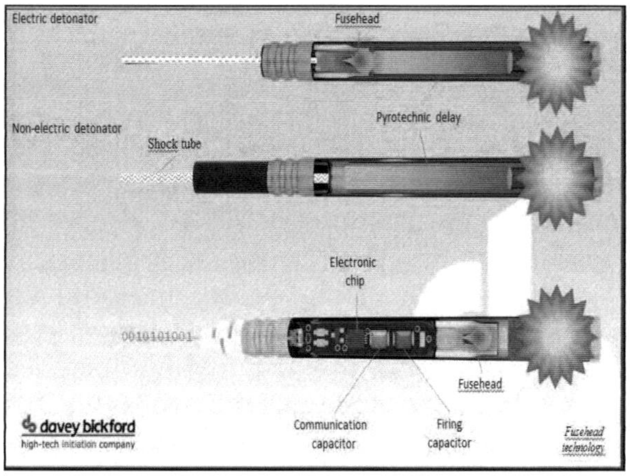

Figure 1. Detonator chronology

2.1 Electronic Blasting Systems

Even if all existing electronic blasting systems are complying with international safety standards and have the same fundamental components as the "digital timing", each of them has unique architecture particularly in the way how the safety constraints have been approached in the design.

Beginning with product development, manufacturers have chosen different fundamental principles for addressing safety issues: one or two capacitor design for the detonator electronic chip, different connector and inherent safety are examples of such basic principles (Fig.2).

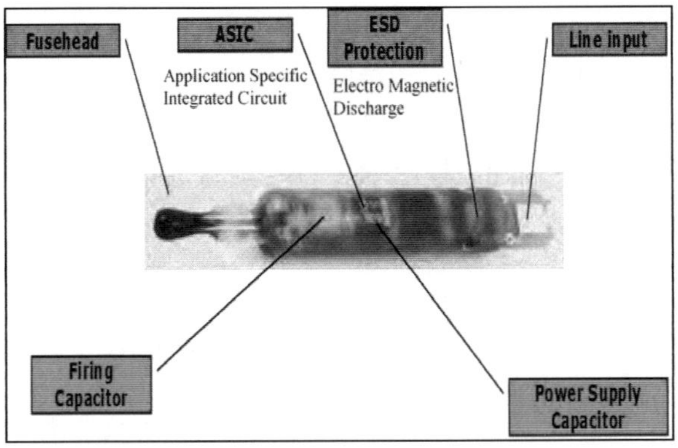

Figure 2. Electronic detonator components (Davey Bickford)

The concept of using a two capacitor technology inside a detonator is done in order to separate the communication circuit from the firing circuit. The first capacitor will be used purely for communication purposes such as reading the detonator ID. The second capacitor will be used purely for the firing as to ensure better insulation from the latter and from external stimuli. In addition to the built-in safety characteristics of the detonators, inherent safety consists in making it physically impossible for the on-bench tester to prematurely detonate a detonator, using both hardware and software to ensure this. This is generally achieved by limiting voltage and/or current output of on-bench testing equipment. Manufacturers' solutions for limiting the consequence of "power" limitation over the testers performance in terms of the number of detonators that can be tested simultaneously, differ significantly from one system to the next.

All of the electronic detonator circuits are protected from external electrical aggression, thanks to passive components such as resistors or transient voltage suppression diodes.

The electronic detonators can be traced by means of the unique identifier embedded into the silicon chip. Anticipated regulations known under the "Track and Trace" standards may improve the traceability of explosives by adding a label/bar code/data matrix on each single detonator, explosive cartridge or piece of detonating cord, but nothing could be more reliable than a permanent tracer programmed into the chip like the unique ID code of each electronic detonator.

The explosives industries don't have the same needs. For example, a quarry which would fire fifty holes each week will not have the same requirements as a large mining operation which might daily detonate five to ten blasts totaling over 2500 holes within a 20 minutes' window between on a daily basis. Therefore manufactures propose different equipment with different features to mines and quarries with different needs. Consequently, to assist with the different needs in the market, we recently observed the integration of high level technology in initiation equipment, such as wireless technology. Even though the basic concept of wireless technology remains the same, there are notable differences between the currently available systems.

The ability to wirelessly fire remotely, is one of the latest technological features improving the safety aspect as well as productivity. This technology, eventually combined with "repeater" devices, not only extends the distance between the actual firing location and the blast area, but also allows parallel firing procedures of several blasts from the same firing point at the same time. Recent developments in features such as multi-blasting, has been welcomed in the mining industry and consists of parallelizing several firing procedures and thus simultaneously controlling several blasts remotely. This has significantly reduced the time the mine is idle during blasting operations.

To challenge the size of the blast using conventional non-electrical detonators which is theoretically infinite, the manufacturers proposed the multiplying of the number of remote controlled firing boxes. This blast synchronization technology again varies from one manufacturer to the next.

The electronic blasting equipment offers the engineers the possibility to design a blast in multiple different ways. The critical feature is not only the programming of the detonators at the desired timing but also the transfer of the blasting information to and from various system components. This feature is not only for transferring blast plans from bench programmers to firing units but also for downloading/uploading history files or software upgrades. Various solutions are proposed, from no transfer at all to total wireless data transfer with Infra-red or RFID technology, a USB port, or cables

2.2 Nonel Vs. Electronic Detonators

2.2.1 Non electric detonators

Non electric detonators use a shock tube detonator designed to initiate explosions, generally for the purpose of demolition of buildings and for use in the blasting of rock in mines and quarries. Instead of electric wires, a hollow plastic tube delivers the firing impulse to the detonator, making it immune to most of the hazards associated with stray electric current. It consists of a small diameter, three-layer plastic tube coated on the innermost wall with a reactive explosive compound, which, when ignited, propagates a low energy signal, similar to a dust explosion. The reaction travels at approximately 2,000 m/s along the length of the tubing with very minimal disturbance outside of the tube.

2.2.2 Electronic detonators

In civil mining, electronic detonators has better precision (0,02%) in regards to delays. Electronic detonators are designed to provide the precise control necessary to produce accurate and consistent blasting results in a variety of blasting applications in the mining, quarrying, and construction industries. Electronic detonators may be programmed in 1-millisecond increments from 1 millisecond to 10 000 milliseconds using the dedicated programing device called the programming unit.

Benefits:

- 100% verification of reliability of connections in initiation network.
- Delay range of 1- 10 000 ms with an increment of 1 ms.
- Precision of 0.02% of nominal delay time.
- Safe and reliable initiation of up to 1500 units in one blast per one blasting machine.
- Unique ID in each detonator.
- Multiple verification of detonators prior each blast.
- Wireless firing.
- On bench safety.

2.3 Effect of Accurate Delays on Blasting

Many tests have been performed to show the different accuracy between the Non electric detonator and the electronic detonator. The accuracy of the non-electrical detonator may reach ±7% or more, which means for a hole supposed to fire at 400 ms, it could fire at 372ms. If we use the 17ms connector between two holes loaded with 400ms downhole detonators, the second hole will be fired before the first one (1st hole 400 ms and the 2nd hole 372ms+17ms = 382ms). This kind of situation can lead to:

- Fly rock
- Poor fragmentation,
- Air blast increases
- Vibration increases

Another benefit from the accuracy is to have what we plan and this will help to have the right correlation between the blast parameters and the blast result.

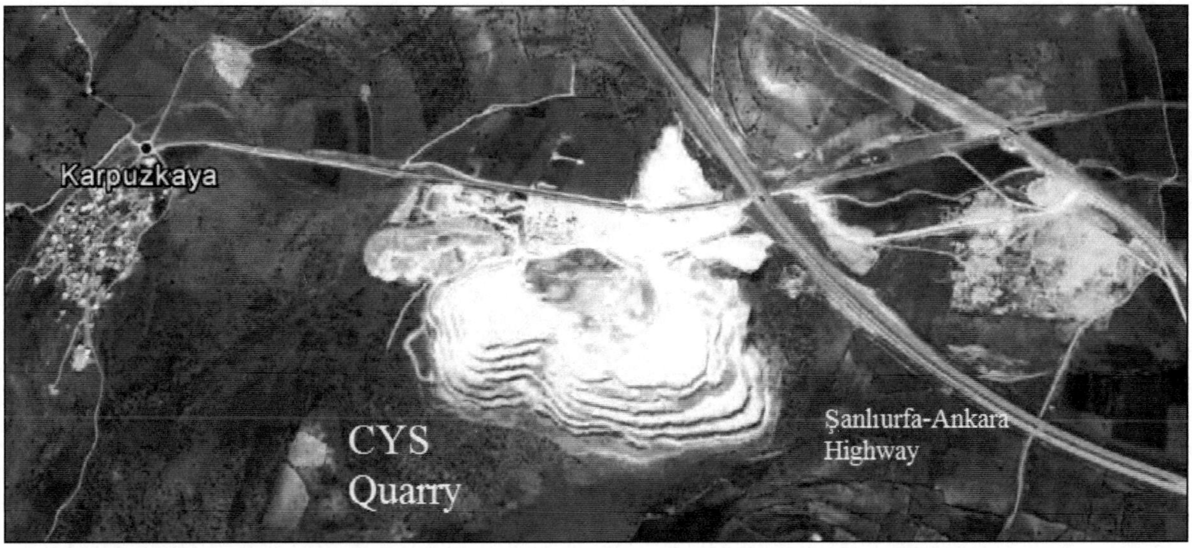

Figure 3. Location of the CYS quarry

3 CASE SITE

3.1 Location

CYS limestone quarry is located in district of Gaziantep in the south-eastern part of Turkey. The Sanlıurfa-Ankara highway passes through the northern part and an urban area called Karpuzkaya is 800m away from the western part of the quarry (Fig.3).

The site uses non-electric detonators for initiation design and ANFO as the main charge in blasting operations. Blast holes are 89 mm and bench height between 13-15 m.

3.2 Mining Issues

CYS initially decided to use electronic initiation systems not only to improve neighbor relations through the control of blast induced vibrations but also increase the efficiency of blast.

As the quarry is closely borders a neighborhood (within 500m) the need to reduce peak particle velocity and also control the neighbors' perception of blast vibration was essential.

3.2.1 Nonel blasts and vibration measurement

Thirty blast holes were initiated by using nonel detonators. The blast parameters are given in Table 1.

Table 1. Nonel blast parameters

#Blastholes	each	30
Hole diameter	mm	89
Bench height	m	13
Hole depth	m	13.5
Burden	m	2-2,5
Spacing	m	2.9
Stemming	m	2.3
Column Charge	kg	56

The most commonly used application for nonel blasts were applied to evaluate both the blast efficiency and vibration results. The 30 Blast holes were divided into six groups and each group of five holes were initiated separately. Only the results of the first and second group are showed in Figure 4.

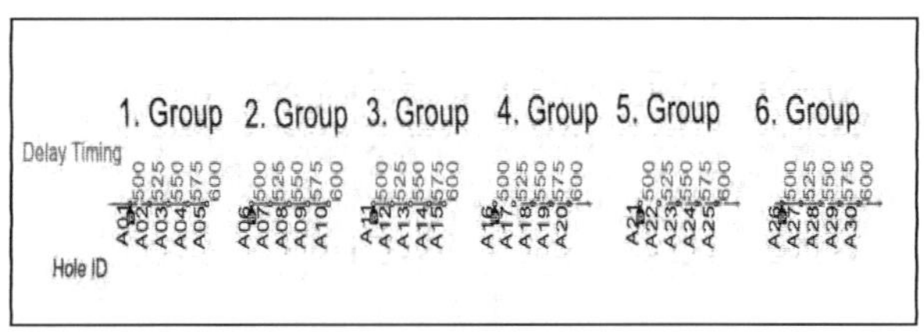

Figure 4. Initiation sequence of Nonel blast

The delay between holes was 25ms and 500ms in hole delay was selected for all holes. The spacing of blast holes is 2.9m but it increases up to 6 m between groups. The accelerometer of the vibration monitor was coupled to the ground in front of the house closest to the blast location as to record the blast induced vibration waveform as well as the wave propagation velocity.

Table 2 summarizes the peak particle velocity recorded from nonel blast which was initiated 820 m away from the monitoring location.

Table 2. Nonel blast induced vibration records

	Tran	Vert	Long	Unit
PPV	0.331	0.292	0.528	mm/s
ZC Frequency	14.0	16.4	16.0	Hz

From the readings, it was determined that the recorded vibration levels, was far above the allowable peak ground vibration levels which is induced in the ground outside the nearest structure due to blasting at mines, quarries and similar sites by Turkish regulations (Tab. 3).

Table 3. Allowable peak ground vibration levels

Vibration Frequency (Hz)	Maximum Allowable Vibration Velocity (Peakvalue-mm/s)
1	5
4-10	19
30-100	50

(Velocity limit lines rise from 5 mm/s to 19 mm/s in the frequency range 1 Hz-4Hz, and from 19 mm/s to 50 mm/s in the frequency range 10 Hz-30Hz in log-log graph)

The blast result was also poor due to the insufficient application. The six-meter spacing between each group resulted in bad fragmentation and toe problems as indicated in Figure 5.

Figure 5. Nonel poor fragmentation and toe problem

3.2.2 Electronic blasts and vibration measurements

To a certain extent, previously single hole firing had been controlled by limiting blast sizes when using nonelectrical initiation systems. However, with the use of electronic detonators, the number of blast holes was no longer a limitation. This was done by ensuring that the timing delay for each hole would not initiate within 8ms of each other.

The initial step was to design each blast hole as to minimize the vibration levels based on the existing parameters but without the additional spacing between the groups. There is no need to divide holes into the groups when using an electronic detonator and initiation system.

The different design parameters indicated in Table 4 were used in the blasts where the electronic initiation systems were applied. Delays of 30ms to 50ms were selected between the holes.

Table 4 Electronic blasting design parameters

Parameters		1st Blast	2nd Blast	3rd Blast
#Blastholes	each	40	47	57
Hole Diam.	mm	89	89	89
Bench height	m	13	13	15
Hole depth	m	13.5	14.3	15
Burden	m	2-2,5	2.9	3.75
Spacing	m	2.9	2.9	3.6
Stemming	m	2.3	2.5	3.25
Column Charge	kg	56	65	67

The recorded values of ground vibration are presented in Table 5 while monitoring from the closest neighbor house during the electronic blasts conducted at CYS quarry.

Table 5 Electronic blasting vibration measurements

# Blast	Particle Velocity (mm/s)			Charge (kg)	Distance (m)	Dominant Frequency (Hz)			Device Location
	Tran	Vert	Long			Tran	Vert	Long	
1	1.348	0.402	0.654	56	790	12.5	21.8	11	In House
2	0.378	0.284	0.473	65	900	13.6	12.2	16	Garden
3	0.762	0.889	0.635	67	790	10	20	20	In House

Blast induced vibration velocity of all blasts recorded at the closest house in the neighborhood district is less than the limit of 5 mm/s as set in Turkish Regulations.

As it is seen from the results, the first electronic blast has higher vibration levels compared to the others. It is known that high levels of vibration within structures are caused by a close match between the ground vibration frequency and some structural elements (Djordjevic et al., 1990). Structures have additional effect on blast induced vibration.

In addition to the positive outcome in terms of vibration, the quarry also recognized the improvement in fragmentation as indicated in Figure 6.

All the blast holes fired in a single shot during the electronic blast while improving the neighbor's perception of the blasting at CYS quarry.

Figure 6 Electronic blast muckpile

3.2.3 Electronic and nonel blast comparison

The lowest levels of ground vibration that can be perceived by humans is about 1.5 mm/s and under special conditions this may be as low as 0.5mm/s (Bilgin et. Al. 2000). Neighbor response to blast induced ground vibrations is the most decisive parameter for the control of ground vibration and blast size at CYS quarry. As it is seen from the vibration records, firing all the blast holes with electronic detonators did not create vibrations higher than the level of the human response limits even inside the house.

The shortest distance between monitoring location and blast site in electronic blasts was 790m (Fig.7). The highest peak particle velocity induced from electronic blast 2 was recorded as 0.473mm/s at 900m in front of the house (garden), whereas it was recorded 0.528 mm/s from the nonel blast in same location. The control of the blast induced vibrations without limiting blast size was achieved by using high accuracy delays with help of electronic detonators.

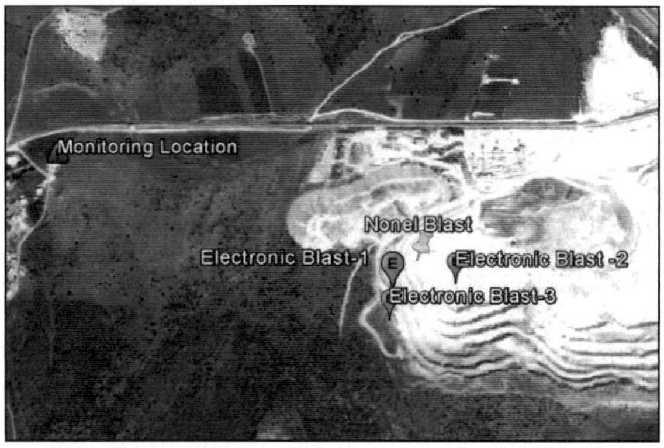

Figure 7 Location of monitoring station and blasting blocks

3.2.4 Electronic blast comparison

The vibration is produced by the radiation of energy from the blast. The blast occurs in geology that has complex structure and is not easily defined by just one or two mechanical parameters.

Current methods of blast vibration prediction may be categorized into five procedures:

- Historical data review
- Charge weight scaling laws
- Waveform superposition
- Scaled charge weight superposition
- Analytical and/or numerical methods such as finite difference or finite element

These five procedures are not distinct, nor mutually exclusive and several of them may be used in any given application.

In our discussion, we will focus on the waveform superposition approach. The radiation of the vibrations from a blast involves waves created at each hole and these combines at the point of interest after traversing the ground (or air). In this process, the superposition of the different waves and nature of the superposition may be characterized as linear or nonlinear. Anderson 2008 provides a recent review of linear waveform superposition and Blair 2008 describes one non-linear superposition case.

If we come back to our trials at CYS, the first test where we performed the blast by group of 5 holes with same distance between group of holes and where we only applied 150ms between a group and 30 to 50ms between holes of the same group. In Figure 8 we can observe that the 150ms between a group help to reduce the maximum amplitude of the vibration and this only by avoiding the waveform superposition.

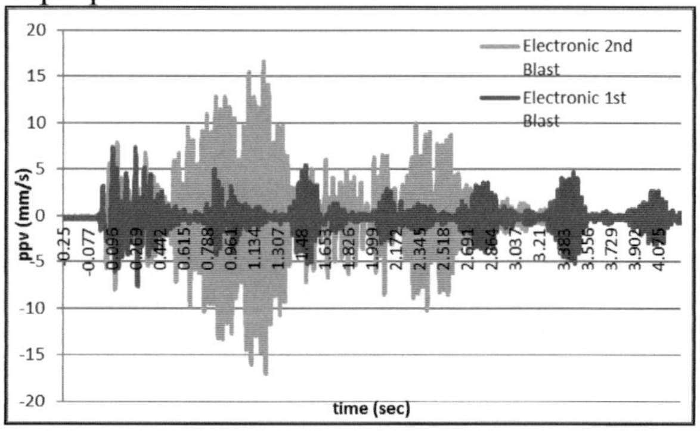

Figure 8. Superposition of waveforms

In this comparison, the monitoring station is located 150m away from the blast site. As it is clearly seen from the Figure 8, the superposition of the waveform can be minimized by applying long delays between holes.

Many quarries have to use a decking blast to reduce the vibration level. This observation can be applied for the quarry where the use of the costly decking can be replaced by the blast using a longest timing between a group of holes and this without losing in the quality of the fragmentation. This method is only possible by using the electronic initiation.

4 CONCLUSION AND RECOMMENDATION

The electronic delay detonator trial at Gaziantep CYS quarry clearly demonstrate that the electronic initiation decrease the vibration level and induce an excellent fragmentation. The size of the blast can be also increased and the shape of the blasted hole can be adapted to improve the loading equipment.

Since the nineties, the Bosman, Bedser and Cunningham (1998) agreed with the results seen by Duniam (1996) which showed vibration levels can be held within a narrow band when electronic detonators are employed. The ability to greatly reduce variability in blast vibration allows a site to 'tune' blasting parameters in order to optimize a blasting without fear of spurious events putting the ground vibrations over the limit.

This way, we recommend for the next step of the study to setup a vibration monitoring in the quarry and after use a numerical model to found out the right timing to apply for each blast to decrease the environmental effect of the blast and increase the production level at a low cost.

ACKNOWLEDGEMENTS

The authors are grateful to CYS Gaziantep carry and to Ideal explosive for their input to this trial. The on-site assistance given by Gaziantep CYS quarry personnel in the trial is also acknowledged.

REFERENCE

Bilgin, H.A., Esen, S., Kılıç, M. and Aldaş, G.G.U., 2000, Investigation of Blast Induced Ground Vibrations at Yeniköy Open Pit lignite Mine, 4th Drilling and Blating Symposium, Ankara, pp. 147-158

Bosman, H, Bedser, G and Cunningham, C, 1998; Production blasting with electronic delay detonators detonators at Peak Quarry in Proceedings Institute of Quarrying South Africa, Annual Conference Durban

Chapot P., Vibrations: Critères de nuisances et prediction. Revue de l'industrie Minérale Mines&Carrières – Les Techniques , 1991.

Davey Bickford, Various User Manuals, Product Literature, Internal Research Documents, 2000 thru 2012

Djordevic N., Kavetsky A., and Scott A., 1990, Blast Design Optimization to Minimize Unduced Vibrations of Structures, International Journal of Blasting and Fragmentation, pp. 373-380

Duniam P, 1999 Electronic Blast Initiation – A System for the New Millennium in Proceedings Institute of Mining and Metallurgy Australia, EXPLO Annual Conference Kalgoorlie.

Kara S, Adamson W, Reisz W, Trousselle R, 2013 The latest generation of the electronic system Enhanced safety and Productivity Procedia Engineering SYNPHO 2013 Agadir.

Paley, N., 2010, Testing Electronic Detonators to Increase SAG Mill Throughput at the Red Dog Mine, Proceedings of the 36th Annual Conference on Explosives and Blasting Technique, Volume II, February 7 – 10, 2010, Orlando, FL

Migairou, P., Case Studies Demonstrating Electronic Initiation Versatility, Proceedings of the 35th Annual Conference on Explosives and Blasting Technique, Volume I, February 8 - 11, 2009, Denver, CO

Proposal of a Multimine Scheduling Approach in the Process of improving economics of a large scale mining copper project. (Case study of KOV pit and Mashamba East pit combined, in the Democratic Republic of the Congo)

P.Mukonki, A.Muhota

University of Lubumbashi,Lubumbashi,Democratic Republic of the Congo)

ABSTRACT: The Kamoto Oliveira Virgule (KOV) pit and Mashamba East open pit mines are both located in the Lualaba province in the Democratic Republic of the Congo, in Kolwezi. They represent a substantial portion of the copper deposit so called "Lambeau Geologique de Kolwezi". When considered separately, KOV pit has a very high strip ratio and 12 Years life, while Mashamba pit, last for 6 years. Optimization scenarios of KOV shows possibility of ore shortage due to the need of increasing the stripping at KOV pit, to achieve Mill target (6.5Mt of ore annually for KOV pit).waste material stripping means lower profit and lower NPV(Net Present Value). This paper outlines, as an alternative, how, by developing Mashamba pit simultaneously with KOV pit, we can balance the KOV waste stripping (improve KOV pit economics or NPV) so that we get to the required ore much quickly to achieve the Kamoto concentrator mill feed target.

1. INTRODUCTION

1.1 Site Location

Mashamba open pit and KOV pit are located in the city of Kolwezi, and are considered as a portion of a sedimentary deposit called "Lambeau Geologique de Kolwezi" located in the major copperbelt inside the Democratic Republic of the Congo. The main geological structures are summarized in the figure 1.

Mashamba East open pit mine has been initially optimized and provided a Life of Mine (LOM) of 6 years at 3.2Mt of ore per year as a target to achieve. The mine has been planned and provided 4 major pushbacks, while KOV pit has been run through the optimizer and provided a life of mine of approximately 12 years, with high strip ratio as mentioned precedently.

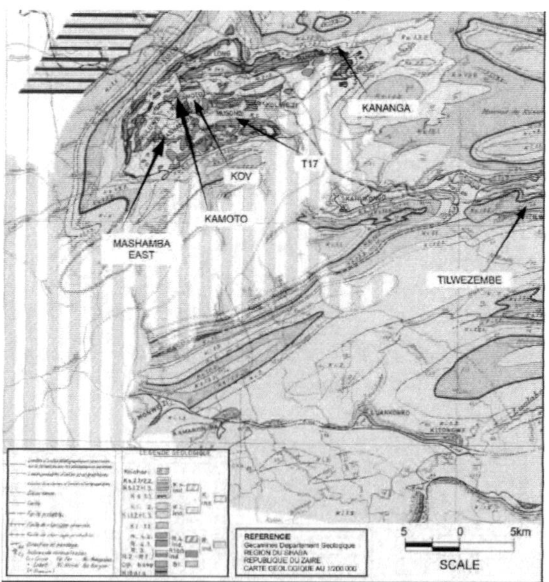

Figure1: Mashamba pit Geological settings by SRK independent technical report-March, 2009

Presented in this context, Mashamba open pit have to be considered as a satellite pit (located on the south of KOV pit (see figure 3) which is to be considered as the main pit). Locations of the two mines are shown on the figure 3.The Katanga Mining Limited property include a major facility called Kamoto Concentrator or KTC as a floatation plant where copper oxide, mixed and sulphide ore are processed before to be sent to the Luilu elecrowining unit for refinery and copper metal production. The property has also an underground mine facility called Kamoto Mine or KTO that we are not going to focus on, for the purpose of this paper.

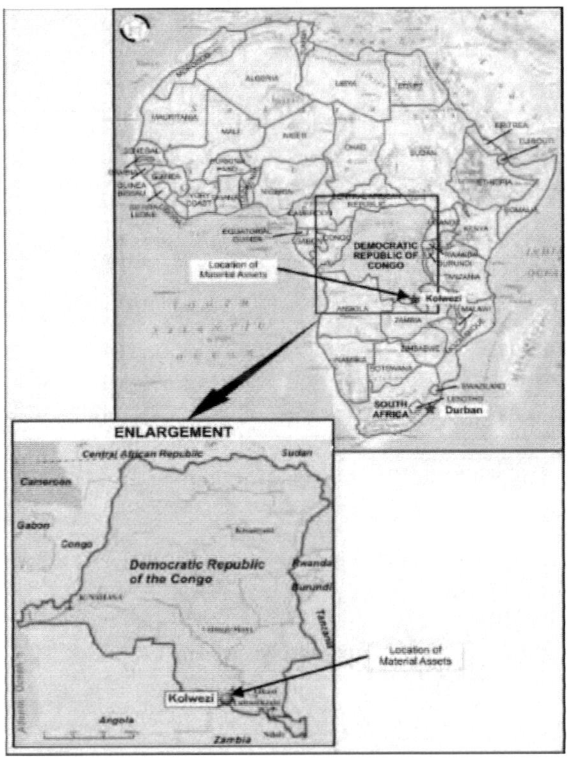

Figure2: Kolwezi town location in the Congo by SRK technical report

Lately in June 2015, the ore mill feed requirement in terms of ore daily throuput was raised from 17,000 tonnes per day to 22,000 tonnes per day, that increase in the ore mill throughput will, consequently , require an important waste material stripping activity, both in KOV pit and Mashamba East open pit, in order to be able to maintain the ore mill feed.

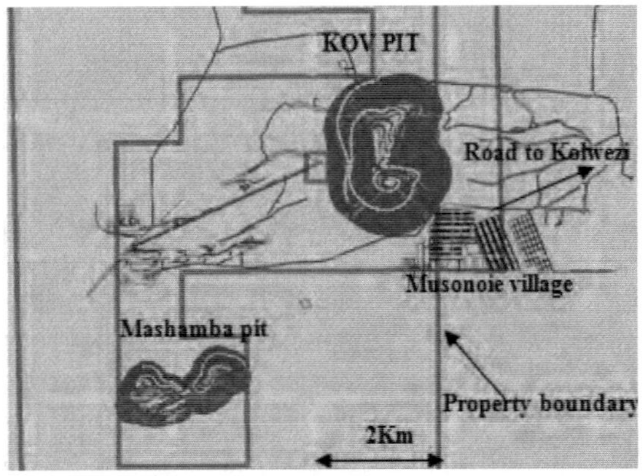

Figure3: Mashamba and KOV pit Location in the Katanga Mining limited property

KOV pit and Mashamba open pit deposits carries with them, copper and cobalt minerals, and, also copper grades that are not homogeneously distributed inside their entire respective mineralization, leading by the way, to challenging optimization process and mining schedules. This study is intending to put in place, or outline, a combined mining schedule that satisfies the mill feed requirement and also satisfies a well-balanced schedule (with an improved net present value), that allows KOV pit and Mashamba pit to be developed much easier while meeting the mill feed requirement.

1.2 Property ownership

Kamoto Copper Company SA (KCC) owns the Material Assets since 2006, including the mining and exploitation rights related to the Mining Assets. Katanga Mining Limited (KML) holds a 75% stake in KCC. La Generale des Carrieres & des Mines (GCM) and La Société Immobilière du Congo (SIMCO), state-owned mining companies in the DRC, own the other 25% of KCC.

2. METHODOLOGY & APPROACH

The approach that will be taken in this paper, will outlines KOV pit mine schedule taken separately, as well as the Mashamba East open pit mine schedule. Analysis and comments will be made regarding the two schedules as separated entities; Then KOV pit, and Mashamba pit, will be combined using the NPV scheduler Multimine tool (MMS, which is a Datamine mine scheduling module that is built to handle multiple block models),and a combined mining schedule that will be including the Mashamba block model and the KOV pit block model will be run ,analyzed, using the Multimine software module, and the results will be subject to comments and suggestions that will be made in regards of the observations. The expected outcome of the Multimine analysis is to find a less aggressive waste stripping ratio (below 12or 11),in order to release the pressure of the waste material stripping in KOV pit, while meeting the Kamoto Concentrator ,daily throuput ore requirement, by using Mashamba pit ore as a contributor to supply the Kamoto Concentrator.

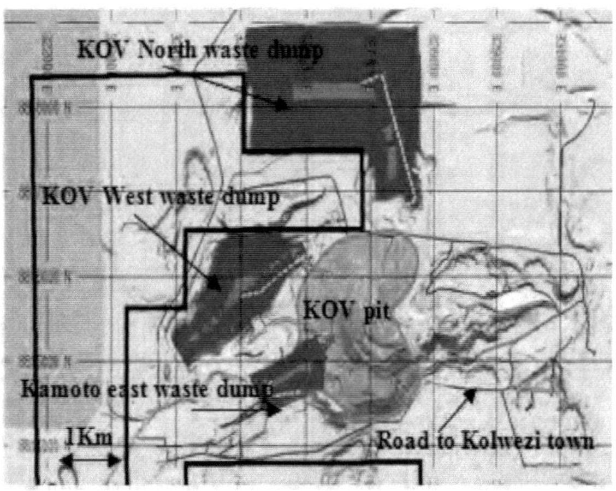

Figure 4: KOV pit and waste dump projected final configuration (sketch from the author)

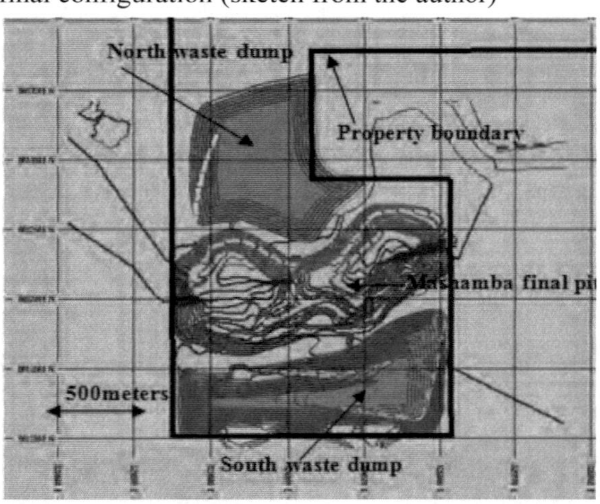

Figure 5: Mashamba pit and waste dump configuration (sketch from the author)

2.1 Kamoto Concentrator (KTC)

In the precedent section, we mentioned that the Kamoto Concentrator or KTC was capable to handle at the same time oxide, mixed and sulphide copper ore from both Mashamba and KOV open pit mines. Although the sulphide ore is likely a small portion of the entire open pit deposits (below 5% of the total open pit ore mass), it is very present in the Kamoto underground mine. The Kamoto Concentrator is designed with fairly old Outokumpu floatation technology, but was subject to several upgrades and it is capable to handle "7.5Million tonnes of ore per annum" (Katanga mining website key facts); at some section of the plant, upgrades have made the facility competitive, although the solvent extraction technology is still preferred. In the diagram below, a summary on how the Kamoto Concentrator is currently handling the run of mine.

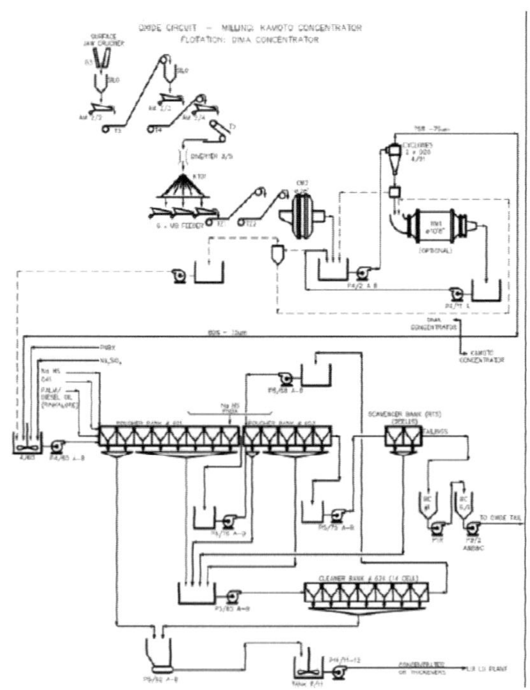

Figure 6: Kamoto Concentrator detailed flow chart from KRP-1june 06rev1-rdye_2 technical report.

2.2 Mashamba Pit Mining Schedule

As we mentioned precedently, the Mashamba East open pit has a mining schedule in which the ore target is 3.2Mt of ore and runs for 6 years ,below (in figure8),a plan view of the scheduler results on how the mine has been developed.

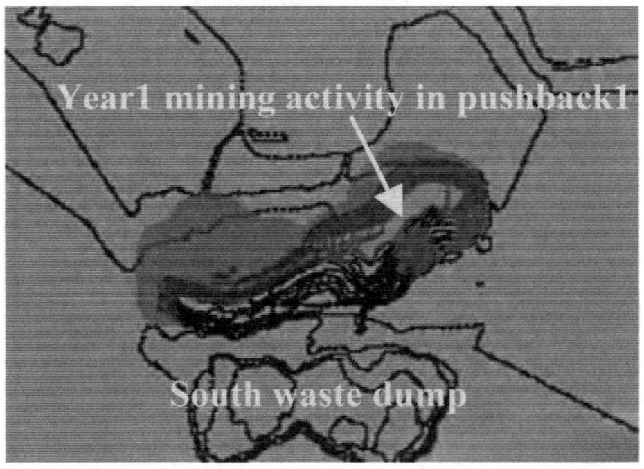

Figure 7: Year 1 of Mashamba schedule graphical sequence, as generated from the NPV software, in red, the mining activity (ore and waste mining), in blue is the topography of the optimized area.

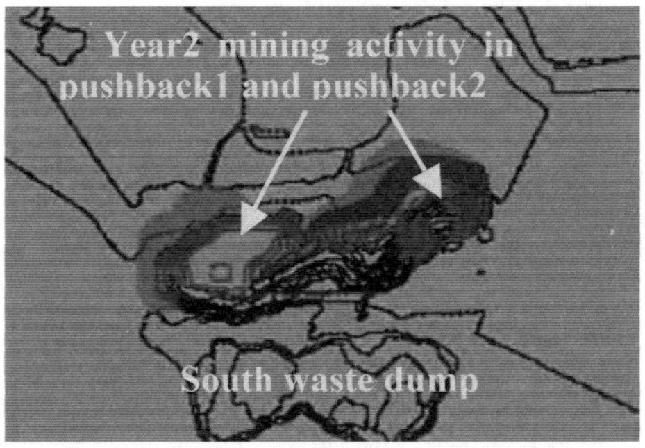

Figure 8: Year 2 of Mashamba schedule graphical sequence, as generated from the NPV software, in red, the mining activity (ore and waste mining), in blue is the topography of the optimized area.

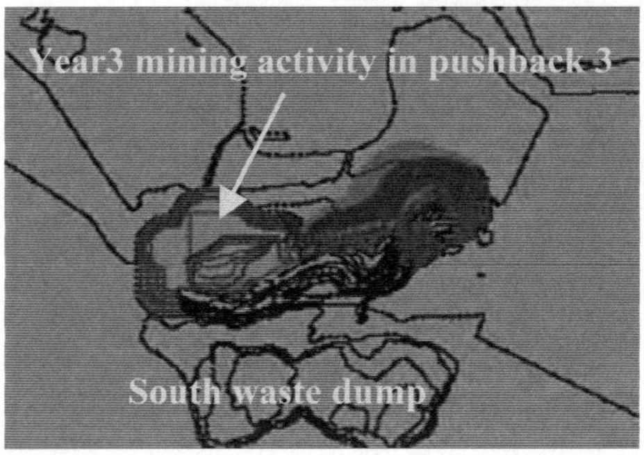

Figure 9: Year 3 of Mashamba schedule graphical sequence, as generated from the NPV software, in red, the mining activity (ore and waste mining), in blue is the topography of the optimized area.

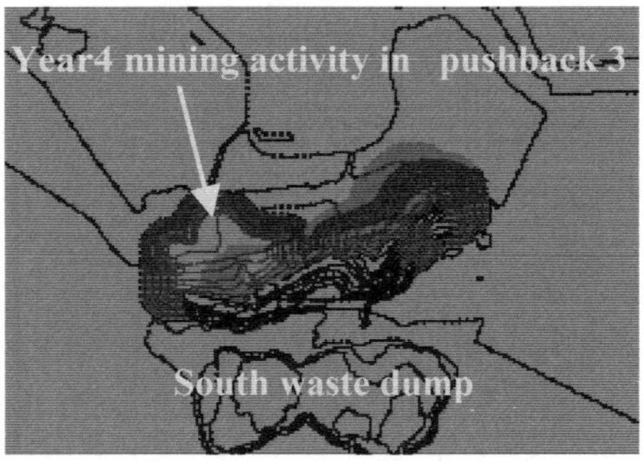

Figure 10: Year 4 of Mashamba schedule graphical sequence, as generated from the NPV software, in red, the mining activity (ore and waste mining), in blue is the topography of the optimized area.

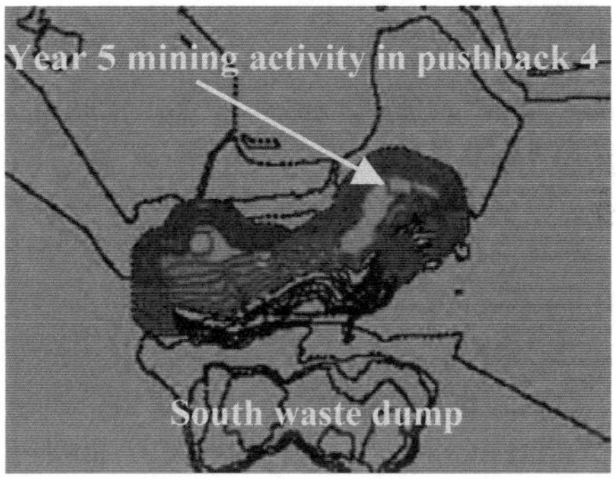

Figure 11: Year 5 of Mashamba schedule graphical sequence, as generated from the NPV software, in red, the mining activity (ore and waste mining), in blue is the topography of the optimized area.

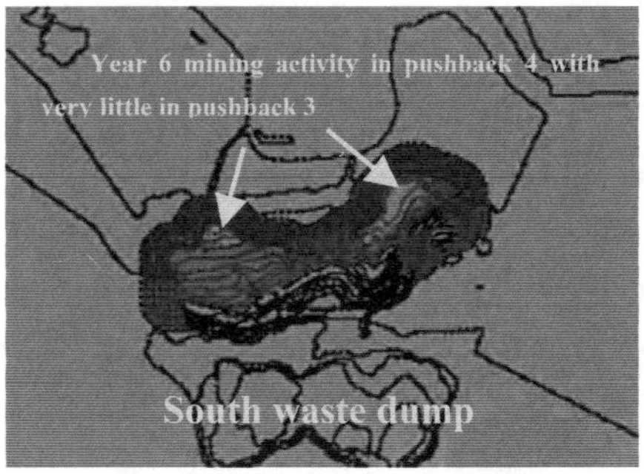

Figure 12: Year 6 of Mashamba schedule graphical sequence, as generated from the NPV software, in red, the mining activity (ore and waste mining), in blue is the topography of the optimized area.

Results of the figure 7-12 have to be homologated to the table 1 that shows the ore and waste quantities associated with the graphical sequence.

Table1.Mashamba east pit 6 years mining Schedule.

Year	Rock (Million Tonnes)	ORE (Million Tonnes)	Waste Million Tonnes)	Strip ratio
YEAR 1	40.3	3.25	37.05	11.4
YEAR 2	37.5	3.25	34.25	10.6
YEAR 3	29.2	3.25	25.95	8.0
YEAR 4	25.9	3.25	22.65	7.0
YEAR 5	26.8	3.25	23.55	7.3
YEAR 6	1.8	1.03	0.77	0.7
TOTAL	162	17.2	144.8	8.4

In order to measure the benefit of conducting the Multimine approach, both Mashamba and KOV pit mine schedules have been associated to the Net present value, that financial notion have the advantage of showing the necessity or not (as a decision maker) of applying such Multimine strategy
The figure 13, illustrates how behave the NPV along the Mashamba pit, 6 years schedule, compared to the yearly amount of ore and waste mined annually.

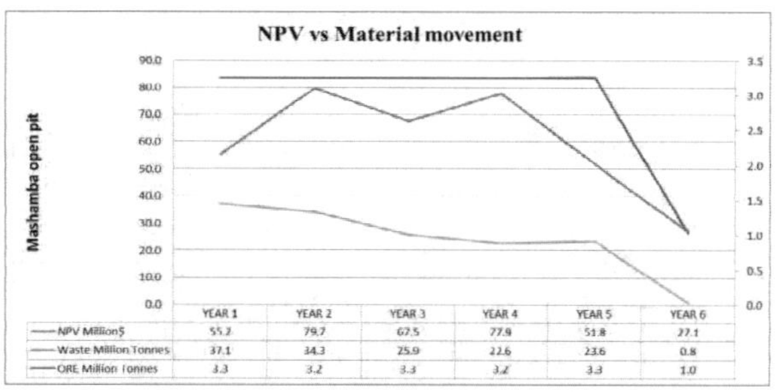

Figure 13: Mashamba pit 6 years ore and waste schedule homologated to the corresponding NPV (in million dollars) per annum.

2.3 KOV Pit Mining Schedule

The KOV long term mining schedule has been produced as per the agreed scenario, to provide a 12 years mining schedule as reported in the table 2 and shown on figure 14-25.

Table2.KOV pit 12 years mining schedule.

Year	Rock (Million Tonnes)	ORE (Million Tonnes)	Waste Million Tonnes)	Strip ratio
YEAR 1	103.17	7.10	96.07	13.52
YEAR 2	83.05	6.70	76.36	11.40
YEAR 3	54.77	6.00	48.77	8.12
YEAR 4	58.72	6.30	52.42	8.31
YEAR 5	68.30	6.70	61.59	9.18
YEAR 6	87.05	6.10	80.95	13.27
YEAR 7	39.80	6.00	33.81	5.6
YEAR 8	37.49	6.00	31.48	5.24
YEAR 9	43.59	6.00	37.59	6.26
YEAR10	49.03	6.20	42.83	6.90
YEAR11	7.22	4.29	2.93	0.68
YEAR12	1.43	1.06	0.38	0.36
TOTAL	633.64	68.46	565.18	8.25

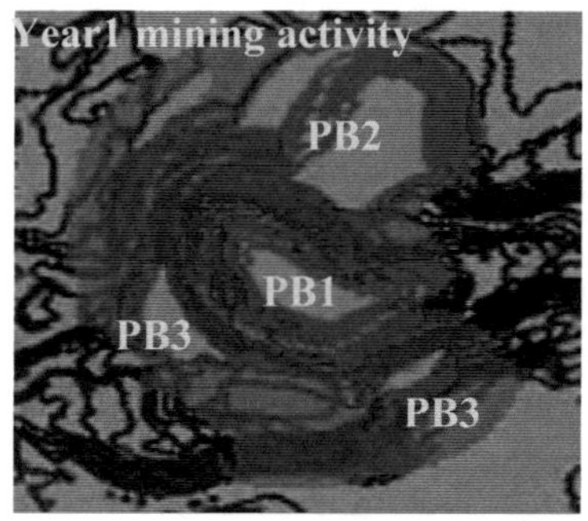

Figure 14: Year 1 of KOV pit schedule graphical sequence, as generated from the NPV software, in red, the mining activity (ore and waste mining), in blue is the topography of the optimized area. PB for pushback 1, 2 and 3.

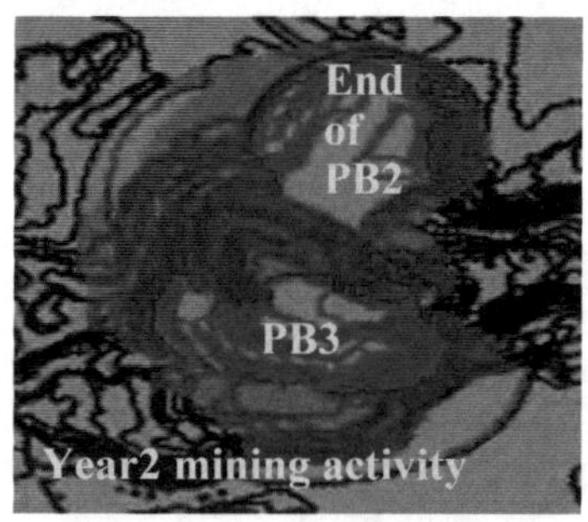

Figure 15: Year 2 of KOV pit schedule graphical sequence, as generated from the NPV software, in red, the mining activity (ore and waste mining), in blue is the topography of the optimized area. PB for pushback 2 and 3.

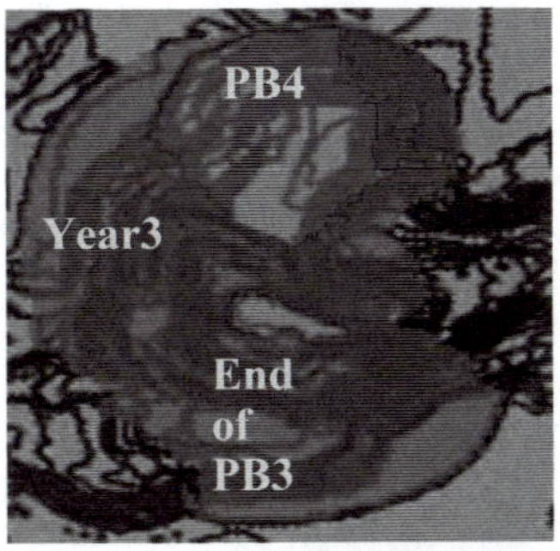

Figure 16: Year 3 of KOV pit schedule graphical sequence, as generated from the NPV software, in red, the mining activity (ore and waste mining), in blue is the topography of the optimized area. PB for pushback 3 and 4.

Figure 17: Year 4 of KOV pit schedule graphical sequence, as generated from the NPV software, in red, the mining activity (ore and waste mining), in blue is the topography of the optimized area. PB for pushback 6.

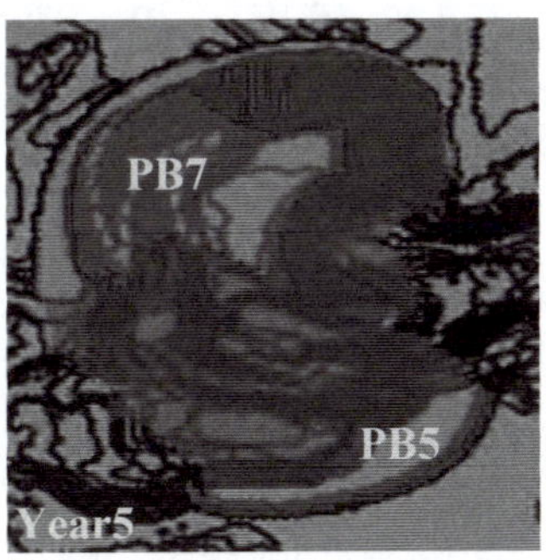

Figure 18: Year 5 of KOV pit schedule graphical sequence, as generated from the NPV software, in red, the mining activity (ore and waste mining), in blue is the topography of the optimized area. PB for pushback 5 and 7.

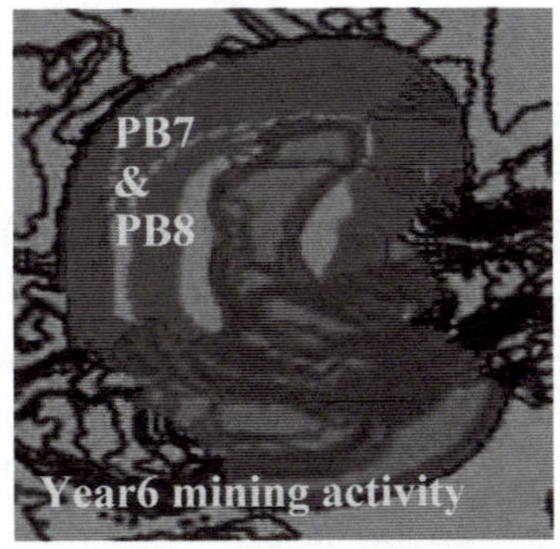

Figure 19: Year 6 of KOV pit schedule graphical sequence, as generated from the NPV software, in red, the mining activity (ore and waste mining), in blue is the topography of the optimized area. PB for pushback 7 and 8.

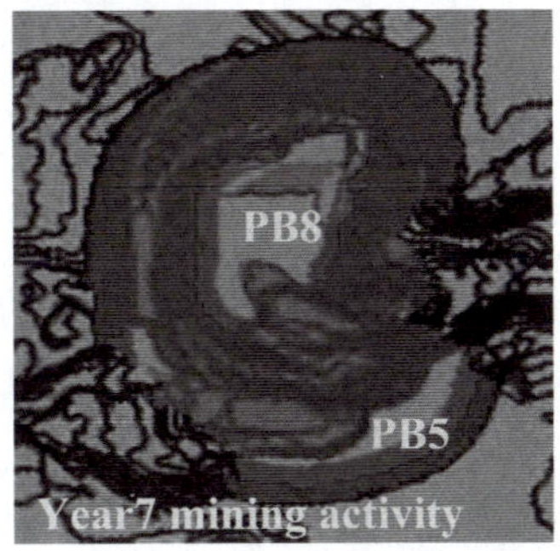

Figure 20: Year 7 of KOV pit schedule graphical sequence, as generated from the NPV software, in red, the mining activity (ore and waste mining), in blue is the topography of the optimized area. PB for pushback 5 and 8.

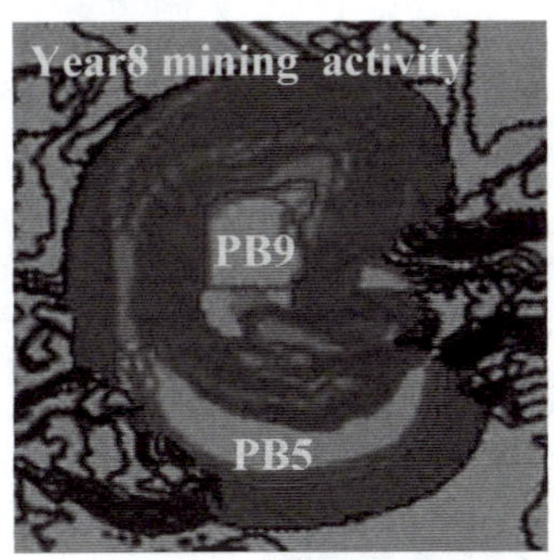

Figure 21: Year 8 of KOV pit schedule graphical sequence, as generated from the NPV software, in red, the mining activity (ore and waste mining), in blue is the topography of the optimized area. PB for pushback 5 and 9.

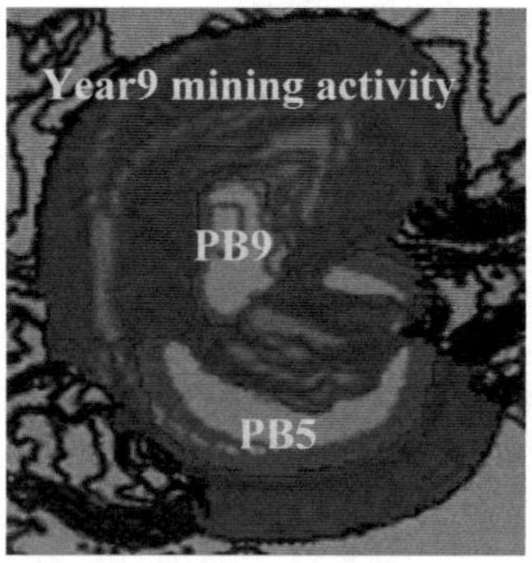

Figure 22: Year 9 of KOV pit schedule graphical sequence, as generated from the NPV software, in red, the mining activity (ore and waste mining), in blue is the topography of the optimized area. PB for pushback 5 and 9.

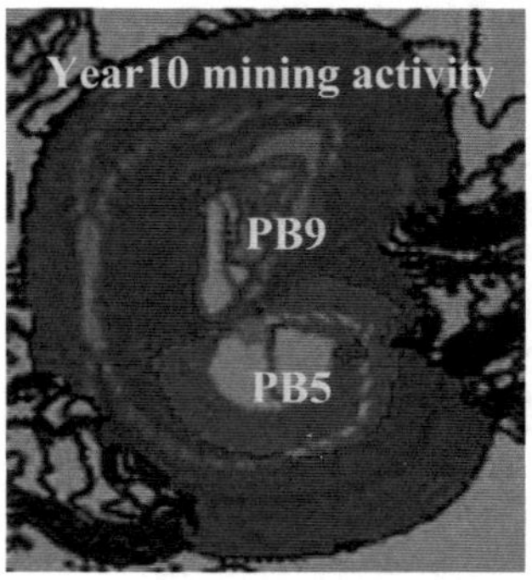

Figure 23: Year 10 of KOV pit schedule graphical sequence, as generated from the NPV software, in red, the mining activity (ore and waste mining), in blue is the topography of the optimized area. PB for pushback 5 and 9.

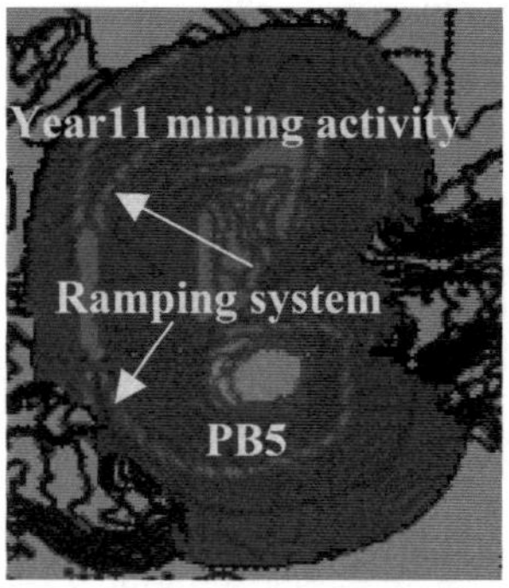

Figure 24: Year 11 of KOV pit schedule graphical sequence, as generated from the NPV software, in red, the mining activity (ore and waste mining), in blue is the topography of the optimized area. PB for pushback 5.

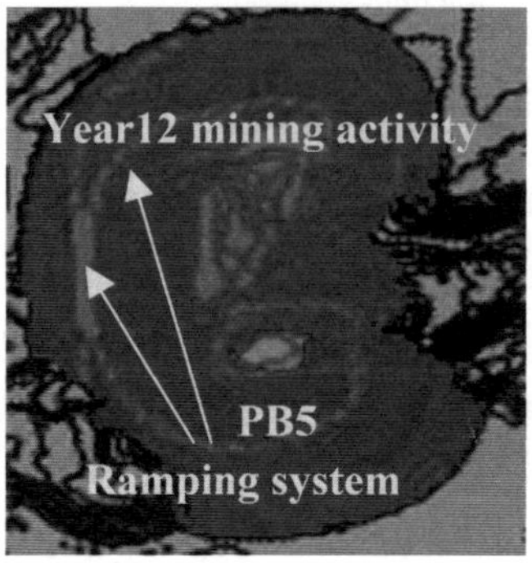

Figure 25: Year 12 of KOV pit schedule graphical sequence, as generated from the NPV software, in red, the mining activity (ore and waste mining), in blue is the topography of the optimized area. PB for pushback 5.

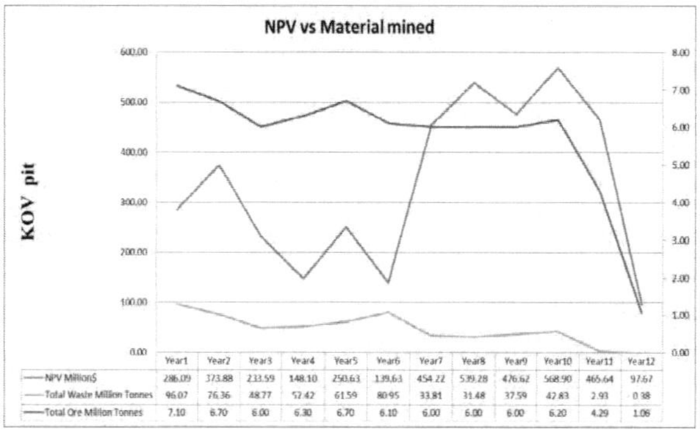

	Year1	Year2	Year3	Year4	Year5	Year6	Year7	Year8	Year9	Year10	Year11	Year12
NPV Million$	286.09	373.88	233.59	148.10	250.63	139.63	454.22	539.28	476.62	568.90	465.64	97.67
Total Waste Million Tonnes	96.07	76.36	48.77	57.42	61.59	80.95	33.81	31.48	37.59	42.83	2.93	0.38
Total Ore Million Tonnes	7.10	6.70	6.00	6.30	6.70	6.10	6.00	6.00	6.00	6.20	4.29	1.06

Figure 26: KOV open pit 12 years ore and waste schedule homologated to the corresponding NPV (in million dollars) per annum.

As mentioned precedently In order to measure the benefit of conducting the Multimine approach, both Mashamba and KOV pit mine schedules have been associated to the Net present value, that financial notion have the advantage of showing the necessity or not (as a decision maker) of applying such Multimine strategy
The figure 26, illustrates how behave the NPV along the KOV pit, 12 years schedule, compared to the yearly amount of ore and waste mined annually.

2.4 KOV Pit Compared to Mashamba Pit

When conducting an NPV comparison between KOV pit and Mashamba pit, taken separately, but assumed to be started at the same time, we realize that when Mashamba is getting better NPV, although, strip ratio is high at the first 3 years, KOV pit has lower NPV for the 6 first years, this is due to lower copper grades from year2 to year 6, and there is a jump from year 7 to the end due to the activity in the bottom of the KOV pit that contains higher copper grades. Mashamba East pit, at the same time had better grades than the first 6 years of KOV; this is an indication that, to sustain a decent mill feed, thus, to improve npv, the combined schedule approach can be an alternative to consider.

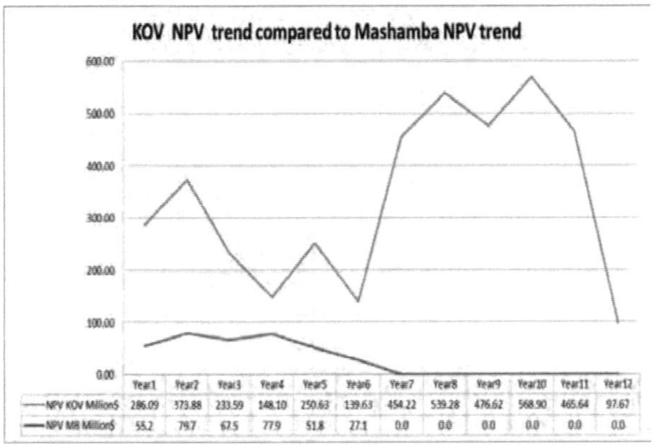

Figure 27: KOV NPV trend compared to Mashamba NPV trend.

Figure 27 illustrates the NPV trends both for KOV pit and Mashamba pit.

In this approach, higher grades means high copper contained in the ore, although, not authorized at this stage to disclose the copper grades, that notion is implicitly expressed in the npv. And this comparison shows that there is room for improvement for the KOV pit, waste material stripping, if we can put to contribution Mashamba East pit grades at the first half of the KOV schedule.

2.5 Combined Mine Schedules

Based on the preceding observations and assumptions, in this section, we are going to combine the KOV bloc model and the Mashamba East pit bloc model, to highlight the best case scenario that will lead to improve the npv for KOV pit. To be able to conduct such schedule, we are going to run the Multipit tool available in the NPV scheduler software. The economics and geotechnical constrains used to generate KOV ultimate pit and Mashamba East ultimate pit will remain the same, but the only difference is that, the tool will be using the two separate bloc models as a whole, to conduct the analysis. For this case, we are going to combine ore target for KOV pit (6.5Mt/year average) and Mashamba East pit ore target (3.2Mt/year) to make the final ore target to achieve at 9.7Mt of ore per year.

2.6 Multimine Tool (MMS) Description

In this paper, we are not going to develop the theoretical approach on the fundamentals of a combined Lersch-Grossman algorithm applied to a Multimine logic, but it is essential to keep in mind that the Multimine Scheduler (MMS) tool helps to determine the optimum limit of one mine and for all the mines considered in the analysis (minimum of two mines).The tool also provides an "optimal extraction sequence (OES)"(NPV scheduler V4 training course ,October 2012,p34) for the mines separately and for all mines combined, with the advantage of providing also an associated npv for all the mines combined.

3. RESULTS & DISCUSSIONS.

The study, as it was conducted, was based on the idea of improving the KOV economics by smoothing its waste stripping, using Mashamba pit as a satellite pit, and by combining KOV pit ore target (6.5Mt) to Mashamba pit ore target (3.2Mt), we realize that the combined life of mine is shortened by 3 years (9years for KOV pit) and also shortened by 3 Years for the Mashamba pit.

This approach tends to show that, to have a chance to improve npv for KOV, by putting Mashamba pit ore to contribution, we have to accelerate Mashamba pit mining by

50% of its total material mined (ore+waste) target every year of its initial life of mine (6 years) and we have to accelerate KOV waste stripping by 25% every year to reach high grades ore zones that are mostly located in the bottom areas of the KOV pit. Table 3 shows the resulting mining schedule, after combining KOV and Mashamba east ore targets.

Table 3.Combined mine schedules

Year	Rock (Million Tonnes)	ORE (Million Tonnes)	Waste Million Tonnes)	Strip ratio
YEAR 1	113.23	9.70	103.53	10.7
YEAR 2	87.28	9.70	77.58	8.0
YEAR 3	119.99	9.71	110.28	11.4
YEAR 4	122.91	9.70	113.21	11.7
YEAR 5	64.33	9.70	54.63	5.7
YEAR 6	109.75	9.70	100.05	10.3
YEAR 7	20.93	9.69	11.23	1.2
YEAR 8	145.23	9.71	135.52	13.9
YEAR 9	12.88	7.85	5.03	0.6
TOTAL	796.52	85.46	711.06	8.3

The red shaded coloured number shows that the combined total ore amount to be mined for KOV pit (68Million tonnes) and Mashamba pit (17Million tonnes) is 85 Million tonnes.

It can also be observed that for the first 3 years, we had very aggressive strip ratio for KOV pit (see table2), but after combining schedules, although, strip ratios still high, they are now relatively affordable to a mining stand point. It can be observed that for the first 3 years , Mashamba open pit is supporting KOV pit in term of copper grades so that, while KOV allows itself to develop lower grades areas, Mashamba East is covering up by developing higher grades zones.

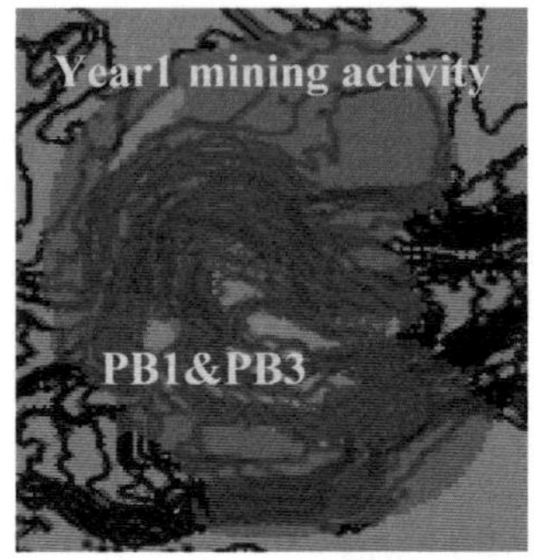

Figure 28: Year 1 of KOV pit accelerated schedule graphical sequence, as generated from the NPV software, in red, the mining activity (ore and waste mining), in blue is the topography of the optimized area. PB for pushback 1 and 3 is combined in a single activity.

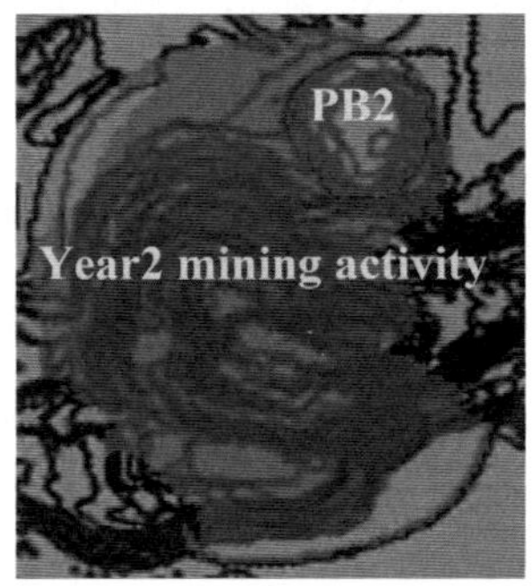

Figure 29: Year 2 of KOV pit accelerated schedule graphical sequence, as generated from the NPV software, in red, the mining activity (ore and waste mining), in blue is the topography of the optimized area. PB for pushback 2.

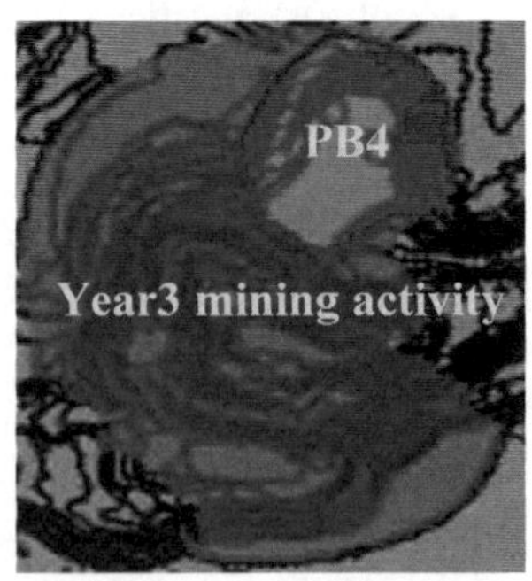

Figure 30: Year 3 of KOV pit accelerated schedule graphical sequence, as generated from the NPV software, in red, the mining activity (ore and waste mining), in blue is the topography of the optimized area. PB for pushback 4.

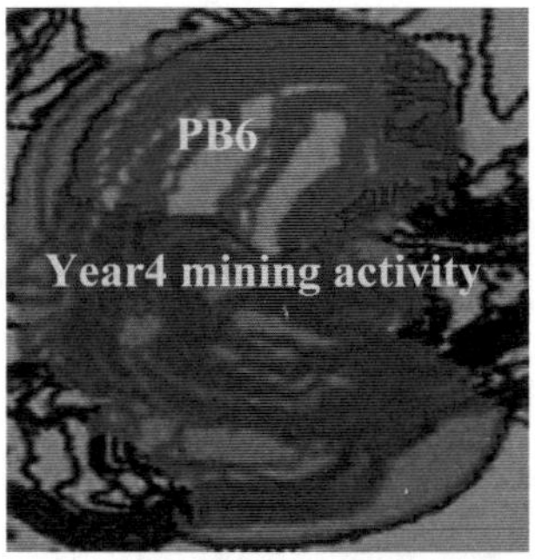

Figure 31: Year 4 of KOV pit accelerated schedule graphical sequence, as generated from the NPV software, in red, the mining activity (ore and waste mining), in blue is the topography of the optimized area. PB for pushback 6 combined with PB7.

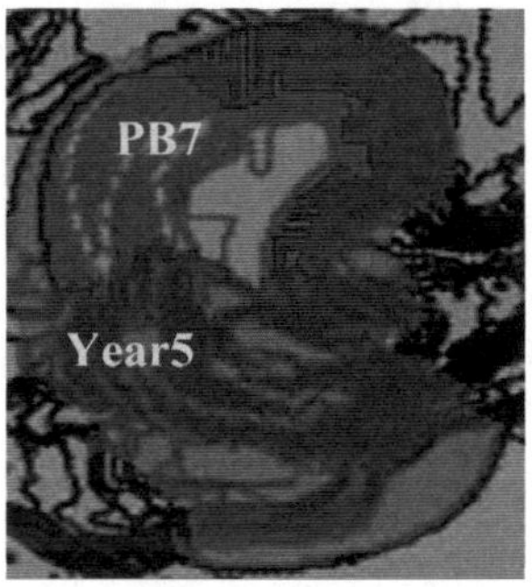

Figure 32: Year 5 of KOV pit accelerated schedule graphical sequence, as generated from the NPV software, in red, the mining activity (ore and waste mining), in blue is the topography of the optimized area. PB for pushback 7.

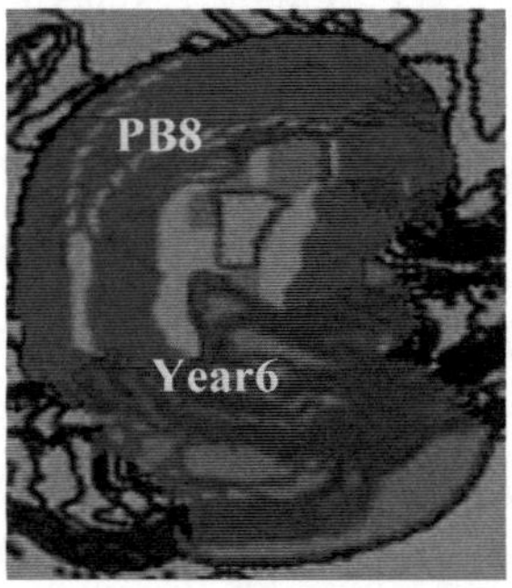

Figure 33: Year 6 of KOV pit accelerated schedule graphical sequence, as generated from the NPV software, in red, the mining activity (ore and waste mining), in blue is the topography of the optimized area. PB for pushback 8.

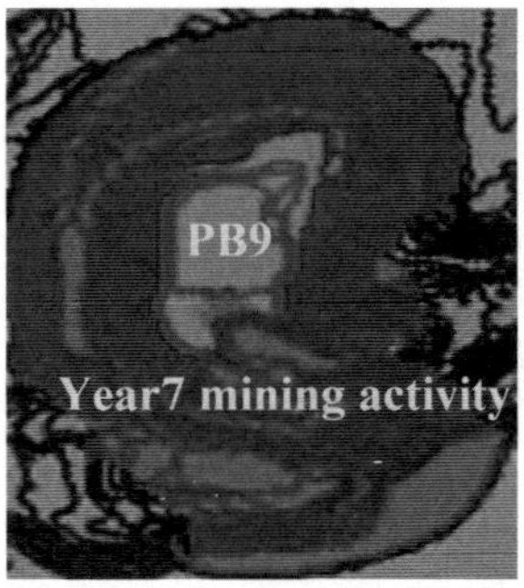

Figure 34: Year 7 of KOV pit accelerated schedule graphical sequence, as generated from the NPV software, in red, the mining activity (ore and waste mining), in blue is the topography of the optimized area. PB for pushback 9.

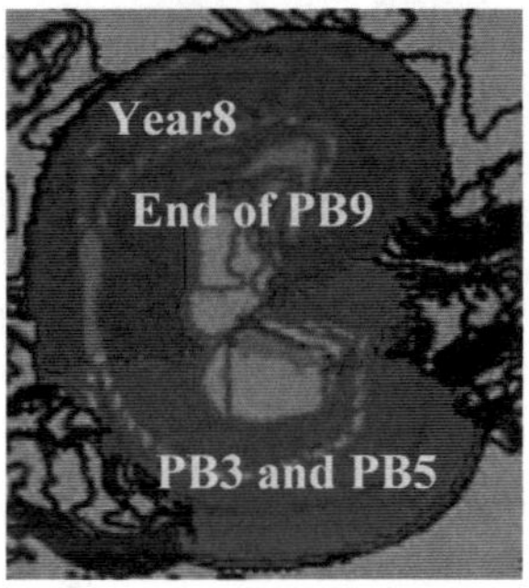

Figure 35: Year 8 of KOV pit accelerated schedule graphical sequence, as generated from the NPV software, in red, the mining activity (ore and waste mining), in blue is the topography of the optimized area. PB for pushback 3, 5 and the bottom of pushback 9.

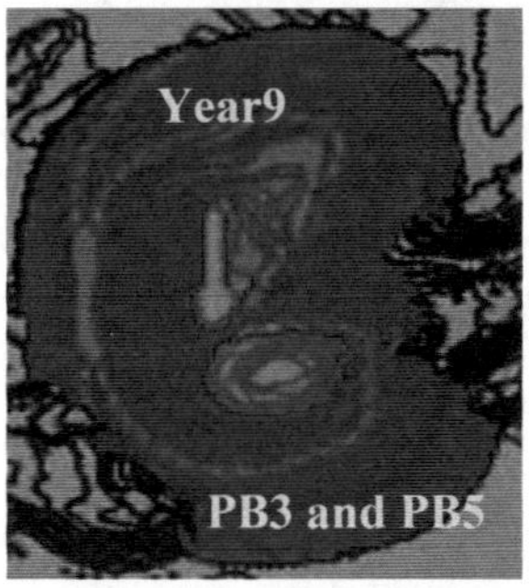

Figure 36: Year 8 of KOV pit accelerated schedule graphical sequence, as generated from the NPV software, in red, the mining activity (ore and waste mining), in blue is the topography of the optimized area. PB for pushback 3, 5 and the bottom of pushback 9.

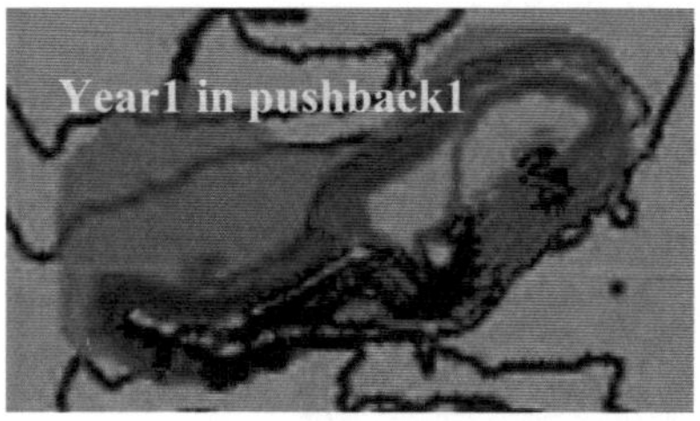

Figure 37: Year 1 of Mashamba pit accelerated schedule graphical sequence, as generated from the NPV software, in red, the mining activity (ore and waste mining), in blue is the topography of the optimized area. PB for pushback 1.

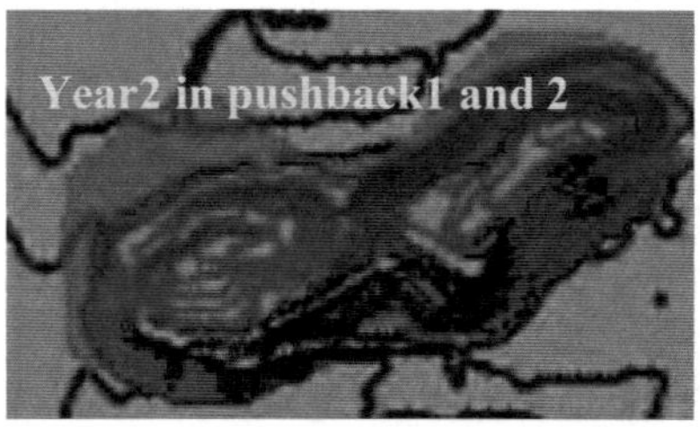

Figure 38: Year 2 of Mashamba pit accelerated schedule graphical sequence, as generated from the NPV software, in red, the mining activity (ore and waste mining), in blue is the topography of the optimized area. PB for pushback 1 and 2.

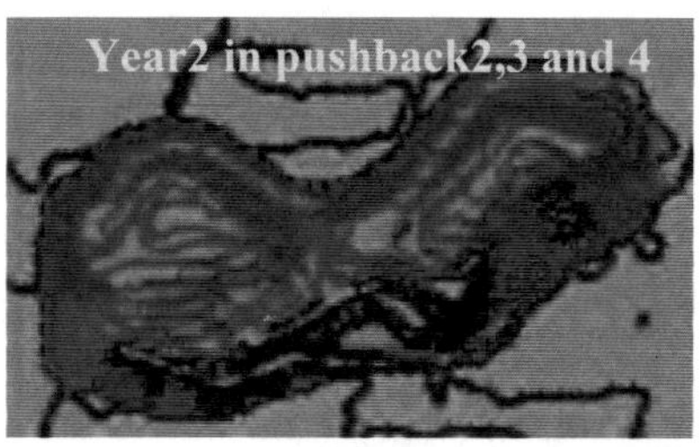

Figure 39: Year 3 of Mashamba pit accelerated schedule graphical sequence, as generated from the NPV software, in red, the mining activity (ore and waste mining), in blue is the topography of the optimized area. PB for pushback 2, 3 and 4.

The graph on figure 40 compares Mashamba npv of the figure 13, to the KOV pit npv of figure 26, with the resulting NPV from the combined schedule.

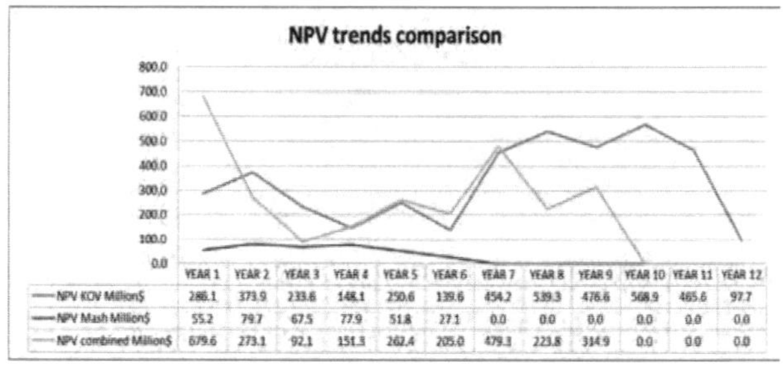

	YEAR 1	YEAR 2	YEAR 3	YEAR 4	YEAR 5	YEAR 6	YEAR 7	YEAR 8	YEAR 9	YEAR 10	YEAR 11	YEAR 12
NPV KOV Million$	286.1	373.9	233.6	148.1	250.6	139.6	454.2	539.3	476.6	568.9	465.6	97.7
NPV Mash Million$	55.2	79.7	67.5	77.9	51.8	27.1	0.0	0.0	0.0	0.0	0.0	0.0
NPV combined Million$	679.6	273.1	92.1	151.3	262.4	205.0	479.3	223.8	314.9	0.0	0.0	0.0

Figure 40: NPV trends comparison (in green, the combined KOV and Mashamba NPV).

When considering the graph on the figure 40, the combined schedule shows a very high npv during the first year, due, certainly to the combined higher grades from the bottom of the KOV pit and the Mashamba ore. Although, the combined npv drops below the KOV pit npv on year 3, it still higher than the Mashamba npv, when Mashamba pit is depleted on Year 6, we can observe that, the combined npv is almost the same as the KOV pit npv alone, the improvement here is mostly observed at the beginning, remember, it is always preferable to have early returns on investments. And the sooner the return is, the better it is for the shareholders. Findings demonstrate that, by combining Mashamba East open pit mine schedule to the KOV schedule, the generated high npv on year 1 allows to strip lower grades areas for KOV, despite the fact that between year 4 to year 6, the combined npv is lower, stripping is occurring at KOV and the npv picks up again from year 6 to the end. Remember again that early returns on investment are always better for investors.

4. CONCLUSIONS

This paper was conducted in order to outline the possibility of improving the schedule of the KOV pit by using Mashamba East open pit as a backup that can help to support the stripping issue at the KOV pit, as outlined along the development of this analysis. Aggressive stripping ratios observed at KOV have been slightly reduced, to an acceptable mining stand point. It is to notice here, that Mashamba pit mining rate have to be increased by 50% rate, in order to expose the ore that will give opportunity to KOV pit to develop and keep a positive NPV. After running several scenarios, the conclusion is that, the npv is good at the beginning (first third form year1 to year3) of the life of mine, and the last third of the combined schedule (from year6 to year9) this is an opportunity for a substantial improvement in the process of improving economics of a such large scale operation as KOV pit. This paper focused on the possibility of implementing the Multimine strategy, we have to agree here that more work has to be conducted in deep analysis, to check for equipments requirements, dumping available areas, and dump designs strategy to fit all the waste material inside the company property. More work still to be done, also on the side of the need of running a complex haulage analysis, to determine exactly, how much dump truck will be required and what will be the ore stockpiling and blending strategy. Deeper analysis will be conducted if possibility arises in the future.

REFERENCES

Hustrulid, W. and Kuchta, (3rd edn), 1998. Open pit Mine planning and design Vol1-Fundamentals, A.A.Balkema, Rotterdam-Brookfield.

Rendu, J.M, (1st edn), 2008. An Introduction to Cut-off grade Estimation, Society for Mining and Metallurgy, Littleton, Colorado.

Juan, P.C, 2002. Management of Mineral Resources, creating value in the Mining Business. Society for Mining and Metallurgy, Littleton, Colorado.

Ian, C.R, 1998. Mining Economics and strategy, Society for Mining and Exploration, Littleton, Colorado.

Elahi, Z.E., Kakaie. And Yousefi, A. 2011. A new algorithm for optimum open pit design: Floating cone method III, journal of mining and environment, Vol N°2, 2011, (pp.118-125).

Dagedelen, K.2001.Open pit optimization-strategies for improving economics of mining projects through Mine planning.*17th International Mining Congress and exhibition of Turkey-IMCET 2001.*

Whittle G. and Burks, S.2010.Simulatneous mining and mineral processing enterprise optimization for the platinum industry. The 4th International platinum Conference, platinum in transition "Boom or Burst". *The South African Institute of Mining and Metallurgy, 2010.*

Madowe, A.C. 2013.The mine planning process for an open pit diamond mine operation-A case study on Letseng diamond mine Lesotho. *The Southern African Institute of Mining and Metallurgy, 2013*

Alexandra, M., Kuchta, M., Martinez, M. Long and short term production scheduling at LKAB`S KIRUNA MINE.*Chapter1*

Dagedelen, K., 1987.Mine Production scheduling Optimization "the state of the Art". *Society for Mining and Metallurgy, Littleton, Colorado.*

Glencore International AG, February 2013.*Golder Associates, Mineral resources and Ore Reserves statement for Kamoto Copper Company. Report number 12614771-11884-3.*

NPV-Scheduler Reference Manual (CAE-Mining 2012).

KRP-1june 06rev1-rdye_2 technical report.

SRK independent technical report-March2009

Kamoto Concentrator Key facts. http://www.katangamining.com/operations/key-assets/kamoto-concentrator.aspx[21 February 2017].

Katanga Mining technical report http://www.katangamining.com/~/media/Files/K/Katanga-mining-v2/operations/reportsoperational/reop2006/technicalreport/techreport10.pdf[21 February 2017].

Functional Dependence of the Moving Forward Dynamics and Production of the Long Wall at the Example of Lignite in the Coal Mine „Kreka" –Bosnia and Herzegovina

Dr.sc. O. Musić[1], Mr. sc. H. Čičkušić[2], Dr. sc. R.Vugdalić[3]

[1]*University of Tuzla, Faculty Mining Geology and Civil engineering, Tuzla, BiH*

[2]*JP „Elektroprivreda BiH" d.d. Sarajevo- Coal Mines „Kreka" Tuzla*

[3]*University of Tuzla, Faculty of Science Tuzla*

ABSTRACT Within the complex mining and geological conditions the Lignite mines „Kreka" Mramor is working on excavation of the main coal layer using the long wall method with the mechanized technology of coal excavation. With selection of the optimal technical-technological parameter of the Long wall excavation such as follows: length and width of long wall, cutting height, production, moving forward dynamic, output act. and defining the functional dependence between the dynamic of moving forward ($n_č$) and production (Q), has been used the manufactories mathematic analyze i.e. least-squares method
The aim of this paper is to use the least-squares method to determine dependence between the dynamic of moving forward ($n_č$) and production (Q) with condition that excavation thickness is d_p = 3.2 m (const.), and Long wall length is $l_č$ = 60 m (const.)

Key words: Long Wall Method, Excavation Parameters, Least-Squares Method

1 LONGWALL TECHNOLOGY WORK

1.1 Preparation of the excavation field

Opening and preparation facilities at the mine „Mramor" can be put in two main groups, as follows:
-Opening facilities (GDVH-m - the main delivery-ventilation corridor, GTVH-m - the main transportation-ventilation corridor, GVU-m - the main ventilation raise, VU-m-ventilation raise and VO-m- ventilation shaft)
-Preparation facilities of TH and VH have been located in the facilities (TH - transportation corridor, VH - ventilation corridor, SU-m1/n..SU-m2/n,.SU-m3/n, - collection raise and ČU- m/n - Long wall raise).

Continuation of opening the excavation field (belt) is made with two parallel corridors: transportation corridor (TH) and the ventilation corridor (VH), with the axial distance of 60 m.
Preparation facilities of TH and VH have been located at the floor of the coal layer, with 1,5 m from the floor consisted of quartz sand, figure 1.

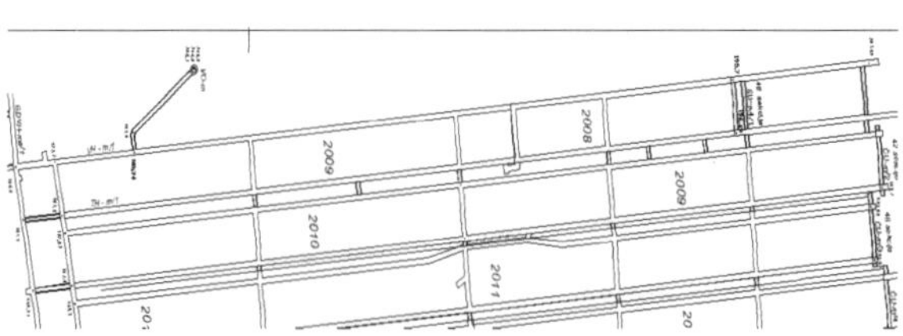

Figure 1. Preparation of the excavation field

1.2. Installation of the Equipment in the Long Wall Raise

Installation of Long wall equipment is being performed in following order.

-Installation of high-pressure pumps

-Assemblage and installation of power roof supports

-Installation of face conveyor along with installation of power roof supports

-Installation of shearer

-Installation of electrical equipment and energetic train

The way of installation of power roof supports in the long wall raise is presented at the figure 2.

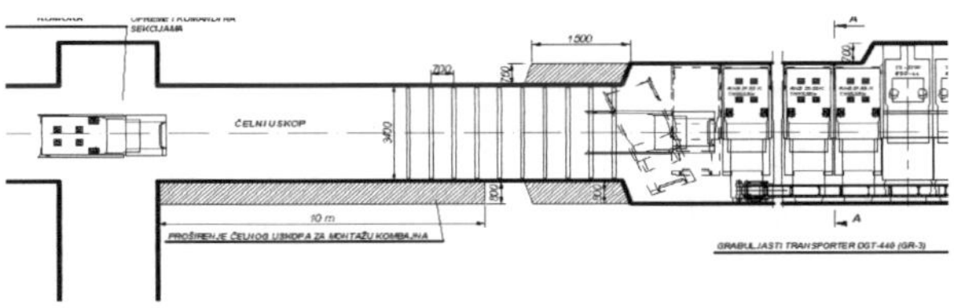

Figure 2. Installation of power roof support in the long wall raise

1.3. Technological Phases of Work

Technological working process at the Long wall implies work in the sub-level part of Long wall (horizontal concentration). As the average thickness of the coal layer is 6,08 m, the height of cutting sub-level part of the Long wall is 3.2 m, thickness of overlying coal strata is 1.3 m and underlying coal strata is 1.58 m.

Technological phases of work are as follows:

-first working operation: "sumping in" of the shearer

-second working operation: cutting of a full cut

-third working operation: cleaning work of shearer

-fourth working operation: cut the wedge shaped portion of coal left while "sumping in"

a) First working operation - "sumping in" of the shearer

First working operation starts in the moment when accomplished all conditions for "sumping in" of the shearer and cutting of a full cut. Work of the shearer at the Long wall (from ventilation towards the transportation corridor) is being performed gradually with "sumping in". In front of the shearer, in the phase of "sumping in", at the safe distance it has to be done indention of the roof parts at the power roof supports, and behind the shearer the roof parts at the power roof supports is to be pull out. After moving forward of the power roof supports the face conveyor has to be moved forward thought entire length, figure 3

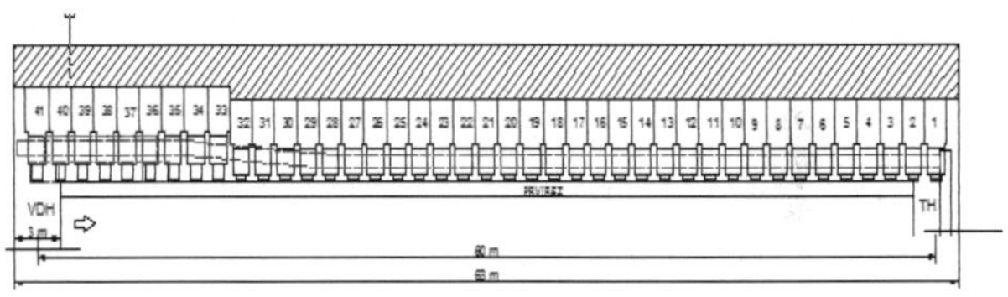

Figure 3. "Sumping in" of the shearer

b) Second working operation - cutting of a full cut

At this stage the work is carried out by the movement of the shearer thought the Long wall from the place of "sumping in" towards the chain scraper conveyor located in the transportation corridor (TH). Simultaneously with the movement of the shearer, when cutting is performed, it has to be done indention of the roof part at the power roof supports which located in front shearer, and behind the shearer the roof parts at the power roof supports is to be pull out. When the shearer come through at least 10 power roof supports, at the safe distance is done moving power roof supports and face conveyor and indention of the movable roof parts at the power roof supports, figure 4.

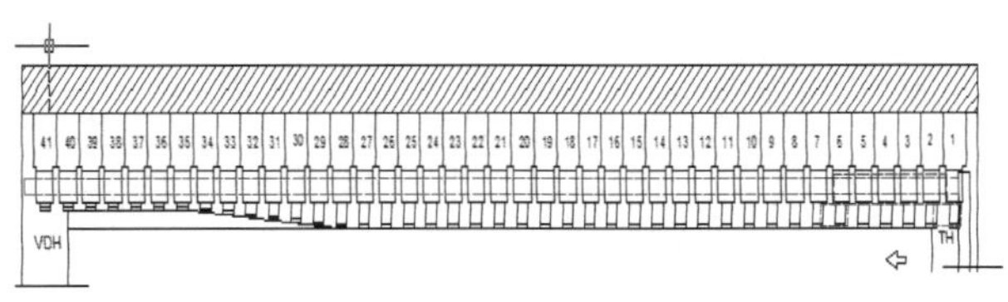

Figure 4. Cutting of a full cut

c) Third working operation - cleaning work of shearer

Before this working operation both shearer cutting drums are taken in position of cutting the down cut (i.e. position of cleaning cut). Conveyor taking position of the straight line, and power roof supports are moved forward next to the conveyer. While moving, the left shearer cutting drum cuts the rest of the down cut from TH, and the right shearer cutting drum cleaning the coal remain from the previous cut.

d) Fourth working operation - cut the wedge shaped portion of coal left while "sumping in"

Fourth phase is cut the wedge shaped portion of coal left while "sumping in". In front of shearer it has to be done indention of the roof parts at the power roof supports, and behind the shearer the floor parts at the power roof supports is to be pull out. With this, it is completed a complete technological cycle of cutting with accompanying working operations, figure 5.

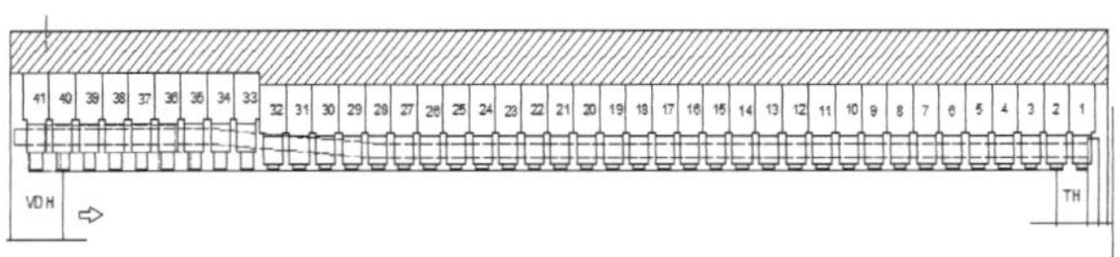

Figure 5. Cut the wedge shaped portion of coal left while sumping in

2.THE CONCEPT OF THE TECHNOLOGICAL PROCESS OF EXCAVATION

According to data from previous research, investment and project documentation it has been obtained the basic parameters for the dimensions and proportions of the excavation fields.

- Lenghtof the mine on the leyor's extend...........................1090 m
- Lenght on the leyor's decline...810 m
- Geodetic heigh (absolte)...79 to 108 m
- Angle of leyer's decline.........2° to 6°
- Thickness of leyer...................6,08 m

The concept of the excevition of the coal layer in the mine „Mramor“ is the following: coal layer thickness from 5.0 – 6,0 m (average 5,50 m) is being excavicted in one operation of thickness 3,2 m with the application of horizontal concentration excavation. The protective plate thickness 1,5-2,0 m is being left at the floor of the coal layer, and protective plate thickness 0,8-1,3 m is being left on the roof of the coal layer. Lenght of the long wall includes:

- 0,75 m from the floor side of the delivery-ventilation corridor - 0,75 m
- 41 power roof support x 1,5 m- 61,50 m
- 0,75 m from the roof side of transportation corridor - 0,75 m

The total lenght of the long wall is 63,00 m and it it presented at the figure 6.

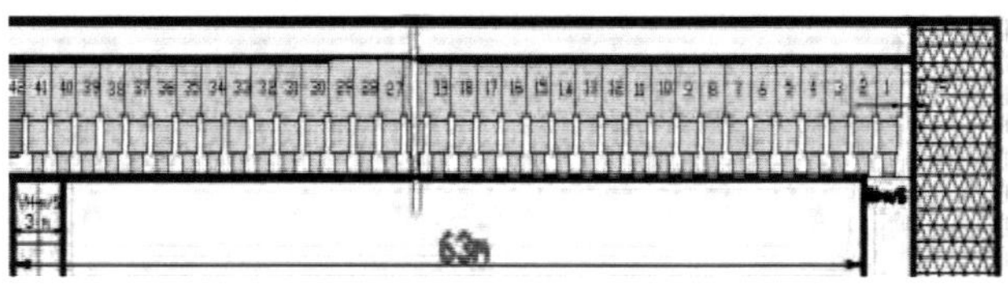

Figure 6. The geometry of the excavation field

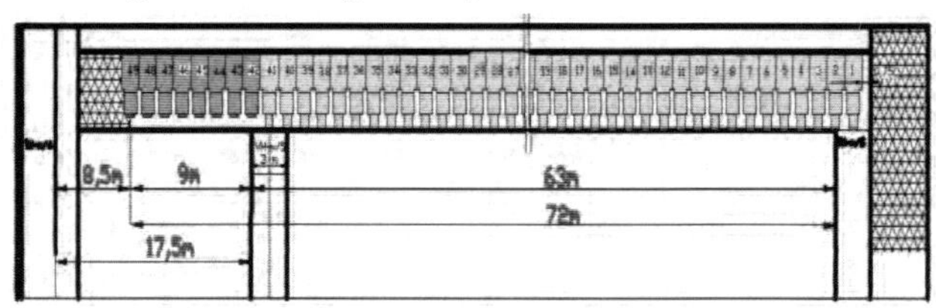

Figure 7. The geometry of the excavation field with protective pillar

In the section of one excavation field can be seen left protective panel of 1.5 m; cutting with the shearer of 3,2 m; roof panel left of 1,38 m. These dimensions are presented at the figure 8.

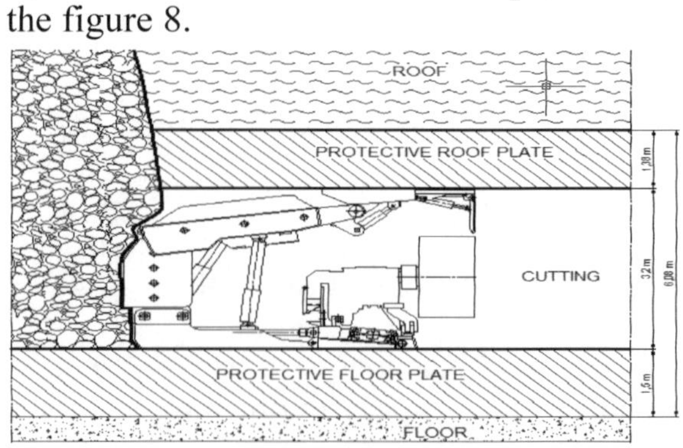

Figure 8. Long wall position in the layer

3. Monofactorial Analysis of the Investigated Parameters of the Long Wall

The task of this work

Find the functional dependence between moving forward dynamic ($v_č$) and production (Q) with condition if:

- thickens of excavation (d_p) = const. = 3,2 m
- length of the long wall ($l_č$) = const. = 60 m

Monitoring the work of the Long wall in five excavation belts led to the indicators where for production (Q) of 1286 t/day of lignite coal the moving forward dynamic ($v_č$) = 4,14 m/day.

For production of Q= 1304 t/day, $v_č$ = 4,20 m/day;

For production of Q = 1455 t/day , $v_č$ = 4,60 m/day

3.1. Application of the Least-Squares Methods – Theoretical Part

If a known function, or some unknown function, we have given points of data (x_i, y_i) $(i = 1, 2, \dots, m)$, then we have to determine the approximate function $y = f(x, a_0, a_1, \dots, a_n)$ which depends on the independent variables x and n+1 parameters $a_0, a_1, \dots, a_n$ which has to be determined from the conditions that is:

$$y_i \approx f(x_i, a_0, a_1, \dots, a_n) \text{ for all } i = 1, 2, \dots, m$$

So we have a system of m approximate equations with n + 1 unknown parameters. $a_0, a_1, \dots, a_n$. In order to solve the system of approximate equations it is necessary that the number of m data points (obtained from well-known functions, or experimental, in the case of unknown function) is greater than or equal to the number of unknown parameters. So, it should be valid $m \geq n + 1$. Parameters $a_0, a_1, \dots, a_n$ seek from requirement that the sum of the squares of values $f(x_i, a_0, a_1, \dots, a_n)$ from real value y_i for $i = 1, 2, \dots, m$, be least. Type of function f(x) (linear, quadratic, cubic, some degree polynomial, logarithmic, exponential, trigonometric, etc.) and the number of parameters it is estimated on the bases of the graphic layout of data points in the plane. If we introduce the function

$$S(a_0, a_1, \dots, a_n) = \sum_{i=1}^{m} [y_i - f(x_i, a_0, a_1, \dots, a_n)]^2$$

then the request of the least-squares method is that the function S has a minimum value as possible.

As the S is function of more changeable, more exactly, function n+1 changeable, for determination of minimum function S (a_0, a_1,...,a_n) a necessary condition of existence of extremes (local maxima or local minima) function S is that all its partial derivatives equal to zero, e.g. $S'_{a_k} = 0$ $(k = 0, 1, 2, \dots, n)$.

In the next calculate it would be applied least-squares method in case of approximation of production depending on the excavation dynamics using quadratic functions that depend on the three parameters a, b and c.

Then the S if the function of three variable, e.g. S=S(a,b,c). The best approximation of our problem of quadratic function will be achieved for those values of real parameters a, b and c for which the value of the partial derivative of the function S with the parameters a, b and c are equal to zero.

3.2. Manufactories Analysis Explored Parameters of a Long Wall with the Least-Squares Method

If in the coordinate system set dynamic progression on the x-axis, and the production on y-axis, such as is shown in Table 1.

Table 1.

X=Vč (m/day)	4,14	4,20	4,60
Y=Q (t/day)	1286	1304	1455

From the given table with the Last-squares method from the form 1.1.

$$y = ax^2 + bx + c, \text{ where is } y = Q; \text{ and } x = v_č \quad (1.1)$$

$$Q = a \cdot v_č^2 + b \cdot v_č + c \quad (1.2)$$

S(a,b,c)= S= $[1286-(4,14^2 a+4,14b + c)]^2 + [1304 - (4,20^2\, a + 4,20\, b + c)]^2 + [1455 - (4,60^2 + 4,60\, b + c)]^2$

(1.3)

Partial excerpt function S per a,b,c, are:

S'_a = - 2· 17,1396·(1286 – 17,1396· a – 4,14· b –c) - 2·17,64 (1304 – 17,64 ·a – 4,2·b –c) -2·21,16·(1455-21,16·a–4,6·b–c)= 151663,7712+2105,362176·a+484,76388+ 111,8792·c (1.4)

S'_b = - 2· 4,14(1286 – 17,139·a – 4,14·b - c) - 2· 4,2 · (1304 – 17,64·a -4,2·b – c) - 2· 4,6 ·(1455 – 21,16·a – 4,6·b – c) =- 34.987,68 + 484,763888·a +111,8792·b + 25,88·c (1.5)

S'_C = -2 (1286 –17,1396·a – 4,14·b – c) - 2· (1304 –17,64·a – 4,2·b - c)-2(1455 – 21,16· a –4,6·b –c) =8090 +111,8792·a + 25,88·b+ 6c (1.6)

From condition that S'_a=S'_b= S'_c= 0 acquire (1.7)

2105,362176·a+484,763888·b+111,8792·c =151663,7712 (1.8)

484,763888·a+111,8792·b+25,88·c =34987,68 (1.9)

111,8792·a+25,88·b+6·c=8090 (1.10)

Multiplying III equation with $\frac{-111,8792}{6}$

and adding into I equation, and

multiplying III equation with $\frac{-25,88}{6}$

and adding into II equation, it will be:

19,202944·a +2,191605·b = 813,3165333 (1.11)

2,191605·a + 0,250133·b = 92,81333333 (1.12)

Determinants of the system are:
D = 0,000157515527; D_a = 0,02713903123; D_b=-0,1793385736 (1.13)

Applying Kramer formula

$$a = \frac{Da}{D} = 172{,}2943239, \quad (1.14)$$

$$b = \frac{Db}{D} = -1138{,}545368 \quad (1.15)$$

From **II** equation of system, in

c =1/6 (8090-111,8792·a – 25,88·b) (1.16)

After the inclusion of the obtained values for a and b it follows:
c=3046,567167 (1.17)

Q=172,2943239·$v_č$–1138,545368·$v_č$+ 3046,567167 (1.18)

$Q_{(4,14)}$=1286,045137 (1.19)

$Q_{(4,2)}$=1303,948495 (1.20)

$Q_{(4,6)}$=1455,006368 (1.21)

Q'(x)=0/y(3,3)=1165,65264·Q'· 344,5886478 · x- 1138,545368=0 (1.22)

For x=3,3 (minimum) y=1165,65264 (1.23)

Max.Nema: y→ ∝ (x → ∝)
(D < 0 → y > 0 ; for all x

Diagram dependency of production from moving forward dynamic would look like this:

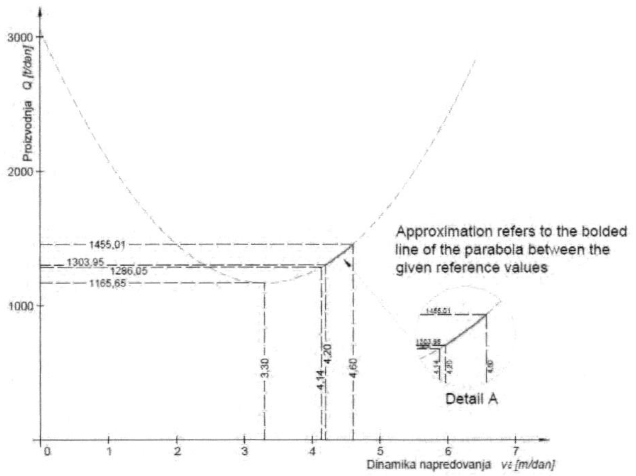

Figure 9. Diagram dependency of production from moving forward dynamic

4. CONCLUSION

Using the least squares method, we conclude that the prognosis dependency of the production Q from the speed of moving forward $v_č$ with the constant long wall length of $l_č$ =60 m and excavation thickness of d_p =3,2 m about behaving on the right half of the parabola obtained for $v_č$ > 3,3 m/day.

For example.

$Q(4{,}5) = 172{,}2943239 \cdot 4{,}5^2 - 138{,}545368 \cdot 4{,}5 + 3046{,}567167 = 1412{,}07307$ t/day

REFERENCES:

1. H. Čičkušić, M. Topalović; Study on the work environment in mine „Marići“ II horizon of the Main layer in the function of installing mechanized way of excavation by the Long wall; Mining company „Kreka“ Tuzla, (March 2016)

2. O. Musić, Shot wall methods of excavation of stratified layers (Tuzla,2010)

3.. Study of classification, categorization and calculation of the coal reserves, Mining company „Mramor“ with situation on the day December 31st 2008

4 H.Čičkušić,O.Musić,K.Gutić; Models of selection of the coal excavation machine in Mining company „Kreka“ Tuzla; VI symposium “ Mining 2015“ –Borsko lake, May 2015

5. O.Musić, J.Kortnik, H. Čičkušić, Š. Sarajlić; Orientation and the way of security of the connection raise when long wall passing through in Mine „Mramor“; II symposium „Mining 2011“, V. Banja 2011.

Tutorial Autodesk – AutoCad, 2013.

https://hr.wikipedia.org/wiki/Interpolacija (April 2016)

The Results of GNSS Observations in the Area of the Underground Disposal of Radioactive Waste (Yenisei Ridge)

V. Tatarinov,
Geophysical Center of RAS, Moscow, Russia
I. Seelev,
Chemical plant Federal State Unitary Enterprise Mining - Chemical Plant
T. Tatarinova
Geophysical Center of RAS, Moscow, Russia

ABSTRACT The paper presents the results of the study of modern movements of the Earth's crust in the southern part of the Yenisei Ridge (Nizhnekansky massif) according to GPS/GLONASS global navigation satellite systems. It is planned to build an underground research laboratory in this area for the study of the ecological safety of disposal of high-level radioactive waste in geological formations. The observations, carried out in 2010-2016, provided data on modern horizontal geodynamic movements on the border of the largest tectonic structures - the Siberian Platform and West Siberian Plate. In 2013-2014 an activation cycle was registered, which manifested itself in the sign change (compression-tension) and increasing the velocity of crustal deformation at the boundary of these structures.

Keywords Disposal Radioactive Waste, Modern Geodynamic Movements, Deformation

1. INTRODUCTION

The Yenisei Ridge is located at the junction of major tectonic structures - the Siberian Platform, the West Siberian Plate and the Altai-Sayan orogenic area. It is an active orogenic zone, i.e. the formation of the Yenisei Ridge is not finished yet. Therefore, prediction of geodynamic movement's velocity is highly relevant for the general fundamental problem of securing the insulation properties of the rocks to ensure the Yenisei Ridge is located at the junction of major tectonic structures - the Siberian Platform, the West Siberian Plate and the Altai-Sayan orogenic area. It is an active orogenic zone, i.e. the formation of the Yenisei Ridge is not finished yet. Therefore, prediction of geodynamic movement's velocity is highly relevant for the general fundamental problem of securing the insulation properties of the rocks to ensure environmental safety of the radioactive waste disposal.

In this regard, experts of GC RAS, JSC "Geolokom" and the ROSATOM Mining and Chemical Combine in Krasnoyarsk (GKHK) created a geodynamic testing ground for modern crustal movements monitoring based on GPS/GLONASS global navigation satellite systems. Over the last five years, 30 stations of the geodynamic polygon carried out regular observations. The results of research are described below.

2. GEOLOGICAL AND TECTONIC CHARACTERISTICS OF THE AREA

The Nizhnekansky rock massif is located in the southern part of the Yenisei Ridge, which is a projection of the crystalline basement of the Siberian platform. In the north it is bordered by the Riphean structures of the Trans Angara area, in the east it is blocked by the Vendian-Cambrian deposits of the Kansk-Taseevsk depression, in the south by the middle Paleozoic and Mesozoic sediments of the Rybinsk depression, and in the west, along the Baikal-Yenisei deep fault borders with the West Siberian Plate [Anderson, 2011].

The massif is formed by an intrusive body, stretched from the southeast to the northwest for 60 km with a width of about 30 km. Its capacity ranges from 5-6 km to 8 km. The age of rock was determined by geologists from 450 to 850 million years. Large faults are of sub-latitudinal and sub-meridional orientation, the latter are considered younger. Despite the complexity of the geological structure and shape of the modern landscape, the area is characterized by the stable tectonic regime, the total amplitude of the territory's uplift in the recent period is estimated at 400-500 m, and the average rate of uplift is 0,2-0,3 mm/year. However, the instrumentally established crustal movement velocities are by orders of magnitude higher due to the scale of the time factor of modern movements.

The Yenisei river divides the area into two parts. The left bank is characterized by predominantly negative vertical movements and the right bank by positive ones [Lobatskaya R.M. 2014, Tatarinov V.N., 2014]. The major tectonic disturbances occur in the areas of the Muratovsky, Atamanovsky, Pravoberezhny and Bolshetelsky submeridional faults [Belov S.V., 2007]. Therefore to create a geodynamic testing ground the observation points layout was arranged as a graph of the profile, crossing all the major tectonic faults from west to east.

3. RESULTS OF RESEARCH

From 2010 to 2015 five cycles of observations were carried out, the dynamics of changes in the lengths of baselines was investigated (ΔL) and maps of crustal deformation velocities were created. For the total period of observations, the average value ΔL (Fig. 1) was +3,8 ±0,8 mm. This suggests that the geodynamic regime of the whole area was characterized by tensile stress (extension).

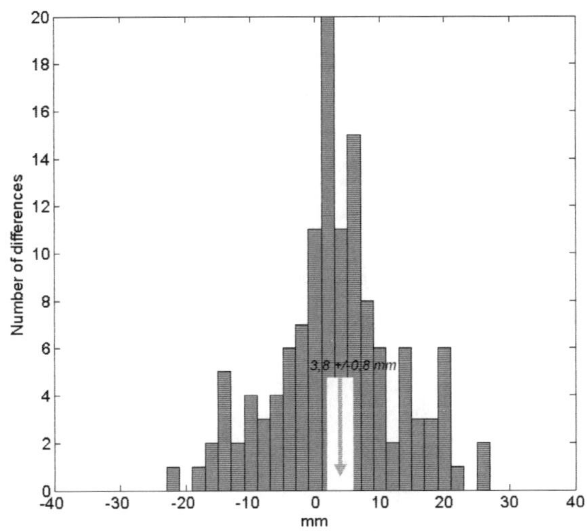

Figure 1. The distribution of length differences over the entire observation period from 2010 to 2015

Figures 2-4 show the changes in the lengths of baselines between observations points for the periods 2010-2013, 2013-2014, 2014-2015, respectively. Their analysis showed that each of the two epochs of observations has its own peculiar features: a) a change in absolute values ΔL and a change of their sign (compression-extension), depending on the epoch of observations. Fig. 2 shows that from 2010 to 2013 the lines shortened on the right bank, i.e., the compression of the upper crust took place, and on the left bank, on the contrary, the length of almost all lines has increased, i.e. the geological medium experienced extension. In 2013 (Fig. 3) the picture has changed to the opposite: lines on the right bank mainly extended, and compressed on the left bank. At that, the absolute values ΔL increased dramatically — by 30-50%.Then, in 2015 (Fig. 4), the left bank has almost completely returned to its original geodynamic regime that existed before 2013, and on the right bank, this return process has just begun.

The reason for this effect, apparently, is the fundamental property of the behavior of the Earth's crust [Batugin A.S., 2016, Lan Tianwei, 2015] which is the cyclical development of geodynamic movements in time, when relatively long periods of accumulation of elastic energy, accompanied by relatively small movements, were followed by periods of intensification of movements, often of the opposite sign.

Geodynamic movements of this nature can be found in other areas. For example, in the Urals, which is tectonically similar to the area of research, the results of measurements of voltages at 25 underground mines at depths of 100 to 800 meters revealed tectonic stress pulsing with a period of 12-18 months. One of the major directions of oscillations is the southeast with azimuths 110-150^0, which corresponds to the direction of maximum compression of the Urals [Zubkov A.W., 2002].

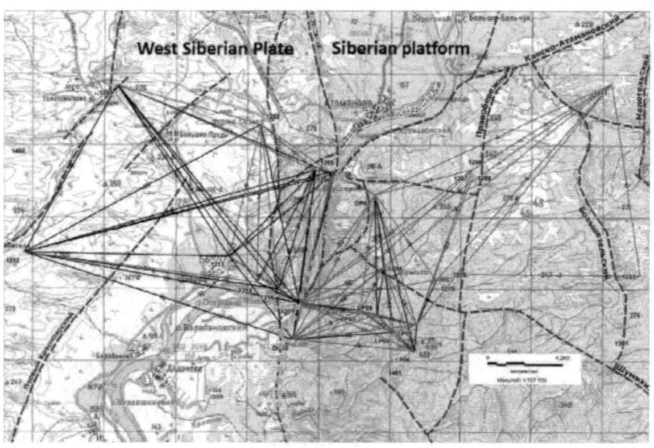

Figure 2. Changing the length of the baselines for 2010-2013, black color - length is not changed or increased (extention), red - length decreased (compression)

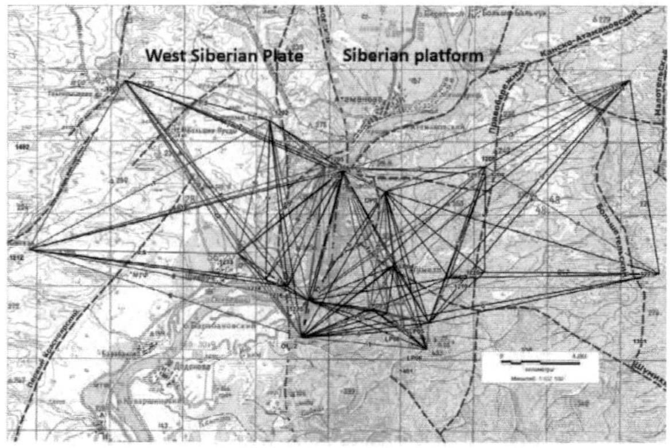

Figure 3. Changing the length of the baselines for 2013-2014, black color - length is not changed or increased (extension), red - length decreased (compression)

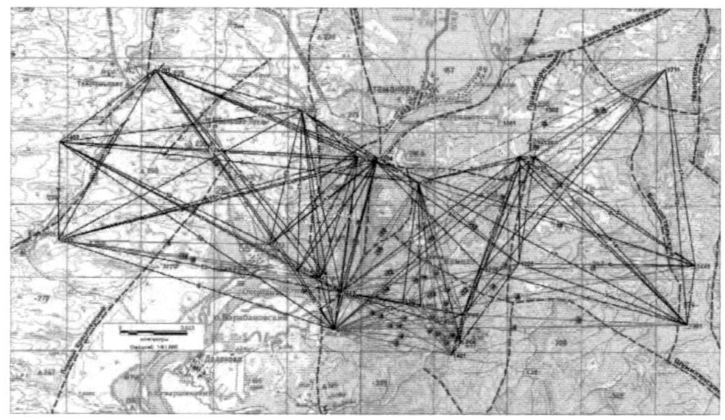

Figure 4. Changing the length of the baselines for 2014-2015, black color - length is not changed or increased (extension), red - length decreased (compression)

There was also an attempt to find out by the changes of absolute values ΔL the main directions of movement associated with the main lines of tectonic forces. This is important to set the boundary conditions for the simulation of stress-strain state of the massif. To this end, we constructed a chart of the lengths of the lines change for separate observation epochs (Fig. 5). The arrows in Fig. 5, d show the azimuths of changes in the maximum absolute values ΔL. Interestingly, these directions coincide with major tectonic fault lines in Figures 2-4. It can be assumed that deformations of the Earth's surface are also associated with shear movements along the faults.

For the quality control and evaluation of the accuracy of the measurement results a network adjustment was performed [Tatarinov V.N., 2006, 2008, 2014]. Fig. 6 shows the estimations of the mean square error (MSE) for the period 2012-2015. The accuracy estimation at the final stage of the equalization calculation provided the most reliable data on the real accuracy of points of the local geodetic network. The MSE of the point's location has shown that the MSE of horizontal and vertical accuracy amounted to 3.0-3.7 and 6.0-6.5 mm, respectively.Thus, it is evident that the greatest ΔL changes during the movements' intensification were caused by geodynamic reasons, and not by noise or possible technical errors in observations or calculations.

Not even rate of movements, but strain rate is essential to forecast the insulating properties of a rock massif. Therefore the Earth's surface dilatation (Δ) map was constructed (the relative change in the area) (Fig. 6). The map showed that the maximum deformations were recorded in the area of Muratovsky and Atamanovsky faults located at the junction of the Siberian Platform and West Siberian plate.

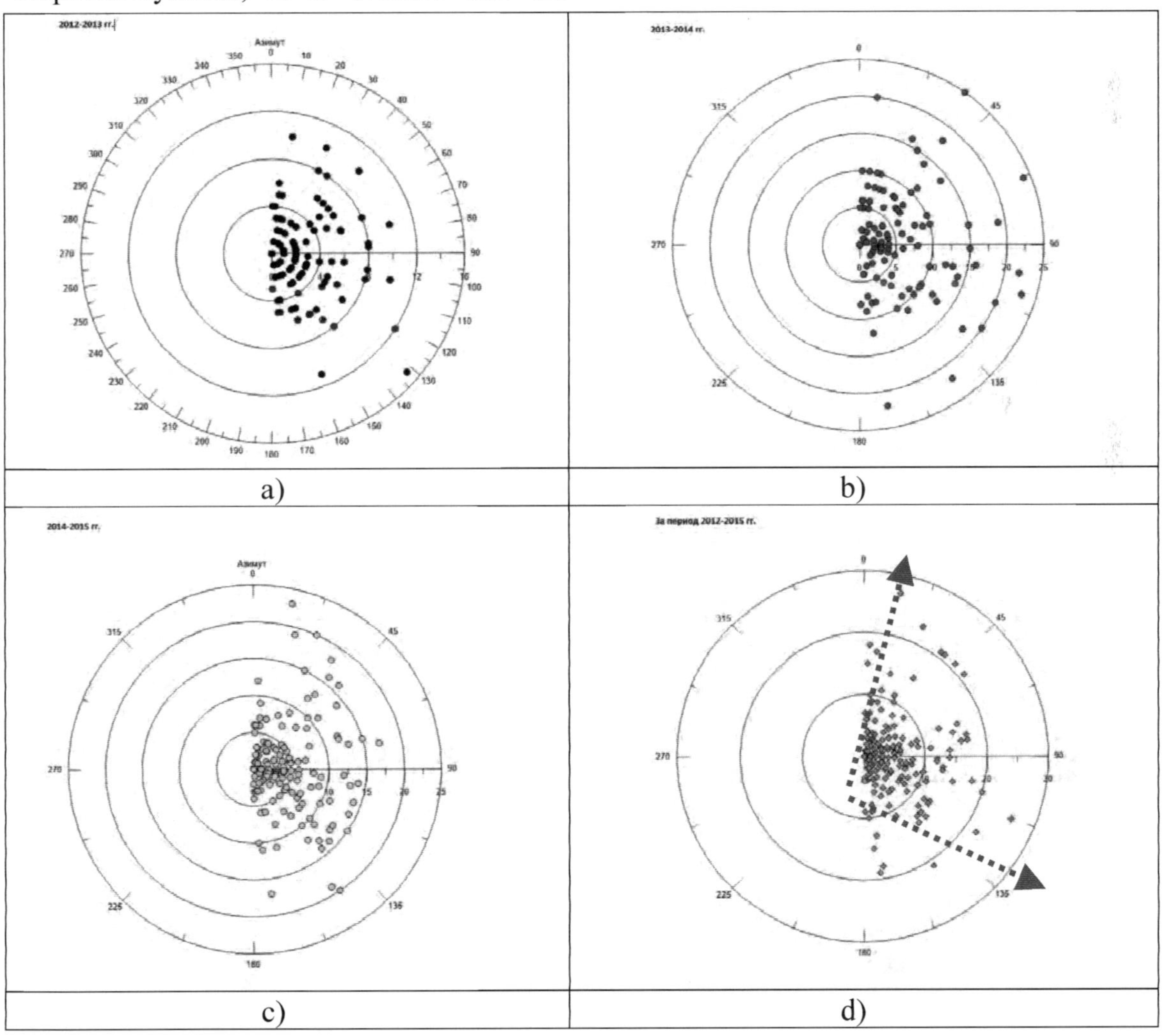

Figure 5. Charts change the lengths of baselines, depending on the azimuth for different observation periods: a) 2012-2013, b) 2013-2014, c) 2014-2015, d) 2012-2015

At the "Yeniseisky" station the strain was significantly lower $\Delta=2 \cdot 10^{-8}$. The work [Kuzmin Y.O., 1999, Tatarinov V.N., 2008] provides the average rate of the relative strain for increased geodynamic hazard areas equal to $5 \cdot 10^{-4}$-$5 \cdot 10^{-5}$. Comparing the magnitude of dilatation with these values, it can be argued that the "Yeniseisky" station in the period of observations was characterized by weak activity.

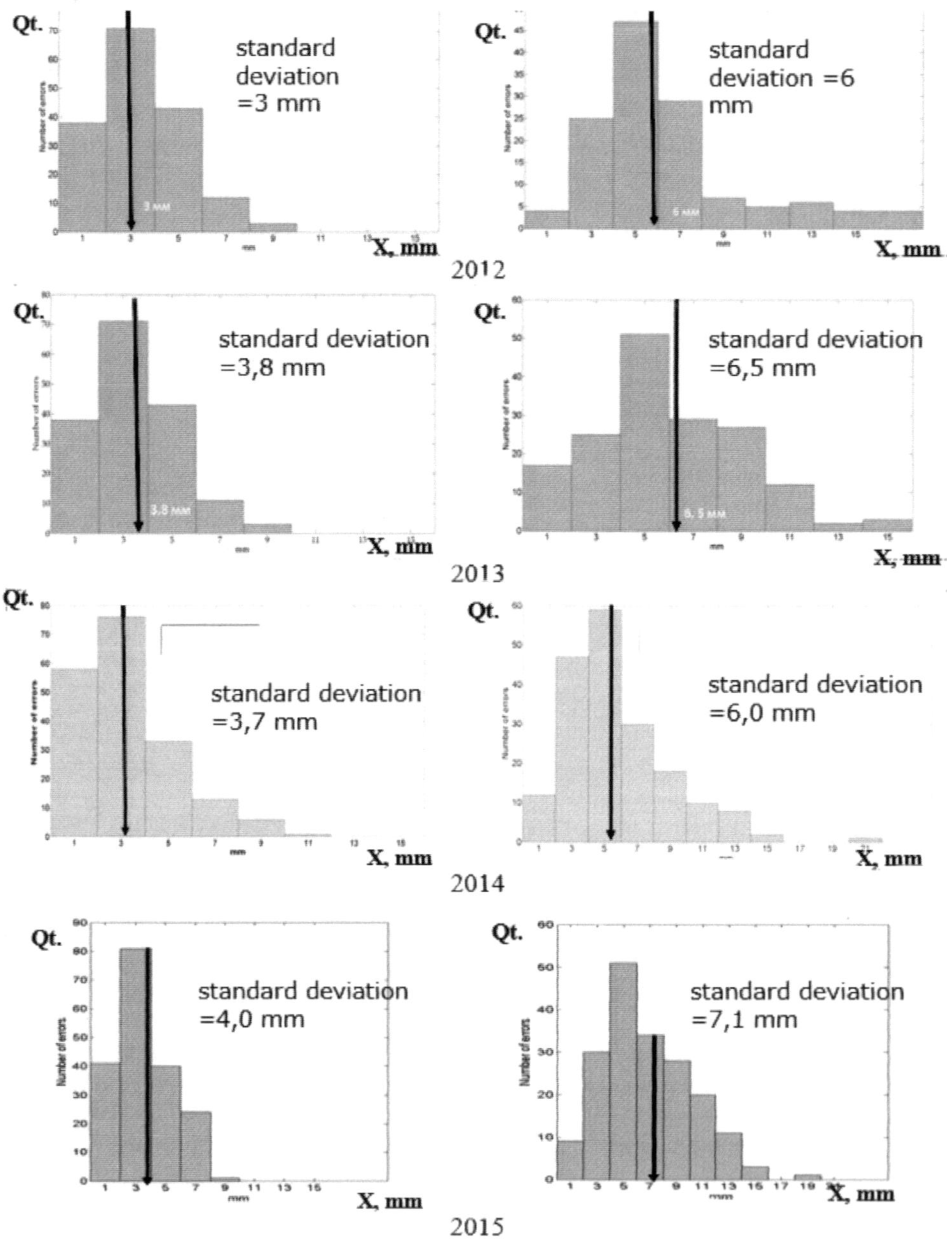

Figure 6. Results of the evaluation of the mean-square error of measurements. Horizontal components - left column, vertical components – right

Fig. 7 shows the region's longitudinal section from east to west as an alternative interpretation of the cyclical development of modern geodynamic processes in the area (the vertical scale is considerably stretched relative to the horizontal scale). Vertical tectonic strain at the Siberian Plate in the southern part of the Yenisei Ridge leads to the formation of the relative extension zones on the Earth's surface and at the same time to compression on the border of the West Siberian platform.

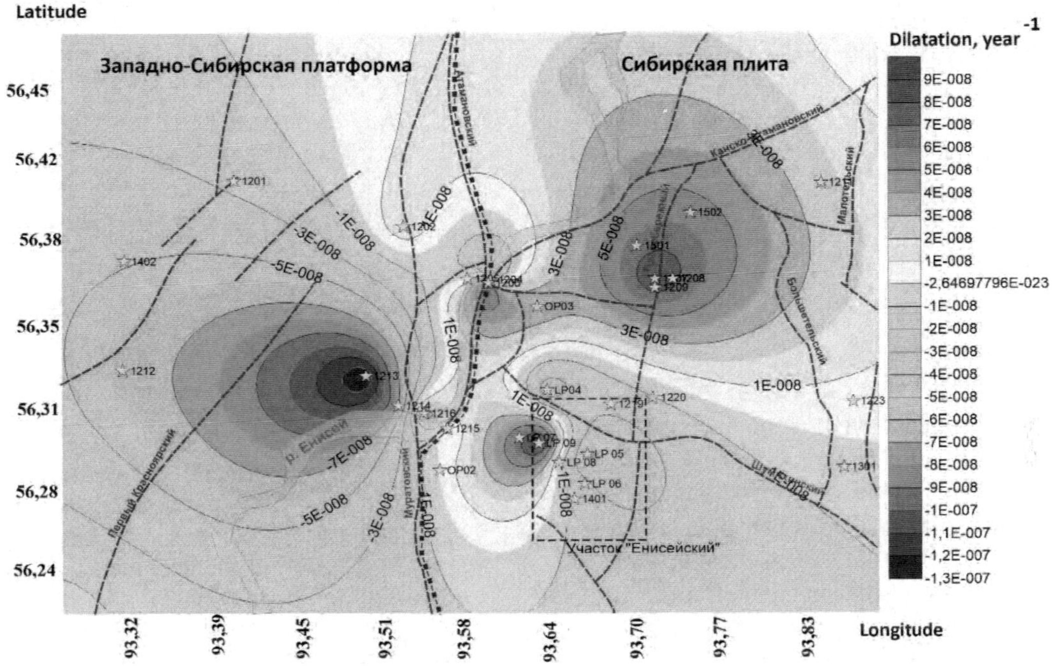

Figure 7. Map of dilatation (strain rate) of the earth's surface for 2010-2015

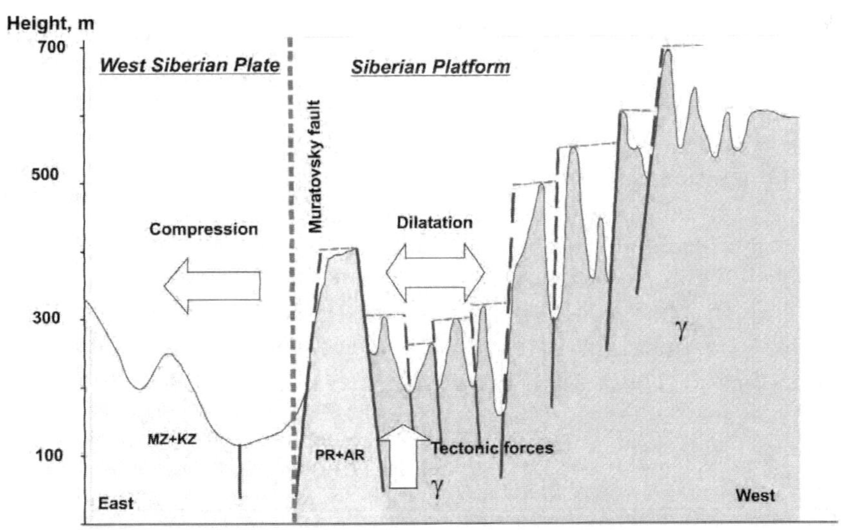

Figure 8. Hypothesis of the geodynamic situation in the southern part of the Yenisei Ridge (the vertical scale is greatly increased relative to the horizontal). The red dotted line - the boundary of two tectonic areas: the West Siberian plate and the Siberian platform

After reaching a certain limit of compression, a relaxation of accumulated stress by a right shift occurs along the Muratovsky and Atamanovsky faults. The rock massif returns to its former condition and after a while the geodynamic cycle repeats. These conclusions are confirmed by the neotectonic diagram of the Baikal-Yenisei fault in [Lobatskaja R.M., 2014], which provides vertical velocities of the Pleistocene period. The scheme is very similar to the dilatation map shown in Fig. 7. It can also be assumed that the modern geodynamic regime of the territory is close to the regime that existed, at least in the Pleistocene (from 2,5 million to 10 000 years ago).

4. CONCLUSIONS

A geodynamic testing ground for observing modern crustal movements based on the use of GNSS was established in the area of construction of an underground research laboratory for the study of the ecological safety of disposal of high-level radioactive waste. For the first time we instrumentally determined the horizontal crustal movements rate for the area, located at the junction of two major tectonic structures — the Siberian Platform and West Siberian Plate, in order to assess the intensity and direction of tectonic processes at the present stage of development of the region and the values of the maximum strain rate.

In 2013-2014 an activation cycle was registered, which manifested itself in an increase in the absolute values of changes in baseline lengths and in the sign change (compression-tension) on the right and left bank of the Yenisei River. The annual rate of baselines' changes during the activation period increased to 15 mm, with pre-existing baselines in the range from 0 to 10 mm in 2010-2013. The probable reason was the cyclical nature of space-time development of geodynamic movements.

As a result of equalization of observations data we equalized baseline vectors' components and evaluated their accuracy. The mean square errors were in the intervals of 3-4 mm and 6-7,4 mm, respectively.

We hypothesized that the modern geodynamic regime of the territory was close to the regime that existed at least in the Pleistocene.

In general, we can say that we didn't record any anomalous values of modern movements of the Earth's crust, which could confirm the existence of tectonic instability of the geological environment in the southern part of the Yenisei Ridge (Nizhnekansky massif).

The work in the geodetic center of the Russian Academy of Sciences and the Institute of Physics of the Earth.

REFERENCES

Anderson E.B., Belov S.V., Tatarinov V.N. and other (2011) Underground disposal of radioactive wastes. M. «Gornaja Kniga», 592p.

Batugin A.S., Kolikov K.S., Batugina I.M., Golovko I.V. (2016) Analysis of Stress Fields and Manifistations of Mining-induced Seismicity During Development of Kusbass Coal Mine Considering the Results of Geodynamic Zoning. Processing of 35th International Conference on Ground Control in Mining. 2016. ISBN 978-7-5020-5491-5. Beijing, 2016. P.60-65.

Belov S.V.., Morozov V.N., Tatarinov V.N., Kamnev E.N., Hamer J. (2007) The study of the structure and geodynamic evolution of the Nizhnekamsk array in connection with the disposal of high level radioactive waste. Geoecology. 2007. No. 2. P. 248-266.

Kuzmin Y.O. (1999) Modern geodynamics and assessment of geodynamic risk in Subsoil Use. M. Economic News Agency. 220 p.

Lan Tianwei, Zhang Hongwei, Li Sheng, Han Jun, Song Weihua, A.C.Batugin, Tang Guoshui (2015) Numerical study on 4-1 coal seam of Xiaoming mine in ascending mining [J]. The Scientific World Journal, 2015.3:4 pages. Article ID 516095. doi:10.1155/2015/516095.

Lobatskaya R.M. (2014) The fault-block structure of the Baikal-Enisei fault in the Region of operating nuclear energy facilities. Geodynamics & Tectonophysics. 2014;5 (2):547–562. DOI:10.5800/GT-2014-5-2-0140.

Morozov V.N., Tatarinov V.N. (2008) Prediction of geological environment stability when selecting placement and operation of the nuclear fuel cycle. Engineering ecology. No.5. C. 10-16.

Tatarinov V.N., Morozov V.N., Kamnev E.N., Kaftan W.I., Kagan A.I. (2014) Geodynamic monitoring as a basis for the conservation of the biosphere in the disposal of radioactive waste. Earth sciences. No. 3. 2014. P.47-60.

Tatarinov V.N., Tatarinova T.A. (2012) Given the magnitude of the effect in observations of land surface deformations satellite navigation systems. Surveying Gazette. No. 5. P. 15-19.

Tatarinov V.N. (2006) Geodynamic safety in nuclear fuel cycle facilities. Use and protection of natural resources in Russia. Bulletin. No. 1 (85). P. 46-51.

Zubkov A.W. (2002) Stress state of the crust of the Urals. Lithosphere. No. 3. P.3-18.

Geophysical Explorations of Vein Copper Deposits Using IP and RS Geophysical Methods in the Southern Part of Sarbisheh City and Its Surroundings, the Exploration Area of Sahl Abad, Eastern Iran

S.Keshavarzipourtafti, M.Saeidnezhad*, M.Dehju

University of Tehran, Tehran, Iran

*University of Tehran, Tehran, Iran**

University of Birjand, Birjand, Iran

ABSTRACT One of the several copper mineralization regions in the eastern part of Iran is considered to be the exploration area of Sahl Abad, which has been studied in detail in this paper. The study area is composed of Ophiolite complexes -related to the late Cretaceous- including peridotite rocks, gabbro, diabase rocks, pillow basalts, and sedimentary rocks especially pelagic limestone, radiolarite and shale. Tectonic structure of the study area and field observations suggests that in this region the mineralization are most likely to have occurred as veins along the existing faults of the area. The collection of data has been done using IP and RS geophysical methods within Sahl Abad exploration area. This study sets out firstly to determine the limits of anomalous zones; secondly to examine depth and thickness of the existing ore deposits and their locations in relation to one another, and finally to determine locations appropriate for exploration drilling. Geoelectrical studies have been carried on the marked area by dipole-dipole array. In this paper, the entire resistivity and chargeability data collected along profiles have been interpreted with the assistance of the statistical analysis as well as forward and direct modeling. Having interpreted the anomalous zones, the structural features of the ore deposits, such as the approximate shape of the mineralization, depth of the mineralization and so forth, were identified; thereby the appropriate locations for drilling were proposed. In the end, the authors recommend collection of magnetic data in the area of interest -if possible- for a better interpretation.
Keywords Sahl Abad, Copper mineralization, IP and RS geophysical methods, Dipole-dipole array.

1 INTRODUCTION

The exploration area of Sahl Abad vain copper is located in the north of Iran, at a distance of 100 km from the south east of Birjand, the center of the Southern Khorasan province, and lied at longitude 59.89° and at latitude 32.26°. This ore deposit is situated in the northern part of Loot desert in terms of structural divisions (Figure1). To explore for the hidden parts of mineralization, geophysical studies has been conducted in the above-mentioned region and its surroundings. Today, geophysical studies are widely used both at a national and international level to explore for copper mineralization (Yugoslav report, 1384).

Given the fact that the geophysical methods of induced polarization and resistivity measurement are considered as the most efficient methods for discovering the hidden parts of copper mineralization in all mineral resources, our study was done based on these methods due to the following reasons:

Firstly, sulphide minerals were abundantly observed in the form of numerous veins that led to the formation of a kind of anomaly considered to be detected more readily by the induced polarization method.

Secondly, the intensity of alteration is so high that the resistivity of rocks decreases.

The purpose of this paper is to explore for hidden parts as well as for the center of mineralization of the vain copper, covered by alluvium, using IP/ RS methods and then interpreting the results with the aid of geological studies, alteration and mineralization.

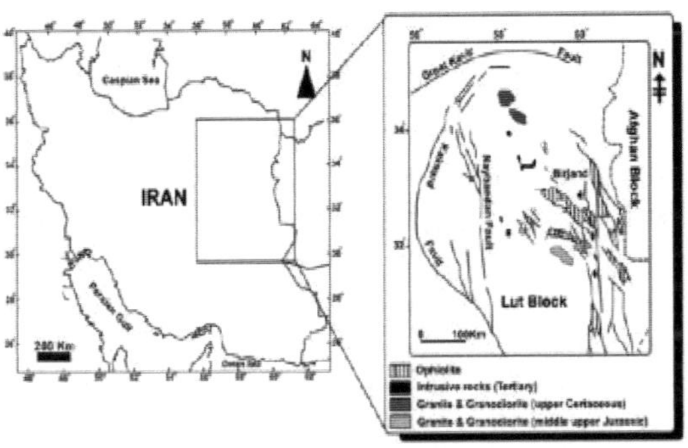

Figure1. This figure indicates the geographical location of Sahl Abad copper exploration region in the east of Iran and in the Loot block (Ra'eisossadat et al., 1391).

2 STRUCTURE AND GEOLOGY OF SAHLABAD

The exploration area of Sahl Abad is regarded as one of the several copper mineralization in eastern Iran, which has been studied in detail. In fact, the east of Iran, especially Loot block, has been turned into a great potential for forming a variety of mineralization, and the reason lies in occurrence of subduction in this area in the past and subsequently in high-magma-production. There is evidence verifying the occurrence of mineralization in different locations of the eastern Iran- such as Sorkhkooh (Tarkian et al., 1983), Rahimi (Khosravi et al., 2006), Chahshalqam (Karimpour et al., 2005), Dehsalam (Ziaee et al., 2003) and so forth. However, until now, no serious action has been taken to explore for the minerals of interest, with no deposit having been exploited.

In geological terms, Sahl Abad's exploration area is located in the north-east of a 1:100000 scale map of Sahl Abad (Vasiq et al., 1975) According to the above-mentioned map, the collection of volcanic rocks exposed on the Earth's surface in the area mainly consists of basaltic rocks, basaltic andesite, pyroxene andesite and dacite related to Eocene epoch (Arjang-Ravesh et al. 1975)

On the basis of the field and experimental studies which have been conducted, the geological units recognized in the Sahl Abad's exploration area are as follows:

A mélange of ophiolite and flysch deposits have been developed in a vast area of the northern Sahl Abad. A mélange of ophiolites consists of rocks such as peridotite, gabbro, diabase dykes, pillow basalts and sedimentary rocks- including pelagic limestone, radiolarite and shale -that are related to the Late Cretaceous period (Figure 2). Flysch sediments consisting of Late Cretaceous and early Cenozoic deposits, including shale, sandstone, limestone, marl and interference of diabase and tuff, have undergone a regional metamorphism, were extended in the east of the region (Zarrinkoob et al., 1993).

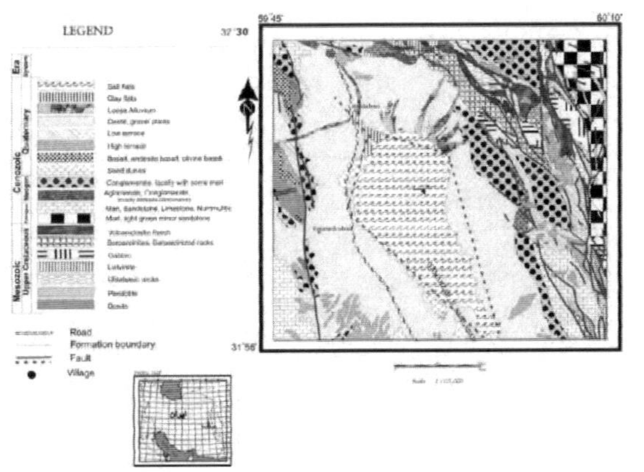

Figure2. The geological map of Sahl Abad region.

3 METHODOLOGY

To explore for hidden parts of mineralization in Sahl-Abad area, the following studies have been conducted:

1. Provision of the digital geological map of the Sahl-Abad exploratory area at the scale of 1:2000;
2. Drawing IP/RS sections and SP-MN curves for each of the three collected profiles;

3. Interpretation of the collected IP/RS data in an approximate area of 3 Km^2 related to Sahl-Abad and its surroundings.

4 THE ANALYSIS OF GEOPHYSICAL COLLECTION

Induced polarization (IP) and resistivity (RS) pseudo-sections were collected by Dipole-Dipole array; for this purpose, the stations of the three collected profiles were selected nearly perpendicular to the mineralization veins - exposed on the Earth's surface – and in the north-south direction. The length of the profiles was 220, 270 and 180 meters, respectively. The electrode spacing was 10 meters. In addition, data was collected, as a whole, in 584 points of the profiles. It is worth noting that length of the profiles were selected based on the existing anomalies in the area where data were gathered.

4.1 Profile 1

At an approximate distance of 50 meters from the beginning of the profile, an increase in the value of the resistivity (RS) and chargeability (IP) was observed when the depth increases to 15 meters. At a depth of 80 to 100 meters, the sections illustrate a simultaneous rise in the resistivity and chargeability values. On the margin of this part, copper mineralization has been observed as an oxide one (Malachite, Azurite) located on the surface, and this oxidation can be generalized to the depth of 40 meters.

At a distance of 60 meters, the values of resistivity and chargeability have dropped markedly, which can be interpreted as a place for the fault's activity where the moisture has increased as a result of fragmentation and clays- arising from alteration- and subsequently caused the resistivity to decrease to nearly 20 ohm-m. In fact, this resistivity value was representative of water-bearing areas. At the end of the profile, at a distance of more than 120 meters as the electrodes entered the margin of the secondary stream and then the primary stream, the moisture content of the underground layers rose which led the resistivity and chargeability values to decline. Figure 3 illustrates the IP and RS pseudo-section of profile 1.

4.2 Profile 2

In profile 2, Like Profile 1, there are two sections where the chargeability (IP) increases, separated by a low IP border at a depth of 80 meters. The first section can be clearly seen at a distance of 50 to 70 meters from the beginning of the profile; the relatively high values of RS and IP in this section can be indicative of mineralization. In the second section, existence of an area with high values of chargeability is more visible at a depth of 80 to 90 meters, and the copper mineralization is observed as an oxide on the surface as well. At the end of the profile, like profile 1, soil the moisture and subsequently the resistivity and chargeability values go up as the major and minor rivers enter the area. Figure 4 indicates the IP and RS pseudo-sections of profile 2.

4.3 Profile 3

Regarding Profile 3, at a distance of 40 to 60 meters, the area of interest shows high values of both resistivity and chargeability. The extension of the RS and IP zones at depth is a good testament to the extension of mineralization at depth. Add to that, the existence of copper oxide mineralization on the surface verifies this claim.

As to profile 3, at a distance of 80 to 120 meters, while the chargeability value increases, there is a decrease in the value of resistivity. In this area, a mineral mass obtained from the past was observed, including a mixture of iron oxides, clay minerals and copper oxide minerals. The reason why the resistivity declined can be due to the existence of clay minerals as well as high rates of water absorption. At a distance of 130 meters, an old furnace and its slag were observed; the increase in the chargeability can be attributable to the above-mentioned furnace slag. At a distance of more than 130 meters, the resistivity values decline as a result of entrance into the primary stream. Figure 5 indicates IP and RS pseudo-sections of profile 3.

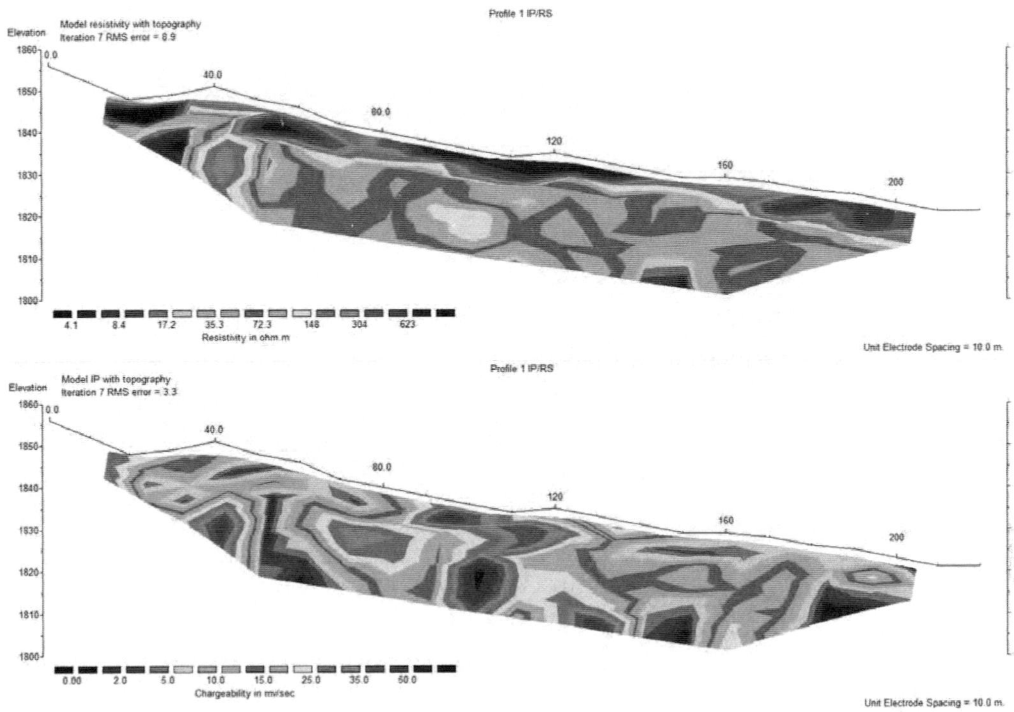

Figure3. The diagram reveals the RS and IP pseudo-section of Profile 1.

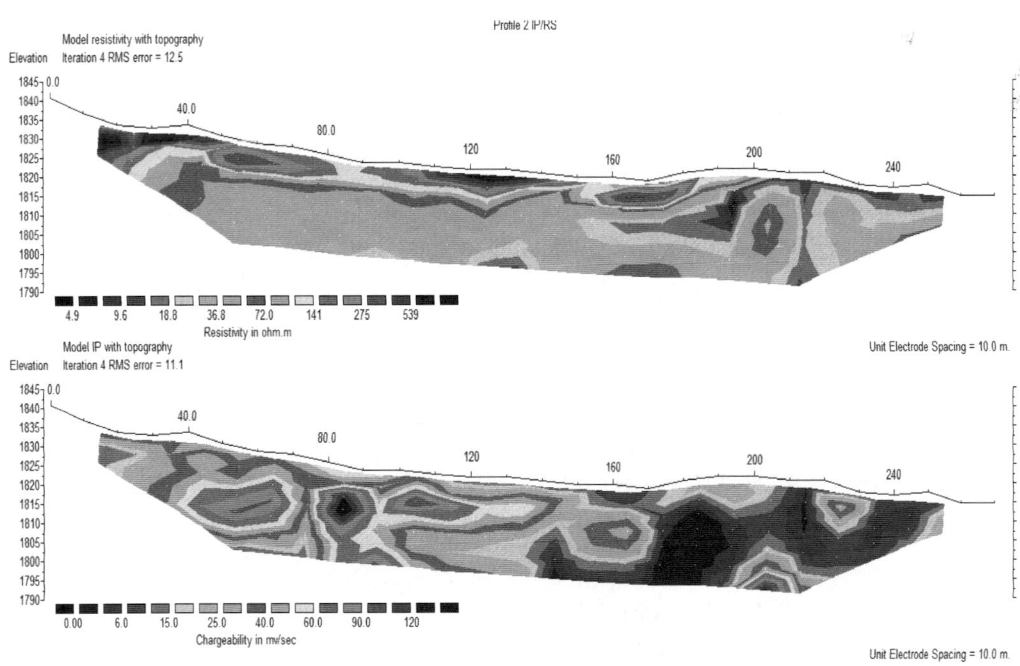

Figure4. The diagram illustrates the RS and IP pseudo-section of Profile 2.

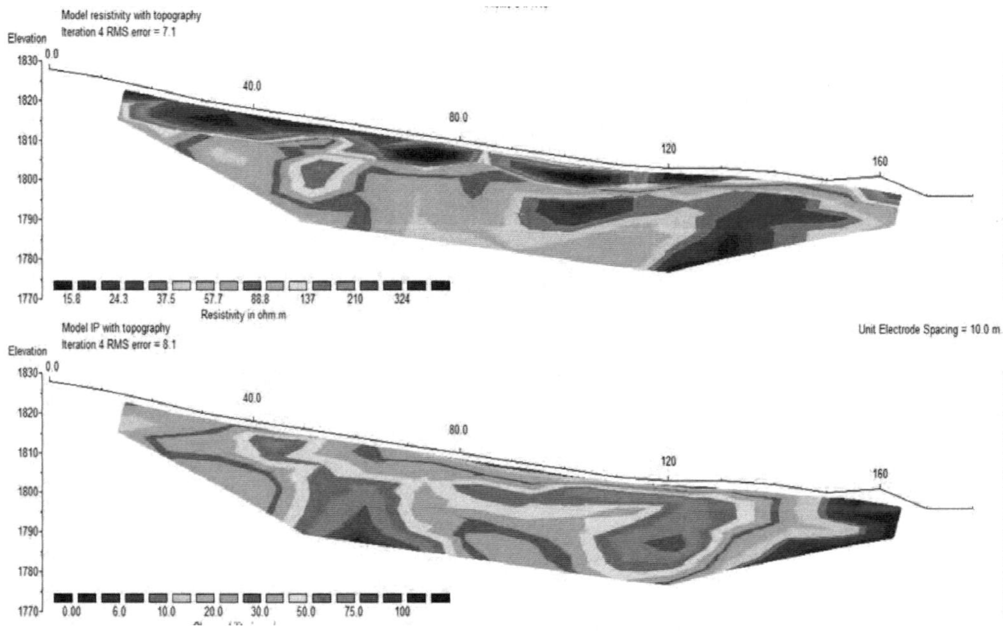

Figure5. The diagram shows the RS and IP pseudo-section of Profile 3.

4.4 The Study of SP-MN Curves

Background potential (SP) arises from fluid flow, the bioelectric activities of plants, varying electrolyte concentrations in groundwater and other chemical actions. The amplitude of these potentials is very variable, but generally less than 100 millivolt. In the far distance, the sum of them, on average, is zero because there are equal numbers of positive and negative SP with identical values (Telford et al., 1989).

In this study, in order to smooth, normalize and remove unnecessary fluctuations of the curves, we have taken arithmetic mean from the SP values of the point of interest and that of the two points located immediately before and after the desired point to calculate the equivalent SP value for each point (MN).

Following plotting SP-MN curves for each of the three profiles, the SP value of each point was compared with the values of IP/ RS in their corresponding sections. Afterward, a distance of MN having the highest value in each of the SP-MN curves as well as in each of the RS and IP sections was introduced as the potential mineralization zone; Besides, the points having the greatest amounts in each of the three methods have been introduced as the best places to dig. Figure 6 shows the SP-MN curves.

5 CONCLUSION

Based on the field observations, the type of mineralization in the area under study, located in the margins of the contact between ultrabasic and acidic volcanic (Rhyolite) units, after considering the sections and the curves plotted and the interpretations mentioned above, can be assessed as sulphide mineralization at depth.

Moreover, all the three parameters of interest- related to each of the three collected profiles- had their maximum values in part of the area located between MN=70 to MN = 110 meters; thus, this part of the area under study is suggested as the mineralization area. Regarding profile 2, given the fact that the depth extension of part of the study area having high IP values is more noticeable, and that copper mineralization is observed as an oxide on its surface, it is suggested that the best place to dig is located at an approximate distance of MN=75 meters from this zone.

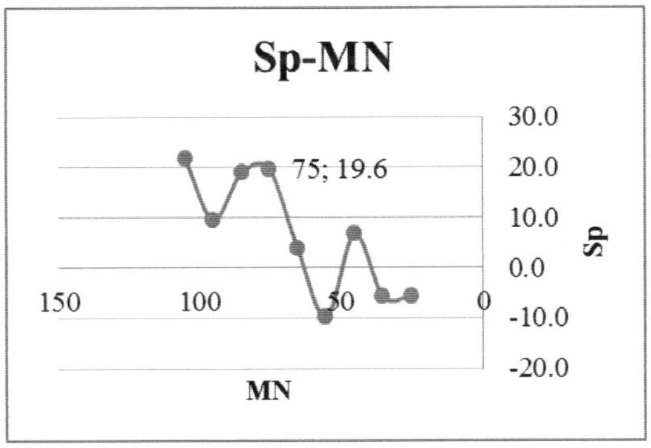

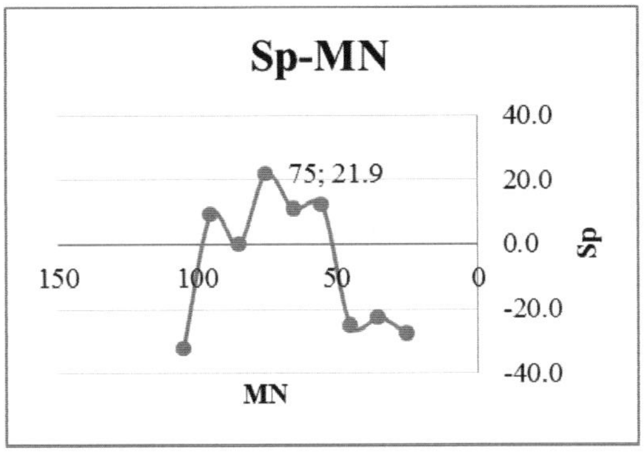

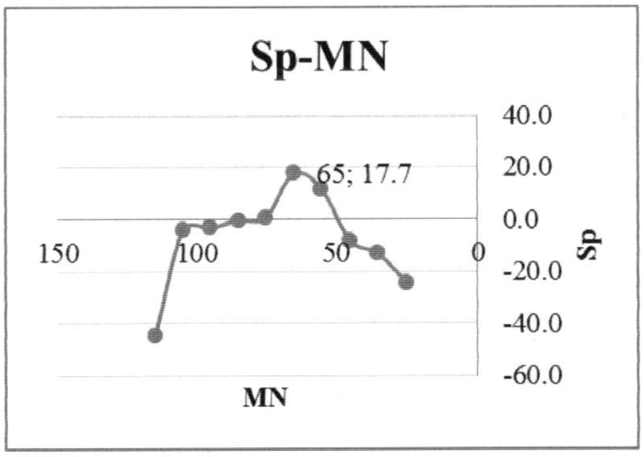

Figure6. The picture indicates SP-MN curves that are painted Profile 1, Profile 2, and Profile 3 respectively.

REFERENCES

Malekzadeh Shafaroudi, A; Heidarian Shahri, M; Karimpour, M. H; 1388; "The Mineralization and Geophysical Exploration by IP/RS and Geomagnetic Methods in MA-*I* and surrounding area, Copper-Gold Porphyria Exploration Area, East of Iran", *Ferdowsi University of Mashhad Economical Geology Journal*, 1, 1, pp.1-17.

Iran Geology Organization; 1384; "Yugoslav Report".

Raeis-Al-Sadat, N; Zarrinkoub, M. H; Khatib, M. M; 1391; "Geomorphology and How to Found Playas Sahl-Abad, East of Iran", *GDI Journal*, 28, pp. 115-132.

Tarkian, M; Lotfi M; Baumann A; 1983; "Tectonic, magmatism and the formation of mineral deposits in the central Lut, east Iran", *Ministry of mines and metals, GSI, Geodynamic project (geotraverse) in Iran*, 51, pp.357-383.

Khosravi, M; 1385; "The Petrology Studies, Alteration, Geochemical Mineralization in Rahimi area (North West of Ferdows)", *The Economical Geology M.Sc. Thesis of Ferdowsi University of Mashhad*, 265.

Karimpour, M. H; 1384; "The Quartz's holes and Quartz-Alunite (high sulphide) Zones of Porphyria Copper High Part of Chah-Shalghami area, South of Birjand", *13th CMSI Conference*, Bahonar University of Kerman, pp.7-11.

Ziaei, M; Abedi, A; 1382; "The Porphyria Copper Mineralization in Metallurgy Belt of Dasht-e Loot Surrounding", *11th CMSI Conference*, Yazd University, pp.57-59.

Vasigh, H; Soheili, V; 1354; "The Sar Chah Shour Geology Map 1:100000 (Sheet 7754)", *The Geology and Mining Explorations of Iran Organisation.*

Arzhang Ravesh, B; Darvishzadeh, A; 1353; "The Geology of South of Birjand (Sahl-Abad area)", *The Sciences Journal of Tehran University,* 104.

Zarrinkoub, M. H; 1372; "The Analyzing of Minerals along with the Causing Water Reactions and the Listvenite Phenomenon in Sahl-Abad area, Birjand", The M.Sc. Thesis of Tehran University, 154.

Telford, W. M; Geldart, L. P; Sheriff, R. E; 1989; "Applied Geophisycs", *Tehran University Publications*, 2, pp.26-33.

Application of Geostatistical Methods in Modeling Coal Resources

Dr.sc. O. Musić[1], Dr.sc. K. Gutić[1], B.sc. D. Osmanović[2], Mr.sc. Š. Sarajlić[3], B.sc. A. Bruljić[2]

[1]*University of Tuzla, Faculty of Mining Geology and Civil engineering, Tuzla, BiH*

[2]*Mining institute Tuzla, Tuzla, BiH*

[3]*Brown coal mine Đurđevik, Đurđevik, BiH*

ABSTRACT Based on exploration work database, the geostatistical model of coal seam is defined with "Surfer" software in the interest zone of RMU "Đurđevik" (RMU – Brown Coal Mine) which fits to be extracted with underground mining methods. Geostatistical model represented coal seam through following parameters: coal seam thickness and quality (NCV- Net Calorific Value, moisture, ash and total sulphur). The result of the analysis had shown that the coal seam is of high quality (with average 14.600 kJ/kg) but with high content of sulphur (average 1,64%). Another conclusion is to conduct more explorations in the areas with less number of exploration drill holes. This 3D representation of the coal seam did not take any tectonic deformations in observation but it will be incorporated in further development of the 3D model.

Keywords: Geostatistics, Database, 3d Modeling, Computer Based Mine Design

INTRODUCTION

RMU „Đurđevik“ is organised within JP Elektroprivreda BiH and provides thermal power plant „Tuzla“ with its coal.

Extraction of coal is organised on two surface mines („Višća II“ and „Potočari“) and on underground mine (Jama „Đurđevik“). In the last ten years average production of coal on RMU „Đurđevik“ is about 500.000 ty.

The remaining amount of coal which will be excavated with surface mining is about 10,5 million of tons. Coal from deeper layers of basin, which can not be extracted via surface mines and favourable economics, will be extracted through underground mining. This paper defines underground mining zone of interest for this coal deposit and coal quality with the use of geostatistical methods.

1. GENERAL PART

1.1 Geographic Location of RMU "Đurđevik"

RMU „Đurđevik“ is placed on North-eastern part of Bosnia and Herzegovina, 15 km southern from Tuzla city, on the left side of the Sarajevo-Županja freeway (N 44°24' E 18°38'), image number 1 Explorations of Đurđevik basin are mainly from the end of the last and from the beginning of this century. The most valuable ones are Kattzer's papers from 1921. They are providing us with details about borders of and coal deposit itself, quantity of coal in Đurđevik basin.

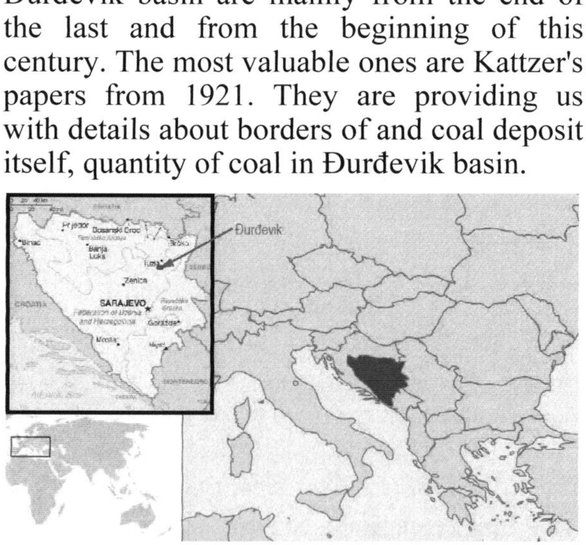

Picture 1. Geographic location of RMU „Đurđevik“

1.2 General Geological Characteristics of Đurđevik's Coal Field

Đurđevik's exploitation filed lies on borders of Gostelja in the east and Oskova rivers on the north. West and south borders of the field are not strictly defined, but the general opinion is that the villages Donja Višća and Požar represent it's southern borders. Foothill relief of the Djedinska mountain is hilly with general inclination towards the north where those hills are gradually disappearing. The exploitation field average height is 350 m a.s.l. Area with the coal seams in the Đurđevik's basin is elongated in the NW-SE direction. The basin is 5,5 km long and its width varies from 1 to 2,5 km. Đurđevik's coal basin is approximately 13 km^2. Picture 2 represents a disposition of the surface and underground mines and overburden dump sites.

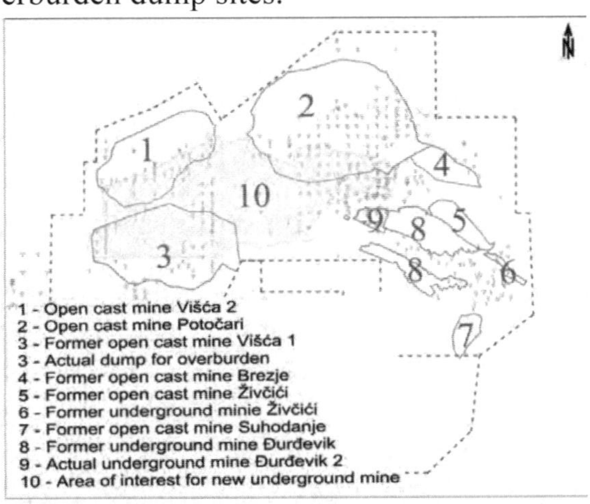

Picture 2. Situation map of former and active mines

The coal is produced in „Đurđevik II" underground mine and surface mines „Višća II" and „Potočari". Coal deposits where the first mines were opened were determinated even before WWII on the outcrops, like „Suhodanj". First data in this area date back to 1948 when the whole Đurđevik basin was explorated for the first time.

Based on this explorations, the earlier information about outcrops, main charachteristis of the coal seams were set up. Until now 100 meter exploration drill hole grid was drilled and examined which led to geological profiles with greater accuracy. All positive drill holes shows different thickness of coal seam, which implies that one existing seam was deranged by vertical and horizontal tectonic movements.

2 SUGGESTION ON ORGANISING EXPLORATION WORK DATA

The traditional methods of analysing and preserving data (raster maps), without any significant use of computer technologies are still being used on the brown coal mine „Đurđevik". Based on this information this paper provides a suggestion to systemize data from exploration drilling.

2.1 Forming Exploration's Work Database

Determining coal seam geometry, coal quality parameters for different exploitation fields, parts of coal deposit and for specific periods of exploitation will lead to the systematisation of the presentation and preservation of numerous data which are inputs in the processing phase. The most suitable solution is the computer made appropriate database.

Database is formed based on the grid of exploration bore holes and current work in underground mining works with quality analysis and determination of exploitation field's dynamics of extraction. Data base can be updated, it can be presented and it can be transferred into other formats as SQL, Paradox, Oracle and others. [1]

In order to define geometry and coal quality in the zone of the interest for this paper a database was formed containing the following:

1. Data on the spatial position of exploration boreholes,
2. Data on the coal seam's hanging and lying wall, respectively lithostratigraphic data.
3. Data on the coal quality parameters,
4. Data on the limits of the mine field, and period of excavation in the limited mine field.

3 GEOSTATISTICAL ANALYSIS OF COAL DEPOSIT

Every mining engineer or geologist is avare the chances for extraction of higher class ore are bigger on the spots of earlier higher class deposits rather on the spots with lower class. Furthermore there is an adequate spatial interdependence. Geostatistics is statistics in which the spatial association was taken into account; there are variables known as regionalized variables. [2]

Development of the geostatistics through the history is very interesting. The name itself comes from Professor Georges Mathéron and his colleagues from Fontainebleau school in France. With that term they described methods developed for problem solving of assessment of ore concentration, primary gold, as well as other types of ores. Idea of geostatistics is developed mainly independent from papers in the field of spatial statistics with its own style and terminology.

Mathéron published the first geostatistical results in his doctoral thesis in 1960' (MATHÉRON, 1962, 1963, 1965). The results still represent capital geostatistical papers. For the first time kriging method was mathematically described, which still stands for the best method for deterministic assessment variable value in the space. The kriging method was for the first time used to assess gold concentration in South Africa's mines. It was conducted by Krige 1951, and this method carries the name in his honor. [3]

3.1 The Choice of Interpolation Method Used for Coal Seam Quality Modeling

The softwares, which are commonly used in mining, are providing different interpolation methods. Some of the provided methods are: inverse distance to a power, kriging, natural neighbor, nearest neighbor, polynomial regression, radial basis function, triangulation with linear interpolation, moving average, local polynomial.

Results of cross validation are main criteria for choosing an appropriate method. The cross validation process was used to determine assessed values in each and every *n* spot in *Surfer* software. The values are in different ranges based on used method of interpolation.

The basin area of interest the analysed method of kriging was adopted it had the smallest value in the process of cross validation. Respectively the model which has to be created with the method of kriging is the most correct one compared to the models created with other method of interpolation. Therefore the kriging wmethod will be used for further coal seam modeling processes and quality assessment.[4] When it comes to model coal seam thickness it inverse distance method is the most suitable. Negative values of coal seam thickness will not be provided with its algorithms.

3.2 Coal Seam's Analysed Data Results in the Zone of the Interest for Underground Mining

When it come to choosing right method for extraction it is crucial to know the coal seam thickness, with its variety, in coal deposit.

Therefore primarily coal seam thickness, with its variety, is geostatistical processed. As a result, a map of thickness has been created with equidistance between isopachs of 2,5 m. Picture 3 represents coal seam thickness map, the data from the geostatistical analysis are shown in table 1.

Picture 3. Coal seam thickness spatial variation [m]

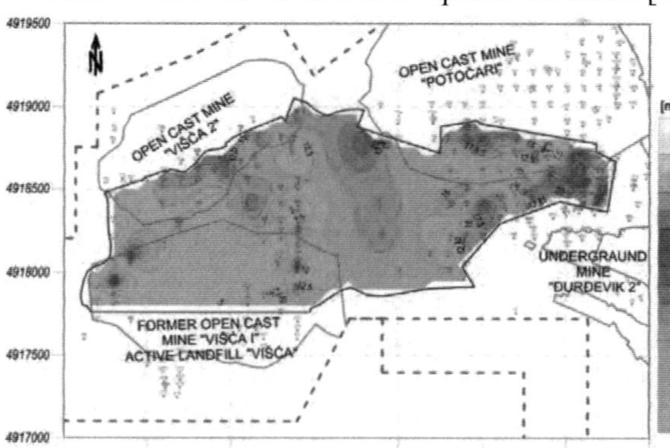

Table 1. Data about coal seam's thickness

Minimal ostensible thickness [m]	0,4
Maximal ostensible thickness [m]	38,34
Mean value [m]	12,28
Standard deviation [m]	4,60
Coefficient of variation	0,3746

- Coal seam thickness

Ostensible coal seam's thickness was determined with geostatistical analysis. It was analysed based on exploration drill holes drilled from the terrain's surface (table 1).

It can be said that the minimum thickness of coal seam is about 0,4 meters in the faults' zones. Maximal thickness of coal seam is approximately 38 m, and mean value circa 12 m. Coal seam's thickness in the interest zone will be in the interval „mean value ± one standard deviation (7,6 ÷ 16,9 m)".

- Coal seam dip

Coal seam angle of dip varies in the interest zone between 11° ÷ 32°. A huge span is caused by the tectonic derangement.

- Zone of interest dimensions for underground exploitation

Zone of interest for underground exploitation in RMU „Đurđevik" covers an area of 260 ha. From west to east the lengts of the coal field is circa 3,2 km, and from north to south 1,3 km. As shown in table 1, mean thickness is around 12 m, so the conclusion is that geological reserves of brown coal in this zone are circa 40 million tons. The commercial reserves, based on current exploitation method in underground exploitation in this mine, are considered as 20 million tons of high quality brown coal.

3.3 Other coal seam quality criteria results in the zone of interest

Based on the structural characteristics and sediment genesis of so far examined coal deposits the assumption of continuous, homogenic quality parameter's distribution is justified.

Primarily it is the matter of quality indicators data as NCV and others.

It implies that it is quite hard to assum greater sudden variations of calorific value between two points of smaller distance between them.

The assumption is taken as continuous change and the greater variations are possible only for bigger distances. [5]

Pictures 3-7 are showing isolines of coal seam's quality parameters in the zone of interest. Table 2 shows geostatistical results of coal quality marks.

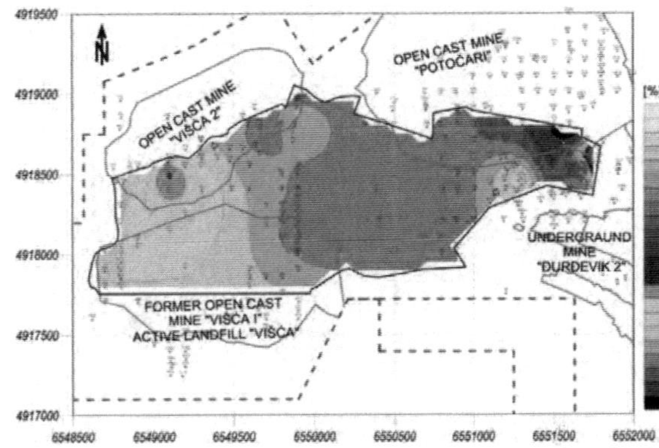

Picture 4. Moisture content spatial variation [%]

Picture 5. NCV spatial variation [kJ/kg]

Picture 6. Ash spatial variation [%]

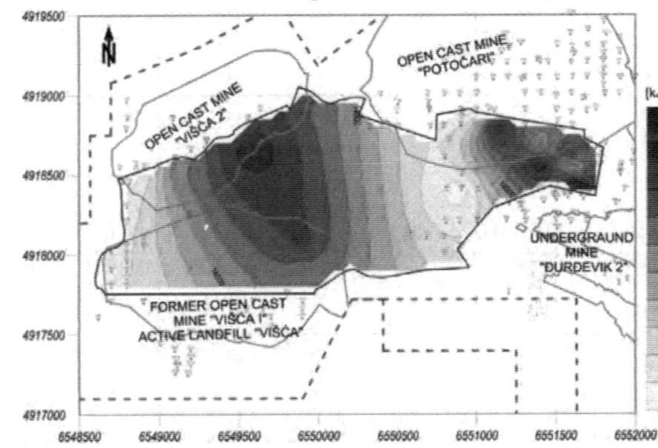

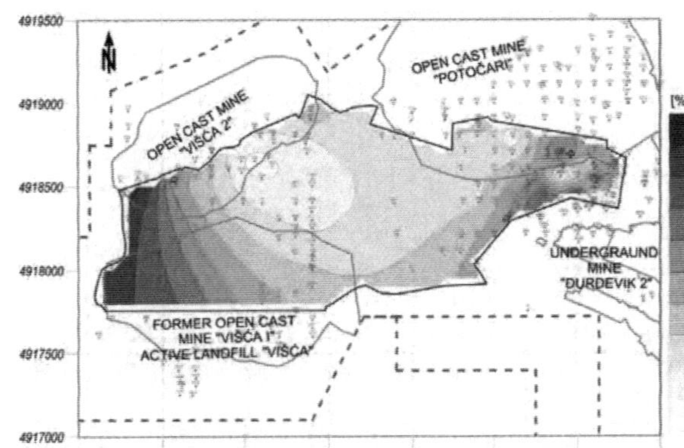

Picture 7. Total sulphur spatial variations [%]

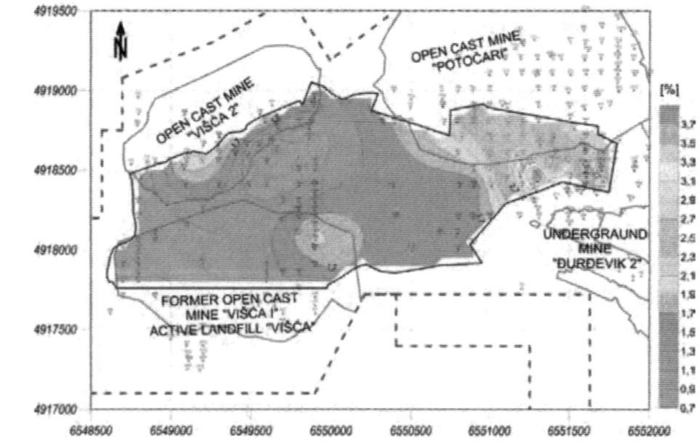

Table 2. Geostatistic results of coal quality marks in the zone of interest for underground exploitation

Parameter	NCV [kJ/kg]	A [%]	S [%]	W [%]
Min.	4862	9,85	0,80	4,7
Max.	23752	54,46	3,65	51,97
Mean	14604	25,97	1,64	10,56
St.deviation	4687	9,23	0,48	3,82
Coef. of variation	0,329	0,355	0,290	0,361

Remark: Data from the table 2 cannot be taken with certainty as there are not enough coal chemistry analysis on the examined area

4 CONCLUSION

The prime goal was to show coal seam interpretation, respectively to define its basic characteristics (mean thickness, angle of dip and strike, quality, etc.).

Through this paper it has been shown that geostatistics is powerful tool which has been used for many years in mining practice.

In RMU „Đurđevik“ interpretation (analysis) of coal deposit like this still has not found a bigger scale application. The possible explanations are as following:

- Current low regulations are not familiar with the present software tools,
- Current topographic maps are not suitable for 3D modeling,
- Current „up to date“ maps of mining works are not suitable for 3D modeling,
- Employees are not well educated for specific software usage,
- Current exploration databases are not providing relevant information necessary for quality modeling

In order to overcome above given problems it is crucial to set up tighter collaboration between mines, institutes and faculties. Imperative is to students and young profesionals a chance to study geostatistics through lectures, workshops, seminars. Moreover participating on other similar professional-scientific gatherings in order to share experiences with other colleagues from the whole region and the rest of the world.

REFERENCES

1. Akif Ibrišimović, Razvoj pravca i fronta napredovanja rudarskih radova na površinskom kopu „Šikulje“ sa aspekta ostvarenja optimalnog kvaliteta uglja; Magistarski rad; RGGF Tuzla septembar 2010.god.
2. Kolegij: Analiza prostornih podataka; Geostatistika; prof.dr.sc. Damir Medak; Geodetski fakultet Zagreb
3. Primijenjena geostatistike u analizi geoloških podataka; Tomislav Malavić; Udžbenici sveučilišta u Zagrebu; INA i RGGF, Zagreb 2008.god.
4. Ante Čubaković, Geostatistički model glinene podine akumulacije HE Tihaljina; Geotehnički fakultet Varaždin 2012.god.
5. A.Ibrišimović; S.Hadžić; Razvoj pravca i fronta napredovanja rudarskih radova sa aspekta ostavrenja optimalne donje toplotne energije uglja na površinskom kopu „Šikulje“ (naučni rad); Rudarstvo 2011; str.49.
6. I.Galić, RGNF Zagreb; Određivanje optimalne točke otvaranja i razvoja površinskog kopa na primjeru ležišta “Kongora”; Rudarsko-geološki glasnik 8; Mostar, prosinac 2004.god.
7. Perišić M; Primjenjena geostatistika – Knjiga I; Rudarski institut Beograd; 1983.god.
8. M. Miladinović, V.Čebašek, N.Gojkovići; Računarski programi za projektovanje i modeliranje u rudarstvu; Podzemni radovi 19; RGF Beograd 2011;Pregledni rad.
9. Pavlić. I.: Statistička teorija i primjena; Tehnička knjiga Zagreb, 1985.god.

Introduction of Quarries Wastes in the Composition of Cement

S.Berdoudi*[1] H.Hebhoub[2] .S.Bentaala[1].Z.Mekti[3]

1. Laboratory of Mining Engineering, National School Polytechnic of Algiers, Algeria
2. LMGHU laboratory, Department of Civil Engineering, Skikda University.
3. LAVMINE laboratory, Mining Department, Annaba University

**Corresponding author: Email: berdoudisaid@yahoo.fr*

ABSTRACT:
The management of mining waste is one of the major environmental problems which are facing the mining professionals. The main objective of this study is to technically show the possibility to recuperate the sand marble wastes as a substituted addition in the cement and mortar. The aggregate used in this study is a sand of marble wastes (excess loads of sand exposed to bad weather conditions) of the quarry derived from Fil-fila marble (Skikda, east of Algeria).

The study focuses the effect of marble waste sand fillers substitution in the cement paste and mortar (5, 10, 15 and 20%) with the same fineness of the cement, to compare the results obtained through control samples (0%) of cement paste properties in the fresh condition and the mechanical performances of mortar in the hardened condition.
The obtained results show that the used material can be technically used as additive in cement.

Keywords: Waste, Quarry, Fillers, Marble, Substitution, Cement.

1 INTRODUCTION

Waste management is currently one of the major encountered problems especially in developing countries. Our study is registered in a politic of recycling and valorizing of wastes that can be used in the cement formulation.

According to Amritpal and Rajwinder (2015), the consumption of cement and cost are more and more increasing , Algeria has a large deficit in building materials, especially cement and sand, the demand rises annually to more than 15 million cubic meters of sand (L.Azzouz and al 2000).

The quarries of granite and marble provided a large amount of scrap masses and waste, also the crushing plants generate a very large amount of waste consisting essentially of powders and slurries, according to Hebhoub and al (2013) the storage of such wastes in repositories promotes the air pollution, contamination of water sources and agricultural land, therefore it is necessary to remove these products, to enhance and reuse them again.

The main objective of this study is to contribute to the re-use of white marble waste (dust) of the Fil-fila career in cement manufacturing.

2 CHARACTERIZATIONS OF THE USED MATERIALS

The used materials for this period are:

- Cement CEMI class 42.5 coming from the cement works of HADJAR SOUD Company,

-Fillers of Marble waste sand are obtained by grinding in a standard normalized ball.

- The mortars are prepared with standard sand CEN according to the norm EN 196-1, this natural siliceous sand with an apparent voluminal mass of 1.63 g/cm^3 and an absolute density mass of $2.5g/cm^3$, this sand is inert from a chemical point of view with fineness module of 2.33.

Table1.Characteristics of used materials

Characteristics	Cement	Marble wastes sand fillers
Absolute density (g/cm^3)	3.33	2.79
Specific surface Blaine) (cm^2/g	**3200**	**3200**
CaO	61.31	55.29
Al_2O_3	5.45	0 .14
Fe_2O_3	3.54	0.09
SiO_2	22.73	0.53
MgO	0.48	0.2
Na_2O	0.19	0.00
K_2O	0,63	0.01
Cl^-	0.035	0.025
SO_3	2.44	0.04
Loss on ignition	2.45	43.40
Insoluble residue	0.85	0.035
CaO (free)	0.4	--
MS	2.52	--
MAF	1.54	--
LSF	0.88	--
MH	1.93	--
C_3S	28.14	--
C_2S	38.71	--
C_3A	8.45	--
C_4AF	10.76	--
$CaCO_3$	--	98.67

With the results listed in Table 1and according to the chemical analysis, we find that the marble waste sand are limestone fillers (98.67 $CaCO_3$). The addition of marble waste sand fillers in cement leads to increase the CaO quantity. This increase gives a lot of C3S; it also reduces the quantity of C2S, C3A and C4AF.

3 EXPERIMENTAL PROGRAM

In the experimental program, we study the substitution of a cement part by the marble wastes sand additions, through varying the substitutions rate (0%, 5%, 10%, 15%, 20%).

3.1 Campaign 1

Manufacturing starting from cement CEM I and marble wastes sand fillers of cement pastes with substitutions rates (0%, 5%, 10%, 15%, 20%).

The constants and variables parameters are water/cement ratio which is equal to 0,27 and the substitution rate respectively. Tests made in this company are consistency, setting and steadiness.

The parameters and tests made for the Campaign 1 are shown in the table 2.

Table 2: Parameters and tests made for the campaign 1.

Scoring	Type of cement paste
-Witness (0%)	-Cement paste made of CEM I
-Pa (5%)	-Cement paste made of 95% of CEM I and 5% of marble wastes sand fillers
-Pa (10%)	Cement paste made of 90% of CEM I and 10% of marble wastes sand fillers
-Pa (15%)	-Cement paste made of 85% of CEM I and 15% of marble wastes sand fillers
-Pa (20%)	-Cement paste made of 80% of CEM I and 20% of marble wastes sand fillers
-Constant parameters	-W / C= 0,27
-Variable parameters	-Substitution Rate
-Tests made	- Consistency - Setting - Steadiness

3.2 Campaign 2

Manufacturing starting from cement CEM I and marble wastes sand fillers of mortars with substitutions rates (0%, 5%, 10%, 15%, 20%).

The constants and variables parameters are water/cement ratio which is equal to 0, 5 and sand, substitution rate respectively. Tests on hardened mortar samples are compressive and tensile strength in flexion at 2, 7, 28, and 90 days.

The parameters and tests made for the Campaign 2 are shown in the table 3.

Table 3: Parameters and tests made for the campaign 2.

Scoring	Type of mortar
-Mo (0%)	Normal Mortar made of CEM I
-Mo (5%)	-Normal mortar made of 95% of CEM I and 5% of marble wastes sand fillers
-Mo (10%)	-Normal mortar made of 90% of CEM I and 10% of marble wastes sand fillers
-Mo (15%)	-Normal mortar made of 85% of CEM I and 15% of marble wastes sand fillers
-Mo (20%)	-Normal mortar made of 80% of CEM I and 20% of marble wastes sand fillers
-Constant parameters	-W /C= 0.5 -Sand
-Variable parameters	-Substitution rate
-Tests made	-Compressive strength in 2, 7, 28, and 90 days. Flexural strength in 2, 7, 28, and 90 days.

4 EFFECT OF RECYCLED FILLERS ON THE CEMENT PASTE CHARACTERISTICS

4.1. Consistency

For specific surface (Blaine) marble waste sand fillers equal to specific surface (Blaine) cement, the penetration of the probe « Vicat » is a decreasing function with the degree of substitution, the bad consistency is obtained for the sample cement (cement rich in C_3A) in accordance with Messan (2006), the addition of marble waste sand fillers led to improved the consistency of the cement paste in accordance with the study executed by Care and al (2002).

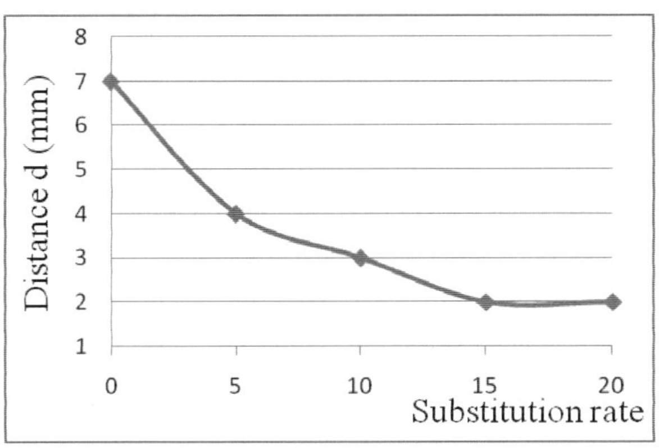

Figure 1: Variation of Consistency according to the substitution rate

4.2 Start and End of Setting

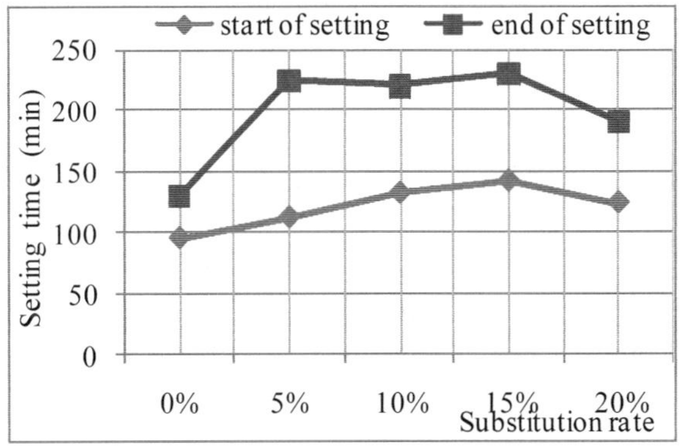

Fig 2: Variation of time of start and end of setting according to the substitution rate.

Figure 2 shows that for W / C constant, the start and end of setting times vary depending on the substitution rate as it was demonstrated by Benia, (2011), minimum values are obtained for the sample cement (presence of large quantities of the C_3A giving a quick time of setting) and the maximum values are obtained for a substitution rate of 15% for marble waste sand as it was proved by Messan (2006).

4.3. Steadiness

Figure 3 show that the introduction of marble waste fillers does not affect the stability of the cement paste whatever the rate of substitution and the fineness of grinding

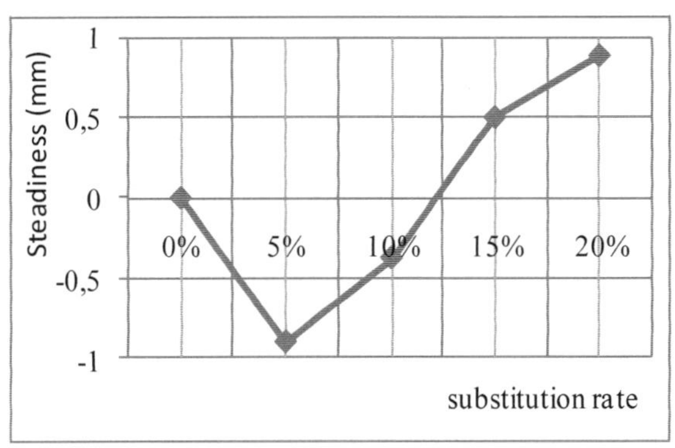

Figure 3: Variation of Steadiness according to the substitution rate

5 EFFECT OF RECYCLED FILLERS ON THE MORTAR CHARACTERISTICS

5.1. Compressive Strength

Up to 7 days, the best performance is obtained for the mortar with 5% substitution rate.

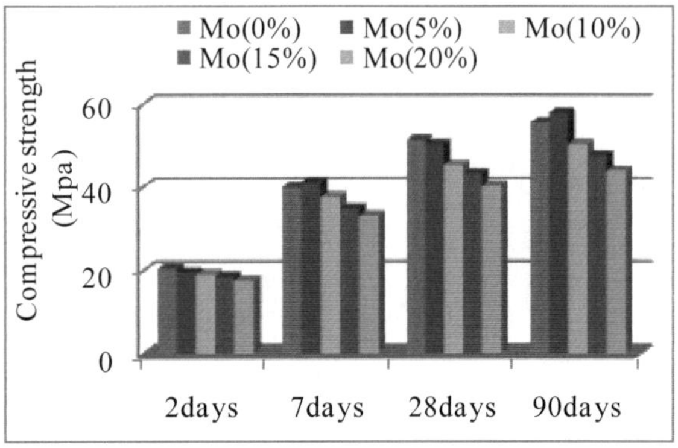

Figure 4: Influence of substitution rate on the compressive strength

The introduction of 5% of marble waste sand leads to increase the compressive strength, this can be explained by the increase of CaO which promotes hydration of the C3S, as it was demonstrated by S.caré and al (2002).

Beyond 5%, the behavior of the different formulations is the same for all ages. Between 2 and 7 days the change in the strength is important compared to 7d, 28d and 90 days, this can be explained by the presence of CaO in large quantities which offers very high strengths at short term. The strength increases with age.

5.2. Flexural Strength

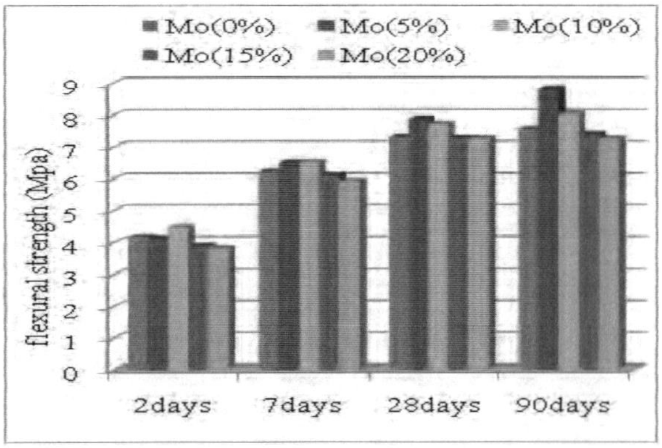

Figure 5: Influence of substitution rate on the flexural strength.

Generally, the behavior of the flexural strength (Figure 5) is the same as in compression; the variation is less between 28 and 90 days compared with 2 and 7 days.

CONCLUSIONS

The study shows that:

-The bad consistency is achieved for the sample cement; adding the marble waste sand, duct improved the consistency of the cement paste; they cause a decrease in the C_3A quantity.

- The introduction of marble waste sand accelerates the mechanical strengths of mortars at a young age.

-The introduction of marble waste leads to increase cohesion.

-The mechanical strength decreases with increasing of the substitution rates (beyond of 5%)

ACKNOWLEDGMENTS

The authors would like to thank the staff of the Civil Engineering Laboratories at the Skikda University and National polytechnic school of Algiers for their assistance and support during the course of this research. Also, the GICA Company (HDJAR EL SOUD), for their generous financial support.

REFERENCES

A.Messan: 2006, Contribution to the study of thin structure out of mortar behavior at very early age. Doctoral thesis, Montpellier.

Er. Amritpal Kaur and Er. Rajwinder Singh. (2015).Strength and duriabilty properties of concrete with partial replacement of cement with metakaolin and marble dust, *Int. J. Eng. Res. Tech.* 04, pp. 1032-1035.

H. Hebhoub, M.Belachia and R. Djebien 2013,Introduction of sand marble wastes in the composition of mortar, *Structural engineering and mechanics*, Vol. 49, No. 4 (2014) 491-498.

L.Azzouz , M.Bouhicha, M , Kenai, S and Hadjoudja, M, 2000, Recyclage des déchets de briques dans le béton de sable de dunes , *Procedings du 1er colloque national de génie civil : Matériaux de Construction*, Université de Mostaganem).

M.Benia, 2011, Influence of specific surface cements with mineral additives on the mechanical behavior of mortar and concrete using local materials, Magister memoy, Msila University, Algeria.

S.Caré, S, linder,R Baroghel Bouny,V De Larrard, F and Charonnat, Y ., 2002, Effect of mineral admixtures on concrete use properties. experience plan and static analysis, *LCPC.* ouvrage d'art OA33.

Maden Projelerinde Yatırım Giderlerinin Analizi
Capital Expenditure Analysis in Mining Projects

C. Hoşgit, V. Aydoğan
Promer Müşavirlik Mühendislik A.Ş., Ankara

T. Özenç
Promer Müşavirlik Mühendislik A.Ş., Ankara

İ. Y. Tonguç
Promer Müşavirlik Mühendislik A.Ş., Ankara

ÖZET Maden projelerinin her safhasında yatırım giderleri (CAPEX) büyük öneme sahiptir. CAPEX çalışmalarının bulgularına göre bir projenin yatırım yapılabilir olup olmadığının kararı verilmektedir.

CAPEX oluşturulmasında yapılan mühendislik çalışmaları sonrasında yerel koşullar detaylı analiz edilerek en uygun yatırım maliyeti belirlenmelidir. Yapılan çalışmalar göstermektedir ki projede uygulanan stratejiler yatırımın CAPEX değerinin ciddi bir oranda yükselmesine veya düşmesine sebep olmaktadır.

NI43-101, JORC, SAMREC veya PERC gibi uluslararası standartlar çerçevesinde hazırlanan teknik raporlar finansal güvenilirlik açısından avantaja sahiptir. Bu avantaj projeye finansman ve itibar kazandırmakta ve böylelikle işveren yatırımı özkaynaklarını kullanarak yapmak zorunda kalmamaktadır.

Bu makalede Türkiye'deki bazı maden projelerinin CAPEX çalışmalarından örnekler sunulacak ve maden yatırımlarınının CAPEX bileşenleri örnekler üzerinden irdelenecektir.

ABSTRACT Capital Expenditure (CAPEX) studies have great importance for mining projects in all phases of the investments. Based on the findings of CAPEX, any project can be identified as feasible or not.

After certain engineering works during CAPEX generation, local conditions shall be analyzed and implemented in detail in order to meet the optimum and accurate investment capital. There are several examples proving the effect of project estimate strategy which may increase/decrease the total CAPEX value significantly.

Any technical report which is prepared according to international standards such as NI43-101, JORC, SAMREC or PERC have the advantage of credit viability. This advantage also brings financial support and reputation to the projects, therefore these projects do not require to be owner financed.

In this article, examples of CAPEX studies for a number of mining projects in Turkey will be presented. Mining investment CAPEX breakdown will be generated based on these sample projects

Keywords: CAPEX, Capital Expenditure, Feasibility, Cost Estimate, Mining

1 ENGINEERING LEVELS IN FEASIBILITY STUDIES

Capital expenditure analysis is the means by which the value creation potential of a project is determined. Correctly determining this potential is critical to the successful operation and expansion on an installation. A cost estimate is produced to approximate the cost of the project. The project budget is a necessity to support feasibility evaluations for project funding requirements and to facilitate the project control.

The evaluation of a mining project from exploration through development and production is a complicated process. Mine development commitment activities for a potential project are initiated when a mineral resource is identified and continue through to the start of construction. The technical feasibility and the economic viability of each project are determined during the phases of mine development with more detailed engineering data required at each stage. (Runge Pincock Minarco, 2015). There are at least three levels of engineering studies during development and those are commonly acknowledged by the mining industry:

1. Preliminary Economic Assessment (PEA), or Scoping study
2. Pre-feasibility Study (PFS)
3. Feasibility Study (FS)

After each study level, investors decide, based on economics, environmental considerations, and market timing, whether to advance to the next level and ultimately whether to build. As the level of engineering detail increases through each study stage, as measured by accuracy, the overall degree of uncertainties decreases. Included in most standards references to definitions of Preliminary Economic Assessment/ Scoping studies, pre-feasibility studies, and feasibility level studies, and the following descriptions are consistent with those definitions.

1.1 Preliminary Economical Assessment (PEA) or Scoping Study

A scoping study is a preliminary evaluation of the project. Although the level of drilling and sampling must be sufficient to define a resource adequetely, flowsheet development, cost estimation and production scheduling are often based on limited test work and engineering design. This level of study is useful for defining subsequent engineering inputs and further required studies, but is not appropriate for economics decision-making.

1.2 Prefeasibility Study

A prefeasibility study represents an intermediate step between the scoping and final feasibility studies. Cost estimates are of the order of ±25-30% accuracy, but not considered sufficiently accurate for final decision making.

1.3 Feasibility Study

A feasibility study, considered the "bankable document", is detailed and accurate enough to be used for positive "go" decisions and financing purposes. Cost estimates are ±15% accurate or better. Mine plans show materials movements and ore grades on an annual basis. Flowhseet development is based on comprehensive test work, material balances and general arrangement drawings. Financial evaluation is based on cash flow calculations for the life of defined reserve. (Lowrie, 2002). Capital cost estimate variations between feasibility studies are given in Table 1. Estimate quality refers to the fullfilment of estimate quality requirements in accordance with the formal quality assurance requirements. An estimate's quality requirements are mostly concerned with accuracy, but they also take into account credibility, confidential level, precision, risk, reliability and validity, throughness, uniformity, consistency, verification and documentation.

Accuracy is the degree to which a measurement or calculation varies from the actual value; thus, estimate accuracy informs us of the degree to which the project final cost outcome may vary from the single point value depicted by the estimated cost. Because the cost estimate is an approximation of the cost of a project, the estimate accuracy is a measure of how closely the estimate is able to predict the actual project expenditures. Unfortunately, this can only be known with certainity after the project is completed (Hickson&Owen, 2015).

2 COST ESTIMATE STANDARDS

Reporting standards provide recognisable benchmarks inside the industry and are recognised by others outside the industry. Standards gain strength and credibility when they are mutually compatible. They reduce confusion, increase understanding, hopefully

Table 1. Miminum report contents for engineering studies (RPM, 2015)

DESCRIPTION	Preliminary Economic Assessment (PEA, Scoping Study)	Prefeasibility Study (PFS)	Feasibility Study (FS)
Civil, Structural, Architectural	Order of magnitude based on historic data or factoring	Estimates from historical factors, percentages and vendor quotes based on materials volumes	Detailed from estimates; engineering 15 to 25% complete; multiple vendor quotes
Piping, HVAC			
Electrical &Instrumentation			
Construction Labor &Productivity			
Material Volumes, Amounts			
Material/Equipment Pricing			
Infrastructure			
Contractors	Included in unit cost or as a percentage of total cost	Percentage of direct cost by cost area for contractor; historic for subcontractors	Written quotes from contractor and subcontractors
EPCM	Percentage of estimated construction cost	Percentage of detailed construction cost	Calculated estimate from EPC(M) firm
Pricing	FOB mine site including all taxes and duties	FOB mine site including all taxes and duties	FOB mine site including all taxes and duties
Owner's	Historic estimate	Estimate from experience factored for similar project	Estimate prepared from detail zero based budget
Environmental Compliance	Factored from historic experience	Estimate from experience factored for similar project	Estimate prepared from detail zero based budget for design engineering and specific permit requirements
Escalation	Typically not considered	Based on company's current budget percentage	Based by cost area with risk
Working Capital	Factored from historic experience	Estimate from experience factored for similar project	Estimate prepared from detail zero based budget
Accuracy	+/- 50%	+/- 25%	+/- 15%
Contingency	25%	15%	10%

keep things as simple as possible and are applicable under all circumstances.

There are many resource/reserve codes worldwide namely:

- CIM GUIDELINES (NI43-101) - Canadian Institute of Mining, Metallurgy and Petroleum (Canada)
- JORC CODE - Joint Ore Reserve Committee (Australia)
- SME GUIDE - Society for Mining, Metallurgy and Exploration (USA)
- SAMREC CODE - South African Mineral Committee (South Africa)
- PERC - Pan-European Reserves & Resources Reporting Committee

When fulfilling public reporting duties, completing an audit, or raising capital, owners may need an evaluation of resources and reserves. An independent reserve evaluation may provide a foundation to develop the project. International banks, stock exchanges, government agencies, investment banks, and funds typically require an Independent Mineral Resources Estimate and Mineral Reserves Estimation.

Resource and reserve reporting may be extremely accurate; however investors usually require a third-party assessment that meets the official standards which also provides credibility and reassurance for investors.

Any feasibility study which is prepared according to national standards have great advantage of providing project funds, in other words owners do not have to finance the project 100% by themselves. Accurate CAPEX analysis in such projects bring the objective approach to the investor which decreases the risk of realization of the investment.

3 ESTIMATE CLASSIFICATIONS

There are two main authorities to classify the cost estimates in industrial investments; US Department of Energy (DOE) and American

Society of Professional Estimators (ASPE). There apppears some confusion in terminology arising from the different classification systems of these organizations. There are also some other organizations which have different levelling systems (AACE International, 2016). In order to minimize the confusion between these classifications a summary table is given in Table 2.

ASPE: American Society of Professional Estimators

DOE: US Department of Energy

ACostE: Association of Cost Engineers

AACE: Association for the Advancement of Cost Engineering International

Table 2. International estimate classifications (Hickson&Owen, 2015)

Estimate Name	ASPE	DOE	ACostE	AACE.
Scoping	Level 1	Class 5	Class III	Class 5
Prefeasibility	Level 2	Class 4		Class 4
Feasibility	Level 3	Class 3	Class II	Class 3
Definitive	Level 4	Class 2	Class I	Class 2
Detailed	Level 5	Class 1		Class 1

4 ELEMENTS OF COST ESTIMATE

The capital cost estimate is a prediction of the total installed cost (TIC) of the facilites described in the scope of work of a project. The principal elements of the estimate are direct costs, indirect costs, EPCM, Owners cost and contingency.

4.1 Direct Costs

Direct costs are overall costs of installation that can be directly attributed to the project. The major items of direct costs are;

Overall site: Bulk earthworks, site control perimeter fence, borrow areas, storm water collection, mine closure.

Pre-operation mining: Pre-stripping/crearing&grubbing, haul roads.

Mine: Power supply and distribution, waste stripping, mine infrastructure, waste water treatment, mine mobile fleet, mining office appliances, furniture, IT and communication, mine survey equipment.

Process facilities: all communition facilities (crushing, screening and grinding), fine ore storage and reclaim, beneficiation (flotation, leaching, gravity separation etc.), dewatering, reagent preparation.

Process plant infrastructure and services: main power supply, switchyard and main substation, power distribution, mine dewatering, process water distribution, fresh water supply and distribution, sewage treatment plant, sewage collection system, potable water system, yard lightning, process CCVT, plant mobile fleet, fire fightning system, storm water collection system.

Tailing management facility: Tailing discharge and distribution, tailings return water system, tailings dam, dry tailings facility.

Ancillary facilities: Administration building, security office, medical building, first aid and emergency services, cafeteria and mine dry, locker building, assay and mill laboratory, mechanical and electrical workshop, warehouse, technical office, generator building, waste management facility, weigh scale, miscellaneous.

4.2 Indirect Costs

Indirect costs are those that can not be directly attributed to the construction of the physical facilities of the plant or associated infrastructure, but are those required to support the construction effort. The major items of indirect costs are;

Indirects: surveying, construction power/lightning, construction sewage/garbage/waste disposal, construction offices, mobilization/demobilization, medical/first aid, IT/communication, miscellaneous cranes/equipment, freight and logistics, vendor representatives, commisioning contractors, QA/QC, safety/security and environment, metalurgical testing and owner R&D cost.

Spares / Warehouse fills: Equipment spares, first fill reagents and consumables.

4.3 EPCM

EPCM costs are significant enough to be in a stand-alone category. EPCM costs include the design and engineering of the project, procurement of all the physical items required and management of the construction contractors. EPCM includes the following activities;

EPCM: EPCM contracts, consultants, owners team salaries, owners team housing, vehicles and expenses, owners team office and admin expenses.

4.4 Owner Cost

Owner cost includes all those costs specifically attributed to the Owner which are not included elsewhere in the estimate. Owner cost includes the following activites;

Owner Cost: Insurance, royalty, duties and taxes, consulting services, finance and marketing, land aquisition and community relations, training and salaries, owners team travel & expenses, working capital, miscellaneous.

4.5 CONTINGENCY

Contingency can be defined as a specific provision added to a base estimate to cover indefinable items that have been historically required but can not be specifically identified in advance. Project Management Institute (PMI) defined contingency as; a provision in the project management plan to mitigate cost risk.

Contingency is always included with an estimate to provide for the unknown costs that are likely occur by experience but are not identifiable. When using an estimate to set the budget, contingency is added to improve the probability that the funding will be adaequate to complete the project. More contingency is needed for early estimates due to the higher uncertainity of accuracy.

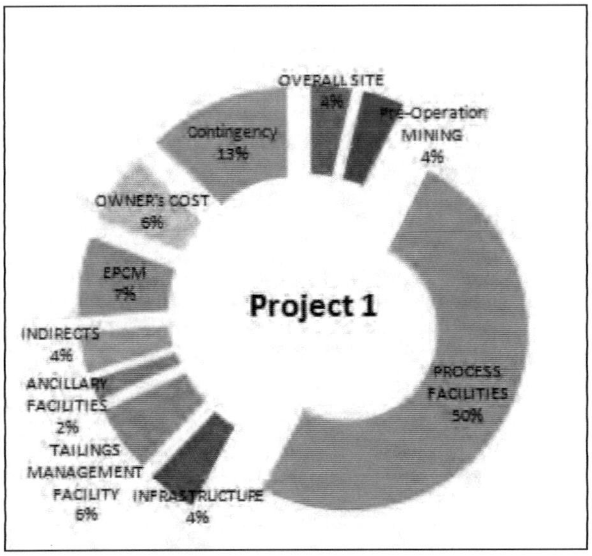

Figure 1. CAPEX elements distribution in example Project

CAPEX elements distribution in two feasiblity studies performed by authors are given in Figure 1 and Figure 2 and comparison between them is given in Figure 3. As given in the examples the elements ratio always vary depending on the mining and beneficiation method, capacity of the plant, contract strategy, local conditions of the mine and plant site and tailing management facilities.

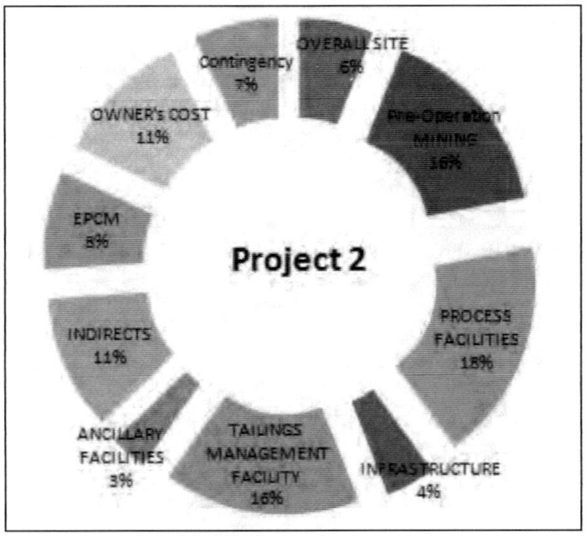

Figure 2. CAPEX elements distribution in example Project2

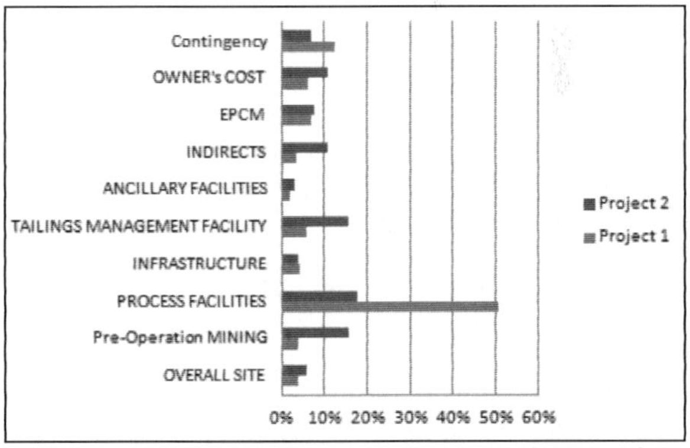

Figure 3. CAPEX elements distribution comparison between Project 1 and Project 2

5 LOCAL CONTRIBUTION AND CONTRACTING STRATEGY IN COST ESTIMATE

Contracting strategy is the most critical item in CAPEX calculations and this strategy has to be determined at the beginning of the project. In countries where mining is well developed like Canada, South Africa,

Australia and United States of America the feasibility studies for such investments are executed by qualified engineering companies. The general approach in those countries is to assign a general contractor for construction works and to assign an EPCM contractor for construction management activities. In EPC alternative method it is observed that the initial capital investment increases relatively. In Turkey usually a general contractor methodology is applied and individual contractors are selected as alternative. In some projects the time schedule is very tight therefore general contractor option is preferred but the cost is higher in these cases. There is a cost saving in the investment when agreed with individual contractors but the importance of management increases in these cases.

In a typical feasibility study, after engineering calculations finalize, budgetary quotations are provided fom vendors and according to calculated bill of quantities, budgetary construction unit costs are determined according to construction contractor budgetary proposals. The accuracy level of these quotations are directly related with the level of engineering completed.

The studies which are prepared by the foreign companies considering the local conditions of that country show significant difference from the studies prepared in Turkey with Turkish local conditions. For example; foreign companies determine the unit cost rates according to their own country engineering man-hour, material supply, workmanship man-hour thus there is a remarkable difference between Turkish all in cost and install prices. Utilization of local talent and embrace local construction techniques bring cost advantage to the project. On the other hand considering the product range of vendors, instead of each equipment supply from different vendors to maximize the supply content from the same vendor brings more negotiation power to the owner and brings cost savings in project budget. This strategy also effects the optimization of supervision and commisioning costs in a positive way. Furthermore foreign companies may fail to notice the local governmental investment incentives and taxing advantages in developing countries like Turkey during the financial studies.

Another point which effects the initial investment is mining method. The initial capitals are higher when owner decides to perform mining activities on their own. Contract mining is an alternative which decreases the initial capital by the application of this cost into years during mine life and as a result increases the operating costs. Owners make decisions according to their engineering company's trade-off studies. In Turkey contract mining is accepted as an advantage considering the low mining cost when compared to international countries.

The feasibility studies which were performed by foreign companies were updated considering the local methodologies explained above in some different case studies by authors. As an example, in one of the gold mine projects located in northwest of Turkey, CAPEX update studies resulted in saving 34% in earthwork cost, 22% in concrete work cost, 32% in steel structure cost considering the local conditions and contracting strategies. The results show that project CAPEX values can be reduced to 25-40% overall. These updated feasibility reports are approved by international organizations according to the standards, presented to the banks for financial support and gained more credibility.

6 CONCLUSION

CAPEX is one of the key advisors for the yes or no decision of a mining project. There is not a certain ratio between the elements of mining initial expenditures. They depend on the mining and beneficiation method, capacity of the plant, contract strategy, local conditions of the mine and plant site and tailing management facilities. Therefore each case must be considered as a new research.

The accuracy of CAPEX relies on the application of right project execution and contracting strategy with the contribution of local conditions of the project country. Several projects approve the right methodology in CAPEX calculations effects the final values signicifantly in a positive way.

The finance support of a project is directly related with the international standards which are applied during engineering studies of the project. Credibility has a great importance in mining since mining projects are sizable investments. The feasibility studies which are prepared according to the international standards have the advantage of financial support hence these projects do not require to be owner financed.

REFERENCES

AACE International, 2016. Cost Estimate Classification System-As applied in Engineering Procurement and Construction for the Process Industries, *AACE International Reccomended Practice No.18R-97*, p3-5.

Hickson R. J., Owen T. L., P., 2015. *Project Management for Mining Handbook for Delivering Project Success*, Society for Mining, Metallurgy and Exploration, Inc (SME) Colorado, p159-185.

Lowrie, R.L., (ed.), 2002. *SME Mining Reference Handbook*, Society for Mining, Metallurgy and Exploration, Inc (SME) United States of America, p1-7.

RungePincockMinarco, 2015. Minimum Engineering Study Requirements Update, *RungePincockMinarco Perspectives*, Issue No. 128, p1-5.

Study of the Effect of the Use of Slag for the Production of Clinker on the CO_2 Emission, Algeria

Z. Mekti, M. Bounouala, A. Boutemedjet, A. Chaib, S. Berdoudi[*]

Mining Resources Valorisation & Environmental Laboratory, Mining Department. Earth Sciences Faculty. Badji Mokhtar University Annaba, Algeria

Mining Engineering Laboratory EMP of Algiers, Algeria[*]

ABSTRACT The main purpose of this work, is to study the possibility of reducing the CO2 emission rate, when substituting the sand by the slag (By-product of the transformation of iron into steel) in raw meal, For the production of composite cement clinker (CPJ CEM II / A 42.5). Two samples of the raw meal were prepared, one with ordinary raw materials (sand), and the other with 9% of slag. The two raw meals were sintered at a temperature of 1450°C. The results of the mineralogical and chemical analyses showed that the use of the blast furnace slag, does not affect the mineralogical characteristics of the clinker. Inter alia, the production of the slag-based clinker, has allowed us to noticeably reduce the annual CO2 emitted by the two processes: decarbonation and combustion, with a decrease of up to 17.56%.

Key words: HADJAR-SOUD Cements Plant, CO_2 Emission, Clinker, Slag, Decarbonation, GEMIS 4.7.

1. INTRODUCTION

Reducing greenhouse gas (GHG) emissions, and mainly CO_2 emissions, which is responsible for 80% of the greenhouse effect in industrialized countries, is now a global issue underlined by the Kyoto agreements, December 1997.

This is what is driving the community to reduce its energy consumption and replace fossil fuels (coal, oil, etc.) with renewable and alternative energy sources; among the activities responsible for this emission is mainly cement production.

The cement, from the Latin "caementum" natural stone, is produced from abundant raw materials, limestone and clay. The cements (Portland cements, blast furnace cements, etc.) are obtained by means of a cooking process which combines these raw materials at 1450°C. The cooking takes place in a rotary furnace, a tube of 50 to 90 meters in length for the (VS) and about 200 meters for wet paths (VH), at the end of which the fuels introduced produce a flame of 2000°C. The cements cannot be obtained without this physicochemical combination at very high temperature. Cement is the basic building block for concrete production.

Cement Production Process

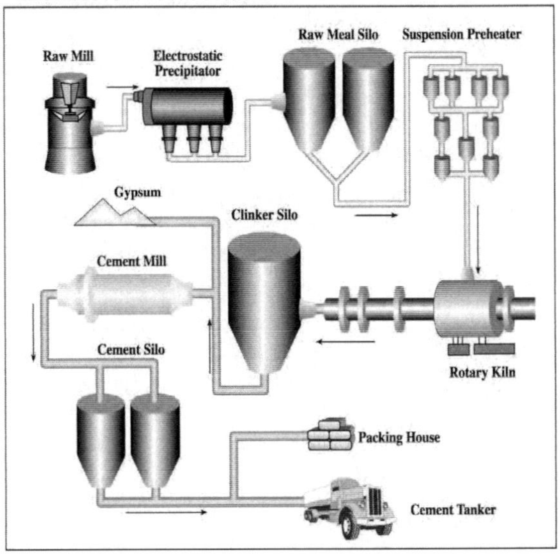

Figure 1: Cement production process
(Source: Lootahgroup)

CO_2 comes firstly, from a process based on a physicochemical combination at very high temperature and energy consumption.

Secondly, a phenomenon of transformation of the limestone ($CaCO_3$), under the effect of heat, into lime (CaO) and carbon dioxide (CO_2).

$$CaCO_3 + \text{heat} \rightarrow CaO + CO_2$$

More than 60% of the CO_2 emissions from cement production come from this "decarbonation".

1.1 Case Study

The cements plant HADJAR-SOUD (SCHS) is a subsidiary of the GICA group. On February 01st, 2008, it started a partnership with the Italian social partner BUZZI-UNICEM at a percentage of 35%, the remaining was hold by GICA group at 65%. HADJAR-SOUD cement plant is located 55km away from Skikda department along national road N°44 leading to Annaba departments. The plant consists of two production lines with a contracted producing capacity over 1 000 000 tons per year:

1st Line: Started on 1973 producing around 1250 T/D of Clinker.

2nd Line: Started on 1975 with a production capacity of 1800 T/D of clinker (After adding of a pre-heating facility on 1993).

1.2 Production and Slag Processing

In the blast furnace, in addition to the iron ore and coke, we add a fondant, generally a base of lime, to lower the melting point of the gangue, this will allow the extraction of the ore at a temperature between 1400°C and 1500°C. The remaining oxides of the load non-transformed to metal will constitute the slag. The slag is evacuated from the blast furnace at a liquid state. According to the corresponding cooling method, we obtain two types of slag; granulated slag (vitrified) at sudden cooling (case study) and the rocky slag (crystallized) result from slow cooling (figure 2)

Figure 2: steel slag production process

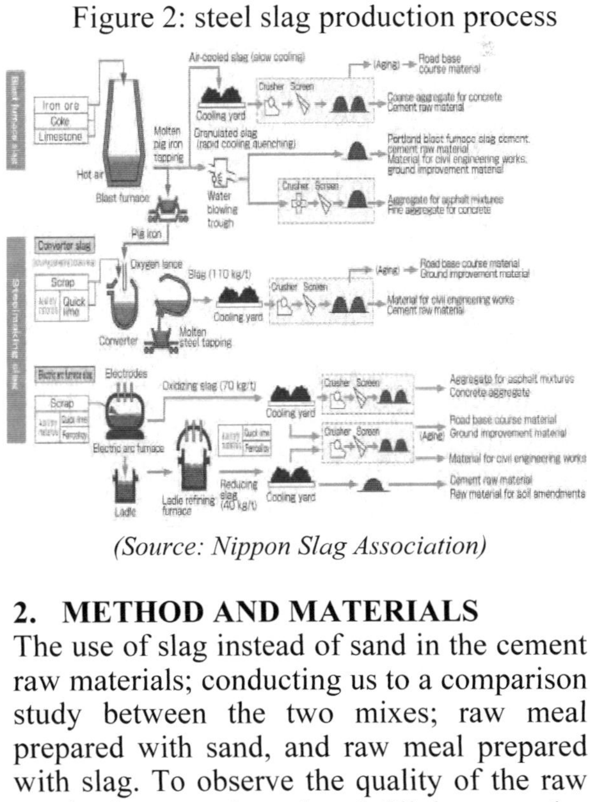

(Source: Nippon Slag Association)

2. METHOD AND MATERIALS

The use of slag instead of sand in the cement raw materials; conducting us to a comparison study between the two mixes; raw meal prepared with sand, and raw meal prepared with slag. To observe the quality of the raw meal mixture and produced Clinker; a series of calculations of the meal's modules were made (MH, MS, MAF ...), then we determine the mineralogical composition of the clinker, such as C_3S, C_2S, C_3A and C_4AF.

The GEMIS 4.7 software, 2011, assesses the impact of products on the environment.

Several trials dosing of raw materials was

made to reach a proper raw mix for the production of clinker. The obtained results during the chemical analysis by fluorescence X, of the various components of the raw material are shown in the following table.

Table 1: Chemical analyses of raw materials

Element	Limestone	Clay	Slag	Iron
CaO	54	5.44	34.04	15.36
Al_2O_3	0.32	17.94	6.61	4.11
Fe_2O_3	0.29	7.57	4.46	40.26
SiO_2	0.61	53.80	32.89	19
MgO	0.36	2.01	8.08	2.07
Na_2O	0.07	0.89	0.42	-
K_2O	0.03	1.75	0.67	0.9
Cl^-	-	0.013	-	-
SO_3	-	0.02	0.42	0.4

In order to perform these tests, the elements are weighed carefully by an electronic balance and then pulverized by a disk mill (type Retesch RS200), the latter acts as a perfect homogenizer of the four elements. With a granulometry of less than 50µm and a moisture content of not more than 5%, analysed by the X-ray fluorescence of the type Bruker AXS S8 LION. The mineralogical phases of the clinker were determined by the DRX type; PANalyticale (X'Pert PRO), the chemical analyses are presented in Table 2.

Table 2: Limitations of elements proportions and dosage of raw materials

Designation	Limestone	Clay	Iron	Slag	Sand	Total
Limit's element (Slag-based meal) (%)	68-72	18.9-19.4	1.6-2	8.5-9.5	-	
Elements limits (meal without Slag or ordinary) (%)	77-83	13.-14	1.5-3	-	2-4	
The amount taken for the test						
Proportion of test elements (with slag) (%)	70	19.2	1.8	9	-	100
Proportion of test elements (without slag) (%)	81.5	13.5	2	-	3	100

Table 3: Chemical analysis of raw meal

Element	CaO	Al_2O_3	Fe_2O_3	SiO_2	MgO	Na_2O	K_2O	Cl^-	SO_3
Meal with slag	42.56	4.09	2.48	13.78	1.36	0.14	0.36	0.005	0.13
Meal without slag	43.97	3.65	2.34	12.39	1.19	0.12	0.32	0.006	0.14

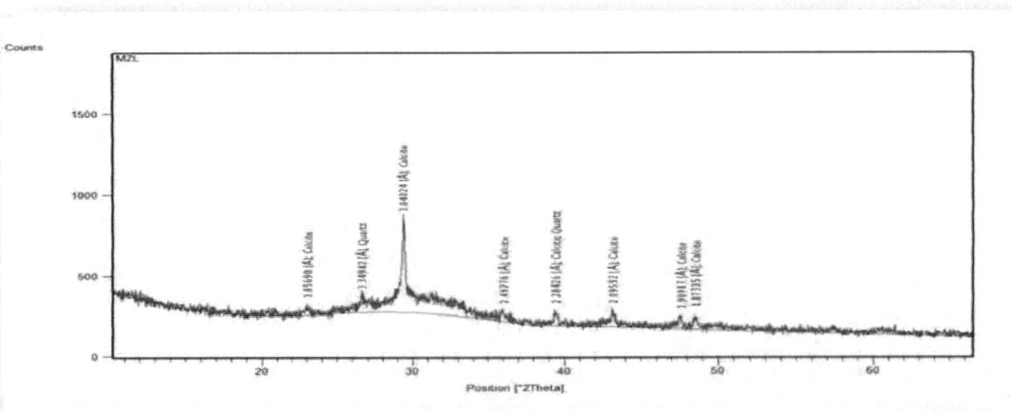

Figure 3: XRD of Slag used

For clinker production with and without slag; a quantity of raw meal was prepared, according to the percentage indicated in the preceding table, and dried at 110°C. They were placed inside a furnace at 500 ° C (Carbolite RHF 16/8, maximum temperature 1600°C) the temperature was increased to 1000°C, at which the samples were kept for 30 minutes, and finally, the temperature was further increased to 1450°C, where the samples remained for 30 min. At the end of the cooking process, the samples were rapidly cooled to avoid the return of C_3S to C_2S. The clinkers produced were analysed by chemical analysis and X-ray diffraction.

3. CHEMICAL AND MINERALOGICAL CHARACTERISTICS OF SLAG USED

As shown in Table 1, Slag consists of 95-98% of its chemical composition of a mixture of four oxides: silica SiO2, lime CaO, alumina Al_2O_3 and magnesia MgO; the complement consists of secondary oxides (FeO, MnO) and sulfur compounds, from a mineralogical point of view (figure 3), its crystallized constituents are essentially silicates or silico-aluminates of lime, but oxides, sulphides, and exceptionally, nitrides are also found.

Table 4: Chemical analyses of Clinker with and without Slag

Element	CaO	Al_2O_3	Fe_2O_3	SiO_2	MgO	Na_2O	K_2O	Cl^-	SO_3	**C_3S**	**C_2S**	**C_3A**	**C_4AF**
Clinker Slag based	65.98	6.34	3.84	21.36	2.11	0.22	0.56	0.007	0.2	**57.61**	**17.80**	**10.30**	**11.69**
Ordinary Clinker, without Slag	66.16	5.64	3.78	22.08	1.01	0.22	0.43	0.017	0.10	**57.30**	**20.10**	**8.35**	**11.5**

4. CHEMICAL AND MINERALOGICAL CHARACTERISTICS OF CLINKERS PRODUCED

The results of the chemical analysis of the clinker produced are given in table4. As shown in the table; the values of the four components of the clinker are in the standards, the addition of 9% blast furnace slag in the raw meal does not appear to affect the chemical composition of the clinker. The chemical analyses by XRF and the XDR of the clinkers produced are given in Table4 and Figure 3. In both types of clinker, the main phases, C_3S, C_2S, C_3A and C_4AF, were well formed. Clinker with Slag contained more C_4AF and less C_2S, a difference that was attributed to the partial replacement of Limestone and Sand with Slag, which is higher in iron than the first. In most clinker, the phases were developed moderately and distributed in an inhomogeneous manner. The clinker phases were well developed and the grains of the Alite and Belite were also observed.

5. EVALUATION OF CO_2 EMISSION, BY GEMIS 4.7, 2011 SOFTWARE

5.1 The Impact Indicators

The Global Warming Potential (GWP), in equivalent of carbon dioxide, is the contribution to atmospheric absorption of infrared radiation by anthropogenic derived gases such as CH_4, CO_2 and N_2O, which contribute to an increase in global temperature. Usually, GWP data refer to a time horizon of 100 years (GEMIS 4.7, 2011; Hauschild and Potting, 2005; Sporyshev and Kattsov, 2006).

5.2 Inventory Inputs (Meal Prepared Based Slag)

The data are provided by the cement plant HADJAR- SOUD (Cetim. Heat balance of the line cooking workshop No.2 (SCHS). July 2015).

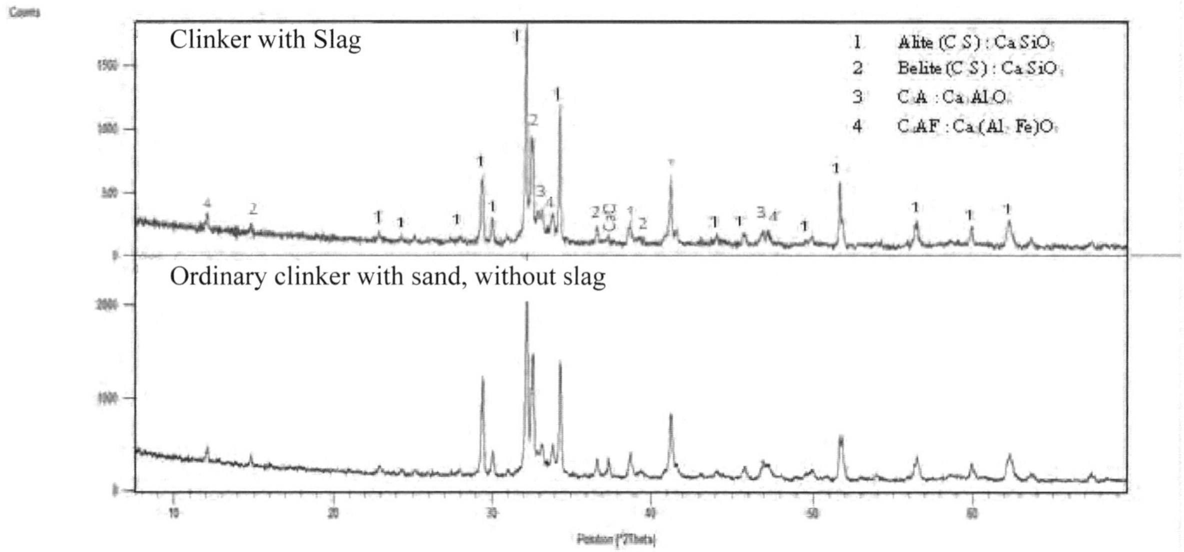

Figure 4: XDR of comparison of the two Clinkers

Consumption related to the production of one ton of clinker;

- Natural gas CH_4: for 01 kg of clinker, Natural gas (CH_4) consumption= 0.1017 Nm^3.
- Or for 01 kg of clinker, the CH_4 fuel consumption = 0.0856 kg.
- Raw material (flour): for 01 kg of clinker we consume 1.89 kg of the raw meal.
- Electricity: For the production of 01 ton of cement, we use 120 kw/h of electricity.
- The factory work at 100% of its capacity.

To prepare one ton of raw meal made from slag, mixing his request 700 kg of limestone, clay 192kg (Marl), iron 18kg and 90kg slag (The chemical composition of these elements are present before).

The values given by the GEMIS 4.7 software are mentioned in Tables 5 and 6, the three main greenhouse gases are: CO2, CH4 and N2O, but what we are interested in is CO2.

The modelling results are shown in Tables 5 and 6.

Table 5: GHG Emission results in tons of CO_2 Equivalent/T of Clinker

Element	Quantity (t)	factor	Equivalent CO_2	Part (%)
CO_2	$802{,}29 \times 10^{-3}$	1,0000000	$802{,}29 \times 10^{-3}$	98,3447
CH_4	$410{,}77 \times 10^{-6}$	25,000000	$10{,}269 \times 10^{-3}$	1,2588
N_2O	$10{,}759 \times 10^{-6}$	298,00000	$3{,}2061 \times 10^{-3}$	0,393
Somme			$815{,}79 \times 10^{-3}$	100

Table 6: Result per Tonne CO_2/t of clinker

Processes	Quantity (t)	Part (%)
Non-metallic minerals\cement clinker-2015	$545{,}00 \times 10^{-3}$	67,9309
heat-process-CaO-gas-2010 (end-energy)	$192{,}51 \times 10^{-3}$	23,9945
gas-CC-DZ-2010-1	$48{,}453 \times 10^{-3}$	06,0393
diesel motor-DE-2010	$7{,}6326 \times 10^{-3}$	$951{,}36 \times 10^{-3}$
compressor-GT-DZ-2010-1	$4{,}7625 \times 10^{-3}$	$593{,}61 \times 10^{-3}$
gas-boiler-DZ-2010-1	$956{,}03 \times 10^{-6}$	$119{,}16 \times 10^{-3}$

5.3 Explanation of the Table 6

- Non-metallic mineral clinker: Decarbonisation.
- Heat-process-CaO-gas-DZ-2010: Natural gas combustion (Algeria).
- Gas-CC-DZ-2010-1: Electricity used.
- Dieselmotor-DE-2010: truck used in quarries iron, limestone and marl.
- Compressor-GT-DZ-2010-1: Compressor used in the transportation of natural gas.
- Gas-boiler-DZ-2010-1: Gas consumed in boilers of the annexes processes.

Natural gas (CH_4) used in the preparation of the clinker contributes up to 23.99% (i.e. 0,192 T of CO_2) of the total GHG emission balance. The main source of CO_2 emission is decarbonisation phase by 67.93% (i.e. 0,545 T of CO_2).

5.4 Comparison of CO_2 Emission Results

If we repeat the modelling of CO_2 for the flour sand-based using the software GEMIS 4.9, 2014, and we keep the same data as for the flour slag-based, except the proportion of elements: limestone 78%, clay 18%, iron 1% and sand 3%. The results obtained are compared with the slag-based meal (Table7).

Table 7: Emission of CO_2 comparison between the two Clinkers

	Decarbonisation T CO_2/T Clinker	Combustion T CO_2/T Clinker	CO_2 T/an
Sand-based Clinker	0.663	0.231	727716
Slag-based Clinker	0.545	0.192	599918
Differences	0,118 T	0,039 T	127798 T/y

The production of raw clinker slag-based by HADJAR-SOUD cement plant may result in a reduction of annual CO_2 emission by the two processes; decarbonisation and combustion, this decrease may reach up to 17.56% (127,798T).

6. CONCLUSION

The use of granulated slag in place of sand for the preparation of raw meal and then the production of Clinker is shown to have very satisfactory results.

The addition of 9% of slag in the raw material does not appear to affect the chemical or mineralogical composition of the clinker, Furthermore, it results a decrease in the rate of limestone utilization, this decrease up to 10%. The chemical composition of the meal remains in the standards.

The results obtained by the software GEMIS 4.7, 2011; showed a decrease in the annual CO_2 emissions from both processes: decarbonation and combustion, with a decrease of 17.56%. This reduction is mainly due to the partial replacement of limestone ($CaCO_3$) in the raw material by already decarbonated products (Slag), knowing that more than 60% of the CO_2 emissions in cement manufacture come from the decarbonation processes (in our case Studied, the emission rate due to decarbonation is 67%).

REFERENCES

Annual Mean Global Carbon Dioxide Growth Rates. Dlugokencky and Tans. Published in December 2012. Retrieved on 18 November 2013

Belboom, S. 2008. Influence de l'utilisation des combustibles de substitution sur le cycle de vie du ciment (Doctoral dissertation, Université de Liège, Liège, Belgium).

Bogue, R.H. 1952. La chimie du ciment portland.

Camille, D. 2004. Chimie du ciment, valorisation des déchets en cimenterie. FDSA, Université libre de Bruxelles.

Cetim. Juillet 2013, Bilan thermique de l'atelier de cuisson de ligne N°2 (SCHS).

J-P. Jacobs, 2007 « la contribution de l'industrie cimentière a la réduction des émissions de CO_2 » syndicat française de l'industrie cimentière.

Morandeau, A. 2013. Carbonatation atmosphérique des systèmes cimentaires à faible teneur en portlandite (Doctoral dissertation, Université Paris-Est).

Makhlouf A., Serradj T., Cheniti H. 2014, Environmental Impact Assessment Review, "Life cycle impact assessment of ammonia production in Algeria: A comparison with previous studies", University Badji Mokhtar Annaba.

Namoulniara, D. K. 2015. Etude expérimentale de la diffusion du CO2 et des cinétiques de carbonatation de matériaux cimentaires à faible dosage en clinker (Doctoral dissertation, Université de La Rochelle).

Pré consultant, 2011. GEMIS 4.7. Life cycle Assessment Software package, version 4,7. Berline office Novalisstr http://www.gemis.de/en/index.htm.

Coagulation/Flocculation Process of Marble Wastewater: A Preliminary (Quadratic) Statistical Model

Mermer Atık Sularının Koagülasyonu ve Flokülasyonu: İlksel Nitelikteki (İkinci Dereceden) İstatistiki Bir Model

M. Çırak, M.A.S.M. Ali
Muğla Sıtkı Koçman University, Mining Engineering Department, Muğla

ABSTRACT In the scope of this preliminary work, the turbidity removal of an aqueous marble waste was investigated. The effect of the Fe-coagulation and PEO-flocculation on the supernatant turbidity was statistically modelled. The obtained mathematical equation from this model adequately explained 89,86% of the variation in the response variable: turbidity. After the confirmation of the model capability, the quadratic equation was used to generate an extensive 3-D surface plot for the dependent and the independent variables. Based on this, the turbidity removal tendencies of the coagulant and the flocculant was identified. When 140-160 mg/L Fe-coagulant and 0,4-0,6 ppm PEO-flocculant was added into the suspension at pH 10-11, it was observed that the marble fines were effectively removed from the supernatant (~ 0 NTU).

ÖZET Bu ön çalışma kapsamında, mermer atık sularındaki bulanıklığın giderilmesi araştırılmıştır. Fe-Koagülantının ve PEO-flokülantının bulanıklık üzerindeki etkileri istatistiksel olarak modellenmiştir. Bu modelden elde edilen matrematiksel denklemin, bağımlı değişken olarak nitelendirilen bulanıklıktaki varyasyonu 89,86% oranında açıklayabilmektedir. Model yeterliliğinin teyit edilmesinin ardından, bu ikinci derece denklem kapsamlı 3-D yüzey grafiklerinin çizilmesinde kullanılmıştır. Bu grafikler temel alınarak, koagülasyon ve flokülasyon işlemlerinin, bulanıklık giderimi üzerindeki etkileri açıklanmiştir. Sonuç olarak, pH 10-11 aralığında ilave edilen 140-160 mg/L Fe-koagülant and 0,4-0,6 ppm PEO-flokülantın, ince mermer parçacıklarını etkili bir şekilde süspansiyondan uzaklaştırdığı gözlemlenmiştir.

1 INTRODUCTION

Turkey has high-quality natural stone reserves with varying physical and chemical properties due to its geological location (Gazi et al., 2012). The mining and the processing of these marbles is one of the most important sectors considering its growing worldwide demand and its economical contribution in Turkey (Domopoulou et al., 2014). For this reason, the sustainability in the marble processing plants is very important and the availability of clean water for the reliable production is the primary factor for the cutting, washing, smoothing and polishing. However, the stone powders generated in the processing stage of the raw marbles contaminates water. As a result, the tailing of these plants contains high amount (2-10%-w/w) of suspended marble fines (Ersoy 2005) (Kavaklı 2003). In general, this wastewater can be recycled back to the plant and it can be reused in the cutting operations. In order to do this, it is essential that the recycled water should not contain any suspended particles. Briefly, the circulated water should have a high quality (Oates, 1998) since the suspended fines wear the equipment and lead to clogging inside the piping systems (Acar 2001).

To prevent the abovementioned problems, the turbidity measurement technique can be used for the determination of the water quality regarding the concentrations of the suspended colloids. According to Ersoy (2005), the turbidity of the wastewater of the marble industry should not exceed 15 NTU. For the elimination of the turbidity generating particles, the physicochemical processes are generally plays an important role (Jiang, 2001). The most commonly used processes are coagulation and flocculation in mining and mineral processing applications. The general aims of the coagulation and flocculation process are the removal of the colloidal particles in an aqueous medium and the agglomeration of the mineral fines into a larger structure (coagula and floc). The principle mechanism of the coagulation is charge neutralization via inorganic electrolytes (Gregory 1989) whereas the principle mechanism of the flocculation is polymer bridging via polymeric macromolecules (Hosten and Çırak, 2013). Nevertheless, these physicochemical processes rely on the different chemical and physical parameters and the tendencies of the operational parameters should be identified and optimized to reduce the turbidity of the wastewater in a cost-effective way (Al Ansari, 2012; Matouq et al., 2014; Ntampegliotis et al, 2006; Gregory, 2005).

In this preliminary work, the turbidity removal of an aqueous marble waste was investigated by using coagulation and flocculation. The effect of solution pH and the reagent dosage on the solid/liquid separation efficiency was analyzed. The performance of the physicochemical processes was evaluated by using nephelometric turbidimeter to determine the most favorable conditions to reduce the turbidity.

2 MATERIAL AND METHODS

The marble waste was collected from a marble company, which is located in Muğla/Turkey. The slurry sample contained 3% (w/w) solid marble and had a natural pH of 10. The suspension was highly turbid (>1000 NTU) with an intense white color. Prior to the solid/liquid separation test, the suspension was intensively mixed at 1500 rpm for complete dispersion.

3.0 g of iron sulfate coagulant was dissolved in 100 ml of distilled water and 0.04 g of polyethylene-oxide flocculant was dissolved in 800 ml of distilled water with help of magnetic stirrer at 1000 rpm for one hour. These feed solutions were prepared on a daily basis to prevent any possible degradation of the coagulant and the flocculant.

The pH of the suspension was adjusted according to the desired acidity/alkalinity level by using HCl and NaOH. This suspension was placed into the Jar test equipment. The completely dispersed suspension was firstly treated with the aimed amount of the coagulant for 2 minutes. Then, the required amount of the flocculant was added into the slurry and it was conditioned for another 2 minutes. Throughout the conditioning stage, the suspensions were mixed at 200 rpm. After the conditioning with the coagulant and the flocculant, the mixers of the jar tests were stopped and the suspensions were rested for 15 minutes. At the end of the 15th minute of the sedimentation, 10 ml sample was drawn from the fixed depth of 2 cm below the water surface. The obtained sample was placed into the turbidimeter and the coagulation-flocculation performance was evaluated in terms of the nephelometric turbidity unit (NTU).

The obtained results from the experiments were evaluated with the help of statistical software (SPSS). During these preliminary statistical tests, some transformation techniques were applied for the response variable to enhance the model.

Consequently, the findings were explained, both mathematically and visually.

3 RESULTS AND DISCUSSION

XRF-based chemical analysis of the sample was carried out for the characterization of the sample and the results were stated in Table 1. As expected, the amount of calcium in the sample was the highest among the other elements. As a matter of course, the surface of the particles can be also considered as calcium-rich surfaces.

Table 1. Chemical analysis of the sample

CaO%	MgO%	Fe_2O_3%	Al_2O_3%
55.8%	0.1%	< 0.01%	< 0.01%
Na_2O%	SiO_2%	MnO%	LOI+ Moisture %
1.2%	0.015%	< 0.01%	49%

The required solid/liquid separation experiments for this Ca-rich sample were carried out in the laboratory. The obtained results were statistically analyzed. The statistical outputs proved that,

- the obtained data was normally distributed (Figure 1),
- the absolute value of the residuals and the standard deviation of them were very small (Figure 2)
- the residuals were randomly distributed throughout the observation without yielding any noticeable pattern. (Figure 2 and 3)
- all data points in the normal probability plot were formed a straight line confirming the normality assumptions (Figure 3).

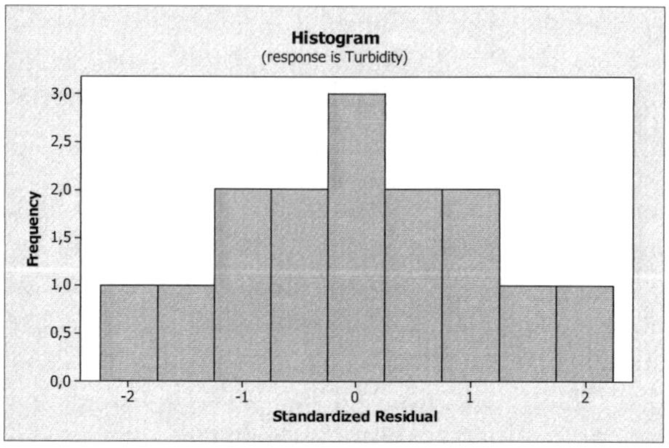

Figure 1. Histogram

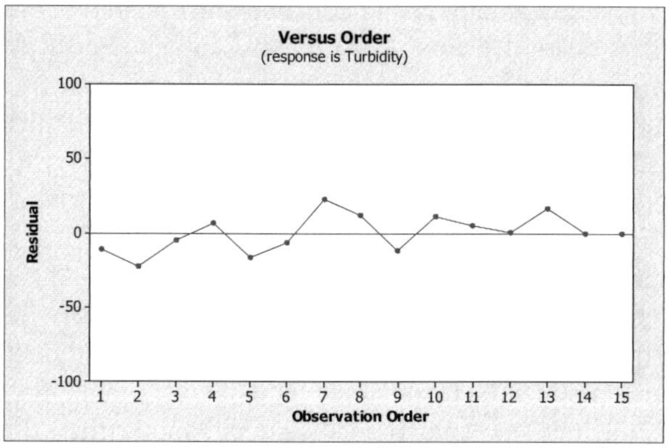

Figure 2. Variation of residuals with observation order.

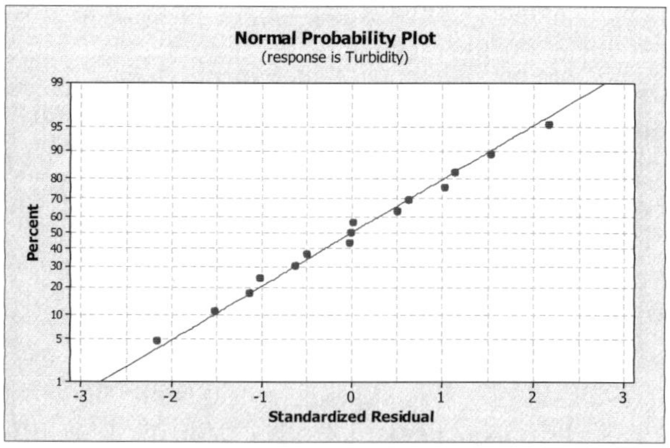

Figure 3. Normal Probability Plot

After the confirmation of the data quality with the previous statistical evidence, the regression coefficients for the obtained model were estimated as stated in Figure 4. Although the p-value of some terms were greater than 0,05, they were not omitted from the model in this preliminary work. The most effective parameter in the model seems to be the solution pH (Figure 5) since their p-values were almost equal to zero. Nevertheless, the third parameter that was called flocculant dosage, seems to be statistically least efficient parameter (p-value > 0,05). When the observed values of the supernatant turbidities were compared with the fitted values according to the sequence suggested by the Box-Behnken Design, only two large standardized residuals were observed for 2nd and 7th runs (Figure 6). The remaining residuals were very close to zero. This statistical information means that the model can explain 89,86% of the variation in the data set. As a result, it was seen that the process of the coagulation-flocculation can be comprehensively identified with this quadratic mathematical equation.

Estimated Regression Coefficients for Turbidity

Term	Coef	SE Coef
Constant	1,28386	0,049978
C1	-0,05358	0,008915
C2	-0,00025	0,000143
C3	-0,06564	0,031456
C1*C1	0,00259	0,000446
C2*C2	0,00000	0,000000
C3*C3	0,01747	0,011144
C1*C2	0,00000	0,000011
C1*C3	0,00346	0,002141
C2*C3	-0,00001	0,000054

Figure 4. Regression Coefficients (C1: pH, C2: Fe-Coagulant, C3: PEO-Flocculant)

Analysis of Variance for Turbidity

Source	DF	Seq SS	Adj SS	Adj MS	F	P
Regression	9	0,001270	0,001270	0,000141	4,92	0,047
Linear	3	0,000026	0,001069	0,000356	12,43	0,009
C1	1	0,000023	0,001035	0,001035	36,12	0,002
C2	1	0,000002	0,000087	0,000087	3,05	0,141
C3	1	0,000001	0,000125	0,000125	4,35	0,091
Square	3	0,001168	0,001168	0,000389	13,58	0,008
C1*C1	1	0,000874	0,000968	0,000968	33,77	0,002
C2*C2	1	0,000223	0,000242	0,000242	8,43	0,034
C3*C3	1	0,000070	0,000070	0,000070	2,46	0,178
Interaction	3	0,000077	0,000077	0,000026	0,89	0,507
C1*C2	1	0,000001	0,000001	0,000001	0,02	0,899
C1*C3	1	0,000075	0,000075	0,000075	2,61	0,167
C2*C3	1	0,000001	0,000001	0,000001	0,04	0,854
Residual Error	5	0,000143	0,000143	0,000029		
Lack-of-Fit	3	0,000143	0,000143	0,000048	1777,48	0,001
Pure Error	2	0,000000	0,000000	0,000000		
Total	14	0,001413				

Figure 5. ANOVA table. (C1: pH, C2: Fe-Coagulant, C3: PEO-Flocculant)

Figure 6. Residuals and standardized

Obs	SE Fit	Residual	St Resid	
1	0,005	-0,003	-1,29	
2	0,005	-0,006	-2,21	R
3	0,005	0,001	0,31	
4	0,005	0,003	1,24	
5	0,005	-0,003	-0,97	
6	0,005	-0,003	-1,24	
7	0,005	0,006	2,21	R
8	0,005	0,002	0,93	
9	0,005	-0,002	-0,93	
10	0,005	0,003	1,29	
11	0,005	-0,001	-0,31	
12	0,003	0,000	0,04	
13	0,005	0,003	0,97	
14	0,003	-0,000	-0,03	
15	0,003	-0,000	-0,01	

R denotes an observation with a large standardized residual.

residuals.

When the efficiency of the statistical model was confirmed with high-R^2 value, the variation of the turbidity as a function of pH and the reagent dosage was analyzed. The performance of the coagulation and the flocculation was separately presented in Figure 7 and 8 by means of 3-D surface plots.

The amount of the coagulant additions resulted in varying efficiencies of the turbidity removal process. At neutral pH, the turbidity firstly increased with increasing amount of the coagulant and then it started to decrease from 250 NTU to 50 NTU. On the contrary to this tendency, at pH 12 the turbidity firstly decreased down to 50 NTU with increasing amount of the coagulant and then it increased up to 100 NTU. When the whole surface plot in Figure 7 was surveyed as a function of the coagulant dosage, it was observed that the lowest turbidity values were obtained at 140-160 mg/L Fe-coagulant.

By comparing Figure 7 and 8, it was proven that the addition of the flocculant into the system significantly improved the efficiency of the process. After the addition of the minimum amount of the flocculant (0.3-0,6 ppm), the supernatant turbidity reached its lowest level throughout the study. According to Figure 8, the increase in the flocculant dosage rather than these optimum values caused highly turbid suspensions.

Figure 7 and 8 indicated that the adjustment of pH in this process was very critical for the turbidity removal. For instance, at pH levels of 7 and 12, the turbidity couldn't decrease below 50 NTU. However, when the solution pH was adjusted to 10-11, the most of the colloidal particles were eliminated from the suspension and the supernatant turbidity reached a value of zero.

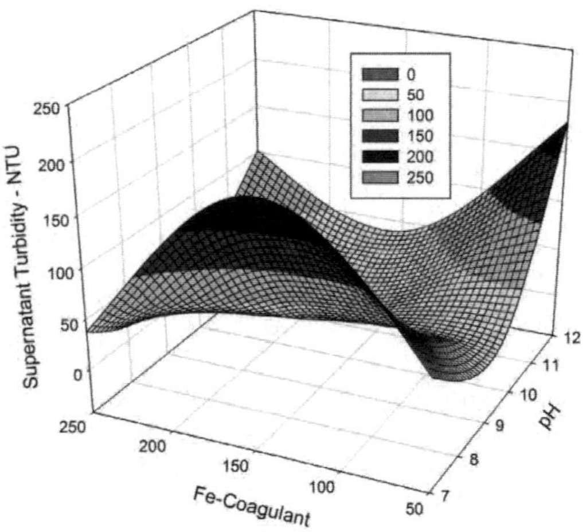

Figure 7. The effect of pH and Fe-coagulant on supernatant turbidity (the amount of PEO-flocculant was fixed at 0.5 ppm.)

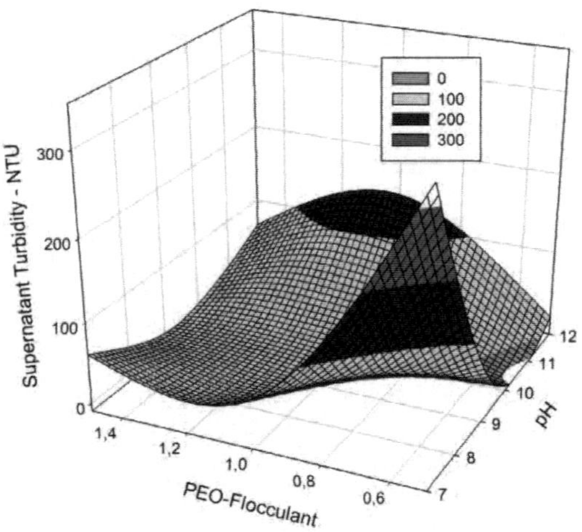

Figure 8. The effect of pH and PEO-flocculant on supernatant turbidity (the amount of Fe-coagulant was fixed at 150 mg/L.)

4 CONCLUSION

In this study, the effect of the Fe-coagulant and the PEO-flocculant on a marble slurry was statistically investigated. Based on the findings of the lab-scale solid/liquid separation experiments, a quadratic mathematical model was formed. The capability of this model was confirmed with high R^2-value of 89,86%. As a result, the statistical outcomes proved that the marble fines in the suspension was completely eliminated from the water when it was treated with 140-160 mg/L Fe-coagulant and 0,4-0,6 ppm PEO-flocculant at pH 10-11.

KAYNAKLAR

Acar, H., 2001, Attention must be paid to matters during the establishment and the running of a wastewater clarity unit for a marble processing plant, The Third Marble Symposium, 289–296, Ankara.

Al Ansari, M. S., 2012. A review of optimal designs in relation to supply chains and sustainable chemical processes, Modern Applied Science, 6(12), 74-85.

Domopoulou, A. E., Spiliotis, X. D., Ntampegliotis, K. I., Gudulas, K. H., Papapolymerou, G. A. P., & Karayannis, V. G., 2014. Compacted ceramics incorporating recycled marble processing residue, Journal of International Scientific Publications: Materials, Methods and Technologies, 8, 786-791.

Ersoy, B., 2005. Effect of pH and polymer charge density on settling rate and turbidity of natural stone suspensions, International Journal of Mineral Processing, 75(3), 207-216.

Gazi, A., Skevis, G., & Founti, M. A., 2012. Energy efficiency and environmental assessment of a typical marble quarry and processing plant, Journal of Cleaner Production, 32, 10-21.

Gregory, J., & O'Melia, C. R., 1989. Fundamentals of flocculation, Critical Reviews in Environmental Science and Technology, 19(3), 185-230.

Gregory, J., 2005. Particles in water: properties and processes, CRC Press.

Hosten, C., & Cırak, M., 2013. Flocculation behavior of clayey dolomites in borax solutions, Powder technology, 235, 263-270.

Jiang, J. Q., 2001. Development of coagulation theory and pre-polymerized coagulants for water treatment, Separation & Purification Reviews, 30(1), 127-141.

Kavaklı, M. Mermer işletme tesisleri proses atıksularının özellikleri, arıtılması ve kontrolü. Türkiye IV. Mermer Sempozyumu (Mersem'2003) bildiriler kitabı 18-19 aralık 2003, 315, 1-2.

Matouq, M., Al-Anber, Z., Susumu, N., Tagawa, T., & Karapanagioti, H., 2014. The kinetic of dyes degradation resulted from food industry in wastewater using high frequency of ultrasound, Separation and Purification Technology, 135, 42-47.

Ntampegliotis, K., Riga, A., Karayannis, V., Bontozoglou, V., & Papapolymerou, G., 2006. Decolorization kinetics of Procion H-exl dyes from textile dyeing using Fenton-like reactions. Journal of Hazardous Materials, 136(1), 75-84.

Oates, J. A. H., 1998. Lime and limestone. Chemistry and Technology, Production and Uses, Weinheim, Wiley VCH.

Variation of Sediment Thickness in Electrocoagulation as a Function of pH, Current and Temperature

pH, Akım ve Sıcaklık gibi Elektrokoagülasyon Parametrelerinin Çamur Kalınlığı Üzerine Etkisi

M. Çırak, T. S. Nurideen
Muğla Sıtkı Koçman University, Mining Engineering Department, Muğla.

ABSTRACT In this electrocoagulation study, the effect of three different parameters on the final sediment thickness was tested. For the analysis of the electrocoagulation parameters, the response surface methodology via Box-Behnken design was used. This statistical analysis suggested a comprehensive quadratic mathematical model with R^2: 87,6% and adjusted-R^2:80,6%. Considering the high-capability of the model, the prediction and interpretation of the electrocoagulation within the studied range of the parameters were reliably carried out. According to the statistical model, the lowest sediment thickness of 0,48cm can be obtained by adjusting pH around 5,0, increasing the solution temperature to 84,19 °C and keeping the electrical current at 0,2A. The lowest value of the electrical current was exclusively suggested by the estimated model since the higher electrical current values resulted in larger agglomerates and higher sediment thickness after the electrocoagulation of the colloidal clay suspension.

ÖZET Bu elektrokoagülasyon çalışması kapsamında, üç farklı parametrenin çamur kalınlığı üzerindeki etkisi test edilmiştir. Elektrokoagülasyon parametrelerinin analizi için, Box-Behnken yöntemi ile oluşturulmuş yüzey etki metodolojisi kullanılmıştır. Bu istatistiksel yöntem, ikincil derece denkleme dayalı kapsamlı bir matematiksel model üretmiştir (R^2: 87,6% and adjusted-R^2:80,6%). Modelin yetkinliği göz önünde bulundurulduğunda, çalışılan parametre sınırları dahilinde prosese ait tahmin ve yorumlamaların güvenilir bir şekilde yapılabildiği görülmüştür. İstatistiksel modele göre, süspansiyonun pH değeri 5,0'a ve sıcaklığı 84,19 °C ayarlanarak elektrik akım şiddeti 0,2A değerine yükseltildiğinde en düşük çamur kalınlığı 0,48cm olarak elde edilebilmektedir. Kolloidal kil süspansiyonunun elektrokoagülasyonunda en düşük elektrik akımının model tarafından tavsiye edilmesinin sebebi yüksek elektrik akımında daha büyük parça boylarının gözlemlenmesi ve buna bağlı olarak çamur yüksekliğinin artması olarak açıklanmaktadır.

1 INTRODUCTION

The presence of ultrafine suspended or colloidal materials such as clay particles, algae, silts, organic particles and soluble substances in water often causes water turbidity.

Most colloidal particles are stable and remain in suspension, and thus lead to pollution in water into which they are discharged or degrade re-circulation water in processing plants (Rubio et al., 2002). The mutual repulsion among colloidal particles owing to the same sign of their surface charges is the main reason for the stability of the system. It is difficult to remove colloidal particles in gravitational sedimentation ponds or devices without any size enlargement treatment. Therefore, agglomeration of particles into a larger floc/coagula is a necessary step for their removal by sedimentation. Size enlargement treatment may involve destabilization of particles or collision of particles to form aggregates.

The conventional treatment method consists of adding metal salts (aluminum, iron etc.), to destabilize the particles either by means of risen the ionic strength of the medium or a neutralizing the surface charge of particles, followed by flocculation and sedimentation.

An alternative to the use of solutions containing the coagulant salts, which causes secondary pollution when added at high concentrations, is the in-situ generation of coagulants by electrolytic oxidation of an appropriate anode material (e.g. iron or aluminum). This process is called electrocoagulation and has many advantages over conventional coagulation (Canizares et al., 2007). In the past decades research on electricity applied directly in water treatment has progressed well, making it an attractive method for coagulation or clarification of water, usually known as the electro-coagulation/electrochemical method.

In the electrocoagulation method, direct current is passed through aluminum/iron plates suspended in water. The process occurs in steps during electrocoagulation are; (i) anode dissolution, (ii) formation of OH^- ions and H_2 at the cathode, (iii) electrolytic reactions at electrode surfaces, (iv) Adsorption of coagulant on colloidal pollutants, and (v) removal by sedimentation or flotation (Can et al., 2006).

A metallic element employed in the anode is oxidized to yield its ions. The metal ions hydrolyze to some extent in water. When aluminum is used, Al^{3+} ions are produced in water, which forms soluble monomeric and polymeric hydroxo-metal complexes. The main reactions occurring at electrodes during electrolysis are as follows:

At the anode, aluminum oxidation occurs,
$Al_{(s)} \rightarrow Al^{3+}_{(aq)} + 3e^-$........................(1)

(Standard electrode potential of aluminum is $Al^{3+}_{(aq)} + 3e^- \rightarrow Al_{(s)}$ $E_1^0 = -1.662$ V)

At the cathode, water reduction occurs,
$3H_2O + 3e^- \rightarrow 3/2H_2 + 3OH^-$ ....(2)

(Standard electrode potential of water is $E_2^0 = -0.828$ V (Szynkarczuk et al., 1994)

At alkaline conditions, $Al^{3+} + 3OH^- \rightarrow Al(OH)_3$......................................(3)

At acidic conditions, $Al^{3+} + 3H_2O \rightarrow Al(OH)_3 + 3H^+$...............................(4)

(Pzhegorlinski et al, 1987) determined the contribution of the individual reactions of equations 1 to 3. Each of these reactions was evaluated by the weight loss of the corresponding electrode and the volume and composition of the collected hydrogen.

The advantages of electrocoagulation method over conventional coagulation include; high particulate removal efficiency, possibility of complete automation, production of less sludge, large and stable flocs, less total dissolved solid (TDS) and gas bubbles at the cathode (Mollah et al., 2001; Chen, 2004). However, it is still not widely accepted owing to high energy costs (Mills, 2000).

The main objective of this study was to conduct an experimental investigation of the effects of operating parameters (initial pH, applied voltage, current density, temperature and time) on the removal efficiency of montmorillonite from suspensions by electrocoagulation.

2 MATERIALS AND METHODS

An electrocoagulation test was conducted with the suspensions of ultrafine montmorillonite in water. The montmorillonite sample was obtained from Death Valley, California.

A montmorillonite suspension used for electrocoagulation was prepared by mixing (800rpm) appropriate amount of powdered material with 200 mL of distilled water so that a solid concentration of 0.20 g/L was obtained.

The pH of suspension was adjusted to the desired values using 0.1 M H_2SO_4 and 0.1 M NaOH solutions and measuring the pH with a pH meter.

The electrocoagulation cell used in the experimental study was constructed with electrode arrangement consisted of two aluminum (99.99%Al) in monopolar mode as stated in Figure 1.

The spacing between the cathode and anode plates was 2cm and the effective plate surface area were $28cm^2$. Before each electrocoagulation run, the aluminum electrodes were changed. The electrodes were connected to a DC power supply.

The current that flows through the cell and the voltage across the electrodes were measured with an ampermeter and a voltmeter, respectively.

A conductivity meter and a digital thermometer were employed to measure electrical conductivity and temperature of the suspensions, respectively. Sodium chloride (0.2g NaCl) was used as a background electrolyte for the electrocoagulation process.

The turbidity of suspensions was measured with a Portable Turbidimeter. The results were expressed in nephelometric turbidity units (NTU).

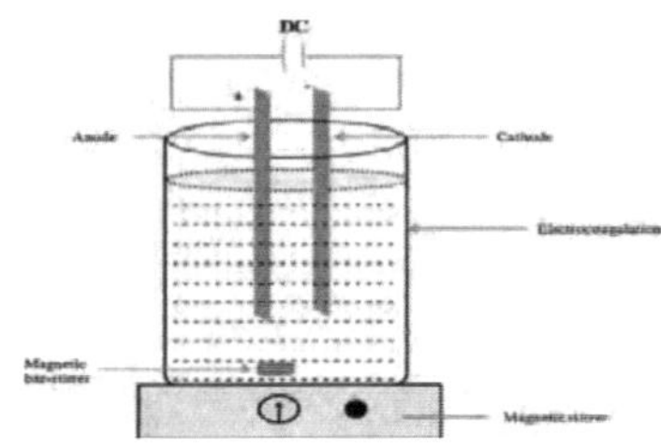

Figure 1. Bench-scale electrocoagulation reactor with monopolar electrodes in parallel connection.

3 RESULTS AND DISCUSSION

After the completion of the experiments, the data was introduced into the statistical software package. Then, each component of the quadratic model was checked with the help of the Pareto Chart. Nonetheless, some components (for instance, B: Temperature, AC: pH x Current, BC: Temperature x Current, CC: Current x Current) were statistically not significant for the model and they were omitted from the model. Once these components were removed, the Pareto Chart proved that all remaining components were significant for the model prediction capability as shown in Figure 2. As a result, the elimination of these non-essential components resulted in a more reliable quadratic model for this study.

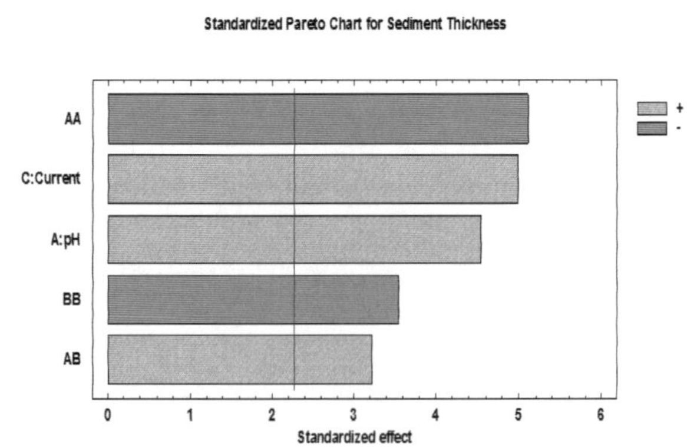

Figure 2. Pareto Chart

For this model the regression coefficients were determined as stated in Table 1.

Table 1. Regression Coefficients for Sediment Thickness

Coefficient	*Estimate*
constant	-5,98602
A:pH	2,19414
C:Current	1,15625
AA	-0,173764
AB	0,00524638
BB	-0,000367509

Based on Table 1, the mathematical expression of the fitted model for this process was summarized as follows,

Sediment Thickness = -5,98602 + 2,19414*pH + 1,15625*Current - 0,173764*pH^2 + 0,00524638*pH*Temperature - 0,000367509*$Temperature^2$

The ANOVA study presented the analysis of variance and the related F-ratios (together with the P-values) of each component of the abovementioned model. As stated in Table 2, all components in the suggested model were found to be significantly different from zero at the 95,0% confidence level. This means that the presence of all five effects in the model was statistically significant for the mathematical expression of the experimental data set (P-values <0,05).

Table 2. Analysis of Variance for Sediment Thickness

Source	*Sum of Squares*	*Df*	*Mean Square*	*F-Ratio*	*P-Value*
A:pH	1,42527	1	1,42527	20,63	0,0014
C:Current	1,71125	1	1,71125	24,77	0,0008
AA	1,79925	1	1,79925	26,05	0,0006
AB	0,720502	1	0,720502	10,43	0,0103
BB	0,867784	1	0,867784	12,56	0,0063
Total error	0,621697	9	0,0690775		
Total (corr.)	4,99733	14			

The R-squared statistics also indicated that the fitted model explained 87,5594% of the variability in the electrocoagulation process considering the sediment thickness. Furthermore, the calculated adjusted R-squared value was determined as 80,648%. This value was also satisfactorily high and close to the R-squared. The standard error of estimate and the mean absolute error were found as 0,262826 and 0,3168297, respectively. These values confirmed that the standard deviation and the average value of the residuals were very low throughout the study. Finally, the Durbin-Watson (DW) statistic (1,73196) tested these residuals to determine whether there is any significant correlation. However, the P-value of the DW statistic was estimated as 0,3965 indicating that there is no serial autocorrelation in the residuals at the 5,0% significance level. For these reasons, it can be claimed that this multi-parameter model was considerable capable to explain the variation in the data set.

The fitted values that obtained from the statistical model was also compared with the actual observed values. The results were presented in the Table 3 together with the lower and upper 95% confidence levels for mean. According to the estimation results for sediment thickness, it was confirmed that the observed values were very close to the fitted values. The lower and the upper confidence levels also formed a very narrow region, which confidingly contains all experimental data points. Moreover, the high prediction performance of the model was also verified with Figure 3 and 4.

Table 3. Estimation Results for Sediment Thickness

Row	*Observed Value*	*Fitted Value*	*Lower 95,0% CL for Mean*	*Upper 95,0% CL for Mean*
1	3,0	2,70322	2,32804	3,07839
2	1,3	1,40288	1,0238	1,78196
3	2,0	2,32788	1,9488	2,70696
4	1,5	1,77822	1,40304	2,15339
5	0,8	0,908798	0,427747	1,38985
6	2,5	2,48113	2,0916	2,87067
7	2,0	2,12788	1,7488	2,50696
8	2,0	1,73836	1,33288	2,14385
9	1,6	1,76044	1,40144	2,11945
10	1,4	1,55613	1,1666	1,94567
11	1,6	1,20288	0,823798	1,58196
12	2,5	2,46043	2,20538	2,71548
13	1,5	1,33088	0,839642	1,82212
14	2,5	2,46043	2,20538	2,71548
15	2,5	2,46043	2,20538	2,71548

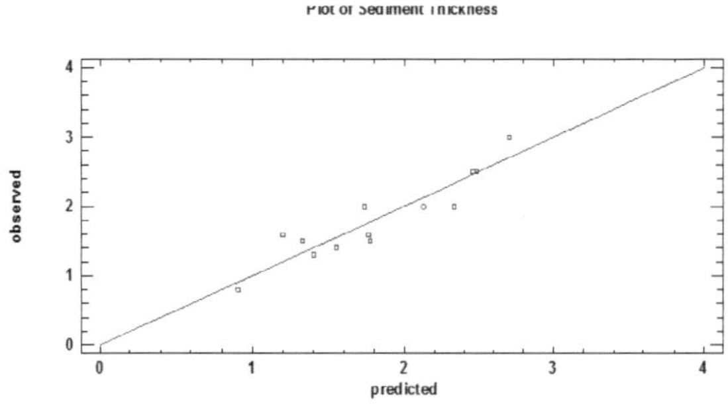

Figure 3. The predicted vs. the observed values

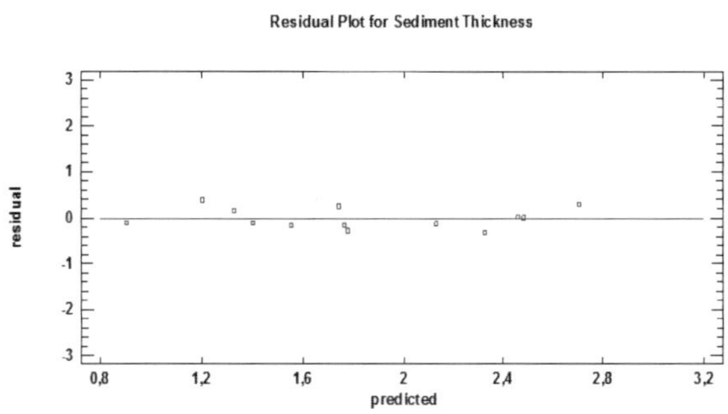

Figure 4. The residual plot

To better comprehend the parameters effect on the electrocoagulation process, the contour plots were given in Figure 5, 6 and 7. The experiments carried out at the acidic pH values between 3-5 produced low sediment thicknesses. On the other hand, the highest sediment thicknesses were obtained between pH of 5 and 7. This variation in the sediment thickness can be explained with the size of the coagula formed in the treatment process. Whereas the size of coagula was very fine and fragile under the acidic pH values, the size of the coagula obtained at pH 5-7 was very strong and large. The sedimentation of these coarse agglomerates led to a less compacted sediment bed rather than the sedimentation of the fine agglomerates.

When the model outputs were visually analyzed with the help of the 3-D counter plots, it was ascertained that the electrical current had a significant effect on the final sediment thickness. In the electrocoagulation process, the higher electrical current increased the amount of the dissolved Al ions from the solid electrodes. These dissolved Al ions specifically hydrolyzed into an aqua-complex form (Mollah et al., 2001; Malakootian and Yousefi, 2009; Kılıç et al., 2009; Mollah et al., 2010; Öztürk et al., 2013; Brahmi et al., 2015; Lu et al., 2015). This hydrolization process was substantiated at pH 5-.7 (Kılıç, 2009) and produced important Al-species like $Al(OH)_2^{+}$, $Al(OH)^{2+}$, $Al_2(OH)_2^{4+}$. These aqueous species were strongly adsorbed on the colloid surfaces and enhanced the size of the individual agglomerated structure. Consequently, the particle-particle interaction was boosted and visible large coagula formation was observed at higher electrical currents (Figure 5,6 and 7). On the contrary, the increasing of the electrical current did not improve the final sediment thickness. Similar relationship between the size of the agglomerated structure and the final sediment thickness was also experienced by the different researchers (Spicer et al., 1998).

The temperature of the solution also affected the final sediment thickness of the electrocoagulation process. However, its effect on the sediment thickness was very limited comparing to the other two parameters, namely electrical current and pH.

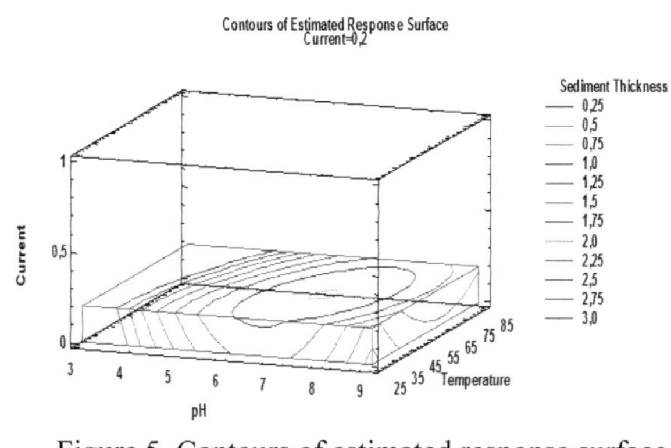

Figure 5. Contours of estimated response surface at current of 0.2 A.

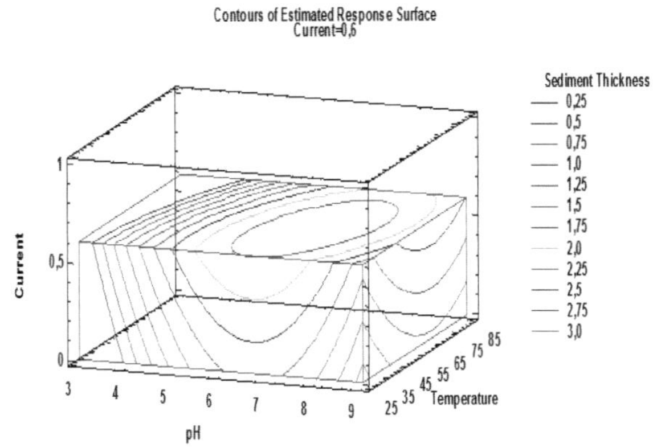

Figure 6. Contours of estimated response surface at current of 0.6 A.

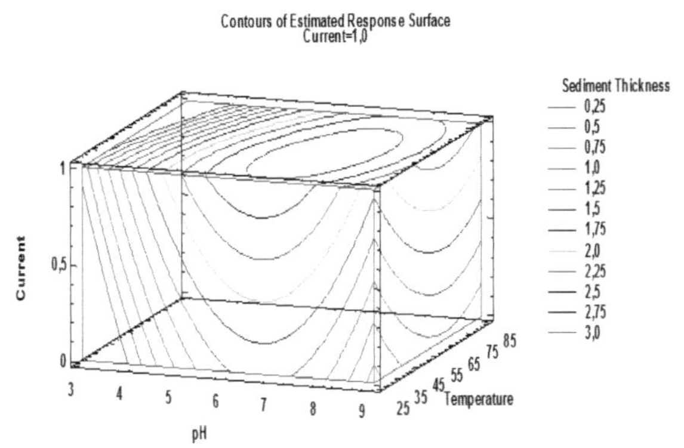

Figure 7. Contours of estimated response surface at current of 1.0 A.

Considering the estimated mathematical model and the related response surface, the physicochemical and electrical parameters of the electrocoagulation process were optimized to obtain minimum sediment thickness. Within the studied range of the parameters stated in Table 3, the optimum values of the pH, temperature and electrical current were found as 5.0, 84.19 and 0.2, respectively. When these values of the parameters were implemented in the electrocoagulation process, the sediment thickness was successfully decreased down to 0.48 cm. The statistical model suggested that the colloidal suspension should be acidic and hot to establish best aqueous chemistry conditions for the electrocoagulation. Nevertheless, the high electrical current was not suggested for the minimization of the sediment thickness due to the formation of the larger agglomerates.

Table 3. The optimum values of the electrocoagulation parameters

Factor	*Low*	*High*	*Optimum*
pH	5,0	9,0	5,0
Temperature	25,0	85,0	84,19
Current	0,2	1,0	0,2

4 CONCLUSION

The response surface methodology, which was constructed with the Box-Behnken design technique, was implemented for the electrocoagulation process of a clay suspension. The variability in the turbidity removal results of the electrocoagulation experiments was successfully identified with a quadratic statistical model (R^2: 87,6%; adjusted-R^2:80,6%; p-values<0,5). With the help of this model, the actual final sediment thickness values were precisely predicted with the attenuated residuals throughout the studied range of the electrocoagulation parameters. This capable model suggested slightly acidic (pH: 5.0) and hot solution (Temperature: 84,19 °C) to obtain the best aqueous geochemical conditions for the hydrolysis of the dissolved aluminum. Furthermore, a low-intensity electrical current (0,2A) was recommended instead of the high-intensity electrical current (1,0A) to minimize the sediment thickness. Following the statistical optimization of these three electrocoagulation parameters, it was found out that the sediment thickness can be diminished from 3,00 cm down to 0,48 cm.

KAYNAKLAR

Brahmi, K., Bouguerra, W., Belhsan, H., Elaloui, E., Loungou, M., Tlili, Z., & Hamrouni, B., 2015. Use of Electrocoagulation with Aluminum Electrodes to Reduce Hardness in Tunisian Phosphate Mining Process Water, Mine Water and the Environment, 1-8.

Can, O. T., Kobya, M., Demirbas, E., Bayramoglu, M., 2006. Treatment of the Textile Wastewater by Combined Electrocoagulation, Chemosphere, 62, 181-187.

Canizares, P., Martinez, F., Jimenez, C., Lobato, J., Rodrigo, M. A., 2007. Coagulation and Electrocoagulation of Wastes Polluted with Colloids, Separation Science and Technology, 42, 2157-2175.

Chen, G., 2004. Electrochemical Technologies in Wastewater Treatment, Separation and Purification Technology, 38, 11-41.

Kılıç, M. G., Hoşten, Ç., & Demirci, Ş., 2009. A parametric comparative study of electrocoagulation and coagulation using ultrafine quartz suspensions, Journal of hazardous materials, 171(1), 247-252.

Kilic, M. G., 2009. A parametric comparative study of electrocoagulation and coagulation of aqueous suspensions of kaolinite and quartz powders (Doctoral dissertation, Ph. D. dissertation, Dept. NAS, Middle East Technical Univ).

Lu, J., Li, Y., Yin, M., Ma, X., & Lin, S., 2015. Removing heavy metal ions with continuous aluminum electrocoagulation: A study on back mixing and utilization rate of electro-generated Al ions, Chemical Engineering Journal, 267, 86-92.

Malakootian, M., & Yousefi, N., 2009. The efficiency of electrocoagulation process using aluminum electrodes in removal of hardness from water.

Mills, D., 2000. A new process for electrocoagulation, American Water Works Association. Journal, 92(6), 34.

Mollah, M. Y. A., Schennach, R., Parga, J. R., & Cocke, D. L., 2001. Electrocoagulation (EC)—science and applications, Journal of hazardous materials, 84(1), 29-41.

Mollah, M. Y. A., Gomes, J. A., Das, K. K., & Cocke, D. L., 2010. Electrochemical treatment of Orange II dye solution—Use of aluminum sacrificial electrodes and floc characterization, Journal of hazardous materials, 174(1), 851-858.

Öztürk, T., Veli, S., & Dimoglo, A., 2013. The effect of seawater conductivity on the treatment of leachate by electrocoagulation, Chemical and Biochemical Engineering Quarterly, 27(3), 347-354.

Przhegorlinskii, 1987. VI. Dissolution of aluminum electrodes in the electrocoagulation treatment of water, Khymiya Tech Vody; 9(2):81-182.

Rubio, J., Souza, M. L., Smith, R.W., 2002. Overview of Flotation as a Wastewater Treatment Technique, Minerals Engineering, 15, 139-155.

Spicer, P. T., Pratsinis, S. E., Raper, J., Amal, R., Bushell, G., & Meesters, G., 1998. Effect of shear schedule on particle size, density, and structure during flocculation in stirred tanks, Powder Technology, 97(1), 26-34.

Physical-chemical Characterization and Enrichment of Orthoclase of Ain Barbar Quarry by Magnetic Separation and Flotation -Est of Algeria-

Chaib [1]; M. Bounouala [1]; M.Chettibi, S. Bouabdallah.[1]; Z. Mekti [1], A. Benselhoub[2].

1. University Badji MokhtarAnnaba, Annaba, Algeria.

2. *State Agrarian and Economic University, Dnipropetrovsk, Ukraine.*

ABSTRACT Feldspars are Common within the earth's crust, but only the potassium and sodium feldspars are highly sought in the ceramic and glass industries. To be utilized, these materials must contain very little harmful impurities especially iron and silica. The Ain Barbar exploits the feldspar material (orthoclase) with respective average grades of 73.26% SiO_2; 14.71% Al_2O_3; 7.78% K_2O; 0.79 % Fe_2O_3 and 0.33%Na_2O. However, there is a very high rate of quartz which decreases the value merchant product and a high iron content which gives to the product coloring aspects making it unmarketable.

For this purpose a mineralogical and chemical characterization, were performed on samples obtained from site extraction of the useful substances. Pre-concentration grain size test, were followed by physical and a physic-chemical concentration (magnetic separation and flotation), by new reagent to provide a valuable quality product which is acceptable for ceramists and glass. During this study, the parameters are examined effect of current intensity the coil, pH, concentration of the different reagent used.

The results collected from the chemical analysis of size fractions, reveal total of SiO_2 contents vary from70.89 to 75.61% in the size fractions. As for the ferriferous inclusions contents are 0.18 to 1.06 % Fe_2O_3 showing excess iron in the raw material, that does not meet the required standard (Fe_2O_3< 0.03 % and quartz < 8%). Also, note that the iron oxide content increases as the particle reduction. The contents of K_2O and Na_2O and SiO_2 are homogeny almost in all size fractions.

1.INTRODUCTION

Feldspar is extensively used in various industrial applications such as glass and ceramics due to its alumina and alkali contents. The mineralogical composition of most feldspar minerals can be expressed in terms of the ternary system: orthoclase ($KAlSi_3O_8$), albite ($NaAlSi_3O_8$) and anorthite ($CaAl_2Si_2O_8$). Chemically, the feldspars are silicates of aluminum, containing sodium, potassium, iron, calcium, barium or combinations of these elements [1].

In industrial applications, the removal of colored impurities and quartz affects the quality, but the ratio of K_2O/Na_2O is more crucial for identifying the quality of the raw material. A ratio of 2/1 to 3/1 is required for ceramic products whereas for high-voltage insulators, abrasives and electrodes production, the ratio must be higher [2]. Feldspar is used in the manufacture of glass as a source of alumina and as a partial replacement for soda ash. It is also used in the production of ceramics.

About 65 % of all feldspar material is used in the glass industry, 30 % in ceramics and 5 % in fillers and other applications. Chemically, the specifications of feldspar products for application in both ceramic and glass industries are nearly the same, total SiO_2 65–68 %, free quartz < 8 %, $A1_2O_3$ 18–19 %, K_2O and Na_2O 11.5–13.5 % and Fe_2O_3 0.08–0.2 %. However, the glass industry requires coarser feldspar concentrates (420 * 74 μm) in comparison with those necessary for ceramic production (98 % < 74 μm). Cengiz Demir [1], conducted differential flotation of Na–K feldspars by fluoride activation in the presence of 15 g/l NaCl and found an increase in the potassium content of the feldspar concentrate.

The best flotation results with Voineasa pegmatites containing equal amount of Na- and K-feldspars using amine collectors were achieved when Na-feldspar was depressed with NaCl [3]. Bayraktar et al studied Demirci-Turkey pegmatite containing 4.8 % K_2O and 2.40 % Na_2O; mica and oxides were first removed followed by separation of Na- and K-feldspars and quartz using NaCl, these researchers achieved a concentrate assaying 3.3 % Na_2O and 13.10 % K_2O.

Gülsoy et al [4] studied the similar feldspar ore from the same region assaying 5.94 % K_2O, 3.14 % Na_2O using magnetic separation followed by flotation with HF and NaCl; a concentrate assaying 3.1 % Na_2O and 12.65 % K2O was achieved. Boulos, and al [5], confirmed that Na+ ions depress Na-feldspar in the presence of amine (G-TAP) at natural pH. Similarly, Na ions were found to be effective in the selective separation of feldspar minerals in HF medium. Gülgönül [6] determined that various feldspar minerals particularly K-feldspars display different surface and floatability properties due to the presence of nano impurities on the mineral surface.

Demir and Gülgönül used mixtures of pure minerals to explain the separation mechanism of feldspar minerals by microflotation experiments. However, it was also observed that pure minerals do not always imitate real ores such as pegmatite, granite, syenite etc.

The formation conditions of feldspar ores particularly the existence of perthitic structure in the ore or rock greatly influence the separation conditions.

2. MATERIALS AND METHODS

2.1. Characterization Studies

The fieldstone (feldspar rhyolites) sample used in these experiments was obtained from Ain Barbar quarry of Annaba (East of Algeria). The ore sample of 90 Kg was reduced to <1 mm in size by a jaw crushers, mixing, quartering and dividing to obtain representative samples of 500 g.

All the ore samples were ground in a ball mill to produce a sample below 500 mm; the particles of approximately 250 mm were separated by particle size analysis using a sieving device of RETSCH type with a diameter of 200 x 50 mm, the particle size measurement range is from 0.045 to 4 mm on a vibratory sieve for 20 min at an amplitude of 60, which amounted to about 80 % by weight of particles in the size range from 250 to 45 mm. The particle size analyses were carried out to determine the optimal mesh release.

The characterization was completed on representative samples by means of optic microscope, diffraction of ray X and X-ray fluorescence in order to better determine the aspect of the qualitative and quantitative characteristics of the feldspar matter.

2.2. Experimental Procedure

The separation of feldspar (orthoclase) from quartz is difficult because they reveal the similar surface properties. The froth flotation is one of the methods for feldspar separation from quartz. But all conditions of this method to get high selectivity are reported to be used in acidic medium with amine as collector and with addition of fluoride (hydrofluoric acid) as modifier; a shortcoming of this reagent system is that it is expensive and involves using environmental pollutants.

In our investigation, we applied physical and physic-chemical processes on samples for the reduction of iron oxides and quartz, the double sided linking by flotation with different collectors including Petroleum sulfonate, Diamine solution with HF and HBr as modifiers in order to obtain a quality product for ceramists.

During this study, the examined parameters were effects of current intensity of the coil, pH, and concentration of different reagents used.

The material of 500 g of the size of fraction 250/45 mm was subjected to magnetic separation tests aimed to remove the ferriferous inclusions contained in the K-feldspar (orthoclase) material.

The range of the current variation in the magnetic separator that was used is from 5 to 12 Amperes, and drum rotation rotor is 60 rpm.

The magnetic separation performance highly depends on the physical particle properties to be separated (and size of the magnetic nature), the quality of the applied magnetic field, and the difference in magnetic susceptibility between the separated particles.

The magnetic separator of high intensity laboratory work dry is composed of three coils surrounding the electromagnet provided with a splined rotor rotating between the pole pieces of a magnetic circuit.The magnetic poles or pole pieces, between which the rotor rotates, are subjected to a magnetic induction [7].

The main magnetic separator parameters are the magnetic flux density which varies from 1.2 to 2 Tesla.

The next step is enrichment by flotation concentrates obtained by magnetic separation.

The first condition to obtain good flotation is to achieve a grinding of the maximum releasing minerals from each other; and a suitable release of the particles to float, in our case the mesh of 250/45 μm is estimated at the release. In some cases, it is necessary to deslime pulp before the float and to avoid the recovery of the mineral particle surfaces. The pulp containing 40 % solids mixture with water is stirred in a conditioning tank to ensure the homogeneity of the medium.

The first time round, dispersants AP 801 and AP 840 are added to avoid any kind of agglomeration before floating the concentrate (feldspar), then the hydrofluoric acid or hydrobromic acid is added to modify the surface feldspar particles during conditioning to be driven to the surface by the air bubbles; this process is repeated every time before flotation concentrates for 5 min. After that, a Diamine or petroleum sulfonate agent is added to the collector, (Table 1).

Table 1. Conditions used in Denver flotation experiments

Paramètres	Conditions
Temperature of medium	20–25°C
Particle size	−250 + 45 μm
Impeller speed	1200 rpm
Solids by wt.%	40%
Collector type	Diamine, Petroleum sulfonate
Amount of collector	150 g/t
Frother type	Pine oïl
Conditioning time for collector	5 min
pH	Acidic pH (1.5–4.0) HF, HBr
Amount of HF	(300 to1200 gr/t) H_2SO_4, as much as needed
Amount of HBr	(300–1200 gr/t) H_2SO_4, as much as needed
Conditioning for frother	3 min
Flotation time	5 min

The flotation operation is repeated for each collector and modifier. After flotation, washing, drying and filtering follows.

The tests were performed in laboratory of valorization of mining resources and environment (LAVAMINE), Mining Department, Badji Mokhtar University, Annaba.

3. RESULTS AND DISCUSSION

3.1. Chemical Analysis and X-ray Diffraction Analysis

The results of chemical and mineralogical analyses are shown in Table 2 and Fig. 1, 2 respectively, these results demonstrate that the ore contains quartz and orthoclase with minor amounts of Albite, Anorthite, and small quantities of clay minerals. The sample has K_2O and Na_2O contents of 7.78 and 0.33 % respectively.

The XRD spectrum and tine section confirm that orthoclase is the principal mineral and other minerals present are very minor to trace their amount.

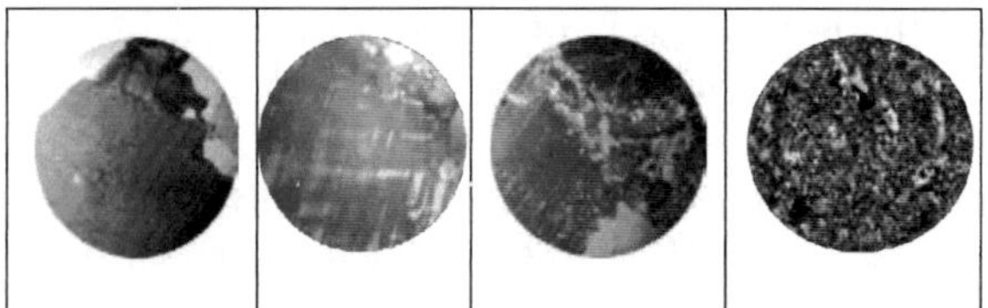

Plagioclase grains in contact with the quartz and orthoclase, +120	Alteration of plagioclase and orthoclase by sericite (mica), +120	Biotite aggregates on the contour of the grain of orthoclase, +120	Pate with feldspar porphyritic quartz and feldspar and mica inclusions, +60

Figure 1. Tine sections

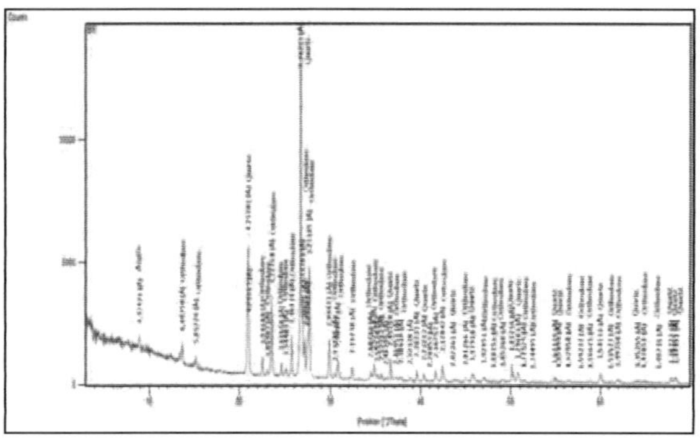

Figure 2. XRD pattern of investigated fieldstone sample

Table 2. Chemical composition of the fieldstone sample

Oxides	SiO_2	Al_2O_3	K_2O	Na_2O	Fe_2O_3	CaO
Contents(%)	73.26	14.71	7.78	0.33	0.79	059

P_2O_5	CaO	P_2O_5	MgO	TiO_2	PbO	ZnO	MnO	PAF
0.31	059	0.31	0.05	0.02	0.02	0.03	0.01	1.04

3.2. Particule Size Analyses

The collected results from chemical analysis of size fractions reveal that the totals of SiO_2 contents vary from 70.89 to 75.61 %. As for the ferriferous inclusions, their contents of Fe_2O_3 are 0.18 to 1.06 % showing an excess of iron in the raw material which does not meet the required standard (Fe_2O_3 < 0.03 % and quartz < 8 %). Besides, we noted that the iron oxide content increases as the particles diminish.

The contents of K_2O and Na_2O and SiO_2 are homogeneous almost in all size fractions. The results of the chemical analysis of size fractions are given in Table 3.

Table 3. Results of chemical analysis of size fractions of *K*-feldspar sample

Fraction, mm	Yield (%) (%)	SiO_2 (%)	Al_2O_3 (%)	K_2O (%)	Na_2O (%)	Fe_2O_3 (%)
brut	100	76.21	13.39	8,40	0.29	0,32
+4	26.02	73.25	15.39	8.45	0.30	0.18
-4 +2	30.54	75.61	13.18	8.81	0.33	0.20
-2 +1	13.00	75.60	12.31	8.48	0.30	0.55
-1 +0.5	07.11	70.89	16.76	8.56	0.34	0.65
-0.5 +0.25	04.04	73.46	13.46	7.98	0.28	1.06
-0.25 + 0.125	03.52	73.56	13.62	8.20	0.36	0.96
-0.125 +0.063	04.03	72.40	15.75	7.60	0.35	0.76
-0.063+ 0.045	01.82	74.61	14.52	7.37	0.27	0.57
-0.045	09.92	74.12	14.46	8.50	0.32	0.70

Fraction, mm	Yield (%) (%)	CaO (%)	MgO (%)	P_2O_5 (%)	TiO_2 (%)	MnO (%)	PAF (%)
brut	100	0,10	0.02	0.29	0.01	0. 01	0.76
+4	26.02	0.52	0.01	0.31	0.01	0.02	0.82
-4 +2	30.54	0.30	0.02	0.29	0.00	0.02	0.61
-2 +1	13.00	0.60	0.02	0.31	0.02	0.01	1.27
-1 +0.5	07.11	0.21	0.03	0.30	0.02	0.01	0.89
-0.5 +0.25	04.04	0.87	0.01	0.32	0.00	0.01	0.98
-0.25 + 0.125	03.52	0.89	0.02	0.28	0.00	0.02	1.24
-0.125 +0.063	04.03	0.59	0.01	0.34	0.07	0.01	1.04
-0.063+ 0.045	01.82	0.05	0.01	0.35	0.05	0.01	1.04
-0.045	09.92	0.20	0.01	0.33	0.03	0.01	1.14

3.3. Evaluation of Efficiency of Iron Removal

The efficiency of iron removal can be calculated by the following equation;

E (%) = [1- (Fe_2O_3 content in concentrate/ Fe_2O_3 content in alimentation)]*100.

Table 4 shows the effect of the current intensity on the effectiveness of iron oxide removal from feldspar, according to obtained results by high intensity magnetic separation (MSHI), we found a significant improvement in feldspar content and a remarkable decrease of impurities such as hematite was obtained in the range between 10 and 12 Amperes. With the increase in the intensity of the electric current at 12 Ampere, it is noted that the iron impurity content decreases from 0.32 to 0.09 %.

The optimum efficiency of removal of iron oxide was obtained in the range of 71.81 %.

Table 4. Reduction of iron content in size fractions of the 250/45 μm material, at different intensities of magnetic field

Intensity (A)	Fe_2O_3 (%)	Fe_2O_3 (%)	E (%)
5		0.28	06.6
7		0.24	25.0
9	0.32	0.19	40.7
10		0.11	65.7
12		0.09	71.81

3.4. Effect of The Ph on Separation of Orthoclase

Separation of feldspar from quartz can be achieved with cationic collectors in acid circuit at a pH value in the vicinity of 1.5 to 2.5. Other works, found that at the pH range of 2.5 to 3, a maximum selectivity could be reached [8].

Fig. 3 and 4 show a maximum selectivity at pH of 1.8 to 2 in both cases, when using the HF or HBr modifier and Diamine as a collector, at concentration of 150 g/t. The contents and recovery of K_2O achieve a maximum, when we use Diamine collectors and hydrobromic acid (HBr) as an activator of orthoclase and quartz depressor (12.65 %; 85 %) respectively.

The results obtained are justified by the point of zero charge of feldspar (1.5 to 2) and quartz varying from 2 to 3.7; at pH 1.5 to 2.5 above the value of the point of zero charge the surface orthoclase takes a positive sign whereas the quartz has a negative sign which allows creating new sites (a better surface activation of orthoclase) to increase the adsorption of collectors (Diamine).

Considerable contents of SiO_2, more than 90.61 %, in tailing signify a better depress of the quartz when using HBr with 150 g/t Diamine at pH = 1.8. The use of HBr increases the quartz depression (increases the speed of sedimentation), compared with the use of HF because the molecular weight of HBr is greater than HF.

From Fig. 5 and 6 it is noted that the use of petroleum sulfonate as a collector in both cases (HF, HBr) reduces the recovery of K_2O, and the content of K_2O is almost constant when the pH ≥ 3, which means that the adsorption of the collector by the quartz surface is started (the start of flotation of quartz in this pH).

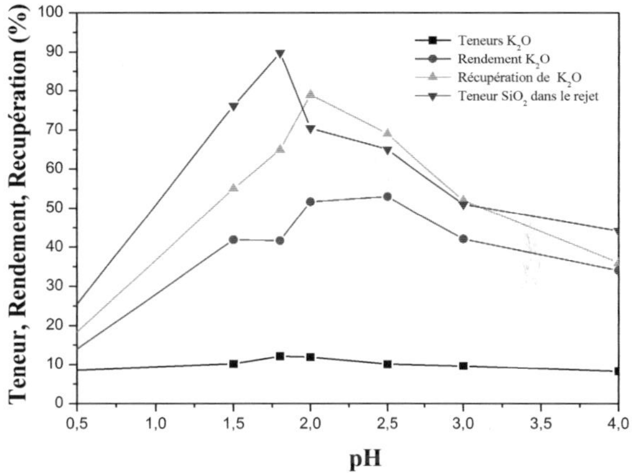

Figure 3. Effect of pH on orthoclase separation with HF and Diamine (150 g/t)

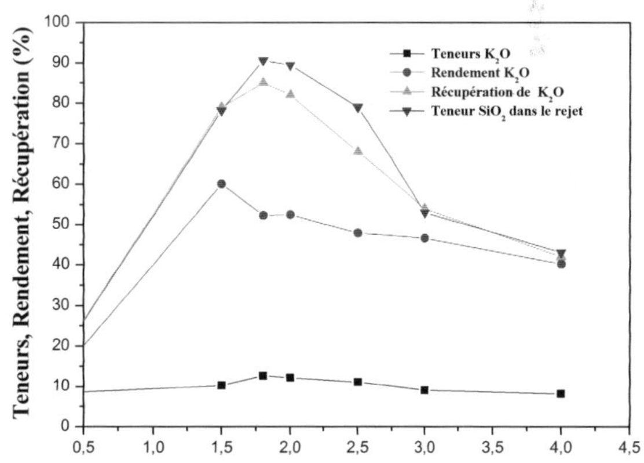

Figure 4. Effect of pH on orthoclase separation with HBr and (150 g/t Diamine)

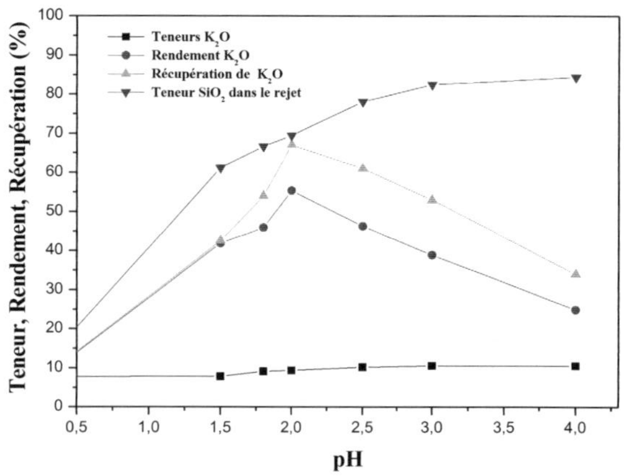

Figure 5. Effect of pH on orthoclase separation with HF and petroleum sulfonate (150 g/t)

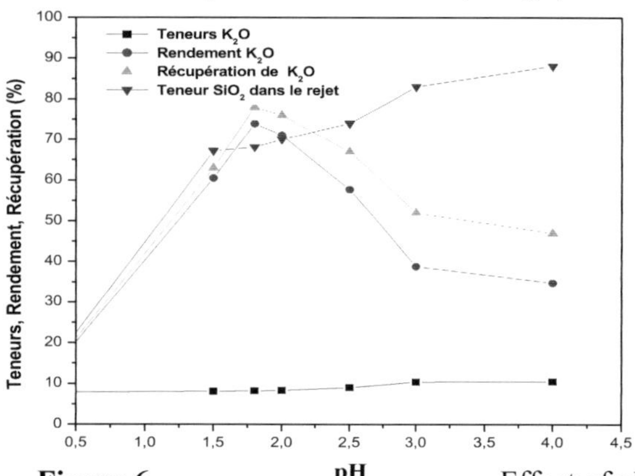

Figure 6. Effect of pH on orthoclase separation with HBr and petroleum sulfonate (150 g/t)

3.5. Effect of the Amount of Hydrofluoric or Hydrobromic Acids on Separation of Orthoclase With Ph = 1.8 to 2 and 150 G/T Diamine or Petroleum Sulfonate

Flotation studies in acidic medium were carried out in two main parts: the amount of F^- and Br^- ions required to activate and depress the minerals (orthoclase) was systematically optimized by using 150 g/t of Diamine or petroleum sulfonate.

All experiments were performed at pH 1.8 to 2 adjusted with H_2SO_4. Tables 7, 8 illustrate that some selective separation between *K*-feldspar (orthoclase) and quartz occurs at substantial recoveries and yield.

It is also found in the present research that conditions of *k*-feldspar formation greatly affect the optimum dosage of HF and HBr. For instance, altered feldspar requires higher HF and HBr dosage for activation. HF and HBr removes altered layers and cleans the surface to facilitate the formation of alumina fluoride complexes.

Pine oil was added to the flotation cell as a frother because HF addition increases the collector consumption and deteriorates the quality of bubble.

The addition of either HF or HBr as pH regulators and modifiers increases the selectivity of the feldspar flotation and the quartz depression as shown in Figure7 and 8. The best results were recorded when we used HBr as a modifier with Diamine as a collector at a concentration of 800 and 150 g/t respectively.

The use of petroleum sulfonate as a collector and HF or HBr as a modifier does not give satisfying results Figure9 and 10.

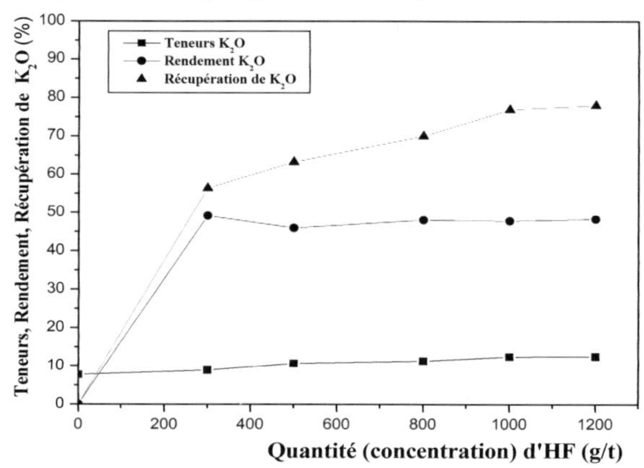

Figure 7. Effect of the concentration of hydrofluoric acid on orthoclase separation with pH = 1.8 to 2 and 150 g/t Diamine

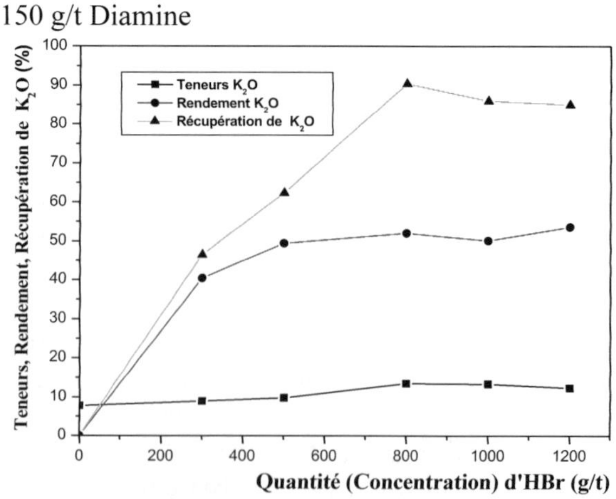

Figure 8. Effect of the concentration of hydrobromic acid on orthoclase separation with pH = 1.8 to 2 and 150 g/t Diamine

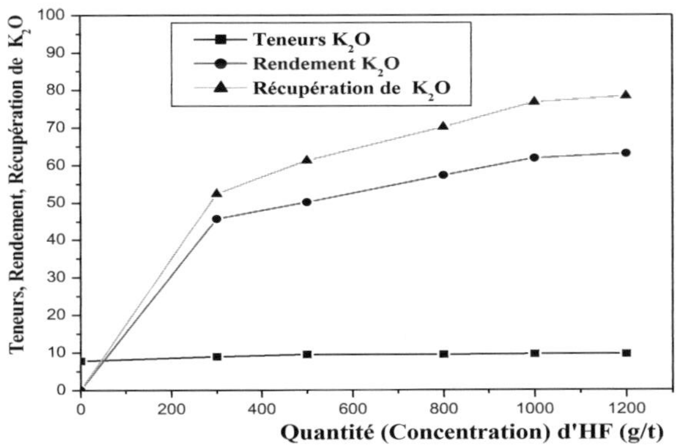

Figure 9. Effect of the concentration of hydrofluoric acid on orthoclase separation with pH =1.8 to 2 and 150 g/t petroleum sulfonate

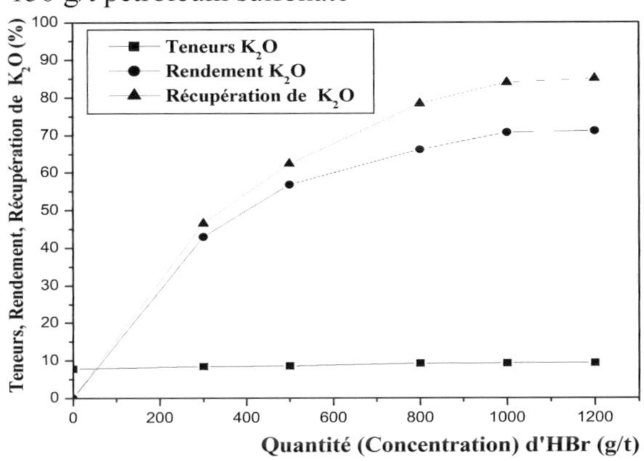

Figure10. Effect of the concentration of hydrobromic acid on orthoclase separation with pH =1.8 to 2 and 150 g/t petroleum sulfonate

3.6. Comparison of Obtained Results with Hydrofluoric or Hydrobromic Acids in Orthoclase Flotation

Comparing the obtained results with the use of hydrofluoric acid or hydrobromic acid in orthoclase flotation, shows that the use of hydrobromic acid (800 g/t of HBr) gives a concentration of 90 % orthoclase and 8 % quartz with 13.51 % K_2O, while in feed it is 56 and 39 % quartz orthoclase with a K_2O content of 7.78. On the other hand, the use of hydrofluoric acid (800 g/t HF) provides a concentrate of 80 % orthoclase and 18 % quartz, with a grade of 9.52 % of K_2O in the same conditions. Then the use 1200 g/t of HF gives concentrates of 86 % orthoclase and 12 % quartz with 12.55 % of K_2O as shown in Figure7 and 8.

3.7. Separation of Orthoclase by Magnetic Separation Followed by Flotation

The operation of orthoclase processing from Ain Barbar quarry is diagrammed in Fig. 11.

CONCLUSION

The beneficiation of feldspar ore (orthoclase) assaying on average 15.16% Al_2O_3, 70.40% SiO_2, 0.03% totals iron oxides, 13.51% K_2O and 0.14%Na_2O, provide several types of products which can be used in the ceramic production and glassmaking.

Attrition by washing operation is necessary for the enrichment of *K*-feldspar, as well as magnetic separation is designed to remove the iron-bearing impurities as presented in our study.

The separation of orthoclase from quartz by flotation with conditions used in this study is very important because it gives good results in terms of economic cost and protection of the environment (the use of HBr is economical and less hazardous than using HF).

Flotation of *K*-feldspar (orthoclase) with pH of 1.8 to 2 by Diamine as a collector and HBr as an activator at dosage of 800 g/t is better than the flotation by petroleum sulfonate and HF in the same conditions.

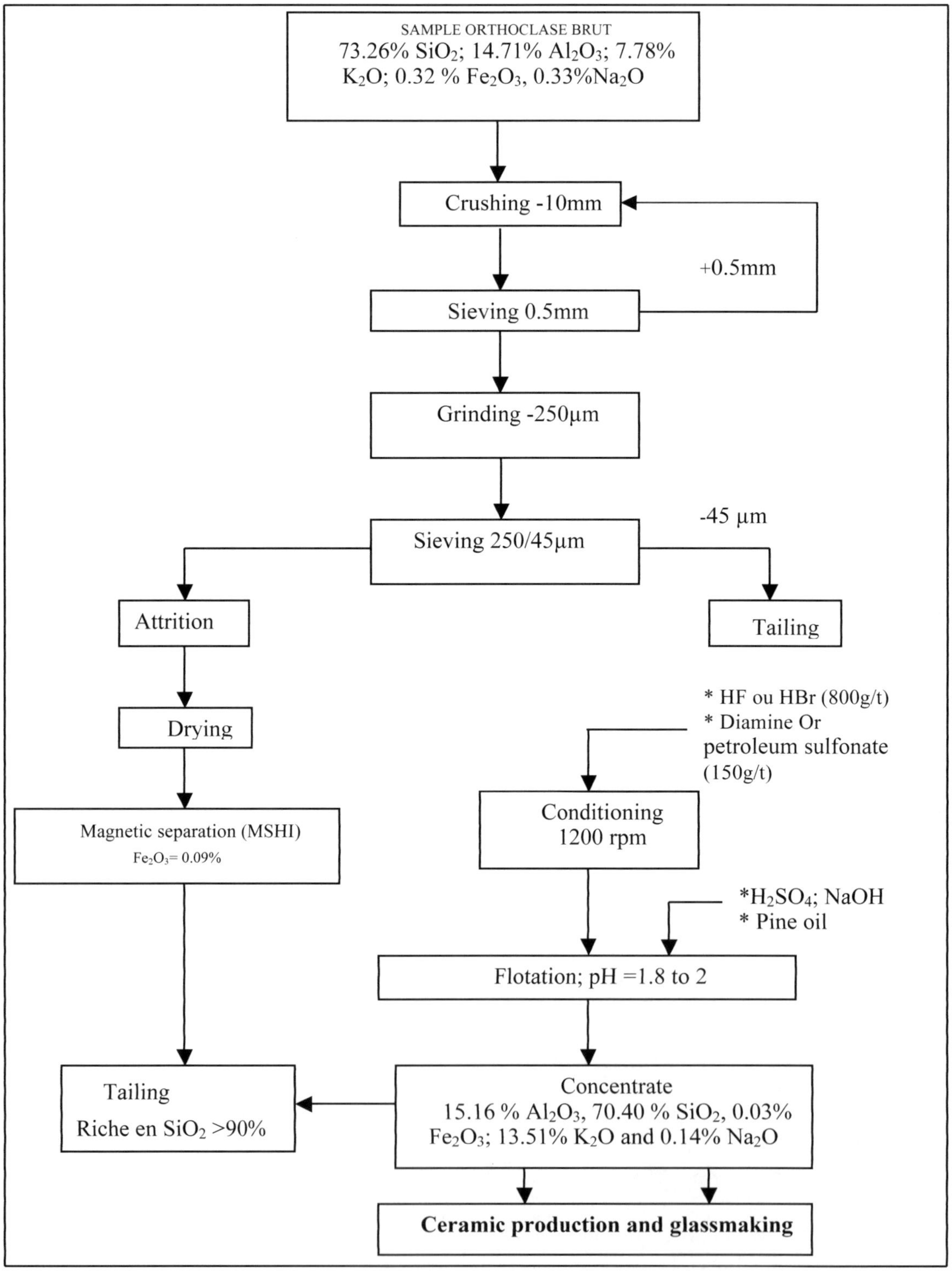

Figure 11. Proposed flow sheet for orthoclase processing from Ain Barbar quarry

REFERENCES

1. Demir, C., 2010. Selective separation of Na- and Kfeldspar from weathered granites by flotation in HF medium. *Ceramics–Silikaty*, Vol. 54, No. 1, pp. 60–64.

2. Lewicka, E., 2010. Conditions of the feldspathic raw materials supply from domestic and foreign sources in Poland. *Gospodarka surowcami mineralnymi,* Vol. 26, pp. 5–19.

3. Heyes, G., Allan, G., Bruckard, W., and Sparrow, G., 2012. Review of flotation of feldspar. Mineral Processing and Extractive Metallurgy (Trans IMM Section C), Vol. 121(2).

4. Gülsoy, O. Y., Can, N. M., and Bayraktar, I., 2005. Production of potassium feldspar concentrate from a

low-grade pegmatitic ore in Turkey. *Mineral Processing and Extractive Metallurgy*, Vol. 114(2), pp. 80–86.

5. Boulos, T. R., Ibrahim, S. S., and Yehia, A., 2015. Differential Flotation of Some Egyptian Feldspars for Separation of Both Silica and Iron Oxides Contaminants. *Journal of Minerals and Materials Characterization and Engineering*, No. 3(06), pp. 435–443.

6. Demir, C., Bentli, I., Gülgönül, I., and Celik, M. S., 2003. Effects of bivalent salts on the flotation separation of Na-feldspar from K-feldspar. *Minerals Engineering*, Vol. 16(6), pp. 551–554.

7. Bouabdallah, S., Bounouala, M., Idres, A., and Chaib, A., 2015. Iron removal process for high purity silica production by leaching and magnetic separation technique. *Naukovyi Visnyk Natsionalnoho Hirnychoho Universytetu*, No. 5, pp. 47–52.

8. Soonthornwiphat, N., Saisinchai, S., and Parinayok, P., 2016. Recovery Slime Waste from Feldspar Flotation Plant at Attanee International Co. Ltd., Tak Province, Thailand. *Engineering Journal*, No. 20(4), pp. 69–78.

Çimento Klinkeri Tanelerinin Tek Tane Darbe Kırılma Dağılımları Benzerliğinin Analizi

Analysis of the Self-Similarity of Single Particle Impact Breakage Distributions of Cement Clinker Particles

Ö. Genç

Muğla Sıtkı Koçman Üniversitesi, Maden Mühendisliği Bölümü, 48000, Kötekli Yerleşkesi, Muğla

ÖZET Bu çalışma darbe kırılması deseninin farklı özgül ufalama enerji seviyelerinde ne şekilde değiştiğini analiz etmektedir. İki farklı çimento klinkeri örneğinin darbe kırılması desenleri, boyuta göre normalleştirilmiş dağılım eğrileri çizilerek incelenmiştir. Bu eğriler, kırılma ürünü tane boyu dağılımlarının dağılımların ortalama tane boyuna (d50) göre normalize edilmesiyle elde edilmiştir. Ağırlık düşürme test tekniği farklı boyut fraksiyonlarındaki tanelerin darbe kırılma dağılımlarının belirlenmesinde kullanılmıştır.

Normalize boyut dağılımları, kırılma deseninin tane boyu ve darbe enerjisi seviyesiyle değiştiğini göstermiştir. 1kWh/t'un üzerindeki enerji seviyelerinde kırılma ürünü boyut dağılımlarının aynı deseni sergilemesi, dağılım benzerliğinin göstergesidir. Bu enerji seviyesinin altında, kırılma dağılımları normalleştirilememiştir. Tane boyuna göre, klinker mikroyapısı, mineralojik ve kimyasal bileşim içeriklerindeki muhtemel değişimler gözlemlenen benzerlikten sapmalar için etkenler olabilir. Darbe mekanizmasının kullanıldığı ufalama ekipmanlarının modellenmesi için, benzer kırılma dağılımlarının elde edildiği (oransal boyut küçültmenin aynı olduğu) enerji seviyeleri kırılma fonksiyonlarının tahmin edilmesi için önerilmiştir.

ABSTRACT This study analysis how the impact breakage pattern changes at different specific comminution energy levels. Impact breakage patterns of two different cement clinker samples were investigated by plotting size-normalized distribution curves. These curves were obtained by normalizing the breakage product particle size distributions according to the median size (d50) of the distributions. Drop-weight test technique was used to determine impact breakage distributions of particles in different size fractions.

Normalized size distributions indicated that, breakage pattern varies with particle size and impact energy level. Breakage product size distributions fall into the same pattern above an energy level of 1kWh/t which indicated a self-similar distribution. Breakage distributions became non-normalizable below this energy level. Probable variations in clinker microstructure, mineralogical and chemical composition on the particle size basis could be the factors for the observed deviations from the self-similarity. Energy levels which provided self-similar breakage distributions (the same proportional size reduction) are proposed to estimate breakage functions for the modelling of comminution equipments in which impact mechanism is used.

1 INTRODUCTION

Portland cement clinker is a nodular, multi-component hydraulic artificial material made by burning, at least to sintering a precisely specified mixture of raw materials containing CaO, SiO_2, Al_2O_3 and Fe_2O_3 (Hewlett,1998). Abbreviations used for CaO, SiO_2, Al_2O_3 and Fe_2O_3 in cement terminology are C, S,A and F respectively. Abbreviations used for clinker minerals are given in Table 1 (Duda, 1985).

Table 1. Abbreviations for clinker minerals

Abbreviation	Composition	Mineral name
C_3S	$3.CaO.SiO_2$	Alite
C_2S	$2CaO.SiO_2$	Belite
C_3A	$3.CaO.Al_2O_3$	Tricalcium Aluminate
C_4AF	$4.CaO.Al_2O_3.Fe_2O_3$	Tetracalcium Aluminoferrite

Clinker consists of at least two-thirds by mass of calcium silicates ($(CaO)_3.SiO_2$) and $(CaO)_2.SiO_2$), the remainder containing aluminium oxide (Al_2O_3), iron oxide (Fe_2O_3) and other oxides. These oxides, first enter in solid-solid reactions in the kiln and later partially melt to form the four major compounds (C_3S, C_2S, C_3A, and C_4AF). The relative and absolute proportions of these compounds determine the character of the resulting cement. Alite (C_3S) and Belite (C_2S) minerals are the strength giving minerals and Tricalcium Aluminate (C_3A), Tetracalcium Aluminoferrite (C_4AF) are the flux minerals of clinker. Clinker is the main material which is ground with different proportions of some additives (i.e. gypsum, trass, limestone, etc.) in different comminution machines to produce various types of cement in the industry (Hewlett,1998). It was found that, as Belite mineral content of the clinker increases, work index also increases (Genç and Benzer, 2009).

Single particle breakage distributions are used to determine breakage functions of materials which describe how particles break inside the comminution machines. This information is required for establishing the crushing and grinding models. Models are used in predicting the throughput, power draw and product size distributions of the equipments so that optimum operational conditions for the comminution equipment can be determined. The widespread applied JKMRC drop-weight test method used in the breakage characterisation of different ore types (Narayanan, 1985; Andersen, 1988;; Man, 2000; Napier Munn et al, 2005) for ball, autogeneous and semi-autogeneous mill modelling studies.

The JKMRC drop-weight tester was modified and the methodology was implemented in clinker breakage characterization tests by Genç (2002). Breakage product size distributions were used to analyse self-similarity of breakage patterns in this study. Self-similarity of breakage patterns of clinker samples provided from different cement plants were compared at different specific energy consumptions using size normalized distribution curves. It was observed that, breakage of each test particle size fraction started to exhibit a self-similar breakage pattern (similar breakage distributions) at energy levels higher than 1kWh/t for both clinker types. This information is useful in selecting energy levels to derive breakage functions to be used in comminution models. Energy levels higher and equal to 1 kWh/t were proposed for breakage function estimation of clinker feed to implement in comminution models of ball mills used in the cement industry.

2 MATERIALS AND METHODS

Clinker samples were provided from two different cement plants. Chemical and mineralogical analysis of the clinker samples belonging to different kiln conditions are given in Table 2 and 3 respectively. Tabulated mineralogical analysis results were estimated by using Bogue Equations (Hewlett, 1998). According to the mineralogical results, strength of clinker-2 is more higher than than clinker-1.

Table 2. Chemical analysis of the clinker samples

Chemical component (%)	Clinker-1	Clinker-2
MgO	1.31	1.54
Al_2O_3	4.56	5.6
SiO_2	21.33	21.53
CaO	66.21	64.79
Fe_2O_3	3.38	2.90
SO_3	1.67	1.64
K_2O	0.54	0.89
Na_2O	0.16	0.16

Cl^-	0.0106	0.0142
Loss of ignition (LOI)	0.25	0.31

Table 3. Mineralogical analysis results

Clinker phases	Clinker-1 (%)	Clinker-2 (%)
C_3S	71.92	58.32
C_2S	6.90	17.73
C_3A	6.36	9.93
C_4AF	10.28	8.82

2.1. Breakage Apparatus

Drop weight apparatus comprises a steel anvil made from steel alloy, a drop weight head, an electromagnet through which an electromagnetic field is formed so that, weights can be hold or released from desired heights. Drop weight apparatus is fitted with a 5.870 kg fixed head mass which can be extended to 46.963 kg with a maximum drop height of 51.50cm which represents a wide energy range. Drop weight head mass has a series of lead weights which can be added or removed when required. Drop weight is raised to a known height through a mechanical arm and then subjected to free fall to achieve an impact force on a particle.
on a root 2 sieve series on a ro-tap sieve shaker for 15 minutes. By this way, breakage product size distributions at the considered impact energy levels were determined. Test

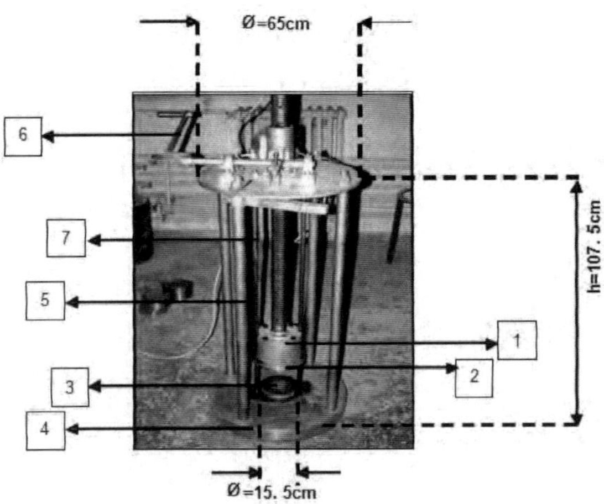

Particle breaks into different sizes at the given impact velocity (Genç, 2002). A photograph of the apparatus is given in Figure 1.

Figure 1. Drop weight apparatus (Genç et.al, 2014)

Drop weight apparatus given in Figure 1 comprises the following parts:

1. Electromagnetic head
2. 5.870 kg lead fixed head mass
3. Steel anvil (Diameter:15.50cm)
4. Steel base
5. Ruler for drop-height adjustment
6. Mechanical arm
7. Drop-weight rail

2.2. Test Conditions

Samples were dry sieved to the required narrow size fractions and breakage tests were conducted on randomly selected particles of each size fraction on at least three levels of specific comminution energy level. Impact energy (kWh/t) supplied by the drop weights are calculated from the equations given by Napier-Munn et al (2005). Broken fragments were collected from each test, and dry sieved

conditions are tabulated for clinker samples in Tables 4 and 5.

Table 4. Test conditions for clinker-1

Test size fractions (mm)	Particle number	Total weight (g)	Drop height (cm)	Dropped weight (kg)	Ecs (kWh/t)
-16+13.2	14	62.71	4.23	10.69464	0.28
	13	57.78	8.08	15.31914	0.76
	12	51.61	13.03	15.31914	1.26
- 13.2+11.2	25	59.92	10.37	15.31914	1.81
	26	63.77	11.47	18.31914	2.33
	23	57.37	16.75	15.31914	2.80
-9.5+8	61	51.99	6.43	15.31914	3.15
	64	52.05	6.43	18.31914	3.95
	63	52.86	9.93	15.31914	4.94

*Particle number is the number of particles broken per energy level in each size fraction

** Ecs is the calculated specific comminution energy level for the test size fraction breakage

Table 5. Test conditions for clinker-2

Test size fractions (mm)	Particle number	Total weight (g)	Drop height (cm)	Dropped weight (kg)	Ecs (kWh/t)
-16+13.2	23	115.73	4.32	10.69464	0.25
	20	91.31	7.98	15.31914	0.73
	22	95.78	12.87	15.31914	1.23
-13.2+11.2	26	79.68	10.50	15.31914	1.43
	26	81.70	11.60	18.31914	1.84
	27	83.58	16.80	15.31914	2.27
-9.5+8	49	50.51	6.43	15.31914	2.60
	50	56.88	6.60	18.31914	2.90
	55	54.64	10.10	15.31914	4.25

3 RESULTS AND DISCUSSION

3.1. Impact Breakage Product Size Distributions

Particle size distribution of the broken size fractions at various energy levels were plotted on log-normal scales in Figure 2 and 3 for clinker samples.

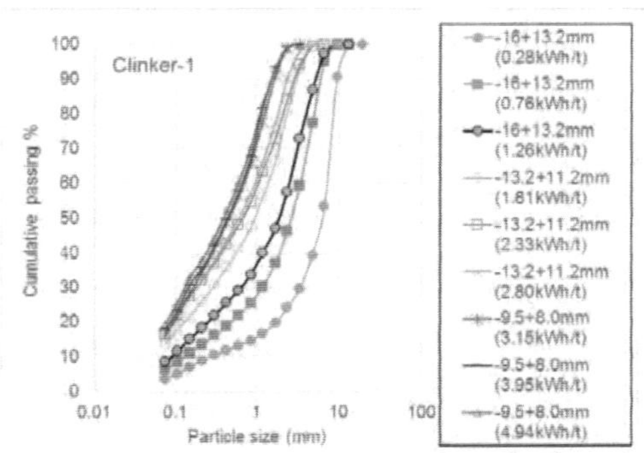

Figure 2. Impact breakage product size distributions of test clinker-1

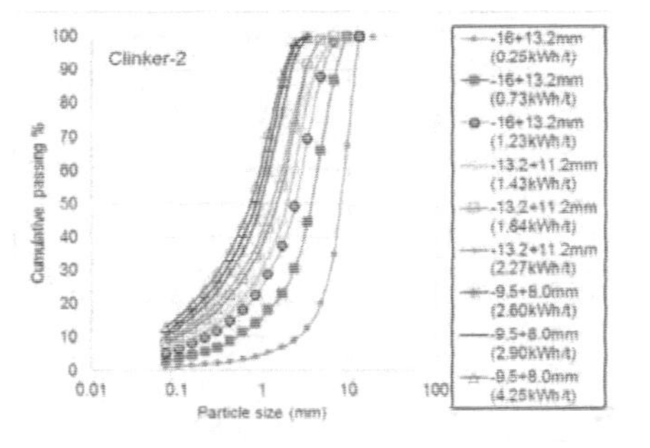

Figure 3. Impact breakage product size distributions of test clinker-2

Particle size distributions of the breakage products indicated that, increase in impact breakage energy level increased the fineness. Breakage size distributions start to become closer above an energy level of 1kWh/t, indicating no more considerable size reduction would occur above this energy level by impact mechanism.

3.1. Size-Normalized Breakage Distributions

Breakage pattern of different particle size fractions broken from different energy levels were investigated on normalized cumulative passing size distributions. In order to construct these curves, particle size was rescaled by dividing each sieve size by the median size (d50) of the broken product size distribution at the considered impact energy level. Self-similarity of breakage products are presented in Figures 4 to 11 for clinker samples 1 and 2. A comparison of normalized curves for test size fraction of -16+13.2mm for clinker-1 and clinker-2 is presented in Figure 12 to show the effect of hardness on fine particle production amount which could be related to the Belite mineral content.

Figure 4. Comparison of normalized curves for different test size fractions for clinker-1

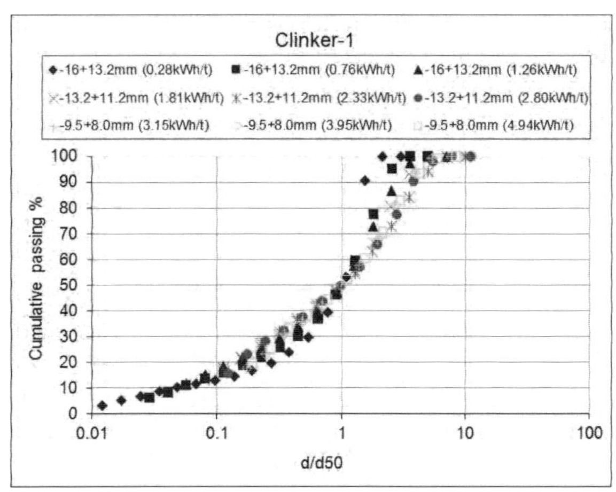

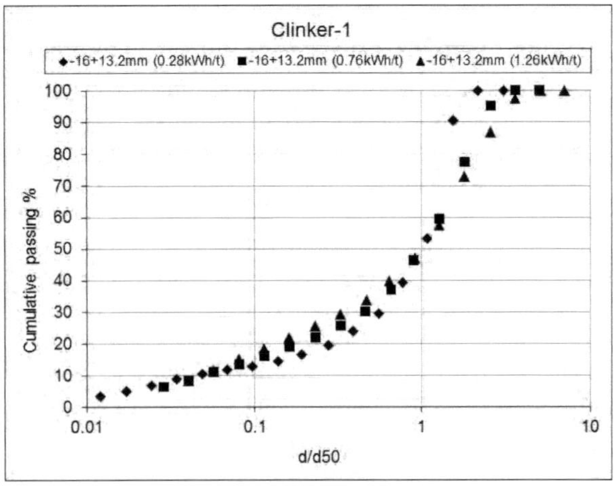

Figure 5. Normalized curves for test size fraction of -16+13.2mm for clinker-1

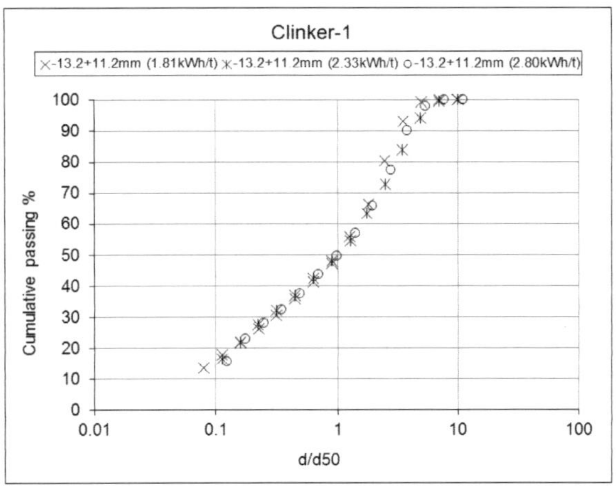

Figure 6. Normalized curves for test size fraction of -13.2+11.2mm for clinker-1

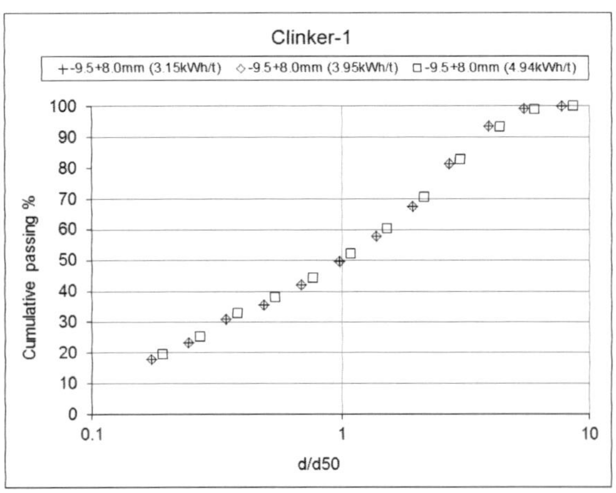

Figure 7. Normalized curves for test size fraction of -9.5+8mm for clinker-1

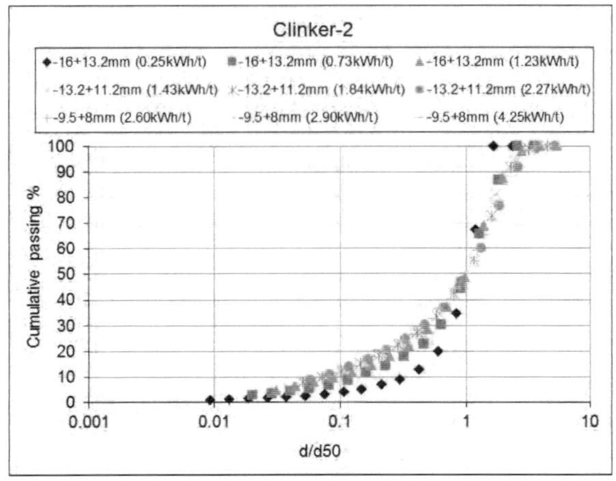

Figure 8. Comparison of normalized curves for different test size fractions for clinker-2

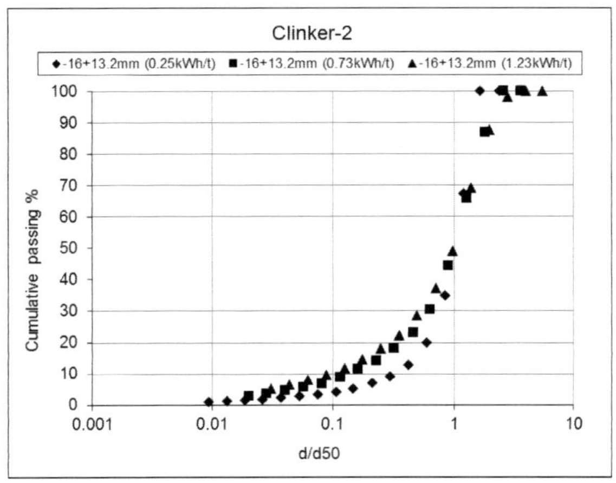

Figure 9. Normalized curves for test size fraction of -16+13.2mm for clinker-2

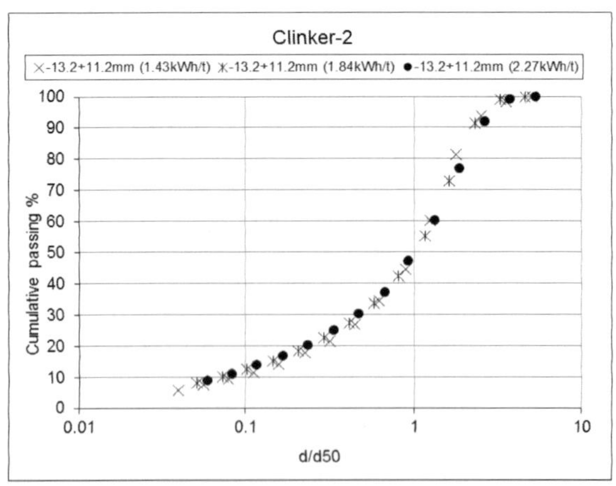

Figure 10. Normalized curves for test size fraction of -13.2+11.2mm for clinker-2

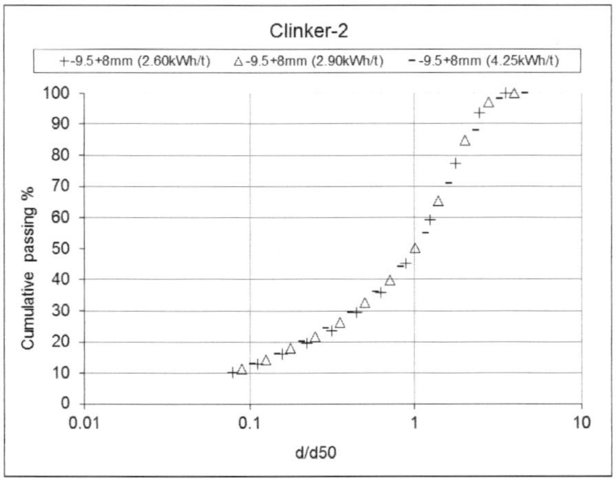

Figure 11. Normalized curves for test size fraction of -9.5+8.0mm for clinker-2

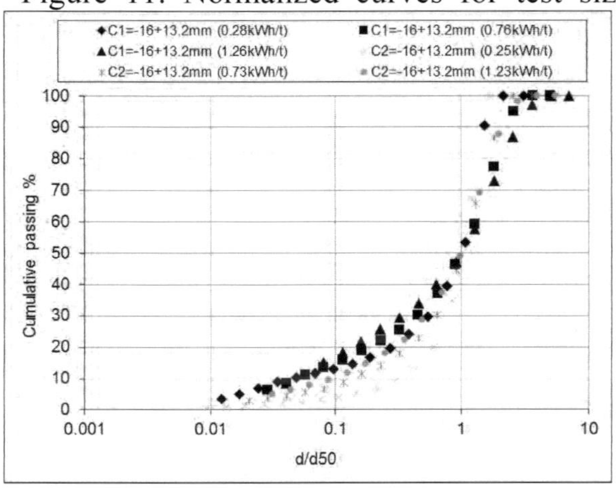

Figure 12. Comparison of normalized curves for test size fraction of -16+13.2mm for clinker-1 and clinker-2 (C1=Clinker-1 and C2=Clinker-2)

Normalized size distributions indicated that;

- Particle breakage pattern varies with size and impact energy level,
- Breakage product size distributions obtained at different specific comminution energy levels do not coincide on a single distribution which means that breakage pattern of each test size fraction is dependent on the impact energy level,
- Breakage of each size fraction start to exhibit a self-similar breakage pattern at energy levels higher than approximately 1 kWh/t.
- Comparison of normalized curves indicated that, clinker-2 produces less fine material as compared to clinker-1 sample which indicated that, clinker-2 is more harder than clinker-1 sample.

4 CONCLUSION

Breakage patterns of different size fractions of two clinker samples were evaluated using size normalized distribution curves. Self similar size distributions were obtained for size fractions broken above an energy level of 1kWh/t whereas distributions were non-normalizable below this energy level. Thus, energy level of 1kWh/t was proposed as the critical energy level for impact breakage of clinker. Findings indicated that, breakage characteristics of clinker change from one plant to another. This could be linked to the different kiln conditions which was expected to affect clinker mineralogy, microstructure and chemical composition on the size fractional base. Thus, deviations from the self-similar breakage occur.

Modified version of the drop weight apparatus was found to be a useful tool for the understanding of the basic impact breakage characteristics of clinker particles such that it can be simply used to simulate the individual breakage process on many breakage energy levels in a mill. Test results can then be used to determine material specific breakage distribution functions to be used in comminution modelling. Energy levels which produce size-normalized breakage distributions could be used to estimate breakage functions which are required in comminution models such as perfect mixing mill model (Napier Munn et.al., 2005) in which step triangular breakage matrices are used to ease the modelling work.

ACKNOWLEDGEMENTS

Authors would like to acknowledge Prof. A.Hakan Benzer from Hacettepe University,

Department of Mining Engineering for the valuable discussions.

REFERENCES

Andersen, J.S., 1988. Development of a cone crusher model., M.Eng.Sc. Thesis, University of Queensland (JKMRC).

Duda, W.H., 1985. Cement-Data-Book, International Process Engineering in the Cement Industry, French & European, 3rd edition, Vol.1.

Genç, Ö., 2002. An Investigation of the Breakage Distribution Functions of Clinker and Additive Materials, MSc Thesis, Mining Engineering Department, University of Hacettepe, Ankara, Turkey.

Genç, Ö., Benzer, A.H., 2009. Single particle impact breakage characteristics of clinkers related to mineral composition and grindability. Minerals Engineering, 22(13), p.1160-1165.

Genç, Ö., Benzer, A.H., Ergün, Ş.L., 2014. Analysis of single particle impact breakage characteristics of raw and HPGR-crushed cement clinkers by drop weight testing. Powder Technology, 259, p.37-45.

Hewlett, C.P.,1998. Lea's Chemistry of Cement and Concrete, Fourth Edition, Butterworth-Heinemann, UK.

Man.,Y.T.,2000. A Model-based scale-up procedure for Wet, Overflow Ball Mills., Julius Kruttschnitt Mineral Research Centre., Department of Mining, Minerals and Materials Engineering, The University of Queensland.

Napier-Munn, T.J., Morrell, S., Morrison, R.D., Kojovic T.,2005. Mineral Comminution Circuits Their Operation and Optimization, JKMRC, Brisbane.

Narayanan, S.S.,1985. Development of a laboratory single particle breakage technique and its application to ball mill modelling and scale-up, PhD thesis, University of Queensland, Australia.

Experimental Investigation of the Effect of Impeller Type and Diameter on the Fluid Flow Hydrodynamics in Copper Solvent Extraction Mixer◊

S. Parvizi[a], E. Kesahvarz Alamdari[b], S. Aosati[b], M. Molaeinasab[c]

[a] *Department of Mechanical engineering, Shahid Rajaee Teacher Training University, P.O. Box. 16785-136, Tehran, Iran.*
[b] *Department of Mining and Metallurgical Engineering, Amirkabir University of Technology, P.O. Box. 15875-4413, Tehran, Iran.*
[c] *National Iranian Copper Industries Co (NICICO)., Kerman, Iran.*

ABSTRACT In this research, laboratory scale of solvent extraction mixer was designed and manufactured. Related works were used to select the final aspect ratios. Organic and aqueous liquids were provided from the solvent extraction unit of Sarcheshmeh copper complex, Iran. A 1000 frame per second camera, 1000 watt projectors and mono chromatic lenses were used for imaging and lighting systems. Rushton, Pitch blade down-flow and up flow turbine with different impeller to tank diameter ratio were studied. According to turbulent and mixing flows, D/T of 0.3 and Rushton impeller are optimal impeller conditions for dynamic mixers of copper industry.

Keywords: Mixer Impeller, Hydrodynamics, Liquid-Liquid Extraction, Solvent Extraction

◊ This work is presented in the 13th Scientific Student Conference of Materials and Metallurgical Engineering of Iran, 2016

1 INTRODUCTION

In many engineering processes like solvent extraction, two immiscible liquids are mixed [1]. Dynamic mixers are broadly used in chemical, mineral and hydrometallurgical industries [2]. Geometry of mixer and impeller have a crucial importance in the efficiency of dynamic mixers [3].

Mixing process is dependent on the impeller geometry because impeller is used to transferring the shear forces to the fluids. In solvent extraction mixers Rushton turbine (RT), pitch blade down-flow turbine (PBDT) and pitch blade up-flow turbine (PBUT) are common impellers [4]. Mechanically agitated contactor (MAC) were used with various impeller type and the result showed that the impeller type is one of the most important parameters that effect the mixing energy [5]. Turbulent kinetic energy of PBDT and 310A were compared. Radial fluctuations for Rushton turbine and axial fluctuations for PBDT and 310A were detected and investigated [6]. In 10% holdup, four different impeller types were compared. The results showed that shear surfaces and turbulent flows of RT is the most. PBDT produce optimal axial fluid flow and RT has a mixing time less than half circle blade disk turbine (HCDT) and more than PBUT [6].

Impeller rotation produce suction zone. This phenomenon is very important factor in efficiency of the dynamic mixers and D/T ratios are typically in the range of 0.25 up to 0.4 [7].

2 MATERIALS AND METHODS

Physical, geometrical and operational parameters are affecting the efficiency of the dynamics mixers [3]. In this research type and diameter of impeller were studied. For this purpose imaging and lighting system were used. Aspect ratios of the mixer parts are set according to the figure 1 [8,9].
Mixer vessel was made of Pyrex with two inlets and one outlet. Four baffles were implemented in the mixer wall vertically and symmetrically. Two magnetic pump were used for transferring aqueous and organic liquids. Casio EX-10 with 1000 fps and shutter speed of 1/12800 was used for direct photography. Two 1000 watt projector with monochromatic lenses were used for lighting system (figure 2).

Organic and aqueous liquid were provided from solvent extraction unit of Sarcheshmeh copper complex, Iran.

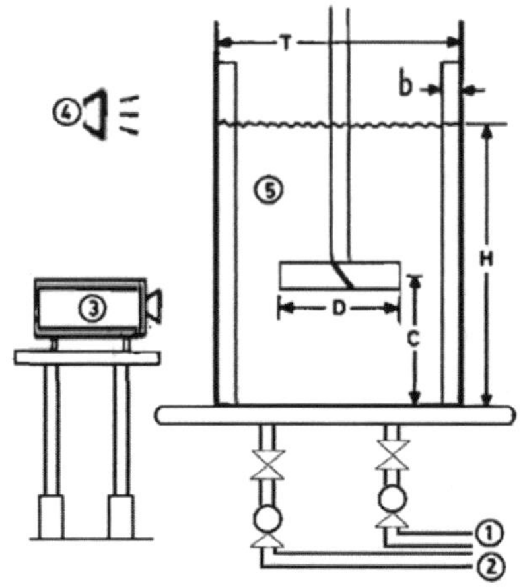

Fig. 1. Setup schematic and mixer vessel geometrical ratios (H=1.2T=10b=3C) (1-organic inlet 2-aqueous inlet 3-imaging system 4-lighting system 5-mixer vessel)

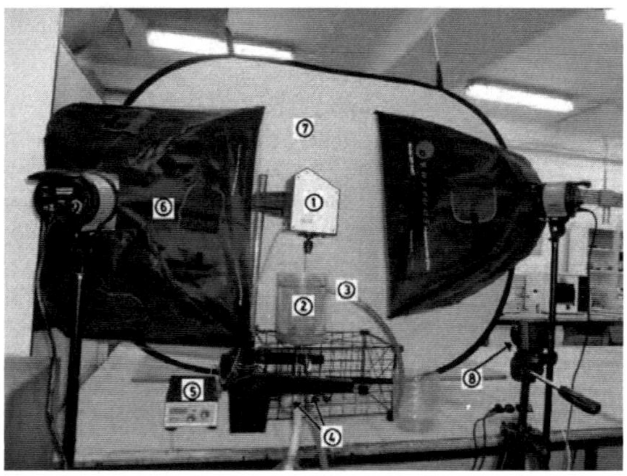

Fig. 2. Experimental equipments (1-mixer motor 2-mixer vessel 3-outlet 4-aqueous and organic inlets 5-impeller speed controller 6-projectors 7-reflecting surface 8-camera)

3 RESULTS AND DISCUSSIONS

3.1 Impeller Type

According to figure 3, low pressure units (suction zone) in RT impeller has a bigger volume that result in increasing impeller efficiency and improvement of mixing condition.

Turbulent flows are very important factor in the mixing process and as shown in the figure 3, RT impeller produce more severe turbulent flows in comparison with PBDT and PBUT impellers, especially in regions near the baffles.

3.1 Impeller Diameter

According to figure 5, with increasing impeller diameter, mixing volume zone is increased while the mixing height was remained fixed approximately. Therefore with increasing impeller diameter mixing efficiency was increased because more fluids participate in mixing process.

The effect of impeller diameter on the baffles function are shown in the figure 6. According to this figure, with increasing impeller diameter dead zone after baffles are tend to vanish and in impeller diameter of 8 cm dead zone around the baffles are omitted approximately.

4. CONCLUSION

RT impeller in comparison with PBUT and PBDT produce more turbulent energy. With increasing impeller diameter dead zone around the baffles are decreased and energy required is increased. Thus RT impeller with 7 cm diameter (D/T=0.3) is recommended for the dynamic mixers of copper solvent extraction units.

Fig 3. Low pressure (suction) zone around the baffle for different impeller typed a)PBDT b)RT c)PBUT (impeller speed: 100 rpm, off-buttom clearance and impeller diameter: 7cm, baffle width=21mm, number of baffles:4)

Fig 4. Turbulent flows at different impeller types a)PBDT b)RT c)PBUT (impeller speed: 100 rpm, off buttom clearance and impeller diameter: 7cm, baffle width=21mm, number of baffles: 4)

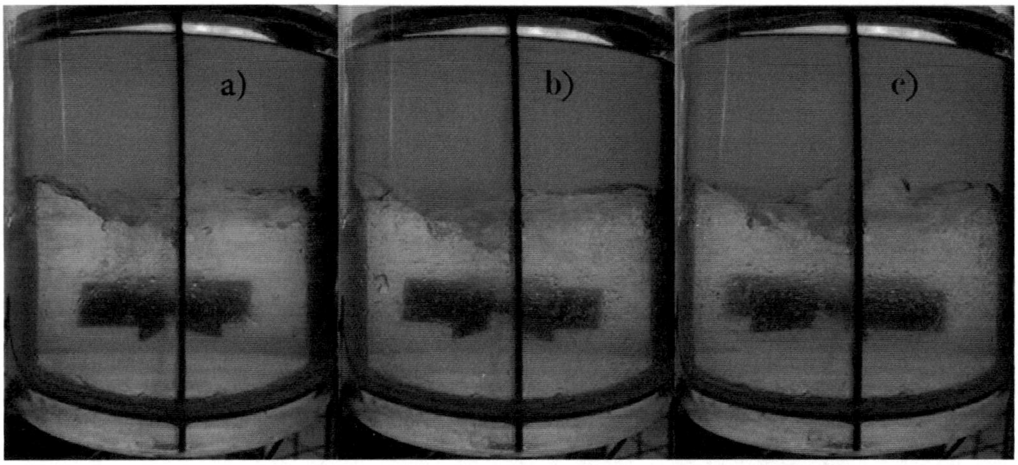

Fig. 5. Mixing volume zone at different impeller diameters a)6 b)7 c)8 cm (impeller speed: 100 rpm, off buttom clearance: 7cm, b=21mm, number of baffle:4, Impeller type: RT)

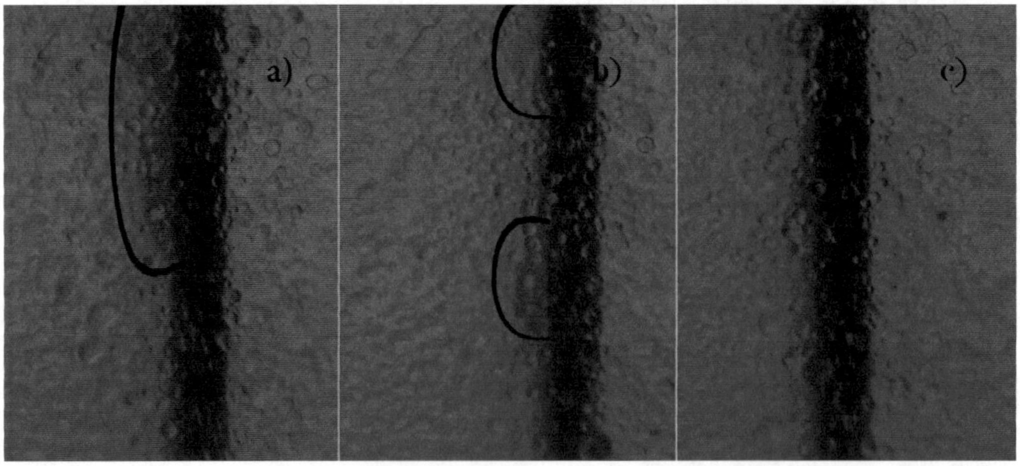

Fig. 6. Dead zones around the baffles at different impeller diameters a)6 b)7 c)8 cm (impeller speed: 100 rpm, off buttom clearance: 7cm, b=21mm, number of baffle:4, Impeller type: RT)

REFERENCES

[1] M. Kraume, A. Gäbler, K. Schulze, Influence of Physical Properties on Drop Size Distribution of Stirred Liquid-Liquid Dispersions, Chemical engineering & technology, 27 (2004) 330-334.

[2] M.O. Shabani, Mazahery, A., Computational Fluid Dynamics (CFD) Simulation of Liquid-Liquid Mixing in Mixer-Settler, Archive of Metallurgy and Materials, 57 (2012) 173-178.

[3] E.L. Paul, V.A. Atiemo-Obeng, S.M. Kresta, Handbook of industrial mixing: science and practice, John Wiley & Sons, 2004.

[4] V. Rewatkar, J. Joshi, Effect of impeller design on liquid phase mixing in mechanically agitated reactors, Chem. Eng. Commun., 102 (1991) 1-33.

[5] K. Raghav Rao, J. Joshi, Liquid phase mixing in mechanically agitated vessels, Chem. Eng. Commun., 74 (1988) 1-25.

[6] G. Zhao, Kresta, S. M.,, Impact of Tank Geometry on the Maximum Turbulence Energy Dissipation Rate for Impellers, AlChE Journal, 42(9) (1996) 2476-2490.

[7] O. Shabani, Alizade, M., Mazahery, A.,, Fluid Flow Characterization of Liquid-Liquid Mixing in Mixer-Settler, Engineering with Computers, 27 (2011) 373-379.

[8] O. Brůha, T. Brůha, I. Fořt, M. Jahoda, Dynamics of the flow pattern in a baffled mixing vessel with an axial impeller, Acta Polytechnica, 47 (2007).

[9] L. Fradette, G. Thomé, P.A. Tanguy, K. Takenaka, Power and Mixing Time Study Involving a Maxblend® Impeller with Viscous Newtonian and Non-Newtonian Fluids, Chem. Eng. Res. Des., 85 (2007) 1514-1523.

Separation of Organic and Non Organic Parts of WEEE by Gravitational Methods

M. Kavousi, A. Sattari, E. Keshavarz Alamdari

Department of Mining and Metallurgical Engineering, Amirkabir University of Technology, Tehran, Iran

ABSTRACT Technological progress has resulted in an increase of waste electric and electronic equipment (WEEE). This waste contains different hazardous materials which are harmful to human health and the environment. In addition, WEEE includes valuable metals and component. Therefore, the recovery of WEEE is important and cost effective. Efficiently physically separating and recovering fine metal particles from WEEE are a key step. This paper will focus on the separation of components by using density separation techniques and an innovative method relevant to leach the existing metals.

Keywords: Recovery, WEEE, The separation techniques.

1 INTRODUCTION

E-waste and Waste Electrical and Electronic Equipment (WEEE) is one of the fastest growing wastes in the world [1]. Also, E-waste is chemically and physically different from other industrial waste; whereas, it contains valuable and base metals, including gold, silver, copper, platinum and toxic materials such as lead, mercury and cadmium. Therefore, the recovery of E-waste is very important from the point of view of environmental problems and economic benefits [1].

Printed circuit board (PCB) is the most significant part of E-waste. The PCBs include nonferrous metals such as Cu, Pb, and Sn [2]. Chemical and mechanical methods have been the traditional methods of recycling PCB waste. Many researcher studied the recovery of metals from PCBs wastes using hydrometallurgical, pyrometallurgical, electrometallurgical, bio-hydrometallurgical processes, and their combinations. Also various mechanical methods were used to separate metals from PCBs, such as shape separation, jigging, and density-based separation [3].

Among of the physically method, the method of traditional air separation shows lower separation efficiency for waste PCBs [4]. One of the well-known physically methods is the density separation techniques. Whereas, the PCB wastes consists different substances with different density, the density separation techniques could be successful to separate these wastes [5].

In this work, a fundamental study has been carried out by using density separation techniques.

2 MATERIALS AND METHODS

Ten pieces of different computer main boards (about 10 kg) were collected and the welded parts were manually removed. About 6 kg WPCBs was obtained from 10 kg waste. Afterward, WPCBs were cut into small pieces of 100 *100 mm via cutting machine. These pieces were shredded by SM-2000 cutting mill (Retsch, Germany) and sieved into different fraction using standard sieves. A RP4 shaking table is used to separate metallic and non-metallic contents. Different kinds of materials are separated into heavy particles and light particles by the difference of densities.

3 RESULTS AND DISCUSSION

The physically separation method focuses on the equipment used for initial processing, or concentration. The object of concentration is to separate the material into two products. Ideally, in recovery of the metals, all of the metals will be in the concentrate, while all other nonmetal material will be in the tailings. In large operation, there is an intermediate step as known scavenging. In fact, whereas the scavenger includes the significant content of variable metals, it should recycle to the first stage.

It means that the final concentrate, usually obtained by repeated processing.

Specific terms are also used to describe the efficiency of the concentration process. Recovery refers to the percentage of metals in the ore that was collected in the concentrate.

Discharged water, feeding capacity, and inclination angle are the major factors that affect the separation efficiency of the shaking table.

The separation efficiency could be calculated according to the Eq 1.[4]

$$\omega=(\alpha-\theta)(\beta-\alpha)/\alpha(\beta-\theta) \quad (1)$$

whereas ω refers to the separation efficiency, α , β , θ refer to the metal contents of feeding materials, metal concentrate, and nonmetallic concentrate, respectively.

The separation results of shaking table tests are shown in Table 1. Experimental results show that the separation efficiency higher than 81 %, recovery rate of more than 76 %, and concentrate grade greater than 57% could be achievable under appropriate conditions. There is a general inverse relationship between recovery and concentrate grade in mineral concentration. Usually, the higher the concentrate grade, the lower the total recovery. The best recovery systems will collect a maximum amount of metals in a minimum amount of concentrate.

In statistics, response surface methodology (RSM) explores the relationships between several explanatory variables and one or more response variables. The analyzing results with the RSM approach are shown in Fig. 1. As it is illustrated in Fig. 1, the recovery rate shows significant interactive relationships with the water discharge and inclination angle. Therefore, establishing the optimum conditions, it is better to maintain the water discharge of 700 cc/10 sec and the inclination angle of 3o.

The results shows that the separation efficiency and concentrate grade gradually

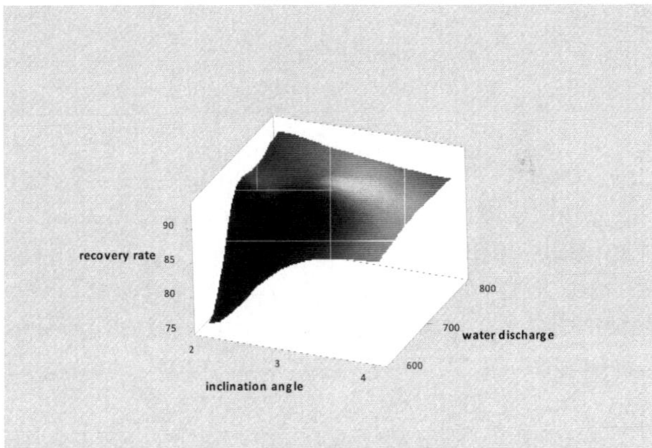

Figure1- The analyzing results with the RSM approach for the variation of the recovery rate versus inclination angle and water discharge (cc/10sec)

Table 1. The results of the separation tests

Water discharge(cc/10sec)	Feeding capacity(gr/min)	Inclination angle(°)	Concentration grade(%)	Recovery rate(%)	Separation efficiency
800	230	3	77.7	91	86
800	230	4	86	86.37	90.09
800	230	2	77	92.8	90.5
700	230	4	74	90	85.6
700	230	2	71.58	91.66	88
700	230	3	76.1	89	88.3
600	230	4	74	79	82
600	230	2	81.8	76	81
600	230	3	83.5	79.09	83.2

Table 2- the chemical analysis of metals

product	Amount (%)	Ag	Cu	Pb	Sn	LOI
concentrate	24	80.5	83.8	89.3	89	0
scavenger	37	11	9.9	5.9	5	12
tail	132	8.5	6.3	4.8	4	78

increase with the increasing of water discharge and feeding capacity should be maintained in a suitable variation extent. The results of the gravity separation of the metallic components from the milled PCBs in optimum condition are shown in Table 2. As shown in table 2, the maximum content of metals were existed in the concentrate. Also, the minimum content of LOI was in the tail. This result shows that our experiments were successful for the separation of metal and nonmetal parts.

4 CONCLUSIONS

In this study, the parameters affecting the separation of metals from other compounds were investigated. The results show that the high separation efficiency could be achievable by optimization of suitable conditions. The optimum condition for separation of waste PCBs is proposed as following: the water discharge of 700 cc/10 sec, the inclination angle of 3° and the feeding capacity range of 200-250 gr/min.

REFERENCES

[1] Kavousi, M., Sattari, A., Alamdari, E.K. and Firozi, S., 2016.. Selective separation of copper over solder alloy from waste printed circuit boards leach solution. Waste Management, In press.

[2] Li, J., Xu, Z. and Zhou, Y., 2007.. Application of corona discharge and electrostatic force to separate metals and nonmetals from crushed particles of waste printed circuit boards. Journal of Electrostatics, 65(4), pp.233-238.

[3] Li, J., Lu, H., Guo, J., Xu, Z. and Zhou, Y., 2007.. Recycle technology for recovering resources and products from waste printed circuit boards. Environmental science & technology, 41(6), pp.1995-2000.

[4] He, J.F., Duan, C.L., He, Y.Q. and Zhang, H.J., 2015.. Recovery of valuable metal concentrate from waste printed circuit boards by a physical beneficiation technology. International Journal of Environmental Science and Technology, 12(8), pp.2603-2612.

[5] Zhang, S. and Forssberg, E., 1997.. Mechanical separation-oriented characterization of electronic scrap. Resources, Conservation and Recycling, 21(4), pp.247-269.

Evaluation of Mixing Efficiency of Turbine Impellers in Stirred Tank Reactors

H. Deveci
Hydromet B&PM Research Group, Div. of Mineral&Coal Processing, Dept. of Mining Engineering, Karadeniz Technical University, Trabzon, Turkey

ABSTRACT Mixing is very important unit operation in hydrometallurgical applications in which suspension of solids, mass and heat transfer are prerequisite. Mixing can be achieved by mechanical agitation using impellers in strirred tank reactors. In this study, the mixing efficiency of Rushton turbine (RT) and 45° pitched blade turbine (PBT) impellers were evaluated based on "just off-bottom suspension" criteria. Effects of solids density (5-30%), air flow rate (0-0.6 L/min), particle size on the critical speed required for suspension of solids were demonstrated. The results have shown that the critical speed increases with increasing solids density and air flow rate. PBT impellers are more suitable for solid-liquid mixing than RT impellers, but, more sensitive to aeration. A minimum tip speed required to achieve satisfactory mixing performance under various conditions was determined for both impeller types. The experimental findings were compared with the estimates based on empirical correlations.

1 INTRODUCTION

Mixing is a physical operation utilised to achieve a degree of uniformity in a given system by eliminating the development of gradients of system properties such as concentration and temperature. It is also provided to facilitate/promote suspension of solids, gas mass transfer and heat transfer to or from the medium. In hydrometallurgical (e.g. leaching) operations, the provision of an optimal environment is pre-requisite or else, the leaching process will be limited by the exposed (available) surface area coupled with poor oxygen transfer, if required.

Mixing can be achieved by mechanical agitation using a suitable impeller (in CSTRs) or by air agitation (in airlift reactors such as pachuca tanks). The CSTRs are most widely utilised within (bio)leaching processes at industrial or laboratory scale due to a number of advantages over pachuca tanks (Rossi, 1999; Acevedo et al., 1988; Fraser, 1984). In a CSTR, mixing is accomplished by displacement and circulation of the material into the most remote parts of the tank. This is performed by means of an impeller which creates turbulence whereby distribution, dispersion and diffusion occur in the system (Doran, 2013). Baffles are also employed to minimise vortex formation and swirling of the slurry.

Impellers can be broadly categorised as radial flow and axial flow based on the direction of pulp leaving the impeller. The typical radial and axial flow impellers are illustrated in Figure 1. Radial flow impellers such as the Rushton turbine are the traditional impellers widely utilised for gas dispersion applications (Oldshue, 1983; Fraser, 1984). The Rushton turbine impellers have also been routinely employed in the early biooxidation plants (Kubera and

Oldshue 1992). However, these impellers had several drawbacks including relatively high shear rate and power draw characteristics, sensitivity to gas flow rate and generation of separate mixing zones where the gradients of pH and temperature may prevail (Weetman and Oldshue, 1988; Fraser, 1992; Hayward et al., 1997). The traditional axial flow impellers such as 45° pitched blade turbine (Figure 1) show moderate shear and higher pumping (flow) capability relative to the Rushton turbine (Weetman and Oldshue, 1988). However, these impellers have a relatively limited capacity for gas dispersion and relatively high sensitivity to gas flow rate (Chapman et al., 1983a; Oldshue et al., 1988).

The limitations of the above impellers have stimulated the development of axial flow hydrofoil impellers (Kubera and Oldshue, 1992) so as to improve efficiency of oxygen and carbon dioxide transfer, reduce shear and provide more efficient suspension of solids and mixing of the slurry with no generation of zones in the reactor (Weetman and Oldshue,1988; Fraser, 1992; Neale and Pinches, 1994). The main rationale behind the adoption of these new design impellers is the achievement of equal process performance with a significantly reduced power consumption of 35-50% compared to the conventional Rushton turbine impellers (Kubera and Oldshue, 1992).

In bioleaching processes, agitation is required to facilitate suspension of solids, transfer of oxygen and carbon dioxide, and heat from or to the medium and also to minimise or eliminate the development of temperature, pH, dissolved oxygen and concentration gradients throughout the bioreactor (Gormerly and Branion, 1989; Bailey and Hansford, 1993). The Continuous Stirred Tank Reactor (CSTR) is currently the most widely utilised type of reactor where the mixing is achieved by an appropriate agitator system (i.e. impeller design). Agitator systems are designed to produce "flow" and "shear" to fulfil the desired mixing duty. However, the hydrodynamic conditions i.e. intensity of shear or turbulence produced to achieve the desired process (mixing) results may affect the leaching performance of the bacteria (Deveci, 2002 and 2004). Therefore, shear conditions and other operating characteristics such as power draw and pumping performance of mixing systems should also be technically evaluated.

In this study, the mixing performance of Rushton turbine (RT) and 45° pitched blade turbine (PBT) impellers for solids suspension were evaluated. A minimum impeller speed required for "just off-bottom suspension" criteria was determined for both impeller types in the presence and absence of air. For comparison, the minimum speed under the test conditions was also estimated using Zwietering and Nienow correlations (Zwietering, 1958; Harnby et al., 1992).

Figure 1. Traditional axial flow 45° pitched blade turbine (top) and radial flow Rushton turbine (bottom) (Oldshue, 1983)

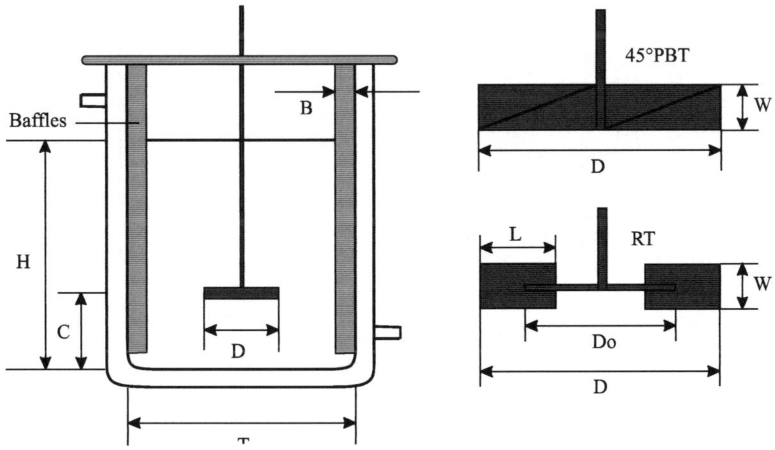

Design Parameter	Dimensions of vessel, impellers and baffles		
	Symbol	Relationship	Actual
Vessel diameter	T	T	95 mm
Slurry height	H	H=T	95 mm
Baffle width	B	B=0.1T	9.5 mm
Impeller clearance from the bottom	C	C=T/3	32 mm
Impeller diameter	D	D=T/3 (or T/2*)	32 mm
Impeller blade width	W	W=D/5	6.4 mm
Rushton turbine blade length	L	L=D/4	8 mm
Rushton turbine disk diameter	Do	Do=2D/3	21.3 mm
Slurry volume	V	$V=\pi T^2H/4$	673 ml

*Note that a 45° pitched blade turbine with a larger diameter (47.5 mm, D=T/2) was also used in the tests

Figure 2. A schematic diagram showing the geometry of impellers, baffles and reactor vessels as well as their design parameters (Deveci, 2002)

2 MATERIAL AND METHOD

A flat-bottomed, jacketed reactor vessel (1.2 L in nominal capacity) was used in this study. Impellers and baffles were sized based on the vessel diameter using the chemical engineering principles for the design and configuration of Stirred Tank Reactors (STRs) (Oldshue, 1983; Neale and Pinches, 1994). The design parameters and geometric relationship between these parameters are presented in Figure 2. In this study, the traditional impeller designs; a six blade Rushton turbine (RT) and two four blade, downward pumping 45° pitched blade turbines (PBT) were used. The vessel was fitted with a "U" shape sparger, which was constructed from titanium tube (1.75 mm in inside diameter) with a hole (0.62 mm in diameter) at its bottom centre to a position underneath the impeller to deliver air into the vessel.

The mixing characteristics of both impeller designs were evaluated with regard to their capability for the suspension of solids and dispersion of air. The vessel was filled with double distilled water (673 ml). Air supply (at 0.2 and 0.5 l/min) was turned on and the impeller speed was gradually increased until no flooding was observed i.e. complete dispersion of air bubbles by the impeller. The corresponding impeller speed was determined using a tachometer.

In a similar manner, the critical (minimum) speed required for complete suspension of solids was determined using size fractions obtained from quartz (Figure 3) at solids concentrations of 5-30% w/w and at airflow rates of 0, 0.2 and 0.5 l/min. A required volume of double distilled water and a known amount of solids corresponding to the solids densities studied were transferred into the reactor equipped with a particular impeller design. The base of the reactor was illuminated and the impeller speed was gradually increased. The speed at which no solids remained resting at the bottom of the vessel was recorded. The determined speed corresponds to "just off-bottom suspension" speed described by Oldsuhe (1983).

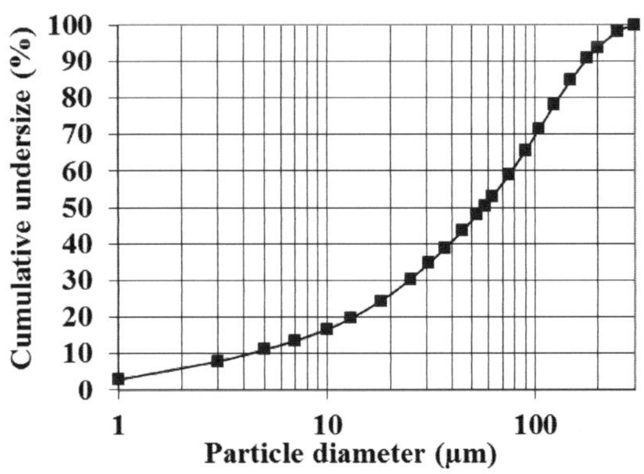

Figure 3. Particle size distribution of quartz material used in this study.

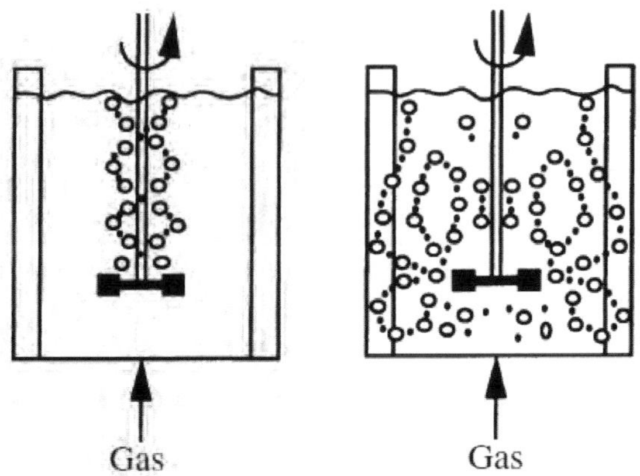

Figure 4. Gas distribution pattern in an aerated stirred tank reactor as a function of impeller speed illustrating flooding (left) and proper dispersion of gas (Doran, 2013)

Table 1. Effect of air flow rate on critical speed (N_g) below which impeller flooding occurs

	Critical Speed, $N_{c,f}$ (m/s)	
Airflow Rate	**0.2 L/min**	**0.6 L/min**
RT	-	2.01
PBT	2.01	2.35

3 RESULTS AND DISCUSSION

3.1 Gas-Handling

Initially flooding conditions at which impeller was unable to perform mixing action effectively was examined for RT and PBT impellers at flow rates of 0.2 and 0.6 L/min. Table 1 indicates the minimum impeller speeds required to avoid flooding of RT and PBT impellers at these air flow rates. The mixing performance of PBT appears to be affected by air even at the lowest flow rate tested (0.2 L/min) at which the RT was stable. Further increase in the air flow rate to 0.6 L/min also led to the flooding of RT at low impeller speeds below 2.01 m/s. For the PBT, the critical impeller speed was determined to increase to 2.35 m/s. These results confirm that axial flow impellers such as pitched blade turbines (PBTs) are more sensitive to aeration than radial flow impellers such as Rushton turbines (RTs).

3.2 Solids Suspension

The complete suspension of solids, which is also known as "just-off-the bottom" suspension, is defined as the criterion that no particles rest on the bottom of vessel for more than few seconds. Figure 4 illustrates the experimental observations for the minimum impeller speed required for complete suspension of solids (-250+53 μm quartz). The critical impeller speed for solids suspension increases with increasing the solids density and air flow rate. For the RT impeller, the critical tip speed is required to be increased from 1.81 m/s at 10% w/w solids to 2.04 m/s at 30% w/w. In a similar manner, the critical impeller speed needed to be increased so as to achieve the suspension of solids in the presence of aeration. To illustrate, a critical speed of 1.98 m/s was sufficient to keep solids in suspension. However, a ≈10% increase in the speed was required under aeration at 0.2 L/min.

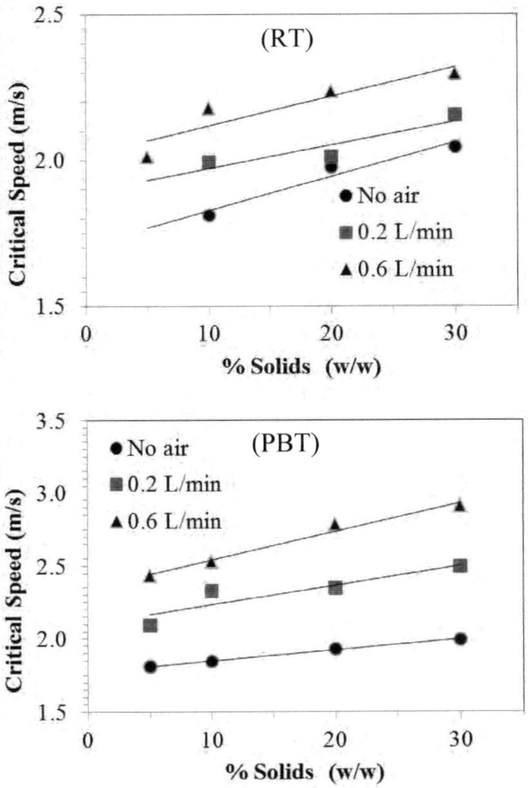

Figure 4. Effect of air flow rate and solids concentration on critical speed for solids suspension by the Rushton turbine (RT) and the pitched blade turbine (PBT) impellers (-250+53 μm quartz)

The PBT impeller appeared to be more effective than the RT impeller for solids suspension in the absence of aeration. In other word, the PBT was able to suspend the solids consistently at lower impeller speeds than the RT. This is consistent with its high pumping capacity compared with the RT (Fraser, 1992; Chapman et al., 1983b). However, the PBT impeller seemed to be more sensitive to aeration (Figure 5), which led to the deterioration in its mixing performance with increasing air flow rate. This is in agreement with its comparatively low gas-handling ability as observed (Table 1). A linear trend of increase in the required impeller speed was observed with increasing the air flow rate in the range tested. At 20% w/w solids, the critical speed was required to be increased by 22% at 0.2 L/min air.

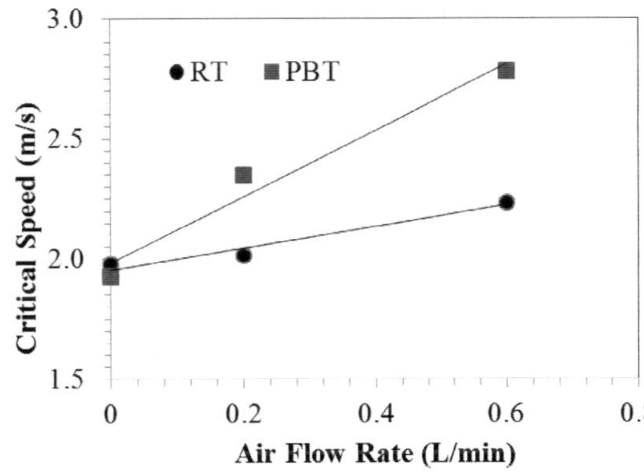

Figure 5. Influence of air flow rate on solids suspension performance of the Rushton turbine (RT) and the pitched blade turbine (PBT) impellers (20% w/w; -250+53 μm quartz)

Table 2 also presents the influence of particle size on the required critical speed for solids suspension by both impellers at different solids loadings and air flow rates.

Table 2. Effect of air flow rate, particle size and solids concentration on the critical speed required for solids suspension by the Rushton turbine (RT) and pitched blade turbine (PBT) impellers

		Critical Speed, N_{js} (m/s)			
Imp.	Air (L/min)	**-125+90 μm (w/w)**		**-90+63 μm (w/w)**	
		10%	**20%**	**10%**	**20%**
RT	0	1.75	1.78	1.71	1.75
	0.2	1.88	2.02	1.83	1.98
	0.6	2.16	2.23	2.12	2.18
PBT	0	1.80	1.88	1.65	1.80
	0.2	2.34	2.36	1.99	2.25
	0.6	2.50	2.70	2.41	2.68
		-53 μm			
		10% w/w		**20% w/w**	
RT	0	1.68		1.73	
	0.2	1.78		1.89	
	0.6	1.96		2.16	
PBT	0	1.49		1.64	
	0.2	1.98		2.18	
	0.6	2.28		2.45	

A similar trend for the effect of aeration and solids density was also observed at different size of solids. It is pertinent to note that these fractions were obtained from the bulk quartz material (Figure 3). Inherently, both impellers were more effective for suspension of solids as the particle size became finer. These findings also suggest the importance of particle size in efficient mixing of slurries in hydrometallurgical operations. For the RT impeller at 10% w/w solids and 0.6 L/min air flow rate, for instance, the minimum impeller speed was determined to be 2.16 m/s for -125+90 μm size fraction compared with 1.96 m/s for -53 μm.

The complete suspension criterion is first described by Zwietering (1958). He developed an empirical equation for critical speed based on more than a thousand tests for different tank and impeller geometries, solids concentrations and liquid/solid densities. The Zwietering correlation for solids suspension is expressed as:

$$N_{JS} = \frac{\text{[illegible]}}{D^{0.85}}$$

Where S is a dimensionless constant depending on impeller type and tank geometry; v is the liquid kinematic viscosity; D_p is the mean particle size; g is the gravitational acceleration; ρ_p and ρ_L are the density of solid and liquid, respectively; X is the solids concentration by weight; and D_i is the impeller diameter. Similar correlations were also proposed by many investigators (Harnby et al., 1992).

The Zwietering correlation indicates that the critical speed depends significantly on impeller diameter. Increasing impeller diameter improves solids suspension. This was confirmed in that adverse effect of aeration on mixing efficiency was significantly mitigated by using a larger PBT impeller (4.75 cm in diameter) than that used in this study. The Zwietering correlation also considers the dependence of critical speed on particle size and solids concentration, which is consistent with the experimental findings in this study (Table 2).

It should be noted that the suspension is probably required to be mixed at speeds higher than the critical speed to achieve a degree of uniform suspension. The Zwietering correlation assumes only two phase, solid-liquid systems. In the presence of aeration, the power draw and pumping capacity of the impeller decrease with the concomitant increase in the required critical speed for solids suspension (Neale and Pinches, 1994; Doran, 2013) as also observed in this study (Figures 4-5 and Table 2). Empirical correlations were also developed for the critical speed in three phase systems (Harnby et al., 1992; Doran, 2013). For Rushton turbine impellers, the following correlation (for D_i/D_T=0.33) was suggested in aerated systems (Chapman et al., 1986b):

$$N_{JSg} = N_{JS} + 2.4F_g$$

where N_{JSg} is the critical speed in the presence of aeration (1/second); and, F_g is the gas flow rate (vvm) i.e. volume of gas per volume of liquid per minute.

Critical speeds required for suspension of solids with and without aeration under the conditions tested were also calculated using these empirical correlations. It was found that these correlations underestimate the critical speed by 10-20%.

4 CONCLUSIONS

This study presents the evaluation of comparative performance of Rushton turbine (RT) and 45° pitched blade turbine (PBT) impellers for suspension of solids in two and three phase systems. The critical speed required for suspension of solids was found to increase with increasing solids loading (5-30% w/w), air flow rate (0-0.6 L/min) and particle size. In the absence of aeration, PBT impellers have higher capacity for solids suspension than RT impellers. However, The PBT was observed to have lower gas-handling capability than the RT. The critical speed for solids suspension in two or three phase systems can be estimated by using empirical correlations; albeit, these estimates are 10-20% lower than the experimental data.

ACKNOWLEDGEMENTS

The author would like to express his sincere thanks and appreciation to the late Dr. CV Phillips, G. Griffits and P. Holman (Camborne School of Mines) for their help and, to the Turkish Ministry of Education for financial support during his studies at Camborne School of Mines.

REFERENCES

Acevedo, F., Cacciuttolo, M.A. and Gentina, J.C. 1988. Comparative performance of stirred and pachuca tanks in the bioleaching of a copper concentrate. *Biohydrometallurgy: Proc. of the Int. Symp.,* Norris P.R. and Kelly D.P. (eds.), Warwick, 1987, p.385-394.

Bailey, A.D. and Hansford, G.S., 1993. Factors affecting the biooxidation of sulphide minerals at high concentrations of solids: A review. *Biotechnology and Bioengineering,* 12(10), p.1164-1174.

Chapman, C.M., Nienow, A.W., Cooke, M., and Middleton, J.C., 1983a. Particle-gas-liquid mixing in stirred vessels: Part II –Gas-liquid mixing. *Chem. Eng. Res. Des.*, 61, p.82-95.

Chapman, C.M., Nienow, A.W., Cooke, M., and Middleton, J.C., 1983b. Particle-gas-liquid mixing in stirred vessels: Part I –Particle-liquid mixing. *Chem. Eng. Res. Des.*, 61, p.71-81.

Deveci, H., 2002. Effect of Solids on Viability of Acidophilic Bacteria. *Minerals Engineering*, 15, p.1181-1189,.

Deveci, H., 2004. Effect of Particle Size and Shape of Solids on the Viability of Acidophilic Bacteria During Mixing in Stirred Tank Reactors. *Hydrometallurgy*, 71(3-4), p.385-396.

Doran, M.P., 2013. *Bioprocess Engineering Principles* (Second Edition). Elsevier, Oxford, 899 p. (ISBN-10: 012220851X).

Fraser, G.M., 1984. Aeration design for gold ore leaching. *Regional Conference on Gold Mining, Metallurgy and Geology*, Melbourne, The Aust. IMM, p.245-255.

Fraser, G.M., 1992. Gas dispersion and mixing for mineral oxidation reactors. *Int. Conference on Extractive Metallurgy of Gold and Base metals*, Misra V.N., Halbe D. and Spottiswood D.J. (eds), Kalgoorlie, W.A., The Aust. I.M.M, pp 293-301.

Gormerly, L.S., and Branion, R.M.R., 1989. Engineering design of microbiological leaching reactors. *Biohydrometallurgy,* Salley J., McCready R.G.L. and Wichlacz P.L (eds.), CANMET SP89-10, pp 499-518.

Harnby, N., Edwards, M.F., and Nienow, A.W., 1992. *Mixing in the process industries.* Butterworth-Heinemann, Oxford, UK, 394-411p.

Hayward, T., Satalic, D.M. and Spencer, P.A., 1997. Engineering, equipment and materials: Developments in the design of a bacterial oxidation reactor. *Minerals Engineering*, 10 (10), p.1047-1055.

Kubera, P.M., and Oldshue, Y., 1992. Advanced impeller technologies to match mixing performance to process needs. *Randol Gold Forum*, Vancouver 92, p.279-285.

Neale, J.W., and Pinches, A., 1994. Determination of gas-liquid mass transfer and solids suspension parameters in mechanically agitated three phase slurry reactor. *Minerals Engineering*, 7(2/3), p.389-403.

Oldshue, J.Y., 1983. *Fluid Mixing Technology.* McGraw-Hill, New York, 547 p.

Oldshue, J.Y., Post, T.A., Weetman, R.J., and Coyle, C.K., 1988. Comparison of mass transfer characteristics of radial and axial flow impellers. *6th European Conference on Mixing*, Pavia, Italy, May 24-26, p.345-350.

Rossi, G., 1999. The design of bioreactors. *Biohydrometallurgy and the Environment toward the Mining of the 21st Century, Part A, IBS`99*, Amils R. and Ballester A. (eds), Madrid, Spain, p.61-80.

Weetman, R.J. and Oldshue, J.Y., 1988. Power, flow and shear characteristics of mixing impellers. *6th European Conference on Mixing*, Pavia, Italy, May 24-26, p.43-50.

Zwietering, T.N., 1958. Suspending of solid particles in liquid by agitators. *Chemical Engineering Science*, 8, p.244-253.

Leaching of Metals from Flotation Tailings of a Copper Smelter Slag in Acidic Solutions

Y. Yiğit, M. Kuzu, E.Y. Yazici, O. Celep, H. Deveci

Hydromet B&PM Research Group, Div. of Mineral&Coal Processing, Dept. of Mining Engineering, Karadeniz Technical University, Trabzon, Turkey

ABSTRACT This study investigated leaching of metals i.e. Cu, Co and Zn from flotation tailings of a copper smelter slag using sulphuric acid solutions in the absence/presence of hydrogen peroxide. The as-received tailings (d_{80}: 70 μm) contained 1.25% Cu, 1.26% Co, 3.80% Zn and 36.5% Fe. The effect of addition of hydrogen peroxide (149 kg H_2O_2 per tonne solids) and particle size (d_{80}: 27-70 μm) was tested on extraction of metals over a period of 24 h. The addition of H_2O_2 slightly improved the leaching of copper by 6.3%. Fine grinding of the tailings from 70 to 27 μm enhanced the extraction of copper, zinc and iron by 18.9%, 8.2% and 22.1%, respectively, over 8 h. of leaching. High extractions for metals (100% Cu and 86.3% Zn) except cobalt (≤10.7%) were achieved under suitable conditions. The addition of H_2O_2 or fine grinding did not contribute to cobalt leaching.

1.INTRODUCTION

Recovery of base and precious metals from urban/industrial/metallurgical waste is of economic and environmental importance (Celep, 2016; Yaylalı, 2016; Yazici and Deveci, 2015). Among various waste streams, copper slag is also an important source for copper as well as other metals e.g., cobalt, nickel, zinc and silver/gold (Fabian et al., 1997; Gbor et al., 2000; Gorai et al., 2003; Sánchez et al., 2004; Shen and Forssberg, 2003). Composition, type and morphology of phases in slag are strongly dependent on smelting and cooling conditions. Copper slag is composed of mainly iron (30-40%) and silica (35-40%) (Sánchez et al., 2004). Iron content may reach up to 50% (Shen and Forssberg, 2003). Copper can be found in metallic, oxide and/or sulphide form (Gbor et al., 2000; Shen and Forssberg, 2003). Common sulphide forms of copper present in slags are covellite (CuS), chalcocite (Cu_2S), bornite (Cu_5FeS_4), stromeyerite (AgCuS) and complex sulphides (Deng and Ling, 2004; Shen and Forssberg, 2003). Cobalt, nickel, zinc and molybdenum in copper slag –if present– are found dominantly as oxides due their relatively higher affinity to oxygen when compared with copper (Gbor et al., 2000; Uzkut and Tunçer, 1981). Cobalt in copper slag can be also present in metallic as well as sulphide, silicate and ferrite form (Deng and Ling, 2007). In addition to its economic potential, disposal of copper slag into environment may result in contamination of surface and/or underground water with hazardous heavy metals (e.g., Cu, Cd, Pb and Zn) through leaching particularly under acidic conditions (Jarošíková et al., 2017; Potysz et al., 2016). Treatment of copper slag for the recovery of metals is therefore required to prevent its possible environmental risk and to exploit its economic potential. Pyrometallurgical methods (carbothermic reduction followed by re-smelting) can be used for further treatment of copper slag for the recovery of metals (Gorai et al., 2003). Other treatment options include flotation, leaching and roasting+leaching (Sánchez et al., 2004; Shen and Forssberg, 2003).

Several researchers investigated leaching of copper slag for recovery of copper, cobalt and other metals in sulphate, chloride, ammonia, cyanide and other lixiviant systems (Banza et al., 2002; Bulut, 2006; Çakır et al., 2013; Deng and Ling, 2007; Kaksonen et al., 2016b; Shen and Forssberg, 2003; Urosevic et al., 2015; Ziyadanogullari, 2000). Bioleaching can be also an alternative for recovery of metals from slags (Kaksonen et al., 2016a; Kaksonen et al., 2016b). Pre-treatment of copper slag using sulphating roasting using different sulphate sources (e.g. ferric sulphate, ammonium sulphate, sulphuric acid, pyrite and SO_2) can be successfully applied to convert sulphide phases to soluble sulphate compounds prior to water or dilute acid leaching (Altundoğan and Tümen, 1997; Arslan and Arslan, 2002; Bulut, 2006; Çakır et al., 2013; Deng and Ling, 2007; Dimitrijevic et al., 2016; Tümen and Bailey, 1990; Ziyadanogullari, 2000). Roasting with ammonium chloride followed by water leaching was also tested by Nadirov et al. (2013).

Flotation of copper slag is similar with sulphide flotation of copper due to the presence of copper largely in the form of sulphide (Roy et al., 2015). In the flotation of copper slag, oxide minerals containing Co and Ni are suppressed and concentrated into the tailings (Shen and Forssberg, 2003; Uzkut and Tunçer, 1981). Flotation tailings of copper slag may contain copper as well as significant amount of cobalt and other metals. Some researchers studied the extraction of metals from flotation tailings of copper slag by sulphuric acid leaching in the presence of oxidants such as H_2O_2 and Fe(III) (Carranza et al., 2009; Muravyov et al., 2012; Urosevic et al., 2015). It is relevant to note that cobalt-bearing tailings is of particular importance due to the fact that cobalt is now classified as a "critical raw material" by European Union (EU) based on its relatively high economic importance and supply risk (EC, 2014). Therefore, the presence of cobalt may add to the economic value of flotation tailings of copper slag.

In this study, extraction of copper, cobalt and zinc from the flotation tailings of a copper smelter slag in sulphuric acid solutions under oxidising conditions was investigated. Effects of addition of hydrogen peroxide (149 kg H_2O_2 per tonne solids) and particle size (d_{80}: 27-32-70 μm) were investigated on the leaching of metals (i.e. Cu, Co, Zn and Fe) over a period of 24 h.

2.MATERIAL AND METHOD

In the leaching tests, a sample of flotation tailings of a copper smelter slag was used. The tailings sample was kindly provided by Eti Bakır Co. (Samsun, Turkey). The as-received sample (d_{80}: 70 μm) was analysed to contain 1.25% Cu, 1.26% Co, 3.80% Zn and 36.5% Fe. Earlier XRD analysis had shown that fayalite (Fe_2SiO_4), magnetite (Fe_3O_4) and SiO_2 are the main phases present in the tailings sample (Alp et al., 2008)

Concentrated sulphuric acid (96% H_2SO_4) and hydrogen peroxide (30% H_2O_2) were used to prepare leach solutions. In the tests, the concentration of H_2SO_4 and solids ratio were kept constant at 1 M and 10% w/v, respectively. The leaching tests were carried out in Erlenmeyer flasks. After the addition of leach solution (200 mL) and tailings sample, the flasks was placed on an orbital shaker operating at 190 rpm and 50°C.

The effect of addition of hydrogen peroxide (E^0=1.78 V) as the oxidant on extraction of metals was examined. Hydrogen peroxide (30% H_2O_2) was dosed into the leach solutions as follows: 5 mL H_2O_2 was added at the start-up with further dosing of 1 mL H_2O_2 at the predetermined intervals of 0.5, 2, 4 and 8 h., amounting to a total H_2O_2 addition of 9 mL, which is equivalent to 149 kg H_2O_2 per tonne solids. A separate leaching test was also performed using 1 M EDTA (ethylenediamine tetraacetic acid) solution (~pH 8) in the presence of hydrogen peroxide (495 kg H_2O_2 per tonne solids) using the as-received sample (d_{80}: 70 μm).

The effect of particle size (d_{80}: 27-70 μm) on the extraction of metals was also investigated. A tema mill was used for grinding the as-received sample down to the required fineness of grind. Analysis of particle size distribution was performed using a Malvern Mastersizer Hydro 2000 MU. During the leaching period of 24 h. samples were taken at predetermined intervals (0.5, 2, 4, 8 and 24 h.) and these samples were centrifuged to obtain clear supernatants for analysis of copper by Atomic Absorption Spectrometer (Perkin Elmer AAnalyst 400). Analysis of cobalt, zinc and iron was also performed using samples taken at 8 and 24 h. Evolution of redox potential (mV *vs.* Ag/ACl) was monitored over the leaching period.

3. RESULTS AND DISCUSSION

Figure 1 shows the effect of hydrogen peroxide (149 kg/t H_2O_2) on the rate and extent of copper leaching over the period of 24 h. The addition of hydrogen peroxide slightly improved the extraction of copper by 6.3 (i.e. from 82.1 to 88.4%) over 24 h. (Fig. 1). High extraction of copper (i.e. 82.1%) in the absence of H_2O_2 could be an indicative of presence of copper mainly in oxide form and, to a less extent, in sulphide form (e.g. Cu_2S). Copper oxide and Cu(I)-sulphide are soluble in acidic solutions (Eq. 1). Dissolution of metallic copper and some secondary copper sulphide minerals (e.g. CuS) is not thermodynamically feasible in the absence of an oxidant (Eqs. 2-7). It is pertinent to note that the previous work (Uzkut and Tunçer, 1981) on a copper slag (originated from the same plant as per the slag sample used in the current study) reported that most of copper (65%) in the slag is in the form of cuprite (Cu_2O) with metallic copper (25%) and chalcocite (Cu_2S, 10%).

$$3Cu_2O + 7H_2SO_4 \rightarrow 6CuSO_4 + 7H_2O + S \quad (\Delta G^0_{(50°C)} = -351 \text{ kJ}) \tag{1}$$

$$Cu + H_2SO_4 \rightarrow CuSO_{4\,(aq)} + H_2(g) \quad (\Delta G^0_{(50°C)} = +25.3 \text{ kJ}) \tag{2}$$

$$CuS + 1.3\ H_2SO_4 \rightarrow CuSO_{4(aq)} + 1.3\ H_2O + 1.3\ S \quad (\Delta G^0_{(50°C)} = +1.2 \text{ kJ}) \tag{3}$$

$$Cu_2S + 2.7\ H_2SO_4 \rightarrow 2CuSO_{4(aq)} + 2.7\ H_2O + 1.7\ S \quad (\Delta G^0_{(50°C)} = -24 \text{ kJ}) \tag{4}$$

$$Cu + H_2O_2 + H_2SO_4 \rightarrow CuSO_{4\,(aq)} + 2H_2O \quad (\Delta G^0_{(50°C)} = -326 \text{ kJ}) \tag{5}$$

$$CuS + H_2O_2 + H_2SO_4 \rightarrow CuSO_{4(aq)} + 2H_2O + S \quad (\Delta G^0_{(50°C)} = -269 \text{ kJ}) \tag{6}$$

$$Cu_2S + 5H_2O_2 + H_2SO_4 \rightarrow 2CuSO_{4(aq)} + 64H_2O \quad (\Delta G^0_{(50°C)} = -1378 \text{ kJ}) \tag{7}$$

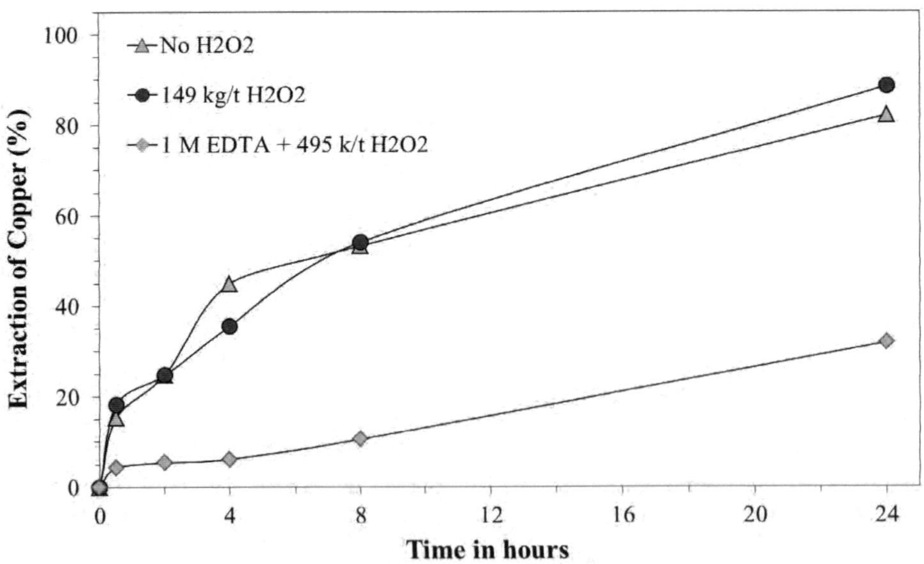

Figure 1. Effect of hydrogen peroxide (149 kg/t H_2O_2) on leaching kinetics of copper using the as-received material (1 M H_2SO_4, d_{80}: 70 µm, 50°C)

The presence of hydrogen peroxide did not contribute significantly (i.e. ≤1%) to cobalt extraction (Fig. 2a,b). It was consistently limited to 10.7% even at a prolonged period of 24 h. (Fig. 2b). Similarly, the addition of hydrogen peroxide did not improve the leaching performance of zinc, which is already soluble in acid (Fig. 2a,b). A major part of zinc with 78.1-81.1% was already extracted at 8 h. (Fig. 2a). In contrast to cobalt and zinc, the dissolution of iron increased by 8.7% (from 63.8 to 72.5% at 8 h) (Fig. 2a) and by 15.3% (from 84.7 to 100% at 24 h) in the presence of hydrogen peroxide (Fig. 2b).

It should be noted that the presence of hydrogen peroxide (Figs. 1-2) would have also led to oxidation of Fe(II) to Fe(III), which is a strong oxidant (E^0=0.77 V), and hence, contributed to leaching by activating Fe(III)/Fe(II) redox couple.

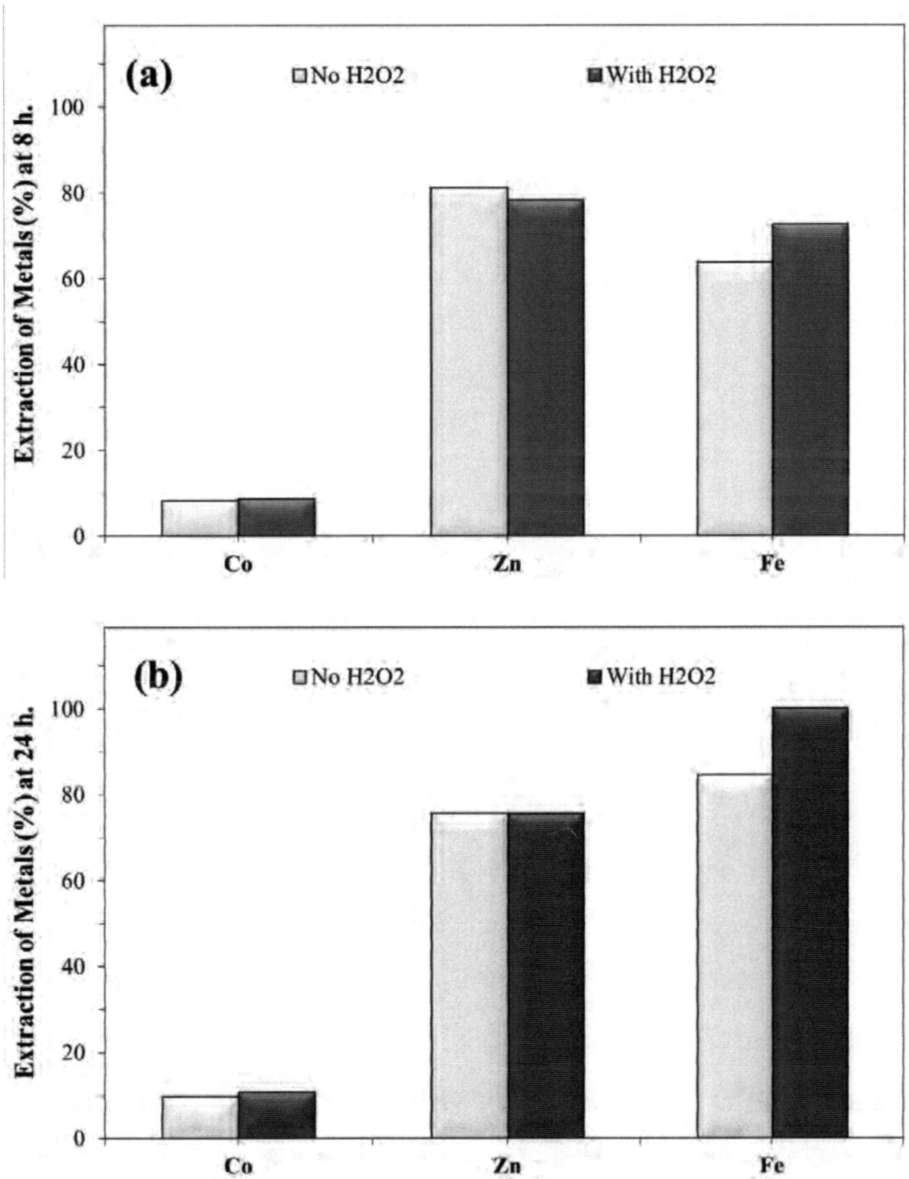

Figure 2. Effect of hydrogen peroxide (149 kg/t H_2O_2) on extraction of cobalt, zinc and iron using the as-received material (d_{80}: 70 μm) at 8 h. (a) and 24 h. (b) (1 M H_2SO_4, 50°C)

Urosevic et al. (2015) studied acidic sulphate leaching of copper and iron from a flotation tailings (-70 μm) of copper slag in the presence of H_2O_2 and $Fe_2(SO_4)_3$ as oxidants. Consistent with the current findings (Figs. 1-2), they found that increasing the concentration of H_2O_2 (0.5-3 M) improved leaching performance of metals. Maximum extraction of metals (copper and iron) was limited to about 60% 2 M H_2O_2 and 50°C (Urosevic et al., 2015). These researchers concluded that H_2O_2 is more effective than Fe(III) under the conditions tested. Çakır et al. (2013) showed sulphuric acid (1 M H_2SO_4) leaching of a flotation tailings of copper slag for 1 h. at 100°C. They reported a limited extraction of copper (≈12%). These investigators also demonstrated that pre-treatment of the flotation tailings by sulphating roasting (using Na_2SO_4) followed by water leaching enhanced the extraction of copper to 60%.

EDTA as a chelating agent forms strong complexes with copper, iron and cobalt. A leaching test using EDTA (1 M) in the presence of 495 kg/t H_2O_2 was also carried out (Fig. 1). Leaching of copper in EDTA solution was determined to be 31.8% Cu over 24 h. (Fig. 1). A limited extraction of cobalt to only 10.4% over 24 h. was also observed. In this test, no dissolution for zinc and iron occurred (data not shown).

Figure 3 shows the effect of particle size on the rate and extent of extraction of copper. Decreasing the particle size (d_{80}) from 70 to 27 μm led to an apparent increase in the rate and extent of extraction of copper particularly after the leaching period of 2 h. The extraction of copper was complete from 27-μm sample over 24 h. compared with 88.4% Cu from coarser sample of 70 μm (Fig. 3).

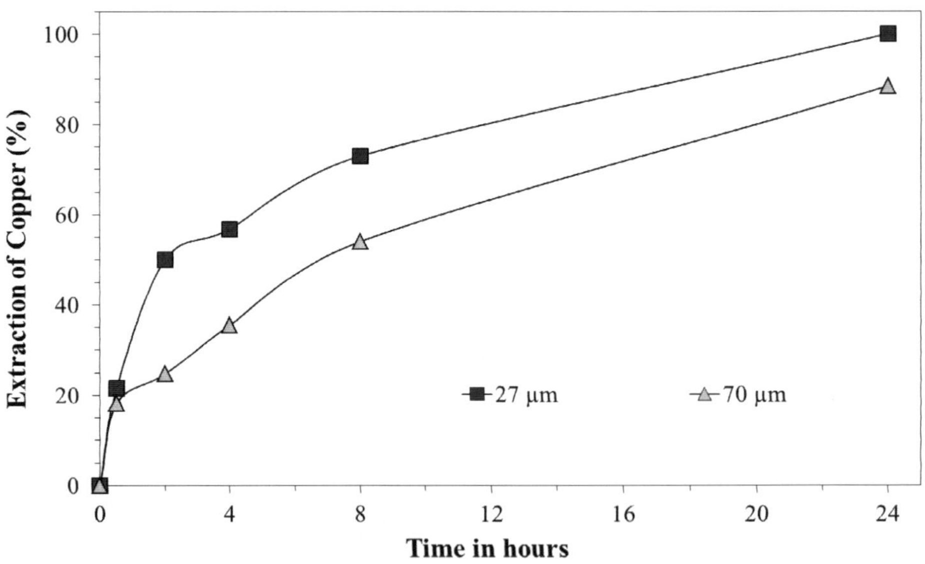

Figure 3. Effect of particle size (d_{80}) on extraction of copper in the presence of hydrogen peroxide (1 M H_2SO_4, 149 kg/t H_2O_2, 50°C)

Figure 4 shows the effect of fineness of the flotation tailings on the extraction of other metals (i.e. Co, Zn and Fe) over the period of 8 and 24 h. Fine grinding of the tailings from 70 to 27 μm did not affect the extraction of cobalt, which did not exceed 10.7% over the leaching period of 24 h. (Fig. 4a,b). An improved extraction of zinc by 8.2% (i.e. from 78.1 to 86.3% over 8 h.) was noted (Fig. 4a). The leaching of iron was also found to increase by 22.1% (Fig. 4a). It can be inferred from these results (Figs. 3-4) that fine grinding down to 27 μm is required to enhance the rate and extent of extraction of metals (except cobalt) from the tailings.

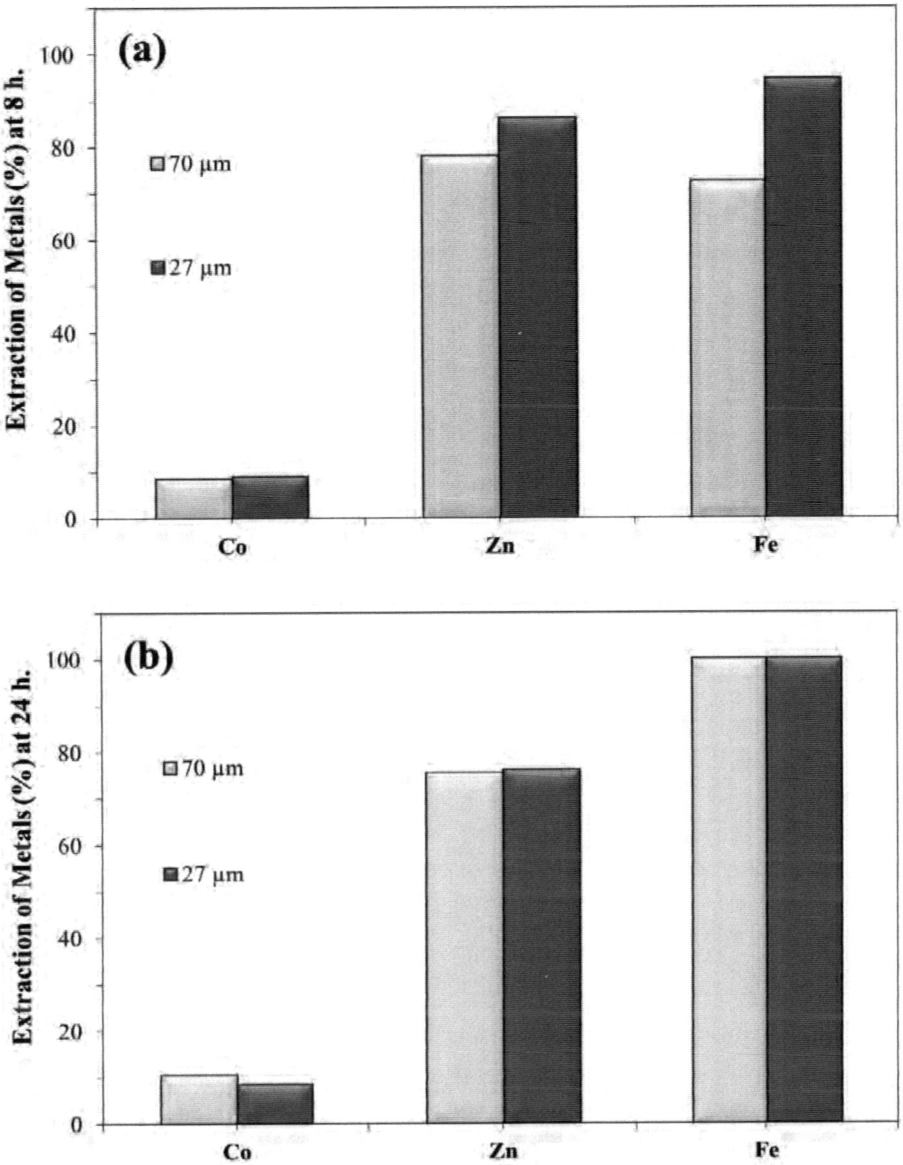

Figure 4. Effect of particle size (d_{80}) on extraction of cobalt, zinc and iron in the presence of hydrogen peroxide (149 kg/t H_2O_2) at 8 h. (a) and 24 h. (b) (1 M H_2SO_4, 50°C)

Cobalt in silicate/ferrite form is commonly found in fayalitic slags like the tailings material used in the current study. Cobalt silicate ($CoO.SiO_2$) and/or ferrite ($CoO.Fe_2O_3$) is soluble in highly concentrated sulphuric acid solutions (Eqs. 8-9) (Arslan and Arslan, 2002; Bulut, 2006). Cobalt oxides containing Co(II) are also soluble in acidic solutions. Low extraction of cobalt (i.e. ≤10.7%) obtained in the current study (Figs. 2-4) could be linked with the presence of cobalt as cobalto-cobaltic oxide (Co_3O_4; $Co^{2+}O.Co^{3+}{}_2O_3$), which needs a reducing agent for conversion of Co(III) to Co(II) for effective dissolution (Hubli et al., 1997). In addition, it has been reported that copper, nickel and cobalt can be physically entrained in slag at fine particle sizes (Sun, 2006). This could also lead to low cobalt extractions by preventing the contact of the metal/mineral with leach solution at coarse particle sizes. Further studies are required to examine low cobalt extractions from the tailings in more detail.

$$(2CoO)SiO_2 + 2H_2SO_4 \rightarrow 2CoSO_{4(aq)} + SiO_2 + 2H_2O \quad (\Delta G^0_{(50°C)} = -205 \text{ kJ}) \tag{8}$$

$$CoOFe_2O_3 + 4H_2SO_4 \rightarrow CoSO_{4(aq)} + Fe_2(SO_4)_3 + 4H_2O \quad (\Delta G^0_{(50°C)} = -212 \text{ kJ}) \tag{9}$$

Redox potentials (mV *vs.* Ag/AgCl) were also monitored in the leaching tests (Figs. 1-2). Figure 5 shows the redox potential profiles in the absence and presence of hydrogen peroxide (149 kg/t H_2O_2) under the conditions of 1 M H_2SO_4 at 50°C. Consistent with the improved extraction of metals, particularly for copper and iron, by the addition of hydrogen peroxide (Figs. 1-2), apparently higher potentials were maintained in the presence of hydrogen peroxide e.g. 357-368 mV (with H_2O_2) *vs.* 296-337 mV (no H_2O_2) between 0.5-8 h. (Fig. 5). It is relevant to note that high temperature (particularly above 50°C) as well as presence of metal ions (e.g. Cu and Fe) even in low concentrations facilitates decomposition of hydrogen peroxide resulting in its excess consumption (Yazıcı and Deveci, 2010). Regarding the high concentration of metals (e.g. ≤1.5 g/L Cu) in the pregnant leach solutions obtained at 50°C in the current study, it can be inferred that a part of hydrogen peroxide was decomposed and hence, wasted by side reactions. This suggests that the addition of H_2O_2 (i.e. redox potential) more than 149 kg/t H_2O_2 should be tested to achieve high extraction of metals particularly for cobalt.

It is interesting to note that, in EDTA leaching, despite relatively higher addition of hydrogen peroxide by 3.3-fold when compared with H_2SO_4+H_2O_2 leaching, negative potentials were noted over the period of 0.5-8 h. (Fig. 5).

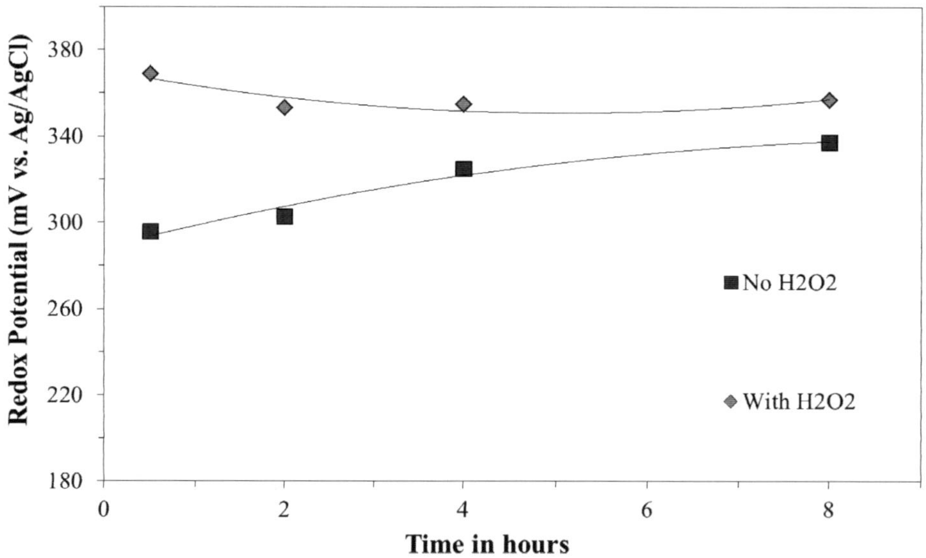

Figure 5. Redox potential (mV *vs.* Ag/AgCl) profiles in the absence and presence of hydrogen peroxide (1 M H_2SO_4, 149 kg/t H_2O_2, 50°C, d_{80}: 70 μm)

4. CONCLUSIONS

Extraction of Cu, Co and Zn from the flotation tailings of a copper smelter slag in sulphuric acid solutions was studied in this study. The influence of addition of hydrogen peroxide (149 kg H_2O_2 per tonne solids) and grinding of the tailings (d_{80}: 27-70 μm) on metal extraction was demonstrated. The presence of hydrogen peroxide improved the rate and extent of extraction of copper from 82.1 to 88.4% over 24 h. using the as-received tailings (d_{80}: 70 μm). No effect of H_2O_2 on leaching of zinc and cobalt was observed. However, the dissolution of iron increased by 15.3% and reached %100 in the presence of H_2O_2.

Extraction of metals increased from 54.1 to 72.9% Cu, from 78.1 to 86.3% Zn and from 72.5 to 94.6% Fe over 8 h. with reducing the particle size from 70 to 27 μm. Complete extraction of copper was achieved from the 27-μm sample over 24 h. Despite high leaching performance of copper (100%) and zinc (86.3%) the extraction of cobalt was consistently limited to ≤10.7% even for the finest particle size tested (d_{80}: 27 μm).

ACKNOWLEDGEMENTS

The authors would like to express their sincere thanks and appreciation to the Eti Bakır Co. for kindly providing the samples.

REFERENCES

Alp, İ., Deveci, H., Süngün, H., 2008. Utilization of flotation wastes of copper slag as raw material in cement production. Journal of Hazardous Materials 159, 390-395.

Altundoğan, H.S., Tümen, F., 1997. Metal recovery from copper converter slag by roasting with ferric sulphate. Hydrometallurgy 44, 261-267.

Arslan, C., Arslan, F., 2002. Recovery of copper, cobalt, and zinc from copper smelter and converter slags. Hydrometallurgy 67, 1-7.

Banza, A.N., Gock, E., Kongolo, K., 2002. Base metals recovery from copper smelter slag by oxidising leaching and solvent extraction. Hydrometallurgy 67, 63-69.

Bulut, G., 2006. Recovery of copper and cobalt from ancient slag. Waste Management & Research 24, 118-124.

Carranza, F., Iglesias, N., Mazuelos, A., Romero, R., Forcat, O., 2009. Ferric leaching of copper slag flotation tailings. Minerals Engineering 22, 107-110.

Celep, O., Altınkaya, P., Yazici, E.Y., Deveci, H., 2016. Extraction of base and precious metals from scrap segments used in marble cutting, in: M.S.Çelik, G.Bulut, F.Karakaş, O.Güven, H.Baştürkçü, Z.Tarsus (Eds.), XV. International Mineral Processing Symposium and Exhibition (IMPS), İstanbul, pp. 880-892.

Çakır, M., Kartal, M., Gül, H., Taşkın, E., Uysal, M., Aydın, A.O., Alp, A., 2013. Bakır Rafinasyon Curufu Flotasyon Atıklarındaki Bakırın Geri Kazanımı, 1st International Symposium on Innovative Technologies in Engineering and Science, Sakarya University, 7-9 June, Sakarya, pp. 200-207.

Deng, T., Ling, Y., 2004. Processing of copper converter slag for metals reclamation: Part II: mineralogical study. Waste Management & Research 22, 376-382.

Deng, T., Ling, Y., 2007. Processing of copper converter slag for metal reclamation. Part I: extraction and recovery of copper and cobalt. Waste Management & Research 25, 440-448.

Dimitrijevic, M.D., Urosevic, D.M., Jankovic, Z.D., Milic, S.M., 2016. Recovery of copper from smelting slag by sulphation roasting and water leaching. Physicochem. Probl. Mineral Pro. 52, 409-421.

EC, 2014. Report on critical raw materials for the EU - Report of the Ad hoc Working Group on defining critical raw materials. European Commission (EC), p. 41 pp.

Fabian, H., Richardson, H.W., Habashi, F., Besold, R., 1997. Copper, in: Habashi, F. (Ed.), Handbook of Extractive Metallurgy. Wiley-VCH, Toronto, pp. 491-579.

Gbor, P.K., Mokri, V., Jia, C.Q., 2000. Characterization of smelter slags. Journal of Environmental Science and Health, Part A 35, 147-167.

Gorai, B., Jana, R.K., Premchand, 2003. Characteristics and utilisation of copper slag—a review. Resources, Conservation and Recycling 39, 299-313.

Hubli, R.C., Mittra, J., Suri, A.K., 1997. Reduction-dissolution of cobalt oxide in acid media: a kinetic study. Hydrometallurgy 44, 125-134.

Jarošíková, A., Ettler, V., Mihaljevič, M., Kříbek, B., Mapani, B., 2017. The pH-dependent leaching behavior of slags from various stages of a copper smelting process: Environmental implications. Journal of Environmental Management 187, 178-186.

Kaksonen, A.H., Särkijärvi, S., Peuraniemi, E., Junnikkala, S., Puhakka, J.A., Tuovinen, O.H., 2016a. Metal biorecovery in acid solutions from a copper smelter slag. Hydrometallurgy.

Kaksonen, A.H., Särkijärvi, S., Puhakka, J.A., Peuraniemi, E., Junnikkala, S., Tuovinen, O.H., 2016b. Chemical and bacterial leaching of metals from a smelter slag in acid solutions. Hydrometallurgy 159, 46-53.

Muravyov, M.I., Fomchenko, N.V., Usoltsev, A.V., Vasilyev, E.A., Kondrat'eva, T.F., 2012. Leaching of copper and zinc from copper converter slag flotation tailings using H2SO4 and biologically generated Fe2(SO4)3. Hydrometallurgy 119–120, 40-46.

Nadirov, R.K., Syzdykova, L.I., Zhussupova, A.K., Usserbaev, M.T., 2013. Recovery of value metals from copper smelter slag by ammonium chloride treatment. International Journal of Mineral Processing 124, 145-149.

Potysz, A., Kierczak, J., Fuchs, Y., Grybos, M., Guibaud, G., Lens, P.N.L., van Hullebusch, E.D., 2016. Characterization and pH-dependent leaching behaviour of historical and modern copper slags. Journal of Geochemical Exploration 160, 1-15.

Roy, S., Datta, A., Rehani, S., 2015. Flotation of copper sulphide from copper smelter slag using multiple collectors and their mixtures. International Journal of Mineral Processing 143, 43-49.

Sánchez, M., Parada, F., Parra, R., Marquez, F., Jara, R., Carrasco, J.C., Palacİos, J., 2004. Management of copper pyrometallurgic slags: giving additional value to copper mining industry, VII. International Conference on Molten Slags Fluxes and Salts, The South African Institute of Mining and Metallurgy, Capte Town.

Shen, H., Forssberg, E., 2003. An overview of recovery of metals from slags. Waste Management 23, 933-949.

Sun, H., 2006. Investigation of physically entrained matte in the flash furnace slag. University of Toronto, Canada, p. 134.

Tumen, F., Bailey, N.T., 1990. Recovery of metal values from copper smelter slags by roasting with pyrite. Hydrometallurgy 25, 317-328.

Urosevic, D., Dimitrijevic, M.D., Janković, Z.D., Antic, D., 2015. Recovery of copper from copper slag and copper slag flotation tailings by oxidative leaching. Physicochem. Probl. Miner. Process. 51, 73-82.

Uzkut, İ., Tunçer, F., 1981. KBİ Samsun izabe tesisi cüruf ve flotasyon artıklarının metal içeriği ve değerlendirilmesi, Türkiye VII. Madencilik Kongresi Ankara, pp. 295-318.

Yaylalı, B., Yazici, E.Y., Celep, O, Deveci, H.,, 2016. Extraction of cobalt from a flotation tailings in different mineral acids under oxidative conditions, in: M.S.Çelik, G.Bulut, F.Karakaş, O.Güven, H.Baştürkçü, Z.Tarsus (Eds.), XV. International Mineral Processing Symposium and Exhibition (IMPS), İstanbul, pp. 726-736.

Yazici, E.Y., Deveci, H., 2015. Cupric chloride leaching (HCl–CuCl2–NaCl) of metals from waste printed circuit boards (WPCBs). International Journal of Mineral Processing 134, 89-96.

Yazıcı, E.Y., Deveci, H., 2010. Factors Affecting Decomposition of Hydrogen Peroxide, XII. International Mineral Processing Symposium (IMPS), Kapadokya, 6-8 October, pp. 609-616.

Ziyadanogullari, B., 2000. Recovery of Copper and Cobalt from Concentrate and Converter Slag. Separation Science and Technology 35, 1963-1971.

Effect of High Concentration of Ammonia in Cyanide Leaching of a Copper-Gold Ore

A. D. Bas [1,2], E.Y. Yazici [1], O. Celep [1], H. Deveci [1]

[1] *Hydromet B&PM Research Group, Div. of Mineral&Coal Processing, Dept. of Mining Engineering, Karadeniz Technical University, Trabzon, Turkey*

[2] *Department of Mining, Metallurgical and Materials Engineering, Laval University, Quebec, Canada*

ABSTRACT Cyanide leaching of gold ores that contain high copper (≥0.5% Cu) is often not economic due to excessively high consumption of cyanide and low gold recoveries. This study investigated the effect of high concentration of ammonia (3-4 M NH_3) at different strengths of cyanide solutions (1.5-2.5 g/L NaCN) to improve the extraction of gold from a copper rich gold-ore (56 g/t Au, 10 g/t Ag, 1.10% Cu). In direct cyanide leaching of the ore at typical cyanide conditions (1.5 g/L NaCN, no ammonia) only 6.7% of gold was extracted over 24 h. However, the addition of ammonia (≥3 M NH_3) at the same concentration of cyanide (1.5 g/L NaCN) induced a substantial increase in gold extraction, which was almost complete over 24 h. Increasing the concentration of ammonia from 3 to 4 M resulted in an increase in the dissolution of copper from 32.4 to 52.3% over the same period.

1.INTRODUCTION

Cyanide, which has been used in gold extraction for 125 years is by far the most suitable and effective reagent for leaching of gold and silver from ores (Adams, 2016; Marsden and House, 2006). Cyanide leaching is not a selective process and some impurities such as copper and arsenic also dissolve and present problems during leaching and downstream processes (Bas et al., 2011; Hayes and Corrans, 1992; Muir, 2011). The presence of copper minerals interferes with cyanide leaching of gold resulting in unacceptably low gold extractions and high consumption of cyanide (Alymore and Muir, 2001; Breuer et al, 2005; Dawson et al, 1997; Sceresini, 2005). Most copper minerals readily react with cyanide, consuming up to 30 kg/tonne NaCN for each 1% copper reagent present (Muir, 2011). Such a high consumption of cyanide can even render the process uneconomic. Thus, alternative process options to direct cyanide leaching should be developed for the treatment of copper bearing gold ores. In this regard, several prosesses have been proposed with the main goal of either removing copper prior to cyanidation or suppressing copper dissolution. These include flotation of copper minerals (as well as gold) to produce a concentrate (Bulatovic, 1998; Forrest et al, 2001), alternative lixiviants such as thiosulphate (Bas et al. 2011), leaching of copper minerals prior to cyanidation (Dawson et al, 1997; Muir et al, 1991).

Proposed by Hunt in 1901, the addition of ammonia to suppress copper dissolution and mitigate copper interferences in cyanide leaching of gold has been known for over 100 years (Jeffrey et al., 2002; Muir, 2011; Sceresini, 2005). The details of the Hunt process and its commercial applications were reviewed in the recent work of Hedjazi and Monhemius (2016). Ammonia is added to stabilize Cu^{2+} (Eq. 1) which can act as an oxidant, thus potentially increases the rate of gold dissolution. In the absence of ammonia, Cu^{+} tends to form strong complexes with cyanide (Eq. 2) thus leading an increase in the consumption of cyanide (Hayes et al, 1992). The enhancing effect of ammonia on the extraction of gold (Eq. 3) and reducing the consumption of cyanide has been reported by different research groups (Breuer et al, 2005; Hayes and Corrans, 1992; Muir, 2011; Dai et al, 2012). It has also been reported that the main role of ammonia in ammonia-cyanide system is to suppress the dissolution of copper by forming a precipitate as $Cu_3(NH_3)_4(CN)_4$ (Drok and Ritchie, 1997; Muir et al. 1993).

$$Cu^{2+} + 4NH_3 \rightarrow Cu(NH_3)_4^{2+} \quad (1)$$
$$Cu^{+} + 3CN^{-} \rightarrow Cu(CN)_2^{-} + 0.5(CN)_2^{-} \quad (2)$$
$$Au^{0} + Cu(NH_3)_2(CN)_2 \rightarrow Cu(NH_3)_2^{+} + Au(CN)_2^{-} \quad (3)$$

In this study, ammoniacal cyanide leaching of gold for the treatment of a copper-rich refractory gold ore was investigated. Effect of high concentrations of ammonia and cyanide on the extraction of gold and dissolution of copper was demonstrated. Direct cyanide leaching of gold was carried out for comparison.

2.MATERIAL AND METHOD

The ore sample used in this study was taken from sulphide-rich zones of Mastra Gold Mine, located in Gümüşhane, Turkey. The sample was crushed down to -4 mm by using jaw and roll crushers in sequence. The crushed ore sample was rod-milled to 80% passing -75 μm (d_{80}). The ground ore was divided into 120 g sub-samples for use in the leaching tests.

The ore sample is the most abundant in quartz (85% SiO_2 as quartz and other silicates). Chalcopyrite and covelline are present as the main copper-bearing minerals (1.1% Cu) in the ore (Bas et al., 2012). The other sulphide phases are reported to be pyrite, galena, and sphalerite in the ore (Bas et al., 2012, 2015; Ozdemir, 2011; Serbest, 2010). The ore sample contains 56 g/tonne Au and 10 g/tonne Ag.

Leaching experiments were conducted in 1000-mL vessels in which leach solutions (480 mL) were prepared at the desired concentrations of cyanide (1.5-2.5 g/L) and ammonia (3-4 M NH_3). The ore sample (120 g, 25% w/v) was placed into the vessel. Agitation of leaching vessels was carried out by overhead stirrers at 600 rpm. The vessels were capped for the duration of the tests. pH was adjusted and maintained at 10.5-11 by the addition of 1 M NaOH or 1 M H_2SO_4. A schematic illustration of the leaching setup could be found elsewhere (Celep et al., 2015). During the tests, the vessels were sampled at predetermined intervals and the collected samples were subjected to centrifugation to produce clear supernatants for subsequent analysis of gold and copper in AAS (Perkin Elmer AAnalyst 400). These samples were also used to determine the concentration of free cyanide by titration with silver nitrate using p-dimethylamino-benzal-rhodanine (0.02% w/w in acetone) as the indicator.

Table 1. Chemical and mineralogical compositions of the ore sample used in the study (Bas et al., 2012, 2015; Ozdemir, 2011; Serbest, 2010)

Chemical Composition					
SiO_2 %	**Fe_2O_3 %**	**Al_2O_3 %**	**CaO %**	**Ba %**	**$S_{(total)}$ %**
85	3.90	2.76	0.07	1.13	3.14
Au g/ton	**Ag g/ton**	**Cu %**	**Pb %**	**Zn %**	**Ni %**
56	10	1.10	0.19	0.36	0.08
Mineralogical Composition					
Quartz as the predominant phase; Chalcopyrite, covellite and chalcocite as copper-bearing minerals; Pyrite, galena and sphalerite as other sulphides					

The leaching tests were carried out for 24 h. and after termination of the tests, leaching residues were separated by filtration using a vacuum filter. Oven dried (at 105°C) residues were then digested in hot modified aqua-regia (Celep et al., 2015) to determine the residual metal content (gold and copper). Metal extraction data presented in the results were calculated based on the calculated head assays using metal content of residues.

3.RESULTS AND DISCUSSION

Ammoniacal cyanide leaching is considered as a promising alternative process for the treatment of copper-rich gold ores (Costello et al., 1992; Dai et al., 2012; Muir et al., 1991; Muir, 2011). Muir (2011) alleged that, in view of reagent costs, ammoniacal cyanide leaching is more economic than ammonia pretreatment despite similar gold extractions. In the earlier studies, 87% gold extraction was obtained using 2 M NH_3 and 2.5 g/L NaCN from the copper sulphide-bearing gold ore used in this study (Bas et al., 2015). In this regard, the influence of excess ammonia concentration on the gold and silver extraction was further investigated in this study.

Figs. 1-2 show the substantial improvement in the extraction of gold by the addition of ammonia. In the absence of ammonia, gold extraction was limited to only 6% at 1.5 g/L NaCN. The addition of ammonia at a concentration of 3 M NH_3, was found to lead to almost complete extraction of gold (Fig. 1). Similar final gold extraction was achieved at 2.5 g/L NaCN (Fig. 2). Zheng et al. (1995) found that the leach rate of pure gold decreased with the increase in ammonia concentration. Consistently, it was observed that (Fig. 1) increasing the ammonia concentration from 3 to 4 M impeded the kinetics of gold leaching. The addition or increasing the concentration of ammonia appeared to increase the availability of cyanide for gold leaching via forming complexes with copper (Fig. 3). Earlier studies (Drok and Ritchie, 1997; Muir et al. 1993) attributed the beneficial effect of ammonia in ammonia-cyanide system to the formation of $Cu_3(NH_3)_4(CN)_4$, thereby suppressing the leaching of copper.

Fig. 2 illustrates the effect of cyanide concentration on leaching of gold at 4 M NH_3. Increasing the concentration of cyanide from 1.5 to 2.5 g/L NaCN was found to improve the gold extraction kinetics. To illustrate, 95.9% extraction of gold at 2.5 g/L NaCN was achieved over an initial period of 8 h., compared with 56.6% gold extraction at 2.5 g/L NaCN.

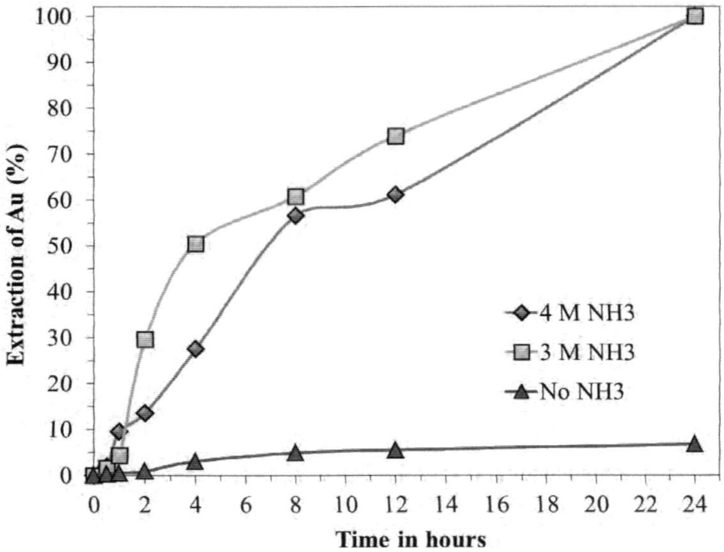

Figure 1. Effect of NH_3 concentration in ammoniacal cyanidation of gold (1.5 g/L NaCN, pH 10.5-11, 25% w/v solids)

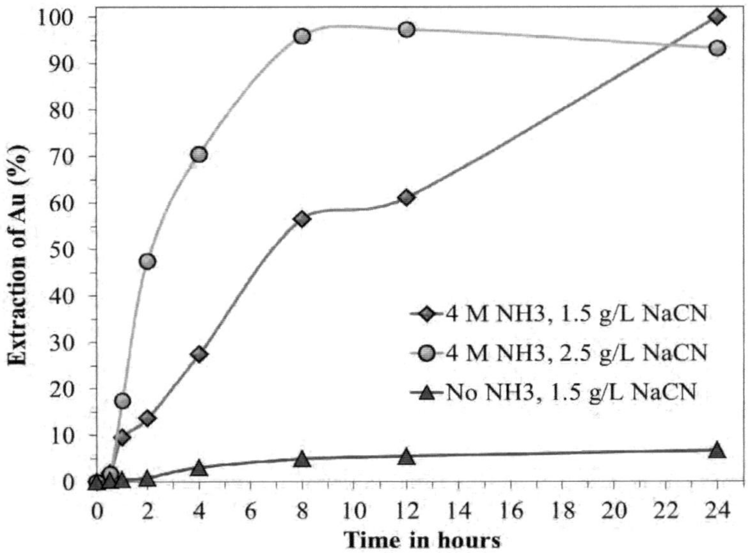

Figure 2. Effect of NH_3 and cyanide concentrations in ammoniacal cyanidation of gold (pH 10.5-11, 25% w/v solids)

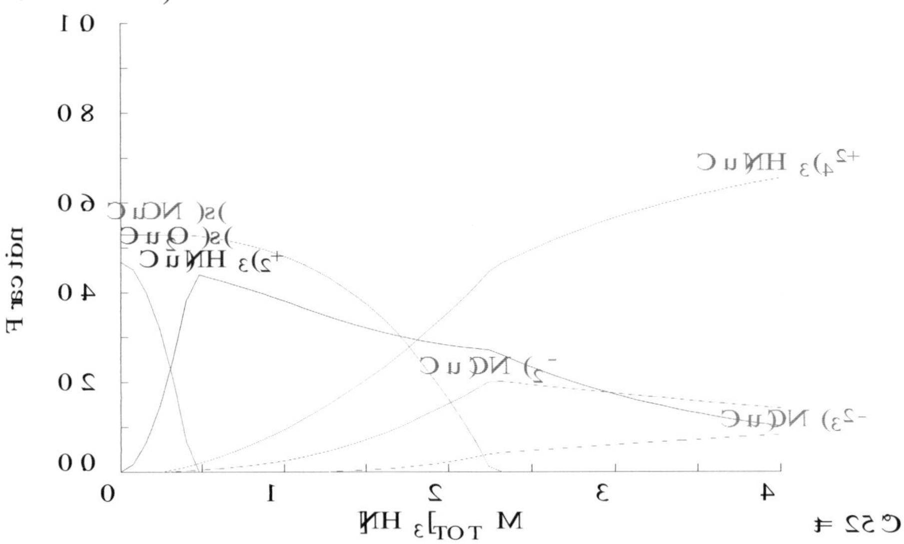

Figure 3. NH_3-dependent speciation of copper in NH_3-CN system ([Cu]: 58 mM, [CN^-]: 31 mM, pH 10.5, 25°C)

Fig. 4 depicts the copper dissolution in ammoniacal cyanide leaching. It was noticed that the leaching of copper was somewhat faster than gold i.e. the release of copper preceded that of gold. Drok and Ritchie (1997) studied the electrochemical behaviour of Au-Ag alloy and demonstrated that ammonia strongly passivates the dissolution of gold, but not in the region of open circuit potential where leaching takes place. The researchers indicated that copper in cyanide solutions is leached more rapidly than gold. They attributed this to the difference in oxygen-reduction mechanism of gold and copper. They suggested that reduction of oxygen on copper to hydroxide is a four-electron process (Eq. 4). On the other hand, oxygen is reduced by gold to hydrogen peroxide essentially via a two-electron process (Eq. 5).

$$4Cu^0 + O_2 + 8CN^- + 2H_2O \rightarrow 4Cu(CN)_2^- + 4OH^- \quad (4)$$

$$2Au^0 + O_2 + 4CN + 2H_2O \rightarrow 2Au(CN)_2^- + H_2O_2 + 2OH^- \quad (5)$$

Muir (2011) also showed that high concentrations of Cu^{2+} (>5 mM) could exert an adverse effect on the ammoniacal cyanide leaching. The researcher proposed that the optimum conditions for oxidised copper-gold ores are 1-2 kg/tonne NH_3 and 1-2 kg/tonne NaCN at pH 10.5-11. Increasing the concentrations of reagents above these levels can adversely influence the extraction of gold and its selectivity over copper. pH is also an important parameter significantly affecting leaching of gold and copper in that high pH promotes the leaching of gold due to the increased availability of free NH_3 for complexation of copper (Muir et al., 1991; Muir et al., 1993; Muir, 2011).

Increasing the concentration of ammonia improved the leaching of copper with the highest dissolution of 60% Cu at 4 M ammonia and 1.5 g/L NaCN (Fig. 4). In comparison, copper extraction was only 5% in the absence of ammonia (Fig. 4). This is consistent with the speciation calculations that also predict the increased solubilisation of copper with increasing NH_3 concentration apparently due to the formation of ammonia complexes (Fig. 3).

The slow rate of dissolution of copper after the initial periods of fast leaching could be attributed to the occurrence of copper. In this regard, copper is also present as chalcopyrite, which has limited solubility in cyanide leaching (Bas et al., 2015).

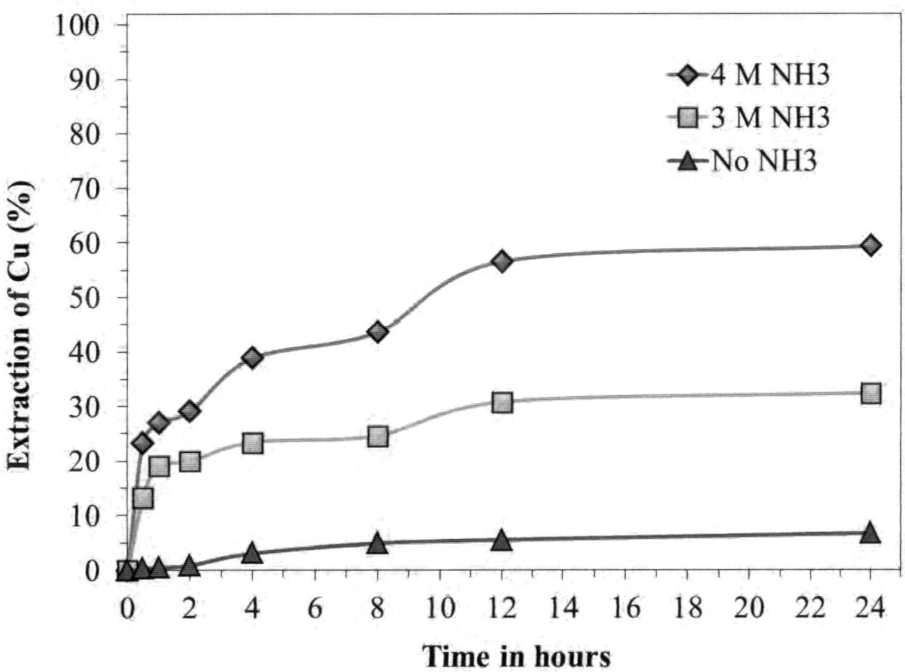

Figure 4. Effect of NH_3 concentrations in ammoniacal cyanidation of copper (1.5 g/L NaCN, pH 10.5-11, 25% w/v solids)

Figure 5 illustrates the effect of ammoniacal cyanidation on the consumption of sodium cyanide. It was found that increasing the concentrations of ammonia and cyanide led to 50% increase in the cyanide consumption. This increase is apparently linked with the improvement in copper and gold extractions (Figs. 1-3). In contrast, lower cyanide consumptions at 0.5-1 M NH_3 than at no NH_3 were reported by Bas et al. (2012) for the same ore. La Brooy et al. (1991) reported that the presence of ammonia (1-2 g/L) in cyanide leach solutions could provide better selectivity for gold over copper with minimising the consumption of cyanide. In an industrial leach plant for the treatment of a copper-gold ore, the addition of ammonia did not contribute to the extraction of gold while the dissolution of copper was reduced by half, thus leading to significant reduction in cyanide consumption (Hadjezi and Monhemius, 2016). It can be inferred from these earlier and current data that cyanide consumption and copper leaching tend to increase with increasing NH_3 concentration which, also results in decrease in selectivity.

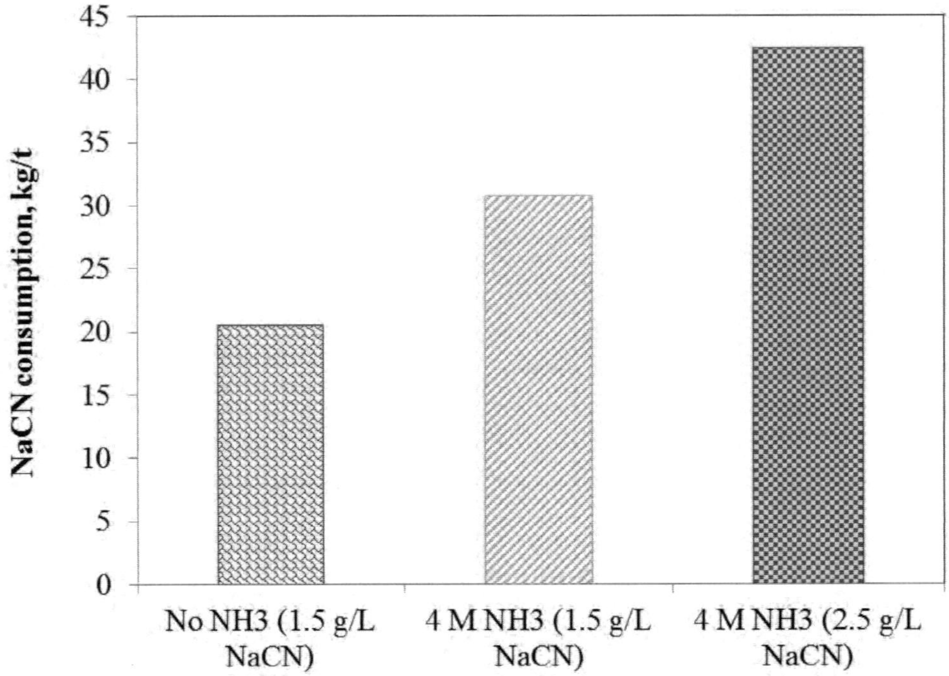

Figure 5. Effect of ammoniacal cyanidation on the consumption of NaCN

4. CONCLUSION

Copper-bearing gold ores prove technically and economically difficult to leach for selective extraction of gold due to the interfering effect of copper present. This study has presented the application of ammoniacal cyanide leaching for such a copper-rich gold ore. Addition of ammonia at 3-4 M was shown to significantly improve the extraction of gold from the ore. Increasing the concentration of cyanide (from 1.5 to 2.5 g/L NaCN) at a fixed ammonia concentration was also observed to enhance the leaching kinetics for gold. Limited copper leaching was presumed to be linked with the occurrence of copper i.e. the presence of chalcopyrite in addition to more cyanide-soluble phases. This could be ascribed to the formation of copper-amine complexes mitigating the interfering effect of copper present. These findings suggest that the ammoniacal cyanide leaching can be considered as a promising option for achieving high gold extractions from refractory copper-bearing gold ores.

ACKNOWLEDGEMENTS

The authors would like to express their sincere thanks to Mastra Gold Mine (Gümüşhane) for kindly providing ore samples.

REFERENCES

Adams, M. D., 2016. *Gold Ore Processing: Project Development and Operations*. Second ed. ISBN 978-0-444-63658-4, Elsevier.

Alymore, M. G. and Muir D., 2001. Thiosulphate leaching of gold - A review, *Minerals Engineering*, 14, 2, 135-174

Bas, A. D., Koç, E., Yazici, E. Y. and Deveci, H., 2015. Treatment of copper-rich gold ore by cyanide leaching, ammonia pretreatment and ammoniacal cyanide leaching, *Trans. Nonferrous Met. Soc. China*, 25, 597–607.

Bas, A. D., Ozdemir, E., Yazici, E. Y., Celep, O. and Deveci, H., 2011. Ammoniacal thiosulphate leaching of a copper-rich gold ore, in Proceedings *Int. Conference on Environmental and Mineral*

Processing (EaMP), Ostrava, Czech Republic, 83-90.

Bas, A. D., Yazici, E. Y., Deveci, H., 2012. Treatment of copper rich gold ores by ammonia assisted cyanide leaching. *Proceedings of XXVI International Mineral Processing Congress (IMPC) 2012*, New Delhi, India, 24-28 September, 356-365.

Breuer, P. L., Dai, X. and Jeffrey, M. I., 2005. Leaching of gold and copper minerals in cyanide deficient coper solutions, *Hydrometallurgy*, 78, 3-4, 156-165.

Bulatovic, S. M., 1997. Flotation behaviour of gold during processing of porphyry copper-gold ores and refractory gold-bearing sulphides, *Minerals Engineering*, 10, 9, 895–908.

Celep, O., Bas, A.D., Yazici, E.Y., Alp, I., Deveci, H., 2015. Improvement of silver extraction by ultrafine grinding prior to cyanide leaching of the plant tailings of a refractory silver ore. *Miner. Process. Extr. Metall. Rev.*, 36, 4, 227-236.

Costello, M. C., Ritchie, I. C. and Lunt, D. J., 1992. Use of the ammonia cyanide leach system for copper gold ores with reference to the retreatment of the Torco tailings, *Minerals Engineering*, 5 (10-12), 1421-1429.

Dai, X., Simons, A. and Breuer, P., 2012. A review of copper cyanide recovery technologies for the cyanidation of copper containing gold ores, *Minerals Engineering*, 25, 1, 1-13.

Dawson, J. N., La Brooy, S. R. and Ritchie, I. M., 1997. Copper-gold ore leaching: A kinetic study on the ammoniacal cyanidation of copper, chalcocite and chalcopyrite, *The AusIMM Annual Conference*, Ballarat, March 12-15, 291-297.

Drok, K., Ritchie, I., 1997. An investigation of the selective leaching of gold over copper using ammoniacal cyanide. *World Gold '97*, Singapore. The Australasian Institute of Mining and Metallurgy, Melbourne, 87-93.

Forrest, K., Yan, D., and Dunne, R., 1997. Optimisation of gold recovery by selective gold flotation for copper-gold-pyrite ores, *Minerals Engineering*, 4, 2, 227-241.

Hayes, G. A. and Corrans, I. J., 1992. Leaching of gold-copper ores using ammoniacal cyanide, *Extractive Metallurgy of Gold and Base Metals*, Kalgoorlie, October 23-25, 349-354.

Hedjazi, F., Monhemius, J., 2016. Industrial application of ammonia assisted cyanide leaching for copper-gold ores, *Emerging Trends in Minerals Engineering*. IMM Transactions, IOM3, London, UK.

Jeffrey, M I, Linda, L, Breuer, P L and Chu, C, 2002. A kinetic and electrochemical study of the ammonia cyanide process for leaching gold in solutions containing copper. *Minerals Engineering*, 15, 1173-1180.

Marsden, J. O. and House, C. I. 2006. *Chemistry of Gold Extraction*, 2nd Edition, Society for Mining, Metallurgy, and Exploration (SME). p.651

Muir, D. M., 2011. A review of the selective leaching of gold from oxidised copper–gold ores with ammonia–cyanide and new insights for plant control and operation, *Minerals Engineering*, 24, 576-582.

Muir, D. M., La Brooy, S. R. and Fenton, K., 1991. Processing copper-gold ores with ammonia or ammonia-cyanide solutions, in Proceedings *World Gold '91*, Cairns, April 21-25, 145-150.

Muir, D. M., La Brooy, S. R., Deng, T. and Sing, P., 1993. The mechanism of the ammonia-cyanide system for leaching copper-gold ores, *Milton Wadsworth Hydrometallurgy Symposium*, 1-5 August, Salt Lake City, 191-204.

Ozdemir, E., 2011. Thiosulphate leaching of gold from copper-rich Mastra gold ore, BSc. Thesis, Karadeniz Technical University, Dept. of Mining Engineering, Trabzon, Turkey. (In Turkish)

Sceresini, B., 2005. Gold-copper ores. In Adams, A.D., (Ed.), *Advances in Gold Ore Processing, Developments in Mineral Processing,* Vol 15 [M]. Amsterdam, The Netherlands, Elsevier Publishers, 789-824.

Serbest V., 2010. Effect of secondary copper minerals on cyanide leaching of gold ores, BSc. Thesis, Karadeniz Technical University, Dept. of Mining Engineering, Trabzon, Turkey. (In Turkish).

Zheng, J., Ritchie, I. M., La Brooy, S. R., Singh, P., 1995. Study of gold leaching in oxygenated solutions containing cyanide-copper-ammonia using a rotating quartz crystal microbalance. *Hydrometallurgy*, 39, 277-292.

Olivin Flotasyonunun Dekstrin Varlığında İstatistiksel Olarak Değerlendirilmesi

Statistical Evaluation of Olivine Flotation in the Presence of Dextrin

E. Polat, T. Güler
Muğla Sıtkı Koçman University, Department of Mining Engineering, Muğla, Turkey

ÖZET Olivin $(Mg,Fe)_2SiO_4$) özellikle dökümhane kumu, cüruf şartlandırma, refrakter demir cevher peleti yapımı gibi metalurjik işlemler için önemli bir endüstriyel hammaddedir. Olivin, zaman içinde yüksek Kızdırma Kaybı (KK) veren ve kristal suyu içeren serpantinlere (örneğin, lizardit), dönüşmektedir. Metalurjik alanlar açısından olivin cevherinin yüksek KK değeri vermesi termal özelliklerini bozduğu için çok önemli bir özelliktir. Bu çalışma, dekstrin ve Na-oleat (toplayıcı) konsantrasyonlarının, pH ve kondisyonlama süreleri ile birlikte olivin flotasyonu ve lizarditin uzaklaştırılabilmesine olan etkilerini aydınlatmak için yapılmıştır. Cevher örneğinin olivin, forsterit gibi cevheri oluşturan ana mineraller ile, yan kayacı oluşturan piroksen, lizardit ve kromit gibi minerallerden oluştuğu belirlenmiştir. Deneysel tasarım Taguchi metodu kullanılarak yapılmış ve sonuçlar ANOVA testi ile değerlendirilmiştir. Köpükte yüksek KK değerine sahip ürünler elde edilmesi, lizarditin olivin ve forsterit gibi minerallere göre yüzebilirliğinin fazla olduğunun bir göstergesidir. Oleat derişiminin flotasyonda test edilen en etkili değişken olduğu bulundu. Lizarditin dekstrin kullanımı ile bastırılması verimli olmamıştır.

ABSTRACT Olivine $(Mg,Fe)_2SiO_4$) is an important industrial raw material especially for metallurgical processes such as foundry sand, slag conditioning, production of refractories and iron ore pellets. Olivine alters throughout the geological ages to serpentines (e.g., lizardite) containing high rate of crystal water, in turn high Loss on Ignition (LOI) value. The LOI value of olivine ore is the major specification for metallurgical areas: thermal properties of olivine deteriorate at high LOI values. This study was performed to elucidate the effect of the concentration of dextrin and Na-oleate together with pH and conditioning time on olivine flotation and rejection of lizardite. Ore sample was found to contain olivine and forsterite as major rock forming minerals together with pyroxenes, lizardite and chromite as accessory minerals. Experimental design was made using Taguchi approach and experimental results were evaluated by ANOVA test. High LOI product was obtained in the froth indicating the higher floatability of lizardite as compared with olivine and forsterite. Collector concentration was found to be most effective tested variable on flotation. Lizardite could not be rejected efficiently by using dextrin as modifying agent.

1 INTRODUCTION

Olivine with the nominal formul $(Mg,Fe)_2SiO_4$, is a naturally occurring orthosilicate mineral group that are typically found in mafic and ultramafic igneous rocks. The mineral group is usually green in color. It is a solid solution series with end members forsterite (Mg_2SiO_4) and fayalite (Fe_2SiO_4). Olivine is a major component in earth forming rocks and is found with varying compositions due to alteration by weathering (Davis 1977; Sarkar, 2002; Krivolutskaya and Bryanchaninova 2011).

Olivine is one of the most important raw material due to its abundancy. Norway, being the largest supplier country of olivine, satisfies the world's olivine production with approximately 60%. Olivine, particularly forsterite, has many favorable properties, so that, it is demanded mostly as a refractory material for certain applications, such as casting and founding processes, foundry sand, production of iron ore pellets, fluidized beds (Manning, 1995; Güler and Aktürk, 2015). It has high melting point around 1760°C, and a low uniform thermal expansivity.

Froth flotation can be applied in order to beneficiate floatability of olivine minerals. Güney and Atak (1997) investigated chromite recovery in olivine containing ore, one of the most encountered mineral in chromite ores, by using different anionic collectors including Na-oleate. It was reported that oleate is not a proper collecting agent to separate chromite from olivine efficiently. Similarly, Gallios (2007) studied the flotation behaviour of chromite and serpentine fine particles in the presence of Na-oleate. It was concluded that moderate oleate consumption may increase the floatability of chromite at alkaline conditions, and CMC, being an organic modifier was found inhibiting its flotation effectively. Liu et al. (2015) evaluated the flotation behaviors of minerals including ilmenite, titanaugite, and forsterite, using sodium oleate in microflotation experiments. Following results were found related with forsterite: its floatability was notable at slightly acid to neutral pH and above the pH 9. Sodium oleate was found to react particularly with Mg elements on the forsterite surface both in alkaline and acidic conditions. Kotlyar et al. (1995) proposed that oxidation of sample by heating did not influence anionic flotation of olivine. They observed slight improvement in chromite flotation at slightly acid to moderate alkaline conditions (pH 6-10). Olivine from Köyceğiz region of Muğla in Turkey was investigated for its flotation behaviour in the presence of collector oleate in a wide pH range (Güler and Aktürk, 2015). Raw sample was first preconcantrated by removing -38 micron lizardite rich fine fraction, to increase the LOI value in the view of industrial applications. According to the flotation experiments: olivine floatability was found very low at highly alkaline conditions, while completely inhibited in acidic pHs. On the other hand, its hydrophobicity was increased at remaining pH conditions. Selectivity between silicate minerals in the tested ore could not be satisfied by Na-oleate flotation because of their similar chemical composition and floatability in flotation processes.

Serpentines are being softer than other silicate minerals, which affect the breakage characteristics of ores (Davis 1977; Velázquez et al. 2008; Tong et al. 2013). According to the characterization tests, ore sample revealed that major constituting minerals might be classified in to two major groups according to their mohs hardness levels: hard minerals including olivine and forsterite, being the major components (Mohs hardness: 5-7) and remaining soft fraction rich in lizardite (Mohs hardness: 2.5). Lizardite is found as a source mineral for most of crystal water content in the ore, playing an important role on determining the LOI (Güler and Aktürk, 2015). Inverse relationship exists between hardness and grindability: breakage rates of soft minerals are higher than those of hard minerals (Davis 1977; Velázquez et al. 2008; Tong et al. 2013). Comminution tests and partitioning behaviour of minerals in the olivine ore from Köyceğiz region, were well documented by Güler et al. (2015): lizardite, being soft and source of crystal water, is concentrated mainly in fine fraction (-38 μm), in which lizardite is structural liberated and concentrated in olivine ore particularly at finer fractions rather than at coarser.

Polysaccharides, such as dextrin, are well known organic reagents for minerals in flotation systems. They are also used in the processes where flocculants/dispersants are required, due to their large molecular sizes. Being non-toxic, made them ecofriendly modifiers more than other widely used inorganic depressants, like cyanide and sulphur dioxide. (Liu and Laskowski, 1988a, 1988b). Dextrin, being a soluble organic polymer, is derived from starch through heating in an acidic conditions. It has a molecular weight ranging from 800 to 79.000. Its basic composition unit is dextrose (a-D-glucopyranose). The only structural difference between starch and dextrin is that dextrin is relatively low in moleculer weight which contains a highly branched structure (Wie and Fuerstenau, 1974; Liu and Laskowski, 1988a; Laskowski et al., 1991; Bulatovic, 1999; Liu et al., 2000; Qin et al, 2013). As a polyhydroxy compound, dextrin is able to participate in a number of chemical reactions through substitution on the hydroxyl groups, or chemical complex formation with the hydroxyl groups (Liu and Laskowski, 1988a; 1988b).

The depressant action of dextrin in many applications have been widely studied, such as in selective flotation of sulfides, oxides, silicates and phosphates. Dextrin was used to depress sulfide minerals: galena was selectively depressed in chalcopyrite-galena complex ores (Liu and Zhang, 2000). Carbonaceous pyrite flotation was inhibited using dextrin at Mount Isa's Hilton and copper concentrator (Grano et al., 1991; Bulatovic, 1999). Mix of dextrin and quebracho was commercially used for depression of pyrite (Bulatovic, 1999). Dextrin was found to be a very effective depressant for the air flotation of molybdenite in the absence of a collector (Wie and Fuerstenau, 1974).

Dextrin adsorption on pyrite surface was evaulated with the presence of metal ions. Contribution of surface metal ions, which act as target sites for dextrin uptake, results in depression of pyrite. According to some studies in the literature type of metal ion is not important for dextrin adsorption on mineral surface (Bogusz et al., 1997). On the other hand, it was reported not reacting directly with metal cations. Coprecipitation between them was not found in the pH ranges where metal cations were present in the pulp. It was concluded that the reaction was possible with dextrin in the case of presence of metal hydroxides (Liu and Laskowski, 1988a, 1988b; Liu, and Laskowski, 1989 Liu et al., 2000). The reason why dextrin forms chemical complexes with metal hydroxides but not with metal cations is accepted to be not clear, yet (Qin et al,_2013; Liu, and Laskowski, 1989).

This study was performed to investigate the floatability of olivine together with separation of lizardite rich fraction at the froth, with the synergic effect of dextrin and Na-oleate consumptions together with pulp pH and conditioning time. Experimental system was designed by applying Taguchi approach. Floatability of lizardite present in the olivine ore was discussed under the light of LOI value.

2 MATERIALS AND METHODS

Representative olivine ore sample was obtained from Köyceğiz, Muğla in Turkey. Mineralogical and chemical characterization of sample was performed by XRD, SEM-EDS, XRF and petrographic analyses. Comminution was made by laboratory type jaw crusher and rod mill. Supplied sample having size of -20 cm was first crushed by a single-toggle blake type jaw crusher down to -1 cm size. Finely sized fraction (-212 µm) of crushed sample was removed prior to further comminution process to avoid over-grinding. Then, sampling of jaw crusher product was performed by conning and quartering followed by sampling using Riffle splitter to get 700 g of crushed ore sample (-1cm+212µm). Each ore sample was dry-ground by a laboratory type rod mill (200x300mm). Grinding time was determined as 10 minutes by preliminary tests. Mill load was adjusted as 40% solid. Mill speed was applied as 60% of critical speed. Laboratory type sieves were used for classification of products.

LOI value of classified sample and flotation products was determined using a muffle furnace. Olivine samples were first dried in an oven at 50°C for approximately 2.5 hours, and then taken to a desiccation. Dried samples weighed and heated in a muffle furnace up to 900°C at a heating rate of 50°C/min, at which temperature it was calcined for 30 minutes (Kleiv and Thorhnhill, 2011). Calcined samples were left in the furnace for cooling down to 150°C. After cooling, samples were taken to desiccator, where it cooled down to room temperature. Prior to each step, weight measurements were made, and LOI value was calculated from weight loss. LOI of raw olivine ore sample was determined as 1.47%.

XRD pattern of olivine sample is given (Fig. 1). Main peaks in the diffractogram overlay to those of olivine and forsterite. Enstatite (a pyroxene group mineral) and lizardite were found as the secondary crystalline phases. Lizardite, which is a serpentine group mineral, was discriminated especially by its high-intensity peak at 12.04° 2θ. Augite,which is a pyroxene group mineral, was hardly discriminated on the diffraction pattern.

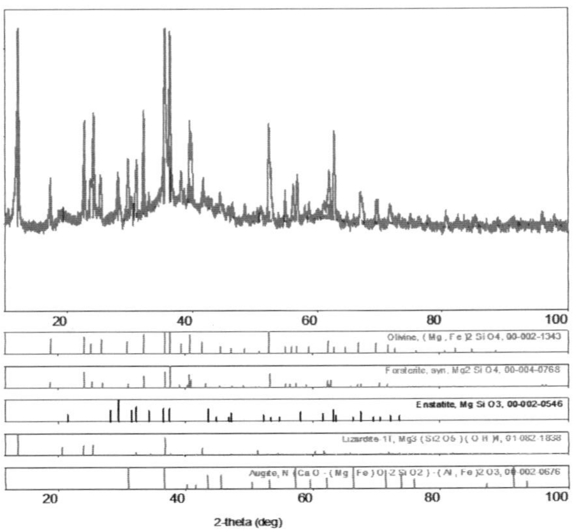

Figure 1. XRD pattern of olivine sample.

Flotation tests were performed by Denver type sub-aerating flotation machine using 50 g olivine sample in a 1 liter cell. Impeller speed was adjusted to 1100 rpm. Na-oleate, as an anionic collecting agent and methyl iso butyl carbinol (MIBC) with a fixed concentration (100 g/t) were used in flotation tests. Dextrin was tested as organic modifying agent. Its stock solution was prepared for daily consumption to minimize the effect of microbiological degradation, as there is evidence that the polymer solution degrades over time (Bogusz et al., 1997; Rath et al., 1997; Qin et al, 2013). The pulp pH was carefully controlled by aqueous solutions of sodium hydroxide (NaOH) or sulphuric acid (H_2SO_4). Conditioning time for collector was applied as 3 minutes followed by skimming the froth into collecting pan for the same duration.

2.1 Experimental Design and Statistical Analysis

Experimental work was designed by applying Taguchi approach. It provides reliable results by conducting relatively less experiments. Number of experiments is reduced by applying Taguchi's orthogonal array, which was derived from full factorial design. For complete analysis, Taguchi suggests to use Signal-to-Noise (S/N) ratio for multiple runs. The S/N ratio is a dimensionless performance measurement, and determines the most robust set of operating conditions from variation within the results. Noise factors are uncontrollable ones that cause variations in the quality property. On the other hand, Signal factors are known as controllable, and decrease the effects of uncontrollable factors (Güler and Polat, 2017). Taguchi's approach was applied for "the larger is better" case to maximize the response, LOI value of the froth products. The S/N ratio was calculated using the following equation, where "y" represents the experimental results.

$$(S/N) = -10\log\left(\frac{1}{n}\sum_{i=1}^{n}\frac{1}{y_i^2}\right) \qquad (1)$$

Four process parameters including concentration ofanionic collector (Na-oleate) and organic reagent dextrin as a depressant, pH and conditioning time were tested at five different levels (Table 1.).

Table 1. Tested factors and their levels.

Factors/Levels	L1	L2	L3	L4	L5
Collector, g/t	0	100	250	1000	2500
pH	7	9	10	11	13
Depressant, g/t	0	20	50	100	500
Cond. time, min.	0	2	5	10	20

Taguchi's $L_{25}(5^4)$ orthogonal array was applied in experimental works (Table 2). A, B, C and D are the tested parameters while 1, 2, 3, 4 and 5 symbolize their levels. The numbers in "Run" column represents the number of each experiment to be conducted to analyze all variables and their levels. Three representative samples taken from both froth and unfloat products, after dewatering and drying stages, were tested for their LOI value. Signal/Noise (S/N) ratio was calculated for froth products, since most of lizardite was concentrated more in floated stream. Increasing the number of LOI tests contributes getting more reliable results for robust design and eliminates man-made errors. Therefore, the LOI tests were repeated three times for each flotation product and their averages were taken as the final result. The S/N ratios was calculated and analysis of variance (ANOVA) table was drawn from the calculated data using Minitab software (version 17.1.0) program (Güler and Polat, 2017).

Table 2. Standard experimental layout for Taguchi's $L_{25}(5^4)$ orthogonal array.

Run	A	B	C	D	Experimental Results			Mean	S/N Ratio
1	1	1	1	1	y_{1-1}	y_{1-2}	y_{1-3}	y_1	S/N_1
2	1	2	2	2	y_{2-1}	y_{2-2}	y_{2-3}	y_2	S/N_2
3	1	3	3	3	y_{3-1}	y_{3-2}	y_{3-3}	y_3	S/N_3
4	1	4	4	4	y_{4-1}	y_{4-2}	y_{4-3}	y_4	S/N_4
5	1	5	5	5	y_{5-1}	y_{5-2}	y_{5-3}	y_5	S/N_5
6	2	1	2	3	y_{6-1}	y_{6-2}	y_{6-3}	y_6	S/N_6
7	2	2	3	4	y_{7-1}	y_{7-2}	y_{7-3}	y_7	S/N_7
8	2	3	4	5	y_{8-1}	y_{8-2}	y_{8-3}	y_8	S/N_8
9	2	4	5	1	y_{9-1}	y_{9-2}	y_{9-3}	y_9	S/N_9
10	2	5	1	2	y_{10-1}	y_{10-2}	y_{10-3}	y_{10}	S/N_{10}
11	3	1	3	5	y_{11-1}	y_{11-2}	y_{11-3}	y_{11}	S/N_{11}
12	3	2	4	1	y_{12-1}	y_{12-2}	y_{12-3}	y_{12}	S/N_{12}
13	3	3	5	2	y_{13-1}	y_{13-2}	y_{13-3}	y_{13}	S/N_{13}
14	3	4	1	3	y_{14-1}	y_{14-2}	y_{14-3}	y_{14}	S/N_{14}
15	3	5	2	4	y_{15-1}	y_{15-2}	y_{15-3}	y_{15}	S/N_{15}
16	4	1	4	2	y_{16-1}	y_{16-2}	y_{16-3}	y_{16}	S/N_{16}
17	4	2	5	3	y_{17-1}	y_{17-2}	y_{17-3}	y_{17}	S/N_{17}
18	4	3	1	4	y_{18-1}	y_{18-2}	y_{18-3}	y_{18}	S/N_{18}
19	4	4	2	5	y_{19-1}	y_{19-2}	y_{19-3}	y_{19}	S/N_{19}
20	4	5	3	1	y_{20-1}	y_{20-2}	y_{20-3}	y_{20}	S/N_{20}
21	5	1	5	4	y_{21-1}	y_{21-2}	y_{21-3}	y_{21}	S/N_{21}
22	5	2	1	5	y_{22-1}	y_{22-2}	y_{22-3}	y_{22}	S/N_{22}
23	5	3	2	1	y_{23-1}	y_{23-2}	y_{23-3}	y_{23}	S/N_{23}
24	5	4	3	2	y_{24-1}	y_{24-2}	y_{24-3}	y_{24}	S/N_{24}
25	5	5	4	3	y_{25-1}	y_{25-2}	y_{25-3}	y_{25}	S/N_{25}

3 RESULTS AND DISCUSSIONS

According to Taguchi approach, each flotation experiments are made randomly, and Taguchi's orthogonal array was constructed (Table 3). Data given in the table was analyzed by applying ANOVA test (Table 4): DF is "degree of freedom", SS is "sum of squares", and MS represents "mean sum of squares". The higher the F value and the lower the P value mean the greater the impact of variable on experimental results. Moreover, if the calculated P value is lower than significance level α, in which case calculated F value is greater than $F_{Table,\alpha}$ value, found elsewhere (Montgomery and Runger, 2010), then the influence of independent variable (experimental parameters) on dependent variable (LOI) is statistically significant, and vice-versa (Güler and Akdemir, 2012; Güler and Polat, 2017; Montgomery and Runger, 2010).

Table 3. All parameters and their levels displayed in L_{25} (5^4) orthogonal array.

Run	Collector g/t	pH	Deprassant g/t	Cond. Time minute	LOI (%) Mean	S/N
1	0	7	0	0	N.D.	N.D.
2	0	9	20	2	N.D.	N.D.
3	0	10	50	5	N.D.	N.D.
4	0	11	100	10	N.D.	N.D.
5	0	13	500	20	N.D.	N.D.
6	100	7	20	5	2.59	8.26
7	100	9	50	10	2.86	9.14
8	100	10	100	20	2.74	8.76
9	100	11	500	0	3.50	10.87
10	100	13	0	2	2.41	7.65
11	250	7	50	20	2.02	6.09
12	250	9	100	0	2.64	8.44
13	250	10	500	2	3.10	9.82
14	250	11	0	5	2.00	6.02
15	250	13	20	10	1.42	3.08
16	1000	7	100	2	1.20	1.59
17	1000	9	500	5	1.78	5.02
18	1000	10	0	10	1.49	3.46
19	1000	11	20	20	1.54	3.76
20	1000	13	50	0	1.59	4.02
21	2500	7	500	10	1.16	1.31
22	2500	9	0	20	1.33	2.46
23	2500	10	20	0	1.42	3.05
24	2500	11	50	2	1.36	2.67
25	2500	13	100	5	1.49	3.49

*N.D: No Data

The main effects plot for S/N ratio was calculated using the LOI data of the froth products (Fig. 2). Shape of line is taken into consideration in the evaluation of interactions of each tested factors. Possible interactions are remarkable when the tested factor slopes are not parallel to each other: curve slopes of both collector and depressant are opposite to each other as expected, giving contradictory results, while their concentrations were increased. Increasing the oleate consumption decreases the LOI value of the froth stream, while not valid in the case of depressant. Similar observations can be done for other tested factors: tested pHs give a parallel trend to collector except for pH 7 while conditioning time seems to slightly detoriates the LOI.

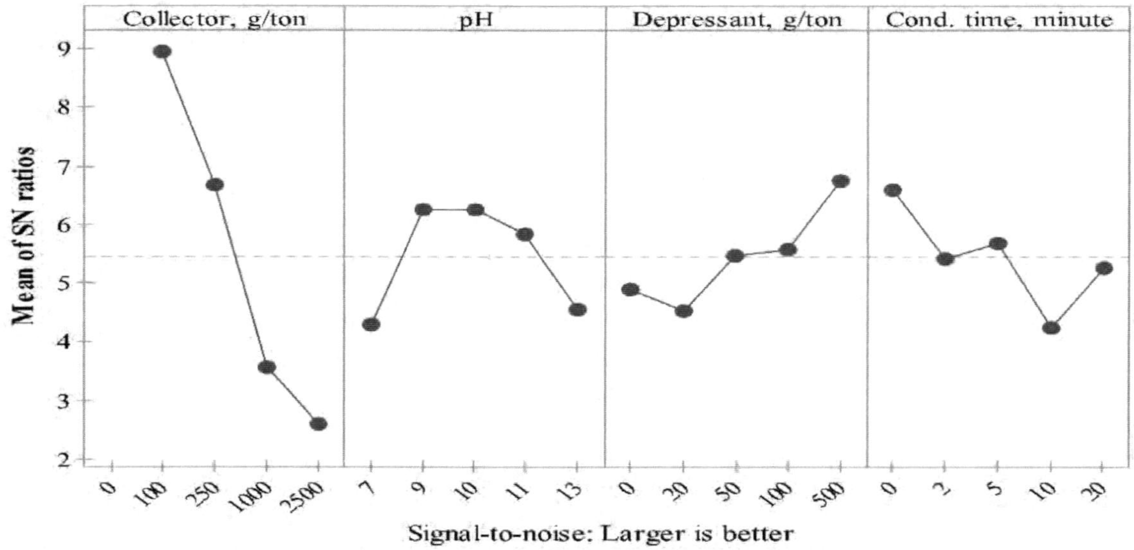

Figure 2. Main effects plot for S/N ratios of factors.

ANOVA test demonstrated that Na-oleate gave the highest F value (19.69). So, Na-oleate consumption was found to be the most effective tested parameter on LOI response data in flotation experiments. Pulp pH, dextrin concentration and conditioning time were recorded as relatively less important ones. Taken the significance level (α-level) as 0.05 (95% confidence), and magnitude of calculated F value as a criterion, then Na-oleate was found to be statistically significant, while the remaining tested variables left at negligible levels. According to applied Taguchi design, maximum hydrophobicity of lizardite rich material was seen to be obtained via LOI value, utilizing the synergic effect of tested factors. Optimized combination of tested factors was determined as (100 g/t Na-oleate) + (500 g/t dextrin) + (alkaline pH conditions) + (no conditioning time) to maximize the S/N ratio, and therefore for maximum hydrophobicity.

Table 4. ANOVA analysis results for all factors.

Source	df	Seq. SS	Adj. SS	Adj. MS	F	P	$F_{Table,\alpha}$=0.05
Collector, g/t	3	126.807	126.807	42.269	19.69	0.007	6.59
pH	4	14.305	10.763	2.691	1.25	0.416	6.39
Depressant, g/t	4	10.563	8.597	2.149	1.00	0.500	6.39
Cond. time, min.	4	8.144	8.144	2.036	0.95	0.520	6.39
Residual Error	4	8.589	8.589	2.147			
Total	19	168.407					

All flotation experiments were performed using -212 μm classified samples in the absence and presence of both Na-oleate and dextrine. Oleate was used as collecter in the tests with the following concentrations: 0, 100, 250, 1000 and 2500 g/t. Similarly, dextrin was consumed in the range of 0-500 g/t. Pulp pH was adjusted in the range of 7-13 and finally five conditioning time was selected: 0, 2, 5, 10 and 20 minutes.

Effect of Na-oleate on the floatability of olivine ore was investigated (Fig. 3). Oleate concentration was found to be significantly effective on olivine flotation: an increase in oleate consumption increased the floatability sharply up to1000 g/t collector concentration, and then slightly at higher concentrations. LOI results showed that lizardite might be hydrophobised by Na-oleate more than other major minerals in the ore, like olivine and forsterite. LOI value of froth product decreased sharply by an increase in oleate concentration up to 1000 g/t in proportion with the rate of froth product. Such a change in the LOI was attributed to the floatability of olivine and fortsterite minerals, not that of lizardite. Sodium oleate reacts mostly with Mg sites in the ore (Liu et al., 2015). Therefore, degree of hydrophobicity of a mineral in the olivine ore will be dependent on the number of Mg containing adsorption sites. Forsterite, having higher number of those sites on the surface relative to lizardite, then, might be readily covered more by Na-oleate. Collector increased the hydrophobicity of Mg rich sites forsterite so that decreased the LOI value of froth. Obtained data clarified that Na-oleate is not so effective on the floatability of lizardite, as expected. Concerning the physical properties of minerals constituting the ore, lizardite has the lowest hardness, therefore during grinding down to predetermined size, lizardite is ground ultrafine sizes while hard minerals, like olivine and forsterite, left at relatively coarser sizes. Lower LOI value at higher collector concentrations, then, is related with the entrainment of lizardite fines: Amount of entrained lizardite may not significantly changed, but its rate decreased due to increase floatability of forsterite. That is, rate of ultrafine entrained lizardite particles decreased by true flotation of olivine and forsterite (Güler and Akdemir, 2012). This study revealed that unwanted lizardite might be rejected from ore to some extent, but specifications of metallurgical application areas, which is lower than 1%, could not be achieved only by oleate flotation.

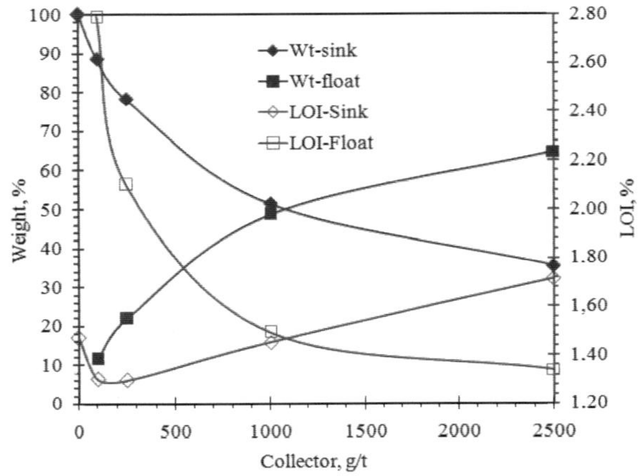

Figure 3. Effect of collector on the floatability and LOI value of olivine ore.

Effect of pH was tested in the range of 7-13 (Fig. 4). Remarkable variation in the floatability was not observed at neutral to moderately alkaline environment. It detoriated and decreased down to a minimum, about 23% at pH 13. Na-oleate adsorbs on negatively charged minerals chemically in neutral to alkaline conditions, where olivine minerals have PZC value lower than pH 7 (Güler and Aktürk, 2015; Liu et al, 2015; Fuerstenau, 2005). Main components of oleate were reported to be $RCOO^-$ and $(RCOO)_2^{2-}$ when pH was above 8.4 (Liu et al, 2015). These sites might be interacted with metal cations and adsorb onto silicates. LOI value of froth increased gradually up to mildly alkaline pH and decreased almost negligibly by increasing pulp pH. Minimum LOI value was obtained at pH 7 as 1.37%. This value became 1.60% in alkaline condition. As mentioned above, rate of lizardite recovery in the froth determines the rate of LOI of the product. Lizardite has positive charge in alkaline condition. Therefore oleate adsorption on it might occur by physical mechanism. Moreover, lizardite recovery in the froth might also be referred to entrainment: lizardite has the lowest hardness in the ore then it accumulates in finer size fractions of ground ores. Finely sized lizardite was thought to be entrained in the froth in addition to true flotation by Na-oleate. (Güler and Akdemir, 2012)

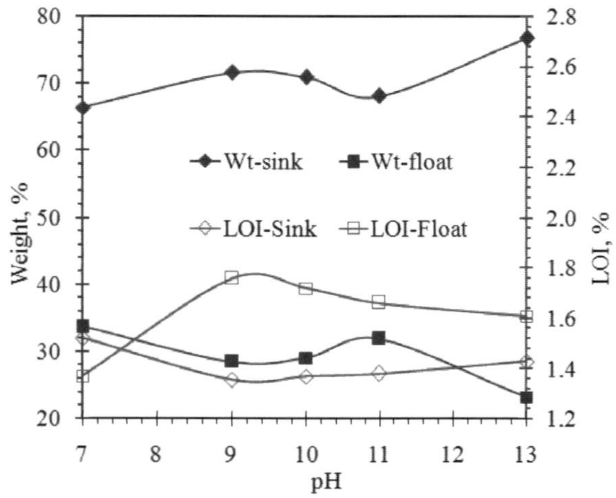

Figure 4. Effect of pH on the floatability and LOI value of olivine ore.

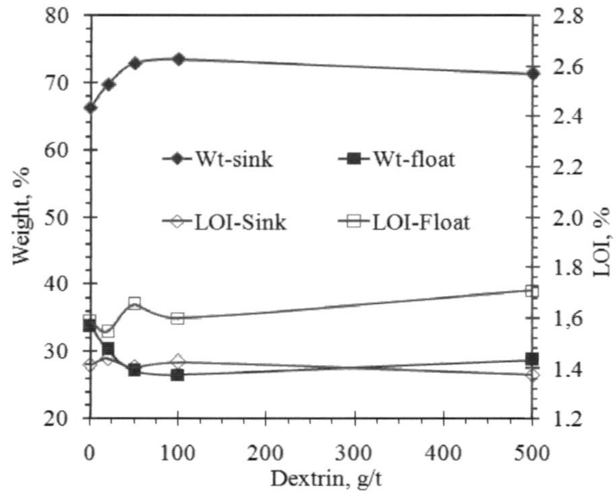

Figure 5. Effect of dextrin on the floatability and LOI value of olivine ore.

Dextrin uptake on silicate minerals as a function of concentration is presented in the Figure 5. The amount of froth product was reduced gradually up to 50 g/t dextrin consumption, while further increase in concentration left it almost at negligible levels. The LOI value of froth was increased in general, except for the fluctuation in the range of 0-100 g/t, while there is no notable change in unfloat stream. Lizardite tends to be concentrated mostly in the froth fraction at higher dextrin consumptions. Mg bearing minerals, such as lizardite and forsterite, release both Mg and other ions, according to the mineral composition in the tested ore, forming oxidation species in the pulp environment to some extent. Forsterite being rich in Mg, then releases more of those ions, forming relatively more hydroxides and their precipitates on mineral surface. Most of dextrin is, then expected to adsorb significantly at forsterite mineral surface due to Mg-hydroxide adsorption sites might present more relative to remaining silicates in the tested ore. It can be concluded that relatively hard minerals, such as forsterite and olivine, might be attract more oleate via adsorption sites (metal-hydroxide speicies) which inhibite their floatability behaviour slightly at higher depressant consumption.

Conditioning time was tested with the following durations: 0, 2, 5, 10 and 20 minutes (Fig. 6). There are peak points, such as at 2nd and 10th minutes of conditioning for the froth stream, giving an inverse relationship with LOI value. These clear fluctuations without any tendency to increase or decrease indicate the negligible importance of tested conditioning time in the flotation experiments. LOI value of froth fraction decreased from about 1.8, down to 1.6 at the maximum tested conditioning time, in general. According to this decrease in LOI, lizardite probably gave a small shift in amount, from froth to unfloat direction.

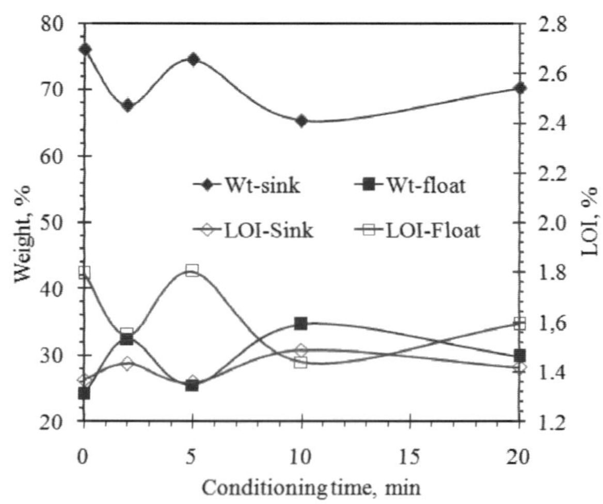

Figure 6. Effect of conditioning time on the floatability and LOI value of olivine ore.

4 CONCLUSION

- Sodium oleate is not a proper collecting agent to separate lizardite efficiently, may be due to similar chemical composition of silicate minerals in the ore.
- Oleate adsorbs on negatively charged minerals chemically in neutral to moderate alkaline conditions, where olivine minerals have PZC value lower than pH 7.
- Dextrin adsorption onto mineral surfaces might be result from its reaction with the metal-hydroxide adsorption species. Lizardite flotation could not be depressed selectively.
- Conditioning time was found as the least significant tested parameter: notable change was not observed as the conditioning time was increased.
- Statistical results support the findings above. Optimized combination of tested factors according to maximizing the LOI value at the froth stream can be summarised as follow: (100 g/t Na-oleate) + (500 g/t dextrin) + (moderate alkaline pH conditions) + (no conditioning time) to maximize the S/N ratio, and therefore maximum lizardite hydrophobicity.

REFERENCES

Bogusz, E., Brienne, S.R., Butler, I., Rao, S.R. and Finch, J.A., 1997. Metal ions and dextrin adsorption on pyrite, *Minerals Engineering,* 10(4), pp.441-445.

Bulatovic, S.M., 1999. Use of organic polymers in the flotation of polymetallic ores: a review, *Minerals Engineering*, 12(4), pp.341-354.

Davis E.G., 1977. Beneficiation of Olivine Foundry Sand by Differential Attrition Grinding, *US Patent No.4039625*, 5 p.

Fuerstenau, D.W., 2005. Zeta potentials in the flotation of oxide and silicate minerals, *Advances in Colloid and Interface Science*, 114, pp.9-26.

Gallios G.P., Deliyanni E.A., Peleka E.N., Matis K.A., 2007. Flotation of chromite and serpentine, *Separation and Purification Technology*, 55, 232-237.

Grano S., Ralston, J. and Smart, R.St. C., 1990. Influence of Electrochemical Environment on the Flotation Behaviour of Mt. Isa Copper and Lead-Zinc Ore, *Int. J. Miner. Process.*, 30, pp.69.

Güler, T., Akdemir, Ü., 2012. Statistical evaluation of flotation and entrainment behavior of an artificial ore, *Transactions of Nonferrous Metals Society of China*, *22*(1), pp.199-205.

Güler, T., Aktürk, S., Özer, A., 2014. Preconcentration of Muğla/Köyceğiz olivines by comminution, 14th International Mineral Processing Symposium, Kuşadası, Turkey, pp.779-785.

Güler, T., Aktürk, S., 2015. Beneficiation of olivine ore by Na-oleate flotation, 16th Balkan Mineral Processing Congress, Belgrade, Serbia, pp.543-548.

Güler, T., Polat, E., 2017. Tannic Acid As a Hydrophobicity Modifier for Galena in the Presence of Metal Ions, *Physicochem. Probl. Miner. Processing*, 53(1), pp.5-16.

Güney A., Atak S., 1997. Separation of chromite from olivine by anionic collectors, *Physicochemical Problems of Mineral Processing*, 31, pp.99-106.

Kleiv R.A., Thorhnhill M., 2011. Dry magnetic separation of olivine sand, *Physicochemical Problems of Mineral Processing*, 47, pp.213-228.

Kotlyar D.G., Tolley W.K., Rice D.A., 1995. The effect of oxidation on the flotation of chromite and associated minerals, *Report of Investigations 9575*, US Bureau of Mines, Washington 10p.

Krivolutskaya N.A., Bryanchaninova N.I., 2011. Olivines of igneous rocks, *Russian Journal of General Chemistry*, 81(6), pp.1302-1314.

Laskowski, J.S., Liu, Q., Bolin, N.J., 1991. Polysaccharides in flotation of sulphides. Part I. Adsorption of polysaccharides onto mineral surfaces, *International journal of mineral processing*, 33(1-4), pp.223-234.

Liu, Q., Laskowski, J.S., 1988a. The role of metal hydroxides at mineral surfaces in dextrin adsorption, I. Studies on modified quartz samples, *International Journal of Mineral Processing*, 26(3-4), pp.297-316.

Liu, Q., Laskowski, J.S., 1988b. The role of metal hydroxides at mineral surfaces in dextrin adsorption, II. Chalcopyrite-galena separations in the presence of dextrin, *International Journal of Mineral Processing*, 27(1-2), pp.147-155.

Liu, Q., Laskowski, J.S., 1989. The interactions between dextrin and metal hydroxides in aqueous solutions, *Journal of Colloid and Interface Science*, 130(1), pp.101-111.

Liu, Q., Zhang, Y., 2000. Effect of calcium ions and citric acid on the flotation separation of chalcopyrite from galena using dextrin, *Minerals Engineering,* 13(13), pp.1405-1416.

Liu, Q., Zhang, Y., Laskowski, J.S., 2000. The adsorption of polysaccharides onto mineral surfaces: an acid/base interaction, *International Journal of Mineral Processing*, 60(3), pp.229-245.

Liu, W., Zhang, J., Wang, W., Deng, J., Chen, B., Yan, W., Xiong, S., Huang, Y. and Liu, J., 2015. Flotation behaviors of ilmenite, titanaugite, and forsterite using sodium oleate as the collector, *Minerals Engineering*, 72, pp.1-9.

Manning, D.A.C., 1995. *Introduction to industrial minerals*, Chapman & HaU, London, 288 p.

Montgomery, D.C., Runger, G.C., 2010. *Applied Statistics and Probability for Engineers*, 5th edition. John Wiley and Sons, Inc., 784 p.

Qin, W., Wei, Q., Jiao, F., Yang, C., Liu, R., Wang, P., Ke, L., 2013. Utilization of polysaccharides as depressants for the flotation separation of copper/lead concentrate, *International Journal of Mining Science and Technology*, 23(2), pp.179-186.

Sarkar, R., 2002. Olivine-The Potential Industrial Mineral: An Overview, *Transactions of the Indian Ceramic Society*, 61(2), pp.80-82.

Tong L., Klein B., Zanin M., Quast K., Skinner W., Addai-Mensah J., Robinson D., 2013. Stirred milling kinetics of siliceous goethitic nickel laterite for selective comminution, *Minerals Engineering*, 49, pp.109-115.

Velázquez A.L.C., Menéndez-Aguado J.M., Brown R.L., 2008. Grindability of lateritic nickel ores in Cuba, *Powder Technology* 182(1), pp.113-115.

Wie, J., Fuerstenau, D.W., 1974. The effect of dextrin on surface properties and the flotation of molybdenite, *International journal of mineral processing*, 1(1), pp.17-32.

Investigation into the Influence of Microwave Pre-treatment on Coal Crushability Using Various Strength Tests

O.Y.Toraman[1,2], M.S.Delibalta[1]

[1]*Ömer Halisdemir University, Faculty of Engineering, Mining Engineering Department, 51245 Nigde, Turkey*

[2]*Ömer Halisdemir University, Industrial Raw Materials & Building Materials Application & Research Center, 51245 Nigde, Turkey*

ABSTRACT In the present study, the effect of microwave energy on crushability of high ash low rank Turkish coal has been investigated. Coal samples were treated by microwave at a frequency of 2.45 GHz with 0.80 kW power level and different exposure times (15-240 s). In order to determine the crushing resistance of lignite coal samples treated by microwave oven, the various strength tests (ISI, CI, $Is_{(50)}$) were applied for each treated and untreated sample and compared with each other. Experimental results have shown that significant increases in crushability were achieved when the coal samples were exposed to microwave radiation.

1 INTRODUCTION

The comminution of coal prior to combustion is an energy intensive process. A significant increase in coal crushability would allow the power requirement to the mills to be reduced significantly. Preheating of coal in an oven has been shown to reduce grind strength (Harrison and Rowson, 1996), although it is unlikely that this would be an economically viable method, at least not through associated energy requirements. Microwave heating is different from conventional heating since microwaves take the form of electromagnetic energy and can penetrate deep in the sample allowing heating to initiate volumetrically (Wang and Forrsberg, 2000; Jones et al., 2002).

Conventional thermal processing heats the sample from the outside inwards through standard heat transfer mechanisms. In addition, microwave heating also has the advantage of being selective and can heat certain phases in a matrix more rapidly than others (Lester and Kingman, 2004). It is assumed that when electromagnetic waves are applied on a material, thermal stresses occur because of moisture in the structure. Coal has a very low loss factor, where as water has very high loss factor (Marland et al., 2000; Marland et al., 2001; Meredith, 1997). Some impurities in coal show different behavior when absorbing electromagnetic irradiation (Chen et al., 1984; McGill et al., 1988; Walkiewicz et al., 1988).

The heating rate of coal and some of its impurities are given in Table 1.

Table 1. The heating rate of coal and its impurities.

Materials		Heating speed (°C/s)	Heating with microwave
Coal		Low ≈ 0.20	Difficult
Water		High ≈ 0.90	Easy
Minerals	Pyrite	High ≈1.89	Easy
	Quartz	Low ≈ 0.07	Difficult
	Feldspar	Low ≈ 0.10	Difficult

Power: 0.65 kW, frequency:2.45 GHz.

Phases consist of different heterogeneous structures because they have different degrees of ability to absorb microwave. Different phases have also different dielectric loss factors. Consequently, heating occurs in coal samples selectively (Marland et al., 2000; Marland et al., 2001; Kingman, 1999).

The effect of microwave treatment on coal was described by Marland et al. (2000). According to author, depending upon the microstructure and geological properties, coals naturally contain water to varying degrees and water is considered to be a good absorber of microwave energy. When microwave are applied, the molecular dipoles of water align and flip around as the applied field is alternating. Their movements produce frictional heating. The average heating rate of water has been calculated at $0.9^{o}C/s$ when exposed to microwave radiation at a power of 0.65 kW and a frequency of 2.45 GHz, mineral matters within coal differ in their ability to absorb microwave radiation. Some minerals readily heat within an applied electric field. Other minerals appear transparent to microwave radiation. Harrison and Rowson (1997) demonstrated that a reduction of 30% in the comparative work index could be achieved using a 0.65 kW 2.45 GHz microwave source. The reduction in relative work index occurring because of cracking initiated around pyrite grains and superheating of water in the porous coal structure. Toraman (2010) investigated the effect of high power (5-20 kW) microwave energy on the grindability of Turkish lignite coal. Microwave treatment increased the Hardgrove Grindability Index (HGI) at high power for a short residance time (2 s.). On the other hand, microwave energy intensity and the exposure time were found as the most important factors affecting crushability. The increase in HGI was up to 125% at a power of 0.90 kW for 60 seconds exposure time according to the non-treated sample (McGill et al., 1998). Recently it was found that by the researchers that the microwave - assisted grinding produced good results particularly for HGI, WI and specific rate of breakage (Ozbayoglu and Depci, 2006). Moreover, it has been obtained that HGI index value increased by approximately 23% after microwave treatment with 0.85 kW (Sahoo, 2011).

On the other hand, the Bond's grindability test can be used to determine the work index (WI) of any type of ore or material. But the process is tedious and long time consumption. The laboratory determination of Hardgrove grindability index (HGI) is less time consuming than that of the selection-for-breakage function. Impact strength test as a criterion of comminution was first developed by Protodyakanov (1950), and then it was used by Evans and Pomeroy (1966) for the classification of coal seams in the UK. The test was then, modified by Paone et al. (1969), Tandanand and Unger (1975), and Rabia and Brook (1980). Various authors Kahraman, 2001; Toraman, et al. 2010; Su, et al., 2010) also used this index value for determination of grindability/crushability of coal and rock samples. This test is simple and very practical. In addition, the apparatus is portable and can be used in the field. As a result, it has been proven that ISI values can be assessed as a criterion of crushability. Moreover, various authors (Kahraman and Toraman, 2008; Toraman et al., 2010) used Crushability Index (CI) value for determination of crushability of rock samples. They stated that CI values have been assessed as a criterion of crushability. The point load test can also be used to determine the strength of the ore samples. This test has two principle advantages: firstly, it can be used with irregular particles smaller than cored specimens required for determination of unconfined compressive strength and, more importantly, the test samples require no preparative treatment. Thus, tests can be carried out more rapidly, giving the requisite statistical validity in a reasonable time (Broch and Franklin, 1972).

The aim of this research is to investigate the influence of microwave pretreatment on the crushability of high-ash low rank Turkish coal. For this, the various strength tests (*ISI, CI,* $Is_{(50)}$) were applied for each treated and untreated sample and compared with each other.

2 MATERIAL AND METHOD

2.1 Material

A low rank lignite coal from Bursa Lignite Enterprise (BLI) (Turkey) was used in the study. The proximate analysis of the coal sample is shown in Table 2. The coal samples were crushed by a jaw crusher and then screened into different size groups (31.5+26.5 mm; 19.0+9.52 mm; 9.52+3.18 mm) for various test studies.

Table 2. Results of proximate analysis of lignite coal.

Analysis	As received wt%	Dry basis wt%
Moisture	31.68	-
Ash	35.19	51.51
Volatile matter (VM)	21.40	31.32
Fixed carbon (FC)	11.73	17.17
Total sulphur	1.97	2.88
Lower calorific value	1840 kcal/kg	2971 kcal/kg

2.2 Method

2.2.1. Microwave test

Microwave treatment was carried out with a 800 W experimental prototype 290 mm(W)×270 mm(D)×190 mm(H) microwave oven with variable power at 2.45 GHz. Coal samples (31.5+26.5 mm; 19.0+9.52 mm; 9.52+3.18 mm) were exposed to microwave radiation at a power of 0.80 kW and a frequency of 2.45 GHz for 15, 30, 60, 90, 120, 150, 180 and 240 seconds. Index tests were determined immediately after the heat treatment to prevent relative humidity affecting the crushing performance.

2.2.2. Point load test

The point load test equipment was used to determine the strength of the coal samples. For testing, the sample (31.5+26.5 mm) is placed between two shaped tips and a compressive force applied. From the dimensions of the particle and the force to failure, an index value, known as Is(50), is calculated from Eq. (1). The index is known as $Is_{(50)}$ since it is intended to correlate to tests performed on 50-mm diameter cores. The relationship is empirical and several have been developed by different authors. The formula used in this work was:

$$Is_{(50)} = [(Depth/50)^{0.45} \text{ x Force} : (4 \text{ x Width x Depth}/\pi)] \quad (1)$$

where all dimensions are in millimeters and the force is in MN. Is(50) has been shown to be approximately 1/14th of the unaxial compressive strength (Hoek and Brown, 1980) but, as only comparative data were required, the Is(50) value itself is reported in this work. The measurement of width and depth and the principle are illustrated in Fig. 1. Five specimens in the single size-range 31.5+26.5 mm were used per test. The average value was reported.

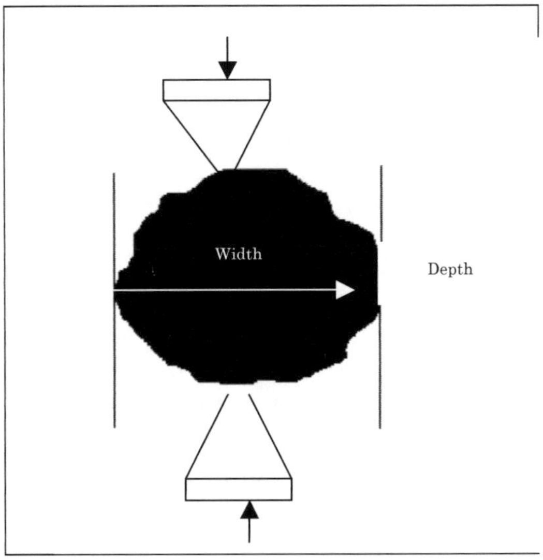

Figure 1. The principle of point load test.

2.2.3. Crushability index test

The crushed samples (19.0-9.52 mm) were sieved to obtain 500 g samples in the selected charging size range. After microwave radiation these samples were charged to the crusher with a outlet gap of 4–8 mm. Crushed materials were passed through the 9.52 mm and the percentage of undersized material was described as crushability index (CI). The test was repeated three times for each coal sample and the average value was recorded as the CI.

Table 3. Microwave heating of coal samples.

Sample	*Microwave power (kW)*	*Exposure time (second)*
Coal x non-treated	0	0
Coal x treated	0.80	15
		30
		60
		90
		120
		150
		180
		240

2.2.4. Impact strength test

Impact strength (ISI) test apparatus consists of a vertical stainless steel cylinder with a diameter of 4.23 cm. Inside this hollow cylinder, there is a 1.8 kg steel plunger positioned vertically (Fig. 2). After microwave radiation, the coal samples with a weight of 100±0.5g and 9.52+3.18 mm range are introduced into the cylinder, and the plunger is dropped freely on the samples 20 times consecutively from 30.48 cm. The Impact Strength Index (ISI) value is equal to the amount of coal retaining in the initial size range after the test. The test was repeated three times for each sample and the average value was recorded as the ISI. In accordance with the impact strength index values, coal samples are classified as in Table 4.

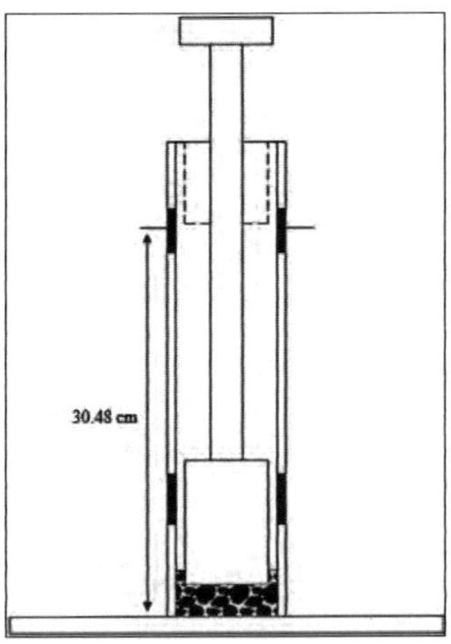

Figure 2. Impact strength index (ISI) test apparatus.

Table 4. Classification of coals in terms of impact strength index test.

ISI	*Classification*
>75	Extremely hard
75-60	Very hard
60-40	Hard
<40	Soft

3. RESULTS AND DISCUSSION

Turkish coal generally has a high ash-moisture content and low calorific value. Lower rank coals are more sensitive to microwave radiation possibly due to increased inherent moisture content. There is evidence to suggest that gaseous evolution (water and volatile matter), as well as gangue mineral expansion, causes cracks and fissure formations that are responsible for the improvement in crushability (Marland et al., 2001). In the study, the exposure time of MW heating has been varied from 15 to 240 seconds. An exposure time of >240 seconds, the coal caught fire (arcing).

The results showed that the increase in exposure time affected the crushability positively, e.g. the change of 0.80 kW caused a 645% decrease in the point load index. Generally, coal samples showed large changes in crushability, with point load index increases of 380-645%. Fig. 3 shows point load index vs. exposure time for coal samples. On the other hand, coal samples also showed large changes in crushability, with increases of 9-42% CI and 2-265% ISI, respectively.

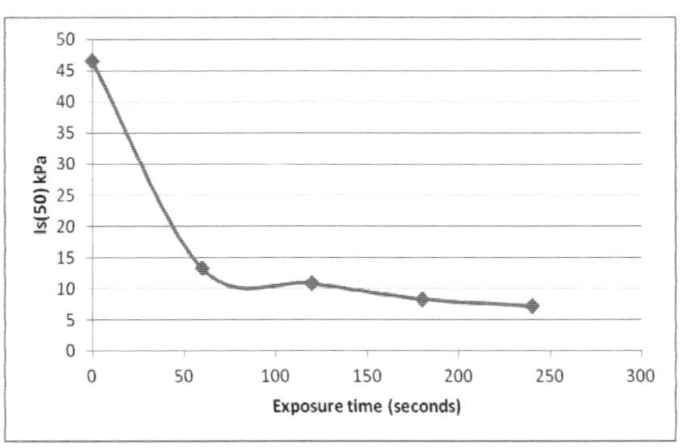

Figure 3. Is(50) vs. exposure time of coal samples (-31.5+26.5 mm).

Fig. 4 shows CI index vs. exposure time and Fig. 5 shows ISI index vs. exposure time for coal samples. In addition, Fig. 6 also shows fractures occurred on coal before (a) and after (b) microwave irradiation.

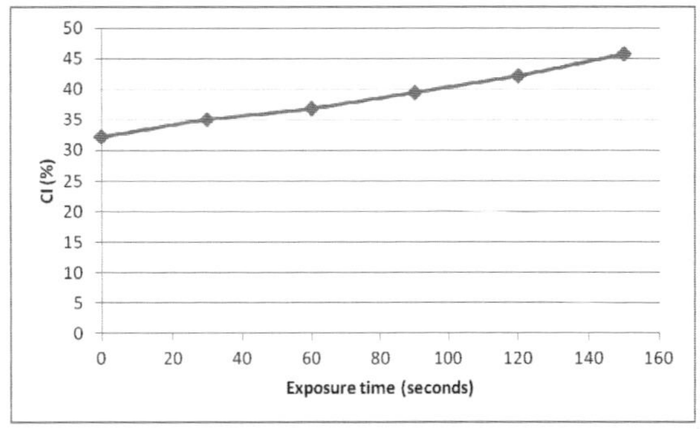

Figure 4. CI vs. exposure time of coal samples (-19+9.52 mm).

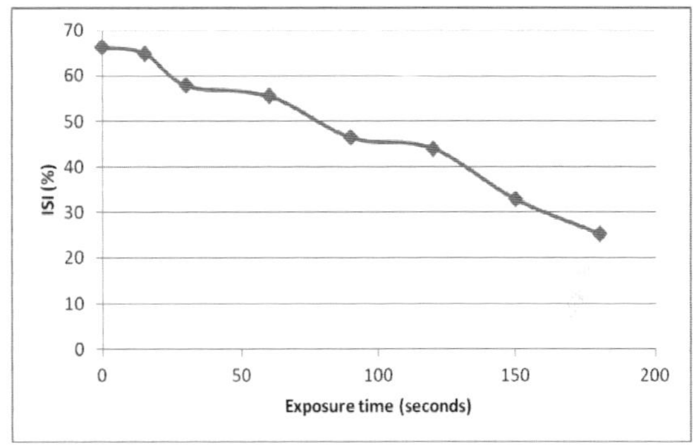

Figure 5. ISI index vs exposure time of coal samples (-9.52+3.18 mm).

(a) (b)

(c) (d)

Figure 6. Fractures occurred on coal before (a) and after (b) microwave irradiation for CI test (90 sec, 0.80 kW); before (c) and after (d) microwave irradiation for ISI test (90 sec, 0.80 kW).

The release of the moisture content (31.68%) of the coal tested might be the major cause of the thermally induced stresses and reduction in resistance to crushability. Most of the mineral matters appear transparent to microwave radiation except pyrite (FeS_2). As the total sulphur content of coal tested was high (2.88% at dry basis), pyritic sulphur effect would also be appreciable. Therefore, the differential expansion of minerals within the coal might be significant during heat treatment. Ozbayoglu and Depci (2006), in a previous study on a bituminous coal, observed a similar trend as above. They also reported that the distribution moisture content in various fractions has more impact on the improvement in crushability than the size.

4 CONCLUSION

In this study we studied the influence of microwave treatment on crushability of low rank Turkish coal by using various strength tests. The results obtained are summarized as follows:

- Experimental results have shown that increases of up to 42% CI index and decreases up to 265% and 645% ISI index and point load index, respectively can be achieved by microwave exposure of coal samples.
- The results were indicated that low rank coals are very sensitive to microwave radiation due to increased inherent and surface moistures. Rapid expansion of moisture within the coal causes some physical changes, such as craks and fissures formations which are responsible for the improvement in crushability. This means that the power requirement to the mills will reduce significantly.
- Microwave treatment had the greatest effect on the crushability of the larger fractions, as indicated by a lower value for the point load index.

References

Broch, E., Franklin, J.A., 1972. The point load strength test, *International Journal of Rock Mechanics and Mining Sciences,* 9, 669-697.

Chen, T.T., Dutrizac, J.E., Haque, K.E., Wyslouzil, W., Kashyap, S., 1984. The relative transparency of minerals to microwave radiation. *Can. Metall. Q.*, 23, 349-351.

Evans, I., and Pomeroy, C.D., 1966. The Strength, Fracture and Workability of Coal. London: *Pergamon Press.*

Harrison, P.C., and Rowson, N.A., 1997. The effect of microwave and conventional heat treatment on the comminution of coal. *IChemE Research Event,* 33-37.

Harrison, P.C., and Rowson, N.A., 1996. The effect of heat treatment on the grindability of coal, *IChemE. Res. Event.*, 268-71.

Hoek, E., Brown, E.T., 1980. The Inst. Min. and Metall. London.

Jones, D.A., Lelyveld T.P., Mavrofidis S.D., Kingman S.W., Miles N.M., 2002. Microwave heating applications in environmental engineering. *Resour. Conserv. Recycl.*, 34:75–90.

Kahraman, S., and Toraman, O.Y., 2008. Predicting Los Angeles abrasion loss of rock aggregates from crushability index, *Bull. Mater. Sci.,* Vol. 31, No. 2, 173–177.

Kahraman, S., 2001. Evaluation of simple methods for assessing the uniaxial compressive strength of rock, *Int. J. Rock Mech. &Mining Sci.*, 38, 981-994.

Kingman, S.W., 1999. The Influence of microwave radiation on the comminution and beneficiation of minerals. Ph.D. thesis. University of Birmingham.

Lester, E, and Kingman, S., 2004. The effect of microwave pre-heating on five different coals, *Fuel,* 83,1941-1947.

Marland, S., Han, B., Merchant, A., Rowson, N., 2000, The effect of microwave radiation on coal grindability. *Fuel,* 79, 1283–1288.

Marland, S., Merchant, A., Rowson, N., 2001. Dielectric properties of coal, *Fuel,* 80, 1839-49.

Marland, S., Han, B., Merchant A., and Rowson, N., 2000. The effect of microwave radiation on coal grindability, *Fuel,* 79, 1283-1288.

McGill, S.L., Walkiewicz, J.W., Smyres, G.A., 1988. The effect of power level on the microwave heating of selected chemicals and minerals. *Mater Res Soc Symp Proc.* Warrendale.

Meredith, R., 1997. Engineers handbook of industrial microwave heating. *Institution of Electrical Engineers.*

Ozbayoglu, G., and Depci, T., 2006. Microwave radiation effect on coal grindability. 15th Int. Coal Prep. Congress, Beijig, China, October.

Paone, J., Madson, D., Bruce, W.E., 1969. Drillability studies – laboratory percussive drilling. USBM RI 7300.

Protodyakanov, E.I., 1950. Determination of coal strength at mines, *Ugol,* 25, 20-24.

Rabia, H., and Brook, W., 1980. An empirical equation for drill performance prediction. In: Proc. 21st US Symp. Rock Mech Univ., Missouri-Rolla, 103–111.

Sahoo, B.K., De, S., Meikap, B.C., 2011. Improvement of grinding characteristics of Indian coal by microwave pre-treatment, *Fuel Processing Technology,* 92, 1920-1928.

Su, O., Toroglu I., and Akcın, N.A., 2010. An evaluation of the impact strength index as a criterion of grindability, *Energy Sources, Part A,* 32, 1671-1678.

Tandanand, S., Unger, H.F., 1975. Drillability determination – a drillability index of percussive drills. USBM RI 8073.

Toraman, O.Y., 2010. The effect of high power microwave energy on the grindability of Turkish Cayirhan lignite, *Energy Sources, Part A,* 32, 1794-1800.

Toraman, O.Y., Kahraman, S., Cayirli, S., 2010. Predicting the crushability of rocks from the impact strength index, *Minerals Engineering,* 23, 752-754.

Walkiewicz, J.W., Kazonich, G., McGill, S.L., 1988. Microwave heating characteristics of selected minerals and compounds, *Miner. Metall. Process.,* 63, 29-44.

Wang, Y., Forrsberg, E., 2000. Microwave assisted comminution and liberation of minerals. Mineral Processing on the Verge of the 21st Century. Rotterdam.